MOLECULAR
MEDICAL
MICROBIOLOGY

MOLECULAR MEDICAL MICROBIOLOGY

Edited by

Max Sussman

Department of Microbiology
University of Newcastle upon Tyne

ACADEMIC PRESS

A Harcourt Science and Technology Company

SAN DIEGO • SAN FRANCISCO • NEW YORK • BOSTON
LONDON • SYDNEY • TOKYO

Academic Press
A Harcourt Science and Technology Company
Harcourt Place, 32 Jamestown Road, London NW1 7BY, UK
http://www.academicpress.com

Academic Press
A Harcourt Science and Technology Company
525 B Street, Suite 1900, San Diego, California 92101-4495, USA
http://www.academicpress.com

ISBN 0-12-677530-3

Library of Congress Catalog Number: 2001089056

A catalogue record for this book is available from the British Library

Typeset by Newgen Imaging Systems (P) Ltd, Chennai, India
Printed and bound in Spain by Grafos SA Arte Sobre Papel, Barcelona
01 02 03 04 05 06 07 GF 9 8 7 6 5 4 3 2 1

Dedication

ונתתי לה בביתי ובחומתי יד וש.....

Isaiah 56:5

Siegmund Hirsch
Hamburg 14.1.1905 - Auschwitz 30.9.1942

Josepha Alexandra Hirsch-Levy
Hamburg 9.9.1905 - Auschwitz 17.7.1942

Ursula Selma Hirsch
Hamburg 11.8.1931 - Auschwitz 17.7.1942

Shulamith Hirsch
Hamburg 28.11.1933 - Auschwitz 17.7.1942

Uncle, Aunt and Cousins

הי״ד

Editorial Board

Preface

The developments of the last almost fifty years suggested that the time might be ripe for a book dedicated to medical microbiology as seen from a molecular vantage point. Such a project faces the problem that the subject is immense and growing at an intimidating rate even while this book was being edited. Nevertheless, I hope that it is at least a faithful representation of the subject.

Molecular medical microbiology cannot, of course, stand in isolation from the 'whole organism' approach of the pre-molecular age that is so well described in many standard texts. It was felt, therefore, that to make the size of the book manageable, contributors could safely assume that readers in search of information about the more classical phenotypic and descriptive aspects of the subject would profitably turn elsewhere.

This book is in a way a *prolegomenon*. Molecular medical microbiology is not a 'settled subject' and authors were given only limited guidance about the content and structure of their contributions. Each chapter, therefore, reflects the particular insights and priorities of its authors.

My Advisory Editors were most helpful, especially during the initial planning stages of the project and I am grateful to them for their advice, which in some cases was sought at very inconvenient times. I wish also to thank all the contributors and particularly those who submitted early and, later in the gestation, readily agreed to revise their chapters.

My successor Professor Carlos Hormaeche has shown his friendship to me in many ways. I am most fortunate, some twelve years after my retirement, to be able to continue to work in the Department of Microbiology and Immunology of the Newcastle Medical School.

Finally, the journey from manuscript to printed page is long, often tortuous and sometimes mysterious and a multitude of hands is involved. The eagle and expert eye of my copy editors has saved me from many a lapse. I should like to express my thanks to the many at Academic Press who assisted me at various stages. Particularly deserving of gratitude amongst these are, Dr Tessa Picknett for her assistance and understanding, Dr Lilian Leung and Soma Mitra for their help with keeping a semblance of order and Sutapas Bhattacharya for his reassuring efficiency and limitless knowledge about book production.

Max Sussman
Newcastle Upon Tyne
July 2001

List of Contributors

Soman N. Abraham
Depts of Pathology and Microbiology, Duke University
Medical Center, Durham NC 27710, USA

Mark Achtman
Max-Planck Institut für Infektionsbiologie and Deutsches
Rheuma-Forschungszentrum, Schumannstr 21/22, 10117
Berlin, Germany

Shin-Ichi Aizawa
Dept of Bioscience, Teikyo University 1-1 Toyosatodai,
Utsunomiya, 320-8551, Japan

David G. Allison
School of Pharmacy and Pharmaceutical Sciences,
University of Manchester, Manchester M13 9PL, UK

Valerie Asche
Menzies School of Health Research, POB 4794, Darwin,
NT 0801, Australia

Larry M. Baddour
Dept of Medicine Section of Infectious Diseases, University
of Tennessee Medical Centre, Knoxville, TN 37920-6999,
USA

Camella C. Bailey
Center for Vaccine Development, University of Maryland
School of Medicine, 685 W Baltimore Street, Baltimore,
MD 21201, USA

Michael R. Barer
Dept of Microbiology and Immunology, The Medical
School, University of Newcastle Upon Tyne, Newcastle
Upon Tyne, NE2 4HH, UK
Present address: Dept of Microbiology and Immunology,
University of Leicester, P.O. Box 138, Medical Sciences
Building, University Road, Leicester LE1 9HN, UK

Arnold S. Bayer
Harbor-UCLA Medical Center, Division of Infectious
Diseases, 1000 W Carson St, Box 466, Torrance CA
90509, USA

Blaine L. Beaman
Depts of Medical Microbiology and Immunology,
University of California School of Medicine, Davis,
CA 95616, USA

Douglas J. Beecher
HMRU, FBI Academy, Building 12, Rm 201, Quantico,
VA 22135, USA

Peter M. Bennett
Depts of Pathology and Microbiology, School of Medical
Sciences, University of Bristol, University Walk, Bristol
BS8 1TD, UK

Sucharit Bhakdi
Insitut für Medizinische Mikrobiologie und Hygiene der
Universität, Hochhaus am Augustusplatz, 55101 Mainz,
Germany

Svend Birkelund
Dept of Medical Microbiology and Immunology,
The Bartholin Building, University of Aarhus, DK-8000
Aarhus C, Denmark

Michael H. Block
AstraZeneca, Alderley Park, Macclesfield SK10 4TG, UK

R.A. Bojar
The Skin Research Centre, Dept of Microbiology,
University of Leeds, Leeds LS2 9JT, UK

Stephen J. Bourke
Dept of Respiratory Medicine, Royal Victoria Infirmary,
Newcastle Upon Tyne NE1 4LP, UK

C. Josephine Brooke
Dept of Microbiology, University of Western Australia,
35 Stirling Highway, Crawley WA 6009, Australia

Robert R. Brubaker
Dept of Microbiology, 57A Giltner Hall, Michigan State
University, East Lansing, MI 48824-1101, USA

Thierry Calandra
Division of Infectious Diseases, BH-19.111, Centre
Hospitalier Universitaire Vaudois, CH-1011 Lausanne,
Switzerland

John A. Chaddock
CAMR, Porton Down, Salisbury, Wilts SP4 0JG, UK

Babara J. Chang
Dept of Microbiology, University of Western Australia,
35 Stirling Highway, Crawley WA 6009, Australia

Matthew R. Chapman
Depts of Molecular Microbiology and Microbial
Pathogenesis, PO Box 8230, Washington University
School of Medicine, 600 S Euclid Avenue St Louis,
MO 63110-1093, USA

Henrik Chart
Division of Enteric Diseases, PHLS Central Public Health
Laboratory, 61 Colindale Avenue, London NW9 5HT, UK

Gunna Christiansen
Dept of Medical Microbiology and Immunology,
The Bartholin Building, University of Aarhus, DK-8000
Aarhus C, Denmark

Steven Clegg
Dept of Microbiology, College of Medicine, University of
Iowa, 3403 Bowen science Building, Iowa City IA 52242-
1109, USA

Frank M. Collins
Laboratory of Mycobacterial Diseases, Bldg 29 Room 505,
CBER/FDA, 1401 Rockville Pike, Rockville MD 20857,
USA

M.J. Colston
Leprosy and Mycobacterial Research, National Institute for
Medical Research, The Ridgeway, London NW7 1AA, UK

Howard N. Cooper
Dept of Medical Microbiology, Imperial College School of
Medicine, St Mary's Campus, Norfolk Place, London W2
1PG, UK

Pascal F. Cossart
Unité des Interactions Bactéries-Cellules, Institut Pasteur,
28 rue de Docteur Roux, Paris 75015, France

Alan S. Cross
Dept of Medicine, Division of Infectious Diseases,
University of Maryland School of Medicine, 22 S. Greene
Street, Baltimore, MD 21201, USA

Sally J. Cutler
Dept of Infectious Diseases, Imperial College School of
Medicine, Hammersmith Hospital Campus, Ducane Road,
London W12 ONN, UK

Simon M. Cutting
School of Biological Sciences, Royal Holloway University
of London, Egham, Surrey TW20 0EX, UK

Jeremy W. Dale
Molecular Microbiology Group, School of Biological
Sciences, University of Surrey, Guildford, Surrey
GU2 5XH, UK

Antoine Danchin
Hong Kong University Research Centre, Dexter Man
Building, 8 Sassoon Road, Pokfulam, Hong Kong
Also at: Regulation of Gene Expression, URA 2171 CNRS,
Institut Pasteur, 28, Rue du Dr Roux, 75724 Paris
cedex 15, France

Rina Das
Dept Molecular Pathology, Walter Reed Army Institute of
Research, Washington DC 20307-5100, USA

Gregory A. Dasch
Viral and Rickettsial Diseases Division, Naval Medical
Research Center, Bethesda MD 20889-5607
Present address: Viral and Rickettsial Zoonoses Branch,
Division of Viral and Rickettsial Diseases, Centers for
Disease Control and Prevention, Atlanta, GA 30333, USA

Charles J. Dorman
Dept of Microbiology, Moyne Institute of Preventitive
Medicine, University of Dublin, Trinity College, Dublin 2,
Republic of Ireland

Ron J. Doyle
Dept of Microbiology, University of Louisville School of
Medicine, Louisville KY 40292, USA

Mette Drasbek
Loke Diagnostics, Science Park, Aarhus, Denmark

Roman Dziarski
Northwest Center for Medical Education, Indiana
University School of Medicine, Gary, IN 46408, USA

Jennifer Eardley
Center for Vaccine Development, University of Maryland
School of Medicine, 685 W Baltimore Street, Baltimore,
MD 21201, USA

Karen L. Elkins
Center for Biologics Evaluation and Research, FDA,
Building 29 Room 428, 1401 Rockville Pike, Bethesda MD
20892, USA

Marina E Eremeeva
Dept of Microbiology and Immunology, University of
Maryland at Baltimore, School of Medicine, 655 West
Baltimore Street, Baltimore MD 21201-1559, USA
Present address: Viral and Rickettsial Zoonoses Branch,
Division of Viral and Rickettsial Diseases, Centers for
Disease Control and Prevention, Atlanta, GA 30333, USA

Paul Everest
Dept of Veterinary Pathology, Glasgow University,
Bearsden, Glasgow G61 1QH, UK

Edward J. Feil
Dept of Biology and Biochemistry, University of Bath,
Bath, BA2 7AY, UK

Sydney M. Finegold
Veterans Affairs Medical Center and UCLA School of Medicine, Los Angeles, California 90073, USA

Brett B. Finlay
Biotechnology Laboratory, University of British Columbia, Room 237 Wesbrook Building, 6174 University Boulevard, Vancouver, BC V6T 1Z3 BC, Canada

James Flexman
Dept of Microbiology and Infectious Diseases, Royal Perth Hospital, Wellington Street, Perth, Western Australia, 6000

John W. Foster
Depts of Microbiology and Immunology, University of South Alabama College of Medicine, Mobile AL 36688-0002, USA

Timothy J. Foster
Dept of Microbiology, Moyne Institute of Preventive Medicine, Trinity College, Dublin 2, Ireland

Laura S. Frost
Dept of Biological Sciences, University of Alberta, Edmonton, Alberta T6G 2E9, Canada

J.S.H. Gaston
Rheumatology Unit, Dept of Medicine, Adenbrook's Hopsital, University of Cambridge, Cambridge CB2 2QQ, UK

Curtis G. Gemmell
University Dept of Bacteriology, Glasgow Royal Infirmary, 84–86 Castle Street, Glasgow G4 0SF, UK

Peter Gilbert
School of Pharmacy and Pharmaceutical Sciences, University of Manchester, Manchester M13 9PL, UK

Michel Pierre Glauser
Division of Infectious Diseases, BH-19 111, Centre Hospitalier Universitaire Vaudois, CH-1011 Lausanne, Switzerland

Tana Green
Dept of Genitourinary Medicine, Royal Hallamshire Hospital, Glossop Road, Sheffield S10 2JF, UK

Elwyn Griffiths
Chief, Biologicals, World Health Organisation, CH-1211 Geneva 27, Switzerland

Antje Haase
Menzies School of Health Research, POB 4794, Darwin NT 0801, Australia

Jorg H. Hacker
Instutut für Molekulare Infektions Biologie, Universität Würzburg, Röntgenring 11, D-97070 Würzburg, Germany

Ian C. Hancock
Dept of Microbiology and Immunology, The Medical School, University of Newcastle Upon Tyne, Newcastle Upon Tyne NE2 4HH, UK

Margaret M. Hannan
Department of Medical Microbiology, Imperial College School of Medicine, St Mary's Campus, Norfolk Place, London W2 1PG, UK *and* National Center for HIV, STD and TB Prevention, Centers for Disease Control and Prevention, Atlanta GA 30333, USA

Colin R. Harwood
Dept of Microbiology, The Medical School, University of Newcastle Upon Tyne, Framlington Place, Newcastle Upon Tyne NE2 4HH, UK

R.J. Hay
Guys, Kings and St Thomas School of Medicine, St John's Institute of Dermatology, Block 7, Basement, KCL, St Thomas's Hospital, London SE1 7EH, UK

Robert A. Heinzen
Dept of Molecular Biology, University of Wyoming, P.O. Box 3944, Laramie WY 82071-3944, USA

I. Henderson
DynPort Vaccine Company, LLC, 60 Thomas Johnson Drive, Frederick, MD 21702, USA

Didier Heumann
Division of Infectious Diseases, BH-19.111, Centre Hospitalier Universitaire Vaudois, CH-1011 Lausanne, Switzerland

Nicola J. High
School of Biological Sciences, Stopford Building, University of Manchester, Oxford Road, Manchester M13 9PT, UK

Jan A. Hobot
Medical Microscopy Sciences, University of Wales College of Medicine, Heath Park, Cardiff CF4 4XN, UK

Keith T. Holland
The Skin Research Centre, Dept of Microbiology, University of Leeds, Leeds LS2 9JT, UK

Lan Hu
Laboratory of Enteric and STDs, FDA Center for Biologic Evaluation and Research, Building 29, Room 420, NIH Campus, Bethesda MD 20892, USA

Scott J. Hultgren
Depts of Molecular Microbiology and Microbial
Pathogenesis, PO Box 8230, Washington University
School of Medicine, 600 S Euclid Avenue, St Louis, MO
63110-1093, USA

Boris Ionin
Dept of Molecular Pathology, Walter Reed Army Institute
of Research, Washington DC 20307-5100, USA

John M. Janda
Microbial Diseases Laboratory, Division of Communicable
Disease Control, California Dept of Health Services, 2151
Berkeley Way – Room 330, Berkeley CA 94704-1011,
USA

Klaus Jann
Max-Planck-Institut für Immunbiologie, Stübweg 51,
Postfach 1169, D-79108 Freiburg-Zähringen, Germany
Present address: Tannenweg 47, D-79183 Waldkirch,
Germany

Barbara Jann
Max-Planck-Institut für Immunbiologie, Stübweg 51,
Postfach 1169, D-79108 Freiburg-Zähringen, Germany
Present address: Tannenweg 47, D-79183 Waldkirch,
Germany

Angela M. Jansen
Division of Infectious Diseases, School of Medicine,
University of Maryland, 10 South Pine Street, Baltimore,
MD 21201-1192, USA

Marti Jett
Dept Molecular Pathology, Walter Reed Army Institute of
Research, 503 Robert Grant Road, Silver Spring, MD
20910, USA

James B. Kaper
Center for Vaccine Development, University of Maryland
School of Medicine, 685 W Baltimore Street, Baltimore,
MD 21201, USA

Alison R. Kerr
Division of Infection and Immunity, Institute of
Biomedical Sciences, Joseph Black Building, University of
Glasgow, Glasgow G12 8QQ, UK

Julian M. Ketley
Dept of Genetics, Adrian Building, University of Leicester,
University Road, Leicester LE1 7RH, UK

Gerald T. Keusch
Associate Director for International Research, Fogarty
International Center, NIH Building 31, Room B2 C02,
Bethesda MA 20892 USA

G.R. Kinghorn
Dept of Genitourinary Medicine, Royal Hallamshire
Hospital, Glossop Road, Sheffield S10 2JF, UK

Dennis J. Kopecko
Laboratory of Enteric and STDs, FDA Center for Biologic
Evaluation and Research, Building29, Room 420, NIH
Campus, Bethesda, MD 20892, USA

Catherine S. Lachenauer
Channing Laboratory, Harvard Medical School, 181
Longwood Avenue, Boston MA 02115-5899, USA

Teresa Lagergård
Dept of Medical Microbiology and Immunology,
University of Göteborg, Guldhedsgatan 10, S-413 46
Göteborg, Sweden

Peter A. Lambert
Depts of Pharmaceutical and Biological Sciences,
University of Aston, Birmingham B4 7ET, UK

Marc Lecuit
Unité des Interactions Bactéries-Cellules, Institut Pasteur,
28 rue de Docteur Roux, Paris 75015, France

Xin Li
Division of infectious Diseases, School of Medicine,
University of Maryland, 10 South Pine Street, Baltimore,
MD 21201-1192, USA

Margaret A. Liu
Bill and Melinda Gates Foundation, PO BOX 23350,
Seattle, WA 98102, USA

Reggie Y. C. Lo
Dept of Microbiology, College of Agricultural Science,
University of Guelph, Guelph, Ontario, Canada

William Lynn
Dept of Infectious Diseases, Level 8, Ealing Hospital,
Uxbridge Road, Southall UB1 3HW, UK

Alastair P. MacMillan
FAO/WHO Collaborating Centre for Brucella Reference
and Research, Dept of Bacterial Diseases, Veterinary
Laboratories Agency, Weybridge, Surrey KT15 3 NB, UK

Angela C. Martin
Department of Biochemistry, University of Oxford, South
Parks Road, Oxford OX1 3QU, UK

Millicent Masters
Institute of Cell and Molecular Biology, University of
Edinburgh, King's Buildings, Mayfield Road, Edinburgh
EH9 3JR, UK

Bruce A. McClane
Dept of Molecular Genetics and Biochemistry, University of Pittsburgh School of Medicine, Pittsburgh PA 15261, USA

Kathleen A. McDonough
Wadsworth Center, New York State Dept of Health and Dept of Biological Sciences, University of Albany, SUNY PO 22002, 120 New Scotland Ave, Albany, NY 12201-2002, USA

Karen F. McGregor
Dept of Microbiology, University of Western Australia, 35 Stirling Highway, Crawley WA 6009, Australia

Brian J. Mee
Dept of Microbiology, University of Western Australia, 35 Stirling Highway, Crawley WA 6009, Australia

Jack Melling
Karl Landsteiner Institute, Rennweg 95B, A1030 Vienna, Austria

Colin Michie
Consultant Paediatrician, Ealing Hospital, Uxbridge Road, Southall UB1 3HW, UK

Michael F. Minnick
Division of Biological Sciences, University of Montana, Missoula MT 59812-1002, USA

Tim J. Mitchell
Division of Infection and Immunity, Institute of Biomedical Sciences, Joseph Black Building, University of Glasgow, Glasgow G12 8QQ, UK

Albert G. Moat
Home: 778 Roslyn Avenue, Glenside, PA 19038-3805, USA
Institutional Affiliation: Emeritus Professor, Dept of Microbiology, Immunology and Molecular Genetics, Marshall University School of Medicine, Huntington, WV, USA

H. L. Mobley
Division of Infectious Diseases School of Medicine, University of Maryland, 10 South Pine St, Baltimore MD 21201-1192, USA

Cesare Montecucco
Centro CNR Biomembrane and, Dipartimento di Scienze Biomediche, Università di Padova, Via G Colombo 3, 35121 Padova, Italy

Sheldon L. Morris
Laboratory of Mycobacterial Diseases, CBER/FDA, Bethesda MD 20852, USA

J. Gareth Morris FRS
Institute of Biological Sciences, University of Wales, Aberystwyth, SY23 3DA, UK

Donald Morrison
Scottish MRSA Reference Laboratory, Dept of Microbiology, Stobhill Hospital, Balornock Road, Glasgow G21 3UW, UK

Inge Muhldorfer
Byk Gulden Pharmaceuticals, Byk-Gulden Straße 2, D-978467 Konstanz, Germany

Matthew A. Mulvey
Depts of Molecular Microbiology and Microbial Pathogenesis, PO Box 8230, Washington University School of Medicine, 600 S Euclid Avenue, St Louis, MO 63110-1093, USA

Francis E. Nano
Department of Biochemistry and Microbiology, University of Victoria, Victoria, British Columbia, Canada

James P. Nataro
Center for Vaccine Development, Depts of Paediatrics and Microbiology and Immunology, University of Maryland School of Medicine, 685 W Baltimore Street, Baltimore MD 21201, USA

Roger Neill
Dept of Molecular Pathology, Walter Reed Army Institute of Research, Washington, DC 20307-5100, USA

Wright W. Nichols
AstraZeneca, Alderley Park, Macclesfield SK10 4TG, UK
Present address: Director of Microbiology, AstraZeneca R&D Boston, 35 Gatehouse Drive, Waltham, MA 02451, USA

David O'Callaghan
INSERM U-431, Faculte de Medicine, Avenue Kennedy, 30900 Nmes, France

Itzhak Ofek
Dept of Human Microbiology, Sackler Faculty of Medicine, University of Tel Aviv, Israel

Iruka N. Okeke
Center for Vaccine Development, University of Maryland School of Medicine, 685 W Baltimore Street, Baltimore, MD 21201, USA

Steven M. Opal
Dept of Medicine, Division of Infectious Diseases, Brown University School of Medicine, Providence, Rhode Island, USA

Roger Parton
Division of Infection and Immunity, Institute of Biomedical and Life Sciences, Joseph Black

Building, University of Glasgow, Glasgow
G12 8QQ, UK

Sheila Patrick
Depts of Microbiology and Immunobiology, School of
Medicine, Queen's University of Belfast, Grosvenor Road,
Belfast BT12 6BN, UK

John H. Pearce
School of Biological Sciences, University of Birmingham,
Birmingham, Birmingham B15 2TT, UK
Present address: 126 Oxford Road, Moseley, Birmingham,
B13 9SH, UK

Charles W. Penn
School of Biological Sciences, University of Birmingham,
Birmingham B15 2TT, UK

T. Hugh Pennington
Dept of Medical Microbiology, Medical School, University
of Aberdeen, Aberdeen AB25 2ZD, UK

Kenneth M. Peterson
Dept of Microbiology, Louisiana State University Health
Sciences Center, 1501 Kings Highway, PO Box 33932,
Shreveport LA 71130-3932, USA

Wolfgang K. Piepersberg
Bergische Universität GH, Faculty of Chemistry, Dept of
Chemical Microbiology, Gauss-Strasse 20, D-42097
Wuppertal, Germany

Carrie A. Poore
Division of Infectious Diseases, School of Medicine,
University of Maryland, 10 South Pine Street, Baltimore,
MD 21201-1192, USA

Rohit S. Prajapati
School of Biological Sciences, Royal Holloway University
of London, Egham, Surrey TW20 0EX, UK

Peter Pujic
Dept of Medical Microbiology and Immunology,
University of California School of Medicine, Davis,
CA 95616, USA

C. P. Quinn
Centre for Applied Microbiology and Research, Porton
Down, Salisbury, UK
Present address: Microbial Pathogenesis and Immune
Response Lab, MSPB, Mailstop D11, Centres for Disease
Control, 1600 Clifton Road NE, Atlanta, GA 30333, USA

Rino Rappuoli
IRIS Chiron SpA, Via Florentina 1, Sienna 53100, Italy

A. Agneta Richter-Dahlfors
Karolinska Institutet, MTC, Box 280, S-171 77
Stockholm, Sweden

Thomas V. Riley
Dept of Microbiology, University of Western Australia and
The Western Australian Centre for Pathology and Medical
Research, Queen Elizabeth II Medical Centre, Nedlands
6009, Western Australia

Mark Roberts
Dept of Veterinary Pathology, Glasgow University
Veterinary School, Bearsden Road, Glasgow
G61 1QH, UK

Roy M. Robins-Browne
Microbiological Research Unit, Royal Children's Hospital,
Flemington Road, Parkville, Victoria 3052, Australia

Julian I. Rood
Bacterial Pathogenesis Research Group, Dept of
Microbiology, Monash University, Clayton 3800, Victoria,
Australia

Ornella Rossetto
Centro CNR Biomembrane and, Dipartimento di Scienze
Biomediche, Università di Padova, Via G Colombo 3,
35121 Padova, Italy

R. R. B. Russell
Dept of Oral Biology, Dental School, Newcastle Upon
Tyne NE2 4BW, UK

James E. Samuel
Dept of Medical Microbiology and Immunology, College
of Medicine, Texas A and M University, College Station,
TX 77843-1114, USA

Chihiro Sasakawa
Institute of Medical Science, University of Tokyo, 4-6-1
Shirokanedai Minato Ku, Tokyo 108-8639, Japan

Giampietro Schiavo
Imperial Cancer Research Fund, 61 Lincoln Inns Fields,
London WC2A 3PX, UK

Carl A. Schnaitman
1902 E Medlock Drive, Phoenix, Arizona 85016-4127,
USA

Ira Schwartz
Dept of Biochemistry and Molecular Biology, New York
Medical College, Valhalla NY 10595, USA

Tricia Ann Scurtz Sebghati
Dept of Microbiology, College of Medicine, University
of Iowa, 3403 Bowen science Building, Iowa City
IA 52242-1109, USA

Nathan Sharon
Dept of Membrane Research and Biophysics, Weizmann Institute of Science , Rehovot, Israel

Richard W.P. Smith
Institut für Molekulare Biotechnologie e.V., Beutenbergstrasse 11, D-07745 Jena, Germany

Noel H. Smith
School of Biological Sciences, University of Sussex, Brighton, BN1 9QG, UK

Rebecca J. Smith
Dept of Microbiology and Immunology, Leicester University, P.O. Box 138, Medical Sciences Building University Road, Leicester LE1 9HN, UK

Valerie A. Snewin
Scientific Programme Officer The Wellcome Trust, The Wellcome Building, 183 Euston Road, London NW1 2BE, UK

Brian G. Spratt
Dept of Infectious Disease Epidemiology, Imperial College School of Medicine, St Mary's Hospital, Norfolk Place, London, W2 1PG, UK

Lola V. Stamm
Program in Infectious Diseases, Dept of Epidemiology, 242 Rosenau Hall, University of North Carolina at Chapel Hill, Chapel Hill NC 27599-7400, USA

Paul M. Sullam
Dept of Infectious Disease, University of California at San Francisco, California, USA

Johnathan D. Sussman
Greater Manchester Neuroscience Centre, Hope Hospital, Eccles Road, Salford, M6 8HD, UK

Max Sussman
Dept of Microbiology, The Medical School, University of Newcastle Upon Tyne, Newcastle Upon Tyne NE2 4HH, UK

Richard W. Titball
Defence Evaluation and Research Agency, CBD Porton Down, Salisbury, Wiltshire SP4 0JG, UK

Kevin J. Towner
Dept of Microbiology and PHLS Laboratory, University Hospital, Queen's Medical Centre, Nottingham NG7 2UH, UK

Peter C. B. Turnbull
Arjemptur Technology Ltd , Porton Down, Science Park, Salisbury SP4 0JQ UK, Contact address: 86 St Mark's Avenue Salisbury, Wiltshire SP1 3DW, UK

Qinning Wang
Dept of Microbiology, University of Western Australia, 35 Stirling Highway, Crawley WA 6009, Australia

Guiqing Wang
Dept of Biochemistry and Molecular Biology, New York Medical College, Valhalla NY 10595, USA

Michael R. Wessels
Channing Laboratory, 181 Longwood Avenue, Boston MA 02115-5899, USA

Brian M. Wilkins
Dept of Genetics, University of Leicester, Leicester LE1 7HN, UK

Anil Wipat
Dept of Microbiology, The Medical School University of Newcastle Upon Tyne, Framlington Place, Newcastle Upon Tyne NE2 4HH, UK

Martin J. Woodward
Dept of Bacterial Diseases, Veterinary Laboratories Agency (Weybridge), Addlestone, Surrey KT15 3NB, UK

Gary P. Wormser
Chief Division of Infectious Disease, New York Medical College, Valhalla NY 10595, USA

Brendan W. Wren
Dept of Infectious and Tropical Diseases, London School of Hygiene and Tropical Medicine, Keppell Street, London WC1E 7HT, UK

Kwok Yung Yuen
Dept of Microbiology, Queen Mary's Hospital, Faculty of Medicine, Hong Kong University, Pokfulam, Hong Kong

Wilma Ziebuhr
Instutut für Molekulare Infektions Biologie, Universität Würzburg, Röntgenring 11 D-97070, Würzburg, Germany

Amy B. Zuppardo
Bureau of Laboratories, Jacksonville, FL, USA
Present address: Dept of Health and Hospitals Office of Public Health, Room 709, 325 Loyola Avenue, New Orleans LA 70112, USA

Contents

Colour plate sections are located as follows

Colour Plates

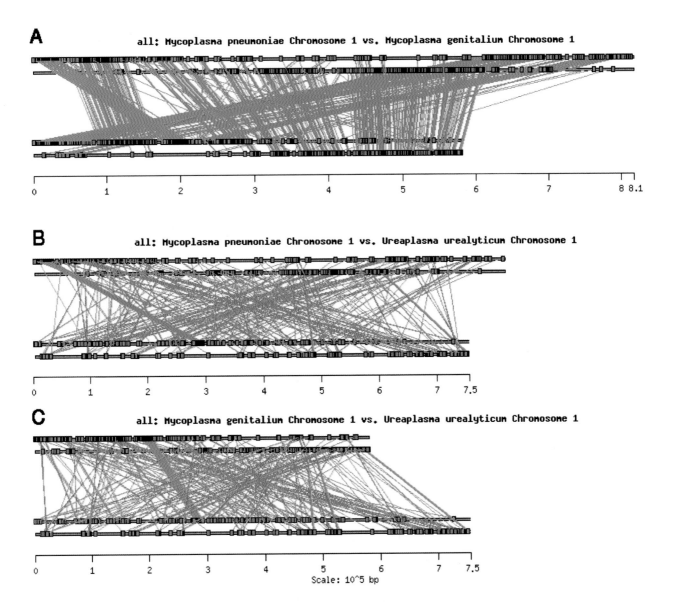

Plate 17 Genome comparison. The genome sequences were aligned with the STD sequence database and the tools genome alignments according to http://www.stdgen.lanl.gov. (a) *Mycoplasma pneumoniae* and *M. genitalium*. (b) *M. pneumoniae* and *U. urealyticum*. (c) *M. genitalium* and *U. urealyticum* genomes are linearised and aligned with the program Bugspray for graphical representation. A green line represents a pair of genes which are found on strands of the same sense, and a red line represents a pair of genes which are found on strands of the opposite sense.

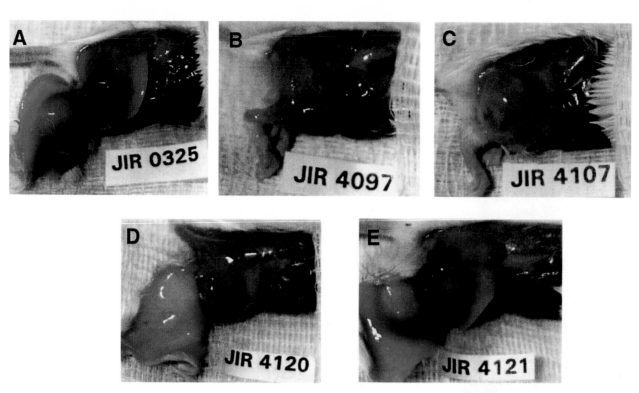

Plate 18 Gross pathology of mice infected with *C. perfringens* strains. The plates show the dissected right hind legs of mice infected with the wild-type strain (a), two independently derived *plc* mutants (b, c), a *plc* mutant carrying a control vector plasmid (d), and the same mutant complemented with a recombinant plasmid carrying a wild-type *plc* gene (e). Note the extensive haemorrhage and necrosis caused by the strains that produce α toxin (a and e), but not by the strains defective in α toxin production (b–d). Reproduced from Awad *et al.* (1995) with the permission of the publishers of *Molecular Microbiology*.

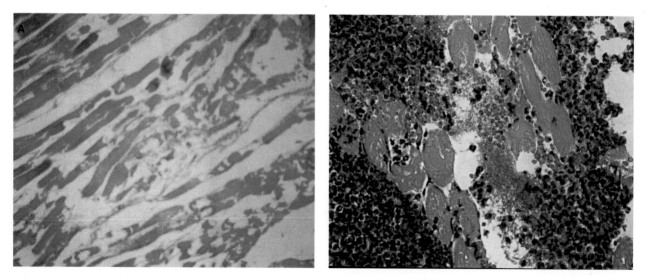

Plate 19 Histopathology of infected muscle tissues. Haemotoxylin and eosin-stained sections of murine tissues taken after infection with A (the wild-type *plc*⁺ strain, JIR325) and B (the isogenic *plc* mutant, JIR4107). The marked influx of polymorphonuclear leukocytes, which is apparent in tissues taken from mice infected with JIR4107, was not apparent in tissues taken from mice infected with the wild-type strain.

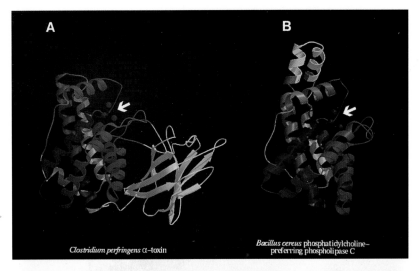

A

B

Clostridium perfringens α-toxin

Bacillus cereus phosphatidylcholine–preferring phospholipase C

Plate 20 (left) Comparison of the crystal structure of *C. perfringens* α toxin (a) with the crystal structure of *B. cereus* PC-*PLC* (b). The traces in regions of structural similarity are colour-coded from **blue** (the N-termini) to **red**. Regions of dissimilarity are coloured in **grey**. The active-site clefts, containing active site zinc ions, are arrowed.

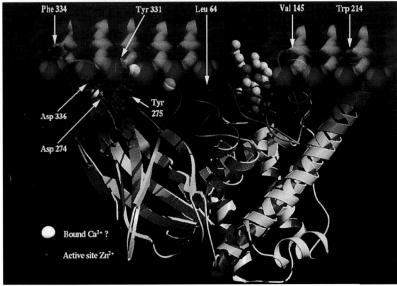

Phe 334   Tyr 331   Leu 64   Val 145   Trp 214

Tyr 275

Asp 336

Asp 274

● Bound Ca²⁺ ?

⦁ Active site Zn²⁺

Plate 21 (left) Model for the interaction of *C. perfringens* α toxin with eukaryotic cell membranes. Key residues which play a role membrane interaction are highlighted. Bound calcium ions are shown in white. A phospholipid molecule is shown partially retracted from the cell membrane into the active site cleft, which also contains zinc ions (**shaded blue**). Reproduced from Naylor *et al.* (1998) with the permission of *Nature Structural Biology*.

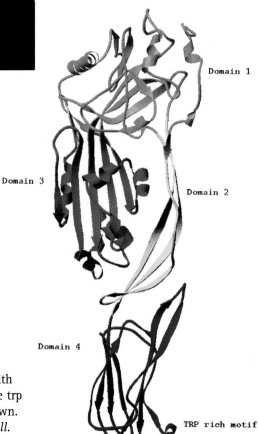

Domain 1

Domain 3

Domain 2

Domain 4

TRP rich motif

Plate 22 (right) Crystal structure of the *C. perfringens* θ toxin, with each domain represented by a different colour. The location of the trp motif, which is conserved in all thiol-activated toxins, is also shown. Reproduced from Rossjohn *et al.* (1997) with the permission of *Cell*.

(a)

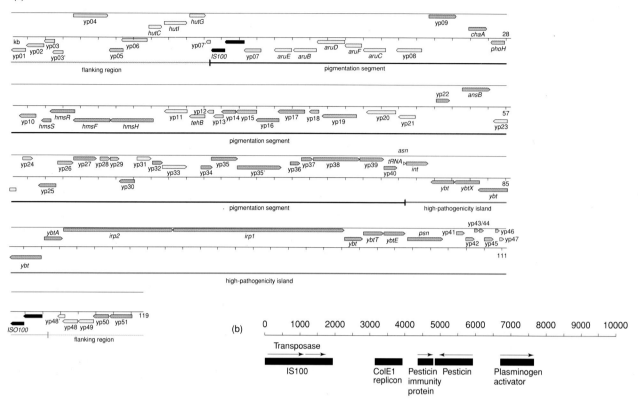

(b)

Plate 23 (a) Map of the *pgm* locus and flanking regions showing the position and orientation of known genes and putative coding sequences: functional categories related to virulence (red), phage-related functions (light blue), transport proteins (dark blue), regulatory proteins (purple), nitrogen and carbon metabolism (brown), miscellaneous (yellow), insertion elements (black), and proteins of unknown function (green). The thick lines below the gene map indicate flanking regions (grey), pigmentation segment (black) and high-pathogenicity island (blue). Numbers at the right end of the map indicate the scale in kilobases. From Buchrieser *et al.* (1999). (b) Structural organisation of the 9610-bp plasmid pPCP1 derived from *Y. pestis* KIM10. The directions of transcription are indicated by the arrows. The single IS*100* element was used to define position 1 of this plasmid. The numbering above the line is the molecular size in base pairs. From Hu *et al.* (1999).

Plate 24 Diagram comparing the organisation of selected genes and elements of pCD1 of *Y. pestis* KIM10 and pYV of *Y. pseudotuberculosis* (pIBI) and *Y. enterocolitica* (pYVe). The relative positions of selected loci with respect to the order of replication of pCD1 are shown: outer circle, pCD1; middle circle, pIBI; inner circle, pYVe. The genes and sequence features of pCD1 and the corresponding regions in pIBI and pYVe are depicted in the same colour to aid in their visualisation. Numbering inside the circles indicates the approximate sizes of the plasmids in nucleotides, measured from the start of their origins of replication. Arrows above each shaded segment representing a gene or gene group point to the direction of transcription. From Hu *et al.* (1999).

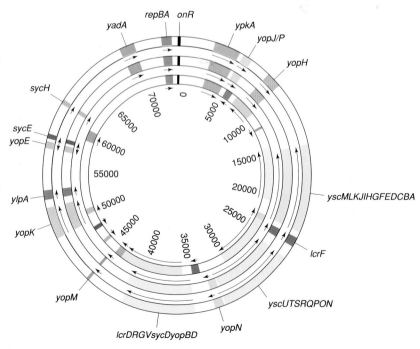

PART
12
RESPIRATORY INFECTIONS

74

Respiratory Tract Infections: A Clinical Overview

Stephen J. Bourke

University of Newcastle upon Tyne, UK

The essential function of the lungs is the exchange of oxygen and carbon dioxide between the blood and the external atmosphere. The respiratory system is therefore designed as an interface between the internal and external environment, such that air from the atmosphere and blood from the circulation are brought into close contact across the alveolar capillary membrane. About eight litres of air are drawn into the lungs each minute and this air is laden with dusts, antigens, toxins, pollutants and pathogenic organisms. It is a remarkable testament to the pulmonary defence systems that in a normal healthy person the lower respiratory tract, below the larynx, is sterile (Cabello *et al.*, 1997).

Respiratory tract infections represent a complex dynamic interplay between pathogen and host (Stockley, 1998). Each of the major respiratory pathogens has evolved its own life cycle, mode of transmission and mechanisms for surviving within the hostile host environment. Conversely the host immune system can mount a carefully regulated response against the invading pathogen. The outcome of the pathogen–host interaction in the respiratory tract depends on the virulence of the pathogen and the vulnerability of the host – virulence versus vulnerability. The dynamic interplay of these factors, and the circumstances of infection, result in a diverse spectrum of clinical disease. In some circumstances the host response successfully eradicates the infection and provides the patient with increased immunity against further infection, as in the case of an acute self-limiting illness such as influenza A virus infection. Conversely, infection with a virulent organism, such as *Streptococcus pneumoniae*, may overwhelm the host defences resulting in acute fulminating pneumonia, even in a previously healthy individual. Frequently the interaction of host and pathogen is inconclusive and a persistent cycle of chronic infection and inflammation ensues resulting in ongoing tissue damage, as in the case of *Pseudomonas aeruginosa* infection in patients with cystic fibrosis.

In some situations infection may be regarded as a primary, exogenous event in which inhalation of a large dose of a virulent pathogen produces a severe infection in a previously healthy person. Thus *Legionella pneumophila* may be inhaled from a contaminated water system (Fraser *et al.*, 1977), or *Chlamydia psittaci* from an infected bird (Bourke *et al.*, 1989), resulting in a severe pneumonia. In these circumstances the key issues are the source of the infection in the environment, the dose inhaled and the virulence of the organism. The patient is in many ways an innocent bystander who has fallen victim to an exogenous pathogen encountered in the environment. In other circumstances infection is a secondary, endogenous event that complicates an underlying disease process. Thus, in patients with smoking-induced chronic bronchitis an organism of low virulence, such as non-capsulated *Streptococcus pneumoniae* or *Haemophilus influenzae*, may chronically colonise the lower respiratory tract, and exacerbate the condition. In these circumstances distortion of the lung architecture and disruption of the mucociliary mechanism allow the organism to gain access to the lower respiratory tract. Infection of the lower respiratory tract may be considered as a process involving several steps, including exposure to the pathogen, inhalation or aspiration of the organism into the lungs, adherence to the respiratory epithelium, failure of clearance, invasion of tissues and initiation of an inflammatory and immune response (Wilson *et al.*, 1996).

Defences of the Respiratory Tract

The respiratory tract has a complex and diverse defence system ranging from physical removal of inhaled particles by coughing and sneezing, to specific local and systemic immune responses to invading pathogens.

Mucociliary Clearance

The upper airway acts as a filter that removes most particles from the inspired air. Particles more than 10 μm in diameter are usually filtered out of the inhaled air stream in the nose, while particles 1–10 μm are mainly deposited in the bronchi and particles less than 1 μm may reach the alveoli. The mucociliary escalator is the main mechanism for clearing particles from the airways (Greenstone and Cole, 1985). The epithelial cells of the bronchi possess cilia which beat in an organised way so as to move particles deposited in the layer of mucus on their surface upwards and out of the lung. The mucus blanket which lies on the cilia consists of a periciliary sol layer of liquid on which a gel layer of viscous mucus floats. Impaired mucociliary clearance may be due to changes in the rheological properties of the mucus, abnormal ciliary function or impaired interaction between mucus and cilia.

Primary ciliary dyskinesia is a relatively rare inherited condition in which abnormalities of ciliary structure and function give rise to chronic infections of the upper and lower respiratory tract such as otitis, sinusitis and progressive bronchiectasis (Bush *et al.*, 1998). In 50% of cases it is associated with dextrocardia and situs inversus (Kartagener's syndrome) (**Fig. 1**). A variety of

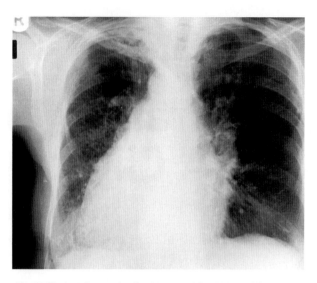

Fig. 1 Chest radiograph of a 78-year-old woman with primary ciliary dyskinesia and dextrocardia (Kartagener's syndrome), which has resulted in bibasal bronchiectasis, progressive lung fibrosis and hypoxia with pulmonary hypertension and cardiac failure.

ciliary abnormalities may occur, including defects in the dynein arms, tubules, radial spokes, ciliary orientation and beat frequency. Transient secondary ciliary defects are common during viral respiratory infections and may facilitate bacterial infection of the lower respiratory tract. Many chronic lung diseases, such as chronic bronchitis or bronchiectasis, are associated with permanent disruption of the mucociliary escalator (De Iongh and Rutland, 1995). This allows bacteria to adhere to the respiratory epithelium and to colonise the lungs. The presence of such bacteria at a normally sterile site stimulates an acute inflammatory response. Failure to clear infection gives rise to a vicious circle of infection and inflammation, which causes further damage to the epithelium, predisposing to further infection and a self-perpetuating cycle of progressive lung damage.

Many of the bacterial pathogens of the respiratory tract have the ability to produce substances that interfere with host defences and facilitate bacterial adherence to the epithelium. *Pseudomonas aeruginosa*, for example, produces pyocyanin and hydroxyphenazine which reduce mucociliary clearance. *H. influenzae* produces various virulence factors which induce ciliary dyskinesia, breakdown of IgA, release of histamine and stimulation of mucus secretion (Wilson *et al.*, 1996). Some of the mechanisms by which bacteria adhere to the respiratory epithelium have been elucidated; they often involve specific interactions between adhesive structures on the bacterial surface and receptors on the mucosal surface. After injury the airway epithelium undergoes a process of repair which involves the spreading, migration and proliferation of epithelial cells. During this process epithelial cells synthesise fibronectin which is required for cell migration and $\alpha5\beta1$ integrin which is required for cell–cell adhesion. These fibronectin and integrin epithelial receptors are used by *Ps. aeruginosa* outer-membrane protein as sites of bacterial adherence (Roger *et al.*, 1999). Thus key elements in the repair process of epithelium are also major receptors for bacterial adherence.

Airway Antimicrobial Peptides

Increasing interest is being paid in the protective effect of antimicrobial peptides present in lung secretions. These include lactoferrin, transferrin, lysozyme, defensins and cathelicidins. Many bacteria, including *H. influenzae*, require co-factors such as iron for proliferation. Lactoferrin and transferrin bind iron and exclude it from the bacteria and thereby impair their proliferation. Lysozyme is capable of attacking bacterial peptidoglycan, thereby disrupting the cell wall and causing bacterial death. Defensins and cathelicidins are epithelially derived peptides with antimicrobial activity against a number of bacterial and fungal pathogens (Goldman *et al.*, 1997). For example, introduction of *Ps. aeruginosa* into the airway results in substantial up-regulation and induction of defensin expression throughout the epithelium of the airway (Bals *et al.*, 1999). The human β defensins, hβD-1 and hβD-2, are salt-sensitive and it is thought that this may be particularly important in patients with cystic fibrosis, since the increased salt concentration of airway mucus in this disease inactivates these peptides thereby contributing to the vulnerability of these patients to infection with organisms such as *Ps. aeruginosa*.

The Immune-Inflammatory Response

Once the initial local lung defences have been breached by an invading pathogen, the body mounts a systemic immune-mediated inflammatory response that involves a complex interplay of humoral and cellular responses (Haslett, 1995). The initial response to bacterial infections involves the activation of alveolar macrophages and other cells resident in the lungs. Inflammatory mediators released from these cells and from the bacteria result in a systemic inflammatory response characterised by fever, increased cardiac output, increased capillary permeability and endothelial cell swelling. The release of cytokines and

chemotaxins, such as tumour necrosis factor (TNFα) and interleukins, such as IL-8, result in the activation and recruitment of other cells. About two hours after an inflammatory stimulus to the lung, neutrophil leucocytes migrate into the alveoli through the endothelium and alveolar epithelium, and monocytes appear about six hours later. Adhesion molecules, such as intercellular and vascular adhesion molecules (ICAM, VCAM), are expressed on the surfaces of endothelial cells, and other adhesion molecules, such as selectins and integrins, are expressed on the surfaces of neutrophils. Neutrophils become sequestered at the site of infection by adhering to the endothelium, rolling along its surface, changing their shape and migrating through junctions into the air spaces (Drost *et al.*, 1999). Leakage of fluid and proteins, including complement, coagulation factors and immunoglobulins, takes place into the alveoli. Neutrophils then begin to ingest and destroy bacteria. The subsequent monocyte influx seems to be important in modifying the later phases of inflammation by amplification or regulation of the response.

Antibody responses to bacterial antigens can be detected in the serum, saliva and pulmonary secretions. Immunoglobulin A (IgA) is the major immunoglobulin of the healthy respiratory tract and it can combine directly with and neutralise pathogenic micro-organisms. The main immunoglobulin produced during the primary immune response is IgM, and IgG is produced during the secondary immune response. In addition to activating complement and the inflammatory cascade of cytokines, immunoglobulins bind to bacterial antigens and this facilitates their phagocytosis. Many respiratory pathogens release enzymes that cleave host immunoglobulins. Where there are major deficits in the host systemic immune response, severe chronic infections, such as severe bronchiectasis in patients with hypogammaglobulinaemia, develop early in life. Deficiency of IgG2 and IgG4 are associated with recurrent respiratory tract infections.

The effectiveness of the host response to infection is exemplified by the fact that, even in the pre-antibiotic era, the majority of patients with pneumococcal pneumonia survived the illness. The disease evolves through four classical stages, characterised by congestion, red hepatisation, grey hepatisation and resolution, which represent the gross appearances of the lung during the various stages of the inflammatory response. The beneficial effects of a carefully regulated inflammatory response are seen as the acute lung infection is eradicated and normal lung architecture is restored. The phase of resolution occurs with successful removal of the inciting stimulus, dissipation of the inflammatory mediators, cessation of neutrophil migration and restoration of normal vascular permeability. This is in contrast to the situation in cystic fibrosis, bronchiectasis or severe chronic obstructive pulmonary disease (COPD). In these diseases the inflammatory response is unable to eradicate the infection and the phase of resolution fails to occur. A persistent uncontrolled inflammatory response becomes deleterious, resulting in tissue damage, which itself disrupts the epithelium and the mucociliary escalator, and encourages the adherence of bacteria to the respiratory mucosa provoking further inflammation and progressive lung damage. Many diseases of the lung, apart from those due to infection, are characterised by damaging inflammation. These, for example, include the acute respiratory distress syndrome (ARDS) in patients in intensive care units, and interstitial lung diseases such as cryptogenic fibrosing alveolitis or extrinsic allergic alveolitis. Once the degree of injury to the epithelium and its basement membrane reach a certain level, excessive scarring occurs. In these circumstances anti-inflammatory therapies, such as with corticosteroids or non-steroidal anti-inflammatory agents, become key elements in treatment. Such anti-inflammatory treatments are also under consideration in an attempt to decrease the deleterious systemic and lung inflammatory responses in mechanically ventilated patients who are receiving antibiotics for severe pneumonia (Monton *et al.*, 1999).

The Clinical Approach to Respiratory Infections

Identification of the micro-organisms responsible for respiratory infections began towards the end of the nineteenth century when bacteria such as *Klebsiella pneumoniae*, *Strep. pneumoniae* and *H. influenzae* were discovered. The list of causative pathogens has increased continually over the last century with the identification, for example, of *Mycoplasma pneumoniae* in 1938 (Reimann, 1938), *Legionella pneumophila* in 1977 (Fraser *et al.*, 1977), and more recently *Chlamydia pneumoniae* (Grayston *et al.*, 1986) and hantaviruses (Duchin *et al.*, 1994). The microbiological approach to respiratory infections focuses particularly on the identification of the pathogen, its characteristics and its susceptibility to antimicrobial therapy. Many of the major respiratory pathogens may, however, also colonise the respiratory tract so that the identification of an organism in respiratory tract secretions may not be sufficient to implicate it as the cause of the

illness. The same microbe may cause various illnesses, such as otitis or bronchitis, and different infecting agents may cause identical clinical syndromes, such as pneumonia.

The clinical manifestations of a respiratory infection often depend more on the site in the respiratory tract involved, the characteristics of the patient and the circumstances of infection than on the nature of the particular pathogen. The term 'chest infection' is an imprecise term often used by patients to refer to nonspecific respiratory symptoms which may or may not be due to infection. The initial clinical approach to respiratory tract infections aims to define as precisely as possible the site, circumstances and severity of infection in an individual patient against the background of the known epidemiology and microbiology of the disease. Many upper respiratory tract infections are self-limiting and may not require specific therapy. Lower respiratory tract infections may be confined to the bronchi as in acute bronchitis or exacerbations of chronic bronchitis, or they may spread to the lung parenchyma (pneumonia), the pleural space (empyema) or bloodstream (septicaemia). The same pathogen may give rise to infections at each of these sites, but the clinical disease process would differ greatly. The circumstances of the illness include the patient's age, the presence of concomitant disease, the severity of the illness, the potential sources of infection in the environment and contact with other infected patients. The severity of the illness depends not only on the virulence of the pathogen but also on the vulnerability of the patient. Recognition of the severity of the illness is crucial in deciding on the choice of antibiotic therapy, the need for supportive care, and whether the patient should be treated in hospital rather than at home or in an intensive care unit rather than on a general ward (see **Fig. 2**).

Diagnosis of the site and severity of respiratory tract infections is notoriously difficult and in an evolving illness careful assessment and review are important. The clinical assessment recognises the complexity of the interactions between the patient, the environment and the pathogen and attempts to move away from an imprecise term such as 'chest infection' to a descriptive statement such as 'severe community-acquired pneumonia in a previously fit adult' or 'probable bronchopneumonia and respiratory failure (hypoxia) in a frail elderly man with COPD'. In both circumstances sputum cultures may yield penicillin-sensitive *Strep. pneumoniae*, but the clinical and microbiological diagnoses are complementary and must be integrated to achieve the best treatment of the patient. Each of the clinical syndromes has its own clinical features, epidemiology and microbiology.

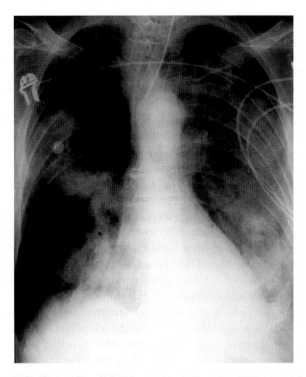

Table 1 A typical spectrum of pathogens implicated in community-acquired pneumonia

Streptococcus pneumoniae	60–70%
Mycoplasma pneumoniae	10%
Haemophilus influenzae	5–10%
Viruses (e.g. influenza)	5–10%
Staphylococcus aureus	3%
Legionella pneumophila	2–5%
Others (e.g. *Chlamydia pneumoniae*)	5%
Gram-negative bacilli	Rare
Anaerobic bacilli	Rare

Fig. 2 Chest radiograph of a previously fit and well 70-year-old man, who developed severe pneumococcal pneumonia with septicaemia, and died despite treatment with antibiotics, mechanical ventilation and full supportive intensive care.

Pneumonia

The term 'pneumonia' is usually used to denote inflammation of the lung parenchyma caused by infection, and the term 'pneumonitis' is best used to denote inflammation due to physical, chemical or allergic stimuli. Pneumonia is an important cause of morbidity and mortality in all age groups. Globally it is estimated that 5 million children die from pneumonia each year, 95% of them in developing countries. In the UK each year about 1 in 1000 of the population are admitted to hospital with pneumonia, and the mortality in these patients is about 10% (Winter, 1993). In the age group 15 to 55 years there are about 3000 deaths from pneumonia each year. About 25% of all deaths in the elderly are related to pneumonia, but this is often the terminal illness in a patient with serious concomitant disease.

The initial clinical approach is to classify pneumonia as community-acquired, hospital-acquired (nosocomial), or pneumonia in the immunocompromised patient.

Community-acquired Pneumonia

Pneumonia acquired in the community may affect a previously healthy individual or it may occur in association with concomitant disease, but a few pathogens – notably *Strep. pneumoniae* – account for the majority of cases (**Table 1**), and Gram-negative or anaerobic organisms are rare (MacFarlane *et al.*, 1982; Winter, 1993).

The spectrum of pathogens implicated varies from study to study and depends on where and when the study was carried out and on the microbiological tests undertaken. In the UK, legionella pneumonia is more common in the summer months, related to foreign travel, and mycoplasma pneumonia tends to occur in epidemics every 3–4 years. In routine clinical practice no pathogen can be identified in at least 50% of patients, but research studies with sensitive tests to detect pneumococcal antigen suggest that *Strep. pneumoniae* accounts for many cases where no pathogen is found (McFarlane *et al.*, 1982). However, many of these studies were carried out before *Chl. pneumoniae* was recognised as a respiratory pathogen (Grayston *et al.*, 1986) and debate continues as to the exact prevalence and role of this organism in pneumonia (Bourke and Lightfoot, 1995). As with *Strep. pneumoniae*, asymptomatic infection and a chronic carriage state have been described for *Chl. Pneumoniae*, so that it may be difficult to define the precise relevance of an identified organism to the patient's illness.

Pneumonia is not a static disease entity, but is constantly changing so that concepts must continually be modified. Not only have new pathogens emerged and antibiotic resistance increased, but the patient characteristics have also changed. Populations are ageing, and increased age is associated with a substantially increased incidence of pneumonia, since factors that favour pathogen aspiration into the lung, such as neurological disease, are more common. Developments in many areas of medicine have resulted in improved survival for patients with chronic complex diseases, and immunosuppressive therapies are more widely used in the fields of

Table 2 Clinical and laboratory features of community-acquired severe pneumonia associated with an increased risk of death

Clinical	Laboratory
Respiratory rate $\geq$ 30/min	Serum urea $\geq$ 7 mmol/L
Diastolic blood pressure < 60 mmHg	Serum albumin < 35 g/L
Age $\geq$ 60 years	Hypoxaemia: $Po_2 \leq$ 8 kPa (60 mmHg)
Underlying disease	White cell count < 4000 $\times$ 10^9/L
Confusion	White cell count > 20 000 $\times$ 10^9/L
Atrial fibrillation Multilobar involvement	Bacteraemia

autoimmune diseases, oncology and organ transplantation. This has led to an increase in the number of patients who in various ways are particularly vulnerable to pneumonia.

Certain clinical and laboratory features (**Table 2**) are associated with an increased mortality or the need for admission to intensive therapy units (Woodhead, 1992; Winter, 1993). Initial treatment of community-acquired pneumonia is based on an assessment of the likely pathogen and the severity and circumstances of the illness. Dual antibiotic therapy, such as with amoxycillin and erythromycin, is commonly used, or cefuroxime and erythromycin in patients with more severe pneumonia.

Hospital-acquired Pneumonia

Hospital-acquired pneumonia is defined as pneumonia developing two or more days after admission to hospital for some other illness (American Thoracic Society, 1995). It is, therefore, a secondary infection in patients compromised by other diseases. Ventilator-associated pneumonia is a particularly important form of hospital-acquired pneumonia. Hospital-acquired pneumonia affects about 5–10 per 1000 hospital admissions with an incidence 20-fold higher in patients who are mechanically ventilated. It is the second most common nosocomial infection, after urinary tract infection, but it has the highest mortality and morbidity.

Although the term 'hospital-acquired pneumonia' often conjures up the idea that infection is exogenous and acquired by aerosol inhalation from the hospital environment or by cross-infection, this is not usually the case (Court and Garrard, 1992). In fact, hospital-acquired pneumonia is mainly a secondary endogenous infection arising from the commensal flora of the

patient as the result of debilitating illness and its treatment which facilitate the colonisation of the oropharynx and stomach by Gram-negative enteric bacteria and predispose to micro-aspiration of oropharyngeal secretions into the lungs. Oropharyngeal colonisation by Gram-negative bacilli is rare (<10%) and of short duration in healthy non-hospitalised individuals; but it increases as patients develop more severe systemic illness, so that about 35% of patients with moderately severe illness and 73% with critical illnesses have oropharyngeal colonisation with Gram-negative bacilli (Johanson *et al.*, 1969). The colonisation rates of healthcare workers in the hospital environment is low, which suggests that patient characteristics are more important than environmental factors. These Gram-negative bacilli seem to be acquired from endogenous sources including the upper gastrointestinal tract, subgingival dental plaque and the periodontal crevices. Severe illness of many different aetiologies is associated with impairment of host defences, and this facilitates the acquisition and adherence of Gram-negative bacilli in the oropharynx. Conditions that predispose to this oropharyngeal colonisation include poor oral hygiene, coma, malnutrition, debility of chronic disease, hypotension, diabetes, alcohol misuse and uraemia. Several medical interventions and procedures also predispose to Gram-negative bacterial oropharyngeal colonisation or micro-aspiration of secretions. These include administration of sedatives, anaesthesia, corticosteroids, broad-spectrum antibiotics, cytotoxic drugs, prolonged surgery, especially thoraco-abdominal procedures, and use of nasogastric tubes. H_2 antagonist drugs used as prophylaxis against stress gastritis may also increase gastric colonisation by Gram-negative bacilli. Patients undergoing mechanical ventilation in intensive therapy units are at particular risk for developing pneumonia since they are critically ill from their underlying illness and because endotracheal tubes bypass the upper airway defences and impair coughing and mucociliary clearance. Endotracheal tubes and the ventilator circuit are often colonised by these organisms as a result of contamination of the equipment by the patients, and spillage of secretions from the tubing into the lungs may pose a hazard.

Since colonisation of the oropharynx by Gram-negative bacilli and micro-aspiration of oropharyngeal secretions into the lungs are the two key stages in the pathogenesis of hospital-acquired pneumonia, it is these organisms that account for most cases of pneumonia in this setting (**Table 3**). On occasion infection may also reach the lung by inhalation of infected aerosols, by gross aspiration of gastric contents, by haematogenous spread from distant sites of infection,

Table 3 A typical spectrum of pathogens implicated in hospital-acquired pneumonia

Enterobacteriaceae	37%
Pseudomonas aeruginosa	17%
Staphylococcus aureus	13%
Fungi	6%
Anaerobic bacteria	2%
Streptococcus pneumoniae	3%
Haemophilus influenzae	3%
Viruses	3%
Others (e.g. *Legionella pneumophila*)	16%

as for example *Staph. aureus* from venous cannulae, or by translocation from the gastrointestinal tract. Infection in these circumstances is often polymicrobial in nature.

Antibiotic therapy must take into account the vulnerability of the critically ill patient and the spectrum of likely pathogens. It frequently consists of a combination of a third-generation cephalosporin (e.g. ceftazidime) and an aminoglycoside (e.g. gentamicin). Various preventative strategies may reduce the risk of hospital or ventilator-acquired pneumonia (Kollef, 1999). These include a formal infection-control programme (e.g. hand washing) to reduce cross-infection, nursing patients semi-upright, maintaining adequate pressures in the cuff of the endotracheal tube, and frequent suction of secretions from the subglottic space to reduce microaspiration, and avoidance of H_2-blocking drugs and broad-spectrum antibiotics, where possible, to reduce oropharyngeal colonisation. Selective decontamination of the oropharynx and upper gastrointestinal tract with topically applied antibiotics has been tried but its effectiveness in reducing mortality from ventilator-associated pneumonia is in doubt. Sucralfate reduces the incidence of stress gastritis without lowering gastric acidity. Maintenance of adequate nutritional status by enteral rather than parenteral feeding may also be important in reducing the risk of nosocomial pneumonia.

Upper Respiratory Tract Infections

Acute upper respiratory infections are a common cause of morbidity, visits to doctors and absence from school or work. They are the commonest respiratory illness and account for about 9% of all consultations in general practice. Children suffer about eight and adults about four respiratory infections each year. An upper respiratory tract infection is a major predisposing factor for secondary bacterial invasion of the lower respiratory tract. Many viruses can produce the same clinical syndrome such as coryza, pharyngitis or otitis for example. They are highly infectious and spread rapidly through close-contact units such as schools, families, workplaces or nursing homes. The multiplicity of strains of rhinovirus, for example, ensures that the state of protection of a community is never sufficient to prevent recurring outbreaks. Although unpleasant, most upper respiratory tract infections are mild and self-limiting, but some may give rise to serious problems, most notably acute epiglottitis in children and influenza A in the elderly, debilitated by chronic underlying disease.

The pattern of illness depends upon the level of immunity in the patient and in the population. In children respiratory syncytial virus (RSV) infection is very common and may cause serious disease, such as bronchiolitis or pneumonia, whereas subsequent infections cause less severe illness usually amounting only to a mild 'cold' in adults. Minor changes in the antigen profile of the influenza A virus, referred to as 'antigenic drift', are sufficient to result in outbreaks of influenza in the winter months each year, whereas major changes, referred to as 'antigenic shift', result in epidemics and pandemics of infection reflecting the lack of immunity in the population to the new strain. In the developed world widespread immunisation with diphtheria, tetanus, pertussis (DTP) and measles, mumps, rubella (MMR) vaccines has revolutionised many aspects of respiratory disease. Diphtheria has virtually been eliminated from the UK, except for a very small number of cases contracted during travel abroad. Similarly, the morbidity and mortality of pertussis in childhood have declined markedly but the level of immunity induced by the vaccine tends to wane after about ten years and outbreaks of pertussis have been reported in young adults who tend to suffer a relatively mild illness characterised by a prolonged and persistent cough (Yaari *et al.*, 1999).

Exposure to respiratory pathogens early in life influences the way in which the immune system develops and responds to antigens. This may be particularly relevant to the pathogenesis of asthma and atopy (Lewis, 1998). T-helper lymphocytes have an important role in the regulation of the inflammatory response. These cells may be divided into two main subsets according to the cytokines they produce. TH2 cells produce interleukin-4 (IL-4), IL-5, IL-6 and IL-10 and up-regulate the specific form of airway inflammation of asthma by enhancing IgE synthesis and eosinophil and mast cell function. In contrast, TH1 cells produce IL-2, interferon β (IFNβ) and lymphotoxin and down-regulate the atopic response. The influence of infection on the pattern of T-helper cell response is complex and incompletely understood and appears to depend on the type and timing of infection. Some infections early in childhood seem to enhance the development of a predominantly TH1 phenotype, which is the natural cell-mediated response to infectious agents, whereas exposure to allergens or some other infections favours a persistent TH2 phenotype, which is associated with allergic responses. A number of factors have led to a reduction in certain childhood infections, including improved sanitation, reduction in family size, introduction of immunisation programmes and increased use of antibiotics. The rising prevalence of asthma has been linked to aspects of a 'westernised, developed-world' lifestyle and to the decline in childhood infections, such as tuberculosis, measles and hepatitis A. Studies in West Africa have shown that measles infection in childhood is associated with a reduced level of atopy (Shaheen *et al.*, 1996) and that tuberculosis down-regulates serum IgE (Adams *et al.*, 1999). Conversely, it appears the RSV infection in infants is associated with a TH2 response, which can enhance allergic airway hyper-responsiveness (Roman *et al.*, 1997). The relationship between infection and asthma is complex and paradoxical, and may be subject to several confounding variables, so that its significance is at present uncertain.

Chronic Obstructive Pulmonary Disease

Chronic obstructive pulmonary disease (COPD) is a form of structural lung damage due to tobacco smoking and is characterised by a combination of chronic bronchitis, airways obstruction and emphysema. Chronic bronchitis is characterised by cough and sputum production associated with mucous gland hypertrophy, hypersecretion and disruption of the mucociliary escalator. In emphysema the terminal air spaces are dilated with destruction of their walls and loss of elastic tissue support to the distal airways, resulting in airway collapse. During exacerbations of COPD, the patient becomes more breathless with an increase in cough, sputum production and sputum purulence. As COPD progresses exacerbations become more frequent and alarming, airways obstruction and gas exchange deteriorate, and mortality and morbidity are substantial. World-wide COPD causes about 3 million deaths each year. In the UK about 30 000 people die and about 30 million working days are lost each year as a result of COPD (British Thoracic Society, 1997).

Disruption of the mucociliary escalator allows colonisation of the lower respiratory tract by bacteria such as *H. influenzae, Strep. pneumoniae* and *Moraxella catarrhalis* (Murphy and Sethi, 1992). Infection promotes inflammation and a vicious circle of infection and inflammation further compromises lung clearance mechanisms. Infection in COPD is a secondary phenomenon in which organisms of low virulence gain a foothold because of compromised lung defences due to structural damage. Extension of bronchial infection into the surrounding lung parenchyma may give rise to bronchopneumonia. Exacerbations may be precipitated by viral infections, which further compromise lung defences and permit secondary bacterial proliferation and invasion. More recently organisms such as *Myc. pneumoniae* and *Chl. pneumoniae* have also been linked to exacerbations (Mogulkoc *et al.*, 1999).

The best use of antibiotics in patients with exacerbations of COPD in clinical practice is uncertain, and it is difficult to distinguish colonisation from active infection. Patients are often given antibiotics in combination with corticosteroids, bronchodilator therapy, oxygen and supportive care so that it is difficult to define the benefit of any one intervention. In general the evidence is that antibiotics are beneficial for patients with more severe exacerbations, whereas the role of antibiotics in milder exacerbations is uncertain. Some studies have shown that an objective measure of airway function, such as the peak expiratory flow rate, improves more rapidly in patients given antibiotics rather than placebo (Anthonisen *et al.*, 1987), whereas other studies have not demonstrated a benefit (Nicotra *et al.*, 1982), and this may reflect the heterogeneous nature of COPD. Long-term prophylactic antibiotic therapy may reduce the frequency of exacerbations in patients with advanced disease (Martinez, 1997), but this strategy is now rarely used, because of concerns about antibiotic resistance, alteration of the normal oropharyngeal flora and unwanted effects such as *Clostridium difficile* colitis. About 15–20% of strains of *H. influenzae* are amoxycillin-resistant and about 90% of strains of *M. catarrhalis* produce β lactamase, so that the optimal choice of antibiotic is unclear. High-dose therapy may be necessary to achieve adequate penetration of antibiotics into the scarred airway mucosa and viscid secretions. Influenza vaccination is recommended for patients with COPD and pneumococcal vaccinations may also be useful.

Bronchiectasis

Bronchiectasis is a chronic disease characterised by irreversible dilatation of bronchi due to bronchial wall damage resulting from infection and inflammation. These morphological changes are usually accompanied by chronic suppurative lung disease with cough productive of purulent sputum. Bronchiectasis represents a particular type of bronchial injury and may result from a number of different underlying disease processes. It may be confined to one area of the lung if there is a local cause, such as bronchial obstruction by a foreign body, or may be diffuse if there is a generalised cause such as immunoglobulin deficiency, ciliary dyskinesia or cystic fibrosis. In many cases the disease evolves through a vicious circle of steps in which an underlying defect or bronchial injury impairs mucociliary clearance, secretions accumulate, infection supervenes, an inflammatory response is provoked, and infection and inflammation cause progressive damage to the bronchial wall with further disruption of mucociliary clearance (Cole, 1995). The prevalence of gross saccular bronchiectasis has declined over the years as the result of childhood vaccination programmes, improved socio-economic conditions and more widespread use of antibiotics. The advent of computed tomography (CT) has, however, led to an increased recognition of less severe forms of cylindrical bronchiectasis. Severe infections in childhood, such as pertussis, measles and tuberculosis, are still important world-wide causes of bronchiectasis. Certain well-defined immunodeficiency states, such as hypogammaglobulinaemia, selective immunoglobulin deficiencies and ciliary dyskinesia give rise to bronchiectasis. It is important to recognise that in patients with HIV infection there is an increased incidence of respiratory tract infections, with bronchiectasis, sinusitis, bronchitis and pneumonia due to common bacterial pathogens at an earlier stage of the disease, before severe immune deficiency leads to infection with opportunistic pathogens. Patients with certain diseases, such as rheumatoid arthritis, ulcerative colitis, Crohn's disease and coeliac disease, have an increased incidence of bronchiectasis, but the mechanisms by which the bronchiectasis arises in these diseases is unclear.

Secondary infection of the bronchial tree is a key element in the initiation and progression of bronchiectasis. Typical organisms isolated are the classical respiratory pathogens, *H. influenzae*, *Strep. pneumoniae* and *M. catarrhalis*, and *Ps. aeruginosa* is particularly common in more advanced disease and in patients with cystic fibrosis. Treatment is multifaceted with particular emphasis on physiotherapeutic clearance procedures, anti-inflammatory drugs, bronchodilator medication and antibiotics. In bronchiectasis the bronchial wall is scarred, thickened and relatively poorly perfused and viscous secretions also constitute a barrier, so that penetration of antibiotics to the site of infection may be poor. In these circumstances high-dose oral antibiotics (e.g. amoxycillin 3 g twice daily) or nebulised antibiotics may be useful and are associated with favourable clinical responses (Hill *et al.*, 1986).

Cystic Fibrosis

Cystic fibrosis is due to a defect in a gene on the long arm of chromosome 7, which codes for a 1480-amino-acid protein, cystic fibrosis transmembrane conductance regulator (CFTR). More than 500 mutations of this gene have been identified, but the most common is designated ΔF508, in which mutation of a single codon of the gene results in the loss of phenylalanine ('delta F') at position 508 of the protein. CFTR functions as a chloride channel in the membrane of epithelial cells. Failure of chloride transport in the bronchial mucosa results in secretions of abnormal viscosity that interfere with mucociliary clearance and permit the adherence of bacteria to the mucosa. Persistent infection and inflammation cause progressive lung damage, bronchiectasis, respiratory failure and death. The airway surface liquid in cystic fibrosis has an abnormally high sodium chloride content, and this may predispose to bacterial infections because the antibacterial activity of airway β-defensins and lysozyme are increasingly inhibited as the salt concentration increases (Goldman *et al.*, 1997). CFTR may also have other functions, including an effect on the sialylation of membrane proteins. This may be relevant to the particular susceptibility of patients with cystic fibrosis to *Ps. aeruginosa* infection, since it results in increased amounts of asialoglycoprotein on the epithelial surface and this appears to act as a binding site for *Ps. aeruginosa* (Welsh and Ramsey, 1998).

Cystic fibrosis patients have recurrent and chronic lung infections from early childhood. At first, typical organisms identified in sputum cultures are *Staph. aureus*, *H. influenzae* and *Strep. pneumoniae*. By the teenage years, many patients have become infected with mucoid strains of *Ps. aeruginosa*, and this is often associated with a decline in lung function and a worse prognosis. Initial infection with *Ps. aeruginosa* is more common during the winter and may be facilitated by viral infections such as RSV. Many other organisms may colonise the respiratory tract in cystic fibrosis, including Enterobacteriaceae, *Stenotrophomonas maltophilia*, atypical mycobacteria and *Aspergillus fumigatus*. In the 1980s, *Burkholderia cepacia* infection emerged as a major clinical problem for patients with cystic fibrosis, and it soon became clear that this organism could be transmitted by person-to-person contact with potentially devastating consequences

(Lipuma *et al.*, 1990). The clinical course of patients with *B. cepacia* infection is variable, but some develop a progressive necrotising pneumonia, which is usually rapidly fatal and is referred to as the dreaded 'Cepacia syndrome'. It appears that certain virulent strains, notably genomovar 3, give rise to epidemic spread of *B. cepacia* infection among patients with cystic fibrosis. Recognition of the transmission of this infection from patient to patient was very traumatic for the cystic fibrosis community and led to the segregation of patients with *B. cepacia* infection from other patients and the abandonment of many support meetings and holiday camps for patients with cystic fibrosis.

The prognosis of patients with cystic fibrosis has improved dramatically over the years as a result of better management of the disease in the first year of life, the development of specialist treatment centres for these patients, and meticulous attention to physiotherapy, antibiotic therapy and nutrition as the crucial elements of a treatment programme. In the 1950s, survival beyond ten years of age was unusual. Now the median survival is about 30 years and it is predicted that it will be at least 40 years for children born in the 1990s. Children with cystic fibrosis should be immunised against pertussis and measles as part of the childhood vaccination programme and they should receive annual influenza vaccination. *Staph. aureus* is a major pathogen in the disease from early childhood and long-term flucloxacillin is often used to suppress this infection. An important strategy is the postponement, for as long as possible, of the colonisation of the airways by *Ps. aeruginosa*. Frequent sputum cultures should be performed and intensive anti-pseudomonas antibiotics should be given when the organism is first isolated. Typically, a prolonged course of oral ciprofloxacin and nebulised colistin is given and, if *Ps. aeruginosa* is not eradicated, intravenous anti-pseudomonas antibiotics should be given (Valerius *et al.*, 1991). Eventually chronic infection with *Ps. aeruginosa* becomes established in most patients. Suppression of this infection involves long-term nebulised antibiotics such as colistin, gentamicin or tobramycin with additional courses of combinations of anti-pseudomonas antibiotics (e.g. ceftazidime and gentamicin), given either regularly at three-monthly intervals, or given for exacerbations of the disease or when there is a reduction in lung function. Most patients are trained in the self-administration of intravenous antibiotics at home, often through a totally implanted central venous device (**Fig. 3**). High doses are required to achieve adequate penetration of antibiotics into scarred bronchial mucosa, and patients with cystic fibrosis have increased renal clearance of many antibiotics.

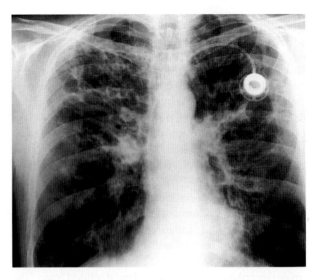

Fig. 3 Chest radiograph of a 50-year-old man with cystic fibrosis who had chronic pulmonary infection with *Pseudomonas aeruginosa* treated by nebulised colistin and frequent courses of self-administered intravenous antibiotics at home by way of an indwelling venous access system (Portacath) situated in his left subclavian vein. The radiograph shows extensive bronchiectasis, peribronchial fibrosis, distortion of lung architecture and areas of consolidation.

Elucidation of the molecular and cellular biology of cystic fibrosis has allowed new treatment strategies to be considered, including gene-replacement therapy, ion-channel drugs which act on sodium and chloride channels (e.g. amiloride), therapies that may facilitate trafficking of mutant CFTR in the cell, anti-inflammatory drugs (e.g. corticosteroids, ibuprofen) and mucolytic drugs (e.g. recombinant human DNase) (Welsh and Ramsey, 1998). It is now clear that there are several different ways in which cystic fibrosis mutations may disrupt CFTR function. The commonest mutation (ΔF508) results in mis-folding of CFTR so that it is degraded in the cell rather than processed to the cell membrane. Experiments *in vitro* show that this process is temperature-sensitive such that a reduction in the incubation temperature results in the mutant CFTR moving to the cell surface where it retains some function. A number of pharmacological interventions may stabilise mis-folded CFTR, so that its trafficking to the cell membrane is improved. Some mutations (e.g. G542X, W1282X) are 'stop mutations' that result in failure to transcribe CFTR. Interestingly, aminoglycosides bind to prokaryotic ribosomes, and permit misreading of stop codons, which allows transcription to go to completion (Howard *et al.*, 1996). The concept is, therefore, emerging that antibiotics may have modes of action over and above their antibacterial effects. Macrolide antibiotics, for example, seem to have important anti-inflammatory effects in some lung diseases. This first came to attention because of the efficacy of low-dose prolonged erythromycin in diffuse panbronchiolitis, a recently described disease, found mainly in Japan and characterised by chronic inflammation of the respiratory bronchioles leading to bronchiectasis and secondary infection with *H. influenzae* and *Ps. aeruginosa* (Koyama and Geddes, 1997). The remarkable efficacy of erythromycin in this disease has led to pilot studies of macrolide therapy in bronchiectasis (Tsang *et al.*, 1999) and cystic fibrosis (Jaffé *et al.*, 1998) with some benefit, which suggests the need for further controlled trials.

Tuberculosis

The World Health Organization estimates that 1720 million people (one-third of the world's population) have latent infection with *Mycobacterium tuberculosis*, 15–20 million people have active infection, and 3 million deaths occur each year from tuberculosis (95% in the developing world) (Sudre *et al.*, 1992). One hundred years ago in the UK about 30 000 people died from tuberculosis each year. Mortality and notification rates declined steadily from 1900 onwards because of improvement in nutrition and socio-economic factors, with a sharper decline in mortality in the late 1940s because of the introduction of effective treatment. Overall the decline in notification rates has levelled off during the last decade, with small increases in some areas. Notification rates in the UK are 20–30 times higher in people who originate from the Indian subcontinent than in the indigenous white population. About 33% of all cases in the UK occur in the white population, and more than 50% of these are over 55 years of age. Infection may have been contracted in childhood and been dormant for years before becoming reactivated. Factors that reduce resistance and precipitate reactivation include ageing, alcohol misuse, poor nutrition, debility from other diseases, use of immunosuppressive drug therapy, and co-infection with human immunodeficiency virus (HIV). In the UK overlap between the population with HIV infection, mainly young white men, and the population with tuberculosis, mainly older white people and younger immigrants from the Indian subcontinent, is limited so that only about 5% of patients with acquired immune deficiency syndrome (AIDS) have tuberculosis and about 2% of patients with tuberculosis are identified as having HIV infection. However 4.5 million people world-wide are estimated to be co-infected with HIV and tuberculosis, and of these 98% are in developing countries (Quinn, 1996).

Control of tuberculosis depends firstly on prompt identification and comprehensive treatment of patients with active tuberculosis, because this limits the spread of infection. In the UK, when a diagnosis of tuberculosis is made, there is a statutory requirement for the doctor to notify the patient to the public health authorities who are then responsible for the screening of contacts. The index patient may have acquired infection from, or transmitted infection to, someone in his or her close environment. In the UK about 10% of close contacts of smear-positive cases are found to have active disease. The current standard treatment of tuberculosis consists of 6 months of rifampicin and isoniazid supplemented by pyrazinamide for the first 2 months. Meticulous supervision of treatment is essential and patients should be seen at least each month for the prescription of medication, to check for compliance with treatment and to monitor for side-effects, such as reduction in liver function. Patients who have difficulty in complying with treatment should have directly observed therapy (DOT), in which the patient is observed to ensure that medication has actually been taken (Morse, 1996). This can sometimes be achieved by the patient attending a hospital or general practice clinic to be given large doses of the medication three times per week under the supervision of a healthcare worker.

At present drug-resistant tuberculosis is rare in the initial treatment of the indigenous population in the UK. About 6% of patients from the Indian subcontinent have tuberculosis that is resistant to isoniazid, and in this situation additional drugs, such as ethambutol, are added to the treatment regimen. Multiple-drug-resistant tuberculosis results from inadequate previous treatment.

In order to focus attention on the threat posed by tuberculosis to global health, the WHO has designated 24 March each year as 'World TB day', marking the anniversary of Koch's discovery of *M. tuberculosis* as the 'cause' of tuberculosis. In spite of the identification of *M. tuberculosis* as long ago as 1882, the development of BCG vaccination in the 1930s and the introduction of effective drug treatment from the 1940s onwards, the fact that 3 million people each year die from tuberculosis represents a tragic political failure. Perhaps it is necessary to rediscover that tuberculosis is 'caused' by poverty, as much as by *M. tuberculosis*, in order to focus attention on the appropriate corrective steps (Grange and Zumla, 1999). Even in developed countries poverty and homelessness are important factors and it is in this population that outbreaks occur of multiple-drug-resistant tuberculosis. The advent of DNA finger-printing techniques makes it possible to distinguish different strains of *M. tuberculosis* and to obtain a more accurate picture of the clinical and molecular epidemiology of the disease in terms of the sources and spread of infection (Van Soolingen and Hermans, 1995).

Respiratory Infections in Immunocompromised Patients

Respiratory tract infections in immunocompromised patients are an increasingly important problem in view of the global impact of HIV infection and the increased use of immunosuppressive drugs in patients with cancer, inflammatory diseases or after organ transplantation. It is estimated that since the first case of AIDS was recognised in 1981, about 5 million people have died from AIDS and about 24 million adults and 1.5 million children have been infected with HIV. About two-thirds of cases have been in Africa where the disease continues to devastate an entire generation of young adults (Quinn, 1996). Pulmonary complications of HIV infection include inflammatory, neoplastic and infectious diseases (Miller, 1996). Although CD4 T-lymphocytes are the main target of HIV infection, the virus also infects pulmonary macrophages and can give rise to a lymphocytic alveolitis and impaired gas diffusion. Late in the disease, neoplastic diseases such as high-grade B-cell lymphoma and Kaposi's sarcoma arise (Miller, 1996). The occurrence of various infections in patients with HIV infection reflects the CD4 T-lymphocyte count and depends on previous and current exposure to pathogens, such as reactivation of tuberculosis or re-infection in areas with a high prevalence. As the CD4 count falls there is at first an increase in the frequency of infection with common pathogens, so that these patients have an increased incidence of respiratory tract infections in the form of sinusitis, bronchitis, bronchiectasis and pneumonia. The clinical features may be unusual with a higher frequency of complications such as bacteraemia, abscess formation, cavitation and empyema. Patients with HIV infection and impaired CD4 lymphocyte function are highly susceptible to the reactivation of latent tuberculosis and also to contracting the disease from an exogenous source with rapid progression to active disease, and transmission of infection to others. Globally it is estimated that 4.5 million people are co-infected with HIV and tuberculosis. Mortality in these patients is as high as 50% and, in some situations, within 18 months. As the CD4 count falls below about $200/\mu L$, infections with opportunistic pathogens develop. These include *Pneumocystis carinii*, *Mycobacterium avium intracellulare* complex, viruses such as cytomegalovirus, and fungi such as *Aspergillus fumigatus*, *Cryptococcus neoformans* and *Candida albicans*.

Increasing numbers of patients are severely immunocompromised by a variety of diseases and by use of

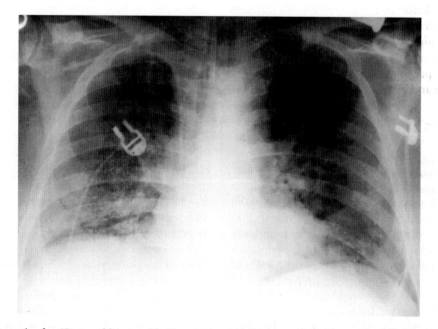

Fig. 4 Chest radiograph of a 45-year-old man with Wegener's granulomatosis. He was treated with prednisolone and cyclophosphamide and developed breathlessness, pyrexia, and diffuse consolidation with progressive respiratory failure that necessitated mechanical ventilation in an intensive care unit. Broncho-alveolar lavage was carried out by way of a bronchoscope. The washings yielded *Pneumocystis carinii*.

immunosuppressive drugs. The clinical problem is that of patients with one of these conditions presenting with pulmonary infiltrates on chest X-ray, with dyspnoea and fever (**Fig. 4**). The lung infiltrates in these circumstances may be due to pulmonary involvement by the underlying disease process, a reaction to drug therapy, or infection resulting from immunosuppression. For example, a patient receiving bleomycin chemotherapy for a testicular carcinoma who presents with fever and lung infiltrates on chest X-ray may have pulmonary toxicity due to chemotherapy ('bleomycin lung'), carcinomatous infiltration of the lung (lymphangitis carcinomatosa) or infection, including bacterial pneumonias and opportunistic infections. Treatment depends crucially on accurate diagnosis often by means of invasive procedures, such as bronchoscopy with bronchoalveolar lavage and transbronchial lung biopsy.

Conclusion

Respiratory tract infections represent a very diverse and disparate group of diseases. They are one of the commonest causes of morbidity and mortality in all age groups, in all countries, and in a range of medical specialities from general practice to oncology, from intensive care medicine to transplant surgery and from paediatrics to geriatrics. The clinical pattern of disease reflects a complex dynamic pathogen–host interaction in terms of virulence and vulnerability. Respiratory tract infections are not a static group of diseases but are constantly changing, so that the approach to these infections must continually be modified to take account of new pathogens, the development of antibiotic resistance and changing patient characteristics.

References

Adams JFA, Scholvinck EH, Gie RP, Potter PC, Beyers N, Beyers AD (1999) Decline in total serum IgE after treatment for tuberculosis. *Lancet* 353: 2030–2033.

American Thoracic Society (1995) Hospital-acquired pneumonia in adults: diagnosis, assessment of severity, initial antimicrobial therapy, and preventative strategies. *Am. J. Respir. Crit. Care Med.* 153: 1711–1725.

Anthonisen NR, Manfreda J, Warren CPW, Hershfield ES, Harding GKM, Nelson NA (1987) Antibiotic therapy in exacerbations of chronic obstructive pulmonary disease. *Ann. Intern. Med.* 106: 196–204.

Bals R, Wang X, Meegalla RL *et al.* (1999) Mouse β-defensin 3 is an inducible antimicrobial peptide expressed in the epithelia of multiple organs. *Infect. Immun.* 67: 3542–3547.

Bourke SJ, Lightfoot NF (1995) *Chlamydia pneumoniae*: defining the clinical spectrum of infection requires precise laboratory diagnosis. *Thorax* 50 (Suppl. 1): S43–S48.

Bourke SJ, Carrington D, Frew CE, Stevenson RD, Banham SW (1989) Serological cross-reactivity among chlamydial strains in a family outbreak of psittacosis. *J. Infect.* 19: 41–45.

British Thoracic Society (1997) Guidelines on the management of chronic obstructive pulmonary disease. *Thorax* 52: Suppl. 5.

Bush A, Cole P, Hariri M *et al.* (1998) Primary ciliary dyskinesia: diagnosis and standards of care. *Eur. Respir. J.* 12: 982–988.

Cabello H, Torres A, Celis R *et al.* (1997) Bacterial colonisation of distal airways in healthy subjects and chronic lung disease: a bronchoscopic study. *Eur. Respir. J.* 10: 1137–1144.

Cole P (1995) Bronchiectasis. In: Brewis RAL, Corrin B, Geddes DM, Gibson GJ (eds) *Respiratory Medicine*. London: WB Saunders, pp. 1286–1316.

Court CA, Garrard CS (1992) Nosocomial pneumonia in the intensive care unit: mechanisms and significance. *Thorax* 47: 465–473.

De Iongh RU, Rutland J (1995) Ciliary defects in healthy subjects, bronchiectasis, and primary ciliary dyskinesia. *Am. J. Respir. Crit. Care Med.* 151: 1559–1567.

Drost EM, Kassabian G, Meiselman HJ, Gelmont D, Fisher TC (1999) Increased rigidity and priming of polymorphonuclear leukocytes in sepsis. *Am. J. Respir. Crit. Care Med.* 159: 1696–1702.

Duchin JS, Koster FT, Peters CJ (1994) Hantavirus pulmonary syndrome: a clinical description of 17 patients with a newly recognised disease. *N. Engl. J. Med.* 14: 949–1005.

Fraser DW, Tsai TR, Orenstein W *et al.* (1977). Legionnaires' disease: description of an epidemic of pneumonia. *N. Engl. J. Med.* 297: 1189–1197.

Goldman MJ, Anderson GM, Stolzenberg ED, Kari UP, Zasloff M, Wilson JM (1997) Human β-defensin-1 is a salt sensitive antibiotic in lung that is inactivated in cystic fibrosis. *Cell* 88: 553–560.

Grange JM, Zumla A (1999) Paradox of the global emergency of tuberculosis. *Lancet* 353: 996.

Greenstone M, Cole PJ (1985) Ciliary function in health and disease. *Br. J. Dis. Chest* 79: 9–26.

Grayston JT, Kuo CC, Wang SP, Altman J (1986) A new *Chlamydia psittaci* strain, TWAR, isolated in acute respiratory tract infections. *N. Engl. J. Med.* 315: 161–168.

Haslett C (1995) Lung inflammation and repair. In: Brewis RAL, Corrin B, Geddes DM, Gibson GJ (eds) *Respiratory Medicine*. London: WB Saunders, pp. 219–237.

Hill SL, Morrison HM, Burnett D, Stockley RA (1986) Short term response of patients with bronchiectasis to treatment with amoxycillin given in standard or high doses orally or by inhalation. *Thorax* 41: 559–565.

Howard M, Frizzell RA, Bedwell DM (1996) Aminoglyco-side antibiotics restore CFTR function by overcoming premature stop mutations. *Nature Med.* 2: 467–469.

Jaffé A, Francis J, Rosenthal M, Bush A (1998) Long-term azithromycin may improve lung function in children with cystic fibrosis. *Lancet* 351: 420.

Johanson WG, Pierce AK, Sanford JP (1969) Changing pharyngeal bacterial flora of hospitalised patients. *N. Engl. J. Med.* 281: 1137–1140.

Kollef MH (1999) The prevention of ventilator-associated pneumonia. *N. Engl. J. Med.* 340: 627–634.

Koyama H, Geddes DM (1997) Erythromycin and diffuse panbronchiolitis. *Thorax* 52: 915–918.

Lewis S (1998) Infections in asthma and allergy. *Thorax* 53: 911–912.

Lipuma JL, Dasen SE, Nielson DW, Stern RC, Stull TL (1990) Person-to-person transmission of *Pseudomonas cepacia* between patients with cystic fibrosis. *Lancet* 336: 1094–1095.

MacFarlane JT, Finch RG, Ward MJ, Macrae AD (1982). Hospital study of adult community-acquired pneumonia. *Lancet* 2: 255–258.

Martinez J (1997) Antibiotics and vaccination therapy in COPD. *Eur. Respir. Rev.* 45: 240–242.

Miller R (1996) HIV-associated respiratory diseases. *Lancet* 348: 307–312.

Mogulkoc N, Karakurt S, Isalska B *et al.* (1999) Acute purulent exacerbations of chronic obstructive pulmonary disease and *Chlamydia pneumoniae* infection. *Am. J. Respir. Crit. Care Med.* 160: 349–353.

Monton C, Ewig S, Torres A *et al.* (1999) Role of gluco-corticoids on inflammatory response in nonimmuno-suppressed patients with pneumonia: a pilot study. *Eur. Respir. J.* 14: 218–220.

Morse D (1996) Directly observed therapy for tuberculosis. *Br. Med. J.* 312: 719–720.

Murphy TF, Sethi S (1992) Bacterial infection in chronic obstructive pulmonary disease. *Am. Rev. Respir. Dis.* 146: 1067–1083.

Nicotra MB, Rivera M, Awe RJ (1982). Antibiotic therapy of acute exacerbations of chronic bronchitis. *Ann. Intern. Med.* 97: 18–21.

Quinn TC (1996) Global burden of the HIV pandemic. *Lancet* 348: 99–106.

Reimann HA (1938) An acute infection of the respiratory tract with atypical pneumonia. *J. Am. Med. Assoc.* 111: 2377–2384.

Roger P, Puchelle E, Bajolet-Laudinat O *et al.* (1999) Fibronectin and α51 integrin mediate binding of *Pseudomonas aeruginosa* to repairing airway epithelium. *Eur. Respir. J.* 13: 1301–1309.

Roman M, Calhoun WJ, Hinton KL *et al.* (1997) Respiratory syncytial virus infection in infants is associated with predominant Th-2-like response. *Am. J. Respir. Crit. Care Med.* 156: 190–195.

Shaheen SO, Aaby P, Hall AJ *et al.* (1996) Measles and atopy in Guinea-Bissau. *Lancet* 347: 1792–1796.

Stockley RA (1998) Role of bacteria in the pathogenesis and progression of acute and chronic lung infection. *Thorax* 53: 58–62.

Sudre P, ten Dam G, Kochi A (1992) Tuberculosis: a global view of the situation today. *Bull. WHO* 70: 149–159.

Tsang KWT, Ho PI, Chan KN *et al.* (1999) A pilot study of low-dose erythromycin in bronchiectasis. *Eur. Respir. J.* 13: 361–364.

Valerius NH, Koch C, Hoiby N (1991) Prevention of chronic *Pseudomonas aeruginosa* colonisation in cystic fibrosis by early treatment. *Lancet* 338: 725–726.

Van Soolingen D, Hermans PWM (1995) Epidemiology of tuberculosis by DNA fingerprinting. *Eur. Respir. J.* 8 (Suppl. 20): 649–656.

Welsh MJ, Ramsey BW (1998) Research on cystic fibrosis. *Am. J. Respir. Crit. Care Med.* 157: S148–S154.

Wilson R, Dowling RB, Jackson AD (1996) The biology of bacterial colonisation and invasion of the respiratory mucosa. *Eur. Respir. J.* 9: 1523–1530.

Winter J (1993) British Thoracic Society guidelines for the management of community-acquired pneumonia in adults admitted to hospital. *Br. J. Hosp. Med.* 49: 346–350.

Woodhead MA (1992) The aetiology, management and outcome of severe community-acquired pneumonia on the intensive care unit. *Respir. Med.* 86: 7–13.

Yaari E, Zimerman YY, Schwartz SB *et al.* (1999) Clinical manifestations of *Bordetella pertussis* infection in immunised children and young adults. *Chest* 115: 1254–1258.

75

Bordetella pertussis

Mark Roberts[1] and Roger Parton[2]

[1]*Glasgow University Veterinary School, Glasgow, UK*
[2]*University of Glasgow, Glasgow, UK*

Bordetella pertussis is a small (0.2–0.5 μm × 0.5–2 μm), Gram-negative, aerobic, non-motile coccobacillus and is the cause of whooping cough (pertussis) in humans. Although effective vaccines have been available and in use for many years, pertussis still causes considerable morbidity and mortality and it is one of the major causes of infant mortality in the developing world. This chapter will discuss the basic biology of *B. pertussis*, the disease it causes, the components of the organism involved in its pathogenicity, the host responses to the pathogen, the epidemiology of pertussis and traditional and contemporary methods for its control. A great deal of information is already available on these subjects. Even so, it is expected that knowledge of this organism will increase greatly in the coming years. This is because the genome of *B. pertussis* strain Tohama 1, a virulent and widely used laboratory strain is being sequenced at the Sanger Centre, Cambridge, UK and the sequence information is publicly available (http://www.sanger.ac.uk/Projects/B_pertussis/).

The genome is 3.88 Mb in size and the majority of the sequencing has been performed but, at the time of writing, the complete annotated genome has not been released. Use of the sequence has already been made, including analysis of the polypeptides produced by the predicted open reading frames. This

information is available at the Pedant web site (http://pedant.mips.biochem.mpg.de/). The current release has 3355 entries, of which ∼46% are reported to have homologues to known proteins. A study utilising the currently available genome sequence has already revealed potential new virulence factors (Antoine *et al.*, 2000).

Sequencing of the genomes of representative strains of *B. bronchiseptica* and *B. parapertussis* is also in progress at the Sanger Centre. Comparison of the genome sequences of these closely related pathogens should yield valuable information on the genetic basis of their different biology.

Bordetella

The seven named species of the genus *Bordetella* appear to be obligatory parasites, mostly inhabiting the surface of the respiratory tract of humans and other warm-blooded animals, including birds (**Table 1**). *B. pertussis* and *B. parapertussis* are important human respiratory tract pathogens. Some species are associated with infections at other sites in humans. The type species is *B. pertussis* and this organism is restricted to humans and the human respiratory tract. According to Lapin

doi:10.1006/bkmm.2001.0075

Table 1 Host specificities of *Bordetella* species

Species	Host	Source of human isolates
B. pertussis	Humans	Respiratory tract
B. parapertussis	Humans, sheep	Respiratory tract
B. bronchiseptica	Mammals, humans	Respiratory tract
B. avium	Birds	—
B. hinzii	Birds, humans	Respiratory tract, blood
B. holmesii	Humans	Blood, respiratory tract
B. trematum	Humans	Wounds, ear

(1943), the small Gram-negative coccobacillus was first clearly described in 1900, by Bordet and Gengou, in the sputum of an infant with whooping cough, although other investigators had noted similar organisms previously. The abundance and purity of organism in the specimen was such that its association with the infection was clear. By 1906, Bordet and Gengou had developed a suitable medium for cultivation of the organism and were then able to establish its morphology, cultural characteristics, virulence and antigenicity (Bordet and Gengou, 1906).

B. parapertussis is found in the respiratory tracts of both humans and sheep but was first isolated by Bradford and Slavin (1937) from cases of mild whooping cough and recognised as being different from *B. pertussis*. Phylogenetic studies based on DNA polymorphisms mediated by insertion sequence (IS) elements (van der Zee *et al.*, 1996a) and arbitrarily primed polymerase chain reaction (PCR)-based fingerprinting (Yuk *et al.*, 1998b) have shown that human and sheep strains form distinct populations. Thus there is no evidence for transmission of *B. parapertussis* between humans and sheep.

B. bronchiseptica was first isolated by Ferry (1910) from the respiratory tract of dogs. It is a common respiratory tract coloniser in wild and domesticated mammals and an important pathogen, for example in kennel cough in dogs, atrophic rhinitis in swine and snuffles in rabbits, but it is seldom isolated from humans. In people it appears to be more of a respiratory tract commensal and opportunist, sometimes associated with cases of septicaemia in compromised patients (Goodnow, 1980; Woolfrey and Moody, 1991). Some human cases have been linked to contact with animals. *B. avium* was first proposed as a species by Kersters *et al.* (1984) to include agents of respiratory disease of turkey poults and other birds. More recently, three new species have been added to the genus. *B. hinzii* was the name given to a *B. avium*-like

group of organisms found mainly in the respiratory tracts of chickens and turkeys but human isolates have been reported (Vandamme *et al.*, 1995). *B. holmesii* describes a group of strains isolated originally from human blood cultures (Weyant *et al.*, 1995) but also from sputum (Tang *et al.*, 1998) and nasopharyngeal specimens from patients with pertussis-like symptoms (Mazengia *et al.*, 2000). The most recently named species, *B. trematum*, contains isolates from human wound and ear infections which were described originally as atypical bordetellae or unidentified (Vandamme *et al.*, 1996). The pathogenic potential of these three species is unclear although *B. hinzii* has been identified as the cause of a case of fatal septicaemia (Kattar *et al.*, 2000). Like *B. bronchiseptica*, they appear to be opportunists in humans. No hosts other than humans have been reported for *B. holmesii* and *B. trematum*.

Currently the genus *Bordetella* is assigned to the $\beta2$ subdivision of Proteobacteria and analysis of 16S rRNA sequences of representatives of five *Bordetella* species (including *B. holmesii*) show they share a common phylogeny with *Alcaligenes* (Weyant *et al.*, 1995). The two genera have a number of characteristics in common and may be difficult to differentiate on the basis of phenotypic criteria. Differential characteristics of these and other phenotypically similar genera are described elsewhere (Johnson and Sneath, 1973; Kersters *et al.*, 1984; Pittman, 1984a; Vandamme *et al.*, 1995, 1996). Some of the properties of *B. pertussis*, namely cultural characteristics, biochemical activities and cellular composition, used to differentiate it from other *Bordetella* species are shown in **Table 2**.

Studies on the relatedness of *B. pertussis*, *B. parapertussis* and *B. bronchiseptica* by multi-locus enzyme electrophoresis, nucleic acid hybridization analyses and other comparisons of their genomes have suggested that their classification as three separate species is unwarranted (Kloos *et al.*, 1981; Kersters *et al.*, 1984; Musser *et al.*, 1986; Muller and Hildebrandt, 1993). On the basis of these and other criteria, it has been concluded that they are, in fact, subtypes of a single genomic species whereas *B. avium*, *B. holmesii*, *B. hinzii* and *B. trematum* each form true (genomic) species (Vandamme *et al.*, 1995, 1996; Weyant *et al.*, 1995). For the sake of convenience in this chapter, however, use of the traditional specific names is appropriate.

An evolutionary history of *B. pertussis*, *B. parapertussis*, *B. bronchiseptica* and *B. avium* has been proposed (Rappuoli, 1994). Their close phylogenetic relationship with *Alcaligenes* and other bacteria widespread in the environment suggests that the ancestral bordetellae were free-living and evolved to

Table 2 Differential characteristics of *Bordetella* species

Characteristic	B. pertussis	B. parapertussis	B. bronchiseptica	B. avium	B. hinzii	B. holmesii	B. trematum
Growth: colonies visible in (days)	3	1–2	1	1	2	2–3	1
Peptone agar:							
growth	−	+	+	+	+	+	+
browning	−	+	−	−	−	+	−
Growth on MacConkey's agar	−	+	+	+	+	+	+
Motile, peritrichous flagella	−	−	+	+	+	−	+
β-like haemolysis	+	+	v	v	?	−	−
Citrate utilisation	−	v	+	v	+	−	−
Nitrate reduction	−	−	+	−	−	−	v
Oxidase	+	−	+	+	+	−	−
Urease	−	+	+	−	v	−	−
Major cellular fatty acids[a]	$C_{16:1w7c}$ $C_{16:0}$	$C_{16:0}$ $C_{17:0cyc}$	$C_{16:0}$ $C_{16:1w7c}$ $C_{17:0cyc}$	$C_{16:0}$ $C_{17:0cyc}$	$C_{16:0}$ $C_{17:0cyc}$	$C_{16:0}$ $C_{17:0cyc}$	$C_{16:0}$ $C_{17:0cyc}$
G + C content of DNA (mol%)	67.7–68.9	68.1–69.0	68.2–69.5	61.6–62.6	65–67	61.5–62.3	64–65

+, positive; −, negative; v, some strains positive, others negative; ?, unknown.

[a] The number before the colon is the number of carbon atoms and the number after the colon is the number of double bonds; w, double bond position; c, *cis* isomer; cyc, cyclopropane ring.

infect warm-blooded animals. Some of their properties reflect this environmental origin: temperature sensing, and regulation of virulence factor expression accordingly (see Regulation of Virulence and Antigenic Variation, p. 1571); the presence of flagella in some species and greater expression of motility at low temperature; the ability of *B. bronchiseptica* and *B. avium* to grow and survive for long periods in low nutrient conditions such as natural waters (Porter *et al.*, 1991; Porter and Wardlaw, 1993; Cotter and Miller, 1994). A phylogenetic tree, based on the presence or absence in the chromosome of two insertion sequences and the nucleotide sequence of the pertussis toxin (PT) gene suggests that the first to diverge from the ancestral line, to infect birds, was *B. avium*. Its distinctiveness is shown by its lower G + C content and lack of genes for PT. Later came a branch leading to *B. pertussis* and *B. parapertussis* that specialised to infect humans, leaving the line to *B. bronchiseptica* with its broad host range (Gross *et al.*, 1989). More recent genotyping studies, by multi-locus

enzyme electrophoresis and the distribution of IS elements, have suggested that *B. pertussis* and *B. parapertussis* evolved at different times from distinct clones of *B. bronchiseptica* (van der Zee *et al.*, 1997). This and other evidence from DNA fingerprinting by pulsed-field gel electrophoresis (Mastrantonio *et al.*, 1998) and arbitrarily primed PCR (Yuk *et al.*, 1998b) shows that *B. pertussis* strains and human strains of *B. parapertussis* are both still very homogeneous clonal populations and supports the suggestion of their relatively recent adaptation to humans. In contrast, sheep strains of *B. parapertussis* are more diverse. Indeed, there is perhaps some historical evidence for a very recent origin of *B. pertussis* since the first descriptions of the highly characteristic disease caused by this organism date from Paris in the mid-sixteenth century (Lapin, 1943; Olson, 1975). It has been suggested that the development of the human population of Europe at that time to a sufficient size, density and mobility could have provided a suitable niche for the emergence of the organism (Wardlaw, 1988).

Growth and Metabolism

Growth of *B. pertussis* was first achieved by Bordet and Gengou (1906) with a glycerol-potato extract medium, without peptone but containing blood. This Bordet–Gengou (BG) medium, with minor modifications, is still in general use. *B. pertussis* is the most fastidious of the bordetellae and fresh isolates will not grow on the usual peptone-containing media. On BG agar after about 3 days, *B. pertussis* gives tiny (0.5 mm diameter), smooth, convex, pearl-like, glistening colonies with entire edges and surrounded by a narrow zone of haemolysis. After 5–6 days, the diameter of the colonies has increased to 2–3 mm. Colonies of *B. parapertussis* develop more quickly and become slightly larger while those of *B. bronchiseptica* are larger still and appear within 24 hours. Changes in colony morphology may occur on repeated subculture, with concomitant loss of antigenic and virulence attributes and alterations in growth requirements. Peppler and Schrumpf (1984) described phenotypic variants of *B. pertussis* that had flat, non-haemolytic colonies on BG agar and some of the variants could grow on nutrient agar.

B. pertussis is a strict aerobe with an optimum growth temperature of 35–37°C, it does not ferment carbohydrates and energy is obtained primarily from the oxidation of amino acids. The nutritional requirements of *B. pertussis* are quite simple and the organism can be cultured in a buffered simple salts medium containing several amino acids, an organic sulphur source, and several 'growth factors' (Rowatt, 1957; Stainer, 1988). The essential amino acids are glutamic acid or proline and one of the sulphur-containing amino acids – cysteine or cystine. Other amino acids can be metabolised but do not substitute for glutamic acid or proline. The growth factors include nicotinic acid (or nicotinamide), glutathione and ascorbic acid. Nicotinic acid (or nicotinamide) is an essential vitamin for *B. pertussis*. Glutathione can serve as an alternative sulphur source but it and ascorbic acid are primarily thought to provide reducing power in defined liquid media.

Although *B. pertussis* has simple nutritional requirements, it can be difficult to culture, particularly in liquid media from a low inoculum (Stainer, 1988). This is because uniquely amongst the bordetellae, *B. pertussis* is exquisitely sensitive to unsaturated fatty acids or residual detergents that may be present on laboratory glassware. *B. pertussis* itself was found to release inhibitory free fatty acids into the medium (Frohlich *et al.*, 1996). For these reasons, media for cultivation of *B. pertussis* are usually supplemented with substances that can absorb these inhibitors, such as blood and starch, as in BG medium, but also charcoal and anion exchange resins (Stainer, 1988).

The difficulty in growth of *B. pertussis* was overcome by Imaizumi *et al.* (1983), who supplemented the simple chemically defined medium of Stainer and Scholte (1971) with heptakis (2,6-*O*-dimethyl) β-cyclodextrin. Cyclodextrins form water-soluble complexes with hydrophobic molecules such as fatty acids. The resulting medium was the first defined medium for *B. pertussis*, in liquid form or when solidified with agarose, to allow reliable growth from low inocula. Supplementation of the medium with casamino acids further improves growth. Dimethyl β-cyclodextrin not only improves the growth of *B. pertussis* but also greatly enhances the yield of the virulence factors pertussis toxin (PT), filamentous haemagglutinin (FHA) and adenylate cyclase toxin (ACT) in culture supernates (Hozbor *et al.*, 1994) and prevents auto-agglutination of *B. pertussis* in defined medium due to surface-associated FHA (Menozzi *et al.*, 1994a). *B. bronchiseptica* and *B. avium* have the remarkable ability to grow in low nutrient conditions, such as in natural waters and unsupplemented phosphate-buffered saline, but this property is not shared by *B. pertussis* or *B. parapertussis* (Porter *et al.*, 1991; Porter and Wardlaw, 1993; Cotter and Miller, 1994).

There is very little information about the metabolic pathways of *B. pertussis* that are important for growth *in vivo*. Inactivation of the *aroA* gene of *B. pertussis* blocks the pre-chorismate (aromatic) pathway that is required for the biosynthesis of the aromatic amino acids, *p*-amino benzoic acid (a precursor of folic acid) and other aromatic compounds. *B. pertussis aroA* mutants are highly attenuated in mice and they can also function as live respiratory *B. pertussis* vaccines (Roberts *et al.*, 1990). Tn5-generated *B. pertussis* leucine and methionine auxotrophs were also attenuated in a lethal infant mouse model (Weiss and Goodwin, 1989).

Iron is an essential nutrient for most bacteria but, *in vivo*, the concentration of free iron is exceedingly small, below that necessary for bacterial growth (Weinberg, 1986). In the extracellular fluids of vertebrates, iron is bound to the iron-binding proteins transferrin (TF) and lactoferrin (LF). Pathogenic bacteria have evolved a number of high-affinity iron uptake systems that can liberate iron from TF or LF or they can acquire iron from haem or haemoglobin released from damaged red cells (Wooldridge and Williams, 1993). Iron can be captured from TF and LF by two different mechanisms. The first involves the secretion of siderophores, which are low-molecular-weight, high-affinity iron chelators. The ferric-siderophore is transported back into the bacterial cell by a cognate outer-membrane protein (OMP) receptor. The second mechanism involves direct binding of TF or LF to a specific TF or LF OMP receptor at the bacterial surface and subsequent transport of iron into the bacterial cytoplasm. The utilisation of iron from other sources, such as haem, also requires specific receptors. There is evidence that *B. pertussis* possesses all three types of iron uptake systems (Menozzi *et al.*, 1991a; Redhead and Hill, 1991; Moore *et al.*, 1995; Brickman and Armstrong, 1999; Pradel *et al.*, 2000).

Usually, in Gram-negative bacteria, the genes for ferric iron uptake are only active under iron-limiting conditions because they are under negative control by the ferric uptake repressor (Fur) (Wooldridge and Williams, 1993). The *B. pertussis fur* gene has been identified (Beall and Sanden, 1995a). Both *B. pertussis* and *B. bronchiseptica* produce the siderophore alcaligin and a ferric-alcaligin receptor, the 79-kDa OMP FauA (Moore *et al.*, 1995; Brickman and Armstrong, 1999). Expression of the genes for alcaligin biosynthesis and FauA require AlcR, an AraC-like activator (Beaumont *et al.*, 1998; Pradel *et al.*, 1998). The alcR gene itself is under negative regulation by iron, probably mediated by Fur (Beaumont *et al.*, 1998). A *B. pertussis alcR* mutant colonised the respiratory tract of mice normally (Pradel *et al.*, 1998), showing that alternative iron uptake mechanisms are available. *B. pertussis*, in common with other Gram-negative bacteria, also has receptors for the siderophores that are produced by other micro-organisms. One such is the ferric-enterobactin receptor BfeA (Beall and Sanden, 1995b).

High-affinity iron-uptake systems of Gram-negative bacteria usually depend on the accessory inner membrane proteins TonB, ExbB and ExbD (the Ton system) (Moeck and Coulton, 1998). TonB interacts directly with the outer membrane receptors to couple them to the proton motive force of the inner membrane to facilitate transport of iron complexes across the outer membrane (Moeck and Coulton, 1998). A *B. pertussis ΔtonB exbB* mutant was unable to utilise haemin, ferrioxamine B and several other iron sources and exhibited a reduced ability to colonise the lungs of mice (Pradel *et al.*, 2000). This indicates that high-affinity iron-uptake systems are important for growth of *B. pertussis in vivo*. However, in other bacteria the Ton system is involved in the uptake of substances in addition to iron, and so further studies are necessary to determine if the effects observed are due solely to disruption of iron supply (Wooldridge and Williams, 1993; Moeck and Coulton, 1998; Pradel *et al.*, 2000).

Pertussis

The clinical features of pertussis tend to vary with age, general health and immune status. For example, adults and older children often show mild or atypical symptoms. Typical pertussis can be divided into the catarrhal, paroxysmal and convalescent stages. After an incubation period of 7–14 days, the catarrhal stage resembles a non-distinctive, upper respiratory tract infection with a mild cough, excessive mucus production and sometimes mild fever. The cough becomes increasingly severe and, after 7–10 days, the patient enters the paroxysmal stage. This is characterised by bouts of uncontrollable coughing, up to 20 or more in 24 hours, when the patient is attempting to clear the tenacious mucus from the airways. The paroxysms are often followed by the distinctive, inspiratory whoop. Vomiting and hypoxia may also ensue. A typical feature of the late catarrhal and early paroxysmal phases of pertussis is a pronounced leucocytosis with a predominant lymphocytosis.

Secondary infections, such as otitis media and pneumonia, are relatively frequent, especially in infants, and may result from impairment of clearance mechanisms by the bacterium. Complications resulting from pressure effects of the coughing paroxysms include subconjunctival haemorrhage, pneumothorax, epistaxis, subdural haematomas, hernias and rectal prolapse. Central nervous system disturbances, in the form of seizures and encephalopathy, may also occur and are thought to be due to the hypoxia or, possibly, to bacterial toxins. The paroxysmal stage may last for 1–6 weeks or more, then the cough gradually subsides over a prolonged convalescent period. A high percentage of all reported cases of pertussis require hospitalisation, especially those in infants <6 months of age. A recent estimate of the case-fatality rate in the US was 0.2% and the majority of these deaths (84%) were in infants <6 months (Centers for Disease Control and Prevention: http://www.cdc.gov/health/diseases.htm). Further information on clinical aspects of pertussis may be found elsewhere (Olson, 1975; Cherry *et al.*, 1988; Walker, 1988; Hodder and Mortimer, 1992; Cherry, 1996; Hewlett, 2000).

Pathogenesis

B. pertussis is believed to enter the respiratory tract in aerosol droplets generated by the coughing of an infected individual. The organism is able to alternate between virulent and avirulent states by switching on and off groups of genes in response to specific environmental cues (see below). It is not known if it travels between human hosts in a fully virulent state, a fully avirulent state or an intermediate phase but conditions within the respiratory tract must provide the stimulus for the expression of the appropriate virulence factors, if they are not already expressed. Mucus is presumably the first point of contact for *B. pertussis* upon entering the respiratory tract but how the organism crosses the mucus layer to reach its site of localisation is unknown. Unlike most bacterial respiratory pathogens, *B. pertussis*, and other bordetellae, are able to adhere very efficiently to healthy, rapidly beating ciliated epithelial cells (**Fig. 1**). The organism then proliferates and produces toxins and adhesins that serve to counteract the host's defences and prevent elimination of *B. pertussis*. One or more toxins counter the muco-ciliary clearance mechanism by paralysing cilia and provoking the extrusion of ciliated and non-ciliated epithelial cells and others impede the actions of phagocytic cells. Other components may stimulate enhanced production of mucus. The net effect of these activities is the accumulation of viscid mucus that contains white cells and epithelial cells and coughing is the only means of clearing the respiratory tract of this material.

Whether there are bacterial factors that directly promote a coughing response is unknown. There is also evidence that *B. pertussis* inhibits specific immune responses within the respiratory tract. The impairment of the respiratory tract defences also predisposes the individual to secondary infections with other bacteria. Prior to the advent of antibacterial chemotherapy, these were serious consequences of pertussis and had a high mortality rate.

B. pertussis does not invade beyond the respiratory tract or, if it does, it is unable to survive. However, at least one *B. pertussis* factor, pertussis toxin, produces systemic effects, such as lymphocytosis and effects on insulin and glucose metabolism, but the role of these effects in the pathogenesis of pertussis remains speculative. *B. pertussis* can invade and survive, but not replicate, within eukaryotic cells. It is proposed that intracellular residence, particularly within alveolar macrophages, may allow prolonged survival within the respiratory tract. Humans are the only natural host of *B. pertussis* and the organism is not known to survive for long outside of its host. Continued existence of *B. pertussis* therefore requires transmission from an infected to a susceptible individual. Almost nothing is known about this process and *B. pertussis* is not naturally transmissible between laboratory animals.

The roles of the known virulence factors of *B. pertussis* in the pathogenesis of whooping cough are discussed below. Our knowledge of this process is clearly incomplete and several caveats should be borne in

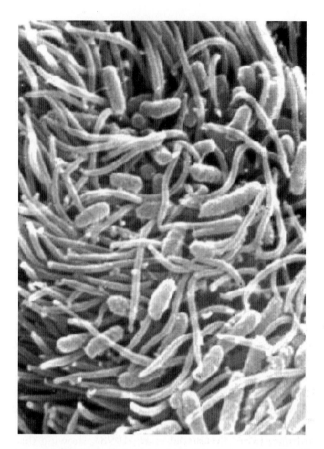

Fig. 1 Adhesion of *B. bronchiseptica* to canine ciliated respiratory epithelia. Scanning electron micrograph of a sample of trachea from a dog infected with *B. bronchiseptica*. The bacteria can be seen tightly adherent to the cilia. Image kindly provided by Dr Hal Thompson.

mind when considering the information presented. It is likely that *B. pertussis* produces additional as yet uncharacterised virulence factors, an assumption supported by recent analysis of the available genome sequence information (Antoine *et al.*, 2000). Secondly, *B. pertussis* naturally only infects humans and there are therefore drawbacks to all of the animal models of infection that are used to study pathogenesis and immunity in pertussis. A more accurate picture of the pathogenic mechanisms in pertussis and the role of individual *B. pertussis* virulence factors can probably be gleaned by including a comparison with the other well-characterised species of *Bordetella*, namely *B. parapertussis*, *B. bronchiseptica* and *B. avium*. The diseases they produce have certain common features in their respective hosts yet they do not have an identical complement of virulence factors (Parton, 1999). The common features include: age-related susceptibility to disease; adherence of organisms to ciliated epithelium and loss of ciliated cells from the respiratory tract; reduced weight gain of the host; excessive mucus production; and some form of cough. The only virulence factors known to be common to these species are tracheal cytotoxin, dermo-necrotic toxin, lipopolysaccharide/lipo-oligosaccharide and fimbriae, indicating their likely importance in pathogenesis.

The molecular basis of *B. bronchiseptica* pathogenicity has been studied extensively recently. This organism can establish respiratory infection in rodents from a low infectious dose and is transmissible between laboratory animals. Therefore, studies on *B. bronchiseptica* infection in its natural host animals will be referred to, where pertinent to *B. pertussis* infection.

Regulation of Virulence and Antigenic Variation

It has long been recognised that *B. pertussis* can alternate between virulent and avirulent forms either by genotypic changes (serotype variation and phase variation) or by phenotypic changes (antigenic modulation) in response to *in vitro* growth conditions. (The mechanism and significance of serotype variation is described under the section on Fimbriae, p. 1578.) Phase variation was first described by Leslie and Gardner (1931) as a stepwise degradative process from phase I, characteristic of fresh isolates, to the avirulent phase IV, with the different phases being distinguished by colony morphology, haemolysis and serology. It is now clear that degradation of *B. pertussis* strains is even more complicated than the above four stages but the spontaneous occurrence of phase IV or 'avirulent-phase' mutants that no longer express any of the above-mentioned virulence factors associated with fresh isolates (except tracheal cytotoxin (TCT) and LPS) is well known (Coote and Brownlie, 1988; Stibitz and Miller, 1994; Cotter and DiRita, 2000). Such mutants occur at a high frequency of 1 per 10^3 to 10^6 organisms (Peppler, 1982; Weiss *et al.*, 1984), due to spontaneous mutation in the *bvg* locus (see below). Both reversible frameshift mutations (Stibitz *et al.*, 1989) and irreversible deletions (Monack *et al.*, 1989) have been found in avirulent phase strains of *Bordetella*. Intermediate-phase variants of *B. pertussis*, expressing different combinations of virulence factors, have also been reported (Goldman *et al.*, 1984; Carbonetti *et al.*, 1994; Stibitz and Miller, 1994; Cotter and DiRita, 2000).

Expression of these same virulence factors is also regulated in response to growth conditions *in vitro* (Robinson *et al.*, 1986; Coote, 1991; Cotter and DiRita, 2000). The term 'antigenic modulation' was introduced by Lacey (1960) to describe this freely reversible phenotypic change which occurs when *B. pertussis* is grown with high levels of certain salts, organic acids or at low temperatures. The so-called C (cyanic) mode or avirulent phenotype occurs after growth, for example, in high concentrations of magnesium sulphate or at 25°C and, in early studies, it was distinguished from the X (xanthic) mode or virulent phenotype by serology and cultural characteristics. A serologically distinct intermediate or I mode, produced under submodulating conditions, was also identified.

The similar loss of virulence factors during antigenic modulation and phase variation suggested that a common regulatory mechanism is involved and that phase variation is the genotypic equivalent of antigenic modulation. Thus, phase I and X mode are essentially the same, as are phase IV and C mode, although the ability of phase IV strains to grow on nutrient agar (Leslie and Gardner, 1931; Peppler and Schrumpf, 1984) suggests that additional changes may occur upon repeated subculture. Weiss *et al.* (1983) introduced the terms 'virulent-phase' for X mode and phase I and 'avirulent-phase' for C mode and phase IV to emphasise this similarity. The main difference is that the X to C change is environmentally regulated, affects all cells in the population and is freely reversible, whereas the phase I to IV change is a relatively infrequent, mutational event which requires subsequent clonal selection for recognition, and reversion is equally infrequent or impossible.

Elucidation of the mechanisms of these processes has been a direct result of the pioneering work of Weiss *et al.* (1983), who developed a system for transposon mutagenesis of *B. pertussis*. They demonstrated that a single Tn5 insertion in the *B. pertussis* chromosome simultaneously abolished the expression of several virulence factors and thus identified a virulence regulatory locus. They suggested that the product of this locus acted as a positive effector to co-ordinately activate the virulence genes. Weiss and Falkow (1984) proposed that modulating conditions could influence the expression of the regulatory locus, which they designated *vir*, that in turn would alter expression of the virulence determinants. Subsequently Knapp and Mekalanos (1988) used Tn*phoA* mutagenesis to identify genes encoding secreted proteins (i.e. potential virulence factors) and which were subject to control by modulating conditions. This work led to the discovery of several virulence-repressed genes (*vrg*) which were regulated in a reciprocal fashion to the virulence-activated genes (*vag*) described above. The products of the *vrg* loci are probably related to the C-mode-specific antigens described by Lacey (1960) but their function remains unknown. The *vir* locus was later re-named *bvg* (*Bordetella* <u>v</u>irulence <u>g</u>ene) to distinguish it from *vir* loci in other organisms (Arico *et al.*, 1989) and the virulent and avirulent phases/modes are now usually referred to as Bvg$^+$ and Bvg$^-$, respectively.

The *bvg* locus encodes two proteins, BvgS and BvgA, which are members of the two-component regulator family which control expression of bacterial genes via phosphorylation/dephosphorylation of component proteins (Arico *et al.*, 1989; Stibitz and Yang, 1991). The two components are usually a sensor kinase (SK) and a response regulator (RR). The sensor kinase is usually an integral cytoplasmic membrane protein with periplasmic and cytoplasmic domains. The RR is usually a cytoplasmic DNA-binding protein that is the actual transcriptional regulator. The SK is responsive to specific environmental cues and transfers this information to the interior of the bacterial cells to effect changes in gene expression by a phospho-relay reaction that involves phosphorylation of the RR.

The SK, BvgS, is a 135-kDa integral protein of the inner (cytoplasmic) membrane. It spans the membrane twice and possesses a periplasmic input (sensor) domain, and cytoplasmic linker, transmitter, receiver and output domains (see **Fig. 2**). The 23-kDa BvgA is a typical cytoplasmic RR (Stibitz and Yang, 1991; Scarlato *et al.*, 1998). Incoming Bvg$^+$ signals are transmitted via a complex phospho-relay system between alternating histidine and aspartate residues (His-Asp-His-Asp), first by autophosphorylation of the His residue at position 729 in the transmitter domain of BvgS, followed by transfer of the phosphate group to the BvgS receiver domain (Asp1023), then to the histidine phospho-transfer output domain (His1172) and finally to the N-terminal receiver domain (Asp54) of BvgA (Uhl and Miller, 1996a, b) (see **Fig. 2**). There is evidence that the phospho-relay involves dimerisation of both BvgS and BvgA (Stibitz and Yang, 1991; Beier *et al.*, 1995, 1996). The BvgS receiver domain contains autophosphatase activity

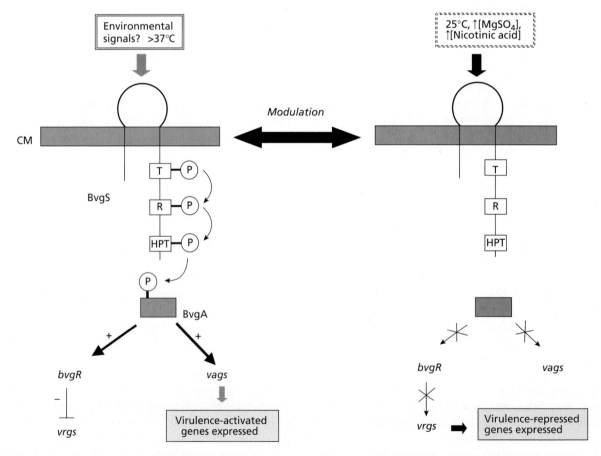

Fig. 2 Bvg-mediated gene regulation in *B. pertussis*. In non-modulating conditions, BvgS autophosphorylates at the transfer domain (T) and the phosphate group is sequentially transferred to the receiver domain (R) and then to the histidine phospho-transfer domain (HPT). The phosphate group is then transferred to BvgA, making it competent to activate transcription of the virulence-activated genes (*vags*) including *bvgR*. BvgR is a repressor that inhibits expression of the virulence-repressed genes (*vrgs*). In modulating conditions, BvgS is inactive, the *vags* are not expressed and the *vrgs* are derepressed. CM, cytoplasmic membrane.

which may modulate the phospho-relay (Uhl and Miller, 1996a).

The *bvgAS* genes form a single operon and are adjacent to the FHA gene, *fhaB*, but are transcribed in the opposite direction (Stibitz *et al.*, 1988; Stibitz and Garletts, 1992). Under modulating (Bvg⁻) growth conditions, BvgA and BvgS are expressed only at a low level, from a weak constitutive promoter (Roy *et al.*, 1990; Scarlato *et al.*, 1990). Transfer of *B. pertussis* to non-modulating (Bvg⁺) growth conditions leads to BvgS-mediated phosphorylation of BvgA. This in turn enables BvgA to bind avidly, via a C-terminal helix-turn-helix domain, to particular DNA sequences upstream of *vag* loci where it interacts with RNA polymerase to allow transcription (Boucher *et al.*, 1994; Kinnear *et al.*, 1999). BvgA also binds to stronger *bvg* promoters to enhance *bvgAS* expression (Scarlato *et al.*, 1990).

Transcription analysis has revealed that the vag genes are turned on sequentially. Activation of the genes that encode some adhesins (fimbriae and filamentous haemagglutinin, FHA) occurs immediately after the shift to Bvg⁺ growth conditions, whereas expression of those that encode pertussis toxin (PT) and adenylate cyclase toxin (ACT) (late *vag* genes) begins after 2 hours (Scarlato *et al.*, 1991). Transcription of the *prn* gene occurs at an intermediate time (Kinnear *et al.*, 1999). The timing of activation of the vag genes appears to depend on the affinity, the number and orientation of BvgA-binding sites upstream of vag genes and the concentration of BvgA within the cell (Cotter and DiRita, 2000).

It is possible that further fine-tuning of the expression of individual virulence genes may occur via environmental influences on, for example, DNA topology and mRNA stability (Scarlato *et al.*, 1993; Graeff-Wohlleben *et al.*, 1995; Coote, 2000). The *vrg* loci are negatively regulated by a virulence-activated gene, *bvgR*, that encodes a 32-kDa repressor protein (Merkel *et al.*, 1998a). Under Bvg⁻ growth conditions, transcription of the *vag*s, including *bvgR*, ceases relieving the inhibition of *vrg* expression. Further details on the mechanism of virulence regulation by the BvgAS system may be obtained from Cotter and DiRita (2000) and Coote (2000).

The significance of Bvg-mediated regulation of virulence expression is not known. Any one of a number of *in vitro* growth conditions, such as high concentrations of magnesium sulphate, nicotinic acid or a temperature of 25°C, will cause modulation, perhaps by interfering in some way with the usual signal-transduction process. However, it is not known when, or even if, this process occurs *in vivo* and what the natural signals are. Likewise, the signals for the organism to adopt the virulent phase (other than growth at 37°C) are unknown but must be present under the usual growth conditions *in vivo* and *in vitro*, although the only known habitat of *B. pertussis* is the human respiratory tract. There has been speculation that *B. pertussis* may undergo modulation to the Bvg⁻ state to evade the host immune response and persist in the host, to aid transmission, to prolong survival between hosts or to colonise some other site (Robinson *et al.*, 1986; Wardlaw and Parton, 1988; Rappuoli, 1994; Stibitz and Miller, 1994). As yet, however, no function has been assigned to any of the *vrg* products. Expression of the *vag* products appears to be temporally controlled so that adhesins are expressed first and toxins later. This would be appropriate if a Bvg⁻ phase organism was transferred to a new host or to a new site within the host where it had to combat host defences.

The idea of a significant role for the co-ordinate regulation of virulence in *B. pertussis* is strengthened by the knowledge that this process also occurs in *B. bronchiseptica*, *B. parapertussis* and *B. avium* and sequences homologous to *bvg* have been shown in all three species (Arico *et al.*, 1991; Gentry-Weeks *et al.*, 1991). However, it is now clear that different sets of genes are controlled by *bvgAS* in these species. For example, in *B. bronchiseptica*, but apparently not in *B. pertussis*, Bvg-activated factors include a type III protein secretion system, possibly involved in down-regulation of the host immune system (Yuk *et al.*, 1998a, 2000) and LPS expression (van den Akker, 1998). Bvg-activated factors unique to *B. pertussis* include pertussis toxin (Gross *et al.*, 1989) and tracheal colonisation factor (Finn and Stevens, 1995). Some factors that are expressed only in the Bvg⁻ phase of *B. bronchiseptica* and not in that of *B. pertussis* include an adhesin (Register and Ackermann, 1997), the siderophore alcaligin (Giardina *et al.*, 1995), flagella (Akerley *et al.*, 1992) and urease (McMillan *et al.*, 1996), whereas the *vrg* products mentioned above are expressed only in *B. pertussis*.

Thus, Bvg regulation appears to control different functions in these two species and this may be related to their different lifestyles. Such differences include severity of disease caused, host range and, possibly, pathways of transmission (Cotter and DiRita, 2000). *B. bronchiseptica*, but not *B. pertussis*, is capable of environmental survival and growth in nutrient-limited conditions (Porter *et al.*, 1991; Porter and Wardlaw, 1993). The Bvg⁻ phase of *B. bronchiseptica* appears to be essential for this (Cotter and Miller, 1994; Akerley *et al.*, 1995). Studies with Bvg⁺ and Bvg⁻ phase-locked mutants in animals have shown that the Bvg⁺ phase of both species is necessary and sufficient for respiratory infection and that ectopic expression of *bvg*-repressed factors in the Bvg$^+$ phase actually hinders the infection process (Cotter and Miller, 1994; Akerley *et al.*, 1995; Martinez de Tejada *et al.*, 1998; Merkel *et al.*, 1998b). Also, antibodies to *vrg* products are not detected in animals infected with wild-type *B. pertussis* or *B. bronchiseptica* (Cotter and Miller, 1994). Such data argue strongly against an *in vivo* role for the Bvg$^-$ phase of *B. pertussis*, at least in rodents, and for the Bvg$^-$ phase of *B. bronchiseptica* in its natural animal hosts.

It has been suggested that the Bvg system may allow the bordetellae to sense their location within or outside of the respiratory tracts of their respective hosts (Cotter and DiRita, 2000). In *B. bronchiseptica*, it could allow the organism to adapt to the very different conditions *in vivo* and *ex vivo* (Cotter and Miller, 1994, 1997). No role for the Bvg$^-$ phase of *B. pertussis* is apparent, as it does not appear to be involved in environmental or intracellular survival. It may be that this phase and the *vrg* products are evolutionary remnants in this organism (Martinez de Tejada *et al.*, 1998; Merkel *et al.*, 1998b). It has been further suggested that an intermediate phase (Bvg^i), between Bvg$^+$ and Bvg$^-$, is important for sensing location within the respiratory tract and perhaps for transmission to new hosts. Such a phase has been reported for *B. bronchiseptica* and *B. pertussis* and is characterised by the absence of *vrg* products, the presence of some but not all *vag* products, and some products unique

to the Bvg^i phase (Cotter and Miller, 1997; Cotter and DiRita, 2000). This phase perhaps corresponds to the I mode of *B. pertussis* induced by growth under sub-modulating conditions, as described by Lacey (1960). Cotter and DiRita (2000) have hypothesised that organisms in the nasal cavity, at a significantly lower temperature than those deeper in the respiratory tract, may be in the Bvg^i phase and primed for transmission to a new host. A Bvg^i phase-specific protein, BipA, has been identified in *B. bronchiseptica* and the *bipA* gene is present in *B. pertussis* (Antoine *et al.*, 2000; Cotter and DiRita, 2000). BipA is a large outer membrane protein that has a region of homology to the intimin protein of enteropathogenic *E. coli* and its function is currently being investigated. For further speculation on the role of Bvg regulation in *Bordetella* species, see Cotter and DiRita (2000) and Coote (2000).

Virulence Factors

A cartoon showing the known virulence factors of *B. pertussis* and their cellular location is shown in **Fig. 3**. The distribution of some of these factors has been confirmed by immuno-electron microscopy with gold-labelled antibodies (Blom *et al.*, 1994). For example, the presence of pertactin was revealed by intense and evenly distributed labelling of the cell surface. Antibody to filamentous haemagglutinin (FHA)

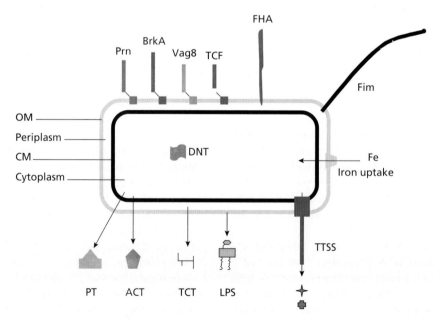

Fig. 3 Virulence factors of *B. pertussis* and their cellular location. OM, outer membrane; CM, cytoplasmic membrane; Prn, pertactin; TCF, tracheal colonisation factor; FHA, filamentous haemagglutinin; Fim, fimbriae; PT, pertussis toxin; ACT, adenylate cyclase toxin; TCT, tracheal cytotoxin; LPS, lipopolysaccharide; DNT, dermonecrotic toxin; TTSS, type III secretion system (it is uncertain if the TTSS is functional in *B. pertussis*).

revealed aggregates of material between or adherent to the cells, which suggests that this component is readily shed, at least during growth *in vitro*. Similarly, antibodies to pertussis toxin and adenylate cyclase toxin labelled amorphous material between or adherent to cells, again indicating that these components are readily dispersed after export. *B. pertussis* is reported to be capsulated when freshly isolated (Lawson, 1940) but the significance of this structure for virulence is unknown and, until recently, there was no information on its composition. A recent report has described the production of a carbohydrate slime-layer when the organism is grown *in vitro* as a biofilm (Bosch *et al.*, 2000). Interestingly, genes for the biosynthesis of a polysaccharide capsule have been identified in the genome sequence of *B. pertussis* but the conditions governing their expression are uncertain (Antoine *et al.*, 2000).

Adhesins

Adhesion to host cells is critical to the initiation of infection by most bacterial pathogens and *B. pertussis* is no exception. *B. pertussis*, *B. bronchiseptica* and *B. parapertussis* exhibit tropism for ciliated epithelia (Matsuyama and Takino, 1980; Bemis and Kennedy, 1981; Tuomanen *et al.*, 1983b; Weiss and Hewlett, 1986; Sekiya *et al.*, 1988; Dugal *et al.*, 1990; Ackermann *et al.*, 1997; Soane *et al.*, 2000). The bacteria have been observed in close apposition to the ciliary shaft, usually located towards the bottom of the cilium (Tuomanen and Hendley, 1983; Soane *et al.*, 2000).

Human nasal turbinate tissue in an air-interface organ culture has been used recently to study *B. pertussis* pathogenesis (Soane *et al.*, 2000). Bacteria bound initially to mucus, cilia and areas of epithelial damage and thereafter mainly to cilia. There was little adherence to non-ciliated cells. Bacterial replication led to great increase in the number of *B. pertussis* attached to the cilia and was associated with the loss of ciliated and non-ciliated cells and hypersecretion. The ability to colonise healthy ciliated respiratory epithelium is unusual among bacterial pathogens and probably accounts for the low infectious dose of the bordetellae in their natural hosts (MacDonald and MacDonald, 1933; Cotter and Miller, 1996). *B. pertussis* is highly host-adapted, naturally infecting only humans, whereas *B. bronchiseptica* has a large host range, causing infection in a wide range of domestic and companion animals and occasionally humans (Goodnow, 1980; Woolfrey and Moody, 1991; Mesnard *et al.*, 1993; Gueirard *et al.*, 1998b). *B. pertussis* adheres less well to ciliated cells from rabbits, mice or hamsters than to human ciliated cells, whereas *B. bronchiseptica* adhered preferentially to non-human ciliated cells (Tuomanen *et al.*, 1983). This may fully or partly explain the difference in species specificity between *B. pertussis* and *B. bronchiseptica* but whether it is due to a different complement of adhesins or to the same basic adhesins with different specificities is not yet clear.

The different *B. pertussis* adhesins and their role in *B. pertussis* pathogenesis are described below. Because of the difficulty of obtaining ciliated cells, particularly from humans, the majority of the *in vitro* studies on *B. pertussis* adhesion have utilised immortal non-differentiated epithelial-like eukaryotic cell lines. Therefore, it must be borne in mind that the results from such studies may not accurately reflect the contribution of a particular virulence factor to *B. pertussis* pathogenesis in humans. This would appear to be the case with many of the studies on *B. pertussis* fimbriae. However, in other cases a good correlation between the contribution of a virulence factor to adhesion to cell lines and ciliated cells (or other relevant cells such as macrophages) has been found, as is the case for filamentous haemagglutinin (FHA) (see below). Also, because experimental infection of humans is not acceptable, *B. pertussis* infection has been studied using small animal models, predominantly mice. Large numbers of *B. pertussis* need to be introduced into the respiratory tract of mice in order to establish infection. This may obviate the requirement for a particular factor that is normally necessary for initiation of infection from a low infectious dose. Conversely, loss of an adhesin may alter the host–pathogen interaction, rendering a *B. pertussis* strain more virulent in animals. An FHA mutant of *B. pertussis* introduced intratracheally into rabbits was not retained in the trachea but rapidly entered the lungs and caused more pathology than a wild-type *B. pertussis* strain (Weiss and Goodwin, 1989). Strains of *B. pertussis* isolated from lethally-infected infant mice were generally afimbrate (Weiss and Goodwin, 1989). We also found that a defined non-fimbrial mutant produced more severe infection in aerosol-infected adult mice (M. Roberts, unpublished observation).

Filamentous haemagglutinin

Filamentous haemagglutinin (FHA) has been shown to be the major adhesin of *B. pertussis* for a variety of cell types. Mature FHA is a polypeptide of 220 kDa that forms filaments of ~ 2–4 nm $\times$ 40–50 nm (Arai and Sato, 1976; Locht *et al.*, 1993; Makhov *et al.*, 1994). It is present on the surface of *B. pertussis* and is also secreted into the medium (Arai and Sato, 1976).

The FHA structural gene, *fhaB*, encodes a precursor (FhaB) of 367 kDa of which the N-terminal ~2000 residues comprise the mature protein (Relman *et al.*, 1989; Delisse-Gathoye *et al.*, 1990; Domenighini *et al.*, 1990). A model of FHA structure has been produced by combining circular dichroism, electron microscopy and computational analyses (Makhov *et al.*, 1994). FHA is predicted to be composed of a shaft ~35 nm in length, formed by the FHA molecule folding back on itself, and a globular head formed from the N- and C-termini producing a horseshoe nail-shaped molecule. The protein contains two large repeat regions: the first (from residue 344 to 1065) consists of 38 copies of a 19-residue sequence and the second (from residue 1440 to 1688) consists of 13 copies of a different 19-residue module. These repeats are predicted to form short β strands connected by two β turns and it is these that make up the bulk of the shaft.

The signals for secretion of FhaB through the inner and outer membrane are contained within the residue at the very N-terminus of the precursor (Jacob-Dubuisson *et al.*, 1996; Lambert-Buisine *et al.*, 1998). The first 71 amino acids of the precursor comprise the signal sequence for directing transport across the inner membrane (Lambert-Buisine *et al.*, 1998). However, the first 22 amino acids of the signal sequence are not necessary for secretion and form what has been called an extension sequence. The signal sequence is removed at a putative Lep signal peptidase cleavage site and the N-terminal glutamine residue is modified to a pyro-glutamine residue. The functional significance of this modification is unknown; it is not required for secretion or functional activity of FHA and it does not occur in *E. coli* expressing FhaB (Lambert-Buisine *et al.*, 1998). Transport of the precursor across the outer membrane requires an accessory outer membrane protein FhaC (Willems *et al.*, 1994; Jacob-Dubuisson *et al.*, 1996). FhaC has homology to ShlB and HmpB of *Serratia marcescens* and *Proteus mirabilis*, OMPs which are required for the secretion and activation of the haemolysins produced by these organisms (Willems *et al.*, 1994). Along from the signal peptide (residues 91–205) FhaB exhibits extensive homology to the *S. marcescens* and *P. mirabilis* haemolysins ShlA and HmpA (Delisse-Gathoye *et al.*, 1990). N-terminal fragments of FHA containing this region are successfully secreted by FhaC (Jacob-Dubuisson *et al.*, 1996). Experimental and structural predictions indicate that FhaC forms a β-barrel pore in the outer membrane but it is unusual in that it has an odd number of β strands (Lambert-Buisine *et al.*, 1998; Guedin *et al.*, 2000). The FhaB precursor is predicted to be translocated through this pore, and proteolytic cleavage at the cell surface would release mature FHA from the C-terminal ~1600 residues of the precursor (Jacob-Dubuisson *et al.*, 1996).

The role of the C-terminal domain of FhaB is uncertain but it is not directly involved in the secretion of FhaB. *B. pertussis* strains with frameshift mutations within the region of *fhaB* encoding the C-terminal domain do not secrete FHA into the culture medium but N-terminal fragments of FhaB are secreted successfully (Locht *et al.*, 1993; Jacob-Dubuisson *et al.*, 1996; Renauld-Mongenie *et al.*, 1996a). It has been proposed that the C-terminus acts as an intramolecular chaperone, interacting with the N-terminal domain in the periplasm to keep it in a secretion-competent form (Renauld-Mongenie *et al.*, 1996a). The C-terminal domain contains a proline-rich region at its C-terminal end and this may function to anchor FhaB to the cell wall (Domenighini *et al.*, 1990; Locht *et al.*, 1993). Release of FHA from the surface of *B. pertussis* may facilitate transmission of the organism within the respiratory tract or from host to host (Arico *et al.*, 1993). It has also been proposed that FHA (and PT) released from *B. pertussis* in the respiratory tract could bind to other bacteria by a process which has been called 'piracy of adhesins', allowing them to efficiently adhere to cilia and promoting secondary bacterial infections (Tuomanen, 1988).

FHA has at least three different-binding specificities for eukaryotic cells. First, in the N-terminus of FHA (residues 1–400) is a region that mediates haem-agglutination, which can be inhibited by sulphated sugars such as heparin and dextran sulphate which are present on glycolipids and proteoglycans but not by non-sulphated sugars such as dextran (Menozzi *et al.*, 1991b, 1994b). Glycolipids and proteoglycans possessing sulphated sugar residues are ubiquitous in the respiratory tract and are bound by *B. pertussis* (Brennan *et al.*, 1991; Hannah *et al.*, 1994). Proteoglycans are widely used by microbes to mediate initial attachment to epithelial surfaces (Rostand and Esko, 1997). Attachment of *B. pertussis* to certain epithelial cell lines (such as WiDr and CHO) cells can be largely inhibited by sulphated sugars (Brennan *et al.*, 1991; Menozzi *et al.*, 1994b).

FHA can mediate binding of *B. pertussis* to the leucocyte integrin complement receptor 3 (CR3, $\alpha_M\beta_2$, CD11b/CD18) both on the surface of macrophages and to purified CR3 (Relman *et al.*, 1990; Van Strijp *et al.*, 1993). FHA possesses a motif, Arg-Gly-Asp (RGD), at positions 1097–1099 (Relman *et al.*, 1989) and RGD sites in eukaryotic proteins such as fibronectin mediate binding to integrin receptors on eukaryotic cells (Ruoslahti and Pierschbacher, 1987). Site-directed mutagenesis of *fhaB* to change the

RGD site in FHA to RAD reduced the binding of a *B. pertussis* strain to macrophages by $\sim 50\%$, implicating the RGD site in direct binding to CR3 (Relman *et al.*, 1990). However, subsequent work has indicated that the FHA RGD site may act indirectly to enhance binding of FHA and *B. pertussis* to CR3. The RGD site binds to the leucocyte response integrin/integrin-associated protein complex, triggering a signal transduction cascade that activates CR3 rendering it receptive for FHA binding by some non-RGD-dependent mechanism (Van Strijp *et al.*, 1993; Ishibashi *et al.*, 1994). FHA RGD-derived synthetic peptides also inhibit CR3-mediated transepithelial migration of neutrophils *in vivo* and *in vitro* (Rozdzinski *et al.*, 1995).

Several other *B. pertussis* molecules also enhance FHA-mediated binding to macrophages. Binding of fimbriae to the very late antigen 5 integrin (β_1/α_5, VLA-5) on macrophages activates CR3 (see below). PT also enhances binding of *B. pertussis* to CR3 and the effect is dependent on the carbohydrate-recognition domains of the S2 and S3 components of the B subunits (van't Wout *et al.*, 1992). Binding of *B. pertussis* to CR3 promotes entry of the bacteria into macrophages *in vivo* in rabbits and mice (Saukkonen *et al.*, 1991; Hellwig *et al.*, 1999). Entry of bacteria into macrophages via CR3 is thought to be advantageous to bacteria because it proceeds without the induction of a respiratory burst (Wright and Silverstein, 1983).

The third adhesion region of FHA maps to residues 1141–1279 and mediates binding to lactosylceramides. This is responsible for attachment of FHA to cilia and for about 50% of the FHA binding to macrophages, although it is not important for uptake of *B. pertussis* into macrophages (Relman *et al.*, 1989, 1990; Saukkonen *et al.*, 1991). Recognition of lactosylceramides is also a property of the S2 subunit of PT and the carbohydrate-recognition domain of FHA and S2 share some sequence similarity (Relman *et al.*, 1990).

FHA also binds C4b-binding protein (C4BP), a plasma protein that inhibits the classical pathway of complement activation (Berggard *et al.*, 1997). However, FHA does not appear to be involved in resistance to serum killing of *B. pertussis* (see below) so binding of C4BP must have some other function, perhaps preventing opsonisation.

FHA can also inhibit the production of IL-12 by macrophages in response to LPS and IFNγ (McGuirk and Mills, 2000b). This is due to FHA stimulating the production of IL-10 by macrophages. *B. pertussis* respiratory infection in mice suppresses the response of pulmonary, but not splenic, T cells to mitogens and antigens (McGuirk *et al.*, 1998). This may be mediated

in part by the action of FHA but the region of the molecule responsible for this activity is unknown. Infection of monocytes by wild-type *B. pertussis*, but not FHA and adenylate cyclase toxin mutants, also inhibits antigen-dependent CD4+ T-cell proliferation (Boschwitz *et al.*, 1997a).

Short regions of FHA also have homology to C3b and coagulation factor X, eukaryotic ligands for CR3 (Sandros and Tuomanen, 1993). Antibodies to FHA are reported to bind to C3b and synthetic peptides derived from the appropriate region of FHA are reported to inhibit coagulation. CR3 is also responsible for mediating attachment of leucocytes to endothelia at sites of inflammation. Some anti-FHA monoclonal antibodies bound to cerebral microvessels in sections of human brains *in vitro* and rabbit brain *in vivo* (Tuomanen *et al.*, 1993). In rabbits this was reported to increase vascular permeability and inhibit leucocyte diapedesis. It should be noted that immunisation with acellular vaccines containing FHA has not been reported to cause cerebral side effects.

A number of *in vitro* adhesion assays using primary cells, cell lines and tissues of various types and from a variety of species has indicated that FHA is the major *B. pertussis* adhesin (Tuomanen, 1988; Relman *et al.*, 1989, 1990; Roberts *et al.*, 1991; Funnell and Robinson, 1993; Bassinet *et al.*, 2000). However, studies in mice using *B. pertussis* FHA mutants have suggested that FHA is far less important *in vivo* (Weiss *et al.*, 1985; Kimura *et al.*, 1990; Roberts *et al.*, 1991; Khelef *et al.*, 1994; Geuijen *et al.*, 1997). In some instances, this may have been because the mutants studied had an internal in-frame deletion in *fhaB* and produced a truncated FHA that may have retained some activity. However, a *B. pertussis fhaC* mutant (which does not have FHA on its surface) exhibited only a slight, transient defect in colonisation of the nasal cavity and tracheas of mice (Geuijen *et al.*, 1997). FHA is absolutely required for the colonisation of the tracheas of conscious but not anaesthetised rats by *B. bronchiseptica* (Cotter *et al.*, 1998). Thus, the use of anaesthesia in mice may account for the relative lack of effect of FHA mutations observed in some, but not all, of the experiments with *B. pertussis*. An advantage of studying *B. bronchiseptica* is that only small numbers of bacteria are necessary to infect laboratory animals, whereas large numbers of *B. pertussis* are needed to produce infection in these animals. As stated above, the introduction of large numbers of *B. pertussis* into the respiratory tract of mice may overcome the need for production of FHA and possibly other adhesins.

FHA, alone or in combination with other antigens, has been demonstrated to be a protective immunogen in rodents and is one of the main components of acellular pertussis vaccines (APVs) in use in humans (see below) (Kimura and Kuno-Sakai, 1990; Novotny *et al.*, 1991; Shahin *et al.*, 1992, 1995; Mills *et al.*, 1993, 1998; Redhead *et al.*, 1993; Roberts *et al.*, 1993; Cahill *et al.*, 1995; Mahon *et al.*, 1996; Guiso *et al.*, 1999).

Fimbriae

B. pertussis produces two related but antigenically distinct types of fimbriae, serotype 2 (Fim2) and serotype 3 (Fim3) fimbriae, whose expression is positively regulated by the *bvg* locus (Zhang *et al.*, 1985; Mooi, 1994). The fimbriae are ~200 μm in length and ~5 nm in diameter (Zhang *et al.*, 1985). The main body of each fimbria is composed of a major subunit protein, Fim2 or Fim3, of 22.5 kDa and 22 kDa, respectively. The same 40-kDa minor fimbrial protein, FimD, is present at the tip of both types of fimbriae (Geuijen *et al.*, 1997). The structural genes for Fim2 and Fim3 are scattered on the chromosome, as is the gene for another fimbrial subunit, FimX (Stibitz and Garletts, 1992). The *fimX* gene is not expressed in *B. pertussis* because of a deletion in its promoter region (Willems *et al.*, 1990). Other genes necessary for the production of Fim2 and Fim3 and the structural gene for the adhesive tip protein (FimD) are located between the structural gene for FHA (*fhaB*) and a gene for FHA secretion (*fhaC*) (Willems *et al.*, 1992). The arrangement of these genes is *fimABCD*, which is typical of fimbrial operons in other organisms, with the first gene encoding the major fimbrial subunit. However, the first gene, *fimA* (which is homologous to the other *B. pertussis* major fimbrial subunit genes) is not expressed because of a deletion at its 5′ end. In contrast, intact *fimA* genes are present in *B. bronchiseptica* and *B. parapertussis*, and *fimA* is expressed in *B. bronchiseptica* (Locht *et al.*, 1992; Willems *et al.*, 1992; Boschwitz *et al.*, 1997b). It has been proposed that *fimA* represents the primordial *Bordetella* fimbrial gene (Mooi, 1994). FimB and FimC respectively encode a periplasmic chaperone and an outer membrane usher/anchor protein necessary for fimbrial biogenesis (Locht *et al.*, 1992; Willems *et al.*, 1992). BvgA can bind upstream of *fimB* to activate transcription and *fimBCDfhaC* form a single transcriptional unit (Willems *et al.*, 1992, 1994; Mooi, 1994). Polar mutations in *fimBCD* also adversely affect FHA production (Locht *et al.*, 1992). Thus, production of two of the major adhesins, FHA and fimbriae, is tightly coupled. This is supported by the kinetics of transcription of the *fha* and *fim* genes upon shift of *B. pertussis* from 22°C to 37°C (Scarlato *et al.*, 1991) (see above).

Fimbrial production is subject to the same control as other *bvg*-activated genes. In addition, fimbrial production may exhibit a distinct form of variation called serotype variation. *B. pertussis* strains may produce none, one type or both types of fimbriae (Robinson *et al.*, 1986; Willems *et al.*, 1990). Production of a particular fimbria can be gained or lost at a frequency of about 1 per 10^3–10^4 cell divisions and is a random event (Willems *et al.*, 1990). This is now known to occur due to the insertion or deletion of cytosines within a run of 15 cytosine residues in the *fim2* and *fim3* promoter regions during chromosome replication by a process called slipped-strand mispairing. Such mutations are thought to interfere with or abolish the interaction between BvgA and RNA polymerase at the *fim* promoters.

Fimbriae are highly immunogenic and stimulate the production of agglutinating antibodies (Robinson *et al.*, 1986; 1989a). Both Fim2 and Fim3 share the same tip protein, FimD, and are thought to be functionally identical (Willems *et al.*, 1993). Therefore fimbrial variation is thought to be a means of counteracting the anti-fimbrial immune response of the host without compromising fimbrial function. There is evidence that fimbrial variation is important *in vivo*. Infection of mice immune to one fimbrial type leads to selection of *B. pertussis* strains that produce the other type (Robinson *et al.*, 1989a). It was thought that fimbriae were the only virulence factors that exhibit serotype variability but other virulence factors such as pertactin and pertussis toxin also exhibit inter-strain variation in their primary sequence (see below).

Early reports suggested that fimbriae were not important for the adhesion of *B. pertussis* to eukaryotic cells (Urisu *et al.*, 1986). More recent work has shown that fimbriae do bind to eukaryotic cells or extracellular matrix proteins. The type of cell or tissue and the species of animal from which the cells or tissues were derived can have major effects on the results of bacterial adhesion assays. For example, *B. pertussis* fim mutants were more adherent to Vero cells than wild-type *B. pertussis* but less adherent to tracheal rings prepared from the primate *Papio anubis* (Funnell and Robinson, 1993). Purified fimbriae bind to sulphated sugars that, as mentioned above, are ubiquitous in the respiratory tract (Geuijen *et al.*, 1996). The binding was attributed to the major subunit because removal of FimD did not significantly affect adhesion (Geuijen *et al.*, 1996). Binding to HEp-2 cells that display sulphated sugars was reduced by alteration of the sulphation state of the sugars (Geuijen *et al.*, 1996). However, it was subsequently found that a maltose-binding protein (MBP)–FimD fusion could bind sulphated sugars, such as heparin, indicating that both major and minor subunits may bind to sulphated sugars (Geuijen *et al.*, 1997). Using a Fim2–MBP fusion, two regions of Fim2, designated H1 and H2, and conserved basic amino acid and tyrosine residues within these regions were shown to be involved in heparin binding (Geuijen *et al.*, 1998). *B. pertussis* thus possesses at least two surface proteins, fimbriae and FHA, that can bind sulphated sugars.

FimD also mediates attachment of fimbriae to monocytes by binding VLA-5 (Hazenbos *et al.*, 1995a, b). Engagement of VLA-5 by FimD activates the macrophage CR3 and significantly enhances the FHA-mediated attachment of *B. pertussis* to monocytes (Hazenbos *et al.*, 1993, b). The binding of another CR3 ligand (C3bi-coated RBCs) to its receptor was also augmented by attachment of FimD to VLA-5 (Hazenbos *et al.*, 1995b). The effect of FimD–VLA-5 ligation on up-regulation of CR3 activity could be inhibited by a protein tyrosine kinase inhibitor. *B. pertussis* fimbriae are thus able to bind to both sulphated sugars (via the major and minor fimbrial subunits) and VLA-5 (via the minor subunit). A eukaryotic extracellular matrix protein, fibronectin, also binds both VLA-5 and sulphated sugars, and two regions of the major fimbrial subunit with similarity to fibronectin peptides that bind sulphated sugars have been identified (Geuijen *et al.*, 1996).

A *B. pertussis* strain unable to produce fimbriae due to a mutation in *fimB* was less able to colonise the nasopharynx, trachea and lungs of intranasally infected mice compared with its wild-type parent and the effect was greatest in the trachea (Geuijen *et al.*, 1997). However, the mutation also reduces the production of FHA by 75%. A strain with a mutation in *fhaC*, which inhibited production of FHA specifically, was less affected for respiratory tract colonisation, as was a strain unable to produce Fim2 or Fim3 (Geuijen *et al.*, 1997). A *B. bronchiseptica* Δ*fimBCD* mutant, unable to produce any fimbriae but unaffected in FHA production, was unaffected in its ability to adhere to different cell lines but was deficient in its ability to colonise the trachea but not the nasal cavity and larynx of rats (Mattoo *et al.*, 2000).

Fimbriae of *B. pertussis* stimulate agglutinating antibodies and there is historical evidence that agglutinating antibodies elicited by whole-cell vaccines correlated with immunity against pertussis in humans (Robinson *et al.*, 1985; Preston, 1988). However, it is uncertain whether the correlation was merely highlighting the general immunogenicity of particular vaccine preparations; it may be that other structures on the surface of *B. pertussis*, including LPS and pertactin, are also agglutinogens (Robinson *et al.*, 1989b). More recent studies using purified preparations have demonstrated that fimbriae are protective immunogens in mice and that a five-component acellular pertussis vaccine containing Fim2 and Fim3 is very efficacious in human infants (Ashworth *et al.*, 1982; Robinson *et al.*, 1989b; Jones *et al.*, 1995, 1996; Gustafsson *et al.*, 1996; Willems *et al.*, 1998).

Pertactin

Pertactin is a Bvg-regulated surface-located and secreted protein. This was originally referred to as P69 or 69-kDa protein which reflected its apparent molecular weight determined by SDS-PAGE (Montaraz *et al.*, 1985; Charles *et al.*, 1989; Makoff *et al.*, 1990). It was subsequently shown that pertactin migrates aberrantly in SDS-PAGE and its true molecular weight is ~ 60 kDa (Capiau *et al.*, 1990; Makoff *et al.*, 1990). Pertactin was first identified in *B. bronchiseptica* as a protein whose presence correlated with the efficacy of anti-*B. bronchiseptica* vaccines in pigs (Novotny *et al.*, 1985a). Homologous proteins were later shown to be present in *B. pertussis* and *B. parapertussis* (Montaraz *et al.*, 1985). The molecular weight of the pertactins produced by these three bordetellae differ slightly because they possess different numbers of one or other of the two repeat units present in pertactin (see below). *B. bronchiseptica* pertactin has an apparent molecular weight of 68 kDa (P68/pertactin) and that of *B. parapertussis* is 70 kDa (P70/pertactin). The *B. pertussis* protein is often referred to as P69/pertactin.

The pertactin gene (*prn*) of *B. pertussis* encodes a large precursor of 93.5 kDa that is processed at both the N- and C-termini to generate mature pertactin (Charles *et al.*, 1989, 1994). The N-terminus of the precursor possesses a conventional (~ 3 kDa) cleavable signal sequence responsible for Sec-mediated transport of pertactin precursor across the inner membrane (Charles *et al.*, 1989, 1994). The C-terminal 30 kDa domain of the precursor is necessary for translocation of pertactin from the periplasm through the outer membrane to the cell surface (Charles *et al.*, 1994). No accessory proteins are known to be involved in the transport of pertactin across the outer membrane. It is now recognised that many proteins use a similar mechanism to cross the outer membrane and they have been collectively called autotransporters (Henderson *et al.*, 1998). This type of protein secretion has been called type IV secretion but confusingly so has the secretion mechanism exemplified by pertussis toxin transport (Finlay and Falkow, 1997; Burns, 1999). In view of this, Henderson *et al.* (2000) have proposed that the autotransporter mechanism should be designated as a type V secretion system. The C-terminal domains of different autotransporters are the regions that exhibit the greatest homology (Henderson *et al.*, 1998). These domains are thought to fold into an amphipathic β barrel composed of an even number of anti-parallel β sheets that form a single channel pore in the outer membranes through which the mature region of the protein (passenger domain) is translocated. Once through the outer membrane, the mature protein may remain at the surface or be released into the external environment. Release from the C-terminal domain is achieved by proteolytic cleavage which, depending on the autotransporter, may be achieved either by self proteolysis or via a separate outer membrane protease. The C-terminal domain of pertactin can be used to deliver heterologous antigens to the surface of Gram-negative bacteria (M. Roberts, unpublished observation).

It is not known how pertactin is released from its 30-kDa transport domain. Proteolytic activity of pertactin has not been demonstrated and examination of the primary and tertiary structure of pertactin does not reveal evidence of a proteolytic domain. A separate outer membrane protease may be required for liberation of pertactin from its autotransporter domain as is the case for some other autotransporters (Henderson *et al.*, 1998). Interestingly, pertactin translocates to the cell surface and is processed normally when expressed in heterologous Gram-negative bacteria and processing also occurs in eukaryotic cells (Charles *et al.*, 1993;, 1994).

Large amounts of the pertactin autotransporter domain are present in the outer membranes of

B. pertussis and *E. coli* expressing the *B. pertussis prn* gene (Charles *et al.*, 1994). Whether the C-terminal domain has any function other than transport of pertactin is unknown, but it also possesses an RGD motif (Charles *et al.*, 1989). *B. pertussis* and other *Bordetella* spp. produce a number of different autotransporter proteins (see below) (see **Fig. 4**). The C-terminal domains of these other proteins are also present in the outer membrane in relatively large amounts and there is evidence that immune responses against these proteins can mediate protection against *B. pertussis* infection in mice (Hamstra *et al.*, 1995).

Mature pertactin as well as the autotransporter domain possesses an RGD motif (Charles *et al.*, 1989) (**Fig. 4**). This provided the first indication that pertactin may function as an adhesin. The role of pertactin in adhesion and the function of the RGD domain are discussed below. Two regions of direct repeats are also present in pertactin and in each case the motif is repeated five times in P69/pertactin (Charles *et al.*, 1989). The first repeat is located towards the N-terminus of the mature protein, adjacent to the RGD motif, and has the sequence Gly-Gly-X-X-Pro (GGXXP). The second repeat is located close to the C-terminus and has the sequence Pro-Gln-Pro (PQP).

The three-dimensional structure of P69/pertactin was determined by X-ray crystallography (Emsley *et al.*, 1996). The molecule is composed of a 16-stranded parallel β helix that is V-shaped in cross-section (**Fig. 5**). A number of loops extend from the helix and these are thought to contribute to the biological

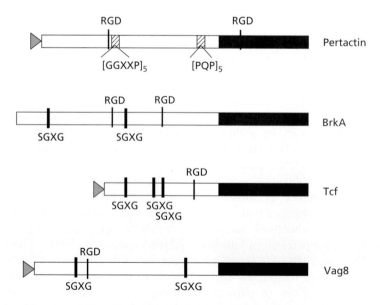

Fig. 4 The autotransporters of *B. pertussis*. The salient features of the structures of the pertactin, BrkA, tracheal colonisation factor (Tcf) and Vag8 precursor proteins are shown. The white bars indicate the mature portions of the proteins and the black bars the autotransporter domains. The triangle represents the Sec-dependent signal sequences. The striped boxes represent the repeat regions in pertactin. The positions of the Arg-Gly-Asp (RGD) and Ser-Gly-X-Gly (SGXG) motifs are indicated.

activity of pertactin, but the actual role of these loops has not been determined. One of the loops is formed from the GGXXP repeats. Mapping of the epitopes recognised by a number of pertactin-specific mono-clonal antibodies (mAbs) revealed that both repeat regions are immunodominant B-cell epitopes, with the majority of the mAbs recognising epitopes in the PQP repeat region (Charles *et al.*, 1991). One mAb, BBO5, recognising an epitope in this region (Pro-Gly-Pro-Gln-Pro-Pro) can confer passive protection against *B. bronchiseptica* infection in mice, and active immun-isation with the BBO5 epitope linked to a carrier protein enhances the clearance of *B. pertussis* from the lungs of mice (Montaraz *et al.*, 1985; Charles *et al.*, 1991).

Pertactin (pertussis and tactus (Latin)) was so named because eukaryotic (CHO) cells bound to

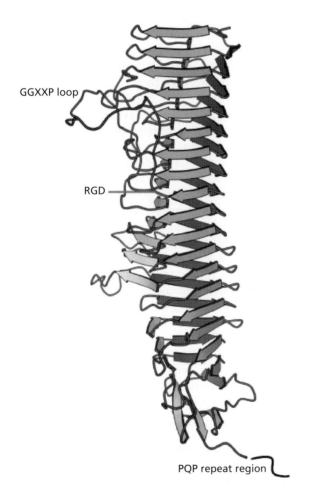

GGXXP loop

RGD

PQP repeat region

Fig. 5 Three-dimensional structure of *B. pertussis* pertactin, determined by X-ray crystallography, showing that pertactin is largely composed of a parallel β helix. The positions of the Gly-Gly-X-X-Pro (GGXXP) and Pro-Gln-Pro (PQP) repeat regions and the Arg-Gly-Asp (RGD) motif are indicated. The diagram was kindly provided by Dr Paul Emsley.

microtitre wells coated with P69/pertactin and because a *B. pertussis prn* mutant exhibited reduced adherence to the same cells (Leininger *et al.*, 1991). Synthetic peptides (SP) based on the region of mature pertactin possessing the RGD motif inhibited binding of CHO cells to purified pertactin whereas SP in which the RGD site had been changed to RGE or which were based on the RGD site in the C-terminal domain were not inhibitory. The role of pertactin in mediating attachment to eukaryotic cells *in vivo* is inferior to that of FHA and in some instances is only significant in the absence of FHA (Roberts *et al.*, 1991). There is some evidence that pertactin and FHA functionally interact (Arico *et al.*, 1993). *B. pertussis prn* mutants also exhibit reduced invasion of cultured epithelial cells and the pertactin RGD SP but not an SP based on the FHA RGD region inhibited invasion of these cells (Roberts *et al.*, 1991; Leininger *et al.*, 1992). Studies with a *B. bronchiseptica prn* mutant show that pertactin is involved in attachment of the organism to macrophages and porcine-ciliated respiratory cells (Forde *et al.*, 1999) (M. Roberts, unpublished observation).

Other work questions the roles of pertactin in invasion and of the pertactin RGD motif in mediating adhesion. Wild-type *prn* or a mutated *prn* which encoded pertactin with an RGE motif rather than RGD were expressed in a non-adhesive, non-invasive laboratory strain of *E. coli* (Everest *et al.*, 1996). The presence of either the wild-type or mutant form of pertactin on the surface of *E. coli* significantly enhanced its attachment to eukaryotic cells. However, expression of either form of pertactin did not improve the invasiveness of the *E. coli* strain. The position of the RGD site in the tertiary structure of pertactin also indicates that it is unlikely to function in eukaryotic cell binding (Emsley *et al.*, 1996). The adhesion to eukaryotic epithelial cells of *E. coli* expressing pertactin has some interesting features. Adhesion required a long incubation period and was reduced at 4°C and by inhibiting host cell protein synthesis (Everest *et al.*, 1996). This indicated that activation of the host cell is required for optimal binding of pertactin and is remi-niscent of the need for activation of the CR3 receptor to render it permissive for FHA binding (see above). Recent evidence suggests that expression of pertactin by *B. pertussis* may actually impede its invasion of human tracheal epithelial cells (Bassinet *et al.*, 2000).

B. pertussis strains with mutations in *prn* or *prn fhaB* colonised the lungs of aerosol-infected mice as effi-ciently as wild-type *B. pertussis* (Roberts *et al.*, 1991). In a separate study using the same strains, the *prn fhaB* mutant but not the *prn* mutant was cleared more rapidly from the lungs of intranasally-infected mice than a wild-type strain (Khelef *et al.*, 1994). In contrast, insertional inactivation of *prn* compromised the ability of *B. bronchiseptica* to colonise the lower respiratory tract but not the nasal cavity of mice (M. Roberts, unpublished observation).

Pertactin is an important protective immunogen of *Bordetella* spp. Immunisation with vaccines containing pertactin, including recombinant pertactin and attenuated *S. typhimurium* expressing pertactin, exerts a protective effect against infection with *B. pertussis* in mice and *B. bronchiseptica* in mice and pigs (Montaraz *et al.*, 1985; Novotny *et al.*, 1985b, c, 1991; Capiau *et al.*, 1990; Shahin *et al.*, 1990, 1995; Charles *et al.*, 1991; Romanos *et al.*, 1991; Roberts *et al.*, 1992, 1993; Strugnell *et al.*, 1992; Gotto *et al.*, 1993; Mills *et al.*, 1993, 1998; Redhead *et al.*, 1993; Boursaux-Eude *et al.*, 1999; Guiso *et al.*, 1999). Acellular pertussis vaccines containing pertactin have a high efficacy in human infants and recent serological analysis has indicated that this adhesin may be the most important component of these vaccines (Cherry *et al.*, 1998; Hewlett and Halperin, 1998; Storsaeter *et al.*, 1998).

BrkA Autotransporter

The second *Bordetella* autotransporter to be described was BrkA (*Bordetella* resistance to killing), which mediates protection of *B. pertussis* against complement killing by non-immune serum (Fernandez and Weiss, 1994). The gene for BrkA was identified by analysis of the site of insertion of Tn*5lac* in a *B. pertussis* mutant that was 10-fold less virulent in intranasally-infected infant mice (Weiss and Goodwin, 1989; Fernandez and Weiss, 1994). Two adjacent but divergently transcribed genes, *brkA* and *brkB*, are required for the production of surface-located BrkA and serum resistance (Fernandez and Weiss, 1994). The intragenic region contains a putative BvgA-binding site. The BrkA precursor lacks a typical N-terminal signal sequence and so is thought to cross the inner membrane by a sec-independent process. BrkB, a predicted inner membrane protein, is assumed to be involved in translocation of BrkA across the inner membrane.

The 103-kDa BrkA precursor is processed to yield a 73-kDa mature protein and a ~30-kDa C-terminal autotransporter domain (Fernandez and Weiss, 1994; Shannon and Fernandez, 1999). Recombinant BrkA C-terminal domain can form pores in lipid bilayers (Shannon and Fernandez, 1999). Two RGD motifs are present in mature BrkA as well as two motifs (SGXG) reported to be involved in glycosaminoglycan binding (see below) (Fernandez and Weiss, 1994) (Fig. 4).

B. pertussis brkA or *brkB* mutants are ~10-fold more sensitive to killing by normal human serum than their wild-type parent strain (Fernandez and Weiss, 1994). A *prn* mutant was also more sensitive to serum killing but less so than *brkA* or *brkB* mutants and the effect was not statistically significant (Fernandez and Weiss, 1994). Mutations affecting the production of a number of other *bvg*-regulated virulence factors (pertussis toxin, adenylate cyclase toxin, filamentous haemagglutinin, dermonecrotic toxin, tracheal colonization factor and Vag8) had no effect on serum sensitivity (Fernandez and Weiss, 1998). Interestingly, BrkA was found to protect against antibody-independent classical pathway complement activation. How this occurs is unknown but it was proposed that it might involve the glycosaminoglycan-binding sites (Bourdon *et al.*, 1987; Fernandez and Weiss, 1994). Two or more SGXG motifs are also present in the autotransporters tracheal colonisation factor and Vag8 (**Fig. 4**). This motif is the recognition signal for the covalent attachment of glycosaminoglycan units to proteins within eukaryotic cells to form proteoglycans but it should be preceded by at least two acidic amino acid residues in the 2–4 residues before the motif (Bourdon *et al.*, 1987). None of the SGXG motifs in the *B. pertussis* autotransporters conform to this rule. It seems unlikely, therefore, that BrkA attaches to glycosaminoglycan residues. However, the presence of the SGXG motif in several *B. pertussis* autotransporters suggests that they may have some function. Expression of BrkA is unable to protect *B. pertussis* from serum killing by immune serum (Weiss *et al.*, 1999). The *brk* genes are present in *B. bronchiseptica* and *B. parapertussis* but, at least in the case of *B. bronchiseptica*, BrkA does not appear to play a role in resistance to complement-mediated killing (Rambow *et al.*, 1998). BrkA may have other functions in *B. pertussis*, including adhesion to epithelial cells and resistance to antimicrobial peptides (Fernandez and Weiss, 1994, 1996). There is no information on the ability of BrkA to function as a protective immunogen and it is not included in any current acellular vaccines.

Tracheal Colonisation Factor

The third autotransporter reported to be produced by *B. pertussis* is tracheal colonisation factor (TCF), so named because a *tcf* mutant had a reduced ability to colonise the trachea but not the lungs of mice (Finn and Stevens, 1995). The TCF precursor has an apparent molecular weight of 90 kDa and contains a typical signal sequence at the N-terminus and a C-terminal autotransporter domain of ~30 kDa (Finn and Stevens, 1995). Unlike pertactin, the majority of the mature N-terminal domain is released from the *B. pertussis* surface (Finn and Stevens, 1995). The mature sequence possesses a RGD motif and three SGXG motifs (Finn and Stevens, 1995) (**Fig. 4**). The ability of purified TCF to protect against *B. pertussis* infection has not been investigated. However, intranasal immunisation of mice with an attenuated *Vibrio cholerae* strain expressing TCF reduced the ability of *B. pertussis* to colonise the trachea (Chen *et al.*, 1998).

Virulence-Activated Gene 8

Virulence-activated gene 8 (Vag8) is the product of a gene identified in a screen, using the transposon Tn*phoA*, for novel genes positively regulated by the *vir* (*bvg*) locus (Knapp and Mekalanos, 1988). The product of *vag8* is a polypeptide of 94.8 kDa. The N-terminus possesses a predicted 37 amino acid signal sequence, the cleavage of which would yield a polypeptide of ~91 kDa (Finn and Amsbaugh, 1998). Vag8 has homology to the other *B. pertussis* autotransporters, pertactin, BrkA and TCF (**Fig. 4**), with homology strongest to pertactin. The greatest homology is between their predicted C-terminal transport domains (33–44% amino acid identity) (Finn and Amsbaugh, 1998).

Sequences similar to *vag8* are present in *B. bronchiseptica* and *B. parapertussis* and antibodies to Vag8 recognise a protein of ~95 kDa in whole-cell lysates of *B. pertussis* and *B. bronchiseptica* but not *B. parapertussis* (Finn and Amsbaugh, 1998). It is not known if this represents a Vag8 precursor that is not processed to remove the C-terminal autotransporter domain or if the mature protein (with a predicted molecular weight of ~60 kDa) migrates anomalously in SDS-PAGE, like pertactin and TCF. The original Tn*phoA*-*vag8* mutant exhibited a reduced ability to colonise the respiratory tract of mice but a non-polar *vag8* mutant colonised the respiratory tract normally (Finn and Amsbaugh, 1998). Vag8 was shown to be identical to the 91–92 kDa virulence-associated surface proteins previously described by others (Armstrong and Parker, 1986; Hamstra *et al.*, 1995). Hamstra *et al.* (1995) demonstrated that, in the presence of non-protective amounts of pertussis toxin, the 92-kDa protein was protective in the intracerebral mouse protection test.

Other *B. pertussis* Autotransporters

The gene sequence for another putative autotransporter in *B. pertussis*, Bap5 (*Bordetella autotransporter protein 5*), has been submitted to GenBank (accession number AF081494). The Bap5 protein is predicted to be 79.5 kDa. No typical N-terminal signal sequence could be detected but it does have a characteristic C-terminal domain (Blackburn, 2000). Removal of this domain from the predicted precursor would yield a mature protein of 49 kDa. The predicted mature protein also has RGD and SGXG motifs. The gene for another potential autotransporter protein, Phg, described as a cold-shock protein of *B. pertussis*, is also listed in the GenBank database (accession number AJ009835).

Invasion

B. pertussis can invade and survive, but not replicate, in both epithelial cells and leucocytes *in vitro* (Ewanowich *et al.*, 1989; Lee *et al.*, 1990; Roberts *et al.*, 1991; Steed *et al.*, 1991; Friedman *et al.*, 1992; Torre *et al.*, 1994). Invasion is a *bvg*- and microfilament-dependent process (Ewanowich *et al.*, 1989; Lee *et al.*, 1990; Mouallem *et al.*, 1990; Roberts *et al.*, 1991). Intracellular survival *in vivo* has been proposed to provide a reservoir of bacteria which are protected from the extracellular defences of the host. There is evidence in both mice and rabbits that *B. pertussis* survives within macrophages *in vivo*, and that the RGD motif of FHA (along with pertussis toxin) is involved in bacterial uptake (Cheers and Gray, 1969; Saukkonen *et al.*, 1991; Hellwig *et al.*, 1999). The importance of cell-mediated immunity in murine pertussis is also evidence that intracellular survival of *B. pertussis* may be significant in animal models (see below) but it is not known if an intracellular phase is important in human *B. pertussis* infection.

The alveolar macrophage is probably the cell in which *B. pertussis* resides *in vivo*, and FHA binding to CR3 mediates uptake of *B. pertussis* into macrophages (Relman *et al.*, 1990; Van Strijp *et al.*, 1993). Entry into macrophages via CR3 is used by a variety of intracellular pathogens, possibly because CR3-mediated phagocytosis does not stimulate a respiratory burst (Wright and Silverstein, 1983). As well as FHA, fimbriae and pertactin participate in attachment of *B. pertussis* to human monocytes (Hazenbos *et al.*, 1994). Binding of FHA to CR3 requires activation of CR3 by the interaction of FHA, fimbriae and PT with other macrophage molecules (see above). *B. bronchiseptica* pertactin is involved in the attachment to and invasion of porcine alveolar macrophages (Forde *et al.*, 1999). *B. pertussis* was found to be associated with alveolar macrophages isolated from several HIV-infected children although it could not be cultured and it is not known if the organisms were viable (Bromberg *et al.*, 1991).

There is contradictory evidence regarding whether or not *B. pertussis* can survive within human polymorphonuclear leucocytes (Steed *et al.*, 1991, 1992; Lenz *et al.*, 2000). Others have found that *B. bronchiseptica* but not *B. pertussis* can survive in a murine alveolar macrophage cell line and that the former but not the latter organism can survive at the low pH values in the acidified phagosome (R. Gross, personal communication).

Mutations in genes encoding a variety of adhesins reduce the ability of *B. pertussis* to enter epithelial cell lines *in vitro* (see above). However, no specific epithelial cell invasin has yet been identified. It is likely that the various adhesins enhance the invasion process by bringing *B. pertussis* cells into close apposition with the eukaryotic cell and this allows an unidentified invasin(s) to function efficiently. Possibly, one or more proteins translocated by the putative type III secretion system may stimulate the uptake of *B. pertussis* into eukaryotic cells, as is the case for the type III secretion systems of other bacteria (Galan and Collmen, 1999).

Although *B. pertussis* can invade and survive within macrophages, it is also able to kill these cells by inducing apoptosis (Khelef *et al.*, 1993b). These results would seem contradictory, but the same is also true for more dedicated intracellular pathogens such as *Salmonella* spp. (Zychlinsky and Sansonetti, 1997). A satisfactory experimental explanation for these seemingly conflicting actions has not been produced. It may be that *B. pertussis* is more efficient at killing activated macrophages or that macrophage killing liberates bacteria that have been residing intracellularly but are unable to replicate there.

Toxins

Pertussis Toxin

Pertussis toxin (PT) is produced exclusively by *B. pertussis*. Although *ptx* and *ptl* genes encoding the toxin subunits and export machinery, respectively, are present in *B. parapertussis* and *B. bronchiseptica*, they are transcriptionally silent due to defective promoters (Arico and Rappuoli, 1987). PT has a remarkable range of biological activities *in vitro* and *in vivo*, resulting from its action on many different cell types and tissues. Activities of PT *in vitro* include: ADP-ribosylation of G proteins; haemagglutination; adhesion/invasion; inhibition of neutrophil oxidative burst; T-cell mitogenicity. Activities of PT demonstrated *in vivo* in animals and humans include: induction of leucocytosis; enhancement of insulin secretion; inhibition of adrenalin hyperglycaemia; sensitisation to histamine, anaphylaxis, anoxia, endotoxin etc.; adjuvanticity;

enhancement of vascular permeability; acute toxicity/lethality; protective antigenicity (in its toxoided form) (Tuomanen, 1988; Wardlaw, 1990; Kaslow and Burns, 1992; Krueger and Barbieri, 1995; Munoz, 1998; Ui, 1988) PT is essential for virulence of *B. pertussis* in a mouse model of infection (Weiss *et al.*, 1984; Weiss and Goodwin, 1989) and one probable target is the circulating cells of the immune system and interference with their normal functioning (Weiss and Hewlett, 1986).

PT is a complex protein of 105 kDa composed of six polypeptides (S1–S5), present in the ratio 1:1:1:2:1. It is a member of the AB family of bacterial toxins, specifically the AB_5 family, and consists of an enzymatic A protomer and a receptor-binding B pentamer. The family includes cholera toxin (CT), the *E. coli* heat-labile enterotoxins (LTs) and the shiga or verotoxins (Burnette, 1994, 1997; Stein *et al.*, 1994a; Merritt and Hol, 1995). The S1 polypeptide or A subunit is the enzymatic moiety and the other polypeptides make up the B subunit (Irons and Robinson, 1988). PT is unique amongst this family in that the B subunit is a heteropentamer whereas the B subunits of the other toxins are all homopentamers. The three-dimensional crystal structure of PT, as an uncomplexed molecule and complexed with ATP and carbohydrate, has been resolved (Stein *et al.*, 1994a, b; Hazes *et al.*, 1996).

Although the different AB_5 toxins lack extensive sequence homology, they do possess remarkable structural homology (Burnette, 1994, 1997; Stein *et al.*, 1994a; Merritt and Hol, 1995). A cartoon of the structure of PT is shown in **Fig. 6**. The S1 (A) subunit is roughly pyramidal in shape and sits on the B subunit. It is an ADP-ribosyltransferase that catalyses the transfer of the ADP-ribose moiety of NAD to target proteins, as do the A subunits of CT and LT. The active site of S1 is formed from the N-terminal amino acid residues and is structurally homologous to

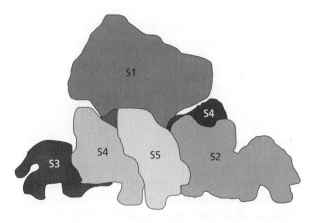

Fig. 6 Schematic diagram of the structure of pertussis toxin.

that of CT, LT and other ADP-ribosylating toxins (exotoxin A of *Pseudomonas aeruginosa* and diphtheria toxin). The orientation of the active site in S1 of PT is rotated 180° relative to those of CT and LT. Two residues, Glu129 and His35, of S1 are conserved in all ADP-ribosylating toxins; the former residue has been shown to be essential for enzymatic activity (Pizza *et al.*, 1988; Locht *et al.*, 1989; Burnette, 1994; 1997; Stein *et al.*, 1994a; Merritt and Hol, 1995). Site-directed mutagenesis to replace certain residues in S1, in particular Arg9 and Glu129, has been used to genetically detoxify PT to produce molecules that are safe and immunogenic (Pizza *et al.*, 1988; Locht *et al.*, 1989; Rappuoli *et al.*, 1992a; Burnette, 1994). The molecular basis of PT activity is discussed more fully below.

The B subunit forms a 'squashed' ring structure formed by the five subunits arranged in the order S5:S2:S4a:S3:S4b (Stein *et al.*, 1994a). S2 and S3 are ~ 70% homologous at the sequence level but the other B subunits exhibit little sequence homology with each other or with S2 and S3 (Locht and Keith, 1986; Nicosia *et al.*, 1986). However, the S4 and S5 subunits and the ~ 100 residues of the C-terminus of S2 and S3 fold almost identically to form a structure called the oligomer binding (OB) fold (Stein *et al.*, 1994a). The OB fold of PT consists of six antiparallel β strands that form a β barrel and an α helix between the fourth and fifth β strands that line the interior of the B subunit ring. The OB fold is also found in the B subunits of the other AB$_5$ toxins (Burnette, 1994, 1997; Stein *et al.*, 1994a; Merritt and Hol, 1995). The N-terminal ~ 100 residues of S2 and S3 form domains that extend beyond the plane of the B-subunit ring and it is these domains that carry the receptor-binding determinants of PT (Stein *et al.*, 1994a, b). These domains consist of five β strands and two α helices and are very similar to a domain in the eukaryotic mannose-binding protein (MBP), a C-type (calcium-dependent) lectin (Stein *et al.*, 1994a). However, there is no evidence that receptor-binding of PT is calcium-dependent and the S2 and S3 domains lack the carbohydrate-recognition site of MBP.

The B subunit targets PT to eukaryotic cells by binding to glycoconjugates, preferentially glycoproteins (Brennan *et al.*, 1988; Tyrrell *et al.*, 1989; Saukkonen *et al.*, 1992; Armstrong *et al.*, 1994). There is some controversy concerning the receptor specificity of S2 and S3 (reviewed by Burnette, 1997) but they appear to have a preference for binding Asp (N)-linked oligosaccharides that contain α (2–6)-linked sialic acid residues. A crystal structure showing binding of S2 and S3 to the terminal sialic acid residue of a branched oligosaccharide has been produced (Stein *et al.*, 1994b). The sialic acid bound to equivalent sites and to conserved Tyr, Ser and Arg residues in the N-terminal regions of S2 and S3. Some of the properties of PT are due solely to binding of the B oligomer to cell surfaces. These include haemagglutination and T-cell mitogenicity (Ui, 1988; Nencioni *et al.*, 1991; Kaslow and Burns, 1992) and at least some of its adjuvant activities (Roberts *et al.*, 1995; Ryan *et al.*, 1997). The other, more harmful effects depend on the activities of the holotoxin.

Upon receptor binding, PT is taken into the eukaryotic cell by receptor-mediated endocytosis. The toxicity of PT is conferred by the enzymatic activity of the S1 subunit which ADP-ribosylates the α subunit of various regulatory G proteins (Krueger and Barbieri, 1995). The modified G proteins are unable to take part in the normal signal transduction across the membrane and, because different cell types and signalling systems are involved, the ensuing biological effects are numerous. In fact, this ability of PT to inactivate G proteins is widely used to probe eukaryotic cell-signalling processes (Ui, 1990).

PT needs to be activated to exert its toxic effects. This requires ATP and the reduction of the disulphide bond formed between the Cys41 and Cys201 residues of S1 (Avigan *et al.*, 1992). Upon entry of PT into a eukaryotic cell it undergoes intracellular trafficking to reach its target G proteins that are located on the inner surface of the plasma membrane and intracellular organelles such as the Golgi (Xu and Barbieri, 1996). Here it binds ATP which destabilises the interaction with S1 and facilitates the reduction of the S1 disulphide bond by a cellular protein, disulphide isomerase. The reduction of the disulphide bond is thought to bring about a conformational change in S1 that enables its substrates to gain access to the active site and may also expose a potential transmembrane domain. S1 is then liberated from the B pentamer and translocates through the membrane and is exposed to the cytoplasm but remains tethered to the membrane by the transmembrane domain (Hazes *et al.*, 1996).

Pertussis toxin is actively secreted by *B. pertussis*. Each polypeptide is synthesised with an N-terminal signal sequence and translocated to the periplasm through the inner membrane by a Sec-mediated process. The holotoxin is assembled here and transported across the outer membrane via a mechanism involving nine Ptl (pertussis toxin liberation) proteins (Covacci and Rappuoli, 1993; Weiss *et al.*, 1993; Burns, 1999; Craig-Mylius and Weiss, 1999; Farizo *et al.*, 2000). Interaction between components of the Ptl system and a region of the S1 subunit of PT appears to be essential for secretion of the assembled toxin (Craig-Mylius *et al.*, 2000) but only the holotoxin and not the S1 subunit in the absence of the B oligomer is exported (Farizo *et al.*, 2000). The *ptl* genes are located downstream from the PT operon and are co-transcribed with the *ptx* genes from the *ptx* promoter (Ricci *et al.*, 1996). The Ptl system is related to the type IV secretion system used by *Agrobacterium tumefaciens* to deliver its tumour-inducing T-DNA–protein complex into plant cells. Other pathogens such as *Legionella pneumophila*, *Brucella* spp. and *Helicobacter pylori* use similar systems to export effector proteins (Burns, 1999).

Although secreted, some PT must remain surface-associated, as demonstrated by immunoelectron microscopy (Ashworth *et al.*, 1985). This surface-located PT may participate in the adhesion of *B. pertussis* to eukaryotic cells (Tuomanen, 1988). A *B. pertussis* PT⁻ mutant bound ∼50% less efficiently to human ciliated respiratory cells than its wild-type parent but a mutation in the FHA structural gene had a far greater effect (Relman *et al.*, 1989). PT appears to play no role in adherence of *B. pertussis* to non-ciliated epithelial cells (Relman *et al.*, 1989; van den Berg *et al.*, 1999; Bassinet *et al.*, 2000). PT also acts co-operatively with FHA in adhesion to and uptake of *B. pertussis* by macrophages and this effect is independent of the enzymatic activity of the toxin (Relman *et al.*, 1990; van't Wout *et al.*, 1992). Binding of the S2 or S3 subunits to macrophages enhances binding of *B. pertussis* to the CR3 integrin mediated by FHA. Activation of CR3 and augmentation of FHA binding is a property that PT shares with *B. pertussis* fimbriae (see above).

Some of the effects of pertussis infection or pertussis vaccination in humans, such as leucocytosis, enhancement of insulin secretion and inhibition of adrenaline hyperglycaemia, can be reproduced in animals by injection of PT. Indeed, some of the effects of PT on experimental animals (Pittman, 1984b) and in humans (Toyota *et al.*, 1980) are long-lasting. In view of these numerous and long-lasting effects of PT, this toxin has been proposed as having a central role in the pathogenesis of pertussis and in some way being responsible for the prolonged course of the disease and for the symptoms that persist after the infection has been cleared (Pittman, 1984b). Its actual role, however, and its main site of action in the host, whether locally in the respiratory tract or systemically, perhaps in the central nervous system, remain obscure (Hewlett, 1997).

Uncertainty over the role of PT in the pathogenesis of pertussis stems from the fact that PT is unique to *B. pertussis* and yet there are common features of *Bordetella* diseases, as referred to above. For example, the clinical characteristics of pertussis and parapertussis can be very similar, even though *B. parapertussis* does not produce PT. *B. parapertussis* is generally considered to cause a milder form of disease than *B. pertussis*, and many parapertussis infections are asymptomatic, but this organism is still capable of causing typical whooping cough (Hoppe, 2000). In two studies of coughing illness in Germany, paroxysmal coughing, whooping and vomiting were all features of parapertussis infections, with frequencies approaching those in pertussis cases (Heininger *et al.*, 1994; Wirsing von Konig and Finger, 1994). The main difference was that lymphocytosis, which is due to PT, was not a feature of parapertussis. Data from animal models are conflicting. In a coughing rat model of pertussis, a wild-type *B. pertussis* strain produced a significant number of coughing paroxysms whereas a PT⁻ mutant, a phase IV strain and a *B. parapertussis* strain, none of which produced PT, did not (Parton *et al.*, 1994). PT is involved in lethal *B. pertussis* infection of infant mice but mutations affecting the production of PT did not affect the growth *B. pertussis* in murine lungs (Weiss *et al.*, 1984; Weiss and Goodwin, 1989). In the adult, non-lethal mouse model, a *B. pertussis* PT⁻ strain caused a greatly reduced pulmonary inflammation compared with a wild-type strain (Khelef *et al.*, 1994). It could be that PT plays a role in exacerbating the disease process, as suggested by Hewlett (1999), possibly by one or more of its known sensitising effects, but other toxins are primarily responsible for causing the symptoms of pertussis.

Despite the uncertainty of the role of PT in pathogenesis, toxoided PT (PTd), is included in all acellular pertussis vaccines (see below). In mice, immunisation with PTd protected against both intracerebral and aerosol challenges with *B. pertussis* (Sato *et al.*, 1984; Sato and Sato, 1988, 1990). In humans, monocomponent acellular vaccines containing toxoided PTd have been shown to be somewhat efficacious in a number of field trials and formalin-detoxified PTd has been a major component of many acellular pertussis vaccines in use since 1981 (Sato *et al.*, 1984). Chemical detoxification of PT, however, is known to reduce its immunogenicity and more immunogenic genetically detoxified recombinant PTd (rPTd) is now available (Rappuoli *et al.*, 1992a, b). One such molecule which has the substitutions Arg9→Lys and Glu129→Gly in S1 (PT-9K/129G) has been tested extensively in mice and humans and been shown to be safe and highly immunogenic. It is a component of an acellular pertussis vaccine that has good efficacy and has been licensed for use in humans (Podda *et al.*, 1992; Rappuoli *et al.*, 1992a, c; Greco *et al.*, 1996). This molecule is also able to function as a mucosal adjuvant (Roberts *et al.*, 1995). Passive immunisation of humans with sera containing high titres of anti-PT antibodies has been shown to have protective effects (see below).

Lipopolysaccharide

The outer leaflet of the outer membrane of *B. pertussis*, in common with that of other Gram-negative bacteria, contains lipopolysaccharide (LPS). The LPS of *B. pertussis*, but not that of *B. bronchiseptica* and *B. parapertussis* lacks an O-polysaccharide side-chain characteristic of the LPS of enteric bacteria and because of this it is sometimes referred to as lipooligosaccharide (Chaby and Caroff, 1988; Martin *et al.*, 1992; Le Blay *et al.*, 1994, 1997; Preston *et al.*, 1996). The lipid As of *Bordetella* spp. are unusual. *B. pertussis* lipid A is hypoacylated compared with enterobacterial lipid As. There are also species-specific differences in the lipid As of the different *Bordetella* spp. (Zarrouk

et al., 1997). Two different forms of LPS are found in *B. pertussis*, LPS-A and LPS-B, which can be resolved electrophoretically (Peppler, 1984). LPS-B is composed of lipid A linked via a single ketodeoxyoctulosonic acid residue to a branched core consisting of glucose, heptose, glucuronic acid, glucosamine and galactosaminuronic acid. Addition of a trisaccharide (*N*-acetyl-*N*-methyl-fucosamine, 2,2-dideoxy-2,3-di-*N*-acetylmannosaminuronic acid and *N*-acetylglucosamine) to LPS-B produces LPS-A (Chaby and Caroff, 1988; Caroff *et al.*, 1990). The chemical composition of *B. pertussis* LPS is shown in **Fig. 7**. Recently, there has been substantial progress in elucidating the genetics of LPS biosynthesis in *B. pertussis*, *B. bronchiseptica* and *B. parapertussis* (Allen and Maskell, 1996; Allen *et al.*, 1998a, b; Middendorf and Gross, 1999). There is evidence that the main regulator of virulence gene expression (the *bvg* locus, see above) influences LPS biosynthesis in *B. bronchiseptica* but not in *B. pertussis* (van den Akker, 1998).

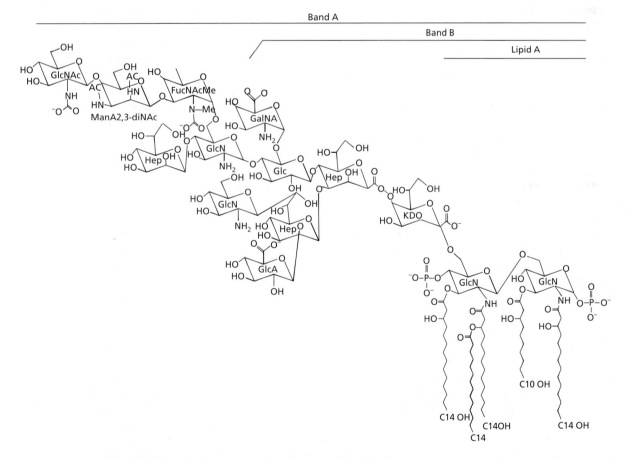

Fig. 7 Chemical composition of the LPS of *B. pertussis*. The structure of the LPS of *B. pertussis* is shown and the parts of the molecule that comprise the lipid A, band B LPS and band A LPS are indicated. GlcN, glucosamine; KDO, ketodeoxyoctulosonic acid; Hep, heptose; Glc, glucose; GlcA, glucuronic acid; GalNA, galactosaminuronic acid; FucNAcMe, *N*-acetyl-*N*-methyl-fucosamine; ManA2,3-diNAc, 2,2-dideoxy-2,3-di-*N*-acetylmannosaminuronic acid; GlcNAc, *N*-acetylglucosamine. The diagram was kindly provided by Dr Andrew Preston and Professor Duncan Maskell.

LPS has the usual properties of endotoxins such as general toxicity, pyrogenicity and adjuvanticity as well as some unusual properties such as ability to induce resistance to mouse adenovirus infection and the ability of the polysaccharide component to induce B-cell mitogenicity and polyclonal B-cell activation in mice (Chaby and Caroff, 1988). Endotoxin mediates most of its diverse biological effects by triggering cytokine production by monocytes. *B. pertussis* LPS has been shown to act synergistically with tracheal cytotoxin to induce inflammation in the respiratory mucosa and damage to ciliated epithelial cells (Flak and Goldman, 1999; Flak *et al.*, 2000) (see below).

The role played by LPS in infection of mice by *B. pertussis*, *B. bronchiseptica* and *B. parapertussis* has been investigated. The *wbl* locus, which is required for expression of LPS-A, was deleted from a strain of each species. The *wbl* mutants of *B. pertussis*, *B. bronchiseptica* and *B. parapertussis* displayed different defects in respiratory tract colonisation in mice (Harvill *et al.*, 2000). All three mutants were defective in colonisation of the trachea, but only the *B. pertussis wbl* mutant was less able to colonise the nasal cavity than the wild-type. This mutant had reduced ability to colonise the lungs as well as the trachea initially but this difference from the wild-type was transient. In contrast, the *B. bronchiseptica wbl* mutant colonised the lungs like the wild-type initially but was cleared from this site by the second week post infection. The *wbl* mutation rendered *B. bronchiseptica* and *B. parapertussis* sensitive to killing by normal rabbit serum, whereas both the *B. pertussis* wild-type and the mutant were killed. Thus the role of LPS in the host–pathogen interaction appears to be different between the different *Bordetella* species.

LPS probably causes much of the reactogenicity of the whole-cell pertussis vaccine and one of the main aims in pertussis vaccine development has been to eliminate this toxin. Nevertheless, LPS may have a role in inducing immunity to pertussis infection. It was shown some time ago that anti-LPS antibodies may be bactericidal in the presence of complement (Ackers and Dolby, 1972). More recently, monoclonal antibodies have been generated that react with either LPS A or LPS B (Archambault *et al.*, 1991). Some of these were agglutinating or bactericidal. Some were able to passively protect mice from lethal (Shahin *et al.*, 1994) and sublethal (Mountzouros *et al.*, 1992) aerosol challenge with *B. pertussis* but *in vitro* bactericidal activity was not a requirement for protection. The presence of smooth LPS in *B. bronchiseptica* correlated with resistance to killing *in vitro* by antibody and complement (Byrd *et al.*, 1991) and antimicrobial peptides (Banemann *et al.*, 1998) whereas *B. pertussis*, lacking the *O*-polysaccharide side-chains, was more sensitive to these agents.

Tracheal Cytotoxin

Tracheal cytotoxin (TCT) was discovered by Goldman and colleagues from its ability to cause ciliostasis and ciliated cell extrusion in hamster tracheal organ cultures and inhibition of DNA synthesis in hamster trachea epithelial (HTE) cell cultures. The selective destruction of ciliated cells is similar to that seen in necropsy material from human pertussis and in turkey coryza (Goldman, 1988). Such effects could well account for some of the pathological events of *Bordetella* respiratory infections such as accumulation of mucus, coughing and pre-disposition to secondary infections. TCT is an unusual, low-molecular-weight (921 Da) toxin consisting of muramyl peptide containing *N*-acetylglucosamine, *N*-acetylmuramic acid, alanine, glutamic acid and diaminopimelic acid in the molar ratio 1:1:2:1:1 (Goldman, 1988; Cookson *et al.*, 1989). It is therefore equivalent to a monomeric subunit of the peptidoglycan component of the Gram-negative cell envelope. This material is generated by most bacteria during the normal turnover of peptidoglycan as the cells grow but, in contrast to most bacteria, *B. pertussis* releases large quantities into the culture supernate (Goldman, 1988; Cookson *et al.*, 1989). An identical substance is released by *Neisseria gonorrhoeae*. Unlike other muramyl peptides, the disaccharide unit of TCT is not required for toxicity but the primary amine group of diaminopimelic acid is necessary (Luker *et al.*, 1993).

TCT and related muramyl peptides have been associated with diverse biological activities such as pyrogenicity, adjuvanticity, arthritogenicity, induction of slow-wave sleep and stimulation of IL-1 production. Nitric oxide, synthesised in response to IL-1, may be the actual cytotoxic factor (Heiss *et al.*, 1993, 1994). In a study of hamster tracheal organ cultures infected with *B. pertussis*, Flak and Goldman (1999) showed that the production of cytokine-inducible NO synthase (iNOS) was induced in non-ciliated cells but not ciliated cells. Moreover, the effect was reproduced by a combination of TCT and *B. pertussis* LPS, but not by either agent alone. A similar synergistic response, resulting in induction of IL-1, iNOS, NO production and inhibition of DNA synthesis was shown in HTE cells (Flak *et al.*, 2000). It has been suggested that the production of inflammatory mediators by mucosal epithelial cells in response to molecules such as muramyl peptides and LPS is part of a normal innate response to infecting bacteria. The abnormal release of large amounts of TCT by *Bordetella* spp. causes an exaggerated response that results in respiratory tract pathology (Flak *et al.*, 2000).

Although the effects of purified TCT and LPS can be demonstrated on ciliated epithelia, Bvg⁻ strains of *B. pertussis*, which produce both of the toxins but not adhesins, cause little damage when incubated with respiratory epithelia (Goldman, 1988). It may be that, during infection, adhesins are required to bring *B. pertussis* into close contact with target cells for their toxic effect to be induced. TCT, even at very low concentrations, also impairs neutrophil functions and may thereby contribute to survival of bordetellae *in vivo* (Cundell *et al.*, 1994). There is no information on the immunogenicity of TCT.

Dermonecrotic Toxin

Dermonecrotic toxin (DNT) is also known as heat-labile toxin. It has potent vasoconstrictive activity and, in experimental animals causes death, reduced weight gain, spleen atrophy and ischaemic lesions or necrosis of skin (Wardlaw and Parton, 1983a; Nakase and Endoh, 1988). These activities are readily destroyed by heating (e.g. 56°C for 10 minutes). DNT is largely responsible for the mouse-lethal toxicity of freshly harvested *B. pertussis* cells. The mouse toxicity test for whole-cell pertussis vaccine was introduced by Pittman primarily to ensure that DNT is destroyed during vaccine manufacture (Pittman, 1952; Pittman and Cox, 1965). The most sensitive *in vivo* assay, however, depends on the production of haemorrhagic lesions after subcutaneous injection of suckling mice (Cowell *et al.*, 1979). Purification of the toxin has been problematical due to its instability and the activities attributed to DNT have been derived mainly from studies with crude or partially purified preparations, before and after heating at 56°C. More recently, recombinant *Bordetella* DNT has become available (Pullinger *et al.*, 1996; Kashimoto *et al.*, 1999).

The genes encoding the DNTs of both *B. pertussis* and *B. bronchiseptica* have been cloned and sequenced (Walker and Weiss, 1994; Pullinger *et al.*, 1996). The products of the *dnt* genes of both organisms are 1464 amino acids in length with a molecular weight of 160 kDa and there are only 11 residues different between the two polypeptides (Walker and Weiss, 1994; Pullinger *et al.*, 1996; Kashimoto *et al.*, 1999). DNT is a member of a family of dermonecrosis-inducing toxins that includes the *P. multocida* toxin (PMT) and *E. coli* cytotoxic necrotising factors 1 and 2 (CNF1 and CNF2) (Walker and Weiss, 1994). The members of this family are large proteins that affect cell division or growth. However, there is little overall homology between the different toxins although DNT and CNF share a small region of homology. The members of this family all stimulate DNA synthesis in target cells but only PMT is mitogenic (Pullinger *et al.*, 1996). They also all stimulate the formation of actin stress fibres and focal adhesions (Horiguchi *et al.*, 1995; Lacerda *et al.*, 1997).

The biological effects of DNT are thought to be due to interference with the activity of small regulatory GTP-binding proteins (GBPs), in particular Rho, Rac1 and Cdc42 (Horiguchi *et al.*, 1995, 1997; Lacerda *et al.*, 1997; Senda *et al.*, 1997; Schmidt *et al.*, 1999; Masuda *et al.*, 2000). DNT has two enzymatic activities that modify GBPs: deamination and trans-glutamination. In the case of Rho, both activities target the Glu residue at position 63 that is critical to GTP hydrolysis. These modifications do not affect its GTP-binding ability but they greatly reduce its capacity to hydrolyse GTP to GDP and thus render Rho constitutively active. One effect of Rho activation is tyrosine phosphorylation of focal adhesion kinase (p125^{fak}) and paxillin (Lacerda *et al.*, 1997). The receptor or cell internalisation domain of DNT resides at the N-terminus of the molecule whereas the catalytically active domain lies at the C-terminus (Pullinger *et al.*, 1996; Lemichez *et al.*, 1997; Schmidt *et al.*, 1999; Kashimoto *et al.*, 1999).

DNT is a cytoplasmic component of the bacterium and does not appear to be released by actively growing cells (Cowell *et al.*, 1979; Nakai *et al.*, 1985; Walker *et al.*, 1994). Despite its potency in various bioassays, *B. pertussis* transposon-insertion mutants deficient in DNT production were unaltered in their ability to cause a lethal intranasal infection in infant mice (Weiss and Goodwin, 1989). However, from studies of other *Bordetella* species in their natural hosts, DNT appears to have a role both in disease and in colonisation. In *B. bronchiseptica*, DNT seems to have a role in producing the turbinate atrophy associated with atrophic rhinitis in pigs. Purified DNT affected an osteoblastic cell line by stimulating protein and DNA synthesis, causing the formation of poly-nucleated cells and inhibiting differentiation. Such degenerative changes in osteoblasts may help to explain how DNT impairs bone formation in atrophic rhinitis (Horiguchi *et al.*, 1994). Significantly, PMT also plays a role in atrophic rhinitis in piglets and severe progressive atrophic rhinitis usually results from dual infection with *B. bronchiseptica* and *P. multocida* (Rutter, 1983; Rhodes *et al.*, 1987; Backstrom *et al.*, 1988; Nielsen *et al.*, 1991; Felix *et al.*, 1992; Gwaltney *et al.*, 1997). Piglets inoculated intranasally with crude DNT showed a disease similar to that observed in the field (Hanada *et al.*, 1979; Nakase and Endoh, 1988). A *B. bronchiseptica* strain deficient in DNT production was associated with considerably less turbinate atrophy and pneumonia in pigs compared with the wild-type parent and also had reduced ability to colonise the nasal cavity (Brockmeier *et al.*, 2000). Similarly, a DNT-deficient mutant strain of *B. avium* was less virulent and persisted for a shorter time than the parent strain in young turkeys, and had reduced ability to adhere to turkey ciliated tracheal cells *in vitro* (Temple *et al.*, 1998).

The question of whether DNT, in toxoided form, could act as a protective antigen in humans or in animal models remains unanswered. However, it is clearly not essential for protection since it is destroyed or removed during preparation of whole-cell and acellular vaccines, respectively, and convalescent sera lack anti-DNT antibodies (Wardlaw and Parton, 1983a).

Adenylate Cyclase Toxin

The adenylate cyclase (AC) of *B. pertussis* was first shown to be associated with toxic activity when Confer and Eaton (1982) found that urea extracts of the bacterium contained an invasive AC that caused accumulation of supraphysiological levels of cyclic AMP in human neutrophils and macrophages. This was associated with phagocytic impotence as shown by attenuation of neutrophil superoxide production and bactericidal capabilities. Subsequently, adenylate cyclase toxin (ACT) was shown to penetrate cells and, because of the intracellular presence of its substrate ATP and its activator, the eukaryotic protein calmodulin, to cause cyclic AMP intoxication and ATP depletion in a wide range of cell types (Hewlett and Goxdon, 1988). Immune effector cells such as neutrophils, monocytes, macrophages and natural killer cells are thought to be the primary targets. An effect on the mucosa to produce the fluid and mucus secretion seen in clinical pertussis has also been suggested. *B. parapertussis* and *B. bronchiseptica* have ACT activity similar to that of *B. pertussis* but the toxins are

antigenically distinct (Gueirard and Guiso, 1993; Khelef *et al.*, 1993a).

ACT or CyaA, the product of the *cyaA* gene, is a bifunctional protein of 1706 amino acids with both AC and weak haemolytic activities and these are associated with the N-terminal 400 amino acids and the C-terminal 1306 amino acids, respectively (Glaser *et al.*, 1988a). The C-terminal domain of ACT is homologous to the RTX (repeats in toxin) family of calcium-dependent, pore-forming cytolysins, of which the prototype is *E. coli* haemolysin (Glaser *et al.*, 1988a; Coote, 1992). These proteins share common structural features, including a series of glycine-rich nonapeptide (GGXGXDXLX) repeat units that bind calcium and are important for interaction of the toxin with target cell membranes (Ladant and Ullmann, 1999). The repeat region and other structural features of ACT are shown in **Fig. 8**.

The RTX toxins also have similar mechanisms of secretion and activation. ACT is exported from *B. pertussis* by a type I secretion pathway that requires a non-processed C-terminal signal sequence on the RTX toxin and the products of the accessory genes *cyaB*, *cyaD* and *cyaE* (Glaser *et al.*, 1988b). It is synthesised as an inactive protoxin and is converted to the active form by palmitoylation of Lys983 catalysed by the product of another accessory gene *cyaC* (Hackett *et al.*, 1994; Basar *et al.*, 2000). The RTX domain of ACT provides a means of entry into the target cell for the N-terminal AC moiety. Once inside the cell, the AC moiety is activated > 1000-fold by the eukaryotic protein calmodulin and causes unregulated synthesis of cyclic AMP. It appears that ACT monomers are sufficient to deliver the AC domain to the target cell interior, whereas haemolysis requires oligomerisation of the monomers in the cell membrane (Gray *et al.*, 1998). The exact mechanism of toxin penetration of target membranes is unknown but a hypothetical

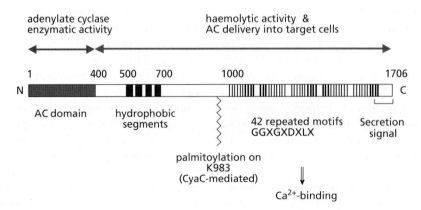

Fig. 8 Structure of the *B. pertussis* adenylate cyclase toxin (CyaA). Numbers represent the amino acid residues. Reproduced from Ladant and Ullmann (1999), with permission. The figure was kindly supplied by Dr Agnes Ullman.

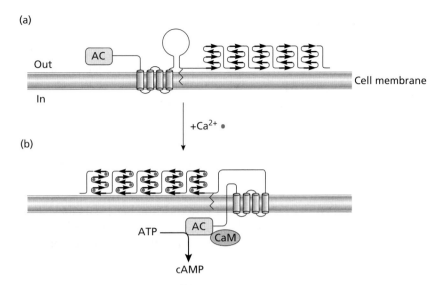

Fig. 9 Hypothetical mechanism for the translocation of *B. pertussis* adenylate cyclase toxin (CyaA) across the target cell membrane. (a) At low calcium concentrations (<0.1 mM), the hydrophobic domain (cylinders) and the palmitoyl chain (zigzag) are first inserted into the plasma membrane. (b) At high calcium concentrations (> 0.1 mM), binding of Ca^{2+} ions (circles) to the repeat motifs (thick arrows in ACT) elicits a conformational change that results in the translocation of the catalytic domain (AC) into the cytosol. Upon binding of calmodulin (CaM), AC catalyses the production of cAMP. Reproduced from Ladant and Ullman (1999) with permission. The figure was kindly supplied by Dr Agnes Ullman.

model for binding and insertion of ACT into the target cell cytoplasmic membrane and translocation of the AC domain into the cytosol is shown in **Fig. 9**.

Studies with *B. pertussis* mutants deficient in ACT activity have confirmed its importance as a virulence factor, at least in mouse models of infection. Such mutants have reduced ability to cause lethal infection in infant mice and to colonise the lungs of older mice (Weiss *et al.*, 1984; Weiss and Goodwin, 1989; Khelef *et al.*, 1992, 1994). A number of recent studies have confirmed that phagocytic cells are a primary target of ACT. For example, ACT is able to induce apoptosis in mouse alveolar macrophages *in vitro* and *in vivo* (Khelef *et al.*, 1993b; Gueirard *et al.*, 1998a). From comparing the effects of wild-type and ACT⁻ *B. bronchiseptica* strains in normal and immunodeficient mice, it was concluded that neutrophils were critical for early defence against infection and that ACT was responsible for overcoming this arm of the innate immune system (Harvill *et al.*, 1999). Other work has shown that expression of ACT inhibits uptake, and possible destruction, of *B. pertussis* by human respiratory epithelial cells (Bassinet *et al.*, 2000) and by human neutrophils after opsonisation (Weingart and Weiss, 2000).

None of the current acellular pertussis vaccines contain ACT but it has been shown to be a protective antigen for mice in a number of studies. The purified AC enzymatic moiety of ACT was reported to protect

mice against *B. pertussis* colonisation of the respiratory tract and brain after sublethal intranasal and lethal intracerebral challenge respectively (Guiso *et al.*, 1990, 1991). Anti-AC antibodies protected mice against lethal intranasal infection with *B. pertussis* (Brezin *et al.*, 1987). Others have shown that immunisation with native ACT and recombinant active ACT promoted more rapid clearance of *B. pertussis* from the respiratory tract than in controls or in mice immunised with inactive (non-acylated) recombinant ACT (Betsou *et al.*, 1993; Hormozi *et al.*, 1999).

Apart from any potential use in new pertussis vaccines, ACT may have other important applications in vaccinology (Ladant and Ullmann, 1999). Genetically detoxified, recombinant ACT has been shown to be an effective vehicle for antigen delivery and induction of cell-mediated immunity. A variety of viral and tumour CD8 T-cell epitopes, recognised by CD8 cytotoxic T lymphocytes, have been inserted into the catalytic domain of AC. The resulting AC toxoids are still cell-invasive and deliver these epitopes into the cytosol of MHC class I antigen-presenting cells. Both protective anti-viral and therapeutic anti-tumour responses have been demonstrated (Saron *et al.*, 1997; Fayolle *et al.*, 1999). The catalytic domain of ACT has also been used as a reporter to study protein targeting and, as a bacterial two-hybrid system, for studying protein–protein interactions (Ladant and Ullmann, 1999).

Type III Secretion System

Type III secretion systems (TTSS) are responsible for translocating effector proteins directly from the bacterial cytoplasm to the cytoplasm of eukaryotic cells (Galan and Collmen, 1999). Genes encoding components of a TTSS and potential translocated proteins have been identified in *B. bronchiseptica*, *B. parapertussis* and *B. pertussis* (Yuk *et al.*, 1998a, 2000; Kerr *et al.*, 1999). However, under the conditions tested, the TTSS genes were expressed only in one of four *B. pertussis* strains examined. It is uncertain, therefore, if the TTSS functions in *B. pertussis* generally (Yuk *et al.*, 1998a). The role of the TTSS has been investigated only in *B. bronchiseptica* where it is Bvg-regulated (Yuk *et al.*, 1998a).

A mutation in *B. bronchiseptica bcsN*, a gene encoding a component of the translocation apparatus, greatly decreased the secretion of at least three polypeptides designated *bopN*, *bsp22* and *bopD* in *B. bronchiseptica* (Yuk *et al.*, 2000). The TTSS appears to have a number of functions, including induction of apoptosis in macrophages *in vitro* and *in vivo;* inhibition of bacterial- or TNFα-induced translocation of the transcription factor NF-κB from the cytoplasm to the nucleus in epithelial cells; and facilitating the persistence of *B. bronchiseptica* in the tracheas of rodents (Yuk *et al.*, 1998a, 2000). Compared with wild-type *B. bronchiseptica*, TTSS mutants were more virulent in immunodeficient (SCID-beige) mice and elicited higher serum antibodies in immunocompetent mice (Yuk *et al.*, 2000). Thus, the TTSS of *B. bronchiseptica* interferes with the action of components of the innate and acquired immune systems, enabling long-term persistence of *B. bronchiseptica*. The role, if any, of a TTSS in *B. pertussis* remains to be determined.

New Virulence Factors

Recently Antoine *et al.* (2000) have searched the genome sequence of *B. pertussis* (http://www.sanger.-ac.uk/Projects/B_pertussis/) for new potential virulence factor genes and determined whether these fusions are regulated positively or negatively by the *bvg* locus using a transcriptional fusion reporter plasmid. The genes identified include putative toxins, adhesins and iron-uptake proteins, e.g. two genes encoding homologues of FHA (*fhaL* and *fhaS*); *bilA*, which encodes a homologue of *E. coli* intimin; capsule biosynthesis and transport genes; *bexB*, which encodes a homologue of *Aeromonas hydrophila* extracellular DNase; two genes, *metC1* and *metC2*, whose products are most homologous to β-cystathionase from *Bordetella avium* and *E. coli* respectively; *bilY*, whose

product is homologous to legiolysin from *Legionella pneumophila*; three genes encoding putative autotransporters with homology to serine proteases, *sphB1–3*; and two genes, *bfrD* and *bfrE*, which encode homologues of hydroxamate siderophore receptors of *Pseudomonas aeruginosa*. It is likely that additional potential virulence genes will be identified when the annotated *B. pertussis* genome sequence is released. The roles of these genes in the biology of *B. pertussis* are likely to be revealed by traditional techniques such as have been described above. Newer techniques such as transcriptome analysis and proteomics, which will be possible with the availability of the complete *B. pertussis* genome sequence, will enable the expression of *B. pertussis* genes to be studied at the whole genome level.

Immune Response

The role of the host innate and acquired immune defences in recovery from and resistance to *B. pertussis* infection has been studied most intensively in mice, in particular the adult mouse model (Sato and Sato, 1988; Shahin and Cowell, 1994). Mice are infected with *B. pertussis* by one of two methods. The first, intranasal infection, involves application of suspensions of *B. pertussis* to the external nares of lightly anaesthetised animals which inhale the inoculum. The second method involves exposure of conscious animals to an aerosol of *B. pertussis*. Several points should be borne in mind when considering the results from such studies. Mice do not develop pertussis, i.e. they do not develop a paroxysmal cough. In immunocompetent adult mice *B. pertussis* infection is generally not fatal. The course of the infection is monitored by determining the number of viable *B. pertussis* in different regions of the respiratory tract (usually the lungs but the nasal cavity and trachea may also be sampled) of mice at different times after challenge. Because of this, there is no generally agreed measure of protection. Usually, *B. pertussis* numbers increase to a peak in the first to second week after infection and will thereafter be cleared from the lungs over the subsequent 2–3 weeks. What is stated as protection following immunisation in these models reflects prevention of the initial increase in bacterial numbers and/or enhanced clearance of *B. pertussis* from the respiratory tract. Different vaccines and immunisation regimes can differ greatly in their effectiveness at promoting clearance of *B. pertussis* from the respiratory tract but they may all be referred to as protective. Despite these caveats, there is a correlation between the protective efficacy of vaccines in humans and mice in both the intranasal and aerosol models (Mills *et al.*, 1998; Guiso *et al.*, 1999).

In mice, both cellular and humoral factors are important for immunity to *B. pertussis*. Much of what follows comes from the seminal work of Mills and co-workers. In naïve mice challenged with *B. pertussis*, the clearance of the organism from the lungs precedes the appearance of high-titre anti-*B. pertussis* serum antibodies by several weeks (Roberts *et al.*, 1990; Redhead *et al.*, 1993). Adoptive transfer of an FHA-specific CD4 T-cell line, spleen cells or CD4 (but not CD8)-enriched splenic T cells into T-cell deficient (nude) mice enhanced clearance of *B. pertussis* from the lungs of recipient mice (Mills *et al.*, 1993). Passive immunisation with convalescent serum had little effect on clearance of *B. pertussis* from these mice. The CD4 T cells had a Th1 phenotype secreting IL-2 and IFNγ but not IL-4 in response to stimulation with whole *B. pertussis* or *B. pertussis* antigens *in vitro*, indicating that cell-mediated immunity, probably involving activation of macrophages, was a major component of immunity to *B. pertussis* (Mills *et al.*, 1993; Redhead *et al.*, 1993). Immunisation of mice with killed whole-cell pertussis vaccine (WCPV) was more effective than immunisation with acellular pertussis vaccine (APV) composed of detoxified PTd, pertactin and FHA at promoting pulmonary clearance of *B. pertussis*. Immunisation with WCPV induced a Th1 CD4 T-cell response, whereas immunisation with APV elicited a Th2 CD4 T-cell response (production of IL-4 and IL-5) (Redhead *et al.*, 1993).

B. pertussis infection in humans, which is known to induce long-lasting immunity to disease, elicits T cells with a Th1 phenotype (Peppoloni *et al.*, 1991; Hafler and Pohl-Koppe, 1998). In human infants, immunisation with APVs which have high efficacy induce a more balanced Th1/Th2 response (production of IFNγ, IL-4 and IL-5) early after immunisation (Ausiello *et al.*, 1997). However, 4 years after immunisation, the response in the infants given APV had become more polarised towards a Th1-type response (Ausiello *et al.*, 1999). Anti-*B. pertussis* antibodies wane relatively quickly, even in infants receiving APVs that have a high efficacy (Giuliano *et al.*, 1998; Salmaso *et al.*, 1998). Moreover, it has been difficult to establish a serological correlate of immunity in APV trials in humans although this may have been because of experimental design (Ad Hoc Group, 1988; Hewlett and Halperin, 1998). Taken together, the studies described above indicate that serum antibodies play a minor role in immunity to *B. pertussis* and that the critical factor in protective immunity was *B. pertussis*-specific CD4 T cells secreting type 1 cytokines.

Other studies, however, have indicated that antibodies and/or B cells have an important role in anti-pertussis immunity. Passive immunisation with mAbs specific for PT, pertactin or *B. pertussis* LPS protects infant mice from lethal *B. pertussis* infection (Sato and Sato, 1990; Shahin *et al.*, 1990, 1994). Administration of immune sera from mice immunised with WCPV or APV significantly enhanced clearance of *B. pertussis* from the lungs of aerosol-challenged mice (Mills *et al.*, 1998). *B. pertussis* numbers in the lungs of passively immunised mice decreased rapidly but thereafter the residual *B. pertussis* population was cleared very slowly over several weeks (Mills *et al.*, 1998). Mice which lacked mature B cells (Igµ-chain knockout mice, BKO) developed a persistent pulmonary infection with *B. pertussis* and this was not affected by immunisation with WCPV or APV. Although this would seem to indicate a critical role of antibodies (or at least B cells) in immunity to *B. pertussis*, the T-cell response in BKO mice was also suppressed. Other studies also indicate a role for B cells in *B. pertussis* immunity other than by the production of antibody (Leef *et al.*, 2000). Recent analysis of the serum antibody response to different pertussis antigens in infants who had received either APV or WCPV in different vaccine trials revealed that protection against clinical pertussis correlated with antibodies to pertactin, fimbriae and, to a lesser extent, PT (Cherry *et al.*, 1998; Hewlett and Halperin, 1998; Storsaeter *et al.*, 1998).

The above pattern of *B. pertussis* clearance in mice passively immunised with anti-APV or anti-WCPV sera resembles that seen in mice actively immunised with APV. Delayed clearance was not seen in mice immunised with WCPV or in mice that had recovered from a previous *B. pertussis* infection (Redhead *et al.*, 1993; Mills *et al.*, 1998). It was thought that the residual population of *B. pertussis* resides intracellularly within macrophages and that cell-mediated immunity requiring activation of macrophages by CD4 Th1 T cells is required to clear the intracellular organisms. IL-12 promotes the development of type 1 (Th1) helper CD4 T-cell responses and mice immunised with APV (PTd + FHA + pertactin) plus recombinant murine IL-12 developed a Th1 response against these antigens and cleared *B. pertussis* from their lungs as efficiently as mice immunised with WCPV (Mahon *et al.*, 1996). The kinetics of *B. pertussis* clearance from the lungs of IFNγ receptor knockout (GKO) mice that had recovered from a previous *B. pertussis* infection supports a role for cell-mediated immunity in the late clearance of *B. pertussis* from murine lungs (Mills *et al.*, 1998).

B. pertussis caused systemic infection and death in naïve GKO mice that received a high-dose challenge of *B. pertussis* by aerosol (Mahon *et al.*, 1997). However, there was no difference in the numbers of *B. pertussis* or the kinetics of clearance of *B. pertussis* from the lungs of GKO and wild-type mice. A separate study in GKO mice did not report that *B. pertussis* infection in these mice was lethal. Higher numbers of *B. pertussis* were recovered from the lungs of GKO than wild-type mice but the kinetics of *B. pertussis* clearance was very similar (Barbic *et al.*, 1997). IL-4 does not seem to play a role in immunity to *B. pertussis* in naïve or immunised mice but it may play a role in controlling the inflammatory response in the lungs (McGuirk and Mills, 2000a).

The fact that circulating antibodies to *B. pertussis* antigens wane quite rapidly after immunisation, even with highly efficacious APVs, without a concomitant rise in susceptibility to whooping cough suggests that the capacity of the vaccines to elicit long-term immunological memory to *B. pertussis* is probably critically important to their effectiveness (Giuliano *et al.*, 1998; Salmaso *et al.*, 1998). Aerosol immunisation with a live attenuated *B. pertussis aroA* strain did not stimulate a detectable serum antibody response but challenge of immunised mice with wild-type *B. pertussis* led to a rapid anamnestic serum anti-*B. pertussis* response that correlated with the clearance of *B. pertussis* from the lungs (Roberts *et al.*, 1990). Likewise, *B. pertussis* challenge of mice that had been immunised with WCPV or APV 44 weeks previously caused a rapid rise in *B. pertussis* antigen-specific serum IgG antibodies (Mahon *et al.*, 2000).

Mucosal immunity would be expected to play a role in the defence against a pathogen such as *B. pertussis* which is located predominantly extracellularly and which remains within the respiratory tract. Anti-*B. pertussis* IgA has been detected in the respiratory tract of mice following mucosal immunisation with a variety of vaccines, delivery systems and vaccination regimes (Molina and Parker, 1990; Guzman *et al.*, 1991; Shahin *et al.*, 1992; Walker *et al.*, 1992; Amsbaugh *et al.*, 1993; Brownlie *et al.*, 1993; Guzman *et al.*, 1993; Roberts *et al.*, 1993; Cahill *et al.*, 1995; Cropley *et al.*, 1995; Mielcarek *et al.*, 1997; Jabbal-Gill *et al.*, 1998; Ryan *et al.*, 1999). However, a definitive role of secretory IgA in the defence against *B. pertussis* has yet to be demonstrated. The problem is the difficulty in differentiating between the contribution of secretory IgA and other components of the immune system in protection against *B. pertussis*. The recent reports of the construction of IgA- and J-chain knockout mice means that this may now be possible (Johansen *et al.*, 1999; Mbawuike *et al.*, 1999). Secretory IgA is likely to be more important for protection of the nasal cavity and the trachea than the lungs because IgA-secreting cells predominate in the upper respiratory tract and serum antibody can transduce into the lungs.

Infection of mice with *B. pertussis* leads to the appearance of *B. pertussis*-specific antibody-secreting cells (ASC), predominantly producing IgA, in the respiratory tract (Shahin *et al.*, 1992; Roberts *et al.*, 1993; Jones *et al.*, 1996). Anti-*B. pertussis* IgA also appears in the nasopharyngeal secretions or sputum of humans with pertussis (Goodman *et al.*, 1981; Granstrom *et al.*, 1988). The enhanced clearance of *B. pertussis* from the lungs of mice immunised intranasally with pertactin correlated better with the presence of pulmonary anti-pertactin ASC rather than serum anti-pertactin antibodies (Roberts *et al.*, 1993). Similarly, clearance of *B. pertussis* from immunised rabbits correlated with nasal anti-FHA IgA not with serum antibodies (Ashworth *et al.*, 1982). Also, the appearance of nasal anti-*B. bronchiseptica* IgA in dogs immunised with a live, attenuated intranasal *B. bronchiseptica* vaccine correlated with immunity to this organism (Bey *et al.*, 1981).

Diagnosis

A clinical diagnosis of pertussis is often possible due to the characteristic and prolonged paroxysmal coughing in the typical disease. However, clinical features tend to vary with age and immune status, with older children and adults, and cases of vaccine failure, often showing less severe or atypical symptoms. Moreover, a proportion of pertussis-like coughs may be caused by *B. parapertussis* (see above) as well as by adenovirus, parainfluenza viruses, respiratory syncytial virus, *Mycoplasma pneumoniae* and *Chlamydia pneumoniae* (Wirsing von Konig *et al.*, 1998b; Hallander *et al.*, 1999). A pronounced leucocytosis is also suggestive of pertussis. The WHO definition of pertussis used in recent vaccine efficacy trials requires a minimum of 21 days of paroxysmal coughing with laboratory confirmation or epidemiological linkage (WHO, 1991). Laboratory methods include direct detection of *B. pertussis* by culture, direct fluorescent antibody and PCR methods, or indirect diagnosis by serology. Each method has its advantages and limitations in terms of sensitivity, specificity and applicability. Recent vaccine efficacy trials have provided a valuable opportunity to evaluate and compare these laboratory methods and results have been reviewed (Muller *et al.*, 1997; Hallander, 1999; Kerr and Matthews, 2000).

Culture of *B. pertussis* from respiratory secretions is often referred to as the gold standard for diagnosis but sensitivity is variable and depends on many factors. Successful isolation is more likely early in the disease and less so as the paroxysmal stage progresses. Age and immunisation status of the patient and any preceding antimicrobial therapy are also important, as are the methods and media for specimen collection, transport and culture. For primary isolation of *B. pertussis*, the medium developed by Regan and Lowe (1977) containing charcoal agar, with defibrinated horse blood 10% and cephalexin (40 mg/L) to suppress the normal nasopharyngeal flora gives better recovery of *B. pertussis* than Bordet–Gengou and other media. Plates should be incubated aerobically in a humid atmosphere at 35°C rather than 37°C and for up to 7 days. The characteristic tiny, smooth, convex, pearl-like colonies of *B. pertussis* usually become visible after 3–4 days whereas those of *B. parapertussis* appear after 2–3 days. Identification and differentiation from related organisms can be confirmed by Gram-staining, urease, oxidase and nitrate reactions (**Table 2**).

A variety of serological tests such as agglutination, complement fixation and ELISA have been used to confirm diagnosis of pertussis and are valuable in epidemiological and vaccine efficacy studies. They are less suitable for rapid diagnosis and generally require seroconversion to distinguish active disease from high titres due to vaccination or maternal antibodies. ELISA tests for IgG antibodies to PT and FHA are widely used and sensitive diagnostic methods and criteria for a positive diagnosis have been suggested by Hallander (1999). Antibody responses to PT are considered to be specific for pertussis whereas seroconversion for FHA may also occur after *B. parapertussis* infection and potentially after exposure to non-encapsulated *H. influenzae*.

The direct detection of *B. pertussis* or its components in clinical specimens is also possible (Onorato and Wassilak, 1987; Friedman, 1988; Gilchrist, 1991). Such methods have the advantages of speed and ability to detect organisms that are no longer viable. The direct fluorescent antibody test uses fluorescein-labelled antibodies to *B. pertussis* and *B. parapertussis* to detect the bacteria in nasopharyngeal secretions. The test can be highly specific, depending on the quality of the reagents used, but is technically demanding and of low sensitivity. It is recommended as an adjunct to, rather than a replacement of, culture. Tests for the detection of *B. pertussis* components such as adenylate cyclase, by its enzymic activity, and pertussis toxin, by immunoassay or by cytotoxicity assay, either directly in respiratory secretions or after enrichment appear to be sensitive and specific but have not found routine use.

The tests likely to become the preferred method in many laboratories for rapid and sensitive detection of *B. pertussis* and other bordetellae are those involving PCR. Such techniques have attracted much interest for the diagnosis of infectious disease, especially with the advent of non-radioactive detection methods, and they are particularly suitable for fastidious, slow-growing pathogens such as *B. pertussis*. With increasing

knowledge of the *Bordetella* genomes, species-specific and genus-specific sequences can be targeted. A PCR assay developed by van der Zee (1993) targeting species-specific insertion sequences allows the simultaneous detection and discrimination of *B. pertussis* and *B. parapertussis* and is sensitive enough to detect a single cell of either pathogen. Reizenstein *et al.* (1993), with primers directed to the PT gene promoter, followed by restriction enzyme analysis of the amplified fragments, was able to discriminate between *B. pertussis*, *B. parapertussis* and *B. bronchiseptica*.

At a meeting summarised by Meade and Bollen (1994), it was predicted that PCR would be comparable with, or more sensitive than, culture for detection of *B. pertussis*, offered a greater possibility than culture or serology for detecting infection in asymptomatic or previously vaccinated individuals and would be more likely to detect infection in antibiotic-treated patients. These predictions have been borne out in many studies (e.g. Edelman *et al.*, 1996; Muller *et al.*, 1997; Schmidt-Schlapfer *et al.*, 1997). Hewlett (2000) agrees that the advantages of PCR methods are clear but has cautioned that efforts to culture *B. pertussis* should continue, to detect strain variation, antibiotic resistance and other properties of the organism that could be missed by a PCR-based system.

Typing

Serotyping is a long-established method for epidemiological investigations in pertussis (Eldering *et al.*, 1957). It depends on the presence or absence of a number of specific agglutinogens on the bacterial cell surface. These are detected by slide agglutination tests with antisera raised in rabbits and made monospecific by absorption with heterologous strains. Nowadays, only agglutinogens 1, 2 and 3 are widely recognised. The antigenic determinants of agglutinogens 2 and 3 are located on fimbriae (Fim2 and Fim3) and can be lost or gained independently, *in vitro* or *in vivo*, due to spontaneous frameshift mutations in the promoter region of the fimbrial subunit genes (Willems *et al.*, 1990) (see above). Agglutinogen 1 is common to all fresh isolates and is located on the LPS. Thus, the four recognised serotypes of *B. pertussis* are 1,2,3; 1,2; 1,3 and 1. Serotyping provides only limited information on *B. pertussis* strains but is still useful as the fimbrial antigens are potentially important in protection.

Nowadays, much more discriminating methods are available that are applicable to both epidemiological and phylogenetic investigations. DNA fingerprinting methods such as ribotyping, RAPD analysis, IS-based methods and PFGE of genomic fragments have been introduced. The latter method in particular has revealed strain differences by the presence of numerous DNA types amongst *B. pertussis* isolates (De Moissac *et al.*, 1994; Beall and Sanden, 1995a; Syedabubakar *et al.*, 1995). More recently, gene typing methods have been introduced and have detected potentially important polymorphisms in the *B. pertussis* population in some of the virulence genes (Mooi *et al.*, 1998, 1999; Mastrantonio *et al.*, 1999; Cassiday *et al.*, 2000) (see below). With advances in sequencing technology, such information is readily obtainable and more easily comparable between laboratories and databases than results of PFGE and other methods based on whole-genome comparison. To facilitate the comparison of *B. pertussis* populations between different countries and for better comparison of the effectiveness of different vaccination regimes used, Mooi and colleagues (2000) have proposed a standardised methodology for epidemiological typing of *B. pertussis*, based on serology, PFGE and gene typing.

Epidemiology

The epidemiology of pertussis has been reviewed recently (Black, 1997; Cherry, 1999; Edwards *et al.*, 1999). The disease has long been responsible for severe morbidity and mortality worldwide. It is still a significant problem even in vaccinated communities and a major problem in unvaccinated populations. Increasing global immunisation coverage by schemes such as the WHO Expanded Programme on Immunisation is steadily reducing the incidence and severity but there is still some way to go. World-wide, *B. pertussis* causes some 20–40 million cases of pertussis, 90% of which occur in developing countries, and an estimated 200 000–300 000 fatalities each year (WHO: http://www.who.int/vaccines-diseases/diseases/pertussisvaccine.htm).

Before widespread vaccination, pertussis was primarily a disease of young children in that the majority of typical cases occurred in the age group 1–5 years, with maternal antibody providing some protection to infants <1 year. All ages are susceptible, however, but when most adults had pertussis as children and then were regularly re-exposed to infection, there was a high level of adult immunity. Pertussis is highly communicable, with attack rates ranging from 50% to 100%, depending on the intensity of exposure. In a recent household study as part of an efficacy trial of acellular pertussis vaccines (APVs), the attack rate was 69% in children and 31% in adults (Wirsing von Konig *et al.*, 1998a). Transmission can mostly be attributed to droplet infection from active cases. Asymptomatic,

culture-positive carriers have occasionally been detected, but their role in transmission of the disease, in the absence of coughing, is unknown. There is no known non-human host, vector or environmental reservoir of *B. pertussis*. There is no major or consistent seasonal pattern for pertussis between different countries, although a recent survey in England and Wales revealed a slight increase above background during late summer and early autumn (Van Buynder *et al.*, 1999).

In countries where vaccination has been introduced, the incidence and severity of pertussis has been greatly reduced, but the age distribution of the disease has also changed significantly. A higher proportion of cases now occur in younger, unvaccinated infants and in adolescents and adults with waning immunity. For example, in the United States between 1992 and 1994, 41% of cases occurred in infants <1 year and 28% in those aged 10 years or older, compared with 20% of cases in those aged <1 year and 3% in individuals >15 years before widespread vaccination (Edwards *et al.*, 1999). The same trend is seen within the highly vaccinated population of England and Wales. Between 1991 and 1997, the proportion of notifications in children under 6 months has increased from 6% to 19% and in those over 15 years from 4.4% to 9.3% (Van Buynder *et al.*, 1999). It is thought that in well-immunised populations the opportunity for boosting vaccine-induced immunity in adults by re-exposure to *B. pertussis* is limited. When these adults are eventually infected, they often develop atypical or asymptomatic disease but can still transmit the disease to others.

There is now much evidence that adults constitute an important reservoir of infection (Black, 1997; Cherry, 1999). For example, a number of studies has confirmed that pertussis is the cause of a high percentage of cases of persistent coughs in adults. The delay in recognition and treatment of such cases provides the opportunity for transmission of infection to those household contacts who are not fully immunised. Reports of re-infection of adults diagnosed with pertussis as children have been documented, suggesting that immunity induced by infection, like that induced by vaccination, is not solid. This conclusion was also reached after a serological study of young adults in the United States, where pertussis is controlled by vaccination, and in Germany, where it is not. The study showed that all had been exposed to *B. pertussis* infection at some time previously and the incidence of recent infection, determined by assaying anti-pertussis IgA antibodies, was similar in the two groups (Cherry, 1999). It appears that whether immunisation is practised or not, the disease is endemic. In most countries, epidemics occur approximately every 3–4 years, presumably due to the number of susceptibles in the population. A lack of effect on this frequency by the introduction of vaccination or when changes in vaccine uptake rates have occurred seems to indicate that the whole-cell pertussis vaccines (WCPVs) in widespread use provide better protection against disease than against infection and that vaccination has little effect on the prevalence of *B. pertussis* in the population (Fine, 1988). This may be partly because the vaccines are administered parenterally and as such are unlikely to stimulate mucosal immunity in the respiratory tract.

In recent years, there has been a resurgence of pertussis in some countries, even though they have maintained high vaccination rates. In the US for example, pertussis is regarded as a re-emerging infection. Infection rates have been rising steadily since the early 1980s. The reasons for this are unclear but the growth of a susceptible adult population, as discussed above, has been suggested as a contributory factor (Black, 1997) and a higher incidence and increasing proportion of cases in older children and adults has been reported (Guris *et al.*, 1999). Improved diagnosis and reporting of cases in these age groups may in part be responsible for the increase. There is also some evidence that one of the two most widely used WCPVs in the US has low efficacy (Edwards *et al.*, 1999). A similar resurgence has been noted in Australia (Andrews *et al.*, 1997), Canada (De Serres *et al.*, 1995) and the Netherlands where the incidence was 5-fold higher than in previous epidemics (de Melker *et al.*, 1997).

In a study with an IS-based fingerprinting method, van der Zee *et al.* (1996b) found evidence that vaccination in the Netherlands had resulted in the selection of strains which differed in DNA type from the vaccine strains and suggested that this could have been due to antigenic differences between the strains. Such antigenic differences in the pertactin and pertussis toxin components of the *B. pertussis* population were discovered by Mooi *et al.* (1998) and provided strong evidence for the vaccine-driven evolution of strains. Sequence analysis of isolates revealed polymorphisms in the genes encoding pertactin and PT that would result in different amino acid sequences for these proteins. Eight *prn* alleles have now been found (Mooi *et al.*, 2000) and the polymorphisms are mainly due to insertions or deletions in regions of *prn* encoding one or other of the repeat sequences in pertactin which are known to be part of immunodominant B-cell epitopes of the adhesin. Polymorphisms in PT were mainly in the S1 subunit, in two regions identified as T-cell epitopes. When the *prn* and *ptx* genotypes of contemporary, historical and vaccine strains were compared, types identical to those included in the Dutch whole-cell vaccine were found in 100% of isolates from the 1950s when the pertussis vaccines were first introduced. Non-vaccine types were found to have gradually replaced the vaccine types and were found in 90% of strains circulating in 1990–1996. A comparison of strains collected from vaccinated and non-vaccinated pertussis patients indicated that the WCPVs protected better against the vaccine strain genotype. Further evidence that antigenic shifts in pertactin and PT have been driven by vaccination has been reported from Finland (Mooi *et al.*, 1999), Italy (Mastrantonio *et al.*, 1999) and the United States (Cassiday *et al.*, 2000).

As suggested by Mooi *et al.* (1998), such findings may have implications for the efficacy of both WCPV and APV and may indicate a future need to adjust the composition of these vaccines to contain several pertactin and PT types. Pertactin and PT are important constituents of APVs and antibodies to them have been shown to correlate with clinical protection (Cherry *et al.*, 1998; Storsaeter *et al.*, 1998). Several of the APVs contain pertactin and PT types found only in a minority of recent Dutch isolates of *B. pertussis*. It is interesting to note that the increase in pertussis in the Netherlands has been associated with an increase in the prevalence of the 1,2 serotype. A similar increase in the proportion of this serotype has been detected in England and Wales and it appears to cause more severe disease, as determined from hospital admissions, than other serotypes (Van Buynder *et al.*, 1999).

Antimicrobial Therapy

B. pertussis is sensitive *in vitro* to a wide range of antibiotics. However, as pointed out by Hoppe (1998), their effectiveness *in vitro* does not necessarily reflect their usefulness against the bacteria localised on the ciliated epithelium of the respiratory tract due to the difficulty in maintaining suitable concentrations of the drugs at this site. In addition, some antibiotics effective *in vitro* are not recommended for use in children. Several antibiotics, including erythromycin, tetracycline, choramphenicol and trimethoprim-sulphamethoxazole, are effective *in vivo* but erythromycin is generally the drug of choice for treatment of pertussis patients and for post-exposure prophylaxis of close contacts. Erythromycin estolate is reported to reach higher concentrations in secretions than other formulations of erythromycin (Hoppe, 1992). Short-term treatment with the newer macrolides, clarithromycin and azithromycin, because of their improved absorption and longer half-life, has been shown in small-scale studies to be at least as effective as standard erythromycin treatment (Aoyama *et al.*, 1996; Bace *et al.*, 1999). These agents, or trimetho-prim-sulphamethoxazole may be appropriate for treatment of adult pertussis patients who cannot tolerate erythromycin (Hoppe, 1996). Routine anti-biotic susceptibility testing of *B. pertussis*, a slow-growing and nutritionally fastidious organism, is cumbersome and is usually unnecessary. However, two reports of erythromycin resistance have been noted and have prompted the development of simpli-fied and standardised testing methods (Hoppe, 1998; Hill *et al.*, 2000) The mechanism of this resistance is at present unknown. Plasmids encoding antibiotic resistance have been described in *B. bronchiseptica* and *B. avium* but not in *B. pertussis*.

Antibiotic treatment in the early (catarrhal or early paroxysmal) stages can ameliorate the symptoms of pertussis and eliminate the infection. Later treat-ment may have no clinical benefit for the patient but may help to limit the spread to susceptible contacts (Olson, 1975; Cherry *et al.*, 1988). Early attempts to modify the course of the illness by passive immunisa-tion were of doubtful efficacy (Edwards *et al.*, 1999) but more recently, new preparations of anti-pertussis immunoglobulin have shown promise. High-titre immunoglobulins prepared from adults immunised with either pertussis toxoid (PTd) or PTd plus FHA gave a significant reduction in the number of whoops when administered to pertussis patients in a double-blind, placebo-controlled trial (Granstrom *et al.*, 1991). Similarly, a high-titre serum from donors immunised with PTd reduced the paroxysmal coughing in children with severe pertussis (Bruss *et al.*, 1999). Further clinical trials of these materials are clearly warranted.

Control

Whole-cell Vaccines

The early history, development and current status of immunisation with whole-cell pertussis vaccines (WCPVs) have been detailed elsewhere (Wardlaw and Parton, 1983b; Cherry *et al.*, 1988; Parton, 1991; Cherry, 1996; Edwards *et al.*, 1999). Pertussis vac-cines in general use throughout the world are whole-cell preparations although the vaccines produced by different manufacturers differ in potency, reactogen-icity and the immune responses that they produce, and immunisation schedules vary from country to country. Where there is good vaccine coverage of the popula-tion, the incidence of pertussis has been reduced to

low levels. Even in those not fully immunised, the disease is less severe. Evidence for the undoubted efficacy of WCPVs has been summarised by Edwards *et al.* (1999) and has been provided, for example, by clinical trials (historical and recent) showing the rapid decline in morbidity and mortality with implementa-tion of vaccination programmes, by the re-emergence of disease in countries where immunisation has been discontinued or where acceptance rates have declined, and by studies showing the inverse correlation between pertussis attack rates and the proportion of children immunised. In recent trials of acellular vac-cines where WCPVs have been used as controls, effi-cacy estimates for WCPVs ranged from 83% to 98%, except for one preparation with a very poor efficacy of 36–47% (Decker and Edwards, 2000).

The current UK vaccines contain less than 2×10^{10} killed *B. pertussis* cells in a 0.5-mL dose, with a potency of more than 4 International Units by comparison with the International Standard for Pertussis Vaccine in the intracerebral mouse protection test (WHO, 1990). The pertussis component is usually combined with diphtheria and tetanus toxoids (DTP) and absorbed to aluminium hydroxide, for greater immunogenicity and less reactogenicity, and with thiomersal as a pre-servative (Department of Health, 1990). Prior to 1974, vaccine uptake in the UK was more than 80% and annual notifications were at the low level of around 2400 (in England and Wales). From 1974 onwards, following widely publicised concerns about the possi-ble harmful effects of WCPVs, there was a rapid decline in vaccine acceptance rates and recurrences of major pertussis epidemics around 1978 and 1982. As vaccine uptake has recovered, control has once again been established. The relationship between whooping cough notifications in England and Wales and vaccine uptake rate are shown in **Fig. 10**, along with the numbers of notifications and deaths since widespread vaccination was introduced in the 1950s.

An accelerated immunisation schedule (at 2, 3 and 4 months of age) was introduced in the UK in 1990 to replace the previous, extended immunisation (at 3, 5, and 10 months of age) in an attempt to give earlier full protection to very young infants in whom pertussis is most severe (Department of Health, 1990). Efficacy, estimated at 94% (White *et al.*, 1996), was not altered by the new schedule although a reduced rate of adverse reactions was achieved (Miller *et al.*, 1997).

Ever since the introduction of pertussis vaccines in the 1930s, their use has been linked with adverse effects, ranging from transient local reactions to per-manent brain damage and death (Cherry *et al.*, 1988; Ross, 1988; Griffith, 1989; Golden, 1990; Hodder and Mortimer, 1992). Whole-cell pertussis vaccines

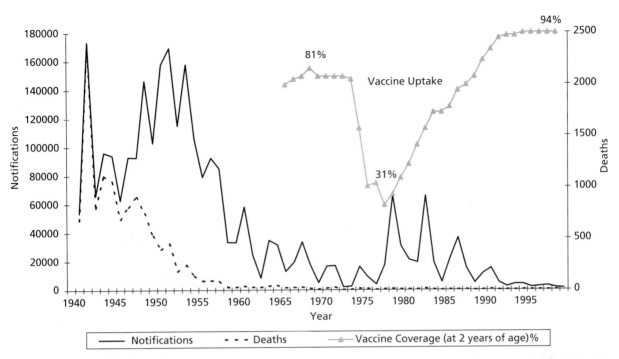

Fig. 10 Whooping cough notifications (cases and deaths) in England and Wales, 1940–1999. The figure also shows when widespread immunisation was introduced and the vaccine uptake rate. Source: PHLS Communicable Diseases Surveillance Centre. The figure was kindly supplied by Dr Elizabeth Miller.

are undoubtedly reactogenic and both local and systemic reactions may occur. Pain, redness or swelling at the site of injection are common, as are mild fever, drowsiness, fretfulness, and anorexia. More severe reactions such as high fever, uncontrollable crying and seizures are less frequent but all are transient. The risk of neurological damage from pertussis immunisation has been the subject of much debate and intensive study. Acute neurological illnesses due to a variety of causes do occur in early childhood, at around the time of pertussis immunisation, and the problem has been to distinguish between causal and temporal relationships to vaccination. After several extensive investigations, it has been concluded that there is no convincing evidence that WCPVs cause permanent neurological damage (Marcuse and Wentz, 1990; Gale *et al.*, 1994; Cherry, 1996).

The situation with WCPVs was aptly described by Manclark and Cowell (1984): 'Present WCPVs are compromises that accept a certain level of toxicity to achieve an acceptable level of potency'. Such vaccines contain at least two active toxins, namely LPS and PT. The former probably contributes much of the reactogenicity of the preparations in terms of fever, local reactions and common systemic reactions. PT is a potent toxin with numerous biological activities and there has been much speculation about its role as a neurotoxin. However, since there is no conclusive link

between pertussis vaccines and permanent neurological illness, the potential hazard of having active PT in a vaccine is unknown. These problems of reactogenicity, both real and perceived, have been the stimulus for the development of APVs, in which the protective components of *B. pertussis* have been retained and the potentially toxic components removed or modified. It has been predicted that acellular vaccines will replace WCPVs in the US and some parts of Europe over the next few years. However, WCPVs, because of their efficacy and low cost will remain widely used for some time to come (Decker and Edwards, 2000).

Acellular Vaccines

The development of pertussis vaccines over the last 2–3 decades (reviewed by Robinson and Ashworth, 1988; Wardlaw, 1992; Edwards *et al.*, 1999; Decker and Edwards, 2000) has been aimed mainly at overcoming the problems of reactogenicity associated with the whole-cell products by preparing defined acellular vaccines, fully characterised in terms of their antigenic components and in which the toxic components of *B. pertussis* are absent or inactivated. A less reactogenic preparation would improve public acceptance of vaccination and could also be used to boost waning immunity in adults. Other aims have been to improve protective efficacy and to protect against infection as well as against disease. To achieve these aims, certain problems have had to be addressed. As emphasised above, the pathogenesis of pertussis is complex and not completely understood. A major hurdle has been the shortcomings of the various animal models available for assessing mechanisms of virulence and immunity, although mouse models, using intracerebral, intranasal and aerosol challenge have provided much useful information (Sato and Sato, 1988). The intracerebral mouse protection test has provided a reliable estimate of the potencies of WCPVs for children but is unreliable for assessing the potency of APVs (Robinson and Funnell, 1992). Thus, the identification and characterisation of the virulence factors of *B. pertussis* and their role as toxins and/or protective antigens has been slow and is still ongoing.

The first of the defined acellular vaccines was developed by Sato and co-workers in Japan in the late 1970s (Sato and Sato, 1984). Such 'first generation' acellular vaccines contained formaldehyde-toxoided PT (PTd) and filamentous haemagglutinin (FHA) as their main components. However, because of the way in which these ingredients were prepared, some vaccines were subsequently found to contain considerable amounts of additional components including pertactin and fimbriae that could contribute to immunity (Hewlett and Cherry, 1997). 'Second-generation' vaccines contain more highly purified ingredients. Two main types of preparations have been used in Japan: the Takeda type, containing PTd and FHA in the ratio 1:9, plus smaller amounts of fimbrial antigens and pertactin; and the Biken type, containing PTd alone or PTd and FHA in equal amounts. Several such vaccines from different manufacturers, and with different compositions, have been used in place of whole-cell vaccines since 1981. At that time, the vaccines were given to children at 2 years of age but, since 1988, the Japanese have reverted to immunisation at 3 months. The available evidence indicates that these vaccines have been effective in preventing clinical disease and are less reactogenic, in terms of causing fever and local reactions, than WCPVs (Noble *et al.*, 1987; Kimura and Kuno-Sakai, 1990). However, they had not been subjected to the most rigorous clinical trials before their widespread use in Japan. Moreover, the important components were not identified and no serological correlates of protection were established (Noble *et al.*, 1987; Mortimer, 1990).

In order to gain more information on the efficacy of APVs, two of the prototype vaccines, a mono-component PTd vaccine and a two component PTd + FHA vaccine were used in a large-scale, double-blind, placebo-controlled field trial in 1986–1987 in Sweden (Ad Hoc Group, 1988). At that time, pertussis was prevalent in Sweden as a result of a policy not to immunise against the disease. The trial involved 3800 children, starting at 5–11 months of age, and only two doses of vaccine were given. The reactogenicity of both vaccines was low, as expected, but efficacy results were disappointing. Their efficacy values against culture-confirmed or serologically confirmed pertussis (69% for the two-component vaccine and 54% for the monocomponent PTd) were lower than would have been expected for a whole-cell vaccine, although such a vaccine was not included in this trial. Both vaccines, however, gave better protection (about 80%) against severe pertussis, i.e. culture-confirmed, with a cough lasting > 30 days, and provided evidence that PTd alone or with other antigens could make effective vaccines. Surprisingly, no correlation was found between serum antibody levels to one or other vaccine component and protection. The serum samples were obtained post-immunisation and not immediately pre-exposure and therefore the antibody levels measured would not necessarily have been relevant to protection. Also, there may be subsets of protective and non-protective antibodies which would not have been distinguished by ELISA (Hewlett and Cherry, 1997).

A further analysis of the trial data showed that the monocomponent vaccine, although protecting against typical pertussis, did not prevent infection. The two-component vaccine gave some protection against infection suggesting that the inclusion of FHA was beneficial (Storsaeter *et al.*, 1990). A long-term follow up showed that clinical immunity after immunisation with the APVs was maintained for at least 4 years, with the two-component vaccine being significantly more efficacious (Storsaeter and Olin, 1992).

As a result of the outcome of the Swedish trials, plans were made for further field trials of other APV formulations and, where possible, to compare them with WCPVs and placebo controls. A number of vaccines were selected, containing from one (PTd alone) to five components (PT, FHA, pertactin and fimbrial antigens 2 and 3). One of the vaccines was a 'third-generation' product containing a highly immunogenic recombinant pertussis toxoid (rPTd) PT-9K/129G developed by Rappuoli and co-workers (Rappuoli *et al.*, 1991). Chemical detoxification with agents such as formaldehyde is known to alter the structure and epitopes of molecules and reduce immunogenicity so that larger amounts of antigen are required to achieve potency. Moreover, the formaldehyde-detoxified PTd shows some reversion to toxicity on storage at 37°C (Kimura *et al.*, 1990), whereas the rPTd is stable.

Large-scale trials of these APVs were carried out in the early to mid-1990s in Sweden (Stockholm 1992 (Gustafsson *et al.*, 1996); Stockholm 1993 (Olin *et al.*,

1997) and Gothenburg (Trollfors *et al.*, 1995)), Germany (Erlangen (Stehr *et al.*, 1998); Mainz (Schmitt *et al.*, 1996) and Munich (Liese *et al.*, 1997)), Italy (Greco *et al.*, 1996) and Senegal (Simondon *et al.*, 1997). In considering the results, it is necessary to bear in mind that the trials were different in design, which makes direct comparison difficult. In addition, a variety of vaccine preparations from different manufacturers were used and so the vaccines differed not only in the number of components but also in the amounts of individual antigens and their method of purification, as well as adjuvant and other additives. All of the vaccines contained PTd but toxoiding was done either with formaldehyde, glutaraldehyde, hydrogen peroxide or by genetic means. Four of these trials (Stockholm 1992, 1993, Italy, Gothenburg) were randomised, prospective and fully blinded and thus had the most rigorous design.

The first three mentioned trials incorporated a WCPV and directly compared APVs of different composition and thus provided the most useful data. The other trials each evaluated only one APV, along with WCPV, and had control groups that were neither randomised nor fully blinded (Decker and Edwards, 2000). The design and outcomes of these trials, and their limitations, have been extensively reviewed (Hewlett and Cherry, 1997; Plotkin and Cadoz, 1997; Edwards *et al.*, 1999; Halperin, 1999; Decker and Edwards, 2000) and only the general conclusions will be summarised here.

All of the acellular preparations were efficacious, with estimates of efficacy against severe pertussis ranging from 59% to 93%. In general, the multi-component vaccines (containing 3–5 antigens) were better than the one- and two-component products. The multi-component vaccines were also more effective against mild disease, providing some optimism that the best APVs have the potential to control the spread of pertussis (Olin, 1997). The similar efficacy (84%) in the Italy trial of two vaccines containing PTd, FHA and pertactin is noteworthy, as one of these contained rPTd at one-fifth of the amount of the formaldehyde-detoxified PTd in the other vaccine, as well as one-tenth of the FHA and one-third of the pertactin. Multi-component vaccines containing pertactin appeared to be amongst the most efficacious. Overall, the results suggested that the number of components, the quantity of each component and the method of production, especially of PTd, all seemed to influence the efficacy of the acellular vaccines (Decker *et al.*, 2000). However, as emphasised by Edwards *et al.* (1999), vaccine efficacy is only one factor in determining the effectiveness of a vaccination programme and even widespread use of the

lower-ranking APVs, as well as WCPVs with moderate efficacy, have been shown to control pertussis well.

In general, the WCPVs were as good or better than the APVs with efficacy estimates from 94% to 98%, although the WCPV used as a control in the Stockholm (1992) trial and in Italy had unexpectedly low efficacy (48% and 36% respectively). In all trials where WCPVs were included, the APVs were significantly less reactogenic in terms of common local (redness, swelling and pain) and systemic (fever, fussiness, drowsiness and anorexia) reactions and also in causing rare but severe events such as persistent crying, hypotonic–hyporesponsive episodes and seizures.

On the basis of the above results, all of the APVs used in the trials have now been licensed for use in various countries. In the US, for example, four APVs containing 1, 2, 3 and 5 components, respectively, together with D and T, are currently licensed and APVs are recommended for all five doses in the US pertussis vaccine schedule. WCPV is no longer recommended (Centers for Disease Control and Prevention: http://www.cdc.gov/health/diseases.htm). However, WCPVs will remain in use in many countries for some time to come (see Control by Whole-cell Vaccines). No pertussis vaccine is currently used in older children or adults. However, in view of the need to control disease in adults and because of the lower reactogenicity of APVs, it is likely that they will soon be used to boost immunity in these groups.

Despite the proven efficacy of APVs and WCPVs, the mechanisms by which they produce immunity in humans remains uncertain. In two separate studies, serological data from two of the trials was subjected to regression analysis to determine the serum antibody levels to the various vaccine components at the time of exposure to infection (Cherry *et al.*, 1998; Storsaeter *et al.*, 1998). It was concluded that antibodies to pertactin, Fim2 and Fim3 and to a lesser extent PT were associated with reduced risk of pertussis. The strongest correlation was with pertactin. The additional presence of antibodies to FHA did not appear to contribute to protection. In one trial, levels of anti-fimbrial antibodies produced in response to WCPV or a five-component APV showed a clear correlation with protection (Olin *et al.*, 1997). However, as explained by Hewlett and Halperin (1998), such data do not prove that the antibodies measured represent the means whereby the vaccines induce protection. Nor do they mean that vaccines lacking, for example, pertactin or fimbriae will not be efficacious or that PTd and FHA are not useful vaccine antigens. Monocomponent PTd vaccines are clearly efficacious and inclusion of FHA gives even better protection (see above). Also, results from studies on the mechanism of immunity to *B. pertussis* in mice and analysis of the cytokine responses of T cells from human infants immunised with WCPVs or APVs or convalescing from *B. pertussis* infection indicate that cell-mediated immunity plays an important role in protection against *B. pertussis* infection (see above).

Although these field trials have provided much valuable information, various questions concerning the use of APVs have still to be answered. Because not all possible combinations of antigens have been tested, or any single antigen other than PTd, it is still not clear which are the most important components or what the optimal composition might be. Until a convenient laboratory method of assessing potency of APVs is devised, or a clear immunological correlate of protection is identified, lengthy and expensive field trials will be necessary. Furthermore, other candidate antigens may become available. For example, ACT is protective in mice and is being developed as a vaccine delivery vehicle for induction of cell-mediated immunity to heterologous epitopes (see above). In addition, BrkA, TCF and other autotransporters, because of their similarity to pertactin, could prove to be useful protective antigens. There is evidence that the C-terminal autotransporter domains themselves are protective antigens in mice (Hamstra *et al.*, 1995). It is also likely that new potential immunogens will be revealed when the annotated genome sequence of *B. pertussis* is released.

The current WCPVs and APVs need to be administered by injection. This is a disadvantage for use of the vaccines in the developing world because of a shortage of sterile needles and the risk of blood-borne infections associated with the re-use of needles. A mucosally delivered pertussis vaccine would avoid this problem and also have the advantage of stimulating mucosal immunity. As mentioned previously, many different types of mucosal pertussis vaccines have been developed and shown to be immunogenic in mice. A live attenuated *B. pertussis* vaccine would present many different *B. pertussis* antigens but could also be used as a vector for mucosal delivery of heterologous antigens. *B. pertussis* has already been demonstrated to be an effective carrier for heterologous antigens (Renauld-Mongenie *et al.*, 1996b; Mielcarek *et al.*, 1997, 1998). It could be envisioned that an attenuated *B. pertussis* strain expressing non-toxic variants of tetanus and diphtheria toxins could be used as a mucosal DTP vaccine. An attenuated *B. bronchiseptica* strain expressing a non-toxic fragment of tetanus toxin has

been shown to induce serum antibodies to the toxin when administered intranasally to mice, demonstrating the utility of this approach (M. Roberts *et al.*, unpublished observation). However, the development and licensing of live vaccines are problematic and the current pertussis vaccines are likely to be in use for some time.

Acknowledgements

We would like to thank Paul Emsley, Duncan Maskell, Andrew Preston, Hal Thompson, Roy Gross, Nichole Guiso, Elizabeth Miller and Agnes Ullman for supplying material for figures and/or sharing unpublished information.

References

Ackermann MR, Register KB, Gentry-Weeks C, Gwaltney SM, Magyar T (1997) A porcine model for the evaluation of virulence of *Bordetella bronchiseptica*. *J. Comp. Pathol.* 116: 55–61.

Ackers JP, Dolby JM (1972) The antigen of *Bordetella pertussis* that induces bactericidal antibody and its relationship to protection of mice. *J. Gen. Microbiol.* 70: 371–382.

Ad Hoc Group (1988) Placebo controlled trial of two acellular pertussis vaccines in Sweden. *Lancet* i: 955–966.

Akerley BJ, Monack DM, Falkow S, Miller JF (1992) The bvgAS locus negatively controls motility and synthesis of flagella in *Bordetella bronchiseptica*. *J. Bacteriol.* 174: 980–990.

Akerley BJ, Cotter PA, Miller JF (1995) Ectopic expression of the flagellar regulon alters development of the *Bordetella*-host interaction. *Cell* 80: 611–620.

Allen A, Maskell D (1996) The identification, cloning and mutagenesis of a genetic locus required for lipopolysaccharide biosynthesis in *Bordetella pertussis*. *Mol. Microbiol.* 19: 37–52.

Allen AG, Isobe T, Maskell DJ (1998a) Identification and cloning of *waaF* (*rfaF*) from *Bordetella pertussis* and use to generate mutants of Bordetella spp. with deep rough lipopolysaccharide. *J. Bacteriol.* 180: 35–40.

Allen AG, Thomas RM, Cadisch JT, Maskell DJ (1998b) Molecular and functional analysis of the lipopolysaccharide biosynthesis locus wlb from *Bordetella pertussis*, *Bordetella parapertussis* and *Bordetella bronchiseptica*. *Mol. Microbiol.* 29: 27–38.

Amsbaugh DF, Li ZM, Shahin RD (1993) Long-lived respiratory immune response to filamentous haemagglutinin following *Bordetella pertussis* infection. *Infect. Immun.* 61: 1447–1452.

Andrews R, Herceg A, Roberts C (1997) Pertussis notifications in Australia, 1991 to 1997. *Commun. Dis. Intell.* 21: 145–148.

Antoine R, Alonso S, Raze D *et al.* (2000) New virulence-activated and virulence-repressed genes identified by systematic gene inactivation and generation of transcriptional fusions in *Bordetella pertussis. J. Bacteriol.* 182: 5902–5905.

Aoyama T, Sunakawa K, Iwata S, Takeuchi Y, Fujii R (1996) Efficacy of short-term treatment of pertussis with clarithromycin and azithromycin. *J. Pediatr.* 129: 761–764.

Arai H, Sato Y (1976) Separation and characterization of two distinct hemagglutinins contained in purified leucocytosis-promoting factor from *Bordetella pertussis. Biochim. Biophys. Acta* 444: 765–782.

Archambault D, Rondeau P, Martin D, Brodeur BR (1991) Characterization and comparative bactericidal activity of monoclonal antibodies to *Bordetella pertussis* lipo-oligosaccharide A. *J. Gen. Microbiol.* 137: 905–911.

Arico B, Rappuoli R (1987) *Bordetella parapertussis* and *Bordetella bronchiseptica* contain transcriptionally silent pertussis toxin genes. *J. Bacteriol.* 169: 2847–2853.

Arico B, Miller JF, Roy C *et al.* (1989) Sequences required for expression of *Bordetella pertussis* virulence factors share homology with prokaryotic signal transduction proteins. *Proc. Natl Acad. Sci. USA* 86: 6671–6675.

Arico B, Scarlato V, Monack DM, Falkow S, Rappuoli R (1991) Structural and genetic analysis of the bvg locus in *Bordetella* species. *Mol. Microbiol.* 5: 2481–2491.

Arico B, Nuti S, Scarlato V, Rappuoli R (1993) Adhesion of *Bordetella pertussis* to eukaryotic cells requires a time-dependent export and maturation of filamentous hemagglutinin. *Proc. Natl Acad. Sci. USA* 90: 9204–9208.

Armstrong SK, Parker CD (1986) Heat-modifiable envelope proteins of *Bordetella pertussis. Infect. Immun.* 54: 109–117.

Armstrong GD, Clark CG, Heerze LD (1994) The 70-kilodalton pertussis toxin-binding protein in Jurkat cells. *Infect. Immun.* 62: 2236–2243.

Ashworth LA, Fitzgeorge RB, Irons LI, Morgan CP, Robinson A (1982) Rabbit nasopharyngeal colonization by *Bordetella pertussis*: the effects of immunization on clearance and on serum and nasal antibody levels. *J. Hyg. Lond.* 88: 475–486.

Ashworth LA, Dowsett AB, Irons LI, Robinson A (1985) The location of surface antigens on the surface of *Bordetella pertussis* by immuno-electron microscopy. *Dev. Biol. Stand.* 61: 143–151.

Ausiello CM, Urbani F, la Sala A, Lande R, Cassone A (1997) Vaccine- and antigen-dependent type 1 and type 2 cytokine induction after primary vaccination of infants with whole-cell or acellular pertussis vaccines. *Infect. Immun.* 65: 2168–2174.

Ausiello CM, Lande R, Urbani F *et al.* (1999) Cell-mediated immune responses in four-year-old children after primary immunization with acellular pertussis vaccines. *Infect. Immun.* 67: 4064–4071.

Avigan J, Murtagh JJJ, Stevens LA, Angus CW, Moss J, Vaughan M (1992) Pertussis toxin-catalyzed ADP-ribosylation of G(o) alpha with mutations at the carboxyl terminus. *Biochemistry* 31: 7736–7740.

Bace A, Zrnic T, Begovac J, Kuzmanovic N, Culig J (1999) Short-term treatment of pertussis with azithromycin in infants and young children. *Eur. J. Clin. Microbiol. Infect. Dis.* 18: 296–298.

Backstrom LR, Brim TA, Collins MT (1988) Development of turbinate lesions and nasal colonization by *Bordetella bronchiseptica* and *Pasteurella multocida* during long-term exposure of healthy pigs to pigs affected by atrophic rhinitis. *Can. J. Vet. Res.* 52: 23–29.

Banemann A, Deppisch H, Gross R (1998) The lipopolysaccharide of *Bordetella bronchiseptica* acts as a protective shield against antimicrobial peptides. *Infect. Immun.* 66: 5607–5612.

Barbic J, Leef MF, Burns DL, Shahin RD (1997) Role of gamma interferon in natural clearance of *Bordetella pertussis* infection. *Infect. Immun.* 65: 4904–4908.

Basar T, Havlicek V, Bezouskova S, Hackett M, Sebo P (2001) Acylation of lysine 983 is sufficient for toxin activity of *Bordetella pertussis* adenylate cyclase: substitutions of alanine 140 modulate acylation site selectivity of the toxin acyltransferase CyaC. *J. Biol. Chem.* 276: 348–354.

Bassinet L, Gueirard P, Maitre B, Housset B, Gounon P, Guiso N (2000) Role of adhesins and toxins in invasion of human tracheal epithelial cells by *Bordetella pertussis. Infect. Immun.* 68: 1934–1941.

Beall BW, Sanden GN (1995a) Cloning and initial characterization of the *Bordetella pertussis* fur gene. *Curr. Microbiol.* 30: 223–226.

Beall B, Sanden GN (1995b) A *Bordetella pertussis* fepA homologue required for utilization of exogenous ferric enterobactin. *Microbiology* 141: 3193–3205.

Beall B, Cassiday PK, Sanden GN (1995) Analysis of *Bordetella pertussis* isolates from an epidemic by pulsed-field gel electrophoresis. *J. Clin. Microbiol.* 33: 3083–3086.

Beaumont FC, Kang HY, Brickman TJ, Armstrong SK (1998) Identification and characterization of alcR, a gene encoding an AraC-like regulator of alcaligin siderophore biosynthesis and transport in *Bordetella pertussis* and *Bordetella bronchiseptica. J. Bacteriol.* 180: 862–870.

Beier D, Schwarz B, Fuchs TM, Gross R (1995) *In vivo* characterisation of the unorthodox BvgS two-component sensor protein of *Bordetella pertussis. J. Mol. Biol.* 248: 596–610.

Beier D, Deppisch H, Gross R (1996) Conserved sequence motifs in the unorthodox BvgS two-component sensor protein of *Bordetella pertussis. Mol. Gen. Genet.* 252: 168–176.

Bemis DA, Kennedy JR (1981) An improved system for studying the effect of *Bordetella bronchiseptica* on the ciliary activity of canine tracheal epithelial cells. *J. Infect. Dis.* 144: 349–357.

Berggard K, Johnsson E, Mooi FR, Lindahl G (1997) *Bordetella pertussis* binds the human complement regulator C4BP: role of filamentous hemagglutinin. *Infect. Immun.* 65: 3638–3643.

Betsou F, Sebo P, Guiso N (1993) CyaC-mediated activation is important not only for toxic but also for protective activities of *Bordetella pertussis* adenylate cyclase-hemolysin. *Infect. Immun.* 61: 3583–3589.

Bey RF, Shade FJ, Goodnow RA, Johnson RC (1981) Intranasal vaccination of dogs with live avirulent *Bordetella bronchiseptica*: correlation of serum agglutination titer and the formation of secretory IgA with protection against experimentally induced infectious tracheobronchitis. *Am. J. Vet. Res.* 42: 1130–1132.

Black S (1997) Epidemiology of pertussis. *Pediatr. Infect. Dis. J.* 16: S85–S89.

Blackburn PE (2000) Characterisation of the virulence-related outer-membrane proteins of *Bordetella pertussis*. PhD Thesis, University of Glasgow.

Blom J, Heron I, Hendley JO (1994) Immunoelectron microscopy of antigens of *Bordetella pertussis* using monoclonal antibodies to agglutinogens 2 and 3, filamentous haemagglutinin, pertussis toxin, pertactin and adenylate cyclase toxin. *APMIS* 102: 681–689.

Bordet J, Gengou O (1906) Le microbe de la coqueluche. *Ann. Inst. Pasteur* 23: 415–419.

Bosch A, Massa NE, Donolo A, Yantorno O (2000) Molecular characterisation by infrared spectroscopy of *Bordetella pertussis* grown as biofilm. *Phys. Status Solidi B-Basic Res.* 220: 635–640.

Boschwitz JS, Batanghari JW, Kedem H, Relman DA (1997a) *Bordetella pertussis* infection of human monocytes inhibits antigen-dependent CD4 T cell proliferation. *J. Infect. Dis.* 176: 678–686.

Boschwitz JS, van der Heide HG, Mooi FR, Relman DA (1997b) *Bordetella bronchiseptica* expresses the fimbrial structural subunit gene *fimA*. *J. Bacteriol.* 179: 7882–7885.

Boucher PE, Menozzi FD, Locht C (1994) The modular architecture of bacterial response regulators. Insights into the activation mechanism of the BvgA transactivator of *Bordetella pertussis*. *J. Mol. Biol.* 241: 363–377.

Bourdon MA, Krusius T, Campbell S, Schwartz NB, Ruoslahti E (1987) Identification and synthesis of a recognition signal for the attachment of glycosaminoglycans to proteins. *Proc. Natl Acad. Sci. USA* 84: 3194–3198.

Boursaux-Eude C, Thiberge S, Carletti G, Guiso N (1999) Intranasal murine model of *Bordetella pertussis* infection: II. Sequence variation and protection induced by a tricomponent acellular vaccine. *Vaccine* 17: 2651–2660.

Bradford WL, Slavin B (1937) An organism resembling *Hemophilus pertussis*. *Am. J. Publ. Health* 27: 1277–1282.

Brennan MJ, David JL, Kenimer JG, Manclark CR (1988) Lectin-like binding of pertussis toxin to a 165-kilodalton Chinese hamster ovary cell glycoprotein. *J. Biol. Chem.* 263: 4895–4899.

Brennan MJ, Hannah JH, Leininger E (1991) Adhesion of *Bordetella pertussis* to sulfatides and to the GalNAc beta 4Gal sequence found in glycosphingolipids. *J. Biol. Chem.* 266: 18827–18831.

Brezin C, Guiso N, Ladant D *et al.* (1987) Protective effects of anti-*Bordetella pertussis* adenylate cyclase antibodies against lethal respiratory infection of the mouse. *FEMS Microbiol. Lett.* 42: 75–80.

Brickman TJ, Armstrong SK (1999) Essential role of the iron-regulated outer membrane receptor FauA in alcaligin siderophore-mediated iron uptake in Bordetella species. *J. Bacteriol.* 181: 5958–5966.

Brockmeier SL, Register KB, Kunkle RA (2000) Evaluation of the virulence of a dermonecrotic toxin mutant of *Bordetella bronchiseptica* in swine. 16th International Pig Veterinary Society Congress, Melbourne, Australia, pp. 480–480.

Bromberg K, Tannis G, Steiner P (1991) Detection of *Bordetella pertussis* associated with the alveolar macrophages of children with human immunodeficiency virus infection. *Infect. Immun.* 59: 4715–4719.

Brownlie RM, Brahmbhatt HN *et al.* (1993) Stimulation of secretory antibodies against *Bordetella pertussis* antigens in the lungs of mice after oral or intranasal administration of liposome-incorporated cell-surface antigens. *Microb. Pathog.* 14: 149–160.

Bruss JB, Malley R, Halperin S *et al.* (1999) Treatment of severe pertussis: a study of the safety and pharmacology of intravenous pertussis immunoglobulin. *Pediatr. Infect. Dis. J.* 18: 505–511.

Burnette WN (1994) AB5 ADP-ribosylating toxins: comparative anatomy and physiology. *Structure.* 2: 151–158.

Burnette WN (1997) Bacterial ADP-ribosylating toxins: form, function, recombinant vaccine development. *Behring. Inst. Mitt.* Feb. No. 98, 434–441.

Burns DL (1999) Biochemistry of type IV secretion. *Curr. Opin. Microbiol.* 2: 25–29.

Byrd DW, Roop RM, Veit HP, Schurig GG (1991) Serum sensitivity and lipopolysaccharide characteristics in *Bordetella bronchiseptica*, *B. pertussis* and *B. parapertussis*. *J. Med. Microbiol.* 34: 159–165.

Cahill ES, O'Hagan DT, Illum L, Barnard A, Mills KH, Redhead K (1995) Immune responses and protection against *Bordetella pertussis* infection after intranasal immunization of mice with filamentous haemagglutinin in solution or incorporated in biodegradable microparticles. *Vaccine* 13: 455–462.

Capiau C, Carr ME, Hemlin D *et al.* (1990) Purification, characterisation, immunological evaluation of the 69-kDa outer membrane protein of *Bordetella pertussis*. In: Manclark CR (ed.) *Proceedings of the Sixth International Symposium on Pertussis*. Bethesda, Maryland: Department of Health and Human Services, United States Public Health Service, pp. 75–86.

Carbonetti NH, Fuchs TM, Patamawenu AA, Irish TJ, Deppisch H, Gross R (1994) Effect of mutations causing overexpression of RNA polymerase alpha subunit on regulation of virulence factors in *Bordetella pertussis*. *J. Bacteriol.* 176: 7267–7273.

Caroff M, Chaby R, Karibian D, Perry J, Deprun C, Szabo L (1990) Variations in the carbohydrate regions of *Bordetella pertussis* lipopolysaccharides: electrophoretic, serological, structural features. *J. Bacteriol.* 172: 1121–1128.

Cassiday P, Sanden G, Heuvelman K, Mooi F, Bisgard KM, Popovic T (2000) Polymorphism in *Bordetella pertussis* pertactin and pertussis toxin virulence factors in the United States, 1935–1999. *J. Infect. Dis.* 182: 1402–1408.

Chaby R, Caroff M (1988) Lipopolysaccharides of *Bordetella pertussis* endotoxin. In: Wardlaw AC, Parton R (eds) *Pathogenesis and Immunity in Pertussis*. Chichester: John Wiley and Sons, pp. 247–271.

Charles IG, Dougan G, Pickard D *et al.* (1989) Molecular cloning and characterization of protective outer membrane protein P.69 from *Bordetella pertussis*. *Proc. Natl Acad. Sci. USA* 86: 3554–3558.

Charles IG, Li JL, Roberts M *et al.* (1991) Identification and characterization of a protective immunodominant B cell epitope of pertactin (P.69) from *Bordetella pertussis*. *Eur. J. Immunol.* 21: 1147–1153.

Charles I, Rodgers B, Musgrave S *et al.* (1993) Expression of P.69/pertactin from *Bordetella pertussis* in a baculovirus/insect cell expression system: protective properties of the recombinant protein. *Res. Microbiol.* 144: 681–690.

Charles I, Fairweather N, Pickard D *et al.* (1994) Expression of the *Bordetella pertussis* P.69 pertactin adhesin in *Escherichia coli*: fate of the carboxy-terminal domain. *Microbiology* 140: 3301–3308.

Cheers C, Gray DF (1969) Macrophage behaviour during the complaisant phase of murine pertussis. *Immunology* 17: 875–887.

Chen I, Finn TM, Yanqing L, Guoming Q, Rappuoli R, Pizza M (1998) A recombinant live attenuated strain of Vibrio cholerae induces immunity against tetanus toxin and *Bordetella pertussis* tracheal colonization factor. *Infect. Immun.* 66: 1648–1653.

Cherry JD (1996) Historical review of pertussis and the classical vaccine. *J. Infect. Dis.* 174 (Suppl. 3): S259–S263.

Cherry JD (1999) Epidemiological, clinical, laboratory aspects of pertussis in adults. *Clin. Infect. Dis.* 28 (Suppl. 2): S112–S117.

Cherry JD, Brunell PA, Golden GS, Karzon DT (1988) Report of the task force on pertussis and pertussis immunization. *Pediatrics* 81 (Suppl.): 939–984.

Cherry JD, Gornbein J, Heininger U, Stehr K (1998) A search for serologic correlates of immunity to *Bordetella pertussis* cough illnesses. *Vaccine* 16: 1901–1906.

Confer DL, Eaton JW (1982) Phagocyte impotence caused by an invasive bacterial adenylate cyclase. *Nature* 217: 948–950.

Cookson BT, Tyler AN, Goldman WE (1989) Primary structure of the peptidoglycan-derived tracheal cytotoxin of *Bordetella pertussis*. *Biochemistry* 28: 1744–1749.

Coote JG (1991) Antigenic switching and pathogenicity: Environmental effects on virulence gene expression in *Bordetella pertussis*. *J. Gen. Microbiol.* 137: 2493–2503.

Coote JG (1992) Structural and functional relationships among the RTX toxin determinants of Gram-negative bacteria. *FEMS Microbiol. Rev.* 88: 137–162.

Coote J (2001) Environmental sensing mechanisms in Bordetella. *Adv. Microb. Physiol.* 44: 141–181

Coote JG, Brownlie RM (1988) Genetics of virulence of *Bordetella pertussis*. In: Wardlaw AC, Parton R (eds) *Pathogenesis and Immunity in Pertussis*. Chichester: John Wiley and Sons, pp. 39–74.

Cotter PA, DiRita VJ (2000) Bacterial virulence gene regulation: an evolutionary perspective. *Annu. Rev. Microbiol.* 54: 519–565.

Cotter PA, Miller JF (1994) BvgAS-mediated signal transduction: analysis of phase-locked regulatory mutants of *Bordetella bronchiseptica* in a rabbit model. *Infect. Immun.* 62: 3381–3390.

Cotter PA, Miller JF (1996) Genetic analysis of the Bordetella-host interaction. *Annu. NY Acad. Sci.* 797: 65–76.

Cotter PA, Miller JF (1997) A mutation in the *Bordetella bronchiseptica* bvgS gene results in reduced virulence and increased resistance to starvation, identifies a new class of Bvg-regulated antigens. *Mol. Microbiol.* 24: 671–685.

Cotter PA, Yuk MH, Mattoo S *et al.* (1998) Filamentous hemagglutinin of *Bordetella bronchiseptica* is required for efficient establishment of tracheal colonization. *Infect. Immun.* 66: 5921–5929.

Covacci A, Rappuoli R (1993) Pertussis toxin export requires accessory genes located downstream from the pertussis toxin operon. *Mol. Microbiol.* 8: 429–434.

Cowell JL, Hewlett EL, Manclark CR (1979) Intracellular localization of the dermonecrotic toxin of *Bordetella pertussis*. *Infect. Immun.* 25: 896–901.

Craig-Mylius KA, Weiss A (1999) Mutants in the *ptlA-H* genes of *Bordetella pertussis* are deficient for pertussis toxin secretion. *FEMS Microbiol. Lett.* 179: 479–484.

Craig-Mylius KA, Stenson TH, Weiss AA (2000) Mutations in the S1 subunit of pertussis toxin that affect secretion. *Infect. Immun.* 68: 1276–1281.

Cropley I, Douce G, Roberts M *et al.* (1995) Mucosal and systemic immunogenicity of a recombinant, non-ADP-ribosylating pertussis toxin: effects of formaldehyde treatment. *Vaccine* 13: 1643–1648.

Cundell DR, Kanthakumar K, Taylor GW *et al.* (1994) Effect of tracheal cytotoxin from *Bordetella pertussis* on human neutrophil function *in vitro*. *Infect. Immun.* 62: 639–643.

de Melker HE, Conyn-van Spaendonck MA, Rumke HC, van Wijngaarden JK, Mooi FR, Schellekens JF (1997) Pertussis in The Netherlands: an outbreak despite high levels of immunization with whole-cell vaccine. *Emerging Infect. Dis.* 3: 175–178.

De Moissac YR, Ronald SL, Peppler MS (1994) Use of pulsed-field gel electrophoresis for epidemiological study of *Bordetella pertussis* in a whooping cough outbreak. *J. Clin. Microbiol.* 32: 398–402.

De Serres G, Boulianne N, Douville FM, Duval B (1995) Pertussis in Quebec: ongoing epidemic since the late 1980s. *Can. Communicable Dis. Rep.* 21: 45–48.

Decker MD, Edwards KM (2000) Acellular pertussis vaccines. *Pediatr. Clin. N. Am.* 47: 309–335.

Delisse-Gathoye AM, Locht C, Jacob F *et al.* (1990) Cloning, partial sequence, expression, antigenic analysis of the filamentous hemagglutinin gene of *Bordetella pertussis*. *Infect. Immun.* 58: 2895–2905.

Department of Health (1990) Whooping cough (pertussis). In: Department of Health (ed.) *Immunisation against Infectious Disease*. London: HMSO, pp. 20–27.

Domenighini M, Relman D, Capiau C *et al.* (1990) Genetic characterization of *Bordetella pertussis* filamentous haemagglutinin: a protein processed from an unusually large precursor. *Mol. Microbiol.* 4: 787–800.

Dugal F, Girard C, Jacques M (1990) Adherence of *Bordetella bronchiseptica* 276 to porcine trachea maintained in organ culture. *Appl. Environ. Microbiol.* 56: 1523–1529.

Edelman K, Nikkari S, Ruuskanen O, He Q, Viljanen M, Mertsola J (1996) Detection of *Bordetella pertussis* by polymerase chain reaction and culture in the nasopharynx of erythromycin-treated infants with pertussis. *Pediatr. Infect. Dis. J.* 15: 54–57.

Edwards KM, Decker MD, Mortimer EA (1999) Pertussis vaccine. In: Plotkin BJ, Orenstein WA (eds) *Vaccines*. Philadelphia: WB Saunders and Co, pp. 293–344.

Eldering G, Hornbeck C, Baker J (1957) Serological study of *Bordetella pertussis* and related species. *J. Bacteriol.* 74: 133–136.

Emsley P, Charles IG, Fairweather NF, Isaacs NW (1996) Structure of *Bordetella pertussis* virulence factor P.69 pertactin. *Nature* 381: 90–92.

Everest P, Li J, Douce G *et al.* (1996) Role of the *Bordetella pertussis* P.69/pertactin protein and the P.69/pertactin RGD motif in the adherence to and invasion of mammalian cells. *Microbiology.* 142: 3261–3268.

Ewanowich CA, Melton AR, Weiss AA, Sherburne RK, Peppler MS (1989) Invasion of HeLa 229 cells by virulent *Bordetella pertussis*. *Infect. Immun.* 57: 2698–2704.

Farizo KM, Huang T, Burns DL (2000) Importance of holotoxin assembly in Ptl-mediated secretion of pertussis toxin from *Bordetella pertussis*. *Infect. Immun.* 68: 4049–4054.

Fayolle C, Ladant D, Karimova G, Ullmann A, Leclerc C (1999) Therapy of murine tumors with recombinant *Bordetella pertussis* adenylate cyclase carrying a cytotoxic T cell epitope. *J. Immunol.* 162: 4157–4162.

Felix R, Fleisch H, Frandsen PL (1992) Effect of *Pasteurella multocida* toxin on bone resorption *in vitro*. *Infect. Immun.* 60: 4984–4988.

Fernandez RC, Weiss AA (1994) Cloning and sequencing of a *Bordetella pertussis* serum resistance locus. *Infect. Immun.* 62: 4727–4738.

Fernandez RC, Weiss AA (1996) Susceptibilities of *Bordetella pertussis* strains to antimicrobial peptides. *Antimicrob. Agents Chemother.* 40: 1041–1043.

Fernandez RC, Weiss AA (1998) Serum resistance in bvg-regulated mutants of *Bordetella pertussis*. *FEMS Microbiol. Lett.* 163: 57–63.

Ferry NS (1910) A preliminary report of the bacterial findings in canine distemper. *Am. Vet. Rev.* 37: 499–504.

Fine PE (1988) Epidemiological considerations for whooping cough eradication. In: Wardlaw AC, Parton R (eds) *Pathogenesis and Immunity in Pertussis*. Chichester: John Wiley and Sons, pp. 451–467.

Finlay BB, Falkow S (1997) Common themes in microbial pathogenicity revisited. *Microbiol. Mol. Biol. Rev.* 61: 136–169.

Finn TM, Amsbaugh DF (1998) Vag8, a *Bordetella pertussis*bvg-regulated protein. *Infect. Immun.* 66: 3985–3989.

Finn TM, Stevens LA (1995) Tracheal colonization factor: a *Bordetella pertussis* secreted virulence determinant. *Mol. Microbiol.* 16: 625–634.

Flak TA, Goldman WE (1999) Signaling and cellular specificity of airway nitric oxide production in pertussis. *Cell. Microbiol.* 1: 51–60.

Flak TA, Heiss LN, Engle JT, Goldman WE (2000) Synergistic epithelial responses to endotoxin and a naturally occurring muramyl peptide. *Infect. Immun.* 68: 1235–1242.

Forde CB, Shi X, Li J, Roberts M (1999) *Bordetella bronchiseptica*-mediated cytotoxicity to macrophages is dependent on bvg-regulated factors, including pertactin. *Infect. Immun.* 67: 5972–5978.

Friedman RL (1988) Pertussis: the disease and new diagnostic methods. *Clin. Microbiol. Rev.* 1: 365–376.

Friedman RL, Nordensson K, Wilson L, Akporiaye ET, Yocum DE (1992) Uptake and intracellular survival of *Bordetella pertussis* in human macrophages. *Infect. Immun.* 60: 4578–4585.

Frohlich BT, d'Alarcao M, Feldberg RS, Nicholson ML, Siber GR, Swartz RW (1996) Formation and cell-medium partitioning of autoinhibitory free fatty acids and cyclodextrin's effect in the cultivation of *Bordetella pertussis*. *J. Biotechnol.* 45: 137–148.

Funnell SG, Robinson A (1993) A novel adherence assay for *Bordetella pertussis* using tracheal organ cultures. *FEMS Microbiol. Lett.* 110: 197–203.

Galan JE, Collmer A (1999) Type III secretion machines: bacterial devices for protein delivery into host cells. *Science* 284: 1322–1328.

Gale JL, Thapa PB, Wassilak SG, Bobo JK, Mendelman PM, Foy HM (1994) Risk of serious acute neurological illness after immunization with diphtheria-tetanus-pertussis vaccine. A population-based case-control study. *JAMA* 271: 37–41.

Gentry-Weeks CR, Provence DL, Keith JM, Curtiss R (1991) Isolation and characterization of *Bordetella avium* phase variants. *Infect. Immun.* 59: 4026–4033.

Geuijen CA, Willems RJ, Mooi FR (1996) The major fimbrial subunit of *Bordetella pertussis* binds to sulfated sugars. *Infect. Immun.* 64: 2657–2665.

Geuijen CA, Willems RJ, Bongaerts M, Top J, Gielen H, Mooi FR (1997) Role of the *Bordetella pertussis* minor fimbrial subunit, FimD, in colonization of the mouse respiratory tract. *Infect. Immun.* 65: 4222–4228.

Geuijen CA, Willems RJ, Hoogerhout P, Puijk WC, Meloen RH, Mooi FR (1998) Identification and characterization of heparin binding regions of the Fim2 subunit of *Bordetella pertussis*. *Infect. Immun.* 66: 2256–2263.

Giardina PC, Foster LA, Musser JM, Akerley BJ, Miller JF, Dyer DW (1995) Bvg repression of alcaligin synthesis in *Bordetella bronchiseptica* is associated with phylogenetic lineage. *J. Bacteriol.* 177: 6058–6063.

Gilchrist MJ (1991) Bordetella. In: Balows A, Hausler WJ (eds) *Manual of Clinical Microbiology*. Washington, DC: American Society for Microbiology, pp. 471–477.

Giuliano M, Mastrantonio P, Giammanco A, Piscitelli A, Salmaso S, Wassilak SG (1998) Antibody responses and persistence in the two years after immunization with two acellular vaccines and one whole-cell vaccine against pertussis. *J. Pediatr.* 132: 983–988.

Glaser P, Ladant D, Sezer O, Pichot F, Ullmann A, Danchin A (1988a) The calmodulin-sensitive adenylate-cyclase of *Bordetella pertussis* – cloning and expression in *Escherichia coli. Mol. Microbiol.* 19–30.

Glaser P, Sakamoto H, Bellalou J, Ullmann A, Danchin A (1988b) Secretion of cyclolysin, the calmodulin-sensitive adenylate cyclase-haemolysin bifunctional protein of *Bordetella pertussis. EMBO J.* 7: 3997–4004.

Golden GS (1990) Pertussis vaccine and injury to the brain. *J. Pediatr.* 116: 854–861.

Goldman S, Hanski E, Fish F (1984) Spontaneous phase variation in *Bordetella pertussis* is a multistep non-random process. *EMBO J.* 3: 1353–1356.

Goldman WE (1988) Tracheal cytotoxin of *Bordetella pertussis*. In: Wardlaw AC, Parton R (eds) *Pathogenesis and Immunity in Pertussis*. Chichester: John Wiley and Sons, pp. 231–246.

Goodman YE, Wort AJ, Jackson FL (1981) Enzyme-linked immunosorbent assay for detection of pertussis immunoglobulin A in nasopharyngeal secretions as an indicator of recent infection. *J. Clin. Microbiol.* 13: 286–292.

Goodnow RA (1980) Biology of *Bordetella bronchiseptica. Microbiol. Rev.* 44: 722–738.

Gotto JW, Eckhardt T, Reilly PA *et al.* (1993) Biochemical and immunological properties of two forms of pertactin, the 69,000-molecular-weight outer membrane protein of *Bordetella pertussis. Infect. Immun.* 61: 2211–2215.

Graeff-Wohlleben H, Deppisch H, Gross R (1995) Global regulatory mechanisms affect virulence gene expression in *Bordetella pertussis. Mol. Gen. Genet.* 247: 86–94.

Granstrom G, Askelof P, Granstrom M (1988) Specific immunoglobulin A to *Bordetella pertussis* antigens in mucosal secretion for rapid diagnosis of whooping cough. *J. Clin. Microbiol.* 26: 869–874.

Granstrom M, Olinder-Nielsen AM, Holmblad P, Mark A, Hanngren K (1991) Specific immunoglobulin for treatment of whooping cough. *Lancet* 338: 1230–1233.

Gray M, Szabo G, Otero AS, Gray L, Hewlett E (1998) Distinct mechanisms for K+ efflux, intoxication, hemolysis by *Bordetella pertussis* AC toxin. *J. Biol. Chem.* 273: 18260–18267.

Greco D, Salmaso S, Mastrantonio P *et al.* (1996) A controlled trial of two acellular vaccines and one whole-cell vaccine against pertussis. Progetto Pertosse Working Group. *N. Engl. J. Med.* 334: 341–348.

Griffith AH (1989) Permanent brain damage and pertussis vaccination: is the end of the saga in sight? *Vaccine* 7: 199–210.

Gross R, Arico B, Rappuoli R (1989) Genetics of pertussis toxin. *Mol. Microbiol.* 3: 119–124.

Guedin S, Willery E, Tommassen J *et al.* (2000) Novel topological features for FhaC, the outer membrane transporter involved in the secretion of the *Bordetella pertussis* filamentous hemagglutinin. *J. Biol. Chem.* 275: 30202–30210.

Gueirard P, Guiso N (1993) Virulence of *Bordetella bronchiseptica*: role of adenylate cyclase-hemolysin. *Infect. Immun.* 61: 4072–4078.

Gueirard P, Druilhe A, Pretolani M, Guiso N (1998a) Role of adenylate cyclase-hemolysin in alveolar macrophage apoptosis during *Bordetella pertussis* infection *in vivo. Infect. Immun.* 66: 1718–1725.

Gueirard P, Le Blay K, Le Coustumier A, Chaby R, Guiso N (1998b) Variation in *Bordetella bronchiseptica* lipopolysaccharide during human infection. *FEMS Microbiol. Lett.* 162: 331–337.

Guiso N, Szatanik M, Rocancourt M (1990) *Bordetella pertussis* adenylate cyclase: a protective antigen against lethality and bacterial colonisation in murine respiratory and intracerebral models. In: Manclark CR (ed.) *Proceedings of the Sixth International Symposium on Pertussis*. Bethesda, Maryland: Department of Health and Human Services, United States Public Health Service, pp. 207–211.

Guiso N, Szatanik M, Rocancourt M (1991) Protective activity of *Bordetella* adenylate cyclase-hemolysin against bacterial colonization. *Microb. Pathog.* 11: 423–431.

Guiso N, Capiau C, Carletti G, Poolman J, Hauser P (1999) Intranasal murine model of *Bordetella pertussis* infection. I. Prediction of protection in human infants by acellular vaccines. *Vaccine* 17: 2366–2376.

Guris D, Strebel PM, Bardenheier B *et al.* (1999) Changing epidemiology of pertussis in the United States: increasing reported incidence among adolescents and adults, 1990–1996. *Clin. Infect. Dis.* 28: 1230–1237.

Gustafsson L, Hallander HO, Olin P, Reizenstein E, Storsaeter J (1996) A controlled trial of a two-component acellular, a five-component acellular, a whole-cell pertussis vaccine. *N. Engl. J. Med.* 334: 349–355.

Guzman CA, Brownlie RM, Kadurugamuwa J, Walker MJ, Timmis KN (1991) Antibody responses in the lungs of mice following oral immunization with *Salmonella typhimurium* aroA and invasive *Escherichia coli* strains expressing the filamentous hemagglutinin of *Bordetella pertussis. Infect. Immun.* 59: 4391–4397.

Guzman CA, Molinari G, Fountain MW, Rohde M, Timmis KN, Walker MJ (1993) Antibody responses in the serum and respiratory tract of mice following oral vaccination with liposomes coated with filamentous hemagglutinin and pertussis toxoid. *Infect. Immun.* 61: 573–579.

Gwaltney SM, Galvin RJ, Register KB, Rimler RB, Ackermann MR (1997) Effects of *Pasteurella multocida* toxin on porcine bone marrow cell differentiation into osteoclasts and osteoblasts. *Vet. Pathol.* 34: 421–430.

Hackett M, Guo L, Shabanowitz J, Hunt DF, Hewlett EL (1994) Internal lysine palmitoylation in adenylate cyclase toxin from *Bordetella pertussis. Science* 266: 433–435.

Hafler JP, Pohl-Koppe A (1998) The cellular immune response to *Bordetella pertussis* in two children with whooping cough. *Eur. J. Med. Res.* 3: 523–526.

Hallander HO (1999) Microbiological and serological diagnosis of pertussis. *Clin. Infect. Dis.* 28 (Suppl. 2): S99–S106.

Hallander HO, Gnarpe J, Gnarpe H, Olin P (1999) *Bordetella pertussis, Bordetella parapertussis, Mycoplasma pneumoniae, Chlamydia pneumoniae* and persistent cough in children. *Scand. J. Infect. Dis.* 31: 281–286.

Halperin SA (1999) Developing better paediatric vaccines. The case of pertussis vaccine. *BioDrugs* 12: 175–191.

Hamstra HJ, Kuipers B, Schijf-Evers D, Loggen HG, Poolman JT (1995) The purification and protective capacity of *Bordetella pertussis* outer membrane proteins. *Vaccine* 13: 747–752.

Hanada M, Shimoda K, Tomita S, Nakase Y, Nishiyama Y (1979) Production of lesions similar to naturally occurring swine atrophic rhinitis by cell-free sonicated extract of *Bordetella bronchiseptica. Nippon. Juigaku. Zasshi.* 41: 1–8.

Hannah JH, Menozzi FD, Renauld G, Locht C, Brennan MJ (1994) Sulfated glycoconjugate receptors for the *Bordetella pertussis* adhesin filamentous hemagglutinin (FHA) and mapping of the heparin-binding domain on FHA. *Infect. Immun.* 62: 5010–5019.

Harvill ET, Cotter PA, Yuk MH, Miller JF (1999) Probing the function of *Bordetella bronchiseptica* adenylate cyclase toxin by manipulating host immunity. *Infect. Immun.* 67: 1493–1500.

Harvill ET, Preston A, Cotter PA, Allen AG, Maskell DJ, Miller JF (2000) Multiple roles for *Bordetella* LPS molecules during respiratory tract infection. *Infect. Immun.* 68: 6720–6728.

Hazenbos WL, van den Berg BM, van Furth R (1993) Very late antigen-5 and complement receptor type 3 cooperatively mediate the interaction between *Bordetella pertussis* and human monocytes. *J. Immunol.* 151: 6274–6282.

Hazenbos W, van den Berg BM, Vant W, Mooi FR, van Furth R (1994) Virulence factors determine attachment and ingestion of nonopsonized and opsonized *Bordetella pertussis* by human monocytes. *Infect. Immun.* 62: 4818–4824.

Hazenbos WL, Geuijen CA, van den Berg BM, Mooi FR, van Furth R (1995a) *Bordetella pertussis* fimbriae bind to human monocytes via the minor fimbrial subunit FimD. *J. Infect. Dis.* 171: 924–929.

Hazenbos WL, van den Berg BM, Geuijen CW, Mooi FR, van Furth R (1995b) Binding of FimD on *Bordetella pertussis* to very late antigen-5 on monocytes activates

complement receptor type 3 via protein tyrosine kinases. *J. Immunol.* 155: 3972–3978.

Hazes B, Boodhoo A, Cockle SA, Read RJ (1996) Crystal structure of the pertussis toxin-ATP complex: a molecular sensor. *J. Mol. Biol.* 258: 661–671.

Heininger U, Stehr K, SchmittGrohe S *et al.* (1994) Clinical characteristics of illness caused by *Bordetella parapertussis* compared with illness caused by *Bordetella pertussis. Pediatr. Infect. Dis. J.* 13: 306–309.

Heiss LN, Moser SA, Unanue ER, Goldman WE (1993) Interleukin-1 is linked to the respiratory epithelial cytopathology of pertussis. *Infect. Immun.* 61: 3123–3128.

Heiss LN, Lancaster JRJ, Corbett JA, Goldman WE (1994) Epithelial autotoxicity of nitric oxide: role in the respiratory cytopathology of pertussis. *Proc. Natl Acad. Sci. USA* 91: 267–270.

Hellwig SM, Hazenbos WL, van de Winkel JG, Mooi FR (1999) Evidence for an intracellular niche for *Bordetella pertussis* in broncho-alveolar lavage cells of mice. *FEMS Immunol. Med. Microbiol.* 26: 203–207.

Henderson IR, Navarro-Garcia F, Nataro JP (1998) The great escape: structure and function of the autotransporter proteins. *Trends Microbiol.* 6: 370–378.

Henderson IR, Nataro JP, Kaper JB *et al.* (2000) Renaming protein secretion in the gram-negative bacteria [letter]. *Trends Microbiol.* 8: 352–352.

Hewlett EL (1997) Pertussis: current concepts of pathogenesis and prevention. *Pediatr. Infect. Dis. J.* 16: S78–S84.

Hewlett EL (1999) A commentary on the pathogenesis of pertussis. *Clin. Infect. Dis.* 28 (Suppl. 2): S94–S98.

Hewlett EL (2000) Bordetella species. In: Mandell GL, Bennett JE, Dolin R (eds) *Principles and Practice of Infectious Diseases.* Philadelphia: Churchill Livingstone, pp. 2414–2422.

Hewlett EL, Cherry JD (1997) New and improved vaccines against pertussis. In: Levine MM, Woodrow GC, Kaper JB, Cobon GS (eds) *New Generation Vaccines.* New York: Marcel Dekker, pp. 367–410.

Hewlett EL, Gordon VM (1988) Adenylate cyclase toxin of *Bordetella pertussis.* In: Wardlaw AC, Parton R (eds) *Pathogenesis and Immunity in Pertussis.* Chichester: John Wiley and Sons, pp. 193–209.

Hewlett EL, Halperin SA (1998) Serological correlates of immunity to *Bordetella pertussis. Vaccine* 16: 1899–1900.

Hill BC, Baker CN, Tenover FC (2000) A simplified method for testing *Bordetella pertussis* for resistance to erythromycin and other antimicrobial agents. *J. Clin. Microbiol.* 38: 1151–1155.

Hodder SL, Mortimer EA (1992) Epidemiology of pertussis and reactions to pertussis vaccine. *Epidemiol. Rev.* 14: 243–268.

Hoppe JE (1992) Comparison of erythromycin estolate and erythromycin ethylsuccinate for treatment of pertussis. The Erythromycin Study Group. *Pediatr. Infect. Dis. J.* 11: 189–193.

Hoppe JE (1996) Update on epidemiology, diagnosis, treatment of pertussis. *Eur. J. Clin. Microbiol. Infect. Dis.* 15: 189–193.

Hoppe JE (1998) State of art in antibacterial susceptibility of *Bordetella pertussis* and antibiotic treatment of pertussis. *Infection* 26: 242–246.

Hoppe JE (2000) Neonatal pertussis. *Pediatr. Infect. Dis. J.* 19: 244–247.

Horiguchi Y, Sugimoto N, Matsuda M (1994) *Bordetella bronchiseptica* dermonecrotizing toxin stimulates protein synthesis in an osteoblastic clone, MC3T3-E1 cells. *FEMS Microbiol. Lett.* 120: 1–2.

Horiguchi Y, Senda T, Sugimoto N, Katahira J, Matsuda M (1995) *Bordetella bronchiseptica* dermonecrotizing toxin stimulates assembly of actin stress fibers and focal adhesions by modifying the small GTP-binding protein rho. *J.Cell Sci.* 108: 3243–3251.

Horiguchi Y, Inoue N, Masuda M et al. (1997) *Bordetella bronchiseptica* dermonecrotizing toxin induces reorganization of actin stress fibers through deamidation of Gln-63 of the GTP-binding protein Rho. *Proc. Natl Acad. Sci. USA* 94: 11623–11626.

Hormozi K, Parton R, Coote J (1999) Adjuvant and protective properties of native and recombinant *Bordetella pertussis* adenylate cyclase toxin preparations in mice. *FEMS Immunol. Med. Microbiol.* 23: 273–282.

Hozbor D, Rodriguez ME, Yantorno O (1994) Use of cyclodextrin as an agent to induce excretion of *Bordetella pertussis* antigens. *FEMS Immunol. Med. Microbiol.* 9: 117–124.

Imaizumi A, Suzuki Y, Ono S et al. (1983) Heptakis(2,6-O-dimethyl)beta-cyclodextrin: a novel growth stimulant for *Bordetella pertussis* phase I. *J. Clin. Microbiol.* 17: 781–786.

Irons LI, Robinson A (1988) Pertussis toxin: production, purification, molecular structure, assay. In: Wardlaw AC, Parton R (eds) *Pathogenesis and Immunity in Pertussis.* Chichester: John Wiley and Sons, pp. 95–120.

Ishibashi Y, Claus S, Relman DA (1994) *Bordetella pertussis* filamentous hemagglutinin interacts with a leucocyte signal transduction complex and stimulates bacterial adherence to monocyte CR3 (CD11b/CD18). *J. Exp. Med.* 180: 1225–1233.

Jabbal-Gill I, Fisher AN, Rappuoli R, Davis SS, Illum L (1998) Stimulation of mucosal and systemic antibody responses against *Bordetella pertussis* filamentous haemagglutinin and recombinant pertussis toxin after nasal administration with chitosan in mice. *Vaccine* 16: 2039–2046.

Jacob-Dubuisson F, Buisine C, Mielcarek N, Clement E, Menozzi FD, Locht C (1996) Amino-terminal maturation of the *Bordetella pertussis* filamentous haemagglutinin. *Mol. Microbiol.* 19: 65–78.

Johansen FE, Pekna M, Norderhaug IN et al. (1999) Absence of epithelial immunoglobulin A transport, with increased mucosal leakiness, in polymeric immunoglobulin receptor/secretory component-deficient mice. *J. Exp. Med.* 190: 915–922.

Johnson R, Sneath PH (1973) Taxonomy of *Bordetella* and related organisms of the Families Achromobacteraceae, Brucellaceae, Neisseraceae. *Int. J. Syst. Bacteriol.* 23: 381–404.

Jones DH, McBride BW, Jeffery H, O'Hagan DT, Robinson A, Farrar GH (1995) Protection of mice from *Bordetella pertussis* respiratory infection using microencapsulated pertussis fimbriae. *Vaccine* 13: 675–681.

Jones DH, McBride BW, Thornton C, O'Hagan DT, Robinson A, Farrar GH (1996) Orally administered microencapsulated *Bordetella pertussis* fimbriae protect mice from *B. pertussis* respiratory infection. *Infect. Immun.* 64: 489–494.

Kashimoto T, Katahira J, Cornejo WR et al. (1999) Identification of functional domains of Bordetella dermonecrotizing toxin. *Infect. Immun.* 67: 3727–3732.

Kaslow HR, Burns DL (1992) Pertussis toxin and target eukaryotic cells: binding, entry, activation. *FASEB J.* 6: 2684–2690.

Kattar MM, Chavez JF, Limaye AP et al. (2000) Application of 16S rRNA gene sequencing to identify *Bordetella hinzii* as the causative agent of fatal septicemia. *J. Clin. Microbiol.* 38: 789–794.

Kerr JR, Matthews RC (2000) *Bordetella pertussis* infection: pathogenesis, diagnosis, management, the role of protective immunity [Review]. *Eur. J. Clin. Microbiol. Infect. Dis.* 19: 77–88.

Kerr JR, Rigg GP, Matthews RC, Burnie JP (1999) The Bpel locus encodes type III secretion machinery in *Bordetella pertussis*. *Microb. Pathog.* 27: 349–367.

Kersters K, Hinz KH, Hertle A et al. (1984) *Bordetella avium* sp. nov, isolated from the respiratory tracts of turkeys and other birds. *Int. J. Syst. Bacteriol.* 34: 56–70.

Khelef N, Sakamoto H, Guiso N (1992) Both adenylate cyclase and hemolytic activities are required by *Bordetella pertussis* to initiate infection. *Microb. Pathog.* 12: 227–235.

Khelef N, Danve B, Quentin-Millet MJ, Guiso N (1993a) *Bordetella pertussis* and *Bordetella parapertussis*: two immunologically distinct species. *Infect. Immun.* 61: 486–490.

Khelef N, Zychlinsky A, Guiso N (1993b) *Bordetella pertussis* induces apoptosis in macrophages: role of adenylate cyclase-hemolysin. *Infect. Immun.* 61: 4064–4071.

Khelef N, Bachelet CM, Vargaftig BB, Guiso N (1994) Characterization of murine lung inflammation after infection with parental *Bordetella pertussis* and mutants deficient in adhesins or toxins. *Infect. Immun.* 62: 2893–2900.

Kimura M, Kuno-Sakai H (1990) Developments in pertussis immunisation in Japan. *Lancet* 336: 30–32.

Kimura A, Mountzouros KT, Relman DA, Falkow S, Cowell JL (1990) *Bordetella pertussis* filamentous hemagglutinin: evaluation as a protective antigen and colonization factor in a mouse respiratory infection model. *Infect. Immun.* 58: 7–16.

Kinnear SM, Boucher PE, Stibitz S, Carbonetti NH (1999) Analysis of BvgA activation of the pertactin gene promoter in *Bordetella pertussis*. *J. Bacteriol.* 181: 5234–5241.

Kloos WE, Mohapatra N, Dobrogosz WJ, Ezzell JW, Manclark CR (1981) Deoxyribonucleotide sequence relationships among bordetella species. *Int. J. Syst. Bacteriol.* 31: 173–176.

Knapp S, Mekalanos JJ (1988) Two trans-acting regulatory genes (vir and mod) control antigenic modulation in *Bordetella pertussis*. *J. Bacteriol.* 170: 5059–5066.

Krueger KM, Barbieri JT (1995) The family of bacterial ADP-ribosylating exotoxins. *Clin. Microbiol. Rev.* 8: 34–47.

Lacerda HM, Pullinger GD, Lax AJ, Rozengurt E (1997) Cytotoxic necrotizing factor 1 from *Escherichia coli* and dermonecrotic toxin from *Bordetella bronchiseptica* induce p21(rho)-dependent tyrosine phosphorylation of focal adhesion kinase and paxillin in Swiss 3T3 cells. *J. Biol. Chem.* 272: 9587–9596.

Lacey BW (1960) Antigenic modulation in *Bordetella pertussis*. *J. Hyg. Camb.* 58: 57–93.

Ladant D, Ullmann A (1999) *Bordetella pertussis* adenylate cyclase: a toxin with multiple talents. *Trends Microbiol.* 7: 172–176.

Lambert-Buisine C, Willery E, Locht C, Jacob-Dubuisson F (1998) N-terminal characterization of the *Bordetella pertussis* filamentous haemagglutinin. *Mol. Microbiol.* 28: 1283–1293.

Lapin JH (1943) *Whooping Cough*. Springfield, Illinois: Charles Thomas, p. 238.

Lawson GM (1940) Modified technique for staining capsules of *Haemophilus pertussis*. *J. Lab. Clin. Med.* 25: 435–438.

Le Blay K, Caroff M, Richards JC, Perry MB, Chaby R (1994) Specific and cross-reacting monoclonal antibodies to *Bordetella parapertussis* and *Bordetella bronchiseptica* lipopolysaccharides. *Microbiology* 140: 2459–2465.

Le Blay K, Gueirard P, Guiso N, Chaby R (1997) Antigenic polymorphism of the lipopolysaccharides from human and animal isolates of *Bordetella bronchiseptica*. *Microbiology* 143: 1433–1441.

Lee CK, Roberts AL, Finn TM, Knapp S, Mekalanos JJ (1990) A new assay for invasion of HeLa 229 cells by *Bordetella pertussis*: effects of inhibitors, phenotypic modulation, genetic alterations. *Infect. Immun.* 58: 2516–2522.

Leef M, Elkins KL, Barbic J, Shahin RD (2000) Protective immunity to *Bordetella pertussis* requires both B cells and CD4(+) T cells for key functions other than specific antibody production. *J. Exp. Med.* 191: 1841–1852.

Leininger E, Roberts M, Kenimer JG *et al.* (1991) Pertactin, an Arg-Gly-Asp-containing *Bordetella pertussis* surface protein that promotes adherence of mammalian cells. *Proc. Natl Acad. Sci. USA* 88: 345–349.

Leininger E, Ewanowich CA, Bhargava A, Peppler MS, Kenimer JG, Brennan MJ (1992) Comparative roles of the Arg-Gly-Asp sequence present in the *Bordetella pertussis* adhesins pertactin and filamentous hemagglutinin. *Infect. Immun.* 60: 2380–2385.

Lemichez E, Flatau G, Bruzzone M, Boquet P, Gauthier M (1997) Molecular localization of the *Escherichia coli* cytotoxic necrotizing factor CNF1 cell-binding and catalytic domains. *Mol. Microbiol.* 24: 1061–1070.

Lenz DH, Weingart CL, Weiss AA (2000) Phagocytosed *Bordetella pertussis* fails to survive in human neutrophils. *Infect. Immun.* 68: 956–959.

Leslie PH, Gardner AD (1931) The phases of *Haemophilus pertussis*. *J. Hyg.* 31: 423–434.

Liese JG, Meschievitz CK, Harzer E *et al.* (1997) Efficacy of a two-component acellular pertussis vaccine in infants. *Pediatr. Infect. Dis. J.* 16: 1038–1044.

Locht C, Keith JM (1986) Pertussis toxin gene: nucleotide sequence and genetic organization. *Science* 232: 1258–1264.

Locht C, Capiau C, Feron C (1989) Identification of amino acid residues essential for the enzymatic activities of pertussis toxin. *Proc. Natl Acad. Sci. USA* 86: 3075–3079.

Locht C, Geoffroy MC, Renauld G (1992) Common accessory genes for the *Bordetella pertussis* filamentous hemagglutinin and fimbriae share sequence similarities with the *papC* and *papD* gene families. *EMBO J.* 11: 3175–3183.

Locht C, Bertin P, Menozzi FD, Renauld G (1993) The filamentous haemagglutinin, a multifaceted adhesion produced by virulent *Bordetella* spp. *Mol. Microbiol.* 9: 653–660.

Luker KE, Collier JL, Kolodziej EW, Marshall GR, Goldman WE (1993) *Bordetella pertussis* tracheal cytotoxin and other muramyl peptides: distinct structure–activity relationships for respiratory epithelial cytopathology. *Proc. Natl Acad. Sci. USA* 90: 2365–2369.

MacDonald H, MacDonald EJ (1933) Experimental pertussis. *J. Infect. Dis.* 53: 328–330.

Mahon BP, Ryan MS, Griffin F, Mills KH (1996) Interleukin-12 is produced by macrophages in response to live or killed *Bordetella pertussis* and enhances the efficacy of an acellular pertussis vaccine by promoting induction of Th1 cells. *Infect. Immun.* 64: 5295–5301.

Mahon BP, Sheahan BJ, Griffin F, Murphy G, Mills KH (1997) Atypical disease after *Bordetella pertussis* respiratory infection of mice with targeted disruptions of interferon-gamma receptor or immunoglobulin mu chain genes. *J. Exp. Med.* 186: 1843–1851.

Mahon BP, Brady MT, Mills KH (2000) Protection against *Bordetella pertussis* in mice in the absence of detectable circulating antibody: implications for long-term immunity in children. *J. Infect. Dis.* 181: 2087–2091.

Makhov AM, Hannah JH, Brennan MJ *et al.* (1994) Filamentous hemagglutinin of *Bordetella pertussis*. A bacterial adhesin formed as a 50-nm monomeric rigid rod based on a 19-residue repeat motif rich in beta strands and turns. *J. Mol. Biol.* 241: 110–124.

Makoff AJ, Oxer MD, Ballantine SP, Fairweather NF, Charles IG (1990) Protective surface antigen P69 of *Bordetella pertussis*: its characterization and very high level expression in *Escherichia coli. Biotechnology* 8: 1030–1033.

Manclark CR, Cowell JL (1984) Pertussis. In: Germanier R (ed.) *Bacterial Vaccines.* New York: Academic Press, pp. 69–106.

Marcuse EK, Wentz KR (1990) The NCES reconsidered: summary of a 1989 workshop. National Childhood Encephalopathy Study. *Vaccine* 8: 531–535.

Martin D, Peppler MS, Brodeur BR (1992) Immunological characterization of the lipooligosaccharide B band of *Bordetella pertussis. Infect. Immun.* 60: 2718–2725.

Martinez de Tejada G, Cotter PA, Heininger U *et al.* (1998) Neither the Bvg- phase nor the vrg6 locus of *Bordetella pertussis* is required for respiratory infection in mice. *Infect. Immun.* 66: 2762–2768.

Mastrantonio P, Stefanelli P, Giuliano M *et al.* (1998) *Bordetella parapertussis* infection in children: epidemiology, clinical symptoms, molecular characteristics of isolates. *J. Clin. Microbiol.* 36: 999–1002.

Mastrantonio P, Spigaglia P, van Oirschot H *et al.* (1999) Antigenic variants in *Bordetella pertussis* strains isolated from vaccinated and unvaccinated children. *Microbiology* 145: 2069–2075.

Masuda M, Betancourt L, Matsuzawa T *et al.* (2000) Activation of rho through a cross-link with polyamines catalyzed by Bordetella dermonecrotizing toxin. *EMBO J.* 19: 521–530.

Matsuyama T, Takino T (1980) Scanning electronmicroscopic studies of *Bordetella bronchiseptica* on the rabbit tracheal mucosa. *J. Med. Microbiol.* 13: 159–161.

Mattoo S, Miller JF, Cotter PA (2000) Role of *Bordetella bronchiseptica* fimbriae in tracheal colonization and development of a humoral immune response. *Infect. Immun.* 68: 2024–2033.

Mazengia E, Silva EA, Peppe JA, Timperi R, George H (2000) Recovery of *Bordetella holmesii* from patients with pertussis-like symptoms: use of pulsed-field gel electrophoresis to characterize circulating strains. *J. Clin. Microbiol.* 38: 2330–2333.

Mbawuike IN, Pacheco S, Acuna CL, Switzer KC, Zhang Y, Harriman GR (1999) Mucosal immunity to influenza without IgA: an IgA knockout mouse model. *J. Immunol.* 162: 2530–2537.

McGuirk P, Mills KH (2000a) A regulatory role for interleukin 4 in differential inflammatory responses in the lung following infection of mice primed with Th1- or Th2-inducing pertussis vaccines. *Infect. Immun.* 68: 1383–1390.

McGuirk P, Mills KH (2000b) Direct anti-inflammatory effect of a bacterial virulence factor: IL-10-dependent suppression of IL-12 production by filamentous hemagglutinin from *Bordetella pertussis. Eur. J. Immunol.* 30: 415–422.

McGuirk P, Mahon BP, Griffin F, Mills KH (1998) Compartmentalization of T cell responses following respiratory infection with *Bordetella pertussis*: hyporesponsiveness of lung T cells is associated with modulated expression of the co-stimulatory molecule CD28. *Eur. J. Immunol.* 28: 153–163.

McMillan DJ, Shojaei M, Chhatwal GS, Guzman CA, Walker MJ (1996) Molecular analysis of the bvg-repressed urease of *Bordetella bronchiseptica. Microb. Pathog.* 21: 379–394.

Meade BD, Bollen A (1994) Recommendations for use of the polymerase chain reaction in the diagnosis of *Bordetella pertussis* infections. *J. Med. Microbiol.* 41: 51–55.

Menozzi FD, Gantiez C, Locht C (1991a) Identification and purification of transferrin- and lactoferrin-binding proteins of *Bordetella pertussis* and *Bordetella bronchiseptica. Infect. Immun.* 59: 3982–3988.

Menozzi FD, Gantiez C, Locht C (1991b) Interaction of the *Bordetella pertussis* filamentous hemagglutinin with heparin. *FEMS Microbiol. Lett.* 62: 59–64.

Menozzi FD, Boucher PE, Riveau G, Gantiez C, Locht C (1994a) Surface-associated filamentous hemagglutinin induces autoagglutination of *Bordetella pertussis. Infect. Immun.* 62: 4261–4269.

Menozzi FD, Mutombo R, Renauld G *et al.* (1994b) Heparin-inhibitable lectin activity of the filamentous hemagglutinin adhesin of *Bordetella pertussis. Infect. Immun.* 62: 769–778.

Merkel TJ, Barros C, Stibitz S (1998a) Characterization of the bvgR locus of *Bordetella pertussis. J. Bacteriol.* 180: 1682–1690.

Merkel TJ, Stibitz S, Keith JM, Leef M, Shahin R (1998b) Contribution of regulation by the bvg locus to respiratory infection of mice by *Bordetella pertussis. Infect. Immun.* 66: 4367–4373.

Merritt EA, Hol WG (1995) AB5 toxins. *Curr. Opin. Struct. Biol.* 5: 165–171.

Mesnard R, Guiso N, Michelet C *et al.* (1993) Isolation of *Bordetella bronchiseptica* from a patient with AIDS. *Eur. J. Clin. Microbiol. Infect. Dis.* 12: 304–306.

Middendorf B, Gross R (1999) Representational difference analysis identifies a strain-specific LPS biosynthesis locus in *Bordetella* spp. *Mol. Gen. Genet.* 262: 189–198.

Mielcarek N, Cornette J, Schacht AM *et al.* (1997) Intranasal priming with recombinant *Bordetella pertussis* for the induction of a systemic immune response against a heterologous antigen. *Infect. Immun.* 65: 544–550.

Mielcarek N, Riveau G, Remoue F, Antoine R, Capron A, Locht C (1998) Homologous and heterologous protection after single intranasal administration of live attenuated recombinant *Bordetella pertussis. Nature Biotechnol.* 16: 454–457.

Miller E, Ashworth LA, Redhead K, Thornton C, Waight PA, Coleman T (1997) Effect of schedule on reactogenicity and antibody persistence of acellular and whole-cell pertussis vaccines: value of laboratory tests as predictors of clinical performance. *Vaccine* 15: 51–60.

Mills KH, Barnard A, Watkins J, Redhead K (1993) Cell-mediated immunity to *Bordetella pertussis*: role of Th1 cells in bacterial clearance in a murine respiratory infection model. *Infect. Immun.* 61: 399–410.

Mills KH, Ryan M, Ryan E, Mahon BP (1998) A murine model in which protection correlates with pertussis vaccine efficacy in children reveals complementary roles for humoral and cell-mediated immunity in protection against *Bordetella pertussis*. *Infect. Immun.* 66: 594–602.

Moeck GS, Coulton JW (1998) TonB-dependent iron acquisition: mechanisms of siderophore-mediated active transport. *Mol. Microbiol.* 28: 675–681.

Molina NC, Parker CD (1990) Murine antibody response to oral infection with live aroA recombinant Salmonella dublin vaccine strains expressing filamentous hemagglutinin antigen from *Bordetella pertussis*. *Infect. Immun.* 58: 2523–2528.

Monack DM, Arico B, Rappuoli R, Falkow S (1989) Phase variants of *Bordetella bronchiseptica* arise by spontaneous deletions in the *vir* locus. *Mol. Microbiol.* 3: 1719–1728.

Montaraz JA, Novotny P, Ivanyi J (1985) Identification of a 68-kilodalton protective protein antigen from *Bordetella bronchiseptica*. *Infect. Immun.* 47: 744–751.

Mooi FR (1994) Genes for the filamentous haemagglutinin and fimbriae of *Bordetella pertussis*: colocation, coregulation and cooperation? In: Miller VL, Kaper JB, Portnoy DA, Isberg RR (eds) *Molecular Genetics of Bacterial Pathogenesis*. Washington DC: American Society for Microbiology, pp. 145–155.

Mooi FR, van Oirschot H, Heuvelman K, van der Heide HG, Gaastra W, Willems RJ (1998) Polymorphism in the *Bordetella pertussis* virulence factors P. 69/pertactin and pertussis toxin in The Netherlands: temporal trends and evidence for vaccine-driven evolution. *Infect. Immun.* 66: 670–675.

Mooi FR, He Q, van Oirschot H, Mertsola J (1999) Variation in the *Bordetella pertussis* virulence factors pertussis toxin and pertactin in vaccine strains and clinical isolates in Finland. *Infect. Immun.* 67: 3133–3134.

Mooi FR, Hallander H, Wirsing vKC, Hoet B, Guiso N (2000) Epidemiological typing of *Bordetella pertussis* isolates: recommendations for a standard methodology. *Eur. J. Clin. Microbiol. Infect. Dis.* 19: 174–181.

Moore CH, Foster LA, Gerbig DGJ, Dyer DW, Gibson BW (1995) Identification of alcaligin as the siderophore produced by *Bordetella pertussis* and *B. bronchiseptica*. *J. Bacteriol.* 177: 1116–1118.

Mortimer E Jr (1990) Pertussis and its prevention: a family affair. *J. Infect. Dis.* 161: 473–479.

Mouallem M, Farfel Z, Hanski E (1990) *Bordetella pertussis* adenylate cyclase toxin: intoxication of host cells by bacterial invasion. *Infect. Immun.* 58: 3759–3764.

Mountzouros KT, Kimura A, Cowell JL (1992) A bactericidal monoclonal antibody specific for the lipooligosaccharide of *Bordetella pertussis* reduces colonization of the respiratory tract of mice after aerosol infection with *B. pertussis*. *Infect. Immun.* 60: 5316–5318.

Muller FM, Hoppe JE, Wirsing vKC (1997) Laboratory diagnosis of pertussis: state of the art in 1997. *J. Clin. Microbiol.* 35: 2435–2443.

Muller M, Hildebrandt A (1993) Nucleotide sequences of the 23S rRNA genes from *Bordetella pertussis*, *B. parapertussis*, *B. bronchiseptica* and *B. avium*, their implications for phylogenetic analysis. *Nucleic Acids Res.* 21: 3320.

Munoz JJ (1988) Action of pertussigen (pertussis toxin) on the host immune system. In: Wardlaw AC, Parton R (eds) *Pathogenesis and Immunity in Pertussis*. Chichester: Wiley, pp. 173–192.

Musser JM, Hewlett EL, Peppler MS, Selander RK (1986) Genetic diversity and relationships in populations of *Bordetella* spp. *J. Bacteriol.* 166: 230–237.

Nakai T, Sawata A, Kume K (1985) Intracellular locations of dermonecrotic toxins in *Pasteurella multocida* and in *Bordetella bronchiseptica*. *Am. J. Vet. Res.* 46: 870–874.

Nakase Y, Endoh M (1988) Heat-labile toxin of *Bordetella pertussis*. In: Wardlaw AC, Parton R (eds) *Pathogenesis and Immunity in Pertussis*. Chichester: John Wiley and Sons, pp. 211–229.

Nencioni L, Pizza MG, Volpini G, De Magistris MT, Giovannoni F, Rappuoli R (1991) Properties of the B oligomer of pertussis toxin. *Infect. Immun.* 59: 4732–4734.

Nicosia A, Perugini M, Franzini C et al. (1986) Cloning and sequencing of the pertussis toxin genes: operon structure and gene duplication. *Proc. Natl Acad. Sci. USA* 83: 4631–4635.

Nielsen JP, Foged NT, Sorensen V, Barfod K, Bording A, Petersen SK (1991) Vaccination against progressive atrophic rhinitis with a recombinant *Pasteurella multocida* toxin derivative. *Can. J. Vet. Res.* 55: 128–138.

Noble GR, Bernier RH, Esber EC et al. (1987) Acellular and whole-cell pertussis vaccines in Japan. Report of a visit by US scientists. *JAMA* 257: 1351–1356.

Novotny P, Chubb AP, Cownley K, Montaraz JA (1985a) Adenylate cyclase activity of a 68,000-molecular-weight protein isolated from the outer membrane of *Bordetella bronchiseptica*. *Infect. Immun.* 50: 199–206.

Novotny P, Chubb AP, Cownley K, Montaraz JA, Beesley JE (1985b) Bordetella adenylate cyclase: a genus specific protective antigen and virulence factor. *Dev. Biol. Standard.* 61: 27–41.

Novotny P, Kobisch M, Cownley K, Chubb AP, Montaraz JA (1985c) Evaluation of *Bordetella bronchiseptica* vaccines in specific-pathogen-free piglets with bacterial cell surface antigens in enzyme-linked immunosorbent assay. *Infect. Immun.* 50: 190–198.

Novotny P, Chubb AP, Cownley K, Charles IG (1991) Biologic and protective properties of the 69-kDa outer membrane protein of *Bordetella pertussis*: a novel formulation for an acellular pertussis vaccine. *J. Infect. Dis.* 164: 114–122.

Olin P (1997) The best acellular vaccines are multicomponent. *Pediatr. Infect. Dis. J.* 16: 517–519.

Olin P, Rasmussen F, Gustafsson L, Hallander HO, Heijbel H (1997) Randomised controlled trial of two-component, three-component, five-component acellular pertussis vaccines compared with whole-cell pertussis vaccine. Ad Hoc Group for the Study of Pertussis Vaccines. *Lancet* 350: 1569–1577.

Olson LC (1975) Pertussis. *Medicine* 54: 427–469.

Onorato IM, Wassilak SG (1987) Laboratory diagnosis of pertussis: the state of the art. *Pediatr. Infect. Dis. J.* 6: 145–151.

Parton R (1991) Changing perspectives on pertussis and pertussis vaccination. *Rev. Med. Microbiol.* 2: 121–128.

Parton R (1999) Review of the biology of *Bordetella pertussis. Biologicals* 27: 71–76.

Parton R, Hall E, Wardlaw AC (1994) Responses to *Bordetella pertussis* mutant strains and to vaccination in the coughing rat model of pertussis. *J. Med. Microbiol.* 40: 307–312.

Peppler MS (1982) Isolation and characterization of isogenic pairs of domed hemolytic and flat nonhemolytic colony types of *Bordetella pertussis. Infect. Immun.* 35: 840–851.

Peppler MS (1984) Two physically and serologically distinct lipopolysaccharide profiles in strains of *Bordetella pertussis* and their phenotype variants. *Infect. Immun.* 43: 224–232.

Peppler MS, Schrumpf ME (1984) Isolation and characterization of *Bordetella pertussis* phenotype variants capable of growing on nutrient agar: Comparison with phases III and IV. *Infect. Immun.* 43: 217–223.

Peppoloni S, Nencioni L, Di Tommaso A *et al.* (1991) Lymphokine secretion and cytotoxic activity of human CD4+ T-cell clones against *Bordetella pertussis. Infect. Immun.* 59: 3768–3773.

Pittman M (1952) Influence of preservatives, of heat, of irradiation on mouse protective activity and detoxification of pertussis vaccine. *J. Immunol.* 69: 201–216.

Pittman M (1984a) Genus *Bordetella.* In: Krieg NR, Holt JG (eds) *Bergey's Manual of Systematic Bacteriology.* Baltimore: Williams & Wilkins, pp. 388–393.

Pittman M (1984b) The concept of pertussis as a toxin-mediated disease. *Pediatr. Infect. Dis.* 3: 467–486.

Pittman M, Cox CB (1965) Pertussis vaccine testing for freedom from toxicity. *J. Appl. Microbiol.* 13: 447–456.

Pizza M, Bartoloni A, Prugnola A, Silvestri S, Rappuoli R (1988) Subunit S1 of pertussis toxin: mapping of the regions essential for ADP-ribosyltransferase activity. *Proc. Natl Acad. Sci. USA* 85: 7521–7525.

Plotkin SA, Cadoz M (1997) The acellular pertussis vaccine trials: an interpretation. *Pediatr. Infect. Dis. J.* 16: 508–517.

Podda A, De Luca EC, Titone L *et al.* (1992) Acellular pertussis vaccine composed of genetically inactivated pertussis toxin: safety and immunogenicity in 12- to 24- and 2- to 4-month-old children. *J. Pediatr.* 120: 680–685.

Porter JF, Wardlaw AC (1993) Long-term survival of *Bordetella bronchiseptica* in lakewater and in buffered saline without added nutrients. *FEMS Microbiol. Lett.* 33–36.

Porter JF, Parton R, Wardlaw AC (1991) Growth and survival of *Bordetella bronchiseptica* in natural-waters and in buffered saline without added nutrients. *Appl. Environ. Microbiol.* 1202–1206.

Pradel E, Guiso N, Locht C (1998) Identification of AlcR, an AraC-type regulator of alcaligin siderophore synthesis in *Bordetella bronchiseptica* and *Bordetella pertussis. J. Bacteriol.* 180: 871–880.

Pradel E, Guiso N, Menozzi FD, Locht C (2000) *Bordetella pertussis* TonB, a Bvg-independent virulence determinant. *Infect. Immun.* 68: 1919–1927.

Preston A, Mandrell RE, Gibson BW, Apicella MA (1996) The lipooligosaccharides of pathogenic gram-negative bacteria. *C. R. Microbiol.* 22: 139–180.

Preston NW (1988) Pertussis today. In: Wardlaw AC, Parton R (eds) *Pathogenesis and Immunity in Pertussis.* Chichester: Wiley, pp. 1–18.

Pullinger GD, Adams TE, Mullan PB, Garrod TI, Lax AJ (1996) Cloning, expression, molecular characterization of the dermonecrotic toxin gene of *Bordetella* spp. *Infect. Immun.* 64: 4163–4171.

Rambow AA, Fernandez RC, Weiss AA (1998) Characterization of BrkA expression in *Bordetella bronchiseptica. Infect. Immun.* 66: 3978–3980.

Rappuoli R (1994) Pathogenicity mechanisms of *Bordetella. Curr. Top. Microbiol. Immunol.* 192: 319–336.

Rappuoli R, Pizza M, Podda A, De Magistris MT, Nencioni L (1991) Towards third-generation whooping cough vaccines. *Trends Biotechnol.* 9: 232–238.

Rappuoli R, Pizza M, Covacci A *et al.* (1992a) Recombinant acellular pertussis vaccine – from the laboratory to the clinic: improving the quality of the immune response. *FEMS Microbiol. Immunol.* 5: 161–170.

Rappuoli R, Pizza M, De Magistris MT *et al.* (1992b) Development and clinical testing of an acellular pertussis vaccine containing genetically detoxified pertussis toxin. *Immunobiology* 184: 230–239.

Rappuoli R, Podda A, Pizza M *et al.* (1992c) Progress towards the development of new vaccines against whooping cough. *Vaccine* 10: 1027–1032.

Redhead K, Hill T (1991) Acquisition of iron from transferrin by *Bordetella pertussis. FEMS Microbiol. Lett.* 61: 303–307.

Redhead K, Watkins J, Barnard A, Mills KH (1993) Effective immunization against *Bordetella pertussis* respiratory infection in mice is dependent on induction of cell-mediated immunity. *Infect. Immun.* 61: 3190–3198.

Regan J, Lowe F (1977) Enrichment medium for the isolation of *Bordetella. J. Clin. Microbiol.* 6: 303–309.

Register KB, Ackermann MR (1997) A highly adherent phenotype associated with virulent Bvg+-phase swine isolates of *Bordetella bronchiseptica* grown under modulating conditions. *Infect. Immun.* 65: 5295–5300.

Reizenstein E, Johansson B, Mardin L, Abens J, Mollby R, Hallander HO (1993) Diagnostic evaluation of polymerase chain reaction discriminative for *Bordetella pertussis, B. parapertussis, B. bronchiseptica. Diagn. Microbiol. Infect. Dis.* 17: 185–191.

Relman DA, Domenighini M, Tuomanen E, Rappuoli R, Falkow S (1989) Filamentous hemagglutinin of

Bordetella pertussis: nucleotide sequence and crucial role in adherence. *Proc. Natl Acad. Sci. USA* 86: 2637–2641.

Relman D, Tuomanen E, Falkow S, Golenbock DT, Saukkonen K, Wright SD (1990) Recognition of a bacterial adhesion by an integrin: macrophage CR3 (alpha M beta 2, CD11b/CD18) binds filamentous hemagglutinin of *Bordetella pertussis*. *Cell* 61: 1375–1382.

Renauld-Mongenie G, Cornette J, Mielcarek N, Menozzi FD, Locht C (1996a) Distinct roles of the N-terminal and C-terminal precursor domains in the biogenesis of the *Bordetella pertussis* filamentous hemagglutinin. *J. Bacteriol.* 178: 1053–1060.

Renauld-Mongenie G, Mielcarek N, Cornette J *et al.* (1996b) Induction of mucosal immune responses against a heterologous antigen fused to filamentous hemagglutinin after intranasal immunization with recombinant *Bordetella pertussis*. *Proc. Natl Acad. Sci. USA* 93: 7944–7949.

Rhodes MB, New CWJ, Baker PK, Hogg A, Underdahl NR (1987) *Bordetella bronchiseptica* and toxigenic type D *Pasteurella multocida* as agents of severe atrophic rhinitis of swine. *Vet. Microbiol.* 13: 179–187.

Ricci S, Rappuoli R, Scarlato V (1996) The pertussis toxin liberation genes of *Bordetella pertussis* are transcriptionally linked to the pertussis toxin operon. *Infect. Immun.* 64: 1458–1460.

Roberts M, Maskell D, Novotny P, Dougan G (1990) Construction and characterization *in vivo* of *Bordetella pertussis aroA* mutants. *Infect. Immun.* 58: 732–739.

Roberts M, Fairweather NF, Leininger E *et al.* (1991) Construction and characterization of *Bordetella pertussis* mutants lacking the vir-regulated P.69 outer membrane protein. *Mol. Microbiol.* 5: 1393–1404.

Roberts M, Tite JP, Fairweather NF, Dougan G, Charles IG (1992) Recombinant P.69/pertactin: immunogenicity and protection of mice against *Bordetella pertussis* infection. *Vaccine* 10: 43–48.

Roberts M, Cropley I, Chatfield S, Dougan G (1993) Protection of mice against respiratory *Bordetella pertussis* infection by intranasal immunization with P.69 and FHA. *Vaccine* 11: 866–872.

Roberts M, Bacon A, Rappuoli R *et al.* (1995) A mutant pertussis toxin molecule that lacks ADP-ribosyltransferase activity, PT-9K/129G, is an effective mucosal adjuvant for intranasally delivered proteins. *Infect. Immun.* 63: 2100–2108.

Robinson A, Ashworth LAE (1988) Acellular and defined-component pertussis vaccines against pertussis. In: Wardlaw AC, Parton R (eds) *Pathogenesis and Immunity in Pertussis*. Chichester: John Wiley and Sons, pp. 399–417.

Robinson A, Funnell SG (1992) Potency testing of acellular pertussis vaccines. *Vaccine* 10: 139–141.

Robinson A, Irons LI, Ashworth LA (1985) Pertussis vaccine: present status and future prospects. *Vaccine* 3: 11–22.

Robinson A, Duggleby CJ, Gorringe AR, Livey I (1986) Antigenic variation in *Bordetella pertussis*. In: Birkbeck TH, Penn CW (eds). *Antigenic Variation in Infectious Diseases*. Oxford: IRL Press, pp. 147–161.

Robinson A, Ashworth LA, Irons LI (1989a) Serotyping *Bordetella pertussis* strains. *Vaccine* 7: 491–494.

Robinson A, Gorringe AR, Funnell SG, Fernandez M (1989b) Serospecific protection of mice against intranasal infection with *Bordetella pertussis*. *Vaccine* 7: 321–324.

Romanos MA, Clare JJ, Beesley KM *et al.* (1991) Recombinant *Bordetella pertussis* pertactin (P69) from the yeast *Pichia pastoris*: high-level production and immunological properties. *Vaccine* 9: 901–906.

Ross EM (1988) Reactions to whole-cell pertussis vaccines. In: Wardlaw AC, Parton R (eds) *Pathogenesis and Immunity in Pertussis*. Chichester: John Wiley and Sons, pp. 375–398.

Rostand KS, Esko JD (1997) Microbial adherence to and invasion through proteoglycans. *Infect. Immun.* 65: 1–8.

Rowatt E (1957) The growth of *Bordetella pertussis*: a review. *J. Gen. Microbiol.* 17: 297–326.

Roy CR, Miller JF, Falkow S (1990) Autogenous regulation of the *Bordetella pertussis bvgABC* operon. *Proc. Natl Acad. Sci. USA* 87: 3763–3767.

Rozdzinski E, Spellerberg B, van der Flier M *et al.* (1995) Peptide from a prokaryotic adhesin blocks leucocyte migration *in vitro* and *in vivo*. *J. Infect. Dis.* 172: 785–793.

Ruoslahti E, Pierschbacher MD (1987) New perspectives in cell adhesion: RGD and integrins. *Science* 238: 491–497.

Rutter JM (1983) Virulence of *Pasteurella multocida* in atrophic rhinitis of gnotobiotic pigs infected with *Bordetella bronchiseptica*. *Res. Vet. Sci.* 34: 287–295.

Ryan EJ, McNeela E, Murphy GA *et al.* (1999) Mutants of *Escherichia coli* heat-labile toxin act as effective mucosal adjuvants for nasal delivery of an acellular pertussis vaccine: differential effects of the nontoxic AB complex and enzyme activity on Th1 and Th2 cells. *Infect. Immun.* 67: 6270–6280.

Ryan MS, Griffin F, Mahon B, Mills KH (1997) The role of the S-1 and B-oligomer components of pertussis toxin in its adjuvant properties for Th1 and Th2 cells. *Biochem. Soc. Trans.* 25: 126S.

Salmaso S, Mastrantonio P, Wassilak SG *et al.* (1998) Persistence of protection through 33 months of age provided by immunization in infancy with two three-component acellular pertussis vaccines. Stage II Working Group. *Vaccine* 16: 1270–1275.

Sandros J, Tuomanen E (1993) Attachment factors of *Bordetella pertussis*: mimicry of eukaryotic cell recognition molecules. *Trends Microbiol.* 1: 192–196.

Saron MF, Fayolle C, Sebo P, Ladant D, Ullmann A, Leclerc C (1997) Anti-viral protection conferred by recombinant adenylate cyclase toxins from *Bordetella pertussis* carrying a CD8+ T cell epitope from lymphocytic choriomeningitis virus. *Proc. Natl Acad. Sci. USA* 94: 3314–3319.

Sato H, Sato Y (1984) *Bordetella pertussis* infection in mice: correlation of specific antibodies against two antigens, pertussis toxin, filamentous hemagglutinin with mouse protectivity in an intracerebral or aerosol challenge system. *Infect. Immun.* 46: 415–421.

Sato Y, Sato H (1988) Animal models of pertussis. In: Wardlaw AC, Parton R (eds) *Pathogenesis and Immunity in Pertussis.* Chichester: John Wiley and Sons, pp. 309–325.

Sato H, Sato Y (1990) Protective activities in mice of monoclonal antibodies against pertussis toxin. *Infect. Immun.* 58: 3369–3374.

Sato Y, Kimura M, Fukumi H (1984) Development of a pertussis component vaccine in Japan. *Lancet* 1: 122–126.

Saukkonen K, Cabellos C, Burroughs M, Prasad S, Tuomanen E (1991) Integrin-mediated localization of *Bordetella pertussis* within macrophages: role in pulmonary colonization. *J. Exp. Med.* 173: 1143–1149.

Saukkonen K, Burnette WN, Mar VL, Masure HR, Tuomanen EI (1992) Pertussis toxin has eukaryotic-like carbohydrate recognition domains. *Proc. Natl Acad. Sci. USA* 89: 118–122.

Scarlato V, Prugnola A, Arico B, Rappuoli R (1990) Positive transcriptional feedback at the bvg locus controls expression of virulence factors in *Bordetella pertussis. Proc. Natl Acad. Sci. USA* 87: 6753–6757.

Scarlato V, Arico B, Prugnola A, Rappuoli R (1991) Sequential activation and environmental regulation of virulence genes in *Bordetella pertussis. EMBO J.* 10: 3971–3975.

Scarlato V, Arico B, Rappuoli R (1993) DNA topology affects transcriptional regulation of the pertussis toxin gene of *Bordetella pertussis* in *Escherichia coli* and *in vitro. J. Bacteriol.* 175: 4764–4771.

Scarlato V, Beier D, Rappuoli R (1998) Molecular genetics of *Bordetella pertussis* virulence. In: Williams PH, Ketley J, Salmond G (eds) *Bacterial Pathogenesis.* London: Academic Press, pp. 395–406.

Schmidt-Schlapfer G, Liese JG, Porter F, Stojanov S, Just M, Belohradsky BH (1997) Polymerase chain reaction (PCR) compared with conventional identification in culture for detection of *Bordetella pertussis* in 7153 children. *Clin. Microbiol. Infect.* 3: 462–467.

Schmidt G, Goehring UM, Schirmer J, Lerm M, Aktories K (1999) Identification of the C-terminal part of *Bordetella* dermonecrotic toxin as a transglutaminase for rho GTPases. *J. Biol. Chem.* 274: 31875–31881.

Schmitt HJ, von Konig CH, Neiss A *et al.* (1996) Efficacy of acellular pertussis vaccine in early childhood after household exposure. *JAMA* 275: 37–41.

Sekiya K, Futaesaku Y, Nakase Y (1988) Electron microscopic observations on tracheal epithelia of mice infected with *Bordetella bronchiseptica. Microbiol. Immunol.* 32: 461–472.

Senda T, Horiguchi Y, Umemoto M, Sugimoto N, Matsuda M (1997) *Bordetella bronchiseptica* dermonecrotizing toxin, which activates a small GTP-binding protein rho, induces membrane organelle proliferation and caveolae formation. *Exp. Cell Res.* 230: 163–168.

Shahin RD, Brennan MJ, Li ZM, Meade BD, Manclark CR (1990) Characterization of the protective capacity and

immunogenicity of the 69-kD outer membrane protein of *Bordetella pertussis. J. Exp. Med.* 171: 63–73.

Shahin RD, Amsbaugh DF, Leef MF (1992) Mucosal immunization with filamentous hemagglutinin protects against *Bordetella pertussis* respiratory infection. *Infect. Immun.* 60: 1482–1488.

Shahin RD, Cowell JL (1994) Mouse respiratory infection models for pertussis. *Meth. Enzymol.* 235: 47–58.

Shahin RD, Hamel J, Leef MF, Brodeur BR (1994) Analysis of protective and nonprotective monoclonal antibodies specific for *Bordetella pertussis* lipooligosaccharide. *Infect. Immun.* 62: 722–725.

Shahin R, Leef M, Eldridge J, Hudson M, Gilley R (1995) Adjuvanticity and protective immunity elicited by *Bordetella pertussis* antigens encapsulated in poly(DL-lactide-co-glycolide) microspheres. *Infect. Immun.* 63: 1195–1200.

Shannon JL, Fernandez RC (1999) The C-terminal domain of the *Bordetella pertussis* autotransporter BrkA forms a pore in lipid bilayer membranes. *J. Bacteriol.* 181: 5838–5842.

Simondon F, Preziosi MP, Yam A *et al.* (1997) A randomized double blind trial comparing a two-component acellular to a whole-cell pertussis vaccine in Senegal. *Vaccine* 15: 1606–1612.

Soane MC, Jackson A, Maskell D *et al.* (2000) Interaction of *Bordetella pertussis* with human respiratory mucosa *in vitro. Respir. Med.* 94: 791–799.

Stainer DW (1988) Growth of *Bordetella pertussis.* In: Wardlaw AC, Parton R (eds) *Pathogenesis and Immunity in Pertussis.* Chichester: John Wiley and Sons, pp. 19–37.

Stainer DW, Scholte MJ (1971) A simple chemically defined medium for the production of phase I *Bordetella pertussis. J. Gen. Microbiol.* 63: 211–220.

Steed LL, Setareh M, Friedman RL (1991) Intracellular survival of virulent *Bordetella pertussis* in human polymorphonuclear leucocytes. *J. Leukoc. Biol.* 50: 321–330.

Steed LL, Akporiaye ET, Friedman RL (1992) *Bordetella pertussis* induces respiratory burst activity in human polymorphonuclear leucocytes. *Infect. Immun.* 60: 2101–2105.

Stehr K, Cherry JD, Heininger U *et al.* (1998) A comparative efficacy trial in Germany in infants who received either the Lederle/Takeda acellular pertussis component DTP (DTaP) vaccine, the Lederle whole-cell component DTP vaccine, or DT vaccine. *Pediatrics* 101: 1–11.

Stein PE, Boodhoo A, Armstrong GD, Cockle SA, Klein MH, Read RJ (1994a) The crystal structure of pertussis toxin. *Structure* 2: 45–57.

Stein PE, Boodhoo A, Armstrong GD *et al.* (1994b) Structure of a pertussis toxin-sugar complex as a model for receptor-binding. *Nature Struct. Biol.* 1: 591–596.

Stibitz S, Garletts TL (1992) Derivation of a physical map of the chromosome of *Bordetella pertussis* tohama I. *J. Bacteriol.* 174: 7770–7777.

Stibitz S, Miller JF (1994) Co-ordinate regulation of virulence in *Bordetella pertussis* mediated by the *vir* (*bvg*) locus. In: Miller VL, Kaper JB, Portnoy DA, Isberg RR (eds) *Molecular Genetics of Bacterial Pathogenesis.* Washington DC: American Society for Microbiology, pp. 407–422.

Stibitz S, Yang MS (1991) Subcellular localization and immunological detection of proteins encoded by the vir locus of *Bordetella pertussis. J. Bacteriol.* 173: 4288–4296.

Stibitz S, Weiss AA, Falkow S (1988) Genetic analysis of a region of the *Bordetella pertussis* chromosome encoding filamentous hemagglutinin and the pleiotropic regulatory locus *vir. J. Bacteriol.* 170: 2904–2913.

Stibitz S, Aaronson W, Monack D, Falkow S (1989) Phase variation in *Bordetella pertussis* by frameshift mutation in a gene for a novel two-component system. *Nature* 338: 266–269.

Storsaeter J, Olin P (1992) Relative efficacy of two acellular pertussis vaccines during three years of passive surveillance. *Vaccine* 10: 142–144.

Storsaeter J, Hallander H, Farrington CP, Olin P, Mollby R, Miller E (1990) Secondary analyses of the efficacy of two acellular pertussis vaccines evaluated in a Swedish phase III trial. *Vaccine* 8: 457–461.

Storsaeter J, Hallander HO, Gustafsson L, Olin P (1998) Levels of anti-pertussis antibodies related to protection after household exposure to *Bordetella pertussis. Vaccine* 16: 1907–1916.

Strugnell R, Dougan G, Chatfield S *et al.* (1992) Characterization of a *Salmonella typhimurium aro* vaccine strain expressing the P.69 antigen of *Bordetella pertussis. Infect. Immun.* 60: 3994–4002.

Syedabubakar SN, Mathews RC, Preston NW, Owen D, Hillier V (1995) Application of pulsed-field gel electrophoresis to the 1993 epidemic of whooping cough in the UK. *Epidemiol. Infect.* 115: 101–113.

Tang YW, Hopkins MK, Kolbert CP, Hartley PA, Severance PJ, Persing DH (1998) *Bordetella holmesii*-like organisms associated with septicemia, endocarditis, respiratory failure. *Clin. Infect. Dis.* 26: 389–392.

Temple LM, Weiss AA, Walker KE *et al.* (1998) *Bordetella avium* virulence measured *in vivo* and *in vitro. Infect. Immun.* 66: 5244–5251.

Torre D, Ferrario G, Bonetta G, Perversi L, Tambini R, Speranza F (1994) Effects of recombinant human gamma interferon on intracellular survival of *Bordetella pertussis* in human phagocytic cells. *FEMS Immunol. Med. Microbiol.* 9: 183–188.

Toyota T, Kai Y, Kakizaki M *et al.* (1980) Effects of islet-activating protein (IAP) on blood glucose and plasma insulin in healthy volunteers (phase 1 studies) Tohoku. *J. Exp. Med.* 130: 105–116.

Trollfors B, Taranger J, Lagergard T *et al.* (1995) A placebo-controlled trial of a pertussis-toxoid vaccine. *N. Engl. J. Med.* 333: 1045–1050.

Tuomanen E (1988) *Bordetella pertussis* adhesins. In: Wardlaw AC, Parton R (eds) *Pathogenesis and Immunity in Pertussis.* Chichester: John Wiley and Sons, pp. 75–94.

Tuomanen EI, Hendley JO (1983) Adherence of *Bordetella pertussis* to human respiratory epithelial cells. *J. Infect. Dis.* 148: 125–130.

Tuomanen EI, Nedelman J, Hendley JO, Hewlett EL (1983) Species specificity of *Bordetella* adherence to human and animal ciliated respiratory epithelial cells. *Infect. Immun.* 42: 692–695.

Tuomanen EI, Prasad SM, George JS *et al.* (1993) Reversible opening of the blood–brain barrier by anti-bacterial antibodies. *Proc. Natl Acad. Sci. USA* 90: 7824–7828.

Tyrrell GJ, Peppler MS, Bonnah RA, Clark CG, Chong P, Armstrong GD (1989) Lectinlike properties of pertussis toxin. *Infect. Immun.* 57: 1854–1857.

Uhl MA, Miller JF (1996a) Central role of the BvgS receiver as a phosphorylated intermediate in a complex two-component phosphorelay. *J. Biol. Chem.* 271: 33176–33180.

Uhl MA, Miller JF (1996b) Integration of multiple domains in a two-component sensor protein: the *Bordetella pertussis* BvgAS phosphorelay. *EMBO J.* 15: 1028–1036.

Ui M (1988) The multiple biological activities of pertussis toxin. In: Wardlaw AC, Parton R (eds) *Pathogenesis and Immunity in Pertussis.* Chichester: John Wiley and Sons, pp. 121–172.

Ui M (1990) Pertussis toxin as a valuable probe for G-protein involvement in signal transduction. In: Moss J, Vaughan M (eds) *ADP-Ribosylating Toxins and G-Proteins. Insights into Signal Transduction.* Washington, DC: American Society for Microbiology.

Urisu A, Cowell JL, Manclark CR (1986) Filamentous hemagglutinin has a major role in mediating adherence of *Bordetella pertussis* to human WiDr cells. *Infect. Immun.* 52: 695–701.

van't Wout J, Burnette WN, Mar VL, Rozdzinski E, Wright SD, Tuomanen EI (1992) Role of carbohydrate recognition domains of pertussis toxin in adherence of *Bordetella pertussis* to human macrophages. *Infect. Immun.* 60: 3303–3308.

Van Buynder PG, Owen D, Vurdien JE, Andrews NJ, Matthews RC, Miller E (1999) *Bordetella pertussis* surveillance in England and Wales: 1995–7. *Epidemiol. Infect.* 123: 403–411.

van den Akker WM (1998) Lipopolysaccharide expression within the genus *Bordetella*: influence of temperature and phase variation. *Microbiology* 144: 1527–1535.

van den Berg BM, Beekhuizen H, Willems RJ, Mooi FR, van Furth R (1999) Role of *Bordetella pertussis* virulence factors in adherence to epithelial cell lines derived from the human respiratory tract. *Infect. Immun.* 67: 1056–1062.

van der Zee A, Agterberg C, Peeters M, Schellekens J, Mooi FR (1993) Polymerase chain reaction assay for pertussis: simultaneous detection and discrimination of *Bordetella pertussis* and *Bordetella parapertussis. J. Clin. Microbiol.* 31: 2134–2140.

van der Zee A, Groenendijk H, Peeters M, Mooi FR (1996a) The differentiation of *Bordetella parapertussis* and *Bordetella bronchiseptica* from humans and animals as determined by DNA polymorphism mediated by two different insertion sequence elements suggests their phylogenetic relationship. *Int. J. Syst. Bacteriol.* 46: 640–647.

van der Zee A, Vernooij S, Peeters M, Van Embden J, Mooi FR (1996b) Dynamics of the population structure of *Bordetella pertussis* as measured by IS1002-associated RFLP: comparison of pre- and post-vaccination strains and global distribution. *Microbiology* 142: 3479–3485.

van der Zee A, Mooi F, Van Embden J, Musser J (1997) Molecular evolution and host adaptation of *Bordetella* spp.: phylogenetic analysis using multilocus enzyme electrophoresis and typing with three insertion sequences. *J. Bacteriol.* 179: 6609–6617.

Van Strijp JA, Russell DG, Tuomanen E, Brown EJ, Wright SD (1993) Ligand specificity of purified complement receptor type three (CD11b/CD18, alpha m beta 2, Mac-1) indirect effects of an Arg-Gly-Asp (RGD) sequence. *J. Immunol.* 151: 3324–3336.

Vandamme P, Hommez J, Vancanneyt M *et al.* (1995) *Bordetella hinzii* sp. nov, isolated from poultry and humans. *Int. J. Syst. Bacteriol.* 45: 37–45.

Vandamme P, Heyndrickx M, Vancanneyt M *et al.* (1996) *Bordetella trematum* sp. nov, isolated from wounds and ear infections in humans, reassessment of *Alcaligenes denitrificans* Ruger and Tan 1983. *Int. J. Syst. Bacteriol.* 46: 849–858.

Walker E (1988) Clinical aspects of pertussis. In: Wardlaw AC, Parton R (eds) *Pathogenesis and Immunity in Pertussis.* Chichester: John Wiley and Sons, pp. 273–282.

Walker KE, Weiss AA (1994) Characterization of the dermonecrotic toxin in members of the genus Bordetella. *Infect. Immun.* 62: 3817–3828.

Walker MJ, Rohde M, Timmis KN, Guzman CA (1992) Specific lung mucosal and systemic immune responses after oral immunization of mice with *Salmonella typhimurium* aroA, *Salmonella typhi* Ty21a, invasive *Escherichia coli* expressing recombinant pertussis toxin S1 subunit. *Infect. Immun.* 60: 4260–4268.

Wardlaw AC (1988) Virulence factors and species specificity in *Bordetella.* In: Donachie W, Griffiths E, Stephen J (eds) *Bacterial Infections of the Respiratory and Gastrointestinal Tract.* Oxford: IRL Press, pp. 41–56.

Wardlaw AC (1990) *Bordetella.* In: Parker MT, Duerden BI (eds) *Topley & Wilson's Principles of Bacteriology, Virology and Immunity,* 8th Edition, Vol. 2. London: Edward Arnold.

Wardlaw AC (1992) Multiple discontinuity as a remarkable feature of the development of acellular pertussis vaccines. *Vaccine* 10: 643–651.

Wardlaw AC, Parton R (1983a) *Bordetella pertussis* toxins. *Pharmacol. Ther.* 19: 1–53.

Wardlaw AC, Parton R (1983b) Pertussis vaccine. In: Easmon CSF, Jeljaszwicz J (eds) *Medical Microbiology,* Vol. 2: *Immunization against Bacterial Disease.* London: Academic Press, pp. 207–253.

Wardlaw AC, Parton R (1988) The host–parasite relationship in pertussis. In: Wardlaw AC, Parton R (eds) *Pathogenesis and Immunity in Pertussis.* Chichester: John Wiley and Sons, pp. 327–352.

Weinberg ED (1986) Iron, infection, neoplasia. *Clin. Physiol. Biochem.* 4: 50–60.

Weingart CL, Weiss AA (2000) *Bordetella pertussis* virulence factors affect phagocytosis by human neutrophils. *Infect. Immun.* 68: 1735–1739.

Weiss AA, Falkow S (1984) Genetic analysis of phase change in *Bordetella pertussis. Infect. Immun.* 43: 263–269.

Weiss AA, Goodwin MS (1989) Lethal infection by *Bordetella pertussis* mutants in the infant mouse model. *Infect. Immun.* 57: 3757–3764.

Weiss AA, Hewlett EL (1986) Virulence factors of *Bordetella pertussis. Annu. Rev. Microbiol.* 40: 661–686.

Weiss AA, Hewlett EL, Myers GA, Falkow S (1983) Tn5-induced mutations affecting virulence factors of *Bordetella pertussis. Infect. Immun.* 42: 33–41.

Weiss AA, Hewlett EL, Myers GA, Falkow S (1984) Pertussis toxin and extracytoplasmic adenylate cyclase as virulence factors of *Bordetella pertussis. J. Infect. Dis.* 150: 219–222.

Weiss AA, Hewlett EL, Myers GA, Falkow S (1985) Genetic studies of the molecular basis of whooping cough. *Dev. Biol. Stand.* 61: 11–19.

Weiss AA, Johnson FD, Burns DL (1993) Molecular characterization of an operon required for pertussis toxin secretion. *Proc. Natl Acad. Sci. USA* 90: 2970–2974.

Weiss AA, Mobberley PS, Fernandez RC, Mink CM (1999) Characterization of human bactericidal antibodies to *Bordetella pertussis. Infect. Immun.* 67: 1424–1431.

Weyant RS, Hollis DG, Weaver RE *et al.* (1995) *Bordetella holmesii* sp. nov, a new gram-negative species associated with septicemia. *J. Clin. Microbiol.* 33: 1–7.

White JM, Fairley CK, Owen D, Mathews RC, Miller E (1996) The effect of an accelerated immunisation schedule on pertussis in England and Wales. *CDR Rev.* 6: R86–R91.

WHO (1990) Requirements for diphtheria, tetanus, pertussis and combined vaccines. WHO Technical Report Series 800, Annex 2, 87–151.

WHO (1991) Meeting on case definitions of pertussis Conference Proceedings MIN/EPI/PERT91.1, 4–5. Geneva: World Health Organization.

Willems R, Paul A, van der Heide HG, ter Avest AR, Mooi FR (1990) Fimbrial phase variation in *Bordetella pertussis*: a novel mechanism for transcriptional regulation. *EMBO J.* 9: 2803–2809.

Willems RJ, van der Heide HG, Mooi FR (1992) Characterization of a *Bordetella pertussis* fimbrial gene cluster which is located directly downstream of the filamentous haemagglutinin gene. *Mol. Microbiol.* 6: 2661–2671.

Willems RJ, Geuijen C, van der Heide HG *et al.* (1993) Isolation of a putative fimbrial adhesin from *Bordetella pertussis* and the identification of its gene. *Mol. Microbiol.* 9: 623–634.

Willems RJ, Geuijen C, van der Heide HG *et al.* (1994) Mutational analysis of the *Bordetella pertussis* fim/fha gene cluster: identification of a gene with sequence similarities to haemolysin accessory genes involved in export of FHA. *Mol. Microbiol.* 11: 337–347.

Willems RJ, Kamerbeek J, Geuijen CA *et al.* (1998) The efficacy of a whole cell pertussis vaccine and fimbriae against *Bordetella pertussis* and *Bordetella parapertussis* infections in a respiratory mouse model. *Vaccine* 16: 410–416.

Wirsing von Konig CH, Finger H (1994) Role of pertussis toxin in causing symptoms of *Bordetella parapertussis* infection. *Eur. J. Clin. Microbiol. Infect. Dis.* 13: 455–458.

Wirsing von Konig CH, Postels-Multani S, Bogaerts H *et al.* (1998a) Factors influencing the spread of pertussis in households. *Eur. J. Pediatr.* 157: 391–394.

Wirsing von Konig CH, Rott H, Bogaerts H, Schmitt HJ (1998b) A serologic study of organisms possibly associated with pertussis-like coughing. *Pediatr. Infect. Dis. J.* 17: 645–649.

Wooldridge KG, Williams PH (1993) Iron uptake mechanisms of pathogenic bacteria. *FEMS Microbiol. Rev.* 12: 325–348.

Woolfrey BF, Moody JA (1991) Human infections associated with *Bordetella bronchiseptica*. *Clin. Microbiol. Rev.* 4: 243–255.

Wright SD, Silverstein SC (1983) Receptors for C3b and C3bi promote phagocytosis but not the release of toxic oxygen from human phagocytes. *J. Exp. Med.* 158: 2016–2023.

Xu Y, Barbieri JT (1996) Pertussis toxin-catalyzed ADP ribosylation of Gi-2 and Gi-3 in CHO cells is modulated by inhibitors of intracellular trafficking. *Infect. Immun.* 64: 593–599.

Yuk MH, Harvill ET, Miller JF (1998a) The BvgAS virulence control system regulates type III secretion in *Bordetella bronchiseptica*. *Mol. Microbiol.* 28: 945–959.

Yuk MH, Heininger U, Martinez dT, Miller JF (1998b) Human but not ovine isolates of *Bordetella parapertussis* are highly clonal as determined by PCR-based RAPD fingerprinting. *Infection* 26: 270–273.

Yuk MH, Harvill ET, Cotter PA, Miller JF (2000) Modulation of host immune responses induction of apoptosis and inhibition of NF-kappaB activation by the *Bordetella* type III secretion system. *Mol. Microbiol.* 35: 991–1004.

Zarrouk H, Karibian D, Bodie S, Perry MB, Richards JC, Caroff M (1997) Structural characterization of the lipids A of three *Bordetella bronchiseptica* strains: variability of fatty acid substitution. *J. Bacteriol.* 179: 3756–3760.

Zhang JM, Cowell JL, Steven AC *et al.* (1985) Purification and characterization of fimbriae isolated from *Bordetella pertussis*. *Infect. Immun.* 48: 422–427.

Zychlinsky A, Sansonetti P (1997) Perspectives series: host/pathogen interactions. Apoptosis in bacterial pathogenesis. *J. Clin. Invest.* 100: 493–495.

76

Streptococcus pneumoniae

Tim J. Mitchell and Alison R. Kerr

University of Glasgow, Glasgow, UK

Streptococcus pneumoniae, the pneumococcus, was simultaneously identified by Pasteur (1881) and Sternberg (1881). Pasteur discovered the organism in saliva from a child who had died of rabies, while Sternberg found pneumococci in his own saliva. Each injected the saliva into rabbits and both recovered diplococci from the blood of the animals. Two years later these bacteria were implicated in disease when Friedlander (1883) noted the association of the organism with lobar pneumonia.

Formerly called *Diplococcus pneumoniae*, *Strep. pneumoniae* is Gram-positive and encapsulated. It is now recognised as a major human pathogen in infancy, childhood and adult life. *Strep. pneumoniae* is capable of causing both non-invasive disease, such as otitis media, and invasive infections, such as pneumonia, meningitis and sepsis.

Bergey's Manual of Systematic Bacteriology describes streptococci as spherical or ovoid, less than 2 μm in diameter, Gram-positive, catalase-negative and facultative anaerobic bacteria (Hardie, 1984). They are further classified by their growth on blood agar, the group-specific carbohydrate (Lancefield) antigen in the bacterial cell wall and biochemical tests. Pneumococci are initially characterised by greenish zones of incomplete haemolysis surrounding the colonies on blood agar (α haemolysis). *Strep. pneumoniae* does not possess Lancefield antigens and thus the further classification of α-haemolytic streptococci has been clarified by molecular studies. These organisms can be classified into three main groups: *Strep. mutans*, *Strep. salivarius* and *Strep. oralis* on the basis of ribosomal RNA and DNA hybridisation. The *Strep. oralis* group can be further divided into the *Strep. milleri*, *Strep. sanguis* and *Strep. oralis* subgroups. *Strep. pneumoniae* belongs to the oralis subgroup, along with *Strep. oralis* and *Strep. mitis*. Susceptibility to optochin (ethyl hydrocuprein) and solubility in bile differentiate pneumococci from the other members of the oralis subgroup.

Structure, Genetics and Virulence Factors of *Streptococcus pneumoniae*

The pneumococcal outer surface comprises three main parts: the capsule, cell wall and the plasma membrane. Virulence factors may be cell associated or be produced as soluble factors (**Fig. 1**).

Capsule

Although *Strep. pneumoniae* exists in encapsulated and non-capsulated forms, the great majority of clinical isolates and strains of *Strep. pneumoniae* recovered from carriage sites have a polysaccharide capsule (Kalin, 1998), of which 90 types are currently known (Henrichsen, 1995). The majority, if not all,

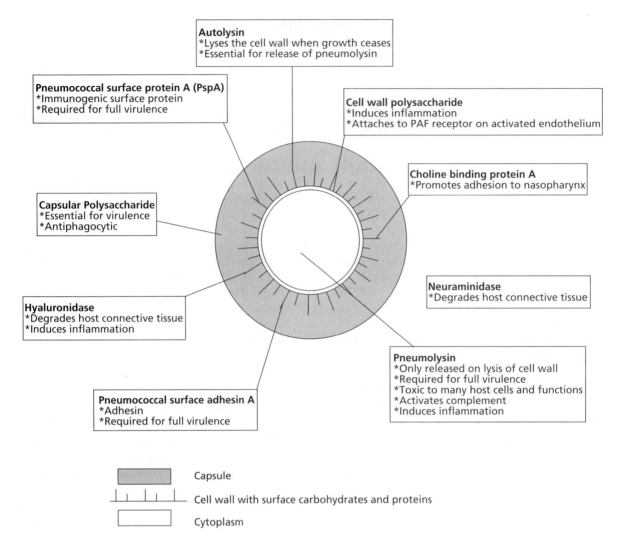

Fig. 1 Schematic diagram of the main virulence factors of *Strep. pneumoniae* including their main functions and cellular location. (Adapted from Catterall, 1999).

of these capsular serotypes are capable of causing serious illness in humans (Kalin, 1998). Few reports have appeared of atypical non-capsulate pneumococci being carried or causing infection in humans (Muller-Graf *et al.*, 1999). Such strains are isolated mainly from immunocompromised individuals, such as HIV-seropositive patients. Thus the majority of non-capsulate pneumococci are only found in the laboratory.

Capsular types were originally differentiated by the capsule swelling (Quellung) reaction in which addition of type-specific antiserum causes macroscopic agglutination and microscopically visible swelling of the capsule (Neufeld, 1902). In the case of pneumococci with small capsules, however, the reaction can be difficult to interpret and the results dubious. More recent methods employ fluorescent antibodies directed against the capsule (Wicher *et al.*, 1982).

Capsular type is classified by one of two systems: the American system assigns numbers in the chronological order of their discovery, whereas the Danish system is based on antigenic similarities. Virulent strains that cause most outbreaks of disease were the first identified and thus first assigned numbers by the American classification system. Accordingly, the lower numbered serotypes are most commonly isolated in human infections. The lowest numbered serotypes such as 1 and 2 are classified identically by both the American and Danish systems. Higher numbered serotypes in the American system that share antigenic similarities with lower numbered types are grouped together by the Danish system. Thus American type 48 becomes Danish 7B, because it is similar to Danish type 7A, American type 7, while serotype 50 became 7C (Lee, 1987). The Danish serotype nomenclature is now generally employed (van Dam *et al.*, 1990).

The capsule forms the outermost layer of all fresh clinical isolates of the pneumococcus. A total of 90 distinct types of capsule are currently known (Henrichsen, 1995). The simplest types of capsular polysaccharide (CPS) are linear polymers composed of repeat units of two or more monosaccharides. More complex capsules with branched backbone structures are composed of 1–6 monosaccharides with additional side-chains. Until recently, very little was known about the genes that encode capsule biosynthesis. Since the recent explosion in the molecular understanding of this process, however, genes encoding the biosynthetic pathways in a range of serotypes have been elucidated (Table 1).

The capsule biosynthesis loci are usually located at the same position in the chromosome, between *dexB* and *aliA*. Major differences between the loci presumably reflect the different complexities of the capsular polysaccharide synthesised. Type 3 CPS is relatively simple and is composed of a disaccharide repeat unit of glucose and glucuronic acid. The *cps3* locus contains three functional genes, *cap3A* (or *cps3D*), *cap3B* (or *cps3S*) and *cap3C* (or *cps3U*). *cap3A* encodes the UDP-Glc dehydrogenase required for the synthesis of UDP-GlcAc (Arrecubieta *et al.*, 1994), while *cps3B* encodes the type 3 synthase, a β-glycosyltransferase that links the alternating Glc and GlcA moieties. The final gene, *cap3C*, encodes a Glc-1-phosphate uridyltransferase. Introduction of the *cap3B* gene alone into *E. coli* is sufficient to drive the synthesis of type 3 CPS (*E. coli* contains UDP-GlcA) and some of the CPS appears in the periplasm. It appears that additional genes are not required for the transport of CPS in these organisms. It has been suggested that the C-terminal portion of the transferase enzyme forms a pore in the cytoplasmic membrane through which the growing polysaccharide chain is extruded (Keenleyside and Whitfield, 1996). The *cps19f* locus is much more complicated than *cps3*. It is 15 kb long and consists of 15 genes designated *cps19fA* to *cps19fO* (Guidolin *et al.*, 1994; Morona *et al.*, 1997b). A summary of the predicted functions of these genes is given in Table 2. Paton and Morona (2000) suggested a mechanism for type 19F CPS biosynthesis. The initial step involves transfer of Glc-1-phosphate to a lipid carrier on the cytoplasmic side of the cell membrane by CPS19fE. Cps19fF, G and H then catalyse the sequential transfer of other component monosaccharide precursors (synthesised by Cps19fK, L, M, N and O) to form the tripeptide repeat unit. These repeat units are then translocated to the extracellular side of the membrane and polymerised in a blockwise manner. These two steps are catalysed by Cap19fJ and I respectively. Cps19fC and D may be involved in regulating chain length.

Cell Wall

The typical Gram-positive cell wall is composed mainly of peptidoglycan (PG): glycan chains of alternating *N*-acetyglucosamine and *N*-acetylmuramic acid. PG is synthesised by a set of enzymes that are the target for the antibiotic penicillin. These enzymes are termed penicillin-binding proteins or PBPs. Penicillin is

Table 1 The capsule loci for which DNA sequence is available

Capsule type	References
1	Munoz *et al.* (1997)
2	Lannelli *et al.* (1999)
3	Arrecubieta *et al.* (1995); Dillard *et al.* (1995)
4	Institute for Genomic Research: http: //www.tigr.org/tdb
14	Kolkman *et al.* (1997)
19A	Morona *et al.* (1999b)
19B	Morona *et al.* (1999b)
19C	Morona *et al.* (1999b)
19F	Morona *et al.* (1997a)
23F	Morona *et al.* (1999a)
33F	Llull *et al.* (1998)

Table 2 Proposed function of the 15 gene products of the capsule biosynthesis locus from serotype 19F *Streptococcus pneumoniae*

Gene product	Function
Cps19fA	Regulation
Cps19fB	Unknown
Cps19fC	Chain length regulation/export
Cps19fD	Chain length regulation/export
Cps19fE	Glucosyltransferase
Cps19fF	*N*-acetylmannosamine transferase
Cps19fG	Homology to LicD of *Haemophilus influenzae*. Function unknown
Cps19fH	Rhamnosyltransferase
Cps19fI	Polysaccharide polymerase
Cps19fJ	Oligosaccharide repeat unit transporter
Cps19fK	UDP-*N*-acetylglucosamine-2-epimerase
Cps19fL	Glucose-1-phosphate thymidylyltransferase
Cps19fM	dTDP-4-keto-6-deoxyglucose-3,5-epimerase
Cps19fN	dTDP-glucose-4,6-dehydratase
Cps19fO	dTDP-L-rhamnose synthase

Information from Paton and Morona (2000).

structurally similar to D-alanyl-D-alanine, a constituent of cell walls and acts as an inhibitor of the cell wall cross-linking process. Changes in the affinity of PBPs for penicillin forms the basis of penicillin resistance. Acidic polysaccharides termed teichoic acids are attached to the cell wall. Teichoic acid (TA) can be attached either directly to the peptidoglycan of the cell wall or may be anchored in the plasma membrane (lipoteichoic acid, LTA). The chains of TA and LTA are made up of identical repeats of D-glucose, 2-acetamido-4-amino-2,4,6-trideoxy-D-galactose, 2 units of N-acetyl-D-galactosaminyl residues and ribitol-5-phosphate. These repeat units are joined by linking the O5 of the ribitol with the O6 of the glucosyl residue by a phosphodiester bond. TA and LTA also contain phosphorylcholine (PC) linked to the O6 of the N-acetyl-D-galactosaminyl residues. Phosphorylcholine is a key molecule in the biology of the pneumococcus because it plays a role as an adhesin and also serves as a site for the attachment of choline-binding proteins. The genes for the synthesis of phosphorylcholine have been located in the pneumococcal genome and mutations in these can be lethal for the pneumococcus (Zhang et al., 1999).

Pneumolysin

The production of pneumolysin, a haemolytic protein toxin, by Strep. pneumoniae was first reported by Libman (1905). Since then, numerous studies have been carried out with crude preparations of the toxin (Cole, 1914; Neill, 1926, 1927; Cohen et al., 1940, 1941; Halbert et al., 1946) which demonstrated that the protein is toxic. It is antigenic, irreversibly inactivated by cholesterol and susceptible to oxidation, a process which could be reversed by treatment with thiol-reducing agents. The toxin belongs to a large family of thiol-activated toxins (Smyth and Duncan, 1978), which are produced by four genera of Gram-positive bacteria. Recent advances in molecular biology and protein purification have allowed large amounts of highly purified pneumolysin to be produced and studied in a range of biological systems both in vitro and in vivo.

Pneumolysin is a lytic toxin able to lyse all cells that have cholesterol in their membranes. The mechanism of action this group of toxins is believed to be similar and has been reviewed (Smyth and Duncan, 1978; Morgan et al., 1996). Two of the major steps in the lytic process are membrane-binding and oligomerisation of the toxin to form pores. These two steps show differences in their temperature dependency. The binding of the toxin to cells is temperature-independent while oligomerisation occurs only at higher temperatures. The lytic activity of the toxin can be inhibited by pre-incubation with cholesterol and this has led to the suggestion that membrane cholesterol is the receptor for this toxin. Once bound to the cell membrane, monomeric toxin inserts into the membrane and undergoes oligomerisation to produce high-molecular-weight pore-like structures composed of up to 50 monomeric units. These pores are believed to mediate the cell lysis by osmotic mechanisms. Inhibition of the lytic process by cholesterol is assumed to represent occupancy of the cholesterol-binding site by free cholesterol that prevents binding of the toxin to target cell membranes. Studies with the related protein toxin listeriolysin O suggest, however, that inhibition of toxin action by cholesterol may not be due to a blockage of binding, but rather to interference with the oligomerisation process (Jacobs et al., 1998). The exact mechanism of cell lysis by pneumolysin and other members of the thiol-activated toxin family still remains to be defined. The latest theories of how this process works will be discussed below.

Pneumolysin consists of a single 53-kDa polypeptide chain and is produced by virtually all clinical isolates of the pneumococcus (Paton et al., 1983; Kanclerski and Mollby, 1987). It is the only member of the family that is not secreted from the cell but remains within the bacterial cytoplasm (Johnson, 1977).

The genes that encode pneumolysin from serotype 1 and serotype 2 pneumococci have been cloned, sequenced and expressed in Escherichia coli (Paton et al., 1986; Walker et al., 1987; Mitchell et al., 1989). Genes for pneumolysin have also been cloned from several other serotypes, including the genome sequence of a type 4 pneumococcus. Comparison of the derived amino acid sequences of these proteins shows a very high level of conservation. Alignment of the amino acid sequence of pneumolysin with the other members of the family shows a high degree of similarity between the proteins. Analysis of the primary amino acid sequence of pneumolysin shows no major areas of hydrophobicity that might be involved in membrane insertion. The hydrophobic nature of pneumolysin (Johnson et al., 1982) must, therefore, reflect the generation of hydrophobic areas during the folding of the primary sequence (see below on the folded form of pneumolysin). In agreement with the cytoplasmic location of the toxin, the predicted amino acid sequence does not contain a signal sequence for secretion.

Since pneumolysin is thiol-activated it was for many years thought that the activation process involved the reduction of an intra-molecular disulphide bond (Smyth and Duncan, 1978). The primary amino acid sequence shows, however, that pneumolysin contains only a single cysteine residue. This lies at the C-terminal end of the molecule (amino acid position 428) in a region of 11 amino acids that is conserved amongst many of the family of toxins (Morgan *et al.*, 1996). This conserved sequence (ECTGLAWEWWR) plays an important role in the activity of the toxin as judged by studies by site-directed mutagenesis. Interestingly, the cysteine residue is not essential for activity and can be replaced with alanine without effect on the lytic activity of the toxin *in vitro* (Saunders *et al.*, 1989). The nature of the residue at this position is important, however, because substitution with glycine or serine results in a reduction in toxin activity. Substitutions in other residues within the motif also affect the activity of the toxin, and changes in the tryptophan (position 433) have a dramatic effect. Attempts to define the nature of the defect in the mutants with reduced activity have been unsuccessful and have only shown that the effect is not due to a gross defect in the ability of the toxins to bind to cells or to oligomerise (Saunders *et al.*, 1989). It seems that the cysteine motif may play an important part in mediating a conformational change in the toxin when it interacts with membranes and allows the insertion of the toxin monomer into the lipid bilayer (see below).

Naturally occurring mutations in the amino acid sequence of pneumolysin can also be informative. Pneumolysin derived from some serotype 8 pneumococci contains several alterations in amino acid sequence when compared with the protein from other serotypes (Lock *et al.*, 1996). The type 8 protein has three substitutions (positions 172, 224 and 265) and two amino acid deletions (residues 270–271). Interestingly, these changes make the protein run on SDS-PAGE with an apparent molecular weight greater than that of the proteins from other more typical strains of the pneumococcus (Lock *et al.*, 1996). Construction of chimeric molecules between the type 8 and the type 2 proteins demonstrated that it is the substitution at position 172 (threonine to isoleucine) that is responsible for the decreased activity of this version of the protein (Lock *et al.*, 1996).

The mechanism by which pneumolysin binds to cells remains unclear. The evidence that cholesterol is the receptor for the toxin is based on studies of the inhibition of toxin activity by incubation of the toxin with cholesterol in solution. The finding that the related toxin listeriolysin O is inactivated by cholesterol but is still able to bind to cells suggests that binding of these toxins to cells may involve a receptor other than cholesterol (Jacobs *et al.*, 1998). Interaction with cholesterol may play a role in the assembly of the functional oligomer. The region of the pneumolysin molecule involved in cell binding has been

inferred from studies using truncated versions of the protein (Owen *et al.*, 1994). Removal of the five C-terminal amino acids of the toxin was sufficient to prevent cell lysis by the protein. A substitution of the proline residue at position 462 with serine causes a 90% reduction in the ability of the toxin to bind to red blood cells.

The mechanism of pore formation by pneumolysin also remains unclear. Site-directed mutations can be made in the conserved cysteine-containing region that have a dramatic effect on the lytic activity of the toxin but have no effect on the ability of the toxin to generate oligomers as observed by electron microscopy. To define the differences in the nature of the pore formed by wild-type pneumolysin and a lysis-deficient mutant (F433) in which the tryptophan at position 433 has been replaced with phenylalanine, the proteins have been studied in their effects on cells and on planar lipid bilayers (Korchev *et al.*, 1998). The mutation at position 433 has no effect, as observed by electron microscopy, on the ability of the toxin to bind to cells or to form oligomeric structures. The mutation does affect the ability of the toxin to induce leakage of markers from Lettre cells and to induce conductance channels in planar lipid bilayers. Pneumolysin-induced leakage from Lettre cells is sensitive to inhibition by divalent cations. The sensitivity to inhibition by divalent cations was much reduced in the mutant form of the toxin. When inserted into planar lipid bilayers, pneumolysin induces a range of conductance channels with small (less than 30 pS), medium (30 pS–1 nS) and large (greater than 1 nS) conductance steps. Small and medium channels are preferentially closed by divalent cations. Wild-type toxin forms mainly small channels, whereas the mutant forms mostly large channels that are insensitive to closure by divalent cations.

Osmotic protection studies show that cells treated with wild-type toxin can be protected from lysis by polysaccharides with a molecular mass of more than 15 kDa, whereas cells treated with the mutant form of the toxin were not protected by polysaccharides with molecular mass over 40 kDa. This also suggests that the functional pores are larger in the membranes treated with the F433 mutant. The cysteine motif therefore probably plays a role in the functionality of the pores generated by pneumolysin. Since differences were not observed by electron microscopy in the structure of the pores, the structural appearance of the pores does not reflect the functional state of the channels.

Purified pneumolysin is also able to activate the classical complement pathway (Paton *et al.*, 1984) and this effect has been linked to the ability of the toxin to bind to the Fc portion of human immunoglobulin (Mitchell *et al.*, 1991). Analysis of the amino acid sequence of pneumolysin showed that it does not contain any homology to known IgG-binding proteins but shows limited homology to C-reactive protein (CRP), a human acute-phase protein (Mitchell *et al.*, 1991). CRP is also able to activate the classical complement pathway, but this effect is mediated by a direct-binding of complement component C1q (Volanakis and Kaplan, 1974). Mutagenesis of the homologous region in pneumolysin revealed that residues 384 and 385 are involved in antibody binding and complement activation by the toxin (Mitchell *et al.*, 1991).

In order fully to understand the mechanism of pore formation by pneumolysin, it would be useful to have a three-dimensional structure for the protein monomer. The 2.7-Å crystal structure of perfringolysin has been solved by Rossjohn *et al.* (1997). Since the pneumolysin and perfringolysin sequences share 48% sequence identity and 60% sequence similarity extending over the full length of the molecule, it has been possible to model a structure for the pneumolysin monomer by homology with that of perfringolysin (Rossjohn *et al.*, 1998). On the basis of the homology model with perfringolysin, the pneumolysin molecule is long and rod-shaped with overall dimensions of $11\,\text{nm} \times 5\,\text{nm} \times 3\,\text{nm}$. These values are in good agreement with those derived for pneumolysin from metal shadow electron microscopy (as discussed above). In agreement with the EM the molecule consists of four domains. The overall structural composition is 41% sheet and 21% helix. The conserved cysteine motif forms a loop at the bottom of domain 4.

Pneumolysin has been crystallised in combination with cholesterol (Kelly and Jedrzejas, 2000). The determination of the three-dimensional structure of the toxin in the presence of cholesterol will be very informative.

Neuraminidase

It has been known for many years that all fresh isolates of pneumococci produce neuraminidase (Kelly *et al.*, 1966; O'Toole *et al.*, 1971). Biochemical studies have revealed that neuraminidases with a range of molecular weights are produced and this led to controversy over the number and molecular size of the enzymes. Genome sequence analysis appears to show that at least three genes code for products related to neuraminidase (*nanA*, *nanB* and *nanC*). NanA is located on the surface of the pneumococcus (Camara *et al.*, 1994). The predicted protein sequence contains an QJ;N-terminal putative signal peptide and a Gram-positive cell surface anchor domain (LPXTG). The predicted protein sequence also contains four aspartic box sequences and a FRIP region that is conserved amongst all sialidases (Camara *et al.*, 1994). The exact size of NanA is unknown as there are three putative translation start codons. The *nanA* gene varies in size because of the presence or absence of three direct repeats of 60 nucleotides towards the 3′-end of the gene (Dowson *et al.*, 1997). Isolates that possess the direct repeats were found to be almost exclusively of serotype 3. Homology searches revealed that the repeats in pneumococcal *nanA* are most similar to a region of the IgA1 protease of *Strep. sanguis* in which there was approximately 50% amino acid homology but identical spacing of conserved PEYTG and EP motifs. The *nanA* gene is a striking example of horizontal gene transfer, with the 5′ end of the recombinant molecule differing from the recipient by approximately 35% at the nucleotide level. This level of divergence is much higher than that previously reported for mosaic genes in the pneumococcus and suggests that a rare recombination event may be driven by a strong selective pressure over a long period of time.

NanB has a predicted size of 74 kDa after cleavage of the 29 amino acid signal peptide (Berry *et al.*, 1996) and is exported from the cell. It lacks a cell wall attachment sequence but contains three aspartic boxes and a FRIP region. NanB has a pH optimum of 4.5 compared with 6–7.5 for NanA. There is negligible sequence homology between NanA and NanB. The *nanB* gene lies about 4.5 kb downstream of *nanA* and appears to be part of an operon consisting of at least six open reading frames. The *nanA* gene is unlikely to be part of this operon as there is a strong terminator at the end of the gene (Camara *et al.*, 1994). Some of the genes in the operon have homology to sugar-binding proteins.

The *nanC* gene has been identified by homology searches of the genome sequence.

Hyaluronidase

Pneumococcal hyaluronidase (hyaluronate lyase) is a spreading factor that may be involved in the disease process and the gene for this enzyme has been cloned and expressed in *E. coli* (Berry *et al.*, 1994). The gene sequence reveals an LPXTGX motif but no secretion signal. The protein was believed to have a molecular weight of 107 kDa (Berry *et al.*, 1994) but analysis of the genomic sequence of a pneumococcus type 4 strain shows a predicted molecular weight of 121 kDa and that the sequence reported by Berry *et al.* (1994) lacks 117 amino acids from the N-terminus, including the secretion signal sequence.

The core of the hyaluronidase enzyme has been expressed for crystallography studies (Jedrzejas *et al.*, 1998a, b), the crystal structure solved (Songlin *et al.*, 2000) and residues involved in the active site of the enzyme identified. This allowed the construction of mutants with reduced or no enzymatic activity, which will be useful in the evaluation of the role hyaluronidase may play in virulence.

Other Surface Proteins

The pneumococcus has several mechanisms to attach proteins to its cell surface. One of these uses the conserved amino acid sequence LPXTGE, which acts as a cleavage site and point of covalent attachment to the cell wall (Schneewind *et al.*, 1995). Examination of the pneumococcus genome sequences shows that there are some 15 of these motifs, including those present in the enzymes IgA protease, neuraminidase and hyaluronidase. Lipoproteins are attached to the cell surface via palmitic acid and the motif LXXC in the N-terminus of the protein serves as the point of lipidation. Examples of these linkages include those found in peptide permease systems. A third means of linking proteins to the pneumococcal cell surface is by non-covalent interactions with cell wall phosphorylcholine, the so-called choline-binding proteins (CBPs). Genomic and biochemical analysis suggests that 12 of these proteins are present in the

pneumococcus (summarised by Gosink *et al.*, 2000) (**Table 3**).

The 12 CBPs all share a common C-terminal choline-binding domain that consists of 2–10 repeats of 20 amino acids (Garcia *et al.*, 1998). This family includes several proteins that play a role in pneumococcal infection, including the autolysin and CbpA (see below). The CBPs are 'modular' in that the C-terminal ends mediate the common function of attachment to the bacterial cell, whereas the more diverse N-terminal ends mediate other biological functions. The biological function of the major autolysin (LytA), *N*-acetylmuramyl-L-alanine amidase, is the degradation of cell wall (Garcia *et al.*, 1986). Two other cell wall hydrolases, LytB and LytC, have been described. LytB plays a role in daughter cell separation after cell division (Garcia *et al.*, 1999a) and LytC has lysozyme activity at 30°C (Garcia *et al.*, 1999b). It has been suggested that PcpA is involved in protein–protein and/or protein–lipid interactions (Sanchez-Beato *et al.*, 1998). PspA and CbpA have been ascribed a role in virulence and their functions are discussed below.

Various predictions can be made concerning the structure of PspA. Sequence analysis of *pspA* suggests that the protein product is composed of several domains (McDaniel *et al.*, 1998; Yother and Briles, 1992). The N-terminal half of the molecule has the features of an α-helical coiled-coil structure. This is followed by a proline-rich domain where the sequence

Table 3 The choline-binding proteins of *Streptococcus pneumoniae*

Choline-binding protein	Role in virulence	References
LytA	Release of pneumolysin and cell wall degradation products	Berry *et al.* (1989a)
LytB	Colonisation of nasopharynx	Gosink *et al.* (2000)
LytC	Colonisation of nasopharynx	Gosink *et al.* (2000)
PcpA	May play a role in adhesion	Sanchez-Beato *et al.* (1998)
PspA	Role in sepsis. Decreases complement deposition. Binds to lactoferrin	Hammerschmidt *et al.* (1999); Neeleman *et al.* (1999); Talkington *et al.* (1992); Tu *et al.* (1999)
CbpA	Colonisation of nasopharynx. Uses polymeric immunoglobulin receptor	Rosenow *et al.* (1997) Zhang *et al.* (2000)
CbpD	Colonisation of nasopharynx	Gosink *et al.* (2000)
CbpE	Reduced colonisation of nasopharynx. Decreased adherence to human cells	Gosink *et al.* (2000)
CbpF	None	Gosink *et al.* (2000)
CbpG	Reduced colonisation of nasopharynx. Reduced binding to human cells. Role in sepsis. Putative serine protease	Gosink *et al.* (2000)
CbpI	None	Gosink *et al.* (2000)
CbpJ	None	Gosink *et al.* (2000)

Pro-Ala-Pro-Ala-Pro is repeated many times. At the C-terminal end of the molecule is a 200-amino-acid choline-binding domain made up of 10 repeats each of 20 amino acids. The final 17 amino acids make up a hydrophilic domain. Several fragments of the PspA molecule have been crystallised for structure determination (Lamani et al., 2000). A 'PspA-like' protein has also been described and variously named PspC, CbpA and SspA by various workers. CbpA was originally reported in the serotype 2 derivative R6x (Rosenow et al., 1997) and in this strain it consists of 663 amino acids with a predicted mass of 75 kDa. The C-terminal choline-binding domain has the same features as those just described for PspA and the N-terminal region is predicted to contain six α-helical structures that form two direct repeats of 107 amino acids designated R1 and R2. CbpA has been reported to function as a pneumococcal adhesin (Rosenow et al., 1997) and that it is involved in interaction with the human polymeric immunoglobulin receptor (Zhang et al., 2000).

Antibiotic Resistance

The standard treatment in the community for pneumococcal pneumonia in the UK is oral benzylpenicillin (Musher, 1992) and when hospital admission is necessary, the antibiotic is given parenterally. In non-industrialised nations treatment is generally with oral amoxycillin, which provides a broader antibacterial coverage than required for pneumococcal infection, while in more serious cases penicillin is administered parenterally. The efficacy of all therapeutic agents is currently under threat by antibiotic resistance (Goldsmith et al., 1997; Hart and Kariuki, 1998).

The first clinically significant pneumococci with reduced penicillin sensitivity were identified during the 1960s in Australia and Papua New Guinea (Hansmann and Bullen, 1967). By 1974 the incidence of penicillin-resistant pneumococci in this area was 12% (Hansmann et al., 1974) and by 1980 it had reached 30% (Gratten et al., 1980). Recent studies have identified a significant problem in western countries, and in some areas of North America 40% of pneumococcal isolates are penicillin-resistant (Butler and Cetron, 1999). In some developing countries, antibiotic-resistant pneumococci are increasing in prevalence dramatically on account of the unregulated use of antibiotics in humans and animals (Hart and Kariuki, 1998).

Pneumococci become resistant to penicillin because of an alteration in the penicillin-binding proteins (PBPs), the target of the antibiotic. The PBPs, which are transpeptidases, are involved in the final stages of peptidoglycan synthesis and are the target of penicillin.

Pneumococcal penicillin resistance develops due to the sequential alteration of three PBPs (1A, 2X and 2B). Alterations in these targets, especially around the active-site serine, reduce their affinity for penicillin (Dowson et al., 1993). The alteration in PBP genes arises by inter-species recombination (Dowson et al., 1989a, b) in which other bacterial species act as donors of low-affinity PBP sequences to generate mosaic genes. For example, Streptococcus mitis, a commensal member of the oral flora has donated PBP2B genes to pneumococci (Dowson et al., 1993). Resistance to cephalosporins is also based on altered PBPs (Coffey et al., 1995).

Erythromycin resistance is mediated by ermAM, which encodes a methylase that modifies 23S rRNA, or by increased efflux of the drug mediated by an efflux pump coded by mefE (Widdowson and Klugman, 1999). Tetracycline resistance is mediated by the genes tetM and tetO, which produce ribosomal protection, and chloramphenicol resistance is due to production of chloramphenicol acetyltransferase by the cat gene (Widdowson and Klugman, 1999). Fluoroquinolone resistance is mediated either by changes in the target DNA gyrase and topoisomerase or by enhanced pump-mediated drug efflux.

Vancomycin-resistant pneumococci have not been reported and this antibiotic is at present the drug of last resort. Vancomycin tolerance, the ability of pneumococci to survive but not grow in the presence of the antibiotic, has recently emerged and is generally believed to be a precursor phenotype to resistance. Vancomycin tolerance is the result of loss of function of the Vnc histidine kinase of a two-component signal transduction system, which, it is suggested, may be critical for bactericidal activity of vancomycin (Novak et al., 1999b). Strains of Strep. pneumoniae resistant to all appropriate antibiotics may emerge in the short term.

Genetic Transformation

Genetic transformation was originally described in Strep. pneumoniae (Griffith, 1928) and may be considered the starting point of molecular biology. This ability may promote the survival of the organism by permitting it to evolve to overcome host-resistance mechanisms, such as the alteration of penicillin-binding proteins in penicillin-resistant isolates of Strep. pneumoniae (Dowson et al., 1989a). Competence for genetic transformation is co-ordinated between pneumococci by a quorum-sensing mechanism that results in most cells in a culture becoming competent simultaneously for DNA uptake during a

10–15 minute window of exponential growth as a critical, but low, cell density is reached.

The state of capsulation is an important factor in determining the transformability. Encapsulated isolates are generally not competent, whilst non-capsulate strains are readily able to take up exogenous DNA. Some encapsulated pneumococci can, however be transformed by the addition of factors from cultures of non-capsulate strains (Yother *et al.*, 1986).

During the period of competence, pneumococcal protein synthesis is altered so that a small group of proteins constitute the vast majority of those transcribed. Many of these proteins have been identified (Morrison and Baker, 1979). Most of these proteins are not transcribed at any other time in the pneumococcal growth cycle and they are thought to be required for transformation to occur or to increase its efficiency.

Two signals for pneumococcal competence have been identified and designated CSP-1 (competence-stimulating peptide-1) and CSP-2 (Havarstein *et al.*, 1995; Pozzi *et al.*, 1996). Genes associated with CSP synthesis and activity have been designated *com* for their involvement in competence regulation. The immature CSP peptide (ComC) contains a 'double glycine' leader peptide that is removed after translation to create mature CSP. An ABC transporter/protease (ComA and ComB) is believed to be responsible for this modification. Mature CSP is sensed by a receptor histidine protein kinase (ComD) linked to a response regulator (ComE). Separate operons encode ComAB and ComCDE and up-regulation of basal levels of both occurs in response to CSP (Pestova *et al.*, 1996; Alloing *et al.*, 1998; Lee and Morrison, 1999) because ComE binds to direct repeat sites adjacent to the promoters of the operons (Ween *et al.*, 1999).

Until recently little was known about the regulatory mechanism activated by CSP that controls the shift in protein synthesis occurring during competence. Most of the indications were that transcription regulation is involved, since transcription of several genes is highly up-regulated at the onset of competence (Alloing *et al.*, 1998). Furthermore, several promoter regions of competence-regulated operons contain a conserved 8 bp sequence 7–20 bp upstream of their 5′ genes termed the combox. A protein (ComX) present in active preparations of pneumococcal RNA polymerase that is capable of transcription from a combox promoter has been identified (Lee and Morrison, 1999); ComX mutants are not transformable. The *comX* gene is transcribed only at the time of competence, and relies on *comE* and CSP for expression, while *comAB* and *comCDE* do not require ComX for expression. Thus, ComX is believed to be the regulator of natural competence in *Strep. pneumoniae*.

Advances in technology such as microarrays should permit the more ready identification of genes whose transcription is induced or repressed during competence. Indeed Rimini *et al.* (2000) have recently utilised such an approach. Many of the genes identified were previously known to be involved in competence but a number not previously known to be induced were also identified. It will now be possible to investigate these newly recognised genes, in order better to understand the regulation of competence.

Addition of CSP-1 to a non-competent growing culture of pneumococci induces competence within 15–20 minutes. This status lasts approximately 20 minutes, after which the cells become refractory to CSP for about one generation. Exogenous DNA taken up in this manner undergoes at least one endonucleolytic cleavage (Morrison and Baker, 1972) followed by reduction into single-strand form (Lacks, 1962). A cut end enters first with uptake following from 3′ to 5′ (Mejean and Claverys, 1993). Recombination then occurs and is mostly completed within 10 minutes at 37°C (Ghei and Lacks, 1967).

Apart from transformation, *Strep. pneumoniae* can also acquire new genetic information by infection with bacteriophage (transfection). Phage gain access to pneumococcal cells via adherence to choline residues in the cell wall (Lopez *et al.*, 1981). Phage-coded pneumococcal lysins and the host autolysin (LytA) described elsewhere in this chapter show levels of homology that suggest that the former amidase is derived directly from the latter (Romero *et al.*, 1990). Thus, genetic interchange between *Strep. pneumoniae* and bacteriophage may be an important factor in the evolutionary adaptation of pneumococci.

Phase Variation

Strep. pneumoniae undergoes spontaneous phase variation which can be observed phenotypically as changes in colony morphology (Weiser *et al.*, 1994). Colony morphology varies between an opaque and transparent phenotype. The variation has a range of effects in terms of the cell surface of the organism and virulence.

In an infant rat model of colonisation only transparent organisms colonise the respiratory tract to any degree (Weiser *et al.*, 1994). When adult mice were challenged by the intra-peritoneal route, animals receiving mostly opaque organisms died of sepsis and transparent organisms were largely avirulent. It therefore appears that transparent forms are adapted for colonisation, whereas opaque forms are adapted for events later in infection.

Saluja and Weiser (1995) isolated a fragment of DNA which, when transformed into a transparent organism, increased the appearance of the opaque phenotype. This 'opacity' operon contains two genes, *glpD* and *glpF*, which have homology to the glycerol regulon genes of other bacteria. Sequences with homology to the BOXA and BOXC intergenic elements and an open reading frame coding for a 126 amino acid protein with no homology to proteins in the database were also present. Strains lacking the BOX elements phase vary at lower rates, suggesting that these elements may be important in the phase-variation process.

Diseases Due to *Streptococcus pneumoniae*

Pneumococcal Pneumonia

Epidemiology
Pneumonia is the sixth leading cause of death in the United States and the most common cause of death from an infectious disease in the non-industrialised world (Garibaldi, 1985; Niederman *et al.*, 1993). Current estimates of mortality rates in children under the age of 5 are 2.7 million per year (World Bank, 1993). Pneumonia is also a major cause of death in the adult population. Every year an estimated 100 000 adults between the ages of 45 and 59 die from this infection, and there are 1.2 million deaths in the group over 60 years of age (Murray and Lopez, 1996). *Strep. pneumoniae* is believed to be responsible for most of these deaths in the industrialised and non-industrialised worlds. Estimates currently stand at 40–60% of cases being due to this one micro-organism (Fang *et al.*, 1990; Mulholland, 1999). One reason for the prevalence of pneumococcal pneumonia is the ubiquity of the bacteria; they colonise the nasopharynx of up to 70% of the community (Kemper and Deresinki, 1994). Such colonisation occurs early in life, and most infants are colonised with pneumococci on at least one occasion before they are 2 years old (Gray *et al.*, 1980). Carriage of one serotype of pneumococci does not protect against colonisation with other serotypes, since up to four serotypes may be carried by an individual at any one time (Tuomanen *et al.*, 1995).

Disease Progression
In the classic description of pneumococcal pneumonia there are four overlapping stages: congestion and oedema, red hepatisation, grey hepatisation and resolution. The arrival of bacteria in the alveoli results in congestion of the alveolar capillaries and an outpouring of serous fluid. This fluid acts as a nutrient source and a mode of transport for the bacteria outward via the pores of Kohn to adjacent alveoli (Harford and Hara, 1950). Capillary engorgement with erythrocytes follows and leads to red hepatisation when neutrophils, a few macrophages and many erythrocytes pass from capillaries via the interstitium to fill the alveolar spaces (Loosli and Baker, 1962). Large amounts of fibrin are present in the exudate. Within a few hours, as exudate continues to accumulate in the alveoli, capillaries are compressed, the red cell content reduced and the leucocyte content increases (McKinsey and Bisno, 1980). Intravascular fibrin deposition causes a decline in perfusion and results in grey hepatisation (Kline and Internitz, 1915). Pneumococci that manage to evade the immune response spread into the interstitial lung tissues and then to the bloodstream, often with serious prognostic implications. Some time may pass before disease progresses to this stage, but most deaths from bacteraemic pneumococcal pneumonia occur within the first 5 days of illness (Sato *et al.*, 1998). Splenectomised individuals may succumb within 18–24 hours without symptoms (Musher, 1992).

The cause of death in pneumococcal pneumonia remains a puzzle. Death may occur even after antibiotic treatment has begun, pneumonia is clearing and the previously infected site is sterile (Austrian and Gold, 1964). A more detailed understanding of the roles played by host factors in the pathogenesis of the disease may be expected to help answer this question and aid the search for more effective control measures.

It is striking that the lungs of patients that recover fully from pneumococcal pneumonia show no permanent damage despite the massive inflammatory response they have endured (Catterall, 1999). Resolution is believed to occur after the appearance of anti-capsular antibodies which facilitate efficient phagocytosis. Later, neutrophils must be removed from the airways and the lungs. The widespread belief is that the neutrophils undergo necrosis before engulfment and removal by macrophages. More recent results indicate that neutrophils become apoptotic rather than necrotic (Haslett, 1992, 1997). With such 'programmed cell death' neutrophils retain their granule contents and membrane integrity but no longer secrete their contents in response to external stimuli (Haslett, 1992). In addition, macrophages that have ingested apoptotic neutrophils do not release pro-inflammatory mediators; these macrophages are then cleared by mucociliary transport. Deposited fibrin acts as a provisional matrix into which fibroblasts migrate in order to repair the damaged lung interstitium (Simon and Paine, 1995). Type II epithelial cells also spread over this matrix before proliferating to repair damaged epithelia.

The inflammatory cytokines IL-1β and TNFα act as primers for alveolar epithelial type II cell chemotaxis, as measured by passage through pored filters, when the cells were pre-incubated with these cytokines before addition of epidermal growth factor, a known type II cell-chemokine. In contrast, IL-6 regulates the chemotaxis in a negative manner. After fibroblast and epithelial cell migration, the fibrin scaffolding is removed by plasminogen activators (Bertozzi *et al.*, 1990). Clearance of fluid, proteins and debris occurs by active transport mechanisms, lymphatic drainage and phagocytosis by macrophages. Since the alveolar walls are not destroyed during pneumococcal pneumonia (McKinsey and Bisno, 1980), the pulmonary architecture returns to normal when the infection has resolved.

Predisposing Factors

Despite the high carriage rates, comparatively few individuals develop pneumococcal pneumonia. Many factors are responsible for determining the outcome of colonisation and these can generally be separated into physiological, non-immunological and immunological (**Table 4**).

Age People at the extremes of age are most likely to suffer from pneumococcal disease. The highest incidences are found in children between 18 and 24 months; infections are rarer in teenagers and young adults but increase with middle age to peak again in individuals in their 70s and 80s (Austrian and Gold, 1964; Burman *et al.*, 1985). This age distribution is also associated with variations in the frequency of capsular types identified in clinical cases. Danish types 6B, 19F and 14 are most likely to cause disease in children whereas 1, 2 and 3 are more common in adults (Leach *et al.*, 1997).

In infants the immature state of B cells, which lack the CD40 surface antigen required for IgG antibody production during T-cell-dependent responses, is responsible for their poor response to pneumococcal polysaccharide (Dullforce *et al.*, 1998). This deficiency also increases their susceptibility to other encapsulated bacteria, such as *Haemophilus influenzae*.

The increased incidence of pneumococcal pneumonia in the elderly population is not understood. Environmental factors, nutritional status and underlying infections undoubtedly play a role but the immune response itself is also likely to be involved. Immune responses are altered in elderly individuals with the major effects of ageing found in the cell-mediated arm of immunity. The most well-known deficiency is the loss of thymic tissue which leads to an elevated number of immature T cells within the circulation (Simons and Reynolds, 1990). The antibody response is also diminished, since antibody levels 2 weeks after immunisation with pneumococcal capsular polysaccharide are significantly lower in the elderly than in young adults (Ammann *et al.*, 1980). The reduced response of the elderly population to pneumococcal vaccination has been studied *in vitro*. If vaccine is added to cultures of spleen cells from young and old mice, cells from young mice generate significantly more vaccine-specific plaque-forming cells (Garg *et al.*, 1996). This was not due to elevated T-cell-mediated suppression but rather to reduced accessory cell function, an effect that could be overcome by the addition of exogenous IL-1. IL-1 secretion diminishes with age, since as assessed by the ability to induce thymocyte proliferation, peritoneal macrophages and T cells from 2-year-old mice have less IL-1 activity than cells from 4-month-old animals (Inamizu *et al.*, 1985). IL-1 stimulates B lymphocytes to proliferate, increases antibody secretion and induces IL-2 production, which in turn causes T lymphocyte proliferation. Thus, with reduced expression of IL-1 by accessory cells, such as macrophages, from aged individuals, a poorer response to vaccination is inevitable.

The innate immune defences also diminish with age. Phagocytosis of zymosan-opsonised yeast particles is significantly impaired in elderly humans as compared with young individuals. Such phagocytic impairment is

Table 4 Host factors that predispose individuals to pneumococcal infection

Type of factor	Factor
Physiological	Age under 2 years or over 65 years
	Male gender
Non-immunological	Disruption of bronchial epithelium, influenza, smoking
	Ethanol, narcotic intoxication
	Decreased vascular perfusion, e.g. sickle cell anaemia or congestive heart failure
	Cold weather
Immunological	Immunocompromised status, e.g. HIV-positive
	Phagocyte abnormalities: neutropenia, hyposplenia
	Hypogammaglobulinaemia
	Specific antibody deficiency
	Complement deficiency, C2, C3

Adapted from Burman *et al.* (1985), Johnston (1981).

evident in subjects in their 30s when compared with those in their teens and 20s. This effect becomes progressively more evident and significant in later decades (Emanuelli *et al.*, 1986).

In contrast to these immune deficiencies, inflammatory responses are elevated in the elderly. This effect has been noted in patients suffering from pneumococcal infection (Bruunsgaard *et al.*, 1999). This effect is likely to represent increased production of inflammatory cytokines such as TNFα (Foster *et al.*, 1992). This indicates that a decrease in neutrophil recruitment and activity or a reduced antibody response may be responsible for the increased susceptibility to pneumococcal pneumonia in older individuals. The pathology may be worse in this population because of inappropriate inflammatory responses, which may result from impaired lymphocyte regulation.

HIV Status Pneumococcal infection occurs between 5- and 17-fold more frequently in HIV-positive populations, making *Strep. pneumoniae* the most commonly isolated respiratory pathogen in HIV-positive individuals (Gilks *et al.*, 1996; Moore *et al.*, 1998). Furthermore, the risk of bacteraemic pneumococcal pneumonia is increased approximately 100-fold (Redd *et al.*, 1990), with a 57% mortality rate in AIDS patients as compared with 22–39% in other populations (Pesola and Charles, 1992). This problem is compounded by the increased incidence of antibiotic-resistant strains of pneumococci in HIV-positive patients (Paul *et al.*, 1995). Since most antibiotic-resistant strains of pneumococci are of the serotypes covered by the present vaccine, it would suggest that the problem could be overcome by widespread vaccination of HIV-positive individuals. Unfortunately, poor vaccine efficacy has frequently been reported. Individuals with CD4 counts below 500 are far less likely to respond to vaccination in terms of anti-pneumococcal IgG titres against serotypes 3, 4, 6A, 8 and 23F (Rodriguez-Barradas *et al.*, 1992). Furthermore, HIV-seropositive individuals who respond well to vaccination may again have very low antibody levels 12 weeks after immunisation (Loeliger *et al.*, 1995).

Sex The incidence of pneumococcal infection in humans is greater in males than in females, with a ratio of 2.1:1 (Burman *et al.*, 1985) but the reason for this difference is unknown. Males do not have a higher pneumococcal carriage rate than females and males do not appear to be colonised by more virulent capsular types than females, since similar capsular types can be isolated from males and females in cases of invasive pneumococcal infection (Scott *et al.*, 1996). This indicates that a defect in the male immune response increases their susceptibility to pneumococcal infection. Indeed it is widely recognised that in most species males have weaker immune responses than females.

Sex differences in antibody levels against pneumococcal capsular polysaccharide have been documented, with males responding less than women to vaccination (Roghmann *et al.*, 1987). Most data in this connection indicate that hormonal control of the immune response is responsible for these differences. By using mouse strains that vary in their sensitivity to, rather than circulating levels of androgen, Cohn (1986) showed that the response to pneumococcal polysaccharides is mediated solely by this one hormone. Male mice with high tissue responses to androgen, as measured by seminal vesicle growth, had lower antibody levels after immunisation than females of the same strain or males from strains with low sensitivity. The reason for the mediation of immune responses by androgen is unclear. It may be that androgen-mediated responses directly affect the immune response. Alternatively, the androgen response may exert its effects via intermediaries such as cytokines. The effects of this sensitivity to androgen on the inflammatory response during pneumococcal pneumonia would therefore be an interesting future direction for research.

Other Genetic Factors It has been noted that adoptees have a 5-fold greater risk of dying from an infectious disease if a biological parent died from the same infection; 29% of the infections were pneumonia (Sorensen *et al.*, 1988). The risk was not increased if an adoptive parent died from an infection. These facts indicate that a strong genetic component determines the outcome of infectious diseases and that environmental factors play a minor role.

Recognition of the above has led to a search for the host genetic factors that might be responsible. A locus on mouse chromosome 1, termed *Bcg*, *Lsh* or *Ity*, is important in controlling infections due to intracellular pathogens, such as *Mycobacterium* spp., *Listeria monocytogenes* and *Salmonella typhimurium* (Skamene *et al.*, 1979; Forget *et al.*, 1981; Plant *et al.*, 1982). *Ity* encodes natural resistance macrophage protein 1 (Nramp1) which affects the ability of macrophages to destroy ingested intracellular parasites early in the infectious process (Vidal *et al.*, 1993). The human homologue NRAMP 1 has been identified and, unlike the murine gene, it is expressed in the lungs, liver and spleen, but it is most abundant in peripheral blood neutrophils (Cellier *et al.*, 1997). Although the function of human *Nramp1* has been far less characterised than the mouse gene, it is also associated with susceptibility to *Mycobacterium* spp. (Bellamy *et al.*, 1998).

More recently, mutations in the human IFNγ receptor I have also been shown to mediate susceptibility to infections with *Mycobacteria* (reviewed by Abel and Dessein, 1997). IFNγ is a potent macrophage activator and appears to be required for development of optimal macrophage anti-mycobacterial activity. In addition, still unidentified host factors influence susceptibility to infections due to *Pseudomonas aeruginosa* (Morissette *et al.*, 1995) and *Candida albicans* (Ashman and Papadimitriou, 1992).

Notably missing from the above list of pathogens are extracellular bacteria. So far, candidate genes for resistance to such organisms have not been identified. During the 1930s it was recognised that the pathology of pneumococcal pneumonia depends on the strain of mouse and the capsular serotype of the pneumococcus, suggesting a host genetic component of disease susceptibility (Schütze *et al.*, 1936). Such findings have also more recently been described by Briles *et al.* (1986). At present the nature of the host factor(s) involved in determining the outcome to pneumococcal infection is unknown. For this reason we established models of pneumococcal pneumonia in a range of inbred strains of mice that are either genetically resistant or susceptible to the infection, with the long-term goal of identifying possible genes involved (Gingles *et al.*, 2001). BALB/c mice were found to be resistant and CBA/Ca mice susceptible to intra-nasal infection with 10^6 cfu *Strep. pneumoniae*. Measurement of bacterial loads after infection showed that BALB/c mice survive the infection by control of the organisms in their lungs and preventing their spread to the bloodstream. In contrast, pneumococci proliferated unchecked in the lungs and circulation of CBA/Ca mice, resulting in a median survival time after challenge of only 27 hours.

Analysis of the host cellular response to infection in whole-lung homogenates revealed that both the total number of leucocytes and the number of neutrophils recruited to BALB/c lungs 12 hours and 24 hours after challenge were significantly higher than those in CBA/Ca mice. This difference was also evident in histological preparations, in which inflammatory lesions were visible earlier and to a greater extent in BALB/c lungs than in CBA/Ca lungs. Thus, the inflammatory response is up-regulated earlier and to a greater extent in mice that are genetically resistant to pneumococcal pneumonia. As will be described below, cell influx is initiated and controlled by a range of host inflammatory mediators secreted by resident lung cells. We, therefore, suggest that differences in the ability of inbred strains of mice to produce these inflammatory mediators may explain their susceptibility or resistance to pneumococcal pneumonia.

Meningitis

Acute bacterial meningitis is a common and much-feared disease with significant mortality. Despite dramatic improvements in antibiotic therapy and supportive care the disease remains a major problem and around 10% of survivors of the infection are left with a neurological deficit, ranging from sensorineural deafness to hemiplegia. *Neisseria meningitidis*, *Streptococcus pneumoniae* and *Haemophilus influenzae* are responsible for the majority of cases of bacterial meningitis, respectively with case fatality rates in the USA of 10%, 26% and 6% (Schlech *et al.*, 1985). The introduction of the new conjugate vaccine has reduced the incidence of meningitis associated with *H. influenzae*.

Common pre-disposing factors for pneumococcal meningitis are pneumonia, sinusitis, middle ear pathology and cerebrospinal fluid leak due to injury. In the UK during the period 1982–1992 there were 22 567 reports of pneumococcal bacteraemia and 3500 reports of pneumococcal meningitis. The annual mean incidence of pneumococcal meningitis in the UK during the period 1989–1992 was 8.7 per 100 000 population (Askenasy *et al.*, 1995). In the US the reported incidence is about 1.1 per 100 000 population but it is 30 per 100 000 in infants under 5 months of age (Schlech *et al.*, 1985). Globally, pneumococcus causes over 1 million deaths per year in children under the age of 5 years (Obaro *et al.*, 1996). Pneumococcal meningitis occurs in epidemics in sub-Saharan Africa and in some areas it is the most common form of meningitis.

The exact pathophysiological mechanisms of pneumococcal meningitis remain unclear, but bacterial and host factors both clearly play a role. Much of the neuronal loss during this disease occurs from the dentate gyrus of the hippocampus, a site associated with cognitive function. The role of the host inflammatory response is illustrated by the finding that down-regulation of inflammation by steroids, cytokine antagonists or inhibitors of leucocyte recruitment decrease neurological complications in both experimental systems and clinically (Tuomanen *et al.*, 1989; Saez-Llorens *et al.*, 1990; Saukkonen *et al.*, 1990). Several factors of both bacterial or host origin are potentially neurotoxic in bacterial meningitis. *Strep. pneumoniae* can be directly toxic to neuronal tissue (Kim *et al.*, 1995) and inflammatory mediators are up-regulated during infection (Quagliarello and Scheld, 1992; Saez-Llorens *et al.*, 1990). Possible mechanisms of toxicity involve excitatory amino acids, reactive oxygen and nitrogen species, cytokines and chemokines, prostaglandins and complement (Braun and Tuomanen, 1998). Apoptosis is the chief mechanism of neuronal loss in pneumococcal meningitis (Braun et al., 1999b) and this can be inhibited by treatment with the caspase inhibitor N-benzyloxycarbonyl-Val-Ala-Asp-fluoromethylketone (z-VAD-fmk). Exposure of human neuronal cells to infected cerebrospinal fluid in vitro also induces apoptosis, whether or not bacteria are present. The potential role of pneumococcal virulence factors in this process is discussed below.

Otitis Media

Otitis media is predominantly a childhood disease in which fluid accumulates in the middle ear space and inflammation occurs in the surrounding mucosa. Almost 95% of children will have experienced at least one attack of acute otitis media by 3 years of age and up to half of cases are due to Strep. pneumoniae (Teele et al., 1989), which makes the pneumococcus the most frequent cause of otitis media in infants and children. Most cases resolve spontaneously in 3–4 weeks but occasionally complications, such as meningitis and bacteraemia, may occur (Poole, 1995).

The fluid that accumulates during otitis media may be serous, purulent or mucoid. In acute otitis media the fluid is purulent in some 60% of cases. In cases of chronic otitis media the fluid is mucous in almost 50%, serous in 10% and purulent in less than 10% (Giebink, 2000). In serous otitis media, fluid accumulates below the middle ear epithelium, while in purulent otitis media large numbers of inflammatory cells are also present. In the mucoid condition metaplasia of goblet cells results in fluid with small numbers of inflammatory cells but high levels of mucous strands.

A chinchilla model has been used extensively to elucidate the pathophysiology of otitis media (Sato et al., 1999). Inoculation of viable pneumococci into the middle ear results in a localised infection. This causes recruitment of polymorphonuclear lymphocytes into the sub-epithelial space of the middle ear mucosa. During the first week of infection this is accompanied by epithelial metaplasia, sub-epithelial oedema and an increase in middle ear fluid volume. This is followed by gradual resolution of inflammation over the next 8–10 weeks. As in the case of other pneumococcal infections, pneumococcal replication in the middle ear is not necessary to initiate the inflammatory response, since inoculation of non-viable pneumococci results in acute middle ear inflammation (Lowell et al., 1980). Various pneumococcal preparations, including capsule, peptidoglycan and teichoic acids, have been

tested for their ability to induce inflammation (Ripley-Petzoldt et al., 1988; Carlsen et al., 1992). Native cell wall results in the highest levels of inflammatory cell recruitment, lysozyme accumulation and epithelial metaplasia. Thus, cell wall appears to be the principal factor that initiates the middle ear inflammatory response during pneumococcal otitis media.

Host factors have also been implicated in the pathogenesis of pneumococcal otitis media. The release of toxic oxygen species, including hydrogen peroxide, from activated neutrophils in the middle ear is thought to affect epithelial cell metabolism. Incubation of chinchilla primary middle ear epithelial cell cultures with activated neutrophils results in reduced epithelial viability as judged by [^{3}H]thymidine incorporation (Kawana et al., 1994). Thus, neutrophil reactive oxygen species may contribute to epithelial metaplasia in otitis media.

Inappropriate production of cytokines may also be detrimental in otitis media. A range of inflammatory cytokines including IL-1β, IL-6, IL-8 and TNFα have been detected in the middle ear fluid of otitis media patients and in the chinchilla model (Himi et al., 1992; Sato et al., 1999). TNFα and IL-1β increase the adherence of Strep. pneumoniae to chinchilla tracheal epithelium (Tong et al., 1999). These cytokines can cause direct disruption to host tissues and induce inflammatory cell influx (Lukacs and Ward, 1996; Stephens et al., 1987). Indeed, TNFα, IL-1β, IL-6 and IL-8 concentrations correlate significantly with middle ear fluid neutrophil numbers in the chinchilla otitis media model (Sato et al., 1999). Furthermore, incubation of primary chinchilla middle ear epithelial cells with TNFα induces mucous glycoprotein secretion (Giebink, 2000) and may therefore be a trigger of mucoid otitis media.

Mechanisms of Pathogenesis

Pneumonia

Pneumococcal pneumonia is believed to occur after spread of the bacteria from the nasopharynx to the lungs by aspiration (Musher, 1992). In the alveoli the organisms are thought to attach preferentially to type II pneumocytes (Cundell and Tuomanen, 1994). Two classes of receptors are expressed on resting pneumocytes: N-acetyl-D-galactosamine linked by either β1,3 or β1,4 to galactose (GalNAcβ1–3Gal and GalNAcβ1–4Gal) (Cundell et al., 1995a). Only 0.1% of pneumococci are capable of attaching to these receptors on resting cells (Tuomanen and Masure, 1997). Activation of pneumocytes leads to the expression of a novel receptor *N*-acetylglucosamine (GalNAc), the platelet-activating factor (PAF) receptor (Cundell *et al.*, 1995a). The appearance of this receptor results in a rapid enhancement of pneumococcal adhesion (Tuomanen and Masure, 1997).

The site of pneumococcal invasion still remains unclear. Pneumococci may directly traverse the respiratory epithelium but the route taken to the bloodstream may be via the lymphatics. Adherence is thought to be followed by bacterial internalisation into host cell vacuoles by receptor-mediated endocytosis (Cundell and Tuomanen, 1994). Pneumococci are then able to pass through these cells and gain access to the bloodstream. The mechanism of cell transcytosis has been investigated *in vitro* by Ring *et al.* (1998). Bacteria enter vacuoles that traverse the cell over a few hours. The number of viable pneumococci in the cells decreases over time because of intracellular killing, transcytosis or recycling of organisms back to their surface of origin. A large proportion of the organisms that enter vacuoles do so by the PAF receptor. Transcytosis is restricted to the transparent phase variants of the organism and is dependent on CbpA. Any opaque variants that enter vacuoles are destroyed. Once it has bound ligand, the PAF receptor is rapidly internalised and is then re-cycled back to the cell surface. It seems the pneumococcus may exploit this pathway to penetrate cell barriers. CbpA also mediates translocation of pneumococci via the polymeric immunoglobulin receptor (Zhang *et al.*, 2000). The mechanism of this process is discussed in relation to CbpA as a virulence factor (see below).

Meningitis

Pneumococcal meningitis causes damage to neuronal tissue mediated by the pneumococcus and the host inflammatory response. Apoptosis appears to be the chief mechanism of neuronal loss during pneumococcal meningitis. Virulence factors that may be involved in the process include pneumolysin and hydrogen peroxide. The role of these virulence factors is discussed below.

Role of Virulence Factors

Pneumococci produce several factors that may be involved in the pathogenesis of disease. These include structural components, surface proteins and toxins (**Fig. 1**). Publication of the DNA sequence of *Strep. pneumoniae* by the Institute for Genomic Research in November 1997 (http://www.tigr.org/tdb) will aid the identification of novel virulence factors by allowing comparison of sequences of known virulence factors from other organisms with the *Strep. pneumoniae* sequence.

Virulence of pneumococci is lost with repeated passage *in vitro*, which makes periodic passage of the bacteria through mice necessary for studies *in vivo* (Johnston, 1991). This might reflect alterations in some of these virulence factors.

Capsule

One of the oldest recognised virulence factors is the polysaccharide capsule and proof of the importance of the capsule to virulence was established in the 1930s. Avery and Dubos (1931) found that an enzyme obtained from a soil bacillus could remove the serotype 3 pneumococcal capsular polysaccharide and that this treatment afforded mice protection from otherwise fatal challenge with *Strep. pneumoniae*.

The role of the capsule in virulence stems from its anti-phagocytic activity (Jonsson *et al.*, 1985). In the non-immune host, antibody to cell wall constituents, which is ubiquitous in adults, becomes attached to the surface of the organisms and in turn binds complement. The presence of the capsule prevents iC3b and the Fc of immunoglobulins on the bacterial cell surface from interacting with receptors on phagocytic cells, with the result that the organisms remain extracellular (Musher, 1992). The role of the capsule in virulence is determined not only by its size but also by its chemical composition. Pneumococci of types 3 and 37 have similar capsules of similar size but the former is highly invasive in animal models whereas the latter is relatively avirulent (Lee *et al.*, 1991).

Pneumococci are the prototypic extracellular pathogens and engulfment of the organisms into a phagocytic vacuole is sufficient to kill them. Granulocytes from patients suffering from chronic granulomatous disease, which are unable to generate hydrogen peroxide and other microbicidal oxidants, retain anti-pneumococcal activity (Johnston and Newman, 1977). The explanation is that since *Strep. pneumoniae* is catalase-negative it is unable to break down its own hydrogen peroxide (Johnston, 1991).

Cell Wall and its Role in Virulence

The cell wall is important in mediating attachment of pneumococci to activated lung cells (Cundell *et al.*, 1995a). The phosphorylcholine of the pneumococcal cell wall binds to the receptor for platelet-activating factor (PAF). The PAF receptor is up-regulated during the inflammatory response and during viral infection,

Table 5 Effects of low levels of pneumolysin relevant to pneumococcal infection

Target	Effect	Reference
Monocytes	Inhibits respiratory burst	Nandoskar *et al.* (1986)
	Decreases anti-pneumococcal activity	Nandoskar *et al.* (1986)
	Stimulates pro-inflammatory cytokine production	Houldsworth *et al.* (1994)
Neutrophils	Inhibits respiratory burst and migration	Paton and Ferrante (1983)
Pulmonary epithelial cells	Disrupts integrity	Rubins *et al.* (1993)
	Inhibits cilial beating	Steinfort *et al.* (1989)
Pulmonary endothelial cells	Disrupts integrity	Rubins *et al.* (1992)
B cells	Decreases antibody production	Boulnois *et al.* (1991)
Peripheral blood lymphocytes	Inhibits proliferation and antibody production	Ferrante *et al.* (1984)

which may explain the increased occurrence of pneumococcal pneumonia following viral infection.

The cell wall plays a significant role in initiating the inflammatory response that occurs during pneumococcal pneumonia. The minimum cell wall component capable of initiating an inflammatory response has been identified as teichoic acid attached to the glycan backbone of peptidoglycan and some small-stem peptides (Tuomanen *et al.*, 1985). By binding to pulmonary epithelia and causing separation of tight junctions, cell wall components are likely to increase the permeability of alveolar epithelia (Tuomanen *et al.*, 1985). This in turn permits influx of inflammatory cells that may be recruited by the ability of cell wall teichoic acids to activate the alternative complement pathway (Winkelstein and Tomasz, 1978). Pro-inflammatory cytokine production, such as TNFα and IL-1, is also implicated in this cell recruitment. Nitric oxide (NO) is another pro-inflammatory mediator released by incubation with cell wall preparations (Orman *et al.*, 1998). NO is also implicated in up-regulation of the inflammatory response and is capable of causing direct damage to the lungs.

Pneumolysin

Pneumolysin is an intracellular protein of the thiol-activated toxin family released by the pneumococci on lysis (Boulnois *et al.*, 1991) and is expressed by all pathogenic strains of *Strep. pneumoniae*. It is released *in vivo*, since anti-pneumolysin antibodies can be detected during infection (Kemper and Deresinki, 1994) and pneumolysin itself can be detected with fluorescent antibodies on lung sections from infected mice (Canvin *et al.*, 1995b).

The ability of pneumolysin to act as a virulence factor is indicated by the reduced virulence of mutant bacteria unable to produce the toxin (Berry *et al.*, 1989b). Mice infected with pneumolysin-deficient *Strep. pneumoniae* survive significantly longer than those infected with pneumolysin-sufficient organisms. These former are less able to survive and multiply in the airways and are incapable of multiplying to the same high levels in the bloodstream as wild-type pneumococci (Kadioglu *et al.*, 2000). This virulence factor is capable of exerting differing effects on the immune response, depending on its concentration. At lower concentrations (~1 ng/mL) pneumolysin is capable of a range of activities (Table 5). At higher concentrations (~10 μg/mL) the toxin binds to cholesterol in the target cell membrane, inserts into the membrane, undergoes oligomerisation and forms transmembrane pores, thus disrupting membrane integrity and causing cell lysis (Paton, 1996).

These higher concentrations of pneumolysin are also implicated in the inflammatory response of pneumococcal pneumonia. Injection of recombinant toxin into the ligated bronchi of rats induces a pattern of inflammation similar to the host response seen during infection with virulent *Strep. pneumoniae*. The lung distal to the site of ligation becomes consolidated with inflammatory cells and there is marked disruption to lung architecture (Feldman *et al.*, 1991). Furthermore, *in vitro* pneumolysin activates the classical complement pathway in the absence of pneumolysin-specific antibody, thus initiating an inflammatory response in the fluid layer, which diverts the response from intact pneumococci (Paton *et al.*, 1984). In addition, opsonisation with pneumolysin-treated human serum significantly reduces phagocytosis of the bacteria by neutrophils *in vitro*. Thus, complement activation correlates with significant inhibition of serum opsonic activity. The mechanism by which pneumolysin exerts this effect is unknown but is believed to involve non-specific-binding of IgG Fc by the toxin (Mitchell *et al.*, 1991).

Pneumolysin stimulates human monocytes to release TNFα and IL-1β, which have potent pro-inflammatory activities (Houldsworth *et al.*, 1994). Pneumolysin also stimulates nitric oxide production and up-regulates TNFα and IL-6 production by macrophages (Braun *et al.*, 1999a). Lower levels of TNFα, IL-6 and IFNγ are released during infection with pneumolysin-deficient *Strep. pneumoniae* mutants than by wild-type bacteria, providing indirect evidence of the inflammatory role played by pneumolysin *in vivo* (Benton *et al.*, 1998). Pneumolysin can also stimulate production of phospholipase A (Rubins *et al.*, 1994) and COX-2 expression (Braun *et al.*, 1999a) by pulmonary artery endothelial cells. Phospholipases release products of the arachidonic acid pathway which are involved in neutrophil recruitment and activation of inflammatory cells during pneumococcal pneumonia (Tuomanen *et al.*, 1987). In addition, phospholipase A degrades phospholipids in cell membranes, which may aid the spread of the pneumococci.

Microscopy of cells treated with pneumolysin *in vitro* indicates that the toxin can also mediate direct damage to host cells. Bronchial and alveolar epithelial cells as well as pulmonary endothelial cells are disrupted (Steinfort *et al.*, 1989; Rubins *et al.*, 1992, 1993). In these inflammatory activities pneumolysin may not act in isolation, since it may also potentiate the inflammatory effects of cell wall products. Such synergy has previously been documented for the α toxin of *Staphylococcus aureus* (Bhakdi *et al.*, 1989).

Finally, infection with pneumococci deficient in the ability to synthesise pneumolysin induces significantly less pulmonary inflammation than infection with wild-type bacteria. A pneumolysin-negative mutant of the pneumococcus termed PLN-A has been constructed (Berry *et al.*, 1989b) and this has been used extensively in the investigation of the role of the toxin in disease. Although the histopathological changes that occur in the lungs of mice infected with pneumolysin-deficient bacteria are similar to those occurring during wild-type infection, the kinetics are delayed and the severity is reduced (Canvin *et al.*, 1995a). In addition, we have characterised the pulmonary cellular response during infection with pneumolysin-deficient bacteria (Kadioglu *et al.*, 2000). After intra-nasal infection of MF1 mice with 10^6 cfu, these pneumococci were unable to attain the same bacterial loads in the lungs and bloodstream. Thus, the presence of pneumolysin is essential for bacterial multiplication in the alveoli, especially during the first 6–8 hours of infection.

It has been shown that instillation of recombinant pneumolysin concurrently with pneumolysin-deficient bacteria returns the pattern of bacteria growth to that of wild-type organisms (Rubins *et al.*, 1995), proving that it is the loss of this single virulence factor that is responsible for reduced pneumococcal proliferation. In addition, mice infected with pneumolysin-deficient

bacteria do not produce symptoms of infection as marked as the wild-type bacteria. Cell recruitment in response to infections with pneumolysin-deficient pneumococci is delayed and reduced in comparison with infection with wild-type bacteria (Kadioglu *et al.*, 2000). Neutrophil responses were the most affected and the re-distribution of T and B lymphocytes in and around inflamed bronchioles was also delayed and reduced. Whether this was due to a poorer inflammatory stimulus by the reduced bacterial loads in the lungs and bloodstream of mice infected with pneumolysin-deficient bacteria or to direct pneumolysin damage by wild-type bacteria is unknown.

The role of pneumolysin in pneumococcal eye infections has been investigated. When a deletion mutant that lacks pneumolysin was constructed and used in a rabbit model of ocular infection virulence was greatly reduced (Johnson *et al.*, 1990). A non-haemolytic pneumococcus produced by chemical mutagenesis, which probably had a point mutation in the pneumolysin gene, had the same virulence characteristics as the parent strain. This suggested that an activity of pneumolysin other than its lytic function was important in the role of the toxin during pathogenesis of ocular infection. This was confirmed with mutants of the toxin devoid of various activities (Johnson *et al.*, 1995).

It is clear, therefore, that pneumolysin contributes to the virulence of the pneumococcus in models of pulmonary, systemic and ocular infections. Its role in some other diseases is less clear. Use of toxin-positive and toxin-negative isogenic pneumococci in the chinchilla otitis media model showed that the amount of inflammation induced was similar (Sato *et al.*, 1996). It should be pointed out that these studies were carried out with a serotype 3 organism, whereas the studies described above with PLN-A were done with a serotype 2 strain. It may be that the contribution of pneumolysin to virulence differs between serotypes. In a study of the role of pneumolysin in pathogenesis of pneumococcal meningitis it was found that although direct intra-cisternal injection of pneumolysin into rabbits causes a rapid inflammatory response, there was no evidence of a contribution of the toxin to the inflammation caused by the whole organism as determined by infection with wild-type and PLN-A pneumococci (Friedland *et al.*, 1995). Therefore, although pneumolysin can stimulate the inflammatory cascade in the central nervous system, this is not necessary for the pathogenesis of meningeal inflammation. It has also been shown that the toxin does not play a role in post-antibiotic enhancement of meningeal inflammation (Friedland *et al.*, 1995). Use of an *in vitro* system has shown that pneumolysin is toxic to brain tissue and that the toxin acts in synergy with hydrogen peroxide (Hirst *et al.*, 2000). The toxin may therefore play a role in direct toxicity to brain tissue.

The findings with regard to inflammation were largely confirmed by Winter *et al.* (1996) who used a guinea-pig model of pneumococcal meningitis. There was no difference in the degree of inflammation induced by wild-type and PLN-A pneumococci in this model. There was, however, less protein influx into the cerebrospinal fluid of animals infected with PLN-A. This may be related to the decreased effect PLN-A is known to have on other cells. Although no difference in inflammation could be detected, PLN-A induced substantially less ultrastructural damage to the cochlea of infected animals and there was less associated hearing loss. The ultrastructural damage after infection with wild-type pneumococci was very similar to that induced by micro-perfusion of the cochlea with purified pneumolysin (Comis *et al.*, 1993) but ultrastructural changes were absent after meningitis due to PLN-A. Thus, in this experimental model, pneumolysin appears not to contribute to inflammation but most of the cochlear damage and hearing loss is due to pneumolysin. This also challenges the dogma that damage to the cochlea is due to bystander damage due to the inflammatory response.

Immunisation with inactivated pneumolysin leads to enhanced survival after subsequent challenge with pneumococci (Paton *et al.*, 1983). Survival times of mice subsequently infected with 5×10^6 cfu *Strep. pneumoniae* are increased from 2.48 days for untreated mice to 5.52 days in immunised animals. Although this effect is not as marked as the protection afforded by anti-capsule antibodies, protection is not serotype-specific. Protection against subsequent infection with at least nine different pneumococcal serotypes had previously been shown in mice immunised with pneumolysin toxoid prepared from type 2 pneumococci (Alexander *et al.*, 1994). For these reasons, pneumolysin has been proposed as a promising antigen for use in novel vaccines.

Neuraminidase

All clinical isolates of *Strep. pneumoniae* produce surface-associated neuraminidase, which cleaves *N*-acetylneuraminic acid from glycolipids, lipoproteins and oligosaccharides on cell surfaces and in body fluids (Camara *et al.*, 1994). This may cause direct damage to the host or it may unmask potential binding sites for the organism. Loss of sialic acid due to neuramidase activity accompanies the advance of pneumococci up the eustachian tube to the middle ear (Linder *et al.*, 1994). Immunisation of mice with

formaldehyde-inactivated neuraminidase significantly increases survival times of mice subsequently challenged with *Strep. pneumoniae* (Lock *et al.*, 1988) but not to the extent seen with pneumolysin or capsular polysaccharide.

The pneumococcus has genes for the production of at least three neuraminidase enzymes (see above) and the role of neuraminidase A in infection has been investigated. It plays a role in nasopharyngeal colonisation and development of otitis media in the chinchilla model (Tong *et al.*, 2000) but does not play a role in meningitis-associated deafness (Winter *et al.*, 1996).

Hyaluronidase

Hyaluronidase, which breaks down the hyaluronic acid component of mammalian connective tissue and extracellular matrix, is secreted by 99% of clinical isolates (Humphrey, 1948). This degradation may aid bacterial spread and colonisation, as shown for hyaluronidase of other micro-organisms. In the case of *Treponema pallidum*, hyaluronidase induces vascular leakage and dissemination of the organism (Fitzgerald and Repesh, 1987). In addition, hyaluronidase may potentiate pulmonary inflammation during pneumococcal pneumonia by complex interactions with pro-inflammatory cytokines and chemokines. TNFα and IL-1β are capable of inducing the production of hyaluronic acid by fibroblasts *in vitro* (Irwin *et al.*, 1994). Hyaluronic acid can promote further cytokine secretion by binding to CD44 on host cells. The system is further complicated by the ability of IL-1 to act in combination with high levels of hyaluronic acid to release hyaluronidase, thereby degrading hyaluronic acid and surrounding connective tissue.

Breakdown products of hyaluronic acid stimulate chemokine expression by macrophages *in vitro* (McKee *et al.*, 1996). Similar expression *in vivo* would induce cell recruitment and potentiate inflammation. Furthermore, these degradation products are also capable of increasing the phagocytosis of IgG-opsonised latex beads and chemotaxis of granulocytes *in vitro* and *in vivo* in response to casein (Hakansson *et al.*, 1980). Thus, the expression of hyaluronidase by *Strep. pneumoniae* during pneumococcal pneumonia might disrupt normal hyaluronic acid metabolism leading to direct tissue disruption, up-regulation of pro-inflammatory cytokines and cell recruitment.

Surface Proteins

Although the receptors to which pneumococci adhere on human cells have been identified, pneumococcal adhesins are largely unknown. Cundell *et al.* (1995b) found that mutations in genes identified as peptide permeases reduce pneumococcal adherence to the receptors expressed on resting lung or endothelial cells. These permeases are believed to bind and transport small peptides across the bacterial membrane, presumably to be used in metabolism. A homologue of these permeases, PsaA, has been ascribed a role in pneumococcal virulence (Talkington *et al.*, 1996), and immunisation of mice with PsaA before intravenous infection with 450 cfu of type 3 pneumococci results in 75% survival, whilst un-immunised mice do not survive. Novak *et al.* (1998) showed that PsaA and three homologues, PsaB, PsaC and PsaD, are manganese transporters. Mutations upstream of the genes for these permeases reduced pneumococcal adherence to lung epithelial cells *in vitro*. Rather than being directly responsible for pneumococcal adhesion, it is now believed that these permeases act in a regulatory manner to alter adhesion by acting on Cbp expression; manganese may be required for the export of Cbp.

An additional range of pneumococcal surface proteins have been identified and proposed as virulence factors. These are the choline-binding proteins described above. Twelve of these proteins are listed in **Table 3**.

PspA is a highly variable protein expressed by all clinically important pneumococcal serotypes (Crain *et al.*, 1990) and is found non-covalently attached to the choline of the cell wall. Mutant *Strep. pneumoniae* variants unable to produce PspA are less virulent in systemic disease because they are more easily cleared from the bloodstream (McDaniel *et al.*, 1987). PspA may act by preventing the deposition of C3b on to the surface of pneumococci. Alternatively, this virulence factor may inhibit the formation of the alternate complement pathway C3 convertase (Tu *et al.*, 1999). PspA inhibits complement activation PspA inhibits complement activation and reduces the effectiveness of complement receptor-mediated clearance mechanisms (Neeleman *et al.*, 1999) and reduces the effectiveness of complement receptor-mediated clearance mechanisms. It has also been suggested that PspA binds to lactoferrin and therefore that it is involved in iron acquisition (Hammerschmidt *et al.*, 1999). The role of this activity in virulence remains to be determined.

Choline-binding protein A (CbpA) is involved in bacterial adhesion to the nasopharynx. Immunisation of mice with this protein, also termed PspC, provides mice with protection against subsequent intravenous challenge with PspC-producing and non-producing strains of *Strep. pneumoniae* (Brooks-Walter *et al.*, 1999). This cross-protection is due to the production cross-reactive antibodies against PspC and PspA, which

display high homology. CbpA-deficient mutants colonise the nasopharynx of rats less well and show reduced binding to human cells (Rosenow *et al.*, 1997). CbpA has activities which may be important in the disease process, including binding to complement component C3 (Smith and Hostetter, 2000). The interaction of CbpA and C3 has been demonstrated in two models of adhesion. CbpA also stimulates the production of IL-8 from pulmonary epithelial cells and may therefore be involved in immune-cell recruitment and chemotaxis (Madsen *et al.*, 2000). It also mediates attachment to the polymeric immunoglobulin receptor (pIgR) and translocation of pneumococci across human nasopharyngeal epithelial cells (Zhang *et al.*, 2000).

pIgR is an integral membrane protein required for the transcytosis of polymeric immunoglobulins (pIg), including IgA and IgM (Mostov and Kaetzel, 1999). Polymeric immunoglobulin binds to pIgR on the basolateral surface of the epithelial membrane and is then transported to the mucosal surface. The extracellular portion of pIgR is cleaved and released with the pIg. The cleaved portion of pIgR is known as secretory component (SC) and the complex is known as secretory immunoglobulin. A pneumococcal surface protein, SpsA, binds to SC and secretory IgA but not monomeric IgA (Hammerschmidt *et al.*, 1997). At the sequence level, SpsA is very similar to CbpA. Zhang *et al.* (2000) showed that the N-terminal repeats of CbpA mediate binding to human pIgR (hpIgR) but not to rabbit pIgR and the level of hpIgR expression in cells correlates with the level of adhesion. There is a gradient of pIgR expression from the upper to lower respiratory tract (Mostov and Kaetzel, 1999) with pIgR expressed at high levels in the nasopharynx. The CbpA–pIgR interaction may, therefore, be important for nasopharyngeal colonisation but be less important for infection processes in the lower respiratory tract where expression is very low. The PAF receptor may act as the receptor in the lower respiratory tract (Cundell *et al.*, 1995a). It has been proposed that pneumococci interact with uncleaved hpIgR and drive the transcytotic process in reverse from apex to base (Zhang *et al.*, 2000).

Other choline-binding proteins known to be involved in virulence include the major autolysin LytA (Berry *et al.*, 1989a) which is believed to mediate its effect by allowing release of other virulence factors such as pneumolysin and cell wall fragments. CbpD, CbpE, CbpG, LytB and LytC play a role in colonisation of the nasopharynx (Gosink *et al.*, 2000) and mutations in CbpE and CbpG reduce adherence to human cells. CbpG, which may be a serine protease, also plays a role in sepsis (Gosink *et al.*, 2000).

Hydrogen Peroxide

Hydrogen peroxide plays a role in the virulence of the pneumococcus. The amounts of hydrogen peroxide produced are similar to those produced by poly-morphonuclear cells. Toxicity has been demonstrated in rat epithelial cells (Duane *et al.*, 1993). Peroxide is produced by the action of pyruvate oxidase, which is important in the virulence of the pneumococcus (Spellerberg *et al.*, 1996). Hydrogen peroxide is also an important factor in toxicity in *ex vivo* rat brain slices (Hirst *et al.*, 2000).

Superoxide Dismutase

The pneumococcus contains two types of superoxide dismutase (SOD), MnSOD and FeSOD. Inactivation of *soda*, the gene for MnSOD, shows that MnSOD plays a role in the virulence of the pneumococcus in animal models (Yesilkaya *et al.*, 1999). The pattern of inflammation in lungs infected with the mutant was different to that seen with wild-type organisms. After infection with the mutant, neutrophils were packed around bronchioles, in contrast to the wild-type infection in which neutrophils are more diffusely localised.

NADH Oxidase

A gene termed *nox* has been identified which encodes a soluble flavoprotein that re-oxidises NADH and reduces molecular oxygen to water (Auzat *et al.*, 1999). Disruption of this gene shows that Nox is involved in genetic competence and virulence. The virulence and persistence in mice of a blood isolate was attenuated by a *nox* insertion mutation. It has been proposed that the enzymatic activity of the NADH oxidase, which is coded by *nox*, is probably involved in transducing a signal related to oxygen availability into the cell and so behaves as an oxygen sensor. Oxygen therefore plays a role in the regulation of pneumo-coccal transformability and virulence by a mechanism involving Nox (Auzat *et al.*, 1999).

Regulation of Virulence

Many important functions in bacteria, such as expression of virulence factors, competence, osmoregulation and chemotaxis, are controlled by two-component systems. A typical two-component system consists of a membrane-spanning sensor and a response regulator located in the cytoplasm. The sensor is a histidine kinase that auto-phosphorylates in response to an external stimulus. The phosphate group is then transferred to the response regulator, which acts as a transcription factor and on phosphorylation changes its DNA-binding properties to alter gene expression (reviewed by Hoch, 2000). Analysis of the pneumo-coccal genome sequence shows that 13 of these systems are present, plus an orphan response regulator (Lange *et al.*, 1999; Throup *et al.*, 2000). Eight of these systems are important for virulence in a mouse pneumonia model (Throup *et al.*, 2000).

The current knowledge of these systems is summarised in **Table 6**. The functionality of six of the systems have been examined, including PnpSR for the sensing of phosphate (Novak *et al.*, 1999a), VncSR for antibiotic-induced autolysis (Novak *et al.*, 1999b), ComDE for development of competence for DNA transformation (Havarstein *et al.*, 1996), CiaSR for competence and penicillin susceptibility (Guenzi *et al.*, 1994), ZmpSR contiguous to a zinc metalloprotease (Novak *et al.*, 2000) and BlpHR for a bacteriocin regulon. The zinc metallo-protease is involved in the targeting of choline-binding proteins for surface expression. The exact mechanisms by which these processes are involved in disease is the subject of current research.

ClpC ATPase is a subfamily of HSP100/Clp molecular chaperones and a *clpC* gene has been identified in the pneumococcus (Charpentier *et al.*, 2000). Disruption of this gene leads to failure of lysis after treatment with antibiotics and increased tolerance to high temperature and decreased transformation. The mutant fails to express pneumolysin and the choline-binding proteins LytA, CbpA, CbpE, CbpF and CbpJ. These defects are associated with reduced adherence of the organism to human type II alveolar cells (Charpentier *et al.*, 2000). Thus, ClpC plays an essential role in the expression of virulence factors.

Host Defence

Structural and Mechanical Defences

The first line of defence against the foreign bacteria is the anatomical design of the respiratory tract. Particles of 10 µm or more are deposited in the nasal passages by impaction and the branching of bronchi also leads particles of 2–10 µm to be removed on impact with the mucus-covered airways. Coughing and sneezing also act to prevent particles passing beyond the respiratory bronchioles (Pison *et al.*, 1994). Only particles between 0.5 µm and 2 µm become deposited in the terminal respiratory units and alveoli (Nelson and Summer, 1998). After contact with the walls of the

Table 6 The 13 two-component systems (TCS) of *Streptococcus pneumoniae*

TCS system[a]	Equivalent system[b]	Role in virulence in respiratory infection model[b]	Putative function	References for putative function
1	480	Attenuated		
2	492	No	Response regulator is essential for viability	Throup *et al.* (2000)
3	474	No		
4	481	Attenuated	PnpS/R. Tested for role in phosphate transport	Novak *et al.* (1999a)
5	494	Attenuated	CiaR/S. Sensitivity to penicillin and induction of competence to take up DNA	Guenzi *et al.* (1994)
6	478	Attenuated		
7	539	Attenuated		
8	484	Attenuated		
9	488	Attenuated	ZmpR/S. Associated with zinc metalloprotease. The protease is involved in targeting of choline-binding proteins. No direct evidence for regulation of the protease by the TCS	Novak *et al.* (2000)
10	491	Attenuated	VncR/S. Sensitivity to antibiotics	Novak *et al.* (1999b)
11	479	No		
12	498	Not tested	ComD/E. Induction of competence to take up DNA	Havarstein *et al.* (1996)
13	486	Attenuated	BlpR/H. Production of bacteriocins	de Saizieu *et al.* (2000)

[a] Lange *et al.* (1999).

[b] Throup *et al.* (2000).

respiratory tract, foreign particles become embedded in the mucous layer. It has been estimated that the beating of the cilia moves the mucous layer upwards at a rate of 0.5–1 mm/min in the small airways and 5–20 mm/min in the larger airways (Wanner, 1977).

Cellular Defences

Macrophages

Alveolar macrophages constitute the first line of antibacterial defence and are ideally situated between the lung tissue and air (Lohmann-Matthes *et al.*, 1994). In a normal resting animal, these cells constitute more than 90% of the cells recovered by broncho-alveolar lavage (Reynolds, 1987).

Alveolar macrophages are believed mainly to originate from monocytes recruited out of the bloodstream to differentiate within the alveoli. The rapid recruitment of macrophages during acute inflammation suggests that local proliferation of the population may also occur (Toews *et al.*, 1979). Additional populations of macrophages in the lungs include

interstitial and intravascular macrophages and dendritic cells. Interstitial macrophages are located in the lung connective tissue and function in innate immunity and in an antigen-presenting capacity (Lohmann-Matthes *et al.*, 1994). Dendritic cells, which are present in the lung interstitium in low numbers, have a low phagocytic function and are specialised for antigen presentation. Intravascular macrophages are seen attached to the endothelium of pulmonary capillaries (Dehring and Wismar, 1989) and, unlike monocytes, are highly phagocytic and can remove bacteria that enter the lungs by way of the circulation (Warner *et al.*, 1987) but they do not appear to play a major role in mice (Brain *et al.*, 1999).

Alveolar macrophages engulf bacteria rapidly after infection. In the case of *Staph. aureus*, 80% of bacteria are ingested within 1 hour (Goldstein *et al.*, 1974). To achieve this, macrophages express a range of surface receptors, including the Fc receptors, complement receptors and lectin receptors (Lohmann-Matthes *et al.*, 1994). Pulmonary macrophages possess several mechanisms to kill ingested bacteria. The phagosomes, which contain acid hydrolases that constitute an anaerobic killing pathway, fuse with lysosomes. Unlike monocytes and neutrophils, airway macrophages do not possess much myeloperoxidase and therefore do not utilise hypochlorous acid as an antimicrobial defence. They do, however, exhibit a 'respiratory burst', which leads to the production of reactive oxygen intermediates and reactive nitrogen intermediates (Poulter, 1997). Airway macrophages also release free oxygen radicals, proteases and acid hydrolases into their environment, which may contribute to extracellular killing of pneumococci.

In addition to their phagocytic role, alveolar macrophages are central to the induction of pulmonary inflammatory responses. These cells secrete a wide range of inflammatory mediators, including cytokines, nitric oxide and metabolites of the arachidonic acid pathways. These mediators act in both a paracrine and autocrine manner to mediate macrophage functions, such as further cytokine release, antigen presentation and phagocytosis (Cavaillon, 1994).

Polymorphonuclear Leucocytes

In the healthy host less than 0.15% of neutrophils are present in the air spaces (Vial *et al.*, 1984) and additional recruitment to the lungs does not occur until the defences of the alveolar macrophages are overcome (Toews *et al.*, 1979). When required, polymorphonuclear leucocytes are sequestered from blood vessels and it is estimated that 40% of neutrophils are marginated in the lung capillaries (Nelson *et al.*, 1995) or are released from the bone marrow.

Normal neutrophil traffic is along the endothelium. Initial tethering is via the L-selectin constitutively expressed on neutrophils and the corresponding receptor on endothelial cells (Hogg, 1994). This interaction is not strong and is easily broken resulting in neutrophils 'rolling' along the endothelial surface. More secure attachment occurs via the endothelial (E) and platelet (P) selectin molecules and their receptors on neutrophils. These interactions are assisted by leucocyte integrins (CD11/CD18) that interact with immunoglobulin-like counter receptors on the endothelium (intercellular adhesion molecules ICAM-1, ICAM-2) and vascular cell adhesion molecule 1 (VCAM-1) (Hogg and Walker, 1995). Neutrophils then migrate from pulmonary capillaries at locations where three endothelial cells meet, since at these sites the tight junctions between cells are incomplete (Walker *et al.*, 1991). Chemotactic stimuli include complement proteins and chemokines. Pneumococci and soluble factors derived from them induce β_2-integrin-dependent neutrophil adhesion to human alveolar epithelial cells (Smith *et al.*, 1998).

Lymphocytes

Lymphocytes in the lungs present aggregates below the mucosal layer of airways (bronchus-associated lymphoid tissue or BALT), in the airspaces and interstitium and in the intra-vascular compartment (Pabst, 1997). The number of lymphocytes in lungs of infected animals increases during pneumococcal pneumonia (Bergeron *et al.*, 1998; Kadioglu *et al.*, 2000) but the role of these recruited lymphocytes is not fully understood. T-cell-deficient nude mice are not more susceptible to pneumococcal infection than wild-type animals (Winkelstein and Swift, 1975), suggesting that T cells are not required for a protective response against *Strep. pneumoniae*. Conversely, lymphocytes are involved in antibody-mediated cytotoxicity against *Strep. pneumoniae in vitro* (Sestini *et al.*, 1987). The lymphocytes responsible possess the phenotype of T helper cells and utilise IgA to mediate their antibacterial activity (Sestini *et al.*, 1988).

Lymphocytes are able to release a range of mediators that can activate and potentiate the effects of other cell populations. Thus, the importance of lymphocytes in pneumococcal pneumonia may be due to their regulatory capacity. Since pulmonary lymphocytes undergo re-distribution in the lung during pneumococcal pneumonia they may play an important role (Kadioglu *et al.*, 2000).

Mast Cells

Mast cells are derived from CD34+ progenitor cells within the bone marrow (Kirschenbaum *et al.*, 1991). These progenitors migrate to common primary sites of infection, such as the genito-urinary, gastro-intestinal and respiratory tracts (Abraham and Malaviya, 1997). The most striking features of mast cells are the large number of densely packed histamine, serotonin and TNFα-containing cytoplasmic granules (Terr, 1994). They are the only cells that store pre-formed TNFα, which on activation is quickly released along with newly synthesised TNFα (Gordon and Galli, 1990, 1991). On degranulation, mast cells also release histamine and serotonin into their environment. Each of these three substances is capable of causing disruption to the surrounding tissue, while the latter two at high concentrations also cause vasodilatation. This increases blood flow and permits a greater influx of inflammatory cells (Cushing and Cohen, 1992; Szarek *et al.*, 1992). Physical contact between the mast cell and bacteria is not required for activation, since the cells can be activated at a distance by bacterial products such as toxins (Abraham and Malaviya, 1997).

In addition to TNFα, mast cells secrete a range of cytokines, such as IL-1, IL-6, IL-8, IL-10 and several chemokines (Mecheri and David, 1997; Lorentz *et al.*,

2000). This implicates mast cells as important first-line defences against pathogenic micro-organisms. Indeed, mast cells can phagocytose and kill Gram-positive and Gram-negative organisms, but uptake of Gram-positive bacteria is higher and is associated with far greater TNFα release than Gram-negative uptake (Arock *et al.*, 1998). However, the phagocytic efficiency of mast cells appears to be far less than that of macrophages and neutrophils.

The importance of mast cells during pulmonary and systemic infections has been highlighted by experiments carried out in mast cell-deficient mice (Malaviya *et al.*, 1996). Such animals are less able to clear invading bacteria, such as *Klebsiella pneumoniae*, because of impaired neutrophil recruitment at systemic and local sites of infection, which reflects a reduction in TNFα levels. In addition, increasing the number of mast cells by treating mice with c-*kit* ligand (mast cell growth factor) results in animals that are more resistant to bacterial peritonitis induced by caecal ligation (Maurer *et al.*, 1998).

Secretory Pulmonary Defences

Proteins in respiratory secretions are produced locally in the airway lumen or they can pass through the alveolar capillaries (Stockley, 1997). The degree of transudation depends on the molecular size of the substance and the integrity of the blood–air barrier.

Surfactant

Surfactant (surface active agent) consists of phospholipid-rich lipoproteins and reduces the surface tension in the lungs that permits their expansion. Apart from this physiological role, surfactant modulates the immune response. The majority of these functions are attributed to the surfactant proteins, SP-A1 and 2, SP-B, SP-C and SP-D. SP-A and SP-D are members of the collectin protein family characterised by the presence of a collagenous region and a lectin domain, which mediates binding to carbohydrates on the surface of micro-organisms. They also contain mannose-binding lectin (MBL) and two bovine serum lectins, conglutinin and CL-43. MBL activates the alternative complement system. Recent data suggest that the lung collectins SP-A and SP-D modulate pulmonary inflammatory responses.

SP-A highly efficiently opsonises bacteria, including *Strep. pneumoniae*, and promotes their phagocytosis by alveolar macrophages. Indeed, addition of SP-A to pneumococci in the presence of rat alveolar macrophages results in higher levels of phagocytosis than IgG (Schagat *et al.*, 1999) and SP-A and IgG act synergistically as opsonins. SP-A and SP-D act as chemoattractants for neutrophils and macrophages *in vitro* and SP-D is more effective than SP-A (Wright and Youmans, 1993; Madan *et al.*, 1997). After phagocytosis, both pulmonary collectins increase the production of reactive oxygen species by neutrophils and macrophages (van Iwaarden *et al.*, 1990, 1992). The effects of SP-A but not those of SP-D can be attenuated by addition of surfactant lipids.

Finally, SP-A modulates the pulmonary inflammatory response by altering cytokine production. It can induce release of immunoglobulins and pro-inflammatory cytokines such as TNFα, IL-1 and IL-6 from peripheral blood mononuclear cells, splenocytes and alveolar macrophages (Kremlev and Phelps, 1994). Conversely, an anti-inflammatory role has been ascribed to SP-A, because it can reduce TNFα activity released from LPS-stimulated alveolar macrophages *in vitro* (McIntosh *et al.*, 1996). Such differences may reflect the local environment, the inflammatory stimulus and levels of other inflammatory mediators.

Lysozyme

The muramidase lysozyme, which degrades peptidoglycan, is the major secretory enzyme of the alveolar macrophage and accounts for some 25% of the protein released (Canto *et al.*, 1994). Although its effects have not been widely studied it is known to be active against *Strep. pneumoniae* (Jacquot *et al.*, 1987). The growth of pneumococci cultured in media containing 1000 µg/mL of human airway lysozyme is delayed. Furthermore, although the addition of lysozyme to a culture during early exponential growth allows the bacteria to multiply, reduced viability counts were recovered as compared with growth in its absence.

Immunoglobulins

IgA, IgG, IgD, IgE and IgM immunoglobulins are detectable in respiratory secretions (Stockley, 1997). Whilst a role in up-regulation of phagocytosis by alveolar macrophages (Richards and Gauldie, 1985) has been identified for IgA and IgG in innate lung defence, this is not the case for IgD, IgE and IgM. The importance of antibody in protection against *Strep. pneumoniae* is highlighted by the increased susceptibility to pneumococcal infections of patients with hypogammaglobulinaemia (Rosen and Janeway, 1966).

Complement Activation

Like other Gram-positive bacteria, pneumococci are resistant to the bactericidal and lytic activities of complement (Kemper and Deresinki, 1994). The role of complement in *Strep. pneumoniae* infection is, therefore, in opsonisation for phagocytosis.

The Classical Pathway Deposition of C3b on the surface of pneumococci by the activation of the classical complement pathway is believed to play a major role in host defence against *Strep. pneumoniae* by increasing efficacy of phagocytosis (Brown *et al.*, 1983). Pneumolysin can also activate the classical pathway.

The Alternative Pathway Pneumococci are capable of direct activation of the alternative pathway of complement. Winkelstein and Tomasz (1978) showed that the teichoic acid of pneumococcal cell walls can deplete serum C3 by activating the alternative complement pathway. Since C3b is bound to the cell, encapsulated pneumococci are not efficiently opsonised for phagocytosis. Thus the classical complement pathway is a more effective mechanism of host defence against *Strep. pneumoniae* (Brown *et al.*, 1983).

Sources of Complement Proteins Most complement components in the lung are derived from plasma during inflammation (Stockley, 1997) but low levels of complement proteins are present within the lungs since the functioning alternative complement pathway can be detected in the lung lavage fluid of normal rats (Coonrod and Yoneda, 1981). This fact may be explained by the ability of pulmonary macrophages and epithelial cells to synthesise and secrete C2, C4, C3 and C5 (Strunk *et al.*, 1988).

Effects of Complement Proteins The most important complement factors in pneumococcal infection are C3 and C5 and both C3a and C5a can stimulate mast cell and neutrophil degranulation and increase vascular permeability (Lukacs and Ward, 1996). C5a appears to be more potent than C3a in these respects and possesses the additional ability of stimulating neutrophil migration and degranulation. In addition to the inflammatory roles played by C3a and C5a, C1q can also mediate inflammatory responses *in vitro* by increasing the production of the pro-inflammatory cytokines IL-6, IL-8 and MCP-1 by human endothelial cells (van den Berg *et al.*, 1998). C3b and C5b opsonise pneumococci, as shown by the depleted C3 and C5 levels in serum after incubation with the bacteria and the agglutination and capsular swelling of opsonised bacteria after the addition of anti-C3 serum opsonins (Shin *et al.*, 1969). Direct evidence for the role of C5 in protection against *Strep. pneumoniae* is highlighted by the inability of C5-deficient mice to clear the organisms unless exogenous C5 is administered, because of impaired neutrophil recruitment (Tomasz, 1981). In addition, humans with complement factor deficiency are more likely to become infected with *Strep. pneumoniae* than immunocompetent individuals (Newman *et al.*, 1978).

The Role of Cytokines

Pulmonary Cytokine Production Several reports have appeared of the production of cytokines in murine models of pneumococcal pneumonia and most of these have concentrated on individual cytokines. The balance between pro- and anti-inflammatory cytokines is poorly understood and the evidence is often conflicting (Takishima *et al.*, 1997; van der Poll *et al.*, 1997a). Indirect support for the importance of pro-inflammatory cytokines in host defence against *Strep. pneumoniae* is provided by the reduced ability of transgenic mice with impaired NF-κB activity to control type 2 pneumococcal infection (Sha *et al.*, 1995). NF-κB is a transcription factor capable of regulating the expression of pro-inflammatory cytokines, which consists of two subunits, p50 and p65. After intra-peritoneal challenge, with 100 cfu of type 2 *Strep. pneumoniae* mice with targeted disruptions in the p50 subunit have reduced survival times as compared with wild-type mice. Moreover, large numbers of viable pneumococci were recovered from the blood and organs of the transgenic mice. Thus, the ability of these mice to contain and kill the bacteria in the peritoneum was reduced.

Direct evidence that cytokines are involved in the host response to *Strep. pneumoniae* is provided by mouse models. Thus, Bergeron *et al.* (1998) studied a model of pneumonia in which female CD1 mice were infected with 10^7 cfu of type 3 pneumococci. The mice died about 72 hours after challenge with acute inflammation with histological changes and cell recruitment similar to that later also reported by Kadioglu *et al.* (2000). There was a rapid influx of neutrophils to the lungs, at first around blood vessels but later around infected bronchioles. Neutrophils were then progressively replaced by monocytes and small numbers of lymphocytes 72–96 hours after challenge.

This cell influx was accompanied by rapid cytokine release. The first cytokine to appear in broncho-alveolar lavage fluid was TNFα. Significantly elevated levels were detected 1 hour after challenge and these continued to increase until 12 hours. By 24 hours post challenge, levels were returning to baseline. A similar transient production of TNFα was evident in lung tissues. Pulmonary IL-1α levels also reached their peak 12 hours after challenge but more slowly and in lower concentrations than TNFα. IL-1α levels remained raised in lung tissues from mid-infection until the end of the experiment. Production of IL-6 in the airways was biphasic, with a peak 4 hours post challenge and again at 48 hours post challenge. This cytokine was detected in the lung tissues throughout the experiment.

In summary, Bergeron *et al.* (1998) found initially ineffective phagocytosis of pneumococci leading to cytokine release over 4–24 hours into the lungs as pneumococci grew in the alveoli. When alveolar injury had occurred and bacteria entered the bloodstream cytokine release in the lungs was downregulated. These kinetics of cytokine production are similar to those found in other models of pneumococcal pneumonia. Intra-nasal infection with type 3 or type 19 pneumococci also results in the rapid production of TNFα in the lungs (Takishima *et al.*, 1997; van der Poll *et al.*, 1997a). van der Poll *et al.* (1996) measured IL-10 production during pneumococcal pneumonia in the mouse and found elevated levels of IL-10 from 12 hours after challenge but peak production did not occur until 72 hours after challenge.

Systemic Cytokine Production Bergeron *et al.* (1998) also investigated the systemic production of cytokines in their model of pneumococcal pneumonia. They found that serum TNFα increased towards the end of the experiment as bacteria gained access to the bloodstream. Systemic levels of IL-1β increased very transiently 12 hours post challenge, at the time when maximum lung levels were detected, which suggests that this was due to overspill from the lungs. The increased IL-6 levels early in the infection (4 hours post challenge) and again increasingly towards the end of the experiment correlated well with bacterial loads. Others have found similar systemic TNFα kinetics but they found very variable amounts (van der Poll *et al.*, 1997a). Similar variations have been found in humans (Marks *et al.*, 1990). The rapidity and sustained nature of IL-6 production in the circulation has also been reported (van der Poll *et al.*, 1997b). Serum IFNγ levels correlate well with the degree of bacteraemia and the severity of illness (Rubins and Pomeroy, 1997). However, mice that survive until 72 hours after challenge with pneumococci of serotypes 1, 2, 3 or 4 showed the lowest IFNγ levels in spite of bacteraemia.

Effects of Cytokines The role of individual cytokines can be determined by altering the levels of the individual cytokines during the course of an infection by antibody neutralisation or the use of transgenic mice. Systemic neutralisation of TNFα results in increased bacterial loads and mortality after infection with *Strep. pneumoniae* (Takishima *et al.*, 1997; van der Poll *et al.*, 1997a). Contradictory explanations for these effects of TNFα have, however, been reported. Neutralisation of TNFα during infection with type 3 pneumococci results in significantly increased bacterial loads in the lungs but bacteraemia is unaffected. Treatment with anti-TNFα also reduces IL-1 levels but has no effect on IL-6, IL-10 or IFNγ levels or neutrophil recruitment to the lungs (van der Poll *et al.*, 1997a). It was suggested, therefore, that TNFα is required to increase the antibacterial activity of resident host cells during pneumococcal pneumonia but that it does not lead to the recruitment of additional cells to the site of infection. In contrast, Takishima *et al.* (1997) found that neutralisation of TNFα before infection with type 19 pneumococci did not affect pulmonary bacterial loads but significantly increased levels of bacteraemia. Furthermore, significantly reduced numbers of circulating neutrophils were found after TNFα neutralisation. This indicates that TNFα is responsible for increasing the numbers of neutrophil during pneumococcal pneumonia and that these prevent overwhelming bacteraemia.

Although most research has been aimed at identifying the role of TNFα during pneumococcal pneumonia, the effects of other cytokines has also been investigated. Intra-nasal infection of IL-6-deficient mice results in a 6-fold increase in bacterial loads and higher mortality during pneumococcal infection (van der Poll *et al.*, 1997b). This was evident in spite of higher levels of pulmonary TNFα, IL-1 and IFNγ at the end of the infection and this highlights the importance of pro-inflammatory cytokines early in the infection but not late in the disease. Interferon γ knockout mice are significantly more susceptible (LD_{50} 2.5×10^6) to infection with type 2 pneumococci than their wild-type counterparts (LD_{50} 10^7) (Rubins and Pomeroy, 1997), probably because IFNγ activates macrophages.

Finally, neutralisation of IL-10 with antibody 2 hours before intra-nasal infection with 10^6 cfu type 3 pneumococci results in increased pulmonary TNFα and IFNγ levels, with a corresponding 6-fold decrease in pulmonary and systemic bacterial loads and increased survival; pre-treatment with re-combinant IL-10 has the opposite effects (van der Poll *et al.*, 1996). This indicates that a high level of IL-10 during pneumococcal pneumonia reduces the levels of pro-inflammatory cytokines and permits higher bacterial loads and ultimately reduced survival of the host.

Vaccines

In spite of successful type-specific protection with capsular polysaccharide preparations in the 1930s and 1940s (Lee *et al.*, 1991), interest in vaccines against *Strep. pneumoniae* infection waned quickly when antibiotics were discovered. Complacency developed in the belief that a vaccine had become unnecessary, even though some 20% of patients with pneumococcal bacteraemia over 65 years of age die in spite of appropriate antibiotic treatment (Butler and Cetron, 1999). Recognition of this fact in the 1970s led to renewed interest in the development of improved pneumococcal vaccines. A vaccine containing capsular polysaccharide against 23 serotypes of pneumococci became available in 1983. This vaccine is effective against the serotypes responsible for more than 90% of cases of pneumococcal pneumonia in the United States and Europe (Alonsodevelasco *et al.*, 1995) but less than 70% of pneumococcal infections in Asia (Lee, 1987).

The vaccine has a low efficacy because some groups of patients has poor antibody responses to polysaccharide-based vaccines (White, 1988).

More effective vaccines may result from conjugation of capsular polysaccharides to a protein carrier as in the case of *H. influenzae* in infants (Ahman *et al.*, 1998). The basis of these conjugate vaccines is that the infant immune system responds to both the protein and polysaccharide antigens (Käyhty and Eskola, 1996). A 7-valent conjugate vaccine has been tested (Black *et al.*, 1999) and licensed for use in the United States. It remains to be seen, however, whether the introduction of this vaccine will shift the disease-causing serotypes to those not included in the vaccine. Natural transformation also allows organisms to switch their capsular type (Coffey *et al.*, 1998).

The possible development of protein-based vaccines is also of interest. Serious candidate proteins are PspA, CbpA, pneumolysin and PsaA and the use of these proteins as vaccines has been reviewed by Briles *et al.* (2000).

References

Abel L, Dessein AJ (1997) The impact of host genetics on susceptibility to human infectious diseases. *Curr. Opin. Immunol.* 9: 509–516.

Abraham SN, Malaviya R (1997) Mast cells in infection and immunity. *Infect. Immun.* 65: 3501–3508.

Ahman H, Kayhty H, Lehtonen H, Leroy O, Froeschle J, Eskola J (1998) Streptococcus pneumonia capsular polysaccharide-diphtheria toxoid conjugate vaccine is immunogenic in early infancy and able to induce immunologic memory. *Pediatr. Infect. Dis. J.* 211–216.

Alexander J, Lock RA, Peeters CCAM *et al.* (1994) Immunization of mice with pneumolysin toxoid confers a significant degree of protection against at least nine serotypes of *Streptococcus pneumoniae. Infect. Immun.* 62: 5683–5688.

Alloing G, Martin B, Granadel C, Claverys JP (1998) Development of competence in *Streptococcus pneumoniae*: phermone autoinduction and control of quorum sensing by the oligopeptide permease. *Mol. Microbiol.* 29: 75–83.

Alonsodevelasco E, Verheul AFM, Verhoef J, Snippe H (1995) *Streptococcus pneumoniae* – virulence factors, pathogenesis, and vaccines. *Microbiol. Rev.* 59: 591.

Ammann AJ, Schiffman G, Austrian R (1980) The antibody responses to pneumococcal capsular polysaccharides in aged individuals. *Proc. Soc. Exp. Biol. Med.* 164: 312–316.

Arock M, Ross E, LaiKuen R, Averlant G, Gao ZM, Abraham SN (1998) Phagocytic and tumor necrosis factor alpha response of human mast cells following exposure to gram-negative and gram-positive bacteria. *Infect. Immun.* 66: 6030–6034.

Arrecubieta C, Garcia E, Lopez R (1994) Molecular characterisation of *cap3A*, a gene from the operon required for the synthesis of the capsule of *Streptococcus pneumoniae* type 3: sequencing of mutations responsible for the unencapsulated phenotype and localisation of the capsular cluster on the pneumococcal chromosome. *J. Bacteriol.* 176: 6375–6383.

Arrecubieta C, Garcia E, Lopez R (1995) Sequence and transcriptional analysis of a DNA region involved in the production of capsular polysaccharide in *Streptococcus pneumoniae. Gene* 167: 1–7.

Ashman RB, Papadimitriou JM (1992) Genetic resistance to *Candida albicans* infection is conferred by cells derived from the bone marrow. *J. Infect. Dis.* 166: 947–948.

Askenasy OM, George RC, Begg NT (1995) Pneumococcal bacteraemia and meningitis in England and Wales 1982–1992. *Communic. Dis. Rep.* 5: R45–R50.

Austrian R, Gold J (1964) Pneumococcal bacteremia with special reference to bacteremic pneumococcal pneumonia. *Ann. Intern. Med.* 60: 759–776.

Auzat I, Chapuy-Regaud S, Le Bras G *et al.* (1999) The NADH oxidase of *Streptococcus pneumoniae*: its involvement in competence and virulence. *Mol. Microbiol.* 34: 1018–1028.

Avery OT, Dubos R (1931) The protective action of a specific enzyme against type iii pneumococcus infection in mice. *J. Exp. Med.* 54: 73–89.

Bellamy R, Ruwende C, Corrah T, McAdam KPWJ, Whittle HC, Hill AVS (1998) Variations in the NRAMP1 gene and susceptibility to tuberculosis in West Africans. *N. Engl. J. Med.* 338: 640–644.

Benton KA, VanCott JL, Briles DE (1998) Role of tumor necrosis factor alpha in the host response of mice to bacteremia caused by pneumolysin-deficient *Streptococcus pneumoniae. Infect. Immun.* 66: 839–842.

Bergeron Y, Ouellet N, Deslauriers AM, Simard M, Olivier M, Bergeron MG (1998) Cytokine kinetics and other host factors in response to pneumococcal pulmonary infection in mice. *Infect. Immun.* 66: 912–922.

Berry AM, Lock RA, Hansman D, Paton JC (1989a) Contribution of autolysin to virulence of *Streptococcus pneumoniae*. *Infect. Immun.* 57: 2324–2330.

Berry AM, Yother J, Briles DE, Hansman D, Paton JC (1989b) Reduced virulence of a defined pneumolysin-negative mutant of *Streptococcus pneumoniae*. *Infect. Immun.* 57: 2037–2042.

Berry AM, Lock RA, Thomas SM, Rajan DP, Hansman D, Paton JC (1994) Cloning and nucleotide sequence of the *Streptococcus pneumoniae* hyaluronidase gene and purification of the enzyme from recombinant *Escherichia coli*. *Infect. Immun.* 62: 1101–1108.

Berry AM, Lock RA, Paton JC (1996) Cloning and characterization on *nanB*, a second *Streptococcus pneumoniae* neuraminidase gene, and purification of the nanB enzyme from recombinant *Escherichia coli*. *J. Bacteriol.* 178: 4854–4860.

Bertozzi P, Astedt B, Zenzius L, AL E (1990) Depressed bronchoalveolar urokinase activity in patients with adult respiratory distress syndrome. *N. Engl. J. Med.* 322: 890–897.

Bhakdi S, Muhly M, Korom S, Hugo F (1989) Release of interleukin-1β associated with potent cytocidal action of staphylococcal alpha toxin on human monocytes. *Infect. Immun.* 57: 3512–3519.

Black S, Shinefield H, Ray P *et al.* (1999) Efficacy of heptavalent conjugate pneumococcal vaccine (Wyeth Lederle) in 7,000 infants and children: results of the northern California kaiser permanente efficacy trial. *Pediatr. Res.* 45: 915.

Boulnois GJ, Paton JC, Mitchell TJ, Andrew PW (1991) Structure and function of pneumolysin, the multifunctional, thiol-activated toxin of *Streptococcus pneumoniae*. *Mol. Microbiol.* 5: 2611–2616.

Brain JD, Molina RM, DeCamp MM, Warner AE (1999) Pulmonary intravascular macrophages: their contribution to the mononuclear phagocyte system in 13 species. *Am. J. Physiol.* 276: L146–L154.

Braun JS, Tuomanen EI (1998) Molecular mechanisms of brain damage in bacterial meningitis. *Adv. Pediatr. Infect. Dis.* 14: 49–72.

Braun JS, Novak R, Gao GL, Murray PJ, Shenep JL (1999a) Pneumolysin, a protein toxin of *Streptococcus pneumoniae*, induces nitric oxide production from macrophages. *Infect. Immun.* 67: 3750–3756.

Braun JS, Novak R, Herzog KH, Bodner SM, Cleveland JL, Tuomanen EI (1999b) Neuroprotection by a caspase inhibitor in acute bacterial meningitis. *Nature Med.* 5: 298–302.

Briles DE, Horowitz J, McDaniel LS *et al.* (1986) Genetic control of the susceptibility to pneumococcal infection. *Curr. Top. Microbiol. Immunol.* 124: 103–120.

Briles DE, Paton JC, Swiatlo E, Nahm M (2000) Pneumococcal vaccines. In: Fischetti VA, Novick RP, Ferretti JJ, Portnoy DA, Rood JI (eds) *Gram-positive Pathogens*. Washington, DC: ASM Press.

Brooks-Walter A, Briles DE, Hollingshead SK (1999) The *pspC* gene of *Streptococcus pneumoniae* encodes a polymorphic protein, PspC, which elicits cross-reactive antibodies to PspA and provides immunity to pneumococcal bacteraemia. *Infect. Immun.* 67: 6533–6542.

Brown EJ, Joiner KA, Cole RM, Berger M (1983) Localisation of complement component C3 on *Streptococcus pneumoniae*: anti-capsular antibody causes complement deposition on the pneumococcal capsule. *Infect. Immun.* 39: 403–409.

Bruunsgaard H, Skinhİj P, Qvist J, Pedersen BK (1999) Elderly humans show prolonged *in vivo* inflammatory activity during pneumococcal infections. *J. Infect. Dis.* 180: 551–554.

Burman LÅ, Norrby R, Trollfors B (1985) Invasive pneumococcal infections: incidence, predisposing factors, and prognosis. *Rev. Infect. Dis.* 7: 133–142.

Butler JC, Cetron MS (1999) Pneumococcal drug resistance: the new 'special enemy of old age'. *Clin. Infect. Dis.* 28: 730–735.

Camara M, Boulnois GJ, Andrew PW, Mitchell TJ (1994) A neuraminidase from *Streptococcus pneumoniae* has the features of a surface protein. *Infect. Immun.* 62: 3688–3695.

Canto RG, Robinson GR II , Reynolds HY (1994) Defense mechanisms of the resipratory tract. In: Chmel H, Bendinelli M, Friedman H (eds) *Pulmonary Infections and Immunity*. New York: Plenum Press, pp. 1–27.

Canvin JR, Marvin AP, Sivakumaran M *et al.* (1995a) The role of pneumolysin and autolysin in the pathology of pneumonia and septicemia in mice infected with a type 2 pneumococcus. *J. Infect. Dis* 172: 119–123.

Canvin JR, Marvin AP, Sivakumaran M *et al.* (1995b) The role of pneumolysin and autolysin in the pathology of pneumonia and septicemia in mice infected with a type-2 pneumococcus. *J. Infect. Dis.* 172: 119–123.

Carlsen BD, Kawana M, Kawana C, Tomasz A, Giebink GS (1992) Role of the bacterial cell wall in middle ear inflammation caused by *Streptococcus pneumoniae*. *Infect. Immun.* 60: 2850–2854.

Catterall JR (1999) *Streptococcus pneumoniae*. *Thorax* 54: 929–937.

Cavaillon JM (1994) Cytokines and macrophages. *Biomed. Pharmacother.* 48: 445–453.

Cellier M, Shustik C, Dalton W *et al.* (1997) Expression of the human NRAMP1 gene in professional primary phagocytes. Studies in blood cells and in HL-60 promyelocytic leukemia. *J. Leukoc. Biol.* 61: 96–105.

Charpentier E, Novak R, Tuomanen E (2000) Regulation of growth inhibition at high temperature, transformation and adherance in *Streptococcus pneumoniae* by ClpC. *Mol. Microbiol.* 37: 717–726.

Coffey TJ, Dowson CG, Daniels M, Spratt BG (1995) Genetics and molecular biology of beta-lactam-resistant pneumococci. *Microb. Drug Resist. Mechan. Epidemiol. Dis.* 1: 29–34.

Coffey TJ, Enright MC, Daniels M *et al.* (1998) Recombinational exchanges at the capsular polysaccharide biosynthetic locus lead to frequent serotype changes among natural isolates of *Streptococcus pneumoniae. Mol. Microbiol.* 27: 73–83.

Cohen B, Perkins ME, Putterman S (1940) The reaction between hemolysin and cholesterol. *J. Bacteriol.* 39: 59–60.

Cohen B, Halbert SP, Perkins ME (1942) Pneumococcal hemolysin: the preparation of concentrates, and their action on red cells. *J. Bacteriol.* 43: 607–627.

Cohn DA (1986) Low immune-responses in high androgen responder C57BR/CDJ males. *Clin. Exp. Immunol.* 63: 210–217.

Cole R (1914) Pneumococcus hemotoxin. *J. Exp. Med.* 20: 346–362.

Comis SD, Osborne MP, Stephen J et al. (1993) Cytotoxic effects on hair cells of guinea pig cochlea produced by pneumolysin, the thiol activated toxin of *Streptococcus pneumoniae. Acta Otolaryngol.* 113: 152–159.

Coonrod JD, Yoneda K (1981) Complement and opsonins in alveolar secretions and serum of rats with pneumonia due to *Streptococcus pneumoniae. Rev. Infect. Dis.* 3: 310–322.

Crain MJ II, Waltman WD, Turner JS et al. (1990) Pneumococcal surface protein A (PspA) is serologically highly variable and is expressed by all clinically important capsular serotypes of *Streptococcus pneumoniae. Infect. Immun.* 58: 3293–3299.

Cundell DR, Tuomanen EI (1994) Receptor specificity of adherence of *Streptococcus pneumoniae* to human type II pneumocytes and vascular endothelial cells *in vitro. Microb. Pathog.* 17: 361–364.

Cundell DR, Gerard NP, Gerard C, Idanpaan-Heikkila I, Tuomanen EI (1995a) *Streptococcus pneumoniae* anchor to activated human cells by the receptor for platelet-activating factor. *Nature* 377: 435–438.

Cundell DR, Pearce BJ, Sandros J, Naughton AM, Masure HR (1995b) Peptide permeases from *Streptococcus pneumoniae* affect adherence to eukaryotic cells. *Infect. Immun.* 63: 2493–2498.

Cushing DJ, Cohen ML (1992) Serotonin-induced relaxation in canine coronary-artery smooth muscle. *J. Pharmacol. Exp. Ther.* 263: 123–129.

de Saizieu A, Gardes C, Flint N et al. (2000) Microarray-based identification of a novel *Streptococcus pneumoniae* regulon controlled by an autoinduced peptide. *J. Bacteriol.* 182: 4696–4703.

Dehring DJ, Wismar BL (1989) Intravascular macrophages in pulmonary capillaries of humans. *Am. Rev. Respir. Dis.* 139: 1027–1029.

Dillard JP, Vandersea MW, Yother J (1995) Characterisation of a cassette containing genes for type 3 capsular polysaccharide biosynthesis in *Streptococus pneumoniae. J. Exp. Med.* 181: 973–983.

Dowson CG, Hutchison A, Spratt BG (1989a) Extensive remodeling of the transpeptidase domain of penicillin-binding protein-2b of a penicillin-resistant South-African isolate of streptococcus-pneumoniae. *Mol. Microbiol.* 3: 95–102.

Dowson CG, Jephcott AE, Gough KR, Spratt BG (1989b) Penicillin-binding protein 2 genes of non-beta-lactamase-producing, penicillin-resistant strains of *Neisseria-gonorrhoeae. Mol. Microbiol.* 3: 35–41.

Dowson CG, Coffey TJ, Kell C, Whiley RA (1993) Evolution of penicillin resistance in *Streptococcus pneumoniae* – the role of *Streptococcus mitis* in the formation of a low-affinity Pbp2b in *Streptococcus pneumoniae. Mol. Microbiol.* 9: 635–643.

Dowson CG, Barcus V, King S, Pickerill P, Whatmore A, Yeo M (1997) Horizontal gene transfer and the evolution of resistance and virulence determinants in *Streptococcus. J. Appl. Microbiol.* 83: S42–S51.

Duane PG, Rubins JB, Weisel HR, Janoff EN (1993) Identification of hydrogen peroxide as a *Streptococcus pneumoniae* toxin for rat alveolar epithelial cells. *Infect. Immun.* 4392–4397.

Dullforce P, Sutton DC, Heath AW (1998) Enhancement of T cell-independent immune responses *in vivo* by CD40 antibodies. *Nature Med.* 4: 88–91.

Emanuelli G, Lanzio M, Anfossi T, Romano S, Anfossi G, Calcamuggi G (1986) Influence of age on polymorphonuclear leukocytes *in vitro* phagocytic activity in healthy human subjects. *Gerontology* 32: 308–316.

Fang G-D, Fine M, Orloff J et al. (1990) New and emerging etiologies for community-acquired pneumonia with implications for therapy. *Medicine* 69: 307–316.

Feldman C, Munro NC, Jeffrey DK et al. (1991) Pneumolysin induces the salient features of pneumococcal infection in the rat lung *in vivo. Am. J. Respir. Cell. Mol. Biol.* 5: 416–423.

Ferrante A, RowanKelly B, Paton JC (1984) Inhibition of *in vitro* human lymphocyte response by the pneumococcal toxin pneumolysin. *Infect. Immun.* 46: 585–589.

Fitzgerald TJ, Repesh LA (1987) The hyaluronidase associated with *Treponema pallidum* facilitates treponemal dissemination. *Infect. Immun.* 55: 1023–1028.

Forget A, Skamene E, Gros P, Miailhe AC, Turcotte R (1981) Differences in response among inbred mouse strains to infection with small doses of *Mycobacterium bovis* BCG. *Infect. Immun.* 32: 42–47.

Foster KD, Conn CA, Kluger MJ (1992) Fever, tumor necrosis factor and interleukin-6 in young, mature and aged Fisher 344 rats. *Am. J. Physiol.* 262: R211–R215.

Friedland IR, Paris MM, Hickey S et al. (1995) The limited role of pneumolysin in the pathogenesis of pneumococcal meningitis. *J. Infect. Dis.* 172: 805–809.

Frielander C (1883) Ueber die Schizomyceten bei der acuten Fibrisen Pneumonie. *Virchow's Arch. Pathol. Anat. Physiol. Klin. Med.* 87: 319–324.

Garcia J, Sanchez-Beato R, Medrano F, Lopez R (1998) Versatility of choline-binding domain. *Microb. Drug Resist.* 4: 25–36.

Garcia P, Garcia JL, Garcia E, Lopez R (1986) Nucleotide sequence and expression of the pneumococcal autolysin gene from its own promoter in *Escherichia coli. Gene* 43: 265–272.

Garcia P, Gonzalez MP, Garcia E, Lopez R, Garcia JL (1999a) LytB, a novel murein hydrolase essential for cell separation. *Mol. Microbiol.* 31: 1275–1281.

Garcia P, Gonzalez MP, Garcia JL, Lopez R (1999b) The molecular characterisation of the first autolytic lysozyme of *Streptococcus pneumoniae* reveals evolutionary mobile domain. *Mol. Microbiol.* 33: 128–138.

Garg M, Luo W, Kaplan AM, Bondada S (1996) Cellular basis of decreased immune responses to pneumococcal vaccines in aged mice. *Infect. Immun.* 64: 4456–4462.

Garibaldi RA (1985) Epidemiology of community-acquired respiratory tract infections in adults: incidence, etiology, and impact. *Am. J. Med.* 78: 32–37.

Ghei OK, Lacks SA (1967) Recovery of donor deoxyribonucleic acid marker activity from eclipse in pneumococcal transformation. *J. Bacteriol.* 93: 816–829.

Giebink GS (2000) Otitis media: the chinchilla model. In: Tomasz A (ed.) *Streptococcus pneumoniae*. New York: Mary Ann Liebert, p. 501.

Gilks CF, Ojoo SA, Ojoo JC *et al.* (1996) Invasive pneumococcal disease in a cohort of predominantly HIV-1 infected female sex-workers in Nairobi, Kenya. *Lancet* 347: 718–723.

Gingles NA, Alexander JE, Kadioglu A *et al.* (2001) The role of genetic susceptibility in invasive pneumococcal infection-identification and study of susceptible and resistant inbred mouse strains. *Infect. Immun.* 69: 426–434.

Goldsmith CE, Moore JE, Murphy PG (1997) Pneumococcal resistance in the UK. *J. Antimicrob. Chemother.* 40 (Suppl. A): 11–18.

Goldstein E, Lippert W, Warshauer D (1974) Pulmonary alveolar macrophage: defender against bacterial infection of the lung. *J. Clin. Invest.* 54: 519–528.

Gordon JR, Galli SJ (1990) Mast cells as a source of both preformed and immunologically inducible TNFα/ cachectin. *Nature* 346: 274–276.

Gordon JR, Galli SJ (1991) Release of both preformed and newly synthesised tumor necrosis factor α (TNFα)/ cachectin by mouse mast cells stimulated via the FcεRI. A mechanism for the sustained action of mast cell-derived TNFα during IgE-dependent biological responses. *J. Exp. Med.* 174: 103–107.

Gosink K, Mann ER, Guglielmo C, Tuomanen EI, Masure HR (2000) Role of novel choline binding proteins in virulence of *Streptococcus pneumoniae*. *Infect. Immun.* 68: 5690–5695.

Gratten M, Naraqi S, Hansmann D (1980) High prevalence of penicillin-insensitive pneumococci in Port Moresby, Papua New Guinea. *Lancet* 2:192–195.

Gray BM, Converse GMI, Dillon HCJ (1980) Epidemiologic studies of *Streptococcus pneumoniae* in infants: acquisition, carriage, and infection during the first 24 months of life. *J. Infect. Dis.* 142: 923–933.

Griffith F (1928) The significance of pneumococcal types. *J. Hyg.* 27: 113–159.

Guenzi E, Gasc A-M, Sicard MA, Hackenbeck R (1994) A two-component signal transducing system is involved in competence and penicillin susceptibility in laboratory mutants of *Streptococcus pneumoniae*. *Mol. Microbiol.* 12: 505–515.

Guidolin A, Morona JK, Morona R, Hansman D, Paton JC (1994) Nucleotide sequence of an operon essential for capsular biosynthesis in *Streptococcus pneumoniae* type 19F. *Infect. Immun.* 62: 5384–5396.

Hakansson L, Hallgren R, Venge P (1980) Regulation of granulocyte function by hyaluronic acid. *In vitro* and *in vivo* effects on phagocytosis. *J. Clin. Invest.* 66: 298.

Halbert SP, Cohen B, Perkins ME (1946) Toxic and immunological properties of pneumococcal hemolysin. *Bull. Johns Hopkins Hosp.* 78: 340–359.

Hammerschmidt S, Talay SR, Brandtzaeg P, Chhatwal GS (1997) SpsA, a novel pneumococcal surface protein with specific binding to secretory Immunoglobulin A and secretory component. *Mol. Microbiol.* 25: 1113–1124.

Hammerschmidt S, Bethe G, Remane PH, Chhatwal GS (1999) Identification of pneumococcal surface protein A as a lactoferrin-binding protein of *Streptococcus pneumoniae*. *Infect. Immun.* 67: 1683–1687.

Hansmann D, Bullen MM (1967) A resistant pneumococcus. *Lancet* 2: 264–265.

Hansmann D, Devitt L, Miles H, Riley I (1974) Pneumococci relatively insensitive to penicillin in Australia and New Guinea. *Med. J. Austr.* 2: 353–356.

Hardie JM (1984) Genus *Streptococcus* Rosenbach 1884: 22[AL]. In: Sneath PHA, Mair NS, Sharpe ME, Holt JG (eds) *Bergey's Manual of Systematic Bacteriology*. Baltimore: Williams & Wilkins, pp. 1043–1071.

Harford CG, Hara M (1950) Pulmonary edema in influenzal pneumonia of the mouse and the relation of fluid in the lung to the inception of pneumococcal pneumonia. *J. Exp. Med.* 91: 245–259.

Hart CA, Kariuki S (1998) Antimicrobial resistance in developing countries. *BMJ* 317: 647–650.

Haslett C (1992) Resolution of acute inflammation and the role of apoptosis in the tissue fate of granulocytes. *Clin. Sci.* 83: 639–648.

Haslett C (1997) Granulocyte apoptosis and inflammatory disease. *Br. Med. Bull.* 53: 669–683.

Havarstein LS, Diep DB, Nes IF (1995) A family of bacteriocin ABC transporters carry out proteolytic processing of their substrates concomitant with export. *Mol. Microbiol.* 16: 229–240.

Havarstein LS, Gaustad P, Nes IF, Morrison DA (1996) Identification of the streptococcal competence-pheromone receptor. *Mol. Microbiol.* 21: 863–869.

Henrichsen J (1995) Six newly recognised types of *Streptococcus pneumoniae*. *J. Clin. Microbiol.* 33: 2759–2762.

Himi T, Suzuki T, Kodama H, Takezawa H, Kataura A (1992) Immunologic characteristics of cytokines in otitis media with effusion. *Ann. Otol. Rhinol. Laryngol.* 101: 21–25.

Hirst RA, Sikand KS, Rutman A, Mitchell TJ, Andrew PW, O'Callaghan C (2000) Relative roles of pneumolysin and hydrogen peroxide from *Streptococcus pneumoniae* in inhibition of ependymal ciliary beat frequency. *Infect. Immun.* 68: 1557–1562.

Hoch JA (2000) Two-component and phosphorelay signal transduction. *Curr. Opin. Microbiol.* 3: 165–170.

Hogg JC (1994) The traffic of polymorphonuclear leukocytes through pulmonary microvessels in health and disease. *Am. J. Respir.* 163: 769–775.

Hogg JC, Walker BA (1995) Polymorphonuclear leukocyte traffic in lung inflammation. *Thorax* 50: 819–820.

Houldsworth S, Andrew PW, Mitchell TJ (1994) Pneumolysin stimulates production of tumor necrosis factor alpha and interleukin-1beta by human mononuclear phagocytes. *Infect. Immun.* 62: 1501–1503.

Humphrey JH (1948) Hyaluronidase production by pneumococci. *J. Pathol. Bacteriol.* 55: 273–275.

Inamizu T, Chang M-P, Makinodan T (1985) Influence of age on the production and regulation of interleukin-1 in mice. *Immunology* 55: 447–455.

Irwin CR, Schor SL, Ferguson MW (1994) Effects of cytokines on gingival fibroblasts *in vitro* are modulated by the extracellular matrix. *J. Periodontal Res.* 29: 309–317.

Jacobs T, Darji A, Frahm N *et al.* (1998) Listeriolysin O: cholesterol inhibits cytolysis but not binding to cellular membranes. *Mol. Microbiol.* 28: 1081–1089.

Jedrzejas MJ, Chantalat L, Mewbourne RB (1998a) Crystallization and preliminary X-ray analysis of *Streptococcus pneumoniae* hyaluronate lyase. *J. Struct. Biol.* 121: 73–75.

Jedrzejas MJ, Mewbourne RB, Chantalat L, McPherson DT (1998b) Expression and purification of *Streptococcus pneumoniae* hyaluronate lyase from *Escherichia coli*. *Protein Exp. Purif.* 13: 83–89.

Johnson MK (1977) Cellular location of pneumolysin. *FEMS Microbiol. Lett.* 2: 243–245.

Johnson MK, Knight RJ, Drew GK (1982) The hydrophobic nature of thiol-activated cytolysins. *Biochem. J.* 207: 557–560.

Johnson MK, Hobden JA, Hagenah M, O'Callaghan RJ, Hill JM, Chen S (1990) The role of pneumolysin in ocular infections with *Streptococcus pneumoniae*. *Curr. Eye Res.* 9: 1107–1114.

Johnson MK, Callegan MC, Engel LS *et al.* (1995) Growth and virulence of a complement-activation-negative mutant of *Streptococcus pneumoniae* in the rabbit cornea. *Curr. Eye Res.* 14: 281–285.

Johnston RB Jr (1981) The host response to invasion by *Streptococcus pneumoniae*: protection and the pathogenesis of tissue damage. *Rev. Infect. Dis.* 3: 282–287.

Johnston RB Jr (1991) Pathogenesis of pneumococcal pneumonia. *Rev. Infect. Dis.* 13 (Suppl. 6): S509–S517.

Johnston RB Jr, Newman SL (1977) Chronic granulomatous disease. *Pediatr. Clin. North Am.* 24: 365–376.

Jonsson S, Musher DM, Chapman A, Goree A, Lawrence EC (1985) Phagocytosis and killing of common bacterial pathogens of the lung by human alveolar macrophages. *J. Infect. Dis.* 152: 4–13.

Kadioglu A, Gingles N, Grattan K, Kerr A, Mitchell TJ, Andrew PW (2000) Host cellular immune response to pneumococcal lung infection in mice. *Infect. Immun.* 68: 492–501.

Kalin M (1998) Pneumococcal serotypes and their clinical relevance. *Thorax* 53: 159–162.

Kanclerski K, Mollby R (1987) Production and purification of *Streptococcus pneumoniae* hemolysin (pneumolysin). *J. Clin. Microbiol.* 25: 222–225.

Kawana M, Kawana C, Amesara R, Juhn SK, Giebink GS (1994) Neutrophil oxygen metabolite inhibition of cultured chinchilla middle ear epithelial cell growth. *Ann. Otol. Rhinol. Laryngol.* 103: 812–816.

Käyhty H, Eskola J (1996) New vaccines for the prevention of pneumococcal infections. *Emerg. Infect. Dis.* 2: 289–298.

Keenleyside WJ, Whitfield C (1996) A novel pathway for O-polysaccharide biosynthesis in *Salmonella enterica* serovar Borreze. *J. Biol. Chem.* 271: 28581–28592.

Kelly RT, Greiff D, Farmer S (1966) Neuraminidase activity in *Diplococcus pneumoniae*. *J. Bacteriol.* 91: 601–603.

Kelly SJ, Jedrzejas MJ (2000) Crystallization and preliminary X-ray diffraction analysis of a functional form of pneumolysin, a virulence factor from *Streptococcus pneumoniae*. *Acta Crystallogr.* D56: 1452–1455.

Kemper CA, Deresinki SC (1994) The immunology of pneumococcal pneumonia. In: Chmel H, Bendinelli M, Friedman H (eds) *Pulmonary Infections and Immunity*. New York: Plenum Press, pp. 29–49.

Kim YS, Kennedy S, Tauber MG (1995) Toxicity of *Streptococcus pneumoniae* in neurones, astrocytes, and microglia *in vitro*. *J. Infect. Dis.* 171: 1363–1368.

Kirschenbaum AS, Goff JP, Metcalfe DD (1991) Demonstration of the origin of human mast cells from CD34+ bone marrow progenitor cells. *J. Immunol.* 146: 1410–1414.

Kline BS, Internitz MC (1915) Studies upon experimental pneumonia in rabbits, VIII. Intra vitam staining in experimental pneumonia and circulation in the pneumonic lung. *J. Exp. Med.* 21: 311–319.

Kolkman MAB, van der Zeijst BAM, Nuijten PJM (1997) Functional analysis of glycosyltransferases encoded by the capsular polysaccharide locus of *Streptococcus pneumoniae* serotype 14. *J. Biol. Chem.* 272: 19502–19508.

Korchev YE, Bashford CL, Pederzolli C *et al.* (1998) A conserved tryptophan in pneumolysin is a determinant of the characteristics of channels formed by pneumolysin in cells and planar lipid bilayers. *Biochem. J.* 329: 571–577.

Kremlev SG, Phelps DS (1994) Surfactant protein A stimulation of inflammatory cytokine and immunoglobulin production. *Am. J. Physiol.* 267: L712–L719.

Lacks S (1962) Molecular fate of DNA in genetic transformation of pneumococcus. *J. Mol. Biol.* 5: 119–131.

Lamani E, McPhearson DT, Hollingshead SK, Jedrzejas MJ (2000) Production, characterisation, and crystallization of truncated forms of pneumococcal surface protein A from *Escherichia coli*. *Protein Exp. Purif.* 20: 379–388.

Lange R, Wagner C, deSaizieu A *et al.* (1999) Domain organisation and molecular characterisation of 13 two-component systems identified by genome sequencing of *Streptococcus pneumoniae. Gene* 237: 223–234.

Leach AJ, Gratten M, Mathews JD (1997) A model to explain the age-dependence of infection with particular pneumococcal serotypes. *Immunol. Cell Biol.* 75: A47.

Lee C-J (1987) Bacterial capsular polysaccharides- biochemistry, immunity and vaccine. *Mol. Immunol.* 24: 1005–1019.

Lee C-J, Banks SD, Lee JP (1991) Virulence, immunity, and vaccine related to *Streptococcus pneumoniae. Crit. Rev. Microbiol.* 18: 89–114.

Lee MS, Morrison DA (1999) Identification of a new regulator in *Streptococcus pneumoniae* linking quorum sensing to competence for genetic transformation. *J. Bacteriol.* 181: 5004–5016.

Libman E (1905) A pneumococcus producing a peculiar form of hemolysis. *Proc. N.Y. Pathol. Soc.* 5: 168.

Linder TE, Daniels RL, Lim DJ, DeMaria TF (1994) Effect of intranasal inoculation of *Streptococcus pneumoniae* on the structure of the surface carbohydrates of the chinchilla eustachian tube and middle ear mucosa. *Microb. Pathog.* 16: 435–441.

Llull R, Lopez R, Garcia E, Munoz R (1998) Molecular structure of the gene cluster responsible for the synthesis of the polysaccharide capsule of *Streptococcus pneumoniae* type 33F. *Biochim. Biophys. Acta* 1449: 217–224.

Lock RA, Paton JC, Hansman D (1988) Comparative efficacy of pneumococcal neuraminidase and pneumolysin as immunogens protective against *Streptococcus pneumoniae. Microb. Pathog.* 5: 461–467.

Lock RA, Zhang QY, Berry AM, Paton JC (1996) Sequence variation in the *Streptococcus pneumoniae* pneumolysin gene affecting haemolytic activity and electrophoretic mobility of the toxin. *Microb. Pathog.* 21: 71–83.

Loeliger AE, Rijkers GT, Aerts P (1995) Deficient anti-pneumococcal polysaccharide responses in HIV seropositive patients. *FEMS Immunol. Med. Microbiol.* 12: 33–41.

Lohmann-Matthes ML, Steinmuller C, Franke-Ullmann G (1994) Pulmonary macrophages. *Eur. Respir. J.* 7: 1678–1689.

Loosli CG, Baker RF (1962) Acute experimental pneumococcal (type I) pneumonia in the mouse: the migration of leucocytes from the pulmonary capillaries into the alveolar spaces as revealed by the electron microscope. *Trans. Am. Clin. Climatol. Assoc.* 74: 15–28.

Lopez R, Garcia E, Ronda C (1981) Bacteriophages of *Streptococcus pneumoniae. Rev. Infect. Dis.* 3: 212–223.

Lorentz A, Schwengberg S, Sellge G, Manns MP, Bischoff SC (2000) Human intestinal mast cells are capable of producing different cytokine profiles: role of IgE receptor cross-linking and IL-4. *J. Immunol.* 164: 43–48.

Lowell SH, Juhn SK, Giebink GS (1980) Experimental otitis media following middle ear inoculation of nonviable *Streptococcus pneumoniae. Ann. Otol. Rhinol. Laryngol.* 89: 479–482.

Lukacs NW, Ward PA (1996) Inflammatory mediators, cytokines, and adhesion molecules in pulmonary inflammation and injury. *Adv. Immunol.* 62: 257–304.

Madan T, Eggleton P, Kishore U *et al.* (1997) Binding of pulmonary surfactant proteins A and D to *Aspergillus fumigatus* conidia enhances phagocytosis and killing by human neutrophils and alveolar macrophages. *Infect. Immun.* 65: 3171–3179.

Madsen M, Lebenthal Y, Cheng Q, Smith BL, Hostetter MK (2000) A pneumococcal protein that elicits interleukin-8 from pulmonary epithelial cells. *J. Infect. Dis.* 181: 1330–1336.

Malaviya R, Ikeda T, Ross E, Abraham SN (1996) Mast cell modulation of neutrophil influx and bacterial clearance at sites of infection through TNF-α. *Nature* 381: 77–80.

Marks JD, Marks CB, Luce JM *et al.* (1990) Plasma tumor necrosis factor in patients with septic shock. *Am. Rev. Respir. Dis.* 141: 94–97.

Maurer M, Echenacher B, Håltner L *et al.* (1998) The c-kit ligand, stem cell factor, can enhance innate immunity through effects on mast cells. *J. Exp. Med.* 188: 2343–2348.

McDaniel LS, Yother J, Vijayakumar M, Mcgarry L, Guild WR, Briles DE (1987) Use of insertional inactivation to facilitate studies of biological properties of pneumococcal surface protein-a (PspA). *J. Exp. Med.* 165: 381–394.

McDaniel LS, McDaniel DO, Hollingshead SK, Briles DE (1998) Comparison of the PspA sequence from *Streptococcus pneumoniae* EF5668 to the previously identified PspA sequence from strain Rx1 and ability of PspA from EF5668 to elicit protection against pneumococci of different capsular types. *Infect. Immun.* 66: 4748–4754.

McIntosh JC, Mervin-Blake S, Conner E, Wright JR (1996) Surfactant protein A protects growing cells and reduces TNFα activity from LPS-stimulated macrophages. *Am. J. Physiol.* 271: L310–L319.

McKee CM, Penno MB, Cowman M *et al.* (1996) Hyaluronan (HA) fragments induce chemokine gene-expression in alveolar macrophages – the role of HA size and CD44. *J. Clin. Invest.* 98: 2403–2413.

McKinsey DS, Bisno AL (1980) Pneumonias caused by gram-positive bacteria. In: Fishman AP (ed.) *Pulmonary Diseases and Disorders,* 2nd edn. New York: McGraw-Hill, p. 1477.

Mecheri S, David B (1997) Unravelling the mast cell dilemma: culprit or victim of its generosity. *Immunol. Today* 18: 212–215.

Mejean V, Claverys JP (1993) DNA processing during entry in transformation of *Streptococcus pneumoniae. J. Biol. Chem.* 268: 5594–5599.

Mitchell TJ, Walker JA, Saunders FK, Andrew PA, Boulnois GJ (1989) Expression of the pneumolysin gene in *Escherichia coli*: rapid purification and biological properties. *Biochim. Biophys. Acta.* 1007: 67–72.

Mitchell TJ, Andrew PW, Saunders FK, Smith AN, Boulnois GJ (1991) Complement activation and antibody binding by pneumolysin via a region of the toxin homologous to a human acute-phase protein. *Mol. Microbiol.* 5: 1883–1888.

Moore D, Nelson M, Henderson D (1998) Pneumococcal vaccination and HIV infection. *Int. J. STD AIDS* 9: 1–7.

Morgan PJ, Andrew PW, Mitchell TJ (1996) Thiol-activated cytolysins. *Rev. Med. Microbiol.* 7: 221–229.

Morissette C, Skamene E, Gervais F (1995) Endobronchial inflammation following *Pseudomonas aeruginosa* infection in resistant and susceptible strains of mice. *Infect. Immun.* 63: 1718–1724.

Morona JK, Morona R, Paton JC (1997a) Characterisation of the locus encoding *Streptococcus pneumoniae* type 19F capsular biosynthetic pathway. *Mol. Microbiol.* 23: 751–763.

Morona JK, Morona R, Paton JC (1997b) Molecular and genetic characterization of the capsule biosynthesis locus of *Streptococcus pneumoniae* type 19B. *J. Bacteriol.* 179: 4953–4958.

Morona JK, Miller DC, Coffey TJ *et al.* (1999a) Molecular and genetic characterization of the capsule biosynthesis locus of *Streptococcus pneumoniae* type 23F. *Microbiology* 145: 781–789.

Morona JK, Morona R, Paton JC (1999b) Comparative genetics of capsular polysaccharide biosynthesis in *Streptococcus pneumoniae* types belonging to serogroup 19. *J. Bacteriol.* 181: 5355–5364.

Morrison DA, Baker M (1972) Transformation and DNA size: extent of degradation on entry varies with size. *J. Bacteriol.* 112: 1157–1168.

Morrison DA, Baker ME (1979) Competence for genetic transformation in pneumococcus depends on synthesis of a small set of proteins. *Nature* 282: 215–217.

Mostov KE, Kaetzel CS (1999) Immunoglobulin transport and the polymeric immunoglobulin receptor. In: Ogra PL, Mestecky J, Stroeber LMEW, Bienenstock J (eds) *Mucosal Immunology*. San Diego: Academic Press, pp. 181–211.

Mulholland K (1999) Strategies for the control of pneumococcal diseases. *Vaccine* 17: S79–S84.

Muller-Graf CDM, Whatmore AM, King SJ *et al.* (1999) Population biology of *Streptococcus pneumoniae* isolated from oropharyngeal carriage and invasive disease. *Microbiology UK* 145: 3283–3293.

Munoz R, Mollerach M, Lopez R, Garcia E (1997) Molecular organization of the genes required for the synthesis of type 1 capsular polysaccharide of *Streptococcus pneumoniae*: formation of binary encapsulated pneumococci and identification of cryptic dTDP-rhamnose biosynthesis genes. *Mol. Microbiol.* 25: 79–92.

Murray CJL, Lopez AD (1996) *Global Health Statistics*. Boston: Harvard School of Public Health.

Musher DM (1992) Infections caused by *Streptococcus pneumoniae*: clinical spectrum, pathogenesis, immunity, and treatment. *Clin. Infect. Dis.* 14: 801–809.

Nandoskar N, Ferrante A, Bates EJ, Hurst N, Paton JC (1986) Inhibition of human monocyte respiratory burst, degranulation, phospholipid methylation and

bactericidal activity by pneumolysin. *Immunology* 59: 515–520.

Neeleman C, Geelen SPM, Aerts PC *et al.* (1999) Resistance to both complement activation and phagocytosis in type 3 pneumococci is mediated by the binding of complement regulatory protein factor H. *Infect. Immun.* 67: 4517–4524.

Neill JM (1926) Studies on the oxidation and reduction of immunological substances. I. Pneumococcus hemotoxin. *J. Exp. Med.* 44: 199–213.

Neill JM (1927) Studies on the oxidation and reduction of immunological substances. V. Production of anti-hemotoxin by immunization with oxidized pneumococcus hemotoxin. *J. Exp. Med.* 45: 105–113.

Nelson S, Summer WR (1998) Innate immunity, cytokines and pulmonary host defense. *Infect. Dis. Clin. North Am.* 12: 555–567.

Nelson S, Mason CM, Kolls J, Summer WR (1995) Pathophysiology of pneumonia. *Clin. Chest Med.* 16: 1–12.

Neufeld F (1902) Ueber die Agglutination der Pneumokokken und aber die Theorien de Agglutination. *Z. Hyg. Infect.* 40: 54–72.

Newman SL, Vogler LB, Feigin RD, Johnston RBJ (1978) Recurrent septicemia associated with congenital deficiency of C2 and partial deficiency of factor B and the alternative complement pathway. *N. Engl. J. Med.* 299: 290–292.

Niederman MS, Bass JH Jr, Campbell GD *et al.* (1993) Guidelines for the initial management of adults with community-acquired pneumonia: diagnosis, assessment of severity, and initial antimicrobial therapy. *Am. Rev. Respir. Dis.* 148: 1418–1426.

Novak R, Braun JS, Charpentier E, Tuomanen E (1998) Penicillin tolerance genes of *Streptococcus pneumoniae*: the ABC-type manganese permease complex Psa. *Mol. Microbiol.* 29: 1285–1296.

Novak R, Cauwels A, Charpentier E, Tuomanen E (1999a) Identification of a *Streptococcus pneumoniae* gene locus encoding proteins of an ABC phosphate transporter and a two-component regulatory system. *J. Bacteriol.* 181: 1126–1133.

Novak R, Henriques B, Charpentier E, Normark S, Tuomanen E (1999b) Emergence of vancomycin tolerance in *Streptococcus pneumoniae*. *Nature* 399: 590–593.

Novak R, Charpentier E, Braun JS *et al.* (2000) Extracellular targeting of choline-binding proteins in *Streptococcus pneumoniae* by a zinc metalloprotease. *Mol. Microbiol.* 36: 366–376.

O'Toole RD, Goode L, Howe C (1971) Neuraminidase activity in bacterial meningitis. *J. Clin. Invest.* 50: 979–985.

Obaro SK, Monteil MA, Henderson DC (1996) The pneumococcal problem. *BMJ* 312: 1521–1525.

Orman KL, Shenep JL, English BK (1998) Pneumococci stimulate the production of the inducible nitric oxide synthase and nitric oxide by murine macrophages. *J. Infect. Dis.* 178: 1649–1657.

Owen RHGO, Boulnois GJ, Andrew PW, Mitchell TJ (1994) A role in cell-binding for the C-terminus of pneumolysin, the thiol-activated toxin of *Streptococcus pneumoniae*. *FEMS Lett.* 121: 217–222.

Pabst R (1997) Localisation and dynamics of lymphoid cells in the different compartments of the lung. In: Stockley RA (ed.) *Pulmonary Defences.* Chichester: John Wiley & Sons, pp. 59–75.

Pasteur L (1881) Note sur la Maladie Nouvelle Provoquçe par la Salive d'un Enfant Mort de la Rage. *Bull. l'Acad. Med. (Paris)* 10: 94–103.

Paton JC (1996) The contribution of pneumolysin to the pathogenicity of *Streptococcus pneumoniae*. *Trends Microbiol.* 4: 103–106.

Paton JC, Ferrante A (1983) Inhibition of human polymorphonuclear leukocyte respiratory burst, bactericidal activity, and migration by pneumolysin. *Infect. Immun.* 41: 1212–1216.

Paton JC, Morona JK (2000) *Streptococcus pneumoniae* capsular polysaccharide. In: Fischetti VA, Novick RP, Ferretti JJ, Portnoy DA, Rood JI. *Gram-positive Pathogens.* Washington: ASM Press, pp. 201–213.

Paton JC, Lock RA, Hansman DJ (1983) Effect of immunization with pneumolysin on survival time of mice challenged with *Streptococcus pneumoniae*. *Infect. Immun.* 40: 548–552.

Paton JC, Rowan-Kelly B, Ferrante A (1984) Activation of human complement by the pneumococcal toxin pneumolysin. *Infect. Immun.* 43: 1085–1087.

Paton JC, Berry AM, Lock RA, Hansman D, Manning PA (1986) Cloning and expression in *Escherichia coli* of the *Streptococcus pneumoniae* gene encoding pneumolysin. *Infect. Immun.* 54: 50–55.

Paul J, Kimari J, Gilks CF (1995) *Streptococcus pneumoniae* resistant to penicillin and tetracycline associated with HIV seropositivity. *Lancet* 346: 1034–1035.

Pesola GR, Charles A (1992) Pneumococcal bacteraemia with pneumonia: mortality in acquired immunodeficiency syndrome. *Chest* 101: 150–155.

Pestova EV, Havarstein LS, Morrison DA (1996) Regulation of competence for genetic transformation in *Streptococcus pneumoniae* by an autoinduced peptide pheromone and a two-component regulatory system. *Mol. Microbiol.* 21: 853–862.

Pison U, Max M, Neuendank A, Weibbach S, Pietschmann S (1994) Host–defense capacities of pulmonary surfactant – evidence for nonsurfactant functions of the surfactant system. *Eur. J. Clin. Invest.* 24: 586–599.

Plant JE, Blackwell JM, O'Brien AD, Bradley DJ, Glynn AA (1982) Are the *Lsh* and *Ity* disease resistance genes at one locus on mouse chromosome 1? *Nature* 297: 510–511.

Poole MD (1995) Otitis media complications and treatment failures: implications of pneumococcal resistance. *Pediatr. Infect. Dis. J.* 14: S23–S26.

Poulter LW (1997) Pulmonary macrophages. In: Stockley RA (ed.) *Pulmonary Defences.* Chichester: John Wiley & Sons, pp. 77–92.

Pozzi G, Masala I, Iannelli F *et al.* (1996) Competence for genetic transformation in encapsulated strains of *Streptococcus pneumoniae*: two allelic variants of the peptide pheromone. *J. Bacteriol.* 178: 6087–6090.

Quagliarello V, Scheld WM (1992) Bacterial meningitis: pathogenesis, pathophysiology and progress. *N. Engl. J. Med.* 327: 864–872.

Redd SC, Rutherford GW, Sande MA *et al.* (1990) The role of human immunodeficiency virus in pneumococcal bacteraemia in San Francisco residents. *J. Infect. Dis.* 162: 1012–1017.

Reynolds HY (1987) Bronchoalveolar lavage. *Am. Rev. Respir. Dis.* 135: 250–263.

Richards CD, Gauldie J (1985) IgA mediated phagocytosis by mouse alveolar macrophages. *Am. Rev. Respir. Dis.* 132: 82–85.

Rimini R, Jansson B, Feger G *et al.* (2000) Global analysis of transcription kinetics during competence development in *Streptococcus pneumoniae* using high density DNA arrays. *Mol. Microbiol.* 36: 1279–1292.

Ring A, Weiser JN, Tuomanen EI (1998) Pneumococcal trafficking across the blood–brain barrier – molecular analysis of a novel bidirectional pathway. *J. Clin. Invest.* 102: 347–360.

Ripley-Petzoldt ML, Giebink GS, Juhn S *et al.* (1988) The contribution of pneumococcal cell wall to the pathogenesis of experimental otitis media. *J. Infect. Dis.* 157: 245–255.

Rodriguez-Barradas MC, Musher DM, Lahart C *et al.* (1992) Antibody to capsular polysaccharides of *Streptococcus pneumoniae* after vaccination of human-immunodeficiency-virus infected subjects with 23-valent pneumococcal vaccine. *J. Infect. Dis.* 165: 553–556.

Roghmann KJ, Tabloski PA, Bentley DW, Schiffman G (1987) Immune response of elderly adults to the pneumococcus: variation by age, sex, and functional impairment. *J. Gerontol.* 42: 265–270.

Romero A, López R, García P (1990) Sequence of the *Streptococcus pneumoniae* bacteriophage HB-3 amidase reveals high homology with the major host autolysin. *J. Bacteriol.* 172: 5064–5070.

Rosen FS, Janeway CA (1966) The gamma globulins. III. The antibody deficiency syndromes. *N. Engl. J. Med.* 275: 709–715.

Rosenow C, Ryan P, Weiser JN *et al.* (1997) Contribution of novel choline-binding proteins to adherence, colonization and immunogenicity of *Streptococcus pneumoniae*. *Mol. Microbiol.* 25: 819–829.

Rossjohn J, Feil SC, McKinstry WJ, Tweten RK, Parker MW (1997) Structure of a cholesterol-binding, thiol-activated cytolysin and a model of its membrane form. *Cell* 89: 685–692.

Rossjohn J, Gilbert RJC, Crane D *et al.* (1998) The molecular mechanism of pneumolysin, a virulence factor from *Streptococcus pneumoniae*. *J. Mol. Biol.* 284: 449–461.

Rubins JB, Pomeroy C (1997) Role of gamma interferon in the pathogenesis of bacteremic pneumococcal pneumonia. *Infect. Immun.* 65: 2975–2977.

Rubins JB, Duane PG, Charboneau D, Janoff EN (1992) Toxicity of pneumolysin to pulmonary endothelial cells *in vitro*. *Infect. Immun.* 60: 1740–1746.

Rubins JB, Duane PG, Clawson D, Charboneau D, Young J, Niewoehner DE (1993) Toxicity of pneumolysin to pulmonary alveolar epithelial cells. *Infect. Immun.* 61: 1352–1358.

Rubins JB, Mitchell TJ, Andrew PW, Niewoehner DE (1994) Pneumolysin activates phospholipase A in pulmonary artery endothelial cells. *Infect. Immun.* 62: 3829–3836.

Rubins JB, Charboneau D, Paton JC, Mitchell TJ, Andrew PW, Janoff EN (1995) Dual function of pneumolysin in the early pathogenesis of murine pneumococcal pneumonia. *J. Clin. Invest.* 95: 142–150.

Saez-Llorens X, Ramilo O, Mustafa MM, Mertsola GHM Jr (1990) Molecular pathophysiology of bacterial meningitis: current concepts and therapeutic interventions. *J. Pediatr.* 116: 671–684.

Saluja SK, Weiser JN (1995) The genetic-basis of colony opacity in *Streptococcus pneumoniae* – evidence for the effect of box elements on the frequency of phenotypic variation. *Mol. Microbiol.* 16: 215–227.

Sanchez-Beato AR, Lopez R, Garcia JL (1998) Molecular characterization of PcpA: a novel choline-binding proteinof *Streptococcus pneumoniae*. *FEMS Microbiol.* 164: 207–214.

Sato K, Quartey MK, Liebeler CL, Le CT, Giebink GS (1996) Roles of autolysin and pneumolysin in middle ear inflammation caused by a type 3 *Streptococcus pneumoniae* strain in the chinchilla otitis media model. *Infect. Immun.* 64: 1140–1145.

Sato Y, van Eeden SF, English D, Hogg JC (1998) Bacteremic pneumococcal pneumonia: bone marrow release and pulmonary sequestration of neutrophils. *Crit. Care Med.* 26: 501–509.

Sato K, Liebeler CL, Quartey M, Le CT, Giebink GS (1999) Middle ear fluid cytokine and inflammatory cell kinetics in the chinchilla otitis media model. *Infect. Immun.* 67: 1943–1946.

Saukkonen K, Sande S, Cioffe C *et al.* (1990) The role of cytokines in the generation of inflammation and tissue-damage in experimental gram-positive meningitis. *J. Exp. Med.* 171: 439–448.

Saunders FK, Mitchell TJ, Walker JA, Andrew PW, Boulnois GJ (1989) Pneumolysin, the thiol-activated toxin of *Streptococcus pneumoniae*, does not require a thiol group for *in vitro* activity. *Infect. Immun.* 57: 2547–2552.

Schagat TL, Tino MJ, Wright JR (1999) Regulation of protein phosphorylation and pathogen phagocytosis by surfactant protein A. *Infect. Immun.* 67: 4693–4699.

Schlech WF, Ward JI, Bard JD, Hightower A, Fraser DW, Broome CV (1985) Bacterial meningitis in the United States, 1978 through 1981. The National Meningitis Surveillance Study. *JAMA* 253: 1749–1754.

Schneewind O, Fowler A, Faull K (1995) Structure of the cell wall anchor of surface proteins in *Staphylococcus aureus*. *Science* 268: 103–106.

Schütze H, Gorer PA, Finlayson MH (1936) The resistance of four mouse lines to bacterial infection. *J. Hyg.* 36: 37–49.

Scott JAG, Hall AJ, Dagan R *et al.* (1996) Serotype-specific epidemiology of *Streptococcus pneumoniae*: associations with age, sex, and geography in 7,000 episodes of invasive disease. *Clin. Infect. Dis.* 22: 973–981.

Sestini P, Nencioni L, Villa L, Boraschi D, Tagliabue A (1987) Antibacterial activity against *Streptococcus pneumoniae* by mouse lung lymphocytes. *Adv. Exp. Med. Biol.* 216: 517–525.

Sestini P, Nencioni L, Villa L, Boraschi D, Tagliabue A (1988) IgA-driven antibacterial activity against *Streptococcus pneumoniae* by mouse lung lymphocytes. *Am. Rev. Respir. Dis.* 137: 138–143.

Sha WC, Hsiou-Chi L, Tuomanen EI, Baltimore D (1995) Targeted disruption of the p50 subunit of NF-κB leads to multifocal defects in immune responses. *Cell* 80: 321–330.

Shin HS, Smith MR, Wood WB Jr (1969) Heat labile opsonins to pneumococcus II. Involvement of C3 and C5. *J. Exp. Med.* 130: 1229–1240.

Simon RH, Paine RI (1995) Participation of pulmonary alveolar epithelial cell in lung inflammation. *J. Lab. Clin. Med.* 126: 108–118.

Simons RJ, Reynolds HY (1990) Altered immune status in the elderly. *Semin. Respir. Infect.* 5: 251–259.

Skamene E, Kongshavn PAL, Sachs DH (1979) Resistance to *Listeria monocytogenes* in mice: genetic control by genes that are not linked to the H-2 complex. *J. Infect. Dis.* 139: 228–231.

Smith BL, Hostetter MK (2000) C3 as a substrate for adhesion of *Streptococcus pneumoniae*. *J. Infect. Dis.* 182: 497–508.

Smith JD, Cortes NJ, Evans GS, Read RC, Finn A (1998) Induction of β_2 integrin-dependent neutrophil adhesion to human alveolar epithelial cells by type 1 *Streptococcus pneumoniae* and derived soluble factors. *J. Infect. Dis.* 177: 977–985.

Smyth CJ, Duncan JL (1978) Thiol-activated (oxygen-labile) cytolysins. In: Jeljaszewicz J, Wadstrom T (eds) *Bacterial Toxins and Cell Membranes*. London: Academic Press, pp. 129–183.

Songlin L, Kelly SJ, Lamani E, Ferraroni M, Jedrzejas MJ (2000) Structural basis of hyaluronan degradation by *Streptococcus pneumoniae* hyaluronate lyase. *EMBO J.* 19: 1228–1240.

Sorensen TIA, Nielsen GG, Andersen PK, Teasdale TW (1988) Genetic and environmental influences on premature death in adult adoptees. *N. Engl. J. Med.* 318: 727–732.

Spellerberg B, Cundell DR, Sandros J *et al.* (1996) Pyruvate oxidase, as a determinant of virulence in *Streptococcus pneumoniae*. *Mol. Microbiol.* 19: 803–813.

Steinfort C, Wilson R, Mitchell T *et al.* (1989) Effect of *Streptococcus pneumoniae* on human respiratory epithelium *in vitro*. *Infect. Immun.* 57: 2006–2013.

Stephens KE, Shizaka A, Larrick JW, Raffin TA (1987) Tumor necrosis factor causes increased pulmonary permeability and edema. *Am. Rev. Respir. Dis.* 137: 1364–1370.

Sternberg GM (1881) A fatal form of septicaemia in the rabbit produced by the subcutaneous injection of human saliva. *Annu. Rep. Natl Board Health* 3: 87–108.

Stockley RA (1997) Soluble proteins in lung defence. In: Stockley RA (ed.) *Pulmonary Defences.* Chichester: John Wiley and Sons, pp. 18–38.

Strunk RC, Eidlen DM, Mason RJ (1988) Pulmonary alveolar type II epithelial cells synthesize and secrete proteins of the classical and alternative complement pathways. *J. Clin. Invest.* 81: 1419–1426.

Szarek JC, Bailly DA, Stewart NL, Gruetter CA (1992) Histamine H(1) receptors mediate endothelium-dependent relaxation of rat isolated pulmonary arteries. *Pulmonary Pharmacol.* 5: 67–74.

Takishima K, Tateda K, Matsumoto T, Iizawa Y, Nakao M, Yamaguchi K (1997) Role of TNF in the pathogenesis of pneumococcal pneumonia in mice. *Infect. Immun.* 65: 257–260.

Talkington DE, Koenig A, Russell H (1992) The 37-kDa protein of *Streptococcus pneumoniae* protects mice against fatal challenge. In: *92nd General Meeting of the American Society for Microbiology.* Washington, DC: American Society for Microbiology, p. 149.

Talkington DF, Brown BG, Tharpe JA, Koenig A, Russell H (1996) Protection of mice against fatal pneumococcal challenge by immunization with pneumococcal surface adhesin A (PsaA) *Microb. Pathog.* 21: 17–22.

Teele DW, Klein JO, Rosner B (1989) Epidemiology of otitis media during the first seven years of life in children in Great Boston: a prospective, cohort study. *J. Infect. Dis.* 160: 83–94.

Terr AI (1994) Inflammation. In: Stites DP, Terr AI, Parslow TG (eds) *Basic and Clinical Immunology,* 8th edn. East Norwalk: Appleton and Lange. pp. 137–150.

Throup JP, Koretke KK, Bryant AP *et al.* (2000) A genomic analysis of two-component signal transduction in *Streptococcus pneumoniae.* *Mol. Microbiol.* 35: 566–576.

Toews GB, Gross GN, Pierce AK (1979) The relationship of inoculum size to lung bacterial clearance and phagocytic cell response in mice. *Am. Rev. Respir. Dis.* 120: 559–566.

Tomasz A (1981) Surface components of *Streptococcus pneumoniae.* *Rev. Infect. Dis.* 3: 190–211.

Tong HH, Fisher LM, Kosunick GM, DeMaria TF (1999) Effect of tumor necrosis factor alpha and interleukin 1-alpha on the adherence of *Streptococcus pneumoniae* to chinchilla tracheal epithelium. *Acta Oto-Laryngol.* 119: 78–82.

Tong HH, Blue LE, James MA, DeMaria TF (2000) Evaluation of the virulence of a *Streptococcus pneumoniae* neuraminidase-deficient mutant in nasopharyngeal colonization and development of otitis media in the chinchilla model. *Infect. Immun.* 68: 921–924.

Tu A-HT, Fulgram RL, McCrory MA, Briles DE, Szalai AJ (1999) Pneumococcal surface protein A inhibits complement activation by *Streptococcus pneumoniae.* *Infect. Immun.* 67: 4720–4724.

Tuomanen EI, Masure HR (1997) Molecular and cellular biology of pneumococcal infection. *Microb. Drug Resist. Mech. Epidemiol. Dis.* 3: 297–308.

Tuomanen E, Liu H, Hengstler B, Zak O, Tomasz A (1985) The induction of meningeal inflammation by components of the pneumococcal cell wall. *J. Infect. Dis.* 151: 859–868.

Tuomanen E, Rich R, Zak O (1987) Induction of pulmonary inflammation by components of the pneumococcal cell surface. *Am. Rev. Respir. Dis.* 135: 868–874.

Tuomanen EI, Saukkonen Sande S, Cioffe C, Wright SD (1989) Reduction of inflammation, tissue damage, and mortality in bacterial meningitis in rabbits treated with monoclonal antibodies against adhesion-promoting receptors of leukocytes. *J. Exp. Med.* 70: 959–969.

Tuomanen EI, Austrian R, Masure HR (1995) Pathogenesis of pneumococcal infection. *N. Engl. J. Med.* 332: 1280–1284.

van den Berg RH, Faber-Krol MC, Sim RB, Daha MR (1998) The first subcomponent of complement, C1q triggers the production of IL-8, IL-6 and monocyte chemoattractant peptide-1 by human umbilical vein endothelial cells. *J. Immunol.* 161: 6924–6930.

van der Poll T, Marchant A, Keogh CV, Goldman M, Lowry SF (1996) Interleukin-10 impairs host defence in murine pneumococcal pneumonia. *J. Infect. Dis.* 174: 994–1000.

van der Poll T, Keogh CV, Buurman WA, Lowry SF (1997a) Passive immunisation against TNF impairs host defence during pneumococcal pneumonia in mice. *Am. J. Respir. Crit. Care Med.* 155: 603–608.

van der Poll T, Keogh CV, Guiro X, Buurman WA, Kopf M, Lowry SF (1997b) Interleukin-6 gene deficient mice show impaired defence against pneumococcal pneumonia. *J. Infect. Dis.* 176: 439–444.

van Iwaarden JF, Shimizu H, Van Golde PHM, Voelker DR, Van Golde LMG (1990) Pulmonary surfactant protein A enhances the host defense mechanism of rat alveolar macrophages. *Am. J. Respir. Cell Mol. Biol.* 2: 91–98.

van Iwaarden JF, Shimizu H, van Golde PHM, Voelker DR, van Golde LMG (1992) Rat surfactant protein D enhances the production of oxygen free radicals by rat alveolar macrophages. *Biochem. J.* 286: 5–8.

van Dam JEG, Fleer A, Snippe H (1990) Immunogenicity and immunochemistry of *Streptococcus pneumoniae* polysaccharides. *Antonie Van Leeuwenhoek* 58: 1–47.

Vial WC, Toews GB, Pierce AK (1984) Early pulmonary granulocyte recruitment in response to *Streptococcus pneumoniae.* *Am. Rev. Respir. Dis.* 129: 87–91.

Vidal SM, Malo D, Vogan K, Skamene E, Gros P (1993) Natural resistance to infection with intracellular parasites: isolation of a candidate for *Bcg.* *Cell* 73: 469–485.

Volanakis JE, Kaplan MH (1974) Interaction of C-reactive protein complexes with the complement system. II. Consumption of guinea pig complement by CRP complexes: requirement for human C1q. *J. Immunol.* 113: 9–17.

Walker DC, Chu F, MacKenzie A (1991) Pathway of leukocyte immigration from the pulmonary vasculature during pneumonia in rabbits. *Am. Rev. Respir. Dis.* 143: A541.

Walker JA, Allen RL, Falmagne P, Johnson MK, Boulnois GJ (1987) Molecular cloning, characterization, and complete nucleotide sequence of the gene for pneumolysin, the sulfhydryl-activated toxin of *Streptococcus pneumoniae*. *Infect. Immun.* 55: 1184–1189.

Wanner A (1977) Clinical aspects of mucociliary transport. *Am. Rev. Respir. Dis.* 116: 73–125.

Warner AE, Molina RM, Brain JD (1987) Uptake of blood-borne bacteria by pulmonary intravascular macrophages and consequent inflammatory responses in sheep. *Am. Rev. Respir. Dis.* 136: 683–690.

Ween O, Gaustad P, Håvarstein LS (1999) Identification of DNA binding sites for ComE, a key regulator of natural competence in *Streptococcus pneumoniae*. *Mol. Microbiol.* 33: 817–827.

Weiser JN, Tuomanen EI, Cundell DR, Sreenivasan PK, Austrian R, Masure HR (1994) The effect of colony opacity variation on pneumococcal colonization and adhesion. *J. Cell. Biochem.* Suppl. 18A: 55.

White RTM (1988) Pneumococcal vaccine. *Thorax* 43: 345–348.

Wicher K, Kalinka C, Mlodozeniec P, Rose NR (1982) Fluorescent antibody technique used for identification and typing of *Streptococcus pneumoniae*. *Am. J. Clin. Pathol.* 77: 72–77.

Widdowson CA, Klugman KP (1999) Molecular mechanisms of commonly used non-beta lactam drugs in *Streptococcus pneumoniae*. *Semin. Respir. Infect.* 14: 225–268.

Winkelstein JA, Swift AJ (1975) Host defense against the pneumococcus in T-lymphocyte-deficient, nude mice. *Infect. Immun.* 12: 1222–1223.

Winkelstein JA, Tomasz A (1977) Activation of the alternative complement pathway by pneumococcal cell walls. *J. Immunol.* 118: 451–454.

Winkelstein JA, Tomasz A (1978) Activation of the alternate complement pathway by pneumococcal cell wall teichoic acid. *J. Immunol.* 120: 174–178.

Winter AJ, Comis SD, Osbourne MP *et al.* (1996) Pneumolysin rather than neuraminidase is chiefly responsible for deafness during experimental pneumococcal meningitis in guinea pigs. *7th International Congress for Infectious Diseases*, Abstract 73.004.

World Bank (1993) *World Development Report.* Washington, DC: The World Bank.

Wright JR, Youmans DC (1993) Pulmonary surfactant protein A stimulates chemotaxis of alveolar macrophage. *Am. J. Physiol.* 264: L338–L344.

Yesilkaya H, Kadioglu A, Gingles N, Alexander JE, Mitchell TJ, Andrew PW (2000) The role of manganese containing superoxide dismutase (MnSOD) in oxidative stress and virulence of *Streptococcus pneumoniae*. *Infect. Immun.* 68: 2819–2826.

Yother J, Briles DE (1992) Structural properties and evolutionary relationships of PspA, a surface protein of *Streptococcus pneumoniae* as revealed by sequence analysis. *J. Bacteriol.* 174: 601–609.

Yother J, McDaniel LS, Briles DE (1986) Transformation of encapsulated *Streptococcus pneumoniae*. *J. Bacteriol.* 168: 1463–1465.

Zhang JR, IdanpaanHeikkila I, Fischer W, Tuomanen EI (1999) Pneumococcal licD2 gene is involved in phosphorylcholine metabolism. *Mol. Microbiol.* 31: 1477–1488.

Zhang J-R, Mostov KE, Lamm ME *et al.* (2000) The polymeric immunoglobulin receptor translocates pneumococci across human nosopharyngeal epithelial cells. *Cell* 102: 827–837.

77
Klebsiella pneumoniae

Steven Clegg and Tricia Ann Schurtz Sebghati

University of Iowa, USA

Klebsiella pneumoniae is an opportunistic pathogen frequently implicated in infections of compromised or debilitated individuals. It is, for example, commonly associated with nosocomially acquired infections and is the cause of respiratory disease in patients with chronic alcoholism (Carpenter, 1990; Craven *et al.*, 1990; Donaldson *et al.*, 1991; Jong *et al.*, 1995). The ability of many *K. pneumoniae* strains to resist the action of multiple antibiotics, including broad-spectrum cephalosporins and *β* lactams, is of great concern particularly in hospital environments where infection due to this organism can present serious problems (Gouby *et al.*, 1994; Jacoby, 1996). *K. pneumoniae* has been implicated in a wide range of hospital-acquired infections including septicaemia (Horan *et al.*, 1988), urinary (Williams and Tomas, 1990), and respiratory tract infections (Duma, 1985). Infections are frequently associated with high mortality rates, since eradication of antibiotic-resistant bacteria in compromised individuals is difficult to achieve. Among the Enterobacteriaceae, *K. pneumoniae* is a leading cause of nosocomially acquired pneumonias (Donaldson *et al.*, 1991), and recent epidemiological evidence suggests that the occurrence of *Klebsiella* infections may be significantly higher than was previously believed (Alphen *et al.*, 1995). For these reasons, the factors responsible for mediating the virulence of the organism have been widely investigated, and a number of bacterial attributes have been implicated as factors that contribute to the pathogenicity of *K. pneumoniae*.

K. pneumoniae has long been known to produce copious amounts of capsular material when cultured *in vitro* (Edwards and Fife, 1952). Since the capsule covers the bacterial surface, it is likely that *in vivo* this substance protects the organism from host defence mechanisms. In addition, the release of excess capsular polysaccharide from the bacterial surface may neutralise the bactericidal effect of specific antibodies. Another surface component that may contribute to the virulence of the organism is the lipopolysaccharide (LPS). The variability of the serum-sensitivity of different strains of *K. pneumoniae* appears to be a function of the LPS, and the binding of complement components is affected by the structure of this molecule (Ciurana and Tomas, 1987). Since *K. pneumoniae* is a mucosal pathogen that colonises epithelial surfaces as a pre-requisite to infection, the production of adherence factors contributes to the virulence of the bacteria. Like many other enteric pathogens, *Klebsiella* express a variety of adherence factors on their surfaces (Old and Adegbola, 1983; Old *et al.*, 1985; Hornick *et al.*, 1991; Podschun *et al.*, 1987). Since growth *in vivo* necessitates competition with the host for essential elemental iron, the bacteria need to scavenge this metal from the host iron-binding proteins. The production of high-affinity iron-binding molecules would, therefore, be advantageous during growth in susceptible hosts, and such iron-binding compounds have been identified in *K. pneumoniae* (Williams *et al.*, 1987; Vernet *et al.*, 1992; Podschun *et al.*, 1993). Finally, many pathogenic bacteria produce extracellular enzymes and toxins that contribute to virulence. Most strains of *K. pneumoniae* can produce urease that has been implicated as a putative virulence factor for isolates that cause urinary tract

 doi:10.1006/bkmm.2001.0077

Table 1 Virulence factors of *Klebsiella pneumoniae*

Factor	Role in virulence
Capsule	Anti-phagocytic
Lipopolysaccharide	Resistance to killing by serum
Fimbrial and non-fimbrial adhesins	Adherence and colonisation
Siderophores	Iron chelation and acquisition
Urease	Growth in the urinary tract

infections (Mobley *et al.*, 1995). These virulence factors are discussed in detail in this chapter and listed in **Table 1**. Current concepts of the molecular biology of these factors are described and how these contribute to the virulence of *K. pneumoniae* is presented.

The pathogenicity of *K. pneumoniae* is attributable to several virulence factors that act in concert during the infective process. In addition, as for all opportunistic pathogens, the status of the host plays a crucial role in development of the infection. Consequently, the mechanisms of *Klebsiella* virulence factor production are diverse and the importance of each factor may depend heavily on the site of its infection.

Capsules

Structure

Most clinical isolates of *K. pneumoniae* produce a hydrophilic polysaccharide capsule that surrounds the bacteria. These molecules are antigenically diverse and at least 77 capsular types of *K. pneumoniae* are recognised (Ørskov and Ørskov, 1978, 1984). This antigenic diversity is the result of an array of biochemical capsular structures and differences in chemical bonding patterns between identical sugar residues (Dutton *et al.*, 1989). The capsules are generally acidic polysaccharides and are homo- or heteropolymers that may be branched or unbranched (Lee, 1987; Dutton *et al.*, 1989). Glucuronic acid is the charged molecule that is most frequently present and uncharged molecules such as galactose, L-rhamnose, fucose, mannose and D-glucose have been identified in various capsular types (Ørskov and Ørskov, 1984; Beurret *et al.*, 1989). Some strains of *K. pneumoniae* also possess non-polysaccharide capsular constituents, such as pyruvate or acetate (Lee, 1987). Although there are large numbers of capsular serotypes, not all of these are frequently associated with infection (Cryz *et al.*, 1986; Podschun, 1990). For example, in a study of bacteraemia isolates

of *K. pneumoniae*, 18 serotypes were identified as responsible for 77% of all *K. pneumoniae* strains isolated from blood (Riser and Noone, 1981). Even though certain capsular serotypes are implicated in clinical infections due to *K. pneumoniae*, there appears to be no relationship between specific serotypes and the type of infection or site of isolation (Riser and Noone, 1981).

In spite of the existence of many serologic types of capsular polysaccharide, structural and biochemical analysis of these molecules has been carried out on only a limited number of serotypes. The K10 capsule of *K. pneumoniae* has been investigated in detail by gas–liquid chromatography and nuclear magnetic resonance techniques (Dutton *et al.*, 1989). The results suggest that the K10 polysaccharide consists of a hexasaccharide repeating unit made up of glucose, mannose and galactose sugars linked by α or β glycosidic bonds. The type I capsules produced by *K. pneumoniae* are characterised by the presence of oligosaccharide repeating subunits.

Genetics

A detailed genetic map of the *K. pneumoniae* chromosome has not been produced, but the genes (*cps*) involved in capsule biosynthesis have been mapped to a region close to the *his* locus. Laakso and colleagues (1988) used R-prime plasmids for the conjugative transfer of the genes that encode the K20 capsule from a strain of *K. pneumoniae* to *Escherichia coli*. His$^+$ transconjugants were isolated that were also phenotypically capsulate and expressed the K20 antigen. These experiments indicate that the *cps* genes are located as a discrete gene cluster on the bacterial chromosome. A second indication that the *cps* genes are clustered was obtained by exchange of *cps* genes of one serotype by those of another (Ofek *et al.*, 1993). In these experiments the *cps* genes that encode the expression of K2 and K21a capsules were exchanged while maintaining the *K. pneumoniae* parental genetic background.

The molecular analysis of the genes involved in *K. pneumoniae* capsular biosynthesis has focused on determinants that encode the K2 capsule, because this antigen has been implicated as an important virulence factor (Mizuta *et al.*, 1983). The successful cloning and expression of the K2 *cps* genes in *E. coli* carried on recombinant plasmids has facilitated the genetic dissection of these determinants. The K2 *cps* genes were on a 29 kb *Bam*HI restriction fragment prepared from genomic DNA of *K. pneumoniae* (**Fig. 1**). DNA sequence analysis of this region showed the presence of 17 open reading frames (ORFs) within the fragment, and Tn5 mutagenesis was used to map the boundaries

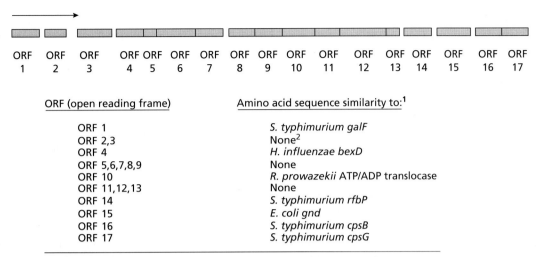

ORF (open reading frame)	Amino acid sequence similarity to:[1]
ORF 1	*S. typhimurium galF*
ORF 2,3	None[2]
ORF 4	*H. influenzae bexD*
ORF 5,6,7,8,9	None
ORF 10	*R. prowazekii* ATP/ADP translocase
ORF 11,12,13	None
ORF 14	*S. typhimurium rfbP*
ORF 15	*E. coli gnd*
ORF 16	*S. typhimurium cpsB*
ORF 17	*S. typhimurium cpsG*

[1] Sequences were compared with those of polypeptides deposited in the databases (Arakawa *et al.*, 1995)
[2] No significant similarity observed

Fig. 1 Genetic organisation of the *Klebsiella pneumoniae cps* gene cluster that encodes the K2 antigen.

of the nucleotide sequences required for K2 capsule expression. Insertion of the transposon into 12 of the 17 ORFs resulted in abrogation of K2 expression. A putative function for some of the gene products that may be encoded by the ORFs was assigned by comparison with proteins from other bacterial systems. Two ORFs could encode polypeptides closely related to the α-D-glucosyl-1-phosphate uridyltransferase of *Salmonella typhimurium* (Jiang *et al.*, 1991) and the 6-phosphogluconate dehydrogenase of *E. coli* (Arakawa *et al.*, 1995). Additionally, one of the putative *cps* gene products showed 81% identity with *S. enterica* phosphomanomutase (Jiang *et al.*, 1991). A 62% identity was observed between one of the ORFs and the mannose-1-phosphate guanyltransferase (Jiang *et al.*, 1991). Two other predicted ORF gene products were found to possess limited identity (26% and 27%, respectively) to the BexD protein of *Haemophilus influenzae* (Kroll *et al.*, 1990) and the RfbP protein of *S. typhimurium*. Both these proteins are involved in the transport of polysaccharides across cell membranes.

Amino acid sequence comparisons are intriguing in that they indicate possible *cps* gene product functions, and these are frequently associated with defined polysaccharide biosynthetic reactions in bacteria other than *K. pneumoniae*. The identification of precise roles for each of the *cps* genes and their products must, however, await a detailed analysis of defined *K. pneumoniae* mutants and the investigation of their phenotypes.

The DNA in the region in which the above ORFs are situated has a G + C content of less than 40%. It has therefore been suggested that this DNA was acquired by lateral transfer into *Klebsiella* from an unrelated organism (Arakawa *et al.*, 1995). Similar lateral gene transfer has also been postulated for the acquisition of certain lipopolysaccharide biosynthetic genes by *Salmonella* (Jiang *et al.*, 1991).

Phenotypic variation of capsule expression by *K. pneumoniae* has been observed *in vitro*. Quantitatively large amounts of capsular material are produced by bacteria grown in the presence of glucose and low nitrogen concentrations (Duguid, 1959). Conversely, in some strains of *K. pneumoniae*, capsule production is inhibited in the presence of salicylate (Domenico *et al.*, 1989). Although phenotypic variation of capsule expression *in vivo* has not been investigated, it can be postulated that capsule production may, under certain environmental conditions, be advantageous for the bacteria, while under other conditions it may be detrimental. It is therefore not surprising that preliminary investigations have indicated that the expression of *cps* genes may be regulated. These investigations have shown that expression of a mucoviscous phenotype by *K. pneumoniae* depends on the presence of a 180-kbp pair plasmid (Nassif and Sansonetti, 1986; Domenico *et al.*, 1989; Nassif *et al.*, 1989b). Loss of the plasmid was associated with the inability of the bacteria to express large amounts of extracellular, soluble viscous material, but it did not affect the expression of a surface-associated capsular polysaccharide. The gene responsible for this phenotype was designated *rmp*A and it was also shown to possess the ability to exert positive control on colanic acid synthesis in *E. coli* (Nassif and Sansonetti, 1986; Domenico *et al.*, 1989; Nassif *et al.*, 1989b). Subsequently, a second *rmp*A allele, designated *rmp*A2, was cloned from the plasmid of a virulent *K. pneumoniae* isolate (Arakawa *et al.*, 1991). This gene was necessary for the production of the K2 capsule by *E. coli* transformed with the *cps* genes. Analysis of the RmpA2 polypeptide suggested that it contains a central region with an amino acid sequence related to NtrC, UhpA and ArcB (Wacharotayankun *et al.*, 1993), which are regulators and act by association with the sigma factor of RNA polymerase. In addition, RmpA2 possesses a helix-turn-helix motif common to DNA-binding proteins. For these reasons, it has been postulated that RmpA2 is a transcriptional regulator and that it may act positively to control *cps* gene expression.

If RmpA is a positive regulator of capsule expression in *K. pneumoniae*, the difference between the mucoviscous phenotype of plasmid-bearing isolates and the non-mucoviscous but capsulate phenotype of strains without the plasmid may represent quantitative differences in the amount of capsular material produced. Comparison of the neutral sugar content of mucoviscous material from *Klebsiella* that possess the cloned *rmp*A gene and capsular polysaccharide from strains without the *rmp*A gene, indicate that there is no difference in composition but simply an excess production of capsule in the strains that possess RmpA (Wacharotayankun *et al.*, 1993). In addition, the introduction of *rmp*A2 into *K. pneumoniae* that produce K9 or K72 capsules, and which are only slightly mucoviscous, results in expression of large amounts of exopolysaccharide. If RmpA is a transcriptional activator of capsular polysaccharide synthesis genes, the precise site of its action has yet to be determined.

Apart from *rmp*A, the 180-kb pair plasmid has also shown to possess a second gene, *rmp*B, that may be involved in capsule expression (Nassif *et al.*, 1989b). The *rmp*B gene controls the expression of *rmp*A, so that loss of RmpB results in reduced *rmp*A expression. Conflicting reports have indicated that *rmp*A can increase capsule expression in the absence of *rmp*B (Nassif *et al.*, 1989a; Wacharotayankun *et al.*, 1993). In this case, however, the *rmp*A gene was carried on a multi-copy plasmid vector so that over-expression of *rmp*A could be achieved without *rmp*B. In wild-type isolates of *K. pneumoniae* the intracellular concentrations, and stoichiometric ratios, of RmpA and RmpB would be expected to differ considerably from those in strains that carry the cloned genes. This may be important, because the activity of regulatory proteins of any one genetic system is likely to be a function of their intracellular concentrations and ratios.

Role in Virulence

Not all capsular types of *K. pneumoniae* are isolated with equal frequency from clinical material (Riser and Noone, 1981; Cryz *et al.*, 1986; Nassif *et al.*, 1989b;

Tarkkanen *et al.*, 1992; Wacharotayankun *et al.*, 1993). This suggests that the mere presence of a capsule is not sufficient to mediate their virulence. Specific K types of *K. pneumoniae* are, however, associated more often with human infections, which suggests that the type of capsule produced by *K. pneumoniae* isolates may contribute to virulence. Moreover, it has been suggested that the amount of capsule produced plays a role in virulence, since there is a correlation between the quantity of capsule expressed by the infecting strain and lethality in an experimental animal model of infection (Domenico *et al.*, 1985). It is believed that, similar to the function of the capsule in *Streptococcus pneumoniae* (Lee, 1987), in the absence of specific antibodies capsules play a role in protecting the bacteria from phagocytosis (Simoons-Smit *et al.*, 1986; Lee, 1987; Williams and Tomas, 1990). This conclusion is based on experimental observations that capsule-specific antibodies increase the efficiency of opsonisation of bacteria by phagocytes (Williams *et al.*, 1983), and that non-capsule derivatives are generally less virulent than their capsulate parents (Williams *et al.*, 1990). Since it is not possible to investigate the role of the *K. pneumoniae* capsule *in vivo* in humans, the mouse has been extensively used as a model of infection (Williams and Tomas, 1990; Ofek *et al.*, 1993; Kabha *et al.*, 1995). *K. pneumoniae* that produces the K2 capsule has been used in most investigations, since this capsular type is frequently isolated from patients and is particularly prevalent among invasive strains associated with bacteraemia (Cryz *et al.*, 1986).

The role of the K2 capsule in virulence has been investigated by exchange of the genes (*cps*K2) that encode this capsule with those (*cps*21a) that encode the K21 capsule (Ofek *et al.*, 1993). These two capsular serotypes were selected since K2$^+$ isolates are frequently lethal after intraperitoneal inoculation into mice, while K21$^+$ isolates are much less virulent (Allen *et al.*, 1987a, b; Athamna *et al.*, 1991; Ofek *et al.*, 1993). The presence of the *his* locus adjacent to the *cps* genes provides a selectable marker for isolating genetically modified bacteria. The ability of the K21a parental strain to bind *in vitro* to macrophages was observed to be approximately five times greater than that of the K2$^+$ strain. This is consistent with reduced phagocytic killing of strains possessing K2 capsule (Mizuta *et al.*, 1983; Ofek *et al.*, 1993; Alberti *et al.*, 1996). The increased interaction of the K21a capsule with phagocytic cells is due to the K21 capsule itself, rather than any other surface-associated molecule or structure, since a recombinant strain in which the *cps*K2 genes had replaced the *cps*K21a determinants showed reduced phagocytic binding activity, which was similar to the wild-type K2$^+$ isolate. The recombinant strain would be expected to differ from the parent only in the production of a different capsular type, while retaining the genotype of the original K21a isolate. Similarly, the reciprocal cross in which the K21a capsule was substituted for the K2 antigen in a K2 background host showed increased binding to phagocytes as compared with the parent. These experiments demonstrated that reduced binding and subsequent uptake of a K2$^+$ strain is a function of the capsular polysaccharide alone.

Mouse lethality has also investigated with the strains described above. In this case, the K2$^+$ isolate killed 100% of mice inoculated with 10^4 bacteria, while the K21a capsule killed only about 33% of inoculated animals. The lethality of the bacteria in these experiments cannot, however, be attributed solely to the type of capsule produced by the bacteria, since 10^4 bacteria of the K2$^+$ recombinant were avirulent. At higher infective doses (10^7 organisms) the K2$^+$ recombinant showed increased lethality as compared with the parental K21a strain. This indicates that the expression of a specific capsular type plays a role in the outcome of disease in the mouse model of infection with *K. pneumoniae*, but other factors also contribute to pathogenesis in this system.

The mechanism of uptake of the bacteria by phagocytic cells has been investigated by Kabha *et al.* (1995), since the type of capsule produced by strains of *K. pneumoniae* can affect the uptake of the bacteria by phagocytic cells (Simoons-Smit *et al.*, 1986). The K21a capsule has the disaccharide sequence Man–α2–Man, which is not present in the K2 capsule (Allen *et al.*, 1987a; Ofek *et al.*, 1993; Kabha *et al.*, 1995). The relatively higher affinity of K21a$^+$ *K. pneumoniae* for macrophages, and the increased bactericidal activity of these cells for K21a$^+$ isolates, can be explained by the binding of the Man–α2–Man to the mannose receptor on macrophages (Athamna *et al.*, 1991). Phagocytosis in which bacterial binding to phagocytic cells with subsequent ingestion is mediated by a receptor–ligand interaction in the absence of specific antibodies has been termed 'lectinophagocytosis' (Ofek and Sharon, 1988). Organisms that possess capsules which avoid this process, such as K2$^+$ isolates, are less likely to be cleared from sites of infection by phagocytic cells. This was experimentally verified by monitoring the clearance of *K. pneumoniae* from the blood of mice that had been intravenously injected with either K2$^+$ or K21a$^+$ *K. pneumoniae* (Kabha *et al.*, 1995). The K21$^+$ strain was cleared more rapidly than the K2$^+$ isolate, and the observed clearance could be inhibited by mannan that would be expected to act as a competitive inhibitor of the Man–α2–Man/ mannose receptor interaction. Derivatives of *K. pneumoniae*, in which the K21a and the K2 antigens had been exchanged, confirmed the role of the capsule in mediating binding to macrophages and their subsequent clearance from the blood (Ofek *et al.*, 1993; Kabha *et al.*, 1995).

As we have seen, the amount of capsule produced by *K. pneumoniae* can affect the ability of the bacteria to resist phagocytosis. Antibody-mediated phagocytosis (opsonophagocytosis) of bacteria is reduced when capsule biosynthesis is inhibited with salicylate (Domenico *et al.*, 1989). It has therefore been suggested that copious production of capsular material and its subsequent release from the bacterial cell may potentiate the virulence of the bacteria *in vivo*, since the cell-free capsular polysaccharide binds circulating *Klebsiella*-specific antibodies (Domenico *et al.*, 1992). This hypothesis has been investigated *in vitro*, and it can be shown that small amounts of cell-free capsule significantly decrease the phagocytosis of whole bacteria (Domenico *et al.*, 1994a, b). A concentration of 120 ng/mL of cell-free capsular polysaccharide inhibits the phagocytosis of a K2$^+$ strain in the presence of anti-K2 opsonins. Whether the amount of capsule produced *in vivo* by *K. pneumoniae* plays a role in the pathogenesis of disease has yet to be determined.

The stimulation of cytokine production by *K. pneumoniae* has not been extensively examined. Since *K. pneumoniae* is particularly implicated in Gram-negative sepsis in debilitated hosts, it may be expected that pathogenicity is influenced by soluble mediators of the inflammatory response produced in response to infection. The K1 and K3 capsular antigens induce tissue necrosis factor α (TNFα) after injection into mice (Choy *et al.*, 1996). Stimulation of TNFα production was independent of LPS, but the pattern of response was similar to that observed with LPS. After intravenous injection, as little as 10 mg of K1 or K3 antigen was sufficient to produce a demonstrable increase in serum TNFα concentration.

Lipopolysaccharide

Structure

Like other Enterobacteriaceae, *K. pneumoniae* has a cell wall structure composed of LPS made up of lipid A, a core oligosaccharide, and an O-antigen specific polysaccharide side-chain (see Chapter 000 [Doyle]). Lipid A is highly conserved in the enterobacteria and consists of a β1,6-linked D-glucosamine disaccharide that possesses phosphomonoester groups. The 2-keto-3-deoxyoctulosonic acid (KDO) is attached to the C6′ through a hydroxyl group, and long chain fatty acids are linked to the disaccharide.

The structure of the core oligosaccharide is also conserved among the enterobacteria, but there is micro-heterogeneity. Early investigations of the core oligosaccharides of different serotypes of *K. pneumoniae* suggested that the core structures were very similar (Nimmich and Korten, 1970). In contrast to many other core structures of enterobacteria, the oligo-saccharides of *K. pneumoniae* O:1 and O:8 do not possess phosphate residues (Severn *et al.*, 1996). It may be predicted that the absence of phosphate groups and the correspondingly weak negative charge weak-ens membrane stability, since the divalent cations required for this stability are most likely to bind to phosphate groups (Vaara, 1992). Consequently, the observed release of LPS as a toxic complex with cap-sular material may be due to the chemical nature of the core oligosaccharide of *Klebsiella*.

Another difference between the cores of *K. pneu-moniae* and *E. coli* resides in the presence of galact-uronic acid in the outer core region of the former and the presence of a hexose containing region in the latter (Severn *et al.*, 1996). Although not present in *E. coli*, galacturonic acid is present in the cores of other enterobacteria, such as *Proteus mirabilis* (Radziejewska-Lebrecht and Mayer, 1989) and *Mor-ganella morganii* (Kotelko *et al.*, 1983) as well as the more distantly related *Vibrio cholerae* (Bock *et al.*, 1994).

The O-antigen-specific polysaccharide side-chains of *E. coli* and *Salmonella* have been used to identify a large number of serogroups because of the structural diversity of these molecules. In *Klebsiella* spp., the analysis and use of O-antigenic typing protocols has not been extensively studied but a limited number of O-serotypes has been identified. The use of an O-sero-typing scheme as an epidemiological tool in *Klebsiella* infections has not been established as a practical tool, because the production of a capsule by most clinical isolates does not allow direct examination for the O-antigen. In addition, only as few as eight *Klebsiella* O-serotypes can be distinguished by conventional sero-typing techniques (Kenne and Lindberg, 1983), but because of the role of the LPS in pathogenicity, the O-antigens of some *K. pneumoniae* serotypes have been investigated and their chemical structure documented.

The O:1 antigen of *K. pneumoniae* consists of two related but distinct galactans (McCallum *et al.*, 1989a, b; Bock *et al.*, 1994). The disaccharide re-peating subunits of these two molecules, referred to as D-galactan I and D-galactan II, are shown below (Kol *et al.*, 1991; Whitfield *et al.*, 1991; Bock *et al.*, 1994). These structures represent antigens found in some

other serotypes of *Klebsiella* spp., and although both forms of LPS are present in purified preparations, the synthesis of D-galactan I and II can occur indepen-dently (Whitfield *et al.*, 1991). The isolation and analysis of *K. pneumoniae* O:1 mutants that synthesise D-galactan I but not D-galactan II led to the hypothesis that in wild-type O:1⁺ isolates D-galactan I is bound to the core of the LPS and D-galactan II is attached to the distal termini of D-galactan I (Kol *et al.*, 1992).

$$\to 3)\text{-}\beta\text{-}\text{D-Gal}f\text{-}(1 \to 3)\text{-}\alpha\text{-}\text{D-Gal}p\text{-}(1 \to$$

D-galactan I

$$\to 3)\text{-}\alpha\text{-}\text{D-Gal}p\text{-}(1 \to 3)\text{-}\beta\text{-}\text{D-Gal}p\text{-}(1 \to$$

D-galactan II

D-galactan I is also present in LPS isolated from serotype O:2 and is most probably the O:2a antigen of this group (Whitfield *et al.*, 1991). In addition, O:2 serotypes that possess the O:2c antigen have the di-saccharide repeating subunit:

$$\to 3)\text{-}\beta\text{-}\text{D-Glc}p\text{NAc-}(1 \to 5)\text{-}\beta\text{-}\text{D-Gal}f\text{-}(1 \to$$

The O8 serotype of *K. pneumoniae* shows antigenic cross-reactivity with the O:1 antigen and this most probably represents a structural relatedness of the two molecules. Biochemical analysis of the O:8 antigen showed that, like the O:1 antigen, it consists of a carbo-hydrate backbone with both the D-galactan I and D-galactan II moieties (Kelly *et al.*, 1993). In the O:8 molecule, however, the D-galactan I contains O-acetyl groups that are not present in the O:1 structure. This acetylation can occur at one of two sites and is found in approximately 40% of the D-galactan I molecules of the O:8 serotype. The structure of the dissacharide of the O:8 D-galactan I is shown below. The D-galactan is also found in serotype O:9 LPS but in this case the subunit backbone is extensively O-acetylated (Whitfield *et al.*, 1991).

$$\begin{array}{c} \text{OAc} \\ | \\ 2/6 \\ \to 3)\text{-}\beta\text{-}\text{D-Gal}f\text{-}(1 \to 3)\text{-}\alpha\text{-}\text{D-Gal}p(1 \to \end{array}$$

O:8 D-galactan I disaccharide

Genetics

Genetic studies of LPS biosynthesis in *K. pneumoniae* have focused on the assembly of the molecules that possess D-galactan I (Clarke and Whitfield, 1992). The polymerisation of sugar residues is believed to be facilitated by a lipid intermediate and polymerisation occurs at the proximal face of the cytoplasmic membrane. Subsequently, the polymerised D-galactan I is transported across the cytoplasmic membrane and is ligated to an acceptor molecule of the lipid A core. The genes necessary for the expression of D-galactan I were initially identified by transfer to *E. coli* K12 (Laakso *et al.*, 1988). Since LPS biosynthetic genes in enteric bacteria are located close to the *his* locus, it was possible to transfer the *rfb* genes of *K. pneumoniae* serotype O:1 to histidine auxotrophs of *E. coli* and screen *his*[+] transconjugants for the expression of D-galactan I with monoclonal antibodies. D-galactan II expression was not observed in these transconjugants and the genes necessary for synthesis of this molecule do not appear to be linked to those that encode expression of D-galactan I. Since D-galactan consists of galacto-pyranose and galactofuranose moieties (see above), UDP-Gal precursors must be synthesised by the *gal*E gene product. This product, UDP-galactose-4-epimerase, is essential for D-galactan I expression and the *gal*E gene is not part of the *rfb* locus.

The genes responsible for the synthesis of the serotype O:1 D-galactan I have been cloned and are designated *rfb*$_{KpO1}$ (Clarke and Whitfield, 1992). Rough strains of *E. coli* with recombinant plasmids that carry the *rfb*$_{KpO1}$ genes express D-galactan I that is structurally identical to that found in O:1 serotypes of *K. pneumoniae*. Since the recombinant *E. coli* can assemble the *K. pneumoniae* O:1 antigen on its surface, the attachment of this polymer to the core region of the LPS does not appear to be very stringent. The *rfb*$_{KpO1}$ gene cluster is shown in **Fig. 2** and possesses six genes, *rfb*A–*rfb*F.

The genes *rfb*A and *rfb*B encode essential components of the polysaccharide cytoplasmic membrane transport machinery. RfbBKpO1 is a 30.8-kDa

ATP-binding protein of the translocation apparatus, and amino acid sequence analysis indicates that it possesses a putative ATP-binding site. In addition, this protein is related to other bacterial gene products implicated as polysaccharide transporters. The proposed ATP binding site of RkbBKpO1 is located near the N-terminus of the polypeptide, but the complete gene product is necessary for the transport function, since deletion of the terminal 14 amino acids from RkbBKpO1 abrogates transport (Bronner *et al.*, 1994). The RfbBKpO1 protein is thought to couple energy derived from the hydrolysis of ATP to an integral membrane protein during transport of the polysaccharide across the membrane. The *RfbA*$_{KpO1}$ gene product may be the integral membrane protein of the transport machinery. This protein possesses a number of possible membrane-spanning domains, but its relatedness to known membrane transporters is limited. *E coli* recombinants that lack the *rfb*A$_{KpO1}$ and *rfb*B$_{KpO1}$ genes, but retain the remaining *rfb*$_{KpO1}$ determinants, accumulate D-galactan I intracellularly, and this is consistent with the proposed function of RfbAKpO1 and RfbBKpO1. The precise mechanism by which the transporter mediates passage of the polysaccharide across the membrane is unknown. The membrane protein may provide the appropriate channel or gate through which the D-galactan I passes or, alternatively, the transport machinery may function to facilitate passage of a separate component that is subsequently required for export of the galactan.

The *rfb*F$_{KpO1}$ gene encodes a 42-kDa protein involved in the intracellular synthesis of the D-galactan I polymer (Clarke *et al.*, 1995). RfbFKpO1 catalyses the following reaction:

$$\text{Glc}p\text{NAc-P-P-lipid} + \text{UDP-Gal}p \rightarrow \text{Gal}p$$
$$\text{-}(1 \rightarrow 3)\text{-Glc}p\text{NAc-P-P-lipid} + \text{UDP}$$

This reaction is one in a series of enzymatically controlled reaction sequences that result in galactan formation, and RfbFKpO1 is, therefore, a galactosyl transferase. A comparison of RfbFKpO1 with other enzymes involved in LPS expression indicates that it is similar to the *Shigella dysenteriae* RfbB protein, which is also a galactosyl transferase. Although RfbFKpO1 is not a membrane-bound protein, it is membrane-associated and presumably functions at the site of polysaccharide transport during LPS biosynthesis. The functions of the *rfb*C–E genes in serotype O:1 expression are not known.

The genetic organisation of *K. pneumoniae rfb* determinants from serotypes other than O:1 has not been determined. The clonality of these gene clusters in various serotypes has, however, been investigated with

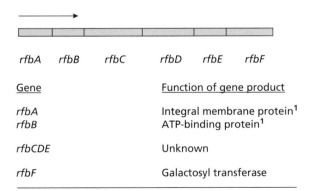

Gene	Function of gene product
rfbA	Integral membrane protein[1]
rfbB	ATP-binding protein[1]
rfbCDE	Unknown
rfbF	Galactosyl transferase

[1] These two gene products are components of an ABC-transporter system (Fath and Kolter, 1993)

Fig. 2 Genetic organisation of the *Klebsiella pneumoniae* O:1 lipopolysaccharide biosynthesis genes.

rfb_{KpO1} gene probes (Kelly *et al.*, 1993). Genomic DNA from serotype O:1 and O:2 strains possesses highly conserved DNA sequences, and the presence of D-galactan I as a component of the LPS in both serotypes may indicate related *rfb* gene clusters. Serotype O:8 strains of *K. pneumoniae* do not, however, possess nucleotide sequences similar to the O:1 gene probes, even though the O:8 strain contains an O-acetylated D-galactan I as part of its LPS. The genes that encode acetylases are unlinked to *rfb* genes, but those involved in synthesis of the D-galactan backbone can be predicted to be highly conserved. Hybridisation of the rfb_{KpO1} gene probes to DNA from an O:8 serotype strain occurs under low stringency conditions, indicating that the diversity between O:8 and O:1 genes is greater than that between O:1 and O:2 serotypes.

Role in Virulence

The bactericidal activity of non-immune serum is mediated by the components of the complement cascade reaction. The pathogenicity of *K. pneumoniae* strains that invade the bloodstream is therefore, in part, a function of their ability to evade the bacteriolytic effect of complement. A number of investigations have shown that the serum sensitivity of *K. pneumoniae* is determined by its LPS molecule (Tomas *et al.*, 1986; Ciurana and Tomas, 1987; Alberti *et al.*, 1993; Kelly *et al.*, 1993). The production of capsular material does not appear to play a role in the degree of serum sensitivity or resistance, because isogenic non-capsulate mutants appear to be equally susceptible to complement-mediated killing as their capsulate parents (Ciurana and Tomas, 1987).

Cell-free LPS from a serum-resistant strain of *K. pneumoniae* inhibits the bactericidal effect of serum on serum-sensitive strain, but the reverse is not the case (Ciurana and Tomas, 1987). The inhibitory effect of the LPS from the serum-resistant strain correlates with its anti-complementary activity. For serotype O:1, the serum resistance was a function of the presence of intact O-antigenic side-chains, while their absence in LPS that contains only a lipidA/core region resulted in incomplete inhibition of serum bactericidal activity. Therefore, one class of serum-resistant *K. pneumoniae* isolates are those that have an intact polysaccharide O:1 side-chain. The smooth LPS inhibits the binding of C1q to target receptors, such as bacterial porins, and so prevent the activation of the classical complement pathway (Alberti *et al.*, 1996).

Although serum-resistant serotype O:1 strains of *K. pneumoniae* do not bind complement component C1q, they do bind C3b (Alberti *et al.*, 1996) but this binding does not result in the formation of a functional C5b-9 membrane attack complex (MAC) and, for this reason, the bacteria are not killed by the alternative complement pathway. Two factors play a role in explaining why smooth strains bind C3b but are not efficiently killed by serum. First, in smooth strains C3b is deposited farther away from the outer membrane than in rough strains, since C3b binds to the LPS O side-chain, whereas in rough strains it is deposited on membrane-associated porins. Second, quantitative differences in the amount of C3b bound by serum-sensitive, as compared with serum-resistant strains, may play a role in bactericidal activity of serum.

LPS mutants of *K. pneumoniae* have been used to investigate the role of the LPS in virulence in the rat urinary tract (Camprubi *et al.*, 1993). The ability of *K. pneumoniae* to cause infection of the urinary tract is a function of LPS independent of the capsule. Strains with an intact O:1 antigen consistently infect the bladder and/or kidneys of experimentally infected rats, whereas strains lacking the O side-chain are responsible for significantly lower infection rates. Even when rough mutants of *K. pneumoniae* are implicated in infection of the urinary tract, they are isolated in smaller numbers than the smooth wild-type strains. The precise function of the intact O:1 side-chain of the LPS in facilitating infection *in vivo* has not been elucidated. In other pathogenic bacteria LPS, or LPS-associated proteins, have been implicated in mediating adherence to epithelial surfaces (McSweegan and Walker, 1986; Dekker *et al.*, 1990; Paerregaard *et al.*, 1991; Ramphal *et al.*, 1991). It is possible, therefore, that LPS mutants of *K. pneumoniae* are altered in the expression of surface components that are necessary for bacterial adherence *in vivo* (Camprubi *et al.*, 1993).

Adherence Factors

Structure

K. pneumoniae expresses numerous types of adherence factors that are associated with the presence of surface appendages called fimbriae or pili (Old and Adegbola, 1983, 1985, Old *et al.*, 1985; see also Chapter 9). Differences in fimbrial types are reflected in their receptor-binding specificities, structure and genetics, but not all types have been implicated as virulence factors. Fimbriae are protein molecules comprised of a major fimbrial subunit that is polymerised to form the shaft of the appendage (Kuehn *et al.*, 1994; Jones *et al.*, 1995). In some cases the protein which confers the ability of fimbriae to bind to their specific receptor is distinct from the major fimbrial subunit and is present in relatively small amounts compared with the major structural subunit (Tewari *et al.*, 1994;

Sokurenko *et al.*, 1995; Johnson and Brown, 1996; Klemm *et al.*, 1996). In the case of *E. coli* P fimbriae, the fimbria-associated adhesin protein is located at the tip of the fimbrial appendage (Hultgren and Normark, 1991). Alternatively, as in the case of type 1 fimbriae, the adhesin molecule may be located at discrete regions throughout the length of the appendage (Ponniah *et al.*, 1991). Since the type 1 fimbriae of *K. pneumoniae* are antigenically and genetically related to those of *E. coli* (Nowotarska and Mulczyk, 1977), it is likely that they are structurally similar. For the remaining *K. pneumoniae* fimbrial types it is unknown whether the adhesin molecule, when present as a distinct gene product, is a tip-associated protein.

The size of *Klebsiella* fimbrial appendages depends on the type of fimbriae examined. Type 1 fimbriae, which confer a mannose-sensitive adherence phenotype on the bacteria, are 6–8 nm wide and can be 1–2 mm long with a characteristic axial channel that can be seen by electron microscopy (Old and Adegbola, 1983). Similar structures have been reported for some of the *K. pneumoniae* fimbriae that mediate a mannose-resistant adherence (Old and Adegbola, 1983). Other *Klebsiella* fimbrial types are slender appendages, approximately 2 nm wide, that can be as long as 6–7 mm (Old *et al.*, 1985). Investigations into the molecular genetics of *K. pneumoniae* fimbriae (see below) have indicated that the assembly of these appendages occurs by a process similar to that reported for *E. coli* type 1 and P fimbriae (Hultgren and Normark, 1991; Hultgren *et al.*, 1991; Kuehn *et al.*, 1993). Structural components of the fimbrial shaft are synthesised in the bacterial cell and then transported through the cytoplasmic membrane. These proteins possess signal peptide regions at their N-termini that facilitate transport via the general secretory machinery of the cell (Collier, 1993; Pohlschroder *et al.*, 1996). After passing through the inner membrane, the fimbrial proteins bind to periplasmic chaperones which prevent degradation, allow ordered assembly, and transport the structural components to the site of fimbrial assembly. At this site an outer membrane protein, referred to as a 'molecular usher', facilitates the assembly and biogenesis of the fimbrial appendage. It has been demonstrated in some fimbrial systems that additional proteins are involved in regulating the length of the fimbriae (Russell and Orndorff, 1992). Experimental evidence also indicates that fimbrial subunits are added to the base of the intact appendage during synthesis (Dodd and Eisenstein, 1984; Eisenstein, 1988).

The different fimbrial types associated with *K. pneumoniae* are classified according to their size,

receptor-binding specificities, and antigenic characteristics. The type 1 fimbriae are representative of the adherence factors that mediate attachment to eukaryotic cells that can be inhibited by soluble D-mannose and its analogues (Nowotarska and Mulczyk, 1977; Eshdat and Sharon, 1984; Firon *et al.*, 1984; Eisenstein, 1988), and therefore the fimbriae are also referred to as mannose-sensitive (MS) adherence factors. Since mannose and its derivatives are the receptors for type 1 fimbriae, they mediate adherence to a wide variety of eukaryotic cells. *K. pneumoniae* type 1 fimbriae are antigenically related to those of *E. coli* but immunologically different from type 1 fimbriae of other enteric bacteria (Nowotarska and Mulczyk, 1977), and they are composed of fimbrial subunits of approximately 18 kDa (Eisenstein, 1988). Type 1 fimbriae have been implicated as colonisation factors of mucosal surfaces (Eisenstein, 1988; Bloch *et al.*, 1992; Gaffney *et al.*, 1995; Venegas *et al.*, 1995).

Type 3 fimbriae are thinner than type 1 fimbriae and adherence, *in vitro*, is not inhibited by D-mannose (Old *et al.*, 1985). The major fimbrial subunit protein has a molecular mass of 20.5 kDa (Korhonen *et al.*, 1983; Old and Adegbola, 1985). These fimbriae mediate the *in vitro* agglutination of erythrocytes that have been treated with tannic acid and they have limited agglutinating activity for fresh, untreated red cells (Old and Adegbola, 1983). The type 3 fimbriae of *Klebsiella* represent a discrete serological group of antigens, and analysis of the amino acid contents of the major fimbrial subunit proteins from different strains indicates a close relatedness (Old and Adegbola, 1985; Old *et al.*, 1985).

A plasmid-encoded fimbria designated KPF-28 has been isolated from a urinary isolate of *K. pneumoniae* (Di Martino *et al.*, 1996). These fimbriae consist of subunit polypeptides (28 kDa) unrelated to the type 1 or type 3 fimbrial subunits. A comparison of the N-terminal amino acid sequence the KPF-28 subunit polypeptide indicates that it is similar to that of the P fimbrial subunit PapA.

Non-fimbrial or afimbrial, adherence factors are also present on the *K. pneumoniae* surface (Di Martino *et al.*, 1995). The CF29K adhesin is associated with the presence of a 29-kDa protein and is not assembled as a discrete fimbrial structure. This adhesin is related to *E. coli* colonisation antigen CS31A, and examination of the gene that encodes CF21K shows that it differs at a single site from that encoding CS31A (Girardeau *et al.*, 1988; Di Martino *et al.*, 1995). Although the CF29K adherence factor is non-fimbriate, its presence on the bacterial surface is associated with a formation of thin and twisted filaments that can be observed by electron microscopy. In this respect, the adhesin may resemble the fibrillar colonisation antigens described on some pathogenic strains of *E. coli* (Casey *et al.*, 1992; Knutton *et al.*, 1992; Svennerholm *et al.*, 1992).

Genetics

The type 1 fimbrial (*fim*) gene cluster of *K. pneumoniae* has been cloned and the genetic organisation is similar to that of *E. coli* (**Fig. 3**). Indeed, the *K. pneumoniae fim* genes can be used to complement deletions of the *E. coli* gene cluster, indicating that the two systems are genetically and functionally related (Clegg *et al.*, 1985a, b; Gerlach *et al.*, 1989c). The *fim*A gene encodes the 18-kDa major fimbrial subunit protein that is polymerised to form the shaft of the protein. The *fim*A gene is highly conserved in fimbriate strains of *Klebsiella* and *E. coli* but does not show a large extent of homology to the *fim*A genes from other members of the Enterobacteriaceae (Buchanan *et al.*, 1985; Clegg *et al.*, 1985b; Gerlach *et al.*, 1989c). This is consistent with serological data that demonstrate that *Klebsiella* and *E. coli* type 1 fimbriae are immunologically cross-reactive but are not recognised by anti-fimbrial sera raised against appendages from other enteric genera (Nowotarska and Mulczyk, 1977). The *fim*C and *fim*D genes encode the periplasmic chaperone and usher (see above) required for the ordered assembly of the fimbriae on the bacterial surface. The *fim*F and *fim*G genes are believed to be involved in controlling the initiation and length of the type 1 fimbriae. FimI is a gene product of unknown function, but may be a minor structural component of the fimbrial shaft. The *fim*H gene encodes the polypeptide that confers the mannose receptor-binding specificity of the type 1 fimbriae. Although the FimH polypeptide is absolutely required for the expression of adhesive fimbriae, its affinity for the binding of mannose derivatives may be influenced by its conformational presentation on the fimbrial shaft (Madison *et al.*, 1994). The *K. pneumoniae* fimH gene product, when assembled by means of the ancillary *E. coli fim* genes, has a binding-specificity similar to that of the *E. coli* product and quantifiably different from that shown by the FimH protein on the *K. pneumoniae* fimbrial shaft. Conversely, an *E. coli fim*H gene can be used to complement a *K. pneumoniae fim* gene cluster, without its homologous fimH gene, and the fimbriae produced are

The *fim* (type 1 fimbriae) gene cluster

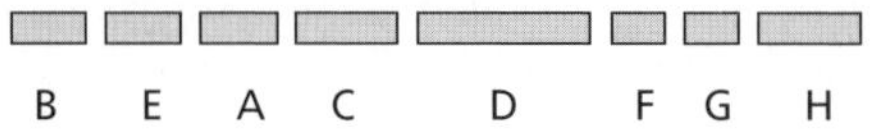

fim gene	Proposed function of gene product[1]
fimB and *E*	Regulation of *fimA* expression
fimA	Major fimbrial stuctural protein
fimC	Periplasmic chaperone
fimD	Periplasmic usher protein
fimF and *G*	Control of fimbrial biogenesis
fimH	Mannose-sensitive adhesin

[1]The proposed function of *FimB, E, C* and *D* are based upon their relatedness to the *E. coli* Fim proteins (Eisenstein, 1988)

The *mrk* (type 3 fimbriae) gene cluster

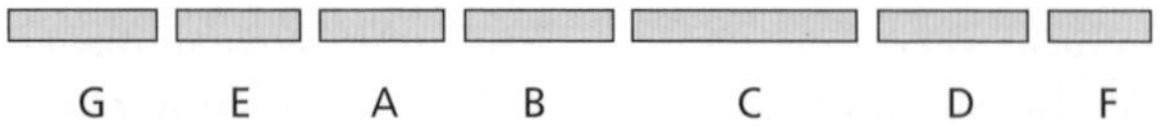

mrk gene	Proposed function of gene product[1]
mrkG and *E*	Two-component regulatory system of *mrkA*[2]
mrkA	Major fimbrial subunit
mrkB	Periplasmic chaperone
mrkC	Periplasmic usher
mrkD	Fimbrial adhesin
mrkF	Fimbrial stability

[1]MrkA-F polypeptides are reviewed in Clegg *et al.* (1994)
[2]MrkG and E polypeptides exhibit relatedness at the amino acid level to two-component regulatory systems

Fig. 3 Type 1 and type 3 *Klebsiella pneumoniae* fimbrial gene clusters.

representative of *K. pneumoniae* rather than *E. coli*. These results have been interpreted to indicate that the assembly and incorporation of the FimH adhesin into the fimbrial appendage results in constraints that may alter the quaternary configuration of the adhesin to modify its binding activity (Madison *et al.*, 1994).

Phenotypic variation of type 1 fimbrial expression has been observed in *K. pneumoniae* under conditions similar to those reported for *E. coli* (Swaney *et al.*, 1977; Clegg and Old, 1979). It is likely, therefore, that the regulation of fimbrial expression in *K. pneumoniae* is subject to molecular controls similar to those described for *E. coli* (McClain *et al.*, 1991, 1993; Gally *et al.*, 1993, 1996). Specific *fim* regulators in *K. pneumoniae* have not been identified.

The *mrk* genes that encode the type 3 fimbriae of *K. pneumoniae* have been cloned and their nucleotide sequence determined (Gerlach and Clegg, 1988; Allen *et al.*, 1991). As in the case of type 1 fimbriae, the *mrk* genes are arranged as a contiguous cluster and at least five genes are required for fimbrial assembly (**Fig. 3**). The major fimbrial subunit gene (*mrk*A) encodes the immunodominant fimbrial shaft protein and is highly conserved among all fimbriate strains of *Klebsiella* (Hornick *et al.*, 1991). Presumably, the biogenesis and assembly of the type 3 fimbriae occurs by a mechanism similar to that for type 1 and Pap fimbriae of *E. coli*, since the *mrk*B and *mrk*C genes encode proteins that have properties characteristic of fimbrial periplasmic chaperones and ushers, respectively (Kuehn *et al.*, 1993). The *mrk*F gene product also plays a role in fimbrial assembly, since mutants that lack this gene can express type 3 fimbriae on their surfaces but they are easily disrupted and removed from the bacteria (Clegg *et al.*, 1994). The receptor-binding specificity of the type 3 fimbriae is determined by the product of the *mrk*D gene (Gerlach *et al.*, 1989a, c), and the MrkD adhesin can be assembled into a Pap pilus lacking its own native PapG adhesin. These hybrid P pili do not bind the galabiose receptor of PapG but mediate the characteristic haemagglutinating and binding properties of type 3 fimbriae (Gerlach *et al.*, 1989b), an observation that confirms the functional relatedness of the two fimbrial assembly pathways.

The use of *mrk*-specific gene probes has shown that the *mrk*A, B, C and F genes are conserved in *Klebsiella* isolates. The *mrk*D determinant, however, exhibits allelic variation between fimbriate strains. One of the genes, $mrkD_{1P}$, is found in most isolates of *K. oxytoca* and rarely in strains of *K. pneumoniae* (Schurtz *et al.*, 1994). This allele is commonly associated with a plasmid-borne *mrk* gene cluster. A second gene, $mrkD_{1C}$, is found most frequently in strains of *K. pneumoniae* and is located within the fimbrial gene cluster on the chromosome. A few isolates of *K. pneumoniae* may possess both a plasmid-borne and a chromosomal *mrk* gene cluster. As will be indicated below, the MrkD1P and MrkD1C adhesins have distinctive receptor-binding specificities, and a comparison of their amino acid sequences shows 58% similarity and 48% identity. Some isolates of *K. pneumoniae* possess a *mrk*D gene related to $mrkD_{1C}$, but the receptor-binding specificity of the type 3 fimbriae is different from that exhibited by MrkD1C-positive fimbriae. Therefore, a third *mrk*D allele, related to $mrkD_{1C2}$, may be present in some *K. pneumoniae* gene clusters.

The expression, *in vitro*, of type 3 fimbriae by some strains of *K. pneumoniae* depends on the cultural conditions and appears to be optimal when the bacteria are nutritionally stressed (Hornick *et al.*, 1991). Therefore, the type 3 fimbriae are not constitutively expressed, and regulation of their expression may, in part, be at the genetic level. The amino acid sequence of the MrkE polypeptide suggests a relatedness to several sensory transcriptional regulators of bacterial two-component systems (Stock *et al.*, 1989). Partial characterisation of the *mrk*G gene immediately upstream of *mrk*E indicates that it is related to the bacterial sensor *lyt*S (Brunskill and Bayles, 1996). These two genes are therefore candidates to act as a molecular sensory mechanism to control type 3 fimbrial expression, possibly at the level of *mrk*A gene expression. The function of *mrk*E and *mrk*G in fimbrial expression has yet to be experimentally demonstrated.

The gene that encodes the KPF-28 fimbrial subunit is present on an R plasmid of *K. pneumoniae* (Di Martino *et al.*, 1996). An oligonucleotide probe for the fimbrial subunit gene, based on N-terminal amino acid sequence data, has been used to show that the gene which encodes the major fimbrial subunit is plasmid-borne. Curing the plasmid from a *K. pneumoniae* isolate resulted in its inability to express KPF-28 fimbriae, and re-introduction of the plasmid into the strain restored fimbrial expression. The phenotypic expression of KPF-28 fimbriae is not, however, due solely to the presence of the R plasmid, since transfer of this DNA molecule to a strain of *E. coli* does not facilitate expression of KPF-28 fimbriae in this strain.

The *cf* 29a gene that encodes the afimbrial CF29K adherence factor has been cloned and its nucleotide sequence determined (Di Martino *et al.*, 1995). The gene is almost identical to that which encodes the related *E. coli* CS31A adhesin and differs by a single nucleotide. Functional and genetic relatedness between these two adherence factors was confirmed by the observation that the CS31A accessory genes could be utilised by the *cf29a* gene to transport and assemble the CF29K adherence factor on to the bacterial surface. The genes responsible for the expression of CF29K are located on the same plasmid as those that encode KPF-28 fimbriae, and therefore single plasmid molecules appear to possess determinants for more than one type of adherence factor.

Role in Virulence

The frequency and occurrence of adhesins on strains of *K. pneumoniae* isolated from clinical samples, and implicated as the aetiological agents of disease, have been reported (Hornick *et al.*, 1991; Tarkkanen *et al.*, 1992; Podschun *et al.*, 1993). Type 1 and type 3 fimbriae are present on urinary isolates of *Klebsiella*, but type 3 fimbriae appear to occur with greater frequency. Similar results were found with strains of *K. pneumoniae* isolated from patients with endotracheal tubes, and implicated as the cause of lower respiratory tract infections. The demonstration that *K. pneumoniae* adhesins can mediate attachment to eukaryotic tissues and cells indicates that these factors may play a role in the initial stages of the infective process. Consequently, there have been numerous investigations to determine the ability of *Klebsiella* adherence factors to bind to epithelial surfaces and connective tissue.

K. pneumoniae type 1 fimbriae can facilitate a mannose-sensitive attachment to a wide range of cells including those from the urinary and respiratory tracts (Maayan *et al.*, 1985; Tomas *et al.*, 1986; Fader *et al.*, 1988). Type 1 fimbria-mediated adherence to freshly isolated human sediment cells from urine has, however, not been detected (Tarkkanen *et al.*, 1997). *K. pneumoniae* type 1 fimbriae bind *in vitro* to urinary slime (Duncan, 1988; Gaffney *et al.*, 1995; Venegas *et al.*, 1995), and it has been suggested that the expression of type 1 fimbriae by uropathogens may not be advantageous, because these fimbriae bind to mannose-containing glycoconjugates in urinary slime and so lead to their elimination from the urinary tract (Gaffney *et al.*, 1995). An investigation of the growth of type 1 fimbriate bacteria *in vivo* indicates, however, that these appendages are important during infection (Maayan *et al.*, 1985). Non-fimbriate bacteria were injected into the bladder of mice and the conversion to expression of type 1 fimbriae during growth was monitored. After 24–48 hours, more than half the animals shed large numbers of type 1 fimbriated *K. pneumoniae*, indicating a selection for fimbriate-phase bacteria. Similarly, inoculation of mice with a mixture of fimbriate (5%) and non-fimbriate bacteria resulted in an infection in which most of the bacteria were

fimbriate. This shows that even when present in relatively small numbers, type 1 fimbriate *K. pneumoniae* can outgrow non-fimbriate strains. Similarly, when relatively small numbers of bacteria were used as the inoculum, the infection rate for type 1 fimbriate *K. pneumoniae* was 85% compared with 14% for the non-fimbriate organisms.

A selective advantage for fimbriate bacteria in infections of the upper urinary tract is not apparent (Maayan *et al.*, 1985). These are cleared as efficiently from the kidney as non-fimbriate bacteria. Since type 1 fimbriae mediate adherence to phagocytic cells, their presence may facilitate increased uptake and killing. Similar observations have been made for *E. coli* that express type 1 fimbriae (Schaeffer *et al.*, 1979; Ofek and Beachey, 1980; Schaeffer *et al.*, 1980; Maayan *et al.*, 1985). These results along with those described above indicate that type 1 fimbriae of *K. pneumoniae* play a role in establishing infection in the lower urinary tract but confer no advantage to the bacteria after migration to the kidney. In infections of the bladder, the fimbriae are believed to mediate adherence to the bladder wall.

Type 3 fimbriae also mediate adherence to epithelial cells but this binding is not inhibited by D-mannose (Hornick *et al.*, 1992). Observations by microscopy of type 3 fimbria-mediated adherence by *K. pneumoniae* to freshly isolated human tracheal epithelial cells indicate that fimbriate bacteria bind to the basolateral margins of the epithelial cells. This is in contrast to P-fimbriated bacteria that adhere to the apical surfaces of the cells. Similarly, incubation of *K. pneumoniae* with thin sections of human lung tissue showed that type 3 fimbriae mediate adherence to the basement membrane but not the luminal margin of the tissue (Hornick *et al.*, 1992). Pre-incubation of the tissue sections with purified type 3 fimbriae inhibits the adherence of fimbriate bacteria. In addition, fimbrial adherence can be inhibited by pre-treatment of the tissues with collagenase but not periodate, which suggests that the fimbriae bind to collagen. Subsequently, it was demonstrated that type 3 fimbriae bind to distinct collagen types of the extracellular matrix (ECM) and that the MrkD adhesin is an ECM-binding protein (Tarkkanen *et al.*, 1990). It can be shown, with a *mrkD*-specific gene probe, that the adhesin gene of type 3 fimbriae is not identical in all strains (Schurtz *et al.*, 1994). One type of MrkD adhesin, associated primarily with *K. oxytoca*, facilitates adherence to type V collagen, while a second MrkD protein mediates attachment to both types V and IV collagen molecules. The latter MrkD adhesin is most frequently found among strains of *K. pneumoniae* (Schurtz, 1997). Some strains of *Klebsiella* express type 3 fimbriae that adhere to eukaryotic cells, but not to collagen, and therefore these fimbriae may possess a third distinct type of MrkD.

Since *K. pneumoniae* is an opportunistic pathogen most frequently associated with infections of compromised individuals, it has been suggested that type 3 fimbriae mediate binding to denuded epithelial surfaces. Such surfaces are likely to be produced in individuals after (1) insertion of catheters or endotracheal tubes, (2) a primary viral or bacterial infection, or (3) therapeutic procedures that may result in an acute inflammatory response. In these cases, receptors that are usually sequestered may become exposed and recognised by the MrkD adhesins with resulting colonisation. Although the role of type 3 fimbriae in mediating attachment to ECMs has been shown *in vitro*, their function *in vivo* has yet to be demonstrated.

A phenomenon termed 'aggregative adherence' has been described for some strains of *K. pneumoniae* that attach to human intestinal cells lines (Favre-Bonte *et al.*, 1995). This pattern of adherence is associated with bacteria in the late exponential phase of growth and is similar to that observed with enteroaggregative *E. coli* and HeLa cells (Scaletsky *et al.*, 1984; Savarino *et al.*, 1994; Hicks *et al.*, 1996; Nataro *et al.*, 1996). Aggregative adherence of *K. pneumoniae* is associated with the formation of a fine fibrillar matrix around the bacteria that is believed to form a capsule-like structure. It has been suggested that induction of the production of this material requires that the bacteria be closely associated with target intestinal cells (Favre-Bonte *et al.*, 1995).

A second pattern of adherence, termed 'diffuse adherence', has been described for CF29K-producing isolates of *K. pneumoniae* (Darfeuille-Michaud *et al.*, 1992; Di Martino *et al.*, 1995; Favre-Bonte *et al.*, 1995). This was observed when strains were incubated with Caco-2, Int407 and Hep-2 intestinal cell lines. The CF29K adhesin most probably binds to a carbohydrate receptor on target cells, since adherence can be inhibited competitively by *N*-acetylneuraminic acid and acetylated chitobiose (Di Martino *et al.*, 1995), while carbohydrates such as glucose, galactose, fucose and their derivatives are not inhibitory (Di Martino *et al.*, 1995). The CF29K antigen is a non-fimbrial adhesin that is not produced by all strains of *K. pneumoniae* (Di Martino *et al.*, 1996). Strains that did not express the CF29K adhesin retained the ability to adhere to intestinal cells *in vitro*. This CF29K-independent adherence was associated with KPF-28 fimbria, since antibodies directed against the fimbriae inhibited adherence of fimbriate bacteria to intestinal cells (Di Martino *et al.*, 1996). Adherence of *K. pneumoniae* isolates to the intestine may, therefore,

be mediated by a number of distinct surface molecules including CF29K and KPF-28. Since nosocomial infections by *K. pneumoniae* are frequently associated with carriage of the organism in the intestinal tract, ability to colonise the intestine is of importance in the development of disease due to *K. pneumoniae*.

Expression of distinct fimbrial types by *K. pneumoniae*, and the ability of these adhesins to recognise a range of receptors facilitates the adherence of bacteria to numerous target cells. Adherence followed by colonisation represents an important first step in the infective process, and it has therefore been suggested that both fimbrial and non-fimbrial adhesins contribute to the virulence of *K. pneumoniae*. It has, however, also been suggested that the expression of *K. pneumoniae* type 1 fimbriae may enable the host more efficiently to clear bacteria from sites of infection.

It has been demonstrated, with mast-cell-deficient mice, that the influx of neutrophils to a site of *K. pneumoniae* infection depends, in part, on the presence of type 1 fimbriae (Darfeuille-Michaud *et al.*, 1992; Malaviya *et al.*, 1994, 1996; Malaviya and Abraham, 1995). In these studies, FimH-negative mutants of *K. pneumoniae* that are fimbriate but non-adhesive did not bind to mast cells and did not stimulate production of TNFα. In contrast, fimbriate bacteria with a functional FimH adhesin interacted with mast cells, and this resulted in the production of TNFα. Since TNFα is a potent mediator of neutrophil influx to sites of bacterial infection, it has been postulated that fimbriate bacteria are more likely to be cleared from such sites. Indeed, it could be demonstrated *in vivo* that fimbriate *K. pneumoniae* are cleared from mucosal surfaces more efficiently than isogenic non-fimbriate strains.

The preceding discussion suggests that the expression of certain types of fimbriae may under some circumstance be beneficial for the bacteria by facilitating attachment to eukaryotic tissues, but that such expression can also result in a heightened immune response by the host. These phenomena emphasise the opportunistic nature of *Klebsiella* infections and serve to illustrate the evolution of bacterial attributes that facilitate the infectious process and the host response to these virulence factors.

Siderophores

Structure and Genetics

Like other Enterobacteriaceae, *Klebsiella* strains synthesise and secrete iron-chelating compounds that efficiently remove iron from the host iron-binding proteins. Since elemental iron is a necessary co-factor for bacterial metabolism, the acquisition of iron *in vivo* represents an important step in *Klebsiella* pathogenesis. These low-molecular-weight iron-binding compounds, termed 'siderophores', are expressed under iron-limiting conditions. Once iron is bound to the siderophore, the complex is recognised by bacterial surface receptors that facilitate its efficient uptake. Two types of siderophores are produced by *K. pneumoniae* (Williams *et al.*, 1987). One, enterochelin, is a catechol-type of molecule that has three phenolic rings which encompass ferric iron (Neilands, 1995). The ferrienterochelin complex is recognised by a 81-kDa outer-membrane protein that is induced under iron-limiting conditions. This receptor appears to be highly conserved among *Klebsiella* isolates and is related to the *E. coli* enterochelin receptor (Williams *et al.*, 1984). Enterochelin expression is de-repressed under conditions of iron starvation. This is a rapid adaptive response to iron limitation, since it can be shown that enterochelin biosynthesis occurs within two generations of the introduction of bacteria grown under iron-replete conditions into media with low levels of iron ($< 1.7 \times 10^{-8}$ M of Fe^{3+}).

Aerobactin, a hydroxamate, is the second type of siderophore produced by *K. pneumoniae*. This molecule is a citrate molecule with N6-hydroxyacetyl lysine substituted for the distal carboxyl groups, and is derived by oxidation of lysine. The aerobactin receptor and the aerobactin biosynthesis genes from the plasmid pColV-K30 have been cloned. Some strains of *K. pneumoniae* have the ability to take up ferriaerobactin without themselves producing the siderophore. These strains, therefore, express the appropriate receptor without producing aerobactin, a phenotype that has also been observed in *E. coli* (Williams *et al.*, 1989). It is possible that *in vivo* these strains may be able to utilise iron by uptake from heterologous ferriaerobactin molecules.

Growth of *Klebsiella* under iron-limiting conditions results in the expression of several outer-membrane proteins that are not observed during growth under iron-replete conditions (Lodge *et al.*, 1986). Although some of these gene products have been identified as putative enterochelin or aerobactin receptors, there are other iron-regulated proteins that have not been characterised (Williams *et al.*, 1989). It has been suggested that some of these proteins may act as uptake mechanisms for exogenous siderophores since *E. coli*, for example, takes up a fungal iron-charged siderophore (Neilands, 1982; Williams *et al.*, 1989; Griffiths, 1991). However, the precise role of these iron-regulated outer-membrane proteins has yet to be elucidated.

Role in Virulence

Growth of *K. pneumoniae in vivo* requires acquisition of iron, which is necessary for metabolic activity and growth. Free-iron concentrations in host tissues are, however, very low, because free iron is toxic and the host produces iron-binding proteins such as transferrin and lactoferrin. Therefore, the ability to accumulate iron contributes to virulence, since it allows for bacterial growth. Indeed, it has been shown that parenteral administration of iron compounds to guinea-pigs enhances their susceptibility to infection by *K. pneumoniae* (Khimji and Miles, 1978). In this case, the concentration of free iron in the tissues is presumably not limiting and the bacteria grow in an iron-replete environment.

Mucosal surfaces and tissues, respectively, possess lactoferrin and transferrin that bind iron with a high affinity (association constant *c*. 10^{22}) (Beeson and Rowley, 1959). These compounds are therefore essentially the only available sources of iron for the bacteria during growth *in vivo*. Consequently, the production of bacterial siderophores with high iron-binding capabilities (*c*. 10^{30}) is necessary for removal of iron from host proteins. An investigation of the type of siderophores produced by clinical isolates of *K. pneumoniae* indicated that most pathogenic strains produce enterochelin (Williams *et al.*, 1989). For example, 32 strains of *K. pneumoniae* from isolated urinary tract infections expressed enterochelin, while none produced detectable amounts of aerobactin (Tarkkanen *et al.*, 1992). Similarly, 14 *Klebsiella* strains from various sources, including blood, urine, and sputum were enterochelin-positive and only three secreted an aerobactin (Williams *et al.*, 1989). This is in contrast to uropathogenic *E. coli* in which both enterochelin and aerobactin expression is frequently observed (Carbonetti *et al.*, 1986). It is possible that aerobactin production by *K. pneumoniae* occurs more frequently in strains associated with septicaemia, since aerobactin expression is found most frequently in strains with the K1 or K2 capsules. These serotypes are most frequently associated with blood-borne infections.

The role of aerobactin production in the virulence of *K. pneumoniae* has been investigated (Nassif and Sansonetti, 1986). K1- and K2-positive strains of *K. pneumoniae* that produce aerobactin were more virulent ($LD_{50} = 10^3$ bacteria) on intraperitoneal injection into mice than two strains of the same serotype that did not express aerobactin ($LD_{50} = 10^6$ bacteria). Moreover, acquisition of an aerobactin-encoding plasmid by a K1-positive isolate of *K. pneumoniae* enhanced its virulence for mice (LD_{50} reduced from 8×10^6 to 5×10^4 bacteria). However, the possible presence on the plasmid of other determinants that encode and/or regulate virulence gene expression was not determined, and it is possible that acquisition or loss of plasmids encoding aerobactin utilisation may also result in an alteration of the ability to produce virulence factors other than aerobactin.

Since most isolates of *K. pneumoniae* implicated in human infections produce enterochelin, it is not clear what advantage is conferred on the bacteria by the ability to synthesise aerobactin. Initial evidence suggests that enterochelin-producing *K. pneumoniae* are not able to grow in media containing transferrin (Nassif and Sansonetti, 1986). These observations were not reproducible, and it is clear that some strains of *K. pneumoniae* utilise transferrin-bound iron in the absence of aerobactin (Williams and Tomas, 1990). Clearly, the construction of isogenic *K. pneumoniae* mutants defined with respect to siderophore production would be useful to investigate further the role of specific iron-binding bacterial factors in bacterial virulence.

Urease

Structure

The urease of *K. pneumoniae* is a nickel-containing enzyme that hydrolyses urea to ammonia and carbamate. The enzyme consists of three subunit polypeptides, α, β and γ (molecular masses 60 304, 11 695 and 11 086, respectively) that are arranged in an $\alpha2\beta4\gamma4$ stoichiometry (Todd and Hausinger, 1989; Jabri *et al.*, 1995; Moncrief *et al.*, 1995). The number of nickel-binding sites per subunit has been determined with phenylphosphorodiamidate, a urease inhibitor that tightly binds to the native enzyme (Todd and Hausinger, 1989). Each catalytic site has two nickel ions and the enzyme molecule has three catalytic sites. The amino acid composition of the urease enzymes from microbial sources, including *K. pneumoniae*, suggests that the enzymes are closely related and differ from those derived from plants (Mobley *et al.*, 1995). Enzymes that do not possess nickel are unable to hydrolyse urea, and bacteria grown in nickel-depleted medium synthesise an inactive enzyme. The ion is tightly bound to the enzyme and resists dissociation in the presence of 1 mmol/L of EDTA but it can be removed by treatment with acids (Duma, 1985). When inactive enzyme was purified from bacteria grown in nickel-depleted medium, it was not possible to reconstitute active enzyme by the simple addition of nickel. This suggested that incorporation of the metal into the active site of the enzyme in the bacterial cell

may be energy-dependent and that it requires other gene products provided by the cell. As we shall see below, part of the urease gene cluster consists of genes that encode polypeptides necessary for the formation of the metalloprotein active centre.

It is believed that urea binds to the nickel-containing catalytic centre of the enzyme, since thiol, a competitive inhibitor of urease, binds directly to nickel (Todd and Hausinger, 1989; Mobley *et al.*, 1995). It has not, however, been determined whether the two nickel ions of one catalytic site act as a single bimetal site or function independently within the same site, but urease consists of a series of complexes associated with non-subunit accessory proteins (Tomas *et al.*, 1986; Lee *et al.*, 1992; Moncrief and Hausinger, 1996). This formation of complexes is thought to be necessary to ensure that the lysine of the catalytic site is carbamylated so that it can act as the metallocentre ligand. Crystallographic analysis of *K. pneumoniae* urease indicates that the two nickel ions of the active centre are separated by 0.35 nm and that the carbamylated lysine bridges these metal ions (Jabri *et al.*, 1995; Moncrief *et al.*, 1995).

Genetics

The urease gene cluster of *K. pneumoniae* has been cloned and plasmids possessing this cluster confer urease activity on heterologous hosts, such as *Salmonella typhimurium* and *E. coli* (Gerlach *et al.*, 1988). The chromosomal urease (*ure*) genes are located on a segment of DNA approximately 12.3 kb in length and the gene cluster consists of seven genes (**Fig. 4**). The three structural subunits of the urease molecule are encoded by *ure*A, *ure*B, and *ure*C, and the protein subunits can be assembled in the absence of nickel to form an inactive apoprotein. In addition, deletions in the remaining *ure* genes also result in the formation of apoprotein even in the presence of nickel, which suggests that these genes may be involved in the formation of the nickel-containing active sites of the enzyme. Three of the genes, *ure*E, *ure*F and *ure*G, are located immediately downstream of the genes that encode the structural proteins, and one gene, *ure*D, is positioned upstream of the subunit genes. All the *ure* genes have the same transcriptional polarity.

The UreE protein is expressed in the bacterial cell at a level equivalent to that of the structural subunits (Lee *et al.*, 1992). The amino acid sequence of this protein indicates the presence of several sites that can be used to bind nickel and it has, therefore, been postulated that UreE is a metal transferase (Mulrooney and Hausinger, 1990; Lee *et al.*, 1992). The predicted sequence of UreG also suggests a function for this

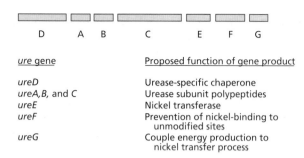

ure gene	Proposed function of gene product
ureD	Urease-specific chaperone
ureA,B, and C	Urease subunit polypeptides
ureE	Nickel transferase
ureF	Prevention of nickel-binding to unmodified sites
ureG	Couple energy production to nickel transfer process

Fig. 4 Genetic organisation of the urease (ure) gene cluster.

protein, since it possesses a nucleotide binding motif (Nassif and Sansonetti, 1986; Saraste *et al.*, 1990; Lee *et al.*, 1992). Although energy is not required for the reconstitution of active protein *in vitro* from apoprotein and nickel, the incorporation of nickel into the enzyme is energy-dependent *in vivo*. Therefore, the function of UreG may be to couple energy production to the metal transfer process from UreE to the active centre.

The UreD protein can be obtained from *K. pneumoniae* cell extracts as a complex associated with the urease subunit proteins (Nassif and Sansonetti, 1986; Saraste *et al.*, 1990; Lee *et al.*, 1992; Park *et al.*, 1994; Moncrief and Hausinger, 1996). These complexes can be purified after the over-expression of ureD, and the urease is able to bind one, two or three UreD molecules, and this is consistent with the trimeric structure of the urease. The UreD–apoprotein urease complex is necessary for nickel activation of the enzyme, which suggests that UreD is a chaperone-like protein that configures the urease enzyme to accept nickel ions. Consistent with this function is the observation that UreD dissociates from the enzyme after nickel incorporation, and that activation by nickel is proportional to the number of UreD molecules complexed with urease. In this respect, the UreD protein would enable the urease to receive nickel from the UreE polypeptide. Similar chaperone proteins are necessary for the activation of other metalloproteins, such as dinitrogenase and tyrosinase, which require iron/molybdenum and copper, respectively (Mobley *et al.*, 1985; Chen *et al.*, 1992, 1993; Homer *et al.*, 1993).

In addition to the UreD-urease complexes described above, UreD–UreF–urease complexes have been isolated from bacterial extracts that show enhanced activation properties (Moncrief and Hausinger, 1996). Since nickel forms a bridge between carbamylated amino acids in the active site of the enzyme, it has been suggested that UreF may modulate activation of urease by preventing nickel binding to the non-carbamylated protein. Consequently, the

efficiency of forming the carbamylated protein *in vivo* is greatly enhanced and reduces the bicarbonate requirement for activation. The precise mechanism by which UreF prevents nickel binding to the non-carbamylated enzyme is not known.

The urease of *K. pneumoniae* is repressed in the presence of ammonia and under nitrogen-rich conditions (Friedrich and Magasanik, 1977). De-repression occurs under nitrogen-limiting or nitrogen-starvation conditions. Regulation of urease expression is effected by a complex regulatory cascade that results in the production of specific factors that mediate transcription of the urease genes. This regulon includes genes that encode factors important for the adaptation of bacteria to the concentration and source of environmental nitrogen. Two of these genes, *ntr*A and *ntr*C, and mutants unable to express these determinants are urease-negative (Friedrich and Magasanik, 1977; Macaluso *et al.*, 1990; Collins *et al.*, 1993). The *ntr*A gene encodes a sigma factor (σ^{54}) that plays a role in the expression of genes involved in the nitrogen regulation system. The *ntr*C gene codes for an important regulator protein of genes that possess promoters that are σ^{54}-dependent.

The production of urease in some *Klebsiella* strains is affected by an additional component of the nitrogen regulation system. Urease activity can be significantly reduced in mutants that are altered in nitrogen assimilatory control (*nac*) (Bender, 1991). The *nac* gene product is a transcriptional regulator and is a member of the LysR family of activators. Expression of the *nac* gene is affected by the presence of *ntr*A and *ntr*C and therefore possesses a σ^{54}-dependent promoter (Magasanik and Neidhardt, 1987). Transcription of *ure*D and *ure*A is controlled by the nitrogen regulatory system, since a nitrogen-regulated promoter has been identified upstream of *ure*D (Collins, 1994). The Nac protein binds to the *ure*D promoter and is required for high-level expression of these *ure* genes. Comparison of the nucleotide sequences in the region of the Nac binding site indicates that *ure*D possesses a σ^{70}-related promoter to which Nac binds (Collins *et al.*, 1993; Collins and Se, 1993). Mutations in this DNA region show that the Nac-binding site is necessary for the expression of *ure*D.

Role in Virulence

The significance of urease as a virulence factor has been suggested for organisms that grow in the urinary tract and may contribute to the formation of infection stones. The hydrolysis of urea leads to an increase of the ammonia concentration in the urine with a resulting increase in pH. This results in the precipitation of the complex inorganic salts, struvite ($MgNH_4PO_4$) and carbonate apatite ($Ca_{10}(PO_4)_6.6H_2O$), which are soluble at pH 6.5 but become insoluble at pH 9.0. The production of urinary stones can contribute to the pathogenesis of infection in two important ways. First, the ability of the host to void urine can be seriously impaired and clearance of the bacteria from the urinary tract reduced. Second, stones may be a nidus of bacterial infection protected from antibiotic action because of reduced permeability at these sites. In the case of *K. pneumoniae*, stone formation is most likely to be associated with the presence of a long-term indwelling catheter (DeVivo *et al.*, 1984). In comparison with *Proteus* spp., however, *K. pneumoniae* produces considerably less urease *in vitro*. Moreover, bacterially induced precipitation of salts on to glass rods from synthetic urine is considerably less for *Klebsiella* than for *Proteus* (Hedelin *et al.*, 1985, 1991; McLean *et al.*, 1991).

Direct experiments to address the role of *Klebsiella* urease as a virulence factor have not been done. The lack of a suitable animal model to mimic the long-term catheterisation with which *Klebsiella* infections are associated in humans has prevented investigation of the role of urease *in vivo*. Consequently, urease activity during infection can only be inferred. It has been speculated that the presence of urease leads to increased levels of an easily assimilated nitrogen source for bacteria and/or that an increase in the pH of the microbial environment favours bacterial growth *in vivo* (McLean *et al.*, 1988). Since ammonia inactivates the fourth component (C4) of complement (Beeson and Rowley, 1959), urease production may act as an anti-complementary factor.

Antibiotic Resistance

Extended-spectrum β Lactamases

The emergence of antibiotic-resistant pathogens is of considerable concern since these bacteria represent a significant reason for treatment failure. Since *K. pneumoniae* is an opportunistic pathogen most likely to infect compromised individuals in the hospital environment, acquisition of antibiotic-resistance factors by these organisms is a serious threat to infected patients. Since antibiotic resistance can facilitate growth *in vivo* in the presence of antimicrobials, antibiotic-resistance factors contribute to the pathogenicity of *K. pneumoniae*. Therefore, when discussing virulence factors produced by *Klebsiella* it is appropriate to consider mechanisms that allow the bacteria to multiply in tissues in the presence of relatively high

concentrations of antibiotics. In fact, the incidence of resistance to antibiotics, such as cefotaxime, ceftriaxone and ceftazidime, designed to be refractory to bacterial β lactamases, is highest in *K. pneumoniae*. The enzymes that mediate this type of resistance are referred to as 'extended-spectrum β lactamases' and are most common in nosocomial isolates of *K. pneumoniae* in which they are present in up to 40% of isolates (Bauernfeind *et al.*, 1989; Reig *et al.*, 1993; Rice *et al.*, 1996).

The extended-spectrum β lactamases have evolved from common plasmid-encoded β lactamases and their activity is primarily due to mutations that give rise to enzymes with high-affinity active sites and greater hydrolytic activity. The most common enzyme in *K. pneumoniae* is a SHV-type β lactamase in which the substitution of serine for glycine at position 238 results in a protein with greater hydrolytic activity for cefotaxime (Jacoby, 1994). A second substitution, of glutamic acid to lysine, at position 240 results in a β lactamase with strong activity for ceftazidime (Jacoby, 1994). These substitutions result in an alteration of the walls of the active sites for each enzyme, with resulting changes in configuration, charge and binding affinities. In some instances, however, resistance to antibiotics cannot be explained by an increased relative affinity for the antimicrobial. In such cases the resistance appears to be a function of increased gene expression resulting in high-level production of the β lactamase. This increase in enzyme production may be due to alterations in the promoter region of the gene, and a single base difference in the promoter region of one family of extended-spectrum β lactamases can result in the creation of dual overlapping promoters. This can result in a 4- to 30-fold increased expression of enzyme (Jacoby, 1994). The promoter strength of the resistance determinant may also be altered by insertion of relatively large DNA sequences upstream of the gene. The insertion of IS15 into *K. pneumoniae* SHV-2-type genes results in a more active -35 promoter region, and this plasmid-borne gene is responsible for high-level resistance to ceftazidime (Podbielski *et al.*, 1991a, b).

A second class of β lactamase is associated with a plasmid isolated from clinical isolates of *K. pneumoniae*. This gene is related to the chromosomal AmpC-type β lactamase found in some Enterobacteriaceae and in *Pseudomonas aeruginosa* (Gonzalez Leiza *et al.*, 1994). The gene encodes an enzyme with two functional isoforms that differ in their pIs. One active fragment is associated with a polypeptide of molecular mass 37 kDa and pI 6.8, while the other is a polypeptide of molecular mass 35 kDa and pI 7.2 (Gonzalez Leiza *et al.*, 1994). Genetic analysis indicates that the differences in size between the two active enzymes may be due to incomplete signal sequence recognition by the bacterial signal peptidase. This low processing efficiency can be predicted to result in two differently sized polypeptides with identical kinetics and reactivity. The ability of *K. pneumoniae* carrying this *Amp*C-type gene to transfer β lactamase resistance to other enteric bacteria is disturbing, because it shows the potential for spread among a range of opportunistic pathogens and reduces the spectrum of antibiotics for the treatment of disease.

The nucleotide sequence of the *K. pneumoniae* *Amp*C-type β lactamase gene suggests that it is most closely related to the chromosomal genes found in *Citrobacter freundii* and *Enterobacter cloacae* and less distantly related to the *Ps. aeruginosa* genes (Gonzalez Leiza *et al.*, 1994). Presumably acquisition of the gene by *K. pneumoniae* was preceded by translocation from the chromosome of one bacterial strain on to a mobilisable plasmid molecule that was eventually transferred into *Klebsiella*. It is unlikely that the gene originated from *Ps. aeruginosa* since, at the nucleotide level, this determinant is only 57% related as compared with a 94% relatedness to that of *C. freundii* (Lindberg and Normark, 1986).

Finally, mention should be made of the observation that carriage of extended spectrum β lactamases in *K. pneumoniae* is frequently associated with large plasmids that carry genes which encode virulence determinants (Jacoby, 1994). Antibiotic resistance determinants have been found on the plasmids that encode siderophore production, mucoid phenotype (Vernet *et al.*, 1992), and the ability to colonise the intestinal tract (Darfeuille-Michaud *et al.*, 1992). It has been suggested that the association of these genetic systems with drug resistance is a reason that *K. pneumoniae*, as compared with *E. coli* and other enteric bacteria, is the predominant host of extended-spectrum β lactamases (Jacoby, 1994).

Role of Outer-membrane Porins

One class of multiple-antibiotic-resistant mutants of *K. pneumoniae* is associated with altered levels of outer-membrane proteins. These mutants may arise spontaneously and show resistance to a range of antimicrobials, including nalidixic acid, trimethoprim, chloramphenicol and β lactams. The phenotype of these mutants is similar to that of multiple antibiotic resistance (Mar) observed in certain *E. coli* mutants (George and Levy, 1983; Cohen *et al.*, 1989), and the *K. pneumoniae* gene responsible for this resistance has been termed *ram*A (resistance antibiotic multiple) (George *et al.*, 1995). The *ram*A gene

from a multiple-drug-resistant (Mdr) mutant of *K. pneumoniae* has been cloned into *E. coli* and the gene has been characterised (George *et al.*, 1995).

K. pneumoniae Mdr mutants show active efflux of tetracycline from the cell and a reduction in active uptake of chloramphenicol in antibiotic transport assays with radiolabelled drugs, and energised or de-energised bacteria. Examination of the membrane-associated proteins in the Mdr mutants, as compared with wild-type bacteria, indicates that antibiotic resistance is associated with loss of the outer-membrane protein of similar size to the *E. coli* OmpF protein. Although the Mdr phenotype is associated with reduced levels of an OmpF-like protein in *K. pneumoniae*, other changes may also play a role in antibiotic resistance, since the introduction of the *ram*A gene from a Mdr mutant into a susceptible strain of *E. coli* resulted in the expression of chloramphenicol resistance that lags behind that to other antibiotics. It has been suggested that the Mdr phenotype is the result of changes in both inner- and outer-membrane composition (George *et al.*, 1995).

Although the cloned *K. pneumoniae ram*A gene confers antibiotic resistance on a *mar* mutant of *E. coli*, the two genes do not appear to be homologues, since DNA probes for each gene hybridise only weakly with the other. In addition, the predicted amino acid sequences of the Mar and Ram proteins have only 40% identity with little similarity at the nucleotide level (George *et al.*, 1995). It is possible that RamA can affect the same set of genes that are controlled by MarA or it may affect MarA expression itself. The identification of an *E. coli soxS* mutation that confers a Mar phenotype shows that multiple-antibiotic resistance may be the result of mutations in several genes. No similarity between the SoxS protein and the *Klebsiella* RamA polypeptide has been observed.

The cloned *ram*A gene was derived from a Mdr mutant of *K. pneumoniae* and, as we have seen, it confers the Mdr phenotype on *E. coli*. The wild-type *ram*A allele will, however, also confer the identical phenotype on an antibiotic-sensitive *E. coli*, and comparison of the RamA protein from the mutant *K. pneumoniae* with its parent shows that they are identical (George *et al.*, 1995). Since investigations into the *E. coli soxS* and *mar*A genes have shown that both are regulated by additional genes, it has been speculated that the mutation responsible for the Mdr phenotype in *K. pneumoniae* is, in fact, within a gene distinct from *ram*A. Transformation of *E. coli* by *ram*A from either the Mdr mutant or the parental strain results in antibiotic resistance in all transformants. It is believed that, in both cases, *ram*A is expressed constitutively from a vector promoter (George *et al.*,

1995). Therefore, the expression of the Mdr phenotype in the *K. pneumoniae* mutant is postulated to be due to a mutation in a *ram*A regulatory gene, which results in altered RamA production.

Additional outer-membrane porin proteins implicated in antibiotic resistance have been identified in clinical isolates of *K. pneumoniae* (Gutmann *et al.*, 1985; Pangon *et al.*, 1989). For example, the resistance of isolates to high levels of cefoxitin is associated with loss of a 35-kDa protein (Martinez-Martinez *et al.*, 1996). However, a detailed analysis of the role of these proteins in mediating antibiotic resistance and their relatedness to other outer-membrane proteins has not been reported.

Conclusion

The incidence of infection by strains of *K. pneumoniae* is low in healthy individuals but it is a serious concern in hospital environments where infection rates can be alarmingly high (Duma, 1985; Emori *et al.*, 1991). Respiratory and urinary tract infections are associated with long-term endotracheal tubes and indwelling catheters. Infections at such sites may lead to septicaemia, or the bloodstream may be directly seeded from infected intravenous lines. Because of advances in supportive care and therapy, large numbers of hospitalised patients are a susceptible group likely to develop nosocomial infections due to *K. pneumoniae* and related species. The high incidence of nosocomially acquired *Klebsiella* infections emphasises the role of the immunological status of the host, particularly of the innate defence mechanisms, in contributing to the development of disease. Under these circumstances, the potential of *K. pneumoniae* to colonise and grow *in vivo* becomes greater. Since debilitated individuals are most susceptible to infection, it has been difficult to investigate the role of various bacterial attributes in mediating virulence. Animal models of infection have been used to mimic respiratory and urinary tract infection due to *K. pneumoniae* (Maayan *et al.*, 1985; Greenberger *et al.*, 1995), and these have provided some insights into the identification of virulence determinants. Clearly, factors such as host specificity and the use of healthy animals in these studies should be borne in mind when interpreting the results of such experiments. Much information has, however, been gained with animal models of infection, and these procedures in conjunction with the use of genetically defined mutants of *K. pneumoniae* should continue to provide insights into *Klebsiella* pathogenicity. Also, the establishment of cellular biological and tissue culture techniques promise to yield interesting information on the molecular biology of *K. pneumoniae*/host-cell interactions.

It is clear that the pathogenicity of *K. pneumoniae* is multi-factorial and a single virulence determinant is not responsible for the clinical manifestations of disease. Indeed, the relative importance of individual virulence factors may vary, depending on the site of infection and the response of the host to infection. Since colonisation represents the initial stage of infection, the ability of bacteria to adhere to host cells is necessary to establish disease. Colonisation factors are therefore important virulence factors and *K. pneumoniae* can express numerous adherence antigens; some of these may be important in the establishment of bacteria in the gastrointestinal tract. Since colonisation of the intestinal lumen frequently occurs after hospitalisation, this is likely to represent a source of bacteria for seeding the respiratory tract by aspiration, and the production of adhesins that mediate attachment to intestinal cells is an important stage in *K. pneumoniae* pathogenesis. The carriage of adhesin genes on large plasmids associated with antibiotic-resistance determinants indicates a selective advantage for bacteria that carry such plasmids. Subsequently, the adherence of bacteria to tissues of the urinary and respiratory epithelium will establish the bacteria at these sites. It is interesting that this type of adherence appears to be mediated by antigens that recognise receptors in the basement membrane that are normally sequestered in healthy tissue. However, exposure as the result of exfoliation of epithelial cells and deposition of matrix proteins on to endotracheal tubes and urinary catheters will provide suitable substrates for colonisation. This is consistent with the high frequency of infection in patients with indwelling devices and localised inflammation (Duma, 1985).

Once bacteria have colonised appropriate tissues, their ability to grow by competing with host cells for important metabolites and overcoming host defence mechanisms is crucial for the further progress of the disease. In these environments siderophore expression is likely to be necessary for acquisition of elemental iron from host cell proteins. Also, because *K. pneumoniae* is not an intracellular pathogen that resists killing by phagocytosis, an ability to interfere with engulfment by phagocytes is required for survival *in vivo*. The relevance of specific capsules in this process has been demonstrated and may be of primary importance in preventing uptake of bacteria by the pulmonary macrophages of the alveoli. Anti-complementary factors such as LPS will also play a role in overcoming the defence mechanisms of the host. Lysis and/or damage of host cells by an exotoxin does not appear to occur in *K. pneumoniae* infections. Sporadic outbreaks of enteritis due to strains of *K. oxytoca* that produce an ST-like enterotoxin related to that found in *E. coli* have been reported (Minami *et al.*, 1992), but most isolates of *K. pneumoniae* that cause extra-intestinal infections do not produce detectable exotoxin. Therefore, toxic activity in these bacteria is due primarily to the lipid A of the LPS molecule.

The increasing population of individuals susceptible to *Klebsiella* infection indicates that the design of therapeutic and immunoprophylactic agents will be important. A basic understanding of the pathogenicity of opportunistic bacteria such as *K. pneumoniae* is fundamental to the development of such agents. The application of molecular genetics and cellular biology to elucidate bacterial pathogenesis and interaction with the host will provide important information about the pathogenetic mechanisms of *K. pneumoniae*. One of the problems associated with the development of molecular genetic techniques in pathogenic isolates of *K. pneumoniae* is the difficulty of using standard recombinant DNA technology in these bacteria. Large amounts of capsule, ill-defined restriction systems, and resistance to multiple antibiotics present practical problems in the construction of defined mutants. During the last few years, however, a number of investigations have focused upon the genetic manipulation of clinical isolates of *K. pneumoniae* (Ofek *et al.*, 1993; Arakawa *et al.*, 1995; Di Martino *et al.*, 1995; Hornick *et al.*, 1995; Kabha *et al.*, 1995; Szabo *et al.*, 1995; Valinluck *et al.*, 1995). These should provide the tools and basic procedures for the examination of the molecular pathogenesis of *K. pneumoniae* and related species.

References

Alberti S, Marques G, Camprubi S *et al.* (1993) C1q binding and activation of the complement classical pathway by *Klebsiella pneumoniae* outer membrane proteins. *Infect. Immun.* 61: 852–860.

Alberti S, Alvarez D, Merino S *et al.* (1996) Analysis of complement C3 deposition and degradation on *Klebsiella pneumoniae*. *Infect. Immun.* 64: 4726–4732.

Allen BL, Gerlach GF, Clegg S (1991) Nucleotide sequence and functions of *mrk* determinants necessary for expression of type 3 fimbriae in *Klebsiella pneumoniae*. *J. Bacteriol.* 173: 916–920.

Allen PM, Fisher D, Saunders JR, Hart CA (1987a) The role of capsular polysaccharide K21b of *Klebsiella* and of the structurally related colanic-acid polysaccharide of *Escherichia coli* in resistance to phagocytosis and serum killing. *J. Med. Microbiol.* 24: 363–370.

Allen PM, Williams JM, Hart CA, Saunders JR (1987b) Identification of two chemical types of K21 capsular polysaccharide from klebsiellae. *J. Gen. Microbiol.* 133: 1365–1370.

Alphen van L, Jansen HM, Dankert J (1995) Virulence factors in the colonization and persistence of bacteria in the airways. *Am. J. Crit. Care Med.* 151: 2094–2100.

Arakawa Y, Ohta M, Wacharotayankun R *et al.* (1991) Biosynthesis of *Klebsiella* K2 capsular polysaccharide in *Escherichia coli* HB101 requires the functions of *rmpA* and the chromosomal *cps* gene cluster of the virulent strain *Klebsiella pneumoniae* Chedid (O1:K2). *Infect. Immun.* 59: 2043–2050.

Arakawa Y, Wacharotayankun R, Nagatsuka T, Ito H, Kato N, Ohta M (1995) Genomic organization of the *Klebsiella pneumoniae cps* region responsible for serotype K2 capsular polysaccharide synthesis in the virulent strain Chedid. *J. Bacteriol.* 177: 1788–1796.

Athamna A, Ofek I, Keisari Y, Markowitz S, Dutton GG, Sharon N (1991) Lectinophagocytosis of encapsulated *Klebsiella pneumoniae* mediated by surface lectins of guinea pig alveolar macrophages and human monocyte-derived macrophages. *Infect. Immun.* 59: 1673–1682.

Bauernfeind A, Chong Y, Schweighart S (1989) Extended broad spectrum beta-lactamase in *Klebsiella pneumoniae* including resistance to cephamycins. *Infection* 17: 316–321.

Beeson PB, Rowley D (1959) The anticomplementary effect of kidney tissue: its association with ammonia production. *J. Exp. Med.* 110: 695–698.

Bender RA (1991) The role of the NAC protein in the nitrogen regulation of *Klebsiella aerogenes*. *Mol. Microbiol.* 5: 2575–2580.

Beurret M, Joseleau JP, Vignon M, Dutton GG, Savage AV (1989) Proof of the occurrence of 5,6-O-(1-carboxyethylidene)-D-galactofuranose units in the capsular polysaccharide of *Klebsiella* K12. *Carbohydr. Res.* 189: 247–260.

Bloch CA, Stocker BA, Orndorff PE (1992) A key role for type 1 pili in enterobacterial communicability. *Mol. Microbiol.* 6: 697–701.

Bock K, Vinogradov EV, Holst O, Brade H (1994) Isolation and structural analysis of oligosaccharide phosphates containing the complete carbohydrate chain of the lipopolysaccharide from *Vibrio cholerae* strain H11 (non-O1) *Eur. J. Biochem.* 225: 1029–1039.

Bronner D, Clarke BR, Whitfield C (1994) Identification of an ATP-binding cassette transport system required for translocation of lipopolysaccharide O-antigen side-chains across the cytoplasmic membrane of *Klebsiella pneumoniae* serotype O:1. *Mol. Microbiol.* 14: 505–519.

Brunskill EW, Bayles KW (1996) Identification of *LytSR*-regulated genes from *Staphylococcus aureus*. *J. Bacteriol.* 178: 5810–5812.

Buchanan K, Falkow S, Hull RA, Hull SI (1985) Frequency among Enterobacteriaceae of the DNA sequences encoding type 1 pili. *J. Bacteriol.* 162: 799–803.

Camprubi S, Merino S, Guillot JF, Tomas JM (1993) The role of the O-antigen lipopolysaccharide on the colonization *in vivo* of the germfree chicken gut by *Klebsiella pneumoniae*. *Microb. Pathogen.* 14: 433–440.

Carbonetti NH, Boonchai S, Parry SH, Vaisanen-Rhen V, Korhonen TK, Williams PH (1986) Aerobactin-mediated iron uptake by *Escherichia coli* isolates from human extraintestinal infections. *Infect. Immun.* 51: 966–968.

Carpenter JL (1990) *Klebsiella* pulmonary infections: occurrence at one medical center and review. *Rev. Infect. Dis.* 12: 672–682.

Casey TA, Nagy B, Moon HW (1992) Pathogenicity of porcine enterotoxigenic *Escherichia coli* that do not express K88, K99, F41 or 987P adhesins. *Am. J. Vet. Res.* 53: 1488–1492.

Chen LY, Chen MY, Leu WM, Tsai TY, Lee YH (1993) Mutational study of *Streptomyces* tyrosinase trans-activator MelC1: MelC1 is likely a chaperone for apotyrosinase. *J. Biol. Chem.* 268: 18710–18716.

Chen LY, Leu WM, Wang KT, Lee YH (1992) Copper transfer and activation of the *Streptomyces* apotyrosinase are mediated through a complex formation between apotyrosinase and its trans-activator MelC1. *J. Biol. Chem.* 267: 20100–20107.

Choy YM, Tsang SF, Kong SK *et al.* (1996) K1 and K3 capsular antigens of *Klebsiella* induce tumor necrosis factor activities. *Life Sci.* 58: PL153–158.

Ciurana B, Tomas JM (1987) Role of lipopolysaccharide and complement in susceptibility of *Klebsiella pneumoniae* to nonimmune serum. *Infect. Immun.* 55: 2741–2746.

Clarke BR, Bronner D, Keenleyside WJ, Severn WB, Richards JC, Whitfield C (1995) Role of *Rfe* and *RfbF* in the initiation of biosynthesis of D-galactan I: the lipopolysaccharide O antigen from *Klebsiella pneumoniae* serotype O:1. *J. Bacteriol.* 177: 5411–5418.

Clarke BR, Whitfield C (1992) Molecular cloning of the *rfb* region of *Klebsiella pneumoniae* serotype O1:K20: the *rfb* gene cluster is responsible for synthesis of the D-galactan I O polysaccharide. *J. Bacteriol.* 174: 4614–4621.

Clegg S, Old DC (1979) Fimbriae of *Escherichia coli* K-12 strain AW405 and related bacteria. *J. Bacteriol.* 137: 1008–1012.

Clegg S, Hull S, Hull R, Pruckler J (1985a) Construction and comparison of recombinant plasmids encoding type 1 fimbriae of members of the family Enterobacteriaceae. *Infect. Immun.* 48: 275–279.

Clegg S, Pruckler J, Purcell BK (1985b) Complementation analyses of recombinant plasmids encoding type 1 fimbriae of members of the family Enterobacteriaceae. *Infect. Immun.* 50: 338–340.

Clegg S, Korhonen KT, Hornick BD, Tarkkanen A-M (1994) Type 3 fimbriae of the Enterobacteriaceae. In: Klemm P (ed.) *Fimbriae: Adhesion, Genetics, Biogenesis, and Vaccines.* Boca Raton, FL: CRC Press, pp. 97–104.

Cohen SP, McMurry LM, Hooper DC, Wolfson JS, Levy SB (1989) Cross-resistance to fluoroquinolones in multiple-antibiotic-resistant (Mar) *Escherichia coli* selected by tetracycline or chloramphenicol: decreased drug accumulation associated with membrane changes in addition to OmpF reduction. *Antimicrob. Agents Chemother.* 33: 1318–1325.

Collier DN (1993) SecB: a molecular chaperone of *Escherichia coli* protein secretion pathway. *Adv. Protein Chem.* 44: 151–193.

Collins CM, Gutman DM, Laman H (1993) Identification of a nitrogen-regulated promoter controlling expression of *Klebsiella pneumoniae* urease genes. *Mol. Microbiol.* 8: 187–198.

Collins CM, Se DO (1993) Bacterial ureases: structure, regulation of expression and role in pathogenesis. *Mol. Microbiol.* 9: 907–913.

Collins CM (1994) The tao of urease. In: Miller VL, Kaper JB, Portnoy DA, Isberg RR (eds) *Molecular Genetics of Bacterial Pathogenesis.* Washington, DC: ASM Press, pp. 437–450.

Craven DE, Barber TW, Steger KA, Montecalvo MA (1990) Nosocomial pneumonia in the 1990s: update of epidemiology and risk factors. *Sem. Respir. Dis.* 5: 157–172.

Cryz SJ, Mortimer PM, Mansfield V, Germanier R (1986) Seroepidemiology of *Klebsiella* bacteremic isolates and implications for vaccine development. *J. Clin. Microbiol.* 23: 687–690.

Darfeuille-Michaud A, Jallat C, Aubel D *et al.* (1992) R-plasmid-encoded adhesive factor in *Klebsiella pneumoniae* strains responsible for human nosocomial infections. *Infect. Immun.* 60: 44–55.

Dekker NP, Lammel CJ, Mandrell RE, Brooks GF (1990) Opa (protein II) influences gonococcal organization in colonies, surface appearance, size and attachment to human fallopian tube tissues. *Microb. Pathogen.* 9: 19–31.

DeVivo MJ, Fine PR, Cutter GR, Maetz HM (1984) The risk of renal calculi in spinal cord injury patients. *J. Urol.* 131: 857–860.

Di Martino P, Bertin Y, Girardeau JP, Livrelli V, Joly B, Darfeuille-Michaud A (1995) Molecular characterization and adhesive properties of CF29K: an adhesin of *Klebsiella pneumoniae* strains involved in nosocomial infections. *Infect. Immun.* 63: 4336–4344.

Di Martino P, Livrelli V, Sirot D, Joly B, Darfeuille-Michaud A (1996) A new fimbrial antigen harbored by CAZ-5/SHV-4-producing *Klebsiella pneumoniae* strains involved in nosocomial infections. *Infect. Immun.* 64: 2266–2273.

Dodd DC, Eisenstein BI (1984) Kinetic analysis of the synthesis and assembly of type 1 fimbriae of *Escherichia coli*. *J. Bacteriol.* 160: 227–232.

Domenico P, Diedrich DL, Straus DC (1985) Extracellular polysaccharide production by *Klebsiella pneumoniae* and its relationship to virulence. *Can. J. Microbiol.* 31: 472–478.

Domenico P, Schwartz S, Cunha BA (1989) Reduction of capsular polysaccharide production in *Klebsiella pneumoniae* by sodium salicylate. *Infect. Immun.* 57: 3778–3782.

Domenico P, Salo RJ, Straus DC, Hutson JC, Cunha BA (1992) Salicylate or bismuth salts enhance opsonophagocytosis of *Klebsiella pneumoniae*. *Infection* 20: 66–72.

Domenico P, Salo RJ, Cross AS, Cunha BA (1994a) Polysaccharide capsule-mediated resistance to opsono-phagocytosis in *Klebsiella pneumoniae. Infect. Immun.* 62: 4495–4499.

Domenico P, Salo RJ, Cross AS, Cunha BA (1994b) Salicylate enhances opsonization of *Klebsiella pneumoniae* with anticapsular antibodies. *Ann. NY Acad. Sci.* 730: 315–317.

Donaldson SG, Aziz SQ, Dal Nogare AR (1991) Characteristics of aerobic Gram-negative bacteria colonizing critically ill patients. *Am. Rev. Respir. Dis.* 144: 202–207.

Duguid JP (1959) Fimbriae and adhesive properties in *Klebsiella* strains. *J. Gen. Microbiol.* 21: 271–286.

Duma RJ (1985) Gram-negative bacillary infections: pathogenic and pathophysiologic correlates. *Am. J. Med.* 78(6A): 154–164.

Duncan JL (1988) Differential effect of Tamm–Horsfall protein on adherence of *Escherichia coli* to transitional epithelial cells. *J. Infect. Dis.* 158: 1379–1382.

Dutton GG, Ng SK, Parolis LA, Parolis H, Chakraborty AK (1989) A re-investigation of the structure of the capsular polysaccharide of *Klebsiella* K10. *Carbohydr. Res.* 193: 147–155.

Edwards PR, Fife MA (1952) Capsule types of *Klebsiella. J. Infect. Dis.* 91: 92–104.

Eisenstein BI (1988) Type 1 fimbriae of *Escherichia coli*: genetic regulation, morphogenesis, and role in pathogenesis. *Rev. Infect. Dis.* 10 (Suppl. 2): S341–344.

Emori TG, Banerjee SN, Culver DH *et al.* (1991) Nosocomial infections in elderly patients in the United States: 1986–1990. National Nosocomial Infections Surveillance System. *Am. J. Med.* 91(3B): 289S–293S.

Eshdat Y, Sharon N (1984) Recognitory bacterial surface lectins which mediate its mannose-specific adherence to eukaryotic cells. *Biol. Cell* 51(2): 259–266.

Fader RC, Gondesen K, Tolley B, Ritchie DG, Moller P (1988) Evidence that *in vitro* adherence of *Klebsiella pneumoniae* to ciliated hamster tracheal cells is mediated by type 1 fimbriae. *Infect. Immun.* 56: 3011–3013.

Favre-Bonte S, Darfeuille-Michaud A, Forestier C (1995) Aggregative adherence of *Klebsiella pneumoniae* to human intestine-407 cells. *Infect. Immun.* 63: 1318–1328.

Firon N, Ofek I, Sharon N (1984) Carbohydrate-binding sites of the mannose-specific fimbrial lectins of enterobacteria. *Infect. Immun.* 43: 1088–1090.

Friedrich B, Magasanik B (1977) Urease of *Klebsiella aerogenes*: control of its synthesis by glutamine synthetase. *J. Bacteriol.* 131: 446–452.

Gaffney RA, Venegas MF, Kanerva C *et al.* (1995) Effect of vaginal fluid on adherence of type 1 piliated *Escherichia coli* to epithelial cells. *J. Infect. Dis.* 172: 1528–1535.

Gally DL, Bogan JA, Eisenstein BI, Blomfield IC (1993) Environmental regulation of the fim switch controlling type 1 fimbrial phase variation in *Escherichia coli* K-12: effects of temperature and media. *J. Bacteriol.* 175: 6186–6193.

Gally DL, Leathart J, Blomfield IC (1996) Interaction of FimB and FimE with the fim switch that controls the phase variation of type 1 fimbriae in *Escherichia coli* K-12. *Mol. Microbiol.* 21: 725–738.

George AM, Levy SB (1983) Amplifiable resistance to tetracycline, chloramphenicol and other antibiotics in *Escherichia coli*: involvement of a non-plasmid-determined efflux of tetracycline. *J. Bacteriol.* 155: 531–540.

George AM, Hall RM, Stokes HW (1995) Multidrug resistance in *Klebsiella pneumoniae*: a novel gene, *ramA*, confers a multidrug resistance phenotype in *Escherichia coli. Microbiology* 141: 1909–1920.

Gerlach GF, Clegg S (1988) Characterization of two genes encoding antigenically distinct type-1 fimbriae of *Klebsiella pneumoniae. Gene* 64: 231–240.

Gerlach GF, Clegg S, Nichols WA (1988) Characterization of the genes encoding urease activity of *Klebsiella pneumoniae. FEMS Microbiol. Lett.* 50: 131–135.

Gerlach GF, Allen BL, Clegg S (1989a) Type 3 fimbriae among enterobacteria and the ability of spermidine to inhibit MR/K hemagglutination. *Infect. Immun.* 57: 219–224.

Gerlach GF, Clegg S, Allen BL (1989b) Identification and characterization of the genes encoding the type 3 and type 1 fimbrial adhesins of *Klebsiella pneumoniae. J. Bacteriol.* 171: 1262–1270.

Gerlach GF, Clegg S, Ness NJ, Swenson DL, Allen BL, Nichols WA (1989c) Expression of type 1 fimbriae and mannose-sensitive hemagglutinin by recombinant plasmids. *Infect. Immun.* 57: 764–770.

Girardeau JP, Der Vartanian M, Ollier JL, Contrepois M (1988) CS31A: a new K88-related fimbrial antigen on bovine enterotoxigenic and septicemic *Escherichia coli* strains. *Infect. Immun.* 56: 2180–2188.

Gonzalez Leiza M, Perez-Diaz JC, Ayala J *et al.* (1994) Gene sequence and biochemical characterization of FOX-1 from *Klebsiella pneumoniae*: a new AmpC-type plasmid-mediated beta-lactamase with two molecular variants. *Antimicrob. Agents Chemother.* 38: 2150–2157.

Gouby A, Neuwirth C, Bourg G *et al.* (1994) Epidemiological study by pulsed-field gel electrophoresis of an outbreak of extended-spectrum beta-lactamase-producing *Klebsiella pneumoniae* in a geriatric hospital. *J. Clin. Microbiol.* 32: 301–305.

Greenberger MJ, Strieter RM, Kunkel SL, Danforth JM, Goodman RE, Standiford TJ (1995) Neutralization of IL-10 increases survival in a murine model of *Klebsiella* pneumonia. *J. Immunol.* 155: 722–729.

Griffiths E (1991) Iron and bacterial virulence: a brief overview. *Biol. Metals* 4(1): 7–13.

Gutmann L, Williamson R, Moreau N *et al.* (1985) Cross-resistance to nalidixic acid, trimethoprim and chloramphenicol associated with alterations in outer membrane proteins of *Klebsiella, Enterobacter* and *Serratia. J. Infect. Dis.* 151: 501–507.

Hedelin H, Grenabo L, Pettersson S (1985) Urease-induced crystallization in synthetic urine. *J. Urol.* 133: 529–532.

Hedelin H, Bratt CG, Eckerdal G, Lincoln K (1991) Relationship between urease-producing bacteria, urinary pH and encrustation on indwelling urinary catheters. *Br. J. Urol.* 67: 527–531.

Hicks S, Candy DC, Phillips AD (1996) Adhesion of enteroaggregative *Escherichia coli* to pediatric intestinal mucosa *in vitro*. *Infect. Immun.* 64: 4751–4760.

Homer MJ, Paustian TD, Shah VK, Roberts GP (1993) The *nifY* product of *Klebsiella pneumoniae* is associated with apodinitrogenase and dissociates upon activation with the iron-molybdenum cofactor. *J. Bacteriol.* 175: 4907–4910.

Hornick DB, Allen BL, Horn MA, Clegg S (1991) Fimbrial types among respiratory isolates belonging to the family Enterobacteriaceae. *J. Clin. Microbiol.* 29: 1795–1800.

Hornick DB, Allen BL, Horn MA, Clegg S (1992) Adherence to respiratory epithelia by recombinant *Escherichia coli* expressing *Klebsiella pneumoniae* type 3 fimbrial gene products. *Infect. Immun.* 60: 1577–1588.

Hornick DB, Thommandru J, Smits W, Clegg S (1995) Adherence properties of an *mrkD*-negative mutant of *Klebsiella pneumoniae*. *Infect. Immun.* 63: 2026–2032.

Hultgren SJ, Normark S (1991) Biogenesis of the bacterial pilus. *Curr. Opin. Genet. Dev.* 1: 313–318.

Hultgren SJ, Normark S, Abraham SN (1991) Chaperone-assisted assembly and molecular architecture of adhesive pili. *Annu. Rev. Microbiol.* 45: 383–415.

Jabri E, Carr MB, Hausinger RP, Karplus PA (1995) The crystal structure of urease from *Klebsiella* aerogenes. *Science* 268: 998–1004.

Jacoby GA (1994) Genetics of extended-spectrum β-lactamases. *Eur. J. Clin. Microbiol. Infect. Dis.* 13 (Suppl. 1): S2–11.

Jacoby GA (1996) Antimicrobial-resistant pathogens in the 1990s. *Annu. Rev. Med.* 47: 169–179.

Jiang XM, Neal B, Santiago F, Lee SJ, Romana LK, Reeves PR (1991) Structure and sequence of the *rfb* (O antigen) gene cluster of *Salmonella* serovar *typhimurium* (strain LT2). *Mol. Microbiol.* 5: 695–713.

Johnson JR, Brown JJ (1996) A novel multiply primed polymerase chain reaction assay for identification of variant papG genes encoding the Gal(alpha 1-4)Gal-binding PapG adhesins of *Escherichia coli*. *J. Infect. Dis.* 173: 920–926.

Jones CH, Pinkner JS, Roth R *et al.* (1995) FimH adhesin of type 1 pili is assembled into a fibrillar tip structure in the Enterobacteriaceae. *Proc. Natl Acad. Sci. USA* 92: 2081–2085.

Jong G-M, Hsuie T-R, Chen C-R, Chang H-Y, Chen C-W (1995) Rapidly fatal outcome of bacteremic *Klebsiella pneumoniae* pneumonia in alcoholics. *Chest* 107: 214–217.

Kabha K, Nissimov L, Athamna A *et al.* (1995) Relationships among capsular structure, phagocytosis and mouse virulence in *Klebsiella pneumoniae*. *Infect. Immun.* 63: 847–852.

Kelly RF, Severn WB, Richards JC *et al.* (1993) Structural variation in the O-specific polysaccharides of *Klebsiella pneumoniae* serotypes O:1 and O:8

lipopolysaccharide: evidence for clonal diversity in *rfb* genes. *Mol. Microbiol.* 10: 615–625.

Kenne L, Lindberg B (1983) Bacterial polysaccharides . In: Aspinall GO (ed.) *Polysaccharides*, New York: Academic Press. Vol. 2. P. 287–363.

Khimji PL, Miles AA (1978) Microbial iron-chelators and their action on *Klebsiella* infections in the skin of guinea-pigs. *Br. J. Exp. Pathol.* 59: 137–147.

Klemm P, Schembri M, Hasty DL (1996) The FimH protein of type 1 fimbriae: an adaptable adhesin. *Adv. Exp. Med. Biol.* 408: 193–195.

Knutton S, Shaw RK, Bhan MK *et al.* (1992) Ability of enteroaggregative *Escherichia coli* strains to adhere *in vitro* to human intestinal mucosa. *Infect. Immun.* 60: 2083–2091.

Kol O, Wieruszeski JM, Strecker G *et al.* (1991) Structure of the O-specific polysaccharide chain from *Klebsiella pneumoniae* O1K2 (NCTC 5055) lipopolysaccharide. *Carbohydr. Res.* 217: 117–125.

Kol O, Wieruszeski JM, Strecker G, Fournet B, Zalisz R, Smets P (1992) Structure of the O-specific polysaccharide chain of *Klebsiella pneumoniae* O1K2 (NCTC 5055) lipopolysaccharide: a complementary elucidation. *Carbohydr. Res.* 236: 339–344.

Korhonen TK, Tarkka E, Ranta H, Haahtela K (1983) Type 3 fimbriae of *Klebsiella* sp.: molecular characterization and role in bacterial adhesion to plant roots. *J. Bacteriol.* 155: 860–865.

Kotelko K, Deka M, Gromska W, Kaca W, Radziejewska-Lebrecht J, Rozalski A (1983) Galacturonic acid as the terminal constituent in the R core polysaccharide of proteus R110 (Ra) mutant. *Arch. Immunol. Therap. Exper.* 31: 833–838.

Kroll JS, Loynds B, Brophy LN, Moxon ER (1990) The *bex* locus in encapsulated *Haemophilus influenzae*: a chromosomal region involved in capsule polysaccharide export. *Mol. Microbiol.* 4: 1853–1862.

Kuehn MJ, Ogg DJ, Kihlberg J *et al.* (1993) Structural basis of pilus subunit recognition by the PapD chaperone. *Science* 262: 1234–1241.

Kuehn MJ, Jacob-Dubuisson F, Dodson K, Slonim L, Striker R, Hultgren SJ (1994) Genetic, biochemical and structural studies of biogenesis of adhesive pili in bacteria. *Meth. Enzymol.* 236: 282–306.

Laakso DH, Homonylo MK, Wilmot SJ, Whitfield C (1988) Transfer and expression of the genetic determinants for O and K antigen synthesis in *Escherichia coli* O9:K(A)30 and *Klebsiella* sp. O1:K20, in *Escherichia coli* K12. *Can. J. Microbiol.* 34: 987–992.

Lee CJ (1987) Bacterial capsular polysaccharides: biochemistry, immunity and vaccine. *Mol. Immunol.* 24: 1005–1019.

Lee MH, Mulrooney SB, Renner MJ, Markowicz Y, Hausinger RP (1992) Klebsiella aerogenes urease gene cluster: sequence of ureD and demonstration that four accessory genes (*ureD*, *ureE*, *ureF* and *ureG*) are involved in nickel metallocenter biosynthesis. *J. Bacteriol.* 174: 4324–4330.

Lindberg F, Normark S (1986) Sequence of the *Citrobacter freundii* OS60 chromosomal ampC β-lactamase gene. *Eur. J. Biochem.* 156: 441–445.

Lodge JM, Williams P, Brown MR (1986) Influence of growth rate and iron limitation on the expression of outer membrane proteins and enterobactin by *Klebsiella pneumoniae* grown in continuous culture. *J. Bacteriol.* 165: 353–356.

Maayan MC, Ofek I, Medalia O, Aronson M (1985) Population shift in mannose-specific fimbriated phase of *Klebsiella pneumoniae* during experimental urinary tract infection in mice. *Infect. Immun.* 49: 785–789.

Macaluso A, Best EA, Bender RA (1990) Role of the *nac* gene product in the nitrogen regulation of some NTR-regulated operons of *Klebsiella* aerogenes. *J. Bacteriol.* 172: 7249–7255.

Madison B, Ofek I, Clegg S, Abraham SN (1994) Type 1 fimbrial shafts of *Escherichia coli* and *Klebsiella pneumoniae* influence sugar-binding specificities of their FimH adhesins. *Infect. Immun.* 62: 843–848.

Magasanik B, Neidhardt FC (1987) Regulation of carbon and nitrogen utilization. In: Neidhardt FC, Ingraham JL, Low KB *et al.* (eds) Escherichia coli *and* Salmonella typhimurium: *Cellular and Molecular Biology.* Washington, DC: ASM Press, pp. 1318–1325.

Malaviya R, Abraham SN (1995) Interaction of bacteria with mast cells. *Meth. Enzymol.* 253: 27–43.

Malaviya R, Ross E, Jakschik BA, Abraham SN (1994) Mast cell degranulation induced by type 1 fimbriated *Escherichia coli* in mice. *J. Clin. Invest.* 93: 1645–1653.

Malaviya R, Ikeda T, Ross E, Abraham SN (1996) Mast cell modulation of neutrophil influx and bacterial clearance at sites of infection through TNF-alpha. *Nature* 381: 77–80.

Martinez-Martinez L, Hernandez-Alles S, Alberti S, Tomas JM, Benedi VJ, Jacoby GA (1996) *In vivo* selection of porin-deficient mutants of *Klebsiella pneumoniae* with increased resistance to cefoxitin and expanded-spectrum cephalosporins. *Antimicrob. Agents Chemother.* 40: 342–348.

McCallum KL, Laakso DH, Whitfield C (1989a) Use of a bacteriophage-encoded glycanase enzyme in the generation of lipopolysaccharide O side-chain deficient mutants of *Escherichia coli* O9:K30 and *Klebsiella* O1:K20: role of O and K antigens in resistance to complement-mediated serum killing. *Can. J. Microbiol.* 35: 994–999.

McCallum KL, Schoenhals G, Laakso D, Clarke B, Whitfield C (1989b) A high-molecular-weight fraction of smooth lipopolysaccharide in *Klebsiella* serotype O1:K20 contains a unique O-antigen epitope and determines resistance to nonspecific serum killing. *Infect. Immun.* 57: 3816–3822.

McClain MS, Blomfield IC, Eisenstein BI (1991) Roles of fimB and fimE in site-specific DNA inversion associated with phase variation of type 1 fimbriae in *Escherichia coli. J. Bacteriol.* 173: 5308–5314.

McClain MS, Blomfield IC, Eberhardt KJ, Eisenstein BI (1993) Inversion-independent phase variation of type 1 fimbriae in *Escherichia coli. J. Bacteriol.* 175: 4335–4344.

McLean RJ, Nickel JC, Cheng KJ, Costerton JW (1988) The ecology and pathogenicity of urease-producing bacteria in the urinary tract. *Crit. Rev. Microbiol.* 16: 37–79.

McLean RJ, Downey J, Clapham L, Wilson JW, Nickel JC (1991) Pyrophosphate inhibition of *Proteus mirabilis*-induced struvite crystallization *in vitro. Clin. Chim. Acta* 200(2/3): 107–117.

McSweegan E, Walker RI (1986) Identification and characterization of two *Campylobacter jejuni* adhesins for cellular and mucous substrates. *Infect. Immun.* 53: 141–148.

Minami J, Saito S, Yoshida T, Uemura T, Okabe A (1992) Biological activities and chemical composition of a cytotoxin of *Klebsiella oxytoca. J. Gen. Microbiol.* 138: 1921–1927.

Mizuta K, Ohta M, Mori M, Hasegawa T, Nakashima I, Kato N (1983) Virulence for mice of *Klebsiella* strains belonging to the O:1 group: relationship to their capsular (K) types. *Infect. Immun.* 40: 56–61.

Mobley HL, Island MD, Hausinger RP (1995) Molecular biology of microbial ureases. *Microbiol. Rev.* 59: 451–480.

Mobley WC, Rutkowski JL, Tennekoon GI, Buchanan K, Johnston MV (1985) Choline acetyltransferase activity in striatum of neonatal rats increased by nerve growth factor. *Science* 229: 284–287.

Moncrief MB, Hausinger RP (1996) Purification and activation properties of UreD–UreF–urease apoprotein complexes. *J. Bacteriol.* 178: 5417–5421.

Moncrief MB, Hom LG, Jabri E, Karplus PA, Hausinger RP (1995) Urease activity in the crystalline state. *Protein Sci.* 4: 2234–2236.

Mulrooney SB, Hausinger RP (1990) Sequence of the *Klebsiella* aerogenes urease genes and evidence for accessory proteins facilitating nickel incorporation. *J. Bacteriol.* 172: 5837–5843.

Nassif X, Sansonetti PJ (1986) Correlation of the virulence of *Klebsiella pneumoniae* K1 and K2 with the presence of a plasmid encoding aerobactin. *Infect. Immun.* 54: 603–608.

Nassif X, Fournier JM, Arondel J, Sansonetti PJ (1989a) Mucoid phenotype of *Klebsiella pneumoniae* is a plasmid-encoded virulence factor. *Infect. Immun.* 57: 546–552.

Nassif X, Honore N, Vasselon T, Cole ST, Sansonetti PJ (1989b) Positive control of colanic acid synthesis in *Escherichia coli* by *rmpA* and *rmpB*, two virulence-plasmid genes of *Klebsiella pneumoniae. Mol. Microbiol.* 3: 1349–1359.

Nataro JP, Hicks S, Phillips AD, Vial PA, Sears CL (1996) T84 cells in culture as a model for enteroaggregative *Escherichia coli* pathogenesis. *Infect. Immun.* 64: 4761–4768.

Neilands JB (1982) Microbial envelope proteins related to iron. *Annu. Rev. Microbiol.* 36: 285–309.

Neilands JB (1995) Siderophores: structure and function of microbial iron transport compounds. *J. Biol. Chem.* 270: 26723–26726.

Nimmich W, Korten G (1970) Die chemische zusammensetzung der *Klebsiella*-lipopolysaccharide (O-antigene). *Pathol. Microbiol.* 36: 179–190.

Nowotarska M, Mulczyk M (1977) Serologic relationship of fimbriae among Enterobacteriaceae. *Arch. Immunol. Therap. Exper.* 25(1): 7–16.

Ofek I, Beachey EH (1980) Bacterial adherence. *Adv. Intern. Med.* 25: 503–532.

Ofek I, Sharon N (1988) Lectinophagocytosis: a molecular mechanism of recognition between cell surface sugars and lectins in the phagocytosis of bacteria. *Infect. Immun.* 56: 539–547.

Ofek I, Kabha K, Athamna A *et al.* (1993) Genetic exchange of determinants for capsular polysaccharide biosynthesis between *Klebsiella pneumoniae* strains expressing serotypes K2 and K21a. *Infect. Immun.* 61: 4208–4216.

Old DC, Adegbola RA (1983) A new mannose-resistant haemagglutinin in *Klebsiella*. *J. Appl. Bacteriol.* 55: 165–172.

Old DC, Adegbola RA (1985) Antigenic relationships among type-3 fimbriae of Enterobacteriaceae revealed by immunoelectronmicroscopy. *J. Med. Microbiol.* 20: 113–121.

Old DC, Tavendale A, Senior BW (1985) A comparative study of the type-3 fimbriae of *Klebsiella* species. *J. Med. Microbiol.* 20: 203–214.

Ørskov F, Ørskov I (1978) Serotyping of Enterobacteriaceae with special emphasis on K antigen determination. In: Norris JR, Bergen T (eds) *Methods in Microbiology*, London: Academic Press, Vol. 11. pp. 37–38.

Ørskov I, Ørskov F (1984) Serotyping of *Klebsiella pneumoniae*. *Meth. Microbiol.* 14: 143–164.

Paerregaard A, Espersen F, Skurnik M (1991) Adhesion of yersiniae to rabbit intestinal constituents: role of outer membrane protein YadA and modulation by intestinal mucus. *Contrib. Microbiol. Immunol.* 12: 171–175.

Pangon B, Bizet C, Bure A *et al.* (1989) *In vivo* selection of a cephamycin-resistant: porin-deficient mutant of *Klebsiella pneumoniae* producing a TEM-3 beta-lactamase [letter]. *J. Infect. Dis.* 159: 1005–1006.

Park IS, Carr MB, Hausinger RP (1994) *In vitro* activation of urease apoprotein and role of UreD as a chaperone required for nickel metallocenter assembly. *Proc. Natl Acad. Sci. USA* 91: 3233–3237.

Podbielski A, Schonling J, Melzer B, Haase G (1991a) Different promoters of SHV-2 and SHV-2a beta-lactamase lead to diverse levels of cefotaxime resistance in their bacterial producers. *J. Gen. Microbiol.* 137: 1667–1675.

Podbielski A, Schonling J, Melzer B, Warnatz K (1991b) Molecular cloning and nucleotide sequence of a new plasmid-coded *Klebsiella pneumoniae* beta-lactamase gene (SHV-2a) responsible for high-level cefotaxime resistance. *Int. J. Med. Microbiol.* 275: 369–373.

Podschun R (1990) Phenotypic properties of *Klebsiella pneumoniae* and *K. oxytoca* isolated from different sources. *Zentral. Hyg. Umweltmed.* 189: 527–535.

Podschun R, Heineken P, Sonntag HG (1987) Haemagglutinatinins and adherence properties to HeLa and intestine 407 cells of *Klebsiella pneumoniae* and *Klebsiella oxytoca* isolates. *Zbl. Bakt. Hyg.* 263A: 585–593.

Podschun R, Sievers D, Fischer A, Ullmann U (1993) Serotypes, hemagglutinins, siderophore synthesis and serum resistance of *Klebsiella* isolates causing human urinary tract infections. *J. Infect. Dis.* 168: 1415–1421.

Pohlschroder M, Murphy C, Beckwith J (1996) *In vivo* analyses of interactions between SecE and SecY, core components of the *Escherichia coli* protein translocation machinery. *J. Biol. Chem.* 271: 19908–19914.

Ponniah S, Endres RO, Hasty DL, Abraham SN (1991) Fragmentation of *Escherichia coli* type 1 fimbriae exposes cryptic D-mannose-binding sites. *J. Bacteriol.* 173: 4195–4202.

Radziejewska-Lebrecht J, Mayer H (1989) The core region of *Proteus mirabilis* R110/1959 lipopolysaccharide. *Eur. J. Biochem.* 183: 573–581.

Ramphal R, Koo L, Ishimoto KS, Totten PA, Lara JC, Lory S (1991) Adhesion of *Pseudomonas aeruginosa* pilin-deficient mutants to mucin. *Infect. Immun.* 59: 1307–1311.

Reig R, Roy C, Hermida M, Teruel D, Coira A (1993) A survey of β-lactamases from 618 isolates of *Klebsiella* spp. *J. Antimicrob. Chemother.* 31: 29–35.

Rice LB, Carias LL, Bonomo RA, Shlaes DM (1996) Molecular genetics of resistance to both ceftazidime and beta-lactam-beta-lactamase inhibitor combinations in *Klebsiella pneumoniae* and *in vivo* response to beta-lactam therapy. *J. Infect. Dis.* 173: 151–158.

Riser E, Noone P (1981) *Klebsiella* capsular type versus site of isolation. *J. Clin. Pathol.* 34: 552–555.

Russell PW, Orndorff PE (1992) Lesions in two *Escherichia coli* type 1 pilus genes alter pilus number and length without affecting receptor binding. *J. Bacteriol.* 174: 5923–5935.

Saraste M, Sibbald PR, Wittinghofer A (1990) The P-loop: a common motif in ATP- and GTP-binding proteins. *Trends Biochem. Sci.* 15: 430–434.

Savarino SJ, Fox P, Deng Y, Nataro JP (1994) Identification and characterization of a gene cluster mediating enteroaggregative *Escherichia coli* aggregative adherence fimbria I biogenesis. *J. Bacteriol.* 176: 4949–4957.

Scaletsky IC, Silva ML, Trabulsi LR (1984) Distinctive patterns of adherence of enteropathogenic *Escherichia coli* to HeLa cells. *Infect. Immun.* 45: 534–536.

Schaeffer AJ, Amundsen SK, Schmidt LN (1979) Adherence of *Escherichia coli* to human urinary tract epithelial cells. *Infect. Immun.* 24: 753–759.

Schaeffer AJ, Amundsen SK, Jones JM (1980) Effect of carbohydrates on adherence of *Escherichia coli* to human urinary tract epithelial cells. *Infect. Immun.* 30: 531–537.

Schurtz TA (1997) *Type 3 Fimbriae of* Klebsiella *Species: Genetic and Functional Analyses of Bacterial Adhesins.* PhD thesis, University of Iowa.

Schurtz TA, Hornick DB, Korhonen TK, Clegg S (1994) The type 3 fimbrial adhesin gene (*mrkD*) of *Klebsiella* species is not conserved among all fimbriate strains. *Infect. Immun.* 62: 4186–4191.

Severn WB, Kelly RF, Richards JC, Whitfield C (1996) Structure of the core oligosaccharide in the serotype O:8 lipopolysaccharide from *Klebsiella pneumoniae. J. Bacteriol.* 178: 1731–1741.

Simoons-Smit AM, Verweij-van Vught AM, MacLaren DM (1986) The role of K antigens as virulence factors in *Klebsiella. J. Med. Microbiol.* 21: 133–137.

Sokurenko EV, Courtney HS, Maslow J, Siitonen A, Hasty DL (1995) Quantitative differences in adhesiveness of type 1 fimbriated *Escherichia coli* due to structural differences in *fimH* genes. *J. Bacteriol.* 177: 3680–3686.

Stock JB, Ninfa AJ, Stock AM (1989) Protein phosphorylation and regulation of adaptive responses in bacteria. *Microbiol. Rev.* 53: 450–490.

Svennerholm AM, McConnell MM, Wiklund G (1992) Roles of different putative colonization factor antigens in colonization of human enterotoxigenic *Escherichia coli* in rabbits. *Microb. Pathogen.* 13: 381–389.

Swaney LM, Liu YP, To CM, To CC, Ippen-Ihler K, Brinton CC (1977) Isolation and characterization of *Escherichia coli* phase variants and mutants deficient in type 1 pilus production. *J. Bacteriol.* 130: 495–505.

Szabo M, Bronner D, Whitfield C (1995) Relationships between *rfb* gene clusters required for biosynthesis of identical D-galactose-containing O antigens in *Klebsiella pneumoniae* serotype O:1 and *Serratia marcescens* serotype O:16. *J. Bacteriol.* 177: 1544–1553.

Tarkkanen AM, Allen BL, Westerlund B *et al.* (1990) Type V collagen as the target for type-3 fimbriae: enterobacterial adherence organelles. *Mol. Microbiol.* 4: 1353–1361.

Tarkkanen AM, Allen BL, Williams PH *et al.* (1992) Fimbriation, capsulation and iron-scavenging systems of *Klebsiella* strains associated with human urinary tract infection. *Infect. Immun.* 60: 1187–1192.

Tarkkanen AM, Virkola R, Clegg S, Korhonen TK (1997) Binding of type 3 fimbriae of *Klebsiella pneumoniae* to human endothelial and urinary bladder cells. *Infect. Immun.* 65: 1546–1549.

Tewari R, Ikeda T, Malaviya R *et al.* (1994) The PapG tip adhesin of P fimbriae protects *Escherichia coli* from neutrophil bactericidal activity. *Infect. Immun.* 62: 5296–5304.

Todd MJ, Hausinger RP (1989) Competitive inhibitors of *Klebsiella* aerogenes urease: mechanisms of interaction with the nickel active site. *J. Biol. Chem.* 264: 15835–15842.

Tomas JM, Benedi VJ, Ciurana B, Jofre J (1986) Role of capsule and O antigen in resistance of *Klebsiella pneumoniae* to serum bactericidal activity. *Infect. Immun.* 54: 85–89.

Vaara M (1992) Agents that increase the permeability of the outer membrane. *Microbiol. Rev.* 56: 395–411.

Valinluck B, Lee NS, Ryu J (1995) A new restriction-modification system: KpnBI: recognized in *Klebsiella pneumoniae. Gene* 167(1/2): 59–62.

Venegas MF, Navas EL, Gaffney RA, Duncan JL, Anderson BE, Schaeffer AJ (1995) Binding of type 1-piliated *Escherichia coli* to vaginal mucus. *Infect. Immun.* 63: 416–422.

Vernet V, Madoulet C, Chippaux C, Philippon A (1992) Incidence of two virulence factors (aerobactin and mucoid phenotype) among 190 clinical isolates of *Klebsiella pneumoniae* producing extended-spectrum β-lactamase. *FEMS Microbiol. Lett.* 75: 1–5.

Wacharotayankun R, Arakawa Y, Ohta M *et al.* (1993) Enhancement of extracapsular polysaccharide synthesis in *Klebsiella pneumoniae* by RmpA2: which shows homology to NtrC and FixJ. *Infect. Immun.* 61: 3164–3174.

Whitfield C, Richards JC, Perry MB, Clarke BR, MacLean LL (1991) Expression of two structurally distinct D-galactan O antigens in the lipopolysaccharide of *Klebsiella pneumoniae* serotype O:1. *J. Bacteriol.* 173: 1420–1431.

Williams P, Lambert PA, Brown MR, Jones RJ (1983) The role of the O and K antigens in determining the resistance of *Klebsiella* aerogenes to serum killing and phagocytosis. *J. Gen. Microbiol.* 129: 2181–2191.

Williams P, Brown MR, Lambert PA (1984) Effect of iron deprivation on the production of siderophores and outer membrane proteins in *Klebsiella* aerogenes. *J. Gen. Microbiol.* 130: 2357–2365.

Williams P, Chart H, Griffiths E, Stevenson P (1987) Expression of high-affinity iron uptake systems by clinical isolates of *Klebsiella. FEMS Microbiol. Lett.* 44: 407–412.

Williams P, Smith MA, Stevenson P, Griffiths E, Tomas JM (1989) Novel aerobactin receptor in *Klebsiella pneumoniae. J. Gen. Microbiol.* 135: 3173–3181.

Williams P, Tomas JM (1990) The pathogenicity of *Klebsiella pneumoniae. Rev. Med. Microbiol.* 1: 196–204.

Williams P, Ciurana B, Camprubi S, Tomas JM (1990) Influence of lipopolysaccharide chemotype on the interaction between *Klebsiella pneumoniae* and human polymorphonuclear leucocytes. *FEMS Microbiol. Lett.* 57: 305–309.

78

Moraxella (Branhamella) catarrhalis

Barbara J. Chang,[1] Brian J. Mee,[1] Karen F. McGregor[1] and Thomas V. Riley[1,2]

[1]*University of Western Australia,*
[2]*Queen Elizabeth II Medical Centre, Nedlands, Western Australia*

Moraxella catarrhalis was for a long time considered a harmless commensal of the upper respiratory tract. In the last 20 years, however, an increasing number of reports of *M. catarrhalis* isolated in a variety of settings has forced a re-evaluation of the clinical significance of this organism. It is unclear whether the increase in the number of reports is due to an increased awareness of the organism by medical investigators or whether there has been an increase in virulence. There is evidence to suggest that both may have occurred.

In the early 1900s, *Micrococcus catarrhalis*, as it was then called, was often isolated in pure culture from patients with respiratory tract infections (reviewed by Berk, 1990). Reports of *Micrococcus catarrhalis* as a pathogen began to dwindle during the 1920s, when the organism was renamed *Neisseria catarrhalis* (Holland, 1920). This classification was based on morphological, cultural and biochemical similarities between *Micrococcus catarrhalis* and *Neisseria meningitidis*. Throughout the middle of the twentieth century, *N. catarrhalis* was regarded as a harmless commensal of the upper respiratory tract and when recovered in sputum culture was often reported as 'normal flora' or 'non-pathogenic' *Neisseria* (Verghese and Berk, 1991).

In the late 1960s, reports of the isolation of *N. catarrhalis* as a pathogen re-emerged (Coffey *et al.*, 1967; Feign *et al.*, 1969) and these continued to increase during the 1970s. The organism came to be recognised as a significant pathogen, commonly associated with otitis media in children and lower respiratory tract infections in adults (Catlin, 1990). While the pathogenic potential of the organism was widely accepted, its nomenclature remained the subject of debate. The role of *Moraxella catarrhalis*, as it is now called, as a pathogen has now been established. Murphy and Sethi (1992) described five criteria that convincingly demonstrate the pathogenicity of *M. catarrhalis* in the respiratory tract. These are, first, that *M. catarrhalis* is found in large numbers in sputum samples evaluated for contamination by strict criteria (McLeod *et al.*, 1986a; Nicotra *et al.*, 1986). Second, the organism can be isolated in pure culture from transtracheal aspirates (Ninane *et al.*, 1977; Aitken and Thornley, 1983; Hager *et al.*, 1987). Third, clinical improvement is observed when patients with *M. catarrhalis* infection are treated with the appropriate antibiotics (McLeod *et al.*, 1986a; Nicotra *et al.*, 1986; Darelid *et al.*, 1993). Fourth, the organism has been recovered from normally sterile sites, such as blood or pleural fluid, in patients with respiratory tract infection (Choo and Gantz, 1989; Wallace and Oldfield, 1990; Collazos *et al.*, 1992). Finally, a bactericidal antibody response to the homologous strain of *M. catarrhalis* develops in patients with chronic bronchitis (Chapman *et al.*, 1985).

Taxonomy

Quite early, evidence accumulated for a genetic difference between *N. catarrhalis* and other neisserias, leading to a re-evaluation of its taxonomic classification. This difference was supported by the findings of nucleic acid hybridisation studies (Kingsbury, 1967), that *N. catarrhalis* genetic transformation with other species of the genus *Neisseria* does not occur (Catlin and Cunningham, 1964), and that their guanine + cytosine (G + C) ratios are different (Catlin and Cunningham, 1961; Bövre *et al.*, 1969). These findings prompted Henriksen and Bövre (1968) to propose the removal of *N. catarrhalis* from the genus *Neisseria*. They suggested either the transfer of *N. catarrhalis* to the genus *Moraxella*, because of genetic similarities to members of this genus; or alternatively, the creation of a new genus for the species. In order to avoid the confusion of including rods and cocci in the same genus, Catlin (1970) supported the latter and proposed the creation of the genus *Branhamella*. The organism became known as *Branhamella catarrhalis*, a name that was readily adopted throughout the 1970s.

There are a number of current proposals for names for this organism. Genetic and physiological studies have demonstrated similarities between *M. catarrhalis* and *Moraxella* species (Bövre, 1970; Henriksen, 1976; Jantzen *et al.*, 1976; Rossau *et al.*, 1986; Veron *et al.*, 1993; Enright *et al.*, 1994) and some investigators believe that *M. catarrhalis* should be classified as a subgenus of *Moraxella* (Bövre, 1979). Rossau *et al.* (1991) proposed that the subgenus *Moraxella* (*Branhamella*) be abandoned and that these species remain part of the *Moraxella* genus in a new family Moraxellaceae. Catlin (1991) argued for the creation of a new family Branhamaceae to accommodate the separate genera *Branhamella* and *Moraxella*.

Currently, the confusing situation is that throughout the literature the organism is alternatively called *Branhamella catarrhalis*, *Moraxella* (*Branhamella*) *catarrhalis* and *Moraxella catarrhalis*. Catlin (1990, 1991) and Murphy (1996) argue that the name *Branhamella catarrhalis* should be maintained and many researchers continue to use the name *Branhamella* (Ahmed *et al.*, 1994; Christensen *et al.*, 1994; Chaibi *et al.*, 1995;

Ishida *et al.*, 1995). To assign the organism to the genus *Moraxella* not only places cocci and rod-shaped organisms in the same genus but also relegates this important pathogen to a genus containing organisms rarely implicated in human infection.

Isolation, Identification and Biochemical Characteristics

The laboratory diagnosis of *M. catarrhalis* respiratory tract infections is not difficult. Usually the microscopy of a good specimen of sputum shows numerous polymorphonuclear leucocytes and many intra- and extracellular Gram-negative diplococci. *M. catarrhalis* is readily cultured on a variety of bacteriological media used for respiratory microbiology, such as blood and heated blood agar. In the past, however, the identification of *M. catarrhalis* has caused some problems. The use of appropriate laboratory tests to distinguish *M. catarrhalis* from non-fermentative, non-pathogenic *Neisseria* species, which may be present in sputum samples, is an important consideration (Doern and Morse, 1980; Enright and McKenzie, 1997). The quality assurance programme of the College of American Pathologists provides an indication of the ability of clinical microbiology laboratories accurately to identify *M. catarrhalis*. In 1983, 74% of 685 participating laboratories correctly identified *M. catarrhalis*, and by 1985, 82% were accurately identifying it (Jones and Sommers, 1986). It is important to note that in the 18% of laboratories that did not identify the organism as *M. catarrhalis*, it was usually reported as *Neisseria* or *Moraxella* species, names that would not have suggested a potential pathogen.

M. catarrhalis is a Gram-negative coccus, commonly arranged in pairs with flattened adjacent sides (Bövre, 1984). It has no special temperature requirements (growth occurs at 22°C and 37°C), or for carbon dioxide or additional nutrients. On blood agar, colonies are small, opaque and grey-white, 1–3 mm diameter, circular and non-haemolytic. The colonies do not adhere to agar and remain intact when pushed over the surface of the agar, resembling a hockey puck on ice (Catlin, 1990). *M. catarrhalis* is oxidase- and catalase-positive, it reduces nitrate and nitrite, fails to produce acid from sugars, and is unable to synthesise polysaccharide from sucrose (Doern and Morse, 1980). *M. catarrhalis* possesses deoxyribonuclease (DNase) activity and it produces a butyric acid esterase, which may be detected with the substrate tributyrin (Riley, 1987).

Epidemiology

Normal Carriage

The natural habitat of *M. catarrhalis* is believed to be exclusively in humans (Henriksen, 1976). It has been isolated from the nasopharynx and pharynx, and occasionally from the conjunctiva and genital tract (Blackwell *et al.*, 1978; Wilhelmus *et al.*, 1980), and there is a strong relationship between age and colonisation rates. In infancy, colonisation of the upper respiratory tract with *M. catarrhalis* is common (Ingvarsson *et al.*, 1982; Vaneechoutte *et al.*, 1990b; Aniansson *et al.*, 1992; Ejlertsen *et al.*, 1994a). Faden *et al.* (1994) showed that 66% of infants in Buffalo, New York, become colonised during the first year of life, with colonisation reaching 78% by the age of 2 years. Colonisation of infants appears to occur at different rates in different populations. In a rural Aboriginal community in Australia, it was reported that 100% of infants were colonised by the age of 3 months (Leach *et al.*, 1994). The factors that account for these differences in colonisation are not clearly defined, but living conditions, environmental factors and genetic background are likely to play a role (Murphy, 1996).

The proportion of children colonised by *M. catarrhalis* tends to decrease with age. Ejlertsen *et al.* (1994a) found the carriage rate of *M. catarrhalis* in 4–15 year olds was only 7% as opposed to 54% in children 1–48 months of age. This reduction in carriage is most likely to be due to increased immunity. An increase in IgG antibodies to *M. catarrhalis* in older children has been reported (Ejlertsen *et al.*, 1994b).

Colonisation is rare in adults, with only between 1% and 5% of healthy adults colonised by *M. catarrhalis* (Pollard *et al.*, 1986; DiGiovanni *et al.*, 1987; Vaneechoutte *et al.*, 1990b; Ejlertsen, 1991; Ejlertsen *et al.*, 1994a). There are few data on colonisation rates in the elderly. Vaneechoutte *et al.* (1990b) reported that 26.5% of people older than 60 years were colonised with *M. catarrhalis*. This suggests an increase in carriage rate as compared with younger adults. In Ireland, however, 694 elderly people were sampled with the colonisation rate varying from 1.7% in the summer to a high of 10.8% in winter (T. Scott, personal communication).

Colonisation with *M. catarrhalis* appears to be a dynamic process. Klingman *et al.* (1995) studied adults over a two-year period and typed strains with molecular methods. Colonisation with a new strain of *M. catarrhalis* occurred frequently, and patients remained colonised with a strain for an average of 2.3 months. Similarly, Faden *et al.* (1994) showed that from birth to 2 years of age children frequently lost and acquired new strains.

A strong seasonal variation, with an increased rate of colonisation in winter months, has been reported (Pollard *et al.*, 1986; Van Hare *et al.*, 1987). A study in Ohio has shown that the nasopharyngeal colonisation rate of healthy children was 46% in winter but only 9% in summer (Van Hare *et al.*, 1987). It has been suggested that the ability of *M. catarrhalis* to adhere to oropharyngeal cells varies in different seasons, with a higher level of adherence in the winter months (Mbaki *et al.*, 1987). This seasonal variation may account for some of the variability seen in the level of respiratory tract colonisation reported by different groups (Vaneechoutte *et al.*, 1990b), and highlights the importance of reporting the time and geographical location of studies that describe colonisation rates.

Respiratory Infections

The most common bacterial causes of respiratory infections are *Streptococcus pneumoniae*, *Haemophilus influenzae* and *M. catarrhalis* (Felmingham *et al.*, 1996). In the Alexander project on lower respiratory pathogens isolated in Europe and the United States from 1992 to 1993, *M. catarrhalis* represented 13.5% of all bacterial isolates (Felmingham *et al.*, 1996). Overall, *M. catarrhalis* was found in approximately 2% of respiratory tract specimens (Christensen *et al.*, 1986; Sarubbi *et al.*, 1990).

Respiratory infections caused by *M. catarrhalis* include pneumonia (West *et al.*, 1982; Wright *et al.*, 1990), bronchitis (Wallace and Musher, 1986), tracheitis (Bodkin and Warde, 1993; Bernstein *et al.*, 1998), laryngitis (Schalen *et al.*, 1980; Hol *et al.*, 1996), sinusitis (Brorson *et al.*, 1976; Wald *et al.*, 1981, 1984; Goldenhersh *et al.*, 1990; Penttila *et al.*, 1997) and persistent cough (Brorson and Malmvall, 1981; Gottfarb and Brauner, 1994), and *M. catarrhalis* is a common bacterial cause in many of these manifestations. It has been isolated from 23.5% of children with long-standing cough (Brorson and Malmvall, 1981) and 18% of cases of sinusitis (Tinkelman and Silk, 1989). *M. catarrhalis* was isolated from more than 50% of cases of acute laryngitis in which it was the dominant causative organism (Schalen *et al.*, 1980).

Underlying cardiopulmonary disease, smoking and other predisposing conditions are common in patients with respiratory tract infection due to *M. catarrhalis* (Slevin *et al.*, 1984; Nicotra *et al.*, 1986; DiGiovanni *et al.*, 1987; Hager *et al.*, 1987; Capewell *et al.*, 1988; Barreiro *et al.*, 1992; Chin *et al.*, 1993). In patients with chronic obstructive pulmonary disease and other chronic lung diseases, purulent exacerbations are provoked by infection with *M. catarrhalis* (Wright *et al.*, 1990; Murphy and Sethi, 1992). Although it is difficult to estimate the proportion of exacerbations due to this organism, one study concluded that approximately one-third of exacerbations are caused by *M. catarrhalis* (Verghese *et al.*, 1990b). Infection with *M. catarrhalis* may be promoted by viral damage to respiratory tract epithelium, for example by respiratory syncytial virus (Arola *et al.*, 1990), or an immunodeficient state (McNeely *et al.*, 1976; Diamond and Lorber, 1984). Although primarily an opportunistic pathogen of people with predisposing conditions, *M. catarrhalis* has been documented as a cause of pneumonia and bronchitis in otherwise healthy adults (Slevin *et al.*, 1984; Davies and Maesen, 1990; Boyle *et al.*, 1991).

M. catarrhalis presents a clinical picture similar to that of other bacterial pneumonias (Slevin *et al.*, 1984; Wright *et al.*, 1990; Chin *et al.*, 1993). Patients experience fever with a cough, leucocytosis is common and pulmonary infiltrates can be seen on chest X-ray. *M. catarrhalis* pneumonia is sometimes considered a mild illness, as it causes a fairly non-invasive infection and the organism is rarely cultured from blood or pleural fluid.

The majority of respiratory isolates have come from the elderly (DiGiovanni *et al.*, 1987; Hager *et al.*, 1987; Wright *et al.*, 1990; Boyle *et al.*, 1991; Chin *et al.*, 1993). Wright *et al.* (1990) analysed respiratory isolates from a hospital in Texas and found that 81% of patients with *M. catarrhalis* infections were over 55 years. They also noted a high short-term mortality rate in the elderly, with 45% of patients dying within 3 months of acquiring *M. catarrhalis* pneumonia. Factors that contribute to the high incidence of respiratory infections in this age group may include immunosuppression (Catlin, 1990) and an increase in the level of adherence to epithelial cells in elderly patients (Carr *et al.*, 1989).

M. catarrhalis is also a cause of lower respiratory tract infections in children. *M. catarrhalis* infections in children often differ from those of adults by the absence of underlying respiratory disease (Mannion, 1987). Boyle *et al.* (1991) found that the majority of *M. catarrhalis* respiratory infections in children occurred in those less than one year of age. *M. catarrhalis* pneumonia may occur in premature infants and in those with pre-existing pulmonary disease (Ohlsson and Bailey, 1985; Haddad *et al.*, 1986; Berg and Bartley, 1987; Dyson *et al.*, 1990). It appeared to be a rare cause of respiratory infection in children in Finland, making up less than 2% of isolates (Korppi *et al.*, 1992). In contrast, Berner *et al.* (1996) isolated *M. catarrhalis* from 38% of children with respiratory

tract infections in Germany. A potential link between asthma and *M. catarrhalis* has been suggested, but the nature of the association is not understood. Seddon *et al.* (1992) found a significantly higher carriage rate in asthmatic children (70–75%) compared with normal children (33%). *M. catarrhalis* synthesises clinically significant amounts of histamine *in vitro* (Devalia *et al.*, 1989), which may contribute to the induction of asthma.

Respiratory infections with *M. catarrhalis* show a distinct seasonal variation, being predominantly a disease of winter and spring months (McLeod *et al.*, 1986b; Pollard *et al.*, 1986; DiGiovanni *et al.*, 1987; Davies and Maesen, 1988; Sarubbi *et al.*, 1990; Boyle *et al.*, 1991). The seasonal variation is similar to that seen for nasopharyngeal colonisation, suggesting an association between colonisation and lung disease. A higher rate of colonisation has been noted in those with symptoms of upper respiratory tract infection, as compared with those without such symptoms (Schalen *et al.*, 1980; Brorson and Malmvall, 1981; Ejlertsen *et al.*, 1994a). Sputum containing *M. catarrhalis* is more likely to be recovered from patients with chronic lung disease than from healthy adults (Pollard *et al.*, 1986; Smith and Lockwood, 1986; Klingman *et al.*, 1995).

Otitis Media

Acute otitis media is common in infants and children. *M. catarrhalis* was recognised as a potential aetiological agent in the 1920s (Marchant, 1990), and Coffey *et al.* (1967) reported its isolation in pure culture from middle ear fluid. Reports of *M. catarrhalis* isolated from patients with otitis media were numerous during the 1980s and the organism has become well-recognised as an important cause of otitis media in children (Catlin, 1990; Murphy, 1996).

M. catarrhalis is responsible for approximately 15% of all otitis media (Kovatch *et al.*, 1983; Bluestone, 1986; Faden *et al.*, 1992; Aspin *et al.*, 1994), making it the third most common cause after *H. influenzae* and *S. pneumoniae*. *M. catarrhalis* was the most common bacterial cause of otitis media in Finland with the organism isolated from 55% of patients (Arola *et al.*, 1990). Murphy (1996) estimated that each year in the United States 3.5 million physician visits are made by children with otitis media due to *M. catarrhalis*.

A strong association between nasopharyngeal colonisation and the development of otitis media has also been described (Faden *et al.*, 1991; Stenfors and Raisanen, 1993). Dickinson *et al.* (1988) used restriction endonuclease analysis to demonstrate that

M. catarrhalis recovered simultaneously from the nasopharynx and middle ear of children with otitis media were identical. A higher rate of nasopharyngeal colonisation of *M. catarrhalis* has been shown in otitis-prone children when compared with normal children (Prellner *et al.*, 1984; Faden *et al.*, 1994). Faden *et al.* (1994) suggested that a high rate of colonisation may be associated with an increased risk of otitis media, with early colonisation a risk factor for first infection (Faden *et al.*, 1997).

Antimicrobial Resistance

For many years *M. catarrhalis* was considered universally susceptible to all antibiotics, including penicillins. By the mid-1980s it became apparent that a dramatic change has occurred in susceptibility to β-lactam antibiotics, with the rapid emergence of resistant strains (Riley, 1988).

Resistance to β Lactams

Almost simultaneously several investigators reported β lactamase-producing strains of *M. catarrhalis* (Malmvall *et al.*, 1977; Ninane *et al.*, 1977; Percival *et al.*, 1977; Buu Hoi-Dang *et al.*, 1978). Credit for the first observation of β lactamase-producing *M. catarrhalis* strains goes to Malmvall *et al.* (1977). They were unable to transfer resistance and noted that further work was needed to study the substrate profile and other biochemical characteristics of the enzyme. Percival *et al.* (1977) showed that the β lactamase from their isolates was biochemically different from the enzyme present in *N. gonorrhoeae*. Ninane *et al.* (1977) in Belgium reported penicillin-resistant strains of *M. catarrhalis* recovered via transtracheal puncture, and Buu Hoi-Dang *et al.* (1978) described a β lactamase that was novel based on substrate profile, isoelectric point and the location of the genetic determinant. Farmer and Reading (1982) analysed β lactamases from Belgian isolates of *M. catarrhalis*, and examined the strains Ravasio, 1646, 1648 and 1908. They showed, by isoelectric focusing, that there were two distinct β lactamases, one represented by the strain Ravasio and the other represented by strain 1908. These two classes were subsequently called BRO-1 and BRO-2 (from BRanhamella and MOraxella). Both enzymes were constitutive, providing low levels of resistance to penicillin and ampicillin, but not to most cephalosporins. Isolates that produce BRO-1 β lactamase are more resistant to ampicillin than those that produce BRO-2 (Luman *et al.*, 1986; Fung *et al.*, 1994a, b). The difference in minimal

inhibitory concentration correlates with the observation that BRO-1 is produced at a level two to three times higher than that of BRO-2 (Wallace *et al.*, 1989, 1990). Strains containing BRO-1 are encountered more frequently than those with BRO-2 (Farmer and Reading, 1982; Kamme *et al.*, 1986; Eliasson *et al.*, 1992; Ikeda *et al.*, 1993). The BRO β lactamases are susceptible to clavulanic acid (Stobberingh *et al.*, 1994; Farmer and Reading, 1986; Labia *et al.*, 1986). Purification of the BRO enzymes requires digestion with papain followed by affinity chromatography and the molecular weight of both enzymes is 28 kDa (Eliasson *et al.*, 1992). The BRO-1 β lactamase is also present in other species of *Moraxella*, such as *M. non-liquefaciens* and *M. lacunata* (Kamme *et al.*, 1986; Wallace *et al.*, 1989).

Beaulieu *et al.* (1989) submitted a 4500-bp sequence containing the BRO-1 DNA gene (accession number U49269); the sequence also contained an amidase gene and a conserved gene of unknown function. Bootsma *et al.* (1996) submitted two sequences, one for the BRO-1 gene of 1080 bp (accession number Z54180) and the other for the BRO-2 gene of 1059 bp (accession number Z54181). The two β lactamases differ by just one amino acid, with a 21-bp deletion in the promoter region of the BRO-2 gene. The $G + C$ content of the *bla* genes was 31%. This is significantly different from flanking DNA and from the *Moraxella* genome (41%), which suggests that horizontal gene transfer is responsible for the appearance of ampicillin resistance in the species. Later, Bootsma *et al.* (1999) showed that the BRO β lactamases are lipoproteins located on the inner part of the outer membrane of *M. catarrhalis*. Previous reports of β-lactamase lipoproteins have been restricted to Gram-positive species, which further supports the role of horizontal transfer as the source and spread of the *bla* gene in this species. The *bla* genes are transmissible by conjugation and by transformation, but neither mode of transfer is related to the presence of plasmid DNA (Wallace *et al.*, 1989; Ikeda *et al.*, 1993; Chaibi *et al.*, 1995). Conjugation and transformation studies by McGregor *et al.* (submitted) showed that when the BRO-1 *bla* gene is transferred to a penicillin-susceptible recipient, or to a BRO-2 recipient, it recombined non-randomly into a sequence. It is always found in a characteristic sequence in the susceptible recipient and replaced the BRO-2 gene in that host.

Isolates of *M. catarrhalis* are intrinsically resistant to vancomycin and trimethoprim (Wallace *et al.*, 1990), but surveys of antibiotic resistance in *M. catarrhalis* indicate that most isolates are susceptible to all other antibiotics, except the penicillins (Doern and Tubert, 1988; Jorgensen *et al.*, 1990; Doern *et al.*, 1999).

In the latter study, β lactamase was detected in every isolate from 13 of 34 participating hospitals in the USA and Canada, with an overall prevalence of 92.2%. Resistance to the other 21 antimicrobials tested was very rare. Resistance to high levels of tetracycline occurs occasionally and is associated with the presence of the *tet*B gene (Roberts *et al.*, 1990). This gene is believed to have originated in Gram-positive species, providing a second example of horizontal gene transfer of antibiotic resistance in *M. catarrhalis*. It is, perhaps, surprising that there is not more resistance in *M. catarrhalis* when the species is naturally transformable (Catlin, 1990) and when it appears with increasing frequency as a nosocomial pathogen.

In addition to the role of *M. catarrhalis* as a pathogen in its own right, β lactamase-producing strains may protect concomitantly infecting more virulent pathogens such as *Strep. pneumoniae* and *H. influenzae* from normally effective antibiotic therapy (Wardle, 1986; Hol *et al.*, 1994).

Structure

Capsule

Early reports of an outer fibrillar layer, described as a capsule (Reyn, 1974), were not confirmed for many years, and the nature and even the existence of a capsular layer on *M. catarrhalis* was in some doubt. Hellio *et al.* (1988) reported 'spicule-like' structures protruding from the cell surface. Ruthenium red staining suggested that these contained polysaccharide, and they were stable after repeated subculture. Ahmed *et al.* (1990, 1991) confirmed a ruthenium red-positive layer in a clinical isolate of *M. catarrhalis*. Later, spicule-like structures were observed on several isolates (Fitzgerald *et al.*, 1999a, b) and they were shown to be trypsin-resistant, and distinguishable from a trypsin-sensitive outer fibrillar coat found on other isolates. Thus, *M. catarrhalis* may potentially produce two surface layers; a proteinaceous diffuse fibrillar layer, and polysaccharide spicules. The capsular nature of the latter layer is still in doubt, since other authors have described apparently similar structures as short fimbriae with knobby ends (Marrs and Weir, 1990) or as peritrichous fimbriae with a knob-like structure at the tip (Ahmed *et al.*, 1992a, 1994).

Pili

In an early report by Wistreich and Baker (1971), one of three *M. catarrhalis* strains was piliated. This strain expressed two pilus types; numerous short pili and sparser, longer filaments. This report was not confirmed for almost 20 years, until Marrs and Weir (1990) detected long, thin pili and short, thick pili with knobby ends. The longer pili appeared morphologically similar to type IV pili, and genomic Southern hybridisation analysis revealed homology to the type-IV Q pilin gene of *M. bovis*. Other reports detailing the two pilus types came from Ahmed *et al.* (1990, 1992a, b, 1994). The pili are trypsin-sensitive (Ahmed *et al.*, 1992a) and often appear to be attached to each other or to form a dense surface layer (Ahmed *et al.*, 1994). The majority of *M. catarrhalis* isolates are piliated (Rikitomi *et al.*, 1991; Ahmed *et al.*, 1992a).

Lipo-oligosaccharides

The outer-membrane glycolipids of *M. catarrhalis* lack the repeating O-antigen polysaccharides of lipopolysaccharides and are termed lipo-oligosaccharides (LOS) (Vaneechoutte *et al.*, 1990a). These occur commonly in non-enteric Gram-negative bacteria, such as those that colonise the mucosal surfaces of the respiratory tract (Griffiss *et al.*, 1988). The LOS of *M. catarrhalis* consists of lipid A plus a core polysaccharide and one oligosaccharide unit (Fomsgaard *et al.*, 1991). The lipid A component is similar to that of other Gram-negative bacteria (Masoud *et al.*, 1994) and cross-reacts with lipid A of the Enterobacteriaceae, but it lacks the 3-hydroxytetradecanoic acid normally present in enteric bacteria (Fomsgaard *et al.*, 1991). Electrophoresis and tandem mass spectrometry reveal a significant level of variability in the lipid A of *M. catarrhalis* (Kelly *et al.*, 1996).

Serological typing of *M. catarrhalis* LOS with hyperimmune rabbit sera demonstrated three major antigenic types among the 302 strains tested; serotype A (61%), B (29%) and C (5%), with 5% untypable strains (Vaneechoutte *et al.*, 1990a). The oligosaccharide structures for all three serotypes are branched with a common inner core (Edelbrink *et al.*, 1994, 1995, 1996). The determinant for serotype A is a terminal α-GlcNAc–(1 → 2)-β-Glc, for serotype B a β-Gal–(1 → 4)-α-Glc, and for serotype C a β-Gal-(1 → 4)–α-GlcNAc. A terminal α-Gal-(1 → 4)–β-Gal-(1 → 4)-Glc trisaccharide is common to the three serotypes and is also found in the LOS of other Gram-negative non-enteric pathogens such as *H. influenzae* and *N. meningitidis* (Weiser, 1992). Tetrameric nucleotide repeat units, which are associated with phase-variable expression of LOS in these pathogens, have also been detected in *M. catarrhalis* (Peak *et al.*, 1996), but there is no evidence of an association in *M. catarrhalis*. Further studies are required to explain why *Neisseria* spp. and

H. influenzae exhibit enormous antigenic heterogeneity in their LOS (Griffiss *et al.*, 1988; Weiser, 1992), while 95% of *M. catarrhalis* isolates belong to just three serotypes.

Outer-membrane Proteins

Techniques for the isolation of the outer membrane (OM) of *M. catarrhalis* were developed by Murphy and Loeb (1989), since the methods widely used for OM isolation from other bacteria were unsuccessful. The protein composition of the *M. catarrhalis* OM is, however, typical of Gram-negative bacteria, with 10–20 OM proteins (OMPs) of which between 6 and 8 predominate. The major OMPs range from 21 kDa to 98 kDa on SDS-PAGE and were designated OMP A to OMP H (Bartos and Murphy, 1988). They are very homogeneous between strains from diverse geographic and clinical sources (Bartos and Murphy, 1988; Murphy, 1990).

The majority of these OMPs have been studied at a molecular and functional level, and their roles in pathogenicity and interaction with the immune system will be discussed below. Characteristics of the best-studied OMPs are summarised in **Table 1**. Some clarification of structure and re-naming of the OMPs has occurred. The original OMP B, an 80-kDa protein, was renamed OMP B2 (also called CopB). This was to distinguish it from a minor OMP of 84 kDa, which was designated OMP B1 after the discovery that it elicited a prominent serum antibody response in patients with bronchiectasis (Sethi *et al.*, 1995). OMP C and OMP D represent a single protein, now referred to as OMP CD, which migrates in SDS-PAGE as a doublet of approximately 60 kDa (Murphy *et al.*, 1993). Sequence analysis of OMP CD predicted a protein of 46 kDa; this is due to a proline-rich region that causes aberrant migration on SDS-PAGE. It shares homology with the Opr F porin of *Pseudomonas* spp., and is, on the basis of this and other evidence – including its abundance in the OM and its high β-sheet content – a porin protein (Murphy *et al.*, 1993). However, definitive proof of this is not yet available. OMP E, a 50-kDa protein on SDS-PAGE, has a molecular mass of 47 kDa predicted from sequence analysis (Bhushan *et al.*, 1994). OMP E has borderline homology with Fad L of *Escherichia coli*, a protein involved in binding and transport of fatty acids, and with OMP F porin proteins (Bhushan *et al.*, 1994). Thus, it is possible that OMP E is a trimeric porin protein, but proof of this awaits functional studies.

In addition to the OMPs described above, a high-molecular-weight protein (HMW-OMP or UspA)

varying in molecular mass from 350 to 720 kDa has been reported (Klingman and Murphy, 1994). Disruption of oligomers with formic acid yielded a single band on SDS-PAGE of 120–140 kDa. The HMP-OMP varied in molecular mass, but was present on all 14 clinical strains tested, and was antigenically conserved (Klingman and Murphy, 1994). UspA was later shown to consist of two related proteins, UspA1 and UspA2 (Aebi *et al.*, 1997). Further nucleotide sequence analysis, N-terminal amino acid sequencing and mass spectrometric analysis indicated sizes of 83 kDa and 60 kDa, respectively (Cope *et al.*, 1999). The anomalous behaviour of UspA1 and UspA2 during SDS-PAGE, and the nature of the very-high-molecular-weight aggregate forms, which are either homoaggregates of UspA1 or UspA2 or hetero-aggregates of both proteins, are not understood. A second type of UspA2 protein, UspA2H, has very recently been identified (Lafontaine *et al.*, 2000). UspA2H is a 'hybrid' of an N-terminal half, which resembles that of UspA1, and a C-terminal half nearly identical to UspA2. Approximately 20% of *M. catarrhalis* isolates possess *usp*A2H rather than a *usp*A2 gene.

Genomic and Metabolic Studies

The initial genomic map of the *M. catarrhalis* genome was reported by Furihata *et al.* (1995). The genome size of the type strain *M. catarrhalis* ATCC 25238 was estimated to be 1940 kb, based on pulsed-field gel electrophoresis of genomic DNA digested by restriction endonucleases *Not*1 (10 fragments) and *Sma*1 (9 fragments). The order of the restriction endonuclease fragments was determined by partial digests, double digests and by using the 'linking' clone method described by Smith and Condemine (1990). With the DNA of a *M. catarrhalis* clinical isolate, SH-5, clones were tested for complementation of *E. coli* auxotrophs. Southern hybridisation revealed the existence of eight amino acid biosynthetic pathways, including those for tryptophan, methionine, cysteine, lysine, leucine, isoleucine and valine, threonine and arginine, and for the purine and pyrimidine biosynthetic pathways. In total, 12 biosynthetic genes were located on the circular *M. catarrhalis* map. Similarly, Nguyen *et al.* (1999) used restriction endonucleases *Not*1, *Sma*1, I-*Ceu*1 (4 fragments) and *Rsr*II (6 fragments) and estimated that the genome of the type strain was 1750 kb. This report had some discrepancies with the earlier map in the relationships of some of the

Table 1 Characteristics of outer membrane proteins of *Moraxella catarrhalis*

OMP	Molecular mass (kDa)	Proposed function	Other properties	ORF (bp)	GenBank no.	References
B1 (TbpB)	80–84	Iron acquisition from transferrin	Bilobed protein; each lobe binds transferrin	~2100	AF039311–AF039316, AF105251	Luke et al. (1999); Yu and Schryvers (1993); Myers et al. (1998); Retzer et al. (1999)
TbpA	115–120	Iron acquisition from transferrin	Binds fibronectin	~3200	AF039312, AF039315	Luke and Campagnari (1999); Myers et al. (1998); Yu and Schryvers (1993)
B2 (CopB)	80–81	Iron acquisition; serum resistance	Amino acid sequence similar to TonB-dependent OMPs	2277–2301	L12346, U69980-2, U83900, U83901	Aebi et al. (1996, 1998a); Helminen et al. (1993a); Sethi et al. (1997)
LbpA	103–105	Iron acquisition from lactoferrin	May act as transmembrane channel	~3000	AF043131, AF043133	Bonnah et al. (1999); Du et al. (1998)
LbpB	95–97	Iron acquisition from lactoferrin	Conserved RGD motif; possible eukaryotic cell attachment site	~2700	AF043131, AF043133	Bonnah et al. (1999); Du et al. (1998)
CD	46	Porin; adhesion to human mucin	55–60 kDa on SDS-PAGE	1362	L10755	Murphy et al. (1993); Reddy et al. (1997); Hsiao et al. (1995)
E	50	Porin	Homology to *E. coli* FadL; fatty acid transport	1377	L31788	Bhushan et al. (1994, 1997)
UspA1 (HMW-OMP)	83	Adhesion to Chang conjunctival cells	> 250 kDa oligomers or aggregates on SDS-PAGE	2496–2823	AF113606, AF113608, AF113610, AF181072, AF181076, U57551, U61725	Aebi et al. (1997, 1998b); Cope et al. (1999); McMichael et al. (1998)
UspA2 (HMW-OMP)	60	Serum resistance	Binds vitronectin	1731–2052	AF113607, AF113609, AF113611, AF181073, U86135	Aebi et al. (1998b); Cope et al. (1999); McMichael et al. (1998)
UspA2H (HMW-OMP)		Adhesion to Chang conjunctival cells	Hybrid protein related to UspA1 and UspA2	2667–2682	AF181074, AF181075	Lafontaine et al. (2000)

fragments, but the authors suggested that the use of four restriction endonucleases eliminated some of the difficulties experienced by Furihata *et al.* (1995). They located four ribosomal RNA operons by hybridisation and by using the restriction endonuclease I-*Ceu*1. A physical map, based on data from Furihata *et al.* (1995) and Nguyen *et al.* (1999), is shown in **Fig. 1**. Juni *et al.* (1986) reported a defined medium for *M. catarrhalis* and showed that the amino acids proline, arginine, glycine and methionine were required for growth, suggesting that these biosynthetic pathways were not operating. It is interesting that, in the cases of methionine and arginine, complementation of genes for *E. coli* auxotrophs does occur (Furihata *et al.*, 1995), indicating that some other gene(s) in these biosynthetic pathways must be absent or inoperative in *M. catarrhalis*.

The study of gene function has been facilitated by cloning and expression of *M. catarrhalis* genes in *E. coli*, followed by sequencing. This has been important for understanding the role of OMPs in virulence or in deciding on the appropriateness of a particular antigen for a vaccine. GenBank sequences are available for several *M. catarrhalis* genes that encode OMPs, and a summary of these genes is included in **Table 1**. Nguyen *et al.* (1999) located genes on the genomic map for five of the OMPs, *usp*A1, *usp*A2, *cop*B, *omp*CD and *omp*E. They are distributed around the chromosome with no linkage between the homologous *usp*A1 and *usp*A2 genes.

Genes that encode putative response regulator proteins have also been cloned from *M. catarrhalis* and partially sequenced (Mibus *et al.*, 1998). Response regulators mediate bacterial responses to environmental stimuli by a signal transduction pathway, and play a central role in the co-ordinate regulation of both virulence, and of metabolic functions, such as the response to phosphate starvation (*pho*B). A *pho*B homologue and other response regulars identified in *M. catarrhalis* should provide a new avenue for investigation of its physiology and pathogenic mechanisms.

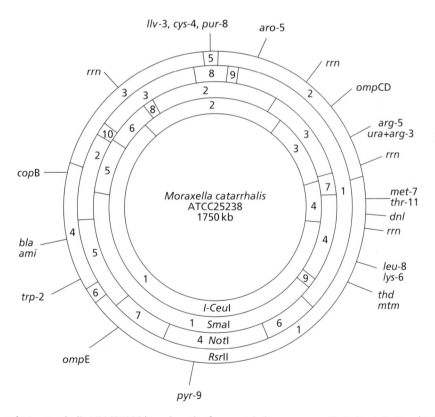

Fig. 1 Physical map of *M. catarrhalis* ATCC25238 based on the four restriction enzymes, *Not*I, *Sma*I, *Rsr*II and I-*Ceu*I. The genes for outer-membrane proteins are described in Table 1. Other symbols include: *ami* – amidase, *bla* – β-lactamase, *dnl* – DNA ligase, *mtm* – DNA methyl transferase, *rrn* – ribosomal RNA operon, *thd* – threonine dehydratase, and the standard amino acid and nucleic acid markers, *arg, aro, cys, ilv, leu, lys, met, thr, trp, pur, pyr* and *ura*. The authors acknowledge the assistance provided by Dr Mark Farinha with this figure, which is based on data from Furihata *et al.* (1995) and Nguyen *et al.* (1999).

Virulence Factors

Iron Uptake

Bacterial pathogens must acquire iron from their hosts and they have evolved a number of mechanisms to do so, including the use of siderophores and iron binding proteins (see Chapter 32). *M. catarrhalis* does not produce siderophores (Campagnari *et al.*, 1994) but utilises transferrin- and lactoferrin-binding proteins (Schryvers and Lee, 1989). A number of *M. catarrhalis* OMPs are iron-repressible and expressed in greater abundance when iron levels in the growth medium are low (Campagnari *et al.*, 1994). These include transferrin-binding proteins A and B (TbpA and TbpB, formerly Tbp1 and 2), lactoferrin-binding proteins A and B (LbpA and LbpB) and CopB (Yu and Schryvers, 1993; Aebi *et al.*, 1996; Bonnah *et al.*, 1998).

The interaction between transferrin and its receptors is conserved among *Neisseria* spp., *H. influenzae* and *M. catarrhalis* (Gray-Owen and Schryvers, 1996). These and members of the Pasteurellaceae produce TbpA proteins that are generally highly conserved within a species, but TbpB proteins that tend to be more variable. TbpA is an integral transmembrane protein that may serve as the channel for the transport of iron across the outer membrane (Gray-Owen and Schryvers, 1996). TbpA proteins from two *M. catarrhalis* strains were 98% identical, while TbpB proteins from six strains were variable, and two potential families of TbpB proteins were identified based on sequence similarities (Myers *et al.*, 1998). TbpB is thought to be a peripheral outer-membrane lipoprotein which, in *M. catarrhalis*, recognises the iron status of transferrin by preferentially binding the iron-saturated form rather than the apo form (Yu and Schryvers, 1993). It consists of two lobes, each with distinct but homologous transferrin-binding regions (Retzer *et al.*, 1999). It has been suggested that TbpB is the same protein as OMP B1, since both proteins bind transferrin, elicit a strong immune response and are approximately 80 kDa OMPs (Campagnari *et al.*, 1996; Mathers *et al.*, 1997; Myers *et al.*, 1998). This was confirmed by the report of 50–71% identity at the amino acid level between OMP B1 and six TbpB proteins (Luke *et al.*, 1999). Isogenic *tbp*A, *tbp*B and double mutants have been constructed and these will be useful tools for future studies of the role of these proteins as virulence factors in *M. catarrhalis* (Luke and Campagnari, 1999).

The genomic organisation of the *M. catarrhalis tbp* genes is unique among bacteria. In general, *tbp*B immediately precedes *tbp*A in a small operon with a promoter upstream of *tbp*B (Gray-Owen and Schryvers, 1996). In *M. catarrhalis*, *tbp*A precedes *tbp*B and a third gene of unknown function is located between them. There are three potential promoter sequences, suggesting independent transcription of the genes (Myers *et al.*, 1998). Thus, expression of the *M. catarrhalis* Tbps may be regulated by a novel mechanism.

A single lactoferrin-binding protein (LbpA) of approximately 103 kDa was originally identified in *M. catarrhalis* (Schryvers and Lee, 1989). More recently, an additional 95-kDa lactoferrin-binding protein, which usually co-migrates with LbpA, was shown to be antigenically distinct from LbpA and designated LbpB (Bonnah *et al.*, 1998). Both genes have been cloned and sequenced (Du *et al.*, 1998). The genes are arranged as *lbp*B followed by *lbp*A, as is found in *N. meningitidis* and *H. influenzae*, but a third gene (*Orf3*) of unknown function downstream of *lbp*A is unique to *M. catarrhalis* (Du *et al.*, 1998). The study of isogenic mutants deficient in LbpA, LbpB or Orf3 protein and RT-PCR revealed that all three genes compose a polycistronic message, and therefore belong to one operon (Bonnah *et al.*, 1999). Insertional inactivation of LbpA demonstrated that LbpA is essential for iron acquisition from lactoferrin, and this is consistent with the proposal that it serves as the transmembrane channel for iron transport (as for TbpA). The LbpB mutant was also impaired in iron acquisition and may facilitate iron uptake in a similar way to TbpB. The *Orf3* mutant was unaffected in iron acquisition, and its function remains a mystery.

A final surprising feature of the iron-acquisition process in *M. catarrhalis* is that insertional inactivation of the gene that encodes the iron-repressible OMP CopB effectively eliminates iron uptake from both transferrin and lactoferrin (Aebi *et al.*, 1996). CopB may play a major role in iron-acquisition that is unique to *M. catarrhalis*, or insertional inactivation of *cop*B may indirectly affect other components of the pathway, such as the permease or energy-transducing complexes (Bonnah *et al.*, 1998, 1999).

Complement Resistance

Resistance to complement-mediated lysis (serum resistance) is an important virulence factor for many Gram-negative bacteria that cause systemic infections, including *Neisseria* spp. (Salyers and Whitt, 1994). Forty-three percent of disease-causing isolates of *M. catarrhalis* were serum-resistant, as compared with 13% of commensal isolates (Jordan *et al.*, 1990), but the numbers were too small for the difference to be of statistical significance. Later reports indicated that complement-resistance is indeed a virulence factor of

M. catarrhalis, since 62% of strains isolated from the sputa of 200 adult patients with bronchopulmonary infections were serum-resistant, significantly more than the 33% of commensal isolates from children (Hol *et al.*, 1993, 1995). In addition, strains associated with respiratory tract infections in the elderly were more often serum resistant than strains from healthy elderly people (Murphy *et al.*, 1997).

Serum resistance in *M. catarrhalis* is likely to be multifactorial, and both CopB and UspA2 have been implicated. The CopB protein was identified as having a role in serum resistance, because an isogenic *cop*B mutant was sensitive to killing by normal human serum, compared with the wild-type parent strain (Helminen *et al.*, 1993b). Verduin *et al.* (1994) showed that complement-resistant strains express a protein capable of binding membrane attack complexes without damaging the bacteria, and suggested that the protein binds to a terminal complement inhibitor such as vitronectin. This protein was not CopB, but UspA2. The evidence is that HMW-OMP (UspA) was found in 10 complement-resistant but not in five complement-sensitive strains (Boel *et al.*, 1998). These authors reported that phage antibodies obtained by competitive panning on complement-resistant *M. catarrhalis* were all directed to the same or closely spaced epitopes on UspA. After their discovery that UspA was comprised of two proteins, UspA1 and UspA2, Aebi *et al.* (1998b) found that an isogenic *usp*A2 mutant was readily killed by normal human serum, while the wild-type parent strain was serum-resistant. UspA2 had previously been found most closely to resemble the YadA OMP involved in serum resistance of *Yersinia* (Aebi *et al.*, 1997). Finally, purified UspA2 binds to vitronectin (McMichael *et al.*, 1998). Further studies on the relative importance of UspA2 and CopB in mediating serum-resistance and their role in the pathogenesis of *M. catarrhalis* disease are clearly required.

Adhesion

Adhesion of bacteria to epithelial cells is an important virulence factor for many pathogens, and allows colonisation of mucosal surfaces such as those of the respiratory tract (Salyers and Whitt, 1994). A number of researchers have studied the adherence properties of *M. catarrhalis* with a variety of model systems *in vitro*, including oropharyngeal cells, bronchial cells, mucin, the HEp-2 laryngeal cell line, the Chang conjunctival cell line and haemagglutination (HA) of human and animal erythrocytes.

Mbaki *et al.* (1987) found that *M. catarrhalis* adheres to human oropharyngeal cells, and that

seasonal variation in the occurrence of *M. catarrhalis* respiratory infections in patients with chronic pulmonary disease (CPD) correlates with the varying efficiency of adhesion to cells taken from these patients. In addition, strains showed increased attachment to oropharyngeal cells from CPD patients as compared with cells from healthy individuals (Rikitomi *et al.*, 1997). An ultrastructural study revealed that *M. catarrhalis* adheres to a presumably polysaccharide granular ruthenium red-positive layer on the microplicae of oropharyngeal cells (Ahmed *et al.*, 1992b). Atomic force microscopy and electron microscopy show that microplicae are positively charged, while the depressions between microplicae are negatively charged, thus explaining why the negatively charged bacterial cells bind to the former domain (Ahmed *et al.*, 2000). The muco-regulating drugs S-carboxymethylcysteine and *N*-acetylcysteine decrease the level of *M. catarrhalis* attachment and also caused loss of the ruthenium-red positive layer on the pharyngeal cell surface (Zheng *et al.*, 1999). A receptor for *M. catarrhalis* on the oropharyngeal epithelium was identified by Ahmed *et al.* (1996) as glycosphingolipid. Varying density of surface receptors may explain the different levels of bacterial adhesion found in oropharyngeal cells from CPD patients, and a further finding that bacterial attachment to bronchial cells was approximately 10 times higher than to oropharyngeal cells (Rikitomi *et al.*, 1997).

Preliminary evidence suggests that *M. catarrhalis* employs both pilus and non-pilus adhesins to bind to cells of the respiratory tract. Piliated and non-piliated strains are able to adhere to oropharyngeal cells (Rikitomi *et al.*, 1991), and there is no correlation between the degree of piliation and adherence (Ahmed *et al.*, 1992a). However, pre-treatment of piliated strains with trypsin, or with antipilus antiserum, significantly decreases adhesion levels (Ahmed, 1992). Piliated strains tend to adhere in higher numbers than a non-piliated strain to bronchial cells (Rikitomi *et al.*, 1997). It will be interesting to determine whether the type IV pili identified in *M. catarrhalis* (Marrs and Weir, 1990), which act as adhesins in a number of genera (Salyers and Whitt, 1994), also play this role in *M. catarrhalis*. Non-pilus adhesins include the 57 kDa OMP CD which was identified by Reddy *et al.* (1997) as binding to purified human middle-ear mucin.

M. catarrhalis isolates agglutinate human, bovine, guinea-pig, dog, rat and rabbit erythrocytes (Soto-Hernandez *et al.*, 1989; Ahmed *et al.*, 1990; Kellens *et al.*, 1995; Fitzgerald *et al.*, 1997). Early studies did not find a correlation between haemagglutination (HA) of human erythrocytes and the source of the isolate (Soto-Hernandez *et al.*, 1989; Jordan *et al.*, 1990).

However, Fitzgerald et al. (1999b) reported that 80% of isolates from the sputum of elderly patients with lower respiratory tract infections were HA positive as compared with 5% of colonising isolates, which suggests that HA may be a virulence factor in this population. HA and adhesion to oropharyngeal cells (Rikitomi et al., 1991) and HEp-2 cells (Fitzgerald et al., 1999b) was not associated, but HA was linked to tracheal cell adhesion (Kellens et al., 1995).

Fitzgerald et al. (1997) identified a 200-kDa surface-expressed protein as a haemagglutinin of M. catarrhalis. HA of both human and rabbit erythrocytes was mediated by a trypsin-sensitive outer fibrillar coat that was absent on non-haemagglutinating isolates (Fitzgerald et al., 1999a). Immuno-electron microscopy with antibodies to the 200-kDa protein showed that it was located on the outer fibrillar layer of the bacteria. The relationship between the 200-kDa protein and previously described OMPs is unknown. The same authors reported that in one isolate of M. catarrhalis trypsin-resistant fimbria-like structures were seen by TEM to mediate attachment to human erythrocytes. A previous study had shown that in strains with a high HA titre, the number of pili was significantly greater than in strains with low HA (Ahmed et al., 1992a). Thus, it is possible that both pilus and non-pilus haemagglutinins exist on M. catarrhalis.

Another potentially useful model for studying M. catarrhalis adhesion in vitro is tissue culture cell lines. M. catarrhalis isolates adhere to HEp-2 cells (Aebi et al., 1998b) and isolates from patients with lower respiratory tract infections have significantly higher adhesion levels than control colonising strains (Fitzgerald et al., 1999b). UspA1 has been identified as a potential HEp-2 cell adhesin, since a uspA1 mutant had 3-fold lower adherence to HEp-2 cells than its isogenic parent (Aebi et al., 1998b). The UspA1 protein was most similar to the hsf gene product of H. influenzae, an adhesin for Chang conjunctival cells (Aebi et al., 1997). When these cells were tested as the target for M. catarrhalis adhesion, a 60-fold reduction in adhesion was observed in the uspA1 mutant (Aebi et al., 1998b). Purified UspA1 binds to HEp-2 cells and anti-UspA1 serum blocks bacterial attachment to the HEp-2 cells. UspA1 also binds to the extracellular matrix protein fibronectin that acts as a receptor for other bacterial pathogens (McMichael et al., 1998). The recently discovered UspA2H protein also functions as an adhesin for Chang conjunctival epithelial cells (Lafontaine et al., 2000).

In contrast to these results, Fitzgerald et al. (1999b) provided evidence that the HEp-2 adhesin is trypsin-resistant and periodate-sensitive, and thus may have a carbohydrate moiety. TEM studies revealed that adhesion was mediated by trypsin-resistant 'tack/spicule-like' structures that resembled the peritrichous fimbriae described by previous workers (Marrs and Weir, 1990; Ahmed et al., 1992a).

The strains studied by the various research groups may indeed possess different HEp-2 adhesins. It should be noted, however, that bacterial growth and assay conditions varied markedly between groups and a standardised system would help to clarify this complex area.

Endotoxin

The lipid A or endotoxin component of LOS may contribute to the pathogenesis of M. catarrhalis disease. DeMaria (1988) summarised the evidence that endotoxin from Gram-negative bacteria, including M. catarrhalis, induces severe inflammatory changes in the middle ear, contributing to the pathogenesis of otitis media. This was confirmed in a chinchilla model in which formalin-killed M. catarrhalis produced effusions in the middle ear (Doyle, 1989). Similarly, heat-killed M. catarrhalis induce middle ear inflammation and mucoperiosteal histopathology in guinea-pigs (Sato, 1997). Fomsgaard et al. (1991) reported that M. catarrhalis LOS is biologically active, causing death in mice and acting as an endotoxin in the Limulus amoebocyte lysate assay.

Histamine Production

Histamine has significant biological effects on the lung, causing smooth muscle contraction, an increase in pulmonary permeability and stimulation of mucus secretion. A number of Gram-negative species found in the respiratory tract, and associated with acute exacerbations of respiratory infections, are capable of histamine production in vitro, and this may contribute to their pathogenic effects (Devalia et al., 1989). Six of eight strains of M. catarrhalis from patients with chronic bronchitis synthesised clinically significant amounts of histamine in vitro (Devalia et al., 1989). M. catarrhalis isolates liberated $^{14}CO_2$ plus histamine from labelled histidine in the presence of the co-factor pyridoxal phosphate, which suggests the presence of a histidine decarboxylase similar to that found in other Gram-negative bacteria (Cundell et al., 1991). Histamine may also have a role in disruption of airway epithelium and it may contribute to airflow obstruction in asthma. Colonisation of the nasopharynx of asthmatic children by M. catarrhalis was significantly higher than in normal children (Seddon et al., 1992), but the role in asthma of histamine or other potential mediators of bronchial hyper-reactivity such as endotoxin is not known.

Immune Response and Vaccine Development

Our understanding of the human immune response to *M. catarrhalis* is complicated by the diversity of strains, antigens and immunoassays used by various research groups. Studies demonstrating a systemic antibody response to *M. catarrhalis* have been summarised by Murphy (1996) and Christensen (1999). Antibodies to *M. catarrhalis* antigens are regularly found in healthy adult sera. The IgG response is low or absent in children under 12 months and it gradually increases during childhood, and IgG3 antibodies predominate (Goldblatt *et al.*, 1990; Rahman *et al.*, 1997). A number of studies have reported increased titres of IgG, IgM and IgA antibodies against different *M. catarrhalis* antigens in convalescent serum from patients with respiratory tract infections and in the middle-ear fluid of children with otitis media (Christensen, 1999).

Little is known about the cell-mediated immune response to *M. catarrhalis* infection. A rapid, predominantly neutrophilic inflammatory response is observed in mouse challenge studies (Verghese *et al.*, 1990a; Kyd *et al.*, 1998). The use of immune-deficient mice suggested that natural killer cells and polymorphonuclear cell functions were important in resolving *M. catarrhalis* challenge (Harkness *et al.*, 1993). Peptidoglycan from *M. catarrhalis* is a potent inducer of the secretion of tumour necrosis factor and nitrite in macrophages, and this may contribute to the very effective triggering of the functional activities of macrophages *in vivo* by *M. catarrhalis* (Keller *et al.*, 1992). *M. catarrhalis*, which is normally susceptible to nitric oxide produced by macrophages, may gain increased resistance to nitric oxide by altered albumin binding (Maluszynska *et al.*, 1998).

Recent work has concentrated on the identification of antigens that may form the basis of a protective vaccine. Vaccine candidates include OMP B1 (TbpB), a major antigen in serum of adults with bronchiectasis (Sethi *et al.*, 1995) and children with otitis media (Campagnari *et al.*, 1996; Mathers *et al.*, 1999). Recombinant TbpB expressed in *E. coli*, and purified, elicits bactericidal antibodies in animals (Myers *et al.*, 1998). The immunogenicity of TbpB was also confirmed in humans (Yu *et al.*, 1999), while Luke *et al.* (1999) identified a surface-exposed epitope of OMP B1 that elicited protective antibodies. Since the epitope was detected on only 31% of clinical isolates, further work is required to define other potentially protective epitopes of OMP B1. Other iron-binding proteins assessed for suitability as vaccine candidates include a 74-kDa transferrin-binding protein (Chen *et al.*, 1999b) and LbpB (Du *et al.*, 1998). CopB is another major antigen in patients with *M. catarrhalis* infections (Helminen *et al.*, 1995; Mathers *et al.*, 1999), and is largely conserved among *M. catarrhalis* strains (Sethi *et al.*, 1997). Similarly, OMP E elicits an IgA response in the sputum of patients with chronic bronchitis, and has at least one surface-exposed, highly conserved epitope (Bhushan *et al.*, 1997).

UspA1 and UspA2 are also considered to be promising vaccine candidates (Helminen *et al.*, 1994; McMichael *et al.*, 1998). Levels of IgG antibodies against these OMPs and serum bactericidal activity correlated with age-dependent resistance to *M. catarrhalis* infections, suggesting that the humoral response to the UspA proteins is critical for protection against *M. catarrhalis* infection (Chen *et al.*, 1999a). Another protein under investigation is OMP CD, which induces bactericidal antibodies in animals (Yang *et al.*, 1997) and enhances pulmonary clearance after intratracheal challenge of mice with *M. catarrhalis* (Murphy *et al.*, 1998). Murphy *et al.* (1999) identified a specific region of OMP CD as important in mediating the human immune response to this protein.

LOS has been considered as a potential vaccine candidate, because serum antibodies to it develop during lower respiratory tract infection and are not serotype-specific (Rahman *et al.*, 1995). Gu *et al.* (1998) constructed conjugates of detoxified LOS coupled to tetanus toxoid or to high-molecular-weight *H. influenzae* proteins. Both conjugates are immunogenic in mice and rabbits, and elicit anti-LOS antibody with complement-dependent bactericidal activity against homologous and heterologous *M. catarrhalis* strains.

Conclusion

It is now generally accepted that *M. catarrhalis* is a significant cause of respiratory disease in adults and children. The rapid appearance of β-lactamase-producing strains still requires further investigation, but resistance to antibiotics other than penicillins is not a problem. Recently, there has been more emphasis on the analysis of putative virulence factors and OMPs driven by the perceived need for a vaccine. Unfortunately, the lack of a suitable animal model has created some difficulties, because the assessment of virulence *in vivo* is crucial for these developments.

References

Aebi C, Stone B, Beucher M *et al.* (1996) Expression of the CopB outer membrane protein by *Moraxella catarrhalis* is regulated by iron and affects iron acquisition from transferrin and lactoferrin. *Infect. Immun.* 64: 2024–2030.

Aebi C, MacIver I, Latimer JL *et al.* (1997) A protective epitope of *Moraxella catarrhalis* is encoded by two different genes. *Infect. Immun.* 65: 4367–4377.

Aebi C, Cope LD, Latimer JL *et al.* (1998a) Mapping of a protective epitope of the CopB outer membrane protein of *Moraxella catarrhalis. Infect. Immun.* 66: 540–548.

Aebi C, Lafontaine ER, Cope LD *et al.* (1998b) Phenotypic effect of isogenic *uspA1* and *uspA2* mutations on *Moraxella catarrhalis* 035E. *Infect. Immun.* 66: 3113–3119.

Ahmed K (1992) Fimbriae of *Branhamella catarrhalis* as possible mediators of adherence to pharyngeal epithelial cells. *APMIS* 100: 1066–1072.

Ahmed K, Rikitomi N, Nagatake T, Matsumoto K (1990) Electron microscopic observation of *Branhamella catarrhalis. Microbiol. Immunol.* 34: 967–975.

Ahmed K, Rikitomi N, Ichinose A, Matsumoto K (1991) Possible presence of a capsule in *Branhamella catarrhalis. Microbiol. Immunology* 35: 361–366.

Ahmed K, Rikitomi N, Matsumoto K (1992a) Fimbriation, hemagglutination and adherence properties of fresh clinical isolates of *Branhamella catarrhalis. Microbiol. Immunol.* 36: 1009–1017.

Ahmed K, Rikitomi N, Nagatake T, Matsumoto K (1992b) Ultrastructural study on the adherence of *Branhamella catarrhalis* to oropharyngeal epithelial cells. *Microbiol. Immunol.* 36: 563–573.

Ahmed K, Masaki H, Dai TC *et al.* (1994) Expression of fimbriae and host response in *Branhamella catarrhalis* respiratory infections. *Microbiol. Immunol.* 38: 767–771.

Ahmed K, Matsumoto K, Rikitomi N, Nagatake T (1996) Attachment of *Moraxella catarrhalis* to pharyngeal epithelial cells is mediated by a glycosphingolipid receptor. *FEMS Microbiol. Lett.* 135: 305–309.

Ahmed K, Nakagawa T, Nakano Y *et al.* (2000) Attachment of *Moraxella catarrhalis* occurs to the positively charged domains of pharyngeal epithelial cells. *Microb. Pathogen.* 28: 203–209.

Aitken JM, Thornley PE (1983) Isolation of *Branhamella catarrhalis* from sputum and tracheal aspirate. *J. Clin. Microbiol.* 18: 1262–1263.

Aniansson G, Alm B, Andersson B *et al.* (1992) Nasopharyngeal colonization during the first year of life. *J. Infect. Dis.* 165 (Suppl. 1): S38–42.

Arola M, Ruuskanen O, Ziegler T *et al.* (1990) Clinical role of respiratory virus infection in acute otitis media. *Pediatrics* 86: 848–855.

Aspin MM, Hoverman A, McCarty J *et al.* (1994) Comparative study of the safety and efficacy of clarithromycin and amoxicillin-clavulanate in the treatment of acute otitis media in children. *J. Pediatr.* 125: 135–141.

Barreiro B, Esteban L, Prats E, Verdaguer E, Dorca J, Manresa F (1992) *Branhamella catarrhalis* respiratory infections. *Eur. Respir. J.* 5: 675–679.

Bartos LC, Murphy TF (1998) Comparison of the outer membrane proteins of 50 strains of *Branhamella catarrhalis. J. Infect. Dis.* 158: 761–765.

Beaulieu D, Piche L, Roy PH (1989) Cloning and characterization of a *Branhamella catarrhalis* β-lactamase gene. Twenty-ninth Interscience Conference on Antimicrobial Agents and Chemotherapy, abstract no. 1120.

Berg RA, Bartley DL (1987) Pneumonia associated with *Branhamella catarrhalis* in infants. *Pediatr. Infect. Dis. J.* 6: 569–573.

Berk SL (1990) The reemergence of *Branhamella catarrhalis. Arch. Intern. Med.* 150: 2254–2257.

Berner R, Schumacher RF, Brandis M, Forster J (1996) Colonization and infection with *Moraxella catarrhalis* in childhood. *Eur. J. Clin. Microbiol. Infect. Dis.* 15: 506–509.

Bernstein T, Brilli R, Jacobs B (1998) Is bacterial tracheitis changing? A 14-month experience in a pediatric intensive care unit. *Clin. Infect. Dis.* 27: 458–462.

Bhushan R, Craigie R, Murphy TF (1994) Molecular cloning and characterization of outer membrane protein E of *Moraxella (Branhamella) catarrhalis. J. Bacteriol.* 176: 6636–6643.

Bhushan R, Kirkham C, Sethi S, Murphy TF (1997) Antigenic characterization and analysis of the human immune response to outer membrane protein E of *Branhamella catarrhalis. Infect. Immun.* 65: 2668–2675.

Blackwell C, Young H, Bain SSR (1978) Isolation of *Neisseria meningitidis* and *Neisseria catarrhalis* from the genitourinary tract and anal canal. *Br. J. Ven. Dis.* 54: 41–44.

Bluestone CD (1986) Otitis media and sinusitis in children Role of *Branhamella catarrhalis. Drugs* 31 (Suppl. 3): 132–141.

Bodkin S, Warde D (1993) *Moraxella catarrhalis*: an unusual pathogen in bacterial tracheitis. *Irish Med. J.* 86: 208–209.

Boel E, Bootsma H, de Kruif J *et al.* (1998) Phage antibodies obtained by competitive selection on complement-resistant *Moraxella (Branhamella) catarrhalis* recognize the high-molecular-weight outer membrane protein. *Infect. Immun.* 66: 83–88.

Bonnah RA, Yu R-H, Wong H, Schryvers AB (1998) Biochemical and immunological properties of lactoferrin binding proteins from *Moraxella (Branhamella) catarrhalis. Microb. Pathogen.* 24: 89–100.

Bonnah RA, Wong H, Loosmore SM, Schryvers AB (1999) Characterization of *Moraxella (Branhamella) catarrhalis lbpB, lbpA* and lactoferrin receptor *orf3* isogenic mutants. *Infect. Immun.* 67: 1517–1520.

Bootsma HJ, van Dijk H, Verhoef J, Fleer A, Mooi FR (1996) Molecular characterization of the BRO β-lactamase of *Moraxella (Branhamella) catarrhalis. Antimicrob. Agents Chemother.* 40: 966–972.

Bootsma HJ, Aerts PC, Posthuma G *et al.* (1999) *Moraxella (Branhamella) catarrhalis* BRO β-lactamase: a lipoprotein of Gram-positive origin? *J. Bacteriol.* 181: 5090–5093.

Bövre K (1970) Pulse-RNA–DNA hybridization between rod-shaped and coccal species of the *Moraxella–Neisseria* groups. *Acta Pathol. Microbiol. Scand. – B: Microbiol. Immunol.* 78: 565–574.

Bövre K (1979) Proposal to divide the genus *Moraxella*, Lwoff 1939 emend. Henriksen and Bövre 1968 into two subgenera: subgenus *Moraxella* (Lwoff 1939) Bövre 1979 and subgenus *Branhamella* (Catlin 1970) Bövre 1979. *Int. J. System. Bacteriol.* 29: 403–406.

Bövre K (1984) Genus II. *Moraxella* Lwoff 1939, emend. Henriksen and Bövre 1968. In: Krieg NR, Holt JG (eds) *Bergey's Manual of Systematic Bacteriology*, Vol. 1. Baltimore: Williams & Wilkins, pp. 296–303.

Bövre K, Fiandt M, Szybalski W (1969) DNA base composition of *Neisseria: Moraxella* and *Acinetobacter* as determined by measurement of buoyant density in CsCl gradients. *Can. J. Microbiol.* 15: 335–338.

Boyle FM, Georghiou PR, Tilse MH, McCormack JG (1991) *Branhamella (Moraxella) catarrhalis*: pathogenic significance in respiratory infections. *Med. J. Austr.* 154: 592–596.

Brorson J, Malmvall B (1981) *Branhamella catarrhalis* and other bacteria in the nasopharynx of children with long standing cough. *Scand. J. Infect. Dis.* 13: 111–113.

Brorson JE, Axelsson A, Holm SE (1976) Studies on *Branhamella catarrhalis* (*Neisseria catarrhalis*) with special reference to maxillary sinusitis. *Scand. J. Infect. Dis.* 8: 151–155.

Buu Hoi-Dang van A, Brive-Le Bouguennce C, Barthelemy M, Labia R (1978) Novel β-lactamase from *Branhamella catarrhalis*. *Ann. Microbiol. (Paris)* 129B: 397–406.

Campagnari AA, Shanks KL, Dyer DW (1994) Growth of *Moraxella catarrhalis* with human transferrin and lactoferrin: expression of iron-repressible proteins without siderophore production. *Infect. Immun.* 62: 4909–4914.

Campagnari AA, Ducey TF, Rebmann CA (1996) Outer membrane protein B1, an iron-repressible protein conserved in the outer membrane of *Moraxella (Branhamella) catarrhalis*, binds human transferrin. *Infect. Immun.* 64: 3920–3924.

Capewell S, McLeod DT, Croughan MJ, Ahmad F, Calder MA, Seaton A (1988) Pneumonia due to *Branhamella catarrhalis*. *Thorax* 43: 929–930.

Carr B, Walsh JB, Coakley D, Scott T, Mulvihill E, Keane C (1989) Effect of age on adherence of *Branhamella catarrhalis* to buccal epithelial cells. *Gerontology* 35: 127–129.

Catlin BW (1970) Transfer of the organism named *Neisseria catarrhalis* to *Branhamella* Gen. Nov. *Int. J. System. Bacteriol.* 20: 155–159.

Catlin BW (1990) *Branhamella catarrhalis*: an organism gaining respect as a pathogen. *Clin. Microbiol. Rev.* 3: 293–320.

Catlin BW (1991) *Branhamaceae* fam. Nov., proposed family to accommodate the genera *Branhamella* and *Moraxella*. *Int. J. System. Bacteriol.* 41: 320–323.

Catlin BW, Cunningham LS (1961) Transforming activities and base contents of deoxyribonucleate preparations from various Neisseria. *J. Gen. Microbiol.* 26: 303–312.

Catlin BW, Cunningham LS (1964) Genetic transformation of *Neisseria catarrhalis* by deoxyribonuclease preparations having different average base compositions. *J. Gen. Microbiol.* 37: 341–352.

Chaibi EB, Mugnier P, Kitzis MD, Goldstein RW, Acar JF (1995) *Branhamella catarrhalis* β-lactamases and their phenotypic implication. *Res. Microbiol.* 146: 761–771.

Chapman AJ, Musher DM, Jonsson S, Calridge JE, Wallace RJ (1985) Development of bactericidal antibody during *Branhamella catarrhalis* infection. *J. Infect. Dis.* 151: 878–882.

Chen D, Barniak V, VanDerMeid KR, McMichael JC (1999a) The levels and bactericidal capacity of antibodies directed against the UspA1 and UspA2 outer membrane proteins of *Moraxella (Branhamella) catarrhalis* in adults and children. *Infect. Immun.* 67: 1310–1316.

Chen D, McMichael JC, VanDerMeid KR *et al.* (1999b) Evaluation of a 74-kDa transferrin-binding protein from *Moraxella (Branhamella) catarrhalis* as a vaccine candidate. *Vaccine* 18: 109–118.

Chin NK, Kumarasinghe G, Lim TK (1993) *Moraxella catarrhalis* respiratory infection in adults. *Sing. Med. J.* 34: 409–411.

Choo PW, Gantz NM (1989) *Branhamella catarrhalis* pneumonia with bacteremia. *Southern Med. J.* 82: 1317–1318.

Christensen JJ (1999) *Moraxella (Branhamella) catarrhalis*: clinical, microbiological and immunological features in lower respiratory tract infections. *APMIS* 107 (Suppl. 88): 5–36.

Christensen JJ, Gadeverg O, Bruun B (1986) *Branhamella catarrhalis*: significance in pulmonary infections and bacteriological features. *APMIS* 94: 89–95.

Christensen JJ, Ursing J, Bruun B (1994) Genotypic and phenotypic relatedness of 80 strains of *Branhamella catarrhalis* of worldwide origin. *FEMS Microbiol. Letters* 119: 155–159.

Coffey JD, Martin AD, Booth HN (1967) *Neisseria catarrhalis* in exudate otitis media. *Arch. Otolaryngol.* 86: 403–406.

Collazos J, de Miguel J, Ayarze R (1992) *Moraxella catarrhalis* bacteremic pneumonia in adults: two cases and review of the literature. *Eur. J. Clin. Microbiol. Infect. Dis.* 11: 237–240.

Cope LD, Lafontaine ER, Slaughter CA *et al.* (1999) Characterization of the *Moraxella catarrhalis* uspA1 and uspA2 genes and their encoded products. *J. Bacteriol.* 181: 4026–4034.

Cundell DR, Devalia JL, Wilks M, Tabaqchali S, Davies RJ (1991) Histidine decarboxylases from bacteria that colonise the human respiratory tract. *J. Med. Microbiol.* 35: 363–366.

Darelid J, Lofgren S, Malmvall BE (1993) Erythromycin treatment is beneficial for longstanding *Moraxella catarrhalis* associated cough in children. *Scand. J. Infect. Dis.* 25: 323–329.

Davies BI, Maesen FPV (1988) The epidemiology of respiratory tract pathogens in Southern Netherlands. *Eur. Respir. J.* 1: 415–420.

Davies BI, Maesen FPV (1990) Treatment of *Branhamella catarrhalis* infections. *J. Antimicrob. Chemother.* 25: 1–7.

DeMaria TF (1988) Endotoxin and otitis media. *Ann. Otol. Rhinol. Laryngol.* 97: 31–33.

Devalia JL, Grady D, Harmanyeri Y, Tabaqchali S, Davies RJ (1989) Histamine synthesis by respiratory tract micro-organisms: possible role in pathogenicity. *J. Clin. Pathol.* 42: 516–522.

Diamond LA, Lorber B (1984) *Branhamella catarrhalis* pneumonia and immunoglobulin abnormalities: a new association? *Am. Rev. Respir. Dis.* 129: 876–878.

DiGiovanni C, Riley TV, Hoyne GF, Yeo R, Cooksey P (1987) Respiratory tract infections due to *Branhamella catarrhalis*: epidemiological data from Western Australia. *Epidemiol. Infect.* 99: 445–453.

Dickinson DP, Loos BG, Dryja DM, Bernstein JM (1988) Restriction fragment mapping of *Branhamella catarrhalis*: a new tool for studying the epidemiology of this middle ear pathogen. *J. Infect. Dis.* 158: 205–208.

Doern GV, Morse SA (1980) *Branhamella (Neisseria) catarrhalis*: criteria for laboratory identification. *J. Clin. Microbiol.* 11: 193–195.

Doern GV, Tubert TA (1988) *In vitro* activities of 39 antimicrobial agents for *Branhamella catarrhalis* and comparison of result with quantitative susceptibility test methods. *Antimicrob. Agents Chemother.* 32: 259–261.

Doern GV, Jones RN, Pfaller MA, Kugler K, and the SENTRY participants group (1999) *Haemophilus influenzae* and *Moraxella catarrhalis* from patients with community-acquired respiratory tract infections: antimicrobial susceptibility patterns from the SENTRY antimicrobial surveillance program (United States and Canada, 1997). *Antimicrob. Agents Chemother.* 43: 385–389.

Doyle WJ (1989) Animal models of otitis media: other pathogens. *Pediatr. Infect. Dis. J.* 81: S45–47.

Du R-P, Wang Q, Yang Y-P *et al.* (1998) Cloning and expression of the *Moraxella catarrhalis* lactoferrin receptor genes. *Infect. Immun.* 66: 3656–3665.

Dyson C, Poonyth HD, Watkinson M, Rose SJ (1990) Life threatening *Branhamella catarrhalis* pneumonia in young infants. *J. Infect.* 21: 305–307.

Edelbrink P, Jansson PE, Rahman MM *et al.* (1994) Structural studies of the O-polysaccharide from the lipopolysaccharide of *Moraxella (Branhamella) catarrhalis* serotype A (strain ATCC 25238). *Carbohydr. Res.* 257: 269–284.

Edelbrink P, Jansson PE, Rahman MM, Widmalm G, Holme T, Rahman M (1995) Structural studies of the O-antigen oligosaccharides from two strains of *Moraxella catarrhalis* serotype C. *Carbohydr. Res.* 266: 237–261.

Edelbrink P, Jansson PE, Widmalm G, Holme T, Rahman M (1996) The structures of oligosaccharides isolated from the lipopolysaccharide of *Moraxella catarrhalis* serotype B: strain CCUG 3292. *Carbohydr. Res.* 295: 127–146.

Ejlertsen T (1991) Pharyngeal carriage of *Moraxella (Branhamella) catarrhalis* in healthy adults. *Eur. J. Clin. Microbiol. Infect. Dis.* 10: 89.

Ejlertsen T, Thisted E, Ebbesen F, Olesen B, Renneberg J (1994a) *Branhamella catarrhalis* children and adults: a study of prevalence, time of colonisation and association with upper and lower respiratory tract infections. *J. Infect.* 29: 23–31.

Ejlertsen T, Thisted E, Ostergaard PA, Renneberg J (1994b) Maternal antibodies and acquired serological response to *Moraxella catarrhalis* in children determined by an enzyme-linked immunosorbent assay. *Clin. Diagn. Lab. Immunol.* 1: 464–468.

Eliasson I, Kamme C, Vang M, Waley SG (1992) Characterization of cell-bound papain-soluble β-lactamases in BRO-1 and BRO-2 producing strains of *Moraxella (Branhamella) catarrhalis* and *Moraxella nonliquefaciens*. *Eur. J. Clin. Microbiol. Infect. Dis.* 11: 313–321.

Enright MC, McKenzie H (1997) *Moraxella (Branhamella) catarrhalis*: clinical and molecular aspects of a rediscovered pathogen. *J. Med. Microbiol.* 46: 360–371.

Enright MC, Carter PE, MacLean IA, McKenzie H (1994) Phylogenetic relationships between some members of the genera *Neisseria, Acinetobacter, Moraxella* and *Kingella* based on partial 16S ribosomal DNA sequence analysis. *Int. J. System. Bacteriol.* 44: 387–391.

Faden H, Brodsky L, Waz MJ, Stanievich J, Bernstein JM, Ogra PL (1991) Nasopharyngeal flora in the first three years of life in normal and otitis prone children. *Ann. Otol. Rhinol. Laryngol.* 100: 612–615.

Faden H, Bernstein J, Stanievich J *et al.* (1992) Effect of prior antibiotic treatment on middle ear disease in children. *Ann. Otol. Rhinol. Laryngol.* 10: 87–91.

Faden H, Harabuchi Y, Hong JJ, Tonawanda/Williamsville Paediatrics (1994) Epidemiology of *Moraxella catarrhalis* in children during the first 2 years of life: relationship to otitis media. *J. Infect. Dis.* 169: 1312–1317.

Faden H, Duffy L, Wasielewdki R, Wolf J, Drystofik D, Tung Y, Tonawanda/Williamsville Pediatrics (1997) Relationship between nasopharyngeal colonization and the development of otitis media in children. *J. Infect. Dis.* 175: 1440–1445.

Farmer T, Reading C (1982) β-Lactamases of *Branhamella catarrhalis* and their inhibition by clavulanic acid. *Antimicrob. Agents Chemother.* 21: 506–508.

Farmer T, Reading C (1986) Inhibition of the β-lactamases of *Branhamella catarrhalis* by clavulanic acid and other inhibitors. *Drugs* 31: 70–78.

Feign RD, San Joaqin V, Middelkamp JN (1969). Purpura fulminans associated with *Neisseria catarrhalis* septicemia and meningitis. *Pediatrics* 44: 120–123.

Felmingham D, Gruneberg RN, for the Alexander Project Group (1996) A multicentre collaborative study of the antimicrobial susceptibility of community-acquired, lower respiratory tract pathogens 1992–93: the Alexander Project. *J. Antimicrob. Chemother.* 38 (Suppl. A): 1–57.

Fitzgerald M, Mulcahy R, Murphy S, Keane C, Coakley D, Scott T (1997) A 200 kDa protein is associated with haemagglutinating isolates of *Moraxella (Branhamella) catarrhalis. FEMS Immunol. Med. Microbiol.* 18: 209–216.

Fitzgerald M, Mulcahy R, Murphy S, Keane C, Coakley D, Scott T (1999a) Transmission electron microscopy studies of *Moraxella (Branhamella) catarrhalis. FEMS Immunol. Med. Microbiol.* 23: 57–66.

Fitzgerald M, Murphy S, Mulcahy R, Keane C, Coakley D, Scott T (1999b) Tissue culture adherence and haemagglutination characteristics of *Moraxella (Branhamella) catarrhalis. FEMS Immunol. Med. Microbiol.* 24: 105–114.

Fomsgaard JS, Fomsgaard A, Hoiby N, Bruun B, Galanos C (1991) Comparative immunochemistry of lipopolysaccharides from *Branhamella catarrhalis* strains. *Infect. Immun.* 59: 3346–3349.

Fung C-P, Yeo S-F, Livermore DM (1994a) Susceptibility of *Moraxella catarrhalis* isolates to β-lactam antibiotics in relation to β-lactamase pattern. *J. Antimicrob. Chemother.* 33: 215–222.

Fung C-P, Yeo S-F, Livermore DM (1994b) Extraction of β-lactamases from *Moraxella catarrhalis. J. Antimicrob. Chemother.* 34: 183–184.

Furihata K, Sato K, Matsumoto H (1995) Construction of a combined *Not*1/*Sma*1 physical and genetic map of *Moraxella (Branhamella) catarrhalis* strain ATCC25238. *Microbiol. Immunol.* 39: 745–751.

Goldblatt D, Turner MW, Levinsky RJ (1990) *Branhamella catarrhalis*: antigenic determinants and the development of the IgG subclass response in childhood. *J. Infect. Dis.* 162: 1128–1135.

Goldenhersh MJ, Rachelefsky GS, Dudley J *et al.* (1990) The microbiology of chronic sinus disease in children with respiratory allergy. *J. Allergy Clin. Immunol.* 85: 1030–1039.

Gottfarb P, Brauner A (1994) Children with persistent cough: outcome with treatment and role of *Moraxella catarrhalis*? *Scand. J. Infect. Dis.* 26: 545–551.

Gray-Owen SD, Schryvers AB (1996) Bacterial transferrin and lactoferrin receptors. *Trends Microbiol.* 4: 185–191.

Griffiss JM, Schneider H, Mandrell RE *et al.* (1988) Lipooligosaccharides: the principal glycolipids of the neisserial outer membrane. *Rev. Infect. Dis.* 10: S287–295.

Gu X-X, Chen J, Barenkamp SJ *et al.* (1998) Synthesis and characterization of lipooligosaccharide-based conjugates as vaccine candidates for *Moraxella (Branhamella) catarrhalis. Infect. Immun.* 66: 1891–1897.

Haddad J, Faou AL, Simconi U, Messer J (1986) Hospital-acquired bronchopulmonary infection in premature infants due to *Branhamella catarrhalis. J. Hosp. Infect.* 7: 301–302.

Hager H, Verghese A, Alvarez S, Berk SL (1987) *Branhamella catarrhalis* respiratory infections. *Rev. Infect. Dis.* 9: 1140–1149.

Harkness RE, Guimond M-J, McBey B-A, Klein MH, Percy DH, Croy BA (1993) *Branhamella catarrhalis* pathogenesis in SCID and SCID/beige mice. *APMIS* 101: 805–810.

Hellio R, Guibourdenche M, Collatz E, Riou JY (1988) The envelope structure of *Branhamella catarrhalis* as studied by transmission electron microscopy. *Ann. Inst. Pasteur Microbiol.* 139: 515–525.

Helminen ME, MacIver I, Latimer JL, Cope LD, McCracken GH, Hansen EJ (1993a) A major outer membrane protein of *Moraxella catarrhalis* is a target for antibodies that enhance pulmonary clearance of the pathogen in an animal model. *Infect. Immun.* 61: 2003–2010.

Helminen ME, MacIver I, Paris M *et al.* (1993b) A mutation affecting expression of a major outer membrane protein of *Moraxella catarrhalis* alters serum resistance and survival *in vivo. J. Infect. Dis.* 168: 1194–1201.

Helminen ME, Maciver I, Latimer JL *et al.* (1994) A large, antigenically conserved protein on the surface of *Moraxella catarrhalis* is a target for protective antibodies. *J. Infect. Dis.* 170: 867–872.

Helminen ME, Beach R, Maciver I, Jarosik G, Hansen EJ, Leinonen M (1995) Human immune response against outer membrane proteins of *Moraxella (Branhamella) catarrhalis* determined by immunoblotting and enzyme immunoassay. *Clin. Diagn. Lab. Immunol.* 2: 35–39.

Henriksen SD (1976) *Moraxella, Neisseria, Branhamella* and *Acinetobacter. Annu. Rev. Microbiol.* 30: 63–83.

Henriksen SD, Bövre K (1968) The taxonomy of the genera *Moraxella* and *Neisseria. J. Gen. Microbiol.* 51: 387–392.

Hol C, Verduin CM, Van Dijk EEA, Verhoef J, van Dijk H (1993) Complement resistance in *Branhamella (Moraxella) catarrhalis. Lancet* 341: 1281.

Hol C, van Dijke EEM, Verduin CM, Verhoef J, van Dijke H (1994) Experimental evidence for *Moraxella*-induced penicillin neutralization in pneumococcal pneumonia. *J. Infect. Dis.* 170: 1613–1616.

Hol C, Verduin CM, Van Dijke EEA, Verhoef J, Fleer A, van Dijk H (1995) Complement resistance is a virulence factor of *Branhamella (Moraxella) catarrhalis. FEMS Immunol. Med. Microbiol.* 11: 207–212.

Hol C, Schalen C, Verduin CM *et al.* (1996) *Moraxella catarrhalis* in acute laryngitis: infection or colonization? *J. Infect. Dis.* 174: 636–638.

Holland DF (1920) Generic index of the commoner forms of bacteria. *J. Bacteriol.* 5: 215–299.

Hsiao CB, Sethi S, Murphy TF (1995) Outer membrane protein CD of *Branhamella catarrhalis*: sequence conservation in strains recovered from human respiratory tract. *Microb. Pathogen.* 19: 215–225.

Ikeda F, Yokota Y, Mine Y, Yamada T (1993) Characterization of BRO enzymes and β-lactamase transfer of *Moraxella (Branhamella) catarrhalis* isolated in Japan. *Chemotherapy* 39: 88–95.

Ingvarsson L, Lundgren K, Ursing J (1982) The bacterial flora in the nasopharynx in healthy children. *Acta Otolaryngol.* 86 (Suppl.): 94–96.

Ishida LK, Ikeda K, Tanno N, Takasaka T, Nishioka K, Tanno Y (1995) Erythromycin inhibits adhesion of *Pseudomonas aeruginosa* and *Branhamella catarrhalis* to human nasal epithelial cells. *Am. J. Rhinol.* 9: 53–55.

Jantzen E, Bryn K, Bøvre K (1976) Cellular monosaccharide patterns of Neisseriaceae. *Acta Pathol. Microbiol. Scand. – B: Microbiology* 84: 177–188.

Jones RN, Sommers HM (1986) Identification and antimicrobial susceptibility testing of *Branhamella catarrhalis* in United States laboratories, 1983–85. *Drugs* 31 (Suppl. 3): 34–37.

Jordan KL, Berk SH, Berk SL (1990) A comparison of serum bactericidal activity and phenotypic characteristics of bacteremic, pneumonia-causing strains: and colonizing strains of *Branhamella catarrhalis*. *Am. J. Medicine* 88(5A): 28S–32S.

Jorgensen JH, Doern GV, Maher LA, Howell AW, Redding JS (1990) Antibiotic resistance among respiratory isolates of *Haemophilus* influenzae, *Moraxella catarrhalis*, and *Streptococcus pneumoniae* in the United States. *Antimicrob. Agents Chemother.* 34: 2075–2080.

Juni E, Heym GA, Avery M (1986) Defined medium for *Moraxella (Branhamella) catarrhalis*. *Appl. Environ. Microbiol.* 52: 546–551.

Kamme C, Eliasson I, Kahl-Knutson B, Vang M (1986) Plasmid-mediated β-lactamase in *Branhamella catarrhalis*. *Drugs* 31 (Suppl. 3): 55–63.

Kellens J, Persoons M, Vaneechoutte M, van Tiel F, Stobberingh E (1995) Evidence of lectin-mediated adherence of *Moraxella catarrhalis*. *Infection* 23: 37–41.

Keller R, Gustafson JE, Keist R (1992) The macrophage response to bacteria: modulation of macrophage functional activity by peptidoglycan from *Moraxella (Branhamella) catarrhalis*. *Clin. Exp. Immunol.* 89: 384–389.

Kelly J, Masoud H, Perry MB, Richards JC, Thibault P (1996) Separation and characterization of O-deacylated lipooligosaccharides and glycans derived from *Moraxella catarrhalis* using capillary electrophoresis–electrospray mass spectrometry and tandem mass spectrometry. *Analyt. Biochem.* 233: 15–30.

Kingsbury DT (1967) Deoxyribonucleic acid homologies among species of the genus *Neisseria*. *J. Bacteriol.* 94: 870–874.

Klingman KL, Murphy TF (1994) Purification and characterization of a high-molecular-weight outer membrane protein of *Moraxella (Branhamella) catarrhalis*. *Infect. Immun.* 62: 1150–1155.

Klingman KL, Pye A, Murphy TF, Hill SL (1995) Dynamics of respiratory tract colonization by *Branhamella catarrhalis* in bronchiectasis. *Am. J. Respir. Crit. Care Med.* 152: 1072–1078.

Korppi M, Katila ML, Jaaskelainen J, Leinonen M (1992) Role of *Moraxella (Branhamella) catarrhalis* as a respiratory pathogen in children. *Acta Paediatr.* 81: 993–996.

Kovatch AL, Wald ER, Michaels RH (1983) Beta-lactamase producing *Branhamella catarrhalis* causing otitis media in children. *J. Pediatr.* 102: 261–264.

Kyd JM, Cripps AW, Murphy TF (1998) Outer-membrane antigen expression by *Moraxella (Branhamella) catarrhalis* influences pulmonary clearance. *J. Med. Microbiol.* 47: 159–168.

Labia R, Barthelemy M, Le Bouguennec CB, Buu Hoi-Dang van A (1986) Classification of β-lactamases from *Branhamella catarrhalis* in relation to penicillinases produced by other bacterial species. *Drugs* 31 (Suppl. 3): 40–47.

Lafontaine ER, Cope LD, Aebi C, Latimer JL, McCracken GH, Hansen EJ (2000) The UspA1 protein and a second type of UspA2 protein mediate adherence of *Moraxella catarrhalis* to human epithelial cells *in vitro*. *J. Bacteriol.* 182: 1364–1373.

Leach AJ, Boswell JB, Asche V, Nienhuys TG, Mathews JD (1994) Bacterial colonization of the nasopharynx predicts very early onset and persistence of otitis media in Australian Aboriginal infants. *Pediatr. Infect. Dis. J.* 13: 983–989.

Luke NR, Campagnari AA (1999) Construction and characterization of *Moraxella catarrhalis* mutants defective in expression of transferrin receptors. *Infect. Immun.* 67: 5815–5819.

Luke NR, Russo TA, Luther N, Campagnari AA (1999) Use of an isogenic mutant constructed in *Moraxella catarrhalis* to identify a protective epitope of outer membrane protein B1 defined by monoclonal antibody. *Infect. Immun.* 67: 681–687.

Luman I, Wilson RW, Wallace RJ, Nash DR (1986) Disk diffusion susceptibility of *Branhamella catarrhalis* and relationship of β-lactam zone size to β-lactamase production. *Antimicrob. Agents Chemother.* 30: 774–776.

Maluszynska GM, Krachler B, Sundqvist T (1998) The ability to bind albumin is correlated with nitric oxide sensitivity in *Moraxella catarrhalis*. *FEMS Microbiol. Lett.* 166: 249–255.

Malmvall BE, Brorsson JE, Johnsson J (1977) *In vitro* sensitivity to penicillin V and β-lactamase production of *Branhamella catarrhalis*. *J. Antimicrob. Chemother.* 3: 374–375.

Mannion PT (1987) Sputum microbiology in a district general hospital: the role of *Branhamella catarrhalis*. *Br. J. Dis. Chest* 81: 391–396.

Marchant CD (1990) Spectrum of disease due to *Branhamella catarrhalis* in children with particular reference to acute otitis media. *Am. J. Med.* 88 (Suppl. 5A): 5S–15S

Marrs CF, Weir S (1990) Pili (fimbriae) of *Branhamella* species. *Am. J. Med.* 88 (Suppl. 5A): 36S–40S.

Masoud H, Perry MB, Richards JC (1994) Characterization of the lipopolysaccharide of *Moraxella catarrhalis*: structural analysis of the lipid A from *M. catarrhalis* serotype A lipopolysaccharide. *Eur. J. Biochem.* 220: 209–216.

Mathers KE, Goldblatt D, Aebi C, Yu R-H, Schryvers AB, Hansen EJ (1997) Characterisation of an outer membrane protein of *Moraxella catarrhalis*. *FEMS Immunol. Med. Microbiol.* 19: 231–236.

Mathers K, Leinonen M, Goldblatt D (1999) Antibody response to outer membrane proteins of *Moraxella catarrhalis* in children with otitis media. *Pediatr. Infect. Dis. J.* 18: 982–988.

Mbaki N, Rikitomi N, Nagatake T, Matsumoto K (1987) Correlation between *Branhamella catarrhalis* adherence to oropharyngeal cells and seasonal incidence of lower respiratory tract infections. *Tohoku J. Exp. Med.* 153: 111–121.

McLeod DT, Ahmad F, Capewell S, Croughan MJ, Calder MA, Seaton A (1986a) Increase in bronchopulmonary infection due to *Branhamella catarrhalis*. *Br. Med. J.* 292: 1103–1105.

McLeod DT, Ahmad F, Croughan MJ, Calder MA (1986b) Bronchopulmonary infection due to *Branhamella catarrhalis*: clinical features and therapeutic response. *Drugs* 31 (Suppl. 3): 109–112.

McMichael JC, Fiske MJ, Fredenburg RA *et al.* (1998) Isolation and characterization of two proteins from *Moraxella catarrhalis* that bear a common epitope. *Infect. Immun.* 66: 4374–4381.

McNeely DJ, Kitchens CS, Kluge RM (1976) Fatal *Neisseria (Branhamella) catarrhalis* pneumonia in an immunodeficient host. *Am. Rev. Respir. Dis.* 114: 399–402.

Mibus DJ, Mee BJ, McGregor KF, Garbin CD, Chang BJ (1998) The identification of response regulators in *Branhamella catarrhalis* using PCR. *FEMS Immunol. Med. Microbiol.* 22: 351–354.

Murphy S, Fitzgerald M, Mulcahy R, Keane C, Coakley D, Scott T (1997) Studies on haemagglutination and serum resistance status of strains of *Moraxella catarrhalis* isolated from the elderly. *Gerontology* 43: 277–282.

Murphy TF (1990) Studies of the outer membrane proteins of *Branhamella catarrhalis*. *Am. J. Med.* 88 (Suppl. 5A): 41S–45S.

Murphy TF (1996) *Branhamella catarrhalis*: epidemiology, surface antigenic structure and immune response. *Microbiol. Rev.* 60: 267–279.

Murphy TF, Loeb MR (1989) Isolation of the outer membrane of *Branhamella catarrhalis*. *Microb. Pathogen.* 6: 159–174.

Murphy TF, Sethi S (1992) Bacterial infection in chronic obstructive pulmonary disease. *Am. Rev. Respir. Dis.* 146: 1067–1083.

Murphy TF, Kirkham C, Lesse AJ (1993) The major heat-modifiable outer membrane protein CD is highly conserved among strains of *Branhamella catarrhalis*. *Mol. Microbiol.* 10: 87–97.

Murphy TF, Kyd JM, John A, Kirkham C, Cripps AW (1998) Enhancement of pulmonary clearance of *Moraxella (Branhamella) catarrhalis* following immunization with outer membrane protein CD in a mouse model. *J. Infect. Dis.* 178: 1667–1675.

Murphy TF, Kirkham C, DeNardin E, Sethi S (1999) Analysis of antigenic structure and human immune response to outer membrane protein CD of *Moraxella catarrhalis*. *Infect. Immun.* 67: 4578–4585.

Myers LE, Yang Y-P, Du R-P *et al.* (1998) The transferrin binding protein B of *Moraxella catarrhalis* elicits bactericidal antibodies and is a potential vaccine antigen. *Infect. Immun.* 66: 4183–4192.

Nguyen KT, Hansen EJ, Farinha MA (1999) Construction of a genomic map of *Moraxella (Branhamella) catarrhalis* ATCC25238 and physical mapping of virulence-associated genes. *Can. J. Microbiol.* 45: 299–303.

Nicotra B, Rivera M, Luman JI, Wallace JR (1986) *Branhamella catarrhalis* as a lower respiratory tract pathogen in patients with chronic lung disease. *Arch. Intern. Med.* 146: 890–893.

Ninane G, Joly J, Piot P, Kraytman M (1977) *Branhamella (Neisseria) catarrhalis* as pathogen. *Lancet* ii: 149.

Ohlsson A, Bailey T (1985) Neonatal pneumonia caused by *Branhamella catarrhalis*. *Scand. J. Infect. Dis.* 17: 225–228.

Peak IRA, Jennings MP, Hood DW, Bisercic M, Moxon ER (1996) Tetrameric repeat units associated with virulence factor phase variation in *Haemophilus* also occur in *Neisseria* spp, *Moraxella catarrhalis*. *FEMS Microbiol. Lett.* 137: 109–114.

Penttila M, Savolainen S, Kiukaanniemi H, Forsblom B, Jousimiessomer H (1997) Bacterial findings in acute maxillary sinusitis – European study. *Acta Otolaryngol.* (Suppl.) 529: 165–168.

Percival A, Corkill JE, Rowlands J, Sykes RB (1977) Pathogenicity of and β-lactamase production by *Branhamella (Neisseria) catarrhalis*. *Lancet* ii: 1175.

Pollard JA, Wallace RJ, Nash DR, Luman JI, Wilson RW (1986) Incidence of *Branhamella catarrhalis* in the sputa of patients with chronic lung disease. *Drugs* 31 (Suppl. 3): 103–108.

Prellner K, Christensen P, Hovelius B, Rosen C (1984) Nasopharyngeal carriage of bacteria in otitis-prone and non-otitis-prone children in day-care centres. *Acta Otolaryngol.* 98: 343–350.

Rahman M, Holme T, Jönsson I, Krook A (1995) Lack of serotype-specific antibody response to lipopolysaccharide antigens of *Moraxella catarrhalis* during lower respiratory tract infection. *Eur. J. Clin. Microbiol. Infect. Dis.* 14: 297–304.

Rahman M, Holme T, Jönsson I, Krook A (1997) Human immunoglobulin isotype and IgG subclass response to different antigens of *Moraxella catarrhalis*. *APMIS* 105: 213–220.

Reddy MS, Murphy TE, Faden HS, Bernstein JM (1997) Middle ear mucin glycoprotein: purification and interaction with nontypable *Haemophilus influenzae* and *Moraxella catarrhalis*. *Otolaryngol. Head Neck Surg.* 116: 175–180.

Retzer MD, Yu R-H, Schryvers AB (1999) Identification of sequences in human transferrin that bind to the bacterial receptor protein, transferrin-binding protein B. *Mol. Microbiol.* 32: 111–121.

Reyn A (1974) Family I. Neisseriaceae Prevot 1993. In: Buchanan RE, Gibbons NE (eds) *Bergey's Manual of Determinative Bacteriology*, 8th edn. Baltimore: Williams & Wilkins, pp. 427–443.

Rikitomi N, Andersson B, Matsumoto K, Lindstedt R, Svanborg C (1991) Mechanism of adherence of *Moraxella (Branhamella) catarrhalis*. Scand. J. Infect. Dis. 23: 559–567.

Rikitomi N, Ahmed K, Nagatake T (1997) *Moraxella (Branhamella) catarrhalis* adherence to human bronchial and oropharyngeal cells: the role of adherence in lower respiratory tract infections. *Microbiol. Immunol.* 41: 487–494.

Riley TV (1987) A note on hydrolysis of tributyrin by *Branhamella* and *Neisseria*. J. Appl. Bacteriol. 62: 539–542.

Riley TV (1988) Antimicrobial resistance in *Branhamella catarrhalis*. J. Antimicrob. Chemother. 20: 147–150.

Roberts MC, Brown BA, Steingrube VA, Wallace RJ (1990) Genetic basis of tetracycline resistance in *Moraxella (Branhamella) catarrhalis*. Antimicrob. Agents Chemother. 34: 1816–1818.

Rossau R, Van Landschoot A, Mannheim W, De Ley J (1986) Inter- and intrageneric similarities of ribosomal ribonucleic acid cistrons of the Neisseriaceae. *Int. J. System. Bacteriol.* 36: 323–332.

Rossau R, Van Landchoot A, Gillis M, De Ley J (1991) Taxonomy of *Moraxellaceae* faM noV, a new bacterial family to accommodate the genera *Moraxella, Acinetobacter* and *Psychrobacter* and related organisms. *Int. J. System. Bacteriol.* 41: 310–319.

Salyers AA, Whitt DD (1994) *Bacterial Pathogenesis: A Molecular Approach*. Washington, DC: ASM Press.

Sarubbi FA, Myers JW, Williams JJ, Shell CG (1990) Respiratory infections caused by *Branhamella catarrhalis*: selected epidemiologic features. *Am. J. Med.* 88 (Suppl. 5A): 5S–9S

Sato K (1997) Experimental otitis media induced by non-viable *Moraxella catarrhalis* in the guinea-pig model. *Auris, Nasus, Larynx* 24: 233–238.

Schalen L, Christensen P, Kamme C, Miorner H, Pettersson KI, Schalen C (1980) High isolation rate of *Branhamella catarrhalis* from the nasopharynx in adults with acute laryngitis. *Scand. J. Infect. Dis.* 12: 277–280.

Schryvers AB, Lee BC (1989) Comparative analysis of the transferrin and lactoferrin binding proteins in the family Neisseriaceae. *Can. J. Microbiol.* 35: 409–415.

Seddon PC, Sunderland D, O'Halloran SM, Hart CA, Heaf DP (1992) *Branhamella catarrhalis* colonization in preschool asthmatics. *Pediatr. Pulmonol.* 13: 133–135.

Sethi S, Hill SL, Murphy TF (1995) Serum antibodies to outer membrane proteins (OMPs) of *Moraxella (Branhamella) catarrhalis* in patients with bronchiectasis: identification of OMP B1 as an important antigen. *Infect. Immun.* 63: 1516–1520.

Sethi S, Surface JM, Murphy TF (1997) Antigenic heterogeneity and molecular analysis of CopB of *Moraxella (Branhamella) catarrhalis*. Infect. Immun. 65: 3666–3671.

Slevin NJ, Aitken J, Thornley PE (1984) Clinical and microbiological features of *Branhamella catarrhalis* bronchopulmonary infections. *Lancet* i: 782–783.

Smith CL, Condemine G (1990) New approaches for physical mapping of small genomes. *J. Bacteriol.* 172: 1167–1172.

Smith JMB, Lockwood BM (1986) A 2-year survey of *Branhamella catarrhalis* in a general hospital. *J. Hosp. Infect.* 7: 277–283.

Soto-Hernandez JL, Holtsclaw-Berk S, Harvill LM, Berk SL (1989) Phenotypic characteristics of *Branhamella catarrhalis* strains. *J. Clin. Microbiol.* 27: 903–908.

Stenfors LE, Raisanen S (1993) Secretory IgA-, IgG- and C3b-coated bacteria in the nasopharynx of otitis-prone and non-otitis-prone children. *Acta Otolaryngol.* 113: 191–195.

Stobberingh EE, Davies BI, van Boven CP (1994) *Branhamella catarrhalis*: antibiotic sensitivities and β-lactamases. *J. Antimicrob. Chemother.* 13: 55–64

Tinkelman DG, Silk HJ (1989) Clinical and bacteriologic features of chronic sinusitis in children. *Am. J. Dis. Child.* 143: 938–941.

Vaneechoutte M, Verschraegen G, Claeys G, Van Den Abeele AM (1990a) Serological typing of *Branhamella catarrhalis* strains on the basis of lipopolysaccharide antigens. *J. Clin. Microbiol.* 28: 182–187.

Vaneechoutte M, Verschraegen G, Claeys G, Weise B, Van Den Abeele AM (1990b) Respiratory tract carrier rates of *Moraxella (Branhamella) catarrhalis* in adults and children and interpretation of the isolation of *M. catarrhalis* from sputum. *J. Clin. Microbiol.* 28: 2674–2680.

Van Hare GF, Shurin PA, Marchant CD *et al.* (1987) Acute otitis media caused by *Branhamella catarrhalis*: biology and therapy. *Rev. Infect. Dis.* 9: 16–27.

Verduin CM, Jansze M, Hol C, Mollnes TE, Verhoef J, van Dijk H (1994) Differences in complement activation between complement-resistant and complement-sensitive *Moraxella (Branhamella) catarrhalis* strains occur at the level of membrane attack complex formation. *Infect. Immun.* 62: 589–595.

Verghese A, Berk SL (1991) *Moraxella (Branhamella) catarrhalis*. Infect. Dis. Clin. N. Amer. 5: 523–538.

Verghese A, Berro E, Berro J, Franzus BW (1990a) Pulmonary clearance and phagocytic cell response in a murine model of *Branhamella catarrhalis* infection. *J. Infect. Dis.* 162: 1189–1192.

Verghese A, Roberson D, Kalbfleisch JH, Sarubbi F (1990b) Randomized comparative study of cefixime versus cephalexin in acute bacterial exacerbations of chronic bronchitis. *Antimicrob. Agents Chemother.* 34: 1041–1044.

Veron M, Lenvoise-Furet A, Coustere C, Ged C, Grimont F (1993) Relatedness of three species of 'False Neisseriae', *Neisseria caviae, Neisseria cuniculi* and *Neisseria ovis*, by DNA–DNA hybridizations and fatty acid analysis. *Int. J. System. Bacteriol.* 43: 210–220.

Wald ER, Milmoe GJ, Bowen A *et al.* (1981) Acute maxillary sinusitis in children. *N. Engl. J. Med.* 304: 749–754.

Wald ER, Reilly JS, Casselbrant M *et al.* (1984) Treatment of acute maxillary sinusitis in childhood: a comparative study of amoxicillin and cefaclor. *J. Pediatr.* 104: 297–302.

Wallace MR, Oldfield EC (1990) *Moraxella (Branhamella) catarrhalis* bacteremia. *Arch. Intern. Med.* 150: 1332–1334.

Wallace RJ, Musher DM (1986) In honor of Dr Sarah Branham: a star is born. The realization of *Branhamella catarrhalis* as a respiratory pathogen. *Chest* 90: 447–450.

Wallace RJ, Steingrube VA, Nash DR *et al.* (1989) BRO β-lactamases of *Branhamella catarrhalis* and *Moraxella* subgenus *Moraxella*, including evidence for chromosomal β-lactamase transfer by conjugation in *B. catarrhalis*, *M. nonliquefaciens* and *M. lacunata*. *Antimicrob. Agents Chemother.* 33: 1845–1854.

Wallace RJ, Nash DR, Steingrube VA (1990) Antibiotic susceptibilities and drug resistance in *Moraxella (Branhamella) catarrhalis*. *Am. J. Med.* 88 (Suppl. 5A): 46S–50S.

Wardle JR (1986) *Branhamella catarrhalis* as an indirect pathogen. *Drugs* 31 (Suppl. 3): 93–96.

Weiser JN (1992) The oligosaccharide of *Haemophilus influenzae*. *Microb. Pathogen.* 13: 335–342.

West M, Berk SL, Smith JK (1982) *Branhamella catarrhalis* pneumonia. *Southern Med. J.* 75: 1021–1023.

Wilhelmus KR, Peacock J, Coster DJ (1980) *Branhamella* keratitis. *Br. J. Ophthalmol.* 64: 892–895.

Wistreich GA, Baker RF (1971). The presence of fimbriae (pili) in three species of *Neisseria*. *J. Gen. Microbiol.* 65: 167–173.

Wright PW, Wallace RJ, Shepherd JR (1990) A descriptive study of 42 cases of *Branhamella catarrhalis* pneumonia. *Am. J. Med.* 88 (Suppl. 5A): 2S–8S.

Yang Y-P, Myers LE, McGuiness U *et al.* (1997) The major outer membrane protein, CD, extracted from *Moraxella (Branhamella) catarrhalis* is a potential vaccine antigen that induces bactericidal antibodies. *FEMS Immunol. Med. Microbiol.* 17: 187–199.

Yu R-H, Schryvers AB (1993) The interaction between human transferrin and transferrin binding protein 2 from *Moraxella (Branhamella) catarrhalis* differs from that of other human pathogens. *Microb. Pathogen.* 15: 433–445.

Yu R-H, Bonnah RA, Ainsworth S, Schryvers AB (1999) Analysis of the immunological responses to transferrin and lactoferrin receptor proteins from *Moraxella catarrhalis*. *Infect. Immun.* 67: 3793–3799.

Zheng CH, Ahmed K, Rikitomi N, Martinez G, Nagatake T (1999) The effects of S-carboxymethylcysteine and N-acetylcysteine on the adherence of *Moraxella catarrhalis* to human pharyngeal epithelial cells. *Microbiol. Immunol.* 43: 107–113.

79

Mycoplasma pneumoniae and other Mycoplasmas

Gunna Christiansen[1], Mette Drasbek[2] and Svend Birkelund[1]

[1]University of Aarhus, Denmark
[2]Loke Diagnostics, Aarhus, Denmark

The genus *Mycoplasma* belongs to a class of bacteria called the Mollicutes (Latin *mollis*: soft, *cutis*: skin), other members of which are the ureaplasmas, spiroplasmas, mesoplasmas, entoplasmas, asteroleplasmas, acholeplasmas and anaeroplasmas. The term mycoplasma has frequently been used to denote any species included in the class Mollicutes. Mycoplasmas are widespread in nature as parasites in humans, mammals, reptiles, fish, arthropods and plants (Razin, 1995). The primary habitats of human and animal mycoplasmas are the mucous surfaces of the respiratory and urogenital tracts, the eyes, alimentary canal and mammary glands (Tully, 1993).

Mycoplasmas are pleiomorphic and are distinguished from other prokaryotic bacteria by their small size, and their total lack of a cell wall. When grown on solid medium they form fried-egg-shaped colonies 0.01–2 mm in diameter (Sub-committee on the Taxonomy of Mollicutes, 1995). They have small and AT-rich genomes and the codon UGA is used as the codon for tryptophan and not as the 'stop' codon as in other organisms. Determination of the $G+C$ content and the genome size (Bak *et al.*, 1969) were major advances in the characterisation of microbes and can be regarded as the beginning of molecular microbiology. Mycoplasmas have a $G+C$ content of 24–39 mol%, which is low compared with that of other bacteria (Bredt, 1979). The Mollicutes have lost a substantial part of their genetic information during the process of evolution and are thought to be the smallest self-replicating prokaryotes. This is reflected in their small genome sizes, which range from 580 kbp (*Mycoplasma genitalium*) (Fraser *et al.*, 1995) to 2220 kbp (*Spiroplasma ixodetis*) (Carle *et al.*, 1995).

Classification

The genus *Mycoplasma* contains more than 100 recognised species. According to the minimum standards for the description of new *Mycoplasma* species (Subcommittee on the Taxonomy of Mollicutes 1995), they comprise non-helical Mollicutes from vertebrates that are not obligate anaerobic, cannot hydrolyse urea, require cholesterol (or sterols) for growth, have a temperature optimum of 37°C or above, and a genome size of 600–1300 kbp.

The sequencing of the genes that encode 16S rRNA has proved to be a valuable taxonomic tool (Weisburg *et al.*, 1989). In this way, the Mollicutes class can be divided into five major phylogenetic groups of which the *Mycoplasma hominis, Mycoplasma pneumoniae* and *Ureaplasma urealyticum* groups are those that contain the members of the human pathogenic mollicutes (Weisburg *et al.*, 1989, 1991).

Identification

Members of the genus *Mycoplasma* are characterised by their very small cell size (diameter 200–300 nm). The cells are often pleiomorphic and the morphology is very much dependent on the age of the culture and medium composition. To cultivate *Mycoplasma* a complex medium containing cholesterol is required (Razin and Tully, 1970). Their special growth requirements and total lack of cell wall makes them readily identifiable from other bacteria, but because of the large number of species in the genus *Mycoplasma* specific identification of each species is necessary. Growth inhibition by homologous and heterologous antiserum is frequently used for this identification (Clyde, 1964, 1983a). DNA amplification by the polymerase chain reaction (PCR) of the gene encoding 16S rRNA and DNA sequencing can also be used as a supplement to species identification. Most *Mycoplasma* genomes contain two rRNA operons with very high homology. Not all 16S rRNA genes have been sequenced but when the sequences become available they will provide excellent complementary tools for species determination (Weisburg *et al.*, 1989).

Structure

The shapes of the mycoplasmas vary from spherical or pear-shaped structures (0.3–0.8 μm in diameter) to branched or helical filaments (Bove, 1993). The helical filamentous shape is found in spiroplasmas. The maintenace of such shapes in the absence of a rigid cell wall has long been thought to indicate the presence of a cytoskeleton in mycoplasmas (Razin, 1978). One group of mycoplasma do have a special well-defined surface protrusion, the tip, that give these cells a characteristic flask-shaped form (**Fig. 1**). The tip-containing species are *Mycoplasma pneumoniae, M. genitalium, M. gallicepticum, M. mobile, M. sualvi, M. alvi* and *M. pulmonis*. Thus, the two important human pathogens *M. pneumoniae* and *M. genitalium* are both members of the tip-bearing *Mycoplasma*. This tip-like organelle plays an important role in the adhesion of the mycoplasmas to the host cells. Adhesion is an important virulence factor, because it is

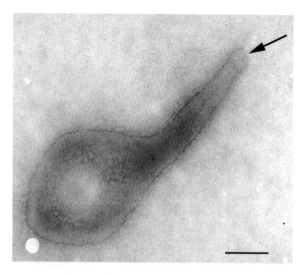

Fig. 1 Electron micrograph of *Mycoplasma pneumoniae*. *M. pneumoniae* was cultivated in SP4 medium in tissue culture flasks. When colour change of the medium was visible the cells were scraped off and directly desorbed to the surface of a carbon coated grid for 30 seconds. The grid was stained with three drops of ammonium molybdate 0.5% and blotted dry. Electron microscopy was done using a JEOL 1010 electron microscope at 40 kV equipped with a Kodak 1.4 slow scan camera. Images were transferred to a SUN Sparc 10 station with an SDV digital camera interface. Arrow point at the tip-structure. Bar 200 nm.

the first step in the colonisation of the host. The tip is also involved in the gliding motility of the mycoplasmas, the tip being positioned in the direction of the movement (Bredt, 1979). Proteins localised at the tip structure and their function are discussed below.

Some *Mycoplasma* are motile and able to glide on solid surfaces. The most motile species is *M. mobile* (Kirchoff *et al.*, 1987; Miyata *et al.*, 2000); *M. pneumoniae* is also capable of gliding motility. Bacterial motility is normally associated with the presence of flagella, but since *Mycoplasma* cells are surrounded by a cytoplasmic cell membrane and contain no cell wall structures, motility must be conferred by other structures. The cytoskeleton-like components are thought to function in modulating both cell shape and cell motility.

Physiology

M. genitalium is the smallest living organism for which the complete genome has been completely sequenced to date (Fraser *et al.*, 1995) and its closest relative is *M. pneumoniae* (Himmelreich *et al.*, 1997), the genome of which was sequenced by Himmelreich *et al.*

(1996). These sequences revealed that both species lacked many of the genes involved in amino acid metabolism (Himmelreich *et al.*, 1997). The small genome sizes of mycoplasmas give them a limited coding capacity. The high A + T content of the mycoplasma genomes is believed to be a result of impaired activity of the enzyme uracil-DNA glycosylase, which removes uracil from DNA arising from spontaneous deamination of cytosine residues (Razin *et al.*, 1998). Other prokaryotes use the UGA codon as a stop codon, but members of the genus *Mycoplasma* translate this to tryptophan (Yamao *et al.*, 1985). This makes it difficult to investigate the mycoplasma genes with expression vectors in *Escherichia coli*.

Both *M. pneumoniae* and *M. genitalium* lack all the genes involved in amino acid synthesis and co-factor biosynthesis, nucleic acids precursors and cholesterol, making them totally dependent on an exogenous supply of the complete spectrum of amino acids and vitamins (Collier, 1983; Fraser *et al.*, 1995). This means that artificial culture media have to be very complex and supplemented with essentially all vitamins. The parasitic life of the mycoplasmas may explain why the loss of coding capacity is tolerated. Investigations of mycoplasmas have demonstrated that they possess only a minimum of organelles for cell growth and replication.

Mycoplasma hominis

Mycoplasma hominis is a commensal of the human genital tract and a facultative pathogen that causes post-partum fever, neonatal infections and infections of immunocompromised individuals. It differs from *M. pneumoniae* and *M. genitalium* in its lack of a tip structure and its genetic variability. The genome has not been sequenced but several genes have been sequenced and compared with those of different isolates. Whereas the 16S rRNA genes and genes encoding housekeeping proteins are similar (Mygind *et al.*, 1998, 2000a), those that encode surface-exposed proteins vary. The variation is based on on/off variation, variation in reaction with monoclonal antibodies and variation in the size of the protein (Boesen *et al.*, 1998; Ladefoged *et al.*, 1995, 1996; Nyvold *et al.*, 1997).

The mode of variation differs between different genes. The gene *lmp* (large membrane protein), which contains multiple repeats of 471 bp, is found in two copies on the genome and the size of the protein differs between isolates (Ladefoged *et al.*, 1995, 1996). A 120-kDa protein, P120, which is not expressed in all isolates, has a hypervariable domain

that differs between isolates (Nyvold *et al.*, 1997) and Vaa (variable adherence-associated) protein, has a modular structure with a variable number of modules. The *vaa* gene is present in only one copy and no reservoir of modules could be detected within the genome (Boesen *et al.*, 1998). P75 is a very conserved protein of 75 kDa with limited genetic variation (Mygind *et al.*, 2000b). Variations in the mode by which the different genes display variability reflects the complicated ways in which *M. hominis* is able to vary its surface. The genetic variation is also reflected in a wide variation in genome size and genome structure (Ladefoged and Christiansen, 1992).

Jensen *et al.* (1998) took advantage of this feature and analysed consecutive isolates of *M. hominis* from pregnant women and their babies. They compared protein profiles and the genome structure by pulsed-field gel electrophoresis of the isolates and found that isolates from a woman and her baby were identical or very close to identical, whereas isolates from different women were different both in protein profile and in genome structure.

Mycoplasma pneumoniae

Mycoplasma pneumoniae was first described as the cause of a clinical syndrome in the late 1930s (Gallagher, 1941). It is an extracellular parasite that colonises mucosal surfaces, and its ability to attach to host epithelial cells is an important characteristic. It infects the upper and lower respiratory tract and is a frequent cause of bronchitis. *M. pneumoniae* is the leading cause of atypical pneumonia in older children and young adults. The patients have flu-like symptoms but characteristically the infection is chronic in onset and recovery. This disease is often called 'walking pneumonia', which reflects the contrast between it and classical pneumonia due to organisms such as *Streptococcus pneumoniae* (Collier, 1983; Clyde, 1983b). *Strep. pneumoniae* is associated with 39% of community-acquired pneumonia cases, while *M. pneumoniae* comes second, causing 16% of the cases (Neill *et al.*, 1996). *M. pneumoniae* infections are often seen as epidemics that occur at intervals of 4–7 years (Ursi *et al.*, 1994).

Because of the lack of a cell wall, *M. pneumoniae* is resistant to the penicillins, as are the other mycoplasmas. The absence of a cell wall probably facilitates close contact between *M. pneumoniae* and the host cell, and this guarantees the exchange of compounds necessary for the growth and proliferation of the bacteria (Dybvig and Voelker, 1996). *M. pneumoniae* bears a tip, which it uses as an attachment organelle. This is a membrane-bound extension of the cell and is characterised by an electron-dense core laterally orientated within the tip (Wilson and Collier, 1976; Tully *et al.*, 1983; Bove, 1993). *M. pneumoniae* is elongated and consists of the frontal tip structure, a thicker body part and sometimes a longer tail-like rear end (Bredt *et al.*, 1983). It moves by gliding motility, and the cells become non-motile when they lose surface contact (Bredt, 1973). The gliding motility is temperature-dependent and the optimal temperature for both cultivation and gliding motility is 37°C.

Proteins

The sequencing of the *M. pneumoniae* genome showed that it is 816 394 bp in length, that the genome is predicted to contain 677 open reading frames (ORFs) and that it encodes 39 RNA species. The G+C content is 40.0 mol%, which is high for the mycoplasmas (Himmelreich *et al.*, 1996). Of the predicted ORFs, 76% show similarity to other gene sequences present in databases, but only 49% could be functionally assigned based on significant sequence similarity to genes or proteins with known function from other organisms. It was possible to make a tentative functional classification of a large number of the gene products in *M. pneumoniae* and to deduce their biochemical and physiological functions. For a classification of the ORFs see Himmelreich *et al.* (1996).

About 8% of the genome is composed of repetitive DNA elements (RepMP1, RepMP2/3, RepMP4 and RepMP5). The genome sequence revealed that the genome had an uneven G+C content. There are regions in which the G+C content is 56% compared with the average of 40% and these regions contain the P1 operon and the repetitive DNA sequences RepMP4, RepMP2/3, RepMP5 and tRNAs. The P1 gene contains a copy of the RepMP2/3 and of RepMP4, while ORF6 contains a copy of RepMP5 (Ruland *et al.*, 1990).

A surprisingly high number of genes, 46 in total, were found to encode putative lipoproteins. The lack of a cell wall may explain the need for a high number of lipid-anchored surface components that are responsible for such cell surface functions as substrate-binding, protein transport and antigenic variation. Whether all the 46 genes give rise to functional proteins and the circumstances under which the proteins are transcribed and translated remain to be determined. It was also surprising that no genes were found to encode proteins responsible for motility, chemotaxis and management of oxidative stress (Himmelreich *et al.*, 1996).

The Tip-structure

The tip-structure of *M. pneumoniae* has been studied extensively because it mediates attachment to the host cells and is involved in gliding motility. *M. pneumoniae* has some highly adhesive membrane components, of which the surface-exposed P1 protein (170 kDa) is the most important. This is densely clustered at the tip organelle, but it is also scattered along the surface (Baseman *et al.*, 1988). The P1 protein is the major component responsible for the attachment of *M. pneumoniae* to host cells, but 10 other proteins may also be involved in this process (Isenberg, 1988; Krause, 1996). These findings suggest that attachment of *M. pneumoniae* is a complex process.

The P1 protein of *M. pneumoniae* is encoded by the P1 gene, which is part of the P1 operon. This consists of three ORFs, ORF4, ORF5 (P1 gene) and ORF6. ORF4 encodes a 28-kDa protein. The product of ORF6 is not, as would be expected, a 130-kDa protein, but rather two proteins of 40 and 90 kDa, which probably arise by proteolytic cleavage of the 130-kDa protein. The 90-kDa (N-terminal part) and 40-kDa (C-terminal part) proteins are both immunogenic and mutants lacking these two proteins are avirulent. The two proteins are involved in cytadherence in an unknown manner (Inamine *et al.*, 1988; Himmelreich *et al.*, 1996).

Other proteins associated with adhesion have been identified; these are P65, P41 and P24, and proteins A, B and C, but their function is not clear (Krause, 1996). Proteins B and C probably correspond to the 130-kDa protein product of ORF6, so that B is the 90-kDa protein and C is the 40-kDa protein (Sperker *et al.*, 1991; Krause, 1996). A cytoskeleton-like filamentous network is associated with the tip-structure, of which the HMW1–3 proteins are the building blocks (Ogle *et al.*, 1992; Krause, 1996, 1998). Some of these proteins are proline-rich and have repeated sequences and other motifs characteristic of eukaryotic cytoskeletal proteins. Protein phosphorylation is a widespread mechanism for regulating intracellular signalling and in the case of P1, HMW1 and HMW3 phosphorylation by an ATP-dependent Ser/Thr kinase has been shown (Dirksen *et al.*, 1994; Krebes *et al.*, 1995). Proteins P65 and P200 share characteristic structural features with HMW1 and HMW3, suggesting that they function as elements in the *M. pneumoniae* cytoskeleton. HMW1–3 are required for P1 to cluster at the tip.

HMW1 and HMW3 are localised in a cytoskeleton-like filamentous network (Stevens and Krause, 1991; Krause, 1997; Hahn *et al.*, 1998). This structure is thought to function by modulating cell shape and by participating in cell division, gliding motility and the proper localisation of adhesion. *M. pneumoniae* cells that lack HMW2 fail to localise the adhesion protein P1 to the attachment organelle, and the levels of HMW1, HMW3 and P65 are also low, because of accelerated turnover. HMW2 may have an effect on the phosphorylation state of HMW1 and HMW3 and this may affect their stability (Dirksen *et al.*, 1994; Fisseha *et al.*, 1999).

P30 is a 30-kDa immunogenic protein. It is, like the P1 protein, an *M. pneumoniae* adhesin, because antibodies raised against P30 inhibit mycoplasma cytadherence (Layh-Schmitt and Hermann, 1994). P30 is clustered at the tip organelle, apparently through an interaction with a complex network of cytadherence accessory proteins (Hedreyda and Krause, 1995). The protein possesses six proline-rich repeated sequences in the C-terminal part (Dallo *et al.*, 1990a), which show 55–67% homology with the C-terminal part of P1 (Dallo, 1996). The proline-rich domains may control the functionality of the P1 and the P30 adhesins (Krause *et al.*, 1983). Further evidence for the existence of this complex structure *in vivo* has come from a study by Layh-Schmitt *et al.* (2000). By means of chemical cross-linking they analysed the interactions of the P1 protein with other membrane proteins and with cytoskeleton-like elements. Proteins were cross-linked *in vivo*, the complexed proteins were isolated by immunoaffinity chromatography, and the cross-linked components were characterised by immunoblotting and by high-mass-accuracy tryptic peptide mapping with matrix-assisted laser desorption mass spectrometry (MALDI MS). In addition to P1 and P30, membrane proteins P90 and P40, the cytoskeleton associated P65 and HMW1 and 3 were identified. These results indicate that *in vivo* these components are in close proximity and they may explain a possible interaction between the surface-localised adhesins and the cytoskeleton elements.

Immunogenic Proteins

The *M. pneumoniae* P1 protein is highly immunogenic and a strong immune response is raised both in patients with *M. pneumoniae* infections and in experimentally infected animals (Hu *et al.*, 1983). *M. pneumoniae* antigens have been characterised by immunoblotting and proteins with molecular masses of 200, 170, 110, 67, 60, 47, 42 and 32 kDa were detected with a polyclonal mouse antiserum. Proteins with molecular masses of 168, 110 and 32 kDa have been shown to be exposed on the surface (Hirschberg *et al.*, 1989) in agreement with the finding of immunogenic proteins

of corresponding size. Immunoblotting has shown that in convalescent serum samples from patients recovering from *M. pneumoniae* infections five bands of 170, 130, 90, 45 and 35 kDa are more often stained than the bands of 110, 92 and 62 kDa (Shepard *et al.*, 1974). It has also shown that the 170-kDa band (P1 protein) is always present in sera from patients with *M. pneumoniae* infection (Baseman *et al.*, 1982; Jacobs *et al.*, 1986).

Subgroups of *Mycoplasma pneumoniae*

Strains of *M. pneumoniae* can be divided into two groups (group I and group II), based on sequence divergence in the repetitive sequences of the P1 gene and the ORF6 gene (Ruland *et al.*, 1994). The DNA sequence of the P1 gene contains two repeated regions (RepMP4 and RepMP2/3), while ORF6 contains one repeated region (RepMP5). The major difference between the two groups is found in the repeated regions of the P1 gene. The P1 protein of group I contains 1627 amino acids whereas the P1 protein of group II contains 1635 amino acid residues (Jacobs *et al.*, 1996). Minor amino acid exchanges and deletions are responsible for these changes (Su *et al.*, 1990a). An aggressive immune response against P1 is found during the human infection, and patients with *M. pneumoniae* infections develop high titres of antiadhesion antibodies. Since some patients have developed antibodies against antigens from both groups, it appears that anti-adhesion antibodies do not protect patients against a new infection, but most patients develop antibodies against only one of the groups.

Variation between group I and group II has also been seen in the 116-kDa protein, which often is denoted the 110-kDa protein (Duffy *et al.*, 1997; Razin and Jacobs, 1992). Epitopes of P1 have been localised to different parts of the protein (Dallo *et al.*, 1988). In ELISA with monoclonal antibodies three regions with epitopes were found, one in the N-terminal region, one in the middle of the protein and one in the C-terminal region. The three regions form a functionally active site that mediates adherence of *M. pneumoniae* to the host cells (Gerstenecker and Jacobs, 1990; Razin and Jacobs, 1992). In western blots with patient sera, however, the epitopes were found in the C-terminal part of the P1 protein (Dallo *et al.*, 1988; Jacobs *et al.*, 1990). A new *M. pneumoniae* P1 gene with a novel nucleotide sequence variation has recently been identified (Kenri *et al.*, 1999).

Major differences in P1 gene sequence between group I and group II *M. pneumoniae* as well as a new P1 gene, named 309 P1, have been found localised between nucleotides 3633 and 3894 in group I and nucleotides 3657 and 3918 in group II. In this region the 309 P1 gene is nine base pairs shorter than P1 gene from group II. The loss of two *Hae*III sites was also found in the 309 P1 sequence. The rest of the 309 P1 gene was almost identical to the group II DNA sequence, except for three point mutations at nucleotides 2539 A–G, 2872 G–A and 2886 A–T. The differences between the amino acid sequences of the group II P1 protein and the 309 P1 protein were found in amino acids 1220–1306. In this region the 309 P1 protein was three amino acids shorter than the P1 protein of group II. It possessed a novel sequence variation of approximately 300 bp in the RepMP2/3 region. Two sequences, closely homologous to the new variable region, have been found in repetitive sequences elsewhere in the *M. pneumoniae* genome. This indicates that it could have been generated by homologous recombination (Kenri *et al.*, 1999).

Comparison between *M. pneumoniae* and *M. genitalium*

The two most closely related mycoplasma species are *M. pneumoniae* and *M. genitalium*. The latter was first isolated in 1980 from urethral specimens of two men with non-gonococcal urethritis (Tully *et al.*, 1983). *M. genitalium* shares many properties with *M. pneumoniae*, such as the flask shape and the terminal tip-like organelle (Boatman, 1979). A consequence of the evolution of the Mollicutes is seen in the genomes of *M. genitalium* and *M. pneumoniae* with the loss of anabolic and catabolic pathways. *M. pneumoniae* and *M. genitalium* show different protein profiles in SDS-PAGE, but they show cross-reaction in serological tests and in western blots (Lind *et al.*, 1984). They also display similar adherence to different types of tissue cells, erythrocytes and inert surfaces such as plastic and glass. Various proteins involved in adhesion have been identified in *M. pneumoniae* and *M. genitalium*: P1 and the accessory proteins in *M. pneumoniae*, and a protein named MgPa and accessory proteins in *M. genitalium*. One of the proteins involved in adhesion in *M. genitalium*, P32, shows 43% homology to the P30 protein from *M. pneumoniae* (Reddy *et al.*, 1995). The MgPa protein shows significant homology (46%) to the P1 protein. They are both large, integral membrane proteins with regions exposed on the mycoplasma cell surface, which attach to receptor sites on the host epithelial cell.

A comparison of the amino acid sequences of the two proteins with maximum homology reveals that they have 747 amino acids in common. These are clustered in extended regions (Opitz and Jacobs, 1992). Cysteine residues are missing in both MgPa

and P1, indicating that the two proteins have a flexible structure, but they both have a proline-rich C-terminal part, which gives them a rigid structure.

Proteins similar to the P1 and the MgPa proteins have been found in other mycoplasma species. A P1 adhesin-like gene has been identified in *Mycoplasma gallicepticum*; it has 28.7% and 26.3% homology at amino acid level to P1 and MgPa. *Mycoplasma pirum* has a gene that codes for a protein showing about 26% amino acid homology to the P1 protein from *M. pneumoniae* (Tham *et al.*, 1994), indicating the importance of P1-like proteins for the tip-containing mycoplasmas.

M. genitalium has the smallest known bacterial genome with a size of 580 kbp, and its gene content must therefore be important for the survival of a minimal bacterial cell. The proteins in *M. pneumoniae* that are functionally similar to those in *M. genitalium* are also important for an understanding of how these micro-organisms function (Hu *et al.*, 1983; Fraser *et al.*, 1995; Peterson *et al.*, 1995). The genome of *M. genitalium* contains 470 ORFs, while that of *M. pneumoniae* contains 677 ORFs, and all the *M. genitalium* ORFs are found in the *M. pneumoniae* genome (Himmelreich *et al.*, 1997). The 470 ORFs of *M. genitalium* are found on six segments of the *M. pneumoniae* genome. The position of the segments in the two genomes is different but the gene order within the segments is identical (**Fig. 2a**). The segments in the *M. pneumoniae* genome are flanked by regions of repeated sequences that may be involved in homologous recombination within the genome. Parts of these repeated sequences, some of which are found in the MgPa operon as in the P1 operon, are also found in the *M. genitalium* genome. The MgPa operon consists of three ORFs: a 29-kDa ORF, the MgPa protein, and a 114-kDa ORF. The MgPa adhesion protein is densely clustered at the tip of *M. genitalium* and elicits a strong immune response in humans and experimentally infected animals. As for the P1 operon, in which multiple extragenic copies of RepMP2/3, RepMP4 and RepMP5 are found spread over the chromosome, repeated elements are also found throughout the genome of *M. genitalium*. The repeats show strong sequence similarity to the repetitive DNA sequences from *M. pneumoniae* (Himmelreich *et al.*, 1997).

The multiple copies of the P1 gene are thought to have evolved by gene conversion and chromosomal re-arrangements in the mycoplasmas (Su *et al.*, 1988). This is indicated by the variation in G + C content, the P1 operon having a G + C content of 56 mol% while the origin of replication of *M. pneumoniae* has a G + C of 26 mol% (Himmelreich *et al.*, 1996). The variable

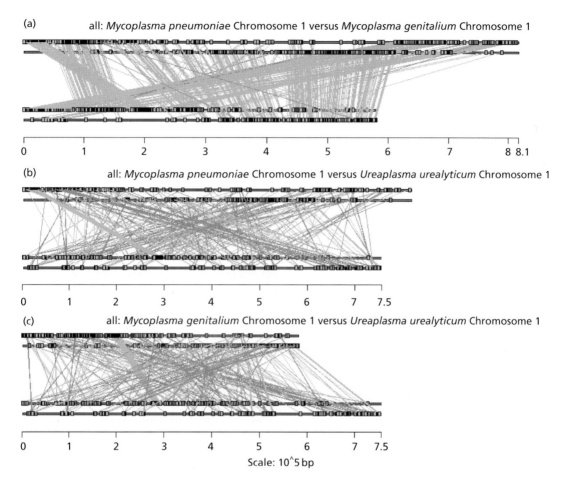

Fig. 2 Genome comparison. The genome sequences were aligned with the STD sequence database and the tools genome alignments according to http://www.stdgen.lanl.gov. (a) *Mycoplasma pneumoniae* and *M. genitalium*. (b) *M. pneumoniae* and *U. urealyticum*. (c) *M. genitalium* and *U. urealyticum* genomes are linearised and aligned with the program Bugspray for graphical representation. A green line represents a pair of genes which are found on strands of the same sense, and a red line represents a pair of genes which are found on strands of the opposite sense (See also Colour Plate 17).

G + C content of the coding regions of the mycoplasma genome has phylogenetic relevance and indicates the possible exogenous origin of adhesion genes. The repeated elements are conducive to homologous recombination and genomic re-arrangements, and may play a role in the induction of antigenic variation of the mycoplasma cell surface and thereby help the parasite to evade the host immune response (Peterson *et al.*, 1995). The sequences found in the MgPa operon and the nine repeats, which are scattered throughout the genome, represent 4.7% of the total genome. Taking into account the small genome size, this means that a substantial part of the genome is used by such repeated elements.

Evidence for recombination between the repetitive elements and the MgPa operon has been seen. This recombination may allow *M. genitalium* to evade the immune response by antigenic variation within the population (Fraser *et al.*, 1995; Peterson *et al.*, 1995). This may also be true for *M. pneumoniae*, because the P1 protein and the MgPa protein resemble each other in structure.

In contrast to the very similar genomes of *M. pneumoniae* and *M. genitalium*, the recently sequenced *Ureaplasma urealyticum* (biovar 1 serovar 3) genome (J.I. Glass, E.J. Lefkowitz, J.C. Glass, C.R. Heimer, E.Y. Chen and G.H. Cassell, unpublished results) has a totally different structure (**Fig. 2b** and **c**). In **Fig. 2b** the *U. urealyticum* biovar 1, serovar 3 genome is compared with that of *M. pneumoniae*, and in **Fig. 2c** it is compared with the genome of *M. genitalium*. It will be seen that the gene order is different and that only a few genes are in the same order (parallel lines). The genome is 751.719 bp in size and has the ability to encode 611 proteins and 39 structural RNAs.

Ureaplasma organisms have features similar to those described for *Mycoplasma*, but in addition they are capable of hydrolysing urea. They do not have a tip-structure.

U. urealyticum can be divided into two biovars and 14 serovars (Shepard *et al.*, 1974), of which serovars 1, 3, 6 and 14 belong to the Parvo biovar, previously biovar 1 (Kong *et al.*, 1999). *U. urealyticum* is a commensal of the human genital tract and a facultative pathogen associated with non-gonococcal urethritis, chorioamnionitis, pre-term membrane rupture and neonatal infections. The biovars can be distinguished by PCR but little is known about the pathogenicity difference between them (Kong *et al.*, 2000). The determined genome sequence and typing by PCR will provide valuable tools for future molecular pathogenicity studies.

The small genome size of *M. genitalium* (580 kbp) and its limited coding capacity of 470 ORFs make this micro-organism attractive for studying the proteome. Wasinger *et al.* (2000) were able to resolve 427 distinct proteins by applying four pH windows in the iso-electrical focusing step followed by SDS-PAGE and silver staining of the two-dimensional gels. By means of peptide mass fingerprinting 158 different proteins were identified, and the results represent the most complete proteome observed. Determination of the *M. genitalium* proteome complements the genome well, because it can provide knowledge about when and under what circumstances individual proteins are expressed, the relative amounts of a given protein, and information on processing and post-translational modification. The published *M. genitalium* proteome can be used as the framework for such determinations.

Products

The parasitic life of *M. pneumoniae*, with its close adherence to the ciliated respiratory endothelial cells, is well described, but so far toxins, enzymes or secreted proteins have not been identified. Attachment of *M. pneumoniae* to these cells is associated with ciliostasis and eventual loss of the ciliated cells; this would indicate the potential effect of an enzyme or toxin. By analysis of the genome many ORFs have been identified with the capacity to encode proteins with unassigned function. Proteome analysis of pulsed-chased *M. pneumoniae*-infected cell cultures may, in combination with transcription analysis, identify gene products produced specifically when *M. pneumoniae* adheres to susceptible cells. Such analyses may identify proteins that are potentially secreted and provide information on their function as enzymes or toxins (Himmelreich *et al.*, 1996).

Pathogenesis

M. pneumoniae causes tracheo-bronchitis and primary atypical pneumonia in older children and young adults. The infection is spread by droplet inhalation. Symptoms are flu-like, with fever, headache and a dry, often prolonged cough. After inhalation, *M. pneumoniae* moves by gliding motility to the ciliated epithelium, where its tip-structure attaches to the cells. It is thought that adherence is necessary for successful colonisation and infection, because *M. pneumoniae* mutants unable to adhere are unable to infect and cause disease in hamsters (Krause *et al.*, 1982, 1983). Adhesion is, therefore, an important pathogenicity factor. Many proteins have been shown to be involved in the adherence process. The major adhesin is the P1 tip protein. Mutants lacking the P1 protein are unable to adhere to cells, and antibodies to this protein can block adherence. Protein P30, which is also localised at the tip, is also involved in adhesion. The characterisation of non-adherent mutants of *M. pneumoniae* further revealed that two groups of proteins, the A, B, and C proteins and the HMW1–5 proteins, are also involved in adherence (Krause, 1997).

After adhering, *M. pneumoniae* causes ciliostasis and desquamation of the ciliated epithelial cells, and loss of ciliary function correlates well with clinical symptoms. The molecular mechanisms that mediate epithelial damage have not been determined, but it has been suggested that hydrogen peroxide and superoxide, which are by-products of *M. pneumoniae* metabolism, may cause cell damage.

Immunity

After a *M. pneumoniae* infection a strong humoral immune response develops. This involves the development of antibodies to several *M. pneumoniae* proteins, of which the P1 protein is the major immunogen, but several other proteins have been described as immunogenic. Vu *et al.* (1987) studied the principal protein antigens of isolates of *M. pneumoniae* by immunoblotting. They analysed the reactivity of human serum samples to *M. pneumoniae* strains collected over a 10-year period and found that even though there was some variation, two major immunogens were dominant. These were the 170-kDa (P1) protein and a 90-kDa protein. There has long been speculation about which gene encodes the 90-kDa protein, but it has been shown that the 116-kDa surface protein encoded by the gene G07-orf1030 (Duffy *et al.*, 1997) is a potential candidate as the antigen for a species-specific serodiagnostic test (Duffy *et al.*, 1999).

The cellular immune response associated with *M. pneumoniae* infections is not well described. It is, however, known that *M. pneumoniae* is capable of inducing transient anergy in patients during the acute phase of infection, although the mechanism has not been clarified (Biberfeld, 1985). It may be speculated that stimulation of IL-10 by *M. pneumoniae* down-regulates and suppresses T-cell proliferation and inhibits production of pro-inflamatory cytokines by mononuclear phagocytes, as in the case of *Mycoplasma arthritidis* infections (Rink *et al.*, 1996).

The clinical picture of mycoplasma infections is more suggestive of damage due to the host immune response and to the inflammatory response rather than to direct toxic effects by mycoplasma components. In this way the ability of mycoplasmas to immunomodulate the host immune response enables them to establish persistent infections and to cause autoimmunity. There is, however, only limited information on the secretion of cytokines and chemokines *in vivo* during autoimmune manifestations and about the molecular mechanisms that cause post-infection sequelae.

Diagnosis

M. pneumoniae infections can be diagnosed by detection of the organisms or the detection of antibodies specific to the organism. Organisms can be detected by cultivation followed by growth inhibition analysis or by colony staining with specific antibodies. Since *M. pneumoniae* is the only *Mycoplasma* that causes respiratory infections in humans, diagnosis by culture is frequently done in the specific *Mycoplasma* medium SP4, in which growth of *M. pneumoniae* is observed by a colour shift of the medium as glucose is fermented (Tully *et al.*, 1979). Identification by cultural methods may take weeks and therefore does not meet the requirements of routine diagnosis.

Detection by DNA amplification methods is useful, but commercial tests are not yet available. Bernet *et al.* (1989) and Jensen *et al.* (1989) first showed the potential of PCR for *M. pneumoniae* diagnosis. Various genes have been used as targets for DNA amplification, including the 16S rRNA gene (Jensen *et al.*, 1989), the *tuf* gene (Luneberg *et al.*, 1993) and the P1 gene (Ursi *et al.*, 1994). In these early studies PCR was compared with cultivation and showed high specificity and good sensitivity. It was pointed out, however, that internal controls for sample inhibitors and for false-positive results were important because of carry-over of target DNA. DNA amplification methods are superior to conventional diagnostic methods, such as cultivation and serological tests, because they are rapid and specific and if

inhibitory substances are not present the sensitivity is also high.

Abele-Horn *et al.* (1998) carried out an extensive study to determine how useful DNA amplification methods are in the diagnosis of pulmonary diseases due to *M. pneumoniae*. Samples were first incubated overnight and then subjected to PCR. Nested PCR was found to be 1000-fold more sensitive than a one-step PCR, and inhibitory substances could often be diluted out. These workers concluded that both one-step PCR and nested PCR are reliable methods for detection of *M. pneumoniae* in respiratory tract samples.

Real-time PCR of the P1 gene was used in the study by Hardegger *et al.* (2000) and compared with a semi-nested PCR of 16S rDNA. They found 97.4% overall agreement and concluded that real-time PCR has the advantage that it is fast and can handle a larger number of samples than conventional PCR. This method will, therefore, probably be used more frequently in the future.

Serodiagnosis of *M. pneumoniae* infection by complement fixation for antibodies against the lipid antigen requires an increase in antibody titre which is not reached until 7–10 days after the onset of clinical symptoms (Kenny *et al.*, 1990). A commercially available test in the form of the IgM enzyme immunoassay (Immuno Well, Gen-Bio, San Diego, CA, USA) (Cimolai and Cheong, 1996) may be valuable, but the accuracy of the test is affected by the prevalence of true illness in the study population.

Antimicrobial Therapy

Since *Mycoplasma* are wall-less bacteria, they are not sensitive to cell wall-specific antibiotics such as the penicillins and cephalosporins. Erythromycin has been the drug of choice for infections due to *M. pneumoniae*, however, even though *M. pneumoniae* shows good sensitivity *in vitro*, clinical symptoms are often unaffected by such a treatment. This suggests either that the concentration of antibiotic is too low to affect *M. pneumoniae* at their site of localisation *in vivo* or that symptoms may be caused by damage due to the host immune response. Other antibiotics, such as fluoroquinolones and macrolides have been used in the treatment of *Mycoplasma* infections, but as is the case with other micro-organisms, antibiotic resistance occurs.

Numerous studies have been carried out to analyse antibiotic resistance in *Mycoplasma* (for a review see Roberts, 1992). The molecular basis for the development of fluoroquinolone-resistance in clinical isolates of *M. hominis* has been studied by Bebear *et al.* (1999), who found that all of five isolates had mutations in the genes encoding DNA gyrase (*GyrA*) and topoisomerase IV (*ParC* and *ParE*). Such analysis may help to elucidate how antibiotic resistance develops.

Epidemiology

Epidemics of *M. pneumoniae* infections occur every 4–7 years (Foy *et al.*, 1979; Noah and Urquhart, 1980; Ponka, 1980; Ghosh and Clements, 1992; Sasaki *et al.*, 1996; Lind *et al.*, 1997; Hauksdottir *et al.*, 1998; Rastawicki *et al.*, 1998). The highest infection rate is among young children and their parents. Infections are spread by droplets with a case-to-case interval of 3 weeks (Foy *et al.*, 1979). Naturally acquired immunity after a *M. pneumoniae* infection lasts for between 2 and 10 years (Foy *et al.*, 1983), and this may explain the periodicity of epidemics.

Clinical isolates can be classified into two groups on the basis of Southern blotting analysis with probes containing fragments of the P1 gene (Hirschberg *et al.*, 1989; Su *et al.*, 1990b). DNA sequence analysis of the P1 gene revealed sequence variation between the two types (Su *et al.*, 1990a) and Ursi *et al.* (1994) took advantage this and typed 24 *M. pneumoniae* isolates by amplifying part of the P1 gene by PCR and cleaving the amplified fragment with restriction endonucleases. The isolates could be divided into the two groups (group I and II); 19 belonged to group I and 5 to group II without variation of the DNA fingerprints. This confirmed the conserved nature of the P1 gene within the two groups.

Even though a new variant of the P1 gene has recently been found (Kenri *et al.*, 1999) and sub-types of the two groups have been described (Wendelein Dorigo-Zestma *et al.*, 2000), there is overall agreement on the existence of two distinct groups of *M. pneumoniae*. The presence of either group I or group II among clinical isolates seems to vary over time. In the most extensive study (Cousin-Allery *et al.*, 2000) 153 *M. pneumoniae* strains from France, Denmark, Belgium and Germany were typed. A shift from group I to group II seemed to have occurred in 1987–88, and almost all strains from Denmark and France during this period belonged to group II. This was in contrast to strains isolated in 1962–86 or in 1991–93, which almost all belonged to group I.

Patients infected with group I *M. pneumoniae* develop adherence-inhibiting antibodies more frequently than patients infected with group II organisms. This may affect the pathogenesis of *M. pneumoniae* diseases and subsequent re-infections. In the period between 1987 and 1989, only group II strains were isolated from patients in Germany (Razin and Jacobs, 1992; Jacobs *et al.*, 1996). The first group I isolate was found in 1990. During the following 4 years the rate of group II incidents decreased (Jacobs *et al.*, 1996). This switch from group II to group I may be explained by the immune status of the population (Ursi *et al.*, 1994). In addition, the adherence-inhibiting antibodies against group II *M. pneumoniae* decreased rapidly after the isolation of group I *M. pneumoniae* (Jacobs *et al.*, 1996).

The change over time between the two types raises the question whether the population may be more susceptible to infection when a change in type occurs. Now that the tools for typing have been developed, it should be possible to determine whether this is the case and also whether subtypes of the two groups are more virulent. Typing can be done directly by PCR amplification, without prior cultivation. This makes it possible to study a large number of samples from clusters of cases and epidemics. Analysis by PCR of a specific DNA fragment only reveals variation within the amplified fragment, however; other variations will not be detected.

However, analysis by other methods, such as sequence analysis of other genes and PCR of long DNA fragments, revealed that existence of the two groups can be confirmed independently of the methods used (Wendelien Dorigo-Zetsma *et al.*, 2000) and indicates the stability of the *M. pneumoniae* genome. This stability is in sharp contrast to that observed in strains of *Mycoplasma hominis*, a commensal of the human genital tract that is only occasionally pathogenic. In the case of *M. hominis*, genotyping by pulsed-field gel electrophoresis of restriction endonuclease-cleaved genomic DNA revealed that different individuals harbour unique genotypes, while isolates taken from the same person were identical or near identical (Jensen *et al.*, 1998). Epidemiological studies of *M. genitalium* have not been carried out but it would be interesting to see whether genetic conservation is another feature that this micro-organism has in common with the closely related *M. pneumoniae*.

Acknowledgements

This study was supported financially by grants from the Danish Health Research Council, the Danish Veterinary and Agricultural Research Council (grants 12-0850-1, 12-0150-1, 20-35031), Aarhus University Research Foundation and the Novo Nordisk Foundation.

References

Abele-Horn M, Busch U, Nitschko H *et al.* (1998) Molecular approaches to diagnosis of pulmonary diseases due to *Mycoplasma pneumoniae*. *J. Clin. Microbiol.* 36: 548–551.

Bak AL, Black FT, Christiansen C, Freundt EA (1969) Genome size of mycoplasma DNA. *Nature* 244: 1209–1210.

Baseman JB, Cole RM, Krause DC, Leith DK (1982) Molecular basis for cytadsorption of *Mycoplasma pneumoniae*. *J. Bacteriol.* 151: 1514–1522.

Baseman JB, Dallo SF, Tully JG, Rose DL (1988) Isolation and characterization of *Mycoplasma genitalium* strains from the human respiratory tract. *J. Clin. Microbiol.* 26: 2266–2269.

Bebear CM, Renaudin J, Charron A *et al.* (1999) Mutations in the gyrA, parC, and parE genes associated with fluoroquinolone resistance in clinical isolates of *Mycoplasma hominis. Antimicrob. Agents Chemother.* 43: 954–956.

Bernet C, Garret M, de Barbeyrac B, Bebear C, Bonnet J (1989) Detection of *Mycoplasma pneumoniae* by using the polymerase chain reaction. *J. Clin. Microbiol.* 27: 2492–2496.

Biberfeld G (1985) Infection sequelae and autoimmune reactions in *Mycoplasma pneumoniae* infection. In: Razin SS, Barile MF (eds) *The Mycoplasmas, Mycoplasma Pathogenicity*. Orlando: Academic Press, Vol. IV: pp. 293–311.

Boatman ES (1979) Morphology and ultrastructure of the Mycoplasmatales. In: Barile MF, Razin S (eds) *The Mycoplasmas*, New York: Academic Press, Vol. 1. pp. 63–102.

Boesen T, Emmersen J, Jensen LT *et al.* (1998) The *Mycoplasma hominis vaa* gene displays a mosaic gene structure. *Mol. Microbiol.* 29: 97–110.

Bove JM (1993) Molecular features of mollicutes. *Clin. Infect. Dis.* 17 (Suppl 1): S10–S31.

Bredt W (1973) Motility of mycoplasmas. *Ann. NY Acad. Sci.* 225: 246–250.

Bredt W (1979) Motility. In: Barile MF, Razin S (eds) *The Mycoplasmas*, Vol. 1. New York: Academic Press, pp. 141–155.

Bredt W, Feldner J, Kahane I, Razin S (1983) Mycoplasma attachment to solid surfaces: a review. *Yale J. Biol. Med.* 56: 653–656.

Carle P, Laigret F, Tully JG, Bove JM (1995) Heterogeneity of genome sizes within the genus *Spiroplasma. Int. J. Syst. Bacteriol.* 45: 178–181.

Cimolai N, Cheong AC (1996) An assessment of a new diagnostic indirect enzyme immunoassay for the detection of anti-*Mycoplasma pneumoniae* IgM. *Am. J. Clin. Pathol.* 105: 205–209.

Clyde WA Jr (1964) Mycoplasma species identification based upon growth inhibition by specific antisera. *J. Immunol.* 92: 958–965.

Clyde WA Jr (1983a) Growth inhibition tests. *Methods Mycoplasmol.* 1: 405–410.

Clyde WA Jr (1983b) *Mycoplasma pneumoniae* respiratory disease symposium: summation and significance. *Yale J. Biol. Med.* 56: 523–527.

Collier AM (1983) Attachment by mycoplasmas and its role in disease. *Rev. Infect. Dis.* 5(Suppl 4): S685–S691.

Cousin-Allery A, Charron A, de Barbeyrac B *et al.* (2000) Molecular typing of *Mycoplasma pneumoniae* strains by PCR-based methods and pulsed-field gel electrophoresis. Application to French and Danish isolates. *Epidemiol. Infect.* 124: 103–111.

Dallo SF (1996) Biofunctional domains of the *Mycoplasma pneumoniae* P30 adhesin. *Infect. Immun.* 64: 2595–2601.

Dallo SF, Su CJ, Horton JR, Baseman JB (1988) Identification of P1 gene domain containing epitope(s) mediating *Mycoplasma pneumoniae* cytoadherence. *J. Exp. Med.* 167: 718–723.

Dallo SF, Chavoya A, Baseman JB (1990a) Characterization of the gene for a 30-kilodalton adhesion-related protein of *Mycoplasma pneumoniae. Infect. Immun.* 58: 4163–4165.

Dallo SF, Horton JR, Su CJ, Baseman JB (1990b) Restriction fragment length polymorphism in the cytadhesin P1 gene of human clinical isolates of *Mycoplasma pneumoniae. Infect. Immun.* 58: 2017–2020.

Dirksen LB, Krebes KA, Krause DC (1994) Phosphorylation of cytadherence-accessory proteins in *Mycoplasma pneumoniae. J. Bacteriol.* 176: 7499–7505.

Duffy MF, Walker ID, Browning GF (1997) The immunoreactive 116 kDa surface protein of *Mycoplasma pneumoniae* is encoded in an operon. *Microbiology* 143: 3391–3402.

Duffy MF, Whithear KG, Noormohammadi AH *et al.* (1999) Indirect enzyme-linked immunosorbent assay for detection of immunoglobulin G reactive with a recombinant protein expressed from the gene encoding the 116-kilodalton protein of *Mycoplasma pneumoniae. J. Clin. Microbiol.* 37: 1024–1029.

Dybvig K, Voelker LL (1996) Molecular biology of mycoplasmas. *Annu. Rev. Microbiol.* 50: 25–57.

Fisseha M, Gohlmann HW, Herrmann R, Krause DC (1999) Identification and complementation of frameshift mutations associated with loss of cytadherence in *Mycoplasma pneumoniae. J. Bacteriol.* 181: 4404–4410.

Foy HM, Kenny GE, Cooney MK (1979) Long term epidemiology of infections with *Mycoplasma pneumoniae. J. Infect. Dis.* 139: 681–687.

Foy HM, Kenny GE, Cooney MK, Allan ID, van Belle G (1983) Naturally aquired immunity to *Mycoplasma pneumoniae. J. Infect. Dis.* 147: 967–973.

Fraser CM, Gocayne JD, White O *et al.* (1995) The minimal gene complement of *Mycoplasma genitalium. Science* 270: 397–403.

Gallagher JR (1941) Acute pneumonitis: a report of 87 cases among adolecents. *Yale J. Biol. Med.* 13: 663–678.

Gerstenecker B, Jacobs E (1990) Topological mapping of the P1-adhesin of *Mycoplasma pneumoniae* with adherence-inhibiting monoclonal antibodies. *J. Gen. Microbiol.* 136: 471–476.

Ghosh K, Clements GB (1992) Surveillance of *Mycoplasma pneumoniae* infections in Scotland 1986-1991. *J. Infect.* 25: 221–227.

Hahn TW, Willby MJ, Krause DC (1998) HMW1 is required for cytadhesin P1 trafficking to the attachment organelle in *Mycoplasma pneumoniae*. *J. Bacteriol.* 180: 1270–1276.

Hardegger D, Nadal D, Bossart W, Altwegg M, Dutly F (2000) Rapid detection of *Mycoplasma pneumoniae* in clinical samples by real-time PCR. *J. Microbiol. Methods* 41: 45–51.

Hauksdottir GS, Jonsson T, Sigurdardottir V, Low A (1998) Seroepidemiology of *Mycoplasma pneumoniae* infections in Iceland. *J. Infect. Dis.* 30: 177–180.

Hedreyda CT, Krause DC (1995) Identification of a possible cytadherence regulatory locus in *Mycoplasma pneumoniae*. *Infect. Immun.* 63: 3479–3483.

Himmelreich R, Hilbert H, Plagens H, Pirkl E, Li BC, Herrmann R (1996) Complete sequence analysis of the genome of the bacterium *Mycoplasma pneumoniae*. *Nucleic Acids Res.* 24: 4420–4449.

Himmelreich R, Plagens H, Hilbert H, Reiner B, Herrmann R (1997) Comparative analysis of the genomes of the bacteria *Mycoplasma pneumoniae* and *Mycoplasma genitalium*. *Nucleic Acids Res.* 25: 701–712.

Hirschberg L, Holme T, Hyden N (1989) Demonstration of membrane association and surface location of *Mycoplasma pneumoniae* antigens using monoclonal antibodies. *J. Gen. Microbiol.* 135: 613–621.

Hu PC, Huang CH, Collier AM, Clyde WA Jr (1983) Demonstration of antibodies to *Mycoplasma pneumoniae* attachment protein in human sera and respiratory secretions. *Infect. Immun.* 41: 437–439.

Inamine JM, Denny TP, Loechel S *et al.* (1988) Nucleotide sequence of the P1 attachment-protein gene of *Mycoplasma pneumoniae*. *Gene* 64: 217–229.

Isenberg HD (1988) Pathogenicity and virulence: another view. *Clin. Microbiol. Rev.* 1: 40–53.

Jacobs E, Bennewitz A, Bredt W (1986) Reaction pattern of human anti-*Mycoplasma pneumoniae* antibodies in enzyme-linked immunosorbent assays and immunoblotting. *J. Clin. Microbiol.* 23: 517–522.

Jacobs E, Pilatschek A, Gerstenecker B, Oberle K, Bredt W (1990) Immunodominant epitopes of the adhesin of *Mycoplasma pneumoniae*. *J. Clin. Microbiol.* 28: 1194–1197.

Jacobs E, Vonski M, Oberle K, Opitz O, Pietsch K (1996) Are outbreaks and sporadic respiratory infections by *Mycoplasma pneumoniae* due to two distinct subtypes? *Eur. J. Clin. Microbiol. Infect. Dis.* 15: 38–44.

Jensen JS, Sondergard-Andersen J, Uldum SA, Lind K (1989) Detection of *Mycoplasma pneumoniae* in simulated clinical samples by polymerase chain reaction. Brief report. *APMIS* 97: 1046–1048.

Jensen LT, Thorsen P, Møller B, Birkelund S, Christiansen G (1998) Antigenic and genomic homogeneity of successive *Mycoplasma hominis* isolates. *J. Med. Microbiol.* 47: 659–666.

Kenny GE, Kaiser GG, Cooney MK *et al.* (1990) Diagnosis of *Mycoplasma pneumoniae* pneumonia: sensitivities and specificities of serology with lipid antigen and isolation of the organism on soy peptone medium for identification of infections. *J. Clin. Microbiol.* 28: 2087–2093.

Kenri T, Taniguchi R, Sasaki Y *et al.* (1999) Identification of a new variable sequence in the P1 cytadhesin gene of *Mycoplasma pneumoniae*: evidence for the generation of antigenic variation by DNA recombination between repetitive sequences. *Infect. Immun.* 67: 4557–4562.

Kirchoff H, Beyene P, Fischer M *et al.* (1987) *Mycoplasma mobile* sp. nov., a new species from fish. *Int. J. Syst. Bacteriol.* 37: 192–197.

Kong F, James G, Ma Z, Gordon S, Bin W, Gilbert GL (1999) Phylogenetic analysis of *Ureaplasma urealyticum* – support for the establishment of a new species, *Ureaplasma parvum*. *Int. J. Syst. Bacteriol.* 49: 1879–1889

Kong F, Ma Z, James G, Gordon S, Gilbert GL (2000) Species identification and subtyping of *Ureaplasma parvum* and *Ureaplasma urealyticum* using PCR-based assays. *J. Clin. Microbiol.* 38: 1175–1179

Krause DC (1996) *Mycoplasma pneumoniae* cytadherence: unravelling the tie that binds. *Mol. Microbiol.* 20: 247–253.

Krause DC (1997) Transposon mutagenesis reinforces the correlation between *Mycoplasma pneumoniae* cytoskeletal protein HMW2 and cytadherence. *J. Bacteriol.* 8: 2668–2677.

Krause DC (1998) *Mycoplasma pneumoniae* cytadherence: organization and assembly of the attachment organelle. *Trends Microbiol.* 6: 15–18.

Krause DC, Leith DK, Baseman JB (1982) Identification of *Mycoplasma pneumoniae* proteins associated with haemadsorption and virulence. *Infect. Immun.* 35: 809–817.

Krause DC, Leith DK, Baseman JB (1983) Reacquisition of specific proteins confers virulence in *Mycoplasma pneumoniae*. *Infect. Immun.* 39: 830–836.

Krebes CD, Dirksen LB, Krause DC (1995) Phosphorylation of *Mycoplasma pneumoniae* cytadherence accessory proteins in cell extract. *J. Bacteriol.* 177: 4571–4574.

Ladefoged SA, Christiansen G (1992) Physical and genetic mapping of the genomes of five *Mycoplasma hominis* strains by pulsed-field gel electrophoresis. *J. Bacteriol.* 174: 2199–2207.

Ladefoged SA, Birkelund S, Hauge S, Brock B, Jensen LT, Christiansen G (1995) A 135 kDa surface antigen of *Mycoplasma hominis* PG21 contains multiple directly repeated sequences. *Infect. Immun.* 63: 212–223.

Ladefoged S, Jensen LT, Brock B, Birkelund S, Christiansen G (1996) Analysis of 0.5-kilobase-pair repeats in the *Mycoplasma hominis* lmp gene system and identification of gene products. *J. Bacteriol.* 178: 2775-2784.

Layh-Schmitt G, Herrmann R (1994) Spatial arrangement of gene products of the P1 operon in the membrane of *Mycoplasma pneumoniae*. *Infect. Immun.* 62: 974–979.

Layh-Schmitt G, Podtelejnikov A, Mann M (2000) Proteins complexed to the P1 adhesin of *Mycoplasma pneumoniae*. *Microbiology* 146: 741–747.

Lind K, Lindhardt BO, Schutten HJ, Blom J, Christiansen C (1984) Serological cross-reactions between *Mycoplasma genitalium* and *Mycoplasma pneumoniae*. *J. Clin. Microbiol.* 20: 1036–1043.

Lind K, Benzon MW, Jensen JS, Clyde WA Jr (1997) A seroepidemiological study of *Mycoplasma pneumoniae* infections in Denmark over a 50-year period 1946–1995. *Eur. J. Epidemiol.* 13: 581–586.

Luneberg E, Jensen JS, Frosch M (1993) Detection of *Mycoplasma pneumoniae* by polymerase chain reaction and nonradioactive hybridization in microtiter plates. *J. Clin. Microbiol.* 31: 1088–1094.

Miyata M, Yamamoto H, Shimizu T, Uenoyama A, Citti C, Rosengarten R (2000) Gliding mutants of *Mycoplasma mobile*: relationships between motility and cell morphology, cell adhesion and microcolony formation. *Microbiology* 146: 1311–1320.

Mygind T, Birkelund S, Christiansen G (1998) DNA sequencing reveals limited heterogeneity in the 16S rRNA gene from the rrnB operon among five *Mycoplasma hominis* isolates. *Int. J. Syst. Bacteriol.* 48: 1067–1071.

Mygind T, Birkelund S, Christiansen G (2000a) Characterization of the variability of a 75-kDa membrane protein in *Mycoplasma hominis*. *FEMS Microbiol. Lett.* 190: 167–176.

Mygind T, Zeuthen Sogaard I, Melkova R, Boesen T, Birkelund S, Christiansen G (2000b) Cloning, sequencing and variability analysis of the gap gene from *Mycoplasma hominis*. *FEMS Microbiol. Lett.* 183: 15–21.

Noah ND, Urquhart AM (1980) Epidemiology of *Mycoplasma pneumoniae* infection in the British Isles. *J. Infect.* 2: 191–194.

Neill AM, Martin IR, Weir R *et al.* (1996) Community-acquired pneumonia: aetiology and usefulness of severity criteria on admission. *Thorax* 51: 1010–1016.

Nyvold C, Birkelund S, Christiansen G (1997) The *Mycoplasma hominis* P120 membrane protein contains a 216 amino acid hypervariable domain that is recognized by the human humoral immune response. *Microbiology* 143: 675–688.

Ogle KF, Lee KK, Krause DC (1992) Nucleotide sequence analysis reveals novel features of the phase-variable cytadherence accessory protein HMW3 of *Mycoplasma pneumoniae*. *Infect. Immun.* 60: 1633–1641.

Opitz O, Jacobs E (1992) Adherence epitopes of *Mycoplasma genitalium* adhesin. *J. Gen. Microbiol.* 138: 1785–1790.

Peterson SN, Bailey CC, Jensen JS *et al.* (1995) Characterization of repetitive DNA in the *Mycoplasma genitalium* genome: possible role in the generation of antigenic variation. *Proc. Natl Acad. Sci. USA* 92: 11 829–11 833.

Ponka A (1980) Occurrence of serologically verified *Mycoplasma pneumoniae* infections in Finland and Scandinavia in 1970–77. *Scand. J. Infect. Dis.* 12: 27–31.

Rastawicki W, Kaluzewski S, Jagielski M (1998) Occurrence of serologically verified *Mycoplasma pneumoniae* in Poland 1970–1995. *Eur. J. Epidemiol.* 14: 37–40.

Razin S (1978) The mycoplasmas. *Microbiol. Rev.* 42: 414–470.

Razin S (1995) Molecular properties of mollicutes: a synopsis. In: *Molecular and Diagnostic Procedures in Mycoplasmology.* New York: Academic Press, Vol. 1. pp. 1–24.

Razin S, Jacobs E (1992) Mycoplasma adhesion. *J. Gen. Microbiol.* 138: 407–422.

Razin S, Tully JG (1970) Cholesterol requirements of mycoplasmas. *J. Bacteriol.* 102: 306–310.

Razin S, Yogev D, Naot Y (1998) Molecular biology and pathogenicity of mycoplasmas. *Microbiol. Mol. Biol. Rev.* 62: 1094–1156.

Reddy SP, Rasmussen WG, Baseman JB (1995) Molecular cloning and characterization of an adherence-related operon of *Mycoplasma genitalium*. *J. Bacteriol.* 177: 5943–5951.

Rink L, Nicklas Luhm WJ, Kruse R, Kirchner H (1996) Induction of a proinflammatory cytokine network by *Mycoplasma arthritidis*-derived superantigen (MAS). *J. Interferon Cytokine Res.* 16: 861–868.

Roberts MC (1992) Antibiotic resistance. In: Maniloff J, McElhany RRN, Finch LR, Baseman JB (eds) *Mycoplasmas: Molecular Biology and Pathogenesis.* Washington, DC: American Society for Microbiology, pp. 513–523.

Ruland K, Wenzel R, Herrmann R (1990) Analysis of three different repeated DNA elements present in the P1 operon of *Mycoplasma pneumoniae*: size, number and distribution on the genome. *Nucleic Acids Res.* 18: 6311–6317.

Ruland K, Himmelreich R, Herrmann R (1994) Sequence divergence in the ORF6 gene of *Mycoplasma pneumonia*. *J. Bacteriol.* 176: 5202–5209.

Sasaki T, Kenri T, Okazaki N *et al.* (1996) Epidemiological study of *Mycoplasma pneumoniae* infections in Japan based on PC restriction fragment length polymorphism of the P1 cytadhesin gene. *J. Clin. Microbiol.* 34: 447–449.

Shepard MC, Lunceford CD, Ford DK *et al.* (1974) *Ureaplasma urealyticum* gen. nov., sp. nov.: proposed nomenclature for the human T (T-strain) mycoplasmas. *Int. J. Syst. Bacteriol.* 24: 160–171.

Sperker B, Hu P, Herrmann R (1991) Identification of gene products of the P1 operon of *Mycoplasma pneumoniae*. *Mol. Microbiol.* 5: 299–306.

Stevens MK, Krause DC (1991) Localization of the *Mycoplasma pneumoniae* cytadherence proteins HMW1 and HMW4 in the cytoskeleton-like shell. *J. Bacteriol.* 3: 1041–1050.

Subcommittee on the Taxonomy of Mollicutes (1995) Mollicutes (International Committee on Systematic Bacteriology) 1995 Revised minimal standards for description of new species of the class mollicutes (division Tenericutes). *Int. J. Syst. Bacteriol.* 45: 605–612.

Su CJ, Chavoya A, Baseman JB (1988) Regions of *Mycoplasma pneumoniae* cytadhesin P1 structural gene exist as multiple copies. *Infect. Immun.* 56: 3157–3161.

Su CJ, Chavoya A, Dallo SF, Baseman JB (1990a) Sequence divergency of the cytadhesin gene of *Mycoplasma pneumoniae*. *Infect. Immun.* 58: 2669–2674.

Su CJ, Dallo SF, Baseman JB (1990b) Molecular distinctions among clinical isolates of *Mycoplasma pneumoniae*. *J. Clin. Microbiol.* 28: 1538–1540.

Tham TN, Ferris S, Bahraoui E, Canarelli S, Montagnier L, Blanchard A (1994) Molecular characterization of the P1-like adhesin gene from *Mycoplasma pirum*. *J. Bacteriol.* 176: 781–788.

Tully JG (1993) Current status of the mollicute flora of humans. *Clin. Infect. Dis.* 17(Suppl 1): S2–S9.

Tully JG, Rose LD, Whitcomb RF, Wenzel RP (1979) Enhanced isolation of *Mycoplasma pneumoniae* from throat washings with a newly modified culture medium. *J. Infect. Dis.* 139: 478–482.

Tully JG, Taylor-Robinson D, Rose DL, Cole RM, Bove JM (1983) *Mycoplasma genitalium*, a new species from the human urogenital tract. *Int. J. Syst. Bacteriol.* 33: 387–396.

Ursi D, Ieven M, van Bever H, Quint W, Niesters HG, Goossens H (1994) Typing of *Mycoplasma pneumoniae* by PCR-mediated DNA fingerprinting. *J. Clin. Microbiol.* 32: 2873–2875.

Vu AC, Foy HM, Cartwright FD, Kenny GE (1987) The principal protein antigens of isolates of *Mycoplasma pneumoniae* as measured by levels of immunoglobulin G in human serum are stable in strains collected over a 10-year period. *Infect. Immun.* 55: 1830–1836.

Weisburg WG, Tully JG, Rose DL *et al.* (1989) A phylogenetic analysis of mycoplasmas: basis for their classification. *J. Bacteriol.* 171: 6455–6467.

Weisburg WG, Barns SM, Pelletier DA, Lane DA (1991) 16S ribosomal DNA amplification for phylogenetic study. *J. Bacteriol.* 173: 697–703.

Wendelien Dorigo-Zetsma J, Dankert J, Zaat SAJ (2000) Genotyping of *Mycoplasma pneumoniae* clinical isolates reveals eight P1 subtypes within two genomic groups. *J. Clin. Microbiol.* 38: 965–970.

Wasinger VC, Pollack JD, Humphery-Smith I (2000) The proteome of *Mycoplasma genitalium*. *Eur. J. Biochem.* 267: 1571–1582.

Wilson MH, Collier AM (1976) Ultrastructural study of *Mycoplasma pneumoniae* in organ culture. *J. Bacteriol.* 125: 332–339.

Yamao F, Muto A, Kawauchi Y *et al.* (1985) UGA is read as tryptophan in *Mycoplasma capricolum*. *Proc. Natl Acad. Sci. USA* 82: 2306–2309.

80

Coxiella burnetii

James E. Samuel[1] and Robert A. Heinzen[2]

[1]*Texas A and M University, Texas, USA*
[2]*University of Wyoming, Laramie, Wyoming, USA*

The aetiological agent of Q fever, *Coxiella burnetii*, is a bacterial obligate intracellular parasite that replicates within the phagolysosome of its eukaryotic host. Edward Derrick, an Australian physician, coined the term 'query (Q) fever' to describe an outbreak of a febrile illness among abattoir workers in 1935 (Derrick, 1937). A few years later, the agent was identified by independent laboratories working in the United States and in Australia. After initial attempts by Derrick to isolate the agent had been unsuccessful, he sent clinical samples in the form of infected guinea-pig liver to MacFarlane Burnet, who reproduced symptoms of the disease in laboratory animals. Impression smears of spleen harvested from infected mice revealed large intracellular vacuoles containing small rod-shaped organisms that resembled typical 'rickettsiae'.

In 1935, a febrile illness was induced in guinea-pigs during a study of tick transmission of Rocky Mountain spotted fever in Hamilton, Montana. The illness did not mimic spotted fever in presentation, and the causative agent was referred to simply as the 'Nine Mile' agent. In 1936, Davis and Cox demonstrated the filterability of the Nine Mile agent (Davis and Cox, 1938). Attempts to cultivate the agent on axenic media failed, but when a laboratory worker investigating the Nine Mile agent developed a febrile illness, a connection was made between the Q fever agent and the Nine Mile agent. Blood from one of the patients was used to inoculate a guinea-pig that subsequently developed a febrile illness. A spleen sample from this animal was infectious for other guinea-pigs and on staining rickettsia-like organisms were observed in spleen samples. Burnet sent a spleen sample from a mouse infected with the Q fever agent to the Montana laboratory. There Davis and Cox demonstrated that convalescent guinea-pigs previously infected with Burnet's spleen samples were immune to challenge by the Nine Mile agent, thus demonstrating antigenic cross-reactivity between the two agents. In 1938, the causative agent was assigned a new genus and species name, *Coxiella burnetti*, in honour of Cox and Burnet. A milestone event in the study of *C. burnetii* and other rickettsia occurred with the successful propagation by Cox of *C. burnetii* in embryonated hen's eggs (Cox, 1941).

Classification

Since *C. burnetii* was unculturable on artificial media and carried by ticks, the organism was taxonomically placed within the tribe *Rickettsiaea*, whose members display similar characteristics. Residing in the $\alpha 1$ subgroup of *Proteobacteria*, the tribe also included the genera *Rickettsia* and *Rochalimae*, though the latter have now been re-classified as the genera *Bartonella* and re-grouped with the $\alpha 2$ sub-group of the

Proteobacteria. Recent phylogenetic analysis, primarily with 16S rRNA sequences, has resulted in the re-classification of *C. burnetii* to the γ subgroup of *Proteobacteria*, where it is most closely related to *Legionella pneumophila* and *Wolbacia persica* (Weisburg *et al.*, 1989). Indeed, while *Rickettsia* species replicate in the host cytoplasm, *C. burnetii* displays both phenotypic and genetic similarities to *L. pneumophila*, including intracellular growth in a membrane-bound vacuole and sequence similarities for a variety of genes (Mo *et al.*, 1995).

Identification

C. burnetii is an obligate intracellular bacterium that displays a prototypic Gram-negative cell wall structure when observed by electron microscopy. It does not, however, consistently stain with the Gram stain, and organisms are typically stained by the Gimenez staining method (Gimenez, 1964). *C. burnetii* is a small (0.3–1.0 μm), highly pleiomorphic coccobacillus. The pleiomorphism results from the production of distinct *C. burnetii* morphological forms that are part of an incompletely defined developmental cycle (see below).

Isolates of *C. burnetii* can been divided into six genomic groups (I to VI) based on restriction-fragment length polymorphisms of DNA (Hendrix *et al.*, 1991; Mallavia, 1991). The genome size of *C. burnetii* isolates ranges from 1600 to 2400 kb (Willems *et al.*, 1996, 1998). Approximately 2% of the coding capacity of *C. burnetii* is carried on a moderately sized plasmid that is maintained at 1–3 copies per cell (Samuel *et al.*, 1983). Four plasmid types ranging from 32.6 to 51 kb have been described as associated with specific genomic groups (Mallavia, 1991). Some isolates do not harbour an autonomously replicating plasmid, but instead have approximately 18 kb of plasmid sequences integrated into the chromosome (Savinelli and Mallavia, 1990; Willems *et al.*, 1997). Interestingly, these isolates have all been obtained from patients suffering from Q fever endocarditis. The absolute maintenance of plasmid sequences in all *C. burnetii* isolates examined to date suggests that they play a critical role in some aspect of *Coxiella* biology and/or virulence.

C. burnetii undergoes a lipopolysaccharide (LPS) phase variation similar to that observed in the *Enterobacteriaceae*. Transition from a smooth (full-length) phase I to a rough (truncated) phase II LPS structure occurs on repeated passage of the organism in an immuno-incompetent host such as embryonated eggs or tissue culture. LPS is the only confirmed virulence factor of *C. burnetii*. Plaque-purified phase II organisms are avirulent for guinea-pigs – a fever animal model of *C. burnetii* infection – while phase I *C. burnetii* causes disease and is always associated with naturally infected mammals and ticks (Hackstadt, 1990). A strain with a semi-rough-type LPS has intermediate virulence for guinea-pigs as shown by fever production but lack of persistence in the spleen (Moos and Hackstadt, 1987). The genetic lesion(s) that account for phase variation has not been precisely determined, but probably involves the loss of chromosomal regions that carry LPS biosynthesis genes (O'Rourke *et al.*, 1985; Vodkin *et al.*, 1986). Phase I and phase II organisms are indistinguishable by electron microscopy, and their intracellular growth characteristics are similar.

Structure

C. burnetii has an impressive ability to survive in the extracellular environment, and to resist physical and chemical disruption, that greatly surpasses that of vegetative bacterial cells. Viable organisms can be recovered after heat treatment at 63°C for 30 minutes, exposure to a 10% salt solution for 180 days at room temperature, exposure to 0.5% formalin for 24 hours, or sonication in distilled water for 30 minutes (Babudieri, 1959; Heinzen, 1997). The remarkable resistance characteristics of *C. burnetii* are attributed to a small, resistant cell form that is part of a developmental cycle (McCaul and Williams, 1981; Heinzen, 1997).

C. burnetii exists as a continuum of morphologically distinct forms termed large-cell variants (LCV) and small-cell variants (SCV). LCV and SCV can be separated on the basis of their different buoyant densities, and preparations highly enriched for SCV can also be obtained by procedures that exploit the sensitivity of LCV to physical disruption (Wiebe *et al.*, 1972; McCaul *et al.*, 1991; Heinzen and Hackstadt, 1996; Heinzen *et al.*, 1996a). The SCV is typically between 0.2 and 0.5 μm in size, rod-shaped, and very compact (**Fig. 1a**). The visible periplasmic space is replaced with an electron-dense region bounded by the cytoplasmic and outer membranes. The most distinctive ultrastructural characteristic of the SCV is the electron-dense, condensed chromatin.

A sub-population of the SCV has been described and termed the small-dense cell (SDC) (McCaul *et al.*, 1991). These display extreme tolerance to breakage by high pressure (20 000 lb/in^2), a procedure that destroys typical SCVs (McCaul *et al.*, 1991).

LCVs can reach a length greater than 1.0 μm and are similar to typical Gram-negative bacteria in possessing

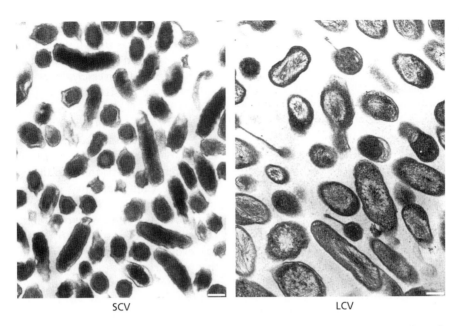

SCV   LCV

Fig. 1 Transmission electron micrograph of *C. burnetii* bacteria separated and purified into SCV and LCV by caesium chloride equilibrium gradient centrifugation. Note the difference between the cytoplasmic regions of the two forms: in LCV it is quite diffuse while SCV has a dense chromatin structure.

a clearly distinguishable outer membrane, periplasmic space and cytoplasmic membrane (**Fig. 1b**). The LCV is more pleiomorphic than the SCV with a thinner cell wall and a dispersed nucleoid. Both LCVs and SCVs divide by binary fission (Heinzen, 1997). In addition to LCVs, SCVs and SDCs, electron-dense, membrane-bound, polar bodies within LCVs have been observed by transmission electron microscopy. This form was termed SLP (spore-like particle), but inclusion of these particles in a current developmental model awaits purification and demonstration of infectivity.

SCVs and LCVs display differences in protein composition. These have been observed by polyacrylamide gel electrophoresis and protein staining of purified SCV and LCV total cell lysates (McCaul *et al.*, 1991; Heinzen and Hackstadt, 1996; Heinzen *et al.*, 1996a, 1999; Heinzen, 1997; Seshadri *et al.*, 1999). Two SCV-specific DNA-binding proteins (Hq1 and ScvA) have been identified and their encoding genes cloned (Heinzen and Hackstadt, 1996; Heinzen *et al.*, 1996a). ScvA is only 30 amino acids in length with a predicted molecular weight of 3610, and has an unusual amino acid composition of 23% arginine, 23% glutamine and 13% proline, with an isoelectric point (pI) of approximately pH 11. Hq1 has a predicted molecular mass of 13 183 Da and a very basic pI (13.1) which is primarily conferred by its high lysine content (29%). Hq1 exhibits 34% and 26% identity with eukaryotic histone H1 and the histone-like protein Hc1 of *Chlamydia trachomatis*, respectively. ScvA has no homologues in the protein database.

Both ScvA and Hq1 bind DNA *in vitro* and are probably integral components of the compact nucleoid structure of the SCV (Heinzen and Hackstadt, 1996; Heinzen *et al.*, 1996a). Chlamydial Hc1 is thought to play a role in the condensation of chromatin during the differentiation of chlamydial reticulate body to elementary body (Hackstadt *et al.*, 1991). Binding of SCV genomic DNA by one or both of these proteins may serve a protective role by stabilising the chromosome, or by inducing topological changes that alter gene expression.

Several proteins that are differentially synthesised by the LCV have also been characterised. Differential expression of the translation elongation factors EF-Tu and EF-Ts between SCV and LCV have been identified by cloning (Seshadri *et al.*, 1999). With specific monoclonal antibodies and immunoblotting, EF-Tu was undetectable in SCV lysates while EF-Ts was present at 4-fold higher concentration in the LCV compared to the SCV. McCaul (1991) and McCaul *et al.* (1991) demonstrated abundant expression of the major outer membrane protein P1 in the LCV, reduced expression in the SCV, and undetectable expression in the SDC. P1 has been purified to near homogeneity, allowing N-terminal and internal amino acid sequence determination and subsequent cloning by a reverse genetic approach. P1 protein has predominantly β[MS1][MS2]-sheet structure and forms pores characteristic of typical bacterial porins in artificial lipid membranes.

Physiology

Growth and Metabolism

C. burnetii is an acidophile that has an absolute requirement for the moderately acidic pH found in the phagolysosome to activate its metabolism (Hackstadt and Williams, 1984). The pH-dependent activation of metabolism by whole cells in axenic medium has been partially explained by the resulting energisation of membranes and substrate transporters such as glutamate (Hackstadt and Williams, 1984). Like more extreme acidophiles, under optimal conditions *C. burnetii* maintains an intracellular pH near neutrality (Hackstadt and Williams, 1984). When host-cell free *C. burnetii* are incubated at a pH level shown to activate metabolism without a metabolisable substrate, the intracellular pH is 5.88, while in the presence of glutamate, the intracellular pH rises to 6.95 (Hackstadt and Williams, 1984).

The organism grows luxuriantly, although it has a slow doubling time of 8–12 hours (Baca and Paretsky, 1983) within vacuoles, in spite of the presence of toxic factors, such as acid hydrolases and defensins, normally regarded as bactericidal. Cytopathic effects are minimal and consequently it is difficult to produce plaques with this organism on a cell monolayer. The mechanism of nutrient acquisition by the organism is undefined. It probably involves trafficking and fusion of nutrient-laden vesicles with the parasite-containing vacuoles where metabolites are hydrolysed to provide precursors for *C. burnetii* metabolism. This model of vesicle-mediated nutrient delivery is supported by the observation that low-molecular-weight fluorescent molecules do not passively diffuse into the *C. burnetii* vacuole when introduced into the cytosol of infected cells by microinjection (Heinzen and Hackstadt, 1997). *C. burnetii* has a growth requirement for iron as shown by the inhibition of replication when desferrioxamine, the intracellular iron chelator, is added to the culture medium (Howe and Mallavia, 1999). Moreover, infection of J774 murine macrophage-like cells by *C. burnetii* results in the up-regulation of transferrin receptor synthesis with a coincident increase in intercellular iron.

Host–Parasite Interaction

The organism plays a passive role in adherence and entry into host cells. *C. burnetii* inactivated by heat or glutaraldehyde are endocytosed via a microfilament-dependent process at rates equal to that observed for viable bacteria (Baca *et al.*, 1993a; Meconi *et al.*, 1998). A recent report indicates that virulent phase I *C. burnetii* are internalised by interacting with a plasma membrane complex that consists of leucocyte response integrin (LRI) $\alpha v \beta 3$ and integrin-associated protein (IAP). Conversely, avirulent phase II organisms enter by the complement CR3 receptor (Capo *et al.*, 1999).

C. burnetii is unique among intracellular bacteria in residing within a vacuole with characteristics of a secondary lysosome. The early parasite-containing phagosome proceeds through the endocytic pathway, eventually acidifying it to a pH of approximately 4.8 (Akporiaye *et al.*, 1983; Maurin *et al.*, 1992a). As described above, *C. burnetii* requires a moderately acidic pH ($\sim$ pH 5) to activate its metabolism and thus has evolved to exploit the only intracellular niche that provides this acidic environment. The *C. burnetii*-containing vacuole fuses with lysosomes as shown by the co-localisation of the lysosomal enzymes $5'$ nucleotidase (Burton *et al.*, 1971), acid phosphatase (Burton *et al.*, 1971, 1978; Akporiaye *et al.*, 1983; Heinzen *et al.*, 1996b), cathepsin D (Heinzen *et al.*, 1996b), vacuolar type H^+-ATPases (Heinzen *et al.*, 1996b), and two predominant lysosomal glycoproteins (LAMP-1 and LAMP-2) (Heinzen *et al.*, 1996b).

The organism undergoes luxurious growth within this vacuole in spite of the presence of factors normally considered to be bactericidal. These can include reactive oxygen species generated during the phagocyte oxidative burst, and antimicrobial agents present in lysosomes such as acid hydrolases and defensins (Reiner, 1994). The *C. burnetii*-containing phagosome may be somewhat atypical, because ingestion of the agent by macrophage-like cell lines results in a greatly diminished respiratory burst with little superoxide anion production (Baca *et al.*, 1994). This phenomenon has also been observed during phagocytosis of *Leishmania*, *Histoplasma*, and some pathogenic mycobacteria (Reiner, 1994). As discussed below, the activity of a *C. burnetii*-encoded acid phosphatase has been implicated in inhibiting the oxidative burst of phagocytes (Baca *et al.*, 1993b).

Developmental Cycle

A model to help an understanding of the roles of the predominant morphological forms of *C. burnetii* (LCV and SCV) is based on the observations that these forms display different protein compositions and metabolic activities (**Fig. 2**). The model suggests aspects of log phase to stationary phase differentiation common to most bacteria, and the generation of resistant extracellular forms required for obligate intracellular organisms to survive in the environment. A few studies support the hypothesis that LCVs are more

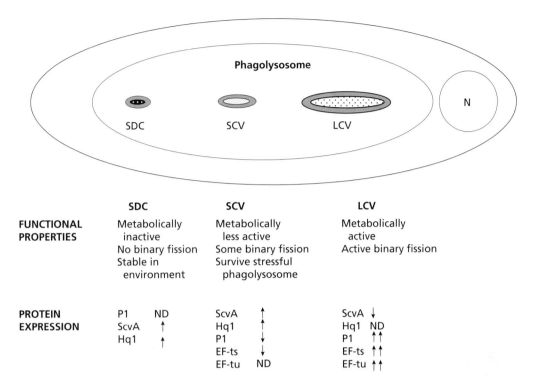

Fig. 2 Model of LCV, SCV and SDC. A working model of *C. burnetii* developmental stages derived from studies summarised in this chapter. A single infected host cell is represented with a phagolysosome that contains all three forms. The ability of each form to develop into one of the alternate forms is unconfirmed. The functional properties assigned to each form and the level of expression of specific proteins is indicated by arrows for high (↑↑), medium (↑), relatively low (↓), or not detected (ND).

active in metabolism and replication than SCVs (McCaul *et al.*, 1981; McCaul, 1991). As such, they may play a more important role than SCV in cell-to-cell spread within the infected host. Conversely, SCVs, especially as SDCs, may for extended periods be able to survive the degradative enzymes and peptides of the phagolysosome. SDCs are probably the cell form responsible for long-term extracellular survival and natural aerosol transmission of the agent. Since *C. burnetii* does not actively lyse host cells and is transmitted by desiccated infected tissues, a sustained supply of resistant cell forms is critical to their survival. In the laboratory, both LCVs and SCVs are infectious for models *in vitro* and *in vivo* (Heggers *et al.*, 1975). Infection by either SCVs or LCVs eventually results in phagosomes harbouring a mixture of cell types. The observation that LCVs are infectious *in vitro* may have little relevance for natural transmission and infection, because the fragile LCVs probably do not persist extracellularly in an infectious form for extended periods.

The environmental conditions that drive *C. burnetii* development are unknown but two obvious candidates for regulators of *C. burnetii* are pH and nutrient availability. Fluctuations in phagolysosomal pH may directly trigger pH-sensitive signal transduction systems in the outer membrane of *C. burnetii*, leading to up- or down-regulation of developmental genes. An adaptive sensory kinase has been identified in *C. burnetii*, but the environmental stimuli to which it responds to are unknown (Mo and Mallavia, 1994). The metabolic status of the host may also drive development. Although not yet tested, the parasitic burden imposed by bacterial growth late in the infectious cycle probably inflicts nutritional stress on the host. Heavily infected, degenerating host cells may reduce trafficking of nutrient-laden vesicles to the *C. burnetii*-containing vacuole (Heinzen *et al.*, 1996b). This, in turn, may drive development of *C. burnetii* to a population dominated by SCVs and SDCs, the cell types most likely to survive extracellularly.

Products

C. burnetii probably secretes molecules into the phagolysosomal vacuole that modify the environment, as has been observed for a number of intracellular parasites that reside in a membrane-bound vacuole (Hackstadt, 1998). Three enzymes have been suggested as playing roles in intracellular survival, based upon their recognised activities in facultative intracellular bacteria. Macrophage infectivity potentiator (Mip) is a peptidylprolyl isomerase (PPIase) localised to the cytoplasm, periplasmic space, and outer surface of *C. burnetii* and *L. pneumophila* (Cianciotto *et al.*, 1995). Mutants of *L. pneumophila* that do not express Mip are attenuated in their ability to infect and survive in macrophages. Interestingly, *C. burnetii* Mip is synthesised as two cytoplasmic products (15 kDa and 15.5 kDa), while a larger form (23.5 kDa) is exported to the periplasmic space and outer membrane by means of a signal sequence (Mo *et al.*, 1998). All three proteins are expressed from a single messenger RNA species, with the smaller forms resulting from alternate translational initiation.

A second periplasmic and outer surface-localised enzyme, Com-1, is a homologue of the disulphide bond-forming enzymes DsbA and DsbC (Hendrix *et al.*, 1993). These enzymes are required for the folding of virulence determinants in *Shigella flexneri* and *Escherichia coli* (Watarai *et al.*, 1995). The *C. burnetii com-1* gene complements an *E. coli dsbA* deletion mutant, and purified re-combinant Com-1 is enzymatically active.

Two groups have shown that the catabolic burst and subsequent superoxide anion production are greatly diminished in phagocytic cells infected with *C. burnetii* (Baca *et al.*, 1984; Ferencik *et al.*, 1984; Akporiaye *et al.*, 1990). Several intracellular pathogens block the oxidative burst, including *Legionella micdadei* (Saha *et al.*, 1985) and *Yersinia pseudotuberculosis* (Bliska and Black, 1995), by expressing an acid phosphatase enzyme. Baca *et al.* (1993b) have shown that an acid phosphatase partially purified from *C. burnetii* sonic extracts blocks superoxide anion production by fMetLeuPhe-stimulated human neutrophils. They also demonstrated, with a series of heteromolybdate complexes, that the *C. burnetii* phosphatase did not have an inhibition pattern characteristic of host cellular acid phosphatase. A *C. burnetii*-specific acid phosphatase inhibitor dramatically reduced the percentage of infected L929 cells in a persistently infected cell line.

Li *et al.* (1996) further purified two major protein species from *C. burnetii* extracts, each of ~ 90 kDa, which expressed acid phosphatase activity. Neutrophils treated with this *C. burnetii* acid phosphatase preparation increased tyrosine phosphorylation of a ~ 44-kDa host protein. Although the identity of this host protein is unknown, it may be involved in regulation of the oxidative burst of neutrophils.

Pathogenesis

The definition of virulence determinants for *C. burnetii* has been hindered by inability to generate and test defined mutants. A different perspective of the nature of *C. burnetii* virulence factors is warranted, because of the persistent but passive parasitic nature of the organism. Studies support the hypothesis that the phagolysosomal vacuole is unmodified – no alterations in trafficking have been noted – and there are no apparent membrane-active toxins. Although *C. burnetii* requires an LPS with a complete O-side-chain (phase I) for virulence, the endotoxin is weakly pyrogenic when compared with other Gram-negative LPS (Hackstadt, 1990).

Acute infection has been modelled as fever development in the guinea-pig (Heggers *et al.*, 1975) and as lethality in several mouse strains (Scott *et al.*, 1987). Guinea-pigs challenged intraperitoneally with as few as 10 virulent organisms develop fever within 5 days. Bacteria can be isolated from a variety of tissues, including the spleen, for several months after infection. Moos and Hackstadt (1987) compared the relative virulence of isolates from acute and chronic infections, and found that as few as 10 inclusion-forming units (IFU) of an acute isolate induce a strong fever response. In contrast, even 10^6 IFU of a chronic isolate do not cause detectable fever in guinea-pigs, although infection could be confirmed by isolation of these organisms from their spleens.

Studies to compare the plasmid profiles (Samuel *et al.*, 1985), chromosomal restriction-fragment length polymorphisms (Hendrix *et al.*, 1991), and LPS banding patterns (Hackstadt, 1986) show that acute isolates are genetically and biochemically distinct from chronic isolates and a grouping of six strains was proposed based on these data (Mallavia, 1991). These studies led to the hypothesis that acute isolates are distinct in virulence potential from chronic isolates, but attempts to model chronic disease have had limited success and have not adequately tested this hypothesis (Atzpodien *et al.*, 1994). The hypothesis that these isolate groups have distinct virulence potentials lost epidemiological support when chronic disease isolates were identified with genetic markers of acute isolates (Thiele and Willems, 1994). Nonetheless, the potential to cause acute or chronic disease may be determined by specific factors encoded by different isolates in combination with host factors such as immune status. The potential for dramatic phenotypic differences between these isolates was underscored by a report that the genome size of different isolates may vary by nearly one megabase (Willems *et al.*, 1996).

Strategies to survive host non-specific defence mechanisms are probably important virulence determinants for *C. burnetii*. Two innate host defence mechanisms that are important in controlling intracellular bacterial pathogens are iron sequestration and the production of reactive oxygen and nitrogen intermediates in the phagolysosome. *C. burnetii* synthesises catalase (Akporiaye and Baca, 1983) and cytoplasmically localised, iron-containing superoxide dismutase (SOD) (Heinzen *et al.*, 1992). These proteins probably do not play a role in detoxifying the phagolysosomal compartment but are important in detoxifying the cytoplasmic compartment of oxygen species produced during aerobic respiration. Periplasmic Cu/Zn SOD or catalase have not been identified, although *L. pneumophila*, the phylogenetically closest pathogen to *C. burnetii* encodes both enzymes and requires them for optimal survival in macrophages (St. John and Steinman, 1996; Bandyopadhyay and Steinman, 1998; Amemura-Maekawa *et al.*, 1999). The molecular mechanisms by which *C. burnetii* acquires iron are undefined, but a ferric iron uptake regulatory protein (Fur) and a periplasmic protein that are transcriptionally regulated by Fur have been reported. Finally, the intensity of the phagocytic cell oxidative burst, as discussed previously, may be affected by a *C. burnetii*-encoded acid phosphatase.

Immunity

Infection of humans with *C. burnetii* induces robust humoral and cell-mediated immune responses. The combined response in patients with acute disease usually abrogates clinical symptoms, results in a self-limiting illness, and typically confers long-lived protection against repeated infection. Occasionally, infection remains sub-clinical and persistent, and patients develop chronic disease. Chronic infection usually involves suppression of components of an effective cell-mediated immune response.

Humoral Response

Antibody develops against *C. burnetii* antigens during the second week of infection and proceeds with typical immunoglobulin class-switching from IgM to IgG, with increasing development of avidity for antigen (Guigno *et al.*, 1992). Antibody that reacts with antigens isolated from phase II organisms develops first, followed several weeks later by antibody reacting with antigens from phase I organisms. The role of antibody in the control of *C. burnetii* replication is not fully understood, but at least two functions have been reported. Specific antibody facilitates opsonised uptake by monocytes/macrophages that may satisfy the requirement of the pathogen to enter host cells. Specific immune serum also lyses infected macrophages by an antibody-dependent cell cytotoxicity

mechanism (Koster *et al.*, 1984). In addition, normal serum kills phase II bacteria, but not phase I cells, by activation of the alternative complement pathway (Vishwanath and Hackstadt, 1988).

Cell-Mediated Response

The requirement for a robust cell-mediated immune response for effective clearance of infection was first highlighted by Kishimoto *et al.* (1978), who showed that athymic mice are less able to clear *C. burnetii* infection from spleen and blood than their euthymic littermates. *C. burnetii* infection and antigens have a potent ability to stimulate a delayed-type hypersensitivity response (DTH) and to activate macrophages (Waag, 1990). In patients with chronic disease T-cell proliferation in response to antigen-specific stimulation is markedly suppressed (Koster *et al.*, 1985a, 1985b). A variety of immunosuppressive disease states, including HIV infection, cancer, lymphoma, chronic renal failure and pregnancy, are associated with a predisposition for the development of chronic Q fever (Raoult *et al.*, 1992; Raoult and Stein, 1993).

Cytokine Response to Infection

Cytokines contribute significantly to the control of *C. burnetii* replication and conversely to the development of chronic disease manifestations. Activated macrophages and monocytes effectively kill intracellular *C. burnetii* (Kishimoto *et al.*, 1977). A key role for interferon γ (IFNγ) in stimulating host cells to inhibit *C. burnetii* replication has been demonstrated in mouse L929 cells and subsequently confirmed in various phagocytic cells (Turco *et al.*, 1984). Recent studies have focused on the interplay between several cytokine signals and the ability effectively to control *C. burnetii* replication.

One model postulates that IFNγ induces tumour necrosis factor α (TNFα) expression by infected peripheral blood mononuclear cells (PBMCs), causing the infected cells to die by apoptosis (Dellacasagrande *et al.*, 1999). This would be a novel pathway for the control of intracellular replication, since monocytes are normally resistant to the induction of apoptotic death by TNFα. Others, using a mouse macrophage cell line (P388D1), demonstrated that phase I *C. burnetii* induced TNFα and IL-1 expression, while infection with phase II *C. burnetii* induced only TNFα (Tujulin *et al.*, 1999). Patients with chronic disease express elevated levels of TNFα and IL-1β compared with uninfected control individuals. PBMCs from patients with chronic infection produce dramatically elevated levels of IL-10 and TGFβ, which probably contributes to the previously reported immune suppression observed in chronic disease (Capo *et al.*, 1996). Monocytes from these patients are unable to control *C. burnetii* replication as compared with those obtained from healthy individuals (Dellacasagrande *et al.*, 2000).

Diagnosis

The signs of Q fever are often sub-clinical or mild. Outbreak surveillance reports indicate that approximately 50% of individuals who sero-convert are asymptomatic (Dupuis *et al.*, 1987). Acute Q fever typically presents as a self-limiting flu-like syndrome with headache, fever and myalgia. Other acute infections include atypical pneumonia, which is clinically mild, and hepatitis with hepatomegaly and fever (Marrie, 1990a, 1990b, 1990c). Non-specific laboratory results associated with acute infection include elevated liver enzyme profiles and auto-antibodies. Leucocyte counts are typically within normal range. Chronic diseases include endocarditis, hepatitis and osteoarthritis. Clinically, endocarditis presents as a sub-acute blood-culture-negative cardiac insufficiency (Raoult *et al.*, 1990).

Specific diagnosis can be made with a variety of assays, but handling of infected material requires biosafety level 3 containment (Fournier *et al.*, 1998). Micro-organisms can be detected in infected tissue samples, which is especially important in patients with chronic disease, by immunoperoxidase staining and immunofluorescence (IFA) (Fournier *et al.*, 1998). Infection has also been demonstrated in tissue samples by PCR amplification with various oligonucleotide primers including primers specific for a *C. burnetii* repetitive element ($\sim$19 copies/genome) (Hoover *et al.*, 1992). Tissue and blood samples from patients with acute or chronic disease, infected animals and arthropods have been used as inoculum for and propagation of *C. burnetii*. Embryonated hen's eggs are very permissive for propagation of large amounts of the organism, while a variety of cloned and primary cell lines also support replication. Guinea-pigs have been used to passage isolates but are able to maintain only virulent organisms. The primary method of microscopic identification and the method used to differentiate *C. burnetii* from other bacteria is Gimenez staining (Gimenez, 1964).

A variety of methods have been used for serological differential diagnosis. Complement fixation was the early standard assay used to identify sero-conversion (Peter *et al.*, 1987). This assay detects phase I and II antigen and is quite specific but less sensitive than other assays. IFA is currently the reference and preferred method with high specificity and sensitivity. ELISA, immunoblotting and micro-agglutination assays have also been used for sero-surveys and diagnostic evaluation of infection.

Antimicrobial Therapy

Acute Q fever is normally a self-limiting infection but intervention with antimicrobial agents effectively minimises clinical symptoms. The sensitivity of *C. burnetii* to selected antimicrobials has been determined in persistently infected cell cultures (Yeaman *et al.*, 1987; Yeaman and Baca, 1991) and in cell cultures by a shell-vial assay (Raoult *et al.*, 1991). Several tetracyclines, especially doxycycline, lincomycin, erythromycin, cotrimoxazole and quinolones are all effective in the treatment of acute Q fever, but chronic infections are more refractory to antibiotic therapy. Mortality in chronically infected individuals ranges from 1% to more than 10% (Raoult *et al.*, 1990).

Since *C. burnetii* replicates in an acidic phagolysosome, and most antimicrobials are either not transported to that compartment or are inactive at low pH, developing an effective course of therapy has been a problem. A combination therapy of doxycycline and chloroquine, a lysosomotrophic agent which neutralises the pH of phagolysosomes, has been devised by Maurin *et al.* (1992a, 1992b). This is effective in controlling infections in cell culture and clinical trials show promising results for chronically infected individuals (Maurin and Raoult, 1996).

Epidemiology and Control

Q fever is a zoonotic disease and *C. burnetii* is maintained in extensive reservoirs in mammal, bird and arthropod species. Wild rodents appear to be a significant wild reservoir, but outbreaks of human disease are most frequently associated with domestic animals. Sheep, goats and cattle are often infected with *C. burnetii* without significant disease except for occasional late-term abortions. Household pets, especially cats during parturition, have been directly associated with several outbreaks (Marrie *et al.*, 1988). Organisms are shed in animal faeces, milk and birth products and remain viable and infectious for long periods. Heavily infected sheep placental tissue, for example, contains up to 10^9 bacteria per gram (Babudieri, 1959).

Transmission of *C. burnetii* to humans occurs primarily by direct inhalation of aerosols from infected material or soil. Workers involved in animal production and abattoir or dairy herd operations are, therefore, at significant risk for infection. Endemic infections are thought to occur worldwide except in New Zealand. In Europe the highest incidence of infection is in spring and summer. Several recent large outbreaks have been reported in the Basque country of Spain (Aguirre Errasti *et al.*, 1984), Switzerland (Dupuis *et al.*, 1987) and Great Britain (Guigno *et al.*, 1992).

In Australia, where Q fever is a particular problem, prevention of infection for populations at risk is provided by vaccination (Marmion *et al.*, 1984). A killed whole-cell vaccine (Q vac) is highly efficacious but skin testing for previous exposure is necessary (Marmion *et al.*, 1990). Vaccinees previously exposed to *C. burnetii* can develop a serious immune reaction at the site of inoculation.

References

Aguirre Errasti C, Montejo Baradna M, Hernandez Almaraz J *et al.* (1984) An outbreak of Q fever in the Basque country. *J. Can. Med. Assoc.* 131: 48–49.

Akporiaye ET, Baca OG (1983) Superoxide anion production and superoxide dismutase and catalase activities in *Coxiella burnetii. J. Bacteriol.* 154: 520–523.

Akporiaye ET, Rowatt JD, Aragon AA, Baca OG (1983) Lysosomal response of a murine macrophage-like cell line persistently infected with *Coxiella burnetii. Infect. Immun.* 40: 1155–1162.

Akporiaye ET, Stefanovich D, Tsosie V, Baca G (1990) *Coxiella burnetii* fails to stimulate human neutrophil superoxide anion production. *Acta Virol.* 34: 64–70.

Amemura-Maekawa J, Mishima-Abe S, Kura F, Takahashi T, Watanabe H (1999) Identification of a novel periplasmic catalase-peroxidase KatA of *Legionella pneumophila. FEMS Microbiol. Lett.* 176: 339–344.

Atzpodien E, Baumgartner W, Artelt A, Thiele D (1994) Valvular endocarditis occurs as a part of a disseminated *Coxiella burnetii* infection in immunocompromised Balb/cj (H-2d) mice infected with the Nine Mile isolate of *C. burnetii. J. Infect. Dis.* 170: 223–226.

Babudieri C (1959) Q fever: a zoonosis. *Adv. Vet. Sci.* 5: 81–84.

Baca OG, Paretsky D (1983) Q fever and *Coxiella burnetii*: a model for host-parasite interactions. *Microbiol. Rev.* 47: 127–149.

Baca O, Akporiaye ET, Rowatt JD (1984) Possible biochemical adaptations of *Coxiella burnetii* for survival within phagocytes: effects of antibody. In: Leive L, Schlessinger D (eds) *Microbiology*, Washington, DC: ASM Press, pp. 269–272.

Baca OG, Klassen DA, Aragon AS (1993a) Entry of *Coxiella burnetii* into host cells. *Acta Virol.* 37: 143–155.

Baca OG, Roman MJ, Glew RH, Christner RF, Buhler JE, Aragon AS (1993b) Acid phosphatase activity in *Coxiella burnetii*: a possible virulence factor. *Infect. Immun.* 61: 4232–4239.

Baca OG, Li Y, Kumar H (1994) Survival of the Q fever agent *Coxiella burnetii* in the phagolysosome. *Trends Microbiol.* 2: 476–480.

Bandyopadhyay P, Steinman HW (1998) *Legionella pneumophila* catalase-peroxidase: cloning of the katB gene and studies of KatB function. *J. Bacteriol.* 180: 5369–5374.

Bliska JB, Black DS (1995) Inhibition of the Fc receptor-mediated oxidative burst in macrophages by the *Yersinia pseudotuberculosis* tyrosine phosphatase. *Infect. Immun.* 6: 681–685.

Burton PR, Kordova N, Paretsky D (1971) Electron microscopic studies of the rickettsia *Coxiella burnetii*: entry, lysosomal response, and fate of rickettsial DNA in L-cells. *Can. J. Microbiol.* 17: 143–150.

Burton PR, Stueckemann J, Welsh RM, Paretsky D (1978) Some ultrastructural effects of persistent infections by the rickettsia *Coxiella burnetii* in mouse L cells and green monkey kidney (Vero) cells. *Infect. Immun.* 21: 556–566.

Capo C, Zaffran Y, Zugan F, Houpikian P, Raoult D, Mege JL (1996) Production of interleukin-10 and transforming growth factor b by peripheral blood mononuclear cells in Q fever endocarditis. *Infect. Immun.* 64: 4143–4150.

Capo C, Lindberg FP, Meconi S *et al.* (1999) Subversion of monocyte functions by *Coxiella burnetii*: impairment of the cross-talk between alpha v beta 3 integrin and CR3. *J. Immunol.* 163: 6078–6085.

Cianciotto NP, O'Connell W, Dasch GA, Mallavia LP (1995) Detection of mip-like sequences and Mip-related proteins within the family *Rickettsiaceae. Curr. Microbiol.* 30: 149–153.

Cox HR (1941) Cultivation of Rickettsiae of the rocky mountain spotted fever, typhus and Q fever groups in the embryonic tissues of developing chicks. *Science* 94: 399–403.

Davis GE, Cox HR (1938) A filter-passing infectious agent isolated from ticks. I. Isolation from *Dermacentor andersonii*, reactions in animals, and filtration. *Public Health Rep.* 53: 2259.

Dellacasagrande J, Capo C, Raoult D, Mege J-L (1999) IFN-γ-mediated control of *Coxiella burnetii* survival in monocytes: the role of cell apoptosis and TNF. *J. Immunol.* 176: 2259–2265.

Dellacasagrande J, Ghigo E, Capo C, Raoult D, Mege JL (2000) *Coxiella burnetii* survives in monocytes from patients with Q fever endocarditis: involvement of tumor necrosis factor. *Infect. Immun.* 68: 160–164.

Derrick EH (1937) 'Q' fever, a new fever entity: clinical features, diagnosis, and laboratory investigation. *Med. J. Aust.* 2: 281–299.

Dupuis G, Petite J, Peter O, Vouilloz M (1987) An important outbreak of human Q fever in a Swiss Alpine valley. *Int. J. Epidemiol.* 16: 282–287.

Ferencik M, Schramek S, Kazar J, Stefanovic J (1984) Effect of *Coxiella burnetii* on the stimulation of hexose monophosphate shunt and superoxide anion production in human polymorphonuclear leukocytes. *Acta Virol.* 28: 246–250.

Fournier P-E, Marrie TJ, Raoult D (1998) Diagnosis of Q Fever. *J. Clin. Microbiol.* 36: 1823–1834.

Gimenez DF (1964) Staining rickettsiae in yolk-sac cultures. *Staining Technol.* 39: 135–140.

Guigno D, Coupland B, Smith EG, Farrell ID, Desselberger U, Caul EO (1992) Primary humoral antibody response to *Coxiella burnetii*, the causative agent of Q fever. *J. Clin. Microbiol.* 30: 1958–1967.

Hackstadt T (1986) Antigenic variation in the phase I lipopolysaccharide of *Coxiella burnetii* isolates. *Infect. Immun.* 52: 337–340.

Hackstadt T (1990) *Rickettiology: Current Issues and Perspectives.* New York: Diamond Point.

Hackstadt T (1998) The diverse habitats of obligate intracellular parasites. *Curr. Opin. Microbiol.* 1: 82–87.

Hackstadt T, Williams JC (1984) Metabolic adaptations of *Coxiella burnetii* to intraphagolysosomal growth. In: Lieve L, Schlessinger D (eds) *Microbiology*, Washington, DC: ASM Press, pp. 266–268.

Hackstadt T, Baehr W, Ying Y (1991) *Chlamydia trachomatis* developmentally regulated protein is homologous to eukaryotic histone H1. *Proc. Natl Acad. Sci. USA* 88: 3937–3941.

Heggers JP, Billups LH, Hinrichs DJ, Mallavia LP (1975) Pathophysiologic features of Q fever-infected guinea pigs. *Am. J. Vet. Res.* 36: 1047–1052.

Heinzen R A (1997) Intracellular development of *Coxiella burnetii*. In: Anderson B, Bendinelli M, Friedman H (eds) *Rickettsial Infection and Immunity.* New York: Plenum Press, pp. 99–129.

Heinzen RA, Hackstadt T (1996) A developmental stage-specific histone H1 homolog of *Coxiella burnetii*. *J. Bacteriol.* 178: 5049–5052.

Heinzen RA, Hackstadt T (1997) The *Chlamydia trachomatis* parasitophorous vacuolar membrane is not passively permeable to low-molecular-weight compounds. *Infect. Immun.* 65: 1088–1094.

Heinzen RA, Frazier ME, Mallavia LP (1992) *Coxiella burnetii* superoxide dismutase gene: cloning, sequencing, and expression in *Escherichia coli*. *Infect. Immun.* 60: 3814–3823.

Heinzen RA, Howe D, Mallavia LP, Rockey DD, Hackstadt T (1996a) Developmentally regulated synthesis of an unusually small, basic peptide by *Coxiella burnetii*. *Mol. Microbiol.* 22: 9–19.

Heinzen RA, Scidmore MA, Rockey DD, Hackstadt T (1996b) Differential interaction with endocytic and exocytic pathways distinguish parasitophorous vacuoles of *Coxiella burnetti* and *Chlamydia trachomatis*. *Infect. Immun.* 64: 796–809.

Heinzen RA, Hackstadt T, Samuel JE (1999) Developmental biology of *Coxiella burnetii*. *Trends Microbiol.* 7: 149–154.

Hendrix LR, Samuel JE, Mallavia LP (1991) Differentiation of *Coxiella burnetii* isolates by analysis of restriction-endonuclease-digested DNA separated by SDS-PAGE. *J. Gen. Microbiol.* 137: 269–276.

Hendrix LR, Mallavia LP, Samuel JE (1993) Cloning and sequencing of *Coxiella burnetii* outer membrane protein gene *com1*. *Infect. Immun.* 61: 470–477.

Hoover TA, Vodkin MH, Williams JC (1992) A *Coxiella burnetii* repeated DNA element resembling a bacterial insertion sequence. *J. Bacteriol.* 174: 5540–5548.

Howe D, Mallavia LP (1999) *Coxiella burnetii* infection increases transferrin receptors on J774A.1 cells. *Infect. Immun.* 67: 3236–3241.

Kishimoto RA, Veltri BJ, Shirley FG, Canonico PG, Walker JS (1977) Fate of *Coxiella burnetii* in macrophages from immune guinea pigs. *Infect. Immun.* 15: 601–607.

Kishimoto RA, Rozmiarek H, Larson RW (1978) Experimental Q fever infection in congenitally athymic nude mice. *Infect. Immun.* 22: 69–71.

Koster FT, Kirkpatrick TL, Rowatt JD, Baca OG (1984) Antibody-dependent cellular cytotoxicity of *Coxiella burnetii* infected J774 macrophage target cells. *Infect. Immun.* 43: 253–256.

Koster FT, Williams JC, Goodwin JS (1985a) Cellular immunity in Q fever: modulation of responsiveness by a suppressor T cell-monocyte circuit. *J. Immunol.* 135: 1067–1072.

Koster FT, Williams JC, Goodwin JS (1985b) Cellular immunity in Q fever: specific unresponsiveness in Q fever endocarditis. *J. Infect. Dis.* 152: 1283–1288.

Li YP, Curley G, Lopez M *et al.* (1996) Protein-tyrosine phosphatase activity of *Coxiella burnetii* that inhibits human neutrophils. *Acta Virol.* 40: 163–272.

Mallavia LP (1991) Genetics of Rickettsiae. *Eur. J. Epidemiol.* 7: 213–221.

Marmion BP, Ormsbee RA, Kyrkou M *et al.* (1984) Vaccine prophylaxis of abattoir-associated Q fever. *Lancet ii*: 1411–1414.

Marmion BP, Ormsbee RA, Kyrkou M *et al.* (1990) Vaccine prophylaxis of abattoir-associated Q fever: eight years' experience in Australian abattoirs. *Epidemiol. Infect.* 104: 275–287.

Marrie TJ (1990a) Acute Q fever. In: Marrie TJ (ed.) *Q fever*, Vol. 1: *The Disease*. Boca Raton, Florida: CRC Press, pp. 125–160.

Marrie TJ (1990b) Epidemiology of Q fever. In: Marrie TJ (ed.) *Q fever*, Vol. 1: *The Disease*. Boca Raton, Florida: CRC Press, pp. 49–70.

Marrie TJ (1990c) Q fever hepatitis. In: Marrie TJ (ed.) *Q fever*, Vol. 1: *The Disease*. Boca Raton, Florida: CRC Press, pp. 171–178.

Marrie TJ, MacDonald A, Durant H, Yates L, McCormick L (1988) An outbreak of Q fever probably due to contact with a parturient cat. *Chest* 93: 98–103.

Maurin M, Raoult D (1996) Optimum treatment of intracellular infection. *Drugs* 2: 45–55.

Maurin M, Benoliel AM, Bongrand P, Raoult D (1992a) Phagolysosomal alkalinization and the bactricidal effect of antibiotics: the *Coxiella burnetii* paradigm. *J. Infect. Dis.* 166: 1097–1102.

Maurin M, Benoliel AM, Bongrand P, Raoult D (1992b) Phagolysosomes of *Coxiella burnetii*-infected cell lines maintain an acidic pH during persistent infection. *Infect. Immun.* 60: 5013–5016.

McCaul TF (1991) The developmental cycle of *Coxiella burnetii*. In: Williams JC, Thompson HA (eds) *Q fever: The Biology of Coxiella burnetii*. Boca Raton, Florida: CRC Press, pp. 223–258.

McCaul TF, Williams JC (1981) Developmental cycle of *Coxiella burnetii*: structure and morphogenesis of vegetative and sporogenic differentiations. *J. Bacteriol.* 147: 1063–1076.

McCaul TF, Hackstadt T, Williams JC (1981) Ultrastructural and biological aspects of *Coxiella burnetii* under physical disruptions. In: Burgdorfer W, Anacker RL (eds) *Rickettsiae and Rickettsial Diseases*. New York: Academic Press, p. 267.

McCaul TF, Banerjee-Bhatnagar N, Williams JC (1991) Antigenic differences between *Coxiella burnetii* cells revealed by postembedding immunoelectron microscopy and immunoblotting. *Infect. Immun.* 59: 3243–3253.

Meconi S, Jacomo V, Boquet P, Raoult D, Mege J, Capo C (1998) *Coxiella burnetii* induces reorganization of the actin cytoskeleton in human monocytes. *Infect. Immun.* 66: 5527–5533.

Mo Y, Mallavia LP (1994) A *Coxiella burnetii* gene encodes a sensor-like protein. *Gene* 151: 185–190.

Mo Y, Cianciotto NP, Mallavia LP (1995) Molecular cloning of a *Coxiella burnetii* gene encoding a macrophage infectivity potentiator (Mip) analogue. *Microbiology* 141: 2861–2871.

Mo YY, Seshu J, Wang D, Mallavia LP (1998) Synthesis in *Escherichia coli* of two smaller enzymically active analogues of *Coxiella burnetii* macrophage infectivity potentiator (CbMip) protein utilizing a single open reading frame from the cbmip gene. *Biochem. J.* 335: 67–77.

Moos A, Hackstadt T (1987) Comparative virulence of intra- and interstrain lipopolysaccharide variants of *Coxiella burnetii* in the guinea pig model. *Infect. Immun.* 55: 1144–1150.

O'Rourke AT, Peacock M, Samuel JE *et al.* (1985) Genomic analysis of phase I and II *Coxiella burnetii* with restriction endonucleases. *J. Gen. Microbiol.* 131: 1543–1546.

Peter O, Dupuis G, Peacock MG, Burfdorfer W (1987) Comparison of enzyme-linked immunosorbent assay and complement fixation and indirect fluorescent-antibody tests for detection of *Coxiella burnetii*. *J. Clin. Microbiol.* 25: 1063–1067.

Raoult D, Stein A (1993) Q fever during pregnancy – a risk for women, fetuses, and obstetricians. *N. Engl. J. Med.* 330: 371.

Raoult D, Raza A, Marrie TJ (1990) Q fever endocarditis and other forms of chronic Q fever. In: Marrie TJ (ed.) *Q fever*, Vol. 1: *The Disease*. Boca Raton, Florida: CRC Press, pp. 1799–2000.

Raoult D, Torres H, Drancourt M (1991) Shell-vial assay: Evaluation of a new technique for determining antibiotic susceptibility, tested in 13 isolates of *Coxiella burnetii*. *Antimicrob. Agents Chemother.* 35: 2070–2077.

Raoult D, Brouqui P, Marchou B, Gastaut J-A (1992) Acute and chronic Q fever patients with cancer. *Clin. Infect. Dis.* 14: 127–130.

Reiner NE (1994) Altered cell signaling and mononuclear phagocyte deactivation during intracellular infection. *Immunol. Today* 15: 374–381.

Saha AK, Dowling JN, LaMarco KI *et al.* (1985) Properties of an acid phosphatase from *Legionella micdadei* which blocks superoxide anion production by human neutrophils. *Arch. Biochem. Biophys.* 243: 150–160.

Samuel JE, Frazier ME, Kahn ML, Thomashow LS, Mallavia LP (1983) Isolation and characterization of a plasmid from phase I *Coxiella burnetii*. *Infect. Immun.* 41: 488–493.

Samuel JE, Frazier ME, Mallavia LP (1985) Correlation of plasmid type and disease caused by *Coxiella burnetii*. *Infect. Immun.* 49: 775–779.

Savinelli EA, Mallavia LP (1990) Comparison of *Coxiella burnetii* plasmids to homologous chromosomal sequences present in a plasmidless endocarditis-causing isolate. *Ann. NY Acad. Sci.* 590: 523–533.

Scott GH, Williams JC, Stephenson EH (1987) Animal models in Q fever: pathological responses of inbred mice to phase I *Coxiella burnetii*. *J. Gen. Microbiol.* 133: 691–700.

Seshadri R, Hendrix LR, Samuel JE (1999) Differential expression of translational elements by lifecycle variants of *Coxiella burnetii*. *Infect. Immun.* 67: 6026–6033.

St. John G, Steinman HM (1996) Periplasmic copper-zinc superoxide dismutase of *Legionella pneumophila*: role in stationary phase survival. *J. Bacteriol.* 178: 1578–1584.

Thiele D, Willems H (1994) Is plasmid based differentiation of *Coxiella burnetii* in 'acute' and 'chronic' isolates still valid? *Eur. J. Epidemiol.* 10: 427–434.

Tujulin E, Lilliehook B, Macellaro A, Sjostedt A, Norlander L (1999) Early cytokine induction in mouse P388D1 macrophages infected by *Coxiella burnetii*. *Vet. Immunol. Immunopathol.* 68: 159–168.

Turco J, Thompson HA, Winkler H (1984) Interferon-γ inhibits growth of *Coxiella burnetii* in mouse fibroblasts. *Infect. Immun.* 45: 781–783.

Vishwanath S, Hackstadt T (1988) Lipopolysaccharide phase variation determines the complement-mediated serum susceptibility of *Coxiella burnetii*. *Infect. Immun.* 56: 40–44.

Vodkin MH, Williams JC, Stephenson EH (1986) Genetic heterogeneity among isolates of *Coxiella burnetii. J. Gen. Microbiol.* 132: 455–463.

Waag DM (1990) Acute Q fever. In: Marrie TJ (ed.) *Q fever*, Vol. 1: *The Disease.* Boca Raton, Florida: CRC Press, pp. 107–123.

Watarai M, Tobe T, Yoskikawa M, Sasakawa C (1995) Disulfide oxidoreductase activity of *Shigella flexneri* is required for release of Ipa proteins and invasion of epithelial cells. *Proc. Natl Acad. Sci. USA* 92: 4927–4931.

Weisburg WG, Dobson ME, Samuel JE *et al.* (1989) Phylogenetic diversity of the *Rickettsiae. J. Bacteriol.* 171: 4202–4206.

Wiebe ME, Burton PR, Shankel DM (1972) Isolation and characterization of two cell types of *Coxiella burnetii. J. Bacteriol.* 110: 368–377.

Willems H, Thiele D, Burger C, Ritter M, Oswald W, Krauss H (1996) *International Society for Rickettsiology and Rickettsial Diseases, Slovakia.* Bratislava: Publishing House of the Slovak Academy of Sciences.

Willems H, Ritter M, Jager C, Thiele D (1997) Plasmid-homologous sequences in the chromosome of plasmidless *Coxiella burnetii* Scurry Q217. *J. Bacteriol.* 179: 3293–3297.

Willems H, Jager C, Bajer G (1998) Physical and genetic map of the obligate intracellular bacterium *Coxiella burnetii. J. Bacteriol.* 180: 3816–3822.

Yeaman MR, Baca OG (1991) Mechanisms that may account for differential antibiotic susceptibilities among *Coxiella burnetii* isolates. *Antimicrob. Agents Chemother.* 35: 948–954.

Yeaman MR, Mitscher LA, Baca OG (1987) *In vitro* susceptibility of *Coxiella burnetii* to antibiotics, including several quinolones. *Antimicrob. Agents Chemother.* 31: 1079–1084.

PART
13

MYCOBACTERIAL INFECTIONS

81

Mycobacterium tuberculosis

Valerie A. Snewin[1], Howard N. Cooper[1] and
Margaret M. Hannan[1,2]

[1]*Imperial College School of Medicine, St. Mary's Campus, London, UK*
[2]*Division of TB Elimination, Centers for Disease Control and Prevention, Atlanta GA, USA*

Mycobacterium tuberculosis, the causative agent of tuberculosis (TB), leads to the death of a human being every 10 minutes. Most cases occur in developing countries where the effect on health provision, the economy and society is so pervasive that it is almost too large to estimate. Sadly, recent years have seen the resurgence of TB, particularly where there is a high prevalence of human immunodeficiency virus (HIV) infection. Approximately 40% of HIV-infected individuals are co-infected with *M. tuberculosis*, which has emerged as a leading opportunistic infection associated with AIDS (Johnson and Ellner, 2000).

Most mycobacteria are saprophytes in soil or water and are harmless to humans, but at some point thousands of years ago *M. tuberculosis* became a significant human pathogen. The origins of tuberculosis are both ancient and obscure. Spinal deformities typical of those resulting from *M. tuberculosis* infection have been found in human remains as far apart as Peru and Egypt and dating from at least 5000 BCE. (Gerszten *et al.*, 2001). It has been suggested that the domestication of cattle led to humans being infected by a novel strain of bovine tuberculosis but, until more molecular evidence is gathered, the origin of the disease will remain a mystery. Tuberculosis did not become truly epidemic until the early seventeenth century.

Increasing urbanisation in western Europe proved to be the ideal environment for these airborne bacteria to spread from host to host with startling rapidity. This epidemic, often called the 'great white plague', spread from England to engulf most of western Europe (Rubin, 1995). Within 200 years most of the European population had been exposed, and perhaps a quarter of the population succumbed to the disease.

Classification and Identification

M. tuberculosis was the type strain of the genus *Mycobacterium* in the classification of Lehmann and Neumann (1896). At that time there was only one other member of the genus: the non-culturable *M. leprae*, the causative agent of leprosy (see Chapter 82). The genus was named after the fungus-like pellicular growth of *M. tuberculosis* in liquid culture.

Mycobacteria are aerobic-to-microaerophilic Gram-positive members of the actinomycete family. *M. tuberculosis* occurs as straight rods which, unlike other members of the actinomycete group, produce neither aerial hyphae nor spores. A marked characteristic of mycobacteria is that they behave anomalously during staining with arylmethane dyes. Once stained the bacteria are highly resistant to de-staining with the weak mineral acids or acid-alcohol solutions that are normally used for this procedure in the Gram stain. This unusual property of acid-resistance led to their being referred to as 'acid-fast bacteria' (AFB). The Ziehl–Neelsen (ZN) stain was developed to exploit this property, and it is still widely used (Bishop and Neumann, 1970; and see the diagnosis section below). Acid-fastness is a distinctive property of mycobacteria, but it is not unique to the genus *Mycobacterium*. *Nocardia* spp. (Chapter 43) are also weakly acid-fast, and in the past this has led to some taxonomic confusion.

For many years the detection of AFB by staining methods such as the ZN stain remained the standard procedure for the identification of mycobacteria. The staining of sputum smears for AFB was a rapid and inexpensive diagnostic method, particularly in the clinical diagnostic laboratory. Recent developments have, however, limited the value of sputum-based AFB staining methods. The rise in the numbers of HIV-infected people in particular has posed significant problems. Many of those who are HIV-infected may become co-infected with normally non-pathogenic mycobacteria such as *M. avium* (see Chapter 83) or they may develop extrapulmonary tuberculosis for which sputum smears would be unhelpful (Foulds and O'Brien, 1998). These considerations have led to the development of automated growth-based detection methods for the identification of *M. tuberculosis*.

Suspected samples are inoculated into selective liquid media, the most common of which is Middlebrook 7H12 broth. This medium, which is available commercially as BACTEC 12B, contains five antimicrobial agents to inhibit the growth organisms other than members of the so-called *M. tuberculosis* complex. The members of this complex are *M. tuberculosis*, *M. bovis*, *M. africanum* and *M. microti*. The growth of the cultures is measured radiometrically and a positive sample can normally be detected within 2 weeks. In developed countries it is rare for a positive diagnosis to arise from any organism other than *M. tuberculosis* (Laverdiere *et al.*, 2000).

The identification of an isolate as *M. tuberculosis* can be confirmed by a number of methods, including the use of nucleic acid probes, inhibition of growth with *M. tuberculosis*-specific inhibitors such as *p*-nitro-α-acetyl-amino-β-hydroxypropiophenone (NAP) or high-performance liquid chromatography of lipids. (Shinnick and Good, 1995; Amsterdam, 1996; Herold *et al.*, 1996). For further details see the Diagnosis section, below.

Structure

The Latin motto *decus et tutamen* (ornament and protection) is most commonly found on coins, but nothing better describes the amazing complexity of the ultrastructure of *M. tuberculosis*. The bacteria must survive in the hostile environment of the interior of the host's phagocytic cells. This may explain the highly unusual cell wall that these bacteria possess. When studied by electron microscopy the mycobacterial cell envelope appears as four distinct layers (Rulong *et al.*, 1991; Paul and Beveridge, 1993). These are the plasma membrane, a layer of moderate electron density, an electron-transparent layer and a layer that is markedly variable in terms of its electron density, thickness and overall homogeneity and granularity.

The plasma membrane appears to be physically and biochemically a typical lipid bilayer as found in Gram-positive bacterial species. The electron density of the second layer suggests that it is a peptidoglycan layer that gives structural rigidity to Gram-positive bacteria. It is in the third layer that *M. tuberculosis* diverges from other Gram-positive organisms. In most of the latter proteins and polysaccharides are attached to the peptidoglycan layer, but *M. tuberculosis* mainly has arabinogalactan mycolate. These molecules are intensely hydrophobic, and their relatively chemically unreactive nature accounts for the failure of this layer to stain well with electron microscopy reagents, thereby leaving an apparently electron-transparent region. The precise nature of the final variable layer is unclear, but it is thought to consist of polar lipids with either short- or long-chain fatty acids attached, glycolipids and the exposed ends of mannose-capped lipoarabinomannans (Brennan and Nikaido, 1995; Lee *et al.*, 1996).

Pathogenicity and Virulence

The pathogenicity of *M. tuberculosis* is complex and remains incompletely understood despite a recent resurgence of interest. Pathogenicity may be defined as the complex of factors required to establish a successful human infection, and includes the damage induced by such an infection (Iseman, 1999). It has been estimated by skin testing that up to one-third of the world population has been exposed to *M. tuberculosis* (Raviglione *et al.*, 1995). What proportion of exposed or infected individuals will eventually succumb to infection is, however, the continued subject of debate amongst TB researchers. In addition, reactivation of disease may occur after a lifetime of latent infection.

Complications such as the impaired immune response of people with HIV/AIDS, which increases by 30-fold the risk of developing active disease, multidrug resistant strains, and increased poverty and drug use in urban life, have led to the increase in TB described by the World Health Organisation as a 'global emergency' (WHO, 1994). *M. tuberculosis* is an obligate intracellular human pathogen and, after inhalation, the bacillus may multiply or be eliminated by alveolar macrophages. If the bacteria establish a foothold, small caseous lesions a few millimetres in size may progress, heal or stabilise. Larger lesions may grow and the bacteria multiply. If the lesions rupture, bacteria are coughed up in infectious sputum, often containing blood and the products of lung tissue breakdown, which is liberated into the bronchial tree and this allows spread to other prospective patients (Bloom, 1994). As an intracellular infection, pathogenesis is intimately linked to the host immune response and, indeed, much of the pathogenic nature of infection has its root in the immune response intended to contain it. Classical investigations into tuberculosis, carried out largely before the antibiotic era, have contributed to our understanding of this pathology (Rich, 1951). The major tissue-damaging response can be produced during delayed hypersensitivity reactions to products of the bacteria.

The genetics of *M. tuberculosis* has been the subject of intensive study. The genome of H37Rv, a laboratory isolate maintained in culture for 80 years, has been completely sequenced (Cole *et al.*, 1998). It consists of 4411 kb, with a G+C content of 65% and at least 3924 predicted open reading frames (ORFs) that potentially encode proteins. Functions have definitively been assigned to about 40% of these, and for another 44% functions have been assigned by homology with genes of other organisms. This leaves a large number of putative proteins about which little or nothing is known (Domenech *et al.*, 2001). Of note is the PE/PPE family of genes, which make up approximately 9% of the genome, the members of which may contribute to a form of 'antigenic variation' (Ramakrishnan *et al.*, 2000). However, 3 years after the genome sequence was determined, relatively few well-defined potential 'virulence factors' have been identified. An example of modern technology shedding light on classic bacteriology is the determination of the genetic basis of 'cording', which was first identified as a morphological difference that correlates with virulence (Glickman *et al.*, 2000).

Host Genetic Factors in Tuberculosis

Since only a minority of those infected with *M. tuberculosis* develop overt disease, many attempts have been made to establish some genetic marker of susceptibility or resistance. Studies on identical and non-identical twins strongly suggest an inherited predisposition or resistance to TB, and racial variations have been suggested (Hermano *et al.*, 1992). Extensive studies in the mouse have revealed that an allele, designated *Bcg*, confers natural macrophage-mediated resistance to a range of intracellular pathogens (Skamene *et al.*, 1982). This allele codes for a protein termed natural-resistance-associated macrophage protein (Nramp) which is involved in the generation of reactive nitrogen intermediates (Vidal *et al.*, 1996). A human homologue of this protein has been found, but its significance for human protection is unknown and it is at present being studied (Cellier *et al.*, 1994).

Many attempts have been made to link susceptibility to TB and leprosy to the class I antigens (HLA-A and HLA-B) of the major histocompatibility complex (Adu *et al.*, 1983; Apt *et al.*, 1993). Although some studies show a low but significant association of a particular HLA type to overt disease, the result may vary from region to region and no definite pattern has emerged. In the case of class II (HLA-D) antigens, it has been shown that HLA-DR2, particularly the DR15 subtype predisposes, to the development of TB, particularly radiologically advanced, smear-positive disease (Bothamley *et al.*, 1989). In future, with the completion of the human genome project new data will be generated, and these should facilitate detailed analysis of the contribution of host genetics in the development of TB.

Genomics

The advance of modern genetic methods and their application to *M. tuberculosis* has led to a plethora of techniques that have been applied to the organism to unravel its mysteries. The complete genome sequence of *M. tuberculosis* H37Rv can now be compared by powerful bioinformatics with an isolate (CDC 1551) involved in a recent tuberculosis outbreak in the USA (Betts *et al.*, 2000), a process which has become known as comparative genomics (Cole, 1998). Thanks to sequencing projects which now include *M. leprae*, *M. bovis* and a non-pathogenic environmental mycobacterium, *M. smegmatis* (see the websites www.tigr.org/tdb and www.sanger.ac.uk/Projects/M.bovis), it is possible to compare the DNA content of these isolates and such studies are also being complemented by the application to *M. tuberculosis* of recent microarray technology (Barry and Schroeder, 2000). Proteomics may provide a link between the genome and the biology of the organism (see Chapter 16).

Other techniques that have been applied to *M. tuberculosis* in attempts to define genes essential for pathogenicity include a variety of molecular genetic screens. Two independent screens have been completed employing signature-tagged mutagenesis (STM), with a mouse model of infection (Camacho *et al.*, 1999; Cox *et al.*, 2000). Both revealed the importance of lipid metabolism, polyketide synthesis, proton-dependent transporters and synthesis and transport of phthiocerol dimycocerosate (PDIM), a complex cell-wall-associated lipid mainly present in pathogenic mycobacteria, interestingly, all located within a 44 kb region of the genome. In-vitro expression technology (IVET), various expression systems, reporter genes, RNA amplification methods including differential display, RNA arbitrarily primed PCR (RAP-PCR) and representational-difference analysis (RDA) have all been applied and are reviewed by DesJardin and Schlesinger (2000).

Epidemiology

Global Epidemiology

M. tuberculosis infection remains the most successful human pathogen worldwide, and more than one-third of the world's population is exposed to the infection every year (Sudre *et al.*, 1992; Dolin *et al.*, 1994; Dye *et al.*, 1999). Of this population more than 10 million develop clinical symptoms, and of those who remain untreated, perhaps 50% will die. In 1992 deaths due to TB were estimated at 2 708 000 (Kochi, 1991).

In 1999 the World Health Organization (WHO) published estimates of the prevalence of TB infection, the incidence of TB disease and the associated morbidity and mortality that occurred worldwide in 1996 (Dye *et al.*, 1999). These data were based on the results of tuberculin skin testing, case notification surveys and TB mortality rates reported by WHO member countries and were calculated according to simple epidemiological models. The prevalence of TB infection is estimated to be 1.7 billion people, or approximately one-third of the world's population. When these figures are compared with earlier reports they indicate a steadily growing problem worldwide.

The recent WHO reports are in striking contrast to the steady downward trends in the incidence of TB that were observed during the 1960s, 1970s and into the 1980s throughout virtually the entire world, although at greatly different rates in different countries. This decline was followed by the levelling off of the downward trends in the mid-1980s (WHO, 1992b).

The geographic distribution of TB has changed considerably over time. In the past, the highest levels of TB were found in the populations of North America and northern Europe. Today the highest annual risk of infection is encountered in the Andes, the Himalayas, sections of Indochina and the Philippines, Haiti and sub-Saharan Africa. Rates in North America and northern Europe are now low. The social distribution of TB, even in the developed world where the disease is uncommon, has always been predominantly at the lower socio-economic level. TB is strikingly associated with poverty, particularly urban poverty. Because TB changes only slowly within a population, when that population moves from one place to another it carries the risk of TB with it for the duration of the lifetime of the people who have moved.

The cause of the global resurgence can be ascribed to the changes in the dynamics of the age-old battle between *M. tuberculosis* and the human host. HIV infection, though an important factor, is not the only contributing influence. Socio-economic poverty, homelessness, overcrowding and malnutrition have historically been associated with TB infection. Much of the decline in mortality from TB began before the availability of effective antituberculosis chemotherapy and had been attributed to improved social conditions, better nutrition and less crowding (McKeown and Record, 1962; Spence *et al.*, 1993). It has been suggested that part of the recent recrudescence of TB may be related to a worsening of the socio-economic circumstances prevailing among certain segments of the world population. One recent example of this is the devastatingly high level of TB and multidrug-resistant (MDR) TB in Russia since the collapse of the political infrastructure of the USSR, which led to breakup of the TB control programmes that had previously been in place. Russia now has one of the highest rates of MDR-TB in the world.

The risk of infection in childhood determines the lifetime experience of TB, both in those who have emigrated and in those who have not. Notification rates of active TB are consistently rising among immigrants to the USA (Talbot *et al.*, 2000), and to a lesser extent this is occurring in Europe also due to the increased number of immigrants from high-prevalence countries. Many immigrants are poor and are obliged to settle in lower-income areas particularly in inner cities where housing is densely crowded. Re-activation of TB in these circumstances results in heavy exposure of household members and so the risk of exposure and reactivation for immigrants remains almost equal to that of their homeland.

Much of the persistence of TB in some parts of the world and its increase in others can be attributed to inadequate regional control programmes (Brudney and Dobkin, 1991).

The Contagious Nature of Tuberculosis

When many sicken and die at once we must look to a single common cause, the air we breathe. Galen (CE 131–201)

A factor that hindered the understanding of TB is the way in which the disease can present a variety of clinical manifestations. The physical wasting became known either as 'consumption' or by the old Greek term 'phthisis', the swollen neck glands as 'scrofula', and deformities of the spine were called 'Pott's disease', after Percival Pott (1713–1788), who first described this condition. It seemed unlikely to early scholars that they were dealing with a single disease, and there was heated debate as to its nature. Most physicians felt that it was humoral condition perhaps caused by abnormal glands, rather than an infective agent, although as early as 1722 Benjamin Martin of London implicated 'animaliculae' carried by the breath of a diseased person (Doetsch, 1978). It is interesting to note that, despite these scholarly reservations, the general public had no illusions about the infective nature of TB. The illustrious Frederick Chopin found that his fame counted for little when an angry mob forced the consumptive composer to flee the island of Majorca, burning his bed and even the plaster from the walls of his room in his wake (Kubba and Young, 1998).

The birth of microbiology is marked by the studies of Robert Koch, who used *M. tuberculosis* as his model organism. Koch's address to the Berlin Physiological Society on 24 March 1882 caused a paradigm shift in scientific thought that has seldom been equalled. He proposed a daring new hypothesis, now known as 'Koch's postulates', which held that not only was disease the product of a micro-organism but that the same micro-organism could be recovered from a diseased animal, isolated and maintained in non-living culture media and then, when it was introduced into a healthy animal, would cause the same disease as had been observed in the original animal. This would have been sufficient in itself to secure Koch's place in history but in reaching these conclusions Koch had developed highly advanced technical methods, such as the use of stains and the first use of solid culture media for the isolation of pure cultures (Sakula, 1982).

Villemin (1827–1892) established the contagious nature of TB years before Koch's isolation of the organism, by inoculating phthisical material into the eye of the rabbit. He suggested in 1868 that TB arose from the inhalation of the infectious material (Riley, 1961). Koch's discovery of the tubercle bacillus encouraged direct tests of this hypothesis, and towards the end of the nineteenth century many workers succeeded in infecting animals by causing them to inhale pulverised or atomised TB. Researchers then realised that for TB to be transmitted it must be converted into some physical form, which is then inhaled. These theories encouraged the search for independent measures of the infectiousness of air.

The situation did not change until the early 1930s when Wells began to re-explore from the beginning the physical and physiological behaviour of organisms suspended in air. Wells' first contribution (1934) was to point out that droplets ejected into the air evaporate while they are falling and the very small ones evaporate completely before they reach the ground. The residues of these evaporated droplets, which he called droplet nuclei, remain suspended for long periods, and it is these that can be carried considerable distances and penetrate to the depths of the lungs. Wells showed that the death rate of all micro-organisms in the transition from droplet to droplet nucleus is high, but that there is a marked difference between the relative survival capacity of airborne organisms, such as the tubercle bacillus, and normally aquatic organisms, such as the typhoid bacillus.

By exposing animals to tubercle bacilli in artificially generated particles of various sizes, Wells (1934) confirmed that the larger particles are trapped by the upper respiratory tract, and most important, showed that the number of lesions developing in the lungs from the inhalation of droplet nuclei closely approximates the number of organisms inhaled. This discovery led to the fundamental concept that infection conveyed in this manner can be induced by the inhalation of a single tubercle bacillus borne on a droplet nucleus; a finding that has since then been directly demonstrated under conditions approximating natural exposure. The importance of this finding is that, since a human being inhales about $14\,m^3$ (500 cubic feet) of air in the course of a day, if infection can be induced by a single droplet nucleus, the nuclei (if evenly distributed) need only be present in a concentration of 1 per $14\,m^3$ to infect everyone continuously breathing that atmosphere for a day. Since this would constitute an explosive outbreak of a severity that is fortunately rare, it is clear that for any infection, which can be induced by a small number of particles, their concentration must generally be less than 1 per $14\,m^3$. The high dilution of organisms, which had been thought to exclude the possibility of truly airborne infection, was thus shown by Wells to be the essence of it.

Molecular Epidemiology and DNA Fingerprinting

The introduction of a molecular typing system for *M. tuberculosis* in 1988 has provided much information on the mode of transmission of TB and made it possible to distinguish re-activation from recent outbreaks of infection in many cases (Eisenach *et al.*, 1988). This is well described in the work of Alland *et al.* (1994) in New York. They showed that, when examined by restriction fragment length polymorphism (RLFP), over 40% of the strains had indistinguishable fingerprint patterns and were possibly derived from recent infection, as compared with previous estimates of 10% for primary infection in the general population.

Most methods of *M. tuberculosis* DNA fingerprinting are based on one of several genetic elements that are repeated at multiple locations in the *M. tuberculosis* genome, the number and position of which are variable. The insertion sequence IS6110 is the most commonly used probe (van Embden *et al.*, 1994). DNA from a cultured isolate is extracted and cleaved by restriction enzymes at many specific sites, including sites within the genetic elements. Because the number and position of genetic elements vary, the enzyme digestion of DNA produces fragments of different sizes. These DNA fragments are separated by agarose gel electrophoresis and transferred on to a nylon membrane. A labelled DNA probe for a selected genetic element is used to visualise the fragments containing that element, and produces banding patterns (fingerprints) on an autoradiogram. Unrelated strains have different numbers and positions of genetic elements and thus different DNA fingerprint patterns (see **Fig. 1**).

Five other genetic elements based on short repetitive DNA sequences that show some degree of genetic diversity have been identified in the *M. tuberculosis* complex and used for DNA fingerprinting research. Three of these, including the polymorphic GC-rich repetitive sequence (PGRS) (Ross *et al.*, 1992; Bifani *et al.*, 1996), a repeat of the triplet GTG (Wilde *et al.*, 1994) and the major polymorphic tandem repeat (MPTR) (Hermans *et al.*, 1992), are present at multiple chromosomal loci. Analysis based on the PGRS element is often referred to by the name of the recombinant plasmid bearing this sequence, pTBN12. Multiple repeats of PGRS and MPTs are now recognised as forming a common domain within members of the PE gene family (van Solingen *et al.*, 1993). The fourth genetic element are the six exact tandem repeat (ETR) loci (Frothingham and Meeker-O'Connell, 1998). In contrast to PGRS or MPTR, these ETR loci have tandem repeats of identical DNA sequences.

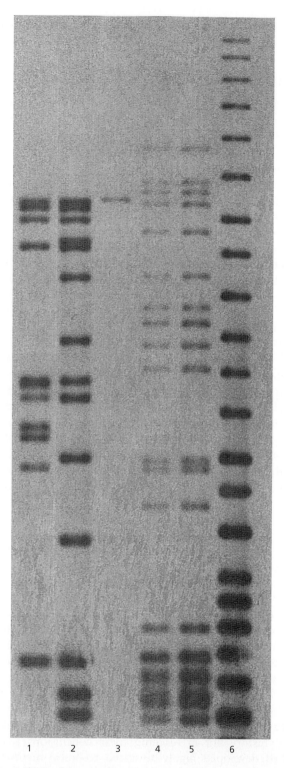

Fig. 1 RFLP gel showing five *M. tuberculosis* isolates in lanes 1–5. Markers are in lane 6. Lanes 1–3 show three epidemiologically unrelated isolates and lanes 4 and 5 represent a cluster of two Beijing isolates from two cases of tuberculosis that were epidemiologically linked. (This gel was kindly provided by Dr C. Woodley, National Centers for Infectious Diseases, Centers for Disease Control and Prevention, Atlanta, Georgia, USA.)

Recently variable numbers of tandem repeats (VNTR) typing was introduced to detect polymorphisms of the MPRT and ETR loci in *M. tuberculosis* complex strains (Frothingham and Meeker-O'Connell, 1998). Finally, the fifth genetic element are the short repetitive direct repeat (DR) elements, which are present at a single genomic locus in *M. tuberculosis* complex strains and are 36 bp in length (Goyal *et al.*, 1994). In the DR locus, the DRs are interspersed by unique DNA spacer sequences of 35–41 bp in length. Clinical isolates differ in the presence of spacers, and this polymorphism can be visualised by spacer oligotyping (spoligotyping) (**Fig. 2**).

Typing methods based on the polymerase chain reaction (PCR) have the advantage of being quicker and easier to perform. The PCR-based typing methods of choice for reproducibility are mixed-linker PCR, VNTR typing and spoligotyping (Kremer *et al.*, 1999).

All of the genetic elements so far identified, including IS*6110*, have limitations for their use in DNA fingerprinting. For instance, in some strains of *M. tuberculosis* insufficient copies of IS*6110* are incorporated in the genome to allow strains to be distinguished accurately, and pTBN12 may have so many small hybridising fragments that the pattern distinction can be difficult. The various methods of DNA fingerprinting also differ in their ability to detect genetic variation or, in other words, their sensitivity and specificity for strain differentiation varies (Chevrel-Dellagi *et al.*, 1993; Godfrey-Faussett *et al.*, 1993; van Soolingen *et al.*, 1993). A very stable genetic element, for instance an insertion sequence which transposes infrequently, may not allow for the differentiation of closely related isolates. For example, IS*1081* and DR have less variability in their numbers and positions in the genome (Kremer *et al.*, 1999; Soini *et al.*, 2000), and thus are less strain specific than IS*6110* (van Soolingen *et al.*, 1993). Further research is required to determine the sensitivity and specificity for strain variation among the methods currently used, and to identify other genetic elements, or methods, that may be better suited for DNA fingerprinting.

The most widely used and standardised method for *M. tuberculosis* DNA fingerprinting employs the insertion sequence IS*6110* as the selected genetic element to visualise the fingerprint patterns (van Embden *et al.*, 1993). IS*6110* appears to have the best balance of attributes of any one marker for DNA fingerprinting so far identified in *M. tuberculosis*. Among unrelated organisms, it is variable in copy number and positions in the genome but stable among related organisms. *M. tuberculosis* isolates can carry up to 20 or even more copies of IS*6110*, with an essentially unlimited number of DNA fingerprint patterns. Rare isolates have been described with one or no IS*6110* copies (Park *et al.*, 2000) and 5–25% of strains contain five or fewer IS*6110* copies, but the majority of strains contain more than five copies (van Soolingen *et al.*, 1991; Chevral-Dellagi *et al.*, 1993; Das *et al.*, 1993). The classification of isolates, which, with the exception of one band, have identical patterns is, however, controversial.

Several studies have evaluated the stability of IS*6110* within strains both *in vitro* and *in vivo* (van Soolingen *et al.*, 1991; Chevrel-Dellagi *et al.*, 1993; Das *et al.*, 1993; Godfrey-Faussett *et al.*, 1993; Cave, 1994). Six *M. tuberculosis* strains passaged weekly for 6 months in liquid culture media and nine strains passaged for 4 weeks in monolayers of cultivated macrophages showed no changes in their DNA fingerprint patterns. In addition, IS*6110* DNA fingerprint patterns remain unchanged in the face of the development of anti-TB drug resistance. Cave *et al.* (1994) examined the stability *in vivo* of 61 isolates from 18 patients. With two exceptions, DNA fingerprints remained identical in serial isolates from individual patients over intervals ranging from 8 months to 4.5 years. In two cases, an additional hybridising fragment was observed approximately 1 year after the original isolate was obtained, a phenomenon occasionally seen in several other studies. Despite increasing experience with IS*6110*, the frequency and timing of its movement in the genome is not known.

Tuberculosis and the HIV Pandemic

The fate of otherwise healthy contacts that become infected with tubercle bacilli has been thoroughly studied. It is now well appreciated that the great majority (possibly as many as 90%) of infected people do not develop clinically significant TB during their lifetime (Horwitz *et al.*, 1973). They can be identified by a positive tuberculin skin test. In the remaining 10%, TB disease does occur, and in about half of these, or 5% of the total, progression to disease occurs during the first few years after exposure. In the other 5% there is a long interval, often several decades, between infection and the onset of disease; this sequence defines what is called *reactivation* or *remotely acquired* TB infection.

By far the most important factor that determines whether infection, new or old, will progress to disease is the adequacy of the host immune response, especially cellular immunity, to the presence of *M. tuberculosis*. Thus TB has been linked to putative cellular immune deficiencies associated with age, malnutrition, genetic factors, the administration of immunosuppressive drugs, and the presence of diseases such as diabetes

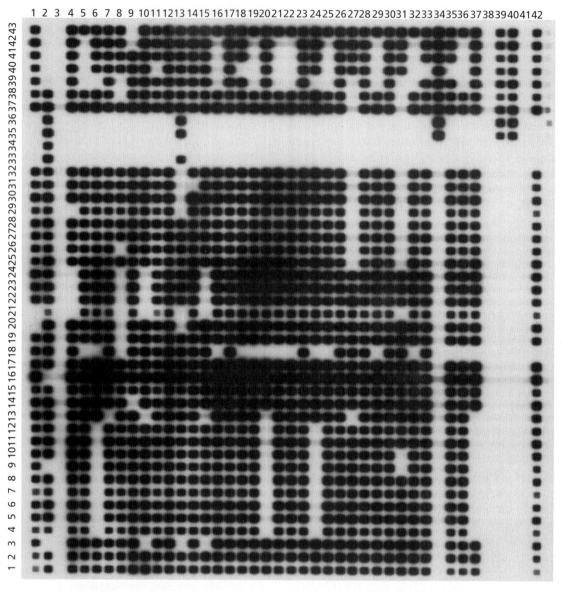

Fig. 2 Spoligotypes identified in *M. tuberculosis* isolates from the USA. Each position represents a spoligotype spacer; black spots indicate hybridisation and empty positions indicate absence of hybridisation. The vertical lanes read from the left are: lane 1, *M. tuberculosis* H37Rv; lane 2, *M. bovis*; lane 3, water as control. Lanes 4–42 represent a collection of unrelated isolates of *M. tuberculosis*. The horizontal rows are 43 DR spacers (see text). (This gel was kindly provided by Lois A. Diems, National Centers for Infectious Diseases, Centers for Disease Control and Prevention, Atlanta, Georgia, USA.)

mellitus, lymphoreticular malignancies, chronic renal insufficiency, silicosis and most notably HIV infection.

The length of time that has elapsed since becoming infected is the most important determinant of the risk of developing TB disease in an immunocompetent person; the risk appears to decrease exponentially during the first few years after infection and then reaches a low level for the rest of the person's lifetime, unless immunological abnormalities supervene. Other modifiers of the risk of disease are age, with greater risk during infancy and adolescence) (Miller *et al.*,

1963; Comstock *et al.*, 1974), and gender, with young women being more likely than young men to develop disease soon after infection. The explanation for the high rates in young people lies in the fact that the greatest likelihood of developing the disease is within the first few years after infection (Frost, 1939; Ferebee, 1969).

The advent of HIV infection is an important new factor in the global resurgence of TB. The acquired immunodeficiency syndrome (AIDS) was first reported by the Centers for Disease Control and Prevention

(CDC, 1981). The pandemic of HIV infection and AIDS has caused marked increases in TB notifications in some countries. By its ability to destroy the immune system, HIV has emerged as the most important risk factor for the progression of dormant TB infection to clinical disease. Nearly 90% of HIV infections have occurred in developing countries, and the majority have occurred in the 15–49-years age group. Using estimates of prevalence of TB and HIV infections in various regions, it has been estimated that by mid-1994 there were 5.6 million people infected with both HIV and TB worldwide, the majority (3.8 million) of whom lived in sub-Saharan Africa. In many developing countries TB has emerged as the most common opportunistic disease associated with HIV infection; up to 54% in Africa, 24% in Latin American countries (Brazil and Mexico), 23% in Haiti, and more than half of patients in India and Thailand (Raviglione *et al.*, 1995).

Immunity

The immune response, sufficient to control the infection in most people, is not generally effective at destroying the organism. *M. tuberculosis* misuses the phagosomal compartment of macrophages, a cell normally able to destroy such intracellular infections (Russell, 1999) (see Chapter 35). The involvement of T cells in controlling mycobacterial infection is underlined by clinical observations; for example, AIDS patients are at greater risk of developing disease. In contrast, patients with impaired humoral responses (multiple myeloma, sickle-cell disease) do not have a predisposition to develop tuberculosis. Through uptake into the macrophage, mycobacterial antigens have access to the antigen-processing machinery of the MHC class II. This results in activation of *M. tuberculosis*-specific CD4+ T cells of T-helper 1 type (Th1). Experiments with knock-out mice have shown aggravation of disease in CD4+-depleted mice and have also provided evidence for participation of CD8+ T cells in protective immunity, restricted by MHC class 1. In addition to these two major T-cell populations, unconventional T cells, $\gamma\delta$ T cells and CD1-restricted $\alpha\beta$ T cells apparently participate in optimum protection (for a review of T-cell-mediated immunity to *M. tuberculosis*, see Stenger and Modlin, 1999). T cells function in protection against disease through activation of antimicrobial activities in macrophages by T-cell cytokines. Interferon γ (IFNγ), a major macrophage-activating cytokine, and other Th1 cytokines, are critical (Bloom, 1994). Direct killing of mycobacteria by T cells through the action of granulysin and perforin has also been demonstrated (Stenger *et al.*, 1999). Infected macrophages are also lysed by T cells, which may release bacteria to be disposed of by more proficient monocytes, or may allow bacterial spread. Initial interaction of *M. tuberculosis* with phagocytes occurs through Toll-like receptors which can induce the adaptive response by inflammatory cytokines and also begin the process of bacterial containment (Flynn and Ernst, 2000).

Animal models are a major help in the understanding of the pathology of an infection, and a number have been developed to aid our understanding of TB. Murine models have advantages, because reagents and knock-out strains are readily available, but mice are more resistant to infection than humans and do not mimic the pathology of TB well. The guinea-pig is extremely susceptible to *M. tuberculosis* and the pathology is more similar to that in humans, so that aerosol infection of guinea-pigs is the standard by which their pathogenicity is compared (Dannenberg *et al.*, 1994; Orme and Collins, 1994). The study of persistence and reactivation of mycobacterial infection is even more complicated, and has been the subject of intense investigation, because an intervention that targets clinically latent infection may become essential in combating this disease.

Diagnosis

Clinical mycobacteriology laboratories play an important role in the control of the spread of tuberculosis through timely detection, isolation, identification and drug susceptibility testing of *M. tuberculosis*. Delay in laboratory diagnosis has been implicated as a major factor in the transmission of disease in many recently reported outbreaks of tuberculosis in Europe and North America (CDC 1990, 1991; Fischl *et al.*, 1992; Valway *et al.*, 1994; Hannan *et al.*, 1995). With delays in identification of disease and further delays in sensitivity reporting, patients remain infectious for longer periods leading to increased risk of the transmission of infection.

Standard laboratory diagnostic evaluation consists of microscopy for detection of the presence of AFB in smears, isolation of organisms by culture, identification of organism species and susceptibility testing (see Classification and Identification, above; Collins *et al.*, 1995). The major obstacle to the provision of timely results is that isolation and growth of sufficient numbers of bacilli for performance of biochemical tests is slow. Most bacteria have a generation time *in vitro* of approximately 20 minutes, but for mycobacteria it is 18–24 hours. As a result, the average culture time on solid medium is 34 days (Morgan et al., 1983) and a further 14 days is required for susceptibility tests.

The second problem is the large number of Mycobacterium species. More than 20 species may cause disease in humans. In the UK the vast majority of isolates are represented by M. tuberculosis and M. avium; others, such as M. kansasaii, are less commonly found. Since M. tuberculosis is the most contagious and pathogenic of the Mycobacterium species, diagnostic laboratories must be able to identify all species to rule out tuberculosis and so allow the prescribing physician to commence appropriate treatment. Identification is important in situations where M. avium infection is more common and pathogenic, as in AIDS patients.

Though less sensitive than culture, the acid-fast smear is an essential adjunct to the diagnosis of tuberculosis. A single smear of a respiratory specimen has a sensitivity of 22–43%, but when multiple specimens are examined the detection rate improves to over 90% (Roberts et al., 1992); the sensitivity of smears of other specimen sources is, however, not as great (Lipsky et al., 1984). Most laboratories in the UK stain Mycobacterium spp. with the ZN stain (see Classification and Identification section, above). The use of a fluorochrome dye such as auramine has replaced the ZN stain in some laboratories, as it is less time consuming and allows easier detection.

Clearly, although smears are important in detection they must be followed by culture. The traditional laboratory media for the cultivation of mycobacteria is Lowenstein Jensen (LJ) medium. In addition, the use of a liquid medium such as Middlebrooks 7H9 is recommended for subculture, tissue specimens and in vitro tests. Substantial improvement in the time to detection and the number of positive cultures can be gained by using a ^{14}C broth-based growth system, such as BACTEC 460 TB (Becton and Dickinson UK Ltd, Oxford). Mycobacteria metabolise ^{14}C substates and liberate ^{14}CO$_2$ which can be detected by the BACTEC 460 instrument. With the BACTEC system the average time to the recovery of M. tuberculosis is 8 days, as compared with 18 days for conventional media. A newer growth detection system is the Mycobacteria Growth Indicator Tube (BBL MGIT 960) which has been developed both as a manual and an automated system. The BBL MGIT system contains the enriched Middlebrooks 7H9 broth medium, with an oxygen-sensitive fluorescent sensor to indicate microbial growth. The mean time to culture is 14.4 days (Hanna et al., 1999). The principal advantages of the MGIT system over the BACTEC 460 system include reduced opportunity for cross-contamination of cultures, no need for needle inoculation,

no radioisotopes, and no need of special instrument other than ultraviolet light. Its limitations include higher contamination rates, masking of fluorescence by blood in specimens, and possible lack of compatibility with some methods of specimen digestion and decontamination of specimens (Hanna, 1999; Kanchana et al., 2000; Whyte et al., 2000). Susceptibility testing methods involving MGIT are still under development. It appears that the MGIT system will become a versatile and rapid system for the detection and characterisation of M. tuberculosis in the future.

Species identification may take a further week after culture has been achieved. Conventional biochemical tests are reaction to niacin, nitrate reduction, catalase and several others, taking an average of 10 days to confirmation. When the BACTEC p-nitro-α-acetylamino-β-hydroxypropiophenone (NAP) test is used for the identification of members of the M. tuberculosis complex directly from 12B medium (Morgan et al., 1985), a diagnosis of M. tuberculosis complex can be ruled out after 5 days in the absence of growth. NAP test results should always be confirmed with nucleic acid probes, chromatography, or other conventional methods of identification.

The introduction in 1985 of the nucleic acid probe kits for the detection of M. tuberculosis complex and M. avium-intercellulare organisms (AccuProbe; Gen-Probe, San Diego, CA) foreshadowed the modern era of clinical microbiology. The revised probe assays utilising acridinium ester have been expanded to include probes for M. avium, M. intercellulare, M. gorgonae and M. kansasii as well as the M. tuberculosis complex and the M. avium complex. In these assays a labelled DNA probe is reacted with the target rRNA, and a DNA–RNA hybrid is formed. Hybridised acridium ester is detected by a luminometer in a chemiluminescence assay. These assays are sensitive and specific for the identification of mycobacteria from a variety of body sites and biological media, and can be carried out in the clinical laboratory in less than 2 hours. Combining a nucleic acid identification system with the BATEC liquid culture system can reduce the time required for detecting and identifying M. tuberculosis to as little as 4–7 days (Evans et al., 1992).

After the introduction of nucleic acid probes in 1985, genetic sequencing was introduced in the early 1990s. Although the method is highly accurate and rapid, it is expensive and labour intensive, and its use is currently restricted to a few research laboratories. These methods rely on the definition of species-specific nucleotide sequences, combined with a procedure for the identification of these sequences. Hypervariable regions of the 16rRNA of mycobacteria have sufficient sequence variability to provide species differentiation, but sequence variability is generally conserved (Rogall *et al.*, 1990a, b). Alternative approaches to DNA sequencing for the genetic characterisation of mycobacterium involve PCR amplification of the 65-kDa heat-shock protein-encoding gene followed by RFLP analysis (Pliyatkis *et al.*, 1992; Telenti *et al.*, 1993b), PCR amplification of the rRNA genes followed by RFLP (De Vaneechoutte *et al.*, 1993); and PCR amplification of the 32-kDa protein-encoding gene (Soini *et al.*, 1994).

Since the introduction of nucleic acid probes to the clinical microbiology laboratory, clinicians have anxiously awaited the possibility of the direct detection of *M. tuberculosis*. Several commercially available kits, based on the target amplification system, where a characteristic component of the organism is amplified to a detectable level, have been evaluated for their ability to offer a rapid diagnosis of tuberculosis. In particular, two commercial amplification assays have been approved by the US Food and Drug Authority (FDA) for use as *in vitro* diagnostic products for the detection of *M. tuberculosis* directly in AFB smear-positive respiratory specimens from untreated patients. The <u>M</u>ycobacterium <u>T</u>uberculosis <u>D</u>irect test (MTD test) (GenProbe) was the first amplification assay for *M. tuberculosis* assay to be approved by the FDA . This test is based on transcription-mediated amplification of the rRNA. The AMPLICOR MTB assay (Roche Molecular Systems, Branchburg, NJ) is a PCR-based assay. The sensitivity of both of these assays for detecting *M. tuberculosis* from respiratory specimens varies between 70% and 100%, and the specificities are generally over 98% (Miller *et al.*, 1994; Chin *et al.*, 1995; Vuorinen *et al.*, 1995; Bradley *et al.*, 1996; Dalovisio *et al.*, 1996; Ichiyama *et al.*, 1997). In the laboratory, each of these procedures can produce a positive signal from specimens that contain as few as 1–10 bacilli, can clearly distinguish *M. tuberculosis* from other species and can be completed in under 1 day. The sensitivities for AFB smear-positive specimens are generally over 95%, but sensitivities as low as 50% have been reported for smear-negative specimens.

Other <u>n</u>ucleic <u>a</u>cid <u>a</u>mplification (NAA) assays based on a variety of different amplification strategies, as well as second-generation versions of the currently available assays, are undergoing clinical evaluation by several manufacturers and will soon be available. These include the ligase chain reaction (Tortoli *et al.*, 1998), the strand displacement amplification reaction (Down *et al.*, 1996), cycling probe technology (Beggs *et al.*, 1996) and nucleic acid sequencing (van der Vliet *et al.*, 1996). At present, NAA for *M. tuberculosis* does not replace conventional diagnostic techniques,

particularly because strains should still be cultured for antimicrobial susceptibility testing.

Antimicrobial Therapy

In his review 'The White Plague' Max Perutz (1994) recounts how Frederic Chopin (1810–1849), Anton Chekhov (1860–1904), George Orwell (1903–1950) and Eleanor Roosevelt (1884–1962) all died of tuberculosis. In the case of some of these famous people we do not know whether the disease was only poorly understood or whether the management of the disease was inappropriate. For some, antimicrobial agents for the treatment of disease came too late. Streptomycin and *p*-aminosalicylic acid were not introduced until the late 1940s and isoniazid, ethambutol, and rifampicin were not used to treat tuberculosis until 1952, 1961 and 1968 respectively. It is possible that some of the more recent of these cases were early victims of antimicrobial resistance; the first report of drug resistance to streptomycin was as early as 1947 (Youmans and Karlson, 1947).

The current approach to treatment is based on several principles that have evolved over the last 40 years (Schluger *et al.*, 1995). Current recommendations for the treatment of tuberculosis have been published by the American Thoracic Society (ATS, 1994) and the Centers for Disease Control (CDC, 1998), and in the UK (BTS, 1998; UK Code of Practice, 2000). The standard recommended treatment is a 6 month regimen of isoniazid, rifampicin and pyrazinamide given for 2 months followed by isoniazid and rifampicin for 4 months. This is the preferred treatment for patients with fully susceptible organisms who adhere to treatment (ATS, 1994). Ethambutol (or streptomycin in children too young to be monitored for visual acuity) should be included in the initial regimen until results of susceptibility are available, unless there is little possibility of drug resistance (i.e. there is less than 4% primary resistance to isoniazid in the community, and the patient has had no previous treatment with antituberculous medications, is not from a country with a high prevalence of drug resistance, and has had no known exposure to drug-resistant cases). Second-line drugs – *p*-aminosalicylic acid, ethionamide, cycloserine, capreomycin, kanamycin, amikacin, capofloxacin, ofloxacin and rifabutin – should be used if resistance or toxicity occurs during administration of first-line drugs (ATS, 1994).

The increased reporting of mycobacterial isolates resistant to one or more drugs has driven renewed interest in sensitivity testing. The problem of drug resistance has been a prominent feature in tuberculosis throughout the USA, Africa and south-east Asia, and more recently in southern Europe (Pablos-Mendez *et al.*, 1998). Reported resistance levels in the UK and the USA remain below 1% (Moore, 1997; Rose *et al.*, 2001). However, with rising numbers of HIV-infected patients and increased population mobility we may expect to encounter more resistant isolates in the future.

The conventional methods for evaluating susceptibility to antituberculous drugs is to grow mycobacteria on solid or liquid media containing various drugs. Four methods have been described: the proportion method, the radiometric or BACTEC method, the absolute concentration method and the resistance ratio method (Inderlied and Salfinger, 2000). In the UK and the USA the agar proportion method is regarded as the standard *M. tuberculosis* reference laboratory susceptibility testing method; isolates are considered to be resistant if the number of colonies on the drug-containing plates exceeds 1% of the number of colonies on the drug-free plates (Woods, 2000). Susceptibility tests are usually carried out after preliminary identification of the organism and it takes up to 3 weeks before results can be interpreted, as the organism grows slowly on solid media. With the BACTEC system, susceptibility test results for all first-line antituberculous drugs – isoniazid, rifampicin, streptomycin, ethambutol and pyrazinamide – can be obtained in 7–14 days, and this is the preferred method in most laboratories.

Increased understanding of the molecular mechanisms of drug resistance and techniques to identify these may eventually replace these time-consuming drug-susceptibility tests. Development of resistance to rifampicin in *M. tuberculosis* follows a 'single step' high-level resistance pattern (Tsukamura, 1972). Mutations arise spontaneously at a rate of one mutation per 10^{-7}–10^{-8} organisms in strains not previously exposed to the antibiotic (Tsukamura, 1972; Mitchison, 1985). Resistance to rifampicin in many *Escherichia coli* strains is known to arise as a result of missense and other mutations that occur in a discrete region of the *rpo*B gene (Ovchinnikov *et al.*, 1981, 1983; Jin and Gross, 1988). Telenti *et al.* (1993b) cloned and sequenced the probe of *M. tuberculosis* and compared this sequence with PCR-generated fragments of the *rpo*B from 122 clinical isolates rifampicin-resistant and -sensitive *M. tuberculosis*. They identified 15 distinct mutations involving 8 conserved amino acids clustered in a 23-amino-acid region (69 bp) in 64 of 66 rifampicin-resistant strains, but none in 56 susceptible organisms. Almost all mutations were missense mutations, and amino acid substitutions at one of two positions (residues 526 and 531) were found in 80% of the resistant strains. Kapur *et al.* (1994)

used automated DNA sequencing to characterise a collection of 128 isolates from the USA, including 121 rifampicin-resistant organisms, for polymorphism in a 350-bp region of *rpo*B containing the 69 bp stretch with mutations associated with rifampicin resistance. They identified 23 distinct *rpo*B alleles. This confirmed that more than 90% of rifampicin-resistant strains have sequence alterations in the 69-bp region, and the great majority of these changes are missense mutations. Molecular mechanisms of resistance to several other antituberculous drugs have also been identified. Zhang *et al.* (1992) reported the detection of the *kat*G gene in *M. tuberculosis* and correlated the deletions with resistance to isoniazid. Mutations in the *pnc*A gene, which encodes the enzyme pyrazinamidase involved in the conversion of inactive pyrazinamide into the active moiety pyrazinoic acid, can be reliably sequenced to detect pyrazinamide resistance (Scorpio and Zhang, 1996). There is evidence that resistance to ethambutol correlates with a specific mutation (at codon 306) in the *emb*B gene (Sreevatsan *et al.*, 1997). The molecular basis of streptomycin resistance has also been elucidated (Finkenet *et al.*, 1993; Nair *et al.*, 1993; Kenney and Churchward, 1994) and shown to result from mutations in the gene that encodes for ribosomal protein S12 or mutations in the 16rRNA region which is structurally linked to the S12 protein in the assembled ribosome.

Mutations in the genes that encode targets of antimicrobial agents can be detected by a variety of molecular methods. A particular focus of such studies has been rifampicin resistance, because of the pivotal role of rifampicin in the treatment of tuberculosis and other mycobacterial infections, the conserved nature of the genetic basis for resistance and the use of rifampicin resistance as a marker for MDR TB. A variety of methods have been use to detect *rpo*B mutations, including PCR amplification of the target sequence and detection by DNA sequencing (Telenti *et al.*, 1993a; Hunt *et al.*, 1994; Kaper *et al.*, 1994), the line probe assay (Cookesey *et al.*, 1997) and single-stranded conformational polymorphism (Telenti *et al.*, 1993c), and similar approaches have been developed to detect mutations involved with resistance to isoniazid and pyrazinamide. These tests are mostly carried out in research laboratories, however, and until they can be optimised for use in the clinical microbiology laboratory their role in clinical diagnosis remains experimental.

Vaccines

According to the World Health Organization (WHO, 1992a), more people alive today have been vaccinated with the live attenuated BCG vaccine (bacille Calmette–Guérin), than have received any other vaccine. In spite of its widespread use and many advantages (it is inexpensive, safe at birth, single shot, and provides some protection against leprosy), BCG vaccination remains controversial because its protective efficacy has varied widely in different parts of the world and its impact on tuberculosis worldwide remains unclear (Rodrigues and Smith, 1990). Although it prevents childhood TB (miliary and meningeal), BCG fails to protect against pulmonary TB in adults, the most prevalent form of the disease. Moreover, once a subject has been vaccinated, the delayed-type hypersensitivity skin (Mantoux) reaction to intradermally administered purified protein derivative (PPD), which is used to define exposed and therefore potentially infected individuals, may become compromised.

Calmette and Guérin, working at the Pasteur Institute in Paris, began the attenuation of a virulent strain of bovine tuberculosis by serial passage on solid media in 1906. Over the following 13 years, they tested the isolate for capacity to revert to virulence and eventually, in 1921, their bacillus was administered orally to a child thought to be at risk from tuberculosis infection. By 1928 the BCG vaccine was recommended by the League of Nations for widespread use to prevent tuberculosis. Many vaccine strains have been derived from this original culture over the years (the original strain is thought to have been lost during World War II) and heterogeneity between vaccines is recognised (Behr *et al.*, 1994), including variations in deleted regions. This vaccine variation, along with other possible explanations, such as exposure to environmental mycobacteria, methodological differences in trails, previous exposure to helminth infection, which may bias the immune response, genetic differences between populations and population differences in susceptibility, may have contributed to the confusing results of BCG efficacy studies (reviewed in Bloom and Fine, 1994).

The development of a better BCG vaccine requires the identification of the immunological mechanisms required for protection and of the antigens recognised by those protective responses. Possible methods include development of recombinant BCG vaccines, genetically attenuated *M. tuberculosis* vaccines, atypical mycobacterial vaccines, auxotroph (transposon mutagenesis) or killed vaccines, subunit vaccine based on immunogenic proteins such as ESAT-6, the Ag85 complex and heat-shock protein Hsp70 (reviewed in Orme, 1999). Such subunit vaccines would not compromise skin-test reactivity. Optimisation of route, dose, adjuvants and improved trials may also be necessary. Whole-genome microarrays have identified

129 *M. tuberculosis*-specific ORFs absent in BCG vaccine strains. These may therefore consist of both virulence factors and protective antigens (Behr *et al.*, 1999). The discovery that plasmids expressing DNA-encoding antigens injected subcutaneously into muscle may provoke an immune response was potentially revolutionary. When this technique was applied to *M. tuberculosis*, many protective antigens were identified, including Ag85, and the use of DNA vaccines as a potential tuberculosis therapy (Lowrie *et al.*, 1999) has opened up a new avenue for research, with the use of a hsp65 DNA construct as a 'post-exposure' vaccine. The gold standard for vaccination efficacy studies is aerosol challenge in the guinea-pig, a time-consuming, expensive and potentially dangerous process that will always be contentious because of its requirement for laboratory animals. Rapid and improved model systems are required to test novel vaccines, such as using luciferase-expressing *M. tuberculosis* as a rapid read-out of viability (Snewin *et al.*, 1999). Although many examples of novel vaccines against tuberculosis have been investigated, so far none has improved on BCG. Whether the new technologies described here will indeed provide the raw materials for a 'better BCG' remains to be seen.

References

Adu HO, Curtis J, Turk JL (1983) Role of the major histocompatibility complex in resistance and granuloma formation in response to *Mycobacterium lepraemurium* infection. *Infect. Immun.* 40(2): 720–725.

Alland D, Kalkut GE, Moss AR *et al.* (1994) Transmission of tuberculosis in New York City. An analysis by DNA fingerprinting and conventional epidemiologic methods. *N. Engl. J. Med.* 330(24): 1710–1716.

Amsterdam D (1996) The laboratory diagnosis of tuberculosis in a period of resurgence: challenge for the laboratory. *Clin. Lab. Sci.* 9(4): 207–212.

Apt AS, Avdienko VG, Nikonenko BV *et al.* (1993) Distinct H-2 complex control of mortality, and immune responses to tuberculosis infection in virgin and BCG-vaccinated mice. *Clin. Exp. Immunol.* 94(2): 322–329.

ATS (1994) Treatment of tuberculosis and tuberculosis infection in adults and children. American Thoracic Society. *Monaldi Arch. Chest Dis.* 49(4): 327–345.

Barnes PF, Modlin RL, Ellner JJ (1994) T cell responses and cytokines. In: Bloom BR (ed.) *Tuberculosis Pathogenesis, Protection and Control.* Washington, DC: ASM Press, pp. 417–435.

Barry CE III, Schroeder BG (2000) DNA micro-arrays: translational tools for understanding the biology of *Mycobacterium tuberculosis*. *Trends Microbiol.* 8: 209–210.

Beenhouwer H De, Lhiang Z, Jannes G *et al.* (1995) Rapid detection of rifampicin resistance in sputum and biopsy specimens from tuberculosis patients by PCR and line probe assay. *Tuber. Lung Dis.* 76(5): 425–430.

Beggs ML, Cave MD, Marlowe C *et al.* (1996) Characterization of *Mycobacterium tuberculosis* complex direct repeat sequence for use in cycling probe reaction. *J. Clin. Microbiol.* 34(12): 2985–2989.

Behr MA *et al.* (1999) Comparative genomics of BCG vaccines by whole-genome DNA microarray. *Science* 284: 1520–1523.

Betts JC, Dodson P, Quan S *et al.* (2000) Comparison of the proteome of *mycobacterium tuberculosis* strain H37Rv with clinical isolate CDC 1551. *Microbiology* 146: 3205–3216.

Bifani PJ, Plikaytis BB, Kapur V *et al.* (1996) Origin and interstate spread of a New York City multidrug-resistant *Mycobacterium tuberculosis* clone family. *JAMA* 275(6): 452–457.

Bishop PJ, Neumann G (1970) The history of the Ziehl–Neelsen stain. *Tubercle* 51(2): 196–206.

Bloom BR (ed.) (1994) *Tuberculosis Pathogenesis, Protection and Control.* Washington, DC: ASM Press, pp. 25–46.

Bloom BR, Fine PEM (1994) The BCG experience: Implications for future vaccines against tuberculosis. In: Bloom BR (ed.) *Tuberculosis Pathogenesis, Protection and Control.* Washington, DC: ASM Press, pp. 531–557.

Bothamley GH, Beck JS, Schreuder GM *et al.* (1989) Association of tuberculosis and M. tuberculosis-specific antibody levels with HLA. *J. Infect. Dis.* 159(3): 549–555.

Bradley SP, Reed SL, Catanzaro A (1996) Clinical efficacy of the amplified *Mycobacterium tuberculosis* direct test for the diagnosis of pulmonary tuberculosis. *Am. J. Resp. Crit. Care Med.* 153(5): 1606–1610.

Brennan PJ, Nikaido H (1995) The envelope of mycobacteria. *Annu. Rev. Biochem.* 64: 29–63.

Brudney K, Dobkin J (1991) Resurgent tuberculosis in New York City. Human immunodeficiency virus, homelessness, and the decline of tuberculosis control programs. *Am. Rev. Respir. Dis.* 9: 745–749.

BTS (1998) Chemotherapy and management of tuberculosis in the United Kingdom: recommendations. Joint Tuberculosis Committee of the British Thoracic Society. *Thorax* 53(7): 536–548.

Camacho LR, Ensergueix D, Perez E, Gicquel B, Guilhot C (1999) Identification of a virulence gene cluster of *Mycobacterium tuberculosis* by signature-tagged mutagenesis. *Mol. Microbiol.* 34: 257–267.

Cave MD, Eisenach KD, Templeton G *et al.* (1994) Stability of DNA fingerprint pattern produced with IS6110 in strains of *Mycobacterium tuberculosis. J. Clin. Microbiol.* 32(1): 262–266.

CDC (1981) Pneumocystis pneumonia – Los Angeles. *MMWR* 30: 250–252.

CDC (1990) Nosocomial transmission of multidrug-resistant tuberculosis to health-care workers and HIV-infected patients in an urban hospital – Florida. *MMWR* 39: 718–722.

CDC (1991) Nosocomial transmission of multidrug resistant tuberculosis among HIV-infected persons – Florida and New York, 1988–1991. *MMWR* 40: 585–591.

CDC (1998) Prevention and treatment of tuberculosis among patients infected with human immunodeficiency virus: principles of therapy and revised recommendations. Centers for Disease Control and Prevention. *MMWR* 47(RR-20): 1–58.

Cellier M, Govoni G, Vidal S *et al.* (1994) Human natural resistance-associated macrophage protein: cDNA cloning, chromosomal mapping, genomic organization, and tissue-specific expression. *J. Exp. Med.* 180(5): 1741–1752.

Chevrel-Dellagi D, Abderrahman A, Haltiti R *et al.* (1993) Large-scale DNA fingerprinting of *Mycobacterium tuberculosis* strains as a tool for epidemiological studies of tuberculosis. *J. Clin. Microbiol.* 31(9): 2446–2450.

Chin DP, Yajko DM, Hadley WK *et al.* (1995) Clinical utility of a commercial test based on the polymerase chain reaction for detecting *Mycobacterium tuberculosis* in respiratory specimens. *Am. J. Resp. Crit. Care Med.* 151(6): 1872–1877.

Cole ST (1998) Comparative mycobacterial genomics. *Curr. Opin. Microbiol.* 1: 567–571.

Cole ST, Brosch R, Parkhill J *et al.* (1998) Deciphering the biology of *Mycobacterium tuberculosis* from the complete genome sequence. *Nature* 393(6685): 537–544.

Collins CH, Grange JM, Yates MD (1985) *Organisation and Practice in Tuberculosis Bacteriology.* London: Butterworths.

Comstock GW, Livesay VT, Woolpert SF (1974) The prognosis of a positive tuberculin reaction in childhood and adolescence. *Am. J. Epidemiol.* 99(2): 131–138.

Cooksey RC, Morlock GP, Glickman S, Crawford JT (1997) Evaluation of a line probe assay kit for characterization of *rpoB* mutations in rifampin-resistant *Mycobacterium tuberculosis* isolates from New York City. *J. Clin. Microbiol.* 35(5): 1281–1283.

Cox JS, Chen B, McNell M, Jacobs WR Jr (2000) Complex lipid determines tissue-specific replication of *Mycobacterium tuberculosis* in mice. *Nature* 402: 79–82.

Dalovisio JR, Montenegro-James S, Kemmerly SA *et al.* (1996) Comparison of the amplified *Mycobacterium tuberculosis* (MTB) direct test, Amplicor MTB PCR, and IS6110-PCR for detection of MTB in respiratory specimens. *Clin. Infect. Dis.* 23(5): 1099–1106; discussion 1107–1108.

Dannenberg AM Jr (1994) Guinea pig model of tuberculosis. In: Bloom BR (ed.) *Tuberculosis Pathogenesis, Protection and Control.* Washington, DC: ASM Press, pp. 135–147.

Das S, Chan SL, Allen BW, Mitchison DA, Lowrie DB (1993) Application of DNA fingerprinting with IS986 to sequential mycobacterial isolates obtained from pulmonary tuberculosis patients in Hong Kong before, during and after short-course chemotherapy [published erratum appears in *Tuber. Lung Dis.* 74(3):217]. *Tuber. Lung Dis.* 74(1): 47–51.

De Vaneechoutte M, Beenhouwer H, Claeys G *et al.* (1993) Identification of Mycobacterium species by using amplified ribosomal DNA restriction analysis [published erratum appears in *J. Clin. Microbiol.* 31(12):3355]. *J. Clin. Microbiol.* 31(8): 2061–2065.

DesJardin LE, Schlesinger LS (2000) Identifying *Mycobacterium tuberculosis* virulence determinants – new technologies for a difficult problem. *Trends Microbiol.* 8: 97–99.

Doetsch RN (1978) Benjamin Marten and his 'New Theory of Consumptions'. *Microbiol. Rev.* 42(3): 521–528.

Dolin PJ, Raviglione MC, Kochi A (1994) Global tuberculosis incidence and mortality during 1990–2000. *Bull. WHO* 72(2): 213–220.

Domenech P, Barry CE III and Cole ST (2001) *Mycobacterium tuberculosis* in the post genomic age. *Curr. Opin. Microbiol.* 4: 28–34.

Down JA, O'Connell MA, Dey MS *et al.* (1996) McLaurin DA 3rd. Cole G. Detection of *Mycobacterium tuberculosis* in respiratory specimens by strand displacement amplification of DNA. *J. Clin. Microbiol.* 34(4): 860–865.

Dye C, Scheele S, Dolin P, Pathania V, Raviglione MC (1999) Consensus statement. Global burden of tuberculosis: estimated incidence, prevalence, and mortality by country. WHO Global Surveillance and Monitoring Project. *JAMA* 282(7): 677–686.

Eisenach KD, Crawford JT, Bates JH (1988) Repetitive DNA sequences as probes for *Mycobacterium tuberculosis*. *J. Clin. Microbiol.* 26(11): 2240–2245.

Evans KD, Nakasone AS, Sutherland PA, de la Maza LM, Peterson EM (1992) Identification of *Mycobacterium tuberculosis* and *Mycobacterium avium-intercellulare* directly from primary BACTEC cultures by using acridine-ester-labelled DNA probes. *J. Clin. Microbiol.* 30: 2427–2431.

Ferebee SH (1969) Controlled chemoprophylaxis trials in tuberculosis. A general review. *Adv. Tuber. Res.* 17: 39–106.

Finken M, Kirschner P, Meier A, Wrede A, Bottger EC (1993) Molecular basis of streptomycin resistance in *Mycobacterium tuberculosis*: alterations of the ribosomal protein S12 gene and point mutations within a functional 16S ribosomal RNA pseudoknot. *Mol. Microbiol.* 9(6): 1239–1246.

Finland M (1942) The present status of the higher types of anti-pneumococcus serums. *JAMA* 120: 1294.

Fischl M, Uttamhandani RB, Daikos GL *et al.* (1992) An outbreak of tuberculosis caused by multidrug-resistant bacilli among patients with HIV infection. *Ann. Intern. Med.* 117: 177–183.

Flynn JA, Ernst JD (2000) Immune responses in tuberculosis. *Curr. Opin. Immunol.* 12: 432–436.

Foulds J, O'Brien R (1998) New tools for the diagnosis of tuberculosis: the perspective of developing countries. *Int. J. Tuber. Lung Dis.* 2(10): 778–783.

Frost WH (1939) The age selection of mortality from tuberculosis in successive decades. *Am. J. Hyg.* 30: 91–96.

Frothingham R, Meeker-O'Connell WA (1998) Genetic diversity in the *Mycobacterium tuberculosis* complex based on variable numbers of tandem DNA repeats. *Microbiology* 144: 1189–1196.

Gerszten PC, Gerszten E, Allison MJ (2001) Diseases of the spine in South American mummies. *Neurosurgery* 48(1): 208–213.

Glickman MS, Jacobs WR Jr (2001) Microbial pathogenesis of *Mycobacterium tuberculosis*: dawn of a discipline. *Cell* 104: 477–485.

Godfrey-Faussett P, Stoker NG, Scott JA *et al.* (1993) DNA fingerprints of *Mycobacterium tuberculosis* do not change during the development of rifampicin resistance. *Tuber. Lung Dis.* 74(4): 240–243.

Goyal M, Young D, Zhang Y, Jenkins PA, Shaw RJ (1994) PCR amplification of variable sequence upstream of *katG* gene to subdivide strains of *Mycobacterium tuberculosis* complex. *J. Clin. Microbiol.* 32: 3070–3071.

Hanna BA, Ebrahimzadeh A, Elliott LB *et al.* (1999) Multicenter evaluation of the BACTEC MGIT 960 system for recovery of mycobacteria. *J. Clin. Microbiol.* 37(3): 748–752.

Hannan MM, Bell A, Easterbrook P *et al.* (1995) An outbreak of multidrug-resistant tuberculosis in an HIV unit and follow up. Federation of Infection Society, Second Biennial Conference, 29 November–1 December 1995, Manchester, p. 26.

Hermans PW, van Soolingen D, van Embden JD (1992) Characterization of a major polymorphic tandem repeat in *Mycobacterium tuberculosis* and its potential use in the epidemiology of *Mycobacterium kansasii* and *Mycobacterium gordonae*. *J. Bacteriol.* 174: 4157–4165.

Herold CD, Fitzgerald RL, Herold DA (1996) Current techniques in mycobacterial detection and speciation. *Crit. Rev. Clin. Lab. Sci.* 33(2): 83–138.

Horwitz O, Edwards PQ, Lowell AM (1973) National tuberculosis control program in Denmark and the United States. *Health Serv. Rep.* 88: 493–498.

Hunt JM, Roberts GD, Stockman L, Felmlee TA, Persing DH (1994) Detection of a genetic locus encoding resistance to rifampin in mycobacterial cultures and in clinical specimens. *Diagn. Microbiol. Infect. Dis.* 18(4): 219–227.

Ichiyama S, Iinuma Y, Yamori S *et al.* (1997) *Mycobacterium* growth indicator tube testing in conjunction with the AccuProbe or the AMPLICOR-PCR assay for detecting and identifying mycobacteria from sputum samples. *J. Clin. Microbiol.* 35(8): 2022–2025.

Inderlied CB, Salfinger M (1999) Antimycobacterial agents and susceptibility tests. In: Murray PR (ed.) *Manual of Clinical Microbiology*. 7th edn. Washington DC: ASM Press, pp. 1601–1623.

Iseman MD (1999) *A Clinicians Guide to Tuberculosis* Philadelphia: Lippincott Williams Wilkins, pp. 63–96.

Jin DJ, Gross CA (1988) Mapping and sequencing of mutations in the *Escherichia coli* rpoB gene that lead to resistance. *J. Mol. Biol.* 202: 45–58.

Johnson JL, Ellner JJ (2000) Adult tuberculosis overview: African versus Western perspectives. *Curr. Opin. Pulm. Med.* 6(3): 180–186.

Kanchana MV, Cheke D, Natyshak I *et al.* (2000) Evaluation of the BACTEC MGIT 960 system for the recovery of mycobacteria. *Diagn. Microbiol. Infect. Dis.* 37(1): 31–36.

Kapur VL, Li S, Iordanescu MR *et al.* (1994) Characterization by automated DNA sequencing of mutations in the gene (*rpoB*) encoding the RNA polymerase B-subunit in Rifampicin-resistant *Mycobacterium tuberculosis* strains from New York City and Texas. *J. Clin. Microbiol.* 32: 1095–1098.

Kenney TJ, Churchward G (1994) Cloning and sequence analysis of the *rpsL* and *rpsG* genes of *Mycobacterium smegmatis* and characterization of mutations causing resistance to streptomycin. *J. Bacteriol.* 176(19): 6153–6156.

Kochi A (1991) The global tuberculosis situation and the new control strategy of the world health organisation. *Tubercle* 72: 1–6.

Kremer K, van Soolingen D, Frothingham R *et al.* (1999) Comparison of methods based on different molecular epidemiological markers for typing of *Mycobacterium tuberculosis* complex strains: interlaboratory study of discriminatory power and reproducibility. *J. Clin. Microbiol.* 37: 2607–2618.

Kubba AK, Young M (1998) The long suffering of Frederic Chopin. *Chest* 113: 210–216.

Laverdiere M, Poirier L, Weiss K *et al.* (2000) Comparative evaluation of the MB/BacT and BACTEC 460 TB systems for the detection of mycobacteria from clinical specimens: clinical relevance of higher recovery rates from broth-based detection systems. *Diagn. Microbiol. Infect. Dis.* 36(1): 1–5.

Lee RE, Brennan PJ, Besra GS (1996) *Mycobacterium tuberculosis* cell envelope. *Curr. Top. Microbiol. Immunol.* 215: 1–27.

Lehmann KB, Neumann R (1896) *Atlas und Grundriss der Bakteriologie und Lehrbuck der speciellen bakteriologischen Diagnostik*, 1st edition. Munich.

Lipsky BJ, Gates J, Tenover FC *et al.* (1984) Factors affecting the clinical value for acid-fast bacilli. *Rev. Infect. Dis.* 6: 214–222.

Lowrie DB *et al.* (1999) Therapy of tuberculosis in mice by DNA vaccination. *Nature* 400: 269–271.

McKeown T, Record RG (1962) Reasons for the decline of mortality in England and Wales during the nineteenth century. *Popul. Stud.* 16: 94–122.

Miller FJW, Seal RME, Taylor MD (1963) *Tuberculosis in Children*. New York: McGraw-Hill.

Miller N, Hernandez SG, Cleary TJ (1994) Evaluation of Gen-Probe amplified *Mycobacterium Tuberculosis* direct test and PCR for direct detection of *Mycobacterium tuberculosis* in clinical specimens. *J. Clin. Microbiol.* 32(2): 393–397.

Mitchison D (1985) Drug resistance in mycobacteria. *Br. Med. Bull.* 40: 84–90.

Moore M, Onorato IM, McCray E, Castro KG (1997) Trends in drug-resistant tuberculosis in the United States, 1993–1996. *JAMA* 278(10): 833–837.

Morgan MA, Doerr KA, Hempel NL *et al.* (1985) Evaluation of the p-nitro-[alpha]-acetylamino-[beta]-hydroxypropiophenone differential test for the identification of *Mycobacterium tuberculosis* complex. *J. Clin. Microbiol.* 21: 634–635.

Morgan MA, Horstmeier CD, DeYoung DR, Roberts GD (1983) Comparison of a radiometric method (BACTEC) and conventional culture media for the recovery of mycobacteria from smear-negative specimens. *J. Clin. Microbiol.* 18: 384–388.

Nair J, Rouse DA, Bai GH, Morris SL (1993) The rpsL gene and streptomycin resistance in single and multiple drug-resistant strains of *Mycobacterium tuberculosis*. *Mol. Microbiol.* 10(3): 521–527.

Orme IM (1999) Beyond BCG: the potential for a more effective TB vaccine. *Molecular Medicine Today* 5: 487–492.

Orme IM, Collins FM (1994) Mouse model of tuberculosis. In: Bloom BR (ed.) *Tuberculosis Pathogenesis, Protection and Control*. ASM Press, Washington, pp. 113–134.

Ovchinnikov YA, Monastyrskaya GS, Gubanov VV *et al.* (1981) Primary structure of *Escherichia coli* RNA polymerase. Nucleotide substitution in the beta subunit gene of the rifampicin resistant rpoB255 mutant. *Mol. Gen. Genet.* 184: 536–538.

Ovchinnikov YA, Monastyrskaya GS, Guriev SO, Kalinina NF, Sverdlov ED *et al.* (1983) RNA polymerase rifampicin resistant mutations in *Escherichia coli*: sequence changes and dominance. *Mol. Gen. Genet.* 190: 344–348.

Pablos-Mendez A, Raviglione MC, Laszlo A *et al.* (1998) Global surveillance for antituberculosis-drug resistance, 1994–1997. World Health Organization-International Union against Tuberculosis and Lung Disease Working Group on Anti-Tuberculosis Drug Resistance Surveillance. *New England Journal of Medicine* 338(23): 1641–1649.

Park YK, Bai GH, Kim SJ (2000) Restriction fragment length polymorphism analysis of *Mycobacterium tuberculosis* isolated from countries in the western pacific region. *J. Clin. Microbiol.* 38(1): 191–197.

Paul TR, Beveridge TJ (1993) Ultrastructure of mycobacterial surfaces by freeze-substitution. *Zentralbl Bakteriol* 279(4): 450–457.

Perutz MF (1994) The white plague. *N Y Rev Books* XLI: 35–39.

Plikaytis BB, Plikaytis BD, Yakrus MA *et al.* (1992) Differentiation of slowly growing Mycobacterium species, including *Mycobacterium tuberculosis*, by gene amplification and restriction fragment length polymorphism analysis. *J. Clin. Microbiol.* 30(7): 1815–1822.

Ramakrishnan L, Federspiel NA, Falkow S (2000) Granuloma-specific expression of Mycobacterium virulence proteins from the glycine-rich PE-RGRS family. *Science* 288: 1436–1439.

Raviglione MC, Snider DE Jr, Kochi A. (1995) Global epidemiology of tuberculosis. Morbidity and mortality of a worldwide epidemic. *JAMA* 273(3): 220–226.

Rich AR (1951) The pathogenesis of tuberculosis. Blackwell Scientific Publications, Oxford.

Riley RLOF (1961) *Airborne Infection Transmission and Control.* Macmillan Press, New York, N.Y., pp. 1–17.

Robert Koch SA (1982) Centenary of the discovery of the tubercle bacillus, 1882. Robert Koch: centenary of the discovery of the tubercle bacillus, (1882). *Thorax* 37(4): 246–251.

Roberts GD, Koneman EW, Kim YK (1992) Mycobacterium. In: Barlows A (ed.) *Manual of Clinical Microbiology*, 5th edn. American Society for Microbiology, Washington DC, pp. 304–309.

Rodrigues LC, Smith PG (1990) Tuberculosis in developing countries and methods for its control. *Trans. R. Soc. Trop. Med. Hyg.* 84: 739–744.

Rogall T, Flohr T, Bottger EC (1990) Differentiation of Mycobacterium species by direct sequencing of amplified DNA. *Journal of General Microbiology* 136(9): 1915–1920.

Rogall T, Wolters J, Flohr T, Bottger EC (1990) Towards a phylogeny and definition of species at the molecular level within the genus Mycobacterium. *International Journal of Systematic Bacteriology* 40(4): 323–330.

Rose AMC, Watson JM, Graham C *et al.* (2001) Tuberculosis at the end of the 20th century in England and Wales: Results of a national survey in 1998. *Thorax* 56: 173–179.

Ross BC, Raios K, Jackson K, Dwyer B (1992) Molecular cloning of a highly repeated DNA element from *Mycobacterium tuberculosis* and its use as an epidemiological tool. *J. Clin. Microbiol.* 30: 942–926.

Rubin SA (1995) Tuberculosis: captain of all these men of death. *Radiol. Clin. N. Am.* 33(4): 619–639.

Rulong S, Aguas AP, da Silva PP (1991) Silva MT Intramacrophagic *Mycobacterium avium* bacilli are coated by a multiple lamellar structure: freeze fracture analysis of infected mouse liver. *Infect. Immun.* 59(11): 3895–3902.

Russell DG (1999) Mycobacterium and the seduction of the macrophage. In: Ratledge C, Dale J (ed.) *Mycobacteria: Molecular Biology and Virulence.* Oxford: Blackwell Science, pp. 371–388.

Schluger NW, Harkin TJ, Rom WN (1995) Principles of therapy of tuberculosis in the modern era. In: Rom WN, Garay S (eds) *Tuberculosis.* New York: Little, Brown, pp. 751–761.

Scorpio A, Zhang Y (1996) Mutations in *pncA*, a gene encoding pyrazinamidase/nicotinamidase, cause resistance to the antituberculous drug pyrazinamide in tubercle bacillus. *Nat. Med.* 2(6): 662–667.

Shinnick TM, Good RC (1995) Diagnostic mycobacteriology laboratory practices. *Clin. Infect. Dis.* 21(2): 291–299.

Skamene E, Gros P, Forget A *et al.* (1982) Genetic regulation of resistance to intracellular pathogens. *Nature* 297(5866): 506–509.

Snewin VA, Gares MP, Gaora PO *et al.* (1999) Assessment of immunity to mycobacterial infection with luciferase reporter constructs. *Infect. Immun.* 67(9): 4586–4593.

Soini H, Bottger EC, Viljanen MK (1994) Identification of mycobacteria by PCR-based sequence determination of the 32-kilodalton protein gene. *J. Clin. Microbiol.* 32(12): 2944–2947.

Soini H, Pan X, Amin A *et al.* (2000) Characterization of *Mycobacterium tuberculosis* isolates from patients in Houston, Texas, by spoligotyping. *J. Clin. Microbiol.* 38(2): 669–676.

Spence DP, Hotchkiss J, Williams CS, Davies PD (1993) Tuberculosis and poverty. *BMJ* 307(6907): 759–761.

Sreevatsan S, Stockbauer KE, Pan X *et al.* (1997) Ethambutol resistance in *Mycobacterium tuberculosis*: critical role of *embB* mutations. *Antimicrob. Agents Chemother.* 41(8): 1677–1681.

Stenger S, Rosat JP, Bloom BR *et al.* (1999) Granulysin: a lethal weapon of cytolytic T cells. *Immunol. Today* 20: 390–394.

Stenger S, Modlin RL (1999) T cell mediated immunity to *Mycobacterium tuberculosis*. *Curr. Opin. Microbiol.* 2: 89–93.

Sudre P, ten Dam G, Kochi A (1992) Tuberculosis: a global overview of the situation today. *Bull. WHO* 70(2): 149–159.

Sultan L, Nyka W, Mills C *et al.* (1960) Tuberculosis disseminators: a study of the variability of aerial infectivity of tuberculosis patients. *Am. Rev. Resp. Dis.* 82: 358.

Talbot EA, Moore M, McCray E, Binkin NJ (2000) Tuberculosis among foreign-born persons in the United States, 1993–1998. *JAMA* 284(22): 2894–2900.

Telenti A, Imboden P, Marchesi F *et al.* (1993a) Detection of rifampicin-resistance mutations in *Mycobacterium tuberculosis*. *Lancet* 341: 647–650.

Telenti A, Marchesi F, Balz M *et al.* (1993b) Rapid identification of mycobacteria to the species level by polymerase chain reaction and restriction enzyme analysis. *J. Clin. Microbiol.* 31(2): 175–178.

Telenti A, Imboden P, Marchesi F, Schmidheini T, Bodmer T (1993c) Direct, automated detection of rifampin-resistant *Mycobacterium tuberculosis* by polymerase chain reaction and single-strand conformation polymorphism analysis. *Antimicrob. Agents Chemother.* 37(10): 2054–2058.

Tortoli E, Lavinia F, Simonetti MT (1998) Early detection of *Mycobacterium tuberculosis* in BACTEC cultures by ligase chain reaction. *J. Clin. Microbiol.* 36(9): 2791–2792.

Tsukamura M (1972) The pattern of resistance development to rifampicin in *Mycobacterium tuberculosis*. *Tubercle* 53: 111–117.

UKCP (2000) Control and prevention of tuberculosis in the United Kingdom: code of practice. *Thorax* 55(11): 887–901.

Valway SE, Greifinger RB, Papania M *et al.* (1994) Multidrug-resistant tuberculosis in the New York State prison system, 1990–1991. *J. Infect. Dis.* 170: 151–156.

Valway SE, Sanchez MP, Shinnick TF *et al.* (1998) An outbreak involving extensive transmission of a virulent strain of *Mycobacterium tuberculosis* [see comments] [published erratum appears in *N. Engl. J. Med.* 338(24):1783]. *N. Engl. J. Med.* 338(10): 633–639.

van Embden JD, Cave MD, Crawford JT *et al.* (1993) Strain identification of *Mycobacterium tuberculosis* by DNA fingerprinting: recommendations for a standardized methodology. *J. Clin. Microbiol.* 31(2): 406–409.

van Soolingen D, Hermans PW, de Haas PE, Soll DR, van Embden JD (1991) Occurrence and stability of insertion sequences in *Mycobacterium tuberculosis* complex strains: evaluation of an insertion sequence-dependent DNA polymorphism as a tool in the epidemiology of tuberculosis. *J. Clin. Microbiol.* 29(11): 2578–2586.

van Soolingen D, de Haas PE, Hermans PW, Groenen PM, van Embden JD (1993) Comparison of various repetitive DNA elements as genetic markers for strain differentiation and epidemiology of *Mycobacterium tuberculosis*. *J. Clin. Microbiol.* 31(8): 1987–1995.

Vidal SM, Pinner E, Lepage P, Gauthier S, Gros P (1996) Natural resistance to intracellular infections: *Nramp1* encodes a membrane phosphoglycoprotein absent in macrophages from susceptible (Nramp1 D169) mouse strains. *J. Immunol.* 157(8): 3559–3568.

van der Vliet GM, Cho SN, Kampirapap K van *et al.* (1996) Use of NASBA RNA amplification for detection of *Mycobacterium leprae* in skin biopsies from untreated and treated leprosy patients. *Int. J. Lepr. Other Mycobact. Dis.* 64(4): 396–403.

Vuorinen P, Miettinen A, Vuento R, Hallstrom O (1995) Direct detection of *Mycobacterium tuberculosis* complex in respiratory specimens by Gen-Probe Amplified *Mycobacterium Tuberculosis* Direct Test and Roche Amplicor *Mycobacterium Tuberculosis* Test. *J. Clin. Microbiol.* 33(7): 1856–1859.

Wells WF (1934) On airborne infection. II. Droplets and droplet nuclei. *Am. J. Hyg.* 20: 611.

WHO (1992a) *Expanded Program for Immunization. Program Report.* Geneva: World Health Organization.

WHO (1992b) Tuberculosis notification update. July. WHO/TB169. Geneva: World Health Organization.

WHO (1994) *TB a global emergency: WHO Report on the TB Epidemic.* Geneva: World Health Organization. Unpublished Document No. WHO/TB/94.177.

Whyte T, Cormican M, Hanahoe B *et al.* (2000) Comparison of BACTEC MGIT 960 and BACTEC 460 for culture of mycobacteria. *Diagn. Microbiol. Infect. Dis.* 38(2): 123–126.

Wilde IJ, Werely C, Beyers N, Donald P, van Helden PD (1994) Oligonucleotide (GTG)5 as a marker for *Mycobacterium tuberculosis* strain identification. *J. Clin. Microbiol.* 32: 1318–1321.

Woods GL (2000) Susceptibility testing for mycobacteria. *Clin. Infect. Dis.* 31(5): 1209–1215.

Woods GL (2001) Molecular techniques in mycobacterial detection. *Arch. Pathol. Lab. Med.* 125(1): 122–126.

Youmans GP, Karlson AG (1947) Streptomycin sensitivity of tubercle bacilli. Studies of recently isolated tubercle bacilli and the development of resistance to streptomycin *in vivo*. *Am. Rev. Tuber.* 55: 529.

Zhang Y, Heym B, Allen B, Young D, Cole S (1992) The catalase-peroxidase gene and isoniazid resistance of *Mycobacterium tuberculosis*. *Nature* 358(6387): 591–593.

82

Mycobacterium leprae

M.J. Colston

National Institute for Medical Research, London, UK

Mycobacterium leprae was one of the first organisms to be established as the cause of disease in humans (Hansen, 1874). In spite of this it is the least understood of any bacterium of medical importance. All mycobacteria are difficult to work with: the pathogens (particularly *M. leprae* and *M. tuberculosis*) have to be handled in high-containment facilities; their unusual cell envelope makes them difficult subjects for genetic and biochemical analysis; and they are generally difficult to grow, many of them having extremely long generation times. In fact *M. leprae* has not been grown at all in culture *in vitro*, so the only source of the organism is infected tissue, either from leprosy patients or experimentally infected animals. It is clear why *M. leprae* has not been an attractive subject for the attention of the microbiologist!

In spite of these difficulties there has seen considerable progress since the 1980s. The availability of large numbers of organisms which can be purified from the tissues of experimentally infected armadillos opened up possibilities for biochemical analysis; equally importantly it enabled the isolation of *M. leprae* DNA, and hence the introduction of molecular biological approaches to studying the organism and its close relatives. Remarkably, by the time this chapter goes to press, the entire genome sequences of both *M. tuberculosis* and *M. leprae* will be known, providing

new insights into the biological behaviour of these unusual and important pathogens.

The Disease

Ten years ago it was estimated that approximately 12 million people suffered from leprosy. However, the rigorous application of vertical control programmes in which multiple drug therapy has been effectively delivered to millions of leprosy patients has drastically reduced the prevalence of the disease. Nevertheless, in many parts of the world, while prevalence has fallen, the incidence (a reflection of the number of *new* patients who are diagnosed) has remained largely unaffected, and it seems likely that, for the foreseeable future, millions of people will continue to suffer the consequences of leprosy.

In common with other pathogenic mycobacteria, *M. leprae* is an intracellular pathogen. As with *M. tuberculosis* it can survive and grow within macrophages, but it also invades, survives and grows within the Schwann cells of peripheral nerves. It is this involvement with nerves that causes the characteristic pathology of the disease. Nerve damage, resulting in sensory loss, secondary tissue damage and a wide range of deforming sequelae results either from uninhibited growth of the bacillus within peripheral nerves, or from immunological damage as a consequence of strong cell-mediated immune responses in and around infected nerve cells. Effective immunity against *M. leprae* is mediated by T cells and macrophages (see below). Most infected individuals do not develop clinical disease, because they are able to mount a successful immune response. Those who do develop clinical leprosy develop symptoms that reflect their ability to mount an effective response. At one end of the spectrum, patients with lepromatous leprosy are unable to restrict growth of *M. leprae* because of a specific immunological defect. These lepromatous patients can harbour large numbers of organisms; they are thought to be the primary source of infection of new cases, they usually require antileprosy treatment for 2 years or longer, and are at greatest risk of developing secondary drug resistance. At the other end of the spectrum, patients with tuberculoid leprosy mount a strong, even exaggerated cell-mediated response to infection; they harbour small numbers of *M. leprae*

and are generally treated for shorter periods, usually 6 months. However, patients who can limit the infection are at high risk of severe nerve- and tissue-damaging immunological reactions. A schematic representation of the possible outcomes of infection with *M. leprae* is shown in **Fig. 1**.

One of the most important characteristics of leprosy is its slow and often insidious development. This is, at least in part, a reflection of the remarkably slow division time of the bacillus. It has been estimated that when growing optimally in experimental animals, *M. leprae* has a generation time of approximately 12 days (Shepard, 1960), compared with optimal generation times of approximately 14 hours for the 'slow-growing' *M. tuberculosis* and 20–30 minutes for many other bacteria. Another important characteristic of the organism is its predilection for cooler body sites. *M. leprae* is thought to have an optimal growth temperature of approximately 32°C. It does not appear to grow significantly in core body tissues such as liver, lungs and spleen, and this low optimal growth temperature has influenced the choice of animal models in which the organism is grown.

Leprosy in Experimental Animals

Experimental animal infections have played a major part in studying *M. leprae* because of the failure to grow the organism in culture medium. There have been attempts to grow *M. leprae* in a wide variety of species. However, laboratory rodents (especially mice) and nine-banded armadillos (*Dasypus novemcinctus Linn.*) have made the biggest contribution to the laboratory study of *M. leprae*, and it is these that are considered in this chapter.

The Mouse Footpad Model

The mouse footpad model was first described by Shepard (1960) and has since been used for a variety of applications. The original studies were carried out by inoculating *M. leprae* from nasal washings of lepromatous leprosy patients into the footpads of laboratory mice. There was limited and localised multiplication of the bacilli over a period of several months. It soon became clear that inoculation of smaller numbers of *M. leprae* was frequently more successful than inoculation of large numbers; this is because the mouse is relatively immune to infection with *M. leprae* and large numbers of organisms provoke an effective immune response which prevents further bacillary multiplication. The standard procedure became to inoculate

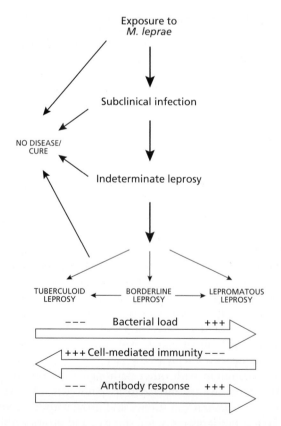

Fig. 1 Schematic of the possible outcomes following exposure to *M. leprae*. The vast majority of exposed individuals never develop clinical symptoms of disease. Those that do develop clinical leprosy exhibit symptoms which reflect their ability to mount an effective cell-mediated immune response against *M. leprae*.

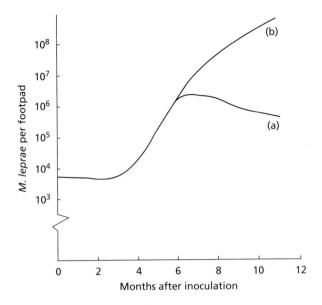

Fig. 2 Schematic of the growth of *M. leprae* in mouse foot-pads. (a) In immunologically normal mice the infection reaches a maximum of approximately 10^6 bacteria per footpad. (b) In immunocompromised mice the infection continues to increase beyond this ceiling and may disseminate to other superficial body sites such as tail and ears.

between 5000 and 10 000 bacteria into the hind footpads of mice. There is an apparent lag phase of about 2 months during which it is difficult to detect bacillary growth, followed by logarithmic growth in which numbers of organisms reach approximately 10^6 per footpad, approximately 6 months after infection (Fig. 2a). This logarithmic growth is followed by a multiplication 'ceiling' in which there is little change in total numbers of *M. leprae*, but loss of viability occurs (Shepard and Chang, 1967; Levy, 1970). The infection is essentially self-limiting and localised to the inoculation site, although some very slow spread of the organisms to other superficial tissues has been reported (Rees *et al.*, 1969).

The mouse footpad model has been extensively used to screen and evaluate potential antileprosy agents, to monitor the infectivity of biopsies taken from patients during therapy, and to screen for drug resistance. Such information has provided the essential rationale for devising the highly successful multi-drug therapies that have so dramatically reduced the prevalence of the disease. The major drawback of the technique is the time required for results to become available. For example, testing for drug sensitivity takes approximately 9 months, which emphasises the importance of designing drug regimens that allow for the possibility that the patients' bacilli are resistant to one of the drugs being used.

Immunosuppressed Rodents

The fact that growth of *M. leprae* is limited in normal mice by the immune response encouraged investigators to achieve higher yields of bacilli with immunocompromised animals. Rees (1966) used thymectomised mice that had received total-body irradiation with 9 Gy (900 rad). Such mice developed enhanced infection (see **Fig. 2b**) as did athymic mice (Colston and Hilson, 1976). Infected athymic mice develop grossly swollen footpads with extensive spread of bacilli to other superficial sites, such as tail skin, forepaws, etc. (Colston and Hilson, 1976). Heavily infected nude mice have been used as a source of bacilli for experiments on the metabolism of *M. leprae* (e.g. Franzblau, 1988). Neonatally thymectomised Lewis rats have also been used (Fieldsteel and Levy, 1976a, b); again these animals developed enhanced infections and generalised spread of bacteria, and have the added advantage of a longer lifespan than mice, making possible long-term chemotherapy experiments (Fieldsteel and Levy, 1976a).

The Nine-banded Armadillo

The mouse footpad model represented a major breakthrough as a tool for studying *M. leprae* growth and drug-sensitivity testing. However, even immunocompromised mice could not produce the large numbers of bacteria necessary to carry out biochemical, physiological and genetic studies. The ability to carry out such basic studies on *M. leprae* was revolutionised by the discovery that the nine-banded armadillo is highly susceptible to infection with *M. leprae* (Kirchheimer and Storrs, 1971; Storrs, 1971, 1978). Armadillos have a low core body temperature; although this does not entirely explain their susceptibility to *M. leprae*, it does mean that it is not only their superficial tissues which become infected, but also core tissues such as liver, spleen and lymph nodes. Such tissues can yield very large numbers of *M. leprae*, often more than 10^{10} bacilli per gram of tissue. This has provided sufficient quantities of bacteria not only to carry out extensive biochemical and genetic studies, but also to produce and test killed *M. leprae* vaccines on a large scale (Immunology of Mycobacterial Diseases Steering Committee, 1995; Gupte *et al.*, 1998).

Classification of *Mycobacterium leprae*

Mycobacteria belong to the high guanosine plus cytosine (G + C) subdivision of Gram-positive bacteria, which also includes such actinomycetes as *Streptomyces*,

Nocardia and corynebacteria. The DNA of most mycobacteria have between 60% and 67% G + C, compared with 72% for streptomycetes and 49% for *Escherichia coli*. However, *M. leprae* has an unusually low G + C content compared with other mycobacteria, with an estimated 56% G + C (Imaeda *et al.*, 1982; Clark-Curtiss *et al.*, 1985); it should be noted that a figure of 57.8% is given on the Sanger Centre website (www.sanger.ac.uk). This, and the inability to grow *M. leprae in vitro*, has meant that its precise taxonomic position has been in doubt; however, the availability of DNA isolated from armadillo-derived *M. leprae*, and the construction of genomic libraries, has made possible a molecular biological approach to classification.

In order to assist in classification, the ribosomal RNA (*rrn*) genes of mycobacteria have been extensively studied (e.g. Rogall *et al.*, 1990a, b). Like *M. tuberculosis*, *M. leprae* has a single *rrn* operon (Bercovier *et al.*, 1986). This has a classical structure with a leader region, 16S rRNA genes, intergene spacer 1, 23S rRNA gene, intergene spacer 2, and 5S rRNA genes. Sequencing of the *rrn* operon of *M. leprae* has revealed that it is closely related to other slow-growing mycobacteria (Smida *et al.*, 1988; Cox *et al.*, 1991; Kempsell *et al.*, 1992), but some unique features have been identified (Cox *et al.*, 1991). The close relationship between *M. leprae* and *M. tuberculosis* is confirmed by sequence comparisons between other genes (e.g. Davis *et al.*, 1994; Fsihi *et al.*, 1996a, b; Ainsa *et al.*, 1997; Azad *et al.*, 1997; Wu *et al.*, 1997). A comparison of the overall structure and organisation of the *M. leprae* and *M. tuberculosis* genomes suggests, however, that there is a complex mosaic of regions of high homology and similar organisation surrounded by non-homologous chromosomal segments (Philipp *et al.*, 1996, 1998; Smith *et al.*, 1997) (see The *M. leprae* Genome, p. 1754).

The Structure of *Mycobacterium leprae*

Much of what is known about the structure of *M. leprae* is derived either from analogy with other mycobacteria, or from studies carried out on armadillo-derived organisms. As far as is known, *M. leprae* has a classical bacterial structure with a single circular chromosome, a cell membrane and a complex cell wall similar to that of other mycobacteria.

The Cell Envelope

The mycobacterial cell envelope is a unique and complex structure which is thought to contribute to the ability of mycobacteria to survive in a variety of environments, some of them extremely hostile. *M. leprae* has to survive and grow within host cells; the cell envelope must act as an interface and barrier between the bacillus and the hostile cell. It is involved in many complex interactions which result in both stimulation and suppression of host cell responses, as well as acting as a permeability barrier.

Mycobacteria are Gram-positive organisms, and like other such bacteria their cell wall has a peptidoglycan backbone and cross-linked polysaccharide. Unlike other Gram-positive bacteria, however, mycobacteria have complex lipid molecules attached to or associated with the cell wall skeleton. It is thought that this forms a second hydrophobic layer analogous to the outer membrane of Gram-negative organisms. Inside the cell wall is a plasma membrane similar to the plasma membranes of other bacteria. Outside the cell wall is a capsule that consists mainly of proteins, polysaccharides and particularly glycolipid (**Fig. 3**).

The Mycobacterial Plasma Membrane

The mycobacterial plasma membrane is similar in principle to other cell membranes, with polar lipids, mainly phospholipids, and proteins assembled into a lipid bilayer. The main phospholipids are phosphatidylinositol mannosides (PIM), phosphatidylglycerol, cardiolipin and phosphatidylethanolamine. Other lipids, such as menaquinones, are also present in mycobacterial plasma membranes (Minnikin, 1982).

In addition to proteins involved in electron transport, two major membrane proteins have been described in *M. leprae* (Hunter *et al.*, 1990). One of these, MMP-II, has been characterised as a bacterioferritin (Pessolani *et al.*, 1994). The other, MMP-I, consists of a 35-kDa protein assembled into a multimeric complex. The function of MMP-I is unknown, but the corresponding gene has been sequenced and the protein has significant homology to a protein that is up-regulated under conditions of sulphur stress in *Synechoccus* spp. and a protein which accumulates during phosphate starvation in *Streptomyces griseus* (Triccas *et al.*, 1998). Iron-regulated envelope proteins have also been detected in membrane fractions of *M. leprae* (Sritharan and Ratledge, 1990), suggesting that nutrient stress is an important part of the *M. leprae* environment.

An important constituent of the mycobacterial cell envelope is lipoarabinomannan (LAM). Although the localisation of LAM within the envelope is unclear, it seems likely that its phosphatidylinositol group anchors it within the plasma membrane (Hunter and Brennan, 1990), although the core structure of mannan backbone, with 6- and 2,6-linked manno-pyranoses and branched arabinofuranosyl-containing chains, probably spans the cell wall (**Fig. 3**). LAM has a wide range of biological activities, which emphasises its importance to the host–parasite interaction. LAM induces synthesis of a range cytokines by host cells, particularly by macrophages (Barnes *et al.*, 1992; Chatterjee *et al.*, 1992; Adams *et al.*, 1993; Bradbury and Moreno, 1993; Roach *et al.*, 1993). It is also thought to be involved in phagocytosis of myco-bacteria (Schlesinger *et al.*, 1994; Stokes and Speert, 1995), and in protecting mycobacteria against oxygen radical intermediates produced by macrophages as part of the defence response (Chan *et al.*, 1991).

The Mycobacterial Cell Wall

The mycobacterial cell-wall skeleton consists of a peptidoglycan backbone covalently linked to arabino-galactan. The mycobacterial peptidoglycan is broadly similar to that of other bacteria, although *M. leprae* has an unusual feature in that L-alanine is replaced by glycine (Draper *et al.*, 1987). Arabinogalactan is composed of D-arabinofuranosyl and D-galactofuranosyl residues (Draper, 1987).

A third constituent of the mycobacterial cell wall are mycolic acids which are attached to the terminal penta-arabinan motifs of the arabinogalactan. Although mycolic acids are found in other bacterial genera, such as *Corynebacterium* and *Nocardia*, mycobacterial mycolic acids are unusual in that they are larger and possess longer side-chains. Mycobacteria generally have a mixture of different types of mycolic acids, and their analysis has been used to identify mycobacterial species (Daffe *et al.*, 1983; Minnikin *et al.*, 1983) *M. leprae* has two types of mycolic acid: α-mycolates and ketomycolates, but appears to lack methoxymycolates which are found in *M. tuberculosis* and other mycobacteria (Draper *et al.*, 1982).

The Mycobacterial Capsule

Many years ago an electron transparent zone was identified as a layer present in many pathogenic mycobacteria when growing inside host cells. The murine leprosy bacillus, *M. lepraemurium*, was seen in electron micrographs to be surrounded by a fibrillar substance inside the phagocytic vacuoles of host macrophages (Draper and Rees, 1973). *M. leprae* in phagosomes was surrounded by a distinctive vacuolar matrix (Nishiura *et al.*, 1977). This substance was subsequently shown to be a unique phenolic glyco-lipid, called PGL-1 (**Fig. 3**), which is produced in large quantities by growing *M. leprae*. It is serologically

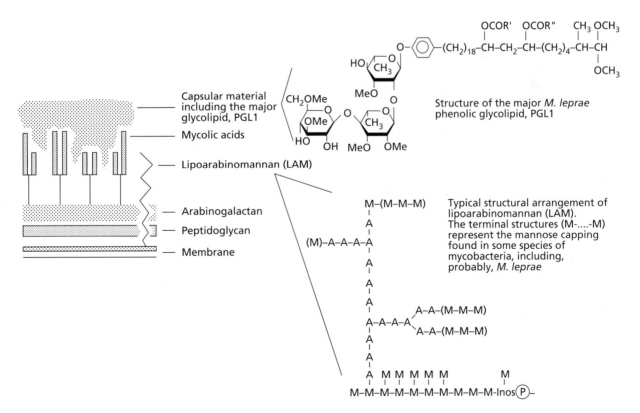

Fig. 3 Schematic of the envelope of *M. leprae*. The general composition is similar to that of other mycobacteria. Two important immunological components are LAM and PGL1 (see right-hand side of the diagram).

highly active (Hunter *et al.*, 1982), and it is thought to protect the intracellular bacilli from the toxicity of radical oxygen intermediates produced by macrophages as part of the host defence mechanism (Chan *et al.*, 1989).

Although PGL-1 is the major constituent of the *M. leprae* capsule, other mycobacteria which do not produce significant amounts of phenolic glycolipids are also seen to be surrounded by an electron-transparent zone. It is thought that in these cases this outer layer consists of proteins and polysaccharides, and it seems likely that such molecules, along with PGL-1, are also constituents of the *M. leprae* capsule.

The *Mycobacterium leprae* Genome

M. leprae appears to have a smaller genome than other eubacteria. The *M. tuberculosis* genome is similar in size to other bacteria, consisting of 4.4 Mb of DNA encoding approximately 4000 genes (Cole *et al.*, 1998). The *M. leprae* genome is approximately 3.27 Mb (see Sanger Centre website at http://www.sanger.ac.uk/Projects/Microbes/). This might suggest that the *M. leprae* genome has shed noncoding DNA; but sequencing of the *M. leprae* genome has revealed that not only is the chromosome small, it also has a low gene density and hence appears to have fewer genes than other bacteria (S.T. Cole, personal communication).

Sequencing of the *M. leprae* cosmid covering the Rif-Str regions revealed that approximately 50% of the potential coding sequence was used (Honore *et al.*, 1993a). It is now clear that this is a general feature of the *M. leprae* genome. Comparison of the *M. leprae* and *M. tuberculosis* genomes has revealed blocks of sequence with a high degree of organisational and sequence homology. In *M. leprae*, however, these are often separated by DNA segments that appear to have degenerated, so that only non-functional gene vestiges can be identified. This is exemplified by the *kat*G locus of *M. leprae* (Eiglmeier *et al.*, 1997). The KatG protein is a haem-containing catalase-peroxidase which is thought to be important for the survival of intracellular bacteria, in that it assists in the detoxication of reactive oxygen intermediates produced by macrophages in response to infection (Chan and Kaufmann, 1994). Mutations in *kat*G result in reduced virulence of *M. bovis* (Wilson *et al.*, 1995) and *M. tuberculosis* (Mitchison *et al.*, 1963). Comparison of the *M. leprae kat*G sequence with those of other mycobacteria reveals, however, that the *M. leprae* gene exists as a vestigial remnant of a gene with large deletions, many frame-shift mutations and a number of in-frame stop codons that disrupt regions of high sequence homology.

Unusual Genetic Structures and Repetitive Elements

The *M. tuberculosis* genome contains a range of genetic elements, repetitive sequences and insertion elements. Some of these – for example, the transposable element-related sequence IS*6110* – have proved extremely valuable as molecular epidemiological markers (Thierry *et al.*, 1990; Hermans *et al.*, 1992; Small *et al.*, 1994). Most of these elements are not present in the *M. leprae* genome. However, two unusual elements are worthy of note. The *M. leprae* chromosome has a large number (*c.* 30) of a variable repeating structure termed RLEP (Clark-Curtiss and Docherty, 1989; Grosskinsky *et al.*, 1989; Woods and Cole, 1990). Each RLEP consists of a conserved 545-bp core with variable flanking sequences (Woods and Cole, 1989; Eiglmeier *et al.*, 1993; Fsihi and Cole, 1995). The function of RLEP is not known. Fsihi and Cole (1995) have, however, reported an unusual arrangement for the *M. leprae* DNA polymerase (*polA*) gene in which *polA* is flanked by two copies of RLEP to form a mobile element-like structure. A second copy of this element was found within the genome of some but not all isolates of *M. leprae*, consistent with the view that this has, at least at some time, been a mobile element.

Another unusual genetic element which appears to be over-represented in the *M. leprae* genome is a protein splicing element, called an 'intein', which was initially found in the *recA* gene of *M. tuberculosis* (Davis *et al.*, 1991, 1992). A similar though unrelated structure was found in the *M. leprae recA* gene (Davis *et al.*, 1994). Subsequently two other protein splicing inteins have been identified in *M. leprae* genes, one in *gyrA* (Fsihi *et al.*, 1996b) and one in a gene of unknown function (Pietrokovski, 1994). Genes with protein splicing elements are transcribed and translated to produce a precursor protein which then undergoes an unusual post-transitional modification in which the central intein region splices out and the N- and C-termini join together to form the mature protein (Colston and Davis, 1994a, b). Although protein splicing inteins appear to be common features of archaebacterial genomes (Bult *et al.*, 1996), they are rare in eubacteria, and so the presence of several such elements in the *M. leprae* genome is surprising – and, as yet, unexplained.

The *rrn* Operon of *Mycobacterium leprae*

Most eubacteria have multiple copies of *rrn* operons, which enable them to respond to different growth conditions. For example, *E. coli* has seven *rrn* operons. Mycobacteria have many fewer operons; most fast-growers have two, while the slow-growers, including *M. leprae*, have a single operon (Bercovier *et al.*, 1986). Thus, *M. leprae* has reduced capacity to produce ribosomes based on rRNA gene dosage. In addition, the single rRNA operon of *M. leprae* (and *M. tuberculosis*) has a reduced number of promoters compared with fast-growing species (Gonzalez-y-Merchand *et al.*, 1996, 1997) and the remaining promoters are weaker (Gonzalez-y-Merchand *et al.*, 1997). Hence the production of ribosomes, and hence the efficiency of protein synthesis, is reduced not only by minimal gene dosage, but also by promoter number and strength, and by the chromosomal location.

Comparison of rRNA nucleotide sequences has also been used to establish phylogenetic relationships in mycobacteria (Rogall *et al.*, 1990a, b). This was of particular interest for *M. leprae*, because of the difference in G + C content as compared with other mycobacteria (see above), plus the inability to grow the organism *in vitro*. It is clear, when phylogenetic analysis of rRNA genes is applied to the mycobacteria, that there is a true phylogenetic division between fast- and slow-growers (Stahl and Urbance, 1990), and that *M. leprae* is correctly positioned in the slow-growing mycobacteria group (Smida *et al.*, 1988; Cox *et al.*, 1991; Kempsell *et al.*, 1992)

Antileprosy Drugs: Modes of Action and Resistance

The three most important antileprosy drugs are dapsone, rifampicin and clofazimine. Used in combination these three drugs have been the cornerstone of leprosy control. Two regimens are commonly used, one for paucibacillary leprosy and one for multibacillary leprosy. In addition, a third regimen of a single dose of rifampicin, ofloxacin and minocycline has recently been introduced for individuals found to have a single lesion. Rifampicin is important because of its very rapid bactericidal activity against *M. leprae*. Dapsone and clofazimine are more weakly bactericidal. The combinations of the three drugs recommended by WHO (see **Fig. 4**) have proved extremely successful in treating and controlling leprosy.

Rifampicin
Rifampicin, unlike either dapsone or clofazimine, has a broad spectrum of activity against many bacterial genera. The mode of action of rifampicin involves blocking transcription by binding to RNA polymerase.

(a) MULTIBACILLARY LEPROSY

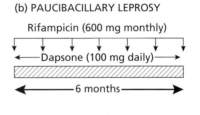

(b) PAUCIBACILLARY LEPROSY

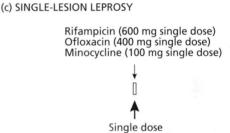

(c) SINGLE-LESION LEPROSY

Rifampicin (600 mg single dose)
Ofloxacin (400 mg single dose)
Minocycline (100 mg single dose)

Single dose

Fig. 4 Schematic of typical chemotherapeutic regimens recommended for the treatment of leprosy. The regimens differ depending on the bacterial loading of the patient: (a) high load, (b) low load, (c) very low load. The regimens combine supervised drug administration (monthly dosage) and unsupervised self-administration (daily dosages).

The study of rifampicin-resistant mutants of *E. coli* has identified a region of approximately 30 amino acids in the β subunit of RNA polymerase encoded by the *rpoB* gene. Mycobacterial rifampicin-resistant isolates also have mutations within this same stretch of 90 nucleotides (Telenti *et al.*, 1993); in *M. leprae* the majority of rifampicin-resistant strains contain a Ser531Leu mutation (Honore and Cole, 1993; Honore *et al.*, 1993a). This concentration of mutations within a small region of DNA has made possible development of a rapid method for detecting rifampicin resistance (Honore *et al.*, 1993b), which is particularly important for the slow-growing *M. tuberculosis* and for *M. leprae* which cannot be grown *in vitro*.

Dapsone
Dapsone (diamino-diphenyl sulphone, DDS), was the first drug to be used against leprosy on a large scale. It was introduced in the early 1950s, and used as monotherapy for the next two or three decades. In the

mid 1960s, however, the first cases of secondary dapsone resistance were reported (Pettit *et al.*, 1966). In the mid 1970s, primary dapsone resistance was found to be widespread (Pearson *et al.*, 1977) and ushered in the use of combined drug regimens.

Dapsone, like other sulphones, acts on the folate biosynthetic pathway. It is a competitive inhibitor of dihydropteroate synthetase and has a remarkably high affinity for the *M. leprae* enzyme, resulting in sensitivity of the organism to very low concentrations of approximately 3 ng/mL (Ellard, 1990). This exquisite sensitivity is unusual even amongst mycobacteria. Only one other species, *Mycobacterium lufu*, shares such a low minimal inhibitory concentration of dapsone (Kulkarni and Seydel, 1983).

Dapsone resistance has been thought to result from over-expression of the target enzyme, dihydropteroate synthetase, rather than structural changes in the enzyme (Kulkarni and Seydel, 1983). Thus, the competitive inhibitor is itself competitively inhibited in resistant strains. Recent studies based on sequence analysis of dapsone-resistant strains of *M. leprae* has, however, revealed mutations in the dihydropteroate synthase gene, *folP* (Kai *et al.*, 1999).

Clofazimine

Clofazimine, the third component of multiple drug regimens used against leprosy, is a rhiminophenozine dye (Gatti, 1975). It is unusual in that it accumulates within cells and adipose tissues, often as crystals (Banerjee *et al.*, 1974), and on prolonged treatment the skin coloration of patients becomes darker, because the dark-red drug accumulates (Levy and Randall, 1970). The antibacterial action of clofazimine is not understood at the molecular level, and it is difficult to measure its minimal inhibitory concentration against *M. leprae*, because of its accumulation in tissue. This tissue accumulation is, however, useful in that it results in slow release of bio-available drug, making intermittent administration a possibility.

Clofazimine has an additional property which makes it an important component of antileprosy regimens. As well as its antibacterial activity, it also has an anti-inflammatory activity which assists in controlling certain immunopathological reactions that are important causes of morbidity in multibacillary leprosy (Karat *et al.*, 1970; Warren, 1970; Helmy *et al.*, 1971).

New Drugs Active Against *Mycobacterium leprae*

Although the rifampicin, dapsone and clofazimine combinations have proved very successful in leprosy control programmes, a number of additional drugs has recently been shown to be active against *M. leprae* (Ji *et al.*, 1991), and their use in therapeutic regimens is currently being investigated.

Ofloxacin, and other fluoroquinolones such as sparfloxacin, are significantly bactericidal against *M. leprae* (Saito *et al.*, 1986; Franzblau and White, 1990). These are broad-spectrum agents, whose molecular target is a subunit of the enzyme DNA gyrase, and are involved in maintaining the tertiary conformational structure of DNA. The tetracycline-related antibiotic minocycline is active against *M. leprae* in experimental animals (Gelber, 1987), and in man (Gelber *et al.*, 1992). In some parts of the world a single dose of combined rifampicin/ofloxacin/minocycline is recommended for the treatment of leprosy patients exhibiting single lesions (see **Fig. 4**; Anonymous, *Indian J. Leprosy*, 69, pp. 121–129).

The macrolide clarithromycin is another drug with significant activity against *M. leprae* (Franzblau and Hastings, 1988). Macrolides interfere with protein synthesis by blocking ribosome function (Malmborg, 1986). Clarithromycin is bactericidal against *M. leprae* in experimentally infected mice (Ji *et al.*, 1991; Walker *et al.*, 1993).

Invasion of Host Cells by *Mycobacterium leprae*

Macrophages

M. leprae is an obligate intracellular pathogen. It invades, survives and grows within host cells, most notably macrophages and Schwann cells of peripheral nerves. Mycobacteria are phagocytosed by macrophages by a receptor–ligand interaction. Complement receptors CR1 (CD35), CR3 (CD11b/CD19), and CR4 (CD11c/CD18) are important (Schlesinger *et al.*, 1990, 1994; Schlesinger and Horwitz, 1990, 1994; Schlesinger, 1993; Stokes *et al.*, 1993), as are mannose receptors (Schlesinger, 1993). The mycobacterial molecules that participate in these interactions are less clearly understood. The macrophage mannose receptor probably interacts with the terminal mannosyl units of mannose-capped LAM (see above). A macrophage invasion mechanism claimed to be specific for pathogenic mycobacteria (*M. leprae* and *M. tuberculosis*), but not used by non-pathogenic mycobacteria (*M. vaccae*, *M. smegmatis* and *M. phlei*) or by other intracellular pathogens (*Leishmania mexicana*, *Listeria monocytogenes*, *Staphylococcus* and *Nocardia asteroides*) has been described (Schorey *et al.*, 1997). In this, a complex is formed between the mycobacteria and C2a, a protein formed by the cleavage of the complement component C2. The mycobacteria-associated c2a is capable of cleaving the C3 component of complement, resulting in opsonisation by C3b, the ligand for complement receptor C31. The mycobacterial component responsible for association with C2a has not been identified.

Schwann Cells

A unique property of *M. leprae* is its ability to invade Schwann cells of peripheral nerves (Mukherjee and Antia, 1986). Recent work has identified the Schwann-cell receptor molecule responsible for this interaction. The Schwann cell is covered by a basal lamina made up of a number of extracellular matrix components including laminins (Cornbrooks *et al.*, 1983). The laminins are glycoproteins that consist of α, β and γ chains, and exist in multiple isoforms. In peripheral nerves the most common isoform is laminin-2 ($\alpha2$, $\beta1$ and $\gamma1$) (Villanova *et al.*, 1997), an isoform not found in the central nervous system. *M. leprae* was found to bind to the G domain of the laminin $\alpha2$ molecule (Rambukkana *et al.*, 1997). The same group (Rambukkana *et al.*, 1998) then went on to show that the *M. leprae*/laminin/$\alpha2$ complex is in turn bound to α-dystroglycan which is expressed on the Schwann-cell surface and can bind to components of the extracellular matrix of the basal lamina (Matsumura *et al.*, 1997). The α-dystroglycan/laminin/$\alpha2$/*M. leprae* complex then associates with the transmembrane protein β-dystroglycan, which in turn binds to

intracellular actin. Thus there is a bridge of four proteins which link the extracellular *M. leprae* bacilli to the intracellular cytoskeleton of the Schwann cell, and ultimately leads to internalisation of the bacteria (**Fig. 5**) The question that then remained was: what is the *M. leprae* component that makes the initial association and binding with laminin-2? Since the invasion of Schwann cells by bacteria is extremely unusual, it seemed reasonable to assume that this binding molecule would be an uncommon surface component of the leprosy bacillus. This has been confirmed (Shimoji *et al.*, 1999). A *M. leprae* cell-wall protein, ML-LBP21, has been identified with a laminin-2 probe. This protein binds strongly to human $\alpha2$ laminins. It is a protein of 21 kDa, but it migrates as a 28-kDa protein in SDS/PAGE, suggesting that it may be post-translationally modified, possibly by glycosylation. Interestingly, when polystryrene beads were coated with recombinant ML-LBP21, they were ingested by cultured primary Schwann cells *in vitro*.

Genes Involved in Intracellular Survival

Intensive current research is aimed at understanding the pathogenicity and intracellular survival of mycobacteria. Most of this research is focused on *M. tuberculosis* in which using a molecular genetic approach it is now possible to address this problem. In the absence of similar techniques for *M. leprae* the approach has been to carry out comparative genomics, with *M. tuberculosis* providing a useful reference point.

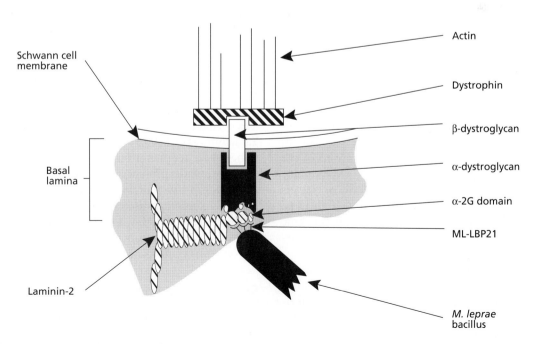

Fig. 5 The proteins involved in the attachment to, and subsequent invasion of, Schwann cells by *M. leprae*.

In addition to invasion of Schwann cells, *M. leprae* has to survive within the hostile environment of the host macrophage. If comparisons with *M. tuberculosis* are appropriate, then this is probably achieved by a combination of mechanisms, including altering the macrophage's response to infection, and mechanisms which involve detoxication of antimicrobial products. The precise mechanisms by which mycobacteria are killed by activated macrophages during the expression of an effective immune response are unknown, but the role of radical oxygen intermediates and radical nitrogen intermediates (ROI and RNI, respectively) is thought to be important.

As mentioned previously, *M. leprae* is surrounded by a thick 'capsule', the major component of which is the specific phenolic glycolipid, PGL-1 (Hunter *et al.*, 1982). PGL-1 is involved in protection against ROI (Chan *et al.*, 1989), and it may well be involved in general protection against other antimicrobial effector mechanisms. The importance of these capsular lipid molecules in mycobacterial pathogenicity has been illustrated recently by means of a genetic approach to studying the growth of *M. tuberculosis* in mice (Cox *et al.*, 1999).

Other molecules which protect against ROI include superoxide dismutase (SOD), thioredoxin, catalases and peroxidases. Most of these enzymes form part of a variety of regulatable responses, induced by exposure of ROI and oxidative stress. SOD is a major *M. leprae* antigen (Thangaraj *et al.*, 1990) and one of the most prominent proteins in the cytoplasm of *M. leprae* isolated from the tissues of experimentally infected armadillo (Hunter *et al.*, 1990), confirming its likely role in intracellular survival.

Thioredoxin is known to scavenge ROI by forming a complex with thioredoxin reductase and NADPH (Fernando *et al.*, 1992). Interestingly, the thioredoxin/thioredoxin reductase system of *M. leprae* is unusual. In other organisms, thioredoxin and thioredoxin reductase are separate proteins encoded by two genes. In *M. leprae*, however, there is a single hybrid protein, encoded by a single gene which is comprised of the active moieties of both proteins linked by a hydrophilic spacer (Wieles *et al.*, 1995). This hybrid protein is more active than the complexes of two individual proteins and, when expressed in the non-pathogenic *Mycobacterium smegmatis*, increases its ability to survive in macrophages (Wieles *et al.*, 1997).

M. leprae appears to lack catalase activity (Eiglmeier *et al.*, 1997), which is rather surprising for an intracellular pathogen. It does, however, have an alkyl-hydroperoxide reductase (*ahpC*) and produces an OxyR protein (Deretic *et al.*, 1995; Sherman *et al.*, 1996; Dhandayuthapani *et al.*, 1997). The OxyR protein is a positive regulator of several genes involved in the oxidative stress response. It is arranged divergently with *ahpC* and binds to an OxyR recognition site. Interestingly the *oxyR* gene of *M. tuberculosis* contains a number of lesions, rendering it inactive (Deretic *et al.*, 1995; Dhandayuthapani *et al.*, 1997). The fact that *M. tuberculosis* is *katG*-positive and *oxyR*-negative, while *M. leprae* is *katG*-negative and *oxyR*-positive, emphasises the caution that needs to be applied in deducing the physiological characteristics of *M. leprae* by analogy with those of *M. tuberculosis*.

Immune Responses to *Mycobacterium leprae*

The Immunological Spectrum

Leprosy has a wide range of clinical manifestations which are a reflection of the type and extent of the immune response. At one extreme, in lepromatous leprosy (see **Fig. 1**), there is a highly selective anergy of the T-cell response to *M. leprae*; T cells do not proliferate and do not produce 'type 1' cytokines such as interferon-gamma (IFNγ) and interleukin 2 (IL-2), in response to *M. leprae*. This lack of a T-cell response results in a failure to activate macrophages, resulting in failure to kill or limit the growth of *M. leprae* and failure to produce the cytokine IL-12 which is important in sustaining a T-cell response. By contrast, patients with tuberculoid leprosy who are at the other end of the spectrum have T cells which proliferate and produce IFNγ in response to *M. leprae*; these patients have strongly activated macrophages and as a consequence have a limited and localised form of infection (see **Fig. 1**). Patients who fall in between these two extremes are immunologically rather unstable and may shift up and down the spectrum as a result of transient changes in immune responsiveness. Interestingly, Yamamura *et al.* (1991) used tissue taken from lesions, rather than peripheral blood cells, and were able to demonstrate the presence of mRNA that encodes IL-2 and IFNγ in tuberculoid lesions, and IL-4, IL-5 and IL-10 in lepromatous lesions. This has led to the concept that the TH-1 and TH-2 types of response are responsible for tuberculoid (paucibacillary) and lepromatous (multibacillary) leprosy respectively.

Reactions and Immunopathology

Between the lepromatous and tuberculoid poles are borderline patients who are immunologically unstable. A significant percentage of these patients experience so-called 'reversal reactions' (or type 1 reactions), in which there is a sudden increase in cell-mediated immunity towards *M. leprae*. This increase in T-cell responsiveness shifts the patients towards the tuberculoid end of the spectrum, but as a consequence they experience local tissue damage (Lockwood *et al.*, 1993; Khanolkar-Young *et al.*, 1995). This tissue damage can take the form of acute inflammation of skin lesions and acute neuritis. Neuritis resulting from reversal reactions can be rapid and severely damaging, resulting in permanent nerve damage and possible deformity. The fact that this severe immunopathology is a consequence of an increased TH-1 type response, with increased expression of TNFα, IFNγ and IL-2, emphasises that protective immunity can be a double-edged sword, on the one hand promoting antibacterial immunity while on the other leading to severe tissue damage and deformity.

In addition to reversal, or type 1 reactions, patients at the lepromatous end of the spectrum may undergo repeated episodes of type 2 reactions, or erythema nodosun leprosum (ENL). The ENL reaction is systemic in nature, with fever, joint pains and dermal nodules. ENL is associated with enhanced B-cell activation (Nath, 1983) and transient peripheral T-cell activation (Laal *et al.*, 1985), with IL-2-producing T cells increasing in the tissues (Modlin *et al.*, 1986).

Thus while the nature of the immune response is responsible for the severity of the level of infection, perturbations in the immune response are also associated with immunopathological consequences.

Conclusion

The way in which leprosy is diagnosed, managed and studied in the laboratory is dictated by the slow growth and lack of cultivability of *M. leprae*. The current genome sequencing project is likely greatly to enhance our understanding of these features of the organism, and lead to new insights into its pathogenicity. The availability of effective antileprosy drugs is likely to reduce the prevalence of the disease still further, although we still know little about its route of transmission, and to date there has been little impact on the incidence of the disease in many parts of the world. The immunological consequences of leprosy, particularly nerve damage, can occur even in the face of effective chemotherapy. This means that leprosy will remain a significant health problem in many parts of the world for the foreseeable future.

Note added in proof:

The sequencing of the *M. leprae* genome has now been completed and published (Cole *et al.*, 2001).

Apart from being smaller than the *M. tuberculosis* genome (3.27 Mb compared with 4.41 Mb) it is estimated that only 49.5% of the *M. leprae* genome contains protein-encoding genes, with 27% containing pseudogenes (inactive 'genes' with recognisable counterparts in *M. tuberculosis*) and 23.5% are non-coding. This gene decay has probably occurred by recombination processes, by which genes that are no longer required for the highly specialised niche in which *M. leprae* lives are destroyed. It appears that *M. leprae* is the most advanced case of this 'reductive evolution' which has been seen to-date.

References

Adams LB, Fukutomi Y, Krahenbuhl JL (1993) Regulation of murine macrophage effector functions by lipoarabinomannan from mycobacterial strains with different degrees of virulence. *Infect. Immun.* 61: 4173–4181.

Ainsa JA, Perez E, Pelicic V, Berthet FX, Gicquel B, Martin C (1997) Aminoglycoside 2'-N-acetyltransferase genes are universally present in mycobacteria: characterization of the aac(2')-Ic gene from *Mycobacterium tuberculosis* and the aac(2')-Id gene from *Mycobacterium smegmatis*. *Mol. Microbiol.* 24: 431–441.

Azad AK, Sirakova TD, Fernandes ND, Kolattukudy PE (1997) Gene knockout reveals a novel gene cluster for the synthesis of a class of cell wall lipids unique to pathogenic mycobacteria. *J. Biol. Chem.* 272: 16741–16745.

Banerjee DK, Ellard GA, Gammon PT, Waters MF (1974) Some observations on the pharmacology of clofazimine (B663). *Am. J. Trop. Med. Hygiene* 23: 1110–1115.

Barnes PF, Chatterjee D, Abrams JS *et al.* (1992) Cytokine production induced by *Mycobacterium tuberculosis* lipoarabinomannan: relationship to chemical structure. *J. Immunol.* 149: 541–547.

Bercovier H, Kafri O, Sela S (1986) Mycobacteria possess a surprisingly small number of ribosomal RNA genes in relation to the size of their genome. *Biochem. Biophys. Res. Comm.* 136: 1136–1141.

Bradbury MG, Moreno C (1993) Effect of lipoarabinomannan and mycobacteria on tumour necrosis factor production by different populations of murine macrophages. *Clin. Exp. Immunol.* 94: 57–63.

Bult CJ, White O, Olsen GJ *et al.* (1996) Complete genome sequence of the methanogenic archaeon, *Methanococcus jannaschii*. *Science* 273: 1058–1073.

Chan J, Kaufmann SHE (1994) Immune mechanisms of protection. In: Bloom B (ed.) *Tuberculosis: Pathogenesis, Protection and Control*. Washington, DC: ASM Press, pp. 389–415.

Chan J, Fujiwara T, Brennan PJ *et al.* (1989) Microbial glycolipids: possible virulence factors that scavenge oxygen radicals. *Proc. Natl Acad. Sci. USA* 86: 2453–2457.

Chan J, Fan XD, Hunter SW, Brennan PJ, Bloom BR (1991) Lipoarabinomannan, a possible virulence factor involved in persistence of *Mycobacterium tuberculosis* within macrophages. *Infect. Immun.* 59: 1755–1761.

Chatterjee D, Roberts AD, Lowell K, Brennan PJ, Orme IM (1992) Structural basis of capacity of lipoarabinomannan to induce secretion of tumor necrosis factor. *Infect. Immun.* 60: 1249–1253.

Clark-Curtiss JE, Docherty MA (1989) A species-specific repetitive sequence in *Mycobacterium leprae* DNA. *J. Infect. Dis.* 159: 7–15.

Clark-Curtiss JE, Jacobs WR, Docherty MA, Ritchie LR, Curtiss RD (1985) Molecular analysis of DNA and construction of genomic libraries of *Mycobacterium leprae*. *J. Bacteriol.* 161: 1093–1102.

Cole ST, Brosch R, Parkhill J *et al.* (1998) Deciphering the biology of *Mycobacterium tuberculosis* from the complete genome sequence. *Nature* 393: 537–544.

Cole ST, Eiglmeier K, Parkhill J (2001) Massive gene decay in the leprosy bacillus. *Nature* 409: 1007–1011.

Colston MJ, Davis EO (1994a) The ins and outs of protein splicing elements. *Mol. Microbiol.* 12: 359–363.

Colston MJ, Davis EO (1994b) In: Bloom B (ed.) *Tuberculosis: Pathogenesis, Protection and Control.* Washington, DC: ASM Press, pp. 217–226.

Colston MJ, Hilson GR (1976) Growth of *Mycobacterium leprae* and *M. marinum* in congenitally athymic (nude) mice. *Nature* 262: 399–401.

Cornbrooks CJ, Carey DJ, McDonald JA, Timpl R, Bunge RP (1983) *In vivo* and *in vitro* observations on laminin production by Schwann cells. *Proc. Natl Acad. Sci. USA* 80: 3850–3854.

Cox JS, Chen B, McNeil M, Jacobs WR (1999) Complex lipid determines tissue-specific replication of *Mycobacterium tuberculosis* in mice. *Nature* 402: 79–83.

Cox RA, Kempsell K, Fairclough L, Colston MJ (1991) The 16S ribosomal RNA of *Mycobacterium leprae* contains a unique sequence which can be used for identification by the polymerase chain reaction. *J. Med. Microbiol.* 35: 284–290.

Daffe M, Laneelle MA, Asselineau C, Levy-Frebault V, David H (1983) Taxonomic value of mycobacterial fatty acids: proposal for a method of analysis. *Ann. Microbiol. (Paris)* 134B: 241–256.

Davis EO, Sedgwick SG, Colston MJ (1991) Novel structure of the *recA* locus of *Mycobacterium tuberculosis* implies processing of the gene product. *J. Bacteriol.* 173: 5653–5662.

Davis EO, Jenner PJ, Brooks PC, Colston MJ, Sedgwick SG (1992) Protein splicing in the maturation of *M. tuberculosis recA* protein: a mechanism for tolerating a novel class of intervening sequence. *Cell* 71: 201–210.

Davis EO, Thangaraj HS, Brooks PC, Colston MJ (1994) Evidence of selection for protein introns in the *recAs* of pathogenic mycobacteria. *EMBO J.* 13: 699–703.

Deretic V, Philipp W, Dhandayuthapani S *et al.* (1995) *Mycobacterium tuberculosis* is a natural mutant with an inactivated oxidative-stress regulatory gene: implications for sensitivity to isoniazid. *Mol. Microbiol.* 17: 889–900.

Dhandayuthapani S, Mudd M, Deretic V (1997) Interactions of OxyR with the promoter region of the *oxyR* and

ahpC genes from *Mycobacterium leprae* and *Mycobacterium tuberculosis*. *J. Bacteriol.* 179: 2401–2409.

Draper P (1987) Mycobacterial wall lipopolysaccharide: aspects of organisation and structure. *Int. J. Leprosy* 55: 781–782.

Draper P, Rees RJ (1973) The nature of the electron-transparent zone that surrounds *Mycobacterium lepraemurium* inside host cells. *J. Gen. Microbiol.* 77: 79–87.

Draper P, Dobson G, Minnikin DE, Minnikin SM (1982) The mycolic acids of *Mycobacterium leprae* harvested from experimentally infected nine-banded armadillos. *Ann. Microbiol. (Paris)* 133: 39–47.

Draper P, Kandler O, Darbre A (1987) Peptidoglycan and arabinogalactan of *Mycobacterium leprae*. *J. Gen. Microbiol.* 133: 1187–1194.

Eiglmeier K, Honore N, Woods SA, Caudron B, Cole ST (1993) Use of an ordered cosmid library to deduce the genomic organization of *Mycobacterium leprae*. *Mol. Microbiol.* 7: 197–206.

Eiglmeier K, Fsihi H, Heym B, Cole ST (1997) On the catalase-peroxidase gene, katG, of *Mycobacterium leprae* and the implications for treatment of leprosy with isoniazid. *FEMS Microbiol. Lett.* 149: 273–278.

Ellard GA (1990) The chemotherapy of leprosy. Part 1 [editorial]. *Int. J. Lepr. Other Mycobact. Dis.* 58: 704–716.

Fernando MR, Nanri H, Yoshitake S, Nagata-Kuno K, Minakami S (1992) Thioredoxin regenerates proteins inactivated by oxidative stress in endothelial cells. *Eur. J. Biochem.* 209: 917–922.

Fieldsteel AH, Levy L (1976a) Dapsone chemotherapy of *Mycobacterium leprae* infection of the neonatally thymectomized Lewis rat. *Am. J. Trop. Med. Hygiene* 25: 854–859.

Fieldsteel AH, Levy L (1976b) Neonatally thymectomized Lewis rats infected with *Mycobacterium leprae*: response to primary infection, secondary challenge, and large inocula. *Infect. Immun.* 14: 736–741.

Franzblau SG (1988) Oxidation of palmitic acid by *Mycobacterium leprae* in an axenic medium. *J. Clin. Microbiol.* 26: 18–21.

Franzblau SG, Hastings RC (1988) *In vitro* and *in vivo* activities of macrolides against *Mycobacterium leprae*. *Antimicrob. Agents Chemother.* 32: 1758–1762.

Franzblau SG, White KE (1990) Comparative *in vitro* activities of 20 fluoroquinolones against *Mycobacterium leprae*. *Antimicrob. Agents Chemother.* 34: 229–231.

Fsihi H, Cole ST (1995) The *Mycobacterium leprae* genome: systematic sequence analysis identifies key catabolic enzymes, ATP-dependent transport systems and a novel *polA* locus associated with genomic variability. *Mol. Microbiol.* 16: 909–919.

Fsihi H, De Rossi E, Salazar L *et al.* (1996a) Gene arrangement and organization in a approximately 76 kb fragment encompassing the *oriC* region of the chromosome of *Mycobacterium leprae*. *Microbiology* 142: 3147–3161.

Fsihi H, Vincent V, Cole ST (1996b) Homing events in the *gyrA* gene of some mycobacteria. *Proc. Natl Acad. Sci. USA* 93: 3410–3415.

Gatti JC (1975) Combined therapy in leprosy. *Leprosy Rev.* 46: 155–160.

Gelber RH (1987) Activity of minocycline in *Mycobacterium leprae*-infected mice. *J. Infect. Dis.* 156: 236–239.

Gelber RH, Murray L, Siu P, Tsang M (1992) Vaccination of mice with a soluble protein fraction of *Mycobacterium leprae* provides consistent and long-term protection against *M. leprae* infection. *Infect. Immun.* 60: 1840–1844.

Gonzalez-y-Merchand JA, Colston MJ, Cox RA (1996) The rRNA operons of *Mycobacterium smegmatis* and *Mycobacterium tuberculosis*: comparison of promoter elements and of neighbouring upstream genes. *Microbiology* 142: 667–674.

Gonzalez-y-Merchand JA, Garcia MJ, Gonzalez-Rico S, Colston MJ, Cox RA (1997) Strategies used by pathogenic and nonpathogenic mycobacteria to synthesize rRNA. *J. Bacteriol.* 179: 6949–6958.

Grosskinsky CM, Jacobs WR, Clark-Curtiss JE, Bloom BR (1989) Genetic relationships among *Mycobacterium leprae, Mycobacterium tuberculosis,* and candidate leprosy vaccine strains determined by DNA hybridization: identification of an *M. leprae*-specific repetitive sequence. *Infect. Immun.* 57: 1535–1541.

Gupte MD, Vallishayee RS, Anantharaman DS *et al.* (1998) Comparative leprosy vaccine trial in south India. *Indian J. Lepr.* 70: 369–388.

Hansen GA (1874) Undersogelser angaaende spedalskhedens aarsager. *Norsk Magazin Laegeviteuskapen.* 4: 1–88.

Helmy HS, Pearson JM, Waters MF (1971) Treatment of moderately severe erythema nodosum leprosum with clofazimine: a controlled trial. *Leprosy Rev.* 42: 167–177.

Hermans PW, van Soolingen D, van Embden JD (1992) Characterization of a major polymorphic tandem repeat in *Mycobacterium tuberculosis* and its potential use in the epidemiology of *Mycobacterium kansasii* and *Mycobacterium gordonae*. *J. Bacteriol.* 174: 4157–4165.

Honore N, Cole ST (1993) Molecular basis of rifampin-resistance in *Mycobacteria leprae*. *Antimicrob. Agents Chemother.* 37: 414–418.

Honore N, Bergh S, Chateau S *et al.* (1993a) Nucleotide sequence of the first cosmid from the *Mycobacterium leprae* genome project: structure and function of the Rif-Str regions. *Mol. Microbiol.* 7: 207–214.

Honore N, Perrani E, Telenti A, Grosset J, Cole T (1993b) A simple and rapid technique for the detection of rifampin resistance in *Mycobacterium leprae*. *Int. J. Lepr. Other Mycobact. Dis.* 61: 600–604.

Hunter SW, Brennan PJ (1990) Evidence for the presence of a phosphatidylinositol anchor on the lipoarabinomannan and lipomannan of *Mycobacterium tuberculosis*. *J. Biolog. Chem.* 265: 9272–9279.

Hunter SW, Fujiwara T, Brennan PJ (1982) Structure and antigenicity of the major specific glycolipid antigen of *Mycobacterium leprae*. *J. Biolog. Chem.* 257: 15072–15078.

Hunter SW, Rivoire B, Mehra V, Bloom BR, Brennan PJ (1990) The major native proteins of the leprosy bacillus. *J. Biolog. Chem.* 265: 14065–14068.

Imaeda T, Kirchheimer WF, Barksdale L (1982) DNA isolated from *Mycobacterium leprae*: genome size, base ratio, and homology with other related bacteria as determined by optical DNA–DNA reassociation. *J. Bacteriol.* 150: 414–417.

Immunology of Mycobacterial Diseases (IMMYC) Steering Committee, WHO (1995) Analysis of vaccines prepared from armadillo-derived *M. leprae*: results of an inter-laboratory study co-ordinated by the World Health Organization. *Int. J. Lepr. Other Mycobact. Dis.* 63: 48–55.

Ji B, Perani EG, Grosset JH (1991) Effectiveness of clarithromycin and minocycline alone and in combination against experimental *Mycobacterium leprae* infection in mice. *Antimicrob. Agents Chemother.* 35: 579–581.

Kai M, Matsuoka M, Nakata N *et al.* (1999) Diaminodiphenylsulfone resistance of *Mycobacterium leprae* due to mutations in the dihydropteroate synthase gene. *FEMS Microbiol. Lett.* 177: 231–235.

Karat AB, Jeevaratnam A, Karat S, Rao PS (1970) Double-blind controlled clinical trial of clofazimine in reactive phases of lepromatous leprosy. *BMJ* 1: 198–200.

Kempsell KE, Ji YE, Estrada IC, Colston MJ, Cox RA (1992) The nucleotide sequence of the promoter, 16S rRNA and spacer region of the ribosomal RNA operon of *Mycobacterium tuberculosis* and comparison with *Mycobacterium leprae* precursor rRNA. *J. Gen. Microbiol.* 138: 1717–1727.

Khanolkar-Young S, Rayment N, Brickell PM *et al.* (1995) Tumour necrosis factor-alpha synthesis is associated with the skin and peripheral nerve pathology of leprosy reversal reactions. *Clin. Exp. Immunol.* 99: 196–202.

Kirchheimer WF, Storrs EE (1971) Attempts to establish the armadillo (*Dasypus novemcinctus* Linn.) as a model for the study of leprosy: I. Report of lepromatoid leprosy in an experimentally infected armadillo. *Int. J. Lepr. Other Mycobact. Dis.* 39: 693–702.

Kulkarni VM, Seydel JK (1983) Inhibitory activity and mode of action of diaminodiphenylsulfone in cell-free folate-synthesizing systems prepared from *Mycobacterium lufu* and *Mycobacterium leprae*: a comparison. *Chemotherapy* 29: 58–67.

Laal S, Bhutani LK, Nath I (1985) Natural emergence of antigen-reactive T cells in lepromatous leprosy patients during erythema nodosum leprosum. *Infect. Immun.* 50: 887–892.

Levy L (1970) Death of *Mycobacterium leprae* in mice, and the additional effect of dapsone administration. *Proc. Soc. Exp. Biol. Med.* 135: 745–749.

Levy L, Randall HP (1970) A study of skin pigmentation by clofazimine. *Int. J. Lepr. Other Mycobact. Dis.* 38: 404–416.

Lockwood DN, Vinayakumar S, Stanley JN, McAdam KP, Colston MJ (1993) Clinical features and outcome of reversal (type 1) reactions in Hyderabad, India. *Int. J. Lepr. Other Mycobact. Dis.* 61: 8–15.

Malmborg AS (1986) The renaissance of erythromycin. *J. Antimicrob. Chemother.* 18: 293–296.

Matsumura K, Chiba A, Yamada H *et al.* (1997) A role of dystroglycan in schwannoma cell adhesion to laminin. *J. Biolog. Chem.* 272: 13904–13910.

Minnikin DE (1982) Lipids: complex lipids, their chemistry, biosynthesis and roles. In: Ratledge C, Stanford J (eds) *The Biology of the Mycobacteria: Vol. 1. Physiology, Identification and Classification*. London: Academic Press, pp. 95–184.

Minnikin DE, Minnikin SM, Dobson G *et al.* (1983) Mycolic acid patterns of four vaccine strains of *Mycobacterium bovis* BCG. *J. Gen. Microbiol.* 129: 889–891.

Mitchison DA, Selkon JB, Lloyd J (1963) Virulence in the guinea pig, susceptibility to hydrogen peroxide, and catalase activity of isoniazid-sensitive tubercle bacilli from South Indian and British patients. *J. Pathol. Bacteriol.* 86: 377–386.

Modlin RL, Ormerod LD, Walsh GP *et al.* (1986) *In situ* characterization of T lymphocyte subpopulations in leprosy in the mangabey monkey. *Clin. Exp. Immunol.* 65: 260–264.

Mukherjee R, Antia NH (1986) Host–parasite interrelationship between *M. leprae* and Schwann cells *in vitro*. *Int. J. Lepr. Other Mycobact. Dis.* 54: 632–638.

Nath I (1983) Immunology of human leprosy: current status. *Leprosy Rev.* (special issue): 31–45.

Nishiura M, Izumi S, Mori T, Takeo K, Nonaka T (1977) Freeze-etching study of human and murine leprosy bacilli. *Int. J. Lepr. Other Mycobact. Dis.* 45: 248–254.

Pearson JM, Haile GS, Rees RJ (1977) Primary dapsone-resistant leprosy. *Leprosy Rev.* 48: 129–132.

Pessolani MC, Smith DR, Rivoire B *et al.* (1994) Purification, characterization, gene sequence, and significance of a bacterioferritin from *Mycobacterium leprae*. *J. Exp. Med.* 180: 319–327.

Pettit JH, Rees RJ, Ridley DS (1966) Studies on sulfone resistance in leprosy: I. Detection of cases. *Int. J. Lepr. Other Mycobact. Dis.* 34: 375–390.

Philipp WJ, Poulet S, Eiglmeier K *et al.* (1996) An integrated map of the genome of the tubercle bacillus, *Mycobacterium tuberculosis* H37Rv, and comparison with *Mycobacterium leprae*. *Proc. Natl Acad. Sci. USA* 93: 3132–3137.

Philipp WJ, Schwartz DC, Telenti A, Cole ST (1998) Mycobacterial genome structure. *Electrophoresis* 19: 573–576.

Pietrokovski S (1994) Conserved sequence features of inteins (protein introns) and their use in identifying new inteins and related proteins. *Protein Sci.* 3: 2340–2350.

Rambukkana A, Salzer JL, Yurchenco PD, Tuomanen EI (1997) Neural targeting of *Mycobacterium leprae* mediated by the G domain of the laminin-α2 chain. *Cell* 88: 811–821.

Rambukkana A, Yamada H, Zanazzi G *et al.* (1998) Role of α-dystroglycan as a Schwann cell receptor for *Mycobacterium leprae*. *Science* 282: 2076–2079.

Rees RJ (1966) Enhanced susceptibility of thymectomized and irradiated mice to infection with *Mycobacterium leprae*. *Nature* 211: 657–658.

Rees RJ, Weddell AG, Palmer E, Pearson JM (1969) Human leprosy in normal mice. *BMJ* 3: 216–217.

Roach TI, Barton CH, Chatterjee D, Blackwell JM (1993) Macrophage activation: lipoarabinomannan from avirulent and virulent strains of *Mycobacterium tuberculosis* differentially induces the early genes c-fos, KC, JE, and tumor necrosis factor-α. *J. Immunol.* 150: 1886–1896.

Rogall T, Flohr T, Bottger C (1990a) Differentiation of *Mycobacterium* species by direct sequencing of amplified DNA. *J. Gen. Microbiol.* 136: 1915–1920.

Rogall T, Wolters J, Flohr T, Bottger EC (1990b) Towards a phylogeny and definition of species at the molecular level within the genus *Mycobacterium*. *Int. J. System. Bacteriol.* 40: 323–330.

Saito H, Tomioka H, Nagashima K (1986) *In vitro* and *in vivo* activities of ofloxacin against *Mycobacterium leprae* infection induced in mice. *Int. J. Lepr. Other Mycobact. Dis.* 54: 560–562.

Schlesinger LS (1993) Macrophage phagocytosis of virulent but not attenuated strains of *Mycobacterium tuberculosis* is mediated by mannose receptors in addition to complement receptors. *J. Immunol.* 150: 2920–2930.

Schlesinger LS, Horwitz MA (1990) Phagocytosis of leprosy bacilli is mediated by complement receptors CR1 and CR3 on human monocytes and complement component C3 in serum. *J. Clin. Invest.* 85: 1304–1314.

Schlesinger LS, Horwitz MA (1994) A role for natural antibody in the pathogenesis of leprosy: antibody in nonimmune serum mediates C3 fixation to the *Mycobacterium leprae* surface and hence phagocytosis by human mononuclear phagocytes. *Infect. Immun.* 62: 280–289.

Schlesinger LS, Bellinger-Kawahara CG, Payne NR, Horwitz MA (1990) Phagocytosis of *Mycobacterium tuberculosis* is mediated by human monocyte complement receptors and complement component C3. *J. Immunol.* 144: 2771–2780.

Schlesinger LS, Hull SR, Kaufman TM (1994) Binding of the terminal mannosyl units of lipoarabinomannan from a virulent strain of *Mycobacterium tuberculosis* to human macrophages. *J. Immunol.* 152: 4070–4079.

Schorey JS, Carroll MC, Brown EJ (1997) A macrophage invasion mechanism of pathogenic mycobacteria. *Science* 277: 1091–1093.

Shepard CC (1960) The experimental disease that follows the injection of human leprosy bacilli into footpads of mice. *J. Exp. Med.* 112: 445–454.

Shepard CC, Chang YT (1967) Effect of DDS on established infections with *Mycobacterium leprae* in mice. *Int. J. Lepr. Other Mycobact. Dis.* 35: 52–57.

Sherman DR, Mdluli K, Hickey MJ *et al.* (1996) Compensatory *ahpC* gene expression in isoniazid-resistant *Mycobacterium tuberculosis*. *Science* 272: 1641–1643.

Shimoji Y, Ng V, Matsumura K, Fischetti VA, Rambukkana A (1999) A 21-kDa surface protein of *Mycobacterium leprae* binds peripheral nerve laminin-2 and mediates Schwann cell invasion. *Proc. Natl Acad. Sci. USA* 96: 9857–9862.

Small PM, Hopewell PC, Singh SP *et al.* (1994) The epidemiology of tuberculosis in San Francisco: a population-based study using conventional and molecular methods. *N. Engl. J. Med.* 330: 1703–1709.

Smida J, Kazda J, Stackebrandt E (1988) Molecular-genetic evidence for the relationship of *Mycobacterium leprae* to slow-growing pathogenic mycobacteria. *Int. J. Lepr. Other Mycobact. Dis.* 56: 449–454.

Smith DR, Richterich P, Rubenfield M *et al.* (1997) Multiplex sequencing of 1.5 Mb of the *Mycobacterium leprae* genome. *Genome Res.* 7: 802–819.

Sritharan M, Ratledge C (1990) Iron-regulated envelope proteins of mycobacteria grown *in vitro* and their occurrence in *Mycobacterium avium* and *Mycobacterium leprae* grown *in vivo*. *Biol. Meth.* 2: 203–208.

Stahl DA, Urbance JW (1990) The division between fast- and slow-growing species corresponds to natural relationships among the mycobacteria. *J. Bacteriol.* 172: 116–124.

Stokes RW, Speert DP (1995) Lipoarabinomannan inhibits nonopsonic binding of *Mycobacterium tuberculosis* to murine macrophages. *J. Immunol.* 155: 1361–1369.

Stokes RW, Haidl ID, Jefferies WA, Speert DP (1993) Mycobacteria–macrophage interactions: macrophage phenotype determines the nonopsonic binding of *Mycobacterium tuberculosis* to murine macrophages. *J. Immunol.* 151: 7067–7076.

Storrs EE (1971) The nine-banded armadillo: a model for leprosy and other biomedical research. *Int. J. Lepr. Other Mycobact. Dis.* 39: 703–714.

Storrs EE (1978) Initiation of armadillo program. *Int. J. Lepr. Other Mycobact. Dis.* 46: 436–438.

Telenti A, Imboden P, Marchesi F *et al.* (1993) Detection of rifampicin-resistance mutations in *Mycobacterium tuberculosis*. *Lancet* 341: 647–650.

Thangaraj HS, Lamb FI, Davis EO, Jenner PJ, Jeyakumar LH, Colston MJ (1990) Identification, sequencing, and expression of *Mycobacterium leprae* superoxide dismutase, a major antigen. *Infect. Immun.* 58: 1937–1942.

Thierry D, Brisson-Noel A, Vincent-Levy-Frebault V, Nguyen S, Guesdon JL, Gicquel B (1990) Characterization of a *Mycobacterium tuberculosis* insertion sequence, IS6110, and its application in diagnosis. *J. Clin. Microbiol.* 28: 2668–2673.

Triccas JA, Winter N, Roche PW, Gilpin A, Kendrick KE, Britton WJ (1998) Molecular and immunological analyses of the *Mycobacterium avium* homolog of the immunodominant *Mycobacterium leprae* 35-kilodalton protein. *Infect. Immun.* 66: 2684–2690.

Villanova M, Sewry C, Malandrini A *et al.* (1997) Immunolocalization of several laminin chains in the normal human central and peripheral nerous system. *J. Submicrosc. Cytol. Pathol.* 29: 409–413.

Walker LL, Van Landingham RM, Shinnick TM (1993) Clarithromycin is bactericidal against strains of *Mycobacterium leprae* resistant and susceptible to dapsone and rifampin. *Int. J. Lepr. Other Mycobact. Dis.* 61: 59–65.

Warren AG (1970) The use of B 663 (clofazimine) in the treatment of Chinese leprosy patients with chronic reaction. *Leprosy Rev.* 41: 74–82.

Wieles B, van Soolingen D, Holmgren A, Offringa R, Ottenhoff T, Thole J (1995) Unique gene organization of thioredoxin and thioredoxin reductase in *Mycobacterium leprae*. *Mol. Microbiol.* 16: 921–929.

Wieles B, Ottenhoff TH, Steenwijk TM, Franken KL, de Vries RR, Langermans JA (1997) Increased intracellular survival of *Mycobacterium smegmatis* containing the *Mycobacterium leprae* thioredoxin–thioredoxin reductase gene. *Infect. Immun.* 65: 2537–2541.

Wilson TM, de Lisle GW, Collins DM (1995) Effect of *inhA* and *katG* on isoniazid resistance and virulence of *Mycobacterium bovis*. *Mol. Microbiol.* 15: 1009–1015.

Woods SA, Cole ST (1989) A rapid method for the detection of potentially viable *Mycobacterium leprae* in human biopsies: a novel application of PCR. *FEMS Microbiol. Lett.* 53: 305–309.

Woods SA, Cole ST (1990) A family of dispersed repeats in *Mycobacterium leprae*. *Mol. Microbiol.* 4: 1745–1751.

Wu QL, Kong D, Lam K, Husson RN (1997) A mycobacterial extracytoplasmic function sigma factor involved in survival following stress. *J. Bacteriol.* 179: 2922–2929.

Yamamura M, Uyemura K, Deans RJ *et al.* (1991) Defining protective responses to pathogens: cytokine profiles in leprosy. *Science* 254: 277–279.

83

The *Mycobacterium avium–intracellulare* Complex

Sheldon L. Morris and Frank M. Collins

CBER, FDA , Bethesda, Maryland, USA

Pulmonary tuberculosis is usually caused by *Mycobacterium tuberculosis*, which was first isolated in pure culture by Robert Koch in 1882 when he demonstrated that this organism satisfied his postulates as the causative agent of this chronic, debilitating disease (Collins, 1982). Later, it was shown that another acid-fast bacillus, *M. avium*, which had previously been shown to infect chickens and swine, could be isolated from the lungs and cervical lymph nodes of some infants and young children (Wolinsky, 1984). It was, however, only after the introduction of anti-tuberculosis chemotherapy, half a century later, that these non-tuberculosis mycobacteria began to be seriously considered as human pathogens (Wayne, 1985).

At first, most mycobacteria other than *M. tuberculosis* were considered to be little more than troublesome laboratory contaminants and their pathogenic significance was generally ignored. In time, it was realized that some strains were being isolated from patients with clinically significant lung disease, and that they were excreting large numbers of acid-fast bacilli in their sputum. As a result, some of these organisms came to be considered as opportunistic human pathogens, able to produce active lung disease whenever the normal host defences became depressed, either iatrogenically or as a result of an intercurrent viral infection (Falkinham, 1996).

Isolation and Identification of the *M. avium* Complex

Initially, these organisms were simply named 'Battey bacilli', after the tuberculosis sanitarium where they were first studied systematically (Runyon, 1965). With the development of suitable diagnostic biochemical and cultural tests, they were classified as members of the *M. avium–intracellulare–scrofulaceum* (MAIS) complex. This consisted of 31 sub-strains which could be distinguished by means of a sero-agglutination test (Schaefer, 1965). The first four serovars were designated as *M. avium*, with 24 serovars classified as *M. intracellulare* and three as *M. scrofulaceum* (McClatchy, 1981). Recent DNA homology studies reclassified 11 of these strains as *M. avium* (serovars 1–6, 8–11 and 21) with the remaining 17 serovars considered to be *M. intracellulare* (Wayne *et al.*, 1993).

M. avium had long been recognised as an avian pathogen, virulent for mice and chickens, whereas the closely related *M. intracellulare* was able to cause disease only in chickens (Schaefer *et al.*, 1970). Both *M. avium* and *M. intracellulare* can, however, be isolated from human clinical material and so should be considered opportunistic human pathogens, able to produce active lung disease in individuals suffering from silicosis, pneumoconiosis and sarcoidosis, as well as in patients who are severely immunosuppressed as a result of prolonged cancer or transplantation chemotherapy (Iseman, 1989). In addition, individuals infected with the human immunodeficiency virus (HIV) are highly susceptible to tuberculosis, as well as to disseminated *M. avium*-complex disease (Kiehn *et al.*, 1985). As many as 30% of patients suffering from acquired immunodeficiency syndrome (AIDS) have been reported to be co-infected with *M. tuberculosis* or one of several serovars in the *M. avium* complex (Young *et al.*, 1986).

Typing of the *M. avium* Complex

The *M. avium* complex comprises a heterogeneous and phenotypically diverse group of opportunistic pathogens. A number of laboratory techniques have been developed to analyse the relatedness of different MAC serovars, including multilocus enzyme electrophoresis, restriction fragment polymorphism analysis (RFLP), nucleotide sequencing of varying genetic regions, and pulsed-field gel electrophoresis (Wasem *et al.*, 1991; Frothingham and Wilson, 1994; Roiz *et al.*, 1995). Among these molecular tools, RFLP analysis with specific insertion sequence probes is being intensively evaluated, and at least five insertion elements have been identified in the *M. avium*–*intracellulare* complex (IS900 in *M. avium* subsp. *paratuberculosis*; IS901, IS1110 and IS1245 in *M. avium*; and IS1141 in *M. intracellulare*). RFLP typing with IS1245 has been used primarily because this *M. avium*-specific element is stable and is present in high copy numbers in many of the *M. avium* strains isolated from human sources (Green *et al.*, 1989; Kunze *et al.*, 1992; Hernandez-Perez *et al.*, 1994; Bauer and Andersen, 1999). Initially, Guerrero *et al.* (1995) described the application of IS1245 hybridisation to strain typing of MAC isolates after determining that the host range was limited to *M. avium* and the element was not even identified in *M. intracellulare*. Subsequently, several investigators in Europe

and the USA confirmed the utility of the IS1245-based RFLP analysis in these molecular epidemiological studies (Lari *et al.*, 1998; Pestel-Caron and Arbeit, 1998; Rittaco *et al.*, 1998; van Soolingen *et al.*, 1998; Bauer *et al.*, 1999).

For most human isolates, highly polymorphic multi-banded IS1245 RFLP patterns are almost invariable. The average number of bands present in human patterns is around 20, similar to the fingerprints for porcine isolates. In contrast, the banding patterns for avian isolates differ significantly from those for the human isolates, showing only a two- or three-band pattern. The presence of these IS1245 low-copy *M. avium* strains highlights an important drawback to this technique. When human strains with the avian IS1245 pattern are detected, the low copy numbers of this genetic element greatly limit the discriminatory power of the RFLP analyses.

Epidemiology of *M. avium*-Complex Infections

Most members of the *M. avium* complex (MAC) can be isolated from the natural environment (mainly soil and water), especially in temperate and subtropical areas of the world (Gruft *et al.*, 1981). Some serovars have been isolated from domestic water supply systems, where they may constitute an infectious hazard for severely immunosuppressed AIDS patients (Yajko *et al.*, 1995). Many of these environmental mycobacteria can, at least temporally, colonise the oropharyngeal and intestinal mucosae of normal people, although the resulting infection does not usually go on to involve the underlying tissues. However, some strains of *M. scrofulaceum* are able to produce self-limiting infections in the cervical lymph nodes of young children (Wolinsky, 1984). Interestingly, these *M. scrofulaceum* serovars do not produce life-threatening, disseminated disease in HIV-infected patients, but the reason for this is not at all clear (Kiehn *et al.*, 1985). However, once these opportunistic pathogens become established within a tissue, they can be very difficult to dislodge, partly because of the ineffectual nature of the cell-mediated immune response by immunosuppressed hosts, and partly because of the innate drug resistance of these organisms, making it very difficult to manage these infections clinically (Ellner *et al.*, 1991).

M. avium is an important avian pathogen, capable of causing severe systemic infections in swine and some domestic animals (Thoen *et al.*, 1984). There is, however, little evidence of direct animal-to-people infections by these organisms (Meissner, 1981; Falkinham, 1996). Some infections may result from the ingestion of MAC-contaminated foods, such as cheese or preserved meat (Horsburgh *et al.*, 1994). Interestingly, avian isolates of *M. avium* (serovar 1) were found to be highly virulent for chickens, compared with isolates of the same serovar obtained from human patients. The latter induced little or no systemic disease when injected intravenously into normal hens (Meissner, 1981). The reason for this striking species difference is unknown (Nel, 1981).

Before the AIDS epidemic, MAC infections usually took the form of a self-limiting infection in young children, usually involving a single lymph node. These same environmental mycobacteria can, however, also cause chronic lung disease in elderly individuals, most of whom have some form of underlying lung disease (Good and Snider, 1982). Systemic MAC disease may also occur in cancer and transplantation patients receiving prolonged radiation treatment and chemotherapy (O'Brien *et al.*, 1987). Far more puzzling are the small number of *M. avium* infections seen in individuals without apparent predisposing factors, and who mostly live in the more temperate regions of Europe, North America and Japan (Good and Snider, 1982). Exact figures for these infections are difficult to obtain, since they are generally not reportable to the Centers for Disease Control or the World Health Organization. In some cases, a lack of suitable diagnostic facilities may make it difficult to obtain good epidemiological data on this disease. Based on these inadequate data, the incidence of MAC disease in the USA before the HIV epidemic was quite low (3 per 100 000). Then, in less than a decade, MAC infections increased to near epidemic proportions in the AIDS population. During the same time period, and for largely unknown reasons, the incidence of MAC disease in the non-AIDS population has also slowly risen, primarily in the most elderly segment of the population (O'Brien *et al.*, 1987). Clearly, much is still not known or understood about this important group of opportunistic pathogens.

Serotyping MAC isolates in the USA indicates striking differences in the dominant MAC strains present within a given community, especially in different parts of the world. For instance, a surprising proportion (as many as one half) of the isolates reported in the USA were *M. intracellulare* serovars, whereas in Europe and Japan the predominant serovars have been identified as *M. avium* (Inderlied *et al.*, 1993). This variability may simply reflect the distribution of environmental mycobacteria in water, dust and food in different parts of the world (Fry *et al.*, 1986). Several studies have implicated drinking water,

and particularly hot water supplies, as a major source of mycobacterial infections other than tuberculosis in HIV-infected individuals, but these isolates do not necessarily reflect the mycobacterial flora found in the natural environment (von Reyn *et al.*, 1993, 1994). There seems little question that the presence of *M. avium* in the hot water supplies of some hospitals can be associated with pulmonary MAC infections in some AIDS patients undergoing treatment (Horsburgh *et al.*, 1994). In most AIDS patients in the USA, an oral route of infection can be inferred from the presence of *M. avium* in biopsy samples taken from individuals who later go on to develop disseminated MAC disease (Fry *et al.*, 1986). It seems likely that we are all exposed from time to time to limited intestinal colonisation by non-tuberculosis mycobacteria, which can be isolated in small numbers from normal faeces with a surprisingly high frequency (Portaels *et al.*, 1988).

Further evidence of widespread exposure to environmental mycobacteria comes from the skin hypersensitivity responses to PPD-A (*M. avium*) or PPD-B (*M. intracellulare*) as reported in surveys of apparently normal adults in the USA (Edwards *et al.*, 1969; Snider, 1982). Such silent infections appear to be surprisingly common, especially in rural areas of the south-eastern and mid-Atlantic states (Fry *et al.*, 1986). Unfortunately, few systematic studies of skin hypersensitivity patterns due to these opportunistic pathogens have been carried out in the United States, but new data should become available from the current American National Health Survey (NHANES).

Low to negligible levels of protection have been reported in several BCG trials carried out in a number of temperate and subtropical countries around the world (Comstock, 1994). Ironically, these countries have the greatest need for a fully protective antituberculosis vaccine. Inapparent MAC infections in normal immunocompetent individuals have been blamed for the poor efficacy of several of these BCG vaccine trials (Grange, 1986; Fine, 1989). The role played by antigenic cross-reactivity between these environmental mycobacteria and BCG vaccine has been discussed extensively (Fine and Rodriguez, 1990; Comstock, 1994). Interestingly, the same cross-reactivity may have an unexpected health benefit, since it has been reported that BCG vaccination reduced *M. avium* infections in Swedish school-children (Romanus *et al.*, 1995). Such a conclusion appears to be consistent with experimental studies in which BCG-vaccinated mice were challenged with virulent *M. avium* (Collins, 1985). Enhanced protection has also been inferred from the resistance to subsequent infection with *M. avium* shown by individuals with latent tuberculosis (tuberculin-positive) (Collins, 1986).

Latently infected people have a small residuum of viable bacilli, remaining from the primary infection, which can persist in the host tissues for long periods (precisely how and where is still controversial; see Grange, 1992). Presumably, antigens released from the dormant lesion induce sufficient immunological memory to prevent the *M. avium* from colonising the GALT organs. Such cross-reactivity may also help to explain the absence of MAC disease in AIDS patients in Uganda and other parts of central Africa (Morrissey *et al.*, 1992). Such a conclusion is not, however, compatible with recent epidemiological data obtained from a group of HIV-infected patients living in the USA and the matter deserves further investigation (Sterling *et al.*, 1998).

Diagnosis of *M. avium*-Complex Infections

A positive diagnosis of infection by a member of the *M. avium*-complex in an apparently immunocompetent patient is usually based on multiple positive sputum cultures, along with the appropriate clinical and radiological evidence (American Thoracic Society, 1997). Despite the prolonged time required to cultivate these organisms on laboratory media, this method remains the 'gold standard' for a positive diagnosis of MAC disease. The efficacy of this approach is often limited because of the small number of viable organisms present in sputum collected from many of these patients. In one retrospective study, 45% of patients with culture-proven MAC disease provided consistently negative sputum cultures (Huang *et al.*, 1999). This is one reason why rapid and direct tests for MAC disease in HIV-negative patients are being sought so actively. Recently, the Roche amplicor assay was reported to have high sensitivity (85%) for the detection of MAC infections in immunocompetent individuals (Matsumoto *et al.*, 1998).

Since pulmonary *M. avium* and *M. tuberculosis* infections often show similar clinical presentations but require different clinical management protocols, prompt and reliable distinction between these two infections is important. In a recent multi-centre, blinded trial, von Reyn *et al.* (1998) demonstrated that dual skin testing with *M. avium* sensitin (PPD-A) and tuberculin (PPD-S) could reliably distinguish between MAC disease and tuberculosis. Similarly, cellular responses to ESAT-6, an *M. tuberculosis*-specific antigen, could discriminate between these two pulmonary pathogens (Lein *et al.*, 1999). Although these tests appear to be promising, both assays will have to be carefully validated in prospective, blinded studies before they can be recommended to differentiate MAC from tuberculosis infections in the clinic.

Diagnosis of Disseminated *M. avium*-Complex Disease

The possible presence of disseminated MAC disease in AIDS patients should always be considered, especially in the face of symptoms such as fever, weight loss, lymphadenopathy and aberrant liver biochemistry (Benson and Ellner, 1993; Inderlied *et al.*, 1993). Such a presumptive diagnosis may be further supported by the response of the infection to drug treatment. A conclusive diagnosis can, however, be made only by culturing the mycobacteria from blood, bone marrow or biopsy material taken from normally sterile body sites. When growth has occurred, nucleic acid probes, HPLC and specific microbiological assays may be useful to identify and type the infecting organism. The ease and rapidity of the commercial AccuProbe culture identification kit (GenProbe, San Diego, CA) makes this test particularly attractive for use in clinical laboratories.

Recently, PCR procedures have been evaluated for their ability to identify cultured MAC strains. Kulski *et al.* (1995) amplified regions of the 16S rRNA gene from *M. intracellulare* and *M. avium* and the MPB70 gene from *M. tuberculosis* in an attempt to determine if the mycobacteria could be detected in BACTEC blood cultures by multiplex PCR. This approach, combined with the sodium iodide–propanol extraction procedure, resulted in sensitive and accurate identification of mycobacterial species when the BACTEC growth index exceeded 20 units. In a related study, Cousins *et al.* (1996) correctly identified 33 of 34 *M. avium* isolates and all of 51 *M. intracellulare* isolates by the multiplex PCR procedure. This again suggests that multiplex PCR can be utilised for the routine identification of cultivated MAC strains.

A significant decrease in the time required for diagnosis of disseminated *M. avium*-complex (DMAC) disease can be achieved if DNA amplification methods could be directly applied to clinical specimens prior to their cultivation on laboratory media (Iralu *et al.*, 1993; De Francesco *et al.*, 1996). In several recent studies, high specificity has been demonstrated using the direct PCR approach. Specificity data suggest that the false-positive PCR results predicted as a result of contamination by environmental MAC strains have not been a confounding problem. However, the sensitivity data have not been as consistent. Bogner *et al.* (1997) and Ninet *et al.* (1997)

reported low sensitivities (26% and 42% respectively) with use of the Amplicor PCR system (16s DNA). In contrast, a PCR sensitivity of greater than 80% has been reported in at least four other studies. Such wide differences are apparently unrelated to the use of different gene targets, because two of these studies also detected amplification of 16S rDNA. Inconsistent results may be due to different blood preparation methods used in the direct PCR analysis. Roth *et al.* (1999) showed that higher sensitivity was obtained in the PCR detection of *M. avium* infections when Ficoll density centrifugation was used instead of simple lysis of the erythrocytes in hypertonic buffer. Alternatively, the variable data may reflect different patient populations, and specifically the extent of the MAC bacteraemia present in these patients. MacGregor *et al.* (1999) reported that direct PCR detection was highly sensitive (94.4%) in specimens with more than 40 organisms per millilitre of blood and that this sensitivity decreased significantly when clinical specimens with less than this threshold were tested.

Virulence of the *M. avium* Complex

A lengthy debate has taken place in the literature about the virulence of many mycobacteria other than *M. tuberculosis*, both for humans and experimental animals (Wolinsky, 1981). Part of the problem stems from the attenuation that occurs whenever these organisms are maintained on laboratory media for any length of time (Barrow, 1991). The decline in virulence has been associated with a shift in the dominant colonial morphology seen when the organisms are grown on Middlebrook 7H11 agar (Kansal *et al.*, 1998). Most primary clinical isolates of *M. avium* produce thin, flat, translucent colonies, designated as SmT, but after growth for a time on laboratory media, an increasing proportion of smooth, domed, opaque (SmD) colonies can be seen and these may eventually come to dominate the culture (Dunbar *et al.*, 1968). When the SmD organisms are grown in the presence of high concentrations of detergent (Tween 80 or Triton X-100), they change to a rough (Rg) colony form (Belisle and Brennan, 1994). The latter variant is completely avirulent for experimental animals, and presumably for people, and passage of the rough-colony variants through mice does not restore virulence, indicating that this change is irreversible. The SmT and SmD colony variants can be maintained as virtually pure cultures by serial transfer on solid media, but they quickly revert to the mixed colony form when transferred back into liquid media (Collins, 1999).

The molecular mediators of these colonial changes are still largely unknown, although the SmD to Rg shift has been associated with a loss of C-mycosides from the cell wall (Barrow and Brennan, 1982). Similarly, the decline in virulence shown by the SmD variant seems to depend on the loss of those cytokines which are normally released in response to specific lipids associated with the SmT cell membrane (Shiratsuchi *et al.*, 1993). Comparison of cell-wall lipids isolated from the translucent and opaque colony variants of *M. intracellulare* suggest that it is the loss of specific acidic surface polysaccharides which causes this change in colonial morphology, along with an increase in drug resistance and reduced mouse virulence (Moulding, 1978).

Determinants of Virulence

The mediators of virulence in the mycobacteria are likely to be multifactoral. In the case of *M. tuberculosis*, disruption of at least three genes (*erp*, *fbp*A and *kat*G) appears to contribute to the poor persistence of tubercle bacilli in the tissues (Berthet *et al.*, 1998; Li *et al.*, 1998). In the case of *M. avium*, the inherent difficulties associated with expressing genes in MAC isolates, and creating gene knockout or gene replacement strains, have been a significant impediment to the identification of the specific factors responsible for virulence. As a result, the bacterial components responsible for MAC pathogenicity are still largely unknown. Recent studies have, however, shown that modification to current gene recombination protocols may facilitate allelic replacement in these mycobacteria. Hinds *et al.* (1999) reported that treatment of transforming DNA, with ultraviolet light or with alkali, significantly enhanced recombination frequencies in *M. tuberculosis* and *M. intracellulare*. Allelic replacement of the 19 kDa antigen gene in *M. intracellulare* was substantially increased with denatured transforming DNA. In another study, Marklund *et al.* (1998) demonstrated the creation of *M. intracellulare* catalase-peroxidase (*kat*G) mutants by new gene replacement methods. Interestingly, the similar growth kinetics of the parental strain and the isogenic *kat*G mutant in mice suggested that KatG was not essential for virulence in *M. intracellulare* and that the intact *oxy*R regulon and the other catalase present in MAC strains could compensate for the loss of KatG; this is in contrast to *M. tuberculosis*, which has only one catalase and a dysfunctional *oxy*R.

In spite of the limited usefulness of current gene replacement methods for the genetic modification of MAC strains, two putative MAC virulence factors have recently been suggested by the use of unique experimental approaches. Plum *et al.* (1997) identified a macrophage-induced gene (*mig*) of *M. avium* which was expressed only when the organism was growing in macrophages. Expression of the MAC *mig* gene in *M. smegmatis* augmented the capacity of the recombinant to proliferate in human monocyte-derived macrophage cultures. A comparison of 20 *M. avium* isolates further indicated the relevance of this gene with respect to virulence (Meyer *et al.*, 1998). In these strains, the presence of an intact *mig* gene correlated with clinical pathogenicity and *in vitro* persistence. In another series of experiments, Maslow *et al.* (1999) investigated the role of the hae-molysin of *M. avium* as a virulence factor associated with the development of a systemic infection. Their evaluation of 28 invasive *M. avium* strains and 32 non-invasive respiratory isolates indicated that only the invasive strains were haemolytic, and the HIV-prevalent serovars 4 and 8 had the greatest activity. These studies suggest that expression of haemolysin by MAC strains may be a necessary factor for the development of disseminated disease in patients.

Pathogenesis of *M. avium*-Complex Disease

Until recently, the incidence of systemic disease due to members of the *M. avium* complex was less than 1000 cases per year in the USA (O'Brien *et al.*, 1987). The AIDS epidemic has changed the incidence of *M. avium*-complex (MAC) disease from a rare, almost exotic infection, to an all too common life-threatening complication of this immunosuppressive disease (Falkinham, 1994). HIV-infected individuals are exquisitely susceptible to systemic *M. avium* infections once the CD4 count drops below 100 cells/mm^3 of blood (Horsburgh, 1999). With the steady rise in the number of AIDS patients worldwide, it is likely that DMAC disease will remain an important public health problem (Piot *et al.*, 1990). The development of the disseminated form of the *M. avium*-complex disease substantially reduces the survival time of AIDS patients, compared with individuals in whom mycobacteria are not involved (Horsburgh and Selik, 1989).

The majority of MAC isolates in the USA are serovar 4 or 8 (Young *et al.*, 1986), with smaller numbers of serovars 1, 6, 9 and 10, mostly reported overseas (Hoffner *et al.*, 1989). These strains have long been regarded as more 'virulent' than other members of the group, and so it might be expected that they would be isolated more frequently from the severely immunosuppressed AIDS patient population than are the other less virulent serotypes. Thus, the virtual absence of several MAC serotypes from AIDS patients, although they occur in cancer and transplant patients in the same communities, is surprising and has yet to be explained (Good and Snider, 1982). Most MAC serovars occur naturally in soil, as well as in seawater and freshwater samples collected in many parts of the world. It is not surprising that they are present in food and drinking water in those areas, where they may serve as a source of infection in the AIDS population (von Reyn *et al.*, 1994). The major route of infection in these patients appears to be intragastric, although there is also evidence for pulmonary infections in some AIDS patients (Yakrus and Good, 1990). A clear demonstration of oral infections caused by these organisms is technically difficult and has, so far, been successfully attempted only in a handful of cases (von Reyn *et al.*, 1993). It is assumed that those MAC strains, which most commonly establish themselves within the intestinal tract, possess adhesins which facilitate their attachment to the mucosal membrane (Sangari *et al.*, 1999). As a result, some of the more virulent strains occasionally gain entry to the sub-mucosa, where they can resist the antibacterial action of the unstimulated macrophages, ultimately reaching the gut-associated lymphoid tissues (GALT). If the population are all subject to occasional exposures to these environmental mycobacteria, it is possible that some organisms establish residence in the lung whenever the normally effective host defences are depressed. It is not clear whether all MAC strains possess this ability, or whether only those serovars that possess a specific virulence factor are able to establish persistent residence *in vivo*. More detailed epidemiological investigation of these infections will be needed before a pathological role can be established for some of these opportunistic pathogens, especially as apparently immunocompetent adults are involved (Iseman, 1989).

Immunopathogenesis

In susceptible strains of mice, C57BL/6 and its beige variant, *M. avium* infections progress through two distinct phases (Cooper *et al.*, 1998). The initial phase is characterised by an almost unrestrained period of growth in the lung, with little or no inflammatory cell response. After 6–8 weeks, a CD4 T-cell response develops, resulting in a reduction of bacterial growth both in the lungs and the spleen, together with a strong granulomatous response. This model has been supported by studies carried out in immunodeficient and knockout mice (Collins and Stokes, 1987;

Doherty and Sher, 1997). Growth in SCID and IFNγ knockout mice is indistinguishable at 4 weeks from that seen in the wild-type controls (Appleberg *et al.*, 1994). After 8 weeks, a time when the cell-mediated responses become most evident in wild-type strains, substantially enhanced growth is seen in the SCID and IFNγ-deficient animals when compared with the immunocompetent controls.

In the primary interaction between the MAC and non-activated macrophages, the binding of the organism to complement receptors facilitates the receptor-mediated mechanisms of phagocytosis. Importantly, uptake via the complement receptor fails to stimulate the respiratory burst, which is an essential step in controlling the MAC infection. Reports of mycobacterial disease in patients with chronic granulomatous disease and the extreme susceptibility to mycobacterial challenge of mice with gp 47 phox gene disruptions, indicate that reactive oxygen intermediates of macrophages, contribute substantially to host defence against the initial infection (Ohga *et al.*, 1997; Cooper *et al.*, 1998). In contrast, the role of reactive nitrogen molecules in controlling early MAC growth appears less certain. Mice lacking the nitric oxide synthase gene and wild-type controls both develop comparable mycobacterial burdens after an intravenous challenge (Doherty and Sher, 1997). In fact, the induction of nitric oxide may actually exacerbate the infection by suppressing the normal immune response to the pathogen (Gomes *et al.*, 1999). Interestingly, Ehlers *et al.* (1999a) recently reported that NOS2-derived nitric oxide down-regulates the size, quantity and quality of granuloma formation by altering the cytokine levels and the cellular composition at the site of infection. This modulation of the *M. avium*-induced granulomas occurred in the absence of any significant effect on the mycobacterial load within the tissues.

Once inside the phagosome, virulent *M. avium* begins to multiply and the vacuole takes on a unique character, which allows the organism to survive within the macrophage. *M. avium* prevents the accumulation of a proton ATPase from entering the vesicle, thereby blocking the acidification of the phagosome. The resulting moderate pH within the infected vacuole (pH 6.5) reduces the hydrolytic enzyme activity and enhances the survival of the bacteria (Strugill-Koszycki *et al.*, 1994). In spite of the limited maturation of the MAC-infected phagosomes, however, these vacuoles remain in a dynamic state and are freely accessible to specific glycolipids and glycoconjugates. Subsequent activation of the macrophage by cytokines released by other cells in the granuloma leads to acidification and maturation of the phagosome (Schaible *et al.*, 1998).

Such changes are concomitant with a reduction in the growth of the mycobacteria during the late stages of the infection, indicating an increased bacteriostatic and bactericidal activity within the phagocytic cell (Stokes *et al.*, 1986).

During the early host response to the MAC infection, the macrophage secretes a number of cytokines, including TNFα, IL-12, IL-6 and IL-10 (Cooper *et al.*, 1998). In addition, neutrophils and dendritic cells may generate TNFα and IL-12 during the innate immune response to the *M. avium* infection (Petrofsky and Bermudez, 1999). These cytokines act together to stimulate IFNγ production by the natural killer (NK) cells. The IFNγ primes the macrophage to produce high levels of IL-12 which, in turn, activates the T-cell mediated antibacterial response.

The role played by the immunosuppressive cytokine IL-10 in the regulation of the early growth of MAC bacilli remains controversial. Some investigators have shown that IL-10 does not modulate the anti-MAC activity of human or murine macrophages, while others have reported that IL-10 down-regulates the antimicrobial activity of macrophages that have taken up MAC bacilli (Shiratsuchi *et al.*, 1996; Sano *et al.*, 1999). In a recent study, Balcewicz-Sablinska *et al.* (1999) demonstrated that IL-10 produced by human monocytes infected with *M. avium* attenuated the mycobacterium-induced apoptosis by inducing the release of TNF type 2 receptor and thus inactivating the TNFα response. Apoptosis is a likely defence mechanism that should lead to reduced bacterial growth and limited spread by the infection. In this circumstance, the bacteria are confined to the apoptotic cells which could easily be taken up by the surrounding macrophages. By reducing the level of MAC-induced apoptosis, IL-10 may decrease the early restriction of *M. avium* growth within the macrophage.

Control of the MAC infection during the second phase of the infection largely depends on the generation of antigen-specific CD4 T cells that synthesise IFNγ. Studies demonstrating that IFNγ knockout mice are highly susceptible to MAC infections prove that IFNγ is an essential component of the acquired immune response (Cooper *et al.*, 1998). The central role of CD4 cells in this process was established by showing that depletion of these cells during a MAC infection greatly enhanced bacterial growth *in vivo*. Clearly, the IL-12 generated during the early phase of the infection induces the production of large numbers of IFNγ-producing CD4 cells, which will activate the macrophages entering the granuloma. Although TNFα does not contribute directly to the bactericidal activity seen during the chronic phase of the infection, this cytokine seems to regulate other important inflammatory responses in the infected host. Ehlers *et al.* (1999b) demonstrated that, in spite of comparable numbers of *M. avium* in TNFα receptor knockout mice and wild-type controls, the developing granulomas in the gene-disrupted animals became progressively necrotic, leading to extremely severe tissue damage and an accelerated mortality. In response to a persistent MAC infection, the absence of the TNF p55 receptor protein led to a dysregulation of the T cell–macrophage interactions and eventually to a widespread necrosis in the granulomatous host.

Immunity to *M. avium*-Complex Infections

Antituberculosis immunity has for nearly 40 years been a paradigm for a cell-mediated immune response to a variety of intracellular pathogens (Mackaness, 1968). Virulent organisms are taken up by resident tissue macrophages, where they multiply with little indication that the normally efficient killing mechanisms have a restraining effect on the intracellular growth of the organism (Xu *et al.*, 1994). Virulent strains of tubercle bacilli are able to reduce the acidity of the phagosome (Crowle *et al.*, 1991), as well as inhibiting the fusion of lysosomal enzymes with the phagosome (Schaible *et al.*, 1998). Although there is evidence that humoral factors, complement and fibronectin are involved in the uptake of MAC cells by monocytes *in vitro* (Bermudez *et al.*, 1991), there is no evidence that convalescent serum is protective when tested in adoptively immunised animals (Reggiardo and Middlebrook, 1974). There is no question that the heavily infected host produces a substantial humoral response to a variety of mycobacterial antigens, some of which serve as useful diagnostic markers in the laboratory (Lee *et al.*, 1991). However, this humoral response usually reaches its peak long after the cellular defences, which consist primarily of CD4+ T cells, have entered the developing lesion from the draining lymph node. In addition, CD8+ cytotoxic T cells, γδ T cells and NK cells can be detected in the lesion, although their precise role in the immune process is still unclear (Inderlied *et al.*, 1993). Relatively little is known about the nature of the cellular response to the more pathogenic members of the *M. avium* complex, especially in apparently normal, immunocompetent adults. In apparently normal adults most clinical isolates of *M. avium* result in a relatively indolent, often fatal, type of lung infection (Huang *et al.*, 1999). On the other hand, only a few *M. intracellulare* serovars are capable of causing progressive lung disease in adults, usually when their lung defences have been depleted.

Experimentally, a slowly progressive disease develops in most inbred strains of mice, as well as in guineapigs, when they are infected with *M. avium* (Thoen *et al.*, 1984). However, mice bearing the *Bcg*^s gene (Goto *et al.*, 1984), beige (*bg/bg*) mice which lack NK cells (Gangadharam *et al.*, 1983), and athymic nude mice which lack CD4+ T cells (Hubbard *et al.*, 1992), rapidly develop a systemic MAC infection. Multiple mechanism(s) are likely to be involved in this increased susceptibility. Thus, the effect can be reversed by the adoptive transfer of immune T cells to nude mouse recipients, while NK cells harvested from *Bcg*^r mice and transferred into *Bcg*^s recipients substantially increased their resistance to a subsequent challenge with *M. avium* (Hubbard *et al.*, 1991).

The primary mediators of the immune response to *M. avium* infection in normal, immunocompetent adults are CD4 T cells. As a result, HIV-infected individuals are unable to resist these opportunistic pathogens, which produce active, life-threatening disease when their CD4 counts fall below a certain threshold (Bermudez, 1994). This is consistent with the finding that disseminated MAC disease regresses sharply as CD4 counts recover after successful antiviral chemotherapy (Shafran, 1998). Presumably, *M. avium* in the lung granulomas is rapidly inactivated as the T-cell defences are restored as a result of the antiviral therapy, which allows activation of the macrophages as they enter the lesion.

Most strains of *M. intracellulare* lack the ability to survive in normal mouse macrophages or even polymorphs and so fail to establish a persistent systemic infection in athymic nude mice (Collins, 1986; Collins and Stokes, 1987). So far, all attempts to transfer 'virulence' genes from *M. avium* to *M. intracellulare* have been unsuccessful (F.M. Collins and S.L. Morris, unpublished data). Part of this problem revolves around an inability to determine which genes are responsible for blocking the killing mechanism(s) in the activated phagocyte. It is possible that different strains of MAC differ in their susceptibility to oxygen and nitric oxide intermediates (Denis, 1991; Bermudez, 1993).

Gamma interferon (IFNγ) is a major cytokine associated with the expression of antibacterial immunity to *M. avium* infection (Bermudez and Young, 1990). The relationship between the amount of IFNγ present in patient serum or tissue biopsy material, and the ability of the host to inactivate this opportunistic pathogen, has not been easy to establish experimentally and it may depend on the release of other modulatory cytokines such as IL-6 and TGFβ

(Inderlied *et al.*, 1993). This seems consistent with the finding that IL-6 stimulates the growth of some strains of *M. avium* complex in macrophage cultures *in vitro* (Shiratsuchi *et al.*, 1993).

Treatment of *M. avium*-Complex Infections

Until recently, treatment of patients with MAC disease has been extremely difficult and often disappointing, largely because of the resistance of this group to most first-line antituberculosis drugs (Wallace *et al.*, 1990). Administration of multi-drug combinations to these patients may reduce the symptoms of the disease, presumably by lowering the number of circulating acid-fast organisms in the blood. These regimens are, however, associated with considerable toxicity, often forcing a cessation to the treatment and recrudescence of active disease. The introduction of the macrolides into this regimen has improved the efficacy of the standard MAC therapeutic protocol dramatically, even for AIDS patients (Griffith, 1999). In the past 5 years, several published trials have provided strong evidence of the clinical benefits to be achieved by inclusion of these agents in the management of MAC disease.

Chemoprophylaxis of HIV-Infected Individuals

The effectiveness of the macrolides in the prevention of MAC disease in AIDS patients has been clearly demonstrated in several recent studies. Pierce *et al.* (1996) assessed the frequency of MAC infections in 667 HIV-infected patients given either clarithromycin or a placebo. A test group of 333 patients received 500 mg of clarithromycin twice daily, while 334 received a placebo. The clarithromycin-treated group showed a significantly lower incidence of MAC bacteraemia (6% in the drug therapy group versus 16% in the placebo group). A significant improvement in patient survival was also reported by these authors. Oldfield *et al.* (1998) also found that the macrolide treatment reduced the incidence of disseminated *M. avium*-complex (DMAC) disease in a group of HIV-infected patients. When azithromycin was administered at 1200 mg once a week, the incidence of DMAC was reduced to 8.2%, compared with 23.3% in the placebo group. To determine whether the azithromycin treatment augmented or simplified the treatment of AIDS patients with DMAC, 1200 mg of azithromycin was administered weekly and the response compared with 300 mg rifabutin given daily or a combination of azithromycin and rifabutin (Havlir

et al., 1996). A significant decline in DMAC was observed in patients treated once a week with azithromycin, relative to daily doses of rifabutin. Although the combined therapy was effective in reducing the number of MAC infections and the emergence of drug-resistant mutants, it was also associated with an increase in dose-limiting adverse events, when compared with either drug given separately. As a result of these trials, the Centers for Disease Control updated the guidelines for the treatment of disease due to these opportunistic pathogens. The current recommendation is that all HIV-infected patients should be treated with clarithromycin (500 mg/day) or 1200 mg of azithromycin given once a week (Centers for Disease Control, 1997).

Treatment of Disseminated *M. Avium*-Complex Disease

Although the optimal regimen for treating DMAC disease has yet to be finalised, recent clinical trials suggest that a macrolide-based treatment may give the best clinical response. Initial studies examined the effectiveness of clarithromycin administered at one of three doses (500, 1000 or 2000 mg given twice daily) to AIDS patients with DMAC disease (Chaisson *et al.*, 1994). A more rapid clearance of organisms from the bloodstream in patients receiving the higher drug doses was reported. Survival of these patients was also greatly enhanced, but monotherapy was associated with a substantial level of clinical relapse and the development of resistance to clarithromycin.

Since the emergence of macrolide resistance in *M. avium* precludes the further use of monotherapy with this drug, its use in combination with other antimycobacterial agents has been evaluated. In a prospective, randomised trial, clarithromycin and clofazimine were administered with or without ethambutol to a group of AIDS patients with DMAC disease (Dube *et al.*, 1997). In both groups, a 100-fold reduction was observed in the number of *M. avium* present in two consecutive blood samples collected from nearly 70% of treated patients. The three-drug regimen, was, however, better at preventing relapse and the emergence of clarithromycin resistance. Further to support the inclusion of ethambutol in this regimen, a group of patients treated with either azithromycin plus ethambutol or with clarithromycin plus ethambutol (Ward *et al.*, 1998). A dose of 600 mg of azithromycin, given once daily with the ethambutol, was compared with 500 mg of clarithromycin given twice daily plus ethambutol. The clarithromycin drug combination resulted in more rapid and effective clearance of the mycobacteria from the bloodstream.

For both regimens, the incidence of macrolide resistance was very low, and only one drug-resistant isolate was identified in the entire study.

Since rifabutin has been recommended for the treatment of MAC disease in AIDS patients, Gordin et al. (1999) tested rifabutin plus clarithromycin and ethambutol in 198 AIDS patients and reported substantially improved responses in patients treated with all three drugs. The addition of rifabutin to the combination had no impact on the bacterial response or on patient survival, but it prevented the development of clarithromycin resistance in patients responding to therapy. Although the macrolide-based combinations were clearly effective, the dosage must be carefully evaluated. In a recent comparison of several drug regimens in MAC-infected AIDS patients, clarithromycin given daily at 500 or 1000 mg and combined with rifabutin or ethambutol appeared to be most effective (Cohn et al., 1999). While an excellent therapeutic outcome was observed in both treatment arms, the high dose of clarithromycin was associated with increased mortality, causing the investigators to conclude that the maximum dosage of clarithromycin should be limited to 500 mg given twice daily to AIDS patients with MAC disease.

The introduction of highly active antiretroviral therapy (HAART) for the treatment of HIV-infected individuals had a dramatic impact on the course of this disease, and on the number of opportunistic life-threatening complications. As a result of the immuno-restoration associated with the increase in CD4 cell numbers in the blood of patients receiving the HAART treatment, management of MAC disease in these patients has been greatly improved. Four patients who had a strong CD4 cell rebound after HAART were treated with combined macrolide therapy and showed complete resolution of their MAC disease to the point where the antimycobacterial therapy could be discontinued (Aberg et al., 1998). It should, however, be noted that continued macrolide treatment may still be indicated, even in the HAART era. Phillips et al. (1999) reported fewer MAC infections, and where they did occur the organisms seemed to be localised in the lymph nodes. It was noted, however, that continued azithromycin prophylaxis, as compared with no treatment, further reduced the incidence of MAC disease in this group.

Treatment of *M. avium*-Complex Lung Disease in Immunocompetent Patients

Macrolide therapy has also improved the prognosis for non-AIDS patients with MAC lung disease. In early studies, patients treated twice daily with 500 mg of clarithromycin tolerated the drug well and showed high response rates (Wallace et al., 1994). However, drug resistance developed rapidly in patients receiving monotherapy. When clarithromycin was given twice daily in combination with ethambutol, rifabutin or rifampin, often with an initial course of streptomycin injections, 92% of the patients underwent sputum conversion, which persisted long-term in 82% of cases. The incidence of drug resistance was low in patients receiving the combined therapy. To minimise costs and reduce the incidence of adverse events, combination therapy with intermittent clarithromycin treatment (Monday, Wednesday and Friday each week) was tested in a group of MAC patients (Griffith et al., 1998). The rate of sputum conversion and the incidence of drug resistance was similar for patients completing 6 months of intermittent therapy compared with a daily clarithromycin treatment. In these patients, 1000 mg of clarithromycin given three times a week was generally well-tolerated.

Mechanisms of Drug Resistance in the *M. avium* Complex

The *M. avium* complex is innately resistant to most first-line antituberculosis drugs. The minimum inhibitory concentration (MIC) values for most of these drugs against *M. avium* isolates are generally 10- to 100-fold higher than those for *M. tuberculosis* (Heifets, 1996). In addition to their innate drug resistance, most MAC strains can develop even higher levels of acquired drug resistance during monotherapy. For instance, after implementation of clarithromycin therapy, breakthrough mutants were detected after only 8 weeks of treatment (Chaisson et al., 1994). Although most of the initial patient isolates were fully susceptible to clarithromycin, resistant strains were recovered from half the patients after 16 weeks of monotherapy.

Cell Envelope Structure

The hydrophobic and relatively impermeable cell envelopes of most mycobacteria contribute substantially to their pathogenicity, as well as to their resistance to a wide spectrum of therapeutic agents (Brennan and Nikaido, 1995; Brennan and Besra, 1997). This substantial lipid-rich barrier consists of a variety of soluble proteins, polysaccharides and lipids. The predominant component of the cell envelope consists of large amounts of C_{60} to C_{90} fatty acids (mycolic acids) which are covalently linked to an arabinogalactan-peptidoglycan cell-wall core. In addition, a number of highly unusual lipids and glycolipids can be extracted from the mycobacterial cell wall, including lipoarabinomannan (LAM), phenolic glycolipids (PGLs) and a number of glycopeptidolipids (GPLs).

Electron microscopic examination of the mycobacterial cell envelope suggests that the cell envelope is composed of an inner layer of moderate electron density, a wide intermediate electron transparent layer and a variably electron-opaque outer layer (Brennan and Nikaido, 1995). The inner layer contains the cell-wall peptidoglycan skeleton which is covalently linked through an L-Rhap(1→3) D-Glc Nac-p linking unit to an arabinogalactan chain. The electron-transparent layer is dominated by mycolic acid residues, which have been esterified to the 5′ position of the arabinogalactan units of the cell-wall core. The outer half of the lipid barrier, the electron-opaque layer, contains a number of unusual polysaccharides and glycolipids. This large mycolyl–arabinogalactan peptidoglycan complex, coupled with the exotic glycolipid and carbohydrates, assemble to form an asymmetric, impermeable bilayer of exceptional thickness and low fluidity.

The *M. avium* complex can be distinguished from most other mycobacteria by the presence of a number of highly antigenic GPL molecules (Aspinall *et al.*, 1995). These major surface antigens are the basis of the 28 MAC serovars and can be divided into non-specific and serovar-specific compounds. All *M. avium*-complex GPLs are comprised of an *N*-acetylated lipopeptide core, which is glycosylated at the terminal alaninol residue with a methyl rhamnose moiety. The non-specific GPLs present in all MAC strains contain a single o-linked 6-deoxytalose residue attached to a D-allo-threonine. The individual MAC serovars possess specific GPLs that have variable oligosaccharides attached to a D-allo-threonine residue. Among the terminal sugars responsible for the serological specificity of the 28 serovar-related GPLs that have been elucidated, there are several unique amidosugars, branched-chain carbohydrates and pyruvate-linked carbohydrates (Brennan and Nikaido, 1995).

The GPL biosynthetic pathways have not yet been completely elucidated for the *M. avium* complex. However, a number of the genes involved have been identified. The ser-2 gene cluster has been shown to encode proteins responsible for the synthesis of the hapten oligosaccharide of the serovar 2-specific GPL (Belisle *et al.*, 1991). Eckstein *et al.* (1998) showed that the rhamnosyltransferase gene known to be involved in GPL biosynthesis was encoded by the ser 2a locus. Further elucidation of these biosynthetic pathways is likely to be greatly facilitated by the expected completion of the MAC genome sequencing project.

Mechanisms of Resistance to Conventional Antituberculosis Drugs

The relative impermeability of the *M. avium*-complex cell envelope is an important factor that contributes to the multiple drug resistance phenotype seen in most MAC strains. Two experimental approaches have been developed to demonstrate that the impermeable cell envelope is largely responsible for the innate drug resistance of MAC bacilli (Rastogi, 1994). First, he showed that inhibitors of cell envelope synthesis, such as ethambutol, significantly enhanced the susceptibility of MAC strains to other antituberculosis drugs. Secondly, amphipathic derivatives of the hydrophobic drug isoniazid (palmitoyl-INH) substantially increased activity against *M. avium*-complex strains, relative to the parent compound. The permeability barrier hypothesis for these MAC strains is further supported by the absence of mutations in genes associated with resistance to many antituberculosis drugs. While evaluating *M. avium* and *M. intracellulare* strains that exhibit varying levels of sensitivity to rifampin, it was found that *rpoB* gene alterations were not associated with the expression of rifampin resistance in a number of clinical MAC isolates (Guerro *et al.*, 1994). Portillo-Gomez *et al.* (1995) extended these findings by showing that mutations in genes responsible for rifampin and streptomycin resistance in *M. tuberculosis* (*rpoB*, *rpsL* and *rrs* genes) were not present in multidrug resistant MAC strains. Moreover, although most MAC isolates are naturally resistant to pyrazinamide (PZA), Zhang and associates found that the MAC *pncA* gene, which encodes the pyrazinamidase that is mutated in most PZA-resistant *M. tuberculosis* strains, conferred PZA susceptibility when expressed in PZA-resistant *M. tuberculosis* (Sun *et al.*, 1997). This suggests that the insensitivity of *M. avium* to PZA was not due to an inefficient pyrazinamidase but was probably due to the ineffectual transport of active metabolites across the MAC cell envelope.

While low drug influx rates undoubtedly contribute to the innate insensitivity of MAC organisms to antibiotics, recent studies suggest that resistance to ethambutol and isoniazid may involve additional mechanisms besides cell-envelope impermeability (Belanger *et al.*, 1996). Over-expression of the *M. avium embAB* genes in *M. smegmatis* rendered this rapid grower resistant to ethambutol. The *embAB* genes encode an arabosyl transferase involved in cell-wall arabinogalactan biosynthesis. In an elegant series of experiments, Mdluli *et al.* (1998) unexpectedly proved that there was no significant difference in the relative permeability to INH between drug-sensitive strains of *M. tuberculosis* and INH-resistant *M. avium*. Furthermore, these investigators compared the rate of activation of INH by KatG (conversion to 4-pyridylmethanol) in *M. tuberculosis* and MAC whole cells and cell lysates to define the role of KatG in isoniazid-resistance. Previously, Rouse *et al.* (1996) had shown that over-expression of the MAC *katG* gene in a *katG* mutant of *M. bovis* (BCG) conferred INH sensitivity. Mdluli *et al.* (1998) showed that the rate of INH activation was 4-fold greater for *M. tuberculosis* when compared with that for *M. avium*. This suggests that the low level of expression of MAC KatG may result in reduced activation of the pro-drug, INH, and this could contribute to the inherent resistance of *M. avium* to isoniazid.

Mechanisms of Resistance to Macrolides

The development of clarithromycin resistance has been associated with a single base mutation in the 23s rRNA gene. Meier *et al.* (1994) examined pairs of susceptible and resistant MAC isolates from patients with MAC disease treated by clarithromycin monotherapy. Nucleotide sequence comparisons of the 23S rRNA genes from these isolate pairs demonstrated that three out of six highly resistant strains had a single mutation at base-pair 2274. Subsequently, Nash and Inderlied (1995) confirmed this finding by demonstrating that seven out of eight macrolide-resistant MAC strains isolated from AIDS patients had a mutation at the same base-pair in the 23S rRNA gene. While rRNA mutations in other organisms may be recessive, this single mutation in *M. avium* had a dominant effect because mycobacteria have only a single rRNA operon. This mutation occurs in the domain V peptidyl-transferase region of the 23S rRNA gene and probably causes a conformational change in the macrolide binding site.

Conclusions

The *M. avium* complex constitutes an important group of potential human and animal pathogens, because they are widely distributed throughout the natural environment and many have the potential to colonise the pharyngeal and intestinal mucosae of individuals exposed to contaminated food and water. While the majority of organisms in this complex do not cause systemic disease in humans, they may induce severe infections in patients with some form of underlying lung disease. When this occurs, the innate drug resistance of these organisms makes it difficult to treat such patients effectively.

The AIDS epidemic had a dramatic effect on the incidence of *M. avium*-complex disease in the USA. The increase in tuberculosis in these patients has raised important questions regarding the mechanisms used by the host defences to limit and control these infections, both in normal and immunodeficient individuals. Despite a great deal of study, surprisingly little is still know about the epidemiology, pathogenicity and immunology of these organisms, or the means by which they subvert or elude these normally effective cellular defences. A great deal of basic research will be needed before new and improved methods can be developed to detect and identify the mycobacteria present in clinical specimens taken from these patients. With the publication of the MAC genome sequence, it should become possible to design more effective drugs and vaccines for use in patients infected with these tenacious opportunistic pathogens.

References

Aberg JA, Yajko DM, Jacobson MA (1998) Eradication of AIDS-related *Mycobacterium avium* complex infection after 12 months of antimycobacterial therapy combined with highly active antiviral therapy. *J. Infect. Dis.* 178: 1446–1449.

American Thoracic Society (1997) Diagnosis and treatment of disease caused by nontuberculous mycobacteria. *Am. Rev. Resp. Crit. Care Med.* 156: S1–25.

Appleberg R, Castro AG, Pedrosa J, Silva RA, Orme IM, Minorprio P (1994) Role of gamma interferon and tumor necrosis factor alpha during T-cell-independent and -dependent phases of *Mycobacterium avium* infection. *Infect. Immun.* 62: 3962–3971.

Aspinall GO, Chattergee D, Brennan PJ (1995) The variable surface glycolipids of mycobacteria: structures, synthesis of epitopes and biological properties. *Adv. Carbohydr. Chem. Biochem.* 51: 169–242.

Balcewicz-Sablinska MK, Gan H, Remold HG (1999) Interleukin-10 produced by macrophages inoculated with *Mycobacterium avium* attenuates mycobacteria-induced apoptosis by reduction of TNF-α activity. *J. Infect. Dis.* 180: 1230–1237.

Barrow WW (1991) Contributing factors of pathogenesis in the *Mycobacterium avium* complex. *Res. Microbiol.* 142: 427–433.

Barrow WW, Brennan PJ (1982) Isolation in high frequency of rough variants of *Mycobacterium intracellulare* lacking C-mycoside glycopeptidolipid antigens. *J. Bacteriol.* 150: 381–384.

Bauer J, Andersen AB (1999) Stability of insertion sequence IS1245, a marker for differentiation of *Mycobacterium avium* strains. *J. Clin. Microbiol.* 37: 442–444.

Bauer J, Andersen AB, Askgaard D, Giese SB, Larsen B (1999) Typing of *Mycobacterium avium* complex strains cultured during a 2-year period in Denmark by using IS1245. *J. Clin. Microbiol.* 37: 600–605.

Belanger AE, Besra GS, Ford ME, Mikusova K *et al.* (1996) The *embAB* genes of *Mycobacterium avium* encode an arabinosyl transferase involved in cell wall arabinan biosynthesis that is the target for the antimycobacterial drug ethambutol. *Proc. Natl Acad. Sci. USA* 93: 11919–11924.

Belisle JT, Brennan PJ (1994) Molecular basis of colony morphology in *Mycobacterium avium*. *Res. Microbiol.* 145: 237–242.

Belisle JT, Pascopella L, Inamine JA, Brennan PJ, Jacobs WR (1991) Isolation and expression of a gene cluster responsible for the glycopeptidolipid antigens of *Mycobacterium avium*. *J. Bacteriol.* 173: 6991–6997.

Benson CA, Ellner JJ (1993) *Mycobacterium avium* complex infection and AIDS: advances in theory and practice. *Clin. Infec. Dis.* 17: 7–20.

Bermudez LE (1993) Differential mechanisms of intracellular killing of *Mycobacterium avium* and *Listeria monocytogenes* by activated human and murine macrophages: the role of nitric oxide. *Cell. Exp. Immunol.* 91: 277–281.

Bermudez LE (1994) Immunobiology of *Mycobacterium avium* infection. *Eur. J. Clin. Microbiol. Infect. Dis.* 13: 1000–1006.

Bermudez LE, Young LS (1990) Killing of *Mycobacterium avium*: insights provided by the use of recombinant cytokines. *Res. Microbiol.* 141: 241–245.

Bermudez LE, Young LS, Enkel H (1991) Interaction of *Mycobacterium avium* complex with human macrophages: roles of membrane receptors and serum proteins. *Infect. Immun.* 59: 1697–1702.

Berthet FX, Langranderie M, Gounon P, Laurent-Winter C *et al.* (1998) Attenuation of virulence by disruption of the *Mycobacterium tuberculosis erp* gene. *Science* 282: 759–762.

Bogner JR, Rusch-Gerdes S, Mertenkotter T, Loch O *et al.* (1997) Patterns of *Mycobacterium avium* culture and PCR positivity in immunodeficient HIV-infected patients: progression from localized to systemic disease. *Scand. J. Infect. Dis.* 29: 579–584.

Brennan PJ, Besra GS (1997) Structure, function and biogenesis of the mycobacterial cell wall. *Biochem. Soc. Trans.* 25: 188–194.

Brennan PJ, Nikaido H (1995) The envelope of mycobacteria. *Ann. Rev. Biochem.* 64: 29–63.

Centers for Disease Control (1997) USPHS/IDSA guidelines for the prevention of opportunistic infections in persons infected with human immunodeficiency virus. *Morbidity and Mortality Wkly Rpt.* 46: 12–13.

Chaisson RE, Benson CA, Dube MP, Heifets LB *et al.* (1994) Clarithromycin therapy for bacteremic *Mycobacterium avium* complex disease. *Ann. Intern. Med.* 121: 905–911.

Cohn DL, Fisher EJ, Peng GT, Hodges JS *et al.* (1999) A prospective randomized trial of four three-drug regimens in the treatment of disseminated *Mycobacterium avium* complex disease in AIDS patients: excess mortality associated with high-dose clarithromycin. *Clin. Infect. Dis.* 29: 125–133.

Collins FM (1982) Immunology of tuberculosis. *Am. Rev. Resp. Dis.* 125: 42–49.

Collins FM (1985) Protection afforded by BCG vaccines against an aerogenic challenge by three mycobacteria of decreasing virulence. *Tubercle* 66: 267–276.

Collins FM (1986) *Mycobacterium avium*-complex infection and the development of acquired immunodeficiency syndrome (AIDS): a casual opportunist or a causal cofactor? *Int. J. Leprosy* 54: 458–474.

Collins FM (1999) *Mycobacterium avium*-complex infection and immunodeficiency. In: Paradise LJ, Friedman H, Bendilli M (eds) *Opportunistic Intracellular Bacteria and Immunity.* New York: Plenum Press, pp. 107–130.

Collins FM, Stokes RW (1987) *Mycobacterium avium*-complex infections in normal and immunosuppressed mice. *Tubercle* 68: 127–136.

Comstock GW (1994) Field trials of tuberculosis vaccines: how could we have done them better? *Contr. Clin. Trials* 15: 247–276.

Cooper AM, Appelberg R, Orme IM (1998) Immunopathogenesis of *Mycobacterium avium* infection. *Front. Biosci.* 3: 141–148.

Cousins D, Francis B, Dawson D (1996) Multiplex PCR provides a low-cost alternative to DNA probe methods for rapid identification of *Mycobacterium avium* and *Mycobacterium intracellulare*. *J. Clin. Microbiol.* 34: 2331–2333.

Crowle AJ, Dahl R, Ross E, May MH (1991) Evidence that vesicles containing living virulent *Mycobacterium tuberculosis* or *Mycobacterium avium* in cultured human macrophages are not acidic. *Infect. Immun.* 59: 1823–1831.

De Francesco MA, Colombrita D, Pinsi G, Gargiulo F *et al.* (1996) Detection and identification of *Mycobacterium avium* in the blood of AIDS patients by the polymerase chain reaction. *Eur. J. Clin. Microbiol. Infect. Dis.* 15: 551–555.

Denis M (1991) Tumor necrosis factor and granulocyte macrophage-colony stimulating factor stimulate human macrophages to restrict the growth of virulent *Mycobacterium avium* and to kill avirulent *M. avium*: killing effector mechanism depends on the generation of reactive nitrogen intermediaries. *J. Leuk. Biol.* 49: 380–387.

Doherty TM, Sher A (1997) Defects in cell-mediated immunity affect chronic, but not innate, resistance of mice to *Mycobacterium avium* infection. *J. Immunol.* 158: 4822–4831.

Dube MP, Sattler FR, Torriani FJ, See D *et al.* (1997) A randomized evaluation of ethambutol for prevention of relapse and drug resistance during treatment of

Mycobacterium avium complex bacteremia with clarithromycin-based combination therapy. *J. Infect. Dis.* 176: 1225–1232.

Dunbar FP, Pejovic I, Cacciatore R, Peric-Golia L, Runyon EH (1968) *Mycobacterium intracellulare* maintenance of pathogenicity in relationship to lyophilization and colony forms. *Scand. J. Resp. Dis.* 49: 153–162.

Eckstein TM, Silbaq FS, Chattergee D, Kelly NJ, Brennan PJ, Belisle JT (1998) Identification and recombinant expression of a *Mycobacterium avium* rhamnosyl-transferase gene (*rtfA*) involved in glycopeptidolipid biosynthesis. *J. Bacteriol.* 180: 5567–5573.

Edwards LB, Acquaviva FA, Livesay VT, Cross FW, Palmer CE (1969) An atlas of sensitivity to tuberculin, PPD-B and histoplasmin in the United States. *Am. Rev. Resp. Dis.* 99: 1–132.

Ehlers S, Kutsch J, Benini J *et al.* (1999a) NOS2-derived nitric oxide regulates the size, quantity and quality of granuloma formation in *Mycobacterium avium*-infected mice without affecting bacterial loads. *Immunology* 98: 313–323.

Ehlers S, Benini J, Kutsch K, Endres R, Rietschel ET, Pfeffer K (1999b). Fatal granuloma necrosis without exacerbated mycobacterial growth in tumor necrosis factor receptor p55 gene-deficient mice intravenously infected with *Mycobacterium avium*. *Infect. Immun.* 67: 3571–3579.

Ellner JJ, Goldberger MJ, Pareti DM (1991) *Mycobacterium avium* infection and AIDS: a therapeutic dilemma in rapid evolution. *J. Infect. Dis.* 163: 1326–1335.

Falkinham JO (1994) Epidemiology of *Mycobacterium avium* infections in the pre- and post-HIV era. *Res. Microbiol.* 145: 169–172.

Falkinham JO (1996) Epidemiology of infection by non-tuberculous mycobacteria. *Clin. Microbiol. Rev.* 9: 177–215.

Fine PEM (1989) The BCG story: lessons from the past and implications for the future. *Rev. Infect. Dis.* 11 (Suppl. 2): S353–S359.

Fine PEM, Rodriguez LC (1990) Modern vaccines: mycobacterial diseases. *Lancet* 335: 1016–1026.

Frothingham R, Wilson KH (1994) Molecular phylogeny of the *Mycobacterium avium* complex demonstrates clinically meaningful divisions. *J. Infect. Dis.* 169: 305–312.

Fry KL, Meissner PS, Falkinham JO (1986) Epidemiology of infection by nontuberculous mycobacteria: VI. Identification and use of epidemiological markers for studies of *Mycobacterium avium*, *M. intracellulare* and *M. scrofulaceum*. *Am. Rev. Resp. Dis.* 134: 39–43.

Gangadharam PR, Edwards CK, Murthy PS, Pratt PF (1983) An acute infection model for *Mycobacterium intracellulare* disease using beige mice: preliminary results. *Am. Rev. Resp. Dis.* 127: 648–649.

Gomes MS, Florido M, Pais TF, Appelberg R (1999) Improved clearance of *Mycobacterium avium* upon disruption of the inducible nitric oxide synthase gene. *J. Immunol.* 162: 6734–6739.

Good RC, Snider DE (1982) Isolation of non-tuberculous mycobacteria in the United States. *J. Infect. Dis.* 146: 829–833.

Gordin FM, Sullam PM, Shafran SD, Cohn DL *et al.* (1999) A randomized placebo-controlled study of rifabutin added to a regimen of clarithromycin and ethambutol for treatment of disseminated infection from *Mycobacterium avium* complex. *Clin. Infect. Dis.* 28: 1080–1085.

Goto Y, Namamura RM, Takahashi H, Tokunaga T (1984) Genetic control of resistance to *Mycobacterium intracellulare* infection in mice. *Infect. Immun.* 46: 135–140.

Grange JM (1986) Environmental mycobacteria and BCG vaccination. *Tubercle* 67: 1–4.

Grange JM (1992) The mystery of the mycobacterial 'persistor'. *Tuberc. Lung Dis.* 73: 249–251.

Green EP, Tizard ML, Moss MT (1989) Sequence and characteristics of IS900, an insertion element identified in a human Crohn's disease isolate of *Mycobacterium paratuberculosis*. *Nucl. Acids Res.* 17: 9063–9073.

Griffith DE (1999) Risk-benefit assessment of therapies for *Mycobacterium avium* complex infections. *Drug Safety* 21: 137–152.

Griffith DE *et al.* (1998) Initial (six months) results of intermittent clarithromycin (CLAR)-containing regimens for *Mycobacterium avium* complex (MAC) lung disease. *Am. J. Respir. Crit. Care Med.* 157: A579.

Gruft H, Falkinham JO, Parker BC (1981) Recent experience in the epidemiology of disease caused by atypical mycobacteria. *Rev. Infect. Dis.* 3: 990–996.

Guerrero C, Stockman L, Marchesi F, Roberts GD, Telenti A (1994) Evaluation of *rpoB* gene in rifampicin-susceptible and -resistant *Mycobacterium avium* and *Mycobacterium intracellulare*. *J. Antimicrob. Chemother.* 33: 661–663.

Guerrero C, Bernasconi C, Burki D, Bodner T, Telenti A (1995) A novel insertion element from *Mycobacterium avium*, IS1245, is a specific target for analysis of strain relatedness. *J. Clin. Microbiol.* 33: 304–307.

Havlir DV, Dube MP, Sattler FR, Forthal DN *et al.* (1996) Prophylaxis against disseminated *Mycobacterium avium* complex with weekly azithromycin, daily rifabutin, or both. *N. Engl. J. Med.* 335: 392–398.

Heifets L (1996) Susceptibility testing of *Mycobacterium avium* complex isolates. *Antimicrob. Agents Chemother.* 40: 1759–1767.

Hernandez-Perez M, Fomunkong NG, Hellyer T, Brown I, Dale JW (1994) Characterization of IS1110, a highly mobile genetic element from *Mycobacterium avium*. *Mol. Microbiol.* 12: 717–724.

Hinds J, Mahenthiralingam E, Kempsell KE *et al.* (1999) Enhanced gene replacement in mycobacteria. *Microbiology* 145: 519–527.

Hoffner SE, Petrini B, Brennan PJ *et al.* (1989) AIDS and *Mycobacterium avium* serotypes in Sweden. *Lancet* 2: 336–337.

Horsburgh CR (1999) The pathophysiology of disseminated *M. avium* complex disease in AIDS. *J. Infect. Dis.* 179 (Suppl. 3): S461–S465.

Horsburgh CR, Selik RM (1989) The epidemiology of disseminated tuberculous mycobacterial infection in acquired immunodeficiency syndrome (AIDS). *Am. Rev. Resp. Dis.* 139: 4–7.

Horsburgh CR, Chin DP, Yajko DM *et al.* (1994) Environmental risk factors for acquisition of *Mycobacterium avium* complex in persons with human immunodeficiency virus infection. *J. Infect. Dis.* 170: 362–367.

Huang JH, Kao PN, Adi V, Ruoss SJ (1999) *Mycobacterium avium–intracellulare* pulmonary infection in HIV-negative patients without pre-existing lung disease. *Chest* 115: 1033–1040.

Hubbard RD, Flory CM, Collins FM (1991) Memory T-cell mediated resistance to *Mycobacterium tuberculosis* infection in innately susceptible and resistant mice. *Infect. Immun.* 59: 2012–2016.

Hubbard RD, Flory CM, Collins FM (1992) T-cell responses in *Mycobacterium avium*-infected mice. *Infect. Immun.* 60: 150–153.

Inderlied CB, Kemper CA, Bermudez LE (1993) The *Mycobacterium avium* complex. *Clin. Microbiol. Rev.* 6: 266–310.

Iralu JV, Sritharan VK, Pieciak WS, Wirth DF, Maguire JH, Barker R (1993) Diagnosis of *Mycobacterium avium* bacteremia by polymerase chain reaction. *J. Clin. Microbiol.* 31: 1811–1814.

Iseman MD (1989) *Mycobacterium avium* complex and the normal host: the other side of the coin. *N. Engl. J. Med.* 321: 896–898.

Kansal RG, Gomez-Flores R, Mehta RT (1998) Changes in colony morphology influences the virulence as well as the biochemical properties of the *Mycobacterium avium* complex. *Microb. Pathogen.* 25: 203–214.

Kiehn TE, Edwards FF, Brannon P, Tsang A *et al.* (1985) Infections caused by *M. avium*-complex in immunocompromised patients: diagnosis by blood culture and fecal examination, antimicrobial susceptibility test and morphological and serological characteristics. *J. Clin. Microbiol.* 21: 168–173.

Kulski JK, Khinsoe C, Pryce T, Christiansen K (1995) Use of mutiplex PCR to detect and identify *Mycobacterium avium* and *Mycobacterium intracellulare* in blood culture fluids of AIDS patients. *J. Clin. Microbiol.* 33: 668–674.

Kunze ZM, Portaels F, McFadden JJ (1992) Biologically distinct subtypes of *Mycobacterium avium* differ in possession of insertion sequence IS901. *J. Clin. Microbiol.* 30: 2366–2372.

Lari N, Cavallini M, Rindi L, Iona E, Fattorini L, Garzelli C (1998) Typing of *Mycobacterium avium* isolates in Italy by IS1245-based restriction fragment length polymorphism analysis. *J. Clin. Microbiol.* 36: 3694–3697.

Lee BY, Chattergee D, Bozic CM, Brennan PJ, Cohn DL *et al.* (1991) Prevalence of serum antibody to the type-specific glycopeptidolipid antigens of *Mycobacterium avium* in human immunodeficiency virus-positive and -negative individuals. *J. Clin. Microbiol.* 29: 1026–1029.

Lein AD, von Reyn CF, Ravn P, Horsburgh CR, Alexander LN, Andersen P (1999) Cellular immune responses to ESAT-6 discriminate between patients with pulmonary disease due to *Mycobacterium avium* complex and those with pulmonary disease due to *Mycobacterium tuberculosis*. *Clin. Diag. Lab. Immunol.* 6: 606–609.

Li Z, Kelley C, Collins FM, Rouse DA, Morris SL (1998) Expression of KatG in *Mycobacterium tuberculosis* is associated with its growth and persistence in mice and guinea pigs. *J. Infect. Dis.* 177: 1030–1035.

MacGregor RR, Dreyer K, Herman S *et al.* (1999) Use of PCR in detection of *Mycobacterium avium* complex (MAC) bacteremia: sensitivity of the assay and effect of treatment for MAC infection on concentrations of human immunodeficiency virus in plasma. *J. Clin. Microbiol.* 37: 90–94.

Mackaness GB (1968) The immunology of antituberculous immunity. *Am. Rev. Resp. Dis.* 97: 337–344.

Marklund B-I, Mahenthiralingam E, Stokes RW (1998) Site-directed mutagenesis and virulence assessment of the *katG* gene of *Mycobacterium intracellulare*. *Mol. Microbiol.* 29: 999–1008.

Maslow JN, Dawson D, Carlin EA, Holland SM (1999) Hemolysin as a virulence factor for systemic infection with isolates of *Mycobacterium avium* complex. *J. Clin. Microbiol.* 37: 445–446.

Matsumoto H, Tsuyuguchi K, Suzuki K, Tanaka E, Amitani R, Kuze F (1998) Evaluation of the Roche Amplicor PCR assay for *Mycobacterium avium* complex in bronchial washings. *Int. J. Tuberc. Lung Dis.* 2: 935–940.

McClatchy JK (1981) The seroagglutination test in the study of nontuberculous mycobacteria. *Rev. Infect. Dis.* 3: 867–870.

Mdluli K, Swanson J, Fischer E, Lee RE, Barry CE (1998) Mechanisms involved in the intrinsic isoniazid resistance of *Mycobacterium avium*. *Mol. Microbiol.* 27: 1223–1233.

Meier A, Kirschner P, Springer B, Steingrube VA *et al.* (1994) Identification of mutations in the 23s rRNA gene of clarithromycin-resistant *Mycobacterium intracellulare*. *Antimicrob. Agents Chemother.* 38: 381–384.

Meissner G (1981) The value of animal models for study of infection due to atypical mycobacteria. *Rev. Infect. Dis.* 3: 953–959.

Meyer M, von Grunberg PW, Knoop T, Hartmann P, Plum G (1998) The macrophage-induced gene *mig* as a marker for clinical pathogenicity and *in vitro* virulence of *Mycobacterium avium* complex strains. *Infect. Immun.* 66: 4549–4552.

Morrissey AB, Aisu TO, Falkinham JO, Ercki PP, Ellner JJ, Daniel TM (1992) Absence of *Mycobacterium avium* complex disease in patients with AIDS in Uganda. *J. AIDS* 5: 477–478.

Moulding T (1978) The relative drug susceptibility of opaque colony forms of *M. intracellulare–avium*: does it affect therapeutic results? *Am. Rev. Resp. Dis.* 117: 1142–1143.

Nash KA, Inderlied CB (1995) Genetic basis of macrolide resistance in *Mycobacterium avium* isolated from patients with disseminated disease. *Antimicrob. Agents Chemother.* 39: 2625–2630.

Nel EE (1981) *Mycobacterium avium–intracellulare* complex serovars isolated in South Africa from humans, swine and the environment. *Rev. Infect. Dis.* 3: 1013–1020.

Ninet B, Auckenthaler R, Rohner P, Delaspre O, Hirschel B (1997) Detection of *Mycobacterium avium–intracellulare* in the blood of HIV-infected patients by a commercial polymerase chain reaction kit. *Eur. J. Clin. Microbiol. Infect. Dis.* 16: 549–551.

O'Brien RJ, Geiter LJ, Snider DE (1987) The epidemiology of non-tuberculous mycobacterial diseases in the United States. *Am. Rev. Resp. Dis.* 135: 1007–1014.

Ohga S, Ikeuchi K, Kadoya K, Miyazaki C, Suita S, Udea K (1997) Intrapulmonary *Mycobacterium avium* infection as a first manifestation of chronic granulomatous disease. *J. Infect.* 34: 147–150.

Oldfield EC, Fessel J, Dunne MW, Dickinson G *et al.* (1998) Once-weekly azithromycin therapy for prevention of *Mycobacterium avium* complex infection in patients with AIDS: a randomized, double-blind, placebo-controlled multicenter trial. *Clin. Infect. Dis.* 26: 611–619.

Pestel-Caron M, Arbeit RD (1998) Characterization of IS1245 for strain typing of *Mycobacterium avium*. *J. Clin. Microbiol.* 36: 1859–1863.

Petrofsky M, Bermudez LE (1999) Neutrophils from *Mycobacterium avium*-infected mice produce TNF-α, IL-12 and IL-1β and have a putative role in early host response. *Clin. Immunol.* 91: 354–358.

Phillips P, Yip B, Hogg R, Bessuille E *et al.* (1999) Azithromycin (AZM) for *Mycobacterium avium* complex (MAC) prophylaxis in HIV patients during the HAART era: evaluation of a provincial program. *Proceedings of the Thirty-Ninth Interscience Conference on Antimicrobial Agents and Chemotherapy*, p. 457.

Pierce M, Crampton S, Henry D, Heifets L *et al.* (1996) A randomized trial of clarithromycin as prophylaxis against *Mycobacterium avium* complex infection in patients with advanced acquired immunodeficiency syndrome. *N. Engl. J. Med.* 335: 384–391.

Piot P, Loga M, Ryder R *et al.* (1990) The global epidemiology of HIV infection: continuity, heterogenicity and change. *J. AIDS* 3: 403–412.

Plum G, Brenden M, Clark-Curtis JE, Pulverer G (1997) Cloning, sequencing and expression of the *mig* gene of *Mycobacterium avium*, which codes for a secreted macrophage-induced protein. *Infect. Immun.* 65: 4548–4557.

Portaels F, Larssen L, Smeets P (1988) Isolation of mycobacteria from healthy persons stools. *Int. J. Leprosy* 56: 468–471.

Portillo-Gomez L, Nair J, Rouse DA, Morris SL (1995) The absence of genetic markers for streptomycin and rifampicin resistance in *Mycobacterium avium* complex strains. *J. Antimicrob. Chemother.* 36: 1049–1053.

Rastogi N (1994) Recent observations concerning structure and function relationships in the mycobacterial cell envelope: elaboration of a model for pathogenicity, virulence and drug-resistance. *Res. Microbiol.* 145: 464–476.

Reggiardo Z, Middlebrook G (1974) Failure of passive serum transfer of immunity against aerogenic tuberculosis in guinea pigs. *Proc. Soc. Exp. Biol. Med.* 145: 173–175.

Rittaco V, Kremer K, van Laan T *et al.* (1998) Use of IS901 and IS1245 in RFLP typing of *Mycobacterium avium* complex: relatedness among serovar reference strains, human and animal isolates. *Int. J. Tuberc. Lung Dis.* 2: 242–251.

Roiz MP, Palenque E, Guerro C, Garcia MJ (1995) Use of restriction fragment polymorphism as a genetic marker for typing *Mycobacterium avium* strains. *J. Clin. Microbiol.* 33: 1389–1391.

Romanus V, Hollander HO, Wahlen P, Olinder-Nielsen AM, Magnusson PH, Juhlin J (1995) Atypical mycobacteria in extrapulmonary disease among children: incidence in Sweden from 1969 to 1990, related to changing BCG-vaccination coverage. *Tuberc. Lung Dis.* 76: 300–310.

Roth A, Fischbach F, Arasteh KN, Futh U, Mauch H (1999) Evaluation of IS-1245-based PCR for detection of *Mycobacterium avium* bacteremia in AIDS patients. *Scand. J. Infect. Dis.* 31: 394–398.

Rouse DA, DeVito JA, Li Z-M, Byer H, Morris SL (1996) Site-directed mutagenesis of the *katG* gene of *Mycobacterium tuberculosis*: effects on catalase-peroxidase activities and isoniazid resistance. *Mol. Microbiol.* 22: 583–592.

Runyon EH (1965) Pathogenic mycobacteria. *Adv. Tuberc. Res.* 14: 235–287.

Sangari FJ, Petrofsky M, Bermudez LE (1999) *Mycobacterium avium* infection of epithelial cells results in inhibition or delay in the release of interleukin-8 and RANTES. *Infect. Immun.* 67: 5069–5075.

Sano C, Sato K, Shimizu T, Kajitana H, Kawauchi H, Tomioka H (1999) The modulating effects of pro-inflammatory cytokines interferon-gamma and tumor necrosis factor and immunoregulating cytokines IL-10 and transforming growth factor on antimicrobial activity of murine peritoneal macrophages against *Mycobacterium avium–intracellulare* complex. *Clin. Exp. Immunol.* 115: 435–442.

Schaefer WB (1965) Serological identification and classification of the atypical mycobacteria by their agglutination. *Am. Rev. Resp. Dis.* 92: 85–93.

Schaefer WF, Davis CL, Cohn ML (1970) Pathogenicity of translucent, opaque and rough variants of *M. avium* in chickens and mice. *Am. Rev. Resp. Dis.* 123: 343–358.

Schaible UE, Sturgill-Koszycki S, Schlesinger PH, Russell DG (1998) Cytokine activation leads to acidification and increases maturation of *Mycobacterium avium*-containing phagosomes in murine macrophages. *J. Immunol.* 160: 1290–1296.

Shafran SD (1998) Prevention and treatment of disseminated *Mycobacterium avium* complex infection in human immunodeficiency virus-infected individuals. *Int. J. Infect. Dis.* 3: 39–47.

Shiratsuchi H, Toosi Z, Mettler MA, Ellner JJ (1993) Colonial morphotype as a determinant of cytokine expression by human monocytes infected with *Mycobacterium avium*. *J. Immunol.* 150: 2945–2954.

Shiratsuchi H, Hamilton B, Toossi Z, Ellner JJ (1996) Evidence against a role for interleukin-10 in the regulation of growth of *M. avium* in human monocytes. *J. Infect. Dis.* 173: 410–417.

Snider DE (1982) The tuberculin skin test. *Am. Rev. Resp. Dis.* 125: S108–S118.

Sterling TR, Moore RD, Graham NM, Astemborski J, Vlahov D, Chaisson RE (1998) *Mycobacterium tuberculosis* infection and disease are not associated with protection against subsequent disseminated *M. avium* complex disease. *AIDS* 12: 1451–1457.

Stokes RW, Orme IM, Collins FM (1986) Role of mononuclear phagocytes in expression of resistance and susceptibility to *Mycobacterium avium* infections in mice. *Infect. Immun.* 54: 811–819.

Sturgill-Koszycki S, Schlesinger P, Chakraborty P, Haddix PL *et al.* (1994) Lack of acidification in mycobacterium containing phagosomes produced by exclusion of the vesicular proton-ATPase. *Science* 263: 678–681.

Sun Z, Scorpio A, Zhang Y (1997) The *pncA* gene from naturally pyrazinamide-resistant *Mycobacterium avium* encodes pyrazinamidase and confers pyrazinamide susceptibility to resistant *Mycobacterium tuberculosis* complex organisms. *Microbiology* 143: 3367–3373.

Thoen CO, Himes EM, Karlson AG (1984) *Mycobacterium avium* complex. In: Kubica GP, Wayne LG (eds) *The Mycobacteria: a Casebook*. New York: Marcel Dekker, pp. 1251–1275.

van Soolingen D, Bauer J, Ritacco V, Leao SC *et al.* (1998) IS1245 restriction fragment polymorphism typing of *Mycobacterium avium* isolates: proposal for standardization. *J. Clin. Microbiol.* 36: 3051–3054.

von Reyn CF, Waddell RD, Eaton T *et al.* (1993) Isolation of *Mycobacterium avium* complex from water in the United States, Finland, Zaire and Kenya. *J. Clin. Microbiol.* 31: 3227–3230.

von Reyn CF, Maslow JN, Barber TW *et al.* (1994) Persistent colonization of potable water as a source of *M. avium* infection in AIDS. *Lancet* 343: 1137–1141.

von Reyn CF, Williams DE, Horsburgh CR, Jaeger AS *et al.* (1998) Dual skin testing with *Mycobacterium avium* sensitin and purified protein derivative to discriminate pulmonary disease due to *Mycobacterium avium* complex from pulmonary disease due to *Mycobacterium tuberculosis*. *J. Infect. Dis.* 177: 730–736.

Wallace RJ, O'Brien R, Glassroth J, Raleigh J, Dutt A. (1990) Diagnosis and treatment of disease caused by non-tuberculous mycobacteria. *Am. Rev. Resp. Dis.* 142: 940–945.

Wallace RJ, Brown BA, Griffith DE, Girard WM *et al.* (1994) Initial clarithromycin monotherapy for *Mycobacterium avium–intracellulare* complex lung disease. *Am. J. Respir. Crit. Care Med.* 149: 1335–1341.

Ward TT, Rimland D, Kauffman C, Huycke M *et al.* (1998) Randomized, open-label trial of azithromycin plus ethambutol as therapy for *Mycobacterium avium* complex bacteremia in patients with human immunodeficiency virus infection. *Clin. Infect. Dis.* 27: 1278–1285.

Wasem CF, McCarthy CM, Murray LW (1991) Multilocus enzyme electrophoresis analysis of the *Mycobacterium avium* complex and other mycobacteria. *J. Clin. Microbiol.* 29: 264–271.

Wayne LG (1985) The 'atypical' mycobacteria: recognition and disease association. *Crit. Rev. Microbiol.* 12: 185–172.

Wayne LG, Good RC, Tsang A *et al.* (1993) Serovar determination and molecular taxonomic correlation in *Mycobacterium avium*, *M. intracellulare* and *M. scrofulaceum*: a cooperative study of the International Working Group on Mycobacterial Taxonomy. *Int. J. Syst. Bacteriol.* 43: 482–489.

Wolinsky E (1981) When is an infection disease? *Rev. Infect. Dis.* 3: 1025–1027.

Wolinsky E (1984) Nontuberculous mycobacteria and associated diseases. In: Kubica GP, Wayne LG (eds) *The Mycobacteria: a Sourcebook*. New York: Marcel Dekker, pp. 1141–1208.

Xu S, Cooper A, Sturgill-Koszycki S *et al.* (1994) Intracellular trafficking in *Mycobacterium tuberculosis-* and *Mycobacterium avium*-infected macrophages. *J. Immunol.* 153: 2568–2578.

Yajko DM, Chin DP, Gonzalez PC *et al.* (1995) *Mycobacterium avium*-complex in water, food and soil samples collected from the environment of HIV-infected individuals. *J. AIDS* 9: 176–182.

Yakrus MA, Good RC (1990) Geographic distribution, frequency and specimen source of *Mycobacterium avium* complex serotypes isolated from patients with acquired immunodeficiency syndrome. *J. Clin. Microbiol.* 28: 926–929.

Young LS, Inderlied CB, Berlin OG, Gottlieb MS (1986) Mycobacterial infections in AIDS patients with an emphasis on the *Mycobacterium avium* complex. *Rev. Infect. Dis.* 8: 1024–1033.

PART
14
SEXUALLY-TRANSMITTED INFECTIONS

84

Sexually Transmitted and Genital Infections: A Clinical Overview

T. Green and G.R. Kinghorn

Royal Hallamshire Hospital, Sheffield, UK

Sexually transmitted infections (STIs) are amongst the most common of all infectious diseases and represent a largely preventable burden of ill-health. The World Health Organization (WHO) has estimated that in the world there are annually more than 330 million new cases of STI, which disproportionately affect the populations of developing countries. The global distribution of STIs is, therefore, very similar to that of infection by the human immunodeficiency virus (HIV). Both STIs and HIV, which favour mutual transmission, are important causes of morbidity and mortality in the developing world. They not only impose enormous economic burdens, but also threaten the social fabric of these countries.

The traditional bacterial infections – syphilis, gonorrhoea and chancroid – remain of paramount importance in the developing world, where increasing antibiotic-resistance is also common and poses threats to the efficacy and cost of treatment. In the UK, as in other developed countries, the most common STIs are bacterial and *Chlamydia trachomatis* infection, and the viral STI genital warts and genital herpes. Whilst the incidence of STIs is highest in people in their late teen years and twenties, recent epidemiological evidence shows rising infection rates in older age groups.

A wide range of other microbial agents can cause a sexually transmitted infection. Globally, the protozoon *Trichomonas vaginalis*, which causes a severe vaginitis in women, but is commonly asymptomatic in men, is thought to be the most frequent agent. It is less common in women in the UK, where vaginal infections are more often due to candidiasis or bacterial vaginosis. Other important sexually transmissible viral infections include hepatitis A, B and C. Arthropod infestations, such as pediculosis pubis or scabies, are amongst the least significant medically, but cause severe distress, and are useful markers for the presence of another STI. It remains an important principle that different STIs occur together, and it is essential that comprehensive investigation be undertaken whenever a STI is diagnosed. Since STIs are frequently asymptomatic, heavy reliance is placed on diagnostic laboratory services for their detection.

In efforts to control STIs, emphasis is placed on the diagnosis and management of those with clinical disease and the detection of asymptomatically infected persons by screening and partner notification. Since the advent of the AIDS pandemic, more resources have been devoted to primary prevention by health education, especially of the young. As yet, the only effective prophylactic vaccines are those for hepatitis B and hepatitis A. Nevertheless, considerable effort is now devoted to improving knowledge of the pathogenesis of HIV and other STIs, in the hope that new prophylactic and therapeutic strategies, including vaccines, will result.

Syphilis

This is due to *Treponema pallidum* (see Chapter 85) and affects more than 12 million people worldwide. It may be acquired sexually, *in utero* or by blood transfusion. The incidence in the developed world is relatively low; in 1996 only 259 cases of infectious syphilis were reported by genitourinary medicine clinics in England and Wales, although an outbreak in the city of Bristol involving upwards of 50 cases was reported in 1997 (Simms *et al.*, 1998). In the USA, syphilis remains an endemic disease, predominantly associated with areas of urban deprivation, drug abuse and prostitution (MMWR, 1998). Numbers of cases also remain relatively low in western Europe, but there is a potential threat from the former Eastern bloc countries where rates of infection are rising (Tichinova *et al.*, 1997). Thus, the majority of cases of syphilis are found in Africa and Asia where co-transmission with, and facilitation of, HIV infection is a growing problem. Fortunately, the treponemes remain sensitive to penicillin, making syphilis a treatable sexually transmitted infection. The key to successful eradication lies in comprehensive screening programmes, since the clinical features of the disease are such that the infectious stages may go unrecognised.

Clinical Features

Traditionally, syphilitic infection has been divided into the infectious primary, secondary, and early latent stages, and the non-infectious late latent and late clinical stages. Pregnant women may also infect their fetus transplacentally early in late latency.

Early Syphilis

Most textbooks give the incubation period as 9–90 days, but the pathognomonic primary chancre usually appears about 3 weeks after infection. This painless indurated ulcer is most commonly sited on the genitals but may be found on other sites that have been the point of sexual contact, including the mouth, anal region, nipples and fingers. Regional lymphadenopathy occurs one week later. The chancre is highly infectious and persists for several weeks; but because it is painless it may easily be overlooked, especially when internal sites such as the cervix are affected.

In untreated patients, a secondary stage occurs 4–6 weeks after the appearance of the chancre. This stage is characterised by a variety of skin manifestations, including a widespread symmetrical coppery rash, initially macular and later papular, that may be seen on the palms and soles as well as on the trunk and face; condylomata lata on moist cutaneous surfaces, such as the axillae and groins; and mucous patches in the mouth and on genital mucosal surfaces. Any or all of these features can occur, and may be florid or scarcely discernible. If untreated, the clinical signs will resolve spontaneously and the disease passes into the early latent stage. This lasts for 2 years, when clinical relapses occur in about a quarter of cases.

Late Syphilis

Patients eventually pass into the late latent stage, when they are no longer infectious to sexual partners but have persistently positive serological tests. About 60% of patients remain asymptomatic and develop no further manifestations of syphilis. The remaining 40% develop complications from the destructive tertiary stages of the disease that may result in chronic debility or death. These syphilitic complications may affect any organ system; they include benign tertiary syphilis, visceral syphilis, cardiovascular disease and neurosyphilis.

Benign Tertiary Syphilis The hallmark lesions, *gummata*, may affect skin, bones and mucosal surfaces. These chronic granulomas have a typical histological appearance, and classically start in the skin as painless raised nodules which erode and ulcerate before healing slowly from the centre to leave a paper-thin depigmented scar surrounded by a region of hyperpigmentation. Mucosal gummas affect the oral cavity and pharynx and are destructive, leading to severe scarring. Syphilitic glossitis causes a swollen smooth-surfaced tongue on which leucoplakia occurs, and malignant epithelial changes may ensue. Bone disease is primarily a periostitis; it may be painful and lead to thickening and deformity mainly of long bones, but lytic lesions of facial and skull bones may occur. Occasionally the inflammatory process may involve adjacent skin to produce a chronic osteomyelitis.

Visceral disease is relatively uncommon, leading to gummata in the liver, stomach, lungs and testes. The eyes may also be involved with uveitis or choroido-retinitis. Haemolytic anaemia occurs rarely.

Cardiovascular Syphilis The cardiovascular lesions of late syphilis predominantly affect the aorta, but ostial stenosis of the openings of coronary arteries can occur. The aortic lesions include asymptomatic aortitis, aortic regurgitation and aortic aneurysm of the ascending aorta. Intimal plaques, vessel dilatation and wall irregularities are macroscopic characteristics, while microscopically endarteritis of the vasa vasorum is the key feature that leads ultimately to destruction of elastic tissue and fibrous replacement. Aortic aneurysm may remain asymptomatic for many years until it either ruptures, leading to collapse and death, or pressure effects supervene. Asymptomatic uncomplicated aortic disease is usually a post-mortem finding.

Neurosyphilis Neurosyphilis may be asymptomatic with abnormalities in the cerebrospinal fluid (CSF) but without neurological signs, or it may be symptomatic with a range of well-recognised neurological deficits and positive CSF serology. The symptomatic complications include meningovascular syphilis, spinal syphilis, general paralysis of the insane (GPI), and tabes dorsalis.

Meningovascular and spinal syphilis result from an endarteritis of the cerebral vessels leading to symptoms and signs that correspond to the area of the brain affected. Cranial nerve palsies may occur. Spinal involvement can lead to spastic paraparesis or syphilitic amyotrophy. GPI was once the most common manifestation of neurosyphilis but it is now relatively rare. It is an encephalitis resulting in cerebral atrophy with a degree of hydrocephalus. Histologically there is perivascular and meningeal infiltration with lymphocytes and plasma cells and treponemes abound. GPI occurs 10–20 years after the primary infection; the onset is usually gradual with memory loss and cognitive defects, including loss of communication and loss of insight, culminating in total dementia.

Tabes dorsalis also has a slow late onset and may take many years to develop. The posterior columns of the spinal cord atrophy through damage to the nerve roots at the lumbosacral levels of the cord, and the autonomic nervous system is also involved. 'Lightning pain' is an early feature, followed by sensory loss, ataxia and hypotonia, and damage to joints may appear (Charcot's joints). Other distressing symptoms such as impotence and incontinence may be present.

Treatment

Treatment of all stages of syphilis is predominantly with penicillin, unless there is a history of allergy, when tetracyclines or erythromycin are substituted. Treatment of primary and secondary syphilis should result in a cure, so that late complications will not develop. The outcome of treating late complications is variable; generally disease progression is halted but earlier damage cannot be reversed. The success of treatment should be monitored serologically at intervals over a period of time (French, 1999; Goh, 1999).

Pathogenesis

Treponema pallidum is a host-specific human pathogen that has not been successfully cultured *in vitro*, but limited propagation and single-passage survival in certain tissue culture systems has been achieved. Thus study of its infective mechanisms at a molecular level has been difficult and most work has been done in experimental animals. Treponemal motility is a key part of the infective process, and allows the organisms to migrate from the primary chancre through the skin to the bloodstream, leading to haematogenous dissemination. Invasion of targeted tissues then occurs leading to secretion of lymphokines that activate tissue macrophages to destroy the invading spirochaetes. The entire treponemal genome has been sequenced in an attempt to identify possible virulence factors (Weinstock *et al.*, 1998), and genes coding for protective immunity and opsonisation to evade the host immune response have been identified. In addition, the outer membrane of *T. pallidum* has been shown to be structurally different from that of other bacteria, a fact that is thought to be related to the chronicity of syphilitic infections.

Cell-mediated delayed-type hypersensitivity (DTH) is the mainstay of the immune response to syphilis infection; antibody production and cytotoxic T lymphocytes are thought to be little involved. The DTH response to syphilis infection leads directly to phagocytosis of the treponemes by tissue macrophages, with resulting necrosis and apoptosis of these cells. Acquired immunity develops slowly after primary infection. It has been postulated that host differences in the extent of cell-mediated responses may account for differences in clinical manifestations (Barbosa-Cesnik *et al.*, 1997).

Diagnostic Tests

Treponema pallidum has a characteristic morphology and motility and can be recognised by dark-field microscopy in serum from abraded lesions in primary and secondary stages.

Serological tests are the mainstay of diagnosis, especially of late syphilis. Tissue damage caused by treponemes releases lipoidal material and lipoidal-like antigens. Serological responses to these antigens are the basis of the quantitative non-treponemal ('reagin') tests for syphilis, which are useful for assessing the response to treatment. Reactivity does not develop until 1–4 weeks after the appearance of the primary chancre. Titres reach a peak in secondary syphilis but then decline, and sometimes disappear altogether in late latent disease and in neurosyphilis. Approximately

2% of sera show the prozone phenomenon, in which undiluted serum gives negative results because of antibody excess or the presence of blocking antibodies. Biological false-positive (BFP) reactions may occur in about 1% of sera; potential causes include pregnancy, genital herpes, alcoholic cirrhosis, connective tissue disorders and Lyme disease.

Specific treponemal antibody tests are used as confirmatory tests and to detect antibodies to antigenic determinants of treponemes. These are qualitative procedures and are not helpful in assessing treatment responses. Once positive, these tests tend to remain positive for life, irrespective of treatment. They are used to differentiate true positives from false positives in the standard non-treponemal antibody tests. The FTA-ABS (fluorescent treponemal antibody absorption) test and double-staining test (FTA-ABS DS) are indirect immunofluorescence tests. MHA-TP (micro-haemagglutination assay for antibodies to *T. pallidum*) detects passive haemagglutination of erythrocytes sensitised with ultrasonicated Nichols strain *T. pallidum*. Commercial treponemal enzyme immunoassay tests (EIAs) have also been designed as confirmatory tests for syphilis. Non-reactive specific tests occur in early primary syphilis and may also occur in the presence of immune suppression associated with HIV infection. It is always essential to confirm the presumptive serological diagnosis of syphilis on a second patient specimen (Kinghorn, 1998).

Gonorrhoea

The causative organism of gonorrhoea, *Neisseria gonorrhoeae*, is a Gram-negative intracellular diplococcus that infects the genital tract mucosa. Other sites that may be infected include the oropharynx and the rectum. Humans are the natural host, and acquisition by adults is through sexual intercourse, although neonates can be infected during vaginal delivery.

Epidemiology

The incidence of gonorrhoea in England and Wales rose by 26% between 1995 and 1997, and figures from genitourinary medicine clinics suggest that this trend is continuing. The greatest rise is in the young sexually active population, with teenage girls increasingly affected, but more gonorrhoea is seen in men than women (a ratio of about 2:1). In addition, homosexual infection in men is rising. Most *N. gonorrhoeae* in the UK are sensitive to penicillin, but a rise in the number of cases of antibiotic-resistant isolates has been reported. Although such resistant strains are more common in developing countries, they have become endemic in the UK. Resistance to a number of commonly used antibiotics, including penicillin, tetracyclines and ciprofloxacin, singly and in combination, has been documented (*CDR Weekly*, 1998).

Clinical Features

Gonorrhoea has a short incubation period of 3–5 days. In men, uncomplicated genital infection manifests itself in 90% as mild dysuria with a purulent urethral discharge, but up to 10% may be asymptomatic. In women with uncomplicated gonorrhoea, up to 50% may be asymptomatic and the remainder have variable symptoms, including vaginal discharge and/or bleeding and dysuria. There is an association with *Trichomonas vaginalis* infection in women with gonorrhoea, and women who present with trichomoniasis should be screened for gonococcal infection. Homosexual men and heterosexual women may acquire rectal infection with gonorrhoea, which may be asymptomatic or may present as a proctitis with soreness and purulent discharge with blood on defaecation. Pharyngeal infection is usually asymptomatic, acquired and transmitted through oral sex.

Complications

Local complications occur in males when various glands in the genital area become infected and form abscesses. Urethral, paraurethral, periurethral and parafrenal gland infections produce penile lesions, and an abscess involving Cowper's glands forms a painful perineal lesion. Urethral stricture is a very rare complication, and post-gonococcal urethritis is likely to be due to co-infection with *Chlamydia*. The most important ascending complication is gonococcal epididymitis, which occurs in 1–2% of untreated infections.

In females, local complications include bartholinitis, and infection in Skene's paraurethral glands. The most serious complication in women is pelvic inflammatory disease (PID), which may occur in 10–15% of women with untreated gonorrhoea. Gonococci pass through the cervical os and colonise the fallopian tubes, leading to salpingitis. An abscess (pyosalpinx) may form and purulent exudate may pass into the peritoneal cavity. Damage to the mucosa of the fallopian tubes leads to adhesions and scarring and may result in decreased fertility or infertility, increased risk of ectopic pregnancy and chronic pelvic pain and dyspareunia. An abscess may rupture, leading to peritonitis, bowel obstruction and shock. Perihepatitis (Fitz–Hugh–Curtis syndrome) may be one of the sequelae of PID but cases in men have been reported.

Disseminated gonococcal infection is rare in the UK and can occur in both sexes, but it is slightly more common in women. Symptoms may range from a mild 'flitting' arthralgia with vague malaise to a severe arthritis with constitutional symptoms. Occasionally a single joint is involved, and associated tenosynovitis is a common feature. Up to 30% of cases have an associated painful rash with sparse lesions containing gonococci, often seen over affected joints. Very rarely meningitis or endocarditis is seen with disseminated infection.

Reiter's disease may follow gonococcal infection, characterised by the triad of urethritis, arthritis and uveitis. This is much commoner in men (ratio 10:1) and in Caucasians, and it is associated with histocompatibility antigen HLA B27. Reiter's disease is also associated with genital *Chlamydia trachomatis* infection, and is discussed further below.

Gonococcal Infection in Children

Neonates may acquire gonococcal ophthalmia neonatorum during vaginal delivery. This is a notifiable disease in the UK and if untreated can cause blindness. Older children with gonorrhoea, which usually presents as a vulvovaginitis in girls, should be investigated for possible sexual abuse.

Treatment

Antimicrobial resistance is an increasing worldwide problem, and regular monitoring of isolate sensitivities is essential. Penicillin-based treatment regimes are no longer recommended. In the UK, treatment of uncomplicated gonorrhoea is currently with a single 500 mg oral dose of ciprofloxacin. For pregnant women, intramuscular spectinomycin is used as an alternative. Gonococcal abscesses are drained if appropriate, and usually treated with a 7-day course of oral or intravenous antibiotics, depending on clinical severity. Epididymitis and PID are also treated with longer courses of antibiotics that include treatment for *Chlamydia*, because of the strong epidemiological link between the two pathogens. Disseminated infection requires intravenous antibiotics, with other supportive measures where appropriate (Bignell, 1999).

Pathogenesis

Neisseria gonorrhoeae is a human pathogen that colonises cuboidal and columnar mucosal epithelial surfaces. Most studies of gonorrhoea pathogenesis have been carried out *in vitro*, as gonococci are relatively easy to culture given the appropriate media and the right laboratory conditions, and there is no simple animal model of infection. Gonococci survive and

multiply after infection by their ability to remain attached to the host mucosal cells and evasion of host mucosal defence systems. The outer cell membrane, which contains phospholipids, lipo-oligosaccharides and multiple proteins surrounded by a loose polyphosphate capsule, mediates the interaction with host cells. Two gonococcal outer-membrane proteins in particular have been the focus of studies into bacterial adhesion to host cells. The heat-modifiable Opa proteins (formerly protein II), named for their presence in opaque colonies in culture, increase adherence between gonococci and between the bacteria and epithelial cells. The second protein, pilin, is a subunit of the helical pili that project from the gonococcal surface through the polyphosphate capsule and show antigenic variation. These subunits, in particular pilE, have been the focus of intensive vaccine research (Barbosa-Cesnik *et al.*, 1997).

Evasion of the host defences is due to gonococcal secretion of catalase, production of sialylated lipo-oligosaccharide that blocks antibody-mediated killing, and production of blocking antibodies against reduction-modifiable protein (Rmp). In addition, a porin channel (por, formerly protein I) appears to be involved in serum-resistance, antibiotic susceptibility and invasiveness. Another outer-membrane protein, protein III, is also immunogenic and may be involved in antimicrobial-resistance.

Antimicrobial resistance may also be plasmid-mediated: it is an increasing problem, especially in developing countries (Dankert, 1998).

Diagnostic Tests

Near-patient testing of Gram-stained urethral or endocervical specimens often provides a presumptive diagnosis of gonorrhoea by the identification of Gram-negative intracellular diplococci. This permits treatment to be commenced immediately. The method is not, however, specific for gonorrhoea, especially in specimens from non-genital sites. The 'gold-standard' specific test is culture, which is also more sensitive. Gonococci are fastidious organisms that require specific growth conditions and specific media, usually made selective by the inclusion of antibiotics. Occasionally vancomycin-sensitive strains fail to grow, and non-selective media may be used in parallel with standard media to detect such cases.

Newer technologies have led to the development of DNA probes and DNA amplification assays for detection of *N. gonorrhoeae*. DNA probes have 93–99% sensitivity and 98–99.5% specificity when compared with culture, and they may be preferred when the optimal conditions for culture specimen transportation to the diagnostic laboratory cannot be guaranteed. The PCR and LCR amplification assays have similar sensitivity and specificity indices, and their main advantage is that, where urethral swabs are unavailable, they can be carried out on first-voided urine samples. The main disadvantage of DNA technologies is that they do not yield bacterial isolates, and antimicrobial susceptibility testing is not possible (Fox and Cohen, 1998).

Genital Chlamydial Infection

Chlamydia trachomatis is an obligate intracellular parasite that causes a number of diseases depending on the serovar involved. Fourteen types of *Chlamydia* are recognised. Serovars A, B and C cause the eye infection, trachoma; types D to K infect the mucosa of the reproductive tract in both sexes and cause genital disease; while serotypes L1, L2 and L3 cause the systemic venereal disease, lymphogranuloma venereum.

Epidemiology

Genital chlamydial infection is the most common curable bacterial STI in the UK, with an incidence of 10–15% amongst the young sexually active female population. It has been estimated that only about 10% of genital chlamydial infections are seen at genito-urinary medicine clinics. Some are seen in primary care and by family planning services and at gynaecological clinics, but the majority are asymptomatic and remain undetected and untreated (*CDR Weekly*, 1997).

Clinical Features

The incubation period of chlamydial infection is difficult to determine because so many patients are asymptomatic, but it is thought to be from about 3 days to 3 weeks. Up to 70% of women and 50% of men with chlamydial infections are asymptomatic, and these subclinical infections may persist for a year or more. Although uncomplicated infection is easy and cheap to treat, the complications of the disease, especially in young women, give cause for concern.

Uncomplicated Infection

Uncomplicated symptomatic infection in men usually produces dysuria and urethral discharge which if untreated will often settle spontaneously. In women it gives rise to symptoms in only about 30% of cases. These may be similar to those seen in gonorrhoea, with an increase in vaginal discharge, which may be offensive or bloody, and there may be vague abdominal pain

or discomfort and rarely dysuria is a feature. In asymptomatic cases there may be no clinical evidence and treatment may not be initiated until laboratory confirmation of chlamydial infection is obtained.

Complications

Chronic prostatitis is a common complication of chlamydial infection in men. It is often difficult to diagnose and even more difficult to treat. Epididymitis is similar in presentation to that seen with gonorrhoea, but the onset is often less acute in chlamydial infection, and occurs in about 1% of untreated cases. Reiter's disease, which has been mentioned above in the context of gonococcal infection, is much more common after genital chlamydial infection. Most patients have urethritis and arthritis, and about one-third have an associated conjunctivitis; other associated features include keratitis and uveitis. The arthritis may affect peripheral joints or the spine and pelvic girdle, and may be a transient polyarthralgia or asymmetrical polyarthritis or a destructive monoarthritis. Females are rarely affected.

Complicated chlamydial infection is much more of a problem in women, since it is often difficult to confirm and harder to treat, and the sequelae have high morbidity and high cost. PID as a consequence of chlamydial infection is often a chronic grumbling illness, and the underlying salpingitis causes problems. Tubal scarring leads ultimately to problems of fertility, ectopic pregnancy and chronic pelvic pain. Another complication of chlamydial infection is neonatal chlamydial infection. As with gonorrhoea, infection is acquired by the infant during vaginal delivery and becomes manifest up to 7 days later as chlamydial ophthalmia neonatorum, or at any time up to 6 months of age as a chlamydial pneumonia, which may be life-threatening.

Lymphogranuloma Venereum

The lymphogranuloma venereum (LGV) serovars are more invasive than those of *Chlamydia trachomatis*, and cause a systemic infection that primarily affects lymphatic tissue. It occurs mainly in tropical regions, and men are affected more than women (ratio 5:1). Two clinical entities are recognised, an inguinal syndrome with painful swelling of the inguinal, femoral and iliac nodes, and an anorectal syndrome with an ulcerative proctitis.

There are three stages of the disease. The primary lesion appears on the external genitalia in less than a quarter of cases, after an incubation period of 3–21 days. The primary sore is often small and may be unnoticed, disappearing spontaneously after a couple of days. The secondary stage is characterised by lymph node involvement with the development of fluctuant inguinal buboes, which are usually unilateral. Constitutional symptoms are also present. Pelvic nodes may also be involved, and in women a purulent cervicitis may develop. Lesions of the tertiary stage are now rare, but can occur after a long latent period. These include genital elephantiasis, rectal stricture and fistula formation.

Treatment

The treatment of choice for uncomplicated adult chlamydial infection is doxycycline. For complicated infections the course is usually 14 days, and for women with PID metronidazole should be added to cover facultative genital tract pathogens that may be involved in the ascending infection (Horner and Caul, 1999).

Tetracyclines are the drugs of choice for LGV and the response is usually good in the primary and secondary stages. Large fluctuant buboes should be aspirated; surgical incision and drainage is not recommended because of the tendency to sinus and fistula formation. Tertiary-stage complications may require surgical repair once the active inflammation has settled with antimicrobial therapy (Maynaud, 1999).

Pathogenesis

Chlamydia are obligate intracellular pathogens, with a unique biphasic lifestyle consisting of morphologically distinct extracellular and intracellular forms. They grow in eukaryotic cells where they parasitise energy systems and utilise host ATP and nucleotide pools for replication. The hallmark of chlamydial infections is that they are of low self-toxicity, and are relatively well shielded from the normal immune responses. Only a minority of infected individuals develop severe disease; most have mild or asymptomatic disease. Humoral immunity plays little or no role in their eradication, whereas T lymphocyte stimulation, macrophage activation with production of interferon γ, and both CD4 and CD8 lymphocytes are of paramount importance (Perry *et al.*, 1997).

The mechanisms of attachment to host epithelial cells and subsequent internalisation may differ between different chlamydial serovars and their biovariants. Serovar-E entry is typically via receptor-mediated endocytosis via clathrin-coated pits. Once inside the cell, the organism is contained within host-derived endosomes which fuse with one another to create a larger vesicle but continue to evade fusion with lysosomes. The elementary bodies undergo re-organisation into reticulate bodies, reproduce by binary fission, and after maturation elementary bodies

escape via the apex of the cell. For slow-growing genital D–K strains, which are predominately luminal pathogens, the development cycle is 48–72 hours. LGV strains differ in their cell-surface receptors, their use of cytoskeletal structures for fusion, and their more rapid maturation. They are transported to the basolateral domain of the cell, and can escape to infect cells in the subepithelial space. LGV strains have a particular predilection for lymphoid cells. Other chlamydial serovars may also appear in the subepithelial spaces and are transported there by macrophages.

Epithelial infection by *Chlamydia* is associated with up-regulation and secretion of pro-inflammatory cytokines such as IL-8, GRO, GM-CSF, and IL-6. In contrast to the rapid cytokine induction caused by extracellular bacteria, the epithelial cytokine response associated with *C. trachomatis* is delayed for 20–24 hours after infection, requires bacterial protein synthesis, and persists throughout its growth cycle. Lysis of epithelial cells is associated with IL-1 release, which stimulates adjacent non-infected epithelial cells to produce additional cytokines. These cytokines are potent chemo-attractants and activators for neutrophils, monocytes, and T lymphocytes. Neutrophils are capable of killing *Chlamydia* by fusion of their lysosomes with ingested elementary bodies. Other cytokines have pleiotropic effects, including the induction of adhesion molecules on endothelial cells, and secretion of cytokines by macrophages and of acute-phase proteins in epithelial cells and macrophages. Chlamydial lipopolysaccharide associated with the bacterial plasma membrane induces release of IL-1 from monocytes and macrophages, which in turn induces the release of TNFα, a potent initiator of the inflammatory process. Nevertheless, invasion of host cells is usually silent; and as host-cell lysis is relatively slow, this minimises the clinical manifestations of the inflammatory response.

During replication, the antigens of *Chlamydia* may be introduced into MHC class 1 processing pathways, present on almost all cell types, and this leads to the activation of CD8 lymphocytes. Major outer membrane protein (MOMP) is the main target for both neutralising antibody and for protective T-cell responses. Since MHC class 2 molecules are not expressed on non-professional phagocytes, they are not recognised by CD4 lymphocytes. Nevertheless, IL-4 from T lymphocytes activates macrophages and promotes their antibacterial actions. This TH1 cell-mediated immune response is characteristic of chlamydial infections and is typified by delayed-type hypersensitivity reactions.

During re-infection or re-activation, the initial inflammatory response occurs and the infiltration of

T lymphocytes occurs more rapidly and in larger numbers. Macrophages are not typically a target of chlamydial infection but they are important for its control and possibly also for its maintenance. Monocytes may also play a role in the extragenital dissemination of *C. trachomatis* and in the establishment of persistent infection. There is *in vitro* evidence that interferon γ, derived from CD8 and CD4 lymphocytes, mediates persistent chlamydial infection in macrophages. This may be mediated, at least in part, by inducing indole-amine-2,3-hydrogenase which limits the supply of an essential amino acid and of iron. In experimental infection of cultured human blood monocytes with serotype K, a persistent non-productive infection occurs in aberrant cytoplasmic inclusions, and primary chlamydial rRNA transcripts, indicative of viable, metabolically active infection, have been demonstrated. The inclusions are characterised by enlarged, abnormal reticulate bodies ultrastructurally similar to those induced by nutrient depletion and the action of β lactam antibiotics. There is a paucity of outer-membrane constituents, such as MOMP and LPS. Nevertheless, levels of heat-shock proteins, which promote intracellular survival, are maintained. The nature of the latent or persistent chlamydial infection *in vivo* remains controversial, but is thought by some investigators to be important in the pathogenesis of the long-term sequelae of chlamydial infections.

Chlamydia heat-shock protein 60 (Chsp60) has immunopathological importance, and its association with PID, ectopic pregnancy, tubal infertility and trachoma is well-documented. Risk factors for antibodies to Chsp60 antibodies include older age and repeated infection. Whether or not antibody to Chsp60 is causally involved in the immunopathogenesis of the long-term sequelae or is merely a marker of persistent chlamydial infections remains controversial. These antibody responses appear to some extent to be genetically restricted, and there may be other host genetic variants that influence the clinical response to chlamydial infections. HLA B27 has long been known for its association with Reiter's syndrome, and the class 1 allele HLA A31 is associated with chlamydial PID. T-cell responses are depressed and inadequate in those with persistent trachoma, compared with those recovering from the infection, whilst humoral responses may be accentuated (Saikku *et al.*, 1998).

Diagnostic Tests

The detection of *Chlamydia* by culture in clinical samples, such as from urethral and cervical swabs, is complex, time-consuming and expensive. In many laboratories, routine diagnosis is with enzyme immunoassay (EIA) or direct immunofluorescence (DIF) techniques, and culture is reserved for cases with medico-legal implications (rape, sexual assault). EIA and DIF have sensitivities of 70–90% and specificities of 97–99% when compared with culture; and since both detect dead as well as living organisms they should not be used for a test of cure within 2–3 weeks of antichlamydial treatment.

DNA probes and DNA amplification tests are available for the detection of *Chlamydia*. The probes have a sensitivity of 86–93% and a specificity that is 97–99% when compared with culture. Signal amplification methods (PCR and LCR) are more sensitive than culture and have a high specificity. These DNA tests also detect dead organisms, and require a delay of at least 3 weeks after the completion of treatment before they are suitable as tests of cure. They also have the advantage over culture in being less demanding in specimen quality, and they can be used with non-invasive samples, such as first-voided urine or self-taken vulval swabs, without significant loss of sensitivity.

A rapid antigen-detection test has been developed for near-patient testing, but its sensitivity is only 50% that of culture, so that its use is limited (Fox and Cohen, 1998; Saikku *et al.*, 1998).

Serological diagnosis is useful in the diagnosis of LGV, and may be of some value in the diagnosis of complicated chlamydial disease.

References

Barbosa-Cesnik CT, Gerbase A, Heymann D (1997) STD vaccines: an overview. *Genitourin. Med.* 73: 336–342.

Bignell C (1999) National guidelines for the management of gonorrhoea in adults. *Sex. Transm. Infect.* 75 (Suppl. 1): S13–S15.

CDR Weekly (1997) Sexually transmitted diseases quarterly report: genital infection with Chlamydia trachomatis in England and Wales. *CDR Weekly* 7: 394–395.

CDR Weekly (1998) Sexually transmitted diseases quarterly report: gonorrhoea in England and Wales. *CDR Weekly* 8: 194–196.

Dankert J (1998) *Neisseria*. In: Armstrong D, Cohen J (eds) *Infectious Diseases*. London: Harcourt, Vol. 2. pp. 16.1–16.14.

Fox KK, Cohen MS (1998) Gonococcal and chlamydial urethritis. In: Armstrong D, Cohen J (eds) *Infectious Diseases*. London: Harcourt, Vol. 1. pp. 63.1–63.10.

French P (1999) National guidelines for the management of late syphilis. *Sex. Transm. Infect.* 75 (Suppl. 1): S34–S37.

Goh B (1999) National guidelines for the management of early syphilis. *Sex. Transm. Infect.* 75 (Suppl. 1): S29–S33.

Horner PJ, Caul EO (1999) National guidelines for the management of *Chlamydia trachomatis* genital infection. *Sex. Transm. Infect.* 75 (Suppl. 1): S4–S8.

Kinghorn GR (1998) Syphilis. In: Armstrong D, Cohen J (eds) *Infectious Diseases*. London: Harcourt, Vol. 1. pp. 64.1–64.10.

Maynaud P (1999) National guidelines for the management of lymphogranuloma venereum. *Sex. Transm. Infect.* 75 (Suppl. 1): S40–S42.

MMWR (1998) Primary and secondary syphilis: United States, 1997. *Morbidity and Mortality Wkly Rpt* 47: 493–497.

Perry LL, Feilzer K, Caldwell HD (1997) Immunity to *Chlamydia trachomatis* is mediated by T helper 1 cells through IFN-γ-dependent and independent pathways. *J. Immunol.* 158: 3344–3352.

Saikku PA (1998) *Chlamydia*. In: Armstrong D, Cohen J (eds) *Infectious Diseases*. London: Harcourt, Vol. 2. pp. 25.1–25.8.

Simms I, Hughes G, Swann AV, Rogers PA, Catchpole M (1998) New cases seen at genitourinary medicine clinics: England 1996. *CDR* Supplement 8: S1–11.

Trichinova L, Borisenko K, Ward H, Meheus A, Gromyko A, Renton A (1997) Epidemic of syphilis in the Russian Federation: trends, origins and priorities for control. *Lancet* 350: 210–213.

Weinstock GM, Hardham JM, McLeod MP, Sodergren EJ, Norris SJ (1998) The genome of *Treponema pallidum*: new light on the agent of syphilis. *FEMS Microbiol. Rev.* 22: 323–332.

85

Treponema pallidum

Lola V. Stamm

University of North Carolina at Chapel Hill, North Carolina, USA

Treponema pallidum subsp. *pallidum* is the causative agent of syphilis, a sexually transmitted infection (STI) with diverse clinical manifestations that occur in distinct stages (Lukehart and Holmes, 1998; Stamm, 1999). Syphilis was first recognised in late fifteenth century Europe, where it rapidly reached epidemic proportions (Singh and Romanowski, 1999). Like many emerging infectious diseases, syphilis was at first highly virulent, but over time it became a more chronic infection.

Although the origin of syphilis remains unknown, two main theories for its origin have been proposed. According to the Columbian theory, syphilis was endemic in the New World and was acquired by Columbus and his sailors who transmitted the infection to a naive European population. The pre-Columbian theory is that syphilis originated in Africa and was introduced into the European population before Columbus' voyage. Regardless of its origin, syphilis spread across the world in the sixteenth century aided by human migrations associated with commerce and conflicts. Despite the advent of effective antibiotic therapies in the mid-twentieth century, syphilis continues to be a worldwide public health problem, particularly because of its association with infection by human immunodeficiency virus (HIV) (Gerbase *et al.*, 1998; Agacfidan and Kohl, 1999; Singh and Romanowski, 1999).

Classification

The genus *Treponema* is a member of the Order Spirochaetales, which contains three other pathogenic genera (*Borrelia*, *Leptospira* and *Brachyspira*) (Smibert, 1984). The treponemal species of human health significance are *Treponema pallidum* subsp. *pallidum* (venereal syphilis), *T. pallidum* subsp. *pertenue* (yaws), *T. pallidum* subsp. *endemicum* (endemic syphilis or bejel), and *T. carateum* (pinta). None of these spirochetes has been serially cultivated *in vitro* for sustained time periods. Although the subspecies of *T. pallidum* are virtually identical on the basis of their morphology, overall antigenic properties and DNA sequence homology, they exhibit differences in geographical distribution, mode and severity of infection, and infectivity for laboratory animals (Smibert, 1984). Until recently, it has not been possible to differentiate the *T. pallidum* subspecies. However, Centurion-Lara *et al.* (1998) and Cameron *et al.* (1999) have reported nucleotide differences in the 5′ region flanking the highly conserved *tpp15* gene and a single nucleotide substitution within the *gpd* gene, respectively. These have permitted differentiation between *T. pallidum* subsp. *pallidum* and other *T. pallidum* subspecies. In addition, Stamm *et al.* (1998) showed significant differences between *T. pallidum* subsp. *pallidum* and *pertenue* in the nucleotide sequence of *tprJ*, a member of the polymorphic multigene *tpr* family.

Several cultivable and non-cultivable *Treponema* species are found in the human oral cavity, gastrointestinal tract and genital tract (Smibert, 1984). By comparative 16S rDNA sequence analysis, Paster *et al.* (1998) showed that there are at least 10 species of cultivable oral treponemes. Many of these spirochetes have limited pathogenic potential, but *T. denticola*,

 doi:10.1006/bkmm.2001.0085

T. vincentii, T. socranskii and *T. pectinovorum* are associated with periodontal disease (Moter *et al.*, 1998) and the non-cultivable pathogen-related oral spirochete (PROS) is associated with chronic period-ontitis and acute necrotising ulcerative gingivitis (Riviere *et al.*, 1991). Analysis of genomic DNA (Greene and Stamm, 1999; Stamm and Bergen, 1999) and 16S rRNA sequence data (Paster *et al.*, 1991) suggests a closer phylogenetic relationship between *T. pallidum* subsp. *pallidum* and some cultivable treponemes than has previously been proposed.

Identification and Detection

Dark-Field Microscopy

T. pallidum subsp. *pallidum* was identified in 1905 by Schaudinn and Hoffman, who demonstrated by modified Giemsa stain the presence of spirochetes in lesion material from syphilis patients. In 1909, Coles described the use of dark-field microscopy to visualise live *T. pallidum* subsp. *pallidum*. The sensitivity and specificity of dark-field microscopy for the detection of *T. pallidum* subsp. *pallidum* in patient materials depends on the experience of the observer and the adequacy of the specimen (Larsen *et al.*, 1995; Young, 1998). Dark-field microscopy is useful for the diagnosis of early syphilis in patients with genital or cutaneous lesions, but this method should not be used to examine material from lesions in the oral cavity or anal region because *T. pallidum* subsp. *pallidum* cannot be distinguished morphologically from other *Treponema* species present at these sites.

Animal Inoculation

Although *in vitro* cultivation of bacterial pathogens from clinical specimens is usually considered the 'gold standard', this is not possible for *T. pallidum* subsp. *pallidum*. The rabbit infectivity test (RIT) is an *in vivo* method that has been used to isolate and identify *T. pallidum* subsp. *pallidum* from clinical samples (Turner *et al.*, 1969; Grimprel *et al.*, 1991; Sanchez *et al.*, 1993). The RIT is sensitive with a theoretical ability to detect one or two viable, virulent trepo-nemes, but it is expensive and requires 3–6 months for completion. These limitations make the RIT imprac-tical for routine clinical use.

Direct Fluorescent Antibody Test

In the mid-1960s, a direct fluorescent antibody test for detection of *T. pallidum* subsp. *pallidum* (DFA-TP) was developed (Larsen *et al.*, 1995) and was later modified for use with monoclonal antibodies. The DFA-TP test differentiates *T. pallidum* subsp. *palli-dum* from cultivable *Treponema* species by a specific antigen-antibody reaction. It is, therefore, applicable to fixed samples from oral, anal, genital and cutaneous lesions, tissues, and body fluids, but the DFA-TP test does not distinguish between *T. pallidum* subsp. *pal-lidum* and other *T. pallidum* subspecies or *T. carateum* because of the presence of cross-reactive antigens.

Molecular Methods

Molecular techniques, such as the polymerase chain reaction (PCR) and reverse transcriptase PCR (RT-PCR), have been used to identify *T. pallidum* subsp. *pallidum* in clinical specimens (blood, cerebrospinal fluid, amniotic fluid, tissues, and lesion exudate) from humans and experimentally infected rabbits (Hay *et al.*, 1990; Grimprel *et al.*, 1991; Sanchez *et al.*, 1993; Larsen *et al.*, 1995; Centurion-Lara *et al.*, 1997). A multiplex PCR is commercially available for the simultaneous detection of *T. pallidum* subsp. *palli-dum* and other genital ulcer disease agents (herpes-simplex virus types 1 and 2 and *Haemophilus ducreyi*) (Orle *et al.*, 1996). PCR assays are more sensitive and specific than direct detection methods (dark-field microscopy and the DFA-TP test) and in conjunction with restriction fragment length polymorphism (RFLP) analysis can differentiate *T. pallidum* subsp. *pallidum* from the non-venereal *T. pallidum* sub-species (Centurion-Lara *et al.*, 1998; Cameron *et al.*, 1999). In addition, Pillay *et al.* (1998) demonstrated that PCR combined with RFLP analysis can be used for the molecular subtyping of clinical isolates of *T. pallidum* subsp. *pallidum* for epidemiological studies.

Structure

Cell Envelope

T. pallidum subsp. *pallidum* is 6–20 μm in length with a diameter of 0.10–0.18 μm, which renders it invisible by light microscopy (Smibert, 1984); but live, un-stained organisms can be observed by dark-field or phase-contrast microscopy. *T. pallidum* subsp. *palli-dum* stains poorly with aniline dyes, but it can be stained by silver impregnation methods. Electron microscopy is useful for special clinical and other investigations (**Fig. 1**).

Like Gram-negative bacteria, *T. pallidum* subsp. *pallidum* has an outer membrane (Johnson *et al.*,

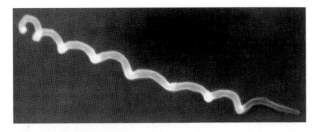

Fig. 1 Scanning electron micrograph of *T. pallidum* subsp. *pallidum* obtained from the testis of an experimentally infected rabbit. Courtesy of S.J. Norris.

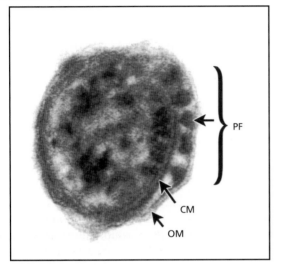

Fig. 2 Transverse section of *T. pallidum* subsp. *pallidum* as seen by transmission electron microscopy. Note the outer membrane (OM), periplasmic flagella (PF), and the cytoplasmic membrane-cell wall complex (CM). Courtesy of S.J. Norris. From Stamm (1999), with permission.

1973), an inner (cytoplasmic) membrane, and a thin peptidoglycan cell wall (Radolf *et al.*, 1989a) (**Fig. 2**).

Unlike Gram-negative bacteria, however, the *T. pallidum* subsp. *pallidum* outer membrane lacks lipopolysaccharide (LPS) (Hardy and Levin, 1983) and it is more susceptible to disruption by routine physical manipulation (centrifugation, washing, incubation) and treatment with low concentrations of detergents (Penn *et al.*, 1985; Stamm *et al.*, 1987; Radolf, 1994). Radolf *et al.* (1989b) and Walker *et al.* (1989) showed by freeze-fracture electron microscopy that the *T. pallidum* subsp. *pallidum* outer membrane contains only rare integral membrane proteins, some of which are exposed at the cell surface (**Fig. 3**).

To facilitate identification of the treponemal rare outer-membrane proteins (TROMPs), Blanco *et al.* (1994) and Radolf *et al.* (1995b) developed methods for the isolation of *T. pallidum* subsp. *pallidum* outer membranes. Recent evidence suggests that the majority of readily identifiable proteins present in the outer-membrane fractions are contaminating periplasmic and cytoplasmic membrane proteins rather than authentic outer-membrane proteins (Shevchenko *et al.*, 1997).

Periplasmic Flagella

T. pallidum subsp. *pallidum* is actively motile (Norris *et al.*, 1993). The spiral shape of the organism and the unique periplasmic location of the flagella enable it to retain motility in viscous fluids, such as those found in

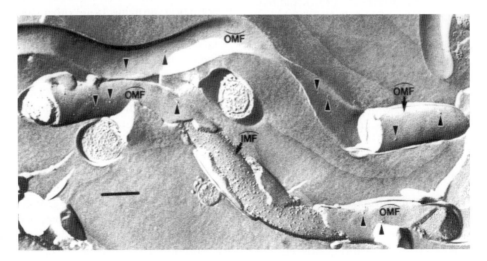

Fig. 3 Membrane architecture of *T. pallidum* subsp. *pallidum* as demonstrated by freeze-fracture electron microscopy. The concave and convex outer-membrane fracture (OMF) faces correspond to the outer and inner leaflets of the outer membrane, respectively, and contain a very low concentration of intra-membranous particles (arrowheads) that represent the TROMPs. In contrast, the convex inner-membrane fracture face (IMF) corresponds to the inner leaflet of the inner membrane, and has a high concentration of intra-membranous particles. Courtesy of E.M. Walker. From Stamm, 1999, with permission.

the joint, eye and extracellular matrix of the skin. Three periplasmic flagella originate at each end of the treponemal cell and entwine the protoplasmic cylinder extending toward the centre of the cell. The periplasmic flagella are composed of a hook–basal body complex and a flagellar filament made up of four proteins that are arranged into an outer sheath (FlaA) and a central core (FlaB1, 2 and 3) (Norris *et al.*, 1993). Genes that encode the protein subunits of the flagellar filament have been cloned and sequenced (Norris *et al.*, 1993) and one of these genes (*fla*A) has been expressed in a heterologous host, *T. denticola* (Chi *et al.*, 1999). A large motility (*fla*) operon that contains 17 genes and encodes proteins associated with flagellar structure, assembly and function has been identified (Hardham *et al.*, 1995; Limberger *et al.*, 1996). *T. pallidum* subsp. *pallidum* also contains an array of 4–6 cytoplasmic filaments which are ribbon-like structures (7.0–7.5 nm wide) of unknown function that run the length of the organism (Norris *et al.*, 1993). The cytoplasmic filaments lie just beneath the inner membrane and parallel to the periplasmic flagella.

Metabolism and Physiology

Cultivation

T. pallidum subsp. *pallidum* is one of the few bacterial pathogens of humans that has not been continuously cultivated *in vitro*. For antigen preparation and experimental studies, the organism is propagated in rabbit testes (Turner and Hollander, 1957). Limited growth of *T. pallidum* subsp. *pallidum* has been obtained in a tissue culture monolayer of Sf1Ep cottontail rabbit epithelial cells under microaerobic conditions (Fieldsteel *et al.*, 1981). The *in vitro* generation time is similar to that estimated for organisms growing *in vivo* (30–33 h). *T. pallidum* subsp. *pallidum* is very sensitive to environmental conditions (temperature, O_2, pH, moisture) and to physical and chemical agents.

Metabolic and Physiological Studies

The fastidious nature of *T. pallidum* subsp. *pallidum* suggests that it has metabolic limitations (Norris *et al.*, 1993). Because of its non-cultivable nature and the lack of systems for genetic manipulation, knowledge of treponemal metabolism and physiology is limited. Previous studies with treponemes extracted from infected rabbit testes demonstrated that *T. pallidum* subsp. *pallidum* is capable of DNA, RNA and protein synthesis (Norris *et al.*, 1993). Glucose is the major energy source, but pyruvate can be metabolised. *T. pallidum* subsp. *pallidum* has a glycolytic system, but lacks a tricarboxylic acid cycle.

Much of what has recently been learned about the metabolism and physiology of *T. pallidum* subsp. *pallidum* has resulted from the cloning, sequence analysis and expression of treponemal genes in *Escherichia coli*. For example, treponemal genes that encode the three enzymes necessary for proline biosynthesis have been identified (Gherardini *et al.*, 1990; Stamm and Barnes, 1997). An *mgl*-like operon that encodes homologues of proteins that putatively facilitate high-affinity transport of glucose/galactose has been cloned and characterised (Porcella *et al.*, 1996; Stamm *et al.*, 1996). The presence of treponemal operons that encode homologues of motility and chemotaxis proteins strongly suggests that *T. pallidum* subsp. *pallidum* has the ability to sense and respond to gradients of nutrients present in host tissues (Hardham *et al.*, 1995; Greene *et al.*, 1997).

Insights from Genome Analysis

The complete sequence of the *T. pallidum* subsp. *pallidum* (Nichols strain) circular chromosome has contributed a wealth of new and confirmatory information about the metabolic and physiological capabilities of this spirochete (Fraser *et al.*, 1998; *T. pallidum* Molecular Genetics Server: http://dpalm.uth.tmc.edu/treponema/tpall.html). The *T. pallidum* subsp. *pallidum* genome is ~1138 kb and contains 1041 predicted open reading frames (ORFs). Predicted biological roles have been assigned to 577 ORFs; 177 ORFs matched hypothetical proteins from other species, and 287 ORFs have no database matches.

Genome analysis confirmed that *T. pallidum* subsp. *pallidum* has limited biosynthetic capabilities and requires multiple nutrients from the host. It is similar to *B. burgdorferi* (the Lyme disease agent), in that *T. pallidum* subsp. *pallidum* is unable to synthesise most amino acids, enzyme cofactors, fatty acids and nucleotides *de novo*. *T. pallidum* subsp. *pallidum* has 18 distinct ATP binding cassette (ABC) transporters with predicted specificities for amino acids, polyamines, carbohydrates, cations and thiamine. Genes encoding all the enzymes of the glycolytic pathway are present, but genes encoding components of the tricarboxylic acid cycle and oxidative phosphorylation are absent. *T. pallidum* subsp. *pallidum* lacks a respiratory electron transport chain, so that ATP production must be accomplished by substrate level phosphorylation. Membrane potential is putatively established by reverse reaction of the V_1V_0 type ATP synthase. Consistent with its microaerophilic nature, *T. pallidum* subsp. *pallidum* lacks genes that encode superoxide dismutase, catalase and peroxidase. As has been proposed by Stamm *et al.* (1991), this spirochete lacks σ_{32} for transcription of heat-shock genes, which probably accounts for its thermal sensitivity. Motility and chemotaxis genes are highly conserved, which indicates the functional importance of their products. *T. pallidum* subsp. *pallidum* lacks a pathway for LPS biosynthesis but contains a complete pathway for peptidoglycan synthesis and complete machinery for protein export and acylation of its 22 lipoproteins. A minimal set of regulatory genes that encode two response-regulator two-component systems, several putative transcriptional repressors, and a potential phosphorylation-based regulatory system are present. Although *T. pallidum* subsp. *pallidum* lacks orthologs for highly conserved bacterial outer-membrane proteins such as porins, it does contain orthologs of *Yersinia pestis* OmpH, *Neisseria gonorrhoeae* Omp85, and *E. coli* OmpA. Of considerable interest is a single multi-gene family that encodes 12 paralogous *T. pallidum* repeat (Tpr) proteins with homology to the major sheath protein (Msp) of *T. denticola* (Stamm *et al.*, 1998; Centurion-Lara *et al.*, 1999). Although the functions and locations of the Tpr proteins have not been definitely determined, some of these proteins may be adhesins or porins.

Pathogenesis

Virulence Attributes

T. pallidum subsp. *pallidum* lacks the 'classical' virulence factors that are produced by many bacterial pathogens, such as a capsule, LPS, potent exotoxins, iron-acquisition mechanisms, etc. (Fraser *et al.*, 1998). Although sequence analysis indicates that the treponemal genome encodes five proteins with similarity to haemolysins/cytotoxins, the significance of these proteins for syphilis pathogenesis is unclear. Putative virulence attributes of *T. pallidum* subsp. *pallidum* include attachment, motility, invasion, and immunoevasion. Attachment of treponemes to a variety of cell types (Fitzgerald *et al.*, 1977; Hayes *et al.*, 1977), by interaction with fibronectin or other host-cell receptors, is thought to be important for the initial establishment of infection (Fitzgerald *et al.*, 1984; Baughn, 1987). Motility facilitates dissemination of the spirochetes throughout the host tissues and is essential for invasion of tight junctions between endothelial cells (Thomas *et al.*, 1988). As previously noted, the paucity of outer-membrane proteins (TROMPs) promotes treponemal immuno-evasion by presenting

fewer targets to the host immune system, thus retarding processes such as antibody binding, complement activation and phagocytosis (Alder *et al.*, 1990; Blanco *et al.*, 1990). Phenotypic (phase and/or antigenic) variation of the TROMPs may account for treponemal persistence in the host and the recrudescent nature of syphilis (Lukehart *et al.*, 1992).

Clinical Manifestations of Untreated Adult Syphilis

Syphilis is a chronic, systemic infection characterised by periods of active clinical disease that are interrupted by periods of latency (Lukehart and Holmes, 1998; Singh and Romanowski, 1999; Stamm, 1999). Factors involved in the pathogenesis of syphilis are poorly understood. After inoculation and penetration of mucosal surfaces or abraded skin, *T. pallidum* subsp. *pallidum* attaches to host cells and multiplies. Within a few hours, significant numbers of treponemes depart from the local site of infection and are carried to the regional lymph nodes. The organisms disseminate to several organs and tissues via the circulation, exit through the tight junctions of vascular endothelial cells, and establish an extracellular residence. The natural course of untreated adult syphilis is divided into four stages: primary, secondary, latent and tertiary.

Primary Stage Syphilis

The primary stage of syphilis occurs after an incubation period of 2–6 weeks. This stage is characterised by the appearance of a lesion (chancre) that develops at the site of inoculation. The chancre begins as a painless indurated papule that necroses to yield a hard-based, well-circumscribed, ulcerated lesion that contains numerous infectious treponemes. Histopathologically, the chancre is characterised by an intense perivascular infiltration of plasma cells, lymphocytes (CD4+ and CD8+), and histiocytes with capillary endothelial proliferation, and eventual obliterative endarteritis (Lukehart *et al.*, 1980). The chancre persists for a few weeks and then heals spontaneously. Resolution and clearance of treponemes is thought to be due to cell-mediated immune mechanisms that involve phagocytosis by macrophages activated by lymphokines released from antigen-specific sensitised T cells. Van Voorhis *et al.* (1996) demonstrated that lesions of primary and secondary syphilis contain mRNA for IL-2, IFNγ, IL-12p40 and IL-10. These findings are consistent with a Th1-predominant local cellular response that leads to macrophage activation resulting in treponemal clearance. In addition, treponemal lipoproteins have been shown to activate monocytes–macrophages, B cells, and endothelial cells *in vitro*, which suggests that these molecules contribute to the immunopathogenesis of syphilis by mediating local inflammatory changes (Riley *et al.*, 1992; Radolf *et al.*, 1995a; Norgard *et al.*, 1996). Clearance of treponemes is also enhanced by low levels of opsonic antibodies that promote both ingestion and killing of organisms by macrophages (Alder *et al.*, 1990; Baker-Zander and Lukehart, 1992). Although the targets of these antibodies have not been identified, they are pathogen-specific treponemal antigens, presumably TROMPs (Shaffer *et al.*, 1993).

Despite destruction of the majority of treponemes by the host immune response, some organisms survive to cause chronic infection. Proposed mechanisms for treponemal persistence include: (1) an antigenically inert treponemal cell surface resulting from a coat of host serum proteins or a paucity of outer-membrane proteins; (2) intracellular localisation or residence within an immunoprotective niche; (3) a sub-population of treponemes resistant to phagocytosis; and (4) premature down-regulation of the local host immune response (Lukehart, 1992). The mechanism with most data to support it is that of the unique treponemal cell surface which presents few targets for the host immune response because of the paucity of the TROMPs (Radolf *et al.*, 1989b; Walker *et al.*, 1989; Blanco *et al.*, 1990; Radolf *et al.*, 1994).

Secondary Stage Syphilis

Secondary syphilis results from the multiplication and dissemination of treponemes throughout the body in spite of the presence of high levels of antitreponemal antibodies (Hanff *et al.*, 1982; Baker-Zander *et al.*, 1985; Stamm and Bassford, 1985). This stage occurs simultaneously with, or up to 6 months after, the healing of the primary lesion. Although clinical manifestations can be subtle in some individuals, the secondary stage is characterised by malaise, low-grade fever, headache, generalised lymphadenopathy, a localised or generalised infectious rash with lesions on the palms and soles, mucous patches in the oral cavity or genital tract, condylomata lata (wart-like lesions), and alopecia. Deposition of immune complexes in the skin and kidneys accounts for the dermal neutrophilic vascular reaction and the glomerulonephritis observed during this stage (Baughn *et al.*, 1986; Jorizzo *et al.*, 1986). Secondary syphilis lasts for several weeks or months, with relapses occurring in approximately a quarter of untreated patients.

Latent Stage Syphilis

Secondary stage lesions usually subside within a few weeks, presumably as the result of macrophage-mediated bacterial clearance and the treponemicidal activity of complement-dependent antibody. The latent stage of syphilis is the period from disappearance of secondary stage manifestations until the occurrence of tertiary stage manifestations. Latent syphilis is arbitrarily divided into early and late stages. Early latent syphilis is defined as occurring within one year of infection, whereas late latent syphilis is defined as occurring after one year of infection. The latter is associated with relative immunity to relapse and increasing resistance to re-infection with the homologous treponemal strain. Although clinical symptoms are not apparent during latency, serological tests for the detection of antibodies to *T. pallidum* subsp. *pallidum* remain positive, indicating that organisms are present, particularly in the spleen and lymph nodes. Absence of exposed mucosal or genital lesions usually precludes venereal transmission. *T. pallidum* subsp. *pallidum* may, however, intermittently seed the bloodstream because syphilis has been transmitted by blood transfusion from patients with latent syphilis of many years duration. Approximately two-thirds of untreated syphilis patients remain in the latent stage for the remainder of their lifetime.

Tertiary Stage Syphilis

The tertiary (late) stage of syphilis occurs in approximately one-third of untreated patients, presumably because of the waning of the host immune response. This stage, which can affect almost any tissue, usually presents within several years to a few decades after the initiation of latency. Treponemes invade the central nervous system, cardiovascular system, eyes, skin, and other internal organs, to produce damage by virtue of their invasive properties, inflammation-provoking cellular components (lipoproteins) and by evoking the delayed-type hypersensitivity response. The manifestations of tertiary syphilis include cardiovascular syphilis, neurosyphilis, and gummatous syphilis. Cardiovascular problems are attributed to local inflammation induced by the multiplication of treponemes within the wall of the aorta. Subsequent aortitis results in complications such as aneurysms and coronary artery stenosis. Neurosyphilis may be symptomatic or asymptomatic. Major clinical categories of symptomatic neurosyphilis are meningeal, meningovascular, and parenchymatous. The last category includes general paresis (destruction of brain parenchyma), tabes dorsalis (destruction of the dorsal roots of the spinal column) or both (taboparesis). Gummas are granulomatous lesions that usually occur in skin, bones or viscera and result from the delayed-type hypersensitivity response to treponemal antigens. Gummas contain few treponemes and may occur singly or multiply, varying in size from microscopic to large tumour-like masses. Although these lesions rarely cause physical incapacity or death, serious complications can occur if the gummas are present in the brain or heart. During the tertiary stage, transmission of syphilis by sexual contact does not occur.

Congenital Syphilis

Congenital syphilis occurs when *T. pallidum* subsp. *pallidum* is transmitted from an infected female to her fetus (Lukehart and Holmes, 1998; Singh and Romanowski, 1999). The rate of transmission, estimated to be 75–95% during untreated primary maternal syphilis, decreases to approximately 40% and 10% during early and late latent syphilis, respectively. Transmission to the fetus rarely occurs during tertiary maternal syphilis. Congenital syphilis can result in spontaneous abortion, stillbirth, premature delivery, perinatal death or non-fatal infection. The infected neonate may be asymptomatic at birth or may have subtle findings or multiple-organ system involvement. Postnatal clinical manifestations are divided into early and late. Early manifestations, which occur at birth or up to two years of age, may include rhinitis (snuffles), mucocutaneous lesions, osteochondritis of the long bones, hepatosplenomegaly, lymphadenopathy, anaemia, jaundice and neurosyphilis. Late manifestations, which occur after two years of age, may include interstitial keratitis, eighth-nerve deafness, interference with secondary tooth development, neurosyphilis, perioral fissures (rhagades), damage to long bones (sabre shins), and perforation of the nasal septum (saddle nose). Congenital syphilis is almost entirely preventable with early prenatal screening and penicillin treatment of infected pregnant females. Infants born to women with untreated or inadequately treated syphilis or infants with physical or laboratory findings consistent with the diagnosis of congenital syphilis require treatment.

Syphilis and the Human Immunodeficiency Virus

There is strong epidemiological synergy between syphilis and HIV infection. Syphilis and other sexually transmitted diseases that produce genital ulcer disease are important risk factors for the acquisition and transmission of HIV (Singh and Romanowski, 1999). Features of the syphilitic chancre that contribute to these processes are: (1) the breach of the epithelial barrier which creates a portal of entry or exit for HIV, (2) the influx of large numbers of macrophages and T cells which provides an environment enriched with receptors for HIV, and (3) the production of

cytokines by macrophages stimulated by treponemal lipoproteins which enhances HIV replication (Norgard *et al.*, 1996).

Several studies have shown that the clinical manifestations of syphilis may be altered in patients with concurrent HIV infection (Schöfer *et al.*, 1996; Singh and Romanowski, 1999). Such patients more often develop syphilitic meningitis, meningovasculitis and ocular syphilis, even after therapy with intramuscular benzathine penicillin for early syphilis. There is also a higher frequency of ulcerating secondary syphilis (malignant syphilis), which is characterised by ulcerating skin lesions and general symptoms such as severe fever and weakness (Schöfer *et al.*, 1996). The higher frequency of ulcerating secondary syphilis is thought to be the consequence of advanced immunodeficiency, which precludes the ability of the host to clear treponemes from skin lesions or the central nervous system. Although atypical serological reactions have been observed in HIV-infected patients with concurrent syphilis, the results of serological tests for syphilis are usually accurate for most patients.

Immunity and Vaccination

Immunity to Re-infection

A degree of immunity to exogenous re-infection eventually develops in humans with syphilis. This so-called 'chancre immunity' is incomplete, since it fails to eradicate the ongoing infection. Magnuson *et al.* (1956) studied male volunteers inoculated intradermally with 100 000 *T. pallidum* subsp. *pallidum* at a single site on the forearm. Control subjects without a previous history of syphilis developed dark-field positive skin lesions at the inoculation site. Five subjects with previously untreated late latent syphilis did not develop lesions or show changes in serological tests for syphilis. All 11 subjects with a previous history of early (primary or secondary) treated syphilis developed lesions and showed increased reactivity in serological tests for syphilis. Among 26 subjects with previously treated latent syphilis, 10 developed lesions (one dark-field positive; nine dark-field negative) and showed increased reactivity in serological tests for syphilis; three developed dark-field negative lesions not associated with changes in serological tests for syphilis, and 13 did not develop any lesions or changes in serological tests for syphilis. These results indicated that a long-term syphilitic infection was necessary to produce immunity to intradermal challenge infection (chancre immunity) and that such immunity waned after effective antibiotic therapy.

Studies conducted with the rabbit model of syphilis confirmed the results observed with human subjects. Turner and Hollander (1957) showed that immunity to re-infection with the homologous *T. pallidum* subsp. *pallidum* strain developed in rabbits 3–6 months after previous treponemal infection. In animals that had been infected with *T. pallidum* subsp. *pallidum* for various time periods and then cured with penicillin, Lewinski *et al.* (1999) showed a positive correlation between the degree of immunity to re-infection and the titre of serum antibody that was treponemicidal and aggregated TROMPs. They concluded that treponemicidal antibody is directed against TROMPs and that the rarity of these proteins accounts for the slow mobilisation of functional immune responses.

Humoral and Cellular Immune Mechanisms

Humoral and cell-mediated immune mechanisms both appear to play a role in immunity to exogenous re-infection by *T. pallidum* subsp. *pallidum*. Passive transfer of serum or IgG antibodies from immune rabbits to naive rabbits partially protected the latter when they were inoculated intradermally with *T. pallidum* subsp. *pallidum* (Turner *et al.*, 1973; Bishop and Miller, 1976; Titus and Weiser, 1979; Blanco *et al.*, 1984). Furthermore, serum or IgG antibodies from immune rabbits blocked attachment to, and invasion of, tissue culture monolayer cells by treponemes (Fitzgerald *et al.*, 1977; Hayes *et al.*, 1977; Thomas *et al.*, 1988), promoted phagocytosis and killing of treponemes by macrophages (Alder *et al.*, 1990; Baker-Zander and Lukehart, 1992; Shaffer *et al.*, 1993), and mediated complement-dependent treponemicidal activity (Blanco *et al.*, 1990; Lewinski *et al.*, 1999). Cellular immunity is critical for the healing of early lesions and the control of infection. Specifically sensitised T cells develop early in infection (Lukehart, 1992; Arroll *et al.*, 1999). Clearance of treponemes is mediated by activated macrophages, presumably in conjunction with opsonic antibody. Partial protection of rabbits from intradermal infection with *T. pallidum* subsp. *pallidum* has been achieved by transfer of lymphocytes from immune rabbits (Schell *et al.*, 1983).

Vaccine Studies

The re-emergence of syphilis in the USA during the late 1980s and early 1990s and the recognition that syphilis is a risk factor for HIV infection has prompted renewed interest in a syphilis vaccine (St Louis and Wasserheit, 1998). Miller (1973) successfully immunised rabbits with *T. pallidum* subsp. *pallidum*

(Nichols strain). The regimen required multiple intravenous injections over a course of 37 weeks with a total inoculum of 3.7×10^9 unwashed γ-irradiated treponemes. Complete immunity to intradermal challenge with the homologous strain persisted for one year. Since this approach is impractical for human immunisation, recent efforts to develop a syphilis vaccine have focused on the use of recombinant-produced treponemal antigens (Norris *et al.*, 1993). Most of these antigens were thought to be cell-surface proteins, but subsequent studies showed that they are subsurface and are unlikely to be primary targets of a treponemicidal response (Radolf, 1994). Although immunisation with some of these proteins did alter the course of syphilis in the rabbit model, none elicited solid immunity to infection.

Several investigators have suggested that an immune response to the TROMPs would prevent infection by *T. pallidum* subsp. *pallidum*. In support of this, Blanco *et al.* (1999a) found that immunisation of a mouse with purified *T. pallidum* subsp. *pallidum* outer-membrane vesicles evoked a significantly greater titre of treponemicidal and TROMP-aggregating activity than that found in immune rabbit serum. Attempts definitively to identify the TROMPS by conventional methods have produced conflicting results (Radolf, 1994). Blanco *et al.* (1995) identified putative cell-surface proteins, Tromp1 and Tromp2 (Champion *et al.*, 1997). Tromp1, subsequently designated TroA (Hardham *et al.*, 1997), contains a cleavable N-terminal signal peptide (Blanco *et al.*, 1999b) and was reported to have porin-like activity (Blanco *et al.*, 1996). However, TroA, which is encoded by a gene that is a component of an ABC transport operon, has extensive sequence homology with metal-ion binding periplasmic transport proteins (Hardham *et al.*, 1997), and lacks structural features consistent with those of bacterial porins (Deka *et al.*, 1999). Immunisation of rabbits with a recombinant expressed TroA fusion protein did not elicit opsonic antibodies (Akins *et al.*, 1997). Akins *et al.* (1997) demonstrated the sub-surface location of native TroA by immunofluorescence analysis of *T. pallidum* subsp. *pallidum* encapsulated in gel micro-droplets.

Genome analysis of *T. pallidum* subsp. *pallidum* has not provided conclusive information about the identities of the TROMPs. A subset of the 12 Tpr proteins encoded by the multicopy polymorphic *tpr* gene family may represent the best TROMP candidates (Fraser *et al.*, 1998). Centurion-Lara *et al.* (1999) showed that rabbits immunised with the recombinant-expressed variable domain of TprK, which is proposed is cell-surface exposed, developed opsonic antibody and showed significant protection against intradermal infection with the homologous *T. pallidum* subsp. *pallidum* strain. The lesions that appeared in the immunised animals were atypical (flat, non-ulcerative, and devoid of treponemes) and healed rapidly as compared with those of the unimmunised control animals. The TprK-immunised animals, however, were not completely protected against infection because they sero-converted after challenge and harboured infectious treponemes in their lymph nodes and testes. Phenotypic variation of TprK may account for the inability of the TprK variable domain to elicit complete protection (Centurion-Lava *et al.*, 2000; Stamm and Bergen, 2000a). The role of TprK and other Tpr proteins as vaccine candidates requires further study.

Diagnosis

The laboratory diagnosis of syphilis usually depends on direct demonstration of *T. pallidum* subsp. *pallidum* in clinical specimens and/or on serology (Larsen *et al.*, 1995; Young, 1998). Serological tests are divided into two categories: non-treponemal and treponemal. It is important to note that infections with other *T. pallidum* subspecies (*T. pallidum* subsp. *endemicum* and *T. pallidum* subsp. *pertenue*) and *T. carateum* also produce a positive reaction in the currently available non-treponemal and treponemal tests.

Non-treponemal Tests

Non-treponemal tests are used as qualitative screening tests and as quantitative tests to monitor treatment efficacy (Larsen *et al.*, 1995). Standard non-treponemal tests include the Venereal Disease Research Laboratory (VDRL) and the rapid plasma reagin (RPR) tests. Serum IgG or IgM antibodies to lipoidal material released from damaged host cells, or potentially associated with treponemes, are detected by microscopic (VDRL) or macroscopic (RPR) flocculation of an antigen suspension that consists of lecithin, cholesterol and purified cardiolipin. The VDRL test can also be used to detect the presence of antibodies in cerebrospinal fluid. Although non-treponemal screening tests are widely available, rapid and relatively inexpensive, they are limited by a lack of sensitivity in primary and tertiary syphilis, in which some 30–40% of sera are non-reactive, and by false-positive reactions. The false-positive reactions are associated with autoimmune or connective tissue diseases, pregnancy, ageing, strong immunological stimulation as in certain bacterial, viral or parasitic infections or immunisations, drug addiction, and chronic liver disease, amongst others.

Treponemal Tests

Treponemal tests are used to confirm reactive non-treponemal tests (Larsen *et al.*, 1995). These tests use *T. pallidum* subsp. *pallidum* as the antigen and are based on the detection of serum IgM or IgG antibodies to treponemal components. Anti-treponemal IgM antibody correlates well with untreated or recently treated early infection, but it is usually absent in late-stage disease. Antitreponemal IgG antibody can remain elevated for years, even after successful treatment, making it difficult to differentiate between current and past infections or persistent infections from re-infections. Standard treponemal tests are the FTA-ABS (fluorescent treponemal antibody absorption) test and the TPPA (*Treponema pallidum* particle agglutination) test. Newer versions of treponemal tests that utilise native or recombinant treponemal antigens include immunoblots (western blots) (Marangoni *et al.*, 1999), enzyme immunoassays (EIA) (Young, 1992), and a latex agglutination test (Young *et al.*, 1998). Treponemal tests are more sensitive and specific than non-treponemal tests for all stages of syphilis. False-positive reactions are rare, but have been associated with autoimmune or connective tissue diseases, pregnancy, viral infections and Lyme disease.

Antimicrobial Therapy

In the pre-antibiotic era, a variety of therapies were used for syphilis treatment. Mercury was one of the first compounds used in 1497. Ehrlich introduced arsphenamine (Salvarsan) in 1909. In 1917, von Jauregg advocated fever (malaria) therapy for neurosyphilis. In the 1920s, bismuth was used in combination with mercurial and arsenical regimens. The major breakthrough occurred when Mahoney *et al.* (1943) used penicillin successfully to treat patients with primary syphilis. Unlike other bacterial pathogens that quickly developed resistance to penicillin, *T. pallidum* subsp. *pallidum* has not developed any documented resistance to this antibiotic. Thus, parenteral penicillin G has remained the drug of choice for treating all stages of syphilis (Centers for Disease Control and Prevention, 1998).

The Centers for Disease Control and Prevention currently recommend that penicillin-allergic pregnant patients should be desensitised and treated with penicillin. Erythromycin and tetracycline or doxycycline are alternatives for the treatment of penicillin-allergic non-pregnant patients whose compliance with therapy and follow-up can be ensured

(Centers for Disease Control and Prevention, 1998). Tetracycline is, however, contraindicated for young children, and erythromycin treatment failures have been documented in some patients with early syphilis. Stamm *et al.* (1988) demonstrated high-level *in vitro* erythromycin resistance in a clinical isolate of *T. pallidum* subsp. *pallidum* that had been obtained from a penicillin-allergic patient in whom erythromycin therapy failed. This isolate, designated Street strain 14, was also cross-resistant to azithromycin (Stamm and Parrish, 1990), a newer macrolide which has been used for syphilis treatment or prophylaxis (Hook *et al.*, 1999). Subsequent genomic analyses have shown that Street strain 14 has a point mutation (A to G transition) in both 23S rRNA genes that is cognate to *E. coli* 23S rDNA A2058 (L. Stamm and H. Bergen, 2000b). Identical point mutations in the 23S rRNA genes of other bacterial species are known to confer constitutive high-level macrolide resistance. Since erythromycin and azithromycin are not commonly used as a first-line treatment for syphilis, the extent of macrolide resistance in *T. pallidum* subsp. *pallidum* 'street strains' is currently unknown.

Epidemiology and Control

Transmission

T. pallidum subsp. *pallidum* is an obligate human parasite. Nearly all cases of syphilis are acquired by direct sexual contact with the lesions of an individual who has active disease. Transmission of syphilis is estimated to occur in approximately one-half of such contacts. Syphilis can be congenitally transmitted from an infected female by transplacental passage of treponemes to the fetus. Rarer modes of syphilis transmission include blood-borne, such as that due to transfusion or sharing of needles, non-sexual personal contact, and accidental inoculation.

Distribution

Syphilis is distributed worldwide. The World Health Organization estimated in 1995 that there were 12 million new cases (Gerbase *et al.*, 1998). Syphilis is a problem in developing countries, particularly in Africa and Asia, where it is a leading cause of genital ulcer disease, which is a risk factor for HIV infection (Agacfidan and Kohl, 1999). In many developed countries syphilis appears to be under control, but it has re-emerged in eastern Europe and in certain recently independent states of the former USSR because economic and social changes have resulted in

a breakdown of public health infrastructure and increased prostitution (Tichonova *et al.*, 1997). In the late 1980s and the early 1990s, syphilis re-emerged in the USA and became concentrated geographically in the urban and rural South and in large urban centres outside the South (Nakashima *et al.*, 1996). At this time the demographics of syphilis also shifted from a disease of Caucasian homosexual males to one affecting predominately African–American heterosexuals. The age distribution of US cases showed that syphilis is usually found in people in their early 20s to early 30s, which is in contrast to other bacterial STIs that mostly affect teenagers and young adults of less than 25 years of age. Rates of primary and secondary syphilis have been declining in the USA since 1990 and are currently the lowest ever reported.

Control

The emergence of the HIV epidemic has re-emphasised the importance of syphilis as a US public health problem. Control of syphilis is currently based on sex education with promotion of condom use, diagnostic screening for blood donors, pregnant women, military recruits, high-risk groups, etc., treatment of infected individuals, and partner notification and treatment (Singh and Romanowski, 1999). The US Public Health Service has targeted syphilis for national elimination (St Louis and Wasserheit, 1998). A major benefit of syphilis elimination would be the eradication of syphilis-related HIV infections and the accompanying treatment costs and lost productivity associated with these infections (Chesson *et al.*, 1999). The nature of the 7–10 year syphilis epidemic cycles presents a window of opportunity to accomplish this. The lack of a vaccine, however, combined with poor surveillance, the limited healthcare access of marginalised populations, and the possibility of re-introduction of syphilis, complicate control and elimination efforts (St Louis and Wasserheit, 1998).

References

Agacfidan A, Kohl P (1999) Sexually transmitted diseases (STDs) in the world. *FEMS Immunol. Med. Microbiol.* 24: 431–435.

Akins DR, Robinson E, Shevchenko D, Elkins C, Cox DL, Radolf JD (1997) Tromp1, a putative rare outer membrane protein, is anchored by an uncleaved signal sequence to the *Treponema pallidum* cytoplasmic membrane. *J. Bacteriol.* 179: 5076–5086.

Alder JD, Friess L, Tengowski M, Schell RF (1990) Phagocytosis of opsonized *Treponema pallidum* subsp.

pallidum proceeds slowly. *Infect. Immun.* 58: 1167–1173.

Arroll TW, Centurion-Lara A, Lukehart SA, Van Voorhis WC (1999) T-cell responses to *Treponema pallidum* subsp. *pallidum* antigens during the course of experimental syphilis infection. *Infect. Immun.* 67: 4757–4763.

Baker-Zander SA, Lukehart SA (1992) Macrophage-mediated killing of opsonized *Treponema pallidum*. *J. Infect Dis.* 165: 69–74.

Baker-Zander SA, Hook EW, Bonin P, Handsfield HH, Lukehart SA (1985) Antigens of *Treponema pallidum* recognised by IgG and IgM antibodies during syphilis in humans. *J. Infect Dis.* 151: 264–272.

Baughn RE (1987) Role of fibronectin in the pathogenesis of syphilis. *Rev. Infect Dis.* 9: S372–S385.

Baughn RE, McNeely MC, Jorizzo JL, Musher DM (1986) Characterization of the antigenic determinants and host components in immune complexes from patients with secondary syphilis. *J. Immunol.* 136: 1406–1414.

Bishop NH, Miller JN (1976) Humoral immunity in experimental syphilis: I. The demonstration of resistance conferred by passive immunisation. *J. Immunol.* 117: 191–196.

Blanco DR, Miller JN, Hanff PA (1984) Humoral immunity in experimental syphilis: the demonstration of IgG as a treponemicidal factor in immune rabbit serum. *J. Immunol.* 133: 2693–2697.

Blanco DR, Walker EM, Haake DA, Champion CI, Miller JN, Lovett MA (1990) Complement activation limits the rate of *in vitro* treponemicidal activity and correlates with antibody-mediated aggregation of *Treponema pallidum* rare outer membrane protein. *J. Immunol.* 144: 1914–1921.

Blanco DR, Reimann K, Skare J *et al.* (1994) Isolation of the outer membranes from *Treponema pallidum* and *Treponema vincentii*. *J. Bacteriol.* 176: 6088–6099.

Blanco DR, Champion CI, Exner MM *et al.* (1995) Porin activity and sequence analysis of a 31-kilodalton *Treponema pallidum* subsp. *pallidum* rare outer membrane protein (TROMP1) *J. Bacteriol.* 177: 3556–3562.

Blanco DR, Champion CI, Exner MM *et al.* (1996) Recombinant *Treponema pallidum* rare outer membrane protein 1 (Tromp1) expressed in *Escherichia coli* has porin activity and surface antigenic exposure. *J. Bacteriol.* 178: 6685–6692.

Blanco DR, Champion CI, Lewinski MA *et al.* (1999a) Immunisation with *Treponema pallidum* outer membrane vesicles induces high-titre complement-dependent treponemicidal activity and aggregation of *T. pallidum* rare outer membrane proteins. *J. Immunol.* 163: 2741–2746.

Blanco DR, Whitelegge JP, Miller JN, Lovett MA (1999b) Demonstration by mass spectrometry that purified native *Treponema pallidum* rare outer membrane protein 1 (Tromp1) has a cleaved signal peptide. *J. Bacteriol.* 181: 5094–5098.

Cameron CE, Castro C, Lukehart SA, Van Voorhis WC (1999) Sequence conservation of glycerophosphodiester

phospho-diesterase among *Treponema pallidum* strains. *Infect. Immun.* 67: 3168–3170.

Centers for Disease Control and Prevention (1998) Guidelines for treatment of sexually transmitted diseases. *Morbidity and Mortality Wkly Rpt* 47: 28–49.

Centurion-Lara A, Castro C, Shaffer JM, Van Voorhis WC, Marra CM, Lukehart SA (1997) Detection of *Treponema pallidum* by a sensitive reverse transcriptase PCR. *J. Clin. Microbiol.* 35: 1348–1352.

Centurion-Lara A, Castro C, Castillo R, Shaffer JM, Van Voorhis WC, Lukehart SA (1998) The flanking region sequences of the 15-kDa lipoprotein gene differentiate pathogenic treponemes. *J. Infect Dis.* 177: 1036–1040.

Centurion-Lara A, Castro C, Barrett L *et al.* (1999) *Treponema pallidum* major sheath protein homologue TprK is a target of opsonic antibody and the protective immune response. *J. Exp. Med.* 189: 647–656.

Centurion-Lara A, Gordornes C, Castro C, Van Voorhis WC, Lukehart SA (2000) The *tprK* gene is heterogeneous among *Treponema pallidum* strains and has multiple alleles. *Infect. Immun.* 68: 824–831.

Champion CI, Blanco DR, Exner MM *et al.* (1997) Sequence analysis and recombinant expression of a 28-kilodalton *Treponema pallidum* subsp. *pallidum* rare outer membrane protein (Tromp2) *J. Bacteriol.* 179: 1230–1238.

Chesson HW, Pinkerton SD, Irwin KL, Rein D, Kassler WJ (1999) New HIV cases attributable to syphilis in the USA: estimates from a simplified transmission model. *AIDS* 13: 1387–1396.

Chi B, Chauhan S, Kuramitsu H (1999) Development of a system for expressing heterologous genes in the oral spirochete *Treponema denticola* and its use in expression of the *Treponema pallidum flaA* gene. *Infect. Immun.* 67: 3653–3656.

Deka RK, Lee Y-H, Hagman KE *et al.* (1999) Physiochemical evidence that *Treponema pallidum* TroA is a zinc-containing metalloprotein that lacks porin-like structure. *J. Bacteriol.* 181: 4420–4423.

Fieldsteel AH, Cox DL, Moeckli RA (1981) Cultivation of virulent *Treponema pallidum* in tissue culture. *Infect. Immun.* 32: 908–915.

Fitzgerald TJ, Johnson RC, Miller JN, Sykes JA (1977) Characterization of the attachment of *Treponema pallidum* (Nichols strain) to cultured mammalian cells and the potential relationship of attachment to pathogenicity. *Infect. Immun.* 18: 467–478.

Fitzgerald TJ, Repesh LA, Blanco DR, Miller JN (1984) Attachment of *Treponema pallidum* to fibronectin, laminin, collagen IV, and collagen I, and blockage of attachment by immune rabbit IgG. *Br. J. Vener. Dis.* 60: 357–363.

Fraser CM, Norris SJ, Weinstock GM *et al.* (1998) Complete genome sequence of *Treponema pallidum*, the syphilis spirochete. *Science* 281: 375–388.

Gerbase AC, Rowley JT, Heymann DHL, Berkley SFB, Piot P (1998) Global prevalence and incidence estimates of selected curable STDs. *Sex. Transm. Infect.* 74 (Suppl. 1): S12–S16.

Gherardini FC, Hobbs MM, Stamm LV, Bassford PJ (1990) Complementation of an *Escherichia coli proC* mutation by a gene cloned from *Treponema pallidum*. *J. Bacteriol.* 172: 2996–3002.

Greene SR, Stamm LV (1999) Molecular characterization of a chemotaxis operon in the oral spirochete, *Treponema denticola*. *Gene* 232: 59–68.

Greene SR, Stamm LV, Hardham JM, Young NR, Frye JG (1997) Identification, sequences, and expression of *Treponema pallidum* chemotaxis genes. *DNA Seq.* 7: 267–284.

Grimprel E, Sanchez PJ, Wendel GD *et al.* (1991) Use of the polymerase chain reaction and rabbit infectivity testing to detect *Treponema pallidum* in amniotic fluid, fetal and neonatal sera, and cerebrospinal fluid. *J. Clin. Microbiol.* 29: 1711–1718.

Hanff PA, Fehniger TE, Miller JN, Lovett MA (1982) Humoral immune response in human syphilis to polypeptides of *Treponema pallidum*. *J. Immunol.* 129: 1287–1291.

Hardham JM, Frye JG, Stamm LV (1995) Identification and sequences of the *Treponema pallidum fliM'*, *fliY*, *fliP*, *fliQ*, *fliR* and *flhB'* genes. *Gene* 166: 57–64.

Hardham JM, Stamm LV, Porcella SF *et al.* (1997) Identification and transcriptional analysis of a *Treponema pallidum* operon encoding a putative ABC transport system, an iron activated repressor protein homolog, and a glycolytic pathway enzyme homolog. *Gene* 197: 47–64.

Hardy PH, Levin J (1983) Lack of endotoxin in *Borrelia hispanica* and *Treponema pallidum*. *Proc. Soc. Exp. Biol. Med.* 174: 47–52.

Hay PE, Clarke JR, Strugnell RA, Taylor-Robinson D, Goldmeier D (1990) Use of the polymerase chain reaction to detect DNA sequences specific to pathogenic treponemes in cerebrospinal fluid. *FEMS Microbiol. Lett.* 68: 233–238.

Hayes NS, Muse KE, Collier AM, Baseman JB (1977) Parasitism by virulent *Treponema pallidum* of host cell surfaces. *Infect. Immun.* 17: 174–186.

Hook EW, Stephens J, Ennis DM (1999) Azithromycin compared to penicillin G benzathine for treatment of incubating syphilis. *Ann. Intern. Med.* 131: 434–437.

Johnson RC, Ritzi DM, Livermore BP (1973) Outer envelope of virulent *Treponema pallidum*. *Infect. Immun.* 8: 291–295.

Jorizzo JL, McNeely MC, Baughn RE, Solomon AR, Cavallo T, Smith EB (1986) Role of circulating immune complexes in human secondary syphilis. *J. Infect Dis.* 153: 1014–1022.

Larsen SA, Steiner BM, Rudolph AH (1995) Laboratory diagnosis and interpretation of tests for syphilis. *Clin. Microbiol. Rev.* 8: 1–21.

Lewinski MA, Miller JN, Lovett MA, Blanco DR (1999) Correlation of immunity in experimental syphilis with serum-mediated aggregation of *Treponema pallidum* rare

outer membrane proteins. *Infect. Immun.* 67: 3631–3636.

Limberger RJ, Slivienski LL, El-Afandi MCT, Dantuono LA (1996) Organization, transcription and expression of the 5′ region of the *fla* operon of *Treponema phagedenis* and *Treponema pallidum*. *J. Bacteriol.* 178: 4628–4634.

Lukehart SA (1992) Immunology and pathogenesis of syphilis. In: Quinn TC, Gallin JI, Fauci AS (eds) *Advances in Host Defense Mechanisms: Vol 8. Sexually Transmitted Diseases*. New York: Raven Press, pp. 141–163.

Lukehart SA, Baker-Zander SA, Lloyd RMC, Sell S (1980) Characterization of lymphocyte responsiveness in early experimental syphilis: II. Nature of cellular infiltration and *Treponema pallidum* distribution in testicular lesions. *J. Immunol.* 124: 461–467.

Lukehart SA, Shaffer JM, Baker-Zander SA (1992) A subpopulation of *Treponema pallidum* is resistant to phagocytosis: possible mechanisms of persistence. *J. Infect Dis.* 166: 1449–1453.

Lukehart SA, Holmes KK (1998) Syphilis. In: Fauci AS, Braunwald E, Isselbacher KJ *et al.* (eds) *Harrison's Principles of Internal Medicine*, 14th edn. New York: McGraw-Hill, pp. 1023–1033.

Magnuson HJ, Thomas EW, Olansky S, Kaplan BI, De Mello L, Cutler JC (1956) Inoculation syphilis in human volunteers. *Medicine* 35: 33–82.

Mahoney JF, Arnold RC, Harris A (1943) Penicillin treatment of early syphilis: a preliminary report. *Vener. Dis. Inform.* 24: 355–357.

Marangoni A, Sambri V, Olmo A, D'Antuono A, Negosanti M, Cevenini R (1999) IgG western blot as a confirmatory test in early syphilis. *Zentralbl. Bakteriol.* 289: 125–133.

Miller JN (1973) Immunity in experimental syphilis: VI. Successful vaccination of rabbits with *Treponema pallidum*, Nichols strain, attenuated by γ-irradiation. *J. Immunol.* 110: 1206–1215.

Moter A, Hoenig C, Choi N-K, Riep B, Gobel UB (1998) Molecular epidemiology of oral treponemes associated with periodontal disease. *J. Clin. Microbiol.* 36: 1399–1403.

Nakashima AK, Rolfs RT, Flock ML, Kilmarx P, Greenspan JR (1996) Epidemiology of syphilis in the United States, 1941–1993. *Sex. Transm. Dis.* 23: 16–23.

Norgard MV, Arndt LL, Akins DR, Curetty LL, Harrich DA, Radolf JD (1996) Activation of human monocytic cells by *Treponema pallidum* and *Borrelia burgdorferi* lipoproteins and synthetic lipopeptides proceeds via a pathway distinct from that of lipopolysaccharide but involves the transcriptional activator NF-kB. *Infect. Immun.* 64: 3845–3852.

Norris SJ, and the *Treponema pallidum* Polypeptide Research Group (1993) Polypeptides of *Treponema pallidum*: progress toward understanding their structural, functional and immunologic roles. *Microbiol. Rev.* 57: 750–779.

Orle KA, Gates CA, Martin DH, Body BA, Weiss JB (1996) Simultaneous PCR detection of *Haemophilus ducreyi*,

Treponema pallidum, and herpes simplex virus types 1 and 2 from genital ulcers. *J. Clin. Microbiol.* 34: 49–54.

Paster BJ, Dewhirst FE, Weisburg WG *et al.* (1991) Phylogenetic analysis of spirochetes. *J. Bacteriol.* 173: 6101–6109.

Paster BJ, Dewhirst FE, Coleman BC, Lau CN, Ericson RL (1998) Phylogenetic analysis of cultivable oral treponemes from the Smibert collection. *J. Syst. Bacteriol.* 48: 713–722.

Penn CW, Cockayne A, Bailey MJ (1985) The outer membrane of *Treponema pallidum*: biological significance and biochemical properties. *J. Gen. Microbiol.* 131: 2349–2357.

Pillay A, Liu H, Chen C *et al.* (1998) Molecular subtyping of *Treponema pallidum* subspecies *pallidum*. *Sex. Transm. Dis.* 25: 408–414.

Porcella SF, Popova TG, Hagman KE, Penn CW, Radolf JD, Norgard MV (1996) A *mgl*-like operon in *Treponema pallidum*, the syphilis spirochete. *Gene* 177: 115–121.

Radolf JD (1994) Role of outer membrane architecture in immune evasion by *Treponema pallidum* and *Borrelia burgdorferi*. *Trends Microbiol.* 2: 307–311.

Radolf JD, Moomaw C, Slaughter CA, Norgard MV (1989a) Penicillin-binding proteins and peptidoglycan of *Treponema pallidum* subsp. *pallidum*. *Infect. Immun.* 57: 1248–1254.

Radolf JD, Norgard MV, Schulz WW (1989b) Outer membrane ultrastructure explains the limited antigenicity of virulent *Treponema pallidum*. *Proc. Natl Acad. Sci. USA* 86: 2051–2055.

Radolf JD, Arndt LL, Akins DR *et al.* (1995a) *Treponema pallidum* and *Borrelia burgdorferi* lipoproteins and synthetic lipopeptides activate monocytes/macrophages. *J. Immunol.* 154: 2866–2877.

Radolf JD, Robinson EJ, Bourell KW *et al.* (1995b) Characterization of outer membranes isolated from *Treponema pallidum*, the syphilis spirochete. *Infect. Immun.* 63: 4244–4252.

Riley BS, Oppenheimer-Marks N, Hansen EJ, Radolf JD, Norgard MV (1992) Virulent *Treponema pallidum* activates human vascular endothelial cells. *J. Infect Dis.* 165: 484–493.

Riviere GR, Wagoner MA, Baker-Zander SA *et al.* (1991) Identification of spirochetes related to *Treponema pallidum* in necrotizing ulcerative gingivitis and chronic periodontitis. *N. Engl. J. Med.* 325: 539–543.

Sanchez PJ, Wendel GD, Grimpel E *et al.* (1993) Evaluation of molecular methodologies and rabbit infectivity testing for the diagnosis of congenital syphilis and neonatal central nervous system invasion by *Treponema pallidum*. *J. Infect Dis.* 167: 148–157.

Schell RF, Chan JK, LeFrock JL (1983) T-cell-mediated resistance. In: Schell RF, Musher DM (eds) *Pathogenesis and Immunology of Treponemal Infection*. New York: Marcel Dekker, pp. 331–348.

Schöfer H, Imhof M, Thoma-Greber E *et al.* (1996) Active syphilis in HIV infection: a multicentre retrospective survey. *Genitourin. Med.* 72: 176–181.

Shevchenko DV, Akins DR, Robinson EJ, Li M, Shevchenko OV, Radolf JD (1997) Identification of homologues for thioredoxin, peptidyl prolyl *cis-trans* isomerase, and glycerophosphodiester phosphodiesterase in outer membrane fractions from *Treponema pallidum*, the syphilis spirochete. *Infect. Immun.* 65: 4179–4189.

Shaffer JM, Baker-Zander SA, Lukehart SA (1993) Opsonization of *Treponema pallidum* is mediated by immunogobulin G antibodies induced only by pathogenic treponemes. *Infect. Immun.* 61: 781–784.

Singh AE, Romanowski B (1999) Syphilis: review with emphasis on clinical, epidemiological, and some biologic features. *Clin. Microbiol. Rev.* 12: 187–209.

Smibert RM (1984) Genus III: *Treponema* Schaudinn 1905, 1728[AL]. In: Kreig NR, Holt JG (eds) *Bergey's Manual of Systematic Bacteriology*, Vol 1. Baltimore: Williams & Williams Co., pp. 49–57.

Stamm LV (1999) Biology of *Treponema pallidum*. In: Holmes KK, Sparling PF, Mardh P-A *et al.* (eds) *Sexually Transmitted Diseases*, 3rd edn. New York: McGraw-Hill, pp. 467–472.

Stamm LV, Barnes NY (1997) Nucleotide sequences of the *proA* and *proB* genes of *Treponema pallidum*, the syphilis agent. *DNA Seq.* 8: 63–70.

Stamm LV, Bassford PJ (1985) Cellular and extracellular protein antigens of *Treponema pallidum* synthesized during in vitro incubation of freshly extracted organisms. *Infect. Immun.* 47: 799–807.

Stamm LV, Bergen HL (1999) Molecular characterization of a flagellar (*fla*) operon in the oral spirochete *Treponema denticola* ATCC 35405. *FEMS Microbiol. Lett.* 179: 31–36.

Stamm LV, Bergen HL (2000a) The sequence-variable, single-copy tprK gene of *Treponema pallidum* Nichols strain UNC and Street Strain 14 encodes heterogeneous TprK proteins. *Infect. Immun.* 68: 6482–6486.

Stamm LV, Bergen HL (2000b) A point mutation associated with bacterial macrolide resistance is present in both 23S rRNA genes of an erythromycin resistant *Treponema pallidum* clinical isolate. *Antimicrob. Agents Chemother.* 44: 806–807.

Stamm LV, Stapleton JT, Bassford PJ (1988) *In vitro* assay to demonstrate high-level erythromycin resistance of a clinical isolate of *Treponema pallidum*. *Antimicrob. Agents Chemother.* 32: 164–169.

Stamm LV, Parrish EA (1990) *In-vitro* activity of azithromycin and CP-63,956 against *Treponema pallidum*. *J. Antimicrob. Chemother.* 25 (Suppl. A): 11–14.

Stamm LV, Hodinka RL, Wyrick PB, Bassford PJ (1987) Changes in cell surface properties of *Treponema pallidum* that occur during in vitro incubation of freshly extracted organisms. *Infect. Immun.* 55: 2255–2261.

Stamm LV, Gherardini FC, Parrish EA, Moomaw CR (1991) Heat shock response of spirochetes. *Infect. Immun.* 59: 1572–1575.

Stamm LV, Young NR, Frye JG, Hardham JM (1996) Identification and sequences of the *Treponema pallidum* mglA and mglC genes. *DNA Seq.* 6: 293–298.

Stamm LV, Greene SR, Bergen HL, Hardham JM, Barnes NY (1998) Identification and sequence analysis of *Treponema pallidum tprJ*, a member of a polymorphic multigene family. *FEMS Microbiol. Lett.* 169: 155–163.

St Louis ME, Wasserheit JN (1998) Elimination of syphilis in the United States. *Science* 281: 353–354.

Thomas DD, Navab M, Haake DA, Fogelman AM, Miller JN, Lovett MA (1988) *Treponema pallidum* invades intercellular junctions of endothelial cell monolayers. *Proc. Natl Acad. Sci. USA* 85: 3608–3612.

Tichonova L, Borisenko K, Ward H, Meheus A, Gromyko A, Retton A (1997) Epidemics of syphilis in the Russian Federation: trends, origins, and priorities for control. *Lancet* 350: 210–213.

Titus RG, Weiser RS (1979) Experimental syphilis in the rabbit: passive transfer of immunity with immunoglobulin G from immune serum. *J. Infect Dis.* 140: 904–913.

Turner TB, Hollander DH (1957) *Biology of the Treponematoses.* Geneva: World Health Organization.

Turner TB, Hardy PH, Newman B (1969) Infectivity tests in syphilis. *Br. J. Vener. Dis.* 45: 183–195.

Turner TB, Hardy PH, Newman B, Nell EE (1973) Effects of passive immunisation on experimental syphilis in the rabbit. *Johns Hopkins Med. J.* 133: 241–251.

Van Voorhis WC, Barrett LK, Koelle DM, Nasio JM, Plummer FA, Lukehart SA (1996) Primary and secondary syphilis lesions contain mRNA for Th1 cytokines. *J. Infect Dis.* 173: 491–495.

Walker EM, Zampighi GA, Blanco DR, Miller JN, Lovett MA (1989) Demonstration of rare protein in the outer membrane of *Treponema pallidum* subsp. *pallidum* by freeze-fracture analysis. *J. Bacteriol.* 171: 5005–5011.

Young H (1992) Syphilis: new diagnostic directions. *Int. J. STD AIDS* 3: 391–413.

Young H (1998) Syphilis serology. *Dermatol. Clin.* 16: 691–698.

Young H, Moyes A, de Ste Croix I, McMillan A (1998) A new recombinant antigen latex agglutination test (Syphilis fast) for the rapid serological diagnosis of syphilis. *Int. J. STD AIDS* 9: 196–200.

86
Haemophilus ducreyi
Teresa Lagergård

University of Göteborg, Göteborg, Sweden

Haemophilus ducreyi is a fastidious, Gram-negative, facultatively anaerobic coccobacillus that causes chancroid, which is also known as soft chancre or *ulcus molle*. This is a sexually transmitted disease (STD) with characteristic mucocutaneous ulcers on the external genitals and often with a unilateral inguinal bubo. The disease is endemic in developing countries, particularly Africa, where an association between genital ulcers and transmission of human immunodeficiency virus (HIV) has been established.

History

The organism was first described by Auguste Ducreyi in 1889 (Ducreyi, 1889). He established the specificity of the infectious agent by serial cutaneous inoculation of the forearm skin of patients with the purulent material from their own genital ulcers. Ducreyi consistently found the same organism in each case and suggested that it was responsible for development of ulcers. He described the bacterium as a short compact streptobacillus, with rounded edges, 1.48 µm in length and 0.5 µm in width. Ducreyi's work was later confirmed by Krefting and Unna in 1892. Lengert in 1898 and Benzacon, Griffon and Le Sourd in 1900 are given credit for the first successful isolation of the organism on a solid culture medium (Sullivan, 1940). Tengue

and Deibert found that, in order to obtain a good growth of *H. ducreyi* serum, red blood cells and peptone should be included in agar medium. Benzancon confirmed that *H. ducreyi* produced ulcers in re-infected humans.

The reactions after intradermal injection of *H. ducreyi* cultured *in vitro* or of sterilised material from chancroid lesions were studied for diagnostic purposes by Ito in 1913 and Reensterna in 1923. The papules that developed between the third and seventh day were interpreted as a positive test. (Kampmeier, 1982; Morse, 1989).

The genus *Haemophilus* was established in 1920, and Ducreyi's bacillus was included in the genus. The designation *Haemophilus ducreyi* was used in the 1923 edition of *Bergey's Manual of Determinative Bacteriology*, but the taxonomic position of *H. ducreyi* has been questioned for a number of years. Its inclusion in the genus *Haemophilus* was based on its requirement for haemin (X factor) and the G + C content of its DNA, which is in the accepted range for the genus. More recent studies, by genetic transformation, DNA hybridisation and S1 nuclease treatment have, however, questioned its position in the genus *Haemophilus*. The results of DNA–RNA hybridisation and 16S rRNA sequencing of three strains, including the reference strain CIP 542, have confirmed that *H. ducreyi* is a member of *Pasteurellaceae* (Trees and Morse, 1995).

Morphology and Growth Conditions

H. ducreyi is a small non-motile, non-spore-forming Gram-negative pleomorphic coccobacillus. It is a fastidious, slow-growing, microaerophilic organism that is difficult to isolate from clinical specimens, and it requires selective, enriched media for growth. A medium consisting of either Muller–Hinton, gonococcal or brain–heart infusion agar, with heated blood, bovine haemoglobin or haemin, serum, vitamins and vancomycin is required for the successful isolation of *H. ducreyi* (Nsanze *et al.*, 1984). Optimal growth is achieved in an humidified and carbon dioxide enriched atmosphere at 30–33°C. After 24–48 hours growth on solid medium, small, non-mucoid, yellow-grey, semi-opaque colonies appear, and a polymorphic colonial morphology may be observed. Colonies can be pushed across the agar surface intact, indicating adherence of cells to each other within colonies.

In liquid medium, a streptobacillary form and the characteristic parallel chains described as 'schools of fish' are commonly observed (**Fig. 1a**); on solid media, more complicated 'fingerprints' are seen (Heyman *et al.*, 1945; Morse, 1989) (**Fig. 1b**). The organisms tend to auto-aggregate and are difficult to disperse.

The electron-microscopic appearance of *H. ducreyi* is typical of Gram-negative bacteria. Close cellular contact between bacteria and formation of blebs are also observed (Morse, 1989; Frisk *et al.*, 1995).

Biochemical Characteristics

H. ducreyi has very limited biochemical activity (Ronald and Albritton, 1990). It is catalase negative and cytochrome-oxidase positive, and it reduces nitrate and produces alkaline phosphatase. *H. ducreyi*

is essentially asaccharolytic, but weekly positive in fermentation tests reactions for glucose, mannose and fructose have been observed. The organisms lack the ability to synthesise porphyrins or porphobilinogen from δ-aminolevulinic acid and they possess a wide range of aminopeptidase activity, some of which shows variability within strains (Morse, 1989).

Clinical Features and Pathology

The stages of ulcer development in chancroid were well described by Sullivan (1940) and Sheldon and Heyman (1946). The portal entry for *H. ducreyi* is thought to be an epithelial break, and the incubation period of chancroid is usually 4–7 days. The infection begins as tender papules surrounded by erythema, and after 24–48 hours the lesions become pustular, and extension of the lesions and rupture of the pustule results in ulceration.

In men, the ulcers of chancroid are often localised on the frenulum, the sulcus, the external or internal surface of the prepuce and the preputial margin. They rarely involve the glans or shaft of the penis or extragenital skin. In women, the majority of ulcers occur on the fourchette, the labia minora and majora, the clitoris, the entrance to the vagina, the vaginal wall or the cervix. Spread by auto-inoculation to other sites, including scrotum, thighs, abdomen and fingers has been reported (Rauschkolb, 1939; Sullivan, 1940). The disease does not, however, spread systemically, even in people with AIDS.

The chancroid ulcers are sharply circumscribed, soft, painful, purulent and have a distinctively foul odour. The edges are ragged and undermined, but there is little inflammation of the surrounding tissue. Genital lesions due to other ulcer-forming pathogens,

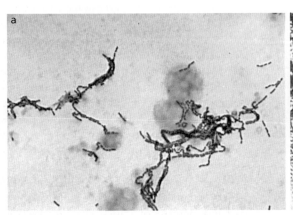 

Fig. 1 Gram stain of *Haemophilus ducreyi* after 48 hours growth: (a) in liquid medium and (b) on solid medium.

such as *Treponema pallidum, Calamytobacterium granulomatis* or herpes simplex virus, and co-infection may mimic chancroid (Sullivan, 1940; Abeck *et al.*, 1992a). Painful inguinal lymphadenopathy (chancroid buboes) are present in 50–87% of cases and may progress and fluctuate until spontaneous rupture occurs. Viable bacteria are seldom recovered from such buboes (Morse, 1989). Ulcer healing is slow and, before the era of antimicrobial therapy, the mean duration of ulcers was 34 days (Rauschkolb, 1939).

The histological picture of chancroid is characteristic and was well described by Sheldon and Heyman (1946). A typical histological picture shows three discrete zones. The narrow, superficial zone consists of necrotic tissue with red cells, fibrin and polymorphonuclear leucocytes, with bacteria present between the cells and within granulocytes. Below this area there is a broader zone of oedematous and inflamed tissue, with numerous small dilated vessels arranged vertically in relation to the surface of the ulcer, and the interstitial connective tissue is oedematous. The lack of significant proliferation of fibroblasts in the presence of the marked endothelial growth is striking, but this zone does not contain micro-organisms. A deep third zone has a dense infiltrate of the plasma cells and fewer lymphocytes. The epidermis at the edge of the ulcer is necrotic, and more peripherally there are indications of oedema and acanthosis with some polymorphonuclear leucocytes.

Recent studies have confirmed the histological observations in patients with chancroid, and have emphasised the epithelial necrosis, intense oedema, acute and chronic inflammation with plasma cells, B lymphocytes, and CD4 and CD8 lymphocytes, macrophages, eosinophils and Russel bodies (Freinkel, 1987; Ortiz-Zepeda *et al.*, 1994; Abeck *et al.*, 1997; King *et al.*, 1998).

Diagnosis

The accurate diagnosis of the aetiological agent responsible for genital ulcers is important because of the need for the specific antimicrobial treatment of different genitourinary diseases.

The diagnosis of chancroid is primarily based on clinical features and character of the lesions. The clinical diagnosis of chancroid and other genital ulcerations can be difficult because of variations in clinical picture, the mimicry of other conditions and co-infection. The positive predictive value of chancroid in men with genital ulcers in endemic areas, based on the clinical picture, can be up to about 84% (Deacon *et al.*, 1956),

but in non-endemic areas the predictive value is considerably lower (Chapel *et al.*, 1977).

Culture Methods

The laboratory confirmation of chancroid is important, but it is difficult in that special media, conditions and expertise are required reliably to culture *H. ducreyi* (Hammond *et al.*, 1978b). The isolation of bacteria from the ulcers or buboes is still regarded as the 'gold standard' for the confirmation of chancroid, although difficulty has been reported in the primary isolation of bacteria from lesions. The isolation of *H. ducreyi* from patients with a clinical diagnosis of chancroid has been evaluated with different selective media with various medium formulations (Morse, 1989), and the sensitivity of culture varies from 56% to 85% in different reports. Many field studies have demonstrated a need to use at least two media to maintain a culture sensitivity of 80–90% (Nsanze *et al.*, 1984; Dangor *et al.*, 1992; Pillay *et al.*, 1998b).

The commonly used medium is Muller–Hinton agar base containing 5% heated horse blood, 5% fetal calf serum, 1% IsoVitalX and 3 mg/L of vancomycin. An alternative culture medium is gonococcal agar base, supplemented with 15% bovine haemoglobin, 5% serum, vitamins and 3 mg/L of vancomycin. The use of a transport medium is also important to maintain the viability of organisms until inoculation into selective media (Dangor *et al.*, 1993).

Direct Examination

Direct examination of the clinical material in Gram-stained smears can be used in some cases as presumptive evidence of *H. ducreyi* infection, but it is not reliable (Sehgal and Shram-Prasad, 1985; Sturm *et al.*, 1987). Such identification is based on characteristics of cell morphology, staining and the arrangement of the cells, but direct microscopy can be misleading because of the presence in ulcers of other bacteria with a morphology similar to that of *H. ducreyi*. For primary identification of the organism, immunostaining with specific monoclonal antibodies, followed by visualisation of binding, for example with fluorescent antibody, has been reported (Karim *et al.*, 1989; Ahmed *et al.*, 1995).

Non-cultural Diagnostic Tests

The introduction of DNA diagnostic methods and the ability to amplify the signal by means of the polymerase chain reaction (PCR) has provided a new possibility to improve and facilitate the diagnosis of chancroid.

DNA probes for the identification of *H. ducreyi* with a specificity of 100% and capacity to detect about 10^4 organisms have been reported. DNA probes do not appear to have the sensitivity necessary for the detection of *H. ducreyi* in clinical specimens, although they have been successfully used to confirm the identification of isolates (Parsons *et al.*, 1989).

Many variant PCR techniques have been developed for diagnosis of chancroid (Chui *et al.*, 1993; Orle *et al.*, 1994; Johnson *et al.*, 1995; West *et al.*, 1995; Gu *et al.*, 1998). Chui *et al.* (1993) used a set of broad-specificity primers based on 16S rRNA genes to amplify a 303-bp sequence from organisms of the Pasteurellaceae and Enterobacteriaceae. A sensitivity of 100% was obtained with two 16-base *H. ducreyi* specific probes, internal for this sequence. The direct detection of bacteria from 100 clinical isolates showed a sensitivity of 83–93 % and a specificity of 51–67%, depending on the number of amplification cycles. Another variant of the PCR assay was developed with a pair of primers selected from sequences of anonymous fragments of *H. ducreyi* DNA. The 1100-bp product was detected by Southern transfer and a ^{32}P-labelled probe consisting of the entire cloned sequence. The assay had a sensitivity of 100% and a specificity of 84% compared with culture (Johnson *et al.*, 1995).

A multiplex PCR assay (M-PCR) with colorimetric detection, which permits simultaneous amplification of DNA from *H. ducreyi*, *Treponema pallidum* and herpes simplex viruses types 1 and 2, was been reported by Orle *et al.* (1994) and evaluated in several studies (Morse *et al.*, 1997; Beyrer *et al.*, 1998; Gu *et al.*, 1998). The results were compared with *H. ducreyi* culture on two different media and confirmatory PCR assay. The sensitivity of M-PCR for *H. ducreyi* was 98.4%, whereas the sensitivity of culture was lower at 74.2%. The concordance between M-PCR and the *H. ducreyi* confirmatory PCR was 99.6%. In general, PCR methods have the potential to become an accurate and easy reference method for detection of *H. ducreyi*.

Attempts have also been made to develop enzyme immunoassays for detection of circulating antibodies to *H. ducreyi*. Comparison of the results is made difficult by the great variability in the tests with respect to different antigen preparations from various strains of *H. ducreyi*, as well as the performance of the tests. The results indicate, however, that specific, circulating antibodies (IgG, IgM and IgA) to outer membrane proteins, lipo-oligosaccharide (LOS) and other undefined antigens of *H. ducreyi* are present in sera from patients with culture-confirmed chancroid (Alfa *et al.*, 1992; 1993a; Desjardins *et al.*, 1992; Roggen *et al.*, 1994; Chen *et al.*, 1997). These assays have a limited diagnostic value, however, because a number of factors affect the sensitivity and specificity of the tests. The results are solely dependent on the specificity of *H. ducreyi* antigens used in the assay. The presence of HIV infection also affects the results of the tests. Enzyme-linked immunoassays (ELISA) are used mainly to evaluate the prevalence of chancroid antibodies in the different populations, as part of intervention strategies to reduce the transmission of HIV.

Models of Pathogenesis

Knowledge of the pathogenesis of chancroid and the host response to *H. ducreyi* is limited. The organism is a strict human pathogen, but some animal models of disease were described in the early literature and have been developed more recently. The experimental infection in humans has also been re-examined. In an effort to elucidate pathogenetic mechanisms in chancroid, various cultured human cells have been employed.

Animal Models

In earlier studies the rabbit was chosen as the most suitable laboratory animal model, and differences in the lesions produced after the intradermal inoculation of live so-called virulent and avirulent strains were reported. The dermal lesions are characterised by an abscess that develops into necrosis and subsequently forms eschars. At that time, however, the number of viable bacteria required to form lesions was not usually determined (Feiner and Mortara, 1945; Dienst, 1948). Later, it was reported that an inoculum of 10^8 colony-forming units (cfu) of bacteria was required to produce dermal lesions in rabbits. The survival of bacteria in lesions has been documented for up to 9 days. Histological examination of lesions showed deep necrosis and an infiltrate of inflammatory cells, especially granulocytes and dilatation of blood vessels. The rabbit model, which is referred to as the 'classical intradermal rabbit test', has its limitations in that the experimental infection is self-limiting and does not mimic the ulcers seen in humans (Hammond *et al.*, 1978a; Campagnari *et al.*, 1991; Lagergård, 1992). Since then, numerous studies have focused on the development of better animal models suitable for the study of *H. ducreyi* pathogenesis.

A mouse model of infection has been investigated, in which the ulcers that developed were similar to those in humans (Tuffrey *et al.*, 1990). Later, however, it was observed that the lesions were also produced by killed *H. ducreyi* or *N. gonorrheae*, as well as by purified LOS

(Campagnari *et al.*, 1991; Lagergård, 1992). A modification of the rabbit model is the temperature-dependent model, in which animals are housed at 15–17°C to achieve a reduction in skin temperature (Purcell *et al.*, 1991). When 10^5 virulent *H. ducreyi* were injected intradermally, the rabbits always developed necrotic lesions. These were found to be dependent on the viability of bacteria, and viable bacteria were recovered from lesions for nearly 2 weeks after the inoculation of the bacteria. Histopathological examination showed acute inflammation, which evolved to abscess formation. Although, as previously observed, the lesions were not similar to human ulcers, the model does reflect the ability of bacteria to multiply in lesions and the model is currently used to study the virulence of isogenic mutants and immunity to *H. ducreyi*.

A swine model of *H. ducreyi* infection, in which bacteria are inoculated into the dorsal surface of pig, was developed by Hobbs *et al.* (1995). Lesions developed 48 hours after the injection of live bacteria, and progress from pustules to ulcers was accompanied by a dermal inflammatory infiltrate containing polymorphonuclear neutrophils, T cells and macrophages. Live bacteria were recovered from lesions for up to 17 days. The ulcers did not resemble those in chancroid patients.

An animal model that resembles the human infection has, however been established in primates. In this model of chancroid in male and female pigtailed macaques, the foreskin or labia were inoculated (Totten *et al.*, 1994b). Ulcers with raised clear margins and a necrotic base developed in males 12 days after injection of live bacteria. The lesions mimicked those seen in humans, but an inoculum of 10^7–10^8 cfu was needed for the production of lesions. In addition, inguinal lymphadenopathy was observed in the majority of primates.

Experimental Infection in Humans

Early studies on chancroid involved isolating bacteria from patient's ulcers and then injecting them into the patient's own skin. The clinical progress of ulcer development was principally reported from these early studies, and the lesions were similar to those observed in patients. The number of bacteria necessary to produce ulcers was not determined (Morse, 1989). A human model of infection has been developed, with the aim of studying the early stages of the infection (Spinola *et al.*, 1994, 1996). Volunteers were injected intradermally on the upper arm with various doses of live bacteria, and for safety reasons they were treated with antibiotics after 14 days. Injection of

approximately 10^3 bacteria resulted in the development of papules that developed into pustules. All doses elicited a skin infiltrate of polymorphonuclear leucocytes, T cells and macrophages, and bacteria were intermittently shed from the papular lesions. In addition, Langerhans cells, CD4+ T cells and expression of HLA-DR by keratinocytes was also noted. There was little evidence of humoral or peripheral blood mononuclear cell responses to bacterial antigens. This model was later standardised (Al-Tawfiq *et al.*, 1998) and seems to be useful for the study of the interaction of bacteria with human skin and involvement of cellular response, but only in the early stages of infection.

Cultured-cell Models

Possible steps in the pathogenesis of chancroid, including adhesion to eukaryotic cells, invasion and cytotoxic damage of cells by *H. ducreyi*, have been investigated *in vitro* with various kinds of cultured human cells and human skin models.

H. ducreyi adheres readily to many transformed and non-transformed cell lines of human and animal origin. Adherence of *H. ducreyi* to the epithelium-like human cell lines as HEp-2; HeLa, A549, human foreskin epithelial cells (HFEC), fibroblasts prepared from neonatal foreskin fibroblasts (HFF, FS2–3) and human keratinocytes has been reported. The capacity for adherence of *H. ducreyi* is high and ranges from 15 to 60% of the inoculum. The process is time, dose and temperature-dependent. 'Virulent' strains manifest significantly higher levels of adherence than 'avirulent' strains (Shah *et al.*, 1992; Alfa *et al.*, 1993a; Lagergård *et al.*, 1993; Lammel *et al.*, 1993; Brentjens *et al.*, 1994; Totten *et al.*, 1994a; Alfa and Degagne, 1997). Binding of *H. ducreyi* to fibrinogen, fibronectin, collagen, gelatine, laminin, and heparin/heparan sulphate, indicates that some of intracellular matrix proteins and proteoglycans may be responsible for adherence of the bacteria to dermis (Abeck *et al.*, 1992b; Frisk and Lagergård, 1998; Bauer and Spinola, 1999).

Electron microscopy indicates that attachment of *H. ducreyi* to eukaryotic cells may result in internalisation or invasion of the cultured cells. Organisms invade HeLa cells better when compared with CHO, MRCC and C16[4] cells. Invagination of the plasma membrane around the bacteria was observed 2 hours after adherence; later HeLa cells were filled with organisms that had invaded, and it was suggested that the rupture of cells was a direct consequence of invasion. Treatment of eukaryotic cells with cytochalasin B reduced the number of internalised bacteria, but did not completely inhibit invasion (Shah *et al.*,

1992). In another study, in which gentamicin was used to kill extracellular bacteria, a much lower rate (<1%) of intracellular localisation of *H. ducreyi* in Hep-2 and HeLa cells was observed as compared with the number of attached bacteria (Lagergård *et al.*, 1993). The penetration of HFFC and HEp-2 cells was reported by Totten *et al.* (1994a), with ruthenium red staining used to distinguish between the intracellular and extracellular localisation of bacteria. About 1% of adhering cells accounted for invasion of the cells and avirulent strain were found to be less invasive than virulent strains. Treatment with cytochalasin B and D inhibited the ability of *H. ducreyi* to invade both cell lines. There are also conflicting results, which demonstrated that bacteria do not clearly enter HeLa cells, although they do enter HFF cells (Lammel *et al.*, 1993). Within HEF cells, bacteria were surrounded by membrane consistent with that of phagocytic vacuoles, and both virulent and avirulent strains displayed intracellular localisation. After incubation, *H. ducreyi* was situated intracellularly and in interstitial spaces between cells, and infection with a virulent strain resulted in cell destruction. The invasion of HFF could not be confirmed by Alfa *et al.* (1993b).

The issue of the intracellular localisation of bacteria in biopsies from chancroid patients has not been well investigated, but bacteria have been seen both inside and outside phagocytes. The biological relevance of the invasion of eukaryotic cells *in vitro* by *H. ducreyi* is not clear.

A cytopathic effect on transformed (A549) and non-transformed (FS2–3; HFF) cells has been reported (Alfa, 1992; Shah *et al.*, 1992; Hollyer *et al.*, 1994). The effect was dose-dependent and for both virulent and avirulent strains was described as depending on the contact of bacteria and eukaryotic cells. In the case of FS2–3 and A549 cells, the effect was detectable

from 3 to 7 days, but only with the virulent strain. Evidence that the destruction of HEp-2 cell monolayers 48 hours after adherence of *H. ducreyi* is due to elaboration and secretion of cytotoxin, but not the result of cell invasion, was reported by Lagergård *et al.* (1993).

A human foreskin fibroblast and keratinocyte model has been used *in vitro* to study *H. ducreyi* cell interactions. Bacteria were observed inside suprabasal keratinocytes, and structural changes in the epidermis were consistent with the pathology of chancroid. An unusual pattern of pro-inflammatory cytokine induction was also observed (Hobbs *et al.*, 1998). A human foreskin fibroblast cell culture and a low-temperature rabbit model were used by Alfa *et al.* (1995) to identify virulence-associated traits of *H. ducreyi*. Slow growth and poor adherence in cell culture correlated with the behaviour of avirulent strains in the rabbit model. The ability to survive, replicate and from microcolonies *in vitro* correlated with virulence as estimated intradermally in rabbit.

Potential Virulence Determinants

In recent years, a number of bacterial components that can directly or indirectly play a role in the pathogenesis of chancroid have been identified, cloned and sequenced, and mutants deficient in these determinants have been produced and tested in experimental infection models.

Toxins

H. ducreyi secretes a potent cytotoxin that can damage cultured human epithelial cell lines, such as HEp-2 and

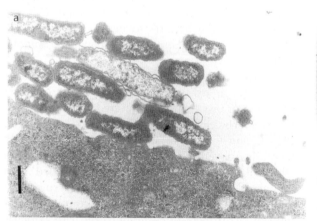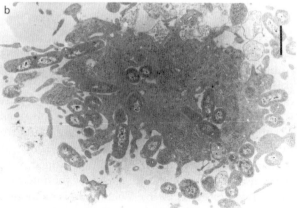

Fig. 2 Effect of *H. ducreyi* on epithelial cells as seen by transmission electron microscopy: (a) Bacteria bound to the surface of HEp-2 cells (bar, 500 nm) and (b) bacteria bound to and entering HeLa cells (bar, 1000 nm).

HeLa after 24–48 hours incubation (Purvén and Lagergård, 1992). Epithelial cells treated with this toxin showed dose-dependent, irreversible cell enlargement, inhibition of proliferation and finally cell death (**Fig. 3**). Later, it was shown that human keratinocytes, fibroblasts, T and B cells are also sensitive (Cortes-Bratti *et al.*, 1999; Svensson *et al.*, 1999; Gelfanova *et al.*, 1999). The majority of *H. ducreyi* isolates produce CDT which is immunologically similar (Lagergård and Purvén, 1993; Purvén *et al.*, 1995). The cluster of chromosomally situated genes that encode for this cytotoxin was identified and cloned. Three linked genes (*cdtABC*) encode proteins with calculated molecular weights of 25, 30 and 20 kDa (Cope *et al.*, 1997), and they most resemble the product of *E. coli*, *Campylobacter jejuni* and *Actinobacillus actinomycetemcomitans cdtABC* genes, which encode the cytolethal distending toxin (CDT) (Pickett and Whitehouse, 1999). The CDT of *H. ducreyi*, like other members of the CDT family, irreversibly blocks mammalian cells proliferation by inducing cell-cycle arrest in the G2 phase, because the p34 cdc2 kinase remains tyrosine phosphorylated and therefore inactive (Cortes-Bratti *et al.*, 1999; Pickett and Whitehouse, 1999). The role of different protein components of CDT in the generation of cell toxicity is not clear, but all three gene products are involved in the action of CDT on cultured cells (Cope *et al.*, 1997, Frisk *et al.*, 2001). An isogenic *cdtC* mutant of *H. ducreyi* was unable to kill HeLa cells or HaCaT keratinocytes in culture. This did not, however, affect the ability of this mutant to cause lesions and survive in the skin of rabbits (Stevens *et al.*, 1999). The *cdtC* gene product plays some role in intoxination of cells, since mouse monoclonal antibodies that bind to cdtC products can neutralise the toxic activity of CDT (Purvén *et al.*, 1997). It has recently been shown that the CdtB protein has DNase activity *in vitro* and this activity may be responsible for the CDT-induced cell cycle arrest (Elwell and Dreyfus, 2000). It has also been shown that it is the *cdtB* gene product of *H. ducreyi* CDT that exerts the DNase activity (Frisk *et al.*, 2001).

A cell-associated haemolysin has been identified, cloned and characterised (Palmer and Munson, 1995; Totten *et al.*, 1995). The toxin is encoded by two adjacent genes, *hhdA* and *hhdB*, which are similar to the pore-forming haemolysins of *Proteus mirabilis* and *Serratia marcescens*. Judging by their homology to *S. marcescens*, one gene probably encodes the haemolysin structural protein and the other is a protein required for secretion and modification of the protein. The haemolysin is immunogenic and expressed *in vivo* by all *H. ducreyi* strains (Dutro *et al.*, 1999). The

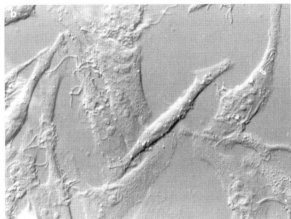

Fig. 3 Effect of cytholethal distending toxin-producing *Haemophilus ducreyi* CCUG 7470 on HEp-2 cells: (a) Monolayer with attached bacteria after 4 hours incubation. (b) Morphological changes in the monolayer after 48 hours incubation. (c) Uninoculated control monolayer after 48 hours incubation. Differential interference microscopy (magnification ×500).

cytopathic effect of *H. ducreyi* bacteria on HFF and the haemolytic activity are due to the same toxin, since an isogenic haemolysin mutant with insertions in *hhdB* failed to produce a cytotoxic effect on HFF cells

(Alfa *et al*, 1996; Palmer *et al.*, 1996). In addition to haemolytic action on erythrocytes and HFF, the haemolysin can destroy macrophages and T and B cells, and enhance the invasion of HEp-2 epithelial cells (Wood *et al.*, 1999). Studies of infection due to a haemolysin-deficient mutant in a human model of early chancroid showed that the haemolysin does not play a role in pustule formation (Palmer *et al.*, 1998b).

Lipo-oligosaccharide

Analysis by electrophoresis of *H. ducreyi* lipopolysaccharide (LPS) shows a pattern of migration similar to that of the LOS of *Haemophilus* and *Neisseria* species, indicating that it lacks the repeating O-side polysaccharide chain characteristic of LPS from most Gram-negative enteric bacteria. The chemical structure of LOS has been reported (Campagnari *et al.*, 1990; Mandrell *et al.*, 1992; Melaugh *et al.*, 1994; Schweda *et al.*, 1995). The LOS of *H. ducreyi* lacks the repeating polysaccharide antigen and contains a more variable branch core oligosaccharide region linked to lipid A. The structures all show a typical constellation of three heptoses linked to Kdo, with a variable oligosaccharide outer core, mostly consisting of hexoses. The majority of strains show a nonasaccharide in the outer core region. The presence of additional glycoforms, which result from the truncation and elongation of the major LOS form, is also indicated. The outer core structure of two strains has a shortened oligosaccharide, making it to hexasaccharide. The terminal Gal1 → 4GlcNAc disaccharide (lactosamine) is partially sialylated in a fashion similar to gonococcal LOS (Mandrell *et al.*, 1992; Melaugh *et al.*, 1996; Ahmed *et al.*, 1997). Furthermore, the LOS of *H. ducreyi* contains an epitope with antigenic and structural similarities to that present in glycosphingolipid, a precursor of the major human blood group antigens, which are present in some gangliosides on human cells. This mimicry may play a role in the evasion of bacteria to certain host immune responses (Mandrell *et al.*, 1992).

The arrangement of genes involved in expression of LOS in *H. ducreyi* has been intensively studied, and mutants deficient in some of the genes have been studied in animal and cell models (Gibson *et al.*, 1997; Stevens *et al.*, 1997; Bauer *et al.*, 1998, 1999). The importance of the *gmhA* gene, which has 87% of homology with that of *H. influenzae*, and the *waaF* gene, which has a gene product similar to that of *Salmonella typhimurium*, are essential for expression of wild-type, virulent LOS in the temperature-dependent rabbit model. An isogenic LOS mutant, constructed by insertion of a *cat* cartridge, was less virulent in the rabbit model.

LOS is also involved in adherence to HFF cells (Alfa and Degagne, 1997), but not to HEp-2 cells (Frisk *et al.*, 1998b). Adherence and invasion of human keratinocytes depends on the gene that encodes D-glycer-D-manno-heptosyltransferase, the enzyme responsible for the terminal portion of the LOS (Gibson *et al.*, 1997). A sialyltransferase gene has been identified in *H. ducreyi*, which represents a novel class of the enzyme responsible for a terminal sialylation of *H. ducreyi* LOS (Bozue *et al.*, 1999). *H. ducreyi* LOS is capable of inducing an inflammatory response in skin, and it has been suggested as the cause of lesions as a result of the toxic action of lipid A. Purified LOS causes ulcer formation in rabbits and in mice after intradermal injection. These lesions are comparable to those produced by the LOS of *N. gonorrheae*, but they are larger than those produced by *E. coli* LPS (Tuffrey *et al.*, 1990; Campagnari *et al.*, 1991; Lagergård, 1992).

Studies of the susceptibility of *H. ducreyi* to the complement-mediated bactericidal effect indicated that avirulent strains are susceptible to the bactericidal action of normal human and rabbit sera, whereas virulent strains are resistant. Moreover, LOS is related to the virulence of *H. ducreyi*, and it has been proposed that certain LOS compositions may make the bacteria more susceptible to complement-mediated serum bactericidal activity (Odumeru *et al.*, 1987). In another study, none of the *H. ducreyi* strains tested showed 100% sensitivity to bactericidal action of high concentrations of human or rabbit serum (Lagergård *et al.*, 1995) and the antibodies specific to LOS were not effective in complement-mediated killing of homologous *H. ducreyi* strains (Frisk *et al.*, 1998a). Similar resistance to normal and hyperimmune sera of a wild strain of *H. ducreyi* and the isogenic LOS mutant has recently been reported (Hiltke *et al.*, 1999).

Adhesins

Spinola *et al.* (1990) demonstrated that *H. ducreyi* expresses fine surface appendages resembling fimbriae. They are morphologically different from the fimbriae found on enteric bacteria and *N. gonorrheae*, and hence were reported as a novel class of fimbriae (Brentjens *et al.*, 1996). The structure was expressed on all strains tested, and was not loose during passages of bacteria *in vitro*. The molecular weight of the fimbrial monomer is 24 kDa, and of whole protein 150–160 kDa. The amino acid sequence of the monomer has been determined, and does not show homology with fimbrial proteins from other bacteria (Frisk *et al.*, 1995). The ability of this structure to mediate adherence to eukaryotic cells, and the receptor on these cells, has not been identified.

Heat-shock Proteins

Heat-shock proteins (Hsp), produced by many bacteria during stress, are highly conserved antigens among different species and eukaryotic cells. They have been identified as contributing to the damage of the host when cross-reactive antibodies and T cells are induced. Parsons *et al.* (1992) reported the molecular analysis of the *H. ducreyi groE* heat-shock operon, and noted a high degree of homology with the *GroES* and *GroEL* genes found in other bacteria. The *H. ducreyi GroEL* genes encode predicted proteins of 57.6 kDa and were the predominant proteins observed during bacterial growth at 33°C in broth. *H. ducreyi* produces unusually high basal levels of GroEL which are involved in the survival, chaining and adherence of bacteria in the presence of the combined stresses of the host environment (Parsons *et al.*, 1997). The GroEL Hsp are associated with the cell surface and to have the capacity to bind to such eukaryotic cells as HEp-2, indicating their possible involvement in the attachment of bacteria to host cells and to each other (Frisk *et al.*, 1998b).

Outer-membrane Proteins

A 28-kDa protein, common and specific for all *H. ducreyi* strains tested, has been cloned (Stewart *et al.*, 1992). In addition, a protein of 18 kDa was described as localised at the surface of the bacteria (Spinola *et al.*, 1992, 1993). A protein, described as a major outer membrane protein, has a molecular mass of 39–42 kDa, depending on the strain. Heat can modify this protein, which is cationic at a pH of 8.0 and contains cysteine residues. These findings indicate that there are differences between this protein and classical porin proteins. The total amino acid content and the N-terminal amino acid sequence showed homology to the outer-membrane proteins of the *Enterobacteriaceae*, *Neisseriaceae* and *Pasteurellaceae*. The function of these proteins as bacterial virulence factors has yet to be determined, but it has been suggested that they may play a role in the serum resistance of *H. ducreyi* (Hiltke *et al.*, 1999).

Other Possible Virulence Factors

A haemoglobin-binding outer membrane protein of *H. ducreyi* has been identified and characterised (Elkins *et al.*, 1995). It has been suggested that the protein is involved in the expression of bacterial

virulence in the low-temperature rabbit model (Stevens *et al.*, 1996).

The importance of superoxide dismutase (SOD) has been investigated with a Cu–Zn SOD-deficient mutant. Such a mutant was significantly more susceptible to killing by extracellular superoxide than the wild strain, indicating that SOD may play a role in bacterial defence against oxidative killing by host immune cells during infection (San-Mateo *et al.*, 1998).

Ward *et al.* (1998) identified two extremely large open reading frames in *H. ducreyi*, 1spA1 and 1spA2, each encoding a predicted protein product whose N-terminal half is approximately 43% similar to the filamentous haemagglutinin of *Bordetella pertussis*, which is responsible for cell attachment of the latter. The role of these proteins in pathogenesis in unknown.

Immunity

Re-infection, auto-inoculation and, before the introduction of antibiotics, the long duration of ulcers are well known. Spontaneous resolution of chancre is, however, also known to occur. As was noted above, the studies of Ito and Reensterna were interpreted as evidence of a delayed hypersensitivity reaction. The results of these early studies were interpreted as implying that protective immunity does not develop during infection with *H. ducreyi* (Morse, 1989).

It is known that the infection with *H. ducreyi* in patients and in experimental animals elicits an antibody response to bacterial antigens such as LOS, Hsp, cytotoxin, fimbrial proteins and outer-membrane proteins (Trees and Morse, 1995). Patients with chancroid have significantly high levels of serum antibodies specific to bacterial components (Alfa *et al.*, 1992, 1993a; Brown *et al.*, 1993; Roggen *et al.*, 1994; Frisk *et al.*, 1995). The role of these antibodies in protection against disease is not known. *H. ducreyi* is relatively resistant to bactericidal action of the complement *in vitro* and to some extent to opsonophagocytic killing, which suggests that antibodies are of limited protective value (Lagergård *et al.*, 1995; Frisk *et al.*, 1998a, Hiltke *et al.*, 1999). Infection with *H. ducreyi* also elicits antibodies that neutralise the cell damaging effect of cytholethal distending toxin, and antibodies against haemolysin, but the role of these antibodies *in vivo* is not known (Lagergård and Purvén, 1993; Dutro *et al.*, 1999).

Results in some animal models have indicated that repeated infection with *H. ducreyi* does not protect against re-infection, and may even contribute to an aggravation of the lesions (Feiner and Morata, 1945; Hobbs *et al.*, 1995). Conflicting results have been reported in other studies. With the temperature-dependent rabbit model of infection, it was reported that immunisation with an accellular or fimbria preparation, but not the LOS of *H. ducreyi*, significantly reduced the severity of lesions and the duration of infection after challenge with live bacteria (Hansen *et al.*, 1994; Desjardins *et al.*, 1995). Since passively transferred antibodies do not protect rabbits in terms of a reduction in the severity of lesions, however, a cell-mediated rather than a humoral response has been proposed as responsible for protection of animals (Desjardins *et al.*, 1996).

In the human model of infection, the host response to intradermal injection of *H. ducreyi* consisted primarily of a cutaneous infiltrate of plymorphonuclear leucocytes, Langerhans cells, macrophages and CD4+ T cells. Expression of HLA-DR by keratinocytes was associated with the presence of interferon γ. There was little evidence for a humoral cell response to bacterial antigens in the early stage of infection (Spinola *et al.*, 1996). Moreover, both papules and pustules were shown to contain a mixed or T helper cell-1 type cytokine mRNA, interleukin 8 and tumour necrosis factor α mRNA, indicating that the immune responses resembled those of a delayed-type hypersensitivity reaction (Palmer *et al.*, 1998a). The experimental infection to the pustular stage of the disease does not provide protective immunity to subsequent challenge in the human experimental model of chancroid (Al-Tawiq *et al.*, 1999).

There are indications that the healing of ulcers in HIV-infected individuals is delayed, as compared with individuals who are HIV-negative; however, the histological and immunohistochemical pictures were identical for HIV-positive patients and HIV-negative individuals (King *et al.*, 1998).

So far, the role of humoral and cell-mediated protective immunity in chancroid is not sufficiently understood, and knowledge about involvement of protective systemic or mucosal immunity is lacking.

Antimicrobial Susceptibility

The antimicrobial susceptibility of *H. ducreyi* reported between 1940 and 1960 showed that the bacterium is sensitive to commonly used antimicrobial drugs (Morse, 1989). Recent studies from various parts of the world have shown that significant antimicrobial resistance has become common to sulphonamides, trimethoprim, tetracycline, chloramphenicol, streptomycin, kanamycin, penicillin, ampicillin and gentamicin. With the development of resistance by *H. ducreyi* to many less costly antimicrobial agents, therapeutic

intervention in chancroid has become increasingly difficult.

Genetic analysis of the antimicrobial resistance of *H. ducreyi* has given an insight into the development of resistance to many commonly used antimicrobial agents. The resistance genes are mainly located on plasmids, but are probably also chromosomally mediated. Many geographical and even temporal differences in antimicrobial susceptibility of *H. ducreyi* depend on the presence or absence of plasmids. Resistance to sulphonamides, aminoglycosides, tetracyclines, chloramphenicol and penicillins is plasmid mediated. The plasmids of *H. ducreyi* display a variability in molecular mass (3.1–34 MDa). They encode resistance separately or in combination, and they can be mobilised by conjugative plasmids, though other mechanisms have also been suggested. Some of those plasmids, such as the class M tetracycline-resistance determinant (*Tet M*), located on a 34-MDa conjugative plasmid, are also present in other microorganisms of the genitourinary tract (Morse, 1989; Trees and Morse, 1995).

Multi-dose therapy with erythromycin or trimoxazole is now regarded as the treatment of choice for the disease in most countries where it is endemic. On the basis of *in vitro* susceptibility testing, the most active drugs against *H. ducreyi* are ceftriaxone, ciprofloxacin and roxithromycin. The first two were effective even in a single dose. Of the new antimicrobial drugs, azithromycin and fleroxacin have been effective in the USA and Kenya respectively (Schulte and Schmid, 1995).

Susceptibility testing of *H. ducreyi* involves some difficulties. In most studies the agar dilution method was used, but there is little experience with the disk diffusion method. The E-test correlates well with the agar dilution method, which may facilitate the estimation of minimum inhibitory concentrations for *H. ducreyi*; however, none of these methods has been standardised (Morse, 1989; Lagergård *et al.*, 1996).

Epidemiology

Strain Typing

The development of a typing method to characterise individual strains is important in epidemiological studies and could address questions about the geographical distributions of strains, their mode of transmission and discrimination between the virulence of strains of this pathogen. Different approaches of strain typing have been proposed, such as aminopeptidase profiles, immunotyping, lectin agglutination, plasmid analysis and ribotyping (Trees and Morse, 1995).

The variability of aminopeptidases between different *H. ducreyi* strains, suggests that these enzymes can be used as useful epidemiological markers (Van Dyck and Piot, 1987). Seven different immunopatterns of *H. ducreyi* have been identified by the immunoblotting technique, and antigenic diversity has been reported in isolates from different geographical locations, and from single areas (Roggen *et al.*, 1992). However, these typing systems have not been further evaluated.

Plasmid analysis of *H. ducreyi* has been studied over a 10-year period in 342 strains isolated from 18 urban areas of the USA (Sarafian and Knapp, 1992). Of these strains, 102 were plasmid free, and among the rest 5 plasmid profiles were identified. The usefulness of this approach in epidemiological studies has obvious limitations, in that many strains may have identical profiles, and plasmids may be lost or transferred by conjugation. This may contribute to the instability of strains tested and to temporary changes in plasmid pattern.

The application of ribotyping to the epidemiological investigation of chancroid is supported by the finding that 4 *Hinc*II and five *Hind*II ribotypes could be distinguished among 14 *H. ducreyi* isolates during a 1-month period in Kenya (Sarafian *et al.*, 1991). Other studies have also indicated that the ribotyping is reproducible, discriminates well between different strains and provides a useful tool for epidemiological studies of chancroid (Brown and Ison, 1993; Pillay *et al.*, 1998a). Ribotyping combined with plasmid analysis was successfully used to investigate the outbreak of chancroid in San Francisco in 1989–91 and to track the epidemiological distribution of the epidemic strain (Flood *et al.*, 1993). A combination of ribotyping, plasmid content and antibiotic susceptibility was used for the epidemiological characterisation of strains isolated from the chancroid outbreak in Mississippi and Louisiana, USA (Haydock *et al.*, 1999).

Transmission

Secretions from chancroid ulcers are highly infectious (Sullivan, 1940), and transmission of chancroid is predominantly by sexual contact and auto-inoculation. It is probably more frequently women than men who transmit the infection, since more women than men are asymptomatic carriers.

Prevalence

The reservoir of *H. ducreyi* is not understood, but it has been suggested that commercial sexual workers are one reservoir (D'Costa *et al.*, 1985). Asymptomatic carriage, especially by women, may contribute to the spread of infection (Hawkes *et al.*, 1995).

Chancroid is particularly common in some parts of eastern and southern Africa (Nsanze *et al.*, 1981; Plummer *et al.*, 1983; Bogaerts *et al.*, 1989; Dangor *et al.*, 1989), south-east Asia (Taylor *et al.*, 1984; Thirumoorthy *et al.*, 1986) and Latin America (Ortiz-Zepeda *et al.*, 1994), where the disease is mostly prevalent in low socio-economic groups who frequent prostitutes. In a number of studies, it was noted that among uncircumcised men the acquisition of chancroid is increased 3-fold. In south-east Asia where chancroid was once endemic, it is now declining as a result of socio-economic development and special programmes to control STDs; genital herpes simplex infection is now more frequently observed (Beyrer *et al.*, 1998). In some African and Latin American countries, a high incidence of chancroid continues to be reported. In general, genital ulcers comprise 18–70% of patients with STDs seen in clinic in endemic countries. Local outbreaks have also been reported in developed countries, including for example the USA and Canada (Hammond *et al.*, 1980; Trees and Morse, 1995) and Europe (Morse, 1989).

Many STDs have been associated with HIV transmission, but the data linking chancroid to HIV transmission is strongest. Genital ulceration, as a clinical syndrome, has in many studies been pointed out as a significant risk factor for HIV sero-conversion (Cameron *et al.*, 1989; Kreiss *et al.*, 1989; Jessamine and Ronald, 1990; Wasserheit, 1992). The presence of chancroid ulcers may increase susceptibility to HIV infection by damage to the epithelial barrier and because inflammation causes an increase of HIV-susceptible cells at the point of entry.

Prevention and Control

Prevention and control of chancroid depend on the treatment of patients with ulcers and their sexual contacts, particularly prostitutes. In addition, conventional STD control programmes, including screening and contact tracing, play a central role.

References

Abeck D, Eckert F, Korting HC (1992a) Atypical presentation of co-existent *Haemophilus ducreyi* and *Treponema pallidum* infection in an HIV-positive male. *Acta Derm. Venereol.* 72: 37–38.

Abeck D, Johnson AP, Mensing H (1992b) Binding of *Haemophilus ducreyi* to extracellular matrix proteins. *Microb. Pathog.* 13: 81–84.

Abeck D, Freinkel AL, Korting HC, Szeimis RM, Ballard RC (1997) Immunohistochemical investigation of genital ulcers caused by *Haemophilus ducreyi*. *Int. J. Sex. Trans. Dis. AIDS* 8: 585–588.

Ahmed HJ, Borrelli S, Jonasson J *et al.* (1995) Monoclonal antibodies against *Haemophilus ducreyi* lipooligosaccharide and their diagnostic usefulness. *Eur. J. Clin. Microbiol. Infect. Dis.* 14: 892–989.

Ahmed HJ, Frisk A, Månsson J-E, Schweda EKH, Lagergård T (1997) Structurally defined epitopes of *Haemophilus ducreyi* lipooligosaccharides recognised by monoclonal antibodies. *Infect. Immun.* 65: 3151–3158.

Al-Tawfiq JA, Thorton AC, Katz BP *et al.* (1998) Standardisation of the experimental model of *Haemophilus ducreyi* infection in human subjects. *J. Infect. Dis.* 178: 1684–1687.

Al-Tawfiq JA, Palmer KL, Chen CY *et al.* (1999) Experimental infection of human volunteers with *Haemophilus ducreyi* does not confer protection against subsequent challenge. *J. Infect. Dis.* 179: 1283–1287.

Alfa MJ (1992) Cytopathic effect of *Haemophilus ducreyi* for human foreskin cell culture. *J. Med. Microbiol.* 37: 43–50.

Alfa MJ, Degagne P (1997) Attachment of *Haemophilus ducreyi* to human foreskin fibroblasts involves LOS and fibronection. *Microb. Pathog.* 22: 39–46.

Alfa MJ, Olson N, Degagne P *et al.* (1992) Use of an adsorption enzyme immunoassay to evaluate the *Haemophilus ducreyi* specific and cross-reactive humoral immune response in humans. *Sex. Transm. Dis.* 19: 309–314.

Alfa MJ, Olsen N, Degagne PP *et al.* (1993a) Humoral immune response of humans to lipooligosaccharide and outer membrane proteins of *Haemophilus ducreyi*. *J. Infect. Dis.* 167: 1206–1210.

Alfa MJ, Degagne P, Hollyer T (1993b) *Haemophilus ducreyi* adheres to but does not invade cultured human foreskin cells. *Infect. Immun.* 61: 1735–1742.

Alfa MJ, Stevens MK, DeGagne P *et al.* (1995) Use of tissue culture and animal models to identify virulence-associated traits of *Haemophilus ducreyi*. *Infect. Immun.* 63: 1754–1761.

Alfa MJ, DeGagne P, Totten PA (1996) *Haemophilus ducreyi* hemolysin acts as a contact cytotoxin and damages human foreskin fibroblasts in cell culture. *Infect. Immun.* 64: 2349–2352.

Bauer BA, Stevens MK, Hansen EJ (1998) Involvement of the *Haemophilus ducreyi gmhA* gene product in lipooligosaccharide expression and virulence. *Infect. Immun.* 66: 4290–4298.

Bauer BA, Lumbley SR, Hansen EJ (1999) Characterisation of a WaaF (RfaF) homolog expressed by *Haemophilus ducreyi*. *Infect. Immun.* 67: 899–907.

Bauer ME, Spinola SM (1999) Binding of *Haemophilus ducreyi* to extracellular matrix proteins. *Infect. Immun.* 67: 2649–2652.

Beyrer C, Jitwatcharanan K, Natpratan C *et al.* (1998) Molecular methods for the diagnosis of genital ulcer disease in a sexually transmitted disease clinic population in northern Thailand: predominance of herpes simplex virus infection. *J. Infect. Dis.* 178: 243–246.

Bogaerts J, Ricart CA, Dyck E Van, Piot P (1989) The etiology of genital ulceration in Rwanda. *Sex. Transm. Dis.* 16: 123–126.

Bozue JA, Tullius MV, Wang J, Gibson BW, Munson RS (1999) *Haemophilus ducreyi* produces a novel sialyltransferase. Identification of sialyltransferase gene and construction of mutants deficient in the production of the sialic acid containing glycoform of the lipooligosaccharide. *J. Biol. Chem.* 274: 4106–4114.

Brentjens RJ, Spinola SM, Campagnari AA (1994) *Haemophilus ducreyi* adheres to human keratinocytes. *Microb. Pathog.* 16: 243–247.

Brentjens RJ, Ketterer M, Apicella MA, Spinola SM (1996) Fine tangled pili expressed by *Haemophilus ducreyi* are a novel class of pili. *J. Bacteriol.* 178: 808–816.

Brown TJ, Ison CA (1993) Non-radioactive ribotyping of *Haemophilus ducreyi* using a digoxigenin labelled cDNA probe. *Epidemiol. Infect.* 110: 289–295.

Brown TJ, Jardine J, Ison CA (1993) Antibodies directed against *Haemophilus ducreyi* heat shock proteins. *Microb. Pathog.* 15: 131–139.

Cameron DW, Simonsen JN, D'Costa LJ *et al.* (1989) Female to male transmission of human immunodeficiency virus type 1: risk factors for seroconversion in men. *Lancet* ii: 403–407.

Campagnari AA, Spinola SM, Lesse AJ *et al.* (1990) Lipooligosaccharide epitopes shared among gram-negative non-enteric mucosal pathogens. *Microb. Pathog.* 8: 353–362.

Campagnari AA, Wild LM, Griffiths GE *et al.* (1991) Role of lipooligosaccharides in experimental dermal lesions caused by *Haemophilus ducreyi*. *Infect. Immun.* 59: 2601–2608.

Chapel TA, Brown WJ, Jeffries C, Stewart J (1977) How reliable is the morphological diagnosis of penile ulcerations? *Sex. Transm. Dis.* 4: 150–152.

Chen CY, Mertz KJ, Spinola SM, Morse SA (1997) Comparison of enzyme immunoassays for antibodies to *Haemophilus ducreyi* in a community outbreak of chancroid in the United States. *J. Infect. Dis.* 175: 1390–1395.

Chui L, Albritton W, Paster B, Maclean I, Marusyk R (1993) Development of the polymerase chain reaction for diagnosis of chancroid. *J. Clin. Microbiol.* 31: 659–664.

Cope LD, Lumbley S, Latimer JL *et al.* (1997) A diffusible cytotoxin of *Haemophilus ducreyi*. *Proc. Natl Acad. Sci. USA* 94: 4056–4061.

Cortez-Bratti X, Chaves-Olarte E, Lagergård T, Thelestam M (1999) A cytotoxin from chancroid bacterium *Haemophilus ducreyi* induces cycle arrest in G2 phase. *J. Clin. Invest.* 103: 107–115.

D'Costa LJ, Plummer FA, Bowmer I *et al.* (1985) Prostitutes are a major reservoir of sexually transmitted diseases in Nairobi, Kenya. *Sex. Transm. Dis.* 12: 64–67.

Dangor Y, Fehler G, Exposto FD, Koornhof HJ (1989) Causes and treatment of sexually acquired genital ulceration in southern Africa. *S. Afr. Med. J.* 76: 339–341.

Dangor Y, Miller SD, Koornhof HJ, Ballard RC (1992) A simple medium for the primary isolation of *Haemophilus ducreyi*. *Eur. J. Clin. Microbiol. Infect. Dis.* 11: 930–934.

Dangor Y, Radebe F, Ballard RC (1993) Transport media for *Haemophilus ducreyi*. *Sex. Transm. Dis.* 20: 5–9.

Deacon WE, Albritton DC, Orlansky S, Kaplan W (1956) V. D. R. L. chancroid studies: A simple procedure for the isolation and identification of *Haemophilus ducreyi*. *J. Invest. Dermatol.* 26: 399–406.

Desjardins M, Thompson CE, Fillon LG *et al.* (1992) Standardization of an enzyme immunoassay for human antibody to *Haemophilus ducreyi*. *J. Clin. Microbiol.* 30: 2019–2024.

Desjardins M, Filion LG, Robertson S, Cameron DW (1995) Inducible immunity with a pilus preparation booster vaccination in an animal model of *Haemophilus ducreyi* in an animal model of chancroid. *Infect. Immun.* 63: 2012–2020.

Desjardins M, Filion LG, Robertson S, Kobylinski L, Cameron DW (1996) Evaluation of humoral and cell-mediated inducible immunity to *Haemophilus ducreyi* infection and disease. *Infect. Immun.* 64: 1778–1788.

Dienst RB (1948) Virulence and antigenicity of *Haemophilus ducreyi*. *Am. J. Syphilis Gonorrhea Vener. Dis.* 32: 289–291.

Ducreyi A (1889) Experimentelle Unterschungen uber den ansteckungssroff des Schancers und uber die Bubonen. *Monatsh. Prakt. Dermatol.* 9: 387–405.

Dutro SM, Wood GE, Totten PA (1999) Prevalence of, antibody response to, and immunity induced by *Haemophilus ducreyi* hemolysis. *Infect. Immun.* 67: 3317–3328.

Elkins C (1995) Identification and purification of a conserved heme-regulated haemoglobin-binding outer membrane protein from *Haemophilus ducreyi*. *Infect. Immun.* 63: 1241–1245.

Elwell AC, Dreyfus LA (2000) DNase I homologous residues in CdtB are critical for cytolethal distending toxin-mediated cell cycle arrest. *Mol. Microbiol.* 37: 952–963.

Feiner RR, Morata F (1945) Infectivity of *Haemophilus ducreyi* for the rabbit and the development of skin hypersensitivity. *Am. J. Syphilis Gonorrhea Vener. Dis.* 29: 71–79.

Flood JM, Sarafian SK, Bolan GA *et al.* (1993) Multistrain outbreak of chancroid in San Francisco, 1989-1991. *J. Infect. Dis.* 167: 1106–1111.

Freinkel AL (1987) Histological aspects of sexually transmitted genital lesions. *Histopathology* 11: 819–831.

Frisk A, Lagergård T (1998) Characterisation of mechanisms involved in adherence of *Haemophilus ducreyi* to eukaryotic cells. *AMPIS* 5: 539–546.

Frisk A, Ahmed H, Dyck E Van, Lagergård T (1998a) Antibodies to surface antigens are not effective in complement-mediated killing of *Haemophilus ducreyi*. *Microb. Pathog.* 25: 67–75.

Frisk A, Ison C, Lagergård T (1998b) Evidence of surface exposure of GroEL heat shock proteins of *Haemophilus ducreyi*. *Infect. Immun.* 66: 1252–1257.

Frisk A, Lebens M, Johansson C *et al.* (2001) The role of different protein components from the *Haemophilus ducreyi* cytolethal distending toxin in generation of cell toxicity. *Microb. Pathog.* (in press).

Gelfanova V, Hansen EJ, Spinola SM (1999) Cytolethal distending toxin of *Haemophilus ducreyi* induces apoptotic death of Jurkat T cells. *Infect. Immun.* 67: 6394–6402.

Gibson GW, Campagnari AA, Melaugh W *et al.* (1997) Characterisation of transpose Tn916-generated mutant of *Haemophilus ducreyi* 35000 defective in lipopolysaccharide biosynthesis. *J. Bacteriol.* 179: 5062–5071.

Gu XX, Rossau R, Jannes G, Ballard R, Van Dyck E, Laga M (1998) The rrs(16S)-rrl(23S)ribosomal intergenic spacer region as a target for the detection of *Haemophilus ducreyi* by a heminested-PCR assay. *Microbiology* 144: 1013–1019.

Hammond GW, Lian CJ, Wilt JC, Ronald AR (1978a) Antimicrobial susceptibility of *Haemophilus ducreyi*. *Antimicrob. Agents Chemother.* 13: 608–612.

Hammond GW, Lian C-J, Wilt JC, Ronald AR (1978b) Comparison of specimen collection and laboratory techniques for isolation of *Haemophilus ducreyi*. *J. Clin. Microbiol.* 7: 39–43.

Hammond GW, Slutchuk M, Scatliff J *et al.* (1980) Epidemiologic, clinical, laboratory, and therapeutic features of an urban outbreak of chancroid in North America. *Rev. Infect. Dis.* 2: 867–879.

Hansen EJ, Lumbley SR, Richardson JA *et al.* (1994) Induction of protective immunity to *Haemophilus ducreyi* in the temperature-dependent rabbit model of experimental chancroid. *J. Immunol.* 152: 184–192.

Hawkes S, West B, Wilson S, Whittle H, Mabey D (1995) Asymptomatic carriage of *Haemophilus ducreyi* confirmed by the polymerase chain reaction. *Genitourinary Med.* 71: 224–227.

Haydock AK, Martin DH, Morse SA *et al.* (1999) Molecular characterisation of *Haemophilus ducreyi* strains from Jackson, Mississippi, and New Orleans, Louisiana. *J. Infect. Dis.* 179: 1423–1432.

Heyman A, Beeson PB, Sheldon WH (1945) Diagnosis of chancroid: The relative efficiency of biopsies, cultures, smears, auto-inoculations and skin tests. *JAMA* 129: 935–938.

Hiltke TJ, Bauer ME, Kleysney-Tait J *et al.* (1999) Effect of normal and immune sera on *Haemophilus ducreyi* 35000HP and its isogenic MOMP and LOS mutants. *Microb. Pathog.* 26: 93–102.

Hobbs MM, San-Mateo LR, Orndorff PE, Almond G, Kawula TH (1995) Swine model of *Haemophilus ducreyi* infection. *Infect. Immun.* 63: 3094–3100.

Hobbs MM, Terry RP, Wyrick PB, Kawula TH (1998) *Haemophilus ducreyi* infection causes basal keratinocyte cytotoxicity and elicits a unique cytokine induction pattern *in vitro* in human skin model. *Infect. Immun.* 66: 2914–2921.

Hollyer TT, DeGagne PA, Alfa MJ (1994) Characterisation of the cytopathic effect of *Haemophilus ducreyi* infection. *Sex. Transm. Dis.* 21: 247–257.

Jessamine PG, Ronald AR (1990) Chancroid and the role of genital ulcer disease in the spread of human retroviruses. *Med. Clin. N. Am.* 74: 1417–1431.

Johnson SR, Martin DH, Cammarata C, Morse SA (1995) Alterations in sample preparation increase sensitivity of PCR assay for diagnosis of chancroid. *J. Clin. Microbiol.* 33: 1036–1038.

Kampmeier RH (1982) The recognition of *Haemophilus ducreyi* as the cause of soft chancre. *Sex. Transm. Dis.* 9: 212–213.

Karim QN, Finn G, Easmon CS *et al.* (1989) Rapid detection of *Haemophilus ducreyi* in clinical and experimental infections using monoclonal antibody: a preliminary evaluation. *Genitourinary Med.* 65: 361–365.

King R, Choudhri SH, Nasio J *et al.* (1998) Clinical and in situ cellular respones to *Haemophilus ducreyi* in the presence or absence of HIV infection. *Int. J. Sex. Trans. Dis. AIDS* 9: 531–536.

Kreiss JK, Coombs R, Plummer F *et al.* (1989) Isolation of human immunodeficiency virus from genital ulcers in Nairobi prostitutes. *J. Infect. Dis.* 160: 380–384.

Lagergård T (1992) The role of *Haemophilus ducreyi* bacteria, cytotoxin, endotoxin and antibodies in animal models for study of chancroid. *Microb. Pathog.* 13: 203–217.

Lagergård T, Purvén M (1993) Neutralizing antibodies to *Haemophilus ducreyi* cytotoxin. *Infect. Immun.* 61: 1589–1592.

Lagergård T, Purvén M, Frisk A (1993) Evidence of *Haemophilus ducreyi* adherence to and destruction of human epithelial cells. *Microb. Pathog.* 14: 417–431.

Lagergård T, Frisk A, Purvén M, Nilsson LA (1995) Serum bactericidal activity and phagocytosis in host defence against *Haemophilus ducreyi*. *Microb. Pathog.* 18: 37–51.

Lagergård T, Frisk A, Trollfors B (1996) Comparison of the E-test with agar dilution for antimicrobial testing of *Haemophilus ducreyi*. *J. Antimicrob. Chemother.* 38: 849–852.

Lammel CJ, Dekker NP, Palefsky J, Brooks GF (1993) *In vitro* model of *Haemophilus ducreyi* adherence to and entry into eukaryotic cells of genital origin. *J. Infect. Dis.* 167: 642–650.

Mandrell RE, McLaughlin R, Aba K-Y *et al.* (1992) Lipooligosaccharides (LOS) of some *Haemophilus* species mimic human glycosphingolipids, and some LOS are sialylated. *Infect. Immun.* 60: 1322–1328.

Melaugh W, Phillips NJ, Campagnari AA, Tullius MV, Gibson BW (1994) Structure of the major oligosaccharide from the lipooligosaccharide of *Haemophilus ducreyi* strain 35000 and evidence for additional glycoforms. *Biochemistry* 33: 13070–13078.

Melaugh W, Campagnari AA, Gibson BW (1996) The lipooligosaccharides of *Haemophilus ducreyi* are highly sialylated. *J. Bacteriol.* 178: 564–570.

Morse SA (1989) Chancroid and *Haemophilus ducreyi*. *Clin. Microbiol. Rev.* 2: 137–157.

Morse SA, Trees DL, Htun YE *et al.* (1997) Comparison of clinical diagnoses and standard laboratory and molecular methods for the diagnosis of genital ulcer diseases in Lesotho: Association with human immunodeficiency virus infection. *J. Infect. Dis.* 175: 583–589.

Nsanze H, Fast MV, D'Costa LJ *et al.* (1981) Genital ulcers in Kenya. Clinical and laboratory study. *Br. J. Vener. Dis.* 57: 378–381.

Nsanze H, Plummer FA, Maggwa AB *et al.* (1984) Comparison of media for the primary isolation of *Haemophilus ducreyi*. *Sex. Transm. Dis.* 11: 6–9.

Odumeru JA, Wiseman GM, Ronald AR (1987) Relationship between lipopolysaccharide composition and virulence of *Haemophilus ducreyi*. *J. Med. Microbiol.* 23: 155–162.

Orle KA, Martin DH, Gates CA *et al.* (1994) Multiple PCR detection of *Haemophilus ducreyi*, *Treponema pallidum*, and herpes simplex virus types-1 and -2 from genital ulcers. Abstr. C-437, 4th Annual Meeting of American Society of Microbiology, p. 568.

Ortís-Zepeda C, Hernandez-Perez E, Marroquín-Burgos R (1994) Gross and microscopic features in chancroid: a study in 200 new culture-proven cases in San Salvador. *Sex. Transm. Dis.* 21: 112–117.

Palmer KL, Goldman WE, Munson RS Jr (1996) An isogenic haemolysin-deficient mutant of *Haemophilus ducreyi* lack the ability to produce cytopathic effects on human foreskin fibroblasts. *Mol. Microbiol.* 21: 13–19.

Palmer KL, Schnizlein-Bick CT, Orazi A *et al.* (1998a) The immune response to *Haemophilus ducreyi* resembles a delayed-type hypersensitivity reaction throughout experimental infection of humans. *J. Infect. Dis.* 178: 1688–1697.

Palmer KL, Thornton AC, Fortney KR *et al.* (1998b) Evaluation of an isogenic hemolysin-deficient mutant in the human model of *Haemophilus ducreyi* infection. *J. Infect. Dis.* 178: 191–199.

Palmer L, Munson RS Jr (1995) Cloning and characterisation of the genes encoding the haemolysin of *Haemophilus ducreyi*. *Mol. Microbiol.* 18: 821–830.

Parsons LM, Shayegani M, Waring AL, Bopp LH (1989) DNA probes for the identification of *Haemophilus ducreyi*. *J. Clin. Microbiol.* 27: 1441–1445.

Parsons LM, Waring AL, Shayegani M (1992) Molecular analysis of the *Haemophilus ducreyi* groE heat shock operon. *Infect. Immun.* 60: 4111–4118.

Parsons LM, Limberger RJ, Shayegani M (1997) Alterations in levels of DnaK and GroEL result in diminished survival and adherence of stressed *Haemophilus ducreyi*. *Infect. Immun.* 65: 2413–2419.

Pickett CL, Whitehouse CA (1999) The cytolethal distending toxin family. *Trends Microbiol.* 7: 292–297.

Pillay A, Hoosen AA, Kiepela P, Sturm AW (1998a) Ribosomal DNA typing of *Haemophilus ducreyi* strains. Proposal for a novel typing scheme. *J. Clin. Microbiol.* 34: 2613–2615.

Pillay A, Hoosen AA, Loykissoonalal D *et al.* (1998b) Comparison of culture media for the laboratory diagnosis of chancroid. *J. Med. Microbiol.* 47: 1023–1026.

Plummer FA, D'Costa LJ, Nsanze H *et al.* (1983) Epidemiology of chancroid and *Haemophilus ducreyi* in Nairobi, Kenya. *Lancet* ii: 1293–1295.

Purcell BK, Richardson JA, Radolf JD, Hansen EJ (1991) A temperature-dependent rabbit model for production of dermal lesions by *Haemophilus ducreyi*. *J. Infect. Dis.* 164: 359–367.

Purvén M, Lagergård T (1992) *Haemophilus ducreyi*, a cytotoxin-producing bacterium. *Infect. Immun.* 60: 1156–1162.

Purvén M, Falsen E, Lagergård T (1995) Cytotoxin production in 100 strains of *Haemophilus ducreyi* from differentgeographic locations. *FEMS Microbiol. Lett.* 129: 221–224.

Purvén M, Frisk A, Lönnroth I, Lagergård T (1997) Purification and identification of *Haemophilus ducreyi* cytotoxin using a neutralizing monoclonal antibody. *Infect. Immun.* 65: 3496–3499.

Rauschkolb JE (1939) Circumcision in treatment of chancroidal lesions of male genitalia. *Arch. Dermatol. Syphilol.* 39: 319–328.

Roggen EL, Breucker S de, Dyck E Van, Piot P (1992) Antigenic diversity in *Haemophilus ducreyi* as shown by western blot (immunoblot) analysis. *Infect. Immun.* 60: 590–595.

Roggen EL, Hoofd G, Dyck E Van, Piot P (1994) Enzyme immunoassays (EIAs) for the detection of anti-*Haemophilus ducreyi* serum IgA, IgG, and IgM antibodies. *Sex. Transm. Dis.* 21: 36–42.

Ronald AR, Albritton WL (1990) Chancroid and *Haemophilus ducreyi*. In: Holmes KK, Mård PA, Sparling PF, Weisner PJ (eds) *Sexually Transmitted Diseases*. New York: McGraw-Hill, pp. 263–271.

San-Mateo LR, Hobbs MM, Kawula TH (1998) Periplasmic copper-zinc superoxide dismutase protects *Haemophilus ducreyi* from exogenous superoxide. *Mol. Microbiol.* 27: 391–404.

Sarafian SK, Knapp JS (1992) Molecular epidemiology, based on plasmid profiles of *Haemophilus ducreyi* infections in the United States. Results of surveillance, 1981-1990. *Sex. Transm. Dis.* 29: 35–38.

Sarafian SK, Woods TC, Knapp JS, Swaminathan B, Morse SA (1991) Molecular characterisation of *Haemophilus ducreyi* by ribosomal DNA fingerprinting. *J. Clin. Microbiol.* 29: 1949–1954.

Schulte JM, Schmid GP (1995) Recommendations for treatment of chancroid, 1993. *Clinical Infectious Diseases* 20: S39–46.

Schweda EKH, Jonasson JA, Jansson P-E (1995) Structural studies of lipooligosaccharides from *Haemophilus ducreyi* ITM 5535, ITM 31 47, and fresh clinical isolate, ACY1: evidence for intrastrain heterogeneity with the production of mutually exclusive sialylated or elongated glycoforms. *J. Bacteriol.* 177: 5316–5321.

Sehgal VN, Shram-Prasad AL (1985) Chancroid or chancroidal ulcers. *Dermatologica* 170: 136–141.

Shah L, Davies HA, Wall RA (1992) Association of *Haemophilus ducreyi* with cell-culture lines. *J. Med. Microbiol.* 37: 268–272.

Sheldon WH, Heyman A (1946) Studies on chancroid. I. Observations on the histology with an evaluation of biopsy as a diagnostic procedure. *Am. J. Pathol.* 22: 415–425.

Spinola SM, Castellazzo A, Shero M, Apicella MA (1990) Characterisation of pili expressed by *Haemophilus ducreyi*. *Microb. Pathog.* 9: 417–426.

Spinola SM, Griffith GE, Bogdan J, Menegus MA (1992) Characterisation of 18,000-molecular-weight outer membrane protein of *Haemophilus ducreyi* that contains a conserved surface-exposed epitope. *Infect. Immun.* 60: 385–391.

Spinola SM, Griffiths GE, Shanks KL, Blake MS (1993) The major outer membrane protein of *Haemophilus ducreyi* is a member of the OmpA family of proteins. *Infect. Immun.* 61: 1346–1351.

Spinola SM, Wild LM, Apicella MA, Gaspari AA, Campagnari AA (1994) Experimental human infection with *Haemophilus ducreyi*. *J. Infect. Dis.* 169: 1146–1150.

Spinola SM, Orazi A, Arno JN *et al.* (1996) *Haemophilus ducreyi* elicits a cutaneous infiltrate of CD4 cells during experimental human infection. *J. Infect. Dis.* 173: 394–402.

Stevens MK, Porcella S, Klesney-Tait J *et al.* (1996) A haemoglobin-binding outer membrane protein is involved in virulence expression by *Haemophilus ducreyi* in animal model. *Infect. Immun.* 64: 1724–1735.

Stevens MK, Kleysney-Tait J, Lumbley S *et al.* (1997) Identification of tandem genes involved in lipooligosaccharide expression by *Haemophilus ducreyi*. *Infect. Immun.* 65: 651–660.

Stevens MK, Latimer SR, Lumbley SR *et al.* (1999) Characterisation of *Haemophilus ducreyi* mutant deficient in expression of cytholethal distending toxin. *Infect. Immun.* 67: 3900–3908.

Stewart S, Ong G, Johanson A, Taylor-Robinson D (1992) Molecular cloning and expression in *Escherichia coli* of a common 28 kD protein antigen of *Haemophilus ducreyi*. *Med. Microbiol. Lett.* 1: 43–48.

Sturm AW, Stolting GJ, Cormane RH, Zanen HC (1987) Clinical and microbiological evaluation of 46 episodes of genital ulceration. *Genitourinary Med.* 63: 98–101.

Sullivan M (1940) Chancroid. *Am. J. Syphilis Gonorrhea Vener. Dis.* 24: 482–521.

Svensson LA, Tarkowski A, Lagergård T (1999) Influence of *Haemophilus ducreyi* cytolethal distending toxin on immune cells. *13th Meeting of ISSTDR 1999*, Abstract 339, p. 192.

Taylor DN, Duangmani C, Suvongse C *et al.* (1984) The role of *Haemophilus ducreyi* in penile ulcers in Bangkok, Thailand. *Sex. Transm. Dis.* 11: 148–151.

Thirumoorthy T, Sng EH, Doraisingham S *et al.* (1986) Purulent penile ulcers of patients in Singapore. *Genitourinary Med.* 62: 253–255.

Totten PA, Lara JC, Norn DV, Stamm WE (1994a) *Haemophilus ducreyi* attaches to and invades human epithelial cells *in vitro*. *Infect. Immun.* 62: 5632–5640.

Totten PA, Morton WR, Knitter GH *et al.* (1994b) A primate model for chancroid. *J. Infect. Dis.* 169: 1284–1290.

Totten PA, Norn DV, Stamm WE (1995) Characterisation of the haemolytic activity of *Haemophilus ducreyi*. *Infect. Immun.* 63: 4409–4416.

Trees DL, Morse SA (1995) Chancroid and *Haemophilus ducreyi*: an update. *Clin. Microbiol. Rev.* 8: 357–375.

Tuffrey M, Alexander F, Ballard RC, Taylor-Robinson D (1990) Characterisation of skin lesions in mice following intradermal inoculation of *Haemophilus ducreyi*. *J. Exp. Med.* 71: 233–244.

Van Dyck E, Piot P (1987) Enzyme profile of *Haemophilus ducreyi* strains isolated on different continents. *Eur. J. Clin. Microbiol.* 6: 40–43.

Ward CK, Lumbley SR, Latimer JL, Cope LD, Hansen EJ (1998) *Haemophilus ducreyi* secretes a filamentous hemagglutinin-like protein. *J. Bacteriol.* 180: 6013–6022.

Wasserheit JN (1992) Epidemiological synergy. Interrelationships between human immunodeficiency virus infection and other sexually transmitted diseases. *Sex. Transm. Dis.* 19: 61–77.

West B, Wilson SM, Changalucha J *et al.* (1995) Simplified PCR for detection of *Haemophilus ducreyi* and diagnosis of chancroid. *J. Clin. Microbiol.* 33: 787–790.

Wood GE, Dutro SM, Totten PA (1999) Target cell range of *Haemophilus ducreyi* hemolysin and its involvement in invasion of human epithelial cells. *Infect. Immun.* 67: 3740–3749.

87

Chlamydia

J.H. Pearce[1] and J.S.H. Gaston[2]

[1]*University of Birmingham, Birmingham, UK*
[2]*Addenbrooke's Hospital, University of Cambridge, Cambridge, UK*

Members of the genus *Chlamydia* are obligate intracellular bacteria that currently comprise the four species, *C. trachomatis*, *C. pneumoniae*, *C. psittaci* and *C. pecorum*. In humans, *C. trachomatis* is responsible for major diseases of the eye (trachoma) and genito-urinary tract, and is now recognised also to cause respiratory infection. *C. psittaci* comprises a highly diverse group responsible for disease in birds and non-primate mammals, and certain strains cause zoonosis in humans. A recently defined small group of strains, *C. pecorum*, cause diseases of cattle and sheep (Fukushi and Hirai, 1992). Major reviews of *C. trachomatis* and *C. psittaci* are those by Storz (1971), Schachter and Dawson (1978), Schachter (1978) and the specialised reviews in Barron (1988). Recent reviews with particular emphasis on *C. trachomatis* and immunological aspects of the subject are those by Ward (1995, 1997), Bavoil *et al.* (1996) and Rank and Bavoil (1996) the mini-review by Raulston (1995) highlights molecular and cellular features.

Progress in understanding chlamydial mechanisms has been greatly handicapped by the lack of a means for gene transfer, by difficulties in production of mutants and by the relative difficulty in growing organisms in quantity. This chapter focuses on cellular and molecular properties of *C. trachomatis* and disease in humans, followed by short sections on human diseases caused by *C. pneumoniae* and *C. psittaci*. Much basic research has been done with certain strains of *C. psittaci* and this has been incorporated into the main section of the chapter.

The species *C. trachomatis* comprises the major trachoma biovar, consisting of serovars A, B, Ba, C, D, E, F, G, H, I, J, K, the lymphogranuloma venereum (LGV) biovar with serovars L1, L2 and L3 – both biovars responsible for disease in humans – and the distinct mouse pathogen, responsible for mouse pneumonitis (MoPn). Members of the trachoma biovar are associated with endemic trachoma (serovars A, B, C) or with oculogenital diseases (serovars D–K) and are essentially pathogens of mucosae. The LGV strains, like serovars D–K, are sexually transmitted and initiate infection at genital mucosal surfaces but are invasive, spreading to local lymph nodes to cause swelling and ulceration. They have other distinctive properties in cell culture.

Transmission of trachoma from humans to orang-utan with conjunctival scrapings was achieved some 90 years ago (Halberstaedter and von Prowazek, 1907). It was accompanied by a description of particles and inclusion bodies in stained smears. The name Chlamydozoa (mantle) was proposed for the

Table 1 Some characteristics of *Chlamydia* species

Characteristics	*C. trachomatis*[a]	*C. pneumoniae*[a]	*C. psittaci*	*C. pecorum*[a]
Natural hosts	Humans	Humans	Birds, mammals	Cattle and sheep
EB morphology	Round	Round or pear shaped	Round	Round
Iodine staining (glycogen presence)	Yes	No	No	No
Sulphonamide sensitivity	Yes	No	No	No
Serovars	At least 15	1	Undefined	3
Characteristic infections	Genital and ocular mucosa	Respiratory tract infections Possible association with heart disease	Pneumonia, abortion	Respiratory, gut and CNS
DNA: Mol % G+C	39.8	40.3	39.6	39.3
Homology % relative to *C. trachomatis*	92			
C. pneumoniae	1–7	94–96		
C. psittaci	1–33	1–8	14–95	
C. pecorum	1–10	10	1–20	88–100

Table adapted from Fukushi and Hirai (1992).

[a] Note that the species *C. trachomatis* includes the distinct strain MoPn, which causes infection in the mouse. Infection of pigs by a *C. trachomatis*-like strain has been described (Zahn *et al.*, 1995), also infection of horses by *C. pneumoniae* (Storey *et al.*, 1993) and koalas (Glassick *et al.*, 1996) and of koalas by *C. pecorum* (Glassick *et al.*, 1996) and pigs (Zahn *et al.*, 1995). Host-range and strain-diversity in the different *Chlamydia* species may be less restricted than previously recognised.

organisms, to describe the appearance of the stained inclusion with its glycogen matrix. Similar organisms were subsequently described in newborns with *ophthalmia neonatorum* but the agent responsible for trachoma was not successfully isolated for another 50 years (T'ang *et al.*, 1957; Collier and Sowa, 1958). Related studies on psittacosis in the 1930s reported similar particles in blood and tissues from birds and human patients (Thygeson, 1934). Subsequent developments led to successful propagation of LGV and psittacosis organisms in the chick embryo and cell culture. Although chlamydiae were isolated from the uterine cervix in 1959, their role in sexually transmitted disease was not defined until the 1970s. By this time the identity of chlamydiae as bacteria had become well accepted as a result of the important contributions of Moulder (1966).

Classification

The order Chlamydiales consists of one family, the Chlamydiaceae, containing one genus, *Chlamydia*, which comprises the four species *C. trachomatis*, *C. psittaci*, *C. pneumoniae* and *C. pecorum*. The characteristics of the four species so far defined indicate considerable strain diversity in *C. psittaci* (**Table 1**), as judged by DNA homology (Fukushi and Hirai, 1989,

1993; Herring, 1992) and the likelihood of further species definition as additional criteria emerge. Analysis of 16S ribosomal RNA sequences (Weisburg *et al.*, 1986) clearly identifies chlamydiae as eubacteria. Taxonomic studies of nucleotide sequences in *ompA*, the gene encoding the common major outer-membrane protein, suggest that chlamydiae have diverged from a common ancestor (Carter *et al.*, 1991; Fitch *et al.*, 1993; Kaltenboeck *et al.*, 1993). It should be noted that a wide range of Chlamydia-like organisms have been reported in invertebrates (Moulder, 1988). Recent studies have revealed the presence of bacterial endosymbionts in acanthamoebae, including organisms that have 86–87% similarity in 16S ribosomal RNA sequences to *Chlamydia* spp. and are believed to constitute a distinct chlamydial genus (Amann *et al.*, 1997).

Identification

Chlamydiae are Gram-negative eubacteria with a genome size of 1045 kb (Birkelund and Stephens, 1992). They are obligate intracellular parasites and exist as spore-like infectious and intracellular metabolically active vegetative forms. The latter undergo successive divisions and differentiate back to infectious forms before release from the host cell. Growth is confined to the so-called 'inclusion membrane', leading to the formation of a micro-colony or inclusion body within the host cell. In *C. trachomatis*, the inclusion contains a glycogen matrix. Chlamydial multiplication depends on the host cell for energy and various nutrients. The envelope structure is unusual for a eubacterium in that muramic acid and conventional peptidoglycan are absent. Non-cultivable chlamydiae of aberrant or bizarre morphology are induced by nutrient deprivation or interferon γ (IFNγ) mediated degradation of intracellular free tryptophan, and these may contribute to pathogenesis.

The Chlamydial Life Cycle and Structures

Chlamydiae have evolved a unique developmental cycle involving two main bacterial forms, the elementary body (EB) and the reticulate body (RB). EBs are the infectious, extracellular, non-replicating form of the organism. In sections viewed in the electron microscope they appear as spherical rigid structures 200–300 nm in diameter. RBs, so-called because of their reticulate structure reflecting ribosomal content, are larger (500–1000 nm) and are the dividing, metabolically active form of the organism; they are non-infectious.

The developmental cycle (**Fig. 1**) in cell culture occurs over approximately 48–72 hours, and usually more rapidly for LGV than for trachoma strains.

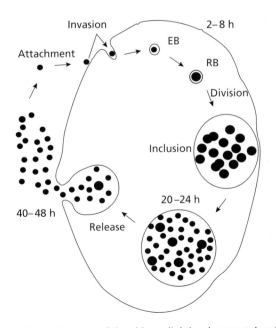

Fig. 1 Schematic view of the chlamydial developmental cycle.

After attachment and entry, the dense-centred EBs are considered to evade the lysosomal compartment, rapidly dissipate their condensed nucleoid (**Fig. 2a**), enlarge and transform to the larger RB form, which then begins to divide. Division is accompanied by the steady expansion of the vacuole (inclusion) membrane (**Fig. 2b**). At about 20 hours, differentiation to EBs begins to be detected (**Fig. 2c**) and EBs steadily accumulate in the inclusion (**Fig. 2d**), although RBs continue to be present. The cycle ends with lysis and the release of organisms. Intermediate condensing forms can often be detected and each RB is generally thought to give rise to a single EB.

Envelope Structure

RB and EB Adaptations

Underlying the developmental cycle is a remarkable alteration in envelope structure when, during RB conversion to EB forms, a novel disulphide-linked protein network is created. Although chlamydiae are Gram-negative eubacteria that possess inner cytoplasmic and outer membranes, they lack muramic acid (Barbour *et al.*, 1982; Fox *et al.*, 1990). They show no evidence of electron-dense peptidoglycan beneath sarkosyl-insoluble outer-membrane preparations that are rich in the 40 kDa major outer-membrane protein (MOMP) (Caldwell *et al.*, 1981). Nevertheless, isolated EBs, unlike RBs, are osmotically stable – in keeping with their capacity for extracellular survival and infectivity. The basis of EB stability appears to reside in the presence in the EB envelope of two cysteine-rich proteins (CRPs) of approximately 12 and 60 kDa, which are cross-linked by disulphide bonds and whose synthesis is initiated at a late cycle stage – about the time RBs differentiate to EBs (Hatch *et al.*, 1984). These findings followed earlier observations that solubilisation of EBs in detergent required the presence of reducing agent (Hatch *et al.*, 1981) and that oligomers of MOMP, a cysteine-containing protein, extracted from EBs undergoes dissociation in the presence of reducing agent (Newhall and Jones, 1983). In addition, restriction of cysteine in growth medium retards development of RB to EB forms (Stirling *et al.*, 1983).

Together with the cysteine-containing MOMP, the 12 kDa (15 kDa in *C. trachomatis*) and 60-kDa CRPs appear to form supramolecular complexes that endow the EB envelope with the rigidity and stability that is absent from RBs. In certain *C. psittaci* and *C. trachomatis* LGV strains the 60-kDa protein appears as a 'doublet' of slightly differing molecular weight, apparently as a result of post-translational

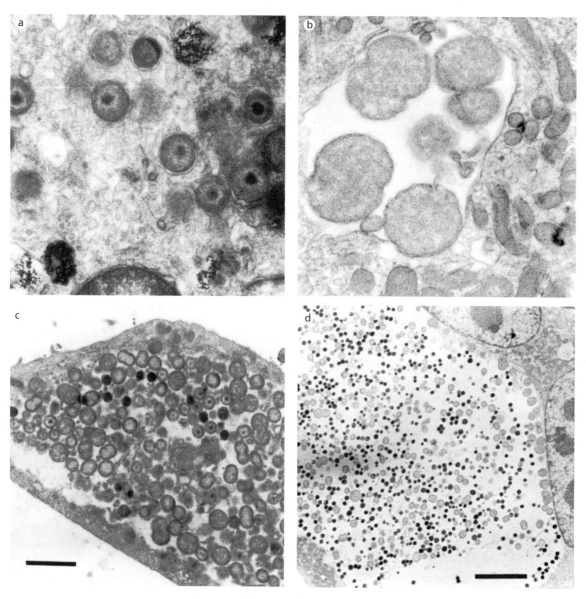

Fig. 2 Cycle stages seen by electron microscopy: (a) Intracellular chlamydiae clustered in the perinuclear area, 4 hours after infection. One EB with characteristic dense nucleoid centre (open arrowhead) is seen together with other chlamydiae (filled arrowheads) which are enlarged and whose nucleoid centres are diminished. The dense granular masses (bent arrows) are thorotrast-labelled secondary lysosomes; these have not associated with the chlamydiae (bar 0.5 μm). (b) Differentiation to large reproductive RBs has occurred, together with division, and the vacuole membrane has expanded (magnification × 12 500, bar marker 1 μm). (c) An inclusion at 28 hours after infection with small numbers of EBs and also dense-centred condensing forms present (bar 2 μm). (d) EBs accumulating within the inclusion (bar 5 μm). Figures 2a, b and d from Pearce, 1986.

modification of the larger form (Allen and Stephens, 1989). Further analysis of cycle events has shown that disulphide cross-linking of the reduced forms of the CRPs and of the cysteine-containing MOMP occurs, either progressively during cycle completion (LGV strain; Newhall, 1987) or at release (*C. psittaci* 6BC; Hatch *et al.*, 1986).

Envelope Architecture

The CRPs were initially thought to be outer-membrane proteins because of their presence in sarkosyl-insoluble extracts. Their exposure on the surface has not, however, been demonstrated with specific antibodies (Collett *et al.*, 1989; Watson *et al.*, 1994). The small CRP has proved to be a lipoprotein and the

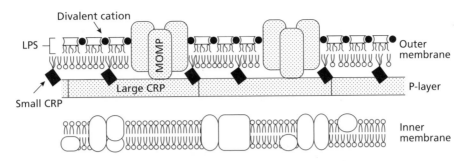

Fig. 3 Model of the EB envelope (Hatch, 1996). Only the MOMP, the large CRP and the small CRP are shown in the outer membrane and the periplasm. The actual shapes of the proteins and the existence of specific interpeptide cross-links have not been established. The P (periplasm) layer, which may consist entirely of cross-linked large CRP, has been observed by electron microscopy (Matsumoto, 1988; Matsumoto and Manire, 1970; Miyashita *et al.*, 1993).

lipid moiety appears to enhance migration in SDS–PAGE, because the protein has a deduced molecular weight of ~ 9000 for both *C. psittaci* and *C. trachomatis* (Everett and Hatch, 1991). Everett and Hatch (1995) showed that an affinity probe which labels protein moieties in contact with hydrophobic environments selectively labels MOMP and the small CRP, but not the large CRP. Hatch (1996) proposed a model of the chlamydial envelope (**Fig. 3**) based on these findings, in which the small CRP is located in the periplasm but is anchored to the outer membrane by its lipid, whereas the large CRP is in the periplasm. This proposal fits with earlier information indicating protein in a regular hexagonal array on the inner surface of the chlamydial outer membrane, which is analogous to the surface S layers found with certain bacteria (Matsumoto and Manire, 1970; Miyashita *et al.*, 1993). The nomenclature EnvA and EnvB, for the small and large CRPs, respectively, has been adopted by Everett and Hatch (1995), to reflect their location and operon order (Lambden *et al.*, 1990), rather than designations as outer-membrane proteins.

The proposed model may not precisely fit the diversity of chlamydial strains present in *C. psittaci*. Data for the GPIC strain indicate that trypsin treatment of EBs yields fragments of EnvB rather than MOMP – as has been found for *C. trachomatis* (see below) – suggesting that at least part of the CRP can be surface-exposed (Ting *et al.*, 1995).

Penicillin-binding Proteins

Despite the absence of muramic acid and peptidoglycan, chlamydiae possess penicillin-binding proteins (PBPs) (Barbour *et al.*, 1982) and are sensitive to penicillin, but not eradicated by it. Treatment of inoculated cell cultures induces the formation of swollen abnormal reticulate forms and various bizarre structures; similar effects occur on treatment with cycloserine, a D-alanine analogue. The paradox of

PBPs and effects of penicillin/cycloserine in the absence of muramic acid or any obvious peptidoglycan target has been reviewed by Moulder (1993). The evident alterations induced by penicillin and cycloserine argue for a functional role of PBPs in envelope biosynthesis. Evidence from the recently sequenced chlamydial genome strongly supports this (Chopra *et al.*, 1998).

Surface Projections

Patches of spike-like projections in hexagonal array have been reported to be clustered on one hemispheric surface of EBs and RBs (Matsumoto, 1988; Miyashita *et al.*, 1993) giving both forms structural polarity (**Fig. 4a**). Matsumoto has presented evidence that the spikes are hollow and pass outwards from anchorages in the cytoplasmic membrane. For organisms attached to the vacuolar inclusion membrane, the spikes appear to project through into the host cytoplasm (**Fig. 4b**), raising the possibility of some form of communication channel. Nothing further is known.

Cell Biology of the Growth Cycle

Although some progress has been made in defining chlamydial ligand structures involved in attachment, little is known of the molecular interactions that mediate entry into host cells and post-entry events. Indeed, nowhere in the study of *Chlamydia* has the lack of a system for gene deletion and exchange, or the ready availability of mutants, been more keenly felt.

Infectivity and Surface Properties

Early biological, structural and physiological studies made use of *C. psittaci* strains 6BC and CAL 10 (Mn). These strains are laboratory adapted and highly infectious in cell culture, and large-scale cultivation is possible in mouse L cell suspension cultures. Trachoma strains of *C. trachomatis*, however, in distinction from

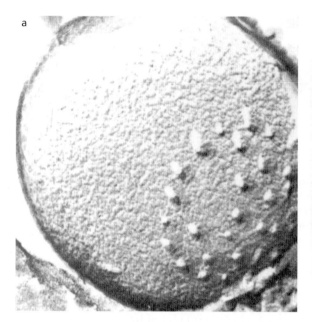
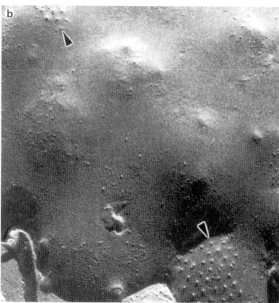

Fig. 4 (a) A localised patch of projections in hexagonal array on the EB surface, seen after freeze etching (magnification × 100 000). From Wyrick and Davis (1984). (b) Freeze-replica of an inclusion-bearing cell after cleavage to expose the convex cytoplasmic face of the inclusion membrane. The particle arrays (arrowheads) from what appear to be RBs underlying the membrane suggest the possibility of connection between the RB and the host cytoplasm. From Miyashita *et al.* (1993).

LGV, are poorly infectious for cell culture systems such as HeLa, and McCoy (mouse L) lines, and this is commonly circumvented by the use of centrifugation of inocula on to monolayers. Although both biovars have a similar net negative surface charge (pI 4.6, Kraaipoel and van Duin, 1979), significant surface differences are evident from the studies of Kuo *et al.* (1973) and Kuo and Grayston (1976). Pre-treatment of cell cultures with polycations, such as DEAE–dextran, selectively enhances attachment and infectivity of trachoma strains. In contrast, heparin (anionic) partially inhibits attachment of both biovars and markedly reduces infectivity of the trachoma strain at a post-attachment stage. Analysis of surface charge and hydrophobicity by phase partitioning (Soderlund and Kihlstrom, 1982) indicates a greater net negative charge on trachoma strains – which may account for the enhancing effect of DEAE-dextran – but similar levels of hydrophobicity. Interestingly, although MOMPS of both biovars have similar weakly acidic isoelectric points (5.3–5.5), the 12 kDa and 60 kDa CRPs, respectively, have a pI of 5.4 and > 8.5 for LGV and 6.9 and 7.3–7.7 for trachoma strains (Batteiger *et al.*, 1985). Present evidence suggests that the CRPs in the two biovars are not surface-exposed, but the 60-kDa CRP may contribute to GPIC surface properties (see above).

The empirically adopted technique of inoculum centrifugation, which is still in use for culture and diagnostic work, enhances trachoma strain infectivity at both attachment and post-attachment stages of cell culture infection (Kuo and Grayston, 1976). Curiously, the guinea-pig inclusion conjunctivitis (GPIC) strain of *C. psittaci*, a useful model of conjunctival and genital tract infection, also resembles trachoma strains in its behaviour in cell culture. Centrifugation increases its infectivity both at attachment and during undefined post-entry events (Prain and Pearce, 1989).

Attachment

The MOMP and a recently defined glycosaminoglycan are the two components so far recognised as most likely to mediate interaction of *Chlamydia* with the host surface. In addition, Joseph and Bose (1991) defined, but did not subsequently pursue, a 38-kDa protein involved in attachment. Also Wyrick and colleagues (Raulston *et al.*, 1993; Schmiel *et al.*, 1994) have used the approach taken by Isberg and Falkow to identify the *Yersinia* invasin protein. A recombinant, with the capacity for attachment to endometrial cultures parasitised by chlamydiae, was isolated from a chlamydial genomic library in *Escherichia coli*. The plasmid insert contained chlamydial proteins of 28 and 82 kDa with homology, respectively, to GrpE- and DnaK-like heat-shock proteins (Hsps). Both proteins were present in sarkosyl-insoluble as well as soluble fractions of chlamydiae and the recombinant. The DnaK-like protein in chlamydiae has been reported to induce neutralising antibody (Zhong and Brunham, 1992) suggesting surface exposure, although not detected by Birkelund *et al.* (1989). The conferring of attachment properties on the *E. coli* recombinant is highly suggestive. Given the chaperone properties of the two proteins, it remains to be seen whether they play direct or indirect modulatory roles in chlamydial attachment.

Glycosaminoglycan Synthesis of this heparan sulphate-like structure by chlamydiae was discovered by Zhang and Stephens (1992) who proposed that, in attachment, the chlamydial glycosaminoglycan (GAG) spans between a chlamydial acceptor molecule and a host-cell receptor. Using both heparin and heparan sulphate glycosaminoglycans they confirmed the earlier findings of Kuo that pre-treatment of host cells resulted in parallel loss of attachment and infectivity of an LGV strain. Pre-treatment of organisms with heparitinase, an enzyme lyase that degrades heparan structures, however, also inhibited attachment and infectivity – implying the presence of a heparan-like structure on the LGV native surface. The heparitinase-dependent loss of attachment and infectivity could be restored by exposure of treated organisms to heparin or heparan but not to other unrelated GAGs, and similar specificity was established for the lyase. Evidence crucial to the proposal was that chlamydiae, despite their ability to bind additional exogenous heparin or heparan, synthesised their own GAG. This was demonstrated by showing that chlamydiae are reactive to monoclonal antibody-specific for heparan sulphate after growth in CHO 761, a host-cell mutant producing only 5% of normal levels of heparan sulphate. Moreover, chlamydial GAG could be isolated from infected mutant cells.

Application of the analysis to trachoma biovar strains (Chen and Stephens, 1994, 1997) and to *C. psittaci* GPIC (Gutierrez-Martin *et al.*, 1997) has given a less clear-cut picture. Heparitinase treatment abolishes the infectivity of trachoma strains and this is restored by heparan, but the lyase does not inhibit all the attachment. This suggests that such strains have both a GAG-dependent and a GAG-independent mode of attachment but that the trachoma strain relies on a GAG structure to mediate a post-attachment step essential for productive infection. Rather similar conclusions were drawn for GPIC from related lines of experimentation. Examination of the effect of pH on GPIC attachment, however, suggested that the GAG-independent mode was favoured by an increase of pH to 7.6–8.0. This intriguing observation would seem to imply that use of one or other attachment mechanism may reflect chlamydial adaptation to different pH environments.

Trial of different GAG ligands for their ability to compete with chlamydiae for binding to host receptors or to compete with heparin for binding to chlamydiae has given information about the likely properties of the chlamydial GAG (Chen *et al.*, 1996). A disaccharide chain length fulfilled both binding requirements but moieties mediating association with the putative chlamydial acceptor and the host-cell receptor appeared to differ.

Major Outer-membrane Protein MOMP, the dominant protein in the outer membrane, is estimated to account for some 60% of membrane protein. It is the major candidate for any chlamydial vaccine and has been studied extensively. Early serological studies indicated serovar, subspecies and species specificities are present in MOMP. This was given meaning once several MOMP biovar and serovar sequences became available. Work with appropriate monoclonal antibodies in the laboratories of Stephens, Caldwell and Ward (Baehr *et al.*, 1988; Conlan *et al.*, 1988; Stephens *et al.*, 1988a), identified the location and epitope properties of all three categories of specificity (**Fig. 5**). The data indicate the presence of five constant and four variable sequences (VS I–IV), visualised as exposed loops stabilised by disulphide bonds. VS I and VS II have regions that determine serovar specificity, VS III appears to be poorly surface-exposed and VS IV, the largest variable region, contains subspecies and species specificity. The latter is represented by an

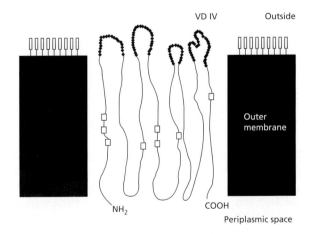

Fig. 5 Schematic diagram of the MOMP within the chlamydial outer membrane. The solid line indicates the membrane-embedded portion of the polypeptide, with the open squares representing conserved cysteine residues. The closed squares represent the exposed variable regions (VD I–IV) with the invariant sequence in VD IV indicated as lighter closed squares. Not to scale. Adapted from Baehr *et al.* (1988).

invariant 9 amino acid sequence, TTLNPTIAG, pictured as a cleft (**Fig. 5**), because of its hydrophobic properties. Substitutions in amino acids in the variable regions of different serovars appear to preserve net negative charge and hydrophilicity, so that changes in these most exposed regions of MOMP do not affect attachment properties.

The key evidence for MOMP as an attachment protein centres on two sets of findings (Su *et al.*, 1988, 1990a). First, that MOMP integrity is essential to attachment and infectivity. Trypsin treatment of trachoma strain B and an LGV strain leads to loss of attachment and infectivity for B but is without effect on LGV. Analysis of products revealed that cleavage at single lysine residues in B at VS II (serovar) and VS IV (subspecies) leads to significant release of MOMP fragments, but not other chlamydial protein. In contrast, only traces of MOMP degradation are detectable for LGV, with cleavage confined to VS II and the VS IV lysine apparently inaccessible. Second, in strain B, monoclonal antibodies to sequences encompassing either the VS II or VS IV cleavage sites inhibits attachment and infectivity, indicating the essential contribution of these regions to binding of chlamydiae to host surfaces. The invariant hydrophobic sequence responsible for species specificity may also contribute to attachment. Heating of organisms at 56°C for 30 minutes results in attachment loss and its concomitant exposure but has little effect on antibody-binding to the serovar and subspecies regions.

Another MOMP contribution to attachment may come from its recently defined glycan substituent (Kuo *et al.*, 1996). This has been identified as an N-linked, predominantly high mannose-containing oligosaccharide with minor complex branched structures and terminal galactose. Inhibition studies resulted in up to 75% reduction in attachment by the most common 9-mannose-residue oligosaccharide. With the identity of the glycan now defined, it will be possible to look directly for glycan exposure on the EB surface. Su *et al.* (1990a) estimate 2.9×10^4 sites per EB for VS II and VS IV, so at one glycan moiety per MOMP molecule there should potentially be comparable surface exposure. A possible glycosylation site for location of the glycan is present in the invariant sequence at NPT within VS IV.

Receptors for Attachment

Early work led to proposals for various carbohydrate receptors on host cells, but these were not sustained in the face of counter-evidence (Moulder, 1991). Data from Su *et al.* (1996) are, however, persuasive that, remarkably, in the light of the GAG findings, heparan

sulphate residues on HeLa cells are receptors for the MOMP. MOMP used as a fusion protein with maltose-binding protein (MBP–MOMP) – a system noted for enhancing conformation of the target protein – was shown to compete with EBs for binding to host cells. Furthermore, attachment is inhibited by heparin or heparan sulphate and by heparitinase treatment of host cells, which implies that host-cell receptors for MBP–MOMP are the heparan sulphate glycans commonly found on mammalian cell surfaces. Confirmation of this was provided by the demonstration that MBP–MOMP (and EBs) showed only low-level attachment to Chinese hamster overay (CHO) cell mutants from which heparan sulphate is almost absent.

A role for integrins as receptors for chlamydial attachment to Hec-1b (human endometrial) cells in non-polarised state has been explored (Wyrick *et al.*, 1994). This protein family, which links a variety of extracellular matrix (cell surface) proteins with the internal cytoskeleton, has been identified as a target for attachment by *Yersinia pseudotuberculosis* during invasion of host cells. Specific antibodies detected integrin of the $\alpha\beta1$ classes but did not inhibit chlamydial attachment or infectivity.

Entry and Post-entry Events

Entry Mechanism and Signalling Detailed study by Ward and Murray (1984) provided the first strong evidence for microfilament-dependent uptake, with 50% uptake inhibition by cytochalasin D, the efficient microfilament inhibitor. Initiation of entry from clathrin-coated pits – a microfilament-independent mechanism – was first recognised by Hodinka and Wyrick (1986) and Hodinka *et al.* (1988). Subsequent analysis has provided evidence that both mechanisms can operate simultaneously to mediate chlamydial entry (Reynolds and Pearce, 1990, 1991). Available data for LGV and the trachoma-like GPIC do not provide any evidence that the entry mechanism is strain-dependent. Intracellular GPIC associates to a significant extent with the host lysosomal compartment, however, except when entry occurs during centrifugation (Prain and Pearce, 1989). These phenomena, initially attributed to differences in entry mechanism, may rather depend on critical interactions with the host surface during centrifugation-induced entry which determine activation of chlamydial metabolism (see below).

Birkelund *et al.* (1994a) found evidence of tyrosine kinase activity during LGV entry into HeLa cells, with three molecular weight classes of protein showing phosphorylation. Phosphorylation sites coincided with the location of organisms and subsequently clustered in the inclusion membrane. Entry did not result in stimulation of epidermal growth factor (EGF) receptor, as found during salmonella entry into Henle cells (Galan *et al.*, 1992). Curiously, LGV entry into McCoy cells is accompanied by massive stimulation of fluid uptake (Reynolds and Pearce, 1990), a phenomenon detected when EGF stimulates receptor-rich cells (West *et al.*, 1989), but phosphorylation has not been studied in the McCoy cell model.

Vacuole Migration It been shown that the early intracellular vacuole does not become acidified (Heinzen *et al.*, 1996; Schramm *et al.*, 1996), a requirement anticipated for chlamydial intracellular survival. Multiple entry of organisms is well recognised to be followed by rapid migration of chlamydial vacuoles to the perinuclear region of the host-cell followed by fusion in the case of *C. trachomatis* strains. Inhibition of migration by cytochalasin has been attributed to interference with actin filament function. Migration and fusion are also dependent on calcium mobilisation (Majeed and Kihlstrom, 1991; Majeed *et al.*, 1993) and involve selective translocation of members of the annexin family of calcium- and membrane-binding proteins (Majeed *et al.*, 1994). Using the same LGV and trachoma strains Schramm and Wyrick (1995) have observed *Chlamydia*-dependent differences in outcome. Inhibitors of both actin filament and microtubule function delayed aggregation and fusion of LGV but not trachoma strain vacuoles. The reason for the different behaviour of LGV remains unclear.

Vacuole Migration and Chlamydial Metabolism A remarkable series of studies by Hackstadt and colleagues has defined key early events that follow LGV strain entry. Endocytosed fluorescent dye markers fail to become associated with newly endocytosed chlamydial vacuoles – indicating that the latter rapidly become isolated from the endocytic pathway normal to all host cells (Heinzen *et al.*, 1996). Consistent with this change, markers of fusion with lysosomes also do not appear in chlamydial vacuole membranes. Apparently critical for the changes in chlamydial vacuole properties is the almost immediate onset of chlamydial metabolism after entry. This implies that some key early event activates metabolism and raises new questions about how metabolic activity is triggered. The presence of chloramphenicol, tetracycline or rifampin results in failure of the vacuole to isolate itself from the endocytic pathway and leads to subsequent association with the lysosomal compartment (Scidmore *et al.*, 1996a). Correspondingly, migration of vacuoles to the

perinuclear region and fusion of multiple vacuoles to form an enlarged inclusion does not take place.

Interaction with the Golgi complex The same workers (Hackstadt *et al.*, 1995, 1996) demonstrated that perinuclear migration leads to a previously unrecognised interaction with the Golgi complex, and interception of ∼50% of newly synthesised Golgi lipid apparently for chlamydial development. Incubation of infected HeLa cells with a fluorescent ceramide probe led to its transfer to the Golgi, the synthesis of fluorescent sphingomyelin and subsequent transfer of Golgi vesicles to the adjacent chlamydial inclusion. Traffic could be demonstrated within 1–2 hours after infection and lipid labelling of organisms occurred rapidly, probably during close contact with the inclusion membrane. Inhibition of lipid transfer occurred on treatment of host cells with brefeldin A – which disrupts the Golgi apparatus – and also by depletion of host ATP pools. Further analysis indicates that routeing of Golgi products to the inclusion is selective for lipid and appears likely to exclude all glycoprotein synthesised by the Golgi (Scidmore *et al.*, 1996b). These novel findings force a revision of the view that the inclusion is essentially an autonomous structure synthesising all its complex molecules from simple nutrients. They also highlight the still unresolved question of how the inclusion receives its energy supply.

Early Changes in Chlamydial Structure and Induction of Metabolism

Dispersal of EB nucleoid and concomitant changes in chromosomal organisation must be initiated early in the cycle – probably after induction of undefined metabolic events which now appear to be crucial for intracellular survival (Scidmore *et al.*, 1996a). Envelope disulphide-bond reduction is detectable for *C. psittaci* 6BC within 1 hour of initiation of infection and is dependent on chlamydial metabolism (Hatch *et al.*, 1986). Relevant here is a remarkable feature of the MOMP-containing outer-membrane complex; it behaves as a porin-like structure that is controlled by its oxidation–reduction state. Test of porin function for outer-membrane complex in liposomes shows that activity is dependent on incorporation in a reduced state; the native oxidised form is inactive (Bavoil *et al.*, 1984). Hence the early reduction detected by Hatch *et al.* (1986) is consistent with the occurrence of nutrient uptake and onset of metabolic activity.

Lundemose *et al.* (1990) used pulse radio-labelling and reported the synthesis of seven proteins within 2–4 hours after infection; one was identified as an S1 ribosomal protein, the another two as Hsps. In a different approach 6BC EBs were isolated 1–2 hours after infection and their newly synthesised RNA was used to probe a genomic library to identify activated genes (Crenshaw *et al.*, 1990; Wichlan and Hatch, 1993). One of several early genes, for which no known sequence homologues could be found, was preferentially and strongly expressed at 1 hour. Another was found to possess a lysine-rich N-terminus and highly hydrophobic C-terminus and was thought to be a membrane-bound protein whose role might be to aid dissociation of DNA-binding proteins (see below) during nucleoid dispersal. A third sequence was homologous to glutamyl t-RNA synthetase.

Mid and Late Cycle Events

As indicated earlier, transformation of EB to RB is complete within about 8 hours after infection and RB division continues uninterrupted until 20–24 hours after infection, when synthesis of the two CRPs is detected. At about the same time synthesis of two histone-like DNA-binding proteins, Hc1 and Hc2 (Hc1 only in *C. psittaci* strains) is also initiated and intermediate forms with 'condensing' nucleoid centres become evident within the inclusion.

During this period the inclusion membrane expands considerably as it maintains separation of the growing micro-colony from the host cytoplasm. Nothing is known of its transport properties other than the finding that it excludes small molecules of 520 Da (Heinzen and Hackstadt, 1997). The expansion occurs in the absence of cycloheximide-inhibitable host protein synthesis, and may receive a contribution of lipid from the Golgi, although this was not detected for sphingomyelin labelled with a fluorescent marker (Hackstadt *et al.*, 1996). Recent work indicates that chlamydial proteins become inserted. For *C. psittaci* GPIC a single protein, Inc A, has been characterised (Rockey *et al.*, 1995). It is found in GPIC RBs but not EBs and is first detected in the inclusion membrane at about 18 hours after infection. Using specific markers, Taraska *et al.* (1996) failed to detect proteins from host endoplasmic reticulum, Golgi or endosomal or lysosomal sources in *C. psittaci* 6BC inclusion membranes. The latter did react with antisera raised against membrane preparations from infected host cells, indicating the presence of unspecified chlamydial antigen, but how extensively is not yet clear. It seems probable that well-defined chlamydial surface antigens are absent, but evidence has been obtained for the presence of chlamydial LPS (see below).

Release

Little is known of this final stage of the infection cycle. In cell culture, release of chlamydiae occurs predominantly by host lysis but nothing is known of the changes that precede the event. Curiously, superficially similar membrane breakdown occurs after ingestion of large numbers of EBs (> 100 inclusion forming units/cell), and is referred to as 'immediate cytotoxicity' (Moulder *et al.*, 1976).

The intact inclusion, which contains organisms, can be released from infected cells, followed by plasma membrane re-sealing and viability retention by the host-cell (Todd and Caldwell, 1985). Additional mechanisms have been described for a bovine *C. psittaci* strain, *in vivo*, where rupture of mucosal cells and release of organisms into the gut lumen, extrusion of intact infected cells or inclusions, or the release of vesicles containing small numbers of chlamydiae was observed (Doughri *et al.*, 1972). Wyrick *et al.* (1993) described release from alternate regions of polarised endometrial cells. A trachoma serovar exited only from apical surfaces, allowing spread of infection to apical regions of adjacent mucosal cells. LGV was released from the basolateral surface, consistent with its ability to pass through to serosal and vascular locations.

It is significant for pathogenesis that direct cell to cell spread has never been reported. Division of inclusions during host cell division has been described, however (Campbell *et al.*, 1989), and may be important for persistence in crypt regions of glandular epithelia or in regenerating tissue surfaces such as the endometrium.

Other Molecular Components

DNA-binding Proteins

Synthesis of Hc1 and Hc2, two basic histone-like proteins, is associated with initiation of chromosomal condensation at the late-cycle stage described earlier. They were first reported by Wagar and Stephens (1988), and were subsequently cloned in several laboratories. The 18-kDa Hc1 is found in all *C. trachomatis* strains. The Hc2 varies from 23 to 32 kDa in *C. trachomatis* and is apparently absent from *C. psittaci*. Expression of Hc1 in *E. coli* causes nucleoid condensation and the fractionation of lysed bacteria revealed DNA sedimenting with Hc1 (Barry *et al.*, 1992). Moreover, Hc1 expression proved to be self-limiting and resulted in down-regulation of transcription, translation and replication (Barry *et al.*, 1993).

The contribution made by each protein has been studied by Christiansen *et al.* (1993), who showed by electron microscopy that DNA preparations complex with Hc1 *in vitro* to form large compact aggregates, with an apparent preference for supercoiled DNA. Inhibition of transcription and translation *in vitro* by Hc1 is also accompanied by the formation of condensed DNA and RNA complexes (Pedersen *et al.*, 1994). Analysis of function revealed domains in the

N- and C-terminal halves of Hc1. Only the C-terminal fragment mediated binding to DNA and RNA; the N-terminal fragment showed high α helix content predicted to contain a putative hydrophobic interface for protein–protein interaction. The N-terminal fragment also contained a site for protein dimerisation, which is common in prokaryote histone-like proteins. Hc1–DNA complex formation was not fully reproduced by replacement with C-terminal fragment, and the N-terminal fragment is proposed also to contribute to binding to DNA (Pedersen *et al.*, 1996a). Similar findings for the C-terminal fragment have been reported by Remacha *et al.* (1996).

Study of Hc2 has been more difficult because of its instability. Since nucleoid condensation occurs in *C. psittaci* with Hc1 alone, Hc2 may have a different role, possibly in gene regulation. Recent findings support this (Pedersen *et al.*, 1996b) with evidence that, although Hc2 is capable of DNA condensation, it is less efficient than Hc1. On the other hand, Hc2 was more efficient than Hc1 in binding RNA and repressing translation and markedly more effective in binding to linearised DNA and inhibiting transcription.

Heat-Shock Proteins

Chlamydial proteins with homology to *E. coli* Hsps have been sequenced from a number of strains. These include GroEL/ES (serovar A: Morrison *et al.*, 1990; serovar L2: Cerrone *et al.*, 1991) and DnaK (serovar L2: Birkelund *et al.*, 1990; serovar D: Danilition *et al.*, 1990) and also the Grp homologue (Schmiel *et al.*, 1994). Major interest in the chlamydial Hsps centres on their role in immune responses – they are highly immunogenic – and, in particular, the significance of the GroEL homologue (Hsp 60) in contributing to host damage by induced T cell responses. As indicated previously the GroEL and DnaK homologues are synthesised early in development and are likely to be important for protein assembly, folding and subsequent stability (Lundemose *et al.*, 1990).

The Chlamydial Mip-like Protein

Discovery of the chlamydial Mip-like protein (Mip) by Lundemose *et al.* (1991) excited interest because of its similarity to the macrophage infectivity potentiator (Mip) protein already defined for legionellae and implicated in their survival within macrophages. It is present in both EBs and RBs and is the second chlamydial lipoprotein to be identified (Lundemose *et al.*, 1993a). It is poorly surface-exposed, being undetectable by immunocytochemical means, although infectivity is neutralised by complement-mediated antibody – suggesting some degree of surface exposure (Lundemose *et al.*, 1992). Its novel

feature is the possession of peptidyl-prolyl-isomerase (PPIase) activity (Lundemose *et al.*, 1993b) similar to that found in the FKBP protein family – so-called because the enzyme activity is inhibited by the immunosuppressive drug FK506. PPIase 'immunophilins' modulate intracellular signalling pathways, of which the best characterised are those affecting T-cell activation (Crabtree and Clipstone, 1994).

Use of immunosuppressive drugs to block PPIase activity during chlamydial infection gave up to 87% inhibition of infectivity in the first 16 hours after infection (Lundemose *et al.*, 1993b). Residual inclusions contained highly abnormal organisms typical of those produced under nutrient deprivation (Coles *et al.*, 1993). Given the presence of a signal peptidase recognition sequence at the N-terminus identical to that in the pullulanase protein of *Klebsiella pneumoniae* (Lundemose *et al.*, 1993a) it was speculated that the chlamydial Mip protein may be inserted into the inner membrane and translocated to the surface. Mammalian PPIases have been speculated to regulate membrane transport channels, and the finding of chlamydial morphology typical of nutrient starvation led to the suggestion that chlamydial PPIase activity may regulate nutrient entry through the inclusion membrane (Lundemose *et al.*, 1993b). A similar protein has been described for *C. psittaci* GPIC (Rockey *et al.*, 1996), but nothing further is known of the mode of action of Mip.

Plasmids

Most chlamydiae carry a plasmid of approximately 7.5 kb that has eight major open reading frames (ORFs). Plasmids of *C. trachomatis* were the first sequenced (Sriprakash and Macavoy, 1987; Hatt *et al.*, 1988; Comanducci *et al.*, 1988) and more than 99% sequence homology was found in two LGV and two trachoma serovar comparisons (Comanducci *et al.*, 1990). Homology within *C. psittaci* strains is lower, with five different restriction site profiles defined (Lusher *et al.*, 1991).

Examination of ORFs 7 and 8 shows that they are homologous and have C-terminal sequences with homology to the integrase family of site-specific recombinases. These integrases are likely to be important for plasmid function, yet appear to be regulated by different sigma factors (Ricci *et al.*, 1993).

Of particular interest is ORF 3 which encodes a 28-kDa polypeptide, pgp3, that has been detected in the EB envelope with specific antisera (Comanducci *et al.*, 1993). Using recombinant ORF 3 protein as antigen, serum antibody responses were detected in some 80% of patients with STD who were seropositive for

C. trachomatis (Comanducci *et al.*, 1994). More recently, sera from HIV-seropositive patients with a 30% seroprevalence of *C. trachomatis* showed high positivity (83%) to pgp3 (Ratti *et al.*, 1995), implying that pgp3 may be unusually immunogenic in comparison with other chlamydial antigens.

Plasmid is present in all *C. trachomatis* serovars with an estimated 10 copies per EB (Palmer and Falkow, 1986), but one isolate (LGV) has been found which lacked plasmid. All avian and most mammalian *C. psittaci* strains contain the plasmid. Not all laboratory stocks of the common *C. psittaci* laboratory strain Cal10 (Mn) have plasmid (Joseph *et al.*, 1986); *C. pneumoniae* strains do not. These findings argue against an essential role in chlamydial survival, but plasmid presence in divergent chlamydial species suggests they make some contribution to host maintenance. The finding that pgp3 is present in EBs provides the first evidence that the plasmid is not cryptic in chlamydial strains.

Lipopolysaccharide

Lipopolysaccharide (LPS) is a major component of the chlamydial outer membrane and carries the heat-stable genus-specific glycolipid antigen. It is phenotypically of the rough form in lacking an O-chain and is similar to the LPS of Re mutant bacteria (Nurminen *et al.*, 1983; Brade *et al.*, 1987a). The saccharide portion is a trisaccharide of 3-deoxy-D-manno-octulosonic acid (Kdo) of sequence αKdo-(2→8)-αKdo-(2→4)-αKdo and of which the α(2→8)-linked saccharide is immunodominant and unique to *Chlamydia*. The trisaccharide has at least two additional antigenic determinants common to other Gram-negative bacteria, in addition to the *Chlamydia* genus-specific determinant (Brade *et al.*, 1987b).

Recombinants from a chlamydial library react with antibody to the genus-specific epitope and are hypothesised to contain the specific Kdo transferase (Nano and Caldwell, 1985). The transferase for *C. trachomatis* was successfully cloned by Belunis *et al.* (1992) and subsequently those for *C. psittaci* (Mamat *et al.*, 1993) and for *C. pneumoniae* (Lobau *et al.*, 1995) were also cloned. The three transferases share only 60% similarity in DNA and deduced amino acid sequences (Lobau *et al.*, 1995). Chlamydiae have been isolated that possess phenotypically smooth LPS that does not react with genus-specific antibody and appears to undergo phase-variation (Lukacova *et al.*, 1994); the extent of their distribution is not yet clear.

LPS is surface-exposed on EBs and RBs and, by cross-linking studies, is in close proximity to the MOMP, probably also stabilising trimeric MOMP complexes (Birkelund *et al.*, 1988) and potentially able to modulate exposure of MOMP antigenic sites (Vretou *et al.*, 1992). Release of LPS from the chlamydial surface appears to occur and is enhanced by exposure to anti-LPS antibody (Birkelund *et al.*, 1989). LPS is also implicated in mediating nitric oxide induction during chlamydial infection of McCoy (mouse) cells (Devitt *et al.*, 1996). Excretion of LPS from the inclusion to the host cell surface during chlamydial infection has been described (Richmond and Stirling, 1981; Campbell *et al.*, 1994) and was confirmed in other work (Karimi *et al.*, 1989; Hearn and McNabb, 1991; Wyrick *et al.*, 1993), but disputed by Baumann *et al.* (1992). The significance of the release is unclear. Recent evidence suggests, however, that chlamydial LPS plays a major role in the induction of pro-inflammatory responses (Ingalls *et al.*, 1995) and LPS excretion from the host cell surface may make an important contribution prior to organism release from infected host cells.

Undefined Glycolipid

This undefined glycolipid (Glxa) has genus specificity and is distinct from LPS (Stuart *et al.*, 1991, 1994). It is shed into supernatants from infected cell cultures and reacts with patient antisera. Its excretion together with LPS (Wyrick *et al.*, 1993a) has been observed in studies on infected endometrial cells and it is inhibited on dissociation of the Golgi complex with brefeldin A. The finding of lipid traffic to the chlamydial inclusion, described earlier, adds support for the possibility of excretion of this material. At present the point of greatest interest is that anti-idiotype antibody to Glxa gives significant protection against chlamydia-induced eye infection in a mouse model (Whittum-Hudson *et al.*, 1996) and may have potential value in a vaccine against trachoma.

Physiology

Metabolism

Studies from the 1960s onwards have succeeded in identifying key limitations of chlamydial metabolism, but progress in this area has been slow, largely because of the inability of chlamydiae to display independent growth and the lack of a system for genetic analysis. A central feature of chlamydial parasitism is the apparent inability of chlamydiae to generate net ATP production and the proposal, unique for bacteria, that they are 'energy parasites' (Moulder, 1962). Studies in this area have focused on properties of host-free chlamydial growth (Hatch, 1988) and on nucleoside metabolism (McClarty, 1994). The reader is referred to these reviews and to the wide-ranging survey of Moulder (1991) for a more detailed treatment.

Obligate intracellular parasitism is generally associated with the absence of unnecessary biosynthetic pathways requiring the expenditure of energy. This, it has been suggested, accounts for the small genome size of *Chlamydia* – about one-quarter that of *E. coli* (Birkelund and Stephens, 1992). McClarty (1994) has, however, pointed out that chlamydiae must retain the ability to synthesise macromolecules, such as nucleic acids and proteins and also low-molecular weight metabolites not made by the host. They must also possess transport systems that allow access to host nutrients. Here, uncertainties about the nature of the inclusion membrane intrude. McClarty (1994) suggested that it may contain non-specific nutrient channels allowing passive transport and resembling those of the malarial parasitophorous vacuole. Recent observations do not support this speculation. Inoculation of fluorescent markers, even as small as 520 Da, into host cytoplasm does not lead to their passage into the inclusion (Heinzen and Hackstadt, 1997).

Synthesis of Macromolecules

It is well established that chlamydiae can synthesise RNA, DNA and protein, as suggested by their sensitivity to prokaryotic inhibitors such as tetracycline, chloramphenicol, rifampicin and nalidixic acid and independence of eukaryotic inhibitors (Hatch, 1988). Other macromolecules include glycogen, present within *C. trachomatis* inclusions and synthesised via a chlamydia-specific glycogen synthetase (Hatch, 1988), LPS with its species-specific transferases (see earlier) and also chlamydia-specific lipids and fatty acids (Newhall, 1988).

Energy Metabolism

Important reactions include provision of precursor molecules such as glucose- and fructose-6-phosphates, ribose-5-phosphate and acetyl-CoA, and also reducing power and energy source. Chlamydiae appear to lack hexokinase and are thought to obtain glucose-6-phosphate from the host cell for metabolism to intermediates and pyruvate via the Embden–Meyerhof–Parnas or Entner–Doudoroff pathways – of which they appear to have components; early studies failed to detect flavoprotein or respiratory cytochromes (Moulder, 1991). Two enzymes of the pentose phosphate pathway have been described (Moulder *et al.*, 1965) and the complete pathway is thought to be present (McClarty, 1994) and provide intermediates for the synthesis of glycan, LPS, fatty acid, phospholipid and folate.

Pyruvate, which may be largely host-derived, is decarboxylated by host-free chlamydiae, raising the possibility that chlamydiae may possess pyruvate dehydrogenase, the enzyme complex required for acetyl-CoA synthesis (Weiss, 1967; Weiss and Wilson, 1969). The reaction is stimulated by addition of lipoic acid, which, as lipoamide, mediates transfer of acetyl groups to CoA. Chlamydial growth requires thiamin and is additionally stimulated by niacin or pantothenate – all vitamin components of co-factors for the complex (Bader and Morgan, 1961).

Positive support for the hypothesis that chlamydiae are energy parasites (see Moulder, 1991), has come from the demonstration of specific ATP–ADP translocase activity in isolated RBs but not in EBs – which are metabolically inactive. Hydrolysis of the transported ATP results in an energised membrane across which lysine transport takes place (Hatch *et al.*, 1982). Chlamydial energy requirements may also be served by other as yet undiscovered energy-yielding catabolic reactions (McClarty, 1994).

Co-factor Metabolism

Chlamydiae are thought to obtain their vitamin requirements from the host cell. Species differences in sulphonamide-resistance have long been recognised, however, with *C. trachomatis* strains sensitive and *C. psittaci* (excepting 6BC) and more recently, *C. pneumoniae* strains, being resistant. Indeed, sulphonamide sensitivity has been considered a key characteristic for species differentiation (Moulder, 1991). Studies by Fan *et al.* (1992) show, however, that all chlamydial strains can probably synthesise folate (required for thymidine synthesis) but prefer to access the host supply. Differences in sulphonamide susceptibility arise because sensitive strains lack the mechanism to transport folate made by the host cell. A significant implication of these findings is that chlamydiae are likely to possess regulatory mechanisms that efficiently inhibit biosynthesis if the co-factor or nutrient is available from the host (McClarty, 1994).

Amino Acid Metabolism

Studies on interference with chlamydial growth by general deprivation of amino acids or sparing by addition of cycloheximide (inhibitor of eukaryotic protein synthesis) suggest that chlamydiae rely on host amino acid pools for most amino acids (Moulder, 1991). Earlier work has provided evidence for both lysine and arginine biosynthesis by *C. psittaci* strains (Moulder *et al.*, 1963; Treuhaft and Moulder, 1968). Apparent strain differences in amino acid requirements in the presence of cycloheximide (Allan and Pearce, 1983) have been suggested to arise from competition for transport into chlamydiae, but may also involve failure of synthesis by chlamydiae (Coles and Pearce, 1987). Deprivation experiments are, however, difficult to interpret because starvation induces increased host protein turnover (McClarty, 1994). Also, selective amino acid deprivation may lead to exiting of other amino acids from the host by counter-transport. Given the observations on suppression of folate biosynthesis by chlamydiae, deprivation studies alone would seem an unreliable guide to capacity for amino acid synthesis.

Non-cultivable Chlamydiae It has long been recognised that chlamydiae can be present *in vivo* in a state sometimes referred to as latent or micro-biologically inapparent, yet capable of reactivation to infectivity (Storz, 1971; Schachter, 1978). Corre-spondingly, culture experiments have suggested that chlamydiae can be non-infectious during deprivation of nutrients, but capable of persistence until re-addition of nutrient leads to restoration of infectivity (Morgan, 1956; Hatch, 1975; Moulder, 1982). It is now clear that under nutrient deprivation chlamydiae undergo an altered developmental pathway which leads to the induction of aberrant non-infectious chlamydiae (Coles *et al.*, 1993). This non-cultivable state appears to be a form of stress or starvation response in which organisms retain metabolic activity and give rise to infectious EBs on restoration of nutrient. Productive chlamydial infection is sensitive to nutrient levels. Reduction of amino acids in the extracellular medium by only 25% is sufficient to trigger intracellular production of non-cultivable organisms with lowered output of infectious EBs (Harper *et al.*, 2000). It may be that the response constitutes a survival mechanism analogous to starvation responses by other bacteria.

Probably the most important mechanism for pro-duction of non-cultivable forms *in vivo* is via local release of IFNγ (Coles *et al.*, 1993; Beatty *et al.*, 1993). This cytokine activates the ubiquitous host cell enzyme, indole-2,3-dioxygenase, whose substrate, tryptophan, is degraded to kynurenines, so that the intracellular environment is depleted in the free amino acid. Importantly, the host cell is not starved of tryptophan, because levels of tryptophanyl synthetase are simultaneously increased, ensuring that tryptophan is sequestered for host use (Kisselev *et al.*, 1993). Intracellular degradation of tryptophan leads both to suppression of chlamydial growth (Byrne *et al.*, 1986) and to the induction of aberrant forms (Shemer and Sarov, 1985a, b; Beatty *et al.*, 1993). Non-cultivable chlamydiae induced by IFNγ synthesise significant levels of Hsp 60, but are reduced levels in MOMP and

are deficient in LPS (Beatty *et al.*, 1993, 1994a); they also show persistence in cell culture (Beatty *et al.*, 1994b, c).

Nucleotide Metabolism

After their success in establishing a host-free system, Hatch and colleagues refined the system to demon-strate that RBs synthesise RNA from supplied nucleotides and to isolate early RNA transcripts during initial EB–RB transformation (Crenshaw *et al.*, 1990; Wichlan and Hatch, 1993).

In an elegant series of studies, characterisation of nucleotide metabolism was achieved in McClarty's laboratory with drug-resistant chlamydiae and mutant host cell lines with well-defined deficiencies. It appears that chlamydiae are unable to synthesise either purines or pyrimidines (McClarty and Fan, 1993; McClarty and Qin, 1993). They receive ribonucleotides ATP, GTP and UTP (but not deoxyribonucleotides) from the host; CTP can also be obtained from the host or it may be synthesised by chlamydiae from UTP via a chlamydial CTP synthetase (Tipples and McClarty, 1993). Chlamydiae also convert nucleotides to deoxy-ribonucleotides via a chlamydia-specific reductase (McClarty and Tipples, 1991; Tipples and McClarty, 1991), and the dTTP needed for DNA synthesis is obtained by the action of a chlamydial thymidylate synthetase which converts dUMP to dTMP (Fan *et al.*, 1991).

An important question arising from these findings is why chlamydiae, which are auxotrophic for ATP, GTP and UTP, possess both CTP synthetase and the capacity to transport CTP from the host. The CTP synthetase has been cloned (Tipples and McClarty, 1995) and its role has been further examined. Synthe-sis occurred only in the mid–late-cycle phase of chla-mydial growth, but synthetase was also stored in EBs – which might spare the need for early production (Wylie *et al.*, 1996a). Co-expressed with CTP synthetase was the enzyme CMP KDO synthetase, required for LPS synthesis, and translated from the same messenger – suggesting that CTP synthetase expression could be important for chlamydial LPS synthesis. To analyse the relationship between CTP pool size and requirement for CTP synthetase activity, a chlamydial strain was selected whose CTP synthetase had lost sensitivity to feedback inhibition by CTP (Wylie *et al.*, 1996b). There was no alteration in expression or inhibition of the transport of CTP to chlamydiae and no increase in levels of CTP in the chlamydial mutant compared with wild-type chlamydiae. Curiously, synthetase-mutated host cells with CTP levels raised above that of the wild-type host did not support chlamydial growth. It appears that maintenance of CTP at the low levels normally found in chlamydiae is in some way critical for chlamydial growth to take place.

Chlamydial Genetics

Difficulties in Gene Transfer and Expression

It has not so far been possible to transfer to or transfect chlamydiae to achieve stable expression of the transferred genes. Both the plasmid and a rare phage found in avian *C. psittaci* (see below) have been considered as possible vectors (Lusher *et al.*, 1991), but the problem of introduction of genetic material into chlamydiae is considerable. With EBs the structure of the envelope may pose a special barrier and extracellular manipulation must not alter subsequent attachment, entry, or avoidance of lysosomes, or affect the properties needed for successful initiation of metabolism. RBs might appear more suitable, but they appear incapable of entry into cultured epithelial or fibroblast cells, although ingestion by macrophages has been reported but leads to fusion with lysosomes (Lawn *et al.*, 1973). Nevertheless, successful electroporation of chlamydiae has been achieved to introduce a plasmid bearing chloramphenicol transferase. Host cells have been successfully infected, with evidence of acquired resistance to chloramphenicol (Tam *et al.*, 1994). A stable population of transformed chlamydiae could not, however, be maintained beyond several passages.

Regulation of Gene Expression

As indicated earlier, the DNA-binding proteins (Hc1 only in *C. psittaci*) mediate major structural changes in the chromosome during both early and late stages of development. Whether they have any additional regulatory role is not known.

With one or two exceptions, successful cloning and expression of chlamydial proteins in *E. coli* has required fusion of inserts to *E. coli* genes so that they come under the control of *E. coli* promoters. From this it has been inferred that the major *E. coli* sigma factor, σ^{66}, does not recognise chlamydial promoters. Yet identification of what appears to be the major chlamydial sigma factor, σ^{66}, has indicated high homology with σ^{70} (Engel and Ganem, 1990; Koehler *et al.*, 1990; Douglas *et al.*, 1994). The existence, for some genes, of tandem promoters has added to the variation in putative promoter sequences and the difficulty in arriving at a consensus sequence. Tandem promoters reported for the MOMP gene of *C. trachomatis* (Stephens *et al.*, 1988b) have not

been confirmed in further analysis (Douglas and Hatch, 1995). Also those for the CRP operon in *C. trachomatis* (Lambden *et al.*, 1990) were not detected in detailed studies extended to *C. psittaci* and *C. pneumoniae* strains, and the earlier finding was attributed to premature chain termination (Watson *et al.*, 1995).

One possible mechanism for regulation of chlamydial development is that it is controlled by stage-specific sigma factors recognising-specific promoter sequences. To investigate the role of sigma factors and promotor diversity, Hatch, Sriprakash and their collaborators developed an *in vitro* transcription system consisting of DNA template isolated at intervals in the cycle, together with substrates and RNA-polymerase-enriched extracts. With this they showed that transcription was blocked by antibody reactive with σ^{66}, that the same sigma factor was required for transcription of the late stage CRP operon (Mathews *et al.*, 1993) and that added recombinant σ^{66} enhanced transcription (Douglas *et al.*, 1994). Moreover, in the same σ^{66}-dependent system, two new late genes, *ltuA* and *ltuB*, were detected in samples taken at 30 hours after infection as well as the expression of one of the histone-like proteins (Hc1) – confirming that σ^{66} is involved in this key stage of development (Fahr *et al.*, 1995). From this work it seems clear that gene expression at different stages during the developmental cycle is mediated by a single sigma factor.

Promoter sequence variation and its effect on recognition by the chlamydial polymerase complex has been explored by analysis of transcription from mutated promotors. For the counter-transcript promoter of the chlamydial plasmid, all single and most multiple mutations in either the -35 or -10 regions were without effect on transcription (Mathews and Sriprakash, 1994), indicating unusual flexibility in polymerase recognition. With the P2 promoter of the MOMP gene, which has an unusual GC-rich -10 region (TATCGC), Douglas and Hatch (1996) observed a similar tolerance of mutations, including substitutions that resembled the consensus sequence for the *E. coli* -10 hexamer.

These unusual promoter recognition properties shown by the chlamydial σ^{66} polymerase may depend on supplementary-binding by transcriptional activator proteins (Busby and Ebright, 1994) that could be present in extracts. An alternative possibility is suggested by work on *Caulobacter crescentus*, a dimorphic Gram-negative bacterium, which is similar to chlamydiae in having promoter sequences not recognised in *E. coli* and a dominant sigma factor, σ^{73}, that shows a broad specificity of promoter recognition (Malakooti and Ely, 1995). The authors speculate that and

N-terminal extension of 20 amino acids, absent from most other sigma factors, may contribute to the unusual recognition properties of σ^{73}. Douglas and Hatch (1996) point out that chlamydial σ^{66} has a comparable extension of 14 amino acids that could fulfil a similar role.

Recombination

It has seemed reasonable to assume that the chlamydial genome is highly stable and subject only to slow change in evolutionary time. Serological and genetic data on genital strains isolated from patients in regions of high-density STD have, however, raised the possibility that recombination between MOMP genes could be taking place (Brunham et al., 1994; Hayes et al., 1994). Complementation analysis by electroporation of recombinant plasmid into an RecA⁻ E. coli was successful in identifying the presence of a chlamydial recombinase, RecA (Zhang et al., 1995). The authors attribute their success in this approach to 'a fortuitous promoter-like sequence capable of recognition by the E. coli RNA polymerase'.

Sequencing of the Chlamydial Genome

The newly initiated programme to sequence the entire chlamydial genome is being funded by NIH. The work is being done as a collaboration between the University of California at Berkeley and Stanford University, under the direction of Dr Richard Stephens (Berkeley). Knowledge of the complete genome sequence will undoubtedly bring considerable benefits and enormously widen the scope of present research. It may also provide information that will allow the production of stable mutants that are essential for certain types of investigations. Information about the programme and available sequence data can be accessed at the website http://violet.berkeley.edu:4231/. The site will remain open after completion of the programme and provide a direct link to the central database.

Pathogenesis

Disease Mechanism

The development of disease after trachoma biovar infections results from immunopathology in which repeated re-infection plays an important part. Immunological mechanisms are considered below. Little is known about disease mechanisms for LGV biovar infections. Trachoma strains are responsible for ophthalmic trachoma (serovars A–C), for ocular and genital infections (serovars D–K) and

for sexually acquired reactive arthritis. The LGV strains (serovars L1, L2, L3) are associated with a distinct sexually acquired disease, lymphogranuloma venereum.

Trachoma, ocular and genital infections involve restricted local multiplication of chlamydiae in mucosal epithelial cells. They are frequently asymptomatic and tend to persist in the absence of treatment. It has been proposed that non-cultivable chlamydiae are responsible for persistence in vivo (Beatty et al., 1994c, d). There is, however, as yet no evidence to suggest that organisms in this state have enhanced survival properties over those of normally replicating organisms (Pearce et al., 1994). Study of this important question is complicated by the likelihood of exogenous re-infection.

Joint damage in reactive arthritis appears to be associated with the presence of non-cultivable chlamydiae in synovial cells, but whether these arise by local multiplication, inward migration of infected macrophages or a combination of the two, is unclear. Lymphogranuloma venereum depends on distinctive properties of the LGV serovars and involves both invasion of non-mucosal tissue and damage to local lymphoid organs. Below we briefly recount the main features of these diseases. The epidemiology of ocular and genital infections is discussed in a later section.

Trachoma

Infection of the conjunctiva is initiated by direct contact and is followed by inflammatory cell responses leading to the development of the characteristic sub-epithelial lymphoid follicles, with germinal centres composed of B and T lymphocytes. The infection can be asymptomatic. Recovery may then take place but if infection persists, and repeated re-infection is usual, fibrosis of the conjunctiva becomes extensive. Follicles contain mainly T cells (Mabey et al., 1992a) and scarring (cicatrisation) leads to infolding of the eyelid (entropion) and trauma to the cornea as a result of abrasion by the inwardly directed lashes (trichiasis). This leads in turn to ulceration, increasing opacity of the cornea and progressive impairment of vision (Dawson et al., 1975; Darougar and Jones, 1983; Schachter et al., 1998).

Oculogenital Infections

Ocular Infection

Infection of the eye (inclusion conjunctivitis) can occur in adults after genital infection, probably as a result of direct transfer, but is usually self-limiting in the absence of re-infection. It is also found in newborn infants (*ophthalmia neonatorum*) as a result of eye infection acquired during birth, and it may persist.

Genital Infections

Infection of the urethra in men leads to a urethritis and early inflammatory response reflected by the presence of inflammatory cells in 'first-catch urine'. The infection can persist in the absence of treatment and may also be asymptomatic. Cervical infection occurs in women, sometimes with mucopurulent discharge, but infection of the urethra is also common (Lee *et al.*, 1995). Again, infection may be asymptomatic, with persistence in the absence of treatment. Chlamydial infection at other sites – the epididymis in men (epididymitis) and rectal mucous membrane (proctitis) in both sexes – can also take place.

In women, ascending infection from the cervix to the uterus and extension to the Fallopian tubes presents the most serious condition, with now well-established evidence that local inflammatory responses can cause salpingitis and irreversible tubal damage (Paavonen and Wolner-Hanssen, 1989; Westrom *et al.*, 1992; Cates *et al.*, 1994). Spread of infection from the cervix leads to the condition of pelvic inflammatory disease (PID) which may be associated with chronic pelvic pain. In a significant proportion of cases ascending infection appears to be asymptomatic (Westrom, 1980; Cates *et al.*, 1994).

The resulting changes in tubal properties are thought to result from the scarring that takes place during recovery from infection. Ectopic pregnancy occurs if entry of the ovum into the tube is prevented. Infertility occurs if sperm or ovum movement within the tube is impaired by damage to the mucosa or cilia or by tubal occlusion (Cates *et al.*, 1994). Damage appears to depend on repeated re-infection, as with trachoma, as indicated by studies with a non-human primate model (Patton, 1985; Patton *et al.*, 1987, 1990).

Reactive Arthritis

This state is triggered by previous infection with any of a number of bacteria. Chlamydial disease occurs after genitourinary tract infection by oculogenital strains in about 1% of cases. It can also follow after LGV, *C. psittaci* (Keat *et al.*, 1983) or *C. pneumoniae* infection. Affected individuals are commonly male and there is a strong association with HLA-B27. Although 'reactive' is intended to imply the presence of microbial antigen at a distant site (Keat *et al.*, 1989), there is now strong evidence that chlamydial antigen is frequently present in the synovial membrane (Keat *et al.*, 1987; Schumacher *et al.*, 1988), probably as IFNγ-induced non-cultivable chlamydiae (Beutler *et al.*, 1995; Branigan *et al.*, 1996). Damage to the joint is believed to depend upon T-lymphocyte-mediated responses to local antigen (Burmester *et al.*, 1995).

Lymphogranuloma Venereum

Sexual transmission of the disease leads initially to local lesions, which are commonly on the penis, vulva or vagina, but have also been found on fingers or tongue or in the rectum. Subsequently, disease progresses to local lymphoid tissue, inguinal nodes or other sites (Schachter and Dawson, 1978). Where the disease is endemic and untreated, the finding of lymphadenopathy, abscess and scarring is common (Bauwens *et al.*, 1995).

Immune Responses to Chlamydiae

Resistance to infection involves both the innate immune system – phagocytes, complement activation – which is not pathogen-specific, and the adaptive immune response whereby components of the pathogen are recognised in an antigen-specific manner. Only the latter is dealt with here, and is considered in five sections:

- Cell-mediated immunity as a means of controlling and/or eradicating primary infection, and the role of different T lymphocyte subsets in this process;
- Chlamydial antigens which are recognised by particular T cell subsets;
- Immune responses associated with protective immunity, i.e. resistance to re-infection, and the role of antibody in this process;
- The relationship between immune responses and pathologies associated with chlamydial infection;
- An account of the possible involvement of chlamydial infection in the pathogenesis of arterial disease.

Phenotype of Cells Participating in the Immune Response to Chlamydiae

Control of an intracellular pathogen such as *C. trachomatis* (CT) depends primarily on cellular immunity. Chlamydiae which have gained access to a cell signal their presence to the immune system by displaying peptide fragments of their antigens on class I and II MHC molecules, where they can be surveyed by CD4+ (helper) and CD8+ (cytotoxic) T lymphocytes respectively. The initial host cell, the epithelial cell, is not, however, competent to initiate cell-mediated immune responses, and it is likely that chlamydiae are taken up by dendritic cells, which transport antigens from epithelial surfaces to lymph nodes where they interact with naive T cells capable of recognising chlamydial antigens (Ojcius *et al.*, 1998; Cyster, 1999). Epithelial cells may assist the recruitment of dendritic cells by production of chemokines such as IL-8 after infection (Rasmussen *et al.*, 1997). In these early stages of infection chlamydial antigens, which may be recognised by specific antibodies, are not accessible, and in any case the production of specific high-affinity antibodies requires help from antigen-sensitised T cells. Thus, it is not surprising that in most models of chlamydia infection, animals which lack T lymphocytes fail to clear infection, whereas those which lack B cells have no such difficulty (Williams *et al.*, 1987a; Ramsey *et al.*, 1988).

Antigen Processing and Presentation to T Cells

In general, antigens or pathogens which enter the cell from the exterior in phagocytic or pinocytic vacuoles are subjected to proteolysis to produce peptides. These peptides then combine with class II MHC molecules, which have been transported from the endoplasmic reticulum via the Golgi apparatus, to the endocytic compartment where peptides are available (Neefjes *et al.*, 1991). In contrast, viruses that replicate within the cytoplasm of infected cells make their antigens available to the proteolytic and transport mechanisms which provide antigenic peptides to complex with newly synthesised class I MHC molecules in the endoplasmic reticulum (Hahn *et al.*, 1996). The peptide/MHC complex is then expressed on the cell surface. Certain bacteria, such as *Listeria monocytogenes*, have the ability to escape from the phagocytic vacuole in which they enter the cell, and by taking up residence in the cytoplasm present their antigens on class I MHC molecules, as described for viruses (Ladel *et al.*, 1994). Chlamydiae do not escape from their membrane-bound intracellular vacuoles (Taraska *et al.*, 1996; Vanooij *et al.*, 1997) until replication is completed, and the rupture of the inclusion body liberates new organisms and destroys the host cell. It may be assumed, therefore, that chlamydial antigens would mostly be presented by class II MHC antigens to CD4+ T cells. In recent years it has been noted, however, that antigens from those intracellular organisms that do not invade the cytoplasm are still made available to class I MHC molecules for presentation to CD8+ T cells (Pfeifer *et al.*, 1993). Although the pathway which permits this has not yet been fully defined, dendritic cells have this property (Svensson *et al.*, 1997). It is likely that it involves

transport of antigen out of the vacuole and into the cytoplasm, where processing and delivery to the endoplasmic reticulum can proceed as for viral antigens (c). Alternatively, since class I MHC molecules are also present in the same intracellular compartments as those where class II MHC molecules meet antigenic peptides, they may gain peptides in the same way. Thus, contrary to initial expectations, CD8+ T cells can also contribute to immunity to intracellular organisms, as has been shown with respect to mycobacterial infection (Ladel *et al.*, 1995a).

Chlamydia-specific CD4+ T Cells

Which of these mechanisms of antigen presentation is critical for protective cellular immune responses to chlamydiae? The most useful studies have been performed in gene targeted mice where one of the presentation pathways has been eliminated (e.g. β_2-microglobulin knockout $(-/-)$ mice, which cannot express class I MHC molecules), or a T-cell subset responsible for responding to antigen is lost. These studies have consistently indicated the importance of class II MHC-restricted CD4+ T cells in clearing chlamydial infection and providing subsequent immunity; class II MHC $-/-$ mice and CD4$-/-$ mice both fail to eliminate the organism (Morrison *et al.*, 1995). A protective role for CD4+ T cells has also been shown by depletion studies *in vivo* (Landers *et al.*, 1991), and by the adoptive transfer of chlamydia-specific CD4+ T-cell clones and lines (Su and Caldwell, 1995). CD4+ T cells, particularly those of the Th1 subset, produce IFNγ (Johansson *et al.*, 1997), which has several antichlamydial functions including stimulation of nitric oxide production (Igietseme, 1996), and induction of tryptophan degradation in the host cells (Beatty *et al.*, 1994a) – tryptophan is essential for normal chlamydial growth. IFNγ also up-regulates class II MHC expression, but chlamydiae are able to counteract this (Zhong *et al.*, 1999).

In humans, a low CD4+ T cell count due to HIV infection is a risk factor for chlamydia-induced PID (Kimani *et al.*, 1996). Patients with ocular infection who recovered had higher proliferative responses to chlamydial antigens mounted by CD4+ T cells than did those with persistent infection (Bailey *et al.*, 1995). Both these reports are consistent with the murine evidence emphasising the critical role for CD4+ T cells.

Chlamydia-specific CD8+ T Cells

Chlamydia-specific CD8+ T cells have been isolated from spleens of mice infected intravenously, or in the upper genital tract, with large numbers of organisms (Buzoni-Gatel *et al.*, 1992; Starnbach *et al.*, 1994,

1995; Beatty *et al.*, 1997); some protection was obtained by transfer of a chlamydia-specific CD8+ clone (Igietseme *et al.*, 1994; Starnbach *et al.*, 1994), but this protection was related to secretion by the clone of large amounts of IFNγ rather than its ability to kill chlamydia-infected cells (Lampe *et al.*, 1998). Generally CD8+ T cells seem to play a much less important role in the overall response to infection, however, since β_2-microglobulin$-/-$ mice clear infection, and depletion of CD8+ cells *in vivo* has minor effects (Su and Caldwell, 1995). In humans MOMP-specific CD8+ T cells have recently been generated with specific peptides *in vitro* from infected individuals, but their role in protection is unknown (Kim *et al.*, 1999).

Other Lymphocyte Subsets

In addition to recognition of chlamydia-derived peptides by T cells which use the $\alpha\beta$ T cell receptor, other chlamydial products may be recognised by different T cell subsets, such as those which use the $\gamma\delta$ receptor (DeLibero, 1997), and possibly the subset of $\alpha\beta$TCR+ T cells which expresses neither CD4 or CD8 (Porcelli *et al.*, 1992). These cells are commonly involved in the immune response to infectious agents, but do not usually perform a critical role – TCR$\delta-/-$ mice can overcome infections by organisms such as mycobacteria (Ladel *et al.*, 1995b). In TCR$\delta-/-$ mice, which develop pneumonia due to infection with MoPn, a large number of organisms was isolated from the lungs at early time points after infection, suggesting a beneficial role of this subset in containing infection. At late time points, however, the role of $\gamma\delta$ T cells appeared deleterious, since the knockout mice had lower mortality and fewer organisms in their lungs (Williams *et al.*, 1996).

Lastly, natural killer (NK) cells can participate in the immune response to infection (Tseng and Rank, 1998). It is unclear whether chlamydial infection renders the host cell more susceptible to lysis by NK cells, but NK activity alone, which is present in T-cell-deficient mice, is clearly incapable of clearing the organism, although the IFNγ which NK cells produce may help to control the infection.

Antigens which Elicit T-cell Responses

Antigens Recognised by CD4+ T Cells

Relatively little is known about the principal targets of the CD4+ T-cell response to chlamydia. In humans, T cell clones specific for chlamydiae were first reported by Quigstad *et al.* (1983), but the nature of the antigens recognised was not known, and awaited the availability of recombinant chlamydial antigens. In both mice and humans, work has been carried out on the MOMP because of its potential as a vaccine, and it is clearly immunogenic. Epitopes have been defined in order to determine whether they are mainly in relatively conserved parts of the MOMP sequence, or in the variable regions which define serovars. Murine T-cell responses to epitopes in the latter regions have been described (Su *et al.*, 1990b), but in human studies a large number of epitopes (presented by different HLA-DR alleles) have been defined in constant regions, using T-cell lines (Ortiz *et al.*, 1996).

In patients who develop reactive arthritis secondary to *C. trachomatis* infection, T cells specific for chlamydial antigens can readily be identified in joint fluid. It has been possible to clone these, and this allowed the first identification of the target antigens of the human CD4+ T-cell response, and the mapping of epitopes. Three such antigens have now been identified; these are the histone-like protein Hc1, the 60-kDa heat-shock protein (Hsp 60) and the 60-kDa outer-membrane protein, OMP2 (Hassell *et al.*, 1993; Gaston *et al.*, 1996; Goodall *et al.*, 1997). In the case of Hc1 and Hsp 60, the specific peptides recognised in the context of HLA-DR1 and HLA-DR4 respectively have been defined (Gaston *et al.*, 1996; Deane *et al.*, 1997). Interestingly, two of these antigens, OMP2 and Hsp 60, have been reported to be recognised by murine chlamydia-specific T-cell lines, but clonal studies are lacking (Beatty and Stephens, 1992). Other T-cell clones have been isolated which respond to chlamydial antigens in a particular molecular weight range when tested by T-cell immunoblotting, but the identity of the stimulating antigens recognised by these clones is not known. As more chlamydial proteins are cloned and sequenced, it is likely that these antigens will be identified. Indeed, now that the chlamydia genome has been sequenced and all ORFs known (Stephens *et al.*, 1998), it will be possible to identify antigens recognised by T-cell clones on the basis of partial sequencing.

Not enough numbers of patients have yet been studied to allow any statement about whether the T cell response to particular chlamydial antigens is associated with uncomplicated genitourinary infection, or with inflammatory sequelae such as reactive arthritis or PID. There are reports of increased responses to hsp 60 by peripheral blood T cells in PID patients, but these responses were relatively modest (Witkin *et al.*, 1993, 1994). In reactive arthritis, responses to chlamydial antigens are generally much more evident within the joint than in peripheral blood, and the same may apply to PID (Gaston *et al.*, 1989).

Responses to chlamydial EBs by peripheral blood mononuclear cells from acute urethritis patients and from healthy controls have also been studied (Shahmanesh *et al.*, 1999). Although the groups could be distinguished statistically there was a large overlap in the responses. This might imply a high level of exposure to *C. trachomatis* in the control population studied, but the result could also be readily explained if T cells recognise antigenic epitopes common to *C. trachomatis* and *C. pneumoniae*, since in patients not known to have had genitourinary infection, previous encounter with *C. pneumoniae* would produce memory T cells able to respond to challenge *in vitro* with *C. trachomatis*. Indeed, mapping the DR4-restricted epitope in *C. trachomatis* hsp 60 (Deane *et al.*, 1997) indicated that it has a sequence which is completely conserved in *C. pneumoniae*, and also in *C. psittaci*. Since respiratory infection with *C. pneumoniae* is common (Gnarpe and Gnarpe, 1993), it is also possible that patients infected with *C. trachomatis* have already been primed for responses to hsp 60 through encounter with *C. pneumoniae*. Priming is important in the pathogenesis of chlamydia-associated inflammation such as trachoma (Grayston *et al.*, 1985), and in experimental models of trachoma challenge with hsp 60 in primed animals can give rise to chronic inflammatory lesions (Morrison *et al.*, 1989b) (see below). Interestingly the sequence of the mapped epitope was also highly conserved in Gram-negative organisms such as *E. coli*, *Salmonella* spp. and *Neisseria gonorrhoeae*, so the possibility of priming by infection with organisms other than chlamydia must also be considered.

Antigens Recognised by CD8+ T Cells

Chlamydia-specific CD8+ T cells have only recently been demonstrated in humans (Kim *et al.*, 1999), and their role in the normal human response to chlamydia infection is unknown. In experiments with *C. trachomatis*-infected macrophages as a stimulus to patient T cells *in vitro*, a cytolytic population which recognised autologous target cells was induced but specificity for chlamydia was not seen (Hassell *et al.*, 1992). In a macaque model of chlamydial infection of the upper genital tract, two-thirds of the infiltrating T cells were CD8+, but their specificity for chlamydia was not directly shown (Vanvoorhis *et al.*, 1996). CD8+ cells are found in urethral mucosa and may recognise chlamydiae at this site (Brunst *et al.*, 1998). If these local CD8+ T cells are important in controlling dissemination of the organisms from the site of infection, inflammatory disease such as PID or reactive arthritis might arise in patients with defective CD8+ responses, and could account for the presence of *C. trachomatis* in distant sites such as the synovial membrane. In an elegant murine study, a chlamydia-specific CD8+ T cell clone was shown to recognise a particular peptide eluted from class I MHC antigens, but the antigen from which this peptide derived was not identified (Starnbach *et al.*, 1994). This approach may, however, eventually identify antigens which elicit a response by CD8+ T cells.

Antigens Recognised by TCRγδ+ T Cells

In the majority of cases γδ T cells are not restricted by MHC antigens. Work on mycobacterial antigen recognition by γδ T cells has shown that non-peptide antigens are recognised, particularly phosphorylated compounds such as isopentenyl pyrophosphate, and widely distributed compounds such as alkylamines (Tanaka *et al.*, 1995; Bukowski *et al.*, 1999). Whether certain chlamydial components are preferentially recogised by γδ T cells is not known. Since γδ T cells are prominent in mucosal sites, such as the genitourinary tract, they too could have a role in local immune responses to chlamydiae, and in the local production of cytokines. Interestingly, 50% of T cell lines derived from the urethral discharge of patients with acute urethritis showed a predominance of γδ T cells, whereas this was not a feature of the lines generated from normal urethra (Brunst *et al.*, 1998).

Immunity to Re-infection

Up to this point the immune mechanisms used to combat established infection have been discussed, as distinct from those that prevent re-infection. Patients frequently present with re-infection, implying that protective immunity is weak or short-lived, or alternatively that infection is persistent and immunity 'non-sterilising'. In the latter case clinical presentation would correspond to reactivation of infection, a process commonly seen with persistent agents such as herpes simplex virus and Epstein–Barr virus. If this analogy were to hold, it would be expected that, like the herpes viruses, chlamydiae would have developed sophisticated strategies to subvert immune responses. As noted, chlamydiae alter expression of MHC antigens (Rodel *et al.*, 1998; Zhong *et al.*, 1999), and also affect the susceptibility of host cells to apoptosis (Fan *et al.*, 1998). Since dendritic cells can acquire antigen by uptake of apoptotic cells (Albert *et al.*, 1998), such inhibition may inhibit generation of immune responses.

IgG and IgA in genital secretions can neutralise infective organisms, and thereby provide a degree of protection from re-infection. The importance of local antibody varies in different species. There is evidence that conjunctival IgA protects guinea-pigs from conjunctival re-challenge with chlamydia (Murray *et al.*, 1973), whereas mice rendered deficient in antibody of all classes showed no difference from controls in the course of infection or immunity to intravaginal challenge (Ramsey *et al.*, 1988). A role for IgA in human genital infection has been suggested by studies showing that there was an inverse relationship between the titre of IgA and the number of organisms isolated from the genital tract of women with established infection (Brunham *et al.*, 1983). No such relationship applied to IgG or serum antibodies.

The formation of high-avidity antibodies of the IgG or IgA class requires specific help from chlamydia-specific T cells. T and B cells need not, however, see the same antigen where the antigens are physically linked, so that a B cell can recognise one component, internalise the complex, and present peptides from the other to a helper T cell (Allen and Stephens, 1993). Antibodies to many chlamydial components have been demonstrated; those which are neutralising include antibodies to MOMP, hsp 70 (Birkelund *et al.*, 1994b) and Mip (Lundemose *et al.*, 1992). Antibodies to MOMP are directed against serovar-specific epitopes in the variable segments of the MOMP sequence (Zhang *et al.*, 1987), but B cells with this specificity may receive help from T cells which see epitopes in the conserved region. Protection mediated by MOMP-specific antibodies of both IgA and IgG class has been tested in mice by the technique of engrafting a hybridoma that produces a monoclonal antibody of the required specificity (Cotter *et al.*, 1995). Protection against vaginal infection was demonstrated but only at relatively low challenge doses of organisms. At higher doses only a marginal effect on organism shedding was seen, and the principal effect of the antibodies was on chlamydial infiltration and consequent inflammation in the upper genital tract. Such an effect of anti-MOMP antibodies would of course be useful if it were duplicated in humans, but there are problems in extrapolating from these experiments which involve relatively short periods of treatment before the mice are overwhelmed by the engrafted hybridoma.

Much research on immunity to re-infection is focused on the question of developing a vaccine (Bavoil *et al.*, 1996; Rank and Bavoil, 1996). In mice, protection requires previous exposure to live organisms, and it has become clear that the route of exposure influences the nature of the immune response which is made and therefore the degree of protection. What is required is an IFNγ-producing cell-mediated response, i.e. one mounted by the Th1 subset. Differentiation of T cells to this phenotype requires IL-12, which is likely to be produced by chlamydia-infected macrophages. When mice were immunised subcutaneously, a Th2 response developed with high levels of antichlamydia antibodies but failed to resolve

subsequent vaginal infection as rapidly as mice immunised orally or intranasally (Kelly *et al.*, 1996; Pal *et al.*, 1996). Both DNA vaccination and the use of infected dendritic cells for immunisation have been tested in mice with encouraging results (Su *et al.*, 1998; Zhang *et al.*, 1999). It remains to be seen whether particular chlamydia antigens, perhaps administered in ways which favour the development of a Th1 immune response, will be able to provide useful protective immunity in humans.

Immune Responses to Chlamydial Antigens and Immunopathology

In diseases associated with chlamydial infection, chronic inflammatory responses are evident – in the upper genital tract in PID, the conjunctivae in trachoma, the joint in reactive arthritis and possibly the arterial wall in atherosclerosis (see below). It is likely that inflammation is maintained by a continuing immune response to chlamydial antigens, or possibly by an autoimmune response triggered as a result of chlamydial infection. An example of the latter mechanism is the finding that T cells which recognise a specific peptide in chlamydial OMP2 cross-react with cardiac myosin and induce inflammatory heart disease (Bachmaier *et al.*, 1999); other examples of antigenic mimicry of potential relevance have been described (Hemmerich *et al.*, 1998). To address the possibility of similar mechanisms in human disease, it is important to define the antigens recognised in chlamydia-associated diseases, and to determine whether immunopathology is associated with responses to particular antigens.

As noted previously, in experimental trachoma in the guinea-pig evidence was produced to implicate T-cell responses to chlamydial hsp 60 in pathogenesis (Morrison *et al.*, 1989a, b), since lesions could be produced in primed animals by challenge with the recombinant protein. Some questions about the role of immune responses to hsp 60 have been raised, however. Using highly purified protein another group has been unable to reproduce experimental trachoma in guinea-pigs (Rank and Bavoil, 1996), and studies in human trachoma have associated lower, rather than higher, T cell responses to hsp 60 with scarring (Holland *et al.*, 1993), although responses in the tissue could not be assessed. When guinea-pigs were deliberately sensitised to hsp 60 before ocular challenge with chlamydiae, an exaggerated inflammatory response was not seen. The latter result may not, however, be surprising; in models of arthritis in which pathogenic responses to hsp 60 have been clearly implicated, it is universally found that previous immunisation with hsp 60 protects against the induction of arthritis (Gaston and Pearce, 1994). It is now clear that prior immunisation can have a number of effects on the response to challenge, including altering the cytokines produced, the dominant epitopes recognised, and inducing regulatory T cells which cross-react with self hsp 60 (Anderton et al., 1995). T cells responsive to hsp 60 have been noted in blood from patients with trachoma (Bailey et al., 1995) and PID (Witkin et al., 1993, 1994), and cloned from affected joints in chlamydia-induced reactive arthritis (Gaston et al., 1996). In the last case these findings have been linked to the demonstration of chlamydiae in the reactive arthritis synovial membrane, as demonstrated by immunofluorescence (Keat et al., 1987) or PCR (Taylor-Robinson et al., 1992; Branigan et al., 1996) even though they cannot be cultivated. It has been suggest that, at this site, the organisms are continuing to produce hsp 60 – certainly hsp 60 mRNA has been detected (Hudson et al., 1996). The state of chlamydiae in the joint could mirror that induced in vitro by various treatments such as nutrient deprivation (Coles et al., 1993) or treatment with IFNγ (Beatty et al., 1993), in which synthesis of major structural proteins such as the major outer-membrane protein is decreased but synthesis of hsp 60 maintained (Beatty et al., 1994a, c). Interferon γ stimulates the degradation of intracellular tryptophan on which chlamydiae are dependent for the synthesis of MOMP – in contrast, hsp 60 does not contain any tryptophan residues. Thus it is possible to postulate that in vivo cells which can traffic to the joint, such as macrophages, could be infected with chlamydiae which are unable to mature into infectious elementary bodies, but which continue to synthesis hsp 60 and stimulate local T cells. Similar mechanisms might be involved in the pathogenesis of PID and trachoma. It is not clear, however, that persistent infection would be established by these means. Using a chlamydia-specific CD8+ T cell clone, it has been shown that host cells infected by chlamydia and treated with IFNγ or tryptophan deprivation remain susceptible to lysis (Rasmussen et al., 1996). Assuming that such cells are normally generated in vivo (the clone was generated from mice given non-physiological intravenous infection), they should be able to eliminate infected cells.

Chlamydial Infection and the Aetiology of Arterial Atheroma

There is evidence to suggest that inflammation may play an important role in the pathogenesis of atheroma, especially in its early stages (Libby and Hansson, 1991). Atheromatous plaques contain inflammatory cells, including significant numbers of macrophages, and T lymphocytes whose surface phenotype is consistent with activation (Stemme et al., 1992). Furthermore, inflammatory plaques are believed to be more unstable than those that do not contain large numbers of inflammatory cells, and to give rise to rupture and thrombosis resulting in vessel blockage. There are two main explanations for the presence of T cells within plaques. First, circulating activated cells may be non-specifically recruited into the evolving plaque by virtue of the increased expression of adhesion molecules which occurs on activation. A second possibility is that at least some of the T lymphocytes are responding to specific antigens within the arterial wall.

Such antigens could be derived either from autologous tissue (autoantigens) or from infective agents. In relation to the former, T cells reactive with oxidised low-density lipoproteins have been described (Stemme et al., 1995). Although it has not been usual to ascribe a role to infectious agents in atherogenesis, however, epidemiological evidence has accumulated to implicate infection by Chlamydia pneumoniae (Saikku et al., 1988; Thom et al., 1992; Melnick et al., 1993; Patel et al., 1995). This has been followed by demonstration of C. pneumoniae in atheromatous tissue, based on immunochemistry, amplification of chlamydial nucleic acids by PCR, and most recently, culture (Kuo et al., 1993, 1995; Ramirez et al., 1996; Jackson et al., 1997). The organism can be shown to disseminate from the respiratory tract within macrophages (Moazed et al., 1998), which could then be recruited to arterial plaques. Recently C. pneumoniae has also been shown to be able to infect the smooth muscle cells present in atheromatous plaques, a property not shared with C. trachomatis (Gaydos et al., 1996; Knoebel et al., 1997). At present it is not possible to decide whether C. pneumoniae has a propensity to infect cells in atheromatous plaques, but does not contribute to the pathological lesion, i.e. it acts as a passenger or innocent bystander. Evidence from animal models of atherosclerosis does, however, suggest that the organism can exacerbate arterial lesions (Muhlestein et al., 1998; Moazed et al., 1999) and it can induce macrophages to form foam cells (Kalayoglu and Byrne, 1998). It also has effects on endothelial cell expression of adhesion molecules which may be relevant to disease progression (Kaukorantatolvanen et al., 1996). Lastly, it is important to consider whether the organism provokes a local T cell-mediated immune response directed against C. pneumoniae antigens, since this could contribute to the inflammatory aspect of atheromatous plaques. The possibility may be clarified by on going studies of the specificity of T cells present in plaque tissue; recent studies suggest that these can recognise chlamydial antigens (Halme *et al.*, 1999; Curry *et al.*, 2000).

Diagnosis

Detection of chlamydiae by culture of infectious organisms has until very recently been the standard laboratory procedure. The availability of monoclonal antibodies in the 1980s led to the development of diagnosis of patient samples by direct immuno-fluorescent staining and followed by methods for enzyme amplification of the immunological reaction. Although they have the advantage that infectious organisms are not required, these alternatives have not supplanted culture as the procedure of choice. All three remain in use, particularly for comparison with mol-ecular genetic methods and for resolving discrepant results. They are considered in detail by Barnes (1989).

Genetic Methods

Gene sequence detection offers increased sensitivity and, as with immunological procedures, the ability to detect samples inactivated during transport or con-taining viable but non-cultivable forms. These tech-niques are under intensive evaluation in the chlamydia field, given the high incidence of genital infection, its frequent asymptomatic character and its association with infertility in women. The aim is to develop detection procedures that can readily be applied to screening patients at risk (Lee *et al.*, 1995; Quinn *et al.*, 1996a; Scholes *et al.*, 1996).

The two main methods under evaluation are PCR and the ligase chain reaction (LCR) (Quinn, 1994; Ridgway *et al.*, 1996) using either plasmid (multiple copies) or MOMP sequences. In PCR (Birkenmeyer and Mushahwar, 1991) two primer sequences are selected which flank the target sequence to be ampli-fied and which are complementary to the opposite strands of the target. The primers are hybridised (annealed) to the target – each strand of which acts as a template along which the primers are extended towards each other by the polymerase. The primer extension products are dissociated by denaturation to give four template strands from the original two of the target. Each serves as a new template to bind fur-ther primer oligonucleotides for extension. In LCR two pairs of primers are used. Each pair member is complementary to the other and the pairs are selected to hybridise adjacently on the target strand. The space or 'nick' between the adjacent primers is sealed by a DNA ligase. Each ligated strand is then separated by denaturation and acts as template for further rounds of primer-binding and template pro-duction (Birkenmeyer and Mushahwar, 1991; Dille *et al.*, 1993).

Epidemiological studies on chlamydial MOMP variation have used primer sequences adjacent to variable regions followed by PCR amplification and sequencing (Dean *et al.*, 1992; Hayes *et al.*, 1992).

Antimicrobials

Drugs capable of inhibiting chlamydiae fall into four groups: those interfering with protein synthesis, with DNA gyrase activity, with folic acid synthesis or with chlamydial envelope formation. The appearance of stable chlamydial mutants resistant to drugs in com-mon use appears to be extremely rare and has not so far posed a problem (Ridgway, 1992). Full resistance to clindamycin, tetracycline and erythromycin has been reported (Jones *et al.*, 1990); partial resistance to erythromycin (Mourad *et al.*, 1980) and to erythro-mycin and tetracycline has also been described (Jones *et al.*, 1990).

Inhibitors of Protein Synthesis

These include macrolides, tetracyclines, lincosamines (clindamycin), rifamycin and chloramphenicol. Tetra-cyclines are in common clinical use against all chla-mydial species. Erythromycin is also in use, but has side effects and requires administration over a longer period. Azithromycin, a macrolide related to erythro-mycin, has unusual pharmacokinetic properties in that it rapidly achieves high tissue levels, penetrates cells readily and shows a prolonged intracellular half-life (Lode *et al.*, 1996). This allows single-dose treatment (Martin *et al.*, 1992; Bailey *et al.*, 1993) and appears likely to revolutionise treatment of *C. trachomatis* and *C. pneumoniae* infections.

Mutation to resistance against the most active rifa-mycin, rifampicin, is readily achieved by single-step mutation *in vitro* (Jones *et al.*, 1983). Chloram-phenicol, of value in studies *in vitro*, is only moderately active against chlamydiae.

DNA Gyrase Inhibitors

Effective clinical activity against *C. trachomatis* by these fluorinated 4-quinolones is shown particularly by ofloxacin and more recently by sparfloxacin and clinafloxacin.

Inhibitors of Folic Acid Synthesis

Differences between chlamydial species in their susceptibility to sulphonamides is well known. The work of Fan *et al.* (1992) suggests, however, that all species are capable of folate synthesis (see section on co-factor metabolism). Strains that are inhibited by sulphonamide appear to lack a mechanism for accessing the folate synthesised by the host cell.

Inhibitors of Envelope Formation

As indicated earlier, penicillin and related compounds interfere with chlamydial growth through effects on PBPs to induce the formation of non-cultivable chlamydiae. These revert to the infectious state once the inhibitor is removed.

Epidemiology

Trachoma

Trachoma has been estimated to affect some 500 million people, of whom several million have greatly impaired vision (Dawson *et al.*, 1981; Thylefors and Negrel, 1995). It is endemic primarily in tropical and subtropical developing countries and is found in rural areas of extreme poverty. An unexplained feature of trachoma is that serovars A–C are almost always the causative strains, in spite of the proximity, in some instances, of genital and neo-natal infections involving serovars D–K. Transmission arises from the reservoir within local populations (Treharne, 1985; Mabey *et al.*, 1992a) with the highest prevalence of active (inflammatory) trachoma in the very young. Recent studies indicate that the prevalence of trachoma shows village-to-village variation with active cases showing clustering within families. Transmission is to a considerable extent intrafamilial, with evidence that sharing a bedroom with an active case increases the incidence of active disease (Mabey *et al.*, 1992b). These findings suggest that the opportunity for prolonged contact is important for transmission and may also favour the repeated re-infection important for disease development (Grayston *et al.*, 1985; Treharne, 1985). A number of possible risk factors are under evaluation for inclusion in control programmes. Unfortunately, the frequently asymptomatic nature of trachoma in its early stages does not encourage the adoption of change in community behaviour (Mabey *et al.*, 1992a).

Analysis of MOMP sequences has provided evidence of allelic variation in endemic trachoma A and B strains (Hayes *et al.*, 1990, 1992) and in retrospective analysis of Tunisian B and Ba strains isolated in 1972 and 1975 (Dean *et al.*, 1992). All these studies indicate retention of serovar properties as defined by immunological reactivity, with similar findings in further studies of A, B variants over a 22-month period (Hayes *et al.*, 1995). More dramatic changes have been seen in genital strains.

Genital Infections

Chlamydial infections of the genitourinary tract are currently the most common sexually acquired bacterial infection. An estimated 50–70 million cases are detected annually worldwide (Piot, 1994) with more than 4 million new cases each year in the US (Quinn, 1996a) and in North America an estimated carriage of asymptomatic infection of 2–5% within the general population (Mahony *et al.*, 1995). National surveillance in the UK is limited to data from genitourinary medicine clinics and voluntary reporting, but treatment of some 40 000 cases was reported in 1995 (Johnson *et al.*, 1996). Surveys in women attending general practices indicate a prevalence of 2–12%, which is highest for ages below 25 years (Oakshott and Hay, 1995). Similar variations in prevalence have been reported elsewhere in Europe and in developing countries (De Schryver and Meheus, 1990; Piot, 1994).

Asymptomatic chlamydial infection can occur in men (Podgore and Holmes, 1982; Stamm and Cole, 1986; Moncada *et al.*, 1994). In women either symptomatic or asymptomatic lower genital tract infection may be followed by ascending infection and PID (Stamm *et al.*, 1984; Westrom, 1980) with the risk of ectopic pregnancy or infertility as an outcome (Cates *et al.*, 1994). Asymptomatic ('silent') infection may be more frequent than the acute form (Westrom and Wolner-Hansson, 1993; Henry-Suchet *et al.*, 1987; Stacey *et al.*, 1990). The extent and seriousness of PID is leading to the adoption of screening programmes, and is discussed below.

MOMP polymorphism has also been examined in genital and LGV strains and evidence obtained of changes arising from recombination (Brunham *et al.*, 1994; Hayes *et al.*, 1994) – as distinct from the behaviour of endemic trachoma strains. Brunham *et al.* (1996) adduced evidence that change in serovar sequence reflects the gradual development of immunity to existing strains, which in turn show a reducing rate of re-infection. In a further development evidence has begun to emerge that different serovars may be associated with distinctive disease patterns. All of seven women with variant F-strain infections were found to have PID, compared with asymptomatic infection in six women with non-variant F infection, whereas 11 of 12 others infected with an E strain displayed asymptomatic infection (Dean *et al.*, 1995).

Control

A vaccine against chlamydial infections is highly desirable to combat both ophthalmic and genitourinary disease. The environment in which endemic trachoma is found strongly pre-disposes to maintenance of disease. A vaccine which prevents the early damaging lesions that lead subsequently to blinding disease appears at present to be the only feasible option for significant improvement. In genital infections in women the serious consequences of salpingitis and infertility – and their frequently asymptomatic nature – have led to increasing debate on the value of control measures designed to monitor individuals most at risk. These problems, and the question of a possible vaccine, well discussed in the literature, are summarised here.

Monitoring of Genital Infection and Screening Programmes

Overall reduction in the number of chlamydial infections has been achieved in Sweden where notification of *C. trachomatis* infection has been in force since 1982 (Mardh, 1997). Effective detection of PID is made difficult, however, by the high frequency of asymptomatic infection. Screening for asymptomatic chlamydial infection was first proposed in the 1980s (Handsfield *et al.*, 1986; Schachter, 1989), but only recently have recommendations for selective screening been released in the USA and not yet in Europe (Johnson *et al.*, 1996; Mardh, 1997).

The first randomised controlled trial of selective screening of young women has been reported (Scholes *et al.*, 1996). This demonstrated that identifying and treating women with early lower-genital-tract infection appeared to reduce by 56% the incidence of subsequent PID over a 1-year period. Recent evaluation, by PCR, of the infection of partners of individuals that were routinely screened showed that bi-directional (male → female and female → male) transmissions were comparable in number, which emphasises the need for routine screening, contact tracing and treatment of all infected partners (Quinn *et al.*, 1996b). This is more similar than previously recognised by culture detection, and emphasises the need for routine

screening, contact tracing and treatment of all infected partners. Paavonen (1997) argues, from a cost–benefit analysis, that screening plus treatment is likely to be of value even where the prevalence is low.

Vaccine Approaches

Much research on immunity to re-infection is focused on the question of developing a vaccine (Brunham and Peeling, 1994; Rank and Bavoil, 1996; Ward, 1997). Two main themes have emerged: the central importance of protective T cell responses involving the CD4+ Th-1 subset (see above), and experimentation with MOMP-derived components to induce neutralising antibodies.

The 'subunit' approach is designed to avoid chlamydial antigens such as hsp 60 that might induce immunopathology, and MOMP is the prime candidate given its role in infection of the host cell, its antigenic properties and the evidence of serovar-specific immunity. Cloned recombinant fragments have been evaluated (Toye *et al.*, 1990; Tuffry *et al.*, 1992), with some protection against heterologous infection by either mucosal or parenteral routes of immunisation in a mouse model (Tuffry *et al.*, 1992). Perhaps more promising has been the use of poliovirus hybrids as live vectors, with the demonstration that a poliovirus hybrid containing a VD I epitope induced antibodies in rabbits that neutralised serovar A infectivity for monkey conjunctivae (Murdin *et al.*, 1993). Hybrids containing serovar, species or sub-species epitopes, generated antisera that neutralised 8 of the 12 trachoma strain serovars (Murdin *et al.*, 1995). Also, as described above, mice bearing IgA- and IgG-engrafted hybridomas released monoclonal antibodies into serum and vaginal secretions that significantly reduced upper-genital-tract infection (Cotter *et al.*, 1995); protection against vaginal infection was also demonstrated, but only at low challenge doses.

In mice it appears that protection requires previous exposure to live organisms, and it has become clear that the route of exposure influences the nature of the immune response and therefore the degree of protection. Thus, when mice were infected subcutaneously a Th-2 response developed, with high levels of anti-chlamydial antibodies. But these animals failed to resolve subsequent vaginal infection as rapidly as mice immunised orally or intranasally (Kelly *et al.*, 1996; Pal *et al.*, 1996), and in which induction of an IFNγ-producing cell-mediated Th-1 response took place.

The MOMP gene in a cytomegalovirus-containing plasmid (i.e. a DNA vaccine) generates a significant delayed-type hypersensitivity Th-1 response in mice immunised by an intramuscular route (Zhang *et al.*, 1997), and this reduced peak chlamydial growth 100-fold after challenge by lung infection, and generated serum antibodies reactive with elementary bodies. These findings suggest that, with appropriate presentation, MOMP may be able to induce both cell-mediated and local mucosal immunity.

Chlamydia pneumoniae

C. pneumoniae was classified as the third species of *Chlamydia* by Grayston *et al.* (1989) after accumulating evidence indicated that chlamydiae of distinctive properties and forming a homogeneous group (TWAR, Taiwan acute respiratory) were associated in humans with acute infections of the lower respiratory tract (see **Table 1**). *C. pneumoniae* strains do not carry a plasmid. Although the 1989 classification defines EBs as pear-shaped it is now clear that some isolates have EB morphology like that of other species (Carter *et al.*, 1991; Popov *et al.*, 1991; Kanamoto *et al.*, 1993).

Structures

Apart from differences in EB morphology the developmental cycle is like that of other species. Chlamydial protein profiles, including MOMP and CRPs appear similar to those of other chlamydiae. Differences in antibody responses and in MOMP solubility in detergent have, however, been described (Iijima *et al.*, 1994) and a cysteine-containing 98-kDa protein in the sarkosyl-insoluble fraction of EBs is thought to contribute to the pear-shaped morphology (Melgosa *et al.*, 1993). The MOMP *ompA* gene shows five conserved segments and four regions corresponding to the variable regions of *C. trachomatis* and *C. psittaci* but without sequence variation between isolates (Carter *et al.*, 1991; Kaltenboeck *et al.*, 1993; Melgosa *et al.*, 1993) or variation in VS IV of 13 isolates (Gaydos *et al.*, 1992).

Epidemiology

Early evidence suggesting novel chlamydial infection was provided by the finding in blood donors of high levels of anti-chlamydial antibody that greatly exceeded those associated with *C. trachomatis* (Forsey *et al.*, 1986). *C. pneumoniae* is now recognised as a common cause of mild pneumonia in children and young adults, often with slow recovery. Bronchitis or pneumonitis can occur, sometimes with early

pharyngitis and often with a persistent cough and disease may be more severe in older patients (Saikku, 1992). Infection may also persist (Falck *et al.*, 1996) or organisms may be present in symptomless carriage (Hyman *et al.*, 1995).

Chlamydia psittaci

Earlier immunotyping and molecular genetic studies revealed major differences within *C. psittaci* (Herring, 1992; Fukushi and Hirae, 1992) and led to the proposal to classify a closely related group of strains that affect cattle and sheep as a distinct species, *C. pecorum* (Fukushi and Hirai, 1992). The remaining strains classed as *C. psittaci* include isolates from more than 130 species of birds and also strains from goats, sheep abortion, feline, musk-rat, guinea-pig and koala bear. The DNA homology within these is extremely variable (**Table 1**) and further subdivision can be expected as additional criteria become available. Here we briefly consider some special features of the species and diseases caused in humans.

Structures

Inclusion Changes During Development

Many *C. psittaci* strains develop as described earlier for *C. trachomatis*, but others are well recognised to form a lobed inclusion structure believed to arise from independent multiple infections by EBs, but Rockey *et al.* (1996) call this into question, because their analysis of GPIC development shows that the lobed structure arises at low multiplicity of infection. Moreover, staining of the inclusion membrane for the presence of IncA antigen clearly delineates its lobed nature and permits stages in its formation to be correlated with EB–RB development. It appears that lobes of inclusion membrane form around individual dividing RBs and expand with RB division. Once differentiation back to EBs is initiated further, lobe formation ceases but individual lobes continue to expand as EBs accumulate within them. These novel findings will undoubtedly be assessed for other *C. psittaci* strains showing a lobed structure. It is possible that the IncA antigen will prove common to all such strains.

Plasmids

Plasmids are found in all avian and many mammalian strains but not in ovine abortion or ocular koala strains. Although these are similar in size to those found in *C. trachomatis*, different plasmid types, as defined by restriction map differences, have been defined for avian, cat, guinea-pig, horse and koala urogenital strains (Lusher *et al.*, 1991). The reader is referred to the review by Lusher *et al.* (1991) for further information.

Chlamydiophage

Phage in chlamydia-like particles was first described by Harshbarger *et al.* (1977) but not investigated in detail. Subsequent detection in *C. psittaci* isolated from ducks, of a phage, Chp1, was described by Richmond *et al.* (1982). It was characterised as an icosahedral phage, 22 nm in diameter, with a single-stranded DNA genome of 4877 bases. Chp1 has five major ORFs, of which three code for structural proteins, VP1, VP2, VP3, of 67, 28 and 17 kDa, respectively (Storey *et al.*, 1989). Although homology studies failed to indicate any relationship with other sequenced phage, homology of VP1 and ORF 4 with two proteins in ΦX174 was detected and other structural and organisational similarities noted. For these reasons Chp1 has been classified with ΦX174 as a member of the Microviridae. Chp1 is transmissible to other avian *C. psittaci* strains but not to mammalian *C. psittaci* or to *C. trachomatis* or *C. pneumoniae* (Lusher *et al.*, 1991).

Pathogenesis

The most common disease in humans is psittacosis or ornithosis by respiratory infection after contact with psittacine or other birds infected with avian *C. psittaci* strains. Infection occurs most commonly in individuals in frequent contact with live birds or their carcasses. In the past this was often as an occupational disease (Meyer and Eddie, 1962; Palmer, 1982). Lower respiratory infection (pneumonitis) develops in 1–2 weeks after exposure, and in more severe cases infection may spread to the spleen and liver, with toxaemia. Death is now uncommon, because of antibiotic therapy.

Occasional infection with other *C. psittaci* strains has been reported (Johnson, 1983). Occasional, but more common, have been reports of infection by the ovine abortion strain (enzootic abortion of ewes). In most instances previous contact with infected flocks had occurred and infection was initiated by the oropharyngeal route. It is most serious in pregnant women, in whom involvement of the placenta and spontaneous abortion is a potential outcome (Flanagan *et al.*, 1996).

References

Albert ML, Sauter B, Bhardwaj N (1998) Dendritic cells acquire antigen from apoptotic cells and induce class I restricted CTLs. *Nature* 392: 86–89.

Allan I, Pearce JH (1983) Amino-acid-requirements of strains of *Chlamydia trachomatis* and *Chlamydia psittaci* growing in McCoy cells – relationship with clinical syndrome and host origin. *J. Gen. Microbiol.* 129: 2001–2007.

Allen JE, Stephens RS (1989) Identification by sequence analysis of two-site post-translational processing of the cysteine-rich outer membrane protein 2 of *Chlamydia trachomatis* serovar L2. *J. Bacteriol.* 171: 285–291.

Allen JE, Stephens RS (1993) An intermolecular mechanism of T-cell help for the production of antibodies to the bacterial pathogen, *Chlamydia trachomatis*. *Eur. J. Immunol.* 23: 1169–1172.

Amann R, Springer N, Schonhuber W *et al.* (1997) Obligate intracellular bacterial parasites of Acanthamoebae related to *Chlamydia* spp. *Appl. Environ. Microbiol.* 63: 115–121.

Anderton SM, Vanderzee R, Prakken B, Noordzij A, Vaneden W (1995) Activation of I cells recognizing self 60-kD heat shock protein can protect against experimental arthritis. *J. Exp. Med.* 181: 943–952.

Bachmaier K, Neu N, delaMaza L (1999) Chlamydia infections and heart disease linked through antigenic mimicry. *Science* 283: 1335–1339.

Bader JP, Morgan HR (1961) Latent viral infections of cells in tissue culture. VII. Role of water-soluble vitamins in psittacosis virus propagation in L cells. *J. Exp. Med.* 113: 271–281.

Baehr W, Zhang YX, Joseph T *et al.* (1988) Mapping antigenic domains expressed by *Chlamydia trachomatis* major outer membrane protein genes. *Proc. Natl Acad. Sci. USA* 85: 4000–4004.

Bailey RL, Arullendran P, Whittle HC, Mabey DCW (1993) Randomised controlled trial of single-dose azithromycin in treatment of trachoma. *Lancet* 342: 453–456.

Bailey RL, Holland MJ, Whittle HC, Mabey DCW (1995) Subjects recovering from human ocular chlamydial infection have enhanced lymphoproliferative responses to chlamydial antigens compared with those of persistently diseased controls. *Infect. Immun.* 63: 389–392.

Barbour AG, Amano KI, Hackstadt T, Perry L, Caidwell HD (1982) *Chlamydia trachomatis* has penicillin-binding proteins but not detectable muramic acid. *J. Bacteriol.* 151: 420–428.

Barnes RC (1989) Laboratory diagnosis of human chlamydial infections. *Clin. Microbiol. Rev.* 2: 119–136.

Barron AL (ed.) (1988) *Microbiology of Chlamydia*. Boca Raton, FL: CRC Press, pp. 250.

Barry CE, Hayes SF, Hackstadt I (1992) Nucleoid condensation in *Escherichia coli* that express a chlamydial histone homolog. *Science* 256: 377–379.

Barry CE, Brickman IJ, Llackstadt T (1993) Hc1-mediated effects on DNA structure – a potential regulator of chlamydial development. *Mol. Microbiol.* 9: 273–283.

Batteiger BE, Newhall WJ, Jones RB (1985) Differences in outer membrane proteins of the lymphogranuloma venereum and trachoma biovars of *Chlamydia trachomatis*. *Infect. Immun.* 50: 488–494.

Baumaun M, Brade L, Fasske E, Brade H (1992) Staining of surface-antigens of *Chlamydia trachomatis* L2 in tissue culture. *Infect. Immun.* 60: 4433–4438.

Bauwens JE, Lampe MF, Suchland RJ, Wong K, Stamm WE (1995) Infection with *Chlamydia trachomatis* Lymphogranuloma venereum serovar L1 in homosexual men with proctitis: molecular analysis of an unusual case cluster. *Clin. Infect. Dis.* 20: 576–581.

Bavoil P, Ohlin A, Schachter J (1984) Role of disulfide bonding in outer membrane structure and permeability in *Chlamydia trachomatis*. *Infect. Immun.* 44: 479–485.

Bavoil PM, Hsia RC, Rank RG (1996) Prospects for a vaccine against chlamydia genital disease 1. Microbiology and pathogenesis. *Bull. Inst. Pasteur* 94: 5–54.

Beatty PR, Stephens RS (1992) Identification of *Chlamydia trachomatis* antigens by use of murine T-cell lines. *Infect. Immun.* 60: 4598–4603.

Beatty PR, Stephens RS (1994) CD8(+) T lymphocyte-mediated lysis of chlamydia-infected L cells using an endogenous antigen pathway. *J. Immunol.* 153: 4588–4595.

Beatty PR, Rasmussen SJ, Stephens RS (1997) Cross-reactive cytotoxic T-lymphocyte-mediated lysis of *Chlamydia trachomatis*- and *Chlamydia psittaci-infected* cells. *Infect. Immun.* 65: 951–956.

Beatty WL, Byrne GI, Morrison RP (1993) Morphologic and antigenic characterization of Interferon gamma-mediated persistent *Chlamydia trachomatis* infection *in vitro*. *Proc. Natl Acad. Sci. USA* 90: 3998–4002.

Beatty WL, Morrison RP, Byrne GI (1994a) Immunoelectron-microscopic quantitation of differential levels of chlamydial proteins in a cell culture model of persistent *Chlamydia trachomatis* infection. *Infect. Immun.* 62: 4059–4062.

Beatty WL, Belanger TA, Desai AA, Morrison RP, Byrne GI (1994b) Tryptophan depletion as a mechanism of gamma interferon-mediated chlamydial persistence. *Infect Immun.* 62: 3705–3711.

Beatty WL, Morrison RP, Byrne GI (1994c) Persistent chlamydiae: From cell culture to a paradigm for chlamydial pathogenesis. *Microbiol. Rev.* 58: 686–699.

Beatty WL, Byrne GI, Morrison RP (1994d) Repeated and persistent infection with *Chlamydia* and the development of chronic inflammation and disease. *Trends Microbiol.* 2: 94–98.

Belunis CJ, Mdluli KE, Raetz CRH, Nano FE (1992) A novel 3-deoxy-D-manno-octulosonic acid transferase from *Chlamydia trachomatis* required for expression of the genus-specific epitope *J. Biol. Chem.* 267: 18702–18707.

Beutler AM, Schumacher HR, Whittum-Hudson JA, Salameh WA, Hudson AP (1995) *In situ* hybridization for detection of inapparent infection with *Chlamydia trachomatis* in synovial tissue of a patient with Reiter's syndrome. *Am. J. Med. Sci.* 310: 206–213.

Birkelund S, Lundemose AG, Christiansen G (1988) Chemical cross-linking of *Chlamydia trachomatis*. *Infect. Immun.* 56: 654–659.

Birkelund S, Lundemose AG, Christiansen G (1989) Characterization of native and recombinant 75-kilodalton immunogens from *Chlamydia trachomatis* serovar L2. *Infect. Immun.* 57: 2683–2690.

Birkelund S, Lundemose AC, Christiansen C (1989) Immunoelectron microscopy of lipopolysaccharide in *Chlamydia trachomatis*. *Infect. Immun.* 57: 3250–3253.

Birkelund S, Lundemose AC, Christiansen G (1990) The 75-kilodalton cytoplasmic *Chlamydia trachomatis* L2 polypeptide is a DnaK-like protein. *Infect. Immun.* 58: 2098–2104.

Birkelund S, Stephens RS (1992) Construction of physical and genetic maps of *Chlamydia trachomatis* serovar-L2 by pulsed-field gel-electrophoresis. *J. Bacteriol.* 174: 2742–2747.

Birkelund S, Johnsen H, Christiansen G (1994a) *Chlamydia trachomatis* serovar L2 induces protein-tyrosine phosphorylation during uptake by HeLa cells. *Infect. Immun.* 62: 4900–4908.

Birkelund S, Larsen B, Holm A, Lundemose AG, Christiansen G (1994b) Characterization of a linear epitope on *Chlamydia trachomatis* serovar L2 DnaK-like protein. *Infect. Immun.* 62: 2051–2057.

Birkenmeyer LC, Mushahwar IK (1991) DNA probe amplification methods. *J. Virol. Methods.* 35: 117–126.

Brade H, Brade L, Nano FE (1987a) Chemical and serological investigations on the genus-specific lipopolysaccharide epitope of chlamydia. *Proc. Natl Acad. Sci. USA* 84(12): 2508–2525.

Brade L, Nano FE, Schlecht S, Schramek S, Brade H (1987b) Antigenic and immunogenic properties of recombinants from *Salmonella typhimurium* and *Salmonella minnesota* rough mutants expressing in their lipopolysaccharide a genus-specific chlamydial epitope. *Infect. Immun.* 55: 482–486.

Branigan PJ, Gerard HC, Hudson AP, Schumacher HR (1996) Comparison of synovial tissue and synovial fluid as the source of nucleic acids for detection of *Chlamydia trachomatis* by polymerase chain reaction. *Arthritis. Rheum* 39: 1740–1746.

Brunham RC, Kuo CC, Cles L, Holmes KK (1983) Correlation of host immune response with quantitative recovery of *Chlamydia trachomatis* from the human endocervix. *Infect. Immun.* 39: 1491–1494.

Brunham RC, Peeling RW (1994) *Chlamydia trachomatis* antigens – role in immunity and pathogenesis. *Infect. Agents Dis. Rev.* 3: 218–233.

Brunham R, Yang CL, Maclean I *et al.* (1994) *Chlamydia trachomatis* from individuals in a sexually transmitted disease core group exhibit frequent sequence variation in the major outer membrane protein (omp1) gene. *J. Clin. Invest.* 94(45): 8–463.

Brunham RC, Kimani J, Bwayo J *et al.* (1996) The epidemiology of *Chlamydia trachomatis* within a sexually-transmitted diseases core group. *J. Infect. Dis.* 173: 950–956.

Brunst M, Shahmanesh M, Sukthankar A, Pearce JH, Gaston JS (1998) Isolation and characterisation of T lymphocytes from the urethra of patients with acute urethritis. *Sex Transm. Infect.* 74: 279–283.

Bukowski J, Morita C, Brenner M (1999) Human γδ T cells recognise alkylamines derived from microbes, edible plants and tea: implications for innate immunity. *Immunity* 11: 57–66.

Burmester CR, Daser A, Kamradt T *et al.* (1995) Immunology of reactive arthritidis. *Annu. Rev. Immunol.* 13: 229–250.

Busby S, Ebright RH (1994) Promoter structure, promoter recognition, and transcription activation in prokaryotes. *Cell* 79: 743–746.

Buzoni-Gatel D, Guilloteau L, Bernard F *et al.* (1992) Protection against *Chlamydia psittaci* in mice conferred by Lyt-2+ T cells. *Immunology* 77: 284–288.

Byrne GI, Lehmann LK, Landry GJ (1986) Induction of tryptophan catabolism is the mechanism for gamma-interferon-mediated inhibition of intracellular *Chlamydia psittaci* replication in t24 cells. *Infect. Immun.* 53: 347–351.

Caldwell HD, Kromhout I, Schachter J (1981) Purification and partial characterization of the major outer membrane protein of *Chlamydia trachomatis*. *Infect. Immun.* 31: 1161–1176.

Campbell S, Richmond SJ, Yates P (1989) The development of *Chlamydia trachomatis* inclusions within the host eukaryotic cell during interphase and mitosis. *J. Gen. Microbiol.* 135: 1153–1165.

Campbell S, Richmond SJ, Yates PS, Storey CC (1994) Lipopolysaccharide in cells infected by *Chlamydia trachomatis*. *Microbiol. UK* 140: 1995–2002.

Carter MW, al-Mahdawi SA, Giles IG, Treharne JD, Ward ME, Clark IN (1991). Nucleotide sequence and taxonomic value of the major outer membrane protein gene of Chlamydia pneumoniae IOL-207. *J. Gen. Microbiol.* 137: 465–475.

Cates W, Wasserheit JN, Marchbanks PA (1994) Pelvic inflammatory disease and tubal infertility – the preventable conditions. *Ann. N Y Acad. Sci.* 709: 179–195.

Cerrone MC, Ma JJ, Stephens HS (1991) Cloning and sequence of the gene for heat shock protein 60 from *Chlamydia trachomatis* and immunological reactivity of the protein. *Infect. Immun.* 59: 79–90.

Chen JCR, Stephens RS (1994) Trachoma and Lgv biovars of *Chlamydia trachomatis* share the same glycosamino-glycan-dependent mechanism for infection of eukaryotic cells. *Mol. Microbiol.* 11: 501–507.

Chen JCR, Stephens RS (1997) *Chlamydia trachomatis* glycosaminoglycan-dependent and independent attachment to eukaryotic cells. *Microb. Pathog.* 22: 23–30.

Chen JCR, Zhang JP, Stephens RS (1996) Structural requirements of heparin-binding to *Chlamydia trachomatis*. *J. Biol. Chem* . 271: 11134–11140.

Chopra I, Storey C, Falla TJ, Pearce JH (1998) Antibiotics, peptidoglycan synthesis and genomics: the chlamydial anomaly revisited. *Microbiol.* 144: 2673-2678.

Christiansen G, Pedersen LB, Koehier JE, Lundemose AG, Birkelund S (1993) Interaction between the *Chlamydia trachomatis* histone hl-like protein (hcl) and DNA. *J. Bacteriol.* 175: 1785–1795.

Coles AM, Pearce JH (1987) Regulation of *Chlamydia psittaci* (strain guinea-pig inclusion conjunctivitis) growth in McCoy cells by amino acid antagonism. *J. Gen. Microbiol.* 133: 701–708.

Coles AM, Reynolds DJ, Harper A, Devitt A, Pearce JH (1993) Low-nutrient induction of abnormal chlamydial development – a novel component of chlamydial pathogenesis? *FEMS Microbiol. Lett.* 106: 193–200.

Collett BA, Newhall WJ, Jersild R Jr, Jones RB (1989). Detection of surface-exposed epitopes on *Chlamydia trachomatis* by immune electron microscopy. *J. Gen. Microbiol.* 135: 85–94.

Collier LH, Sowa J (1958) Isolation of trachoma virus in embryonate eggs. *Lancet* i: 993–996.

Comanducci M, Ricci S, Ratti G (1988) The structure of a plasmid of *Chlamydia trachomatis* believed to be required for growth within mammalian cells. *Mol. Microbiol.* 2: 531–538.

Comanducci M, Ricci S, Cevenini R, Ratti C (1990) Diversity of the *Chlamydia trachomatis* common plasmid in biovars with different pathogenicity. *Plasmid* 23: 149–154.

Comanducci M, Cevenini R, Moroni A *et al.* (1993) Expression of a plasmid gene of *Chlamydia trachomatis* encoding a novel 28 kDa antigen. *J. Gen. Microbiol.* 139: 1083–1092.

Comanducci M, Manetti R, Bini H *et al.* (1994) Humoral immune response to plasmid protein pgp3 in patients with *Chlamydia trachomatis* infection. *Infect. Immun.* 62: 5491–5497.

Conlan JW, Clarke IN, Ward ME (1988) Epitope mapping with solid phase peptides – identification of type-reactive, subspecies-reactive, species-reactive and genus-reactive antibody binding domains on the major outer membrane protein of *Chlamydia trachomatis*. *Mol. Microbiol.* 2: 673–679.

Cotter TW, Meng Q, Shen ZL *et al.* (1995) Protective efficacy of major outer membrane protein specific immunoglobulin-A (IgA) and IgG monoclonal-antibodies in a murine model of *Chlamydia trachomatis* genital tract infection. *Infect. Immun.* 63: 4704–4714.

Crabtree GR, Clipstone NA (1994) Signal transmission between the plasma membrane and nucleus of T lymphocytes. *Annu. Rev. Biochem.* 63: 1045–1083.

Crenshaw RW, Fahr MJ, Wichlan DC, Hatch TP (1990) Developmental cycle-specific host-free RNA synthesis in *Chlamydia* spp. *Infect. Immun.* 58: 3194–3201.

Curry AJ, Portig I, Goodall JC Kirkpatrick PJ, Gaston JSH (2000) T lymphocyte lines isolated from atheromatous plaque contain cells capable of responding to chlamydial antigens. *Clin. Exp. Immunol.* 121: 261–269

Cyster JG (1999) Chemokines and the homing of dendritic cells to the T cell areas of lymphoid organs. *J. Exp. Med.* 189: 447–450.

Danilition SL, Maclean IW, Peeling R, Winston S, Brunham RC (1990) The 75-kilodalton protein of *Chlamydia trachomatis*: A member of the heat shock protein 70 family? *Infect. Immun.* 58: 189–196.

Darougar S, Jones BR (1983) Trachoma. *Br. Med. Bull.* 39: 117–122.

Dawson CR, Jones BR, Darougar S (1975) Blinding and non-blinding trachoma: assessment of intensity of upper tarsal inflammatory disease and disabling lesions. *Bull. WHO* 52: 279–282.

Dawson CR, Jones BR, Tarizzo ML (1981) *A Guide to Trachoma Control*. Geneva: World Health Organization pp. 56.

De Schryver A, Meheus A (1990) Epidemiology of sexually transmitted diseases: the global picture. *Bull. WHO* 68: 639–654.

Dean D, Schachter J, Dawson CR, Stephens RS (1992) Comparison of the major outer membrane protein variant sequence regions of B/Ba isolates: a molecular epidemiologic approach to *Chlamydia trachomatis* infections. *J. Infect. Dis.* 166: 383–392.

Dean D, Oudens E, Bolan G, Padian N, Schachter J (1995) Major outer membrane protein variants of *Chlamydia trachomatis* are associated with severe upper genital tract infections and histopathology in San Francisco. *J. Infect. Dis.* 172: 1013–1022.

Deane K, Jecock R, Pearce J, Gaston J (1997) Identification and characterization of a DR4-restricted T cell epitope within chlamydia hsp60. *Clin. Exp. Immunol.* 109: 439–445.

DeLibero G (1997) Sentinel function of broadly reactive human gamma delta T cells. *Immunol. Today* 18: 22–26.

Devitt A, Lund PA, Morris AG, Pearce JH (1996) Induction of alpha/beta interferon and dependent nitric oxide synthesis during *Chlamydia trachomatis* infection of McCoy cells in the absence of exogenous cytokine. *Infect. Immun.* 64: 3951–3956.

Dille BJ, Butzen CC, Birkenmeyer LG (1993) Amplification of *Chlamydia trachomatis* DNA by ligase chain reaction. *J. Clin. Microbiol.* 31: 729–731.

Doughri AM, Storz J, Altera KP (1972) Mode of entry and release of chlamydiae in infections of epithelial intestinal cells. *J. Infect. Dis.* 126: 652–657.

Douglas AL, Saxena NK, Hatch TP (1994) Enhancement of *in vitro* transcription by addition of cloned, overexpressed major sigma factor of *Chlamydia psittaci* 6BC. *J. Bacteriol.* 176: 3033–3039.

Douglas AL, Hatch TP (1995) Functional analysis of the major outer membrane protein gene promoters of *Chlamydia trachomatis*. *J. Bacteriol.* 177: 6286–6289.

Douglas AL, Hatch TP (1996) Mutagenesis of the p2 promoter of the major outer membrane protein gene of *Chlamydia trachomatis*. *J. Bacteriol.* 178: 5573–5578.

Engel JN, Ganem D (1990) A polymerase chain reaction-based approach to cloning sigma factors from eubacteria and its application to the isolation of a sigma 70 homolog from *Chlamydia trachomatis*. *J. Bacteriol.* 172: 2447–2455.

Everett KD, Hatch TP (1991) Sequence analysis and lipid modification of the cysteine-rich envelope proteins of *Chlamydia psittaci* 6BC. *J. Bacteriol.* 173: 3821–3830.

Everett KD, Hatch TP (1995) Architecture of the cell envelope of *Chlamydia psittaci* 6BC. *J. Bacteriol.* 177: 877–882.

Fahr MJ, Douglas AL, Xia W, Hatch TP (1995) Characterization of late gene promoters of *Chlamydia trachomatis*. *J. Bacteriol.* 177: 4252–4260.

Falck G, Gnarpe J, Gnarpe H (1996) Persistent *Chlamydia pneumoniae* infection in a Swedish family. *Scand J. Infect. Dis.* 28: 271–273.

Fan HZ, McClarty G, Brunham RC (1991) Biochemical evidence for the existence of thymidylate synthase in the obligate intracellular parasite *Chlamydia trachomatis*. *J. Bacteriol.* 173: 6670–6677.

Fan HZ, Brunham RC, McClarty G (1992) Acquisition and synthesis of folates by obligate intracellular bacteria of the genus *Chlamydia*. *J. Clin. Invest.* 90: 1803–1811.

Fan T, Lu H, Hu H et al. (1998) Inhibition of apoptosis in Chlamydia-infected cells: Blockade of mitochondrial cytochrome c release and caspase activation. *J. Exp. Med.* 187: 487–496.

Fitch WM, Peterson EM, Delamaza LM (1993) Phylogenetic analysis of the outer membrane protein genes of chlamydiae, and its implication for vaccine development. *Mol. Biol. Evol.* 10: 892–899.

Flanagan PG, Westmoreland D, Stallard N, Stokes IM, Evans J (1996) Ovine chlamydiosis in pregnancy. *Br. J. Obstet. Gynaecol.* 103: 382–385.

Forsey T, Darougar S, Treharne JD (1986) Prevalence in human beings of antibodies to *Chlamydia* IOL-207, an atypical strain of *Chlamydia*. *J. Infect.* 12: 145–152.

Fox A, Rogers JC, Gilbert J et al. (1990) Muramic acid is not detectable in *Chlamydia psittaci* or *Chlamydia trachomatis* by gas chromatography-mass spectrometry. *Infect. Immun.* 58: 835–837.

Fukushi H, Hirai K (1989) Genetic diversity of avian and mammalian *Chlamydia psittaci* strains and relation to host origin. *J. Bacteriol.* 171: 2850–2855.

Fukushi H, Hirai K (1992) Proposal of *Chlamydia pecorum* sp. nov. for chlamydia strains derived from ruminants. *Int. J. Syst. Bacteriol.* 42: 306–308.

Fukushi H, Hirai K (1993) Restriction fragment length polymorphisms of ribosomal RNA as genetic markers to differentiate *Chlamydia* spp. *Int. J. Syst. Bacteriol.* 43: 613–617.

Galan JE, Pace J, Hayman MJ (1992) Involvement of the epidermal growth factor receptor in the invasion of the epithelial cells by *Salmonella typhimurium*. *Nature* 357: 588–589.

Gaston JSH, Life PF, MerilahtiPalo R *et al.* (1989) Synovial T lymphocyte recognition of organisms that trigger reactive arthritis. *Clin. Exp. Immunol.* 76: 348–353.

Gaston JSH, Pearce JH (1994) Immune responses to heat shock proteins in reactive arthritis. In: van Eden W (ed.) *Stress Proteins in Medicine*. New York: Marcel Dekker, pp. 103–117.

Gaston JSH, Deane KHO, Jecock RM, Pearce JH (1996) Identification of 2 *Chlamydia trachomatis* antigens recognised by synovial fluid T cells from patients with chlamydia-induced reactive arthritis. *J. Rheumatol.* 23: 130–136.

Gaydos CA, Quinn TC, Bobo LD, Eiden JJ (1992) Similarity of *Chlamydia pneumoniae* strains in the variable domain IV region of the major outer membrane protein gene. *Infect. Immun.* 60: 5319–5323.

Gaydos CA, Summersgill JT, Sahney NN, Ramirez JA, Quinn TC (1996) Replication of *Chlamydia pneumoniae in vitro* in human macrophages, endothelial cells and aortic artery smooth muscle cells. *Infect. Immun.* 64: 1614–1620.

Glassick T, Giffard P, Timms P (1996) Outer membrane protein 2 gene sequences indicate that *Chlamydia pecorum* and *Chlamydia pneumoniae* cause infections in koalas. *Syst. Appl. Microbiol.* 19: 457–464.

Gnarpe H, Gnarpe J (1993) Increasing prevalence of specific antibodies to *Chlamydia pneumoniae* in Sweden. *Lancet* 341: 381.

Goodall J, Clements S, Deane K, Hassell A, Gaston J (1997) Recognition of *chlamydia trachomatis* (CT) outer membrane protein 2 (OMP2) by a reactive arthritis synovial T cell clone. *Arthritis Rheum.* 41: S143.

Grayston JT, Wang S-P, Yeh L-J, Kuo C-C (1985) Importance of reinfection in the pathogenesis of trachoma. *Rev. Infect. Dis.* 7: 717–725.

Grayston JT, Kuo C-C, Campbell LA, Wang S-P (1989) *Chlamydia pneumoniae* sp. nov. for *Chlamydia* sp. strain TWAR. *Int. J. Syst. Bacteriol.* 39: 88–90.

Gutierrez-Martin CB, Ojcius DM, Hsia RC *et al.* (1997) Heparin-mediated inhibition of *Chlamydia psittaci* adherence to HeLa cells. *Microb. Pathog.* 22: 47–57.

Hackstadt T, Scidmore MA, Rockey DD (1995) Lipid metabolism in *Chlamydia trachomatis* infected cells – directed trafficking of golgi-derived sphingolipids to the chlamydial inclusion. *Proc. Natl Acad. Sci. USA* 92: 4877–4881.

Hackstadt T, Rockey DD, Heinzen RA, Scidmore MA (1996) *Chlamydia trachomatis* interrupts an exocytic pathway to acquire endogenously synthesised sphingomyelin in transit from the golgi apparatus to the plasma membrane. *EMBO J.* 15: 964–977.

Hahn YS, Yang B, Braciale TJ (1996) Regulation of antigen processing and presentation to class I MHC restricted CD8(+)T lymphocytes. *Immunol. Rev.* 15: 131–149.

Halberstaedter L, Prowazek S von (1907) Uber Zelleinschlusse parasitärer Natur beim Trachom. *Arb. Kaiserlichen Gesund.* 26: 44–47.

Halme S, Juvonen T, Laurila A *et al.* (1999) *Chlamydia pneumoniae* reactive T lymphocytes in the walls of abdominal aortic aneurysms. *Eur. J. Clin. Invest.* 29: 546–552.

Handsfield HH, Jasman LL, Roberts PL *et al.* (1986) Criteria for selective screening for *Chlamydia trachomatis* infection in women attending family-planning clinics. *JAMA* 255: 1730–1734.

Harper A, Pogson CI, Jones ML, Pearce JH (2000) Chlamydial development is adversely affected by minor changes in amino acid supply, blood plasma amino acid levels and glucose deprivation. *Infect. Immun.* 68: 1457–1464.

Harshbarger JC, Chang SC, Otto SV (1977) Chlamydiae (with phage), Mycoplasma and Rickettsiae in Chesapeake Bay bivalves. *Science* 196: 666–668.

Hassell AB, Pilling D, Reynolds D *et al.* (1992) MHC restriction of synovial fluid lymphocyte responses to the triggering organism in reactive arthritis. Absence of a class I restricted response. *Clin. Exp. Immunol.* 88: 442–447.

Hassell AB, Reynolds DJ, Deacon M, Gaston JSH, Pearce JH (1993) Identification of T-cell stimulatory antigens of *Chlamydia trachomatis* using synovial fluid-derived T Cell clones. *Immunology* 79: 513–519.

Hatch TP (1975) Competition between *Chlamydia psittaci* and L cells for host isoleucine pools: a limiting factor in chlamydial multiplication. *Infect. Immun.* 12: 211–220.

Hatch TP, Vance DW, Al Hossainy K (1981) Identification of a major envelope protein in *Chlamydia* spp. *J. Bacteriol.* 146: 426–429.

Hatch TP, Al-Hossainy E, Silverman JA (1982) Adenine nucleotide and lysine transport in *Chlamydia psittaci*. *J. Bacteriol.* 150: 662–670.

Hatch TP, Allan I, Pearce JH (1984) Structural and polypeptide differences between envelopes of infective and reproductive life-cycle forms of *Chlamydia* spp. *J. Bacteriol.* 157: 13–20.

Hatch TP, Miceli M, Sublett JE (1986) Synthesis of disulfide-bonded outer membrane proteins during the developmental cycle of *Chlamydia psittaci* and *Chlamydia trachomatis*. *J. Bacteriol.* 165: 379–385.

Hatch TP (1988) Metabolism of *Chlamydia*. In: Barron AL (ed.) *Microbiology of Chlamydia*. Boca Raton FL: CRC Press, pp. 97–110.

Hatch TP (1996) Disulfide cross-linked envelope proteins – the functional equivalent of peptidoglycan in chlamydiae. *J. Bacteriol.* 178: 1–5.

Hatt C, Ward ME, Clarke IN (1988) Analysis of the entire nucleotide sequence of the cryptic plasmid of *Chlamydia trachomatis* serovar L1 – evidence for involvement in DNA replication. *Nucl. Acids Res.* 16: 4053–4067.

Hayes LJ, Pickett MA, Conlan JW *et al.* (1990) The major outer membrane proteins of *Chlamydia trachomatis* serovar A and serovar B: intra-serovar amino acid changes do not alter specificities of serovar- and C subspecies-reactive antibody-binding domains. *J. Gen. Microbiol.* 136: 1559–1566.

Hayes LJ, Bailey RL, Mabey DC *et al.* (1992) Genotyping of *Chlamydia trachomatis* from a trachoma-endemic village in the Gambia by a nested polymerase chain reaction: identification of strain variants. *J. Infect. Dis.* 166: 1173–1177.

Hayes LJ, Yearsley P, Treharne JD *et al.* (1994) Evidence for naturally occurring recombination in the gene encoding the major outer membrane protein of lymphogranuloma venereum isolates of *Chlamydia trachomatis. Infect. Immun.* 62: 5659–5663.

Hayes LJ, Pecharatana S, Bailey RL *et al.* (1995) Extent and kinetics of genetic change in the ompl gene of *Chlamydia trachomatis* in 2 villages with endemic trachoma. *J. Infect. Dis.* 172: 268–272.

Hearn SA, McNabb GL (1991) Immunoelectron microscopic localization of chlamydial lipopolysaccharide (LPS) in McCoy cells inoculated with *Chlamydia trachomatis. J. Histochem. Cytochem.* 39: 1067–1075.

Heinzen RA, Scidmore MA, Rockey DD, Hackstadt T (1996) Differential interaction with endocytic and exocytic pathways distinguish parasitophorous vacuoles of *Coxiella burnetii* and *Chlamydia trachomatis. Infect. Immun.* 64: 796–809.

Heinzen RA, Hackstadt T (1997) The *Chlamydia trachomatis* parasitophorous vacuolar membrane is not passively permeable to low molecular weight compounds. *Infect. Immun.* 65: 1088–1094.

Hemmerich P, Neu E, Macht M *et al.* (1998) Correlation between chlamydial infection and autoimmune response: molecular mimicry between RNA polymerase major sigma subunit from *Chlamydia trachomatis* and human L7. *Eur. J. Immunol.* 28: 3857–3866.

Henry-Suchet J, Utzmann C, Debrux J, Ardoin P, Catalan F (1987) Microbiologic study of chronic inflammation associated with tubal factor infertility – role of *Chlamydia trachomatis. Fertil. Steril.* 47: 274–277.

Herring AJ (1992) The molecular biology of *Chlamydia* – a brief overview. *J. Infect.* 25: 1–10.

Hodinka RL, Wyrick PB (1986) Ultrastructural study of mode of entry of *Chlamydia psittaci* into L 929 cells. *Infect. Immun.* 54: 855–863.

Hodinka RL, Davis CH, Choong J, Wyrick PB (1988) Ultrastructural study of endocytosis of *Chlamydia trachomatis* by McCoy cells. *Infect. Immun.* 56: 1456–1463.

Holland MJ, Bailey RL, Hayes LJ, Whittle HC, Mabey DCW (1993) Conjunctival scarring in trachoma is associated with depressed cell-mediated immune responses to chlamydial antigens. *J. Infect. Dis.* 168: 1528–1531.

Hudson AP, Gerard HC, Branigan PJ, Schumacher HR (1996) Differential gene expression in *Chlamydia trachomatis* during inapparent synovial infection. *Arthritis Rheum.* 39: 184.

Hyman CL, Roblin PM, Gaydos CA *et al.* (1995) Prevalence of asymptomatic nasopharyngeal carriage of *Chlamydia pneumoniae* in subjectively healthy adults – assessment by polymerase chain reaction-enzyme immunoassay and culture. *Clin. Infect. Dis.* 20: 1174–1178.

Igietseme JU (1996) Molecular mechanism of T-cell control of *Chlamydia* in mice: Role of nitric oxide *in vivo. Immunology* 88: 1–5.

Igietseme JU, Magee DM, Williams DM, Rank RG (1994) Role for CD8(+) T cells in antichlamydial immunity defined by *Chlamydia-specific* T-lymphocyte clones. *Infect. Immun.* 62: 5195–5197.

Iijima V, Miyashita N, Kishimoto T *et al.* (1994) Characterization of *Chlamydia pneumoniae* species-specific proteins immunodormant in humans. *J. Clin. Microbiol.* 32: 583–588.

Ingalls RR, Rice PA, Qureshi N *et al.* (1995) The inflammatory cytokine response to *Chlamydia trachomatis* infection is endotoxin mediated. *Infect. Immun.* 63: 3125–3130.

Jackson LA, Campbell LA, Kuo CC *et al.* (1997) Isolation of *Chlamydia pneumoniae* from a carotid endarterectomy specimen. *J. Infect. Dis.* 176: 292–295.

Johansson M, Schon K, Ward ME, Lycke N (1997) Genital tract infection with *Chlamydia trachomatis* fails to induce protective immunity in gamma interferon receptor-deficient mice despite a strong local immunoglobulin A response. *Infect. Immun.* 65: 1032–1044.

Johnson FWA (1983) Chlamydiosis. *Br. Vet. J.* 139: 93–101.

Johnson AM, Grun L, Haines A (1996) Controlling genital chlamydial infection. *BMJ* 313: 1160–1161.

Jones RB, Ridgway GL, Boulding S, Hunley KL (1983) *In vitro* activity of rifamycins alone and in combination with other antibiotics against *Chlamydia trachomatis. Rev. Infect. Dis.* 5: S556–S561.

Jones RB, Van der Pol B, Martin DH, Shepard MK (1990) Partial characterization of *Chlamydia trachomatis* isolates resistant to multiple antibiotics. *J. Infect. Dis.* 162: 1309–1315.

Joseph TD, Bose SK (1991) Further characterization of an outer membrane protein of *Chlamydia trachomatis* with cytadherence properties. *FEMS Microbiol. Lett.* 68: 167–171.

Joseph T, Nano FE, Garon CF, Caldwell HD (1986) Molecular characterization of *Chlamydia trachomatis* and *Chlamydia psittaci* plasmids. *Infect. Immun.* 51: 699–703.

Kalayoglu MV, Byrne GI (1998) A *Chlamydia pneumoniae* component that induces macrophage foam cell formation is chlamydial lipopolysaccharide. *Infect. Immun.* 66: 5067–5072.

Kaltenboeck B, Kousoulas KC, Storz J (1993) Structures of and allelic diversity and relationships among the major outer membrane protein (ompa) genes of the 4 chlamydial species. *J. Bacteriol.* 175: 487–502.

Kanamoto Y, Iijima Y, Miyashita N, Matsumoto A, Sakano T (1993) Antigenic characterization of *Chlamydia pneumoniae* isolated in Hiroshima, Japan. *Microbiol. Immunol.* 37: 495–498.

Karimi ST, Schloemer RH, Wilde CE 3rd (1989). Accumulation of chlamydial lipopolysaccharide antigen in the plasma membranes of infected cells. *Infect. Immun.* 57: 1780–1785.

Kaukorantatolvanen SSE, Ronni T, Leinonen M, Saikku P, Laitinen K (1996) Expression of adhesion molecules on endothelial cells stimulated by *Chlamydia pneumoniae*. *Microb. Pathog.* 21: 407–411.

Keat A, Thomas BJ, Taylor-Robinson D (1983) Chlamydial infection in the aetiology of arthritis. *Br. Med. Bull.* 39: 168–174.

Keat A, Thomas B, Dixey J *et al.* (1987) *Chlamydia trachomatis* and reactive arthritis – the missing link. *Lancet* i: 72–74.

Keat A, Thomas B, Hughes R, Taylor-Robinson D (1989) *Chlamydia trachomatis* in reactive arthritis. *Rheumatol. Int.* 9: 197–200.

Kelly KA, Robinson EA, Rank RG (1996) Initial route of antigen administration alters the T-cell cytokine profile produced in response to the mouse pneumonitis biovar of *Chlamydia trachomatis* following genital infection. *Infect. Immun.* 64: 4976–4983.

Kim SK, Angevine M, Demick K *et al.* (1999) Induction of HLA class I-restricted CD8(+) CTLs specific for the major outer membrane protein of *Chlamydia trachomatis* in human genital tract infections. *J. Immunol.* 162: 6855–6866.

Kimani J, Maclean IW, Bwayo JJ *et al.* (1996) Risk factors for *Chlamydia trachomatis* pelvic inflammatory disease among sex workers in Nairobi, Kenya. *J. Infect. Dis.* 173: 1437–1444.

Kisselev L, Frolova F, Haenni A-L (1993) Interferon inducibility of mammalian tryptophanyl-tRNA synthetase: new perspectives. *Trends Biochem. Sci.* 18: 263–267.

Knoebel E, Vijayagopal P, Figueroa JE, Martin DH (1997) *In vitro* infection of smooth muscle cells by *Chlamydia pneumoniae*. *Infect. Immun.* 65: 503–506.

Koehler JE, Burgess RR, Thompson NE, Stephens RS (1990) *Chlamydia trachomatis* RNA polymerase major sigma subunit. Sequence and structural comparison of conserved and unique regions with *Escherichia coli* sigma 70 and *Bacillus subtilis* sigma 43. *J. Biol. Chem.* 265: 13206–13214.

Kraaipoel RJ, Duin AM van (1979) Isoelectric focusing of *Chlamydia trachomatis*. *Infect. Immun.* 26: 773–778.

Kuo C-C, Wang S-P, Grayston JT (1973) Effect of polycations, polyanions, and neuraminidase on the infectivity of trachoma-inclusion conjunctivitis and lymphogranuloma venereum organisms in HeLa cells: sialic acid residues as possible receptors for trachoma-inclusion conjunctivitis. *Infect. Immun.* 8: 313–317.

Kuo C-C, Grayston JC (1976) Interaction of *Chlamydia trachomatis* with HeLa 229 cells. *Infect. Immun.* 13: 1103–1109.

Kuo CC, Shor A, Campbell LA *et al.* (1993) Demonstration of *Chlamydia pneumoniae* in atherosclerotic lesions of coronary arteries. *J. Infect. Dis.* 167: 841–849.

Kuo CC, Grayston JT, Campbell LA *et al.* (1995) *Chlamydia pneumoniae* (TWAR) in coronary arteries of young adults (15–34 years old). *Proc. Natl Acad. Sci. USA* 92: 6911–6914.

Kuo CC, Takahashi N, Swanson AF, Ozeki Y, Hakomori SI (1996) An N-linked high-mannose type oligosaccharide, expressed at the major outer membrane protein of *Chlamydia trachomatis*, mediates attachment and infectivity of the microorganism to HeLa cells. *J. Clin. Invest.* 98: 2813–2818.

Ladel CH, Flesch IEA, Arnoldi J, Kaufmann SHE (1994) Studies with MHC-deficient knock-out mice reveal impact of both MHC I- and MHC II-dependent T cell responses on *Listeria monocytogenes* infection. *J. Immunol.* 153: 3116–3122.

Ladel CH, Daugelat S, Kaufmann SHE (1995a) Immune response to *Mycobacterium bovis* bacille Calmette Guerin infection in major histocompatibility complex class I- and II-deficient knock-out mice: Contribution of CD4 and CD8 T cells to acquired resistance. *Eur. J. Immunol.* 25: 377–384.

Ladel CH, Blum C, Dreher A, Reifenberg K, Kaufmann SHE (1995b) Protective role of gamma/delta T cells and alpha/beta T cells in tuberculosis. *Eur. J. Immunol.* 25: 2877–2881.

Lambden PR, Everson JS, Ward ME, Clarke IN (1990) Sulfur-rich proteins of *Chlamydia trachomatis*: developmentally regulated transcription of polycistronic mRNA from tandem promoters. *Gene* 87: 105–112.

Lampe MF, Wilson CB, Bevan MJ, Starnbach MN (1998) Gamma interferon production by cytotoxic T lymphocytes is required for resolution of Chlamydia trachomatis infection. *Infect. Immun.* 66: 5457–5461.

Landers DV, Erlich K, Sung M, Schachter J (1991) Role of L3T4-bearing T-cell populations in experimental murine chlamydial salpingitis. *Infect. Immun.* 59: 3774–3777.

Lawn AM, Blyth WA, Taverne J (1973) Interaction of TRIC agents with macrophages and BHK-2l cells observed by electron microscopy. *J. Hygiene* 71: 515–528.

Lee HH, Chernesky MA, Schachter J *et al.* (1995) Diagnosis of *Chlamydia trachomatis* genitourinary infection in women by ligase chain reaction assay of urine. *Lancet* 345: 213–216.

Libby P, Hansson GK (1991) Biology of disease. Involvement of the immune system in human atherogenesis: current knowledge and unanswered questions. *Lab. Invest.* 64: 5–15.

Lobau S, Marnat U, Brabetz W, Brade H (1995) Molecular cloning, sequence analysis, and functional characterization of the lipopolysaccharide biosynthetic gene *kdtA* encoding 3-deoxy-a-D-manno-octulosonic acid transferase of *Chlamydia pneumoniae* strain TW-183. *Mol. Microbiol.* 18: 391–399.

Lode H, Borner K, Koeppe P, Schaberg T (1996) Azithromycin – review of key chemical, pharmacokinetic and microbiological features. *J. Antimicrob. Chemother.* 37: 1–8.

Lukacova M, Baumann M, Brade L, Mamat U, Brade H (1994) Lipopolysaccharide smooth-rough phase variation in bacteria of the genus *Chlamydia*. *Infect. Immun.* 62: 2270–2276.

Lundemose AG, Birkelund S, Larsen PM, Fey SJ, Christiansen G (1990) Characterization and identification of early proteins in *Chlamydia trachomatis* serovar L2 by two-dimensional gel electrophoresis. *Infect. Immun.* 58: 2478–2486.

Lundemose AG, Birkelund S, Fey SJ, Larsen PM, Christiansen G (1991) *Chlamydia trachomatis* contains a protein similar to the *Legionella pneumophila* Mip gene product. *Mol. Microbiol.* 5: 109–115.

Lundemose AG, Rouch DA, Birkelund S, Christiansen G, Pearce JH (1992) *Chlamydia trachomatis* Mip-like protein. *Mol. Microbiol.* 6: 2539–2548.

Lundemose AG, Rouch DA, Penn CW, Pearce JH (1993a) The *Chlamydia trachomatis* mip-like protein is a lipoprotein. *J. Bacteriol.* 175: 3669–3671.

Lundemose AG, Kay JE, Pearce JH (1993b) *Chlamydia trachomatis* mip-like protein has peptidyl-prolyl cis trans isomerase activity that is inhibited by FK506 and rapamycin and is implicated in initiation of chlamydial infection. *Mol. Microbiol.* 7: 777–783.

Lusher M, Storey CC, Richmond SJ (1991) Extrachromosomal elements of the genus *Chlamydia*. *Adv. Gene Technol.* 226: 1–285.

Mabey DCW, Bailey RL, Hutin YJF (1992a) The epidemiology and pathogenesis of trachoma. *Rev. Med. Microbiol.* 3: 112–119.

Mabey DCW, Bailey RL, Ward MK, Whittle HC (1992b) A longitudinal study of trachoma in a Gambian village: implications concerning the pathogenesis of chlamydial infection. *Epidemiol. Infect.* 108: 343–351.

Mahony JB, Luinstra KE (1995) Multiplex PCR for detection of *Chlamydia trachomatis* and *Neisseria gonorrhoeae* in genitourinary specimens. *J. Clin. Microbiol.* 33: 3049–3053.

Majeed M, Kihistrom K (1991) Mobilization of F-actin and clathrin during redistribution of *Chlamydia trachomatis* to an intracellular site in eukaryotic cells. *Infect. Immun.* 59: 4465–4472.

Majeed M, Gustafsson M, Kihistrom E, Stendahl O (1993) Roles of Ca^{2+} and F-actin in intracellular aggregation of *Chlamydia trachomatis* in eukaryotic cells. *Infect. Immun.* 61: 1406–1414.

Majeed M, Ernst JD, Magnusson KE, Kihistrom E, Stendahi O (1994) Selective translocation of annexins during intracellular redistribution of *Chlamydia trachomatis* in HeLa and McCoy cells. *Infect. Immun.* 62: 126–134.

Malakooti J, Ely B (1995) The principle sigma subunit of the *Caulobacter crescentus* RNA polymerase. *J. Bacteriol.* 177: 6854–6860.

Mamat U, Baumann M, Schmidt G, Brade H (1993) The genus-specific lipopolysaccharide epitope of *Chlamydia* is assembled in *C. psittaci* and *C. trachomatis* by glycosyltransferases of low homology. *Mol. Microbiol.* 10: 935–941.

Mardh P-A (1997) Is Europe ready for STD screening? *Genitourin. Med.* 73: 96–98.

Martin DH, Mroczkowski TF, Dalu ZA *et al.* (1992) A controlled trial of a single dose of azithromycin for the treatment of chlamydial urethritis and cervicitis. *New Engl. J. Med.* 327: 921–925.

Mathews SA, Douglas A, Sriprakash KS, Hatch TP (1993) In vitro transcription in *Chlamydia psittaci* and *Chlamydia trachomatis*. *Mol. Microbiol.* 7: 937–946.

Mathews SA, Sriprakash KS (1994) The RNA polymerase of *Chlamydia trachomatis* has a flexible sequence requirement at the –10 and –35 boxes of its promoters. *J. Bacteriol.* 176: 3785–3789.

Matsumoto A, Manire GP (1970) Electron microscopic observations on the fine stricture of cell walls of *Chlamydia psittaci*. *J. Bacteriol.* 104: 1332–1337.

Matsumoto A (1988) Structural characteristics of chlamydial bodies. In: Barron AL (ed.) *Microbiology of Chlamydiae.* Boca Raton, FL: CRC Press, pp. 21–45.

McClarty G, Tipples G (1991) *In situ* studies on incorporation of nucleic acid precursors into *Chlamydia trachomatis* DNA. *J. Bacteriol.* 173: 4922–4931.

McClarty G, Fan HZ (1993) Purine metabolism by intracellular *Chlamydia psittaci*. *J. Bacteriol.* 175: 4662–4669.

McClarty G, Qin B (1993) Pyrimidine metabolism by intracellular *Chlamydia psittaci*. *J. Bacteriol.* 175: 4652–4661.

McClarty G (1994) Chlamydiae and the biochemistry of intracellular parasitism. *Trends Microbiol.* 2: 157–164.

Melgosa MP, Kuo CC, Campbell LA (1994) Isolation and characterization of a gene encoding a *Chlamydia pneumoniae* 76-kilodalton protein containing a species-specific epitope. *Infect. Immun.* 62: 880–886.

Melnick SL, Sharar E, Folsom AR *et al.* (1993) Past infection with *Chlamydia pneumoniae* strain TWAR and asymptomatic carotid atherosclerosis. *Am. J. Med.* 95: 499–504.

Meyer KF, Eddie B (1962) Immunity against some *Bedsonia* in man resulting from infection and in animals from infection or vaccination. *Ann. N Y Acad. Sci.* 98: 288–313.

Miyashita N, Kanamoto Y, Matsumoto A (1993) The morphology of *Chlamydia pneumoniae.* *J. Med. Microbiol.* 38: 418–425.

Moazed TC, Kuo CC, Grayston JT, Campbell LA (1998) Evidence of systemic dissemination of *Chlamydia pneumoniae* via macrophages in the mouse. *J. Infect. Dis.* 177: 1322–1325.

Moazed TC, Campbell LA, Rosenfeld ME, Grayston JT, Kuo CC (1999) *Chlamydia pneumoniae* infection accelerates the progression of atherosclerosis in apolipoprotein E-deficient mice. *J. Infect. Dis.* 180: 238–241.

Moncada J, Schachter J, Shafer MA *et al.* (1994) Detection of *Chlamydia trachomatis* in first catch urine samples from symptomatic and asymptomatic males. *Sex. Transm. Dis.* 21: 8–12.

Morgan HR (1956) Latent viral infection of cells in tissue culture. I. Studies on latent infection of chick embryo tissues with psittacosis virus. *J. Exp. Med.* 103: 34–47.

Morrison RP, Lyng K, Caidwell HD (1989a) Chlamydial disease pathogenesis. Ocular hypersensitivity elicited by a genus-specific 57-kD protein. *J. Exp. Med.* 169: 663–675.

Morrison RP, Belland RJ, Lyng K, Caidwell HD (1989b) Chlamydial disease pathogenesis. The 57 kD chlamydial hypersensitivity antigen is a stress response protein. *J. Exp. Med.* 170: 1271–1283.

Morrison RP, Su H, Lyng K, Yuan Y (1990) The *Chlamydia trachomatis* hyp operon is homologous to the groE stress response operon of *Escherichia coli*. *Infect. Immun.* 58: 2701–2705.

Morrison RP, Feilzer K, Tumas DB (1995) Gene knockout mice establish a primary protective role for major histocompatibility complex class IT-restricted responses in *Chlamydia trachomatis* genital tract infection. *Infect. Immun.* 63: 4661–4668.

Moulder JW (1962) *The Biochemistry of Intracellular Parasitism.* Chicago: University of Chicago Press.

Moulder JW, Novosel DL, Tribby IC (1963) Diaminopimelic acid decarboxylase of the agent of meningopneumonitis. *J. Bacteriol.* 85: 701–706.

Moulder JW, Grisso DL, Brubaker RB (1965) Enzymes of glucose catabolism in a member of the psittacosis group. *J. Bacteriol.* 89: 810–812.

Moulder JW (1966) The relation of the psittacosis group (chlamydiae) to bacteria and viruses. *Annu. Rev. Microbiol.* 20: 107–130.

Moulder JW, Hatch TP, Byrne GI, Kellogg KR (1976) Immediate toxicity of high multiplicities of *Chlamydia psittaci* for mouse fibroblasts (L cells). *Infect. Immun.* 14: 277–289.

Moulder JW (1982) The relation of basic biology to pathogenic potential in the genus *Chlamydia*. *Infection* 10: S10–S18.

Moulder JW (1988) Characteristics of chlamydiae. In: Barron AL (ed.) *Microbiology of Chlamydia.* Boca Raton, FL: CRC Press, pp. 3–19.

Moulder JW (1991) Interaction of chlamydiae and host cells *in vitro*. *Microbiol. Rev.* 55: 143–190.

Moulder JW (1993) Why is *Chlamydia* sensitive to penicillin in the absence of peptidoglycan? *Infect. Agents Dis. Rev. Issues Comm.* 2: 87–99.

Mourad A, Sweet RL, Sugg N, Schachter J (1980) Relative resistance to erythromycin in *Chlamydia trachomatis*. *Antimicrob. Agents Chemother.* 18: 696–698.

Muhlestein JB, Anderson JL, Hammond EH *et al.* (1998) Infection with *Chlamydia pneumoniae* accelerates the development of atherosclerosis and treatment with azithromycin prevents it in a rabbit model. *Circulation* 97: 633–636.

Murdin AD, Su H, Manning OS, Klein MH, Parnell MJ, Caldwell HD (1993) A poliovirus hybrid expressing a neutralization epitope from the major outer membrane protein of *Chlamydia trachomatis* is highly immunogenic. *Infect. Immun.* 61: 4406–4414.

Murdin AD, Su H, Klein MH, Caldwell HD (1995) Poliovirus hybrids expressing neutralization epitopes from variable domain-i and domain-iv of the major outer membrane protein of *Chlamydia trachomatis* elicit broadly cross-reactive *C. trachomatis* neutralizing antibodies. *Infect. Immun.* 63: 1116–1121.

Murray ES, Charbonnet LT, MacDonald AB (1973) Immunity to chlamydial infections of the eye. 1. The role of circulatory and secretory antibodies in resistance to reinfection with guinea pig conjunctivitis. *J. Immunol.* 110: 1518–1525.

Nano FE, Caldwell HD (1985) Expression of the chlamydial genus-specific lipopolysaccharide epitope in *Escherichia coli*. *Science* 228: 742–744.

Neefjes BJ, Schumacher TN, Ploegh HL (1991) Assembly and intracellular transport of major histocompatibility complex molecules. *Curr. Opin. Cell Biol.* 3: 601–609.

Newhall WJ, Jones RB (1983) Disulphide-linked oligomers of the major outer membrane protein of chlamydiae. *J. Bacteriol.* 154: 998–1001.

Newhall WJ (1987) Biosynthesis and disulfide cross-linking of outer membrane components during the growth-cycle of *Chlamydia trachomatis*. *Infect. Immun.* 55: 162–168.

Newhall WJ (1988) Macromolecular and antigenic composition of chlamydiae. In: Barron AL (ed.) *Microbiology of Chlamydia.* Boca Raton, FL: CRC Press, pp. 47–70.

Nurminen M, Leinonen M, Saikku P, Makela PH (1983) The genus-specific antigen of *Chlamydia* – resemblance to the lipopolysaccharide of enteric bacteria. *Science* 220: 1279–1281.

Oakeshott P, Hay P (1995) General practice update: Chlamydia infection in women. *Br. Gen. Pract.* 45: 615–620.

Ojcius DM, Dealba YB, Kanellopoulos JM *et al.* (1998) Internalization of Chlamydia by dendritic cells and stimulation of Chlamydia-specific T cells. *J. Immunol.* 160: 1297–1303.

Ortiz L, Demick KP, Petersen JW *et al.* (1996) *Chlamydia trachomatis* major outer membrane protein (MOMP) epitopes that activate HLA class II-restricted I cells from infected humans. *J. Immunol.* 157: 4554–4567.

Paavonen J, Wolner-Hanssen P (1989) *Chlamydia trachomatis*: a major threat to reproduction. *Hum. Reprod.* 4: 111–124.

Paavonen J (1997) Is screening for *Chlamydia trachomatis* infection cost-effective? *Genitourin. Med.* 73: 103–104.

Pal S, Peterson EM, Delamaza LM (1996) Intranasal immunization induces long-term protection in mice against a *Chlamydia trachomatis* genital challenge. *Infect. Immun.* 64: 5341–5348.

Palmer L, Falkow S (1986) A common plasmid of *Chlamydia trachomatis*. *Plasmid* 16: 52–62.

Palmer SR (1982) Psittacosis in man. Recent developments in the UK: a review. *J. R. Soc. Med* 75: 262.

Patel P, Mendall MA, Carrington D *et al.* (1995) Association of *Helicobacter pylori* and *Chlamydia pneumoniae* infections with coronary heart disease and cardiovascular risk factors. *BMJ* 311: 711–714.

Patton DL (1985) Immunopathology and histopathology of experimental chlamydial salpingitis. *Rev. Infect. Dis.* 7: 746–753.

Patton DL, Kuo CC, Wang SP, Halbert SA (1987) Distal tubal obstruction induced by repeated *Chlamydia trachomatis* salpingeal infections in pig-tailed macaques. *J. Infect. Dis.* 155: 1292–1299.

Patton DL, Wolner-Hanssen P, Cosgrove SJ, Holmes KK (1990) The effects of *Chlamydia trachomatis* on the female reproductive tract of the *Macaca nemestrina* after a single tubal challenge following repeated cervical inoculations. *Obstet. Gynecol.* 76: 643–650.

Pearce JH (1986) Early events in chlamydial infection. *Ann. Inst. Pasteur Microbiol.* 137A: 325–332.

Pearce J, Gaston H, Deane K *et al.* (1994) 'Persistent' forms and persistence of *Chlamydia. Trends Microbiol.* 2: 257–258.

Pedersen LB, Birkelund S, Christiansen G (1994) Interaction of the *Chlamydia trachomatis* histone h1-like protein (hc1) with DNA and RNA causes repression of transcription and translation *in vitro. Mol. Microbiol.* 11: 1085–1098.

Pedersen LB, Birkelund S, Hoirn A, Ostergaard S, Christiansen G (1996a) The 18-kilodalton *Chlamydia trachomatis* histone hi-like protein (hcl) contains a potential N-terminal dimerization site and a C-terminal nucleic acid-binding domain. *J. Bacteriol.* 178: 994–1002.

Pedersen LB, Birkelund S, Christiansen G (1996b) Purification of recombinant *Chlamydia trachomatis* histone hi-like protein hc2, and comparative functional-analysis of hc2 and hcl. *Mol. Microbiol.* 20: 295–311.

Pfeifer JD, Wick MJ, Roberts RL *et al.* (1993) Phagocytic processing of bacterial antigens for class-I MHC presentation to T-cells. *Nature* 361: 359–362.

Piot P (1994) Epidemiology and control of genital chlamydial infection. In: Orfila J, Byrne GT, Chemesky MA *et al.* (eds) *Chlamydial Infections, Proceedings of the 8th International Symposium on Human Chlamydial Infections,* Bologna, Italy. pp. 7–16.

Podgore JK, Holmes KK, Alexander ER (1982) Asymptomatic urethral infections due to *Chlamydia trachomatis* in male U.S. military personnel. *J. Infect. Dis.* 146: 828.

Popov VL, Shatkin AA, Pankratova VN, Smirnova NS *et al.* (1991) Ultrastructure of *Chlamydia pneumoniae* in cell culture. *FEMS Microbiol. Lett.* 68: 129–134.

Porcelli S, Morita CT, Brenner MB (1992) CD1b restricts the response of human CD4-8-lymphocytes-T to a microbial antigen. *Nature* 360: 593–597.

Prain CJ, Pearce JH (1989) Ultrastructural studies on the intracellular fate of *Chlamydia psittaci* (strain guinea pig inclusion conjunctivitis) and *Chlamydia trachomatis* (strain lymphogranuloma venereum 434): modulation of intracellular events and relationship with endocytic mechanism. *J. Gen. Microbiol.* 135: 2107–2123.

Quigstad E, Digranes S, Thorsby E (1983) Antigen-specific proliferative human T lymphocyte clones with specificity for *Chlamydia trachomatis. Scand. J. Immunol.* 18: 291–297.

Quinn TC (1994) Recent advances in diagnosis of sexually transmitted diseases. *Sex Transm. Dis* 21: S19–S27.

Quinn TC, Welsh L, Lentz A *et al.* (1996a) Diagnosis by amplicor PCR of *Chlamydia trachomatis* infection in urine samples from women and men attending sexually-transmitted disease clinics. *J. Clin. Microbiol.* 34: 1401–1406.

Quinn TC, Gaydos C, Shepherd M *et al.* (1996b) Epidemiologic and microbiologic correlates of *Chlamydia trachomatis* infection in sexual partnerships. *JAMA* 276: 1737–1742.

Ramirez JA, Ahkee S, Summersgill JT *et al.* (1996) Isolation of *Chlamydia pneumoniae* from the coronary artery of a patient with coronary atherosclerosis. *Ann. Intern. Med.* 125: 979–982.

Ramsey KH, Soderberg LSF, Rank RG (1988) Resolution of chlamydial genital infection in B-cell-deficient mice and immunity to reinfection. *Infect. Immun.* 56: 1320–1325.

Rank RG, Bavoil PM (1996) Prospects for a vaccine against *Chlamydia* genital disease. II. Immunity and vaccine development. *Bull. Inst. Pasteur* 94: 55–82.

Rasmussen SJ, Timms P, Beatty PR, Stephens RS (1996) Cytotoxic T-lymphocyte-mediated cytolysis of L cells persistently infected with *Chlamydia* spp. *Infect. Immun.* 64: 1944–1949.

Rasmussen SJ, Eckmann L, Quayle AJ *et al.* (1997) Secretion of proinflammatory cytokines by epithelial cells in response to Chlamydia infection suggests a central role for epithelial cells in chlamydial pathogenesis. *J. Clin. Invest.* 99: 77–87.

Ratti C, Comanducci M, Orfila J, Sueur JM, Gommeaux A (1995) New chlamydial antigen as a serological marker in HIV infection. *Lancet* 346(9): 12.

Raulston JE (1995) Chlamydial envelope components and pathogen host-cell interactions. *Mol. Microbiol.* 15: 607–616.

Raulston JE, Davis CH, Schmiel DH, Morgan MW, Wyrick PB (1993) Molecular characterization and outer membrane association of a *Chlamydia trachomatis* protein related to the hsp70 family of proteins. *J. Biol. Chem.* 268: 268: 23139–23147.

Remacha M, Kaul R, Sherburne R, Wenman WM (1996) Functional domains of chlamydial histone hi-like protein. *Biochem. J.* 315: 481–486.

Reynolds DJ, Pearce JH (1990) Characterization of the cytochalasin D-resistant (pinocytic) mechanisms of endocytosis utilised by chlamydiae. *Infect. Immun.* 58: 3208–3216.

Reynolds DJ, Pearce JH (1991) Endocytic mechanisms utilised by chlamydiae and their influence on induction of productive infection. *Infect. Immun.* 59: 3033–3039.

Ricci S, Cevenini R, Cosco E *et al.* (1993) Transcriptional analysis of the *Chlamydia trachomatis* plasmid pct identifies temporally regulated transcripts, antisense RNA and sigma 70-selected promoters. *Mol. Gen. Genet.* 237: 318–326.

Richmond SJ, Stirling P (1981) Localization of chlamydial group antigen in McCoy cell monolayers infected with *Chlamydia trachomatis* or *Chlamydia psittaci*. *Infect. Immun.* 34: 561–570.

Richmond SJ, Stirling P, Ashley CR (1982) Virus infecting the reticulate bodies of an avian strain of *Chlamydia psittaci*. *FEMS Microbiol. Lett.* 14: 31–36.

Ridgway GL (1992) Advances in the antimicrobial therapy of chlamydial genital infections. *J. Infect.* 25: 51–59.

Ridgway GL, Mumtaz G, Robinson AJ *et al.* (1996) Comparison of the ligase chain-reaction with cell culture for the diagnosis of *Chlamydia trachomatis* infection in women. *J. Clin. Pathol.* 49: 116–119.

Rockey DD, Heinzen RA, Hackstadt T (1995) Cloning and characterization of a *Chlamydia psittaci* gene coding for a protein localised in the inclusion membrane of infected cells. *Mol. Microbiol.* 15: 617–626.

Rockey DD, Fischer ER, Hackstadt T (1996) Temporal analysis of the developing *Chlamydia psittaci* inclusion by use of fluorescence and electron-microscopy. *Infect. Immun.* 64: 4269–4278.

Rockey DD, Chesebro BB, Heinzen RA, Hackstadt T (1996) A 28 kDa major immunogen of *Chlamydia psittaci* shares identity with mip proteins of *Legionella* spp and *Chlamydia trachomatis*–cloning and characterization of the *Chlamydia psittaci* mip-like gene. *Microbiol. UK* 142: 945–953.

Rodel J, Groh A, Vogelsang H *et al.* (1998) Beta interferon is produced by *Chlamydia trachomatis*-infected fibroblast-like synoviocytes and inhibits gamma interferon-induced HLA-DR expression. *Infect. Immun.* 66: 4491–4495.

Saikku P, Leinonen K, Mattila K *et al.* (1988) Serological evidence of an association of a novel *chlamydia*, TWAR, with coronary heart disease and acute myocardial infarction. *Lancet* 339: 883.

Saikku P (1992) The epidemiology and significance of *Chlamydia pneumoniae*. *J. Infect.* 25: S27–S34.

Schachter J, Dawson CR (1978) *Human Chlamydial Infections*. Littleton, MA: PSG Publishing.

Schachter J (1978) Chlamydial infections. *N. Engl. J. Med.* 298: 428–435; 490–495; 540–549.

Schachter J, Ridgway GL, Collier L (1998) Chlamydial diseases. In Hausler WJ Jr, Sussman M (eds) *Topley & Wilson's Microbiology and Microbial Infections*. 9th edn. London: Arnold, pp. 977–994.

Schachter J (1989) Why we need a program for the control of *Chlamydia trachomatis*. *N. Engl. J. Med.* 320: 802–804.

Schmiel DH, Wyrick PB (1994) Another putative heat-shock gene and aminoacyl-transfer-RNA synthetase gene are located upstream from the Grpe-like and Dnak-like genes in *Chlamydia trachomatis*. *Gene* 145: 57–63.

Scholes D, Stergachis A, Heidrich FE *et al.* (1996) Prevention of pelvic inflammatory disease by screening for cervical chlamydial infection. *N. Engl. J. Med.* 334: 1362–1366.

Schramm N, Wyrick PB (1995) Cytoskeletal requirements in *Chlamydia trachomatis* infection of host cells. *Infect. Immun.* 63: 324–332.

Schramm N, Bagnell CR, Wyrick PB (1996) Vesicles containing *Chlamydia trachomatis* serovar L2 remain above pH 6 within Hec-1b cells. *Infect. Immun.* 64: 1208–1214.

Schumacher HR, Magge S, Cherian PV *et al.* (1988) Light and electron microscopic studies on the synovial membrane in Reiter's syndrome – immunocytochemical identification of chlamydial antigen in patients with early disease. *Arthritis Rheum.* 31: 937–946.

Scidmore MA, Rockey DD, Fischer ER, Heinzen RA, Hackstadt T (1996a) Vesicular interactions of the *Chlamydia trachomatis* inclusion are determined by chlamydial early protein synthesis rather than route of entry. *Infect. Immun.* 64: 5366–5372.

Scidmore MA, Fischer ER, Hackstadt T (1996b) Sphingolipids and glycoproteins are differentially trafficked to the *Chlamydia trachomatis* inclusion. *J. Cell Biol.* 134: 363–374.

Shahmanesh M, Brunst M, Sukthankar A, Pearce J, Gaston J (1999) Peripheral blood T-cell proliferative response to chlamydial organisms in gonococcal and non-gonococcal urethritis and presumed pelvic inflammatory disease. *Sex. Trans. Infect.* 75: 327–331.

Shemer Y, Sarov I (1985a) Inhibition of growth of *Chlamydia trachomatis* by human gamma interferon. *Infect. Immun.* 48: 592–596.

Shemer Y, Sarov I (1985b) Active vs latent *Chlamydia trachomatis* infection modulated by gamma interferon. *Isr. J. Med. Sci.* 21: 178–179.

Soderlund G, Kihlstrom E (1982) Physico-chemical surface properties of elementary bodies from different serotypes of *Chlamydia trachomatis* and their interaction with mouse fibroblasts. *Infect. Immun.* 36: 893–899.

Sriprakash KS, Macavoy ES (1987) Characterization and sequence of a plasmid from the trachoma biovar of *Chlamydia trachomatis*. *Plasmid* 18: 205–214.

Stacey C, Munday P, Thomas B *et al.* (1990) *Chlamydia trachomatis* in the fallopian tubes of women without laparoscopic evidence of salpingitis. *Lancet* 336: 960–963.

Stamm WE, Harrison HR, Alexander ER *et al.* (1984) Diagnosis of *Chlamydia trachomatis* infections by direct immunofluorescence staining of genital secretions – a multicenter trial. *Ann. Intern. Med.* 101: 638–641.

Stamm WE, Cole B (1986) Asymptomatic *Chlamydia trachomatis* urethritis in men. *Sex. Transm. Dis.* 13: 163–165.

Starnbach MN, Bevan MJ, Lampe MF (1994) Protective cytotoxic T lymphocytes are induced during murine infection with *Chlamydia trachomatis*. *J. Immunol.* 153: 5183–5189.

Starnbach MN, Bevan MJ, Lampe MF (1995) Murine cytotoxic T lymphocytes induced following *Chlamydia trachomatis* intraperitoneal or genital tract infection respond to cells infected with multiple serovars. *Infect. Immun.* 63: 3527–3530.

Stemme S, Hoim J, Hansson GK (1992) T lymphocytes in human atherosclerotic plaques are memory cells expressing CD45RO and the integrin VLA- 1. *Arterioscler. Thromb.* 12: 206–211.

Stemme S, Faber B, Hoim J, Wikiund O, Witztum JL, Hansson GK (1995) T lymphocytes from human atherosclerotic plaques recognise oxidised low density lipoprotein. *Proc. Natl Acad. Sci. USA* 92: 3893–3897.

Stephens RS, Wagar EA, Schoolnik GK (1988a) High resolution mapping of serovar-specific and common antigenic determinants of the major outer membrane protein of *Chlamydia trachomatis*. *J. Exp. Med.* 167: 817–831.

Stephens RS, Wagar EA, Edinan U (1988b) Developmental regulation of tandem promoters for the major outer membrane protein gene of *Chlamydia trachomatis*. *J. Bacteriol.* 170: 744–750.

Stephens RS, Kalman S, Lammel C et al. (1998) Genome sequence of an obligate intracellular pathogen of humans: *Chlamydia trachomatis*. *Science* 282: 754–759.

Stirling P, Allan I, Pearce JH (1983) Interference with transformation of chlamydiae from reproductive to infective body forms by deprivation of cysteine. *FEMS Microbiol. Lett.* 19: 133–136.

Storey CC, Lusher M, Richmond SJ (1989) Analysis of the complete nucleotide sequence of Chp 1, a phage which infects *Chlamydia psittaci*. *J. Gen. Virol.* 70: 3381–3390.

Storey C, Lusher M, Yates P, Richmond SJ (1993) Evidence for *Chlamydia pneumoniae* of nonhuman origin. *J. Gen. Microbiol.* 139: 2621–2626.

Storz J (1971) *Chlamydia and Chlamydia-induced Diseases*. Springfield, IL: Charles C Thomas.

Stuart ES, Wyrick PB, Choong J, Stoler SB, MacDonald AB (1991) Examination of chlamydial glycolipid with monoclonal antibodies: cellular distribution and epitope-binding. *Immunology* 74: 740–747.

Stuart ES, Troidle KM, MacDonald AB (1994) Chlamydial glycolipid antigen – extracellular accumulation, biological activity, and antibody recognition. *Curr. Microbiol.* 28: 85–90.

Su H, Zhang YX, Barrera O, Watkins NG, Caldwell HD (1988) Differential effect of trypsin on infectivity of *Chlamydia trachomatis* – loss of infectivity requires cleavage of major outer membrane protein variable domain ii and domain iv. *Infect. Immun.* 56: 2094–2100.

Su H, Watkins NG, Zhang YX, Caldwell HD (1990a) *Chlamydia trachomatis*-host cell interactions: role of the chlamydial major outer membrane protein as an adhesin. *Infect. Immun.* 58: 1017–1025.

Su H, Morrison RP, Watkins NG, Caldwell HD (1990b) Identification and characterization of i-helper cell epitopes on the major outer membrane protein of *Chlamydia trachomatis*. *J. Exp. Med.* 172: 203–202.

Su H, Caldwell HD (1995) CD4(+) T cells play a significant role in adoptive immunity to *Chlamydia trachomatis*

infection of the mouse genital tract. *Infect. Immun.* 63: 3302–3308.

Su H, Raymond L, Rockey DD, Fischer E, Hackstadt T, Caldwell HD (1996) A recombinant *Chlamydia trachomatis* major outer membrane protein binds to heparan sulfate receptors on epithelial cells. *Proc. Natl Acad. Sci. USA* 93: 11143–11148.

Su H, Messer R, Whitmire W et al. (1998) Vaccination against chlamydial genital tract infection after immunization with dendritic cells pulsed *ex vivo* with nonviable Chlamydiae. *J. Exp. Med.* 188: 809–818.

Svensson M, Stockinger B, Wick MJ (1997) Bone marrow-derived dendritic cells can process bacteria for MHC-I and MHC-II presentation to T cells. *J. Immunol.* 158: 4229–4236.

Tam JE, Davis CH, Wyrick PB (1994) Expression of recombinant DNA introduced into *Chlamydia trachomatis* by electroporation. *Can. J. Microbiol.* 40: 583–591.

T'ang K-F, Chang H-L, Huang Y-T, Wang K-C (1957) Trachoma virus in chick embryo. *Natl Med. J. China* 43: 81–86.

Tanaka Y, Morita CT, Tanaka Y et al. (1995) Natural and synthetic non-peptide antigens recognised by human gamma delta T cells. *Nature* 375: 155–158.

Taraska T, Ward DM, Ajioka RS et al. (1996) The late chlamydial inclusion membrane is not derived from the endocytic pathway and is relatively deficient in host proteins. *Infect. Immun.* 64: 3713–3727.

Taylor-Robinson D, Gilroy CB, Thomas BJ, Keat AC (1992) Detection of *Chlamydia trachomatis* DNA in joints of reactive arthritis patients by polymerase chain reaction. *Lancet* 340: 81–82.

Thom DH, Grayston JT, Siscovick OS et al. (1992) Association of prior infection with *Chlamydia pneumoniae* and angiographically demonstrated coronary artery disease. *JAMA* 268: 68–72.

Thygeson P (1934) The etiology of inclusion blenorrhea. *Am. J. Ophthalmol.* 17: 1019–1035.

Thylefors AD, Negrel R (1995) Global data on blindness – an update. *Bull. WHO* 73: 115–121.

Ting LM, Hsia RC, Haidaris CG, Bayou PM (1995) Interaction of outer envelope proteins of *Chlamydia psittaci* gpic with the HeLa cell surface. *Infect. Immun.* 63: 3600–3608.

Tipples G, McClarty G (1991) Isolation and initial characterization of a series of *Chlamydia trachomatis* isolates selected for hydroxyurea resistance by a stepwise procedure. *J. Bacteriol.* 173: 4932–4940.

Tipples G, McClarty G (1993) The obligate intracellular bacterium *Chlamydia trachomatis* is auxotrophic for 3 of the 4 ribonucleoside triphosphates. *Mol. Microbiol.* 8: 1105–1114.

Tipples G, McClarty G (1995) Cloning and expression of the *Chlamydia trachomatis* gene for ctp synthetase. *J. Biol. Chem.* 270: 7908–7914.

Todd WJ, Caldwell HD (1985) The interaction of *Chlamydia trachomatis* with host cells – ultrastructural studies of the mechanism of release of a biovar ii strain from HeLa 229 cells. *J. Infect. Dis.* 151: 1037–1044.

Toye B, Zhong GM, Peeling R, Brunham RC (1990) Immunologic characterization of a cloned fragment containing the species-specific epitope from the major outer membrane protein of *Chlamydia trachomatis.* *Infect. Immun.* 58: 3909–3913.

Treharne JD (1985) The community epidemiology of trachoma. *Rev. Infect. Dis.* 7: 760–764.

Treuhaft MW, Moulder JW (1968) Biosynthesis of arginine in L cells infected with chlamydiae. *J. Bacteriol.* 96: 2004–2011.

Tseng CTK, Rank RG (1998) Role of NK cells in early host response to chlamydial genital infection. *Infect. Immun.* 66: 5867–5875.

Tuffrey M, Alexander F, Conlan W, Woods C, Ward ME (1992) Heterotypic protection of mice against chlamydial salpingitis and colonization of the lower genital tract with a human serovar F isolate of *Chlamydia trachomatis* by prior immunization with recombinant serovar L1 major outer membrane protein. *J. Gen. Microbiol.* 138: 1707–1715.

Vanooij C, Apodaca G, Engel J (1997) Characterization of the *Chlamydia trachomatis* vacuole and its interaction with the host endocytic pathway in HeLa cells. *Infect. Immun.* 65: 758–766.

Vanvoorhis WC, Barrett IK, Sweeney YTC, Kno CC, Patton DL (1996) Analysis of lymphocyte phenotype and cytokine activity in the inflammatory infiltrates of the upper genital tract of female macaques infected with *Chlamydia trachomatis.* *J. Infect. Dis.* 174: 647–650.

Vretou E, Psarrou E, Spiliopoulou D (1992) The role of lipopolysaccharide in the exposure of protective antigenic sites on the major outer membrane protein of *Chlamydia trachomatis.* *J. Gen. Microbiol.* 138: 1221–1227.

Wagar EA, Stephens RS (1988) Developmental form-specific DNA-binding proteins in *Chlamydia* spp. *Infect. Immun.* 56: 1678–1684.

Ward ME, Murray A (1984) Control mechanisms governing the infectivity of *Chlamydia trachomatis* for HeLa cells – mechanisms of endocytosis. *J. Gen. Microbiol.* 130: 1765–1780.

Ward ME (1995) The immunobiology and immunopathology of chlamydial infections. *APMIS* 103: 769–796.

Ward ME (1997) *Chlamydia* host and host cell interactions. *Symp. Soc. Gen. Microbiol.* 55: 153–189.

Watson MW, Lambden PR, Everson JS, Clarke IN (1994) Immunoreactivity of the 60 kDa cysteine-rich proteins of *Chlamydia trachomatis,* *Chlamydia psittaci* and *Chlamydia pneumoniae* expressed in *Escherichia coli.* *Microbiol. UK* 140: 2003–2011.

Watson MW, Clarke IN, Everson JS, Lambden PR (1995) The crp operon of *Chlamydia psittaci* and *Chlamydia pneumoniae.* *Microbiol. UK* 141: 2489–2497.

Weisburg WG, Hatch TP, Woese CR (1986) Eubacterial origin of chlamydiae. *J. Bacteriol.* 167: 570–574.

Weiss E (1967) Transaminase and other enzymatic reactions involving pyruvate and glutamate in *Chlamydia* (psittacosis-trachoma group). *J. Bacteriol.* 93: 177–184.

Weiss E, Wilson NN (1969) Role of exogenous adenosine triphosphate in catabolic and synthetic activities of *Chlamydia psittaci.* *J. Bacteriol.* 97: 719–724.

West MA, Bretscher MS, Watts C (1989) Distinct endocytotic pathways in epidermal growth factor-stimulated human carcinoma. *J. Cell. Biol.* 109: 2731–2739.

Westrom LR (1980) Incidence, prevalence and trends of acute pelvic inflammatory disease and its consequences in industrialised countries. *Am. J. Obstet. Gynecol.* 138: 880–892.

Westrom LR, Joesoef R, Reynolds G, Hadgu A, Thompson SE (1992) Pelvic inflammatory disease and infertility: A cohort study of 1844 women with laparascopically verified disease and 657 control women with normal laparascopic results. *Sex Transm. Dis.* 19: 185–192.

Westrom L, Wolner-Hanssen P (1993) Pathogenesis of pelvic inflammatory disease. *Genitourin. Med.* 69: 9–17.

Whittum-Hudson JA, An LL, Saltzman WM, Prendergast RA, MacDonald AB (1996) Oral immunization with an anti-idiotypic antibody to the exoglycolipid antigen protects against experimental *Chlamydia trachomatis* infection. *Nat. Med.* 2: 1116–1121.

Wichlan DG, Hatch TP (1993) Identification of an early stage gene of *Chlamydia psittaci* 6BC. *J. Bacteriol.* 175: 2936–2942.

Williams D, Grubbs B, Schachter J (1987a) Primary murine *Chlamydia trachomatis* pneumonia in B-cell-deficient mice. *Infect. Immun.* 55: 2387–2390.

Williams DM, Schachter J, Grubbs B (1987b) Role of natural-killer cells in infection with the mouse pneumonitis agent (murine *Chlamydia trachomatis).* *Infect. Immun.* 55: 223–226.

Williams DM, Grubbs BG, Kelly KW, Pack E, Rank RG (1996) Role of gamma-delta T cells in murine *Chlamydia trachomatis* infection. *Infect. Immun.* 64: 3916–3919.

Witkin SS, Jeremias J, Toth M, Ledger WJ (1993) Cell-mediated immune response to the recombinant 57-kDa heat-shock protein of *Chlamydia trachomatis* in women with salpingitis. *J. Infect. Dis.* 167: 1379–1383.

Witkin SS, Jeremias J, Toth M, Ledger WJ (1994) Proliferative response to conserved epitopes of the *Chlamydia trachomatis* and human 60-kilodalton heat-shock proteins by lymphocytes from women with salpingitis. *Am. J. Obstet. Gynecol.* 171: 455–460.

Wylie JL, Berry JD, McClarty C (1996a) *Chlamydia trachomatis* ctp synthetase – molecular characterization and developmental regulation of expression. *Mol. Microbiol.* 22: 631–642.

Wylie JL, Wang LL, Tipples G, McClarty G (1996b) A single-point mutation in ctp synthetase of *Chlamydia trachomatis* confers resistance to cyclopentenyl cytosine. *J. Biol. Chem .* 271: 15393–15400.

Wyrick PB, Davis CH (1984) Elementary body envelopes from *Chlamydia psittaci* can induce immediate cytotoxicity in resident mouse macrophages and L-cells. *Infect. Immun.* 45: 297–298.

Wyrick PB, Davis CH, Knight ST *et al.* (1993) An *in vitro* human epithelial cell culture system for studying the pathogenesis of *Chlamydia trachomatis. Sex. Transm. Dis.* 20: 248–256.

Wyrick PB, Davis CH, Wayner EA (1994) *Chlamydia trachomatis* does not bind to alpha-beta-i integrins to colonise a human endometrial epithelial-cell line cultured *in vitro. Microb. Pathog.* 17: 159–166.

Zahn I, Szeredi L, Schiller I *et al.* (1995) Immunohistological determination of *Chlamydia psittaci/C. pecorum* and *C. trachomatis* in the piglet gut. *J. Vet. Med. Ser. B Infect. Dis. Vet. Publ. Health* 42: 266–276.

Zhang DJ, Fan H, McClarty G, Brunharn RC (1995) Identification of the *Chlamydia trachomatis* RecA-encoding gene. *Infect. Immun.* 63: 676–680.

Zhang DJ, Yang X, Berry J *et al.* (1997) DNA vaccination with the MOMP gene induces acquired immunity to *Chlamydia trachomatis* (mouse pneumonitis) infection. *J. Infect. Dis.* 176: 1035–1040.

Zhang DJ, Yang X, Shen C, Brunham RC (1999) Characterization of immune responses following intramuscular DNA immunization with the MOMP gene of *Chlamydia trachomatis* mouse pneumonitis strain. *Immunology* 96: 314–321.

Zhang JP, Stephens RS (1992) Mechanism of *C. trachomatis* attachment to eukaryotic host cells. *Cell* 69: 861–869.

Zhang YX, Stewart S, Joseph T, Taylor HR, Caldwell HD (1987) Protective monoclonal antibodies recognise epitopes located on the major outer membrane protein of *Chlamydia trachomatis. J. Immunol.* 138: 575–581.

Zhong GM, Brunham RC (1992) Antibody responses to the chlamydial heat-shock proteins hsp60 and hsp70 are h-2 linked. *Infect. Immun.* 60: 3143–3149.

Zhong GM, Fan T, Liu L (1999) *Chlamydia* inhibits interferon gamma-inducible major histocompatibility complex class II expression by degradation of upstream stimulatory factor 1. *J. Exp. Med.* 189: 1931–1937.

PART

15

ANAEROBIC INFECTIONS

88

Anaerobic Infections: A Clinical Overview

Sydney M. Finegold[1] and Max Sussman[2]

[1]Veterans Administration Medical Center and UCLA School of Medicine,
Los Angeles, California, USA
[2]Medical School, Newcastle upon Tyne, UK

Infections due to anaerobes are common and associated with considerable morbidity and potential mortality. The anaerobic bacteria responsible for disease in humans are to be found in the normal flora of humans and animals and in the environment. Since, in many cases, they are not immediately clinically distinguishable from many other infections, anaerobic infections may readily be overlooked and their diagnosis depends in the first place on a high level of clinical suspicion. If many or most of the anaerobes present are to be cultivated successfully, special precautions are necessary for specimen collection and transport. Though, in principle, the culture and identification of anaerobes is straightforward, the practice requires great attention to technical detail. Anaerobic infections may also be difficult to treat.

Anaerobic Bacteria in Infection

The most important non-sporing anaerobes from the clinical point of view are six genera of Gram-negative rods. Of these *Bacteroides* spp. (see Chapter 91) are particularly important, especially the species of the *B. fragilis* group. The other medically important Gram-negative anaerobic genera are *Prevotella*, *Porphyromonas*, *Fusobacterium*, *Bilophila* and *Sutterella*. Among the Gram-positive non-sporing anaerobes,

there are cocci, primarily *Peptostreptococcus*, and bacilli of the genera *Actinomyces* (see Chapter 43), *Eubacterium* and *Propionibacterium* and, finally, the Gram-positive spore-forming anaerobic bacilli of the genus *Clostridium* (see Chapters 54, 89 and 90). A wide variety of Gram-negative anaerobes (**Table 1**) and Gram-positive anaerobes (**Table 2**) are encountered in

Table 1 Major anaerobic Gram-negative bacilli encountered in infections

B. fragilis group
  Especially *B. fragilis*, *B. thetaiotaomicron*, *B. distasonis*,
   B. ovatus, *B. vulgatus*
Other *Bacteroides* spp.
  B. splanchnicus, *B. ureolyticus*
Porphyromonas spp.
  Especially *P. asaccharolytica*, *P. gingivalis*, *P. endodontalis*
Pigmented *Prevotella* spp.
  P. corporis, *P. denticola*, *P. intermedia*, *P. loescheii*,
   P. melaninogenica, *P. nigrescens*, *P. pallens*, *P. tannerae*
Other *Prevotella* spp.
  P. oris, *P. buccae*, *P. oralis* group, *P. bivia*, *P. disiens*
Fusobacterium spp.
  F. nucleatum, *F. necrophorum*, *F. mortiferum*, *F. varium*
Bilophila wadsworthia
Sutterella wadsworthensis
Campylobacter spp.
  C. curvus, *C. gracilis*, *C. rectus*

Table 2 Major anaerobic Gram-positive bacteria encountered clinically

Cocci
Peptostreptococcus spp.
    Especially *P. anaerobius, P. intermedius,*
      P. micros, P. magnus, P. asaccharolyticus,
      P. prevotii
Microaerophilic streptococci
    Especially *Strep. anginosus, Strep. constellatus,*
      Strep. intermedius, Strep. milleri group

Spore-forming bacilli
Clostridium spp.
    Especially *Cl. perfringens, Cl. ramosum,*
      Cl. septicum, Cl. novyi, Cl. histolyticum,
      Cl. sporogenes, Cl. sordellii, Cl. bifermentans,
      Cl. fallax, Cl. difficile, Cl. innocuum,
      Cl. botulinum, Cl. tetani, Cl. clostridioforme

Spore non-forming bacilli
Actinomyces spp.
    A. israelii, A. meyerii, A. naeslundii,
    A. odontolyticus, A. viscosus,
    A. neuii, A. radingae, A. turicensis
Propionibacterium propionicum, P. acnes
Eubacterium lentum, E. nodatum
Bifidobacterium dentium

Table 3 Some anaerobic bacteria commonly associated with infections[a]

Non-spore-formers
    B. fragilis group (especially *B. fragilis*)
    Pigmented *Prevotella* and *Porphyromonas*
    Fusobacterium nucleatum
    Peptostreptococcus

Spore-formers
    Cl. perfringens, C. ramosum

[a] These five groups, together, account for about two-thirds of anaerobes isolated from clinically significant infections that involve anaerobes.

clinical specimens, and those most commonly encountered are listed in **Table 3**. All these organisms are present in the normal flora of humans and animals, and members of the genus *Clostridium* are also present in soils and dust. The bacteriology of the anaerobic bacteria involved in human diseases is described in the standard texts including the following: Shah *et al.* (1998) for *Bacteroides, Prevotella* and *Porphyromonas*; Hofstad (1998) for *Fusobacterium* and *Leptotrichia*; Murdoch (1998) for the Gram-positive anaerobic cocci; and Hatheway and Johnson (1998) for *Clostridium*.

Pathogenesis of Anaerobic Infection

Infections Caused by Non-sporing Anaerobes

The sources of non-sporing anaerobes that cause infection are the indigenous flora of mucosal surfaces and, to a much lesser extent, that of the skin. Non-sporing anaerobes outnumber aerobes and facultative anaerobes by a ratio of 10 : 1 in the oral and vaginal floras and by a factor of 1000 : 1 in the colon. The factors that predispose to anaerobic infection include disruption of normal mucosal or cutaneous barriers by malignant or other disease, surgery, trauma, obstruction of a hollow viscus and presence of a foreign body.

Types of Infection

In terms of overall frequency, the four major sites of anaerobic infection are pleuro-pulmonary, intra-abdominal, female genital tract, and skin and soft tissue infections with or without involvement of underlying bone. Examples of other less common infections that primarily involve anaerobic bacteria are brain abscesses and human or animal bite-wound infections. Virtually all types of infection that occur in humans may involve anaerobic bacteria, and no organ or tissue of the body is immune to infection with these organisms. Some common infections that involve non-sporing anaerobic bacteria and the source of infection are listed in **Table 4**, and their prevalence in some common infections is shown in **Table 5**. Common characteristics of anaerobic infection are abscess formation and tissue destruction and a tendency to spread to secondary sites by 'natural routes', such as in the bloodstream from pelvic sites of infection to the lung or from the pharynx to the middle ear. Septicaemia is a serious complication and may occur at any stage but particularly the late stages of anaerobic infections.

Virulence Factors

The variety of different genera and species of non-sporing anaerobes at sites of colonisation far exceeds those isolated from sites of infection. Thus, *Fusobacterium* spp. and of the *Bacteroides* spp. *B. fragilis* are more frequently isolated than other anaerobes. This suggests that specific virulence factors are involved. Indeed, *Prevotella melaninogenica* possesses an antiphagocytic capsule and *B. fragilis* also has a capsule (Kasper, 1976), which acts as a virulence factor (Onderdonk *et al.*, 1977).

Synergy between various anaerobes or between anaerobes and non-anaerobes is often important in mixed anaerobic infections. An example is

Table 4 Sites of human anaerobic bacterial colonisation and some infections derived from these

Route of spread	Site of colonisation		
	Oral cavity	Bowel	Skin
Direct	Human and animal bite infections	Appendicitis	Infected foot ulcers
	Periodontal disease	Pyogenic liver abscess	Infected bed-sores
	Root canal infection	Peritonitis	Necrotising fasciitis
	Odontogenic infections	Intra-abdominal abscess	
	Chronic sinusitis	Post-operative wound infection	
	Peri-tonsillar abscess	Antibiotic-induced colitis	
	Chronic otitis media	Pseudomembranous colitis	
	Aspiration pneumonia		
	Lung abscess		
	Actinomycosis		
Indirect	Cerebral abscess	Lung abscess	
	Subdural empyema	Endometritis	
	Mastoiditis	Salpingitis, tubo-ovarian abscess	
	Neck-space infections		
	Pleural empyema		

Table 5 The prevalence of anaerobic bacteria in some common infections

Chronic sinusitis and otitis media	> 50%
Dental and oral infections	90–100%
Pleuro-pulmonary infections	> 75%
Intra-abdominal infections	50–90%
Female genital tract infections	50–75%
Non-puerperal breast abscesses	50–80%
Diabetic foot ulcers	85–95%

post-operative abdominal wound infection after operations in which the bowel has been opened, such as appendicectomy. The bacterial synergy in these infections appears to be a real phenomenon, rather than merely due to inoculum size. Thus, acute ulcerative gingivitis has long been known to be due to synergy between spirochaetes and *Fusobacterium* spp. When both organisms are inoculated into animals, lung abscesses can be produced, but each organism alone does not do so (Smith, 1930). More recently evidence has accumulated to show that non-sporing anaerobes in combination with facultative anaerobes are more effective in producing intra-abdominal abscesses than each organism alone (Onderdonk *et al.*, 1976). This may be due to the inhibition of phagocytosis by *B. fragilis* (Ingham *et al.*, 1977). The subject has been reviewed by MacLaren (1997).

As in the case of many other bacteria, some Gram-negative anaerobes adhere to cells by way of fimbriae (Brook and Myhal, 1991). Some other recognised virulence factors are listed in **Table 6**.

Acute anaerobic infections may become chronic if they are at sites contiguous with anaerobic bacterial colonisation: for example, chronic sinusitis, otitis media, cholecystitis and peritonitis. Acute osteomyelitis, which is usually due to common pyogenic organisms, particularly *Staphylococcus aureus*, has a tendency to become chronic and if anaerobic bacteria are present, for example because of wound contamination, these become involved in the chronic inflammatory process.

Some anaerobic infections, such as lung abscess and actinomycosis, are unique and are readily suspected clinically. Most anaerobic infections are, however, of mixed aerobic and anaerobic aetiology and are not clinically distinctive. Only the foul or putrid odour of a lesion or its discharge is specific, but other clues may nevertheless be highly suggestive. A simple Gram stain is useful because many anaerobes are morphologically unique. Information obtained in this way about the relative numbers of various organisms may be very useful in directing empirical antimicrobial therapy.

Infections Caused by Spore-forming Anaerobes

All the spore-forming anaerobic pathogens belong to the genus *Clostridium*, which also includes many non-pathogenic species that may on occasion be present at sites of infection. The clostridia are principally found in the soil, but some species are present in the bowel flora of humans and animals. From their main habitat in the soil and from faeces the clostridia may be dispersed to other sites and they are present in house dust.

The pathogenic clostridia are responsible for two types of diseases. First the now uncommon histo-toxic infections, such as gas gangrene (clostridial

Table 6 Some virulence factors of anaerobic non-sporing Gram-negative bacteria (after Lorber, 1995)

Virulence factor	Organism	Activity
Capsular polysaccharide	*Bacteroides fragilis*	Cell adherence
	Prevotella melaninogenica	Abscess formation
		Inhibition of phagocytosis
Fimbriae	*B. fragilis* group	Adherence to cells and mucus
	Porphyromonas gingivalis	
Lipopolysaccharide	*Bacteroides* spp.	Lacks Lipid A – low endotoxicity
	Fusobacterium	Potent endotoxic action
Succinic acid	Many species	Inhibition of phagocytosis and intracellular killing
Enzymes		
Hyaluronidase	*Bacteroides* spp.	Spread in tissues
Collagenase	*Bacteroides* spp.	Tissue damage
Prevotella melaninogenica	Tissue damage	
Phospholipase A	*Prevotella melaninogenica*	Cell membrane damage

myonecrosis) (see Chapter 89), which occurs after traumatic implantation of the organisms into ischaemic tissues and, second, toxin-induced disease in various target tissues, including the bowel (see below and Chapter 54) and the nervous system (botulism, tetanus) (see Chapters 55 and 90).

Wound Infection

Clostridial wound infections occurred in antiquity (Sussman, 1958). Though they are now rare they deserve attention here as a classical example of how the pathogenesis of an infection became understood. Indeed, the mode of action of the α toxin (lecithinase) of *Cl. perfringens*, which is important in the pathogenesis of gas gangrene, was the first of any exotoxin to be understood in any detail at the molecular level.

Clostridial wound infections first came into prominence during World War I, when the incidence of gas gangrene was extremely high in battle wounds. Since it had been uncommon in earlier wars and was always uncommon in civilian injuries, it came to be regarded as a disease of modern warfare associated with extensive and heavily contaminated wounds. Major wounds of the kind seen in war are uncommon in civilian practice, but gas gangrene can occur in major neglected injuries of any kind if they become contaminated with soil or dust that contains *Clostridium* spp. Prevention by appropriate surgical treatment and antibiotic prophylaxis is easily available in the developed world, but this is not so in other parts of the world where gas gangrene after accidental trauma is said still to be common.

Three types of clostridial wound infection are recognised. In increasing order of severity these are: (1) *simple contamination* in which the organisms are present in the tissues but there is no clinical evidence of infection; (2) *clostridial cellulitis* in which the

infection is localised to skin and soft tissue, the fascia and muscle are not involved and there is little evidence of toxaemia, and (3) *clostridial myonecrosis* in which muscle is involved and there is severe toxaemia. If for any reason the tissue oxygen supply is compromised, simple wound contamination may rapidly progress to cellulitis and then to myonecrosis.

Gas gangrene is characteristically associated with serious injuries, but it may follow clean elective surgery and it may rarely follow the injection of adrenaline, which causes vasoconstriction, or insulin into the lower parts of the body, where the skin may be contaminated with the *Cl. perfringens* derived from the bowel. Criminal instrument-induced abortion used not uncommonly to lead to clostridial infection of the uterus but it is now rare, because of the widespread availability of the legal pregnancy termination. Accounts of clostridial soft-tissue infections have been provided by Lorber (1995), and gas gangrene and related infections have been reviewed by Finegold and George (1989) and Willis (1991). The pathology of gas gangrene has been described by Aikat and Dible (1956, 1960) .

Bacteriology The distinction between pathogenic and non-pathogenic *Clostridium* spp. is that in experimental animals the former can by themselves give rise to clostridial infection, whereas the latter cannot do so, though they render more severe infections by pathogenic species. The principal pathogenic clostridial species responsible for gas gangrene include *Cl. perfringens*, *Cl. novyi*, *Cl. septicum* and *Cl. histolyticum*, but other species may also be involved; *Cl. sordelli*, *Cl. sporogenes*, *Cl. bifermentans* and *Cl. tertium* may also often be present. Since anaerobic infections are often due to wound contamination, they are almost always polymicrobial. In the course of injury, soil and clothing contaminated with the patient's own bacterial flora may enter the wound. The clostridia present on clothing consist mainly of *Cl. perfringens* and *Cl. sporogenes*, derived from bowel flora. Surveys during World War I showed that there was a dense and diverse distribution of clostridia in the soil of the Somme and Ypres battlefields, and this was associated with a high incidence of gas gangrene. Similar observations during World War II showed that numbers of anaerobes in desert sands were negligible and gas gangrene was rare (MacLennan, 1943). The clostridial flora of gas gangrene in battle casualties consists mainly of *C. perfringens* and *C. novyi*, but non-clostridial anaerobes and various facultative anaerobes may also be present and facilitate tissue invasion by synergy with histotoxic clostridia.

Wounds may become contaminated with *Cl. botulinum* but this rarely gives rise to clinical botulism (wound botulism), which when it occurs is usually due to *Cl. botulinum* types A or B. Wound botulism has recently been observed in injecting drug abusers (Athwal *et al.*, 2000, 2001; Jensenius *et al.*, 2000).

Pathogenesis   Clostridia do not multiply and cannot produce disease in normal tissues, because the high oxidation–reduction potential (Eh 126–246 mV) of the circulating blood and of the tissues is higher than that necessary for anaerobic bacterial growth (Oakley, 1954). Broth cultures of *Cl. perfringens*, *Cl. septicum* or *Cl. novyi* injected into guinea-pigs produce a disease similar to gas gangrene, but washed bacilli free of toxins do so only in large numbers. Mixtures of small doses of toxin-free bacilli and sub-lethal doses of culture filtrate are, however, highly virulent. The filtrate allows the bacilli to proliferate in the tissues, produce fresh toxin, and finally kill the animal. The clostridial toxins and other substances in the filtrate interfere with tissue defences and allow the potentially toxigenic bacilli to multiply. The evidence that biologically active clostridial products participate in the pathogenesis of gas gangrene is that: (1) the pathogenicity of clostridia correlates with their ability to produce these substances; (2) their injection into tissues mimics the disease and (3) antibodies to these substances is protective. The pathogenesis of experimental gas gangrene has been reviewed by Sussman *et al.* (1998).

Since clostridia are ubiquitous, most accidental wounds are exposed to the risk of contamination; but since anaerobic conditions do not usually exist in the lesion organisms cannot multiply and gas gangrene does not occur. Tissue anoxia is the central factor that allows anaerobes to grow in wounds. High-velocity missiles cause destructive 'cavitation' that reduces tissue perfusion and the shock waves cause damage to distant blood vessels, and this predisposes to anaerobic infections. Missiles may suck in soil, clothing and skin, which carry potential pathogens and bring about deep contamination. Facultative anaerobes, such as *Escherichia coli* and *Proteus* spp., which may also be present in the wound, contribute to the reduction of the local Eh by utilising any remaining oxygen.

Infections that complicate clean elective surgery, such as mid-thigh amputations, are similar to those after accidental trauma. The infecting *Cl. perfringens* is present on the skin and is implanted during surgery.

Damaged and anoxic tissues are subject to a rapidly falling Eh, which creates an ideal environment for clostridial growth (Oakley, 1954; Willis, 1969). Bacterial growth is promoted by the production of toxins and products of bacterial metabolism so that gas gangrene becomes established. Under such conditions neither phagocytes nor antibodies can reach the site and the lack of perfusion may prevent antimicrobial agents from reaching adequate concentrations in the affected tissue.

Clostridial Bacteraemia

Late in the natural history of gas gangrene clostridia involved may invade the bloodstream, and *C. perfringens* bacteraemia may rarely occur after surgery on the gastrointestinal tract, or after perforations of stomach or bowel. Entry of *Cl. perfringens* or *Cl. septicum* into the circulation through a malignant lesion of the colon, probably through an ulcer, is sometimes associated with spontaneous gas gangrene and *Cl. septicum* bacteraemia may spontaneously complicate malignant disease of the colon. *C. septicum* is relatively sparse in the human intestine, and it is not known why it, more often than *C. perfringens*, behaves in this way. Clostridial bacteraemia has also been observed in patients with leukaemia.

Enteric Infections

Mild diarrhoea associated with the consumption of re-heated foods may be due to *Cl. perfringens* present in the food; this is due to certain strains of *Cl. perfringens* type A. This and other clostridial enteric infections are considered in Chapter 54.

The association between *Cl. difficile* and antibiotic-associated diarrhoea and pseudomembranous colitis is discussed in Chapter 56.

Necrotising jejunitis (enteritis necroticans; pigbel) is a severe and often fatal disease due to *Cl. perfringens* type C. It was first recorded in 1946 in north-west Germany. Severe lower abdominal pain and diarrhoea developed some hours after the patients had eaten

Table 7 Incidence of various Gram-positive anaerobes of the normal flora in humans

	Clostridium	Actinomyces	Bifido-bacterium	Eubacterium	Lactobacillus	Cocci	Propionibacterium
Skin	0	0	0	±	0	2	1
Upper respiratory tract[a]	0	1	0	±	0	1	1
Mouth	±	1	1	1	1	±	2
Bowel	2	±	2	2	1–2	±	2
External genitalia	0	0	0	U	0	U	1
Urethra	±	0	0	U	±	0	±
Vagina	±	±	±	±	2	±	2
Endocervix	±	0	0	±	1	±	2

[a] Including nasal passages, nasopharynx, oropharynx and tonsils.
U, unknown; 0, not found or rare; ±, irregular; 1, usually present; 2, usually present in large numbers.

Table 8 Incidence of various Gram-negative anaerobes of the normal flora in humans

	B. fragilis group	Fusobacterium	Other Gram-negative bacilli	Cocci
Skin	0	0	0	0
Upper respiratory tract	0	1	2	1
Mouth	0	2	2	2
Bowel	2	1	2	1
External genitalia	±	±	1	0
Urethra	±	±	1	U
Vagina	±	±	1	±
Endocervix	±	±	1	±

U, unknown; 0, not found or rare; ±, irregular; 1, usually present; 2, usually present in large numbers.

rabbit, tinned meat or fish paste and their deaths were due to peripheral circulatory collapse or intestinal obstruction due to massive jejunal necrosis and mucosal oedema. The disease is referred to as the 'pigbel' syndrome, because of its relationship to the widespread practice of pork feasting in New Guinea. It is predominantly an affliction of children and is due to the β toxin of *Cl. perfringens* type C. The organism proliferates in the intestine and releases its β toxin, which is normally destroyed by intestinal proteinases. The local staple diet of sweet potato contains heat-stable trypsin inhibitors that prevent the destruction of β toxin and allow it to damage the bowel. Active immunisation with *C. perfringens* type C β toxoid confers a high degree of lasting protection.

Prevalence

The prevalence of infections involving anaerobes varies considerably according to the nature of the infection. As many as 5% of bacteraemias may be due to these

organisms and more than 80% of cerebral abscesses involve anaerobes. Similarly, anaerobic bacteria are involved in the vast majority of neck space infections and infections after head and neck surgery. The prevalence of anaerobes in some sites of normal carriage are given in **Tables 7** and **8**.

Specimen Collection and Transport

The proper collection and transport of specimens is crucial for the recovery of anaerobes in the laboratory. Since anaerobes are part of the normal flora, one must be certain not to contaminate the specimens with such flora. This may at times be difficult. A good example of the problem is the patient with suspected aspiration pneumonia. Expectorated sputum would not be suitable because of the large numbers of anaerobes and other organisms present in saliva as indigenous flora. It is, therefore, necessary to 'bypass' the normal flora. When there is a collection of pus in a body cavity, such as the chest (empyema), percutaneous needle aspiration provides a good specimen. If pleural fluid is absent, a suitable fluid can be introduced into the lower respiratory tract through a special double lumen catheter and then removed (bronchoalveolar lavage) for laboratory examination. Alternatively, fluid can be aspirated through a needle introduced into the trachea (transtracheal aspiration).

Since many anaerobic bacteria are extremely sensitive to oxygen, proper transport involves placing the specimen into an oxygen-free glass tube or vial under anaerobic conditions in a non-nutritive holding medium for transport to the laboratory.

Treatment

The two key approaches to the treatment of all anaerobic infections are surgery and antimicrobial therapy. Typically it is essential to remove infected material surgically (debridement) and to drainage collections of pus or infected exudate. Failure to do so promptly and thoroughly may lead to failure of response to appropriate antimicrobial agents.

Hyperbaric oxygen therapy has revolutionised the treatment of gas gangrene. It consists of exposing the patient to oxygen at a pressure of 2.5–3 atmospheres (25–30 kPa) in specially designed chambers, which had originally been designed to treat deep-sea divers with 'bends' (decompression sickness). It does not reverse the myonecrosis but it reduces or even stops its progress, so that surgery can be far less radical than would otherwise be necessary. The therapy may also promote the viability of tissues to which the blood supply is impaired, and it is often possible to save limbs that would otherwise have had to be amputated. Hyperbaric oxygen directly inhibits the growth of clostridia. It also prevents production of some *Cl. perfringens* toxins and kills vegetative bacilli (Gottlieb, 1971).

Initial antimicrobial therapy is necessarily empirical, since it takes some time to obtain definitive information about the susceptibility of the infecting flora to antimicrobial agents. Antimicrobial resistance is an increasing problem with anaerobic bacteria, and the mechanisms for this resistance are similar to the mechanisms that are involved with non-anaerobes. One of the most common mechanisms of such resistance is β lactamase production, but this can to some extent be overcome by the use of combinations of β lactam drugs with various β lactamase inhibitors. Hyper-production of β lactamases and production of metalloenzyme β lactamases may render some of the otherwise better drugs inactive.

The four groups of drugs active against almost all anaerobic bacteria are nitroimidazoles, carbapenems, chloramphenicol and thiamphenicol, and combinations of β lactam drugs with a β lactamase inhibitor. Non-sporing anaerobic Gram-positive bacilli are commonly resistant to nitroimidazoles. A small numbers of strains of the *B. fragilis* group may be resistant to all of the above agents except chloramphenicol.

References

Aikat BK, Dible JH (1956) Pathology of *Clostridium welchii* infection. *J. Pathol. Bacteriol.* 71: 461–476.

Aikat BK, Dible JH (1960) Local and general effects of cultures and culture filtrates of *Clostridium oedematiens, Cl. septicum, Cl. sporogenes* and *Cl. histolyticum. J. Pathol. Bacteriol.* 79: 227–241.

Athwal BS, Gale AN, Brett MM, Youl BD (2000) Wound botulism in UK. *Lancet* 356: 2011–2012.

Athwal BS, Gale AN, Brett MM, Youl BD (2001) Wound botulism in the UK. *Lancet* 357: 234.

Brook I, Myhal ML (1991) Adherence of *Bacteroides fragilis* group species. *Infect. Immun.* 59: 742–744.

Finegold SM, George WL (1989) *Anaerobic Infections in Humans.* San Diego, CA: Academic Press.

Gottlieb SF (1971) Effect of hyperbaric oxygen on microorganisms. *Annu. Rev. Microbiol.* 25: 111–152.

Hatheway CL, Johnson EA (1998) *Clostridium*: The spore-bearing anaerobes. In: Balows A, Duerden BI (eds) *Topley and Wilson's Microbiology and Microbial Infections* 9th edn. London: Arnold, Vol. 2. pp. 731–782.

Hofstad T (1998) *Fusobacterium* and *Leptotrichia*. In: Balows A, Duerden BI (eds) *Topley and Wilson's Microbiology and Microbial Infections* 9th edn. London: Arnold, Vol. 2. pp. 1355–1364.

Ingham HR, Sisson PR, Tharagonnet D (1977) Inhibition of phagocytosis *in vitro* by obligate anaerobes. *Lancet* 2: 1252–1254.

Jensenius M, Løvstad RZ, Dhaenens G, Rørvik LM (2000) A heroin user with a wobbly head. *Lancet* 356: 1160.

Kasper DL (1976) The polysaccharide capsule of *Bacteroides fragilis* subspecies *fragilis.* Immunochemical and morphologic definition. *J. Infect. Dis.* 133: 79–87.

Lorber B (1995) Bacteroides, Prevotella and Fusobacterium species (and other medically important anaerobic Gram-negative bacilli). In: Mandell GL, Bennett JE, Dolin R (eds) *Principles and Practice of Infectious Diseases.* New York: Churchill Livingstone, pp. 2195–2204.

MacLennan JD (1943) Anaerobic infections of war wounds in the Middle East. *Lancet* 2: 63–66; 94–99; 123–126.

MacLaren DM (1997) Soft tissue infection and septicaemia. In: Sussman M (ed.) *Escherichia coli: Mechanisms of virulence.* Cambridge: Cambridge University Press, pp. 469–493.

Murdoch DA (1998) Gram-positive anaerobic cocci. In: Balows A, Duerden BI (eds) *Topley and Wilson's*

Microbiology and Microbial Infections 9th edn. London: Arnold, Vol. 2. pp. 783–797.

Oakley CL (1954) Gas gangrene. *Br. Med. Bull.* 10: 52–58.

Onderdonk AB, Bartlett JG, Louie T *et al.* (1976) Microbial synergy in experimental intra-abdominal abscess. *Infect. Immun.* 13: 22–26.

Onderdonk AB, Kasper DL, Cisneros RL (1977) The capsular polysaccharide of *Bacteroides fragilis* as a virulence factor: Comparison of the pathogenic potential of encapsulated and unencapsulated strains. *J. Infect. Dis.* 136: 82–89.

Shah HN, Gharbia SE, Duerden BI (1998) *Bacteroides, Prevotella* and *Porphyromonas.* In: Balows A, Duerden BI (eds) *Topley and Wilson's Microbiology and Microbial Infections* 9th edn. London: Arnold, Vol. 2. pp. 1305–1330.

Smith DT (1930) Fusospirochetal disease of the lungs produced with cultures from Vincent's angina. *J. Infect. Dis.* 46: 303–310.

Sussman M (1958) A description of *Clostridium histolyticum* gas gangrene in the *Epidemics* of Hippocrates. *Med. Hist.* 2: 226.

Sussman M, Borriello SP, Taylor DJ (1998) Gas gangrene and other clostridial infections. In: Hausler WJ Jr, Sussman M (eds) *Topley and Wilson's Microbiology and Microbial Infections* 9th edn. London: Arnold, Vol. 3. pp. 669–691.

Willis AT (1969) *Clostridia of Wound Infection.* London: Butterworths.

Willis AT (1991) Gas gangrene and clostridial cellulitis. In: Duerden BI, Drasar BS (eds) *Anaerobes in Human Disease.* New York: Wiley-Liss, pp. 299–323.

Further Reading

Allen SD, Duerden BI (1998) Infections due to non-sporing anaerobic bacilli and cocci. In: Hausler WJ Jr, Sussman M (eds) *Topley and Wilson's Microbiology and Microbial Infections* 9th edn. London: Arnold, Vol. 3. pp. 743–776.

Borriello SP (ed.) (1990) *Clinical and Molecular Aspects of Anaerobes.* Petersfield, UK: Wrightson Biomedical Publishers.

Brook I (1989) *Pediatric Anaerobic Infection. Diagnosis and Management,* 2nd edn. St Louis: Mosby.

Duerden BI, Drasar BS (eds) (1991) *Anaerobes in Human Disease.* New York: Wiley Liss.

Finegold SM (1977) *Anaerobic Bacteria in Human Disease* New York: Academic Press.

Finegold SM, George WL, Rolfe RD (1984) International Symposium on Anaerobic Bacteria and Their Role in Disease. *Rev. Infect. Dis.* 6 Suppl. 1.

Finegold SM, Goldstein EJC (1993) Proceedings of the First North American Congress on Anaerobic Bacteria and Anaerobic Infections. *Clin. Infect. Dis.* 16 Suppl. 4: S159–S457.

Finegold SM, Mulligan ME (1994) Centennial Symposium on Anaerobes: A Memorial to Andre Veillon. *Clin. Infect. Dis.* 18 Suppl. 4: S245–S320.

Finegold SM (1998) Tetanus. In: Collier L, Balows A, Sussman M (eds) *Topley and Wilson's Microbiology and Microbial Infections.* London: Arnold, Vol. 3. pp. 693–722.

Kasper DL, Onderdonk AB (1990) International Symposium on Anaerobic Bacteria and Bacterial Infections. In: Kasper DL, Onderdonk AB (eds) *Rev. Infect. Dis.* 12 Suppl. 2: S121–S261.

National Committee for Clinical Laboratory Standards (1993) *Methods for Antimicrobial Susceptibility Testing of Anaerobic Bacteria.* 3rd edn; Approved Standard NCCLS document M11-A3. National Committee for Clinical Laboratory Standards, USA.

Summanen P, Baron EJ, Citron DM *et al.* (1993) *Wadsworth Anaerobic Bacteriology Manual,* 5th edn. Belmont, CA: Star Publishing.

89

Clostridium perfringens: Wound Infections

Richard W. Titball[1] and Julian I. Rood[2]

[1] Defence Evaluation and Research Agency, Salisbury, Wiltshire, UK
[2] Bacterial Pathogenesis Research Group, Monash University, Clayton, Australia

Descriptions of gangrenous infections can be found in the medical literature from the Middle Ages onwards, but it was not until the world wars of 1914–18 and 1939–1945 that the disease was more frequently encountered. The bacteria most frequently associated with gas gangrene (**Table 1**) are *Clostridium perfringens*, *Clostridium novyi* and *Clostridium septicum*. *Clostridium histolyticum*, *Clostridium bifermentans*, *Clostridium sordellii* or *Clostridium fallax* are rarely isolated from gangrenous lesions. Often more than one species of clostridium can be isolated from gas gangrenous wounds (Stock, 1947). Of the clostridia associated with gas gangrene, *C. perfringens* has received most attention, and this chapter deals with the molecular biology of *C. perfringens* with particular reference to the pathogenesis of gas gangrene.

Although the pathogenic *Clostridia* have achieved notoriety for their association with gas gangrene, it is increasingly apparent that they may also be associated with non-gangrenous diseases. Rheumatoid arthritis (Olhagen, 1976), Crohn's disease (Gustafson and Tagesson, 1990), carcinoma of the lower intestinal tract (Parkinson, 1987), sudden infant death syndrome (Murrell *et al.*, 1993) and primary septicaemia (Hübl *et al.*, 1993) have all been associated with *C. perfringens* infection. The pathogenesis of *C. perfringens*-mediated food poisoning is dealt with in Chapter 89.

The ability of *C. perfringens* to cause disease is associated with the production of a variety of extracellular toxins. Different isolates of *C. perfringens* can be assigned to one of five types (A–E) on the basis of the differential production of these toxins (**Table 1**). Of these, only type A isolates are associated with gas gangrene; types B–E isolates are associated with various diseases of livestock (McDonel, 1986; Songer, 1996).

Entry into the Host

Most cases of gas gangrene are associated with an initial focus of infection which develops into a rapidly spreading disease with active invasion of the surrounding connective or muscle tissue. The disease has been reported after hypodermic injections (Bishop and Marshall, 1960), and even after an insect bite (Barry, 1935), but gas gangrene is usually associated with severe traumatic injury (MacLennan, 1962). Such injuries are invariably associated with the entry of organic material which, because of the ubiquitous nature of *C. perfringens* in the soil or faecal matter, means that the wound is likely to become contaminated

Table 1 Toxins and exoenzymes produced by clostridial species associated with gas gangrene

Species	Type	Gas gangrene associated	Phospholipase C	Thiol-activated cytolysin	Protease	Hyaluronidase	Collagenase (gelatinase)	Sialidase (neuraminidase)
C. perfringens	A	+	α toxin	θ toxin (perfringolysin O)	–	μ toxin	κ toxin	Large enzyme, Small enzyme
	B	–	α toxin	θ toxin	λ toxin	μ toxin	κ toxin	Large enzyme, Small enzyme
	C	–	α toxin	θ toxin	–	μ toxin	κ toxin	Large enzyme, Small enzyme
	D	–	α toxin	θ toxin	λ toxin	μ toxin	κ toxin	Large enzyme, Small enzyme
	E	–	α toxin	θ toxin	–	–	κ toxin	Large enzyme, Small enzyme
C. novyi	A	+	γ toxin	δ toxin	η toxin			
	B	–	β toxin	–	–			
	C	–		–	–			
	D	–	β toxin	–	η toxin			
C. septicum		+		δ toxin		γ toxin		Sialidase
C. histolyticum		+		ε toxin		γ toxin	β toxin	
C. sordellii		+	Phospholipase C	O$_2$-labile cytolysin	γ toxin			Sialidase
C. bifermentans								

with vegetative cells or more probably with spores of the bacterium. The incubation period of the disease varies according to the aetiological agent. In the case of *C. perfringens* it is usually short, often less than 24 hours after wound infection, but cases are recorded of incubation periods of only a few hours (MacLennan, 1962). In contrast, infections with *C. novyi* have a much longer incubation period, the average time being 5.25 days (MacLennan, 1943). The incidence of disease ranged from 1% or less of wounded personnel during World War II to 10% of wounded personnel during World War I. In addition, the frequency of gas gangrene in different geographical areas shows wide variation (MacLennan, 1943), for unknown reasons.

The disease is also associated with other traumatic injuries, and in the UK at least 100 cases of gas gangrene are reported each year. Between 10% and 30% of all wounds become infected with *C. perfringens* but only a small proportion of these infections develop into gas gangrene (MacLennan, 1962). Road traffic or industrial accidents may result in the severe tissue damage required to establish infection. Abdominal surgery, especially for the treatment of gangrenous appendicitis (Keighley, 1992), may lead to tissue damage with infection by bacteria which are normally members of the gut flora and antibiotic prophylaxis is usually adopted after such surgery. In women, gas gangrene is occasionally associated with infections of the uterus after criminal abortion with non-sterile equipment. Even when careful procedures are adopted, uterine gas gangrene may follow a difficult birth, with resulting damage to the uterine tissues, which is a devastating and often fatal disease (Duerden, 1992). In many developing countries, gas gangrene after childbirth is associated with the practice of covering the umbilical cord with bovine faeces – a practice that is being strongly discouraged by international health agencies.

Invasion of Host Tissues

The entry of *C. perfringens* into host tissues can lead to several outcomes (**Fig. 1**). Bacterial growth may not occur and therefore no disease results. Anaerobic cellulitis results when there is infection without any effects on the muscle tissues and the disease develops slowly. Involvement of muscle tissues leads to clostridial myonecrosis, an acute infection which develops rapidly. The term gas gangrene is usually reserved to describe clostridial myonecrosis, and this form of the disease is the main subject of this chapter.

One of the characteristic features of gas gangrene is the rapid spread of infection, mainly along the longitudinal axis of the infected muscle. The muscle fibres

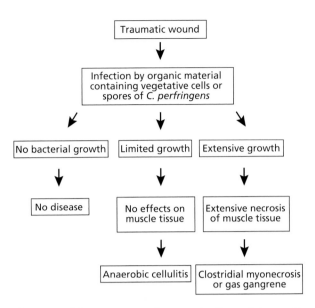

Fig. 1 Possible outcomes of contamination of traumatic wound infections with *C. perfringens*.

undergo degenerative changes, eventually losing all structure (McNee and Dunn, 1917). Early histological studies showed that few bacteria are present close to the advancing edge of muscle damage, which led McNee and Dunn (1917) to suggest that damage to muscle fibres is due to a toxic fluid that spreads between the muscle fibres. Many workers over the past 40 years have suggested that the phospholipase C (or α toxin) produced by *C. perfringens* was the main virulence factor associated with gas gangrene (Evans, 1945). Early studies showed that injection of α toxin into animals results in the reproduction of many of the symptoms of the disease, including muscle necrosis, but the purity of the toxin preparations used in these studies is open to debate, making interpretation of these studies difficult.

Only recently has the role of the α toxin in disease been proved, by different but complementary approaches which relied on the availability of the cloned and sequenced *plc* gene that encodes the α toxin (Leslie *et al.*, 1989; Okabe *et al.*, 1989; Saint-Joanis *et al.*, 1989; Titball *et al.*, 1989; Tso and Siebel, 1989). In a study with a genetically-engineered α toxoid produced in *Escherichia coli*, protection was demonstrated against experimental gas gangrene in the mouse (Williamson and Titball, 1993). Although this study revealed the overriding role of the toxin in disease the construction and virulence testing of a defined *plc* mutant (Awad *et al.*, 1995) was required to allow a more precise determination of the role of the toxin in disease. In these experiments a suicide vector was constructed that was suitable to replace the chromosomal *plc* gene by homologous recombination or allelic exchange. The plasmid pJIR919 was able to replicate in *E. coli* but not in *C. perfringens* and contained a *plc* gene that had been inactivated by the insertion of the erythromycin resistance gene *ermBP*. Transformation of *C. perfringens* cells with pJIR919 and screening of the resultant derivatives led to the isolation of chromosomal mutants that contained the insertionally inactivated *plc* gene (**Fig. 2**). The *plc* mutants could be complemented by providing a copy of a functional *plc* gene on a recombinant plasmid (Awad *et al.*, 1995).

These strains were tested for virulence in a mouse myonecrosis model. Extensive tissue destruction and muscle necrosis was observed after inoculation of the wild-type strain directly into the hind muscle, but not after injection of the *plc* mutants (Awad *et al.*, 1995). These effects were reversed (**Fig. 3**) when one of the mutants was complemented with a recombinant *plc* gene, thereby providing definitive genetic proof of the essential role of α toxin in cell and tissue damage and in the spread of the disease.

The α toxin possesses membrane-active phospholipase C activity and the preferential substrates for the enzyme are phosphatidylcholine and sphingomyelin (Titball, 1993), which are major components of the outer leaflet of most mammalian cell membranes. Although liposomes constructed from only phosphatidylcholine or sphingomyelin are lysed by the α toxin (Nagahama *et al.*, 1996, 1998), α toxin-mediated lysis of erythrocytes, or other cells (Thelestam and Möllby, 1975; Titball *et al.*, 1991) is thought to depend on the degradation of both of these membrane phospholipids (Titball, 1993). It seems likely that α toxin is produced within the initial focus of infection and then diffuses into the adjacent healthy tissues, where it causes cell membrane damage and cell death. At more distant sites, sub-lytic concentrations of α toxin appear to cause cell damage by activating a variety of biochemical pathways in cells, such as the arachidonic acid cascade. The mechanisms by which these pathways might be activated are discussed later in this chapter.

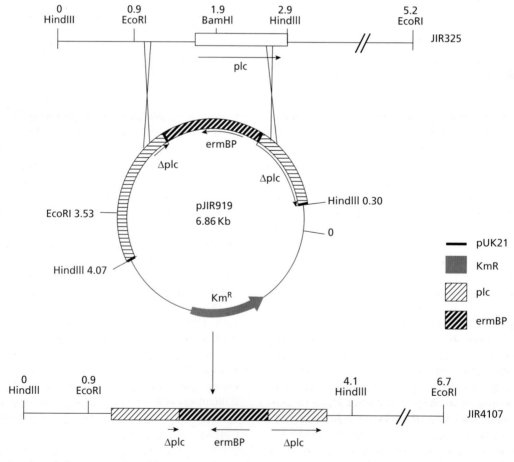

Fig. 2 Construction of *plc* mutants. The wild-type *C. perfringens* strain JIR325 was transformed to erythromycin-resistance with the suicide plasmid pJIR919, and *plc* mutants, such as JIR4107 (Awad *et al.*, 1995), were detected by their inability to produce a zone of precipitation on egg yolk agar. The diagram shows the genetic organisation of the *plc* gene region in the wild-type strain (light hatching), the suicide plasmid, and the resulting mutant. The scale is shown in kilobasepairs.

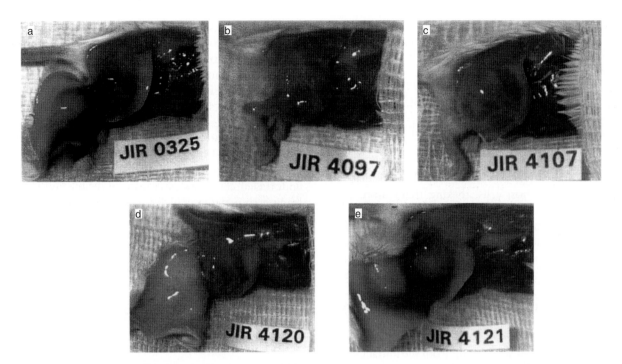

Fig. 3 Gross pathology of mice infected with *C. perfringens* strains. The plates show the dissected right hind legs of mice infected with the wild-type strain (a), two independently derived *plc* mutants (b, c), a *plc* mutant carrying a control vector plasmid (d), and the same mutant complemented with a recombinant plasmid carrying a wild-type *plc* gene (e). Note the extensive haemorrhage and necrosis caused by the strains that produce α toxin (a and e), but not by the strains defective in α toxin production (b–d). Reproduced from Awad *et al.* (1995) with the permission of the publishers of *Molecular Microbiology* (See also Colour Plate 18).

Although α toxin appears to play a key role in cell damage, the destruction of connective tissues, which are rich in collagen, proteoglycan and various complex polysaccharides such as hyaluronic acid, is also a characteristic feature of gas gangrene. It is likely that a battery of *C. perfringens* enzymes, such as the collagenase (κ toxin), sialidases (neuraminidases), hyaluronidase (μ toxin), protease (λ toxin), and endoglycosidases, contribute to this process (Okabe and Cole, 1997). The hyaluronidase enzyme appears to potentiate the cytolytic effect of α toxin by facilitating its diffusion through tissues (Smith, 1979), but virulence studies with toxin mutants are required to confirm this suggestion. Other extracellular toxins and enzymes produced by *C. perfringens* may also have a direct effect on connective tissues. In particular, collagenase degrades collagen (Hatheway, 1990), the most abundant matrix protein. Intriguingly, studies with the collagenase of *C. histolyticum* have shown that proteolytic processing of the collagenase yields a gelatinase (Yoshihara *et al.*, 1994). Close to the site of infection the enzyme may bind to collagen and hydrolyse it, whereas the proteolytically processed enzyme may be able to digest other host proteins. Studies that involved the construction and virulence testing of a collagenase mutant of *C. perfringens*

have indicated, however, that collagenase does not play a major role in the pathogenesis of clostridial myonecrosis (Awad *et al.*, 2000). A second protease produced by *C. perfringens* is λ toxin (Jin *et al.*, 1996). This is able to hydrolyse a variety of substrates including casein, gelatine, azocoll and silk (Jin *et al.*, 1996). Although many workers have suggested that λ toxin plays a role in the pathogenesis of clostridial myonecrosis it is difficult to reconcile this suggestion with the finding that the toxin is not produced by type A strains of *C. perfringens*.

The combined effect of this array of extracellular enzymes and toxins is to cause tissue damage and to enable the infection to spread. It is likely that the degradative products of these enzymes also provide essential nutrients for bacterial growth. *C. perfringens* requires at least 14 amino acids, and other growth factors such as iron, which are not present in appreciable quantities in healthy tissues (Smith, 1979). In soil and decaying organic matter – a rich source of *C. perfringens* – the function of these enzymes may be to release essential nutrients. For example, the sialidases may provide a source of sialic acid, which can be converted to pyruvic acid by *N*-acetylneuraminate lyase (Roggentin and Schauer, 1997). Proteases may provide essential amino acids, whilst phospholipases might provide a source of phosphate.

Tissue Anoxia

Traumatic damage to the tissues often results in a reduced blood supply, which generates the anoxic conditions necessary for the growth of pathogenic clostridial cells. The vulnerability of arm or leg muscles to reductions in blood supply explains why these muscles are most frequently affected by the disease. The anoxic conditions which are established around the site of infection are not only necessary for the growth of *C. perfringens*, but may also modulate the metabolism of host cells, making them more sensitive to the effects of α toxin. Anoxic conditions are associated with decreased levels of glucose pyrophosphorylase within cells, and cells deficient in their ability to produce this enzyme are hypersensitive to the effects of α toxin (Flores-Diaz *et al.*, 1997, 1998).

In addition to the direct effect of traumatic injury on blood supply, there is an increasing awareness of the roles toxins play in reducing the blood supply to infected tissues. At sites distant from the initial focus of infection, sub-lytic concentrations of α toxin appear to mimic the effects of endogenous eukaryotic phospholipases C; the toxin is able to induce contraction of smooth muscle in a wide variety of tissues including the aorta (Fujii and Sakurai, 1989), ileum (Sakurai *et al.*, 1990) and vas deferens (Sakurai *et al.*, 1985), and calcium channel-blocking agents such as verapamil and nifedipine are able to abolish the effects of the toxin (Fujii *et al.*, 1986).

The molecular events leading to muscle contraction appear to involve activation of the arachidonic acid cascade, with the resultant generation of prostaglandins and thromboxanes (Fujii and Sakurai, 1989) and an increase in the level of inositol triphosphate (Sakurai *et al.*, 1990), which in turn stimulates release of calcium stored in the endoplasmic reticulum (**Fig. 4**) leading to the activation of calcium gates and the consequent contraction of smooth muscle (Berridge, 1987). The anoxic conditions required for bacterial growth may, therefore, partly be established as a result of the contraction of the blood vessels serving the tissues surrounding the site of infection and partly established as the direct result of physical injury to these blood vessels (McNee and Dunn, 1917).

C. perfringens toxins may also cause a reduction in blood supply to infected tissues by other mechanisms. One effect of α toxin is to cause platelet aggregation. It is known that platelet aggregation can be induced by exposure to arachidonic acid metabolites, such as thromboxane A_2 (Samuelsson, 1983). The ability of α toxin to induce platelet aggregation factor (PAF) formation may further potentiate these effects (Kald *et al.*, 1994). Aggregates of platelets formed in this way might well block the microvasculature. This effect may be potentiated by the ability of α toxin to cause profound changes in endothelial cell shape and the direct cytotoxic effect of θ toxin on endothelial cells. These combined effects of α and θ toxins would result in the exposure of the sub-endothelium, which would provide sites for platelet aggregation. A further reduction in arterial blood flow may result from the leakage of albumin, electrolytes and water across the damaged endothelial cell surfaces (Bryant and Stevens, 1996) leading to oedema, which is one of the earliest signs of gas gangrene. Both α toxin (Sugahara *et al.*, 1977) and θ toxin (Stevens and Bryant, 1993) have been reported to enhance vascular permeability and may therefore contribute to this process.

Evasion of Host Defence Mechanisms

Early studies by McNee and Dunn (1917) showed that polymorphonuclear leukocytes (PMNLs) are generally absent from muscle tissues affected by the disease. In much later studies with purified *C. perfringens* toxins, however, large numbers of leukocytes could be observed between fascial planes and close to the border between diseased and healthy tissues (Bryant *et al.*, 1993). This contrasts markedly with that seen in other bacterial soft tissue infections in which a massive influx of phagocytic cells occurs. The direct cytotoxic activity of *C. perfringens* α and θ toxins may provide one explanation for the lack of leukocytes. Certainly the direct cytotoxic activity of α toxin on leukocytes has been demonstrated (Titball *et al.*, 1993) and θ toxin disrupts the PMNL cytoskeleton to cause degranulation and priming (Bryant *et al.*, 1993; Stevens *et al.*, 1987a; Stevens and Bryant, 1993). It is now known, however, that the *C. perfringens* toxins may perturb phagocyte function in more subtle ways. The increase in leukotriene release after exposure of PMNLs to low levels of α toxin would compromise the ability of these cells to infiltrate the site of infection (Bremm *et al.*, 1985). In addition, exposure of endothelial cells to α toxin and θ toxin leads to the up-regulation of PAF synthesis, the intercellular adhesion molecule 1 (ICAM) and endothelial leukocyte adhesion molecule 1 (ELAM), cell adherence markers and the production of interleukin 8 (IL-8) (Bryant and Stevens, 1996; Stevens and Bryant, 1993). PMNLs treated with sub-lethal concentrations of θ toxin show up-regulation of CD11b and CD18 adherence glycoproteins (Bryant *et al.*, 1993; Stevens and Bryant,

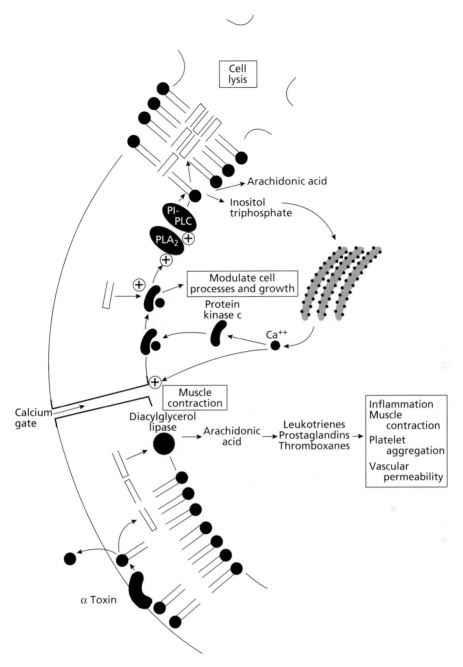

Fig. 4 Activation of eukaryotic cell pathways by *C. perfringens* α toxin (shown partially inserted into the outer leaflet of the eukaryotic cell membrane). The α toxin may cleave outer leaflet phospholipids, leading to the production of diacylglycerol which is then converted to arachidonic acid and then to leukotrienes, prostaglandins or thromboxanes. The diacyglycerol may also activate protein kinase C which can modulate cell growth or activate eukaryotic phospholipases – phospholipase A_2; PLA_2 and phosphatidylinositol phospholipase C; PI-PLC). The activated phospholipases are able to cause a variety of effects, including further activation of the arachidonic acid cascade. Inositol triphosphate generated by the PI-PLC can further activate protein kinase C and cause the release of calcium ions from the endoplasmic reticulum (ER) leading to calcium-gate activation and muscle contraction. Modified from Titball (1993) with the permission of the American Society of Microbiology.

1993). These changes would influence the ability of phagocytes to become correctly activated and trafficked into intervascular spaces further to compromise the host response.

α Toxin is the major toxin involved in modulating the inflammatory response to infection (Stevens *et al.*, 1997). Tissue sections of mice infected with a wild-type strain of *C. perfringens* reveal very little neutrophil influx into the lesion. By contrast, infection with a *plc*-mutant, which does not produce α toxin (Awad *et al.*, 1995), leads to a marked inflammatory response and a very significant influx of PMNLs to the site of infection (**Fig. 5**). When the *plc* mutation is complemented with a recombinant plasmid carrying a wild-type *plc* gene, these effects are reversed and very little PMNL influx is observed. These data clearly indicate the importance of α toxin in modulating the host inflammatory response. Similar experiments were also carried using an isogenic *pfoA* mutant, which does not produce any θ toxin. Although some effects on neutrophil influx were observed, it is clear that θ toxin is not the major toxin involved in attenuating the host inflammatory response (Stevens *et al.*, 1997).

Some other *C. perfringens* exoenzymes may inactivate other components of the host defence system. The λ toxin is capable of degrading immunoglobulin A,

complement component C3 and $α_2$ macroglobulin (Jin *et al.*, 1996) but the significance of this enzyme in the pathogenesis of gas gangrene is questionable on two counts. First, the enzyme is not produced by type A strains (McDonel, 1986) and, second, the enzyme is inhibited by non-immune serum (Smith, 1979; McDonel, 1986). Other *C. perfringens* enzymes may perturb the ability of the host immune cells to communicate with each other. For example, the sialidases produced by the bacterium (Okabe and Cole, 1997; Roggentin and Schauer, 1997) may cleave cell surface glycoproteins involved in recognition. The inability of the host immune cells to respond to appropriate signals would severely restrict the ability of the host to eliminate the bacterium from infected tissues.

Inflammation and Damage to the Host

The initial symptom of gas gangrene is often pain in the region of the wound. Soon after this, the patient complains of local swelling and oedema, and the pulse rate and temperature rise (MacLennan, 1962). In the mouse myonecrosis model, studies with a *plc⁻* mutant showed that inflammation and limb swelling are markedly reduced in comparison with the wild-type strain or the mutant strain carrying a recombinant plasmid encoding the α toxin (Awad *et al.*, 1995). This result might be expected if one of the effects of α toxin on host cells is to activate the arachidonic acid cascade and thereby produce inflammatory mediators such as leukotrienes, thromboxanes and prostaglandins (Fujii and Sakurai, 1989; Diener *et al.*, 1991). The θ toxin may also contribute to this process by inducing the production and release of cytokines such as tumour necrosis factor (TNFα) and IL-1β (Hackett and Stevens, 1992; Asmuth *et al.*, 1995).

Central to any of the effects of α toxin on eukaryotic cells is the generation of diacylglycerol (**Fig. 4**). The mechanism by which α toxin activates the arachidonic acid cascade has been the subject of speculation. The hydrolysis of membrane phosphatidylcholine by α toxin would yield diacylglycerol, but this would require further processing by diacyglycerol lipase to generate substrate for the arachidonic acid cascade (Samuelsson, 1983). While this may occur, it has also been shown that endogenous membrane phospholipases, and especially phospholipase A_2; C and D, are activated in cells treated with α toxin (Gustafson and Tagesson, 1990; Gustafson *et al.*, 1990; Sakurai *et al.*, 1993, 1994). It is thought that these activated endogenous phospholipases are responsible for much of

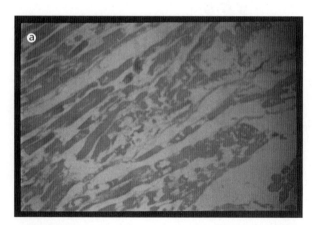

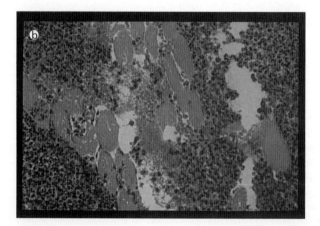

Fig. 5 Histopathology of infected muscle tissues. Haemotoxylin and eosin-stained sections of murine tissues taken after infection with A (the wild-type *plc⁺* strain, JIR325) and B (the isogenic *plc* mutant, JIR4107). The marked influx of polymorphonuclear leukocytes, which is apparent in tissues taken from mice infected with JIR4107, was not apparent in tissues taken from mice infected with the wild-type strain (See also Colour Plate 19).

the membrane phospholipid hydrolysis seen after treatment of cells with α toxin (Sakurai *et al.*, 1993, 1994). The mechanism of activation of these enzymes is not fully clarified, but it is suggested that diacylglycerol-activated protein kinase C plays a key role in this process (**Fig. 4**).

Inositol triphosphate also plays a key role in modulating host cell metabolism but α toxin is not active against phosphatidylinositol-4,5-bisphosphate which, in any case, is found within the inner leaflet of the membrane. The phosphatidylinositol-specific enzymes are one class of membrane phospholipases C activated by protein kinase C. These activated enzymes would generate IP$_3$, which would have a variety of effects on host cells, including the release of calcium from endoplasmic reticulum and the activation of calcium gates.

During the later stages of gas gangrene the patient displays symptoms of profound shock (MacLennan, 1962). Both α toxin and θ toxin are active against human erythrocytes (Hübl *et al.*, 1993) but haemoglobinuria and haemoglobinemia are rarely seen in wound gas gangrene. Although haemolysis does occur after the administration of purified toxins (Asmuth *et al.*, 1995), haemolysis does not appear to play an important role in shock in humans. In spite of the lack of extensive haemolysis (Asmuth *et al.*, 1995) death almost certainly results from the leakage of *C. perfringens* toxins into the arterial circulation and the resultant haemodynamic collapse. The α toxin appears to play a key role in this process by reducing cardiac contractility, heart rate and arterial and venous pressures, finally causing death (Asmuth *et al.*, 1995). The effects of the α toxin on the release of intracellular calcium ion and on muscle contraction may provide a molecular explanation for these effects.

Treatment and Prevention of Disease

Antibiotics

The major problem associated with the treatment of gas gangrene is that the blood supply to the affected tissue is disrupted, making it difficult to deliver drugs to the site of infection. The most commonly used therapy for gas gangrene (Muhvich *et al.*, 1994) involves the surgical debridement of wounds and the use of antibiotics and hyperbaric oxygen (Brummelkamp and Boerema, 1963), but, even with this treatment regime, amputation of the infected limb is often necessary. Penicillin has a microbicidal effect on *C. perfringens* cultured *in vitro*, and has often been the antibiotic of choice for treatment of gas gangrene. Some studies have shown, however, that penicillin does not cause bacteriolysis or cessation of α toxin activity (Stevens *et al.*, 1987b), and is ineffective in controlling an experimental infection in the mouse model when the challenge inoculum is high (Stevens *et al.*, 1987c).

Antibiotic Resistance

Antibiotic-resistant strains of *C. perfringens* are relatively easily isolated from human and animal sources, but penicillin-resistant isolates have never been reported. The isolation of antibiotic-resistant strains from animal faeces may well be related to the use of macrolide-lincosamide and tetracycline antibiotics in feedstuffs (Rood *et al.*, 1978, 1985). Several studies have examined the molecular basis of resistance and the transfer of the antibiotic-resistance determinants (Rood, 1992). Several of these determinants are carried on plasmids, which may or may not be conjugative. All of the conjugative R-plasmids are identical to, or closely related to, the 47-kb plasmid pCW3 and contain the *TetP* antibiotic-resistance determinant. This determinant is unique because it consists of two overlapping genes, *tetA(P)* and *tetB(P)*, that encode tetracycline-resistance by different mechanisms (Sloan *et al.*, 1994). The *tetA*(P) gene encodes a tetracycline efflux protein only found in *C. perfringens*, whereas the *tetB(P)* gene encodes a ribosomal modification protein that belongs to the Tet M family. Recent studies have also revealed a separate *tet(M)*-like gene in some strains of *C. perfringens* (Lyras and Rood, 1996). The worldwide distribution of pCW3-related plasmids suggests that frequent exchange of genetic information occurs between human and animal strains of *C. perfringens*. It is perhaps surprising that only one type of conjugative R-plasmid is found in a bacterium which colonises the gastrointestinal tract, a niche that is often associated with frequent transfer of genetic information between different species of bacteria.

Two chloramphenicol-resistance genes are found in *C. perfringens*. The chromosomal *catQ* gene appears to be unique to *C. perfringens*, but the plasmid-determined *catP* determinant is also found in *C. difficile*, suggesting that cross-species exchange of genetic information can occur (Rood, 1992). The *catP* gene is found on the conjugative R-plasmid pIP401 as part of the 6.3 kb transposable genetic element Tn4451 (Bannam *et al.*, 1995). A very similar transposon, Tn4453, is found in *C. difficile* (Lyras *et al.*, 1998).

Two distinct erythromycin-resistance genes have also been detected in *C. perfringens* (Rood, 1992). One of these, *ermQ*, is unique to *C. perfringens* and is chromosomal (Berryman *et al.*, 1994). The other, *erm(B)*, on the non-conjugative plasmid pIP402, is identical to similar genes from *C. difficile* (Farrow *et al.*, 2000), and is bounded by two directed repeated sequences (Berryman and Rood, 1995). These *erm(B)* genes belong to the Erm(B) family of erythromycin-resistance determinants, which are found in many different bacterial species, and are identical to the *erm* gene of the conjugative plasmid pAMβ1 from *Enterococcus faecalis* (Berryman and Rood, 1995). This provides the strongest evidence that ready exchange of genetic information occurs between *C. perfringens* and other Gram-positive bacteria in the gastrointestinal tract.

Vaccines

Although newer antibiotics may provide improved therapy of gas gangrene (Stevens *et al.*, 1987c), there has also been an intermittent interest in vaccines against gas gangrene, with most effort during the World Wars I and II devoted to the therapeutic use of antisera. Although the conclusions from these studies have been varied, it seems on balance that there is benefit from the therapeutic use of antisera, especially if these are raised against toxoids from all of the five species of clostridia associated with gas gangrene, and if the antiserum is given soon after trauma (Hall, 1945). Active immunisation against the disease has received less attention, but several studies have demonstrated the feasibility of this approach for preventing gas gangrene (Boyd *et al.*, 1972; Kameyama *et al.*, 1975; Williamson and Titball, 1993).

Many of the early studies used formaldehyde-toxoids, but a recurring problem has been to devise a method for the reliable preparation of *C. perfringens* α toxoid that retains full immunogenicity (Ito, 1968). The solution might involve devising a genetically engineered, non-toxic form of the α toxin, and two approaches have been proposed. These are site-directed mutants with greatly reduced phospholipase C and haemolytic activities, such as $Asp_{56}Asn$ (Guillouard *et al.*, 1996), or mutants in which zinc-co-ordinating histidine residues were replaced with alanine or glycine (Nagahama *et al.*, 1995), both of which have reduced lethal activity. All of these mutants retain a low level of enzymatic activity, however, which suggests that they would not be entirely safe as vaccines.

An alternative approach arose from studies with the isolated N- and C-domains of α toxin: this has yielded a vaccine which is easily produced, non-toxic and protective against gas gangrene in an animal model. Immunisation with the isolated N-domain (CPA_{1-249}) induces high levels of circulating antibody to α toxin (Williamson and Titball, 1993). This antibody did not, however, protect mice from challenge with α toxin. In contrast, immunisation with the C-domain ($CPA_{247-370}$) both induced high levels of circulating antibody and protected mice against at least 50 LD_{50} doses of the toxin (Williamson and Titball, 1993; Titball *et al.*, 1998). When immunised mice were challenged intramuscularly with 10 LD_{50} doses of *C. perfringens* type A, they were protected against an otherwise fatal infection. Recombinant vaccinia viruses that express the non-toxic C-terminal domain can also be used to protect mice against α toxin (Bennett *et al.*, 1999). Since $CPA_{247-370}$ is devoid of enzymatic and toxic activity (Titball *et al.*, 1993), a vaccine based on this polypeptide should be considered for further study. The advantage of this vaccine over a formaldehyde-toxoid lies both with the ease of production of $CPA_{247-370}$, which can be produced in high yields from *E. coli*, and the avoidance of formaldehyde toxoiding steps which may lead to batch-to-batch variation of the vaccine. This vaccine might be of use in the prevention of disease in 'at risk' populations such as elderly people, those with obliterative arterial disease and those about to undergo surgery of the lower gastrointestinal tract.

The findings that both the N- and C-domains of α toxin are immunogenic, yet only antibody against the C-domain is protective, may reflect the functions of these domains. It has been suggested that the C-domain is involved in the initial binding of α toxin to cell membranes (Titball, 1993). It is possible that the protein then undergoes a conformational change that allows insertion of the enzymatically active N-domain into the membrane. Antibody reactive with the C-domain may prevent binding of toxin to the membrane, whereas antibody reactive with the N-domain would not prevent binding, and the subsequent conformational change would reduce the affinity of antibody for the protein. In support of this suggestion it has been shown that antibody against the C-domain is able to neutralise all biological activities of the toxin, whereas antibody against the N-domain neutralises only the phospholipase C activity (Williamson and Titball, 1993). This proposal is also supported by the finding that a monoclonal antibody capable of neutralising phospholipase C, which binds to a peptide (ARGFAK) located in the N-domain, is over 50 times less effective in neutralising the haemolytic and lethal activities of the toxin (Logan *et al.*, 1991).

The Molecular Biology of Virulence Determinants

Genome Organisation

The structure and genetic organisation of virulence genes of *C. perfringens* has relatively recently been reviewed (Rood, 1998), and detailed physical and genetic maps of seven strains of *C. perfringens* have been reported (Canard *et al.*, 1992). All of these strains contain nine ribosomal RNA operons, which is consistent with the rapid growth rate of the organism; a physiological property that contributes to the rapid spread of the infection. The *rrnA* operon has been identified near the putative origin of replication on

the basis of linkage with the DNA gyrase genes, a pattern that is found in organisms such as *Bacillus subtilis* and *Staphylococcus aureus* (Rood and Cole, 1991). The genes that encode tRNA molecules are clustered around the origin of replication, and it has been suggested that this arrangement ensures that there is an immediate supply of translational precursors after the spore germination. Many of the virulence determinants of *C. perfringens* are concentrated in localised regions of its 3.6-Mb circular chromosome (Canard and Cole, 1989), often in regions which are somewhat variable, but the significance of this finding is unknown (Canard *et al.*, 1992). Notwithstanding these minor variations, the chromosomal organisation (**Fig. 6**) appears to be broadly similar in different

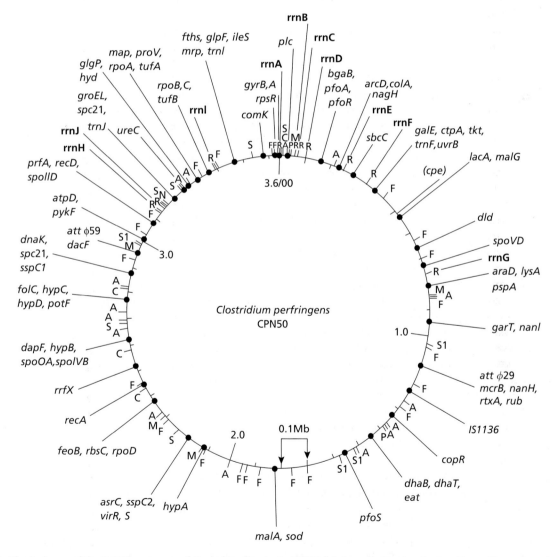

Fig. 6. Physical map of the 3.6-Mb genome of *C. perfringens* strain CPN50 (biotype A) showing the locations of housekeeping and virulence genes. Where several genes map to the same site, their order is arbitrary. The enterotoxin (*cpe*) gene is not found in strain CPN50. The location of the chromosomal *cpe* gene in other strains of *C. perfringens* is shown. Reproduced from Katayama *et al.* (1995) with the permission of the American Society of Microbiology.

strains, with as much variation seen between strains of the same type as between strains in different types (Canard *et al.*, 1992). Many of the mapped genes that encode potential virulence factors (e.g. *plc*, *pfoA*, *nagH* and *colA*) are found in a 250-kb region close to the origin of replication (Katayama *et al.*, 1995). It is possible that this region constitutes a pathogenicity island. In type B and D strains the ε toxin is carried on plasmids of between 3.2 kb and 140 kb in size (Okabe and Cole, 1997). The genes that encode the typing toxins (β, ε, and ι toxins) are all plasmid encoded (Canard *et al.*, 1992; Katayama *et al.*, 1996). Many of these toxin-encoding genes are flanked by IS or IS-like elements, suggesting that movement of genes between plasmids occurs. It is, therefore, tempting to speculate that type A strains represent the simplest *C. perfringens* genetic background, with conjugal transfer of additional genetic information resulting in the generation of other biotypes. The reasons why types B–E strains are not associated with gas gangrene requires additional investigation, however. It is possible that differential regulation of α toxin production may modulate the ability of these strains to cause myonecrosis (Katayama *et al.*, 1993; Bullifent *et al.*, 1996).

Phospholipases C (α Toxin)

C. perfringens α toxin, *C. bifermentans* phospholipase C, *Listeria monocytogenes PLC*-B, *C. novyi* γ toxin and *B. cereus PC-PLC* are all C-phospholipases that are inactivated by treatment with zinc-chelating compounds, such as EDTA and *o*-phenanthroline (Sato *et al.*, 1978; Krug and Kent, 1984; Hough *et al.*, 1989; Titball and Rubidge, 1990; Geoffroy *et al.*, 1991). They have therefore been grouped as zinc-metallophospholipases C (Titball, 1993). These enzymes have different substrate-specificities, but this has often been determined with monodispersed substrates that do not reflect the organisation of membrane-bound phospholipids. The problem has been addressed by some workers who have used liposomes of different compositions as the enzyme substrate. Using this methodology Nagahama *et al.* (1996) showed that *C. perfringens* α toxin hydrolyses phosphatidylcholine and sphingomyelin, but not phosphatidylethanolamine, phosphatidylserine or phosphatidylglycerol.

The genes that encode *C. perfringens* α toxin, *C. bifermentans* phospholipase C, *C. novyi* γ toxin, *L. monocytogenes PLC*-B, and the *B. cereus PC-PLC* phospholipases C have been cloned and sequenced. Each of these genes encodes a protein with a typical N-terminal signal peptide that directs the export of the protein from the cytoplasm (Johansen *et al.*, 1988;

Gilmore *et al.*, 1989; Leslie *et al.*, 1989; Okabe *et al.*, 1989; Saint-Joanis *et al.*, 1989; Titball *et al.*, 1989; Tso and Siebel, 1989; Vazquez-Boland *et al.*, 1992; Tsutsui *et al.*, 1995).

There is some evidence that there is variation in the properties of the enzyme produced by different strains of each species. Miles and Miles (1950) reported that enzymes isolated from different strains of *C. bifermentans* had similar phosphatidylcholine hydrolysing activity, but had up to an 8-fold difference in haemolytic and lethal activities. Although there are amino acid sequence differences between the α toxins produced by different strains of *C. perfringens* (Titball *et al.*, 1999), the specific activities of these phospholipases are similar (Tsutsui *et al.*, 1995; Ginter *et al.*, 1996). α Toxins produced by different strains of *C. perfringens* vary with respect to chymotrypsin resistance, however, which could be correlated with differences in the primary sequence of the enzymes (Ginter *et al.*, 1996).

All of the enzymes have significant amino acid sequence identity, especially within the first 250 amino acids (**Fig. 7**), and the clostridial enzymes possess an additional C-terminal polypeptide (approximately 120 amino acids). This domain is present in the *C. perfringens* α toxin, the *C. novyi* γ toxin and the *C. bifermentans* phospholipase C, but not in the nontoxic *L. monocytogenes PLC*-B and *B. cereus PC-PLC* enzymes (**Fig. 7**) (Titball, 1993). This suggests that the clostridial enzymes may be two-domain enzymes, with the active site located in the N-terminal domain. In support of this suggestion, it was shown that a truncated form of α toxin (residues 1–249) behaved as a functional homologue of the *B. cereus PC-PLC* and retained phospholipase C activity but was devoid of haemolytic and lethal activities (Titball *et al.*, 1991). The crystallisation of α toxin (Basak *et al.*, 1994, 1996, 1998) allowed the structure of the protein to be determined (Naylor *et al.*, 1998) and compared with the reported structure of the *B. cereus PC-PLC* (Hough *et al.*, 1989). This comparison confirmed that α toxin was a two-domain protein.

The crystal structure confirmed that the α-helical N-terminal domain of α toxin has an overall topology similar to the entire *B. cereus PC-PLC* (**Fig. 8**) and, like the *B. cereus PC-PLC*, contains three zinc ions in the putative active-site cleft. These ions may well help to restrain the conformation of the molecule and explain the ability of these enzymes to survive heating for short periods to 100°C (Otnaess *et al.*, 1977; Krug and Kent, 1984). Site-directed mutagenesis of residues in α toxin which are involved in zinc-coordination (Nagahama *et al.*, 1995; Guillouard *et al.*, 1996) or residues involved in the formation of

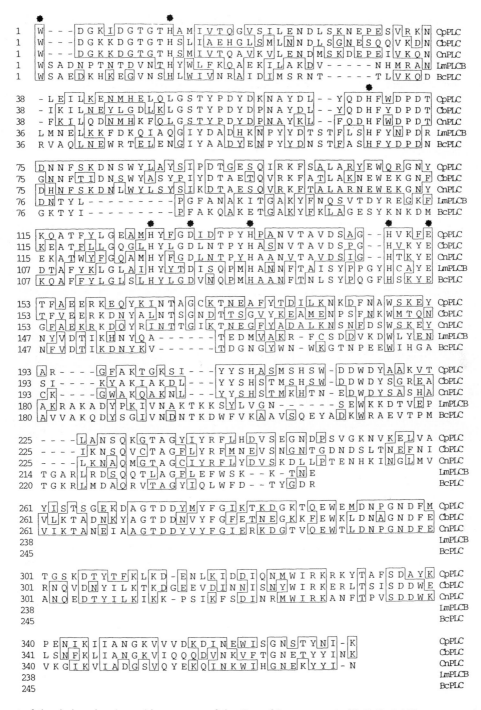

Fig. 7 Alignment of the deduced amino acid sequences of the *C. perfringens* α toxin (CpPLC), *C. bifermentans* phospholipase C (CbPLC), *C. novyi* γ toxin (CnPLC), *B. cereus* PC-PLC (BcPLC) and the *L. monocytogenes* PLC-B (LmPLCB). Consensus residues are shown boxed. Residues involved in the co-ordination of zinc ions in the *B. cereus* PC-PLC are shown starred. Modified from Titball (1993) with the permission of the American Society of Microbiology.

an acidic pocket in the active site cleft (Asp56; Asp130 and Glu152; Nagahama *et al.*, 1997) reduce the phospholipase C, sphingomyelinase, haemolytic and lethal activities to below 3% of activities of the wild-type toxin (**Table 2**).

The removal of the C-terminal domain from α toxin abolishes haemolytic and lethal activity, and indicates that this domain plays a key role in these activities. The isolated C-terminal domain is devoid of enzymatic or cytotoxic activity (Titball *et al.*, 1993), however,

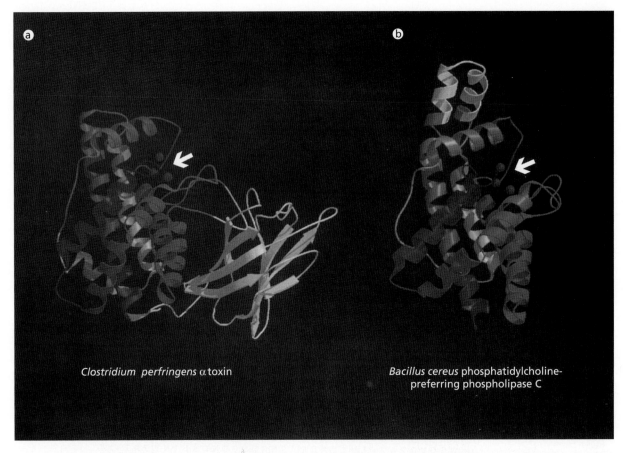

Fig. 8 Comparison of the crystal structure of *C. perfringens* α toxin (a) with the crystal structure of *B. cereus* PC-*PLC* (b). The traces in regions of structural similarity are colour-coded from blue (the N-termini) to red. Regions of dissimilarity are coloured in grey. The active-site clefts, containing active site zinc ions, are arrowed (See also Colour Plate 20).

indicating that this domain is not responsible for these effects. Further, the mutation of residues in the active site cleft in the N-terminal domain (**Table 2**) results in proteins devoid not only of phospholipase C activity, but also of haemolytic and lethal activities. These findings indicate that the phospholipase activity is essential for all biological activities of α toxin.

The precise function of the C-terminal domain of α toxin has yet to be determined. In *B. cereus*, the PC-*PLC* is encoded in an operon that also encodes a sphingomyelinase and, although these enzymes are individually non-haemolytic, a mixture of both enzymes causes erythrocyte lysis (Gilmore *et al.*, 1989). It was suggested, therefore, that the C-domain of α toxin may have sphingomyelinase activity. As discussed above, however, the isolated C-domain does not possess any detectable enzymatic activity, and several lines of investigation now support the suggestion that this domain plays a key role in the recognition of membrane phospholipids. This hypothesis is supported by the finding that mixing the individual N- and C-terminal domains in solution restores

haemolytic activity (Titball *et al.*, 1993). More importantly, the fold of the C-domain is similar to the 'C2' and 'C2-like' domains of eukaryotic phospholipid-binding proteins such as synaptotagmin and rat phosphatidylinositol phospholipase C (Guillouard *et al.*, 1997; Naylor *et al.*, 1998). The eukaryotic C2 and C2-like domains are often calcium-dependent for phospholipid recognition – and it is known that calcium ions are required for α toxin activity. Recent studies have shown that calcium ions can be identified bound to the C-terminal domain of α toxin, and that the C-terminal domain undergoes a conformational change on calcium-binding (Naylor *et al.*, 1999). Calcium-binding to the C-domain also appears to trigger the opening of the active site cleft of α toxin (Eaton *et al.*, 1999).

Some of the calcium-binding residues in the C-domain have been changed by site-directed mutagenesis, and the mutated proteins ($Asp_{269}Asn$, $Asp_{336}Asn$) have reduced activity under calcium-limited conditions (Guillouard *et al.*, 1997). It is thought that these residues, in addition to others, play a role in the

Table 2 Biological activities of site-directed mutants of *C. perfringens* α toxin

Mutation	Domain	Proposed function of target amino acid	Activity					Binding to RBC membranes		Bound zinc (mol)
			pc	sph	hly	let	plt	$+Ca^{2+}$	$+Co^{2+}/Mn^{2+}$	
Wild-type			+	+	+	+	+	+	ND	2
EDTA-treated wild-type										2
W1S[a]	N	Zn^{2+} binding	−	−	−	−	ND	ND	ND	ND
H11S[a]	N	Zn^{2+} binding	−	−	−	ND	ND	ND	ND	ND
H68S[a]	N	Zn^{2+} binding	−	−	−	ND	ND	ND	ND	ND
H68G[b]	N	Zn^{2+} binding	−	−	−	−	ND	−	+	2
H126S[a]	N	Zn^{2+} binding	−	−	−	ND	ND	ND	ND	ND
H126G[b]	N	Zn^{2+} binding	−	−	−	−	ND	−	+	2
H136S[a]	N	Zn^{2+} binding	−	−	−	ND	ND	ND	ND	ND
H136G[b]	N	Zn^{2+} binding	−	−	−	−	ND	−	+	2
H136A[b]	N	Zn^{2+} binding	−	−	−	−	ND	ND	ND	ND
H148S[a]	N	Zn^{2+} binding	−	−	−	ND	ND	ND	ND	ND
H148G[b]	N	Zn^{2+} binding	−	−	−	−	ND	+	ND	1
H148L[b]	N	Zn^{2+} binding	−	−	−	−	ND	ND	ND	ND
E152D[c]	N	Zn^{2+} binding	−	−	−	ND	ND	+	ND	1.5
E152Q[c]	N	Zn^{2+} binding	−	−	−	ND	ND	ND	ND	1
E152G[c]	N	Zn^{2+} binding	−	−	−	ND	ND	+	ND	1
F69C[a]	N	Substrate-binding	−	−	−	ND	ND	ND	ND	ND
D71N[a]	N	Substrate-binding	+	+	+	ND	ND	ND	ND	ND
E152Q[a]	N	Active site acidic pocket	−	−	−	ND	ND	ND	ND	ND
D56N[a]	N	Active site acidic pocket	−	−	−	−	−	ND	ND	ND
D56N[c]	N	Active site acidic pocket	−	−	−	ND	ND	ND	ND	2
D56E[c]	N	Active site acidic pocket	−	−	−	ND	ND	ND	ND	2
D56G[c]	N	Active site acidic pocket	−	−	−	ND	ND	+	ND	2
D56S[c]	N	Active site acidic pocket	−	−	−	ND	ND	ND	ND	2
D130N[a]	N	Active site acidic pocket	−	−	−	ND	ND	ND	ND	ND
D130E[c]	N	Active site acidic pocket	−	−	−	ND	ND	ND	ND	2
D130G[c]	N	Active site acidic pocket	−	−	−	ND	ND	+	ND	2
D130A[c]	N	Active site acidic pocket	−	−	−	ND	ND	ND	ND	2
H46G[b]	N		+	+	+	+	ND	ND	ND	ND
H207G[b]	N		+	+	+	+	ND	ND	ND	ND
H212G[b]	N		+	+	+	+	ND	ND	ND	ND
H241G[b]	N		+	+	+	+	ND	ND	ND	ND

[a] Guillouard *et al.* (1996)
[b] Nagahama *et al.* (1995)
[c] Nagahama *et al.* (1997)

+, greater than 85% retention of the activity of the wild-type toxin; −, less than 3% retention of activity of the wild-type toxin; ND; not determined. pc, hydrolysis of phosphatidylcholine; sph, hydrolysis of sphingomyelin; hly, haemolytic activity; let, lethality in the mouse; plt, platelet aggregation; binding to RBC membranes, ability to bind to rabbit erythrocyte membranes in the presence of 3 mmol/l cation.

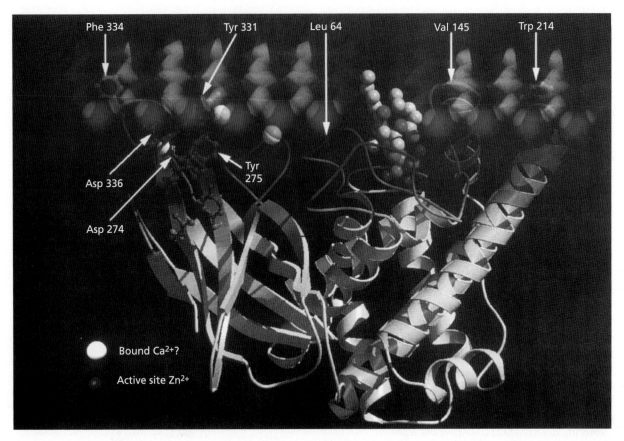

Fig. 9 Model for the interaction of *C. perfringens* α toxin with eukaryotic cell membranes. Key residues which play a role membrane interaction are highlighted. Bound calcium ions are shown in white. A phospholipid molecule is shown partially retracted from the cell membrane into the active site cleft, which also contains zinc ions (shaded blue). Reproduced from Naylor *et al.* (1998) with the permission of *Nature Structural Biology* (See also Colour Plate 21).

co-ordination of calcium ions that act as bridges between the phospholipid phosphate group and aspartate residues in α toxin. With all of this information it is possible to devise a model of the interaction between α toxin and membrane phospholipids (**Fig. 9**). This model suggests that, in addition to calcium-mediated-binding of the C-domain to the head group of phospholipids, several surface-exposed hydrophobic amino acid hydrophobic loops are appropriately positioned to interact with the hydrophobic tail groups.

These findings suggest that all the clostridial enzymes that possess a C-terminal domain should be potent cytolysins and toxins. *C. bifermentans PLC* is only weakly haemolytic, however, and is considered to be non-toxic. The replacement of the C-terminal domain of *C. bifermentans PLC* with the C-terminal domain of α toxin yields a hybrid protein with increased haemolytic and lethal activities (Jepson *et al.*, 1999). Therefore, it seems that the C-terminal domains of the clostridial enzymes do not have identical functions and have different abilities to recognise

membrane phospholipids. The molecular basis of these differences is under investigation.

Thiol-activated Cytolysin (θ Toxin)

The *C. perfringens* θ toxin (perfringolysin O) is one of the pore-forming cytolysins, a group which includes toxins produced by *Streptococcus*, *Listeria* and *Bacillus* spp. (Morgan *et al.*, 1996). These toxins share the properties of inactivation by cholesterol and, at least in an impure form, of requiring treatment with a reducing agent for activation. The gene that encodes θ toxin has been cloned and sequenced (Tweten, 1988a, b; Shimizu *et al.*, 1991). The deduced amino acid sequence of the protein has 42–65% identity with other pore-forming toxins such as pneumolysin and streptolysin O. These different pore-forming cytolysins appear to have evolved to match the lifestyle of the pathogen. For example, listeriolysin, which is produced by *L. monocytogenes*, is a primary virulence factor which plays an essential role in lysis of the phagocytic vacuole and the escape of the bacterium into the cytoplasm (Cossart *et al.*, 1989). When the listeriolysin structural gene (*hly*) was replaced with the *C. perfringens* θ toxin structural gene (*pfoA*) the resulting bacteria were still able to escape from the phagosome but were avirulent, primarily because the more cytotoxic θ toxin resulted in the lysis of the host cell (Jones and Portnoy, 1994). In a subsequent study it was shown that single mutations in *pfoA* modulated the cytolytic activity to allow the intracellular growth of the recombinant *L. mono-cytogenes* (Jones *et al.*, 1996). These findings have significant implications for the evolution of bacterial toxins, as they demonstrate that a single base change is sufficient to convert a cytolysin that is primarily extracellular in nature into a vacuole-specific lysin.

Mice infected with a defined θ toxin mutant of *C. perfringens* strain 13 had an observable reduction in gross pathology but still showed significant muscle destruction and displayed signs of systemic illness (Awad *et al.*, 1995). Subsequent studies with this mutant indicated that θ toxin plays a role in modulating the inflammatory responses, in particular by acting to induce vascular leukostasis (Stevens *et al.*, 1997; Ellemor *et al.*, 1999). The current evidence is, therefore, that θ toxin plays a minor but important role in gas gangrene.

Significant insight into the mode of action of θ toxin has been provided by the crystal structure of the protein (Rossjohn *et al.*, 1997), which appears to have an architecture similar to that of *Streptococcus pneumoniae* pneumolysin (Rossjohn *et al.*, 1998). The four domains of θ toxin (**Fig. 10**) and pneumolysin fold to form elongated rod-shaped molecules which are rich in β sheet. The region of greatest amino acid sequence similarity between θ toxin and the other cytolysins is a stretch of 12 amino acids (ECTGLAWEWWR) termed the trp-rich motif, which includes the single cysteine found in all of the pore-forming toxins. The trp motif forms a surface-exposed loop at the tip of domain 4 with the cysteine residue sandwiched between this loop and one of the β sheets in this domain. The role of this cysteine residue has been the subject of much debate in recent years. Chemical modification of the thiol group generally abolishes the membrane-binding and cytolytic activity of the pore-forming toxins (Iwamoto *et al.*, 1987), but substitution of the cysteine with alanine did not affect the activity of pneumolysin (Saunders *et al.*, 1989) or *Strep. pyogenes* streptolysin O (Pinkney *et al.*, 1989). Replacement of the cysteine with glycine reduces the haemolytic activity of the pneumolysin (Saunders *et al.*, 1989), however, the crystal structure of θ toxin provides an explanation for these contradictory findings by showing that conformational changes in the trp-rich region

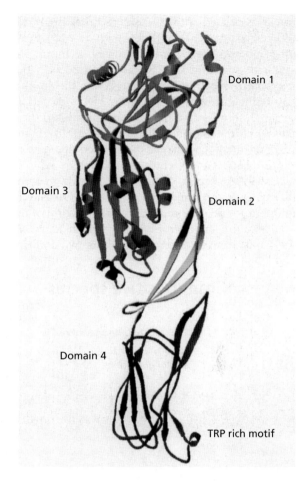

Fig. 10 Crystal structure of the *C. perfringens* θ toxin, with each domain represented by a different colour. The location of the trp motif, which is conserved in all thiol-activated toxins, is also shown. Reproduced from Rossjohn *et al.* (1997) with the permission of *Cell*. (See also Colour Plate 22).

would modulate cholesterol-binding. The chemical modification of the cysteine with a bulky thiol reactive reagent would profoundly affect the conformation of this region, to prevent cholesterol-binding, whereas some amino acid replacements would have a less profound effect on the structure in this region (Rossjohn *et al.*, 1997, 1998).

After the binding of θ toxin to membrane cholesterol, there is good evidence that 40–50 molecules of θ toxin oligomerise within the target cell membrane (Rossjohn *et al.*, 1997) to form large pores with diameters up to 15 nm, resulting in cell lysis (Tweten, 1995). Arc- and ring-shaped structures of θ toxin can be seen with negative stain electron microscopy (Mitsui *et al.*, 1979; Olofsson *et al.*, 1993), and these data have been used to construct a consensus model of the oligomer (Rossjohn *et al.*, 1997). The exact nature of the pore has yet to be determined, but all models of pore formation propose structural changes in the protein associated with this event (Rossjohn *et al.*, 1997; Shepard *et al.*, 1998; Gilbert *et al.*, 1999). Some models suggest that domain 4 forms the membrane-spanning pore (Rossjohn *et al.*, 1997), but fluorescence spectroscopy analysis, has shown that a 30-residue stretch (Lys189–Asn218), located in domain 3, forms a two-stranded amphipathic β sheet that spans the membrane (Shepard *et al.*, 1998). In the case of pneumolysin it has been proposed that regions from domains 3 and 4 enter the membrane (Gilbert *et al.*, 1999). Recent studies have shown that the C-terminal amino acids of domain 4 of θ toxin are important for the maintenance of the structure of the toxin and for cholesterol-binding (Shimada *et al.*, 1999) and recent evidence suggests that two domains

of θ toxin are involved in membrane insertion (Shatursky *et al.*, 1999).

Proteases (λ Toxin)

C. perfringens produces several enzymes that might be classified as proteases, but the term is usually applied to the caseinase or λ toxin produced by this organism (McDonel, 1986). The toxin is produced only by biotype B and D strains, which are rarely associated with gas gangrene, and the role of this toxin in the pathogenesis of gas gangrene is therefore questionable. The λ toxin structural gene, *lam*, encodes a 533-amino acid protein. Analysis of the protein indicates that the mature protein starts at residue 236,

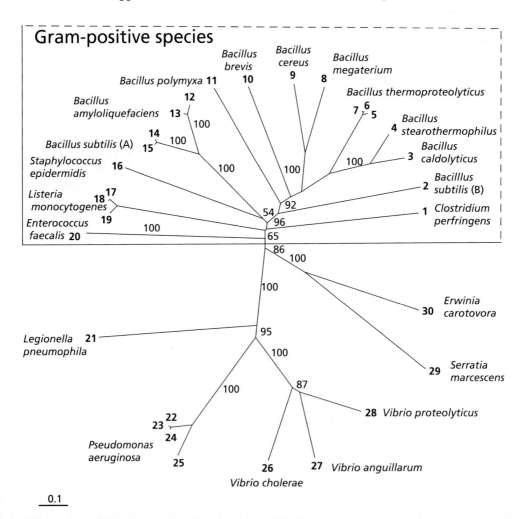

Fig. 11 Phylogenetic relationship of the *C. perfringens* λ toxin with other thermolysin-like proteases, constructed using the neighbour-joining method. The scale bar corresponds to 0.1 estimated fixed mutation per amino acid position. Bootstrap (confidence) values shown adjacent to a node were calculated from 100 trees. Accession numbers for the sequences used in this analysis were: 1, D45904; 2, A41042; 3, P23384; 4, P06874; 5, M21663; 6, B36706; 7, 441267; 8, S21934; 9, S22690; 10, PN0114; 11, P29148; 12, M36723; 13, P06832; 14, D10773; 15, P06142; 16, 396259; 17, C60280; 18, P23224; 19, B60280; 20, A43580; 21, P21347; 22, C38166; 23, B38166; 24, P14756; 25, A38166; 26, P24153; 27, L02528; 28, Q00971; 29, M59854; 30, A41048. Reproduced from Okabe and Cole (1997) with the permission of Academic Press.

suggesting that the N-terminal 253 amino acids are removed during its export and maturation (Jin *et al.*, 1996). The mature protein contains the HEXXH motif of the zinc-metalloprotease family and the consensus sequence involved in binding of the three zinc ions in the thermolysin proteases (Jongeneel *et al.*, 1989). Examination of a phylogenetic tree of these proteases (Okabe and Cole, 1997) indicates that λ toxin is closely related to the thermolysin family (**Fig. 11**) but the finding that the G + C content of the *lam* gene (30%) is similar to that of chromosomal *C. perfringens* DNA suggests that the gene has not been introduced recently into *C. perfringens* (Okabe and Cole, 1997). By analogy with the related Gram-positive proteases, it is likely that the N-terminal propeptide of λ toxin is involved in secretion and folding of the enzyme (Shinde *et al.*, 1993; Li and Inouye, 1994), and in ensuring that the enzyme is active only when in the appropriate cellular (or extra-cellular) compartment (Simonen and Palva, 1993).

Collagenases (κ Toxin)

Early studies with the collagenase from *C. histolyticum* suggested that there were a number of different forms of the collagenase (Peterkofsky, 1982), but it is now known that proteolytic processing of the 116-kDa collagenase enzyme results in the removal of a C-terminal peptide, yielding a 98-kDa gelatinase enzyme (Yoshihara *et al.*, 1994). The 120-kDa *C. perfringens* collagenase or κ toxin is encoded by the *colA* gene and has extensive amino acid sequence similarity to collagenases from *C. histolyticum* and *Vibrio alginolyticus* (Matsushita *et al.*, 1994). It is proposed that these collagenases consist of three domains (**Fig. 12**). The N-terminal domain (domain 1)

contains the HEXXH motif of zinc-metalloproteases, and replacement of the histidine residues in this motif with phenylalanine and arginine, respectively, abolishes the ability of the protein to hydrolyse Pz peptide – a synthetic substrate for collagenase (Okabe and Cole, 1997). It seems likely, therefore, that the collagenase catalytic site lies within domain 1 of the enzyme. Domains 2 and 3 are repeated in the *C. perfringens* and *C. histolyticum* enzymes and are thought to be involved in collagen-binding. This proposal is supported by the finding that a segment of the *C. histolyticum* protein containing domains 2b and 3, fused with glutathione-*S*-transferase (GST), is able to bind to collagen fibres (Matsushita *et al.*, 1995), and by the finding that the domain 1 and 2a segment, although it contains the putative active site, is unable to hydrolyse collagen (Matsushita *et al.*, 1995; Okabe and Cole, 1997). The finding that this polypeptide is able to hydrolyse gelatine, confirms the findings that proteolytic processing of the 116-kDa collagenase yields a gelatinase and also suggests that the C-domain does not play a role in the recognition of gelatine.

The general structural organisation of the *C. perfringens* and *C. histolyticum* collagenases into three domains is also seen in other bacterial enzymes, such as the *V. alginolyticus* collagenase (Yoshihara *et al.*, 1994), but there is significant variation in the primary structures of domain 3, suggesting that this region of the protein plays a key role in determining the differences in substrate-specificities between these enzymes (Okabe and Cole, 1997).

Collagen is the predominant protein in connective tissue, where it is involved in determining the structure and strength of the individual tissues. Collagen degradation would, therefore, have significant consequences for the functional integrity of any tissue exposed to infection by bacteria such as *C. perfringens* that have the ability to produce an extracellular collagenase (Harrington, 1996). The construction and virulence of isogenic *colA* mutants of *C. perfringens* indicates, however, that collagenase does not play a major role in the pathogenesis of clostridial myonecrosis (Awad *et al.*, 2000).

Hyaluronidase (μ Toxin)

Hyaluronidase (μ toxin) is an *N*-acetylglucosaminidase and is encoded by the *nagH* gene (Canard *et al.*, 1994). Export of the protein is directed by a signal sequence and the 114-kDa mature protein appears to be organised into two domains. Since a fragment of the protein comprising residues 1–370 retains enzymatic activity, it appears that the active site of the enzyme is located within the N-terminal domain

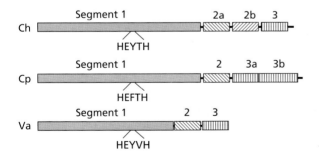

Fig. 12 Proposed molecular architecture of the *C. histolyticum* (Ch), *C. perfringens* (Cp) and *V. alginolyticus* (Va) collagenases. Segment (domain) 1 of the proteins contains the HEXXH motif found in zinc metalloproteases. Domains 2 and 3, which are thought to be involved in collagen-binding, are repeated in the *C. histolyticum* and *C. perfringens* proteins. Reproduced from Okabe and Cole (1997) with the permission of Academic Press.

(residues 1–700) of the enzyme (Canard *et al.*, 1994). The C-terminal domain contains a triple repeat of a 45-amino acid motif and, although the function of this domain is yet to be determined, it is known that such motifs are associated with ability to bind complex carbohydrate substrates (Okabe and Cole, 1997).

Hyaluronic acid consists of repeated β-(1–3)-linked *N*-acetyl-D-glucosamine-β-(1–4)-D-glucuronic acid subunits and is also a major component of connective tissue. Production of hyaluronidase by an invading bacterium could provide a selective advantage by making possible the spread of the focus of infection throughout the tissues. There is no direct evidence, however, that hyaluronidase has any role in *C. perfringens* infections. Again, this evidence must await the isolation and virulence testing of *nagH* mutants.

Sialidases

Glycoproteins that contain terminal sialic acid or *N*-acetylneuraminic acid residues are important components of mammalian cell membranes and play a role in mediating cell-to-cell interactions. Sialidases are exoglycosidases that cleave these glycoproteins to release sialic acid and could therefore either prevent cell–cell interactions, or expose new cell surface receptors as binding-sites for bacterial exotoxins. Many different sialidases are produced by the clostridia associated with gas gangrene, but there is no direct evidence that these enzymes play a role in infection or disease.

Two sialidases have been identified in *C. perfringens* (Roggentin and Schauer, 1997). The NanH enzyme

(73 kDa) is encoded by the *nanH* gene and is exported from the cell (Traving *et al.*, 1994). This is probably the enzyme which is often referred to as the neuraminidase of *C. perfringens* (McDonel, 1986), and it is a member of the group of 'large' sialidases, which include the *V. cholerae* and *C. septicum* enzymes (Roggentin *et al.*, 1993). The NanI enzyme (43 kDa), which is found within bacterial cells, is encoded by the *nanI* gene (Roggentin *et al.*, 1988) and is a member of the group of 'small' sialidases, which include the *S. typhimurium*, influenza virus and *C. sordellii* enzymes (Roggentin *et al.*, 1993). All the large and small enzymes contain a single 'FRIP box' (X-Arg-X-Pro) and several repeated 'Asp-boxes', in which the central asparagine residue is the most highly conserved in a Ser-x-Asp-x-Gly-x-Thr-Trp motif (Roggentin *et al.*, 1989). These motifs are located within the active-site domains of the large enzymes and they are repeated four or five times in different bacterial, viral, and animal sialidases (**Fig. 13**), suggesting that the genes encoding these enzymes have a common evolutionary origin (Roggentin *et al.*, 1993; Saito and Yu, 1995).

Considerable insight into the possible structure of the enzymes of *C. perfringens*, *C. sordellii* and *C. septicum* can be gained from the structures of the *S. typhimurium* (Taylor *et al.*, 1992; Crennell *et al.*, 1993) and *V. cholerae* (Taylor *et al.*, 1992; Crennell *et al.*, 1994). The active site domains of these enzymes (**Fig. 13**) show the canonical neuraminidase fold of six four-stranded antiparallel β sheets arranged in the form of a propeller, with the active site located along the central axis (Crennell *et al.*, 1993). On the basis of

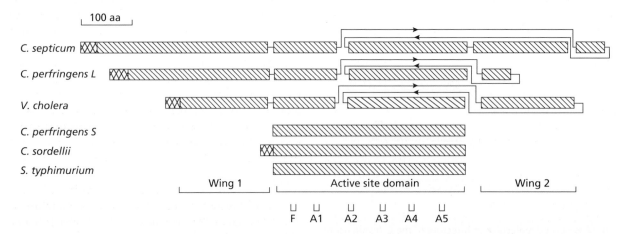

Fig. 13 Comparison of the molecular architecture of the *S. typhimurium* and *V. cholerae* sialidases with the deduced amino acid sequences of the *C. perfringens* NanH (L) or NanI (S) enzymes and the *C. septicum* and *C. sordellii* sialidases. Regions corresponding to signal sequences are shown double hatched. The continuity of the domains of the *V. cholerae* enzyme (determined from the crystal structure of the enzyme; Crennell *et al.*, 1994) is indicated with single lines and, on the basis of amino acid sequence alignments, the possible organisation of the *C. septicum* and *C. sordellii* enzymes is similarly indicated. The locations of the 'FRIP' boxes (F) and 'Asp' boxes (A1–A5) within the active site domain are depicted.

the structures of sialidases complexed with the enzyme inhibitor 2-deoxy-2,3-deoxy-N-acetylneuraminic acid (DANA), a number of key active site residues has been identified (Crennell *et al.*, 1994) and many of these residues are conserved in the *C. perfringens* enzymes, suggesting that the active site architectures are similar.

Although the small enzymes appear to contain only the active-site domain, the large enzymes may well have structures similar to the *V. cholerae* sialidase, in which the active-site domain is flanked by two wing domains with lectin-like structures (Crenell *et al.*, 1994). The C-terminal wing domain is formed from a region of β sheet structure which arises from within the active-site domain (**Fig. 13**). The substrate range of the large enzymes is much greater than that of the small enzymes, which may be related to the N- and C-terminal domains of these enzymes. In the case of the *V. cholera* enzyme the lectin-like domains are thought to enable the enzyme to recognise sialic acids on the surface of epithelial cells. This suggestion would be in accordance with export of the *C. perfringens* NanH enzyme from the bacterial cell, when it would then be able to interact with eukaryotic cells.

Regulation of Virulence

Gene-specific Regulatory Mechanisms

The relationship between the ability to produce α toxin and to cause disease has been the subject of considerable debate. An early study by Evans (1945) showed that in general there is a correlation between the level of production of α toxin *in vitro* and virulence. Strains of *C. perfringens* that produce low levels of α toxin are unable to cause disease. The requirement for the production of a threshold level of α toxin was been confirmed by Ninomiya *et al.* (1994), who showed that α toxin production is increased 10-fold in a strain of *C. perfringens* carrying the *plc* gene on a multicopy plasmid, with an associated conversion of the strain from an avirulent to virulent phenotype. Although these studies indicate that the level of *plc* gene expression is important in determining the pathogenic potential of different strains, why different strains produce different levels of toxin have only recently been investigated.

The level of α toxin production by type A strains is generally much higher than by type B–E strains. This is in agreement with the demonstration that a *plc* gene from a type A strain is expressed at a lower level in a type B genetic background than in a type A background (Bullifent *et al.*, 1996). Several studies have shown that the levels of *plc* mRNA are generally lower in biotype B–E strains, suggesting that *plc* gene expression is regulated at the level of transcription (Katayama *et al.*, 1993; Tsutsui *et al.*, 1995; Bullifent *et al.*, 1996). Comparative gel retardation studies have provided preliminary evidence that type A strains produce a *plc*-specific activator that is not produced in at least one biotype C strain (Katayama *et al.*, 1993). Neither the regulatory gene nor its protein product has yet been identified or characterised.

Examination of the sequence immediately upstream of the -35 box of the *plc* promoter has revealed the presence of an AT-rich region that contains three repeated $(dA)_{5-6}$ tracts that impart localised DNA bending (Toyonaga *et al.*, 1992). Deletion of this region in *E. coli* increased the level of α toxin production 10-fold, suggesting that the localised DNA bending had a negative regulatory effect on transcription of the *plc* gene. This region includes a directly repeated sequence (TCATTCAAAAAT) that may be involved in DNA–protein interactions (Katayama *et al.*, 1993). Studies which utilised deletions of the poly(A) tracts and insertion derivatives that increased the spacing between the tracts and the promoter, showed that in *C. perfringens* the region of DNA bending stimulates the transcription of the *plc* gene, perhaps through binding to RNA polymerase (Matsushita *et al.*, 1996). Deletion of the poly(A) tracts significantly decreased transcription of the *plc* gene (Matsushita *et al.*, 1996), a result opposite to that previously observed in *E. coli* (Toyonaga *et al.*, 1992). These results highlight the importance of carrying out functional genetic analysis in the original host organism.

Recent studies *in vitro* have shown that at lower temperatures the poly(A) tracts facilitate the formation of a complex between RNA polymerase and the *plc* promoter by extending their region of contact and stabilising the RNA polymerase-promoter complex (Katayama *et al.*, 1999). These data provide evidence that suggests that although α toxin is the major virulence determinant in disease this is not its major function. Since *plc* expression is up-regulated at lower temperatures, it appears that *C. perfringens* primarily uses its phospholipase C to scavenge organic material and energy from decaying eukaryotic cells in its normal soil environment.

A putative regulatory gene, *pfoR*, has been identified 591 bp upstream of the θ toxin structural gene *pfoA* (Shimizu *et al.*, 1991). *E. coli* strains that carry both *pfoA* and *pfoR* produce significantly more θ toxin than strains carrying only *pfoA*, and deletion analysis confirmed that the *pfoR* gene product is involved in the activation of *pfoA*. Analysis of the amino acid sequence of the putative PfoR protein revealed SPXX and helix–turn–helix (HTH) motifs that are common in regulatory proteins. Deletion of an 87-bp fragment that encodes the latter motif eliminated the ability to activate *pfoA*, providing evidence in support of the hypothesis that HTH motifs are involved in binding to regions upstream of *pfoA*. Finally, complementation studies show that activation by PfoR is more effective in *cis* than it is in *trans* (Shimizu *et al.*, 1991). These studies provide good evidence that *pfoR* encodes a *cis*-acting regulatory protein involved in the activation of *pfoA* expression. All these studies have, however, been carried out in *E. coli*. Confirmation of the importance of the *pfoR* gene in the regulation of θ toxin production depends on the isolation and complementation of chromosomal *pfoR* mutants of *C. perfringens*.

Random sequencing has identified a homologue of *pfoR*, *pfoS*, which is located at a distinct site elsewhere on the *C. perfringens* chromosome (Katayama *et al.*, 1995). It is not known whether this gene has any role in the regulation of virulence genes.

The Two-component VirR/VirS Global Regulatory System

In recent years significant advances have been made in the understanding of the global regulation of extracellular toxin production in *C. perfringens* (Rood and Lyristis, 1995). These advances have largely been brought about by the application of newly developed methods for the genetic manipulation of *C. perfringens*. In particular, the isolation and genetic analysis of pleiotropic mutants, altered in their ability to produce more than one extracellular toxin, has led to the isolation and characterisation of a two-component global regulatory system (Lyristis *et al.*, 1994; Shimizu *et al.*, 1994).

This system consists of an operon (Ba-Thein *et al.*, 1996) containing two genes, *virR* and *virS*, which regulate the expression of the genes that encode α toxin, θ toxin, κ toxin, protease and sialidase (Lyristis *et al.*, 1994; Shimizu *et al.*, 1994). Analysis of RNA transcripts has shown that regulation occurs at the level of transcription (Ba-Thein *et al.*, 1996). The *virRS* locus is located diametrically opposite the *plc*, *pfoA* and *colA* genes on the chromosome (Katayama *et al.*, 1995), which is consistent with the conclusion that this regulatory system acts in *trans* to regulate gene expression.

The VirR and VirS proteins have significant sequence similarity (Lyristis *et al.*, 1994) to members of the family of bacterial two-component regulatory proteins (Stock *et al.*, 1989). In particular, the C-terminal domain of VirS has similarity to sensor histidine kinases (Lyristis *et al.*, 1994), and the N-terminal domain of VirR has similarity to response

regulators (Lyristis *et al.*, 1994; Shimizu *et al.*, 1994). It is proposed (Rood and Lyristis, 1995) that this two-component system is activated by a specific environmental or growth-phase stimulus, perhaps involving a dialysable compound, substance A, produced by *C. perfringens* cells (Imagawa and Higashi, 1992). The stimulus is detected by the VirS sensor kinase, the N-terminal domain of which contains six transmembrane domains that are presumably located in the cell membrane (**Fig. 14**). Detection of the stimulus leads to the autophosphorylation of VirS, presumably at the histidine residue located in the cytoplasmic domain of VirS, and which is conserved in all histidine sensor kinases. The phosphorylated VirS protein then acts as a specific phosphodonor, with the phosphoryl group transferred to a conserved aspartate residue of the cytoplasmic VirR response regulator. Phosphorylation of VirR results in its activation, enabling it in turn to activate the transcription of the toxin structural genes.

Site-directed mutagenesis of the VirS sensor kinase has shown that the conserved C-terminal domains, including the histidine residue that is the predicted site of auto-phosphorylation, are essential for functional integrity of the VirS protein. In addition, two glutamate residues located in putative transmembrane domains are also required for VirS to be functional. These residues may be required for the transmission of an activation signal from the external environment to the cytoplasmic C-terminal kinase domain (Cheung and Rood, 2000b).

The toxin structural genes are differentially activated by VirR (Ba-Thein *et al.*, 1996). The *pfoA* gene is the gene most stringently controlled by the VirR/VirS system, with the expression of θ toxin totally-dependent on the presence of a functional VirR protein. Both *virR* (Shimizu *et al.*, 1994) and *virS* (Lyristis *et al.*, 1994) mutants fail to produce any θ toxin or *pfoA*-specific mRNA (Ba-Thein *et al.*, 1996). By contrast, expression of both the *colA* and *plc* genes depends only partially on activated VirR. There are two major promoters upstream of *colA*, only one of which is VirR/VirS-dependent. Therefore, in either *virR* or *virS* mutants constitutive lower level expression of *colA* occurs from the VirR/VirS-independent promoter. Only one promoter is located upstream of the *plc* gene. This promoter is partially regulated by VirR, with decreased levels of *plc* transcription occurring in the absence of activated VirR (Ba-Thein *et al.*, 1996).

The VirR protein has now been purified and shown to bind to the *pfoA* promoter but not to the *plc* or *colA* promoters (Cheung and Rood, 2000a). The VirR-binding site encompasses a 52-bp core region located immediately upstream of the *pfoA* promoter. The response regulator binds specifically to

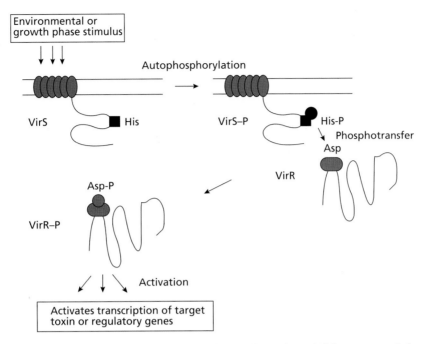

Fig. 14 Mode of action of the two-component VirR/VirS system. The membrane-bound VirS sensor protein is auto-phosphorylated in response to environmental signals and then phosphorylates an asparagine residue in the VirR response regulator. The activated VirR protein is able to activate a variety of target genes within the bacterial cell.

two 12-bp imperfect directly repeated sequences (5′ CCAGTNTNCAC 3′) located 6 bp from the −35 box. Site-directed mutagenesis shows that alteration of the CCA residues to TAG eliminates VirR-binding. In addition, binding to the two direct repeats is non-co-operative (Cheung and Rood, 2000a).

These studies indicate that VirR directly regulates the transcription of the *pfoA* gene. The precise mechanism by which VirR activates transcription is not known. However, it is likely that it either binds to RNA polymerase, and therefore enables it to bind to the promoter, or that it interacts with an RNA polymerase-promoter complex so that transcription can occur. The latter presumably involves changing the conformation of the complex so that it becomes optimal for the initiation of transcription.

It is also clear that VirR must activate the transcription of the *plc* and *colA* genes by a different mechanism. The 12-bp VirR-binding sequence is not found within the promoter or coding domains of these genes, and VirR does not bind specifically to these gene regions (Cheung and Rood, 2000a). It is postulated that VirR activates the expression of one or more as yet unknown regulatory genes, which in turn specifically activate the transcription of the toxin genes (Ba-Thein *et al.*, 1996; Cheung and Rood, 2000a). Further studies that involve the identification of these regulatory genes, and biochemical studies on their putative protein products would be needed to verify

this hypothesis. Irrespective of the outcome of these experiments it is clear that the regulation of toxin production in *C. perfringens* is a complex process. The identification and analysis of the VirRS system is just beginning to lead to an understanding of these mechanisms.

Acknowledgements

Research at Monash University was supported by grants from the Australian National Health and Medical Research Council. The authors thank Dr Claire Naylor, Prof David Moss and Dr Ajit Basak for kindly providing pictures of the crystal structure of *C. perfringens* α toxin, Dr Dennis L. Stevens for providing colour plates of the histopathology of infected tissues and Dr Michael Parker for kindly providing pictures of the crystal structure of *C. perfringens* θ toxin.

References

Asmuth DM, Olson RD, Hackett SP *et al.* (1995) Effects of *Clostridium perfringens* recombinant and crude phospholipase C and θ toxin on rabbit hemodynamic parameters. *J. Infect. Dis.* 172: 1317–1323.

Awad M, Ellemor D, Bryant A *et al.* (2000) Construction and testing of a collagenase mutant of *Clostridium perfringens. Microb. Pathog.* 28: 107–117.

Awad MM, Bryant AE, Stevens DL, Rood JI (1995) Virulence studies on chromosomal α toxin and θ toxin mutants constructed by allelic exchange provide genetic evidence for the essential role of α toxin in *Clostridium perfringens*-mediated gas gangrene. *Mol. Microbiol.* 15: 191–202.

Ba-Thein W, Lyristis M, Ohtani K *et al.* (1996) The *virR/virS* locus regulates the transcription of genes encoding extracellular toxin production in *Clostridium perfringens. J. Bacteriol.* 178: 2514–2520.

Bannam TL, Crellin PK, Rood JI (1995) Molecular genetics of the chloramphenicol resistance transposon Tn4451 from *Clostridium perfringens*: the TnpX site-specific recombinase excises a circular transposon molecule. *Mol. Microbiol.* 16: 535–551.

Barry JR (1935) Insect bite followed by gas gangrene in a diabetic; report of a case. *N. Eng. J. Med.* 212: 198.

Basak A, Fearn AM, Kelly DC *et al.* (1994) Purification and crystallisation of the alpha toxin (phospholipase C) of *Clostridium perfringens. J. Mol. Biol.* 244: 648–650.

Basak A, Eaton J, Moss D, Titball RW (1996) Crystallographic studies of alpha toxin (phospholipase C) from *Clostridium perfringens. Acta Crystallogr.* A 52: C110.

Basak AK, Howells A, Eaton JT *et al.* (1998) Crystallisation and preliminary X-ray diffraction studies of α toxin from two different strains (NCTC8237 & CER89L43) of *Clostridium perfringens. Acta Crystallogr.* D 54: 1425–1428.

Bennett AM, Lescott T, Phillpotts RJ, Mackett M, Titball RW (1999) Recombinant vaccinia viruses protect against *Clostridium perfringens* α toxin. *Viral Immunol.* 12: 97–105.

Berridge MJ (1987) Inositol triphosphate and diacylglycerol: two interacting secondary messengers. *Annu. Rev. Biochem.* 56: 159–193.

Berryman DI, Lyristis M, Rood JI (1994) Cloning and sequence analysis of *ermQ*, the predominant macrolide-lincosamide-streptogramin B resistance gene in *Clostridium perfringens. Antimicrob. Agents Chemother.* 38: 1041–1046.

Berryman DI, Rood JI (1995) The closely related *ermB/AM* genes from *Clostridium perfringens, Enterococcus faecalis* (pAMβ1), and *Streptococcus agalactiae* (pIP501), are flanked by variants of a directly repeated sequence. *Antimicrob. Agents Chemother.* 39: 1830–1834.

Bishop RF, Marshall V (1960) The enhancement of *Cl. welchii* infection by adrenaline-in-oil. *Med. J. Aust.* 47: 656–657.

Boyd NA, Thomson RO, Walker PD (1972) The prevention of experimental *Clostridium novyi* and *Cl. perfringens* gas gangrene in high-velocity missile wounds by active immunisation. *J. Med. Microbiol.* 5: 467–472.

Bremm KD, König W, Pfeiffer P *et al.* (1985) Effect of thiol-activated toxins (streptolysin O, alveolysin, and theta toxin) on the generation of leukotrienes and leukotriene-inducing and -metabolizing enzymes from human polymorphonucleargranulocytes. *Infect. Immun.* 50: 844–851.

Brummelkamp WH, Boerema I (1963) Treatment of clostridial infections with hyperbaric oxygen drenching. A report on twenty-six cases. *Lancet* i: 235–238.

Bryant AE, Bergstrom R, Zimmerman GA *et al.* (1993) *Clostridium perfringens* invasivenss is enhanced by effects of theta toxin upon PMNL structure and function:the roles of leukocytotoxicity and expression of CD11/CD18 adherence glycoprotein. *FEMS Immunol. Med. Microbiol.* 7: 321–336.

Bryant AE, Stevens DL (1996) Phospholipase C and perfringolysin O from *Clostridium perfringens* up-regulate endothelial cell-leukocyte adherence molecule 1 and intercellular leukocyte adherence molecule 1 expression and induce interleukin-8 synthesis in cultured human umbilical vein endothelial cells. *Infect. Immun.* 64: 358–362.

Bullifent HL, Moir A, Awad MM *et al.* (1996) The level of expression of α toxin by different strains of *Clostridium perfringens* is dependent on differences in promoter structure and genetic background. *Anaerobe* 2: 365–371.

Canard B, Cole ST (1989) Genome organisation of the anaerobic pathogen *Clostridium perfringens. Proc. Natl Acad. Sci. USA* 86: 6676–6680.

Canard B, Saint-Joanis B, Cole ST (1992) Genomic diversity and organisation of virulence genes in the pathogenic anaerobe *Clostridium perfringens. Mol. Microbiol.* 6: 1421–1429.

Canard B, Garnier T, Saint-Joanis B, Cole ST (1994) Molecular genetic analysis of the *nagH* gene encoding a hyaluronidase of *Clostridium perfringens. Mol. Gen. Genet.* 243: 215–224.

Cheung JK, Rood JI (2000a) The VirR response regulator from *Clostridium perfringens* binds independently to two imperfect direct repeats located upstream of the pfoA promoter. *J. Bacteriol.* 182: 57–66.

Cheung JK, Rood JI (2000b) Glutamate residues in the putative transmembrane region are required for the function of the VirS sensor histidine kinase from *Clostridium Microbiology* 146: 517–525.

Cossart P, Vicente MF, Mengaud J *et al.* (1989) Listeriolysin O is essential for virulence of *Listeria monocytogenes*: direct evidence obtained by gene complementation. *Infect. Immun.* 57: 3629–3636.

Crennell SJ, Garman EF, Laver WG, Vimr ER, Taylor GL (1993) Crystal structure of a bacterial sialidase from *Salmonella typhimurium* shows the same fold as an influenza virus neuraminidase. *Proc. Natl Acad. Sci. USA* 90: 9852–9856.

Crennell SS, Garman EF, Laver WG, Vimr ER, Taylor GL (1994) Crystal structure of the *Vibrio cholerae* neuraminidase reveals dual lectin-like domains in addition to the catalytic domain. *Structure* 2: 535–544.

Diener M, Egleme C, Rummel W (1991) Phospholipase C-induced anion secretion and its interaction with carbachol in the rat colonic mucosa. *Eur. J. Pharmacol.* 200: 267–276.

Duerden BI (1992) Anaerobic genito-urinary infections. In: Duerden BI, Brazier JS, Seddon SV, Wade WG (eds) *Medical and Environmental Aspects of Anaerobes.* Petersfield: Wrightson Biomedical Publishing, pp. 31–37.

Eaton JT, Naylor CE, Howells A *et al.* (1999) Loop movements open and close the active site of *C. perfringens* alpha toxin. *Structure.* Submitted.

Ellemor D, Baird R, Awad M *et al.* (1999) Use of genetically manipulated strains of *Clostridium perfringens* reveals that both alpha toxin and theta-toxin are required for vascular leukostasis to occur in experimental gas gangrene. *Infect. Immun.* 67: 4902–4907.

Evans DG (1945) The *in-vitro* production of α toxin, θ haemolysin and hyaluronidase by strains of *Cl. welchii* type A, and the relationship of *in-vitro* properties to virulence for guinea-pigs. *J. Pathol. Bacteriol.* 57: 77–85.

Farrow KA, Lyras D, Rood JI (2000) The macrolide-lincosamide-streptogramin B resistance determinant from *Clostridium difficile* 630 contains two *erm*(B) genes. *Antimicrob. Agents Chemother.* 44: 411–413.

Flores-Díaz M, Alape-Girón A, Persson B *et al.* (1997) Cellular UDP-glucose deficiency caused by a single point mutation in the UDP-glucose pyrophosphorylase gene. *J. Biol. Chem.* 272: 23784–23791.

Flores-Díaz M, Alape-Girón A, Titball RW *et al.* (1998) Cellular UDP-Glucose deficiency causes hypersensitivity to Clostridium *perfringens phospholipase C. J. Biol. Chem.* 273: 24433–24438.

Fujii Y, Nomura S, Oshita Y, Sakurai J (1986) Excitatory effect of *Clostridium perfringens* alpha toxin on the rat isolated aorta. *Br. J. Pharmacol.* 88: 531–539.

Fujii Y, Sakurai J (1989) Contraction of the rat isolated aorta caused by *Clostridium perfringens* alpha toxin (phospholipase C): evidence for the involvement of arachidonic acid metabolism. *Br. J. Pharmacol.* 97: 119–124.

Geoffroy C, Raveneau J, Beretti J-L *et al.* (1991) Purification and characterisation of an extracellular 29-kilodalton phospholipase C from *Listeria monocytogenes. Infect. Immun.* 59: 2382–2388.

Gilbert RJC, Jiménez JL, Chen S *et al.* (1999) Two structural transitions in membrane pore formation by pneumolysin, the pore-forming toxin of *Streptococus pneumoniae. Cell* 97: 647–655.

Gilmore MS, Cruz-Rodz AL, Leimeister-Wachter M, Kreft J, Goebbel W (1989) A *Bacillus cereus* cytolytic determinant, cereolysin AB, which comprises the phospholipase C and sphingomyelinase genes; nucleotide sequence and genetic linkage. *J. Bacteriol.* 171: 744–753.

Ginter A, Williamson ED, Dessy F *et al.* (1996) Molecular variation between the α-toxins from the type strain (NCTC8237) and clinical isolates of *Clostridium perfringens* associated with disease in man and animals. *Microbiology* 142: 191–198.

Guillouard I, Garnier T, Cole ST (1996) Use of site-directed mutagenesis to probe structure-function relationships of alpha toxin from *Clostridium perfringens. Infect. Immun.* 64: 2440–2444.

Guillouard I, Alzari PM, Saliou B, Cole ST (1997) The carboxy-terminal C2-like domain of the α toxin from *Clostridium perfringens* mediates calcium-dependent membrane recognition. *Mol. Microbiol.* 26: 867–876.

Gustafson C, Tagesson C (1990) Phospholipase C from *Clostridium perfringens* stimulates phospholipase A2 – mediated arachidonic acid release in cultured intestinal epithelial cells (INT 407). *Scand. J. Gastroenterol.* 25: 363–371.

Gustafson C, Sjodahl R, Tagesson C (1990) Phospholipase C activation and arachidonic acid release in intestinal epithelial cells from patients with Crohn's disease. *Scand. J. Gastroenterol.* 25: 1151–1160.

Hackett SP, Stevens DL (1992) Streptococcal toxic shock syndrome:synthesis and tumor necrosis factor and interleukin 1 by monocytes stimulated with pyrogenic exotoxin A and streptolysin O. *J. Infect. Dis.* 165: 879–885.

Hall IC (1945) An experimental evaluation of American commercial bivalent and pentavalent gas gangrene antitoxins. *Surg. Gynecol. Obstet.* 81: 487–499.

Harrington DJ (1996) Bacterial collagenases and collagen-degrading enzymes and their role in human disease. *Infect. Immun.* 64: 1885–1891.

Hatheway CL (1990) Toxigenic clostridia. *Clin. Microbiol. Rev.* 3: 66–98.

Hough E, Hansen LK, Birkness B *et al.* (1989) High resolution (1.5A) crystal structure of phospholipase C from *Bacillus cereus. Nature* 338: 357–360.

Hübl W, Mostbeck B, Hartleb H *et al.* (1993) Investigation of the pathogenesis of massive hemolysis in a case of *Clostridium perfringens* septicemia. *Ann. Hematol.* 67: 145–147.

Imagawa T, Higashi Y (1992) An activity which restores theta toxin activity in some theta-toxin deficient mutants of *C. perfringens. Microbiol. Immunol.* 36: 523–527.

Ito A (1968) Purification and toxoiding of α toxin of *Cl. perfringens (welchii). Jpn J. Med. Sci. Biol.* 21: 379–391.

Iwamoto M, Ohno-Iwashita Y, Ando S (1987) Role of the essential thiol group in the thiol-activated cytolysin from *Clostridium perfringens. Eur. J. Biochem.* 167: 425–430.

Jepson M, Howells AM, Bullifent HL *et al.* (1999) Differences in the carboxy-terminal (putative phospholipid-binding) domains of the *Clostridium perfringens* and *Clostridium bifermentans* phospholipases C influence the haemolytic and lethal properties of these enzymes. *Infect. Immun.* 67: 3297–3301.

Jin F, Matsushita O, Katayama S-I *et al.* (1996) Purification, characterisation, and primary structure of *Clostridium perfringens* lambda-toxin, a thermolysin-like metalloprotease. *Infect. Immun.* 64: 230–237.

Johansen T, Holm T, Guddal PH *et al.* (1988) Cloning and sequencing of the gene encoding the phosphatidylcholine-preferring phospholipase C of *Bacillus cereus. Gene* 65: 293–304.

Jones S, Portnoy DA (1994) Characterization of *Listeria monocytogenes* pathogenesis in a strain expressing perfringolysin O in place of listeriolysin O. *Infect. Immun.* 62: 5608–5613.

Jones S, Preiter K, Portnoy DA (1996) Conversion of an extracellular cytotoxin into a phagosome-specific lysin which supports the growth of an intracellular pathogen. *Mol. Microbiol.* 21: 1219–1225.

Jongeneel CV, Bouvier J, Bairoch A (1989) A unique signature identifies a family of zinc-dependent metallopeptidases. *FEBS Lett.* 242: 211–214.

Kald B, Boll R-M, Gustaffson-Svard C, Sjodahl R, Tagesson C (1994) Phospholipase C from *Clostridium perfringens* stimulates acetyltransferase-dependent formation of platelet-activating factor in cultured intestinal epithelial cells (INT 407). *Scand. J. Gastroenterol.* 29: 243–247.

Kameyama S, Sato H, Murata R (1975) The role of α-toxin of *Clostridium perfringens* in experimental gas gangrene in guinea pigs. *Jpn J. Med. Sci. Biol.* 25: 200.

Katayama S, Matsushita O, Jung C-M, Minami J, Okabe A (1999) Promoter upstream bent DNA activates the transcription of the *Clostridium perfringens* phospholipase C gene in a low temperature-dependent manner. *EMBO J.* 18: 3442–3450.

Katayama S-I, Matsushita O, Minami J, Mizobuchi S, Okabe A (1993) Comparison of the alpha toxin genes of *Clostridium perfringens* type A and C strains: evidence for extragenic regulation of transcription. *Infect. Immun.* 61: 457–463.

Katayama S-I, Dupuy B, Garner T, Cole ST (1995) Rapid expansion of the physical and genetic map of *Clostridium perfringens* CPN50. *J. Bacteriol.* 177: 5680–5685.

Katayama S-I, Dupuy B, Daube G, China B, Cole ST (1996) Genome mapping of *Clostridium perfringens* strains with I-*CeuI* shows may virulence genes to be plasmid-borne. *Mol. Gen. Genet.* 251: 720–726.

Keighley MRB (1992) Anaerobes in abdominal surgery. In: Duerden BI, Brazier JS, Seddon SV, Wade WG (eds) *Medical and Environmental Aspects of Anaerobes*. Petersfield: Wrightson Biomedical Publishing, pp. 24–30.

Krug EL, Kent C (1984) Phospholipase C from *Clostridium perfringens*: preparation and characterisation of homogenous enzyme. *Arch. Biochem. Biophys.* 231: 400–410.

Leslie D, Fairweather N, Pickard D, Dougan G, Kehoe M (1989) Phospholipase C and haemolytic activities of *Clostridium perfringens* alpha toxin cloned in *Escherichia coli;* sequence and homology with a *Bacillus cereus* phospholipase C. *Mol. Microbiol.* 3: 383–392.

Li Y, Inouye M (1994) Autoprocessing of prothiolsubtilisin E in which active-site serine 221 is altered to cysteine. *J. Biol. Chem.* 269: 4169–4174.

Logan AJ, Williamson ED, Titball RW *et al.* (1991) Epitope mapping of the alpha toxin of *Clostridium perfringens. Infect. Immun.* 59: 4338–4342.

Lyras D, Rood JI (1996) Genetic organisation and distribution of tetracycline resistance determinants in *Clostridium perfringens. Antimicrob. Agents Chemother.* 40: 2500–2504.

Lyras D, Storie C, Huggins A *et al.* (1998) Chloramphenicol resistance in *Clostridium difficile* is encoded on Tn4453 transposons that are closely related to Tn4451 from *Clostridium perfringens. Antimicrob. Agents Chemother.* 42: 1563–1567.

Lyristis M, Bryant AE, Sloan J *et al.* (1994) Identification and molecular analysis of a locus that regulates extracellular toxin production in *Clostridium perfringens. Mol. Microbiol.* 12: 761–777.

MacLennan JD (1943) Anaerobic infections of war wounds in the Middle East. *Lancet* ii: 123–126.

MacLennan JD (1962) The histotoxic clostridial infections of man. *Bacteriol. Rev.* 26: 177–276.

Matsushita C, Matsushita O, Katayama S *et al.* (1996) An upstream activating sequence containing curved DNA involved in activation of the *Clostridium perfringens plc* promoter. *Microbiology* 142: 2561–2566.

Matsushita O, Yoshihara K, Katayama S-I, Minami J, Okabe A (1994) Purification and characterisation of a *Clostridium perfringens* 120-kilodalton collagenase and nucleotide sequence of the corresponding gene. *J. Bacteriol.* 176: 149–156.

Matsushita O, Yoshihara K, Minami J, Okabe A (1995) Clostridial collagenases. *Abstracts of the ASM First International Conference on the Molecular Genetics and Pathogenesis of the Clostridia,* Rio Rico, AZ, Abstract 18.

McDonel JL (1986) Toxins of *Clostridium perfringens* types A,B,C,D and E, p. In: Dorner F, Drews J (eds) *Pharmacology of Bacterial Toxins*. Oxford: Pergamon Press, pp. 477–517.

McNee JW, Dunn JS (1917) The method of spread of gas gangrene into living muscle. *BMJ* 1: 727–729.

Miles EM, Miles AA (1950) The relation of toxicity and enzyme activity in the lecithinases of *Clostridium bifermentans* and *Clostridium welchii. J. Gen. Microbiol.* 4: 22–35.

Mitsui K, Sekiya T, Okamura S, Nozawa Y, Hase J (1979) Ring formation of perfringolysin O as revealed by negative strain electron microscopy. *Biochim. Biophys. Acta* 558: 307–313.

Morgan PJ, Andrew PW, Mitchell TJ (1996) Thiol-activated cytolysins. *Rev. Med. Microbiol.* 7: 221–229.

Muhvich KH, Anderson LH, Mehm WJ (1994) Evaluation of antimicrobials combined with hyperbaric oxygen in a mouse model of clostridial myonecrosis. *J. Trauma* 36: 7–10.

Murrell WG, Stewart BJ, O'Neil C, Siarkas S, Kariks S (1993) Enterotoxigenic bacteria in the sudden infant death syndrome. *J. Med. Microbiol.* 39: 114–127.

Nagahama M, Okagawa Y, Nakayama T, Nishioka E, Sakurai J (1995) Site-directed mutagenesis of histidine residues in *Clostridium perfringens* alpha toxin. *J. Bacteriol.* 177: 1179–1185.

Nagahama M, Michiue K, Sakurai J (1996) Membrane-damaging action of *Clostridium perfringens* alpha toxin on phospholipid liposomes. *Biochim. Biophys. Acta* 1280: 120–126.

Nagahama M, Nakayama T, Michiue K, Sakurai J (1997) Site-specific mutagenesis of *Clostridium perfringens* alpha toxin: replacement of Asp-56, Asp-130 or Glu-152 causes loss of enzymatic and hemolytic activities. *Infect. Immun.* 65: 3489–3492.

Nagahama M, Michiue K, Mukai M, Ochi S, Sakurai J (1998) Mechanism of membrane damage by *Clostridium perfringens* alpha toxin. *Microbiol. Immunol.* 42: 533–538.

Naylor C, Eaton JT, Howells A *et al.* (1998) Structure of the key toxin in gas gangrene. *Nat. Struct. Biol.* 5: 738–746.

Naylor CE, Crane D, Jepson M *et al.* (1999) Characterisation of the calcium-binding C-terminal domain of *Clostridium perfringens* alpha toxin. *J. Mol. Biol.* 294: 757–770.

Ninomiya M, Matsushita O, Minami J *et al.* (1994) Role of alpha-toxin in *Clostridium perfringens* infection determined by using recombinants of *C. perfringens* and *Bacillus subtilis. Infect. Immun.* 62: 5032–5039.

Okabe A, Shimizu T, Hayashi H (1989) Cloning and sequencing of a phospholipase C gene of *Clostridium perfringens. Biochem. Biophys. Res. Comm.* 160: 33–39.

Okabe A, Cole ST (1997) Extracellular enzymes from *Clostridium perfringens* and *Clostridium histolyticum* that damage connective tissue. In: Rood JI, McClane BA, Songer JG, Titball RW (eds) *The Clostridia: Molecular Biology and Pathogenesis.* London: Academic Press, pp. 411–422

Olhagen B (1976) Intestinal *Clostridium perfringens* in arthritis and allied conditions. In: Dummonde DC (ed.) *Infection and Immunology in the Rheumatic Diseases.* Oxford: Blackwell Scientific Publications, pp. 141–145.

Olofsson A, Herbert H, Thelestam M (1993) The projection structure of perfringolysin O. *FEBS Lett.* 319: 125–127.

Otnaess A-B, Little C, Sletten K *et al.* (1977) Some characteristics of phospholipase C from *Bacillus cereus. Eur. J. Biochem.* 79: 459–468.

Parkinson EK (1987) Phospholipase C mimics the differential effects of phorbol-12-myristate-13-acetate on the colony formation and cornification of cultured normal and transformed human keratinocytes. *Carcinogenesis* 8: 857–860.

Peterkofsky B (1982) Bacterial collagenase. *Meth. Enzymol.* 82: 453–471.

Pinkney M, Beachey E, Kehoe M (1989) The thiol-activated toxin streptolysin O does not require a thiol group for activity. *Infect. Immun.* 57: 2553–2558.

Roggentin P, Rothe B, Lottspeich F, Schauer R (1988) Cloning and sequencing of a *Clostridium perfringens* sialidase gene. *FEBS Lett.* 238: 31–34.

Roggentin P, Rothe B, Caper JB *et al.* (1989) Conserved sequences in bacterial and viral sialidases. *Glycoconjugate J.* 6: 349–353.

Roggentin P, Schauer R, Hoyer LL, Vimr ER (1993) The sialidase superfamily and its spread by horizontal gene transfer. *Mol. Microbiol.* 9: 915–921.

Roggentin P, Schauer R (1997) Clostridial sialidases. In: Rood JI, McClane BA, Songer JG, Titball RW (eds) *The Clostridia: Molecular Biology and Pathogenesis.* London: Academic Press, pp. 423–437.

Rood JI, Maher EA, Somers EB, Campos E, Duncan CL (1978) Isolation and characterisation of multiply antibiotic-resistant *Clostridium perfringens* strains from porcine feces. *Antimicrob. Agents Chemother.* 13: 871–880.

Rood JI, Buddle JR, Wales AJ, Sidhu R (1985) The occurrence of antibiotic resistance in *Clostridium perfringens* from pigs. *Aust. Vet. J.* 62: 276–279.

Rood JI, Cole ST (1991) Molecular genetics and pathogenesis of *Clostridium perfringens. Microbiol. Rev.* 55: 621–648.

Rood JI (1992) Antibiotic resistance determinants of *Clostridium perfringens.* In: Sebald M (ed.) *Genetics and Molecular Biology of Anaerobic Bacteria.* New York: Springer , pp. 141–155

Rood JI, Lyristis M (1995) Regulation of extracellular toxin production in *Clostridium perfringens. Trends Microbiol.* 3: 192–196.

Rood JI (1998) Virulence genes of *Clostridium perfringens. Annu. Rev. Microbiol.* 52: 333–360.

Rossjohn J, Feil SC, McKinstry WJ, Tweten RK, Parker MW (1997) Structure of a cholesterol-binding, thiol-activated cytolysin and a model of its membrane form. *Cell* 89: 685–692.

Rossjohn J, Gilbert RJC, Crane D *et al.* (1998) The molecular mechanism of pneumolysin, a virulence factor from *Streptococcus pneumoniae. J. Mol. Biol.* 284: 449–461.

Saint-Joanis B, Garnier T, Cole ST (1989) Gene cloning shows the alpha-toxin of *Clostridium perfringens* to contain both sphingomyelinase and lecithinase activities. *Mol. Gen. Genet.* 219: 453–460.

Saito M, Yu RK (1995) Biochemistry and function of sialidases. In: Rosenberg A (ed.) *Biology of the Sialic Acids.* New York: Plenum Press, pp. 261–313.

Sakurai J, Nomura S, Fujii Y, Oshita Y (1985) Effect of *Clostridium perfringens* alpha toxin on the isolated rat vas deferens. *Toxicon* 23: 449–455.

Sakurai J, Fujii Y, Shirotani M (1990) Contraction induced by *Clostridium perfringens* alpha-toxin in the isolated rat ileum. *Toxicon* 28: 411–418.

Sakurai J, Ochi S, Tanaka H (1993) Evidence for coupling of *Clostridium perfringens* alpha-toxin-induced hemolysis to stimulated phosphatidic acid formation in rabbit erythrocytes. *Infect. Immun.* 61: 3711–3718.

Sakurai J, Ochi S, Tanaka H (1994) Regulation of *Clostridium perfringens* alpha-toxin-activated phospholipase C in rabbit erythrocyte membranes. *Infect. Immun.* 62: 717–721.

Samuelsson B (1983) Leukotrienes: mediators of immediate hypersensitivity reactions and inflammation. *Science* 220: 568–575.

Sato H, Yamakawa Y, Ito A, Murata R (1978) Effect of zinc and calcium ions on the production of alpha-toxin and proteases by *Clostridium perfringens. Infect. Immun.* 20: 325–333.

Saunders KF, Mitchell TJ, Walker JA, Andrew PW, Boulnois GJ (1989) Pneumolysin, the thiol-activated toxin from *Streptococcus pneumoniae* does not require a thiol group for *in vitro* activity. *Infect. Immun.* 57: 2547–2552.

Shatursky O, Heuck AP, Shepard LA *et al.* (1999) The mechanism of membrane insertion of a cholesterol-dependent cytolysin: a novel paradigm for poreforming toxins. *Cell* 99: 293–299.

Shepard LA, Heuck AP, Hamman BD *et al.* (1998) Identification of a membrane-spanning domain of the thiol-activated pore-forming toxin *Clostridium perfringens* perfringolysin O: an α helical to β sheet transition identified by fluorescence spectroscopy. *Biochemistry* 37: 14563–14574.

Shimada Y, Nakamura M, Naito Y, Nomura K, Ohno-Iwashita Y (1999) C-terminal amino acids are required for the folding and cholesterol-binding property of perfringolysin O. *J. Biol. Chem.* 274: 18536–18542.

Shimizu T, Okabe A, Minami J, Hayashi H (1991) An upstream regulatory sequence stimulates expression of the perfringolysin O gene of *Clostridium perfringens*. *Infect. Immun.* 59: 137–142.

Shimizu T, Ba-Thein W, Tamaki M, Hayashi H (1994) The *virR* gene, a member of a class of two-component response regulators, regulates the production of perfringolysin O, collagenase, and haemagglutinin in *Clostridium perfringens*. *J. Bacteriol.* 176: 1616–1623.

Shinde U, Li Y, Chatterjee S, Inouye M (1993) Folding pathway mediated by an intramolecular chaperone. *Proc. Natl Acad. Sci. USA* 90: 6924–6928.

Simonen M, Palva I (1993) Protein secretion in *Bacillus* species. *Microbiol. Rev.* 57: 109–137.

Sloan J, McMurry LM, Lyras D, Levy SB, Rood JI (1994) The *Clostridium perfringens* Tet P determinant comprises two overlapping genes: *tetA*(P), which mediates active tetracycline efflux and *tetB*(P), which is related to the ribosomal protection family of tetracycline-resistance determinants. *Mol. Microbiol.* 11: 403–415.

Smith LDS (1979) Virulence factors of *Clostridium perfringens*. *Rev. Infect. Dis.* 1: 254–260.

Songer JG (1996) Clostridial enteric diseases of domestic animals. *Clin. Microbiol. Rev.* 9: 216–234.

Stevens DL, Mitten J, Henry C (1987a) Effects of α and θ toxins from *Clostridium perfringens* on human polymorphonuclear leukocytes. *J. Infect. Dis.* 156: 324–333.

Stevens DL, Maier KA, Mitten JE (1987b) Effect of antibiotics on toxin production and viability of *Clostridium perfringens*. *Antimicrob. Agents Chemother.* 31: 213–218.

Stevens DL, Maier KA, Laine BM, Mitten JE (1987c) Comparison of clindamycin, rifampin, tetracycline, metronidazole, and penicillin for efficacy in prevention of experimental gas gangrene due to *Clostridium perfringens*. *J. Infect. Dis.* 155: 220–228.

Stevens DL, Bryant AE (1993) Role of θ toxin, a sulfhydyl-activated cytolysin, in the pathogenesis of clostridial gas gangrene. *Clin. Infect. Dis.* 16: S195–S199.

Stevens DL, Tweten RK, Awad MM, Rood JI, Bryant AE (1997) Clostridial gas gangrene: evidence that alpha and theta toxins differentially modulate the immune response and induce acute tissue necrosis. *J. Infect. Dis.* 176: 189–195.

Stock AH (1947) Clostridia in gas gangrene and local anaerobic infections during Italian Campaign. *J. Bacteriol.* 54: 169–174.

Stock JB, Ninfa AJ, Stock AM (1989) Protein phosphorylation and regulation of adaptive responses in bacteria. *Microbiol. Rev.* 53: 450–490.

Sugahara T, Takahashi T, Yamaya S, Ohsaka A (1977) Vascular permeability increase by α toxin (phospholipase C) of *Clostridium perfringens*. *Toxicon* 15: 81–87.

Taylor G, Vimr ER, Garman E, Laver G (1992) Purification, crystallisation and preliminary crystallographic study of neuraminidase from *Vibrio cholerae* and *Salmonella typhimurium* LT2. *J. Mol. Biol.* 226: 1287–1290.

Thelestam M, Möllby R (1975) Sensitive assay for detection of toxin-induced damage to the cytoplasmic membrane of human diploid fibroblasts. *Infect. Immun.* 12: 225–232.

Titball RW, Hunter SEC, Martin KL *et al.* (1989) Molecular cloning and nucleotide sequence of the alpha-toxin (phospholipase C) of *Clostridium perfringens*. *Infect. Immun.* 57: 367–376.

Titball RW, Rubidge T (1990) The role of histidine residues in the alpha-toxin of *Clostridium perfringens*. *FEMS Microbiol. Lett.* 68: 261–266.

Titball RW, Leslie DL, Harvey S, Kelly DC (1991) Haemolytic and sphingomyelinase activities of *Clostridium perfringens* alpha-toxin are dependent on a domain homologous to that of an enzyme from the human arachidonic acid pathway. *Infect. Immun.* 59: 1872–1874.

Titball RW (1993) Bacterial phospholipases C. *Microbiol. Rev.* 57: 347–366.

Titball RW, Fearn AM, Williamson ED (1993) Biochemical and Immunological properties of the C-terminal domain of the alpha-toxin of *Clostridium perfringens*. *FEMS Microbiol. Lett.* 110: 45–50.

Titball RW, Naylor CE, Moss D, Williamson ED, Basak AK (1998) Mechanisms of protection against disease caused by *Clostridium perfringens*. *Immunology* 95: 34.

Titball RW, Naylor CE, Basak AK (1999) The *Clostridium perfringens* α toxin. *Anaerobe* 5: 51–64.

Toyonaga T, Matsushita O, Katayama S, Minami J, Okabe A (1992) Role of the upstream region containing an intrinisic DNA curvature in the negative regulation of the phospholipase C gene of *Clostridium perfringens*. *Microbiol. Immunol.* 36: 603–613.

Traving C, Schauer R, Roggentin P (1994) Gene structure of the 'large' sialidase isoenzyme from *Clostridium perfringens* A99 and its relationship with other clostridial *nanH* proteins. *Glycoconjugate J.* 11: 141–151.

Tso JY, Siebel C (1989) Cloning and expression of the phospholipase C gene from *Clostridium perfringens* and *Clostridium bifermentans*. *Infect. Immun.* 57: 468–476.

Tsutsui K, Minami J, Matsushita O *et al.* (1995) Phylogenetic analysis of phospholipase C genes from *Clostridium perfringens* types A to E and *Clostridium novyi*. *J. Bacteriol.* 177: 7164–7170.

Tweten RK (1988a) Cloning and expression in *Escherichia coli* of the perfringolysin O (theta-toxin) gene from *Clostridium perfringens* and characterisation of the gene product. *Infect. Immun.* 56: 3228–3234.

Tweten RK (1988b) Nucleotide sequence of the gene for perfringolysin O (theta-toxin) from *Clostridium perfringens*: significant homology with the genes for streptolysin O and pneumolysin. *Infect. Immun.* 56: 3235–3240.

Tweten RK (1995) Pore-forming toxins in gram-positive bacteria. In: Roth JA, Bolin CA, Brogden KA, Minion C, Wannemuehler MJ (eds) *Virulence Mechanisms of Bacterial Pathogens*. Washington, DC: American Society for Microbiology Press, pp. 207–229.

Vazquez-Boland J-A, Kocks C, Dramsi S *et al.* (1992) Nucleotide sequence of the lecithinase operon of *Listeria monocytogenes* and possible role of lecithinase in cell-cell spread. *Infect. Immun.* 60: 219–230.

Williamson ED, Titball RW (1993) A genetically engineered vaccine against the alpha-toxin of *Clostridium perfringens* also protects mice against experimental gas gangrene. *Vaccine* 11: 1253–1258.

Yoshihara K, Matsushita O, Minami J, Okabe A (1994) Cloning and nucleotide sequence analysis of the *colH* gene from *Clostridium histolyticum* encoding a collagenase and a gelatinase. *J. Bacteriol.* 176: 6489–6496.

90

Clostridium tetani

Ornella Rossetto, Giampietro Schiavo[1] and Cesare Montecucco

Università di Padova, Padova, Italy
[1]Imperial Cancer Research Fund, London

Tetanus was first recorded in the very earliest medical literature. Twenty-five centuries ago Hippocrates described a paralysed sailor who developed generalised spasmodic contractions of his muscles. He termed this spastic paralysis tetanus (Greek $\tau\varepsilon\tau\alpha\nu o\sigma$, muscle spasm). The most common form of the disease is *generalised tetanus*, in which the portal of entry is a minor wound in 80% of cases. It usually begins with a characteristic facial trismus (lockjaw or *risus sardonicus*) (Mayor, 1945; Weinstein, 1973). Subsequently, neck stiffness develops and later spreads to the vertebral muscles and those of the abdomen and limbs. The typical tetanic seizure is characterised by a sudden burst of tonic muscle contraction that causes arching of the back (opisthotonos), flexion and adduction of the arms, and extension of the lower extremities. The patient is completely conscious during such episodes and experiences intense pain. Later, autonomic symptoms develop with alterations of blood pressure and of the cardiac rhythm, and sweating. Dysphagia may occur and lead to hydrophobia. Spasm of the glottis and larynx may also develop and cause cyanosis and asphyxia if they are not promptly relieved by medical or surgical means. Dysuria or urinary retention may develop. In milder cases, a local form of tetanus develops with rigidity of the group of muscles close to the site of injury. This *local tetanus* may persist for a considerable time without further developments, or it may proceed to generalised tetanus. Local tetanus is due to dysfunction of interneurons that inhibit the α motor neurons of the affected muscles, without further spread through the central nervous system (CNS).

Since immunisation was introduced, tetanus has almost disappeared from more developed countries, but it still costs hundreds of thousand of lives where vaccination is not practised (Galazka and Gasse, 1995). In these areas *tetanus neonatorum* is common. In babies born to non-immunised mothers, this form of the disease develops from the site where the umbilical cord is cut. Indeed, instruments contaminated with *Clostridium tetani* spores are often employed to sever the cord at birth, and old rags, often soiled with faeces, are used as dressings.

Tetanus is often fatal because of exhaustion and usually occurs as the result of respiratory failure or heart failure (Bleck, 1989). Mortality rates have decreased considerably in the developed world, thanks to modern intensive care techniques, but it remains high because it is often the elderly who get tetanus and in such patients the mechanically assisted ventilation necessary for long periods carries the risk of pulmonary infections.

Clostridium tetani

The name *Clostridium tetani*, given to the bacterium that produces tetanus neurotoxin, refers to its elongated shape $(0.5–1.7 \times 2.1–18.1 \, \mu m)$, which frequently harbours a terminal spore. Most strains possess peritrichous flagella that convey active motility to the organism (Cato *et al.*, 1986). *Cl. tetani* is widespread in nature in the form of spores that contaminate most soil samples, particularly those that contain animal faeces.

Cl. tetani spores germinate under conditions of very low oxygen tension and slight acidity, when nutrients are available (Popoff, 1995). Such conditions may be present in ischaemic wounds, skin injuries or abrasions, where spores can germinate and give rise to limited bacterial multiplication without significant spread and inflammatory reaction. Toxigenic strains of *Cl. tetani* produce a protein toxin that is released by autolysis. This toxin, known as tetanus neurotoxin (TeNT), is entirely responsible for all the symptoms of tetanus (Faber, 1890; Tizzoni and Cattani, 1890; Kitasato, 1891).

The spore of *Cl. tetani* endows the organism with considerable resistance to various disinfectants and to heat. To ensure destruction of the spores, contaminated samples must be heated at 100°C for 4 hours. *Cl. tetani* is strictly anaerobic because it does not possess the redox enzymes necessary to reduce oxygen. Oxygen is toxic to it, because peroxides, superoxide and radicals accumulate and cause cell death. The optimal pH for growth is 7.4 and the optimal temperature is 37°C; little or no growth takes place at 25°C or 42°C. *Cl. tetani* can be cultivated on various solid agar media under strictly anaerobic conditions, but in liquid media, growth does not require strictly anaerobic conditions, provided that the medium has been degassed by boiling, followed by cooling under a nitrogen atmosphere, before inoculation (Hatheway, 1990).

Tetanus Neurotoxin

For hundreds of years the cause of tetanus was puzzling because the symptoms developed long after the initial injury, which might be of any kind. It is now known that even very small skin lesions, such as those caused during tattooing or body-piercing, may allow contamination. Tetanus was believed to be a nervous disease with elements of epilepsy, even though tetanus patients remain conscious, until it was shown to be caused by a bacterium (Carle and Rattone, 1884), which was identified by Nicolaier (1884) and then isolated in pure culture and characterised by Kitasato (1891) and Tizzoni and Cattani (1889). In the following year (Faber, 1890; Tizzoni and Cattani, 1890) it was shown that culture filtrates contain a protein toxin capable of inducing tetanus on injection into various animals.

Toxicity

The lethal dose of TeNT varies in different animal species: in most mammalian species it is between 0.1 and 2 ng injected intramuscularly. This is an extremely low value, and it is expected to be even lower in the wild, where even a very small deficit in mobility may be sufficient to impair survival. Absolute neurospecificity and intracellular catalytic activity (see below) are at the basis of such high toxicity. Different animal species show a great range of sensitivity to the toxin. Horses and humans are among the most sensitive species, whereas cats and dogs are less sensitive, but all of them are killed by much lower doses if injected intracerebrally. Doses thousands of times higher are required to kill birds, and amphibians and reptiles are highly resistant to TeNT.

The time of onset of paralysis of animals injected with TeNT is variable depending on species, dose and route of injection. There is always a lag phase, ranging from several hours up to days, between the time of injection and the appearance of symptoms. The lag phase is much longer when the disease is caused by contamination of wounds with spores of *Cl. tetani*, because of the extra time required for germination and bacterial proliferation. In humans, lag phases longer than 1 month have been recorded. The temporal development of symptoms in each form of tetanus is important, since the rapidity of progression largely determines prognosis. The more rapidly this period passes, the worse the prognosis, regardless of the clinical form in which the disease is first manifest (Bleck, 1989).

Electrophysiological Activity and Site of Action of Tetanus Neurotoxin

Tetanus neurotoxin binds specifically to the presynaptic membrane of motoneuron nerve endings and of sensory and adrenergic neurons (Habermann and Dreyer, 1986). The presynaptic receptor has not been identified, but the toxin is expected to bind with high-affinity to account for the low lethal doses, which correspond to sub-picomolar concentrations. TeNT has little or no peripheral action, unless very high doses are injected (Matsuda *et al.*, 1982). Unlike botulinum neurotoxin (BoNT), TeNT is transported in a retrograde direction within the motoneuron axon up to the spinal cord. Here, it localises mainly in the ventral horn of the grey matter (Price *et al.*, 1975; Stockel *et al.*, 1975; Erdmann *et al.*, 1981; Halpern and Neale, 1995). An intra-axonal ascent rate of 7.5 mm/h has been estimated (Stockel *et al.*, 1975), and neuromuscular stimulation enhances the uptake of the toxin (Ponomarev, 1928; Wellhoner *et al.*, 1973).

Within the spinal cord, TeNT migrates trans-synaptically across the synaptic cleft from the dendrites of peripheral motoneurons into coupled inhibitory interneurons (Schwab and Thoenen, 1976; Schwab *et al.*, 1979), where it blocks the release of inhibitory neurotransmitters (Brooks *et al.*, 1955; Benecke *et al.*, 1977; Mellanby and Green, 1981; Bergey *et al.*, 1983). Excitatory synapses appear not to be affected in the early stages (Brooks *et al.*, 1955; Mellanby and Green, 1981; Bergey *et al.*, 1983, 1987), but they may be inhibited at later stages (Takano *et al.*, 1983). This specificity of TeNT for inhibitory versus excitatory synapses is maintained when the toxin is applied to hippocampal slices (Calabresi *et al.*, 1989) or injected into the hippocampus (Mellanby and Thompson, 1977; Mellanby *et al.*, 1977). Such specificity for inhibitory synapses of the CNS also accounts for the neurodegenerative and epileptogenic effects of TeNT injected intracerebrally, which mainly result from unopposed release of glutamate from excitatory synapses (Bagetta *et al.*, 1991). The selective action of the toxin on inhibitory synapses in the spinal cord may at least in part be due to the anatomical organisation of

the tissue; it is not preserved in spinal cord neurons in culture (Bergey *et al.*, 1983). During trans-synaptic migration, TeNT can be neutralised by antitoxin antibodies injected into the spinal fluid (Erdmann *et al.*, 1981).

The blockade of inhibitory synapses brought about by TeNT at the spinal cord impairs the neuronal circuit that ensures balanced voluntary muscle contraction, and causes the spastic paralysis characteristic of tetanus (Mellanby and Green, 1981; Wellhoner, 1982).

The amount of toxin that reaches the CNS, after uptake by the peripheral nervous system (PNS) is clearly an important parameter that determines the severity of the disease and may partly account for the different toxicity of TeNT in different vertebrates (Payling-Wright, 1955). In the laboratory hundreds of mouse lethal doses are frequently used in order to obtain consistent effects rapidly, particularly when insensitive animals such as birds or fish are studied, or *in vitro* with cultured cells or isolated hemidiaphragm muscle preparations. Under such conditions TeNT also inhibits peripheral synapses, causing a botulism-like flaccid paralysis (Matsuda *et al.*, 1982).

Morphological examinations of synapses intoxinated (or intoxicated with toxin) *in vivo* or *in vitro* with TeNT reveal only one significant and consistent change: an increased number of synaptic vesicles close to the cytosolic face of the presynaptic membrane (Pozdniakov *et al.*, 1972; Duchen, 1973; Mellanby *et al.*, 1988; Hunt *et al.*, 1994; Neale *et al.*, 1999) (**Fig. 1**). This observation parallels the large and persistent inhibition of end-plate potentials (EPPs) and

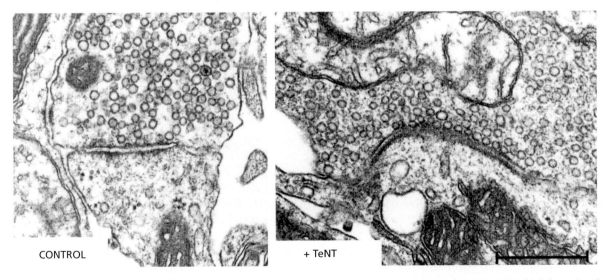

CONTROL   + TeNT

Fig. 1 A nerve poisoned by TeNT. Electron micrographs of control (left) and TeNT-treated spinal cord neurons (right). Notice the increase of synaptic vesicles next to the presynaptic membrane of the intoxicated synapse. (Photographs courtesy of Dr. E. A. Neale, National Institutes of Health, Bethesda, Maryland.)

the large reduction of the frequency, but not of the amplitude, of evoked miniature end-plate potentials (MEPPs). Hence, TeNT lowers the number of vesicles capable of undergoing fusion and release, without affecting the size of the quantum of neurotransmitter, the propagation of the nerve impulse or calcium homeostasis at the synapse (Dreyer *et al.*, 1983).

Genetics and Structure of Tetanus Neurotoxin

Both toxigenic and non-toxigenic strains of *Cl. tetani* are known. The main difference between them is the presence in the toxigenic strain of a 75-kb plasmid (Laird *et al.*, 1980; Eisel *et al.*, 1986). The plasmid was characterised by partial digestion with endonucleases, and the sequencing of the 3945-nucleotide of the TeNT-encoding gene allowed the determination of the 1315 amino acid sequence of the protein. TeNT is synthesised as a single inactive chain without a leader sequence, which is in keeping with the fact that it is released by bacterial autolysis. The toxin is activated by specific proteolysis within a surface-exposed loop subtended by a disulphide bridge, generating a heavy chain (H, 100 kDa) and a light chain (L, 50 kDa), which remain associated via non-covalent protein–protein interactions and via the interchain S–S bond, the integrity of which is essential for neurotoxicity (Schiavo *et al.*, 1990) (**Fig. 2a**).

The central region of the L chain of TeNT contains the His-Glu-Xaa-Xaa-His-binding motif of zinc endopeptidases and this led to the identification of its intracellular catalytic activity (Eisel *et al.*, 1986; Fairweather and Lyness, 1986; Schiavo *et al.*, 1992a, b). TeNT is inactivated by heavy metal chelators which remove bound zinc and generate an inactive apotoxin (Schiavo *et al.*, 1992b), without appreciable changes in the secondary structure of the L chain. The active-site metal atom can be re-acquired on incubation in zinc-containing buffers to re-form the active holotoxin (Schiavo *et al.*, 1992b). The same procedure allows the active-site zinc atom to be exchanged with other divalent transition metal ions, forming active metal-substituted TeNT (Tonello *et al.*, 1997).

The crystallographic structures of the TeNT-related botulinum neurotoxins type A (BoNT/A) and type B (BoNT/B) (Lacy and Stevens, 1999; Swaminathan and Eswaramoorthy, 2000) and of the C-terminal domain of TeNT (H_C) (Umland *et al.*, 1997) have been determined. The toxin structure reveals three distinct functional domains, a unique hybrid of previously characterised structural motifs, and new insight into the mechanism of toxicity of this protein (**Fig. 2b**).

BoNT/A consists of three ~ 50-kDa domains: an N-terminal domain (L chain) endowed with zinc-endopeptidase activity; a membrane translocation domain H_N characterised by the presence of two 10-nm α helices, which are reminiscent of similar elements present in colicin and in the influenza virus haemagglutinin; and a binding domain (H_C) composed of two unique sub-domains similar to the legume lectins and Kunitz inhibitor (Lacy *et al.*, 1998). On basis of the similar structure of the C-terminal domains and of the sequence homology, (Lacy and Stevens, 1999), it is likely that the overall structure of TeNT is very similar to those of the BoNTs (see Chapter 55).

Such structural organisation is functionally related to the fact that the tetanus and botulinum neurotoxins intoxinate neurons by a four-step mechanism: (1) binding, (2) internalisation, (3) membrane translocation and (4) enzymatic target modification. The L chain is responsible for the intracellular catalytic activity; the N-terminal 50-kDa domain of the H chain (H_N) is implicated in membrane translocation, whereas the C-terminal part (H_C) is mainly responsible for the neurospecific binding (Schiavo *et al.*, 2000; Rossetto *et al.*, 2001).

Neurospecific Binding

The H_C or binding domain of TeNT has an overall elongated shape similar to that of BoNT/A and B (**Fig. 2b**). The H_C domains of the two BoNTs appears to be very flexible with respect to the H_N domain. The binding domains of these three clostridial neurotoxins consist of two distinct sub-domains, the N-terminal half (H_{CN}) and the C-terminal half (H_{CC}), with few protein–protein contacts between them. H_{CN} has two seven-strand β sheets arranged in a jelly-roll motif similar to that of the legume lectin carbohydrate-binding proteins. The amino acid sequence of this sub-domain is highly conserved among clostridial neurotoxins, suggesting a closely similar three-dimensional structure. In contrast, the sequence of the H_{CC} sub-domain is poorly conserved. It contains a modified β trefoil folding motif present in several proteins involved in recognition and binding functions such as interleukin 1 (IL-1), fibroblast growth factor and Kunitz-type trypsin inhibitors. Removal of H_{CN} from H_C does not reduce H_C nerve membrane-binding, whereas deletion of only 10 residues from the C-terminus of H_{CC} abolishes its binding to spinal cord neurons (Halpern and Loftus, 1993). The last 34 residues of H_{CC}, and in particular His1293 of TeNT, are very important for binding the oligosaccharide portion of polysialogangliosides (Halpern and Loftus, 1993; Shapiro *et al.*, 1997). The major difference

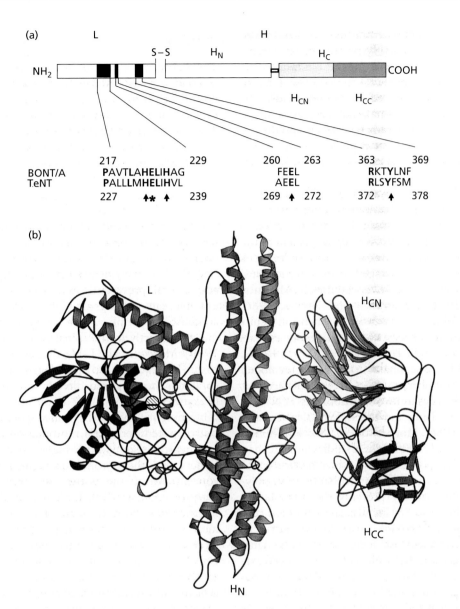

(a)

L   H

S-S H_N H_C

NH₂   COOH

H_{CN} H_{CC}

| | 217 | | 229 | | 260 | 263 | | 363 | | 369 |
BONT/A PAVTLAHELIHAG   FEEL   RKTYLNF
TeNT   PALLMHELIHVL   AEEL   RLSYFSM

227   ↑* ↑   239   269 ↑ 272   372 ↑ 378

(b)

L

H_{CN}

H_{CC}

H_N

Fig. 2 The three functional domain structure of clostridial neurotoxins: (a) Schematic structure of active two-chain clostridial neurotoxins. TeNT and BoNT are composed of two polypeptide chains held together by a single disulphide bridge, which must be reduced intracellularly for fully activity. The C-terminal portion of the heavy chain (H, 100 kDa) is responsible for neuro-specific-binding (domain H_C), and the N-terminus (H_N) is implicated in the translocation of the light chain in cytosol and pore formation. Structurally H_C can be further subdivided into two 25-kDa portions, H_{CN} and H_{CC}. The light chain (L, 50 kDa) is a zinc endopeptidase responsible for the intracellular activity. The segments of high homology between TeNT and BoNT/A are in **bold**. A short α helix (217–229 in BoNT/A and 227–239 in TeNT), in the central part of the L chain, shows the highest homology and contains the zinc-binding motif of metallo-endopeptidases. Amino acids involved in the co-ordination of zinc or in the hydrolysis of the substrate are indicated by an arrow. The glutamic acid co-ordinating a water molecule responsible for the target hydrolysis is indicated by an asterisk. (b) The crystallographic structure of BoNT/A highlights the three functional domains. The L chain contains both α helix and β strand secondary structures with the zinc (⊘) and the zinc-binding motif (contained in the black α helix) in the centre. The presence of two long α helices and a long loop interacting with L chain characterises the translocation domain. H_{CN} has two seven-strand β sheets arranged in a jelly-roll motif, whereas the H_{CC} contains a modified β trefoil folding. On the basis of the similarity of the C-terminal domain structures and the sequence homology, it is likely that the overall structure of TeNT is very similar to that of BoNT/A.

between H_{CC} of TeNT and BoNT/A lies in the structure of the loops, suggesting that these external segments may be responsible for the binding to the different protein receptors. Additional regions of the toxins may be involved in binding, however, because H_C shows only a partial protection from intoxication with the intact toxin molecule and the H_C fragment of TeNT does not prevent retro-axonal transport of the holotoxin (Bizzini *et al.*, 1977; Weller *et al.*, 1991).

Extensive studies of the binding of ^{125}I-TeNT to brain synaptosomes, membrane preparations of various tissues, cells in culture and lipids have not so far led to the identification of the TeNT receptor present at the peripheral nerve endings, nor that responsible for its entry into the spinal cord inhibitory interneurons. These studies have been reviewed extensively (Mellanby and Green, 1981; Habermann and Dreyer, 1986; Montecucco, 1986). Here we will only mention results that are relevant to the recent identification of a sugar-binding sub-domain in TeNT (Umland *et al.*, 1998). Beginning with the seminal work of Van Heyningen (1974), a large number of studies have established that polysialogangliosides are involved in binding TeNT (for references see Mellanby and Green, 1981; Habermann and Dreyer, 1986; Montecucco, 1986; Wellhoner, 1992; Schiavo *et al.*, 2000). These studies indicate that: (1) TeNT binds to polysialogangliosides, particularly to G_{D1b}, G_{T1b} and G_{Q1b}; (2) pre-incubation with polysialogangliosides partially prevents the retro-axonal transport of the toxin; (3) incubation of cultured cells with polysialogangliosides increases their sensitivity to TeNT; (4) treatment of membranes with neuraminidase, which removes sialic acid residues, decreases toxin-binding. Binding to polysialogangliosides accounts for an unsaturable low-affinity-binding of TeNT to nerve cells and to nerve tissue membranes. As previously discussed in detail by Mellanby and Green (1981) and Montecucco (1986), however, it is unlikely that polysialogangliosides are the sole receptors of these neurotoxins. Experiments carried out with cells in culture have indicated that cell-surface proteins may be involved in toxin-binding (Pierce *et al.*, 1986; Yavin and Nathan, 1986; Parton *et al.*, 1988; Schiavo *et al.*, 1991b; Herreros *et al.*, 2000a, b). The sugar-binding and protein-binding sub-domains present in the H_C domain of TeNT and BoNT/A and the protection experiments mentioned above support the suggestion that clostridial neurotoxins may bind strongly and specifically to the presynaptic membrane because they display multiple interactions with sugar- and protein-binding sites.

Generally, receptors for toxins and viruses are cell-surface molecules essential for the life of the cell, and

their study has led to important progresses in cell biology and neurosciences. The identification of the TeNT receptor(s) is particularly relevant because this molecule is transported in a retrograde direction inside motoneurons from the periphery to the spinal cord, thus revealing a pathway from the PNS to the CNS. This should prove helpful in devising novel routes to deliver biological agents, such as analgesics and anaesthetics, into the spinal cord.

Entry into Nerves and Axonal Transport

To reach its final site of action in the spinal cord, TeNT has to enter two different neurons: a peripheral motoneuron and an inhibitory interneuron of the spinal cord (**Fig. 3**). There is evidence that its binding to peripheral and central presynaptic terminals is different. For example, cats and dogs are highly resistant to the toxin administered peripherally, but very sensitive when it is injected directly in the spinal cord (Shumaker *et al.*, 1939). Moreover, the L-H_N fragment of TeNT is not toxic when injected into the cat leg, but it causes a spastic paralysis when injected directly into the spinal cord (Takano *et al.*, 1983). It is possible that the concentration of TeNT in the limited space of the synaptic cleft between peripheral motoneuron and inhibitory interneuron is significantly higher than that at the periphery, with the motoneuron acting as a sort of toxin concentrator. If this is the case, even a low-affinity receptor could mediate the entry of TeNT into the interneurons, because of the anatomically restricted location within the intersynaptic space. Lipid monolayer studies have documented the ability of 10^{-8} mol/L TeNT to interact with acidic lipids (Schiavo *et al.*, 1991a). Similar concentrations are routinely used with cells in culture and in hippocampal injections *in vivo* (Mellanby *et al.*, 1984) or in the experimental induction of flaccid paralysis in mice treated with 1000 times the mouse LD_{50} (Matsuda *et al.*, 1982). On the other hand, in clinical tetanus, the peripheral concentration of TeNT is sub-picomolar. A possible scenario that reconciles the presently available data is summarised in **Fig. 3**. Glycoprotein and glycolipid binding sites are implicated in the peripheral binding of the neurotoxins, which is characterised by high affinity and high specificity. The protein receptor of TeNT is responsible for its inclusion in an endocytic vesicle that moves in a retrograde direction along and into the axon, whereas the protein receptors of BoNTs guide them inside vesicles that acidify within the neuro-muscular junction. The vesicles carrying TeNT reach the cell body and then move to dendritic terminals to release the toxin in the intersynaptic space. TeNT equilibrates between pre- and post-synaptic membranes and then binds to and

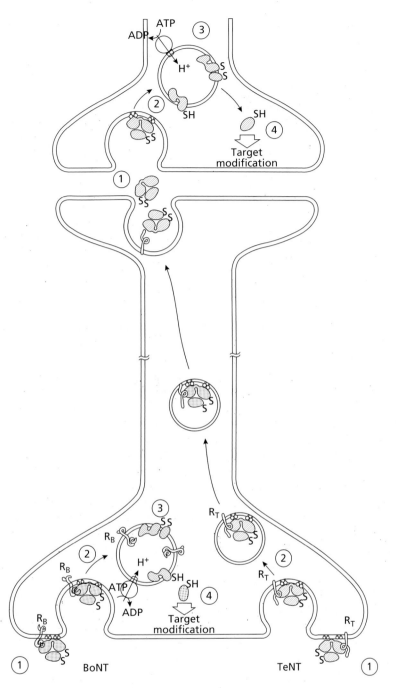

Fig. 3 Entry of TeNT and BoNTs into nerve terminals: (1) TeNT and BoNTs bind to the presynaptic membrane at as-yet-unidentified receptors (R_T and R_B respectively) of peripheral nerve terminals. (2) The protein receptor of TeNT is thought to be responsible for its inclusion in an endocytic vesicle that moves in a retrograde direction along the axon to the inhibitory interneurons of the spinal cord, whereas the BoNT protein receptors guide them to enter vesicles that acidify within the neuromuscular junction. (3) At low pH, the BoNTs and TeNT change conformation, insert into the lipid bilayer of the vesicle membrane and translocate the L chain into the cytosol of peripheral and central neurons respectively. (4) In the cytosol the L chain catalyses the proteolysis of one of the three SNARE proteins.

enters the inhibitory interneurons via synaptic vesicle endocytosis.

Since the L chains of TeNT block neuro-exocytosis by acting in the cytosol (Penner *et al.*, 1986; Ahnert

Hilger *et al.*, 1989), at least this toxin domain must reach the cell cytosol. All available evidence indicates that TeNT does not enter the cell directly from the plasma membrane, rather it is endocytosed into acidic cellular compartments. Electron microscopy studies have shown that, after binding, the toxin enters the lumen of vesicular structures by a temperature- and energy-dependent process (Critchley *et al.*, 1985; Staub *et al.*, 1986; Parton *et al.*, 1987; Matteoli *et al.*, 1996). The H_C domains appears to be sufficient for internalisation into murine spinal cord neurons (Lalli *et al.*, 1999). Gold-labelled TeNT was internalised by spinal cord neurons inside a variety of vesicular structures and only a minority was found in the lumen of small synaptic vesicles (Parton *et al.*, 1987). In contrast, Matteoli *et al.* (1996) found TeNT almost exclusively inside small synaptic vesicles of hippocampal neurons after a 5-minute membrane depolarisation. It has long been known that nerve stimulation facilitates intoxication (Ponomarev, 1928; Wellhoner *et al.*, 1973). A prominent neuro-exocytosis correlates with a high rate of synaptic vesicle recycling by endocytosis and refilling with neurotransmitter (Cremona and De Camilli, 1997). The simplest way to account for the shorter onset of paralysis induced by TeNT under conditions of nerve stimulation is that the neurotoxin enters the synaptic terminal by endocytosis within the lumen of small synaptic vesicles (SSV). Indeed, studies in hippocampal neurons and granular cells of the cerebellum (Matteoli *et al.*, 1996; Verderio *et al.*, 1999; Rossetto *et al.* unpublished results) indicate that TeNT uses SSVs as 'Trojan horses' to enter CNS neurons.

Membrane Translocation

Once internalised in the lumen of an endocytic vesicle, the L chain must cross the hydrophobic barrier of the vesicle membrane to reach the cytosol where it demonstrates its activity (**Fig. 3**). The different trafficking of tetanus and botulinum toxins at the neuromuscular junction clearly indicates that internalisation is not necessarily linked to, and followed by, membrane translocation into the cytosol, i.e. internalisation and membrane translocation are clearly distinct steps of the process of cell intoxination (**Fig. 3**). There is indirect but compelling evidence that TeNT has to be exposed to low pH for nerve intoxication to occur (Williamson and Neale, 1994; Matteoli *et al.*, 1996; Verderio *et al.*, 1999). Acidic pH does not activate TeNT directly, since the introduction of a non-acid-treated L chain in the cytosol is sufficient to block exocytosis (see Humeau *et al.*, 2000). Hence, low pH is instrumental in the process of membrane translocation of the L chain from the vesicle lumen into the cytosol. TeNT has been shown to undergo a structural change, driven by low pH, from a water-soluble 'neutral' form to an 'acid' form capable of inserting

both the H and L chains in the hydrocarbon core of the lipid bilayer (Boquet and Duflot, 1982; Montecucco *et al.*, 1986). After this membrane insertion, TeNT forms transmembrane ion channels which open with high frequency at low, but not at neutral, pH, with low conductance and permeability to molecules smaller than 700 Da (Boquet and Duflot, 1982; Hoch *et al.*, 1985; Menestrina *et al.*, 1989; Beise *et al.*, 1994).

There is evidence that these channels are formed by the oligomerisation of the H_N domain (Menestrina *et al.*, 1989). The H_N domains are highly homologous among the various clostridial neurotoxins (Lacy and Stevens, 1999). The membrane translocation domains of BoNT/A and BoNT/B have a cylindrical shape determined by the presence of a pair of unusually long and twisted 10 nm α helices (see **Fig. 2b**) (Wiener *et al.*, 1997; Swaminathan and Eswaramoorthy, 2000). At each end of the pair, there is a shorter α helix which lies parallel to the main helices and, in addition, several strands pack along the two core helices. It is difficult to identify the residues and segments involved in the formation of ion channels at low pH, but the overall structure of H_N resembles that of some viral proteins which undergo an acid-driven conformational change. After exposure to the neutral pH of the cytosol, the L chain re-folds and is released from the vesicle by reduction of the interchain disulphide bond. In this 'tunnel' model, the formation of a transmembrane ion-conducting pore is a prerequisite for translocation. This model does not explain the fact that the L chain of TeNT penetrates the lipid bilayer in such a way as to be exposed to the fatty acid chains of phospholipids, i.e. it is not shielded from lipids inside the H chain tunnel (Montecucco *et al.*, 1986). Moreover, the low channel conductance does not account for the dimensions expected for a protein channel that has to accommodate a polypeptide chain with lateral groups of different volume, charge and hydrophilicity. The protein-conducting channels of the endoplasmic reticulum, of *E. coli* and of mitochondrial membranes, characterised in planar lipid bilayers, have a conductance of 220 pS (Martoglio and Dobberstein, 1998). These channels are closed when plugged by a transversing polypeptide chain.

A second model (Beise *et al.*, 1994) envisages that, as the vesicle internal pH decreases following the operation of the vacuolar-type ATPase proton pump, TeNT inserts into the lipid bilayer, forming ion channels that grossly alter electrochemical gradients and cause an osmotic lysis of the toxin-containing acidic vesicle. The membrane barrier is broken and the cargo of toxin molecules is released into the cytosol. This model does not, however, explain the fact that TeNT does not lyse the plasmalemma of neuronal cells at pH 5.0.

An alternative hypothesis proposes that the L chain translocates across the vesicle membrane within a channel open laterally to lipids, rather than inside a proteinaceous pore (Montecucco *et al.*, 1994). The two polypeptide chains of the toxin are supposed to change conformation at low pH in a concerted fashion, such that both of them expose hydrophobic surfaces and enter into contact with the hydrophobic core of the lipid bilayer. The H chain is suggested to form a transmembrane hydrophilic cleft that nests the passage of the partially unfolded L chain with its hydrophobic segments facing the lipids. Facing the cytosolic neutral pH, the L chain re-folds and regains its water-soluble neutral conformation. It is possible that cytosolic chaperones are involved in treadmilling the L chain out of the vesicle membrane and assisting its cytosolic re-folding, but as yet there is no supporting evidence. As the L chain is released from the vesicle membrane, after hydrolysis of the interchain S–S bond, the trans-membrane hydrophilic cleft of the H chain is supposed to tighten up to reduce the amount of hydrophilic protein surface exposed to the membrane hydrophobic core. This, however, leaves a peculiarly shaped channel across the membrane, with two rigid protein walls and a flexible lipid seal on one side. This is proposed to be the structure responsible for the ion-conducting properties of TeNT. In this 'cleft' model, the ion channel is a consequence of membrane translocation, rather than a pre-requisite for it. Moreover, ion transport is mediated by a transmembrane structure that derives from but is physically different from the one involved in the L chain translocation.

Zinc Endopeptidase Activity

The catalytic activity of TeNT was discovered after the identification of the zinc-binding motif of zinc endo-peptidases in the corresponding gene (Eisel *et al.*, 1986; Fairweather and Lyness, 1986) and the demonstration that TeNT inhibited release of acetyl-choline (ACh) at synapses of the buccal ganglion of *Aplysia californica* via a zinc-dependent protease activity (Schiavo *et al.*, 1992b). The intracellular sub-strate of TeNT was identified by proteolysis assays, working on the hypothesis that it had to be located either on neurotransmitter-containing vesicles or on the plasma membrane, to account for blockade of exocytosis (Schiavo *et al.*, 1992a).

The catalytic metalloprotease domain contains both α helix and β strand secondary structures and has little similarity to related enzymes of known structure, apart from the helical segment including the zinc-binding motif (see **Fig. 2b**) (Lacy *et al.*, 1998). In BoNT/A and BoNT/B the zinc atom is co-ordinated by the two

histidines and the glutamate of the motif, and by another glutamate which is completely conserved among clostridial neurotoxins (**Fig. 2a**). Moreover, the phenolic ring of a tyrosine residue, essential for activity (Rossetto *et al.*, 2001) is directly opposite the metal atom.

TeNT is a very specific protease: among the many proteins and synthetic substrates so far assayed, only vesicle-associated membrane protein (VAMP, synaptobrevin), a small membrane protein of the SSV, is cleaved at the Gln76–Phe77 peptide bond. Strikingly, TeNT and BoNT/B cleave VAMP at the same peptide bond, and yet when injected into an animal they cause the very different symptoms of tetanus and botulism respectively (Schiavo *et al.*, 1992a). This clearly demonstrates that the different symptoms derive from different sites of intoxination rather than from a different molecular mechanism of action of the two neurotoxins.

Recombinant VAMP is cleaved at the same peptide bond as the corresponding cellular protein, thus indicating that no additional endogenous factors are involved in determining the specificity of TeNT, and making it possible to assay its metalloprotease activity *in vitro* (Schiavo and Montecucco, 1995). Continuous assays based on the use of fluorescent substrates (Soleilhac *et al.*, 1996) will prove particularly useful. The proteolytic activity of TeNT can be detected in living neurons in culture and in tissues with anti-VAMP specific antibodies that recognise epitopes present in the 1–76 fragment of VAMP, which are released into the cytosol following the action of the toxin. Thus, a highly sensitive single-cell assay can be performed by following the progressive loss of staining of the nerve terminals (Matteoli *et al.*, 1996; Verderio *et al.*, 1999).

Vesicle-associated Membrane Protein, the Cytosolic Substrate of Tetanus Neurotoxin

VAMP is a 13-kDa membrane protein of synaptic vesicles, dense core granules and synaptic-like micro-vesicles and is the prototype of the vesicular SNAREs (v-SNARE) (Rothman, 1994).

VAMP forms a coiled-coil heterotrimer with SNAP-25 and with syntaxin, the two partner SNARE proteins that reside on the cytosolic face of the plasma membrane (**Fig. 4a**) (Brunger, 2000). The formation of this complex brings the neurotransmitter-containing vesicle in close contact with the plasma membrane, ready to fuse and to deliver its content. The cleavage of VAMP by TeNT releases a 76-residue fragment into the cytosol. As a consequence the heterotrimeric

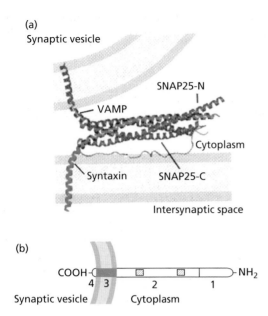

(a)
Synaptic vesicle

SNAP25-N

VAMP

Cytoplasm

Syntaxin   SNAP25-C

Intersynaptic space

(b)

COOH—   —NH₂
4 3   2   1

Synaptic vesicle   Cytoplasm

Fig. 4 (a) Ribbon diagram of the four-helix bundle of the neuronal core SNARE complex with syntaxin, VAMP, SNAP25 C-terminal coil (SNAP25-C) and SNAP25 N-terminal coil (SNAP25-N). (b) VAMP has a short C-terminal tail protruding into the vesicle lumen (4) and a transmembrane segment (3), followed by a 95-residue cytosolic part (2). This portion is highly conserved among isoforms and species, whereas the N-terminal portion (1) is poorly conserved and rich in prolines. The two segments required for TeNT–substrate interaction, characterised by the presence of three carboxyl residues alternating with hydrophobic and hydrophilic ones, are shown by small squares.

complex cannot assemble and the vesicle is prevented from exocytosing its load.

Four functional domains can be distinguished in the VAMP molecule (**Fig. 4b**). The N-terminal 33-residue part is proline-rich and isoform-specific. The adjacent region (residues 33–96), which is highly conserved, is directly implicated in the formation of the coiled-coil SNARE heterotrimeric complex. It is followed by a transmembrane segment and then by a small C-terminal tail that faces the vesicle lumen.

Of the many VAMP isoforms identified by database searches, four isotypes have been extensively characterised in terms of TeNT sensitivity: VAMP-1, VAMP-2, VAMP-3 (cellubrevin) and TI-VAMP (tetanus toxin-insensitive VAMP) (Schiavo *et al.*, 1992a; McMahon *et al.*, 1993; Galli *et al.*, 1998; Coco *et al.*, 1999). VAMP-1 and VAMP-2 are present in many different type of secretory vesicles and granules, and their cleavage always causes impairment of membrane fusion (Trimble, 1993; Rossetto *et al.*, 1996). The lack of activity of TeNT on non-nervous tissues is therefore due not to lack of substrate, but to insufficient toxin-binding or internalisation.

In rats, VAMP-2 is sensitive to TeNT, but VAMP-1 is not cleaved because of an amino acid substitution at the cleaved peptide bond. Since VAMP-1 is the predominant VAMP isoform in the spinal cord, this could account for the resistance of rats to tetanus (Patarnello *et al.*, 1993). Though similar data are not available for other species, this effect could also be implicated in the resistance of other animal species to this toxin and suggests the possibility of a TeNT-driven natural selection.

The structural basis of the remarkable specificity of TeNT for VAMP is not completely understood, but two regions of VAMP have been identified as essential: a segment characterised by the presence of three carboxyl residues alternating with hydrophobic and hydrophilic ones, and a positively charged cluster of residues. The two segments are located in N-terminal and C-terminal positions, respectively, with respect to the cleavage site (Rossetto *et al.*, 1994; Pellizzari *et al.*, 1996; Cornille *et al.*, 1997). This clearly indicates that TeNT recognises the tertiary, rather than the primary, structure of its substrate. This remarkable specificity and the widespread presence of VAMP makes this toxin a very useful tool in the study of cell biology (Hayashi *et al.*, 1994; Schiavo *et al.*, 2000).

Recovery of Synaptic Function

The biochemical and cellular events responsible for the long duration of tetanus are unknown. Several factors may contribute: (1) the lifetime of the TeNT L chain in the cytosol of spinal cord inhibitory interneurons; (2) the turnover of the truncated VAMP and VAMP fragment 1–76, which may interfere with neuroexocytosis; (3) secondary biochemical events triggered by the production of truncated VAMP and/or the released fragment 1–76. In rat spinal cord and in cultured cells ^{125}I-TeNT has a half-life of several days (Habermann and Dimpfel, 1973; Habig *et al.*, 1986; Marxen and Bigalke, 1991). What determines the lifetime of the L chain and the kinetics of its inactivation are not known, but it is not surprising that it may persist for many days because bacterial toxins are generally very stable molecules. It should be recalled that until the last molecule of L chain has entered a spinal cord inhibitory interneuron, the newly-synthesised VAMP molecules would be fragmented by the toxin. The recovery of synaptic function will be greatly speeded up when specific inhibitors of the TeNT metalloprotease activity, capable of crossing biological membranes, become available. Considerable progresses has recently been made with the identification of pseudotripeptides containing ethylene sulphonamide or *m*-sulphonamidophenyl moieties (Martin *et al.*, 1999).

Tetanus Neurotoxin and the Ecology of *Clostridium tetani*

A successful bacterium is one that can multiply effectively and spread so that it is present in nature in large numbers (Mims, 1995). During evolution, such a bacterial species attains a state of balanced pathogenicity that causes the smallest alteration to host physiology compatible with the need to enter and multiply in the host body and to spread to other individuals. As noted earlier, genes coding for clostridial neurotoxins are episomic. It has been suggested that chromosomal genes code for proteins essential for the bacterial life cycle, whereas episomal genes are important for related functions, such as growth and spreading under unusual environmental conditions. Within the bodies of living vertebrates there are only very small anaerobic niches where clostridia can survive. The production and release of a neurotoxin makes a bacterial population capable of killing its host, which then becomes an anaerobic fermentor able to support the growth of extremely large numbers of clostridia of endogenous as well as exogenous origin. Clostridia are known to produce a variety of hydrolases that facilitate the breakdown of tissues. In this simplified view, the production of clostridial neurotoxins is functional in relation to multiplication. During the massive growth of anaerobic bacteria, including clostridia, that takes place on cadavers, the plasmid that carry the TeNT-encoding gene can be exchanged. Of course, a cadaver cannot support bacterial spreading to other hosts. Clostridia overcome this limitation by sporulation, which takes place when nutrients are used up. The spores are dispersed in the environment by natural forces even over long distances.

Our knowledge of the ecology of clostridia in general, and of the toxigenic clostridia in particular, is currently quite poor, but on the basis of the outline given here, they do not appear to be balanced pathogens. However, if one takes into account their anaerobic nature, killing the host, followed by sporulation when nutrients become limiting, is almost an obligatory step to allow bacterial proliferation.

Immunological Properties of Tetanus Neurotoxin

Since a single protein is responsible for all the clinical symptoms of tetanus, the disease can be completely prevented by specific antitoxin antibodies (Galazka and Gasse, 1995; Middlebrook and Brown, 1995). Toxin-neutralising antibodies can be acquired passively, by injection of immunoglobulins isolated from immunised donors, or actively, as a result of vaccination with tetanus toxoid. The toxoid is obtained by treating TeNT with paraformaldheyde (Ramon and Descombey, 1925; Galazka and Gasse, 1995). Tetanus toxoid is highly immunogenic and it is used as a standard immunogen in a variety of immunological studies (Corradin and Watts, 1995). The current vaccination schedule, consisting of 3 inoculations at various time intervals from 2–3 months after birth, is extremely effective in providing complete and long-lasting immunity. A booster injection 20–30 years after the last injection is advisable to ensure almost lifelong protection. The vaccine is well known as to be inexpensive, safe and efficacious but in developing countries even a relatively low-cost vaccine may be unaffordable. The main problem, however, is the difficulty of carrying out the complete schedule of injections. A major development would be a single-dose vaccine, or one that can be introduced into a carrier micro-organism. To reach this goal, antitetanus vaccine has been developed by genetic engineering techniques employing the C-terminal third of the TeNT (Halpern *et al.*, 1990). Antibodies against the receptor-binding domain are immunoprotective (Middlebrook and Brown, 1995). Tetanus toxoid has been used for decades to immunise animals to produce high-titre antisera for therapeutic use. Heterologous antibodies may, however, cause severe allergic reactions (Bardenwarper, 1962), and they have therefore been replaced by anti-TeNT immunoglobulins of human origin. Monoclonal antibodies for immunotherapy have been prepared, but so far no efficacious neutralising monoclonal antibody against tetanus toxin is available (Middlebrook and Brown, 1995). It is likely that antibody binding to multiple epitopes is required for efficient neutralisation.

Several methods have been developed over the years to detect anti-TeNT properties. In general terms, these assays can be divided in two main groups: (1) immunochemical methods which detect the formation of an antigen–antibody complex and (2) biological tests of the ability of an antibody or an antiserum to prevent toxicity of native TeNT. Immunochemical assays detect protective and non-protective anti-TeNT antibodies simultaneously, thus overestimating the antibody titre. Moreover, some of these tests have poor sensitivity. On the other hand, biological assays detect only protective antibodies and do provide a functional response. Traditionally, biological assays are performed by injecting groups of mice with the toxin or with varying ratios of toxin–antisera mixtures, and following the development of tetanus symptoms. These tests are time-consuming and there is a growing concern about the use of laboratory animals for toxicity tests. Alternatively, isolated nerve–muscle preparations can be used, with a reduction in the use of animals and a more quantitative output, similar to that for the botulinum neurotoxins (Goschel *et al.*, 1997; Lalli *et al.*, 1999). *In vitro* methods using neurons in culture are being developed in several laboratories, but such tests have not yet reached the reliability and significance of biological assays.

Acknowledgements

The work described in the authors' laboratory is supported by Telethon-Italia grant 1068 and by EU contract BMH4-97 2410.

References

Ahnert Hilger G, Weller U, Dauzenroth ME, Habermann E, Gratzl M (1989) The tetanus toxin light chain inhibits exocytosis. *FEBS Lett.* 242: 245–248.

Bagetta G, Nistico G, Bowery NG (1991) Characteristics of tetanus toxin and its exploitation in neurodegenerative studies. *Trends Pharmacol. Sci.* 12: 285–289.

Bardenwarper HW (1962) Serum neuritis from tetanus antitoxin. *JAMA* 179: 763–766.

Beise J, Hahnen J, Andersen Beckh B, Dreyer F (1994) Pore formation by tetanus toxin, its chain and fragments in neuronal membranes and evaluation of the underlying motifs in the structure of the toxin molecule. *Naunyn Schmiedebergs Arch. Pharmacol.* 349: 66–73.

Benecke R, Takano K, Schmidt J, Henatsch HD (1977) Tetanus toxin induced actions on spinal Renshaw cells and Ia-inhibitory interneurones during development of local tetanus in the cat. *Exp. Brain Res.* 27: 271–286.

Bergey GK, MacDonald RL, Habig WH, Hardegree MC, Nelson PG (1983) Tetanus toxin: convulsant action on mouse spinal cord neurons in culture. *J. Neurosci.* 3: 2310–2323.

Bergey GK, Bigalke H, Nelson PG (1987) Differential effects of tetanus toxin on inhibitory and excitatory synaptic transmission in mammalian spinal cord neurons in culture: a presynaptic locus of action for tetanus toxin. *J. Neurophysiol.* 57: 121–131.

Bizzini B, Stoeckel K, Schwab M (1977) An antigenic polypeptide fragment isolated from tetanus toxin: chemical characterization, binding to gangliosides and retrograde axonal transport in various neuron systems. *J. Neurochem.* 28: 529–542.

Bleck TP (1989) Clinical aspects of tetanus. In: Simpson LL (ed.) *Botulinum Neurotoxin and Tetanus Toxin.* San Diego CA: Academic Press, pp. 379–398.

Boquet P, Duflot E (1982) Tetanus toxin fragment forms channels in lipid vesicles at low pH. *Proc. Natl Acad. Sci. USA* 79: 7614–7618.

Brooks VB, Curtis DR, Eccles JC (1955) Mode of action of tetanus toxin. *Nature* 175: 120–121.

Brunger AT (2000) Structural insights into the molecular mechanism of Ca^{2+}-dependent exocytosis. *Curr. Opin. Neurobiol.* 10: 293–302.

Calabresi P, Benedetti M, Mercuri NB, Bernardi G (1989) Selective depression of synaptic transmission by tetanus toxin: a comparative study on hippocampal and neostriatal slices. *Neuroscience* 30: 663–670.

Carle A, Rattone G (1884) Studio sperimentale sull'eziologia del tetano. *Giorn. Accad. Med. Torino* 32: 174–179.

Cato EP, George WL, Finegold SM (1986) Genus *Clostridium* Prazmowski 1880. In: Sneath PH, Mair NS, Sharpe MEH (eds) *Bergey's Manual of Systemic Bacteriology,* Baltimore: Williams & Wilkins, Vol. 1, pp. 1141–1200.

Coco S, Raposo G, Martinez S *et al.* (1999) Subcellular localization of tetanus neurotoxin-insensitive vesicle-associated membrane protein (VAMP)/VAMP7 in neuronal cells: evidence for a novel membrane compartment. *J. Neurosci.* 19: 9803–9812.

Cornille F, Martin L, Lenoir C *et al.* (1997) Cooperative exosite dependent cleavage of synaptobrevin by tetanus toxin light chain. *J. Biol. Chem.* 272: 3459–3464.

Corradin G, Watts C (1995) Cellular immunology of tetanus toxoid. *Curr. Top. Microbiol. Immunol.* 195: 77–87.

Cremona O, De Camilli P (1997) Synaptic vesicle endocytosis. *Curr. Opin. Neurobiol.* 7: 323–330.

Critchley DR, Nelson PG, Habig WH, Fishman PH (1985) Fate of tetanus toxin bound to the surface of primary neurons in culture: evidence for rapid internalization. *J. Cell. Biol.* 100: 1499–1507.

Dreyer F, Mallart A, Brigant JL (1983) Botulinum A toxin and tetanus toxin do not affect presynaptic membrane currents in mammalian motor nerve endings. *Brain Res.* 270: 373–375.

Duchen LW (1973) The effects of tetanus toxin on the motor end-plates of the mouse. An electron microscopic study. *J. Neurol. Sci.* 19: 153–167.

Eisel U, Jarausch W, Goretzki K *et al.* (1986) Tetanus toxin: primary structure, expression in *E. coli*, and homology with botulinum toxins. *EMBO J.* 5: 2495–2502.

Erdmann G, Hanauske A, Wellhoner HH (1981) Intraspinal distribution and reaction in the grey matter with tetanus toxin of intracisternally injected anti-tetanus toxoid F(ab')2 fragments. *Brain Res.* 211: 367–377.

Faber K (1890) Die Pathogenie des Tetanus. *Berl. Klin. Wochenschr.* 27: 717–720.

Fairweather NF, Lyness VA (1986) The complete nucleotide sequence of tetanus toxin. *Nucleic Acids Res.* 14: 7809–7812.

Galazka A, Gasse F (1995) The present status of tetanus and tetanus vaccination. *Curr. Top. Microbiol. Immunol.* 195: 31–53.

Galli T, Zahraoui A, Vaidyanathan VV *et al.* (1998) A novel tetanus neurotoxin-insensitive vesicle-associated membrane protein in SNARE complexes of the apical plasma membrane of epithelial cells. *Mol. Biol. Cell* 9: 1437–1448.

Goschel H, Wohlfarth K, Frevert J, Dengler R, Bigalke H (1997) Botulinum A toxin therapy: Neutralizing and nonneutralizing antibodies-therapeutic consequences. *Exp. Neurol.* 147: 96–102.

Habermann E, Dimpfel W (1973) Distribution of 125 I-tetanus toxin and 125 I-toxoid in rats with generalized tetanus, as influenced by antitoxin. *Naunyn Schmiedebergs Arch. Pharmacol.* 276: 327–340.

Habermann E, Dreyer F (1986) Clostridial neurotoxins: handling and action at the cellular and molecular level. *Curr. Top. Microbiol. Immunol.* 129: 93–179.

Habig WH, Bigalke H, Bergey GK *et al.* (1986) Tetanus toxin in dissociated spinal cord cultures: long-term characterization of form and action. *J. Neurochem.* 47: 930–937.

Halpern JL, Loftus A (1993) Characterization of the receptor-binding domain of tetanus toxin. *J. Biol. Chem.* 268: 11188–11192.

Halpern JL, Neale EA (1995) Neurospecific-binding, internalization, and retrograde axonal transport. *Curr. Top. Microbiol. Immunol.* 195: 221–241.

Halpern JL, Habig WH, Neale EA, Stibitz S (1990) Cloning and expression of functional fragment C of tetanus toxin. *Infect. Immun.* 58: 1004–1009.

Hatheway CL (1990) Toxigenic clostridia. *Clin. Microbiol. Rev.* 3: 66–98.

Hayashi T, McMahon H, Yamasaki S *et al.* (1994) Synaptic vesicle membrane fusion complex: action of clostridial neurotoxins on assembly. *EMBO J.* 13: 5051–5061.

Herreros J, Lalli G, Montecucco C, Schiavo G (2000a) Tetanus toxin fragment C binds to a protein present in neuronal cell lines and motoneurons. *J. Neurochem.* 74: 1941–1950.

Herreros J, Lalli G, Schiavo G (2000b) C-terminal half of tetanus toxin fragment C is sufficient for neuronal binding and interaction with a putative protein receptor. *Biochem. J.* 1: 199–204.

Heyningen WE Van (1974) Gangliosides as membrane receptors for tetanus toxin. *Nature* 249: 415–417.

Hoch DH, Romero Mira M, Ehrlich BE *et al.* (1985) Channels formed by botulinum, tetanus, and diphtheria toxins in planar lipid bilayers: relevance to translocation of proteins across membranes. *Proc. Natl Acad. Sci. USA* 82: 1692–1696.

Humeau Y, Doussau F, Grant NJ, Poulain B (2000) How botulinum and tetanus neurotoxins block neurotransmitter release. *Biochimie* 82: 427–446.

Hunt JM, Bommert K, Charlton MP *et al.* (1994) A post-docking role for synaptobrevin in synaptic vesicle fusion. *Neuron* 12: 1269–1279.

Kitasato S (1891) Experimentelle Untersuchungen uber das Tetanusgift. *Z. Hyg. Infekt.* 10: 267–305.

Lacy DB, Stevens RC (1999) Sequence homology and structural analysis of the clostridial neurotoxins. *J. Mol. Biol.* 291: 1091–1104.

Lacy DB, Tepp W, Cohen AC, DasGupta BR, Stevens RC (1998) Crystal structure of botulinum neurotoxin type A and implications for toxicity. *Nat. Struct. Biol.* 5: 898–902.

Laird WJ, Aaronson W, Silver RP, Habig WH, Hardegree MC (1980) Plasmid-associated toxigenicity in *Clostridium tetani*. *J. Infect. Dis.* 142: 623.

Lalli G, Herreros J, Osborne SL *et al.* (1999) Functional characterisation of tetanus and botulinum neurotoxins binding domains. *J. Cell. Sci.* 112: 2715–2724.

Martin L, Cornille F, Turcaud S *et al.* (1999) Metallopeptidase inhibitors of tetanus toxin: A combinatorial approach. *J. Med. Chem.* 42: 515–525.

Martoglio B, Dobberstein B (1998) Signal sequences: more than just greasy peptides. *Trends Cell Biol.* 8: 410–415.

Marxen P, Bigalke H (1991) Tetanus and botulinum A toxins inhibit stimulated F-actin rearrangement in chromaffin cells. *Neuroreport* 2: 33–36.

Matsuda M, Sugimoto N, Ozutsumi K, Hirai T (1982) Acute botulinum-like intoxication by tetanus neurotoxin in mice. *Biochem. Biophys. Res. Commun.* 104: 799–805.

Matteoli M, Verderio C, Rossetto O *et al.* (1996) Synaptic vesicle endocytosis mediates the entry of tetanus neurotoxin into hippocampal neurons. *Proc. Natl Acad. Sci. USA* 93: 13310–13315.

Mayor RH (1945) *Classic Description of Disease*. Springfield, IL: Thomas.

McMahon HT, Ushkaryov YA, Edelmann L *et al.* (1993) Cellubrevin is a ubiquitous tetanus-toxin substrate homologous to a putative synaptic vesicle fusion protein. *Nature* 364: 346–349.

Mellanby J, Thompson PA (1977) Tetanus toxin in the rat hippocampus. *J. Physiol. Lond.* 269: 44p–45p.

Mellanby J, Green J (1981) How does tetanus toxin act? *Neuroscience* 6: 281–300.

Mellanby J, George G, Robinson A, Thompson P (1977) Epileptiform syndrome in rats produced by injecting tetanus toxin into the hippocampus. *J. Neurol. Neurosurg. Psychiatr.* 40: 404–414.

Mellanby J, Hawkins C, Mellanby H, Rawlins JN, Impey ME (1984) Tetanus toxin as a tool for studying epilepsy. *J. Physiol. Paris* 79: 207–215.

Mellanby J, Beaumont MA, Thompson PA (1988) The effect of lanthanum on nerve terminals in goldfish muscle after paralysis with tetanus toxin. *Neuroscience* 25: 1095–1106.

Menestrina G, Forti S, Gambale F (1989) Interaction of tetanus toxin with lipid vesicles. Effects of pH, surface charge, and transmembrane potential on the kinetics of channel formation. *Biophys. J.* 55: 393–405.

Middlebrook JL, Brown JE (1995) Immunodiagnosis and immunotherapy of tetanus and botulinum neurotoxins. *Curr. Top. Microbiol. Immunol.* 195: 89–122.

Mims CA (1995) *The Pathogenesis of Infectious Disease*. London: Academic Press.

Montecucco C (1986) How do tetanus and *Botulinum* neurotoxins bind to neuronal membranes? *Trends Biochem. Sci.* 11: 314–317.

Montecucco C, Schiavo G, Brunner J *et al.* (1986) Tetanus toxin is labeled with photoactivatable phospholipids at low pH. *Biochemistry* 25: 919–924.

Montecucco C, Papini E, Schiavo G (1994) Bacterial protein toxins penetrate cells via a four-step mechanism. *FEBS Lett.* 346: 92–98.

Neale EA, Bowers LM, Jia M, Bateman KE, Williamson LC (1999) Botulinum neurotoxin A blocks synaptic vesicle exocytosis but not endocytosis at the nerve terminal. *J. Cell. Biol.* 147: 1249–1260.

Nicolaier A (1884) Ueber infectiosen Tetanus. *Dtsche Med. Wchnschr.* 10: 842–844.

Parton RG, Ockleford CD, Critchley DR (1987) A study of the mechanism of internalisation of tetanus toxin by primary mouse spinal cord cultures. *J. Neurochem.* 49: 1057–1068.

Parton RG, Ockleford CD, Critchley DR (1988) Tetanus toxin binding to mouse spinal cord cells: An evaluation of the role of gangliosides in toxic internalization. *Brain Res.* 475: 118–127.

Patarnello T, Bargelloni L, Rossetto O, Schiavo G, Montecucco C (1993) Neurotransmission and secretion. *Nature* 364: 581–582.

Payling-Wright G (1955) The neurotoxins of *Clostridium botulinum* and *Clostridium tetani*. *Pharmacol. Rev.* 7: 413–465.

Pellizzari R, Rossetto O, Lozzi L *et al.* (1996) Structural determinants of the specificity for synaptic vesicle-associated membrane protein/synaptobrevin of tetanus and botulinum type B and G neurotoxins. *J. Biol. Chem.* 271: 20353–20358.

Penner R, Neher E, Dreyer F (1986) Intracellularly injected tetanus toxin inhibits exocytosis in bovine adrenal chromaffin cells. *Nature* 324: 76–78.

Pierce EJ, Davison MD, Parton RG (1986) Characterization of tetanus toxin binding to rat brain membranes. Evidence for a high-affinity proteinase-sensitive receptor. *Biochem. J.* 236: 845–852.

Ponomarev AW (1928) Zur frage der pathogenese des tetanus und des fortbewegungsmechanismus des tetanus-toxin entlang der nerven. *Z. Ges. Exp. Med.* 61: 93–106.

Popoff MR (1995) Ecology of neurotoxigenic strains of clostridia. *Curr. Top. Microbiol. Immunol.* 195: 1–29.

Pozdniakov OM, Polgar AA, Smirnova VS, Kryzhanovskii GN (1972) Change in the ultrastructure of neuromuscular junctions due to tetanus toxin. *Biull. Eksp. Biol. Med.* 73: 113–116.

Price DL, Griffin J, Young A, Peck K, Stocks A (1975) Tetanus toxin: direct evidence for retrograde intraaxonal transport. *Science* 188: 945–947.

Ramon G, Descombey PA (1925) Sur l'immunization antitetanique et sur la production del'antitoxine tetanique. *C. R. Soc. Biol.* 93: 508–598.

Rossetto O, Schiavo G, Montecucco C *et al.* (1994) SNARE motif and neurotoxins. *Nature* 372: 415–416.

Rossetto O, Gorza L, Schiavo G *et al.* (1996) VAMP/synaptobrevin isoforms 1 and 2 are widely and differentially expressed in nonneuronal tissues. *J. Cell. Biol.* 132: 167–179.

Rossetto O, Seveso M, Caccin P, Schiavo G, Montecucco C (2001) Tetanus and botulinum neurotoxins: turning bad guys into good by research. *Toxicon* 39: 27–41.

Rothman JE (1994) Mechanisms of intracellular protein transport. *Nature* 372: 55–63.

Schiavo G, Montecucco C (1995) Tetanus and botulism neurotoxins: isolation and assay. *Meth. Enzymol* 248: 643–652.

Schiavo G, Papini E, Genna G, Montecucco C (1990) An intact interchain disulfide bond is required for the neurotoxicity of tetanus toxin. *Infect. Immun.* 58: 4136–4141.

Schiavo G, Ferrari G, Rossetto O, Montecucco C (1991) Tetanus toxin receptor. Specific cross-linking of tetanus toxin to a protein of NGF-differentiated PC 12 cells. *FEBS Lett.* 290: 227–230.

Schiavo G, Benfenati F, Poulain B *et al.* (1992a) Tetanus and botulinum-B neurotoxins block neurotransmitter release by proteolytic cleavage of synaptobrevin. *Nature* 359: 832–835.

Schiavo G, Poulain B, Rossetto O *et al.* (1992b) Tetanus toxin is a zinc protein and its inhibition of neurotransmitter release and protease activity depend on zinc. *EMBO J.* 11: 3577–3583.

Schiavo G, Demel R, Montecucco C (1990) On the role of polysialoglycosphingolipids as tetanus toxin receptors. A study with lipid monolayers. *Eur. J. Biochem.* 199: 705–711.

Schiavo G, Matteoli M, Montecucco C (2000) Neurotoxins affecting neuroexocytosis. *Physiol. Rev.* 80: 717–766.

Schwab ME, Thoenen H (1976) Electron microscopic evidence for a transsynaptic migration of tetanus toxin in spinal cord motoneurons: an autoradiographic and morphometric study. *Brain Res.* 105: 213–227.

Schwab ME, Suda K, Thoenen H (1979) Selective retrograde transsynaptic transfer of a protein, tetanus toxin, subsequent to its retrograde axonal transport. *J. Cell. Biol.* 82: 798–810.

Shapiro RE, Specht CD, Collins BE *et al.* (1997) Identification of a ganglioside recognition domain of tetanus toxin using a novel ganglioside photoaffinity ligand. *J. Biol. Chem.* 272: 30380–30386.

Shumaker HB, Lamont A, Firor WM (1939) The reaction of 'tetanus sensitive' and 'tetanus resistant' animals to the injection of tetanal toxin into the spinal cord. *J. Immunol.* 37: 425–433.

Soleilhac JM, Cornille F, Martin L *et al.* (1996) A sensitive and rapid fluorescence-based assay for determination of tetanus toxin peptidase activity. *Anal. Biochem.* 241: 120–127.

Staub GC, Walton KM, Schnaar RL *et al.* (1986) Characterization of the binding and internalization of tetanus toxin in a neuroblastoma hybrid cell line. *J. Neurosci.* 6: 1443–1451.

Stockel K, Schwab M, Thoenen H (1975) Comparison between the retrograde axonal transport of nerve growth factor and tetanus toxin in motor, sensory and adrenergic neurons. *Brain Res.* 99: 1–16.

Swaminathan S, Eswaramoorthy S (2000) Structural analysis of the catalytic and binding sites of *Clostridium botulinum* neurotoxin B. *Nat. Struct. Biol.* 7: 693–699.

Takano K, Kirchner F, Terhaar P, Tiebert B (1983) Effect of tetanus toxin on the monosynaptic reflex. *Naunyn Schmiedebergs Arch. Pharmacol.* 323: 217–220.

Tizzoni G, Cattani G (1889) Ricerche batteriologiche sul tetano. *Riforma Med.* 5: 512–513.

Tizzoni G, Cattani G (1890) Uber das Tetanusgift. *Zbl. Bakteriol.* 8: 69–73.

Tonello F, Schiavo G, Montecucco C (1997) Metal substitution of tetanus neurotoxin. *Biochem. J.* 322: 507–510.

Trimble WS (1993) Analysis of the structure and expression of the VAMP family of synaptic vesicle proteins. *J. Physiol. Paris* 87: 107–115.

Umland TC, Wingert LM, Swaminathan S et al. (1997) Structure of the receptor binding fragment HC of tetanus neurotoxin. *Nat. Struct. Biol.* 4: 788–792.

Umland TC, Wingert L, Swaminathan S, Schmidt JJ, Sax M (1998) Crystallization and preliminary X-ray analysis of tetanus neurotoxin C fragment. *Acta Crystallogr. D Biol. Crystallogr.* 54: 273–275.

Verderio C, Coco S, Bacci A et al. (1999) Tetanus toxin blocks the exocytosis of synaptic vesicles clustered at synapses but not of synaptic vesicles in isolated axons. *J. Neurosci.* 19: 6723–6732.

Weinstein L (1973) Tetanus. *N. Engl. J. Med.* 289: 1293–1296.

Weller U, Dauzenroth ME, Gansel M, Dreyer F (1991) Cooperative action of the light chain of tetanus toxin and the heavy chain of botulinum toxin type A on the transmitter release of mammalian motor endplates. *Neurosci. Lett.* 122: 132–134.

Wellhoner NH (1982) Tetanus neurotoxin. *Rev. Physiol. Biochem. Pharmacol.* 93: 1–68.

Wellhoner HH (1992) Tetanus and botulinum neurotoxins. In: Herken H and Hucho F (eds) *Handbook of Experimental Pharmacology.* Berlin: Springer, pp. 357–417.

Wellhoner HH, Seib UC, Hensel B (1973) Local tetanus in cats: the influence of neuromuscular activity on spinal distribution of ^{125}I labelled tetanus toxin. *Naunyn Schmiedebergs Arch. Pharmacol.* 276: 387–394.

Wiener M, Freymann D, Ghosh P, Stroud RM (1997) Crystal structure of colicin Ia. *Nature* 385: 461–464.

Williamson LC, Neale EA (1994) Bafilomycin A1 inhibits the action of tetanus toxin in spinal cord neurons in cell culture. *J. Neurochem.* 63: 2342–2345.

Yavin E, Nathan A (1986) Tetanus toxin receptors on nerve cells contain a trypsin-sensitive component. *Eur. J. Biochem.* 154: 403–407.

91

Bacteroides

Sheila Patrick

The Queen's University of Belfast, Belfast, UK

Bacteroides are present in the faeces and are the predominant bacteria of the open-ended culture system of the human intestinal tract. It has been estimated that *Bacteroides* spp. are one of the major groups of commensals that inhabit the human body. They and other related bacteria are also found in the upper respiratory tract, mouth and urogenital tract of humans and animals. This chapter deals mainly with the *Bacteroides sensu stricto* (**Table 1**). The oral Gram-negative obligate anaerobes *Porphyromonas*, *Prevotella* and *Fusobacterium* are considered in Chapter 51. Certain *Bacteroides* species, in particular *Bacteroides fragilis*

and *B. thetaiotaomicron*, are the strict anaerobes most frequently isolated from clinical infection (Johnson, 1978). Such infections generally arise from the contamination by the commensal microbiota of normally non-colonised body sites. In this sense, *Bacteroides* spp. are opportunistic pathogens.

Though Louis Pasteur is credited with the first cultivation of strictly anaerobic bacteria, it was Veillon and Zuber (1897, 1898) who demonstrated clearly the involvement of Gram-negative non-spore-forming strictly anaerobic bacteria in infections such as brain and lung abscesses, peritonitis, bartholinitis and otitis media. They named the most abundant and frequently observed bacterium in appendicitis and peritonitis *Bacillus* (now *Bacteroides*) *fragilis*.

B. fragilis is the obligately anaerobic Gram-negative bacterium most frequently isolated from clinical infection and *B. thetaiotaomicron* the second most frequent isolate (**Table 2**; Duerden, 1980; Brook, 1989; Willis, 1991). These infections include intra-abdominal, vaginal, pilonidal, perianal and brain abscesses and soft-tissue infections. *B. fragilis* is also the most common cause of anaerobic septicaemia, with a potential mortality of up to 19% (Redondo *et al.*, 1995). The risk of *B. fragilis* infection after surgical operations of the lower intestinal tract makes prophylactic antibiotic cover necessary. Before the introduction of antibiotics, the mortality from infection arising particularly from large-bowel perforations was

Table 1 Members of *Bacteroides sensu stricto*[a]

Species	Genome size[b] (Mbp)
B. fragilis (type species)	5.3
B. caccae	
B. distasonis	4.8
B. eggerthii	4.4
B. merdae	
B. ovatus	6.9
B. stercoris	
B. thetaiotaomicron	4.8
B. uniformis	4.6
B. vulgatus	5.1

[a] As defined by Shah and Garbia (1991).
[b] Data from Shaheduzzaman *et al.* (1997).

Table 2 Incidence of *Bacteroides* spp. in faeces and clinical samples

	[a]Faeces (%)	[a]Clinical samples (%)
B. vulgatus	43–45	2–3
B. thetaiotaomicron	15–29	13–17
B. distasonis	9	3–6
B. fragilis	4–13	63–81
B. ovatus	4	0–7

[a] Compiled from Duerden (1980), Brook (1989), Namavar *et al.* (1989), Willis (1991).

30–100%, for example after rupture of an inflamed appendix. In the 1970s intra-abdominal abscesses were associated with a mortality of up to 30%, largely because the role of anaerobes in these infections was not appreciated and treatment was therefore not given (Tally and Ho, 1987), in spite of the early observations of Veillon and Zuber in the 1890s.

Knowledge and understanding of *B. fragilis* is about to blossom as a result of the provision of funding by the Wellcome Trust for the Sanger Centre (UK) to sequence the complete genome of the *B. fragilis* type strain NCTC 9343 (ATCC 25285) and a partial shotgun sequence of the rifampicin-resistant strain 638R. This sequence information is in the public domain and available for view on the web sites of Wellcome Trust Beowulf Genomics (http://www.beowulf.org.uk/home.htm) and the Sanger Centre (http://www.sanger.ac.uk/Projects/Microbes/).

Classification and Taxonomic Position

The genus *Bacteroides* was at first a repository for any Gram-negative strictly anaerobic non-spore forming bacillus that was neither a *Fusobacterium* nor a *Leptotrichia*, but it has been refined over the years. Many of the more than 50 species listed in the ninth edition of *Bergey's Manual of Systematic Bacteriology* have been assigned to new genera (**Table 3**). The proposal formally to limit the genus *Bacteroides* to the species listed in **Table 1** was made by Shah and Collins (1989). The type species is *Bacteroides fragilis* and, therefore, the 10 species listed in **Table 1** are sometimes referred to as the '*B. fragilis* group'. The phylogenetic tree based on whole genome analyses of 7 species of the *B. fragilis* group was reported to be broadly similar to that obtained from 16S rRNA sequences, with the exceptions of *B. thetaiotaomicron* and *B. ovatus*. The sizes of the genomes analysed are included in **Table 1** (Shaheduzzaman *et al.*, 1997).

Table 3 Former members of the genus *Bacteroides*

Dichelobacter nodosus
Prevotella
Porphyromonas
Ruminobacter
Campylobacter gracilis
Bacteroides ureolyticus[a]

[a] Designated a member of the *Campylobacteraceae* but not yet re-named.

DNA homology studies indicate that there is a good correlation between DNA homology group and phenotypic characteristics of the designated species (Johnson and Ault, 1978). Although *B. fragilis*, *B. ovatus* and *B. thetaiotaomicron* could each be subdivided into more than one homology group, the phenotypic tests used identified the different homology groups as the relevant species. Of a group of 60 strains of DNA homology group I, for which the source was known, 76.6% were clinical isolates. Of the 11 group II isolates whose source was known, approximately half were from faeces and the other half from infections. More recent analysis of the two DNA homology groups of *B. fragilis* indicate that group I and group II *B. fragilis* can also be distinguished on the basis of ribotyping, analysis of PCR-generated fragment patterns, insertion sequence content (Podglajen *et al.*, 1995) and small-subunit rDNA sequencing (Ruimy *et al.*, 1996). It was noted that *B. fragilis* DNA homology group I lack the gene *cfiA* (also called *ccrA*) which encodes a metallo-β-lactamase (Ambler class B enzyme). This enzyme confers resistance to the majority of β-lactam antibiotics, such as carbapenems, imipenem and meropenem, and is not susceptible to inhibitors such as clavulanic acid, sulbactam and tazobactam. This gene was identified in about 3% of isolates. Only about one-third of the strains examined expressed this β lactamase, however; the gene is silent in the remainder of these strains. The data suggest that DNA homology group II strains relate to the *cfiA*-gene positive group. Interestingly, three insertion sequences (IS4351, 942 and 1186), which provided promoter sequences for the transcription of *cfiA*, were also identified. This will be further considered below. On the other hand group I strains, which represent the majority of clinical isolates, lack the *cfiA* gene but carry the *cepA* gene, which encodes an active-site serine β lactamase, related to the Ambler Class A enzymes. The *B. fragilis* type strain NCTC 9343 (ATCC 25285) and strain 638R (also known as AIP 638R, TM4000, IB101 and 638rfm), which has been used extensively in genetic studies, are members of the DNA homology group I.

Table 4 Position of *Bacteroides* within the Phylum *Bacteroides–Flavobacterium*

Phylum *Bacteroides–Flavobacterium*	
Bacteroides subdivision	Group
Bacteroides	
Bacteroides	fragilis
Prevotella	melaninogenicus–oralis
Porphyromonas	saccharolytic pigmented
Dichelobacter	
Fusobacterium	fusiform
Leptotrichia	
Flavobacterium subdivision	
Flavobacterium	
Cytophoga	
Flexibacter	

Table 5 Major characteristics of the genus *Bacteroides*

Microscopical
  Gram negative,
  Non-sporing
  Non-motile
  Pleomorphic rod-shaped
  Cell size $0.5–1.3 \times 1.6–11\ \mu m$

Components
  Principal respiratory quinones: menaquinones
    (with 10 and/or 11 isoprene units)
  Dibasic amino acid of peptidoglycan:
    meso-diaminopimelic acid
  DNA G + C ratio: [a]40–48 mol%
  Lipid content : sphingolipids
  Predominant cell fatty acid: 3-hydroxylated and
    non-hydroxylated (straight chain saturated,
    anteiso- and *iso*-methyl branched-chain types)

Growth and metabolism
  Strictly anaerobic
  Bile tolerant
  Esculin hydrolysed
  Fermentation: saccharolytic, major metabolic
    end-products acetate and succinate
  Metabolic enzymes: enzymes for pentose phosphate
    pathway e.g. glucose-6-phosphate dehydrogenase,
    6-phosphogluconate dehydrogenase and also
    malate dehydrogenase and glutamate dehydrogenase

Compiled from Shah and Garbia (1991).

[a] Lower figure variously quoted as 39, 40 and 41.

In the wider taxonomic picture, on the basis of rRNA sequence comparisons, the *Bacteroides* fall into the *Bacteroides–Flavobacterium* phylum (**Table 4**; Woese, 1987). Bearing in mind the diverse phenotypes of these two groups of bacteria, this is an unexpected taxonomic association, since *Flavobacterium*, which includes *Cytophaga* and *Flexibacter*, are aerobic. It may be that as more is learned of the characteristics of these two bacterial groups, their similarities will become more apparent. Interestingly, it is thought that this phylum diverged from other eubacteria early in evolutionary terms. This is thought to have occurred well before the divergence of the Gram-positive bacteria from the phylum, which contains the majority of Gram-negative bacteria, such as the enteric bacteria *Escherichia coli* and the pseudomonads (Phylum, Proteobacteria, Purple bacteria). Indeed, the sequences of some antibiotic-resistance genes bear remarkable homology with those of Gram-positive bacteria. This taxonomic divergence may also explain the difficulties that have been encountered in carrying out genetic experiments in *Bacteroides*, since the methods rely heavily on *E. coli* (see below).

For the taxonomic history of the genus *Bacteroides* the reader is referred to Shah and Garbia (1991) and Shah *et al.* (1998).

Identification

The key distinguishing feature of the *Bacteroides* spp. and related anaerobic Gram-negative bacteria is their anaerobic nature, but their ability to tolerate oxygen may be variable (see below). They are non-sporing and are generally non-motile. When grown on anaerobic blood agar, they lack the black pigmentation and fluorescence that may be associated with some of the oral Gram-negative anaerobes, such as *Porphyromonas* or *Prevotella* spp., and they form circular, entire and smooth colonies. They are generally sensitive to metronidazole and resistant to kanamycin, vancomycin and colistin. Haemin and vitamin K are stimulatory for the growth of many *Bacteroides* spp. and they are an essential requirement for some. The major characteristics are detailed in **Table 5**.

Further details of the current techniques used for the clinical diagnosis of *Bacteroides* spp. are presented below. The reader is also directed to the *Wadsworth Anaerobic Bacteriology Manual* (Summanen *et al.*, 1993) and relevant chapters in Cowan and Steel's *Manual for the Identification of Medical Bacteria* (Brown *et al.*, 1989; Barrow and Felthan, 1993; Jousimies-Somer *et al.*, 1995). Although it is not yet part of the routine identification of *Bacteroides* spp., there is scope for the extension to clinical practice of identification methods based on 16S rRNA sequences that have been used for the identification of *Bacteroides* in human faeces (e.g. Manz *et al.*, 1996; Bonnett *et al.*, 1999).

Structures

Fimbriae and extracellular polysaccharide capsules are two surface structures that have been studied in detail in *Bacteroides*. The vast majority of structural investigations have centred on *B. fragilis*, which is considered to be the most frequent clinical isolate. Some strains of *B. fragilis* release quantities of outer membrane vesicles that carry with them the surface polysaccharides (**Fig.1**). The potential role of these structures in virulence is discussed below.

Polysaccharides

Encapsulating structures have been implicated in resistance to complement-mediated killing, uptake and killing by phagocytes (Reid and Patrick, 1984), and in abscess formation in an animal model (Tzianabos *et al.*, 1993). These aspects are discussed in detail below in relation to virulence.

Capsules are discrete clear areas around negatively-stained *B. fragilis*, *B. ovatus*, *B. vulgatus* and *B. thetaiotaomicron*, but they were not present in five strains of *B. distasonis* (Babb and Cummins, 1978). With the exception of *B. ovatus*, which were all capsulate, 10% or less of the bacteria of other species had capsules. Capsule size varied between strains and also, notably, within populations of a given strain. It is now clear that *B. fragilis* exhibits not only within-strain phase variation in capsule production, but also between- and within-strain antigenic variation of different types of capsules, which continues not to be taken into account. In individual strains of *B. fragilis* it is possible by electron microscopy to identify large (LC) or small capsules (SC) that are both fibrous in appearance but are antigenically different, and bacteria with an encapsulating electron-dense-layer (EDL) adjacent to the outer membrane (**Fig. 2**) (Patrick and Reid, 1983; Patrick *et al.*, 1986). Bacteria with an EDL are non-capsulate (NC) by light microscopy, but small and large capsules are visible with negative staining. Expression of the different capsular types is inheritable, since populations can be enriched by subculture from different interfaces of Percoll step density gradients. EDL-enriched populations also produce polysaccharide slime (**Fig. 3**) which shares epitopes with the discrete LC. Observation by microscopy of the populations enriched for the three capsular types with monoclonal antibodies specific for surface polysaccharides shows that non-capsulate bacteria are antigenically different from SC bacteria, but share epitopes with LC bacteria. In addition, immunofluorescent and immunogold labelling for fluorescence and electron microscopy, respectively, reveal antigenic variation in populations

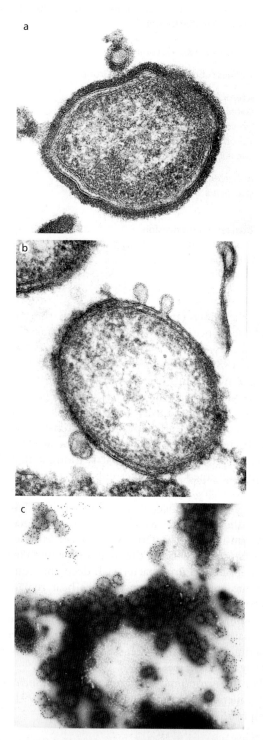

Fig. 1 Transmission electron micrographs of outer-membrane vesicles of *B. fragilis*: (a) Ultrathin section of bacterium (non-capsulate by light microscopy) with an EDL adjacent to the outer membrane. Note the outer membrane vesicle with associated electron dense material. (b) Ultrathin section of bacterium to illustrate vesicle being released from the outer membrane. (c) Negatively stained outer membrane vesicles immunogold-labelled with monoclonal antibody QUBF5. From Lutton *et al.* (1991), with permission.

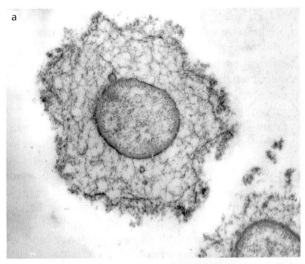

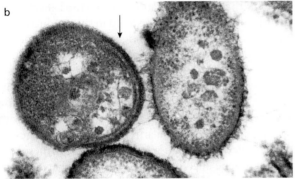

Fig. 2 Transmission electron micrograph of ultrathin sections to illustrate encapsulating structures of *B. fragilis*: (a) Bacterium with large fibrous network, which equates to a LC visible by light microscopy. (b) Bacterium with small fibrous network, which equates to a SC visible by light microscopy (right-hand side) and bacterium with marginal EDL, non-capsulate by light microscopy (left-hand side and arrow).

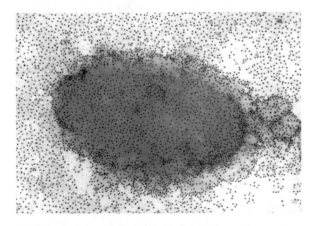

Fig. 3 Transmission electron micrograph of negatively stained bacterium from non-capsulate population of *B. fragilis* immunogold-labelled with monoclonal antibody QUBF7 to illustrate extracellular material. From Lutton *et al.* (1991), with permission.

that appear to be structurally homogeneous (Reid *et al.*, 1985, 1987; Patrick and Lutton, 1990a; Lutton *et al.*, 1991). This phenomenon is observed in recent clinical isolates from a variety of anatomical sites, in isolates from different geographical locations and in culture collection type cultures (Patrick *et al.*, 1995b). If a broth culture with a mixture of epitopes is plated on agar and the resulting colonies are labelled with surface polysaccharide-specific monoclonal antibody and a fluorescent dye and examined by microscopy, two types of colonies are observed; in one of these 90% or more, and in the other 10% or less, of the bacteria express the same epitope. Subculture of single colonies in which the majority of the bacteria express that particular epitope into broth culture, results in enrichment for this epitope, and failure to enrich for other epitopes (Patrick *et al.*, 1999). This strongly suggests that a reversible switching mechanism is in operation.

The genetic mechanism that generates the observed phase and antigenic variation is not known, but multiple mechanisms for the generation of phase and antigenic variation are well characterised in other bacteria (Patrick and Larkin, 1995). Examples which may be relevant to *Bacteroides* include variation in *Neisseria meningitidis* (Jennings *et al.*, 1999) and *Haemophilus influenzae* lipopolysaccharide (LPS).

The precise nature of the biochemical differences that generate the antigenic variation is unknown. It is possible that polysaccharides are biochemically similar in terms of sugar moieties, for example, but for a wide variety of antigenic variation to be generated by alteration of the linkage of the substituent moieties, their chemical substitution or both. Antigenic variation generated by these means is well documented in the polysaccharides of other pathogenic bacteria such as *H. influenzae*, *E. coli* and *N. meningitidis* (Patrick and Larkin, 1995).

Immunochemical analysis of these antigenically variable polysaccharides after polyacrylamide gel electrophoretic (PAGE) separation reveals diffuse bands of high molecular mass. In the non-capsulate, slime-producing populations, associated ladder patterns of lower molecular mass are also observed. This type of banding pattern is characteristic of different chain lengths of heteropolymeric polysaccharides with repeating subunits, each 'step' in the ladder corresponding to the addition of another subunit. It may be that polysaccharides with different chain lengths are more characteristic of slime as opposed to discrete capsular structures. Ladders with a fine step size and a large step size have been observed (**Fig. 4**). The larger step size resembles the pattern observed with the O antigens of enteric bacteria. The controversy surrounding the O antigen of *B. fragilis* is further discussed below.

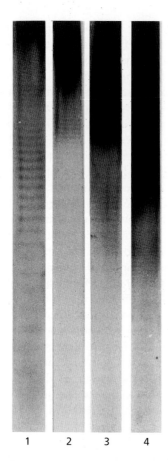

Fig. 4 Immunoblots of aqueous phenol extracts from *B. fragilis* strain NCTC 9343 after PAGE reacted with QUBF5 (track 1), QUBF6 (track 2), QUBF7 (track 3) and QUBF8 (track 4). Note high-molecular-mass material and associated ladder patterns. From Patrick *et al.* (1999), with permission.

Two *B. fragilis* polysaccharides, extracted from populations of strain NCTC 9343 which were not defined for capsule expression by microscopy, have been chemically characterised and designated A and B (Pantosti *et al.*, 1991, 1995; Tzianabos *et al.*, 1992). These two entities can be separated by iso-electro-focusing from extracts obtained by hot-phenol–water extraction followed by ethanol precipitation and separation on a Sephacryl S-300 column (Pantosti *et al.*, 1991), and high-resolution NMR spectroscopy was used to analyse the two polysaccharides (Baumann *et al.*, 1992). Polysaccharide A (PSA) is zwitterionic with a positive and a negative charge. It consists of tetrasaccharide repeating units of three sugars. Polysaccharide B (PSB) possesses one positive and two negative charges and has a hexasaccharide repeating unit. See **Fig. 5** for structures of PSA and PSB.

Although it was originally reported that monoclonal antibodies specific for either polysaccharide had been produced (Pantosti *et al.*, 1991; Tzianabos *et al.*,

1992), it was subsequently reported that antibodies thought to be specific for PSB cross-reacted with PSA (Comstock *et al.*, 1999). The immunochemistry of these entities and whether they are co-expressed on the same bacterial cell is therefore unknown. The evidence for the involvement of these polysaccharides in abscess formation is discussed below.

Many strains of *B. fragilis* are refractory to the introduction of foreign DNA, which presents difficulties for genetic studies that involve the use of *E. coli*-based vectors. A spontaneous rifampicin-resistant mutant, 638R, seems to lack the restriction enzymes of most *B. fragilis* strains, since it can be conjugated with *E. coli* (Privitera *et al.*, 1979). It also lacks a visible capsule when negatively stained for light microscopy (unpublished). Mating of this strain with an *E. coli* strain that contains the *B. fragilis* transposon Tn4351 located on the suicide vector pNJR6 resulted in the production of transposon mutants that did not react with monoclonal antibody 4D5, which is specific for the PSA of *B. fragilis* NCTC 2429 (Pantosti *et al.*, 1995) and cross-reacts with 638R but not with NCTC 9343. Sequencing of the DNA at the junction of the transposon insertion revealed an open reading frame (ORF) with approximately 80% similarity to the *rmlA* genes of *Shigella flexneri* and *Salmonella enterica*. In these bacteria this gene encodes a glucose-1-phosphate thymidylate-transferase enzyme known to be involved in polysaccharide biosynthesis. A 638R *rmlA* probe identified a similar gene on the chromosome of NCTC 9343. Sequencing of the region downstream resulted in the identification of a 15 379-bp locus with 16 ORFs, designated *wcf*. Seven ORFs were identified, which probably encode sugar transferase enzymes and two *O*-acetyltransferases. A further two ORFs were thought to be involved in polysaccharide transport and polymerisation. Interestingly, a deletion mutant from which the last 10 of the 16 ORFs had been removed appeared to lack PSB but not PSA, in spite of the specificity of the 4D5 monoclonal antibody used to locate the biosynthesis region. When compared with the wild type, there was no apparent difference in the ability of the mutants to induce abscesses in a rat model of intra-abdominal infection (Comstock *et al.*, 1999).

Lipopolysaccharide and Common Antigen

The LPS of *B. fragilis* differs in composition from that of enteric bacteria in a number of respects (Lindberg *et al.*, 1990). Enterobacterial lipid A diglucosamine is bisphosphorylated, whereas in *B. fragilis* it is monophosphorylated, since the distal glucosamine residue of the lipid A molecule lacks a phosphate group. *E. coli*

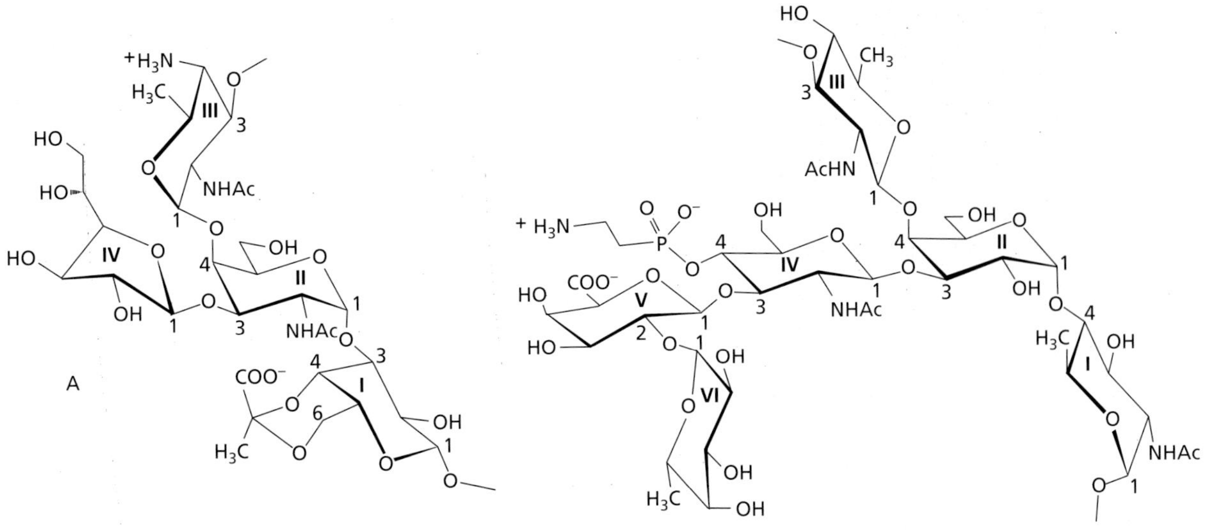

PSA Sugars

		Trivial Name
I	β-galactopyranose-4,6 pyruvate	
II	2-acetamido-2-deoxy-α-D-galactopyranose	*N*-acetyl-D-galactosamine
III	2-acetamido-4-amino-2,4,6-trideoxy-α-D-galactopyranose	*N*-acetyl-D-trideoxyfucosamine
IV	β-D-galactofuranose	

PSB Sugars

I	2-acetamido-2,6-dideoxy--α-L-glucopyranose	*N*-acetyl-L-quinovosamine
II	α-D-galactopyranose	D-galactose
III	2-acetamido-2,6-dideoxy-β-D-glucopyranose	*N*-acetyl-D-quinovosamine
IV	2-acetamido-2-deoxy-β-D-glucopyranose-4(2-aminoethyl-phosphate)	
V	β-D-glucopyranuronic acid	D-galacturonic acid
VI	6-deoxy-α-L-galactopyranose	L-fucose

Fig. 5 Structure of PSA and PSB (see Baumann *et al.*, 1992).

has six fatty acid chains, or acyl groups, per diglucos-amine backbone, which may have a chain length of 12–14 carbon atoms, whereas *B. fragilis* has 4–5 fatty acids of chain lengths 15–17 carbon atoms, with branched 3-hydroxylated and non-hydroxylated fatty acids. One of the predominant fatty acids in enterobacterial lipid A, 3-hydroxytetradecanoic acid, is thought to be lacking in *B. fragilis*. The differences in structure of the *Bacteroides* and the enterobacterial lipid A almost certainly relate to the different immunomodulatory properties of these molecules (Patrick and Larkin, 1995). The potential role of *Bacteroides* LPS in systemic inflammatory response syndrome is discussed later.

Reports are conflicting about the presence or absence of *Bacteroides* LPS of L-glycero-D-manno-heptose and keto-deoxyoctonate (KDO), both of which are present in enterobacterial LPS. KDO is almost certainly present in *Bacteroides*, but in a phosphorylated form that renders it undetectable in the standard thiobarbituric acid assay (Beckmann *et al.*, 1989; Fujiwara *et al.*, 1990). In addition, rhamnose, galactose and glucose have been reported in the core region (Lindberg *et al.*, 1990). Whether the LPS of *Bacteroides* spp. possesses an O antigen similar to that

found in the enterobacteria is subject to controversy. The literature contains emphatic statements that extended repeating O antigen is absent from *Bacteroides* (Lindberg *et al.*, 1990; Comstock *et al.*, 1999). It is suggested that *Bacteroides* LPS is more like the lipooligosaccharide of, for example, *N. meningitidis* (Jennings *et al.*, 1999) or 'rough'-type mutants of enterobacteria. Other publications, however, clearly illustrate silver-stained LPS PAGE profiles with ladder patterns characteristic of the smooth LPS of *B. fragilis* (Poxton and Brown, 1986) and *B. vulgatus* (Delahooke *et al.*, 1995a: **Fig. 6**). A number of possible factors may explain the inability of some workers to isolate the repeating polysaccharide chains of smooth LPS. The most likely is variation within and between *Bacteroides* strains and possible loss of the ability to synthesise O antigen after repeated subculture *in vitro*. A monoclonal antibody, QUBF5, has been described, which labels a carbohydrate component with a PAGE profile similar to that of the apparent O antigen (see **Fig 4**: Lutton *et al.*, 1991), but it labels only a proportion of the bacteria in a population. Enrichment for two different antigenic types of the high-molecular-mass polysaccharides did not co-enrich for the QUBF5

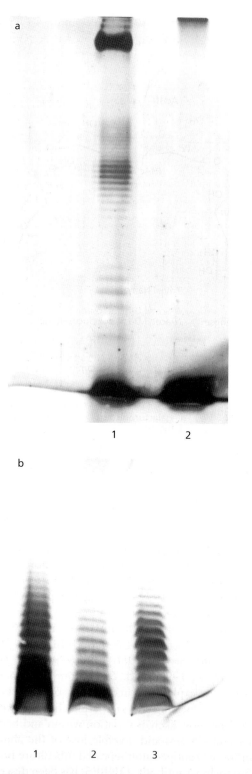

epitope, as determined by immunofluorescence microscopy (Patrick *et al.*, 1999). Whether this represents within-strain antigenic variation of this component or phase variation is not known, as only one monoclonal antibody specific for this type of structure has been described. Further discussion of the conflicting data and inaccurate literature citation since the middle of the 1980s with respect to *Bacteroides* LPS (Weintraub *et al.*, 1985; Poxton and Brown, 1986) can be found in Patrick (1993). It has been suggested that some strains of *E. coli* produce not only an LPS, but also a LOS distinct from rough LPS, by a separate biosynthetic pathway (Heinrichs *et al.*, 1999). Also, the K30 capsular polysaccharide of *E. coli* may be covalently linked to core lipid A, but this is not essential either for export of this high-molecular-mass polysaccharide or for formation of a capsular structure at the cell surface (MacLachlan *et al.*, 1993). The distinction between an O antigen and a capsular or extracellular polysaccharide is therefore somewhat blurred and may finally only be resolved for *Bacteroides* when the genetic loci and pathways for biosynthesis have been identified in a range of strains.

The common antigen described by Poxton and Brown (1986) is extractable by phenol–water, runs behind the rough LPS on SDS PAGE and is distinct from the O-antigen ladder. Polyclonal mono-specific antiserum raised against immunoblot-purified material detects the antigen in *B. fragilis* strains from culture collections and recent clinical isolates from a range of geographical locations. The antiserum also labels strains of *B. ovatus* and *B. thetaiotaomicron*, by immunofluorescence microscopy, but with less intensity. This antiserum has also been used for the direct detection by immunofluorescence microscopy of *Bacteroides* in pus samples and blood culture bottles (Patrick *et al.*, 1995b). The nature of this common antigen is not known.

Fimbriae

Two types of fimbriae have been identified on *B. fragilis*, but in comparison with other bacterial species the fimbriae are not well characterised. This is probably due to poor knowledge of the precise growth conditions required for the assembly of the fimbriae. Pruzzo *et al.* (1984) described filamentous fimbriae of approximately 30 nm in diameter, and subsequently fimbriae of 4–5 nm in diameter were described (van Doorn *et al.*, 1987). Fimbria were observed by electron microscopy and immunogold labelling with fimbria-specific polyclonal antiserum. Immunoblotting after polyacrylamide gel electrophoresis revealed a fimbrial subunit of 40–42 kDa, depending on the

Fig. 6 Silver stained PAGE profiles: (a) *B. fragilis* strain NCTC 9344 aqueous phenol extract (track 1), phenol chloroform petroleum ether extract (track 2); (b) *B. vulgatus* strain MRPL 1985 aqueous phenol extract (track 1), phenol chloroform petroleum ether extract (track 2), triton-magnesium extract (track 3). Photographs courtesy of I.R. Poxton and D.M. Delahooke, Department of Medical Microbiology, University of Edinburgh).

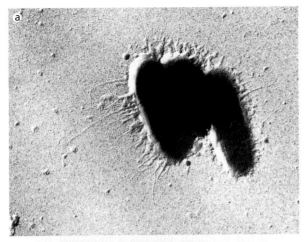

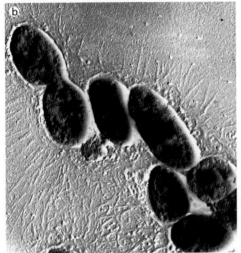

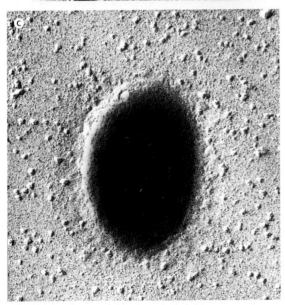

Fig. 7 Transmission electron micrographs of platinum gold-shadowed *B. fragilis*: (a) bacteria with a SC by light microscopy; (b) bacteria with a LC by light microscopy; (c) non-capsulate by light microscopy.

strain. Expression of the fimbrial subunit was reduced at low iron concentrations and low temperature. In a variety of strains of *B. fragilis*, grown in broth or on agar, and in a model of peritoneal infection *in vivo*, fimbrial subunits can be detected by immunoblotting. Intact fimbriae, however, have consistently been observed only by transmission electron microscopy after negative staining in agar cultures. One of the difficulties in observing fimbriae by electron microscopy relates to the interference by polysaccharide. Long strands of material are observed under the electron microscope when *B. fragilis* from broth cultures and producing a small capsule are platinum-gold shadowed or coated. Shorter strands are also observed with the LC population (**Fig. 7**). It remains to be determined whether these strands are purely polysaccharide or cover underlying fimbrial appendages. If the latter is the case, the polysaccharides would mask epitopes on the fimbriae, and account for erratic immunolabelling. Another possibility is that fimbrial subunits are expressed, but not always assembled into intact fimbriae (Lutton *et al.*, 1989). N-Terminal amino acid sequencing of the 40-kDa subunit of *B. fragilis* strain BE1 did not reveal any similarity with the sequences of *Porphyromonas gingivalis*, *Dichelobacter* (formerly *Bacteroides*) *nodosus* or *E. coli* type 1, K88 or CFA/1 fimbrial subunits (van Doorn *et al.*, 1992). The relationship between *B. fragilis* fimbriae and microbial attachment to host cells is considered below.

Physiology

Iron Uptake Mechanisms

The success of the majority of bacteria that grow in human hosts depends on their ability to obtain iron, which is strictly limited in availability. In the host iron is bound to transferrin and lactoferrin, and concentration of freely available iron in body fluids is about 10^{-18} mol/L (Patrick and Larkin, 1995). *B. fragilis* has not only a requirement for iron, but also an absolute requirement for haem, which is necessary for cytochrome and catalase synthesis (Morris, 1991). Although *B. fragilis* can use free haem, it is unlikely that *in vivo* there would be sufficient of this to sustain growth; it is, however, likely that *B. fragilis* is able to release haem from haemoglobin, haptoglobin–haemoglobin complexes or haemopexin–haemoglobin complexes *in vivo*. In defined medium depleted of iron by precipitation with calcium chloride, *B. fragilis* can use haemoglobin or haptoglobin–haemoglobin as a source of both iron and protoporphyrin (Otto *et al.*, 1990, 1994). This is related to the expression of a number of iron-repressible outer-membrane proteins (OMP). Four haem-binding proteins have been isolated by haemin–agarose affinity chromatography and a further 44-kDa iron-repressible OMP is required for the haem-binding protein complex in *B. fragilis* strain BE1. *B. vulgatus*, grown under similar conditions, does not produce the same range of OMP; in particular, it lacks the 44-kDa OMP (Otto *et al.*, 1988). This may be a key factor in the predominance of *B. fragilis* over *B. vulgatus* in clinical infections (**Table 2**). An iron-repressible OMP of approximately 40 kDa associated with increased binding of Congo red has also been described in the type strain *B. fragilis* NCTC 9343 (Larkin *et al.*, 1988). The presence of specific antibody to the 44-kDa OMP in experimentally infected rats and also in sera from patients with *B. fragilis* infection, indicates that it is produced *in vivo* (Otto *et al.*, 1991). There was no detectable response in patients infected with other *Bacteroides* spp., such as *B. ovatus*, *B. distasonis* or *B. vulgatus*. The gene encoding the 44-kDa protein, *hupA*, from *B. fragilis* strain BE1 has been cloned and sequenced. The deduced amino acid sequence indicates that there are haem-binding motifs characteristic of haem lyase enzymes, but little similarity with other putative bacterial haem receptors (Otto *et al.*, 1996). There is no evidence that *Bacteroides* spp. produce their own iron-chelating agents such as the siderophores found in other bacteria, but the possibility cannot be ruled out. A comparison of the OMP profiles of *B. fragilis* growing *in vivo* and *in vitro* did not, however, reveal the induction of the high-molecular-mass OMP which is characteristically produced by *E. coli* grown under iron limitation (Patrick and Lutton, 1990b) and are associated with siderophore uptake.

Oxygen Sensitivity and Redox Potential

Bacteroides spp. are obligate anaerobes that, for their isolation and cultivation, require reducing conditions and, therefore, the absence of dissolved oxygen in media. To achieve this, media are usually boiled to drive off dissolved oxygen and a reducing agent, such as cysteine, is included. A redox potential of less than −42 mV at pH 7, as monitored by the redox dye resazurin, is suitable for the isolation and growth of *Bacteroides* spp. (Poxton *et al.*, 1989; Morris, 1991). Pure cultures of some strains will, however, grow in oxygen-free defined medium at an initial redox potential of +100 mV (Goldner *et al.*, 1993). Separate from this need for reducing conditions, oxygen may be directly toxic to *Bacteroides* spp., but this sensitivity is variable and some survive exposure to oxygen for quite long periods. The relative aerotolerance of some *Bacteroides* spp. probably relates to

the production of superoxide dismutase and catalase. *B. fragilis* and *B. distasonis* produce catalase, but only some strains of *B. thetaiotaomicron*, *B. ovatus* and *B. eggerthi* do so, and *B. vulgatus* and *B. uniformis* are catalase negative. It has been proposed that the relative aerotolerance of *B. fragilis* contributes to its virulence, and recent clinical isolates have been reported as more aerotolerant than faecal isolates. The catalase of *B. fragilis*, KatB, is induced in late exponential phase cultures and in cultures exposed to oxidative stress when moved from anaerobic to aerobic incubation conditions. The catalase is a haemoprotein, composed of two identical subunits, each with an approximate molecular mass of 60 kDa. The catalase of *B. thetaiotaomicron* is similar but has a molecular mass of approximately 250 kDa. Cloning and sequencing of the *katB* gene of *B. fragilis* 638R has indicated amino acid sequence similarities with Gram-positive bacterial and mammalian catalases. This may reflect the taxonomic position of the *Bacteroides–Flavobacterium* phylum, but interestingly the *B. fragilis* catalase shares 71% amino acid similarity and 66% nucleotide identity with the catalase of *H. influenzae* (HktE). It has been suggested that this reflects transfer of the catalase gene into both *B. fragilis* and *H. influenzae* from a common source in the evolutionary past (Rocha and Smith, 1995). A strain with a mutation in the *katB* gene survived as well as the parent strain when exposed to oxygen, but it was more sensitive to the toxic effects of hydrogen peroxide (Rocha *et al.*, 1996). Apart from catalase, oxidative stress or exposure to hydrogen peroxide induces a further 28 proteins ranging in molecular mass from 12 to 79 kDa. If protein synthesis is inhibited by chloramphenicol, the survival of *B. fragilis* under aerobic conditions is limited (Rocha *et al.*, 1996). A number of genes are, therefore, associated with survival to oxidative stress. Some of these may also be involved in the ability of *B. fragilis* to penetrate HeLa cells (Goldner *et al.*, 1993) and in observed changes to the fermentation pathways after growth under oxidising conditions (Goldner *et al.*, 1997). Another mutant of *B. fragilis* 638R, generated by transposon mutagenesis, grows poorly in association with murine fibroblast or Chinese hamster ovary (CHO) cell line when compared with the parent strain (Tang *et al.*, 1999). This mutant was also less tolerant of exposure to oxygen than the parent strain. A 6.6 kbp fragment complemented the mutant, and sequencing of this region revealed five open reading frames which have been designated the *Bacteroides* aerotolerance operon (bat A–E). The precise function of the proteins is unknown, but they may form a membrane-associated complex involved in the generation or export of components necessary for aerotolerance.

Products

Degradative Enzymes and Outer-membrane Vesicles

The release from *Bacteroides* spp. of extracellular enzymes with the potential to degrade components of the host has been long recognised (Rudek and Haque, 1976; von Nicolai *et al.*, 1983) and, indeed, a scheme for the rapid identification of *Bacteroides* spp. based on enzymatic activities has been proposed (Hofstad, 1980). Enzymes so far described include those capable of degrading components of the host extracellular matrix, host cells and tissue such as hyaluronidase, chondroitin sulphatase, fibrinolysin, DNAase, lipases, proteases and neuraminidase. These may play a role in virulence, but in many cases their precise contribution is not proven. It may be that some of these enzymatic activities are related to the growth of *Bacteroides* spp. in the intestinal environment as part of the normal microbiota, rather than for virulence.

Proteolytic activity that may be related to either or both of virulence and survival in the intestine has been reported in *B. fragilis*. Arylamidases for the amino acids leucine, valyl-alanine and glycyl-proline are periplasmic or outer-membrane-associated in exponential phase of batch cultures. Proteolytic activity increased in the culture medium in stationary phase and was either in a soluble form or associated with particulate matter (Macfarlane *et al.*, 1992a, b). It is probable that the particle-associated enzymatic activity relates to the release of outer membrane vesicles (see below). These proteases are active against casein and gelatin (collagen type IV), but not elastin, collagen types I, II, III, VI,

ovalbumin or bovine serum albumin. A protease with similar hydrolytic activity for valyl-alanine and glycyl-proline hydrolyses fibrinogen, casein and gelatin (Chen *et al.*, 1995). The purified enzyme has a molecular mass of approximately 100 kDa and the characteristics of a serine-thiol-like protease. It completely hydrolysed the α chain of fibrinogen. When the fibrinolytic activity of crude extracts of *B. fragilis*, *B. ovatus*, *B. eggerthii*, *B. uniformis* and *B. thetaiotaomicron* were compared, *B. fragilis* had the greatest activity. Since soluble fibrinogen is converted to insoluble fibrin as part of the normal blood coagulation and wound healing, this enzyme has the potential to slow clot formation, and this may play a role in virulence.

Similarly, a clear distinction between glycosidase activities of *Bacteroides* spp. that may play a role in virulence, and those that relate to degradation of intestinal components, has yet to be made. A range of glycosidases has been identified in *B. fragilis* (**Table 6**). For example, an endo-β-galactosidase releases oligosaccharides from the human blood group O erythrocytes (Scudder *et al.*, 1987), whereas some glycosidases degrade the carbohydrate moieties of gastric mucin. The latter may, therefore, be involved in survival within the intestine (Macfarlane and Gibson, 1991). Glycosidase, esterase, lipase (Patrick *et al.*, 1996) and neuraminidase (Domingues *et al.*, 1997) activities are associated with purified outer-membrane vesicles of *B. fragilis*. In culture, *B. fragilis* produces large amounts of single membrane-bound outer-membrane vesicles. These can be observed, by electron microscopy of ultrathin sections, and are released from the outer membrane (**Fig. 1b**). Immunogold labelling

Table 6 Major bacteria of the adult human faecal microbiota

Bacteria	Gram reaction	Morphology	Total viable count[a] (per g of faeces)
Bacteroides	−	Rod	10^9–10^{14}
Eubacterium	+	Rod	10^5–10^{13}
Bifidobacterium	+	Rod	10^5–10^{13}
Clostridium	+	Rod	10^3–10^{13}
Lactobacillus	+	Rod	10^4–10^{13}
Peptostreptococcus	+	Coccus	10^4–10^{13}
Ruminococcus	+	Coccus	10^5–10^{13}
Streptococcus	+	Coccus	10^7–10^{12}
Methanobrevibacter	+	Cocco–bacillus	10^7–10^{11}
Desulfovibrio	−	Rod	10^5–10^{11}
Fusobacterium	−	Rod	10^9
Enterococcus	+	Coccus	10^7
Escherichia coli	−	Rod	10^7
Prevotella/Porphyromonas	−	Rod	10^4

[a] Compiled from Willis (1991) and Macfarlane and Gibson (1994).

indicates that these vesicles carry epitopes associated with the outer membrane (**Fig. 1c**). Purified outer-membrane vesicles cause erythrocytes to agglutinate (Patrick *et al.*, 1996). It is, therefore, interesting to speculate that this targets, and also possibly limits, the destructive enzymatic activity to particular host cell types. It may also partly explain why *B. fragilis* does not act as a progressive rapidly acting histolytic pathogen, like other bacteria that release enzymes, such as *Clostridium perfringens*.

Polysaccharide-degrading enzymes of *B. thetaiotaomicron* have been studied in relation to their potential role in the intestinal environment. *B. thetaiotaomicron* produces two enzymes capable of degrading chondroitin sulphate and related mucopolysaccharides into disaccharides. These two enzymes have a similar molecular mass of 104–108 kDa, but they differ in their affinity for heparin. The gene for one of these chondroitin lyase enzymes has been cloned in *E. coli* (Guthrie *et al.*, 1985). Studies of starch breakdown by *B. thetaiotaomicron* indicate that the enzymes involved are not extracellular and that initial binding of the starch to an OMP complex and translocation into the periplasmic space is necessary. An operon of five genes that encode OMPs (*susC–G*) is involved in starch utilisation. At least four of the OMPs are involved in binding of starch (Reeves *et al.*, 1997).

Neuraminidases, which cleave sialic acid from oligosaccharides on host cell glycoproteins and glycolipids, may have a more subtle effect on the host and could play a key role in virulence. More than 20 naturally occurring sialic acids are known; they are formed by various substitutions and additions to neuraminic acid (Schauer, 1985). There is growing evidence that these sugar residues are involved in the biological activities of the host cells and molecules (Rademacher *et al.*, 1988). A prime example of the role of sialic acid in normal host function is its role in the movement of lymphocytes (Imai *et al.*, 1991). Microbial neuraminidases may subvert the normal function of the immune system or other systems of the host (Patrick and Larkin, 1995). The neuraminidase gene of *B. fragilis* 638R, *nanH*, has been cloned. A mutant, produced by an insertion in *nanH*, was able to revert to *nanH*[+], whereas another in which part of the gene had been deleted could not. The mutant capable of reversion grew as well as the parent strain in association with CHO cells and in a rat granuloma pouch model of infection, and reverted to *nanH*[+]. Glucose limitation favoured reversion to *nanH*[+]. The deletion mutant also grew as well as the parent strain with the CHO cell line, but grew more slowly in the rat model of infection (Godoy *et al.*, 1993). These results indicate that neuraminidase may play a role in virulence.

Enterotoxin

Enterotoxigenic *B. fragilis* were first described as a cause of acute watery diarrhoeal disease in newborn lambs (Myers *et al.*, 1984). Subsequently they were also implicated as a cause of diarrhoea in calves (Border *et al.*, 1985) and humans –particularly children (Myers *et al.*, 1987). The toxin can be detected in culture supernatants by its stimulation of secretion in ligated intestinal loops in animals and alteration of the morphology of the colonic epithelial cell line HT29/C$_1$ (Weikel *et al.*, 1992). The enterotoxin is a zinc-dependent metalloprotease of approximately 20 kDa. Purified toxin undergoes spontaneous breakdown by autodigestion at 37°C. The enterotoxin is active against type IV collagen (gelatin), actin, tropomyosin and fibrinogen (Moncrief *et al.*, 1995). Three isoforms of the protein have been identified. BFT-1 and BFT-2 have 92% amino acid sequence similarity. The third isoform, Korea-BFT, was identified in extra-intestinal isolates obtained in Korea and was related to BFT-2 (Chung *et al.*, 1999). The enterotoxin gene is contained in a small genetic element, termed 'pathogenicity islet', of about 6 kpb, which is bounded by 12 bp nearly perfect direct repeats. In non-toxigenic strains, a putative chromosomal integration site has been identified in a 17 bp G + C-rich region. The toxin gene encodes a 44 kDa pre-pro-toxin. It is thought that removal of a signal peptide produces a 22 kDa pro-toxin which is then cleaved to produce the 20 kDa protease. A second ORF in the pathogenicity islet that also encodes a predicted metalloprotease was identified and has a deduced amino acid sequence of 28.5% identity to the enterotoxin (Moncrief *et al.*, 1998).

In spite of the association of enterotoxin-producing strains of *B. fragilis* with diarrhoea, its importance as a cause of diarrhoea as compared with other bacterial types is a matter for debate. A study in Italy indicates that the carriage rate of enterotoxigenic *B. fragilis* in healthy adults is 15% and in adults with diarrhoea it is only 9.4%. In children it was identified in 17% of children with diarrhoea, but also in 12% of healthy children (Pantosti *et al.*, 1997). Polymerase chain reaction (PCR) assays based on the enterotoxin gene have been developed and correlate reliably with enterotoxin production in isolates (Kato *et al.*, 1996; Leszczynski *et al.*, 1997). Of 188 *B. fragilis* not associated with diarrhoea isolated in a Japanese clinical laboratory 18.6% carried the enterotoxin gene. Included in these were 64 bacteraemia isolates, of which approximately 28% carried the enterotoxin gene (Kato *et al.*, 1996). The precise role of the enterotoxin in *B. fragilis* virulence therefore remains unclear.

Bacteroides as Part of the Resident Microbiota

In normal humans *Bacteroides* spp. may colonise a number of sites, where generally their interaction with the host is considered to be either benign or potentially beneficial. Comprehensive information pertaining to aspects of the normal microbiota of humans and animals can be found in the books by Gibson (1994) and Mackie *et al*. (1997).

Although *Bacteroides* spp., including *B. fragilis* (Leszczynski *et al*., 1997), may be isolated from the female genital tract and also transiently from skin in the anal region, the major reservoir for *Bacteroides* spp. in the human body is the large intestine (Drasar and Duerden, 1991). It is estimated that in humans there are at least 10^{11} bacteria per gram of faeces, with probably as many as 500 different bacterial types, and many of these remain to be cultured. The application of nucleic-acid-based analysis of the human microbiota, such as 16S rRNA-specific oligonucleotide probes, in combination with the detection of expression of particular components, will undoubtedly provide a comprehensive picture of the complexity of the ecology of the normal microbiota which is currently not possible with cultural techniques (Raskin *et al*., 1997).

By far the predominant group that can be isolated from faeces are *Bacteroides* spp. of the '*B. fragilis* group', with an estimated average of 10^{11}/g of faeces. The prevalence of different bacterial types is presented in **Table 6**. *Bacteroides* spp. can be isolated from infants in the first day of life, but they do not begin to predominate in numbers until the introduction of solid foods (Conway, 1997). The relative incidence of different *Bacteroides* spp. in the faecal microbiota does not reflect the frequency with which these species are isolated from clinical infections, such as intra-abdominal abscesses (**Table 2**). *B. fragilis* is the predominant clinical isolate, whereas *B. vulgatus* may account for 40% or more of *Bacteroides* in faeces. This has led to the assumption that *B. fragilis* has determinants of virulence lacking in *B. vulgatus*, which may be the case, but in their virulence studies few workers have compared *B. vulgatus* with *B. fragilis*. Only a few studies of the mucosa-associated microbiota of the human intestine have been carried out, probably because of the difficulty in obtaining such samples, in particular from healthy individuals. Those that have been done indicate that the mucosa-adherent *Bacteroides* spp. distribution may differ considerably from that in faeces, with *B. fragilis* found in greater numbers in faeces than *B. vulgatus* (Namavar *et al*., 1989; Poxton *et al*., 1997).

Given that there are estimated to be more bacterial cells in the normal microbiota than there are mammalian cells in the human, it is highly likely that the metabolic activity of these bacteria has a major impact on the host. It has been suggested that in terms of metabolic activity and biochemical transformations, the human large intestine rivals the liver (Macfarlane and Gibson, 1994). In the gut *Bacteroides* spp. degrade ingested material, in particular heterologous polysaccharides in plant material, that the mammalian system is incapable of degrading. This process results in the production of short-chain fatty acids, which, once absorbed, are potential substrates for energy metabolism in the intestinal mucosal epithelium. There is also evidence that bacteria in the intestine produce vitamins that can be used by the host (Drasar and Duerden, 1991). Other positive interactions of the host with the microbiota include, development of the immune system, which does not reach its full repertoire in 'germ-free' animals, and colonisation resistance to enteric pathogens provided by the resident microbiota. A detailed account of current knowledge of the complex relationship between the mucosal immune system and the intestinal microbiota is provided by Gaskins (1997).

Negative aspects of the resident microbiota include the possibility that bacterial metabolism in the intestine produces potential carcinogens (Gibson and Macfarlane, 1994). The role of anaerobic bacteria, such as *Bacteroides*, in chronic inflammatory bowel disease (IBD) of humans is controversial. There are two major types of IBD: ulcerative colitis, mainly of the colon, and Crohn's disease, which affects both the small and large intestine. A range of bacteria, including *B. vulgatus*, has been suggested as playing a role in Crohn's disease. Since the lesions in Crohn's disease are granulomatous, attention has been focused on possible mycobacterial involvement, in particular *M. paratuberculosis*. A similar situation arises with ulcerative colitis where a conclusive relationship with one type of bacterium has yet to be made. Evidence for anaerobe involvement includes the response of the condition to treatment with metronidazole (Gibson and Macfarlane, 1994). An allele of the major histocompatability complex (MHC) class I, HLA-B27, is associated with ankylosing spondylitis and reactive arthritis, and there is evidence of a relationship between patients with ankylosing spondylitis and IBD. Rats transgenic for human HLA-B27 spontaneously develop colitis, followed by arthritis and spondylitis. This does not occur, however, if the rats are raised under germ-free conditions. *B. vulgatus* was found to be a critical component of the microbiota introduced into germ-free rats for the induction of colitis (Rath *et al*., 1996,

1999). This strongly suggests that at least one of the members of the resident microbiota could play a key role in IBD.

Although few molecular studies have been carried out concerning the fine detail of molecular interaction between the normal microbiota and humans, studies of *B. thetaiotaomicron* hint at the potential for intimate interactions and bacterial adaptation to different ecological niches along the length of the intestine. In the developing intestinal epithelium, multipotent stem cells proliferate in the crypts of Lieberkühn, and several crypts 'supply' cells for each villus. The stem cells differentiate into Paneth cells (antimicrobial-peptide-producing cells) as they migrate to the base of the crypt, and into absorptive enterocytes, enteroendocrine cells and mucus-producing goblet cells as they move up to the apex of the villus. During this differentiation, alterations occur in the composition of cell surface oligosaccharides (glycoconjugates) which can be related to the stage of differentiation of different cell types. In the mouse ileal epithelium differences in the degree of fucosylation of the glycoconjugates can be observed with the age of the germ-free mice. At 21 days, Paneth cell glycoconjugates became fucosylated in some crypts and this gradually spreads by 28 days to Paneth cells in all crypts. In contrast, when mice are raised with a conventional microbiota, not only Paneth cells, but also enterocyte and goblet-cell lineages became fucosylated. Similar development of fucosylation was noted if the germ-free mice were colonised with the normal microbiota at a later date. For example, 70-day-old germ-free mice developed fucosylation of some villi 7 days after introduction of the normal microbiota and full fucosylation 14 days later. Colonisation of the germ-free mice with a pure culture of

B. thetaiotaomicron also restored full glycosylation, whereas an isogenic transposon insertion mutant incapable of using L-fucose as a carbon source did not. There was no difference between fucosylation of large intestinal epithelium of germ-free and colonised mice, which were both fucose positive (Bry *et al.*, 1996). Since the major site of colonisation by *Bacteroides* is the large intestine, rather than the small intestine (Drasar and Duerden, 1991; Wilson, 1997), this appears to represent specific adaptation to the small-intestinal environment. The suggestion is that the bacterium induces fucosylation of the villus glycoconjugates, then cleaves the fucose with secreted α-fucosidase, which *B. thetaiotaomicron* are known to produce, and uses the fucose as a source of carbon and energy. Further studies of *B. thetaiotaomicron* have identified FucR, a molecular sensor of L-fucose concentration, that represses the genes in the fucose utilisation operon. FucR binds fucose and it is thought that this reduces its interaction with the promoter for the L-fucose utilisation operon (**Fig. 8**). This is very different from the control of the fucose-utilisation gene cluster in *E. coli* which is regulated by an activator. It is not known how *B. thetaiotaomicron* causes the epithelial cells to become fucosylated, but it is postulated that there is a separate 'control of signal production' locus which is also regulated by FucR (Hooper *et al.*, 1999).

It is possible that in future the normal microbiota will no longer be seen as simple colonisers but more as an integral part of the mammalian host, inextricably linked with the normal development and functioning of the eukaryotic component. A better understanding of the normal microbiota should also allow a better understanding of the problems that arise when these

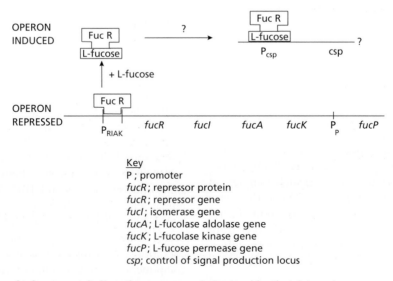

Fig. 8 Fucose induction of L-fucose metabolic pathway enzymes in *Bacteroides thetaiotaomicron*.

organisms cause opportunistic infection at body sites that are not normally colonised.

Bacteroides as Pathogens

Pathogenesis by *Bacteroides* results from opportunistic infection by the normal microbiota, usually manifested by abscess formation or soft-tissue infection, which may lead to septicaemia. The possible involvement of large-scale leakage of *Bacteroides* from the intestine, in conditions such as systemic inflammatory response syndrome, has recently been examined. Enterotoxigenic *B. fragilis* associated with diarrhoea have been described (see above). The possible role of *Bacteroides* in the gastrointestinal microbiota in chronic inflammation of the bowel was also considered above. In this section, attachment mechanisms, avoidance of the host defences and immunomodulatory activities are considered. Most studies have centred on *B. fragilis*, since it is the most common clinical isolate, and the potential virulence determinants of *B. fragilis* are summarised in **Table 7**.

Attachment Mechanisms

B. fragilis can attach to host cells and components of the extracellular matrix. Studies of the attachment of *B. fragilis* to host cells have yielded apparently conflicting results. Polysaccharides (Riley and Mee, 1984;

Vel *et al.*, 1986; Patrick *et al.*, 1996) and fimbriae (Pruzzo *et al.*, 1984) have been reported to be involved in attachment. It appears that *B. fragilis* may produce more than one type of ligand to mediate host cell attachment, and that expression of these ligands is subject to within-strain variation. *B. fragilis* populations enriched for the EDL (see above) cause haemagglutination and this is reduced by pre-treatment of the bacteria with sodium periodate, which suggests that saccharides are involved. Populations enriched for the LC do not cause haemagglutination, and it is possible to mix different proportions of these populations and correlate the degree of haemagglutination with the proportion of EDL bacteria present (Patrick *et al.*, 1988). Since recent clinical isolates vary considerably in the proportion of bacteria that express the LC, it is not surprising that there is confusion in the literature. Furthermore, the EDL population releases extracellular vesicles (see above and **Fig. 1**) which by themselves will cause haemagglutination (Patrick *et al.*, 1996) and also extracellular polysaccharide (**Fig. 3**). If quantities of vesicles or polysaccharides are present, they may block attachment sites on tissue cell lines and therefore exclude bacteria, and in assays that rely on quantification of the attached bacteria, erroneously negative results may be obtained. In assessing attachment potential it is therefore essential to have bacterial populations that are clearly defined for surface structure expression and the presence of vesicles or polysaccharide. Despite the confusion in identifying the structures involved in attachment, it is clear that

Table 7 Potential virulence determinants of *B. fragilis*

Characteristic	Bacterial component/gene[a] implicated
Attachment to host cells and extracellular matrix	Surface polysaccharides
	Fimbriae
Aerotolerance	Catalase (*KatB*)
Haem-binding	44-kDa OMP (*HupA*)
Tissue breakdown	
Casein hydrolysis	Protease
Gelatin hydrolysis	Protease
Fibrinogen hydrolysis	100-kDa serine-thiol-like protease
Degradation of extracellular matrix/oligosaccharides	Glycosidases, esterase lipase, neuraminidase (*nanH*)
Enterotoxicity/diarrhoea	20 kDa metalloprotease
Endotoxicity/immunomodulation	LPS
Avoidance of host defence	(Antigenically variable surface polysaccharides?)[b]
Resistance to phagocytic uptake and killing	LC
	(Outer membrane vesicles/extracellular polysaccharide?)
Resistance to complement-mediated killing	(Electron dense layer/extracellular polysaccharide/ outer membrane vesicles?)
Abscess formation	Extracellular polysaccharide

[a] If identified.
[b] Brackets indicate speculative involvement.

B. fragilis can attach to host cells. This may relate to both the virulence of *B. fragilis* and to its ability to colonise the intestine as part of the normal microbiota. In a comparative study of a number of isolates from faeces, abscesses and blood culture, in which the attachment ligand was not characterised, strains from all three sources caused haemagglutination and attached to human cheek epithelial cells, the intestinal cell line Intestine 407 and human polymorphonuclear leucocytes (PMNL) (Guzman *et al.*, 1997). Sialic-acid-containing glycoproteins have been implicated as potential adhesin receptors, since sialic acid and compounds that contain this sugar inhibit haemagglutination. Treatment of erythrocytes with sialidase, however, increases haemagglutination in some strains (Domingues *et al.*, 1992). This suggests that there is more than one attachment ligand in *B. fragilis* and also more than one cell receptor. There is evidence that, when grown under less reduced conditions, *B. fragilis* can penetrate tissue culture cells (Goldner *et al.*, 1993). The mechanism of penetration is unknown, but clearly this may be important for the virulence of *B. fragilis*, particularly in bacteraemia where the route of spread from the initial focus of infection to the bloodstream is unclear.

Studies of the attachment to components of the extracellular matrix are subject to the same criticism – a lack of appreciation of variability within strains, and the potential attachment of bacterial extracellular components. There is, however, clear evidence that *Bacteroides* can attach to some components of the human extracellular matrix. Binding by *B. fragilis* and *B. vulgatus* to fibronectin, collagen type I and vitro-nectin has been demonstrated. The binding was heat- and protease-sensitive, suggesting the involvement of a bacterial protein ligand, but fimbriae were not observed by electron microscopy (Szoke *et al.*, 1996). *B. fragilis* also binds to laminin, a major structural component of basement membranes (Eiring *et al.*, 1995).

Avoidance of the Host Defences

Resistance to complement-mediated cell lysis by *B. fragilis* in normal human serum can be observed in bacteria non-capsulate by light microscopy but with an EDL adjacent to the outer membrane (**Fig. 2**). The relationship between encapsulation and complement resistance is not clear, since a capsulate strain may be sensitive to complement, whereas non-capsulate populations of the same strain are resistant (Reid and Patrick, 1984; Allan and Poxton, 1994). Other *Bacteroides* spp., including *B. thetaiotaomicron*, *B. caccae*, *B. ovatus* and *B. vulgatus*, may also resist killing by

serum (Bjornson, 1987; Allan and Poxton, 1994), but the proportion of bacteria that resist killing by serum can vary with the growth conditions (Allan and Poxton, 1994). Clearly, different *Bacteroides* spp. can potentially survive killing by the alternative complement pathway. Comprehensive studies of the opsonisation of *B. fragilis* with isolated components of the alternative pathway, indicate that C3 deposition occurred with C3, properdin, factors H, I, B and D. This indicates that resistance is not due to a lack of initial activation of the complement cascade, and may be due to interference with the regulation of cascade at another point in the pathway, as is the case with other bacteria that resist complement-mediated lysis (Patrick and Larkin, 1995).

Populations of *Bacteroides* enriched for the large capsule are not phagocytosed by isolated PMNL in normal human serum, but those that are non-capsulate are phagocytosed (Reid and Patrick, 1984). Though this appears to confirm the well-recognised potential antiphagocytic properties of the capsules, its relevance to infection can be questioned in the light of experiments *in vivo*. Bacteria enriched for the LC and grown *in vivo* in a chamber implanted in the mouse peritoneal cavity, not only no longer produce the LC, but on subculture they are enriched for a non-capsulate population (Patrick *et al.*, 1984). This is observable both in the presence and absence of apparently active mouse PMNL which contain phagocytosed bacteria (Patrick, 1988; Patrick *et al.*, 1995a). The small capsule-enriched population and non-capsulate bacteria do not alter capsule expression during growth *in vivo* and survive equally well in the implanted chambers. Interestingly, the influx of PMNL has no impact on the total viable count of the bacteria in the mouse peritoneal cavity model over periods of 20 days or more. Similarly, in a mouse peritoneal abscess model induced by injection of a mixture of *B. fragilis* and *E. coli* with bran, bacteria remain viable for 10 weeks or more, despite the infiltration of PMNL (Kocher *et al.*, 1996). In these abscesses, high levels of CP-10, a murine chemo-attractant which recruits neutrophils and monocytes–macrophages, and migration inhibition factor-related protein (MRP-14) which retains the recruited cells were detected. These high levels were measurable in the early acute stages of abscess formation and in chronic abscesses after 30–55 days. The release of factors such as succinic acid by *B. fragilis*, which inhibit PMNL chemotaxis (Rotstein *et al.*, 1989b), has been suggested to play a role in virulence, but this appears to be contrary to the evidence for high levels of neutrophils in models of abscess infection. Pathogenic synergy, whereby *B. fragilis* protects and enhances the survival of other facultatively anaerobic bacteria, such as *E. coli*, has been recognised (Rotstein *et al.*, 1989a). Interestingly, abscess-derived neutrophils harbour viable bacteria and are less efficient at killing bacteria than neutrophils derived from peritoneal aspirates or peripheral blood *in vitro* (Finlay-Jones *et al.*, 1991). It is highly likely that factors other than capsules are involved in intraperitoneal survival of *B. fragilis* (Patrick *et al.*, 1995a). The potential for excreted extracellular polysaccharide and extracellular vesicles to mop up opsonins, activate complement and interact with phagocytes, to divert host defences from bacteria, should not be overlooked.

The primary opsonising molecules for attachment to PMNL have been identified as C3b and iC3b, which interact with CR3 and CR1. Purified IgM from normal humans doubles the deposition of opsonising C3 on *B. fragilis*, which suggests a potential role for IgM in alternative complement pathway activation for opsonisation of *B. fragilis* (Foreman and Bjornson, 1994). Opsonisation of *B. thetaiotaomicron* with pentameric, but not monomeric, IgM and complement components renders the capsule visible by electron microscopy after paraformaldehyde and glutaraldehyde fixation (Bjornson and Detmers, 1995). This was interpreted as evidence that the opsonins modified the capsular structure, and prevented it from being destroyed during processing for electron microscopy. The inclusion of osmium tetroxide and ruthenium red is necessary to make *Bacteroides* capsules sufficiently electron dense for observation by electron microscopy (Patrick *et al.*, 1986). Fixation with paraformaldehyde and glutaraldehyde, although necessary to retain sufficient antigenicity for immunolabelling, does not resolve the capsule even if it is present (Reid *et al.*, 1987). A more probable explanation is that the opsonising proteins render the capsule sufficiently electron dense to be visible with the electron microscope. Nevertheless, pentameric IgM enhances alternative complement pathway opsonisation of *B. thetaiotaomicron* and its subsequent adherence to PMNL.

When the extracted capsular complex of *B. fragilis* NCTC 9343, which contains PSA and PSB, is injected into the rat peritoneal cavity along with sterile caecal contents, abscesses are formed in approximately 80% of rats. The presence of both carboxyl (negative charge) and amino (positive charge) groups on the PSA is required for abscess induction in this model (Tzianabos *et al.*, 1993), and a 5-week regime of subcutaneous inoculation with the polysaccharide protects against abscess formation in at least 70% of the rats. The presence of the positive and negatively charged groups is essential for this protection. Interestingly, inoculation with *Strep. pneumoniae* capsule type 1 protects against challenge with the *B. fragilis* PS

and vice versa, and indicates that this phenomenon is not restricted to *B. fragilis* (Tzianabos *et al.*, 1994). Immunity to challenge with bacteria and polysaccharide can be passively transferred by splenic lymphocytes, and T cells appear to be involved in the process (e.g. Tzianabos *et al.*, 1999). The precise role of individual T-cell types is unclear, but CD4+8+, CD4+ and CD8+ T lymphocytes may all be involved. Early reports indicated a role for CD4+8+ in abscess induction and CD8+ T cells in protection (Crabb *et al.*, 1990). More recent reports implicate CD4+ T cells in protection against abscess formation (Tzianabos *et al.*, 1999) and also conversely in an increase in abscess formation (Sawyer *et al.*, 1995). A serum IgG response to *B. fragilis* protein and polysaccharide surface antigens is detectable in a mouse chamber model of intra-peritoneal infection (Patrick *et al.*, 1995a) but it does not alter the numbers of viable bacteria in the chambers. Although the humoral immune response may not affect abscess formation, it may be an important limiting factor in bacteraemia and the systemic inflammatory response syndrome (see below). Interestingly, IgA specific for *Bacteroides* has also been detected in whole-gut lavage fluid (Poxton *et al.*, 1995). Perhaps a clearer understanding of how the *Bacteroides* spp. in the normal microbiota interact with the immune system will help to resolve the problem of the clonal immune recognition of *Bacteroides*.

Immunomodulation and Systemic Inflammatory Response Syndrome

Systemic inflammatory response Syndrome (SIRS) or 'sepsis' is the most common cause of death in patients who are critically ill in intensive care units. LPS is known to play a central role. In patients who do not have obvious septicaemia, bacteria or bacterial components appear to translocate across the ischaemic intestinal wall, and the normal intestinal microbiota are thought to be the source of the endotoxin. This sets in motion a cascade of events involving cytokines, complement and clotting factors; essentially, nonclonal immune recognition events occur. The extent of the immune response is concentration-dependent; high concentrations of bacterial immunomodulatory molecules result in what has been termed immunological 'panic'. This may ultimately result in shock, multiple organ failure and a mortality of up to 90%. It has generally been considered that the major cause of Gram-negative SIRS is the LPS of enterobacteria, but the possible role of immunopotentiation by components of the *B. fragilis* cell envelope, such as the LPS, has not been much considered in the literature. This is largely because of the low toxicity of *B. fragilis* LPS in mouse lethality studies as compared with *E. coli* LPS. *Bacteroides* LPS is reported to be up to 10 000-fold less active than that of *E. coli* (Lindberg *et al.*, 1990). More recent investigations indicate that the extraction procedure used to prepare *Bacteroides* LPS is critical in studies of its immunomodulatory activity. Although *B. fragilis* LPS is 5000-fold less toxic in a mouse lethality model, it was 7-fold more active than phenol–water extracted *E. coli* LPS in the *Limulus* amoebocyte assay. It was also able to induce tumour necrosis factor at levels similar to *E. coli* LPS, but independently of CD14 expression. Given the high numbers of *Bacteroides* spp. in the intestinal tract, where enterobacteria may be outnumbered by up to 1000 to 1, and even allowing for the lower toxicity, it is likely that the quantitative dominance of *Bacteroides* LPS plays a role in endotoxic shock and SIRS (Delahooke *et al.*, 1995a, b). A decrease in the level of serum IgG specific for enterobacterial LPS correlates well with endotoxaemia; non-survivors of sepsis had low levels of enterobacteria-specific IgG. In a small study of 6 survivors and 6 non-survivors of sepsis, levels of IgG specific for *B. fragilis* endotoxin fluctuated more than in the healthy controls, but it is difficult to draw conclusions from such small numbers of patients; exposure to *Bacteroides* LPS had occurred in these patients (Allan *et al.*, 1995). A study of 55 patients undergoing cardiac surgery provided evidence for a decrease in anti-*B. fragilis* immunoglobulin levels during surgery, which indicates that the patients were exposed to *B. fragilis* endotoxin (Bennett-Guerrero *et al.*, 2000). In the light of these results, the potential role of *Bacteroides* in SIRS warrants further consideration.

Diagnosis

The diagnosis of *Bacteroides* and related bacteria depends largely on classical cultural methods, and in the clinical setting is usually taken no further than the genus level. For details of clinical diagnostic methods the reader is referred to the *Wadsworth Anaerobic Bacteriology Manual* (Summanen *et al.*, 1993), Brown *et al.* (1989), Wren (1991), and Jousimies-Somer *et al.* (1995). *Bacteroides* spp. can be detected in pus by direct immunofluorescence microscopy, providing the target antigen is carefully chosen and not antigenically variable (Patrick *et al.*, 1995b).

Molecular detection methods for *Bacteroides* are not as yet in general use in clinical diagnostic laboratories. PCR finger printing identification of pure culture isolates has been examined and resulted in good identification and differentiation of members of the fragilis group (Claros *et al.*, 1997). PCR-based assays for the detection of enterotoxigenic *B. fragilis* have proved reliable (e.g. Leszczynski *et al.*, 1997). Fluorescence *in situ* hybridisation (FISH) of 16S rRNA is possible with pure cultures of *B. fragilis*, but such techniques must be approached with caution when applied directly to clinical samples, as the penetration of the probe into the bacterial cell depends on the nature of encapsulating surface structures (Ramage *et al.*, 1998). Reliable direct identification of *Bacteroides* spp. and other obligate anaerobes in clinical samples is attractive, because in spite of efforts to ensure the handling and transit of samples under anaerobic conditions, a high proportion of *Bacteroides* infections may be missed because of loss of viability (Patrick *et al.*, 1995b).

Antimicrobial Agents

Antibiotics are used for the treatment of anaerobic infections and as prophylaxis for patients undergoing intestinal surgery. *Bacteroides* are intrinsically resistant to aminoglycoside antibiotics, as they are unable to transport these antibiotics. This is exploited for the selection of *Bacteroides* spp. from clinical samples that contain enterobacteria. Current understanding of the genetic systems of *Bacteroides* has come largely from studies of antibiotic resistance.

Metronidazole

Since the 1970s (e.g. Willis *et al.*, 1976) metronidazole has frequently been used for prophylaxis and for the treatment of non-clostridial anaerobic infection. Resistance to metronidazole in anaerobic bacteria is extremely low, and it has excellent tissue penetrability (Freeman *et al.*, 1997). The activity of metronidazole is restricted to anaerobes and organisms during anaerobic metabolism, as reduction of the nitro group of the drug is necessary for its antibacterial action. Reduction of the drug results in the production of radicals with the potential to accept electrons from DNA, resulting in strand breaks and helix destabilisation. In the presence of oxygen, DNA accepts electrons from the reduced drug and converts it back to the oxidised form, in a futile cycle (Edwards, 1997). False resistance to metronidazole is sometimes reported as

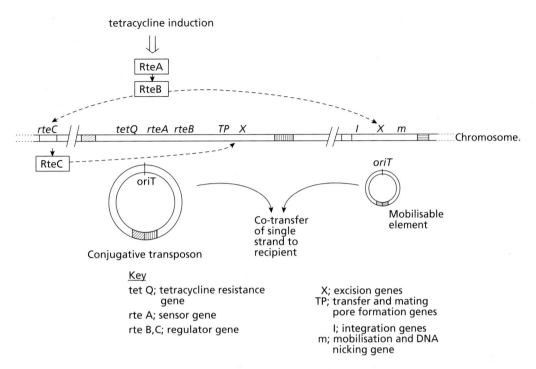

Fig. 9 Mechanism of tetracycline-inducible transfer of conjugative transposons and mobilisable elements in *Bacteroides* spp.

a result of this futile cycle if the *Bacteroides* are not cultured under strictly anaerobic conditions.

Resistance to metronidazole is rare, particularly in the USA (Dubreuil, 1996; Snydman *et al.*, 1996). It is also used for the treatment of peptic ulcer disease (Freeman *et al.*, 1997) and this wider use may lead to the development of resistance in *Bacteroides* of the normal microbiota. 5-Nitroimidazole resistance determinants (*nim* genes) have been identified on the chromosome of some strains and small plasmids (e.g. pIP417, 7.7 kbp; pIP419, 10 kpb) of *B. fragilis*. The plasmids can be transferred by a conjugation-like process in *B. fragilis*, *B. vulgatus*, *B. thetaiotaomicron* and *Prevotella* spp. (Breuil *et al.*, 1989; Sebald, 1994; Trinh *et al.*, 1996). The plasmids are not thought to be self-transmissible, but rely on conjugal helper elements, which may be chromosomal (see below and **Fig. 9**). Three *nim* genes that encode proteins with 85% similarity have been cloned and sequenced from plasmids and chromosomal sites (Sebald, 1994). Insertion sequences (IS) elements have been identified in the 5′ upstream region of *nim* genes, and these are involved in the regulation of expression of the antibiotic resistance gene (Smith *et al.*, 1998: **Table 8**). IS element involvement in the regulation of antibiotic resistance determinant expression in *Bacteroides* has also been reported for other antibiotics (see below). The mechanism of

resistance to the activity of metronidazole is not fully understood.

β Lactams

The *B. fragilis* group are usually resistant to benzylpenicillin, other penicillins and many cephalosporins, with the exception of cephamycins such as cefoxitin. Carbapenems such as imipenem and meropenem are active, but some monobactams are inactive, because they lack affinity for the penicillin-binding proteins (PBP) of *B. fragilis* (Greenwood and Edwards, 1997).

Resistance to penicillin in *B. fragilis* homology group I is associated with the production of active-site serine-β lactamases distantly related to Ambler's class A. Serine-β lactamases attack the carbonyl (C = O) group of the β-lactam ring with the nucleophilic hydroxyl group of an active site serine, which results in hydrolysis of the C–N bond (Wang *et al.*, 1999). These β lactamases are susceptible to the inhibitory effects of compounds such as clavulanic acid, sulbactam and tazobactam. Cephalosporinase genes of this type – *cepA* from *B. fragilis*, *cfxA* from *B. vulgatus* and *cblA* from *B. uniformis* have been cloned and sequenced (Rogers *et al.*, 1993). In terms of amino acid sequence, these three β lactamases are more similar to each other than to all other β lactamases so far identified in other bacteria. From phylogenetic analyses, it appears that

Table 8 Examples of genetic elements of *Bacteroides* spp.

Genetic element	Size (kbp)	Resistance marker	Gene	Gene product action
Self-transmissible				
Plasmids				
pBF4	41	[a]MLS	*ermF*	RNA-methyltransferase
(associated transposon Tn4351)	(5)		[b]*tetX*	
PBI136	80	MLS	*ermFS*	
(associated transposon Tn4551)	(8.5)		[c]*aadS*	
Chromosomal conjugative transposons				
Tc[r] DOT and related elements e.g. Tc[r]	70–150	Tetracycline	*tetQ*	Ribosome protection
ERL, V479, 12256 (also termed Tn5030)		MLS	*ermF*	
Tc[r]7853	70–80	Tetracycline	*tetQ*	
		MLS	*ermG*	
XBU4422	65	?	?	
Mobilisable by self-transmissible plasmid or conjugative transposons				
Plasmids				
PBFTM1O (also termed pCP1)	15			
Tn4404	5	MLS	*ermF*	
			tetX	
PBFKW1	40	Ampicillin		
Nm[r] plasmids				
e.g. pIP417	7.7	5-Nitroimidazole	*nimA*	?
PIP419	10		*nimC*	?
Cryptic plasmids				
e.g. pB8–51	<8	?	?	
pBI143 (also termed pBFTM2006)				
Chromosomal integrated elements				
Transposons				
Tn4555	12	Cephalosporin,	*cfxA*	Serine protease
Tn4399	9.6	Penicillin, cefoxitin		
NBU 1, 2, 3	10–12	?	?	
Insertion sequences involved in gene activation				
IS942	1.6	β Lactam	*ccrA*	Metallo β lactamase
IS1224	>1.7	β Lactam	*cepA*	Serine protease
IS1186	1.3	β Lactam	*cfiA*	Metallo β lactamase
IS1168	1.3	5-Nitroimidazole	*nimA, B*	?
IS1169	1.3	5-Nitroimidazole	*nimD*	?
IS1170	1.6	5-Nitroimidazole	*nimC*	?
IS4351	1.2	MLS	*ermF/FS*	RNA-methyltransferase

[a] Macrolide, lincosamide, streptogramin.
[b] Silent tetracycline-resistance gene.
[c] Silent streptomycin-resistance gene.
Compiled from Sebald (1994), Salyers and Shoemaker (1996), Smith *et al.* (1998).

these *Bacteroides* enzymes evolved at an earlier stage, which is consistent with the early evolutionary divergence of the *Bacteroides*.

Determinants for these β lactamases may be chromosomal, but can be transferred between species of the *B. fragilis* 'group'. The co-transfer of ampicillin resistance with a conjugative transposon that carries a tetracycline resistance marker has been demonstrated. Similarly, the *cfxA* gene, which encodes a

cephalosporinase that degrades cephalosporins, penicillins and cefoxitin, is located on a mobilisable transposon, Tn4555 (see below; Sebald, 1994). Ampicillin resistance has also been located on a large transferable plasmid, but plasmid-mediated resistance to extended β lactams such as carbapenem and cefoxitin has not been reported.

B. fragilis strains that express either high or low levels of the CepA β lactam, which does not degrade cefoxitin or imipenem, have been identified. A 40-fold increase in expression has been linked with the insertion of IS1224 upstream of the *cepA* gene. It has been suggested that the up-regulation arises from the formation of a promoter partly encoded by the IS element (Smith *et al.*, 1998).

A minority of *B. fragilis* strains produce a zinc-requiring metallo-β-lactamase of Ambler's class B, which is associated with resistance to cephamycins and carbapenems. These enzymes are resistance to the β-lactam inhibitors (clavulanic acid, sulbactam and tazobactam) and sensitive to EDTA, which chelates zinc. The proposed mechanism for the hydrolysis of the C-N bond involves an initial interaction of the nuclophile zinc hydroxide in the active site of the enzyme. This polarises the carbonyl group of the β lactam, and a second zinc interacts with the nitrogen, resulting in bond cleavage (Wang *et al.*, 1999). The chromosomal genes *cfiA* and *ccrA* have been identified. With the *cfiA* as a probe, the gene was identified in 3% of *B. fragilis* clinical isolates, which relates to the *B. fragilis* DNA homology group II (Ruimy *et al.*, 1996). These strains were lacking in the *cepA* gene that encodes the more common β lactamase of *B. fragilis* found in DNA homology group I. Differences in the composition of the OMPs and LPS of the two groups have also been noted (Edwards and Greenwood, 1997). The *cfiA* gene-positive group also carried one or more of the three IS elements IS4351, IS1186 or IS942. These are involved in the regulation of gene expression and appear to encode promoters for the antibiotic resistance genes. A promoter region has been identified in IS1186, and IS4351 encodes three promoter-like sequences. These insertion sequences were found in only 1% of the *cfiA*-negative group (Podglajen *et al.*, 1994). It has been suggested that the low incidence of the *cfiA* gene in *B. fragilis* isolates is indicative of recent acquisition of the gene (Ruimy *et al.*, 1996). Alteration of penicillin-binding proteins or outer membrane permeability may also play a role in *Bacteroides* spp. resistance to penicillin, but this remains to be fully determined (Greenwood and Edwards, 1997).

Macrolide, Lincosamide and Streptogramin

Macrolide (e.g. erythromycin), lincosamide (e.g. clindamycin) and streptogramin (e.g. pristinamycin, virginiamycin) (MLS) resistance is transferable within and between *Bacteroides* spp. by conjugative plasmid transfer (Privitera *et al.*, 1979; Sebald, 1994). MLS resistance can also be located on self-transmissible conjugative elements which also carry tetracycline resistance genes. Plasmids that have been studied include pBF4 (41 kb), pBFTM10 or pCP1 (15 kb) and pBI136 (80 kb). Plasmids pBF4 and pBI136 are self-transmissible, whereas pBFTM10 relies on mobilisation by conjugal helper elements (see below). MLS resistance is conferred by the genes *ermF* or *ermFS* (the latter differs by a single base) present on compound transposons (TN 4351, 4551 and 4400). The *ermF* gene encodes a rRNA methylase. Expression of the *erm* genes can be up-regulated by the presence of IS elements (**Table 8**). These *erm* genes are similar in sequence to the *erm* genes of Gram-positive bacteria. The G + C content of these genes is 32%, which is considerably less than the rest of the *B. fragilis* (42%). Tn 4351 and 4400 also carry the *tetX* gene, which is silent in *B. fragilis*, but confers tetracycline resistance in aerobically grown *E. coli*, as resistance is mediated by an NADP-requiring oxidoreductase enzyme. Tn4551 carries a silent streptomycin-resistance gene, *aadS*. The equivalent protein sequence for the *aadS* is 30% identical to the protein produced by the aadK gene of *Bacillus subtilis*. The intrinsic resistance of *Bacteroides* spp. to aminoglycoside antibiotics, because of lack of uptake, adds weight to the hypothesis that these genes have been acquired from a Gram-positive source. The IS elements that flank these transposons are either identical (IS4351 in Tn 4351 and 4551) or almost identical (IS4400 in Tn 4400). IS4351 encodes a putative transposase and is similar in G + C content to the *B. fragilis* genome (Smith *et al.*, 1998). It seems, therefore, that although the transposon genes may be of Gram-positive origin, the flanking IS elements are not.

Tetracyclines

Resistance to tetracycline is widespread in clinical and bowel *Bacteroides* spp., and mediated by the *tetQ* gene product which has a 45% amino acid identity with the *tetM* and *tetO* products in Gram-positive bacteria. TetQ and the related TetM, O, B (P) and S of other bacteria have a molecular mass of about 72 kDa and protect the ribosome from the action of tetracyclines. They have N-terminal sequence similarities with elongation factors Tu and G. The mechanism of action of TetQ is not known, but TetM increases the rate of dissociation of tetracycline from the ribosome when GTP is present (Leng *et al.*, 1997). The *tetQ* gene is carried on a large conjugative element (70 to >150 kb) which is capable of self-transfer and may also carry the MLS resistance *erm* gene. These elements are sometimes referred to as Tet elements, Tc^r elements or conjugative transposons (CT or CTn), but it has been proposed that CTn should be used as the convention (Salyers *et al.*, 1999). They are similar to the conjugative transposons found in Gram-positive bacteria. They differ from standard transposons in that they are larger and are not flanked by insertion sequences, but imperfect repeats of 24 bp have been identified at the ends of the element and these may be involved in the binding of excision proteins. A model has been proposed for the mechanism of conjugation in which the CTn is excised from the chromosome and forms a non-replicating plasmid-like covalently closed circle. This is transferred in a manner similar to a plasmid, with the centre of origin located near the centre of the element. A single-stranded copy enters the recipient, the second strand is synthesised to form a double-stranded circular element which integrates into the recipient chromosome, without duplication of the target site. This process may result in the mobilisation of other plasmids which are not self-transmissible and other unlinked chromosomal regions.

Resistance to tetracycline has increased in recent years as result of the spread of the *tetQ* gene (Leng *et al.*, 1997). TetQ can be transferred by conjugative transposition not only among *Bacteroides* such as *B. fragilis*, *B. uniformis* and *B. thetaiotaomicron*, but also between *Bacteroides* spp. and bacteria from the bovine rumen, such as *Prevotella* (*Bacteroides*) *ruminicola* (Shoemaker *et al.*, 1992) and *Enterococcus faecalis* (Leng *et al.*, 1997). Given the apparent genetic mobility of the *Bacteroides* antibiotic-resistance genes, there is the potential for the development of multiple-drug-resistant *B. fragilis*.

Genetic Systems

The development of genetic systems in *B. fragilis* has been hampered by the general lack of genetic compatibility between *E. coli* and the majority of strains of *B. fragilis* (Salyers and Shoemaker, 1996; Salyers *et al.*, 1999). *E. coli* promoters do not function in *Bacteroides*, and *E. coli* plasmids do not replicate in *Bacteroides* spp. The restriction endonucleases of *B. fragilis* preclude the introduction of *E. coli* DNA by electroporation and in many strains also by conjugation. Also, there are so far no reports of *Bacteroides* plasmids that replicate in *E. coli* (Smith *et al.*, 1998). The lack of replication of *E. coli* plasmids in *Bacteroides* has, however, been used to some advantage in the construction of suicide vectors. *Bacteroides* also appear to lack GATC sites, which are methylated at the adenine residue by the Dam methylase in *E. coli*. This process is important in *E. coli* for the repair of damaged DNA, regulation of gene expression and some gene re-arrangements (Holdman *et al.*, 1976; Patrick and Larkin, 1995). It has been suggested that this may explain the instability of *Bacteroides* DNA in cosmid clones in *E. coli* hosts

(Shoemaker *et al.*, 2000). The apparent general genetic incompatibility probably reflects the taxonomic distance between the two genera. *Bacteroides* spp. fall into the *Bacteroides–Flavobacterium* phylum (Woese, 1987). It is thought that this phylum, which includes the majority of Gram-negative organisms, such as the enteric bacteria, *E. coli* and the pseudomonads, diverged from other eubacteria early in evolutionary terms, well before the divergence of the Gram-positive bacteria from the Purple bacteria.

In spite of this, the study of the potential spread of antibiotic-resistance determinants has lead to the recognition of two key features in *Bacteroides*. First, they possess a number of different types of potentially mobile genetic elements. These include the unusual chromosomal transposons that are mobilisable by other genetic elements and have not yet been described in other genera (see **Table 8**, **Fig. 8** and below). Second, there is a relationship between insertion sequences and up-regulation of antibiotic gene expression, but whether this up-regulation occurs with other genes remains to be determined. Interestingly in *D. nodosus*, the causative agent of ovine footroot, chromosomal genetic elements (*int*A, B and C elements) that up-regulate virulence determinants have been described (Whittle *et al.*, 1999).

Mobile Genetic Elements

A small number of self-transmissible plasmids that carry antibiotic resistance genes have been identified (e.g. pBF4 which carries MLS resistance), but much of the movement of genetic elements among *Bacteroides* appears to be driven by large chromosomal conjugative transposons. The most extensively studied of these are the Tc^r elements that carry the tetracycline resistance gene, *tetQ*, but other 'cryptic' elements have also been identified (e.g. XBU4422 in *B. uniformis*). The chromosomal insertion site of these elements seems to be non-random and restricted to 3–7 chromosomal sites. These elements carry the genes necessary for self-transmission by conjugation, and the mechanism of transfer is thought to involve a circular, non-replicating intermediate form (Salyers and Shoemaker, 1996). Unusually, the frequency of conjugation increases more than 100-fold at low concentrations of tetracycline (about 1 μg/mL). The ORFs of two genes downstream of *tetQ* are similar to the two-component sensor/regulator genes described in a wide range of other bacteria. These have been designated *rte* (regulation of TC^r elements) A (sensor kinase) and B (regulator) (Stevens *et al.*, 1992). A third unlinked gene, *rteC*, is also involved in regulation. These conjugative transposons mediate the conjugative transfer of other genetic elements that are not capable of transferring by themselves. They act as conjugative helpers for plasmids and also a novel class of mobile genetic elements, chromosomal mobilisable transposons, which includes the non-replicating *Bacteroides* units (NBU). *RteB* activates the excision and integration genes of the mobilisable transposons. The conjugative transposons provide the mating pore proteins. Interestingly, many of these mobilisable elements are also mobilised by the IncP plasmids of *E. coli* (Salyers and Shoemaker, 1996; Smith *et al.*, 1998).

It has been estimated that up to 50% of *Bacteroides* faecal and clinical isolates carry cryptic plasmids for which a phenotype has yet to be detected. All the cryptic plasmids so far studied carry an origin of transfer site (*oriT*), regions essential for replication and a mobilisation gene (*mob*). Although they are not self-transmissible, these genes enable them to be mobilised in *trans* by the chromosomal conjugative transposons. Antibiotic-resistance carrying plasmids that encode 5-nitroimidazole resistance and the pBFTM10 plasmid which carries MLS resistance are thought to be mobilised by the Tc^r elements.

The mobilisable transposons are 9–12 kb in size and carry the genes necessary for their own excision from the chromosome, nicking of the circular intermediate and subsequent re-integration, but not conjugation (Murphy and Malamy, 1995; Smith *et al.*, 1998). Excision may be induced by *RteB* encoded on a conjugative transposon and is therefore also up-regulated by tetracycline concentration. After excision, the mobilisable transposon forms into a non-replicating circular element, hence the assignation of the name 'non-replicating *Bacteroides* unit' (NBU) to some of these elements. The circular element is nicked by the Mob protein at the transfer origin and a single strand of the element is transferred to the recipient through the mating pore provided by the conjugative transposon. The elements identified to date include Tn4399 (*B. fragilis*), Tn4555 which encodes the *cfxA* cefoxitin resistance gene (*B. vulgatus*), NBU1 (*B. thetaiotaomicron*), NBU1 and 2 (*B. uniformis*). The NBU1 element has been sequenced. The *intN1* gene and the region of the closed ends of the circularised transposon, *attN1*, are required for the integration of NBU1 at the target chromosomal site. The *intN1* gene encodes the integrase, which has some amino acid similarity with the λ-integrase family. The mobilisation gene, *mobN1*, has been located upstream of the origin of transfer site, *oriT*. The process of excision requires, in addition to the integrase region and *oriT*, two-thirds of the *mobN1* gene region (*prmN1*) which encodes

a protein involved in excision, but which is not a primase, since the circular transposon does not replicate nor is it self-transmissible. Three additional ORFs are also necessary for the excision process (Shoemaker *et al.*, 2000). The excision process therefore appears to be complex, and this may relate to the subsequent conjugation mediated by the Tc^r elements. Unlike the Tc^r elements, NBU1 integrates at a specific site located at the 3′ end of the leucine-tRNA gene.

Studies of a range of mobilisable plasmids and transposons suggest that a common feature is the cassette of genes necessary for DNA processing, and which with adjacent *oriT* form a 'mobilisation region'. Although the number and content of genes in these mobilisation regions may vary, where they are incorporated into *Bacteroides* plasmids and transposons, it enables their co-transfer with self-transmissible plasmids and Tc^r elements. Smith and colleagues (1998) provide a comprehensive review of mobile genetic elements of *Bacteroides*.

References

Allan E, Poxton IR (1994) The influence of growth medium on serum sensitivity of *Bacteroides* species. *J. Med. Microbiol.* 41: 45–50.

Allan E, Poxton IR, Barclay GR (1995) Anti-bacteroides lipopolysaccharide IgG levels in healthy adults and sepsis patients. *FEMS Immunol. Med. Microbiol.* 11: 5–12.

Babb JL, Cummins CS (1978) Encapsulation of *Bacteroides* species. *Infection* 19: 1088–1091.

Barrow GI, Felthan RKA (1993) *Cowan and Steel's Manual for the Identification of Medicial Bacteria* 3rd edn. Cambridge: Cambridge University Press.

Baumann H, Tzianabos AO, Brisson J, Kasper DL, Jennings HJ (1992) Structural elucidation of two capsular polysaccharides from one strain of *Bacteroides fragilis* using high-resolution NMR spectroscopy. *Biochemistry* 31: 4081–4089.

Beckmann I, Eijk HG van, Meisel-Mikolajczyk F, Wallenburg HC (1989) Detection of 2-keto-3-deoxyoctonate in endotoxins isolated from six reference strains of the *Bacteroides fragilis* group. *Int. J. Biochem.* 21: 661–666.

Bennett-Guerrero E, Barclay GR, Youssef ME *et al.* (2000) Exposure to *Bacteroides fragilis* endotoxin during cardiac surgery. *Anesth. Analg.* 90: 819–823.

Bjornson AB, Detmers PA (1995) The pentameric structure of IgM is necessary to enhance opsonization of *Bacteroides thetaiotaomicron* and *Bacteroides fragilis* via the alternative complement pathway. *Microb. Pathog.* 19: 117–128.

Bjornson AB (1987) Opsonization of bacteroides by the alternative complement pathway reconstructed from isolated plasma proteins. *J. Exp. Med.* 164: 777–798.

Bonnett SA, Sutren M, Godon JJ *et al.* (1999) Direct analysis of genes encoding 16S rRNA from complex communities reveals many novel molecular species within the human gut. *Appl. Environ. Microbiol.* 65: 4799–4807.

Border M, Firehammer BD, Shoop DS, Myers LL (1985) Isolation of *Bacteroides fragilis* from the feces of diarrheic calves and lambs. *J. Clin. Microbiol.* 21: 472–473.

Breuil J, Dublanchet A, Truffaut N, Sebald M (1989) Transferable 5-nitroimidazole resistance in the *Bacteroides fragilis* group. *Plasmid* 21: 151–154.

Brook I (1989) Pathogenicity of the *Bacteroides fragilis* group. *Ann. Clin. Lab. Sci.* 19: 360–376.

Brown R, Collee JG, Poxton IR, Fraser AG (1989) Bacteroides, Fusobacterium and related organisms: anaerobic cocci: identification of anaerobes. In: Collee JG, Duguid JP, Fraser AG, Marmion BP (eds) *Mackie and McCartney Practical Medical Microbiology*, 13th edn. Edinburgh: Churchill Livingstone, pp. 553–570.

Bry L, Falk PG, Midtvedt T, Gordon JI (1996) A model of host-microbial interactions in an open mammalian ecosystem. *Science* 273: 1380–1383.

Chen Y, Kinouchi T, Kataoka K, Akimoto S, Ohnishi Y (1995) Purification and characterization of a fibrinogen-degrading protease in *Bacteroides fragilis* strain YCH46. *Microbiol. Immunol.* 39: 967–977.

Chung G, Franco AA, Wu B *et al.* (1999) Identification of a third metalloprotease toxin gene in extraintestinal isolates of *Bacteroides fragilis*. *Infect. Immun.* 67: 4945–4949.

Claros MC, Citron DM, Schonian G, Hunt-Gerardo S, Goldstein EJC (1997) Characterization of *Bacteroides* species using PCR fingerprinting. In: Eley AR, Bennett KW (eds) *Anaerobic Pathogens*. Sheffield: Sheffield Academic Press, pp. 359–369.

Comstock LE, Coyne MJ, Tzianabos AO *et al.* (1999) Analysis of a capsular polysaccharide biosynthesis locus of *Bacteroides fragilis*. *Infect. Immun.* 67: 3525–3532.

Conway PL (1997) Development of the intestinal microbiota. In: Mackie RI, White BA, Isaacson RE (eds) *Gastrointestinal Microbiology*, Vol. 2. London: Chapman & Hall, pp. 3–38.

Crabb JH, Finberg R, Onderdonk AB, Kasper DL (1990) T cell regulation of *Bacteroides fragilis*-induced intraabdominal abscesses. *Rev. Infect. Dis.* 12 Suppl. 2: S178–184.

Delahooke DM, Barclay GR, Poxton IR (1995a) A reappraisal of the biological activity of bacteroides LPS. *J. Med. Microbiol.* 42: 102–112.

Delahooke DM, Barclay GR, Poxton IR (1995b) Tumor necrosis factor induction by an aqueous phenol-extracted lipopolysaccharide complex from *Bacteroides* species. *Infect. Immun.* 63: 840–846.

Domingues RMCP, Cavalcanti SMB, Andrade AFB, Ferreira MCS (1992) Sialic acid as receptor of *Bacteroides fragilis* lectin-like adhesin. *Zlb. Bakteriol.* 277: 340–344.

Domingues RMCP, Souza WGSS, Moraes SR *et al.* (1997) Surface vesicles: a possible function in commensal relations of *Bacteroides fragilis*. *Zlb. Bakteriol.* 285: 509–517.

Doorn J van, Mooi J, Verweij-van Vught AMJJ, MacLaren DMA (1987) Characterization of fimbriae from *Bacteroides fragilis*. *Microb. Pathog.* 3: 87–95.

Doorn J van, Oudega B, MacLaren DM (1992) Characterization and detection of the 40-kDa fimbrial subunit of *Bacteroides fragilis* BE1. *Microb. Pathog.* 13: 75–79.

Drasar BS, Duerden BI (1991) Anaerobes in the normal flora of man. In: Duerden BI, Drasar BS (eds) *Anaerobes in Human Disease*. London: Edward Arnold, pp. 162–179.

Dubreuil L (1996) Antibiotic susceptibility patterns of the *Bacteroides fragilis* group. *Med. Mal. Infect.* 26: 196–207.

Duerden BI (1980) The identification of gram-negative anaerobic bacilli isolated from clinical infections. *J. Hygiene* 84: 301–313.

Edwards DI (1997) Resistance to nitroimidazoles in anaerobes. In: Eley AR, Bennett KW (eds) *Anaerobic Pathogens*. Sheffield: Sheffield Academic Press, pp. 437–448.

Edwards R, Greenwood D (1997) Distinctive outer membrane protein and lipopolysaccharide composition of *Bacteroides fragilis* strains that produce metallo-beta-lactamase. *Anaerobe* 3: 233–236.

Eiring P, Manncke B, Gerbracht K, Werner H (1995) *Bacteroides fragilis* adheres to laminin significantly stronger than *Bacteroides thetaiotaomicron* and other species of the genus. *Zlb. Bakteriol.* 282: 279–286.

Finlay-Jones JJ, Hart PH, Spencer LK *et al.* (1991) Bacterial killing *in vitro* by abscess-derived neutrophils. *J. Med. Microbiol.* 34: 73–81.

Foreman KE, Bjornson AB (1994) The alternative complement pathway promotes IgM antibody-dependent and -independent adherence of *Bacteroides* to polymorphonuclear leukocytes through CR3 and CR1. *J. Leukocyte Biol.* 55: 603–611.

Freeman CD, Klutman NE, Lamp KC (1997) Metronidazole – a therapeutic review and update. *Drugs* 54: 679–708.

Fujiwara T, Ogawa T, Sobue S, Hamada S (1990) Chemical, immunobiological and antigenic characterizations of lipopolysaccharides from *Bacteroides gingivalis* strains. *J. Gen. Microbiol.* 136: 319–326.

Gaskins HR (1997) Immunological aspects of host/microbiota interactions at the intestinal epithelium. In: Mackie RI, White BA, Isaacson RE (eds) *Gastrointestinal Microbiology*, London: Chapman & Hall, Vol. 2, pp. 537–587.

Gibson GR, Macfarlane GT (1994) Intestinal bacteria and disease. In: Gibson SAW (ed.) *Human Health: the Contribution of Microorganisms*. Berlin: Springer, pp. 53–62.

Gibson SAW (ed.) (1994) *Human Health: the Contribution of Microorganisms*. Berlin: Springer.

Godoy VG, Dallas MM, Russo TA, Malamy MH (1993) A role for *Bacteroides fragilis* neuraminidase in bacterial growth in two model systems. *Infect. Immun.* 61: 4415–4426.

Goldner M, Coquis-Rondon M, Carlier J-P (1993) Effect of growth of *Bacteroides fragilis* at different redox levels on potential pathogenicity in a HeLa cell system: demonstration by confocal laser scanning microscopy. *Zlb. Bakteriol.* 278: 529–540.

Goldner M, Mingot N, Emond JP, Dublanchet A (1997) Influence of different levels of redox potential on fermentative products formed by *Bacteroides fragilis. Clin. Infect. Dis.* 25 (Suppl)(2): S147–S150.

Greenwood D, Edwards R (1997) Mechanisms of resistance to beta-lactam antibiotics in *Bacteroides fragilis.* In: Eley AR, Bennett KW (eds) *Anaerobic Pathogens.* Sheffield: Sheffield Academic Press, pp. 449–456.

Guthrie EP, Shoemaker NB, Salyers AB (1985) Cloning and expression in *Escherichia coli* of a gene coding for a chondroitin lyase from *Bacteroides thetaiotaomicron. J. Bacteriol.* 164: 510–515.

Guzman CA, Biavasco F, Pruzzo C (1997) Adhesiveness of *Bacteroides fragilis* strains isolated from feces of healthy donors, abscesses and blood. *Curr. Microbiol.* 34: 332–334.

Heinrichs DE, Yethon JA, Whitfield C (1999) Molecular basis for structural diversity in the core regions of the lipopolysaccharides of *Escherichia coli* and *Salmonella enterica. Mol. Microbiol.* 30: 221–232.

Hofstad T (1980) Evaluation of the API ZYM system for identification of *Bacteroides* and *Fusobacterium* species. *Med. Microbiol. Immunol.* 168: 173–177.

Holdman LV, Good IJ, Moore WEC (1976) Human fecal flora: variation in bacterial composition within individuals and possible effect of emotional stress. *Appl. Environ. Microbiol.* 31: 359–375.

Hooper LV, Xu J, Falk PG, Midtvedt T, Gordon JI (1999) A molecular sensor that allows a gut commensal to control its nutrient foundation in a competitive ecosystem. *Proc. Natl Acad. Sci. USA* 96: 9833–9838.

Imai Y, Singer MS, Fennie C, Lasky LA, Rosen SD (1991) Identification of a carbohydrate-based endothelial ligand for a lymphocyte homing receptor. *J. Cell Biol.* 113: 1213–1221.

Jennings MP, Srikhanta YN, Moxon ER *et al.* (1999) The genetic basis of the phase variation repertoire of lipopolysaccharide immunotypes in *Neisseria meningitidis. Microbiology* 145: 3013–3021.

Johnson JL (1978) Taxonomy of the Bacteroides I. Deoxyribonucleic acid homologies among *Bacteroides fragilis* and other saccharolytic *Bacteroides* species. *Int. J. Syst. Bacteriol.* 28: 245–256.

Johnson JL, Ault DA (1978) Taxonomy of Bacteroides II. Correlation of phenotypic characteristics with deoxyribonucleic acid homology groupings for *Bacteroides fragilis* and other saccharolytic *Bacteroides* species. *Int. J. Syst. Bacteriol.* 28: 257–268.

Jousimies-Somer HR, Summanen PH, Finegold SM (1995) *Bacteroides, Porphyromonas, Prevotella, Fusobacterium,* and other anaerobic Gram-negative bacteria. In: Murray PR (ed.) *Manual of Clinical Microbiology,* 6th edn. Washington, DC: American Society for Microbiology, pp. 603–620.

Kato N, Kato H, Watanabe K, Ueno K (1996) Association of enterotoxigenic *Bacteroides fragilis* with bacteraemia. *Clin. Infect. Dis.* 23 (Suppl)(1): S83–S86.

Kocher M, Kenny PA, Farram E *et al.* (1996) Functional chemotactic factor CP-10 and MRP-14 are abundant in murine abscesses. *Infect. Immun.* 64: 1342–1350.

Larkin MJ, McGuigan J, Patrick S (1988) Iron limitation and induction of congo red binding in *Bacteroides fragilis* NCTC9343. In: Hardie JM, Boriello SP (eds) *Anaerobes Today.* Chichester: Wiley, pp. 216–217.

Leng Z, Riley DR, Berger RE, Krieger JN, Roberts MC (1997) Distribution and mobility of the tetracycline resistance determinant *tetQ. J. Antimicrob. Chemother.* 40: 551–559.

Leszczynski P, Belkum A van, Pituch H, Verbrugh H, Meisel-Mikolajczyk F (1997) Vaginal carriage of enterotoxigenic *Bacteroides fragilis* in pregnant women. *J. Clin. Microbiol.* 35: 2899–2903.

Lindberg AA, Weintraub A, Zahringer U, Rietschel ET (1990) Structure–activity relationships in lipopolysaccharides of *Bacteroides fragilis. Rev. Infect. Dis.* 12 Suppl. 2: S133–S141.

Lutton DA, Patrick S, Doorn J van, Emmerson M, Clarke G (1989) Expression of *Bacteroides fragilis* fimbrial antigen *in vitro* and *in vivo. Biochem. Soc. Trans.* 17: 758–759.

Lutton DA, Patrick S, Crockard AD *et al.* (1991) Flow cytometric analysis of within-strain variation in polysaccharide expression by *Bacteroides fragilis* by use of murine monoclonal antibodies. *J. Med. Microbiol.* 35: 229–237.

MacLachlan PR, Keenleyside WJ, Dodgson C, Whitfield C (1993) Formation of the K30 (Group I) capsule of *Escherichia coli* O9:K30 does not require attachment to lipopolysaccharide lipid A-core. *J. Bacteriol.* 175: 7515–7522.

Macfarlane GT, Gibson GR (1991) Formation of glycoprotein degrading enzymes by *Bacteroides fragilis. FEMS Microbiol. Lett.* 77: 289–293.

Macfarlane GT, Gibson GR (1994) Metabolic activities of the normal colonic flora. In: Gibson SAW (ed.) *Human Health: The Contribution of Microorganisms.* Berlin: Springer, pp. 17–52.

Macfarlane GT, Gibson SAW, Gibson GR (1992a) Proteolytic activities of the fragilis Group of *Bacteroides* spp. In: Duerden BI, Brazier JS, Seddon SV, Wade WG (eds) *Medical and Environmental Aspects of Anaerobes.* Petersfield: Wrightson Biomedical Publishing, pp. 130–143.

Macfarlane GT, Macfarlane S, Gibson GR (1992b) Synthesis and release of proteases by *Bacteroides fragilis. Curr. Microbiol.* 24: 55–59.

Mackie RI, White BA, Isaacson RE (eds) (1997) *Gastrointestinal Microbiology,* Vol. 2 New York: Chapman & Hall.

Manz W, Amman R, Ludwig W, Vancanney M, Schleifer K-H (1996) Application of a suite of 16S rRNA-specific oligonucleotide probes designed to investigate bacteria of the phylum cytophaga-flavobacter-bacteroides in the natural environment. *Microbiology* 142: 1097–1106.

Moncrief JS, Obiso R, Barroso LA *et al.* (1995) The enterotoxin of *Bacteroides fragilis* is a metalloprotease. *Infect. Immun.* 63: 175–181.

Moncrief JS, Duncan AJ, Wright RL, Barroso LA, Wilkins TD (1998) Molecular characterization of the fragilysin pathogenicity islet of enterotoxigenic *Bacteroides fragilis. Infect. Immun.* 66: 1735–1739.

Morris JG (1991) Characteristics of anaerobic metabolism. In: Duerden BI, Drasar BS (eds) *Anaerobes in Human Disease.* London: Edward Arnold, pp. 16–37.

Murphy CG, Malamy MM (1995) Requirements for strand- and site-specific cleavage within the *oriT* region of the Tn(4399), a mobilizing transposon from *Bacteroides fragilis. J. Bacteriol.* 177: 3158–3165.

Myers LL, Firehammer BD, Shoop DS, Border MM (1984) *Bacteroides fragilis:* a possible cause of acute diarrheal disease in newborn lambs. *Infect. Immun.* 44: 241–244.

Myers LL, Shoop DS, Stackhouse LL *et al.* (1987) Isolation of enterotoxigenic *Bacteroides fragilis* from humans with diarrhea. *J. Clin. Microbiol.* 25: 2330–2333.

Namavar F, Theunissen EB, Verweij-Van Vught AMJJ *et al.* (1989) Epidemiology of the *Bacteroides fragilis* group in the colonic flora of patients with colonic cancer. *J. Med. Microbiol.* 29: 171–176.

Nicolai H von, Hammann R, Werner H, Zilliken F (1983) Isolation and characterization of sialidase from *Bacteroides fragilis. FEMS Microbiol. Lett.* 17: 217–220.

Otto BR, Verweij-van Vught AMJJ, van Doorn J, MacLaren DM (1988) Outer membrane proteins of *Bacteroides fragilis* and *Bacteroides vulgatus* in relation to iron uptake and virulence. *Microb. Pathog.* 4: 279–287.

Otto BR, Sparrius M, Verweij-van Vught AMJJ, MacLaren DM (1990) Iron-regulated outer membrane protein of *Bacteroides fragilis* involved in heme uptake. *Infect. Immun.* 58: 3954–3958.

Otto BR, Verweij WR, Sparrius M, Verweij-van Vught AMJJ, Nord CE, Maclaren DM (1991) Human immune-response to an iron-repressible outer-membrane protein. *Infect. Immun.* 59: 2999–3003.

Otto BR, Sparrius M, Wors DJ, Graaf FK de, MacLaren DM (1994) Utilization of haem from the haptoglobin– haemoglobin complex by *Bacteroides fragilis. Microb. Pathog.* 17: 137–147.

Otto BR, Kusters JG, Luirink J, Graaf FK de, Oudega B (1996) Molecular characterization of a heme-binding protein of *Bacteroides fragilis* BE1. *Infect. Immun.* 64: 4345–4350.

Pantosti A, Tzianabos AO, Onderdonk AB, Kasper DL (1991) Immunochemical characterization of two surface polysaccharides of *Bacteroides fragilis. Infect. Immun.* 59: 2075–2082.

Pantosti A, Colangeli R, Tzianabos AO, Kasper DL (1995) Monoclonal antibodies to detect capsular diversity among *Bacteroides fragilis* isolates. *J. Clin. Microbiol.* 33: 2647–2652.

Pantosti A, Menozzi MG, Frate A *et al.* (1997) Detection of enterotoxigenic *Bacteroides fragilis* and its toxin in stool samples from adults and children in Italy. *Clin. Infect. Dis.* 24: 12–16.

Patrick S (1988) Phagocytosis of *Bacteroides fragilis in vitro.* In: Hardie JM, Boriello SP (eds) *Anaerobes Today.* Chichester: Wiley, pp. 31–41.

Patrick S (1993) The virulence of *Bacteroides fragilis. Rev. Med. Microbiol.* 4: 40–49.

Patrick S, Reid JH (1983) Separation of capsulate and non-capsulate *Bacteroides fragilis* on a discontinuous density gradient. *J. Med. Microbiol.* 16: 239–241.

Patrick S, Lutton DA (1990a) *Bacteroides fragilis* surface structure expression in relation to virulence. *Med. Malad. Infect.* 20(hors serie): 19–25.

Patrick S, Lutton DA (1990b) Outer membrane proteins of *Bacteroides fragilis* grown *in vivo. FEMS Microbiol. Lett.* 71: 1–4.

Patrick S, Larkin MJ (1995) *Immunological and Molecular Aspects of Bacterial Virulence.* Chichester: Wiley.

Patrick S, Reid JH, Larkin MJ (1984) The growth and survival of capsulate and non-capsulate *Bacteroides fragilis in vivo* and *in vitro. J. Med. Microbiol.* 17: 237–246.

Patrick S, Reid JH, Coffey A (1986) Capsulation of *in vitro* and *in vivo* grown *Bacteroides* species. *J. Gen. Microbiol.* 132: 1099–1109.

Patrick S, Coffey A, Emmerson AM, Larkin MJ (1988) The relationship between cell surface structure expression and haemagglutination in *Bacteroides fragilis. FEMS Microbiol. Lett.* 50: 67–71.

Patrick S, Lutton SA, Crockard AD (1995a) Immune reactions to *Bacteroides fragilis* populations with three different types of capsule in a model of infection. *Microbiology* 141: 1969–1976.

Patrick S, Stewart LD, Damani N *et al.* (1995b) Immuno-logical detection of *Bacteroides fragilis* in clinical samples. *J. Med. Microbiol.* 43: 99–109.

Patrick S, McKenna JP, O'Hagan S, Dermott E (1996) A comparison of the haemagglutinating and enzymic activities of *Bacteroides fragilis* whole cells and outer membrane vesicles. *Microb. Pathog.* 20: 191–202.

Patrick S, Gilpin D, Stevenson L (1999) Detection of intrastrain antigenic variation of *Bacteroides fragilis* surface polysaccharides by monoclonal antibody labelling. *Infect. Immun.* 67: 4346–4351.

Podglajen I, Breuil J, Collatz E (1994) Insertion of a novel DNA sequence, IS1186, upstream of the silent carbapenemase gene *cfiA*, promotes expression of carbapenem resistance in clinical isolates of *Bacteroides fragilis. Mol. Microbiol.* 12: 105–114.

Podglajen I, Breuil J, Casin I, Collatz E (1995) Genotypic identification of two groups within the species *Bacteroides fragilis* by ribotyping and by analysis of PCR-generated fragment patterns and insertion sequence content. *J. Bacteriol.* 177: 5270–5275.

Poxton IR, Myers CJ, Johnstone A, Drudy TA, Ferguson A (1995) An ELISA to measure mucosal IgA specific for

Bacteroides surface antigens in whole gut lavage fluid. *Microb. Ecol. Health Dis.* 8: 129–136.

Poxton IR, Brown R (1986) Immunochemistry of the surface carbohydrate antigens of *Bacteroides fragilis* and definition of a common antigen. *J. Gen. Microbiol.* 132: 2475–2481.

Poxton IR, Brown R, Wilkinson JF (1989) pH measurements and buffers, oxidation-reduction potentials, suspension fluids and preparation of glassware. In: Collee JG, Duguid JP, Fraser AG, Marmion BP (eds) *Mackie & McCartney Practical Medical Microbiology*, 13th edn., Edinburgh: Churchill Livingstone, Vol. 2, pp. 90–99.

Poxton IR, Brown R, Sawyerr A, Ferguson A (1997) Mucosa-associated bacterial flora of the human colon. *J. Med. Microbiol.* 46: 85–91.

Privitera G, Dublanchet A, Sebald M (1979) Transfer of multiple antibiotic resistance between subspecies of *Bacteroides fragilis. J. Infect. Dis.* 139: 97–101.

Pruzzo C, Dainelli B, Ricchetti M (1984) Piliated *Bacteroides fragilis* strains adhere to epithelial cells and are more sensitive to phagocytosis by human neutrophils than nonpiliated strains. *Infect. Immun.* 34: 189–194.

Rademacher TW, Parekh RB, Dwek RA (1988) Glycobiology. *Annu. Rev. Biochem.* 57: 785–838.

Ramage G, Patrick S, Houston S (1998) Combined fluorescent *in situ* hybridisation and immunolabelling of *Bacteroides fragilis. J. Immunol. Meth.* 212: 139–147.

Raskin L, Capman WC, Sharp R, Poulsen LK, Stahl DA (1997) Molecular ecology of gastrointestinal ecosystems. In: Mackie RI, White BA, Isaacson RE (eds) *Gastrointestinal Microbiology.* New York: Chapman & Hall, pp. 243–297.

Rath HC, Herfarth HH, Ikeda JS *et al.* (1996) Normal luminal bacteria, especially Bacteroides species, mediate chronic colitis, gastritis and arthritis in HLA-B27/human beta2 microglobulin transgenic rats. *J. Clin. Invest.* 98: 945–953.

Rath HC, Wilson KH, Sortor RB (1999) Differential induction of colitis and gastritis in HLA-B27 transgenic rats selectively colonized with *Bacteroides vulgatus* or *Escherichia coli. Infect. Immun.* 67: 2969–2974.

Redondo MC, Arbo MDJ, Grindlinger J, Snydman DR (1995) Attributable mortality of bacteraemia associated with the *Bacteroides fragilis* group. *Clin. Infect. Dis.* 20: 1492–1496.

Reeves AR, Wang G, Salyers AA (1997) Characterization of four outer membrane proteins that play a role in utilization of starch by *Bacteroides thetaiotaomicron. J. Bacteriol.* 179: 643–649.

Reid JH, Patrick S (1984) Phagocytic and serum killing of capsulate and non-capsulate *Bacteroides fragilis. J. Med. Microbiol.* 17: 247–257.

Reid JH, Patrick S, Dermott E, Trudgett A, Tabaqchali S (1985) Investigation of antigenic expression of *Bacteroides fragilis* by immunogold labelling and immunoblotting with a monoclonal antibody. *FEMS Microbiol. Lett.* 30: 289–293.

Reid JH, Patrick S, Tabaqchali S (1987) Immunochemical characterization of a polysaccharide antigen of *Bacteroides fragilis* with an IgM monoclonal antibody. *J. Gen. Microbiol.* 133: 171–179.

Riley TV, Mee BJ (1984) Haemagglutination of *Bacteroides fragilis. FEMS Microbiol. Lett.* 25: 229–232.

Rocha ER, Smith CJ (1995) Biochemical and genetic analyses of a catalase from the anaerobic bacterium *Bacteroides fragilis. J. Bacteriol.* 177: 3111–3119.

Rocha ER, Selby T, Coleman JP, Smith CJ (1996) Oxidative stress response in an anaerobe, *Bacteroides fragilis*: a role for catalase in protection against hydrogen peroxide. *J. Bacteriol.* 178: 6895–6903.

Rogers MB, Parker AC, Smith DJ (1993) Cloning and characterization of the endogenous cephalosporinase gene, *cepA*, from *Bacteroides fragilis* reveals a new subgroup of Ambler class A beta-lactamases. *Antimicrob. Agents Chemother.* 37: 2391–2400.

Rotstein OD, Kao J, Houston K (1989a) Reciprocal synergy between *Escherichia coli* and *Bacteroides fragilis* in an intraabdominal infection model. *J. Med. Microbiol.* 29: 269–276.

Rotstein OD, Vittorini T, Kao J *et al.* (1989b) A soluble Bacteroides by-product impairs phagocytic killing of *Escherichia coli* by neutrophils. *Infect. Immun.* 57: 745–753.

Rudek W, Haque R (1976) Extracellular enzymes of the genus *Bacteroides. J. Clin. Microbiol.* 4: 458–460.

Ruimy R, Podgaljen I, Breuil J, Christen R, Collatz E (1996) A recent fixation of *cfiA* genes in a monophyletic cluster of *Bacteroides fragilis* is correlated with the presence of multiple insertion elements. *J. Bacteriol.* 178: 1914–1918.

Salyers AA, Shoemaker N (1996) Genetics of human colonic bacteroides. In: Mackie RI, White BA, Isaacson RE (eds) *Gastrointestinal Microbiology.* New York: Chapman & Hall, pp. 299–320.

Salyers AA, Shoemaker N, Cooper A, D'Elia J, Shipman JA (1999) Genetic methods for *Bacteroides* species. In: Smith MCM, Socket RE (eds) *Genetic Methods for Diverse Prokaryotes.* Methods in Microbiology, Vol. 29. London: Academic Press, pp. 229–249.

Sawyer RG, Adams RB, May AK, Rosenlof LK, Pruett TL (1995) CD4 + T cells mediate preexposure-induced increases in murine intraabdominal abscess formation. *Clin. Immunol. Immunopathol.* 77: 82–88.

Schauer R (1985) Sialic acids and their role as biological masks. *Trends Biochem. Sci.* 10: 357–360.

Scudder P, Lawson AM, Hounsell EF *et al.* (1987) Characterisation of oligosaccharides released from human-blood-group O erythrocyte glycopeptides by the endo-beta-galactosidase of *Bacteroides fragilis. Eur. J. Biochem.* 168: 585–593.

Sebald M (1994) Genetic basis for antibiotic resistance in anaerobes. *Clin. Infect. Dis.* 18 (Suppl)(4): S297–304.

Shah HN, Collins MD (1989) Proposal to restrict the genus *Bacteroides* (Castellani and Chalmers) to *Bacteroides fragilis* and closely related species. *Int. J. Syst. Bacteriol.* 39: 85–87.

Shah HN, Gharbia SE (1991) *Bacteroides* and *Fusobacterium*: classification and relationships to other bacteria. In: Duerden BI, Drasar BS (eds) *Anaerobes in Human Disease*. London: Edward Arnold, pp. 62–84.

Shah HN, Gharbia SE, Duerden BI (1998) *Bacteroides, Prevotella, Porphyromonas*. In: Collier L, Balows A, Sussman M (eds) *Topley and Wilson's Microbiology and Microbial Infections*, 9th edn, London: Edward Arnold, Vol. 2, pp. 1305–1330.

Shaheduzzaman SM, Akimoto S, Kuwahara T, Kinouchi T, Ohnishi Y (1997) Genome analysis of *Bacteroides* by pulsed-field gel electrophoresis: chromosome sizes and restriction patterns. *DNA Res.* 4: 19–25.

Shoemaker NB, Wang G, Salyers AA (1992) Evidence for natural transfer of a tetracycline resistance gene between bacteria from the human colon and bacteria from the bovine rumen. *Appl. Environ. Microbiol.* 58: 1313–1320.

Shoemaker NB, Wang G, Salyers AA (2000) Multiple gene products and sequences required for excision of the mobilizable integrated *Bacteroides* element NBU1. *J. Bacteriol.* 182: 928–936.

Smith CJ, Tribble GD, Bayley DP (1998) Genetic elements of *Bacteroides* species: a moving story. *Plasmid* 40: 19–29.

Snydman DR, McDermott L, Cuchural GJ *et al.* (1996) Analysis of trends in antimicrobial resistance patterns among clinical isolates of *Bacteroides fragilis* group species from 1990 to 1994. *Clin. Infect. Dis.* 23 Suppl. 1: S54–S65.

Stevens AM, Sanders JM, Shoemaker NB, Salyers AA (1992) Genes involved in production of plasmid-like forms by a *Bacteroides* conjugal chromosomal element share amino acid homology with two-component regulatory systems. *J. Bacteriol.* 174: 2935–2942.

Summanen P, Baron EJ, Citron DM *et al.* (1993) *Wadsworth Anaerobic Bacteriology Manual*, 5th edn. Belmont, CA: Star Publishing Company.

Szoke I, Pascu C, Nagy E, Ljung A, Wadstrom T (1996) Binding of extracellular matrix proteins to the surface of anaerobic bacteria. *J. Med. Microbiol.* 45: 338–343.

Tally FP, Ho JL (1987) Management of patients with intraabdominal infection due to colonic perforation. *Curr. Clin. Topics Infect. Dis.* 8: 266–295.

Tang YP, Dallas MM, Malamy MH (1999) Characterization of the *BatI* (Bacteroides aerotolerance) operon in *Bacteroides fragilis*: isolation of a *B. fragilis* mutant with reduced aerotolerance and impaired growth in *in vivo* model systems. *Mol. Microbiol.* 32: 139–149.

Trinh S, Haggoud A, Reysset G (1996) Conjugal transfer of the 5-nitroimidazole resistance plasmid pIP417 from *Bacteroides vulgatus* BV-17: characterization and nucleotide sequence analysis of the mobilization region. *J. Bacteriol.* 178: 6671–6676.

Tzianabos AO, Pantosti A, Baumann H *et al.* (1992) The capsular polysaccharide of *Bacteroides fragilis* comprises two ionically linked polysaccharides. *J. Biol. Chem.* 267: 18230–18235.

Tzianabos AO, Onderdonk AB, Rosner B, Cisneros RL, Kasper DL (1993) Structural features of polysaccharides that induce intra-abdominal abscesses. *Science* 262: 416–419.

Tzianabos AO, Onderdonk AB, Smith RS, Kasper DL (1994) Structure–function relationships for polysaccharide-induced intra-abdominal abscesses. *Infect. Immun.* 62: 3590–3593.

Tzianabos AO, Russell PR, Onderdonk AB *et al.* (1999) IL-2 mediates protection against abscess formation in an experimental model of sepsis. *J. Immunol.* 163: 893–987.

Veillon A, Zuber A (1897) Sur quelques microbes strictement anaerobies et leur role dans la pathologie humaine. *C R Soc. Biol. (Paris)* 49: 253–255.

Veillon A, Zuber A (1898) Sur quelques microbes strictement anaerobies et leur role en pathologie. *Arch. Med. Exp. Anat.* 10: 517–545.

Vel WAC, Namavar F, Marian A *et al.* (1986) Haemagglutination by the *Bacteroides fragilis* group. *J. Med. Microbiol.* 21: 105–107.

Wang Z, Fast W, Benkovic SJ (1999) On the mechanism of the metallo-beta-lactamase from *Bacteroides fragilis*. *Biochemistry* 38: 10013–10023.

Weikel CS, Grieco FD, Reuben J, Myers LL, Sack RB (1992) Human colonic epithelial cells, HT29/C1, treated with crude *Bacteroides fragilis* enterotoxin dramatically alter their morphology. *Infect. Immun.* 60: 321–327.

Weintraub A, Larsson BE, Lindberg AA (1985) Chemical and immunochemical analyses of *Bacteroides fragilis* lipopolysaccharides. *Infect. Immun.* 49: 197–201.

Whittle G, Bloomfield GA, Katz ME, Cheetham BF (1999) The site-specific integration of genetic elements may modulate thermostable protease production, a virulence factor in *Dichelobacter nodosus*, a causative agent of ovine footrot. *Microbiology* 145: 2845–2855.

Willis AT (1991) Abdominal sepsis. In: Duerden BI, Drasar BS (eds) *Anaerobes in Human Disease*. London: Edward Arnold, pp. 197–223.

Willis AT, Ferguson JR, Jones PH *et al.* (1976) Metronidazole in prevention and treatment of bacteroides infections after appendicectomy. *BMJ* 1: 318–321.

Wilson KH (1997) Biota of the human gastrointestinal tract. In: Mackie RI, White BA, Isaacson RE (eds) *Gastrointestinal Microbiology*. New York: Chapman & Hall, Vol. 2, pp. 39–58.

Woese CR (1987) Bacterial evolution. *Microbiol. Rev.* 51: 221–271.

Wren MD (1991) Laboratory diagnosis of anaerobic infection. In: Duerden BI, Drasar BS (eds) *Anaerobes in Human Disease*. London: Edward Arnold, pp. 180–196.

PART

16

CENTRAL NERVOUS SYSTEM INFECTIONS

92

Central Nervous System Infections: A Clinical Overview

Jonathan D. Sussman

Greater Manchester Neuroscience Centre, Salford, UK

The central nervous system may be infected by a wide variety of bacteria, often as part of what is at first a bacteraemia but may later become a septicaemia, and sometimes as a result of extension from adjacent tissues. This chapter considers the pathophysiology of CNS infections and some of the host responses to these infections. In some cases an inadequate immune response may allow chronic infection to develop. Under other circumstances, notably meningitis, it is the host response that gives rise to tissue injury.

Bacterial Meningitis

Meningitis is an inflammatory disease of the meninges with involvement of the endothelial, ependymal, glial and neuronal cells. Injury to the nervous system arises from the infecting bacteria and also from the inflammatory mediators produced by lymphocytes, monocytes and macrophages as well as astrocytes, microglia and the endothelial cells in response to the infection. In the case of Gram-positive bacterial meningitis, the inflammatory response is triggered by teichoic acid-containing fragments of the cell wall. In the case of Gram-negative bacteria, such as *Haemophilus*

influenzae and *Escherichia coli*, lipopolysaccharide (endotoxin) induces an inflammatory response. In spite of evidence that an initial paucity of polymorphonuclear leucocyte (PMN) activity may permit bacterial growth to become established, PMNs are responsible for many of the adverse affects of the disease. In fatal cases, infiltration in the subarachnoid space by PMNs may extend into cranial nerves and invade the brain surface to produce foci of tissue destruction, ranging from inflammation of the brain (cerebritis) to abscess formation resulting in thrombosis in small and large arteries with resulting infarction.

Clinical Features

The early clinical manifestations of meningitis include lethargy, irritability, fever, muscle pain (myalgia), headache and vomiting. As the disease progresses, photophobia, neck stiffness and drowsiness often develop. Presentation as a pure meningitis is found in only 15–30% of patients, with the remainder showing features of septicaemia, which can progress over as little as a few hours to severe shock or cerebral oedema; death occurs in up to 40% of cases. Long-term

 doi:10.1006/bkmm.2001.0092

neurological sequelae are found in 30% of survivors and include hearing loss, focal neurological deficits, impaired cognitive function and hydrocephalus. Whereas many cases of *Neisseria meningitidis* infection respond well to short courses of antibiotics and recover completely, meningitis due to *Streptococcus pneumoniae*, and neonatal meningitis due to group B streptococci and *E. coli* give rise to significant morbidity and mortality.

Mechanisms of Bacterial Invasion

The development of meningitis requires that bacteria enter the cerebrospinal fluid (CSF) by way of the circulation or by direct spread from infection in adjacent tissues. In the case of meningitis arising from the upper respiratory tract, bacteria attach to and colonise the host nasopharyngeal mucosal epithelium from where they invade adjacent vascular spaces. Here they may be exposed to the host complement-mediated defences against bacteraemia. If these defences fail, the organisms must then cross the blood–brain barrier and enter the CSF where they must survive and replicate. At each step of this process, specific host defences must be overcome; bacterial IgA proteases offer resistance against mucosal IgA, and once in the circulation, capsular polysaccharide protects against complement. Bacteria must adhere to and cross the blood–brain barrier in order to enter the subarachnoid space. Studies have shown that *in vitro Strep. pneumoniae* can anchor to and invade activated vascular endothelial cells through the platelet-activating factor receptor (Cundell *et al.*, 1995). In the case of *E. coli*, outer membrane protein A contributes to endothelial cell invasion by interacting with epitopes on endothelial cell glycoproteins (Prasadarao *et al.*, 1996a, b).

Cerebrospinal fluid offers only limited initial resistance against infection, since the ability of PMNs to restrict bacterial growth is limited by low levels of immunoglobulin and complement which results in insufficient opsonisation of the pathogen (Ernst *et al.*, 1983). Normal CSF flow then results in a rapid spread of bacteria over the surface of the brain into the paravascular spaces.

Inflammatory Responses

After CSF colonisation, host responses eventually generate an immune response, which may lead to bacterial killing, but it also results in inflammation, oedema and brain cell injury.

The mere presence of bacteria does not of itself generate inflammation, and extensive evidence from experimental meningitis indicates that suppression of the host responses to infection may diminish inflammation and tissue injury. The role of host responses in neurological damage is reflected in the severity of *Strep. pneumoniae* meningitis, which is related to the amount of bacterial product released per generation time, and this correlates with the magnitude of the inflammatory response (Täuber *et al.*, 1991).

Bacterial cell wall material and lipopolysaccharide stimulate meningeal inflammation by inducing the secretion of interferon γ (IFNγ), IL-1β, IL-6, tumour necrosis factor α (TNFα) and prostaglandins. This results in increased permeability of the blood–brain barrier. Bacterial replication or lysis in the CSF results in an initial phase that lasts for a few hours, during which the intrathecal release of the pro-inflammatory cytokines IL-1 and TNFα induce up-regulation of the expression of adhesion molecules that enhance neutrophil-binding to the blood–brain barrier. It is thought that the selectins ELAM-1 and CD62 are responsible for the initial phase of leucocyte entry to the CNS (Quagliarello and Scheld, 1992; Fassbender *et al.*, 1997). During the early phase of experimental pneumococcal meningitis in rats, histamine, possibly released by activated mast cells, initiates P-selectin up-regulation and subsequent leucocyte rolling, but not firm adhesion (Weber *et al.*, 1997b). This initial weak adhesion is important in capturing leucocytes, which is a prerequisite for subsequent firm adhesion and diapedesis through the blood–brain barrier. In a rat model of bacterial meningitis, inflammatory changes are attenuated by heparin, which interferes with leucocyte rolling (Weber *et al.*, 1997a).

Serum concentrations of the endothelium-derived adhesion molecules sELAM-1 and sICAM-1, are significantly increased in patients with bacterial meningitis, and the serum level of sELAM-1 is associated with the extent of CSF pleocytosis and with the concentrations of the potentially harmful pro-inflammatory cytokines IL-1β and TNFα in the CSF. The importance of adhesion molecules in enhancing leucocyte entry into the CNS is demonstrated by the ability of anti-ICAM-1 monoclonal antibody to reduce the inflammatory changes in experimental bacterial meningitis (Weber *et al.*, 1995).

Cerebral vascular endothelial cells may regulate the critical steps in inflammatory blood–brain barrier disruption. In response to cell walls of *Strep. pneumoniae*, they express mRNA for ICAM-1, TNFα, and also iNOS. NO production is mediated in part by an autocrine pathway that involves TNFα, whereas ICAM-1 expression is entirely mediated by this autocrine loop (Freyer *et al.*, 1999). Vascular endothelial cells also activate the alternative complement pathway (Winkelstein and Tomasz, 1977), express pro-coagulant factors (Geelen *et al.*, 1992) and release platelet activating factor (Cabellos *et al.*, 1992).

The low concentration of complement in normal CSF offers little defence against infection, but this is up-regulated in response to infection (Tuomanen *et al.*, 1986; Stahel *et al.*, 1997; Gasque *et al.*, 1998). Activated complement in the CSF is chemotactic and contributes to CSF leucocytosis, as well as forming membrane attack complexes that damage neurones.

Chemoattractant cytokines (chemokines) also play a role in attracting neutrophils, monocytes and T cells to sites of developing inflammation (Spanaus *et al.*, 1997). In patients with bacterial meningitis, MCP-1, MIP-1α and MIP-1β are detectable in the CSF. In experimental *Listeria monocytogenes* meningitis CXC and CC chemokines, MIP-1α, MIP-1β and MIP-2 are produced intrathecally by macrophages and leucocytes (Lahrtz *et al.*, 1998).

Matrix metallo-proteinases may also be involved in altering the blood–brain barrier permeability during meningitis. In a rat model of meningococcal meningitis, intracisternal injection of heat-killed organisms disrupts the blood–brain barrier, resulting in increased intracranial pressure, and a CSF pleocytosis that parallels gelatinase B (MMP-9) activity (Paul *et al.*, 1998). Elevated concentrations of MMP-9, collagenase-3, stromelysin-1 and tissue inhibitor of metalloproteinase-1 (TIMP-1) have been detected in the CSF of adults with bacterial meningitis (Kieseier *et al.*, 1999).

The cell surface receptor CD14 of mononuclear phagocytes may be responsible for the transcription of pro-inflammatory cytokines in response to bacterial endotoxin. A variety of cells of myeloid origin can express CD14 in response to systemic endotoxin. These include microglia, choroid plexus and leptomeningeal macrophages, and parenchymal and perivascular-associated microglial cells (Lacroix *et al.*, 1998). A positive correlation between CSF levels of IL-12p40 and IFNγ, TNFα, IL-6 and IL-10 suggests that they are induced by a common stimulus. There is evidence that IL-12, together with TNFα as a co-stimulator, are probably responsible for the production of IFNγ and contribute to local host defence in the CSF compartment (Kornelisse *et al.*, 1997). When TNFα alone is injected intrathecally, only minor inflammatory changes result, whereas it dramatically augments experimental meningitis (Angstwurm *et al.*, 1998).

In response to TNFα and IL-1 there is induction of phospholipase A2, which hydrolyses precursors of cell membrane arachidonic acid and intracellular platelet-activating factor (PAF). This results in the synthesis of arachidonate and PAF. Arachidonate is converted to the pro-inflammatory prostaglandins, thromboxanes and leukotrienes. Cyclo-oxygenase (COX) is required for the synthesis of prostaglandins and is induced in microglia in response to lipopolysaccharide. When COX is inhibited in experimental pneumococcal meningitis, CSF leucocytosis is reduced by more than 90% (Tuomanen *et al.*, 1987). The role of resident brain cells in generating injurious inflammation is clear. Platelet-activating factor, which increases vascular permeability and activates neutrophils, is an inflammatory mediator derived from endothelial cells, macrophages, neutrophils and platelets. Pneumococcal cell wall material can bind to the PAF receptor on microglia, where it contributes to inflammation and leucocytosis (Cabellos *et al.*, 1992).

The contribution individual cytokines make to injury is unclear. The concentration of IL-1 in the CSF of patients with bacterial meningitis correlates both with the concentration of TNFα, and with the risk of neurological sequelae (Mustafa *et al.*, 1989; Arditi *et al.*, 1990). TNFα causes increased leakiness of the blood–brain barrier, leucocyte passage into the CNS, brain oedema, raised intracranial pressure and CSF lactate, and it promotes neuronal apoptosis (Braun and Toumanen, 1999; Bogdan *et al.*, 1997). Nevertheless, inhibition of TNFα or IL-1 does not significantly diminish experimental lipopolysaccharide-induced CSF inflammation (Paris *et al.*, 1995).

Protective anti-inflammatory cytokines are also secreted during the course of meningitis. IL-10 inhibits the release of pro-inflammatory cytokines, including TNFα, IL-1 and IL-6 (de Waal-Malefyt *et al.*, 1991). It also inhibits the release of free oxygen radicals by macrophages (Fiorentino *et al.*, 1991). When IL-10 is given systemically in experimental pneumococcal meningitis, CSF blood flow increases, oedema is reduced and CSF leucocytosis is diminished (Koedel *et al.*, 1996). Transforming growth factor β (TGFβ) suppresses macrophage hydrogen peroxide release, IFNγ-induced release of nitrogen species, macrophage adhesion to endothelium and the production of pro-inflammatory cytokines (Braun and Tuomanen, 1999). In experimental pneumococcal meningitis TGFβ reverses the reduction in CNS perfusion, and reduces cerebral oedema and intracranial pressure (Pfister *et al.*, 1992).

Effectors of Cell Injury

Neither bacterial components, toxins nor the cytokines and chemokines they induce give rise to significant cytotoxicity. At concentrations up to 5 μg/mL endotoxin is not toxic for cultured neurones, astrocytes or microglia (Kim *et al.*, 1995).

Pneumococcal cell wall is not toxic for neurones, but it has some toxicity for microglia and low toxicity for endothelial cells and astrocytes (Geelen *et al.*, 1993; Kim *et al.*, 1995). The substantial neuronal toxicity of these components when injected intracisternally in mice demonstrates that host-derived inflammatory mediators are the neurotoxic mediators (Täuber *et al.*, 1992; Kim and Tauber, 1996). The final step in cell injury arises from reactive oxygen and nitrogen species, excitotoxicity and alterations in cerebral blood flow. The marked increase in CSF lactic acid and the reduction in CSF glucose are thought in part to reflect a shift to glycolysis in the brain with impaired oxygen utilisation, which does not arise simply from reduced cerebral perfusion (Guerra-Romero *et al.*, 1992).

Superoxides are generated in meningeal, glial and CSF inflammatory cells and are highly neurotoxic, and induce necrosis and apoptosis (Freyer *et al.*, 1996, Leib *et al.*, 1996). Resident brain microglia and invading leucocytes are capable of producing nitric oxide in response to *Strep. pneumoniae*, *H. influenzae* and *E. coli* meningitis (Mustafa *et al.*, 1989), and there is evidence *in vitro* that astrocytes may produce nitrite in response to pneumococci (Bernatowicz *et al.*, 1995). Nitric oxide is cytotoxic for oligodendrocytes (Mitrovic *et al.*, 1995), and reacts with superoxide radicals to produce peroxynitrite, a strong oxidant that may be the final effector of neuronal injury (Pfister *et al.*, 1993). Nitric oxide is a powerful vasodilator that contributes to the pathophysiology in early meningitis, and in later stages of experimental meningitis it may counteract cerebral ischaemia (Täuber *et al.*, 1997).

Excitotoxicity is a mechanism of neuronal injury that arises through over-stimulation by excitatory amino acids. CSF macrophages and microglia may be stimulated to express excitatory amino acids (Spranger *et al.*, 1996a, b; Wood, 1995), and peroxynitrite induces release of γ-aminobutyric acid from cortical neurones (Ohkuma *et al.*, 1995). In meningitis, glutamate levels are elevated and correlate with disease severity and outcome (Spranger *et al.*, 1996a, b).

Alterations in blood flow are highly variable in meningitis. Cortical blood flow tends at first to be increased, but it falls as the disease progresses (Paulson *et al.*, 1974; Pfister *et al.*, 1990). Eventual loss of auto-regulation results in changes of cerebral perfusion that correlate with the difference between the arterial pressure and intracranial pressure, so that hypotensive shock may result in reduced brain perfusion that may exacerbate the ischaemia that results from cortical thrombophlebitis. A poor clinical outcome is significantly related to increased cerebral arterial blood flow velocity, probably because it reflects diffuse vasospasm (Ries *et al.*, 1997).

Cytotoxic, vasogenic and interstitial cerebral oedema have been described in meningitis. Vasogenic oedema, which arises from increased blood–brain barrier pinocytosis and separation of vascular endothelial tight junctions, is probably the major source of oedema (Quagliarello *et al.*, 1986). Bacterial cell wall components, in particular disaccharide tetrapeptide, can open the blood–brain barrier *in vivo*, and disrupt tight junctions *in vitro* (Spellerberg *et al.*, 1995). Increased blood–brain barrier permeability may increase CSF volume, and bacterial toxins and brain-derived cytokines increase intracellular water. Experimental neutralisation of TNFα (Saukkonen *et al.*, 1990) and elimination of granulocytes both result in a reduction of oedema in experimental models (Tuomanen *et al.*, 1989). A role for arachidonic acid metabolites, particularly PGE_2, is suggested by a variety of experimental animal models (Chan and Fishman, 1984; Tureen *et al.*, 1991). Loss of cerebral blood flow auto-regulation may lead to an increase in cerebral blood volume, and the syndrome of inappropriate secretion of antidiuretic hormone is frequently seen in patients with meningitis, and both may contribute to brain oedema (Garcia *et al.*, 1981). The drainage of CSF through the arachnoid villi into the cerebral sinus system is also perturbed in meningitis and contributes to increased CSF pressure (Scheld *et al.*, 1980). The resulting increase in intracranial volume varies in degree, but when associated with major infarction may result in rapid herniation of the brain through the foramen magnum with resulting death (Klatzo, 1987).

Host Susceptibility Factors

Young adults and children are at risk of acquiring *N. meningitidis* infection, whereas older individuals are more likely to be infected by *Strep. pneumoniae*. Staphylococcal infection is seen in individuals with endocarditis or after head injury, neurosurgery or in those with indwelling intravascular catheters or shunts. Patients who are being ventilated are susceptible to meningitis due to *Pseudomonas*, *Flavobacterium* and *Proteus*. Gram-negative meningitis may be associated with head injury or neurosurgery, and sepsis including that of infective endocarditis, ruptured brain abscesses and systemic *Strongyloides* infection, and in conditions of immunocompromise, including that after splenectomy. Gram-negative anaerobic meningitis often arises as a result of rupture of a cerebral abscess. *Strep. pneumoniae* is the commonest cause of meningitis in adults, and is particularly associated with pneumonia, sickle cell disease, cirrhosis and myeloma, and dural tears after head injury.

Neisseria meningitidis

Neisseria meningitidis commonly colonises the naso-pharynx without producing symptoms. Asplenic individuals are at increased risk of developing meningococcal disease, and certain inherited complement pathway defects also increase the risk of meningococcal disease. Inherited deficiencies in the terminal common complement pathway proteins (C5–C9) are a risk factor for meningococcal disease. Properdin deficiency also increases the risk of meningococcal disease, and factor D deficiency is associated with recurrent *Neisseria* infection.

In meningococcal septic shock, endotoxin causes intravascular fibrin deposition with microthrombus formation that results in infarction of the skin and extremities and multi-organ failure, with a mortality of up to 50%. Some patients, however, develop bacteraemia or meningitis without septic shock, and this runs a benign course with a mortality rate of less than 5% (Kilpi *et al.*, 1991). The severity of meningococcal disease is related to the plasma concentration of endotoxin, which is related to higher concentrations of inflammatory mediators such as TNFα and interleukin-1β (Waage *et al.*, 1989; Brandtzaeg *et al.*, 1989b). In patients with fulminating meningococcal sepsis, neutrophils, monocytes and the coagulation and complement cascades become activated, and in almost all patients coagulation is deranged, but this is most marked in those with severe septic shock (Brandtzaeg *et al.*, 1989a).

A number of host factors that influence the severity of the disease have been investigated. At presentation with meningococcal disease, a high concentration of plasminogen activator inhibitor-1 (PAI-1), a key inhibitor of the fibrinolytic system, is associated with an adverse outcome in patients with septic shock (Kornelisse *et al.*, 1996; Hermans *et al.*, 1999). However, the 4G/5G deletion/insertion in the promoter region of the PAI-1 gene, which results in elevated plasma concentrations of PAI-1 in meningococcal meningitis, is not a risk factor for meningococcal infection. Patients whose relatives are carriers of the 4G/4G genotype have a 6-fold increased risk of developing septic shock rather than meningitis, but mortality is not affected (Westendorp *et al.*, 1999). In children, the homozygous 4G phenotype may be associated with a 2- to 4.8-fold increased risk of death (Hermans *et al.*, 1999). Inherited prothrombotic abnormalities are not related to the development of septic shock (Westendorp *et al.*, 1996).

Polymorphisms in inflammatory cytokines correlate with outcome in meningococcal disease (Westendorp *et al.*, 1994; Van Dissel *et al.*, 1998). High TNFα levels are associated with a poor outcome, rather than a greater severity of disease (Tracey, 1995). Both worse outcome and greater disease severity are associated with a G-to-A substitution at position 308 in the TNFα gene promoter region, which is associated with higher inducible levels of TNFα (Booy *et al.*, 1997), and IL-1β levels also contribute to outcome (Westendorp *et al.*, 1995).

The bactericidal/permeability-increasing protein (BPI) in the azurophilic granules of polymorphonuclear leucocytes that binds to bacteria and their endotoxins, inhibits endotoxin-induced inflammatory responses and kills a broad range of Gram-negative bacteria. Recombinant BPI inhibits endotoxin-induced cytokine release and circulatory changes, endotoxin-dependent activation of neutrophils, and the coagulation and fibrinolytic pathways. Evidence for the importance of these pathways is the significantly reduced mortality and morbidity when recombinant BPI is administered to children with severe meningococcal sepsis (Giroir *et al.*, 1997).

Listeria monocytogenes

The highest age-specific rates for *Listeria monocytogenes* meningitis are amongst neonates and adults over 60. Neonates acquire early-onset infection *in utero*, or late-onset within a few weeks of birth, when infection is more likely to lead to meningitis. In adults, pregnancy is the commonest risk factor. In other adults, risk factors include advanced age, immunosuppression, malignancy, renal failure, diabetes, alcohol abuse, HIV infection, organ transplantation and administration of corticosteroids or cytotoxic chemotherapy. Since the organism is intracellular, defects of cell-mediated immunity, but not complement or immunoglobulin deficiency, are risk factors for infection.

Listeria monocytogenes expresses surface proteins termed internalins that promote its entry into epithelial cells and hepatocytes. After phagocytosis it becomes enclosed in a sub-cellular organelle, a phagolysosome, the low pH of which activates listeriolysin O, an exotoxin that leads to rapid lysis of the organelle membrane and release of the organism. Once in the cytoplasm, the organism proliferates, and becomes encased in host cell actin filaments that enable the organism to form a filopod, which may be ingested by adjacent cells, so allowing transmission of the disease without bacterial exposure to the extracellular environment. The organism crosses the meninges and blood-brain barrier by endothelial cell or macrophage phagocytosis, and the use of the host-cell contractile system to migrate to and grow within the brain (Southwick and Purich, 1996).

Unlike most forms of bacterial meningitis, in which neutrophils account for more than 80% of the cellular response, in *Listeria* meningitis the percentage of neutrophils is lower (Hansen *et al.*, 1987), and organisms are rarely found in the CSF. In the case of *Mycobacterium tuberculosis*, monocytes exposed to the organisms release monocyte chemoattractant protein-1 (MCP-1), which attracts monocytes and lymphocytes (Kasahara *et al.*, 1994), but whether a similar mechanism operates in *Listeria* infection of the CNS is unknown.

The greater degree of invasion of the CNS in *Listeria* meningitis than in other forms of meningitis is reflected in more common focal and generalised seizures. *Listeria* may invade the brainstem and the cerebral cortex; this does not occur in other forms of bacterial meningitis.

Streptococcus pneumoniae

Pneumococcal meningitis is the commonest cause of meningitis between the ages of 1 and 23 months, and above the age of 19 (Schuchat *et al.*, 1997). The nasopharynx is the primary site of colonisation, and the vast majority of pneumococcal isolates are capsulated (Tuomanen *et al.*, 1995).

The ability of clinical isolates of pneumococci to invade and migrate through brain microvascular endothelial cell monolayers is variable, and loss of the capsule markedly increases invasion. A two-directional pathway of pneumococcal trafficking across the blood–brain barrier has been described in which interaction of pneumococci with the platelet-activating factor receptor results in the transcytosis of bacteria across the cell. However, non-PAF receptor entry shunts bacteria for exit and re-entry on the apical surface in a novel recycling pathway (Ring *et al.*, 1998).

Invasive pneumococcal disease occurs with increased frequency in individuals with defective antibody production and those with defects in the ability to clear bacteria. In children, sinus or ear infection results in a transient bacteraemia with haematogenous seeding of the choroid plexus. In adults the major risk factors include splenectomy, diabetes mellitus, alcohol abuse, liver disease and HIV infection. Patients with pneumococcal pneumonia may also develop meningitis.

Haemophilus influenzae

Risk factors for the development of *H. influenzae* meningitis in adults include recent or past head injury or neurosurgery, sinusitis, otitis media and dural tears with CSF leakage.

Streptococcus agalactiae

Most neonatal infections with *Strep. agalactiae* (group B *Streptococcus*) are acquired during birth. Risk factors include maternal colonisation, *Strep. agalactiae* bacteriuria during pregnancy, premature birth, low birthweight, and prolonged rupture of the membranes. In adults, pregnancy-associated group B streptococcal infection commonly becomes manifest during labour, but meningitis is rare. Meningitis in non-pregnant adults is associated with advanced age, diabetes, cirrhosis of the liver and malignancy.

Infections with Spirochaetes

Neurosyphilis

Up to one-third of patients infected with *Treponema pallidum* may develop tertiary syphilis, which consists of meningeal invasion, an endarteritis and parenchymal invasion. This is manifested as meningitic or meningovascular disease 5–10 years after, or parenchymal disease 10–30 years after the primary infection. Late neurosyphilis primarily affects CNS parenchyma. An estimated 4–9% of patients with untreated syphilis develop symptomatic neurosyphilis, with meningovascular syphilis in 2–3%, general paresis in 2–5% and tabes dorsalis in 1–5%. Since early diagnosis of syphilis is now usual, the classical forms of tertiary syphilitic disease are now uncommon.

Acute syphilitic meningitis occurs 2 months to 26 years after the primary infection (Merritt and Moore, 1935). It is characterised mainly by neck stiffness, headache and cranial neuropathies, and deafness is common. Meningovascular disease that gives rise to large or small vessel arteritis with occlusion and infarction, may give rise to focal neurological disease, and cognitive or psychiatric symptoms. The disease occurs an average of 7 years after the primary infection (Merritt *et al.*, 1946).

General paresis ('general paralysis of the insane', GPI) is a parenchymal disorder that arises from a progressive encephalitis 5–25 years after the primary infection (Merritt and Moore, 1935; Merritt *et al.*, 1946). The presentation may include prominent psychiatric symptoms, and the CSF shows a lymphocytic meningitis. Tabes dorsalis, a spinal form of the disease has the longest latency, with an interval of up to 47 years (Merritt *et al.*, 1946). It is characterised by ataxia.

Cell-mediated immunity is necessary for clearance of *T. pallidum*. In HIV infection or after inadequate antibiotic therapy, systemic infection is cleared but neurological or ocular infection may persist with an increased likelihood of relapse (Marra *et al.*, 1992).

Sonicated *T. pallidum* injected into rabbits induces the expression of mRNA for IL-2 and IFNγ, but not for IL-10, which indicates that the T cell response to *T. pallidum* antigens is biased towards the TH1 phenotype. The endoflagellar sheath protein TpN37 contributes most to this response and may play a key role in the clearance of *T. pallidum* from lesions (Arroll *et al.*, 1999). Primary and secondary syphilitic lesions contain mRNA that encodes IL-2, IFNγ, IL-12p40 and IL-10. Messenger RNA for IL-4 is not present and IL-5 and IL-13 are found in only 25% of lesions. This suggests a local TH1-predominant cellular response activates macrophages, with IFNγ-activated macrophages as the primary effectors of treponemal clearance (Van Voorhis *et al.*, 1996a). Spirochaetal lipoprotein is the principal component of intact bacteria responsible for monocyte activation by a CD14-dependent pathway, and is an important determinant of the pro-inflammatory capacity of the spirochaete (Sellati *et al.*, 1998, 1999). Early in infection by intratesticular infections of rabbits by *T. pallidum* there is a peak of serum opsonic activity, but antibodies that promote optimal macrophage-mediated killing develop much later. The early antibody response that augments phagocytosis and killing corresponds with the *in vivo* clearance of treponemes from the primary site of infection, suggesting that macrophages are the major effector mechanism for the elimination of *T. pallidum* during early infection (Baker-Zander *et al.*, 1993a). VDRL antibodies appear to enhance phagocytosis of *T. pallidum*, but they do not enhance macrophage killing of *T. pallidum*, and rabbits passively immunised with VDRL antibodies are partially protected against *T. pallidum* infection, suggesting a functional role for VDRL antibodies in syphilis (Baker-Zander *et al.*, 1993b). A sub-population of *T. pallidum* that persists after the majority of bacteria have been cleared appears to be resistant to opsonisation and phagocytosis and may provide a mechanism for bacterial persistence (Lukehart *et al.*, 1992).

Circulating immune complexes have been described in experimental syphilis in rabbits. IgG complexes in chronic syphilis significantly decrease the immunological responsiveness of lymphocytes in the macrophage migration inhibition test, and may facilitate the multiplication of treponemes in the host (Podwinska, 1991). In experimental syphilis in rabbits, production of IL-2 by spleen cells is enhanced after 4 days but it is subsequently suppressed. The premature down-regulation of IL-2 secretion may explain why some treponemes persist and why the secondary phase of the disease occurs. It has been proposed that organisms that cause chronic infection may have evolved mechanisms by which a TH1 to TH2 switching occurs, so permitting the establishment of chronicity (Fitzgerald, 1992).

Immunohistochemical examination of secondary syphilis lesions in human skin has shown that most CD4$^+$ cells have histiocytic, rather than lymphocytic, morphological characteristics and that they are CD14$^+$, which suggests that they have a monocytic origin. The lymphocytes are mainly CD8$^+$, activated and express granzyme B and perforin (Van Voorhis *et al.*, 1996b). The presence of HIV-1 infection does not affect the cutaneous response to syphilitic infection (McBroom *et al.*, 1999).

The Jarisch Herxheimer reaction is a systemic response to the release of pyrogens that occurs soon after the first adequate dose of antibiotic treatment. The symptoms comprise rigors, fever, hypotension, leucopenia and exacerbation of meningovascular symptoms and signs that may not be reversible, and death may occur. Studies of this reaction in relapsing fever due to *Borrelia recurrentis* suggest that it is associated with transient elevation of plasma TNFα, IL-6 and IL-8 (Negussie *et al.*, 1992).

Lyme Neuroborreliosis

Borrelia burgdorferi, a spirochaete, is transmitted by *Ixodes* ticks that require deer for their survival, though these are not the reservoir of the organism (see Chapter 97). The illness usually commences with a spreading red rash (erythema chronicum migrans), followed by general dissemination of the organism, including to the CNS. Facial nerve palsy may occur, sometimes before sero-conversion has taken place, but usually the patients already have systemic symptoms.

Early neuroborreliosis includes a combination of lymphocytic meningitis, cranial neuropathies and painful radiculoneuritis. Less frequent manifestations include mononeuritis multipex, optic neuritis and Guillain-Barré syndrome. More advanced disease includes peripheral neuropathy and encephalomyelitis.

Abscesses

Cerebral abscesses are focal areas of suppuration most commonly arising by spread of infection from local structures such as ear, sinuses or teeth. They may also arise from brain injury, surgery or craniofacial osteomyelitis. Cerebral abscesses may also arise by haematogenous spread, particularly from lung abscesses and bronchiectasis. Metastatic abscesses, which are often multiple, are commonly found at the grey-white matter border in the distribution of the middle cerebral artery, where local perfusion is poor. Solitary abscesses are most common in the frontal and temporal lobes. A source of infection cannot be found in approximately one-fifth of abscesses (Anderson, 1993). Bacteria require damaged brain tissue, such as areas of necrosis resulting from ischaemia, thrombophlebitis or embolisation in order to establish infection. Once infection is established, cerebritis develops followed by central necrosis and capsule formation (Britt et al., 1981). Cerebral abscesses present with focal neurological signs appropriate to the site of the lesions combined with symptoms of raised intracranial pressure.

The microbiology of the abscess depends on the source of the primary infection. In 30–60% of cases mixed infections are present, and these often include non-sporing anaerobes. Patients with defects in T cell immunity are predisposed to infections with intracellular pathogens.

Epidural abscesses are those in the space between the skull and the dura mater, and they commence as cranial osteomyelitis that spreads from ear, sinus or orbital infections, or nasopharyngeal malignancy. Spinal epidural abscess is associated with intravenous drug abuse, diabetes mellitus and previous spinal surgery.

Subdural empyemas are collections of pus in the space between dura mater and the arachnoid. They commonly arise from ear or sinus infections, penetrating injury, or infection of subdural effusions.

Tuberculosis

Tuberculosis is primarily a pulmonary infection, but many organs may become involved by blood-borne spread. Involvement of the CNS is found in up to 10% of immunocompetent patients. During initial infection with *Mycobacterium tuberculosis*, the bacteria that reach the distal air spaces of the lung are phagocytosed by alveolar macrophages. *M. tuberculosis* has evolved multiple mechanisms to promote its efficient entry into macrophages, and the passage of the organism through macrophages may be an essential early step in the pathogenesis of tuberculosis (Zimmerli et al., 1996).

Phagocytosis initiates the innate immune response, which in turn orchestrates the adaptive response. This interaction may occur by opsonisation with C3 by the alternative complement pathway, by way of complement receptors CR1, CR3 and CR4. Interaction of mannose-containing cell wall glycolipids with macrophage lectins is also recognised, in which surfactant-associated protein A up-regulates macrophage mannose receptor activity and enhances the potential for phagocytosis of *M. tuberculosis* (Schlesinger et al., 1994, Gaynor et al., 1995). Neither the C3 or mannose mechanisms are specific for pathogenic mycobacteria and are also used by other intracellular organisms (Schlesinger, 1993, Stokes et al., 1993). Monocyte-derived macrophages from tuberculous patients have a lesser ability to adhere to and ingest *M. tuberculosis* than those from healthy controls. This does not arise from defects in complement receptors, but suggests the existence of other molecules in the interaction between bacteria and macrophage (Zabaleta et al., 1998).

A further mechanism of phagocytosis found only for pathogenic mycobacteria requires the association of the complement cleavage product C2a with mycobacteria, resulting in the formation of a C3 convertase. This results in C3b opsonisation of the mycobacteria and phagocytosis. This mechanism requires the cleavage of less than 1% of the serum concentration of C2, suggesting that the local concentrations of C2a at the site of a mycobacterial infection is likely to be in the biologically active range (Schorey et al., 1997). Patients with active pulmonary tuberculosis have high levels of circulating immune complexes with activation of the classical complement pathway resulting in increased serum levels of C2a (Sai-Baba et al., 1990).

The mannose and complement receptors induce rearrangements in the actin cytoskeleton of the macrophage, which leads to the internalisation of the bacterium; the extent of phagocytosis is tightly coupled to activation of phospholipase D (Kusner et al., 1996; Aderem and Underhill, 1999). The choice of host cell receptor and the mechanism of binding may influence the intracellular fate of the tubercle bacilli, and this may be of importance in the outcome of primary infection, where the number of bacilli is presumed to be very low (Cywes et al., 1997; Schlesinger, 1998).

The first requirement for infection is intracellular survival in the phagocytic cells, which may be regulated by bacterial surface glycolipids that may also influence pathogenicity (Fujiwara, 1997). After phagocytosis, the organism modifies vacuolar maturation in a manner that favours their survival and replication within human macrophages. Tubercle bacilli induce macrophage fusion, to form multi-nucleated giant cells that are a histopathological feature of tuberculosis. Studies of swine microglia in cultures containing *M. bovis* suggest that two microglial cell receptors, CD14 and a β-2 integrin, and TNFα participate in their formation (Peterson *et al.*, 1996).

There are a number of host responses to the early stages of *M. tuberculosis* infection. Alveolar macrophages play a prominent and efficient role in the primary defence of the lung by complement receptor-mediated uptake, predominantly by way of CR4, and TNFα-mediated killing of bacteria (Hirsch *et al.*, 1994). There is evidence for the involvement of a range of cytokines in pulmonary disease. IFNγ activates human alveolar macrophages and induces antimycobacterial activity by increasing the proportion of human alveolar macrophages that ingest bacteria, increasing the number of ingested BCG in individual alveolar macrophages, and increasing the killing activity of macrophages (Shimokata, 1996).

A further mechanism of host defence arises from macrophage apoptosis induced by *M. tuberculosis*. This may contribute to protective immunity by removing the environment in which bacteria replicate, but it may also enhance microbial virulence by reduced bacterial killing (Kornfeld *et al.*, 1999).

Once pulmonary infection is established, monocytes express a range of chemokines including monocyte chemotactic and activating factor/monocyte attractant protein-1, which further enhances the inflammatory response (Kasahara *et al.*, 1994).

Classical complement activation is induced by mycobacteria, and this may be mediated by antilipoarabinomannan IgG on the bacilli. Classical complement activation may be important for the extent of phagocytosis of *M. tuberculosis* by mononuclear phagocytes, which may affect the course of the infection (Hetland *et al.*, 1998). In pulmonary disease, pleural SC5b-9, an activation product of complement common pathway, is significantly raised and its level correlates significantly with factor Bb levels, and with lactate dehydrogenase, a marker of tissue damage, which suggests that complement activation plays a significant role in tuberculosis (Hidaka *et al.*, 1995).

T cells play a complex role in protection against tuberculosis. CD4+ T cells remain the dominant subset, though $\gamma\delta$ and CD8+ T cells probably have important complementary roles. All three subsets are sources of IFNγ and competent cytotoxic effector cells (Boom, 1996). Continuing recirculation of T lymphocytes enables the repeated contacts of sensitised CD4+ T cells with specific antigen. TH1 and TH2 cells are activated, leading to the secretion of many cytokines. The optimal development of cellular response is usually accompanied by a lack of humoral response and vice versa. Continuing stimulation by *M. tuberculosis* antigens leads not only to T-lymphocyte differentiation towards TH1–TH2 population,

but also causes the activation of cells to form tuberculous granulation because *M. tuberculosis* is not eliminated. *M. tuberculosis*-specific cytolytic activity is mostly mediated by CD4$^+$ cytotoxic lymphocytes that kill infected target cells by inducing Fas (APO-1/CD95)-mediated apoptosis. Apoptosis of infected macrophages induced through receptors of the TNF family may be an immune effector mechanism that deprives mycobacteria of their growth environment and also reduces viable bacterial counts, but the mechanism is unknown. Interference by *M. tuberculosis* with the FasL system may represent an escape mechanism for bacteria attempting to evade the effect of apoptosis (Oddo *et al.*, 1998).

Under the stimulation of mycobacterial antigens and cytokines, B lymphocytes mature and secrete immunoglobulin, and immune complexes are formed. There is a correlation between the increased level of immune complexes in severe tuberculosis and the decrease of T-cell activity and tissue damage. Immune complexes activate complement by classical or alternative pathways, resulting in increased concentrations of C3 and C4 in active disease. The role of the humoral response in the development and course of tuberculosis remains unknown (Dubaniewitz, 1997)

Raised concentrations of TNFα, IFNγ, sTNFR-75, sTNFR-55, and IL-10 are found in the CSF of patients with tuberculous meningitis. The levels of TNFα do not decrease over 16 months after treatment is commenced. Cytokine release appears unrelated to the stage of tuberculous meningitis or the clinical outcome (Mastroianni *et al.*, 1997).

Only about one in ten of those who become infected by tuberculosis develop clinical disease (Murray *et al.*, 1990), and in most cases there is no identifiable risk factor. Comparison of the incidence of TB in siblings of an index case identifies a higher concordance between monozygotic than dizygotic twins, which demonstrates a genetic element to host susceptibility (Bellamy, 1998a). Resistance to infection with intracellular pathogens in mice has been attributed to the natural resistance-associated macrophage protein (*Nramp*) gene on chromosome 1. Variations in the equivalent human gene are associated with tuberculosis in West Africans, and it is likely that other genes also contribute to disease resistance or susceptibility (Bellamy, 1998b).

A number of risk factors are associated with acquisition of tuberculosis. In HIV-infected persons, immune suppression, as indicated by low CD4$^+$ cell count, is an independent risk factor for tuberculosis (Sudre *et al.*, 1996). CNS tuberculosis is associated with immunosuppression following gastrectomy. Host factors appear to influence susceptibility to CNS infection. A retrospective Canadian study of tuberculosis in non-HIV patients demonstrated that patients with CNS tuberculosis are significantly more likely to be under 40 years old, female, and of Aboriginal origin than those without CNS involvement (Arvanitakis *et al.*, 1998). In children, the development of tuberculous meningitis is associated with significantly low numbers of CD4 T lymphocyte counts as compared with children with a primary pulmonary complex only (Rajajee and Narayanan, 1992). Studies of experimental tuberculous meningitis in rabbits suggests that TNFα is a determinant of pathogenesis and disease progression in mycobacterial infection in the central nervous system, where virulence parallels the expression of TNFα, which in turn correlates with high CSF leucocytosis, high protein accumulation, severe meningeal inflammation, persistent bacillary load and progressive clinical deterioration (Tsenova *et al.*, 1999). Analysis by RFLP of the molecular diversity of *M. tuberculosis* isolates from patients with CNS tuberculosis has revealed that, although several strains of *M. tuberculosis* cause CNS tuberculosis, the predominance of one strain suggests that its occurrence may be strain-dependent (Arvanitakis *et al.*, 1998).

Initially, mycobacterial replication occurs in the alveoli. After a few weeks an asymptomatic phase of haematogenous spread ensues, during which CNS tubercles develop. Patients with miliary pulmonary infection without symptoms of CNS infection may have CNS granulomas on imaging (Gupta *et al.*, 1997), but in those with evolving brain tuberculosis a lesion may only be visible on the chest radiograph in 30% of cases (Garg, 1999). Initially, small tuberculous lesions consisting of mononuclear cells surrounding a necrotic centre develop in the CNS (Rich's foci). In some individuals, mononuclear cells invade the caseous lesions, resulting in lesion expansion and these may eventually rupture. The site of the expanding tubercle determines whether rupture gives rise to meningitis, a parenchymal abscess or a tuberculoma. Meningitis with or without tuberculoma is three times as common as tuberculoma alone. Tuberculosis of the CNS has a morbidity rate of about 29% and a mortality rate of about 26%. In tuberculosis, adverse outcome in terms of morbidity or mortality is significantly more common in those with meningitis (Arvanitakis *et al.*, 1998).

In tuberculous meningitis, an exudate of neutrophils, mononuclear cells, erythrocytes and bacilli occupies the subarachnoid space. As the infection becomes chronic, lymphocytes and connective tissue cells predominate with resulting involvement of the cranial nerves. Persisting neutrophilic meningitis may occur in HIV-infected patients with multi-drug resistant mycobacteria. This may also be found in immunocompromised individuals without HIV infection. An arteritis eventually develops with vessel occlusion giving rise to focal ischaemia of the brain, brainstem or spinal cord. In the cord, myelitis, radiculitis or sometimes chronic transverse myelitis may be found. Obstruction of CSF pathways may give rise to obstructive hydrocephalus. Localised brain inflammation adjacent to areas of exudate may also give rise to tissue injury (Udani *et al.*, 1971).

The symptoms in tuberculous meningitis depend on the stage of the disease and the site of the lesions in the CNS. In the prodromal period patients have nonspecific fatigue, malaise, myalgia and fever. Infection of the upper respiratory tract is often present, but the fever, lethargy and irritability are out of proportion to it. In adults with tuberculous meningitis, fever, headache and vomiting are the commonest features with variable cognitive disturbance and neck stiffness. Cranial neuropathies are found in a quarter of patients. Seizures and paralysis are found in a quarter of children but are less common in adults. Hydrocephalus affects over 80% of children, but only 50% of adults. The clinical features of tuberculous meningitis are not modified by concurrent HIV infection, but there may be a greater incidence of tuberculous intracerebral mass lesions in HIV-positive intravenous drug users (Berenguer *et al.*, 1992; Dube *et al.*, 1992).

Tuberculous granulomas (tuberculomas) consist of a zone of glial scarring and oedema surrounding a capsule of collagenous tissue, multi-nucleated giant cells and monocytes that contains a necrotic centre that contains a few bacilli. They may be multiple in up to one-third of patients and may co-exist with meningitis, but they are uncommon in the spinal cord. When the core of a tuberculoma liquefies, a tuberculous abscess results. These are larger than tuberculomas, contain many more bacteria, and lack the granulomatous reaction that surrounds tuberculomas.

Infection of the intervertebral disc at the interface of the disc vertebrae, usually in the thoraco-lumbar region, may result in vertebral collapse and spinal cord or radicular compression. This may arise by *M. tuberculosis* infection leading to expression of chaperonin 10, which stimulates bone resorption (Meghji *et al.*, 1997).

References

Aderem A, Underhill DM (1999) Mechanisms of phagocytosis in macrophages. *Ann. Rev. Immunol.* 17: 593–623.

Anderson M (1993) Management of cerebral infection. *J. Neurol. Neurosurg. Psychiatry* 56: 1243–1258.

Angstwurm K, Freyer D, Dirnagl U *et al.* (1998) Tumour necrosis factor alpha induces only minor inflammatory changes in the central nervous system, but augments experimental meningitis. *Neuroscience* 86: 627–634.

Arditi M, Manogue KR, Caplan M *et al.* (1990) Cerebrospinal fluid cachectin/tumor necrosis factor-alpha and platelet-activating factor concentrations and secerity of bacterial meningitis in children. *J. Infect. Dis.* 162: 139–147.

Arroll TW, Centurion-Lara A, Lukehart SA *et al.* (1999) T-cell responses to *Treponema pallidum* subsp, pallidum antigens during the course of experimental infection. *Infect. Immun.* 67: 4757–4763.

Arvanitakis Z, Long RL, Hershfield ES *et al.* (1998) *M. tuberculosis* molecular variation in CNS infection: Evidence for strain-dependent neurovirulence. *Neurology* 50: 1827–1832.

Baker-Zander SA, Shaffer JM, Lukehart SA (1993a) Characteristics of the serum requirement for macrophage killing of *Treponema pallidum* spp. Pallidum: Relationship to the development of opsonizing antibodies. *FEMS Immunol. Med. Microbiol.* 6: 273–279.

Baker-Zander SA, Shaffer JM, Lukehard SA (1993b) VDRL antibodies enhance phagocytosis of *Treponema pallidum* by macrophages. *J. Infect. Dis.* 167: 1100–1105.

Bellamy R (1998a) Genetic susceptibility to tuberculosis in human populations. *Thorax* 53: 588–593.

Bellamy R (1998b) Variations in the NRAMP1 gene and susceptibility to tuberculosis in West Africans. *N. Engl. J. Med.* 338: 640–644.

Berenguer J, Moreno S, Laguna F *et al.* (1992) Tuberculous meningitis in patients infected with the human immunodeficiency virus. *N. Engl. J. Med.* 326: 668–672.

Bernatowicz A, Kodel U, Frei K *et al.* (1995) Production of nitrite by primary rat astrocytes in response to pneumococci. *J. Neuroimmunol.* 60: 53–61.

Bogdan I, Leib SL, Bergeron M *et al.* (1997) Tumor necrosis factor-alpha contributes to apoptosis in hippocampal neurons during experimental group B streptococcal meningitis. *J. Infect. Dis.* 176: 693–697.

Boom WH (1996) The role of T-cell subsets in *Mycobacterium tuberculosis* infection. *Infect. Agents Dis.* 5: 73–81.

Booy R, Nadel S, Hibbert M *et al.* (1997) Genetic influence on cytokine production in meningococcal disease. *Lancet* 349: 1176.

Brandtzaeg P, Kierulf P, Gaustad P *et al.* (1989a) Plasma endotoxin as a predictor of multiple organ failure and death in systemic meningococcal disease. *J. Infect. Dis.* 159: 195–204.

Brandtzaeg P, Sandset PM, Joo GB *et al.* (1989b) The quantitative association of plasma endotoxin, antithrombin, protein C, extrinsic pathway inhibitor and fibrinopeptide A in systemic meningococcal disease. *Thromb. Res.* 55: 459–70.

Braun JS, Tuomanen EI (1999) Molecular mechanisms of brain damage in bacterial meningitis. *Adv. Pediatr. Infect. Dis.* 14: 49–71.

Britt R, Enzmann D, Yeager A *et al.* (1981) Neuropathological and computed tomographic findings in experimental brain abscess. *J. Neurosurg.* 55: 590–603.

Cabellos C, MacIntyre DE, Forrest M *et al.* (1992) Differing roles for platelet-activating factor during inflammation of the lung and subarachnoid space. The special case of *Streptococcus pneumoniae*. *J. Clin. Invest.* 90: 612–618.

Chan PH, Fishman RA (1984) The role of arachidonic acid in vasogenic brain edema. *Fedn Proc.* 43: 210–213.

Cundell DR, Gerard NP, Gerard C *et al.* (1995) *Streptococcus pneumoniae* anchor to activated human cells by the receptor for platelet-activating factor. *Nature* 377: 435–438.

Cywes C, Hoppe HC, Daffe M *et al.* (1997) Nonopsonic binding of *Mycobacterium tuberculosis* to complement receptor type 3 is mediated by capsular polysaccharides and is strain dependent. *Infect. Immun.* 65: 4258–4266.

de Waal-Malefyt R, Abrams J, Bennett B *et al.* (1991) Interleukin 10(IL-10) inhibits cytokine synthesis by human monocytes: an autoregulatory role of IL-10 produced by monocytes. *J. Exp. Med.* 174: 1209–1220.

Dubaniewiz A (1997) Humoral response in pulmonary tuberculosis. *Med. Sci. Monitor* 3: 956–960.

Dube MP, Holtom PD, Larsen RA (1992) Tuberculous meningitis in patients with and without human immunodeficiency virus infection. *Am. J. Med.* 93: 520–524.

Ernst JD, Decazes JM, Sande MA (1983) Experimental pneumococcal meningitis: role of leukocytes in pathogenesis. *Infect. Immun.* 41: 275–279.

Fassbender K, Schminke U, Ries S *et al.* (1997) Endothelial-derived adhesion molecules in bacterial meningitis: association to cytokine release and intrathecal leukocyte-recruitment. *J. Neuroimmunol.* 74: 130–134.

Fiorentino DF, Zlotnik A, Mosmann TR *et al.* (1991) IL-10 inhibits cytokine production by activated macrophages. *J. Immunol.* 147: 3815–3822.

Fitzgerald TJ (1992) The Th-1/Th-2-like switch in syphilitic infection: Is it detrimental? *Infect. Immun.* 60: 3475–3479.

Freyer D, Weih M, Weber JR *et al.* (1996) Pneumococcal cell wall components induce nitric oxide synthase and TNF-alpha in astroglial-enriched cultures. *Glia* 16: 1–6.

Freyer D, Manz R, Ziegenhorn A *et al.* (1999) Cerebral endothelial cells release TNF-alpha after stimulation with cell walls of *Streptococcus pneumoniae* and regulate inducible nitric oxide synthase and ICAM-1 expression via autocrine loops. *J. Immunol.* 163: 4308–4314.

Fujiwara N (1997) Distribution of antigenic glycolipids among *Mycobacterium tuberculosis* strains and their contribution to virulence. *Kekkaku* 72: 193–205.

Garg RK (1999) Tuberculosis of the central nervous system. *Postgrad. Med. J.* 75: 133–140.

Garcia H, Kaplan SL, Feigin RD (1981) Cerebrospinal fluid concentration of arginine vasopressin in children with bacterial meningitis. *J. Pediatr.* 98: 67–70.

Gasque P, Singharao SK, Neal JW *et al.* (1998) The receptor for complement anaphylotoxin C3a is expressed by myeloid cells and nonmyeloid cells in inflamed human central nervous system: analysis in multiple sclerosis and bacterial meningitis. *J. Immunol.* 160: 3543–3554.

Gaynor CD, McCormack FX, Voelker DR *et al.* Pulmonary surfactant protein A mediates enhanced phagocytosis of *Mycobacterium tuberculosis* by a direct interaction with human macrophages. *J. Immunol.* 155: 5343–5351.

Geelen S, Bhattacharyya C, Tuomanen E (1992) Induction of procoagulant activity on human endothelial cells by *Streptococcus pneumoniae. Infect. Immun.* 60: 4179–4183.

Geelen S, Bhattacharyya C, Tuomanen E (1993) The cell wall mediates pneumococcal attachment to and cytopathology in human endothelial cells. *Infect. Immun.* 61: 1538–1543.

Giroir BP, Quint PA, Barton P *et al.* (1997) Preliminary evaluation of recombinant amino-terminal fragment of human bactericidal/permeability-increasing protein in children with severe meningococcal sepsis. *Lancet* 350: 1439–1443.

Guerra-Romero L, Tureen JH, Täuber MG (1992) Pathogenesis of central nervous system injury in bacterial meningitis. *Antibiot. Chemother.* 45: 18–29.

Gupta RK, Kohli A, Guar V *et al.* (1997) MRI of the brain in patients with miliary pulmonary tuberculosis without symptoms or signs of central nervous system involvement. *Neuroradiology* 39: 699–704.

Hansen PB, Jensen TH, Lykkegaard S *et al.* (1987) *Listeria monocytogenes* meningitis in adults: sixteen consecutive cases 1973–1982. *Scand. J. Infect. Dis.* 19: 55–60.

Hermans PWM, Hibbert ML, Booy R (1999) 4G/5G promotor polymorphism in the plasminogen-activator-inhibitor-1 gene and outcome of meningococcal disease. *Lancet* 354: 556–560.

Hetland G, Wiker HG, Hogasen K *et al.* (1998) Involvement of antilipoarabinomannan antibodies in classical complement activation in tuberculosis. *Clin. Diagn. Lab. Immunol.* 5: 211–218.

Hidaka K, Abe M, Tanaka T *et al.* (1995) Comparison of complement activation between tuberculous and malignant pleuritis. *Jpn J. Thoracic Dis.* 33: 379–383.

Hirsch CS, Ellner JJ, Russell DG (1994) Complement receptor-mediated uptake and tumor necrosis factor-alpha-mediated growth inhibition of *Mycobacterium tuberculosis* by human alveolar macrophages. *J. Immunol.* 152: 743–753.

Kasahara K, Tobe T, Tomita M *et al.* (1994) Selective expression of monocyte chemotactic and activating factor/monocyte attractant protein 1 in human blood monocytes by *Mycobacterium tuberculosis. J. Infect. Dis.* 170: 1238–1247.

Kieseier BC, Paul R, Koedel U *et al.* (1999) Differential expression of matrix metalloproteinases in bacterial meningitis. *Brain* 122: 1579–1587.

Kilpi T, Anttila M, Kallio MJT, Peltola H (1991) Severity of childhood bacterial meningitis and duration of illness before diagnosis. *Lancet* 338: 406–409.

Kim YS, Täuber MG (1996) Neurotoxicity of glia activated by gram-positive bacterial products depends on nitric acid production. *Infect. Immun.* 64: 3148–3153.

Kim YS, Kennedy S, Täuber MG (1995) Toxicity of *Streptococcus pneumonia* in neurons, astrocytes and microglia *in vitro. J. Infect. Dis.* 171: 1363–1368.

Klatzo I (1987) Pathophysiological aspects of brain edema. *Acta Neuropathol.* 72: 236–239.

Koedel U, Bernatowicz A, Frei K *et al.* (1996) Systemically (but not intrathecally) administered IL-10 attenuates pathophysiologic alterations in experimental pneumococcal meningitis. *J. Immunol.* 157: 5185–5191.

Kornelisse RF, Hazelzet JA, Savelkoul HFJ *et al.* (1996) The relationship between plasminogen activator inhibitor-1 and proinflammatory and counterinflammatory mediators in children with meningococcal septic shock. *J. Infect. Dis.* 173: 1148–1156.

Kornelisse RF, Hack EC, Savelkoul HFJ *et al.* (1997) Intrathecal production of interleukin-12 and gamma interferon in patients with bacterial meningitis. *Infect. Immun.* 65: 877–881.

Kornfeld H, Mancino G, Colizzi V (1999) The role of macrophage cell death in tuberculosis. *Cell Death Different.* 6: 71–78.

Kusner DJ, Hall CF, Schlesinger LS (1996) Activation of phospholipase D is tightly coupled to the phagocytosis of *Mycobacterium tuberculosis* or opsonized zymosan by human macrophages. *J. Exp. Med.* 184: 585–95.

Lacroix S, Feinstein D, Rivest S (1998) The bacterial endotoxin lipopolysaccharide has the ability to target the brain in upregulating its membrane CD14 receptor within specific cellular populations. *Brain Pathol.* 8: 625–640.

Lahrtz F, Piali L, Spanaus KS *et al.* (1998) Chemokines and chemotaxis of leukocytes in infectious meningitis. *J. Neuroimmunol.* 85: 33–43.

Leib SL, Kim YS, Chow LL *et al.* (1996) Reactive oxygen intermediates contribute to necrotic and apoptotic neuronal injury in an infant rat model of bacterial meningitis due to group B streptococci. *J. Clin. Invest.* 98: 2632–2639.

Lukehart SA, Shaffer JM, Baker-Zander SA (1992) A subpopulation of *Treponema pallidum* is resistant to phagocytosis: possible mechanism of persistence. *J. Infect Dis.* 166: 1449–1453.

Marra CM, Handsfield HH, Kuller L *et al.* (1992) Alterations in the course of experimental syphilis associated with concurrent simian immunodeficiency virus infection. *J. Infect. Dis.* 165: 1020–1025.

Mastroianni CM, Paoletti F, Lichtner M (1997) Cerebral fluid cytokines in patients with tuberculous meningitis. *Clin. Immunol. Immunopathol.* 84: 171–176.

McBroom RL, Styles AR, Chiu MJ *et al.* (1999) Secondary syphilis in persons infected with and not infected with HIV-1: a comparative immunohistologic study. *Am. J. Dermatopathol.* 21: 432–441.

Meghji S, White PA, Nair P *et al.* (1997) *Mycobacterium tuberculosis* chaperonin 10 stimulates bone resorption: a potential contributory factor in Pott's disease. *J. Exp. Med.* 186: 1241–1246.

Merritt HH, Moore M (1935) Acute syphilitic meningitis. *Medicine* 14: 119–183.

Merritt HH, Adams RD, Solomon HC (1946) *Neurosyphilis.* New York: Oxford University Press.

Mitrovic B, Ignarro LJ, Vinters HV *et al.* (1995) Nitric oxide induces necrotic but not apoptotic cell death in oligodendrocytes. *Neuroscience* 65: 531–539.

Murray CJL, Stybo K, Rouillon A (1990) Tuberculosis in developing countries: burden, intervention and cost. *Bull. Int. Union Tuberculosis Lung Dis.* 65: 6–24.

Mustafa MM, Lebel MH, Ramilio O *et al.* (1989) Correlation of interleukin-1 beta and cachexin concentrations in cerebrospinal fluid and outcome from baterial meningitis. *J. Pediatr.* 115: 208–213.

Negussie Y, Remick DG, DeForge LE *et al.* (1992) Detection of plasma tumor necrosis factor, interleukins 6, and 8 during the Jarisch-Herxheimer reaction of relapsing fever. *J. Exp. Med.* 175: 1207–1212.

Oddo M, Renno T, Attinger A *et al.* (1998) Fas ligand-induced apoptosis of infected human macrophages reduces the viability of intracellular *Mycobacterium tuberculosis. J. Immunol.* 160: 5448–5454.

Ohkuma S, Narihara H, Katsura M *et al.* (1995) Nitric oxide-induced [3H] GABA release from cerebral cortical neurons is mediated by peroxynitrite. *J. Neurochem.* 65: 1109–1114.

Paris MM, Friedland IR, Ehrett S *et al.* (1995) Effect of interleukin-1 receptor antagonist and soluble tumor necrosis factor receptor in animal models of infection. *J. Infect. Dis.* 171: 161–169.

Paul R, Lorenzl S, Koedel U *et al.* (1998) Matrix metalloproteinases contribute to the blood–brain barrier disruption during bacterial meningitis. *Ann. Neurol.* 44: 592–600.

Paulson OB, Brodersen P, Hansen EL *et al.* (1974) Regional cerebral blood flow, cerebral metabolic rate of oxygen, and cerebrospinal fluid acid-base variables in patients with acute meningitis and with acute encephalitis. *Acta Med. Scand.* 196: 191–198.

Peterson PK, Gekker G, Hu S *et al.* (1996) Multinucleated giant cell formation of swine microglia induced by *Mycobacterium bovis. J. Infect. Dis.* 173: 1194–1201.

Pfister HW, Koedel U, Haberl RL *et al.* (1990) Microvascular changes during the early phase of experimental bacterial meningitis. *J. Cerebral Blood Flow Metab.* 10: 914–922.

Pfister HW, Frei K, Ottnad B *et al.* (1992) Transforming growth factor beta 2 inhibits cerebrovascular changes and brain oedema formation in the tumor necrosis factor alpha-independent early phase of experimental pneumococcal meningitis. *J. Exp. Med.* 176: 265–268.

Pfister HW, Fontana A, Täuber MG *et al.* (1993) Mechanisms of brain injury in bacterial meningitis: workshop summary. *Clin. Infect. Dis.* 19: 463–479.

Podwinska J (1991) Circulating immune complexes in experimental syphilis and their relationship to immunological response against *Treponema pallidum. FEMS Microbiol. Immunol.* 3: 83–91.

Prasadarao NV, Wass CA, Kim KS (1996a) Endothelial cell GlcNAc□1–4GlcNAc epitopes for outer membrane protein A enhance traversal of *Escherichia coli* across the blood-brain barrier. *Infect. Immun.* 64: 154–160.

Prasadarao NV, Wass CA, Weiser JN *et al.* (1996b) Outer membrane protein A of *Escherichia coli* contributes to invasion of brain microvsacular cells. *Infect. Immun.* 64: 146–153.

Quagliariello V, Scheld WM (1992) Bacterial meningitis: pathogenesis, pathophysiology, and progress. *N. Engl. J. Med.* 327: 864–872.

Quagliariello VJ, Long WJ, Scheld WM (1986) Morphological alterations in the blood–brain-barrier with experimental meningitis in the rat. *J. Clin. Invest.* 77: 1085–1095.

Rajajee S, Narayanan PR (1992) Immunological spectrum of childhood tuberculosis. *J. Trop. Pediatr.* 21: 490–496.

Ries S, Schminke U, Fassbender K *et al.* (1997) Cerebrovascular involvement in the acute phase of bacterial meningitis. *J. Neurol.* 244: 51–55.

Ring A, Weiser JN, Tuomanen EI (1998) Pneumococcal trafficking across the blood-brain barrier molecular analysis of a novel bidirectional pathway. *J. Clin. Invest.* 102: 347–360.

Sai-Baba KSS, Moudgil KD, Jain RC *et al.* (1990) Complement activation in pulmonary tuberculosis. *Tubercle* 71: 103–107.

Saukkonen K, Sande S, Cioffe *et al.* (1990) The role of cytokines in the generation of inflammation and tissue damage in experimental gram-positive meningitis. *J. Exp. Med.* 171: 439–448.

Scheld WM, Dacey RG, Winn HR *et al.* (1980) Cerebrospinal fluid outflow resistence in rabbits with experimental meningitis. *J. Clin. Invest.* 66: 243–253.

Schlesinger LS (1993) Macrophage phagocytosis of virulent but not attenuated strains of *Mycobacterium tuberculosis* is mediated by mannose receptors in addition to complement receptors. *J. Immunol.* 150: 2920–2930.

Schlesinger LS (1998) *Mycobacterium tuberculosis* and the complement system. *Trends Microbiol.* 6: 47–49.

Schlesinger LS, Hull SR, Kaufman TM (1994) Binding of the terminal mannosyl units of lipoarabinomannan from a virulent strain of *Mycobacterium tuberculosis* to human macrophages. *J. Immunol.* 152: 4070–4079.

Schorey JS, Carroll MC, Brown EJ (1997) A macrophage invasion mechanism of pathogenic mycobacteria. *Science* 277: 1091–1093.

Schuchat A, Robinson K, Wenger JD *et al.* (1997) Bacterial meningitis in the United States in 1995. *N. Engl. J. Med.* 337: 970–976.

Sellati TJ, Bouis DA, Kitchens RL *et al.* (1998) *Treponema pallidum* and *Borrelia burgdorferi* lipoproteins and synthetic lipopeptides activate monocytic cells via a CD-14 dependent pathway distinct from that used by lipopolysaccharide. *J. Immunol.* 160: 5455–5464.

Sellati TJ, Bouis DA, Caimano MJ *et al.* (1999) Activation of human monocytic cells by *Borrelia burgdorferi* and *Treponema pallidum* is facilitated by CD14 and correlates with surface exposure of spirochaetal lipoproteins. *J. Immunol.* 163: 2049–2056.

Shimokata K (1996) Analysis of cellular immunity against tuberculosis in man with special reference to tuberculous pleurisy and cytokines. *Kekkaku* 71: 591–596.

Southwick FS, Purich DL (1996) Intracellular pathogenesis of listeriosis. *N. Engl. J. Med.* 334: 770–776.

Spellerberg B, Prasad S, Cabellos C *et al.* (1995) Penetration of the blood–brain-barrier: Enhancement of drug delivery and imaging by bacterial glycopeptides. *J. Exp. Med.* 182: 1037–1044.

Spanaus KS, Nadal D, Pfister HW *et al.* (1997) C-X-C and C-C chemokines are expressed in the cerebrospinal fluid in bacterial meningitis and mediate chemotactic activity on peripheral blood-derived polymorphonuclear and mononuclear cells *in vitro. J. Immunol.* 158: 1956–1964.

Spranger M, Krempien S, Schwab S *et al.* (1996a) Excess glutamate in the cerebrospinal fluid in bacterial meningitis. *J. Neurol. Sci.* 143: 126–131.

Spranger M, Schwab S, Krempien S *et al.* (1996b) Excess glutamate levels in the cerebrospinal fluid predict clinical outcome of bacterial meningitis. *Arch. Neurol.* 53: 992–996.

Stahel PF, Frei K, Fontana A *et al.* (1997) Evidence for intrathecal synthesis of alternative pathway complement activation proteins in experimental meningitis. *Am. J. Pathol.* 151: 897–904.

Stokes RW, Haidl ID, Jefferies WA *et al.* (1993) Mycobacteria-macrophage interactions: macrophage phenotype determines the nonopsonic binding of *Mycobacterium tuberculosis* to murine macrophages. *J. Immunol.* 151: 7067–7076.

Sudre P, Hirschel B, Toscani L *et al.* (1996) Risk factors for tuberculosis among HIV-infected patients. *Eur. Respir. J.* 9: 279–283.

Täuber MG, Burroughs M, Niemoller UM *et al.* (1991) Differences of pathophysiology in experimental meningitis caused by three strains of *Streptococcus pneumoniae. J. Infect. Dis.* 163: 806–811.

Täuber MG, Sachdeva M, Kennedy SL *et al.* (1992) Toxicity in neuronal cells caused by cerebrospinal fluid from pneumococcal and Gram-negative meningitis. *J. Infect. Dis.* 166: 1045–1050.

Täuber MG, Kim YS, Leib SL (1997) Neuronal injury in meningitis. In: Peterson PK, Remington JS (eds) *In Defence of the Brain: Current Concepts in the Immunopathogenesis and Clinical Aspects of CNS Infections.* Malden, MA: Blackwell Science, pp. 125–143.

Tracey KJ (1995) TNF and Mae West or; death from too much of a good thing. *Lancet* 345: 75–76.

Tsenova L, Bergtold A, Freedman VH *et al.* (1999) Tumor necrosis factor alpha is a determinant of pathogenesis and disease progression in mycobacterial infection in the central nervous system. *Proc. Natl Acad. Sci. USA* 96: 5657–5662.

Tuomanen EI, Hengstler B, Zak O *et al.* (1986) The role of complement in inflammation during experimental pneumococcal meningitis. *Microb. Pathogen.* 1: 15–32.

Tuomanen EI, Hengstler B, Rich R *et al.* (1987) Nonsteroidal anti-inflammatory agents in the therapy of experimental pneumococcal meningitis. *J. Infect. Dis.* 55: 483–509.

Tuomanen EI, Saukkonen K, Sande S (1989) Reduction in inflammation, tissue damage, and mortality in bacterial meningitis in rabbits treated with monoclonal antibodies against adhesion-promoting receptors of leukocytes. *J. Exp. Med.* 170: 959–968.

Tuomanen EI, Austrian R, Masure HR (1995) Pathogenesis of pneumococcal infection. *N. Engl. J. Med.* 332: 1280–1284.

Tureen JH, Täuber MG, Sande MA (1991) Effects of indomethacin on the pathophysiology of experimental meningitis in rabbits. *J. Infect. Dis.* 163: 647–649.

Udani PM, Parekh UC, Dastur DK (1971) Neurological and related syndromes in CNS tuberculosis. Clinical features and pathogenesis. *J. Neurol. Sci.* 14: 341 .

Van Dissel , van Langevelde P, Westendorp RGJ *et al.* (1998) Anti-inflammatory cytokine profile and mortality in febrile patients. *Lancet* 351: 950–953.

Van Voorhis WC, Barrett LK, Koelle DM *et al.* (1996a) Primary and secondary lesions contain mRNA for Th1 cytokines. *J. Infect. Dis.* 173: 491–495.

Van Voorhis WC, Barrett LK, Nasio JM *et al.* (1996b) Lesions of primary and secondary syphilis contain activated cytolytic T cells. *Infect. Immun.* 64: 1048–1050.

Waage A, Brandtzaeg P, Halstensen A *et al.* (1989) The complex pattern of cytokines in serum from patients with meningococcal septic shock: association between interleukin 6, interleukin 1, and fatal outcome. *J. Exp. Med.* 169: 333–338.

Weber JR, Angstwurm K, Burger W *et al.* (1995) Anti ICAM-12 (CD54) monoclonal antibody reduces inflammatory changes in experimental bacterial meningitis. *J. Neuroimmunol.* 63: 63–68.

Weber JR, Angstwurm K, Rosenkranz T *et al.* (1997a) Heparin inhibits leukocyte rolling in pial vessels and attenuates inflammatory changes in a rat model of bacterial meningitis. *J. Cerebral Blood Flow Metab.* 17: 1221–1229.

Weber JR, Angstwurm K, Rosenkranz T *et al.* (1997b) Histamine (H1) receptor antagonist inhibits leukocyte rolling in pial vessels in the early phase of bacterial meningitis in rats. *Neurosci. Lett.* 226: 17–20.

Westendorp RGJ, Langermans JAM, Huizinga TWJ *et al.* (1994) Genetic influence on cytokine production and fatal meningococcal disease. *Lancet* 349: 170–173.

Westendorp RGJ, Langermans JAM, de Bel CE *et al.* (1995) Release of tumour necrosis factor: an innate host characteristic that contributes to the outcome of meningococcal disease. *J. Infect. Dis.* 71: 1057–1060.

Westendorp RGJ, Reitsma PH, Bertina RM (1996) Inherited prothrombotic disorders and infectious purpura. *Thromb. Haemost.* 75: 899–901.

Westendorp RGJ, Hottenga JJ, Slagboom PE (1999) Variation in plasminogen-activator-inhibitor-1 gene and risk of meningococal septic shock. *Lancet* 354: 561–563.

Winkelstein JA, Tomasz A (1977) Activation of the alternative pathway by pneumococcal cell walls. *J. Immunol.* 118: 451–454.

Wood PL (1995) Microglia as a unique target in the treatment of stroke. Potential neurotoxic mediators produced by activated microglia. *Neurol. Res.* 17: 242–248.

Zabaleta J, Arias M, Maya JR *et al.* (1998) Diminished adherence and/or ingestion of virulent *Mycobacterium tuberculosis* by monocyte-derived macrophages from patients with tuberculosis. *Clin. Diagn. Lab. Immunol.* 5: 690–694.

Zimmerli S, Edwards S, Ernst JD (1996) Selective receptor blockade during phagocytosis does not alter the survival and growth of *Mycobacterium tuberculosis* in human macrophages. *Am. J. Resp. Cell Mol. Biol.* 15: 760–770.

93

Haemophilus influenzae

Nicola J. High

School of Biological Science, University of Manchester, Manchester, UK

Haemophilus influenzae is a facultative anaerobe that has an absolute requirement for NAD (factor V) and a source of haem (factor X). It was first described by Pfeiffer, during the influenza pandemic of 1889–1892. At the time, a Gram-negative coccobacillus was frequently isolated from the sputum of individuals with influenzal pneumonia. The frequency of this observation led to the conclusion that this organism, named *Haemophilus influenzae*, was the cause of influenza. Forty years later, it was finally discovered that the causative agent of influenza was in fact a virus, but by then the name *H. influenzae* had become firmly established in the literature (Smith *et al.*, 1933). After this discovery, *H. influenzae* was more accurately defined as a commensal organism of the human respiratory tract and it was shown to be a common cause of upper and lower respiratory tract infections. Occasionally, it also causes more serious, life-threatening invasive disease such as meningitis (reviewed by Turk, 1982).

In 1931, Dr Margaret Pittman showed that the species could be divided into capsulated and non-encapsulated strains. The former could be further subdivided into six distinct serotypes, designated a–f, based upon the chemical structures of the capsular polysaccharide. The spectrum of disease caused by non-encapsulated and capsulated *H. influenzae* differs. Type b strains alone are responsible for 95% of invasive infections, including cellulitis, septic arthritis, epiglottitis, pneumonia and meningitis (Turk, 1984). These diseases, in particular meningitis, are predominant amongst young children between the ages of 4 months and 4 years. The remaining serotypes and non-encapsulated strains cause localised upper and lower respiratory tract infections, including sinusitis and pneumonia. Non-typable strains are also a common cause of otitis media and conjunctivitis (Turk, 1984).

Of the infections caused by *H. influenzae*, meningitis is of particular importance because of its potential to cause lasting neurological damage, even with appropriate supportive and antibiotic treatment (Rodriguez *et al.*, 1972). Before the introduction in 1992 of the *H. influenzae* (Hib) type b vaccine, it was a leading cause of meningitis and was responsible for approximately 45% of cases (Wenger *et al.*, 1990). The peak incidence of *H. influenzae* meningitis is in children between 9 and 12 months of age. The new-born and children over 5 and adults rarely contract this infection. In 1933 Fothergill and Wright demonstrated that the incidence of *H. influenzae* coincided with the age when bactericidal antibodies specific for *H. influenzae* are absent. New-born babies were shown to be protected against *H. influenzae* infection by bactericidal antibodies passively acquired from the maternal circulation, whilst in older children and adults a functional T-independent immune response efficiently generated *H. influenzae*-specific bactericidal antibodies (Rodrigues *et al.*, 1971; Schneerson *et al.*, 1971; Anderson *et al.*, 1972). Consequently, children are most at risk during the period when maternal antibodies have waned and their own immune system has yet to reach maturity.

 doi:10.1006/bkmm.2001.0093

Haemophilus influenzae Disease

Epidemiology

Haemophilus influenzae is present in the respiratory tract of many individuals (Pfeiffer, 1892) and, for the greater part of the time, it is a harmless commensal. More than 90% of individuals aged over 1 year carry one or more *Haemophilus* species in their nasopharynx. Of these, 49% are *H. influenzae*, and a significant minority (6.6%) are encapsulated (Turk, 1982). Colonisation is transient, and characterised by the acquisition and elimination of a large number of different strains over a given period of time. Carriage of *H. influenzae* by children is an important source of disease transmission, particularly in day-care centres, where a single isolate can be passed to several children simultaneously.

Vaccination has dramatically reduced *H. influenzae* carriage rates, decreasing the chances of transmission of infection from one child to another (herd effect) (Barbour *et al.*, 1995). However, infections caused by other capsular serotypes and non-typable organisms remain an important cause of disease.

Colonisation of the nasopharynx is for all *H. influenzae* strains a critical first step in pathogenesis. In the case of type b strains penetration of the nasopharyngeal mucosa and subsequent entry into the bloodstream results in systemic infection, whilst in the case of non-typable strains contiguous spread within the respiratory tract can occur, resulting in infection of the middle ear, sinuses or lungs. Analysis of the stages in pathogenesis with human nasopharyngeal organ cultures has delineated many of the steps that follow the initial colonisation event and lead to the development of invasive disease. In contrast, the events that precede the onset of disease caused by non-typable organisms are not as clearly defined.

Invasive Disease

The four major stages in the development of invasive disease by *H. influenzae* type b are depicted in **Fig. 1**. These events depend absolutely on the successful colonisation of the nasopharynx, during which *H. influenzae* initially attaches to the mucus layer overlying the nasopharyngeal epithelium. Nasopharyngeal mucus provides a first line of defence against invading organisms, facilitating their removal from the respiratory tract. The cilia on the surface of epithelial cells wave backwards and forwards in a co-ordinated manner and in so doing surface mucus and any bacteria

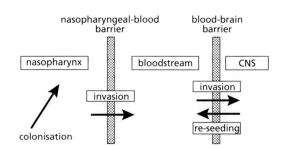

Fig. 1 Stages in the development of *Haemophilus influenzae* type b meningitis.

trapped within it are gradually propelled out of the respiratory tract by the so-called 'mucociliary escalator'. To persist within the nasopharynx *H. influenzae* must, therefore, escape the adhesive mucus matrix and inactivate the mucociliary escalator. The inherent cytotoxic nature of *H. influenzae* plays a key role in this process, bringing about paralysis of the cilia (ciliostasis), the sloughing of ciliated cells and the disruption of epithelial cell tight junctions. Having penetrated the mucus layer, *H. influenzae* attaches selectively to non-ciliated cells, passes between tight junctions and gains access to the submucosa (Farley *et al.*, 1986). Once the mucosal barrier has been crossed, the bacterium is exposed to the full array of host immune defence mechanisms. Survival in the bloodstream and the progress of invasive disease, therefore, depends on the ability of the organism to protect itself from the bactericidal activity of serum.

The mechanisms by which *H. influenzae* enters the subarachnoid space, and thence the central nervous system, are unclear. The port of entry into the CNS is believed to be the choroid plexus, a region of exceptionally high blood flow, which ensures delivery of a maximal numbers of organisms (Tunkel and Scheld, 1993). The ability of *H. influenzae* to invade and transcytose endothelial cells is also predicted to play a crucial role in facilitating translocation across the blood–brain barrier, particularly during the early stages of infection (Virji *et al.*, 1992). The continuing escalation of the inflammatory response, however, in response to developing bacteraemia, also results in the destruction of the tight junction between endothelial cells, allowing recruitment of serum factors and entry of additional organisms into the CNS (Wispelway *et al.*, 1988; Patrick *et al.*, 1992).

Negligible immune defence mechanisms are present in the cerebrospinal fluid (CSF), which allows rapid proliferation of bacteria after translocation of the blood–brain barrier (Zwahlen *et al.*, 1982). For example, the concentration of immunoglobulin G

(IgG) is 800 times greater in the bloodstream than it is in the CSF. Even during the later stages of infection, complement components and immunoglobulins are present at much lower concentrations than in the bloodstream. The presence of endotoxin-containing bacterial cell wall fragments and peptidoglycan in the CSF induces the release of cytokines such as tumour necrosis factor α (TNFα) and interleukin 1β (IL-1β) by glial and endothelial cells. This triggers an inflammatory cascade, which in turn results in an influx of leucocytes into the subarachnoid space, causing cerebral oedema, raised intracranial pressure and neuronal damage. Oxygen radicals and excitatory amino acids have also been shown to contribute to neuronal cell damage (Tuomanen, 1995).

The clinical features of *H. influenzae* infection induced by this sequence of events are non-specific in new-born and young infants, and hamper diagnosis. They include abnormal temperature, somnolence, irritability and poor feeding. The presence of a bulging fontanelle is also a relatively characteristic sign in this age group, but it may not be present early in disease (Tunkel and Scheld, 1995).

Investigation of the pathophysiology of *H. influenzae* meningitis has been greatly aided by the *in vivo* rat experimental model of AL Smith *et al.* (1973). *H. influenzae* administered to infant rats by the intraperitoneal route leads to a progressive disease, in a fashion similar to that observed in humans. The susceptibility of the rat model to infection is influenced by age, with 5-day old rats more prone to infection than 20-day old rats. Further to mimic the events during invasive disease in humans, the infant rat model was modified to allow intranasal inoculation (Moxon *et al.*, 1974). With this model it was demonstrated that after nasopharyngeal colonisation, bacteraemia can arise from inoculation with a single organism, which crosses the nasal mucosa and undergoes clonal expansion in the bloodstream (Moxon and Murphy, 1978).

Disease due to Non-type B and Non-serotypable Strains

Capsular serotypes a and c–f rarely cause systemic infections, which are more commonly associated with type b stains. Although it has been suggested that serotype a and c strains have a particular relationship with sinusitis, encapsulated strains other than type b can generally be regarded as non-pathogenic (Turk, 1984). Upper respiratory tract carriage of non-encapsulated *H. influenzae* is common, occurring in approximately 50–80% of the population (Turk, 1984). These serotypes can therefore be regarded as

a normal component of the human nasopharyngeal flora, nevertheless they are responsible for far more illness than type b strains. The pathogenesis of disease due to non-typable *H. influenzae* involves spread in the respiratory tract, and this is often a consequence of non-specific or specific abnormalities of host immune defences. In particular, viral infections impair mucociliary clearance, and disrupt the integrity of the mucosa, predisposing to infection. Exposure to cigarette smoke also disrupts ciliary function and can increase the likelihood of non-typable *H. influenzae* infection. These cause a broad spectrum of upper and lower respiratory tract infections and are responsible for infection of the middle ear (otitis media) and conjunctivitis.

Lower respiratory tract infections associated with non-typable *H. influenzae* are a major cause of mortality in infants and children in developing countries. Such infections are also prevalent in individuals with chronic obstructive airways disease and cystic fibrosis, where infection can result in impairment of respiratory function and pneumonia (Kilbourn *et al.*, 1983; Gilligan, 1991). Non-typable *H. influenzae* are also responsible for one-third of all cases of acute otitis media, the most common bacterial infection of pre-school children, and they are the primary bacterial cause of chronic otitis media with effusion (Teele *et al.*, 1989). The age of peak incidence is between 6 and 24 months, and by 3 years of age, two-thirds of children may have had one or more episodes of acute otitis media. Although not a life-threatening disease, hearing damage caused by otitis media can result in deficits in language acquisition, speech development and cognitive function (Teele *et al.*, 1989) The consequences on education, employment prospects and quality of life are substantial. The impact of otitis media in world population terms is substantial and it is estimated to be the cause of disabling hearing impairment in 60 million people.

Non-typable *H. influenzae* are also recognised as a neonatal and post-partum pathogen, due to colonisation of the genito-urinary tract. Whilst post-partum sepsis due to non-typable *H. influenzae* produces a relatively mild illness, neonatal infections, in contrast, are severe and manifest as respiratory distress. The onset of symptoms usually occurs within 24 hours of birth, suggesting maternal genital–neonatal transmission (Wallace *et al.*, 1983).

Non-typable *H. influenzae* are increasingly recognised as a significant cause of morbidity and mortality. The associated economic cost is also substantial, prescription drugs and physician visits resulting from non-typable *H. influenzae* infection are estimated, in the United States alone, to total nearly one billion dollars a year (Stool and Field, 1989). The need for a vaccine to protect against infection by these organisms is therefore becoming increasingly clear. Successful vaccination for prevention of disease caused by serotype b strains has been well demonstrated, but the phenotypic diversity inherent in non-typable strains and the absence of an antigenic capsule make the development of an effective vaccine much less straightforward.

Pathogenesis of *Haemophilus influenzae*

Colonisation of Respiratory Mucosa

A critical step in the pathogenesis of *H. influenzae* disease depends upon the ability of the organisms to colonise the nasopharynx. To establish itself on the nasal mucosa the organism must first overcome local immune defences, including the mucociliary escalator and the presence of secretory IgA. To achieve this *H. influenzae* has evolved mechanisms to abolish ciliary function and inactivate IgA. In addition, to persist in respiratory tract *H. influenzae* expresses an array of fimbrial and non-fimbrial adhesins, which promote binding of the organism to the epithelium that underlies the dense, protective layer of respiratory mucus.

Inactivation of IgA1

IgA is the predominant immunoglobulin present on mucosal surfaces, and IgA1 is the dominant antibody in the upper respiratory tract (Brandtzaeg, 1992). IgA plays a major role in the defence of mucosal surfaces by promoting humoral immunity, inactivating bacterial toxins and inhibiting microbial adhesion (Kilian *et al.*, 1996). It prevents colonisation of the respiratory tract by binding bacteria via their antigen-binding sites and simultaneously anchoring itself via its Fc domain into the mucus layer. In this way bacteria are retained within the respiratory mucus, facilitating their removal by the mucociliary escalator. To evade this defence mechanism and persist within the respiratory tract, both capsulate and non-encapsulated *H. influenzae* and other mucosal pathogens, including *Streptococcus pneumoniae* and *N. gonorrhoeae*, secrete an IgA protease (Kilian *et al.*, 1980). The IgA proteases derived from *H. influenzae* are antigenically diverse, and more than 30 antigenic types have been recognised (Lomholt *et al.*, 1993). This antigenic variability is more pronounced in non-typable stains than in encapsulated *H. influenzae*.

The function of IgA protease is, as its name suggests, to disrupt IgA and prevent it from trapping the

organism in the nasopharyngeal mucus. To achieve this it cleaves the heavy chain of human IgA1 in the hinge region (Plaut, 1978). Cleavage of the heavy chain results in the formation of a monomeric Fab fragment, which retains antigen-binding ability, and a monomeric or dimeric functionally inactive Fc fragment. Although the cut IgA molecule can no longer anchor *H. influenzae* into the mucus layer, the antigen-binding domains are still capable of binding to the bacterial cell surface. In so doing they can still, by steric hindrance, prevent adhesion of the organism to the underlying epithelium. The ability of IgA protease-producing bacteria to cleave IgA can be inhibited by human milk (Qiu *et al.*, 1998). This may, in part, explain the protective effect of milk in preventing nasopharyngeal colonisation of *H. influenzae* and other IgA protease-producing respiratory tract pathogens. Inhibition of IgA protease activity is catalysed by lactoferrin, an iron-binding protein present in milk. Although its mode of action is unclear, lactoferrin specifically extracts the IgA protease from the bacterial cell wall, making it susceptible to attack by milk anti-IgA protease antibodies and thereby abolishing its activity.

Two distinct IgA1 proteases have been described for *H. influenzae*, termed type 1 and type 2. Although each cleaves IgA1 at the hinge region, the exact peptide bond forming the cleavage site differs for each enzyme. Type 1 proteases cleave IgA at a proline–serine bond, while type 2 cleave at a proline–threonine bond, four amino acids away (Kilian *et al.*, 1980). The type of IgA protease secreted by *H. influenzae* is dictated by the serotype of the strain. Thus, type 1 enzymes are produced by serotypes a, b, d and f strains, while serotypes c and e strains produce type 2 enzymes (Mulks *et al.*, 1982). IgA2 is not cleaved by either class of enzyme since it lacks the cleavage site located in the hinge region of the molecule (Plautt *et al.*, 1974; Frangione and Wolfenstein-Todel, 1972).

In order to cleave IgA at mucosal surfaces, IgA proteases must be transported to the bacterial cell surface and released extracellularly. This is achieved by a process of autosecretion, the mechanism of which has was first characterised in *N. gonorrhoeae* (Pohlner *et al.*, 1987). Secretion of *H. influenzae* IgA protease was subsequently shown to occur in an identical manner (Poulsen *et al.*, 1989). The *H. influenzae* Rd protease is synthesised as a 185-kDa protein comprising four domains, an N-terminal signal sequence, an internal serine protease domain, a highly basic and α-helical domain and a C-terminal β or helper-domain (Poulsen *et al.*, 1989). The signal sequence directs the export of the protein across the inner membrane and is then cleaved from the protein. The remainder of the protein is subsequently anchored into the outer membrane by its C-terminal domain, which is predicted to form a β-barrel-like structure, containing a hydrophilic channel. This then allows translocation of the α and IgA protease domains of the molecule to the cell surface. Finally, after cleavage within the α domain, the mature IgA protease is released into the extracellular space (Pohlner *et al.*, 1987).

Fimbrial Adhesins

Fimbriae are one of the virulence factors implicated in colonisation in both encapsulated and non-typable strains of *H. influenzae*. These were first identified in type b strains by Guerina *et al.* (1982) and Pichichero *et al.* (1982), who reported that their expression correlated with ability of the organism to haemagglutinate and to adhere to buccal epithelial cells. Four distinct families of fimbriae have been distinguished, but only the long, thick, haemagglutination-positive (LKP) fimbriae mediate adherence to human mucosal epithelial cells. LKP fimbriae also confer binding to mucus and agglutinate human erythrocytes. The latter property facilitated the identification of the receptor recognised by this class of fimbriae, which was shown to comprise a sialic acid-containing lactosylceramide structure which forms part of the human AnWj blood group antigen. More than 14 different strain-specific types of LKP fimbriae have been observed serologically which are composed of subunits ranging from 22 kDa to 27 kDa (Brinton *et al.*, 1989). Of these, only serotype LKP4 is expressed by type b strains.

The amino acid sequence of the LKP fimbrial subunit shows homology to that of *E. coli* P, type 1 and F17 fimbriae (Guerina *et al.*, 1985). In spite of this homology, however, the structure of LKP fimbriae is different. Both fimbriae are composite structures composed of a thick, supporting shaft joined to a thin, adhesive tip fibrillum (Kuehn *et al.*, 1992; St. Geme *et al.*, 1996a). The supporting shaft of the *E. coli* fimbriae all comprise a single strand composed of subunits arranged in a helical structure. In contrast, LKP fimbriae comprise a double helix with a structure analogous to filamentous actin (St. Geme *et al.*, 1996a). In spite of these structural differences, assembly of *E. coli* and *H. influenzae* fimbriae occurs via a similar mechanism (Hultgren *et al.*, 1993; St. Geme *et al.*, 1996b).

Although LKP fimbriae exhibit antigenic variation between strains, unlike *Neisseria gonorrhoeae* pili, intra-strain antigenic variation has not been observed. The expression of LKP fimbriae is, however, subject to phase variation such that non-fimbriated and fimbriated organisms can occur in a given population at

any one time (Guerina *et al.*, 1982; Pichichero *et al.*, 1982; Connor and Loeb, 1983). Organisms isolated from the nasopharynx are often fimbriated, whilst their counterparts isolated from the bloodstream during systemic infection are non-fimbriated. This indicates that, although expression of fimbriae is desirable during initial colonisation of the host, at more advance stages in the disease process during invasion of the bloodstream, it becomes detrimental.

The Molecular Basis of Fimbrial Phase Variation

Biosynthesis of LKP fimbriae involves five genes designated *hifA*–*hifE*, where *hifA* encodes the major structural subunit and *hifB* encodes a chaperone (van Ham *et al.*, 1994). The *hifB*–*E* genes are transcribed as a polycistronic message, while *hifA* is transcribed divergently. To investigate the molecular basis of fimbrial switching in *H. influenzae*, van Ham *et al.* (1993) analysed the organisation of the fimbrial gene cluster in fimbriate and non-fimbriate *H. influenzae* phase variants. Attention was focused on the two divergently transcribed genes, *hifA* and *hifB*, the promotors for which are located on opposite strands in the intervening inter-genic region. Analysis of this region in fimbriate and non-fimbriate phase variants revealed that there were no discernible differences in genetic organisation. For example, no large deletions or re-arrangements, which might be attributable to a re-combination events were observed. Subsequently, however, with northern blot analysis, it was shown that in non-fimbriate strains transcription of neither the fimbrial subunit gene *hifA*, nor its chaperone *hifB* occurred. To investigate why transcription of *hifA* and *hifB* had been inhibited in such strains, the nucleotide sequence of the region between the two genes, which housed each promoter, was determined for both fimbriate and non-fimbriate phase variants. In both cases the −10 and −35 RNA polymerase-binding sites were shown to be separated by dinucleotide repeats of 5′-TA-3′. However, the number of repeats present differed, depending on whether fimbriae were or were not being expressed (see **Fig. 4**). This indicated that transcription of *hifA* and *hifB*, and hence expression of fimbriae, was dictated by the number of copies of 5′-TA-3′ present. By varying the number of copies of 5′-TA-3′, the spacing between the −10 and −35 RNA polymerase-binding sites would be altered.

For transcription to occur under normal circumstances, the optimal spacing between the −10 and −35 regions is 17 nucleotides. Variation in this distance is known to impede docking of the polymerase and would therefore influence the efficiency of transcription. Accordingly, in non-fimbriate variants nine copies of 5′-TA-3′ repeats were consistently detected, reducing the distance between the −10 and −35 sites to just 14 nucleotides, which abolished binding of RNA polymerase and thereby inhibited transcription and the expression of fimbriae. In contrast, fimbriate phase variants contained either 10 or 11 copies of 5′-TA-3′ separating the −10 and −35 sites respectively, by either 16 or 18 bp. Thus, by the loss or gain of one or more copies of 5′-TA-3′, transition between a fimbriate and a non-fimbriate state and vice versa can readily be achieved to enable organisms expressing the most appropriate phenotype for a given stage in infection to be generated.

Fimbriae are not the only phase-variable virulence determinants expressed by *H. influenzae*, further examples and alternative mechanisms for generating this phenomenon and their significance in the pathogenesis of *H. influenzae* disease will be discussed below.

High-molecular-weight Surface Proteins

In addition to the LKP family of fimbriae, non-typable strains of *H. influenzae* also express two unique surface-exposed high-molecular-weight proteins, HMW1 and HMW2, which play an important role in mediating adhesion to epithelial cells (St. Geme *et al.*, 1993). A survey of clinical isolates has indicated that 75% of all non-typable strains express proteins that cross-react with sera raised against HMW1 or HMW2. In contrast, encapsulated strains of *H. influenzae* uniformly fail to express either of these proteins and, moreover, they have been shown, by Southern hybridisation analysis, to lack the genes that encode them (St. Geme *et al.*, 1994b). The presence of the non-fimbriate HMW adhesins was first indicated by the observation that non-typable *H. influenzae*, which lacked fimbriae, were still capable of binding to epithelial cells (St. Geme and Falkow, 1990). Some time later, however, the identity of the non-fimbriate adhesin itself was determined.

The genes that encode HMW1 and HMW2 were originally isolated because both of these proteins are the major targets of human serum antibody generated during the course of non-typable *H. influenzae* infection and they were considered to be potential vaccine candidates (Barenkamp and Bodor, 1990). The cloned genes were 4.4 and 4.7 kb in length and their nucleotide sequences were 100% identical over the first 1259 bp. They encoded, respectively, a 125-kDa and a 120-kDa protein, which showed 71% identity to each other at the amino acid level. Comparison of the amino acid sequence of HMW1 and HMW2 revealed that they shared sequence similarity to the filamentous haemagglutinin expressed by *Bordetella pertussis* (Barenkamp and Leininger, 1992). This provided the first indication that these two proteins might both function as adhesins. The subsequent generation of *H. influenzae* HMW-deficient mutants revealed that, in the absence of these proteins, the ability of non-typable *H. influenzae* to bind to epithelial cells was markedly reduced, confirming the adhesive nature of the HMW proteins. Adhesion was not, however, completely abrogated, indicating that the HMW proteins were not the whole story and that another, as yet unidentified, adhesin was also expressed by non-typable *H. influenzae* (St. Geme *et al.*, 1993).

Down-modulation of High-molecular-weight Proteins

The high-molecular-weight proteins HMW1 and HMW2 are the predominant targets of serum antibody response to natural infection by non-typable *H. influenzae*. In addition, these two proteins play a critical role in adhesion of non-typable *H. influenzae* to epithelial cells. Although the expression of HMW1 and HMW2 does not switch on and off as described for LKP fimbriae or LPS, it can be down-modulated.

This phenomenon was first observed following intrabulbar challenge with a wild-type strain of non-typable *H. influenzae* of chinchillas, previously immunised with purified HMW protein preparations. Isolates from middle ear fluid uniformly expressed decreased amounts of HMW1 and HMW2, suggesting that the expression of these two proteins had been down-modulated. Analysis of the promoter sequence upstream of the *hmw1* and *hmw2* revealed the presence of a series of 7 bp direct repeats arranged in a tandem array. Comparison of this region in the chinchilla variants that expressed low levels of HMW with that of organisms that expressed high levels of HMW, demonstrated an inverse relationship between the number of repeats and the level of mRNA (see **Fig. 4**). As the number of repeats increased, the level of mRNA production and protein expression decreased. In a manner similar to other highly repetitive sequences described for *H. influenzae*, variation in the number of repeats present in the *hmw* promoter region occurs spontaneously and is independent of RecA. By varying the levels of expression of HMW proteins in this manner it has been suggested that non-typable *H. influenzae* can oscillate between efficient colonisation of mucosal surfaces and effective immune evasion (Barenkamp and St. Geme, 1999).

Haemophilus influenzae Adhesin

Approximately 25% of non-typable *H. influenzae* lack HMW1/2-like adhesins, but they are, nevertheless, still capable of adherence to epithelial cells that is independent of fimbriae. To identify this adhesin these strains were further analysed and this revealed a further class of surface-exposed high-molecular-weight proteins which, like HMW1 and HMW2, were a major target for serum antibodies. A 3.3 kb gene encoding a 115-kDa protein termed *H. influenzae* adhesin (Hia) was subsequently isolated (Barenkamp and St. Geme, 1996). The introduction of this gene into *E. coli* conferred the ability to adhere to epithelial cells, confirming the role of Hia as an adhesin. Analysis of the amino acid sequence revealed that Hia was 72% identical to the Hsf adhesin expressed by Hib strains (Barenkamp and St. Geme, 1996). Moreover, the genes that encode these two proteins were located at the same position on the chromosome, suggesting that they were alleles of the same locus.

Expression of Hsf by type b strains is associated with the presence of fine fibril-like structures on the bacterial surface. Analogous structures have not, however, been observed on non-typable strains that express Hia, possibly indicating a subtle difference between the structure of these two organelles. Analysis of a set of non-typable clinical isolates has revealed that Hia is expressed by 80% of these strains, which lack both HMW1 and HMW2. In contrast, it is universally absent in strains that express HMW adhesins (St. Geme *et al.*, 1998). This observation, in conjunction with the observation of an Hia homologue in *H. influenzae* type b strains, has led to the suggestion that non-typable *H. influenzae* that express Hia have diverged more recently from the encapsulated strains than non-typable *H. influenzae* that express HMW proteins.

Hap Adhesin

H. influenzae mutants deficient in the expression of HMW1, HMW2 and Hia can surprisingly still promote low-level adhesion to cultured epithelial cells. To identify the molecule responsible for this adhesion event, a gene library was constructed and a gene *hap* was identified. Nucleotide sequence analysis of this gene revealed that it has considerable homology to the serine-type IgA1 protease expressed by *H. influenzae* (St. Geme *et al.*, 1994). Similar to the IgA protease family, Hap contains a serine protease catalytic domain and is subject to auto-secretion from the cell. In spite of its homology to IgA protease, Hap is unable to cleave purified IgA1. Instead, its homology seems to stem from a similar mechanism of processing and secretion. The concept that a secreted protein functions as an adhesin is intriguing and has led to the suggestion that Hap may, to some extent, remain

surface-associated, as has been observed for the secreted filamentous haemagglutinin expressed by *Bordetella pertussis*.

In addition to promoting interaction between *H. influenzae* and epithelial cells, Hap also promotes bacterial aggregation and the formation of microcolonies on the surface of epithelial cells (Hendrixson and St. Geme III, 1998). Secretory leucocyte protease inhibitor (SLPI), a natural component of respiratory mucus, increases the adhesive properties of Hap. It achieves this by inhibiting auto-proteolysis of Hap so that, instead of being secreted, it remains intimately attached to the bacterial cell surface. It has been suggested that when concentrations of SLPI are low auto-proteolysis of Hap can recur leading to the breakdown of aggregated clumps of bacteria. Dispersal of *H. influenzae* within the respiratory tract, in this manner, may be the trigger that initiates contiguous spread and the onset of disease. The combination of its proteolytic and adhesive properties have also led to the suggestion that Hap may perform a dual function. It has been proposed that proteolytic digestion of structures expressed on either the host or bacterial cell surface may serve to modify potential receptors or ligands, thereby facilitating subsequent adhesion events intrinsic to colonisation. The relative role of Hap in relation to the other adhesins expressed by non-typable strains of *H. influenzae* has yet to be determined.

Invasion of Eukaryotic Cells

Although *H. influenzae* strains are not traditionally considered to be intracellular pathogens, it has been demonstrated than non-typable strains can invade host cells, and in an intracellular location the organism would be protected from the action of local host immune defences. Entry into host cells may, therefore, provide an additional mechanism for promoting colonisation of the nasopharynx after fimbrial or HMW-mediated adhesion to mucosal epithelial cells. Wild-type non-typable strains have been found to be able to penetrate and survive in organ cultures of epithelial monolayers and primate respiratory tissue (Roberts *et al.*, 1984; St. Geme and Falkow, 1990).

Internalisation of non-typable *H. influenzae* by epithelial cells is abolished by cytochalasin D, an inhibitor of actin polymerisation (St. Geme and Falkow, 1990). This indicates that uptake of the organism is an active process triggered by the bacteria, since under normal circumstances epithelial cells are non-phagocytic. *In vitro* models of *H. influenzae* adherence have indicated that expression of the Hap adhesin may promote cell entry. However, the efficiency of invasion mediated by this protein is low, indicating that other factors are likely to be essential (St. Geme et al., 1994a).

In addition to epithelial cell invasion, *H. influenzae* can also gain access to the submucosa by penetrating the mucosal surface at tight junctions, between cells and points of cellular necrosis (van Scilfgaarde et al., 1995). This phenomenon has been termed paracytosis and has been observed in tissue sections from patients with acute and chronic muco-purulent bronchitis, where passage of *H. influenzae* from the mucosal surface, between epithelial cells and down to and beyond the basement membrane appears to occur (Hers and Mulder, 1953). Analysis of adenoid organ cultures infected with non-typable *H. influenzae* by scanning electron microscopy has shown similar results, which indicates that the organism itself is responsible for the disruption of epithelial tight junctions. As the infection progresses, clusters of organisms begin to accumulate between adjacent epithelial cells. Occasionally, some organisms are also found intracellularly in mononuclear cells located in the epithelial layer and below the basement membrane (Farley et al., 1986). It was not, however, possible from these experiments to determine whether these organisms were alive, or had been killed after phagocytosis.

In a later study, non-typable *H. influenzae* were again identified in macrophage-like cells in adenoid tissue obtained from children persistently infected with this organism but who, at the time of surgery, were clinically free from infection (Forsgren et al., 1994). Critically, these intracellular organisms were shown to be viable, suggesting that non-typable *H. influenzae* could survive macrophage-mediated killing. It therefore appeared that, in addition to epithelial cells, non-typable *H. influenzae* may also be able to survive in professional phagocytes. Since epithelial cells are short-lived, constantly sloughed off and expelled from the respiratory tract, survival in macrophages, which live for several weeks, may promote longer term persistence in the respiratory tract. In addition to facilitating colonisation of the respiratory tract, the ability to enter and survive in host cells may also have important clinical significance. A feature of non-typable *H. influenzae* otitis media is that it often recurs when antibiotic treatment has ceased. Such recurrent infections are often caused by the original infecting strain, suggesting that in these individuals clearance of non-typable *H. influenzae* infections is in some way impaired (Barenkamp et al., 1984). By the same token, in otitis-prone children, intermittent nasopharyngeal carriage of the same strain can occur for periods of at least 5 months, regardless of any intervening antibiotic treatment (Samuelson et al., 1995).

Similar observations have also been reported for respiratory tract infections caused by non-typable *H. influenzae* in individuals with chronic, obstructive pulmonary disease and cystic fibrosis, suggesting that this attribute is a common feature amongst this group of organisms (Groenveld et al., 1988). The ability of non-typable *H. influenzae* to survive in macrophages may explain in part why infections caused by this organisms so frequently recur. By providing a niche where the organism is protected from the action of antibiotics, macrophages may allow a reservoir of non-typable *H. influenzae* to develop in the respiratory tract. After release of organisms, on death of the macrophages, further episodes of otitis media may be triggered on termination of antimicrobial chemotherapy. This hypothesis awaits further clarification.

Iron and Haem Acquisition

Iron is a co-factor for several metabolic enzymes and is used as a prosthetic group in components of the electron transport chain. A source of iron is therefore critical for the survival of living cells. In humans, extracellular iron is bound to transferrin in serum, or lactoferrin in mucosal secretions. As a consequence, the concentration of free iron is below that required for bacterial survival (Otto et al., 1992). To survive in its human host, *H. influenzae* has therefore evolved a complex process for extracting iron from lactoferrin and transferrin. This mechanism is applicable for persistence on mucosal surfaces and intravascular survival. Under conditions of iron limitation, *H. influenzae* binds human transferrin via two transferrin-binding proteins, Tbp1 and Tbp2 (Holland et al., 1992). It has been proposed that transferrin-bound iron initially binds to the bacterial cell surface by interaction with Tbp2. The resulting complex then binds to Tbp2, which releases iron into the periplasmic space through a TonB-dependent transport mechanism (Jarosik et al., 1995). TonB is a cytoplasmic membrane-bound protein that spans the periplasmic space and can interact directly with outer membrane proteins. Once released into the periplasmic space, the iron binds to a high-affinity iron-binding protein, FbpA, which facilitates its translocation into the cytoplasmic compartment of the cell. This last step involves two proteins: HitB, a cytoplasmic permease, and HitC, a nucleotide-binding protein that facilitates active transport of iron across the inner membrane (Sanders et al., 1994; Adhikari et al., 1995).

Haem is also another vital requirement for the growth of *H. influenzae*, both *in vivo* and *in vitro*. The enzymatic machinery required to convert δ amino levulinic acid to proto-porphyrin, a precursor for the synthesis of haem, is absent in *H. influenzae*. Therefore, to grow under aerobic conditions *H. influenzae* has an absolute requirement for haem. As in the case of iron, free haem and haemoglobin are, respectively, sequestered in the body by haemopexin and haptoglobin. *H. influenzae* has, therefore, evolved mechanisms to extract haem from these storage proteins and to acquire iron directly from haemoglobin. Haem acquisition is complex and involves outer membrane proteins that bind haemoglobin–haptoglobin complexes and those that mediate the transfer of haem to the cytoplasm.

Three proteins, HuxA, HuxB and HuxC, have been implicated in this process. HuxA binds haem–haemopexin complexes and is released from growing cells into the culture medium. HuxB is involved in the release of HuxA, while HuxC is believed to be involved in the transport of haem into the cytoplasmic compartment of the cell (Hanson *et al.*, 1992; Cope *et al.*, 1995).

Virulence Determinants of Invasive Disease Due to *H. influenzae* Type b

Whilst many of the strategies and the virulence determinants involved in colonisation of the nasopharynx are common to encapsulated and non-encapsulated strains of *H. influenzae*, the mechanisms by which disease due to these two groups of organisms subsequently progresses are very different and involve a distinct repertoire of virulence determinants. In the case of serotype b strains, the expression of a polysaccharide capsule and lipopolysaccharide plays a critical role in protecting *H. influenzae* against the bactericidal activity of serum, which is fundamental for the development of systemic infection.

Translocation from the nasopharyngeal mucosa to the bloodstream presents additional challenges to *H. influenzae*. In these two anatomical sites the environmental conditions to which the organism is exposed differ considerably. Thus, after translocation to the bloodstream, *H. influenzae* must be able rapidly to adapt to its new surroundings and express the virulence determinants required for protection from serum-mediated killing. To achieve this, *H. influenzae* has evolved the ability to phase vary the expression of a diverse array of virulence determinants that enable it successfully to oscillate between colonisation of the nasopharynx and intravascular survival.

Capsular Polysaccharide

Role in Pathogenesis

Arguably the most important virulence factor expressed by *H. influenzae* type b is its capsular polysaccharide, which is polyribosylribitol phosphate (PRP), a polymer of 3)-αDRib$f$-(1-1)-ribitol-5-PO$_4$ (ribosylribitol phosphate) (Crisel *et al.*, 1975). In general terms, expression of a polysaccharide capsule prevents bacterial phagocytosis and acts as an effective shield from the action of complement (reviewed by Moxon and Kroll, 1990). It reduces opsonisation of the cell surface by C3b and masks C3b deposited on the underlying cell wall, reducing the action of phagocytic cells on the organism. Capsule material is also continuously shed from the cell surface and provides an effective mechanism to jettison attached host factors, or to nullify the role of circulating antibodies.

The reason why type b strains, of all the *H. influenzae* capsule serotypes, are so virulent and promote invasive disease is not fully understood. One possibility is that the enhanced shielding effect provided by the type b capsule may be due to the high level of PRP expression. However, since types a, d and f express relatively more capsular polysaccharide than type b strains, this hypothesis has proved incorrect (Sutton *et al.*, 1982).

To investigate the role of PRP in the pathogenesis of meningitis and bacteraemia, a set of *H. influenzae* transformants were generated, each of which expressed a different capsule serotype, but which were otherwise genetically identical. The ability of these transformants to cause invasive disease was assessed by intranasal inoculation of infant rats. Only the serotype a and b transformants consistently caused bacteraemia, and of these the type b strain was far more virulent and uniformly caused higher levels of bacteraemia (Zwahlen *et al.*, 1982). Further confirmation of the protective effect of the PRP capsule was obtained by transforming a capsule-deficient variant of a serotype d strain (Rd) with chromosomal DNA from a type b strain as donor. The resulting encapsulated transformant was as virulent as the wild-type serotype b strain and demonstrated comparable intravascular survival (Moxon and Vaughn, 1981).

It is noteworthy that the type a capsule, which can also promote limited intravascular survival, also contains ribitol phosphate, suggesting that it is the chemical composition of the saccharide that provides resistance to the bactericidal activity of serum. A further feature of the PRP capsule implicated in the enhanced virulence of serotype b strains is that, of the six *H. influenzae* capsule serotypes, only type b requires serum antibody to initiate the protective action of complement. In contrast, the remaining capsular serotypes are susceptible to the alternative complement pathway and are killed after introduction into the bloodstream. As a consequence, babies with an immature immune system who cannot generate a T-independent immune response are highly susceptible to infection because they cannot produce bactericidal antibody to *H. influenzae* type b. The protective effect of serum antibodies against the type b capsule provided further circumstantial evidence that PRP is a critical determinant of virulence and indicated that this polysaccharide would form the ideal basis of a vaccine against *H. influenzae* type b infection (Pietola *et al.*, 1977).

Genetics of Capsule Biosynthesis

The genes involved in the biosynthesis of capsular polysaccharide are chromosomal and are clustered in a region of approximately 36 kb, designated *cap*. The *cap* locus comprises two contiguous 17-kb repeated regions, separated by a smaller region (1–2 kb) of non-repeated DNA (Hoiseth *et al.*, 1986). A survey of more than 100 type b strains isolated from systemic infections and commensal organisms has shown that this duplication is typical of more than 98% of the strains examined (Musser *et al.*, 1988). A similar arrangement is also found in the *cap* loci of other serotypes of *H. influenzae*. Each 17-kb region contains a 4–5 kb serotype-specific locus (region 2). This region encodes enzymes required for the synthesis of capsular polysaccharide and is unique to each serotype.

Region 2 is flanked by DNA common to all serotypes (regions 1 and 3) (Kroll *et al.*, 1989). Region 1 contains the *bex* genes (*bexA*–*bexD*), which encode an ATP-driven export apparatus essential for translocation of the polysaccharide to the cell surface (Kroll *et al.*, 1990). Region 3 contains two ORFs of uncertain function, although they may also be involved in polysaccharide export. A similar tripartite arrangement has also been observed in the capsulation loci of *E. coli* (Boulnois *et al.*, 1987). Moreover, close sequence similarity between BexA and BexB and products of the *kpsT* and *kpsM* genes in *E. coli* K5 suggests that capsulation genes in these organisms may have a common ancestry.

Although the *cap* locus is duplicated in more than 98% of strains, the *bexA* gene, which is crucial for polysaccharide export, is present only in one of the two copies, the second copy having suffered a 1.2 kb deletion. The functional copy of *cap*-containing *bexA* is flanked by direct repeats of the insertion sequence IS*1016* and has been described as a compound transposon (Kroll *et al.*, 1991). The duplication of *cap* and the presence of IS*1016* facilitates a high frequency,

spontaneous, irreversible *rec*-dependent recombination between these two sequences. This results in the loss of one copy of *cap* repeats, the functional copy of *bexA* and part of the insertion element (**Fig. 2**). This molecular event leads to the generation of class I mutants, which appear at a frequency of approximately 20% in late-exponential liquid cultures (Kroll *et al.*, 1988). The majority die, but some remain viable (0.1–0.3%) and form small, iridescent, slow-growing colony variants. These colonies, which have an irregular surface, contain pleiomorphic, often filamentous organisms filled with large cytoplasmic vacuoles that are believed to contain capsular material.

Class I mutants may undergo a point mutation in region 2 of the remaining copy of *cap*. This generates healthy non-iridescent colonies that do not produce PRP (Brophy, 1991). Given the importance of the serotype b capsule for virulence, it is intriguing to speculate why a mechanism that promotes irreversible loss of capsule expression has evolved at such high frequency. One possibility is that capsule loss may facilitate colonisation of the nasopharynx. In support of this hypothesis it has been demonstrated that non-encapsulated mutants show increase levels of adhesion to cultured epithelial cells as compared with wild-type *H. influenzae*. However, although the majority of commensal isolates of *H. influenzae* in the nasopharynx are non-encapsulated, they are only rarely genetic variants of encapsulated strains.

In addition to the spontaneous deletion of a functional copy of *cap*, duplication can also facilitate the amplification of the *cap* locus. A survey of clinical isolates revealed that some strains can possess three, four or even five copies of the *cap* repeat (Corn, 1993). Amplification of the *cap* repeat region in this manner results in a concomitant increase in the production of PRP; strains that carry five copies produce as much as six times more capsular polysaccharide (Kroll and Moxon, 1988; Corn, 1993). The biological significance of *cap* amplification remains a matter for speculation. It has been suggested that such organisms may be at a selective advantage where the expression of capsule is a prerequisite for survival. However, an alternative hypothesis is that increased amounts of PRP may provide more protection against desiccation and increase chances of the survival of the organism between hosts.

Haemophilus influenzae Vaccine

The observation that antibodies directed against the type b capsule are bactericidal and protective provided the first indication that PRP might be an effective vaccine candidate. Accordingly, the first vaccine developed against *H. influenzae* type b consisted only

Phenotype

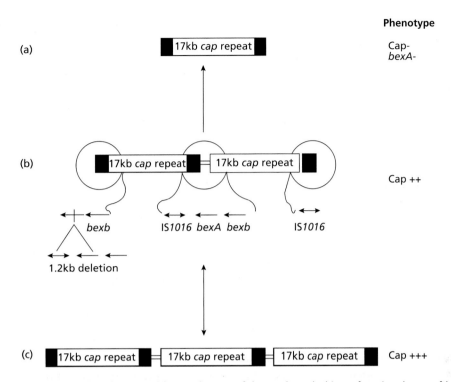

Fig. 2 The *cap* locus of *Haemophilus influenzae*. (a) A single copy of the *cap* locus lacking a functional copy of *bexA* as a result of recombination. (b) The duplicated *cap* locus of *H. influenzae* type b. The duplication is bounded by direct repeats of IS*1016* (black boxes). The bridge region containing *bexA* is shown as constrictions. A unique 1.2-kb region, consisting of *bexA* and the right-hand end of IS*1016* has been lost from the left-hand end of the duplicated locus. (c) The presence of the insertion element also facilitates amplification of the *cap* copy number.

of purified PRP. This was, however, only capable of eliciting a T cell-independent immune response without immunological memory. In addition, it was poorly immunogenic in children under the age of 2 years, the very children most susceptible to *H. influenzae* type b infection (DH Smith *et al.*, 1973). To improve the immunogenicity of PRP with the aim of generating a T-dependent immune response, the polysaccharide was covalently linked to protein. The choice of carrier protein varied from one pharmaceutical company to another. Common examples included tetanus toxoid, a non-toxic mutant form of diphtheria toxin and meningococcal outer membrane protein (Robbins *et al.*, 1996). When these conjugates were given as vaccines they induced a T cell-dependent immune response to the polysaccharide with a substantial improvement in immunogenicity, such that adequate protection is provided in children aged 2 years or less (Robbins *et al.*, 1996). Since the vaccine elicits immunological memory, a booster injection, or subsequent exposure to *H. influenzae*, raises the titre of anti-PRP antibodies.

With this in mind, a series of Hib vaccinations given at the age of 2, 3 and 4 months has been routine in the UK since 1992 (reviewed by Booy and Kroll, 1997). The success of the vaccine programme has been considerable. Within 2 years of its introduction, the incidence of *H. influenzae* infection declined by more than 90%. A further benefit derived from vaccination was that the nasopharyngeal carriage of *H. influenzae* in vaccinated infants was abolished. This reduced transmission of the disease, so that even in communities where vaccination was not complete, the incidence of disease was reduced as a consequence of herd immunity (Barbour *et al.*, 1995).

Since only humans are infected by *H. influenzae* and children are mainly responsible for transmission, world-wide introduction of the vaccine may, theoretically, eradicate this organism. Given the expense of vaccination programmes, however, this seems unlikely, particularly in developing countries. However, if it can be achieved, it is interesting to speculate whether the niche previously filled by *H. influenzae* type b might be repopulated by other serotypes of *H. influenzae* or lead to an increase in colonisation by other virulent pathogens, such as *Streptococcus pneumoniae* and *N. meningitidis*, but this has yet to be substantiated.

Genetic Relationship Between Capsulated and Non-encapsulated Strains

A major question in the evolution of *H. influenzae* is whether non-encapsulated strains are derived from encapsulated organisms. Multi-locus enzyme electrophoresis (MLEE) has been used to examine the genetic relationship between these two groups of organisms (Porra *et al.*, 1986). MLEE is based upon analysis of polymorphisms in essential metabolic enzymes encoded by chromosomally located genes, which, over time, accumulate neutral mutations that affect charged amino acids. These variant polypeptides can be distinguished by gel electrophoresis, based upon changes in mobility caused by variation in the net electrostatic charge generated by such mutations. By analysing a sufficiently large group of polypeptides, the genetic relationship between strains can accurately be predicted (Selander *et al.*, 1986).

With this technique it was determined that the population structure of encapsulated stains is clonal and can be segregated into genetically related clusters, grouped into two major phylogenetic classes (Musser *et al.*, 1988). Division I includes serotypes a, b, c, d and f strains and division II includes a, b and f serotypes. In contrast, considerable genetic diversity is apparent in the non-typable strains. In a survey of 65 epidemiologically distinct isolates no two electrophoretic-types were the same. Moreover, non-typable strains appear to be genetically distinct from type b strains, since they do not share any electrophoretic-types, which indicates that they are not phenotypic variants of type b strains (Musser *et al.*, 1986).

To address the relationship between encapsulated and non-encapsulated strains of *H. influenzae* in greater detail, 123 pharyngeal isolates of non-typable *H. influenzae* were collected from healthy 3-year-old Finnish children (St. Geme *et al.*, 1994b). Based upon Southern hybridisation analysis it was determined that one of these isolates was in fact a capsule-deficient type b strain, which had lost a functional copy of *cap*. Of the remaining isolates 33% hybridised with IS*1016* sequences contained in the *cap* b locus, indicating that a subset of non-typable strains was derived from, or closely related to, an encapsulated ancestor. These strains lack the high-molecular-weight adhesins characteristic of non-typable strains of *H. influenzae*, but they express Hia, the homologue of the major non-pilus adhesin expressed by encapsulated strains.

Taken as a whole, this indicates that some non-typable strains of *H. influenzae* are more closely related to encapsulated strains than others. It seems most probable, therefore, that the common ancestor of *H. influenzae* was a non-encapsulated organism which diverged to form two overlapping lineages.

One lineage comprises organisms that acquired genes involved in encapsulation and associated colonisation factors, while the other contains the non-typable organisms, which acquired genes to promote a different series of interactions with the host, giving rise to a different spectrum of disease.

Lipopolysaccharide

Lipopolysaccharide Structure

Unlike the lipopolysaccharide (LPS) of the *Enterobacteriaceae*, *H. influenzae* LPS lacks structures equivalent to the long polysaccharide O-antigen side-chains and expresses a so-called 'rough' LPS (Flescher and Insel, 1978). LPS of a similar composition is also expressed by the genera *Neisseria*, *Branhamella* and *Bordetella* which, like *H. influenzae*, are respiratory tract pathogens and not subject to the membrane dispersive effects of bile salts found in the intestinal tract. The LPS of *H. influenzae* consists of lipid A, in which the endotoxic properties of the LPS reside and which is embedded in the outer membrane. This part of the molecule is linked, via a single phosphorylated ketodeoxyoctulosonic acid (KDO), to a heterogeneous polymer of neutral sugars, including heptose, glucose and galactose (Zamze and Moxon, 1987). This core-type region is more complex than the analogous region in enterobacterial LPS, and contains variable branched chain structures.

The structure of the core varies from strain to strain and is characterised by different linkages and molar ratios of sugars (**Fig. 3**) (Phillips *et al.*, 1992, 1993; Masoud *et al.*, 1997; Risberg *et al.*, 1999). In non-typable strains predominantly sialylated terminal lactosamine structures can also be detected (Hood *et al.*, 1999), and it has been demonstrated that *H. influenzae* LPS can also be substituted with phosphorylcholine (Weiser *et al.*, 1997; Risberg *et al.*, 1999). Several *H. influenzae* LPS oligosaccharide structures, including the terminal digalactoside αGal(1−4)βGal and sialylated lactosamine, have also been shown to be components of the LPS expressed by pathogenic *Neisseria* spp. (Virji *et al.*, 1989; Campagnari *et al.*, 1990; Di Fabio *et al.*, 1990). αGal(1−4)βGal is also a component of the globoseries glycolipids expressed by human epithelial cells and is integral to the structure of the P blood group antigen. In this context it acts as a receptor for *E. coli* P-fimbriae (Bock *et al.*, 1985), the B subunit of Shiga toxin (Lindberg *et al.*, 1987) and parvovirus B19 (Brown *et al.*, 1993).

It has been suggested that by molecular mimicry of host cell surface antigens the αGal(1−4)βGal structure of *H. influenzae* LPS may play a critical role in

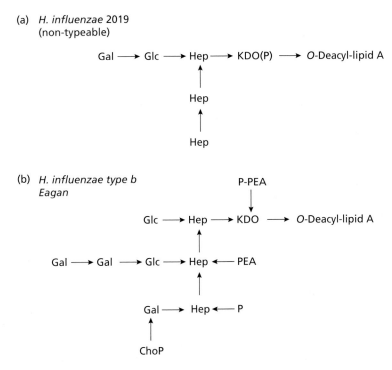

Fig. 3 Structural model showing the LPS of (a) non-typable *Haemophilus influenzae* 2019 and (b) *H. influenzae* type b strain Eagan. Glc, glucose; Gal, galactose; Hep, heptose, KDO, 2-keto-3-deoxyoctulosonic acid; PEA, pyrophosphoethanolamine; P, phosphate; ChoP, phosphorylcholine. From Masoud *et al.* (1997) and Weiser *et al.* (1998).

shielding *H. influenzae* from antigen-specific host immune defences. In support of this hypothesis, LPS mutants that lack this structure have been shown to be more susceptible to the bactericidal activity of human serum (Erwin *et al.*, 2000).

Lipopolysaccharide Phase Variation

A characteristic feature of *H. influenzae* LPS is that it undergoes high-frequency phase variation (10^{-2}/cell per generation), characterised by the spontaneous loss or gain of oligosaccharide structures situated within the outer core. This was first observed by Kimura and Hansen (1987) with monoclonal antibodies (mAbs) directed against structures, including αGal(1–4)βGal present in the outer core. By colony immunoblotting it was demonstrated that a single colony can give rise to daughter colonies with a pattern of reactivity to LPS-specific mAbs different from the parent colony. LPS phase variation can also be detected in individual colonies and is indicated by the formation of sectored patterns after colony immunoblotting (Fig. 3).

In a given strain, multiple LPS structures are subject to phase variation. Some of these vary independently, whilst others vary in a co-ordinate fashion in conjunction with one, or more other structures (Weiser *et al.*, 1989a). As a consequence of phase

variation, the structure of *H. influenzae* LPS exhibits considerable heterogeneity. For example, at any time a given population of *H. influenzae* comprises individual organisms expressing LPS molecules with distinct chemical structures generated by independent phase-variation events. This diversity ensures that at each stage in pathogenesis, individual organisms expressing the most appropriate LPS structure are constantly available to make possible efficient colonisation of the diverse anatomical niches encountered during invasive disease. The ability constantly to change the structural composition of its cell surface antigens also provides *H. influenzae* with an efficient mechanism for the evasion of antigen-specific host immune defence mechanisms.

Genetics of Lipopolysaccharide Synthesis and Phase Variation

Mechanisms that contribute to LPS phase variation were first identified in the *lic1* locus, which was cloned from *H. influenzae* strain RM7004 (Weiser *et al.*, 1989a). This locus was able to confer on *H. influenzae* strain Rd, a non-encapsulated variant of a serotype d strain, the ability to express phase-variable LPS structures recognised by the mAbs 12D9 and 6A2. Nucleotide sequence analysis of the cloned region showed that it consisted of four ORFs in close juxtaposition, such that the start and stop codons of the first three ORFs overlap. Of these four ORFs, ORFs 1, 3 and 4 were essential for LPS synthesis, while the function of ORF2 has yet to be determined. ORF3 and ORF4 were required for the expression of the LPS structures defined by mAb 6A2 and 12D9, respectively. The synthesis of both of these structures, however, ultimately depends on the first ORF, *lic1A*.

On the basis of amino acid sequence homology, it has been shown that *lic1A* encodes the phosphorylcholine kinase, which promotes the choline substitution of LPS (Weiser *et al.*, 1997). Analysis of the nucleotide sequence of this gene revealed that it contains 29 tandem repeats of the tetrameric sequence 5′-CAAT-3′, which encodes the tetrapeptide SINQ, in its 5′-coding region (Weiser *et al.*, 1989b). Such highly repetitive DNA sequences are subject to slipped strand mis-pairing, a non-recombinational mechanism that causes the number of repeat units either to increase or decrease. A 5′-AT-3′-rich region upstream of the 5′-CAAT-3′ repeats easily becomes denatured and is thought to facilitate mis-pairing. As the number of repeats varies upstream, translation initiation codons would be moved in and out of frame with the complete ORF, causing translational frame shifting (**Fig. 4**). As a consequence, only when the number of repeats is consistent with translation of the complete ORF will the product of *lic1A* be expressed.

To investigate the correlation between the number of 5′-CAAT-3′ repeats in *licA* and expression of LPS epitopes the number of 5′-CAAT-3′ repeats present in cultures derived from a single colony either expressing or not expressing 12D9 and 6A2 was determined by the polymerase chain reaction. These experiments demonstrated that, as predicted by the model, the number of 5′-CAAT-3′ repeats in *lic1A* varied in size by 4 bp. In addition, LPS structures which reacted with 12D9 and 6A2 were expressed only when the number of 5′-CAAT-3′ repeats present was consistent with translation of a full-length ORF. The level of reactivity with mAbs 12D9 and 6A2 was also dictated by which of the two potential translation initiation codons, situated in different ORFs, were placed in frame by the 5′-CAAT-3′ repeats. Thus, not only do the number of 5′-CAAT-3′ repeats dictate whether LPS structures recognised by 12D9 are expressed, but they also modulate the level of expression.

Of the battery of monoclonal antibodies that recognise phase-variable LPS structures, it was clear that *lic1A* was only responsible for the phase variation of structures recognised by 12D9 and 6A2. To identify the other genetic loci involved in LPS phase variation, a 20-mer oligonucleotide of five copies of 5′-CAAT-3′ was used to probe the chromosome of *H.*

influenzae strain RM7004 (Weiser *et al.*, 1990a). This revealed the presence of two other loci, termed *lic2* and *lic3*, that contain this motif. Deletion and insertion mutagenesis of *lic2* showed that this locus is a pre-requisite for the synthesis of four phase-variable LPS structures, including αGal(1−4)βGal. It comprises four ORFs termed *lic2A*, *ksgA*, ORF3 and *lic2B* (High *et al.*, 1995). By amino sequence homology *lic2A*, *lic2B* and *orf3* were shown to encode glycosyltransferases, while *ksgA* encodes a 16S rRNA methyltransferase. Similarly to *lic1A* in *lic1*, *lic2A* contains multiple tandem repeats of 5′-CAAT-3′ in the 5′ end of its coding region. With PCR to amplify the repeat region from *lic2A* directly from immunostained colonies, it was determined that variation in the number of 5′-CAAT-3′ repeats correlates directly with the phase-variable expression of αGal(1−4)βGal (High *et al.*, 1995).

Similarly to *lic1A* and *lic2*, *lic3* also comprises four ORFs. The first of these contains multiple copies of 5′-CAAT-3′, but the function of the protein that it encodes is unknown. Of the three remaining ORFs, ORF3 also has no known function, and ORF4 has homology to adenylate kinase. Only ORF2 appears to play a defined role in LPS biosynthesis and shows extensive sequence similarity to UDP galactose-4-epimease (GalE), a pivotal enzyme in LPS biosynthesis (Maskell *et al.*, 1992a). GalE enables cells to interconvert UDP-glucose and UDP-galactose, the building blocks used to synthesis polysaccharides. It is also an intrinsic component of the Leloir pathway, enabling bacteria to use galactose as a sole carbon source. In the *Enterobacteriaceae*, *galE* is located in a gene cluster along with the other Leloir pathway genes, *galT*, *glaK*, *glaM* and *glaR*. In *H. influenzae*, however, these genes are located in a different region of the chromosome.

In the absence of a functional copy of *galE*, *H. influenzae* depends on exogenous galactose to make UDP-galactose. If this sugar is not present in growth media, the *galE* mutant expresses a truncated LPS molecule deficient in galactose, which no longer reacts with mAb 4C4 that recognises αGal(1−4)βGal (Maskell *et al.*, 1992b). The subsequent addition of galactose to growth media restores the wild-type phenotype. These experiments provide strong evidence that *galE* is intrinsic to LPS biosynthesis.

In addition to *lic1A*, *lic2A* and *lic3A*, two further phase-variable genes required for LPS biosynthesis have been defined. These include *lex2A*, which contains multiple copies of the tetrameric sequence 5′-GACA-3′ in its coding sequence (Jarosik and Hansen, 1994; Hood *et al.*, 1996), and *lgtC*, which contains the tetrameric repeat 5′-CACA-3′.

Lipopolysaccharide and Virulence

In addition to the capsule, LPS has also been implicated as a major contributory factor to the pathogenicity of *H. influenzae*. It is now clear that *H. influenzae* LPS can enhance bacterial survival in the nasopharynx, facilitate invasion of organisms across cellular barriers, promote intravascular survival and cause damage to tissues, particularly those in the respiratory tract and meninges. The broad spectrum of pathophysiological effects attributable to LPS reflects the amphipathic nature of this molecule. The hydrophobic lipid A moiety is responsible for tissue damage, while the oligosaccharide core structures are intrinsic to bacterial survival in the host. As with endotoxin, purified LPS causes loss of ciliary activity when applied to guinea-pig and rat trachea organ cultures (Johnson

and Inzana, 1986). This is consistent with the injury to epithelium observed in cultures of human respiratory mucosa (Farley *et al.*, 1986; Read *et al.*, 1991). As a consequence, *H. influenzae* inhibits the mucociliary clearance mechanism, so prolonging and facilitating colonisation of the nasopharynx.

Application of purified LPS directly into the subarachnoid space induces blood–brain barrier permeability and meningeal inflammation, which indicates a critical role for this molecule in the pathophysiology of bacterially induced CNS injury during meningitis (Syrogiannopoulos *et al.*, 1988; Wispelwey *et al.*, 1988). The mechanisms by which LPS mediates these responses and generates an inflammatory response is unclear, but it is believed to result from the induction of the cytokines IL-1 and TNF α (Beutler *et al.*, 1986;

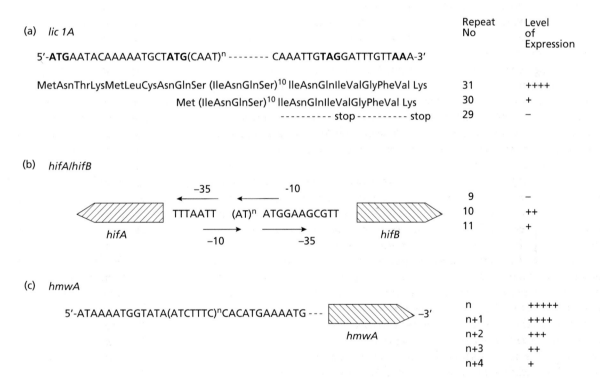

Fig. 4 Mechanisms of phase variation in *Haemophilus influenzae*. (a) The role of tetrameric repeats in promoting phase variation illustrated by the 5'-CAAT-3' repeats in *lic1A*. When 31 copies of 5'-CAAT-3' are present ATG_1 is in-frame with the complete ORF, resulting in full expression of 12D9 reactive LPS structures. When a copy of 5'-CAAT-3' is lost through slipped strand mis-pairing the resulting 30 copies of 5'-CAAT-3' place ATG_2 in-frame with the complete ORF, but in this case expression of the 12D9 reactive LPS structures is lower. When 29 copies of 5'-CAAT-3' are present, translation initiated from either ATG terminates prematurely at stop codons situated downstream of the 5'-CAAT-3' repeats. (b) The hifA and hifB genes involved in fimbriae biosynthesis contain overlapping promoters. In both cases the −10 and −35 regions are separated by 5'-AT-3' repeats. Variation in the number of repeats alters the distance between the RNA polymerase-binding sites and in so doing dictate whether transcription can occur. When 9 copies of 5'-AT-3' repeats are present, the distance between the −10 and −35 sites is non-permissive for RNA polymerase binding. When 10 copies of 5'-AT-3' are present, transcription can occur and high-level expression of fimbriae is observed. When 11 copies of 5'-AT-3' are present, transcription occurs, but at lower levels causing a concomitant reduction in the expression of fimbriae. (c) The *hmwA* genes which encode the biosynthesis of high-molecular-weight proteins in non-typable *Haemophilus influenzae* contain tandem 7-bp repeats of 5'-ATCTTTC-3' upstream of the translation initiation codon. As the number of repeats increases, the level of transcription of both genes and therefore protein production decreases.

Camussi *et al.*, 1991). Complement activation with subsequent releases of the chemotactic component C5a may also be a contributory factor (Ernst *et al.*, 1984).

A critical role for the oligosaccharide moiety of the LPS was first indicated by the isolation of an avirulent mutant of a type b strain (Zwhalen *et al.*, 1985). This mutant expresses a minimal LPS structure, comprising lipid A and heptose, which provides a strong indication that the oligosaccharide moiety of this molecule is a prerequisite for disease production. Subsequently core oligosaccharide phase variants were shown to display differing levels of virulence, indicating that specific structures in the outer core were essential for virulence (Kimura and Hansen, 1987).

The cloning of genes required for the expression of phase-variable oligosaccharide structures has made possible the construction of defined mutants, demonstrating the intrinsic role of LPS in the pathogenesis of invasive disease. The generation of deletion mutations in *galE* and *galK* generates a mutant unable to synthesis galactose containing LPS structures. After intraperitoneal challenge of infant rats, this mutant does not induce detectable bacteraemia in comparison with wild-type *H. influenzae* (Maskell *et al.*, 1992b). The degree of attenuation observed in this mutant is equivalent to that of *cap* mutants, indicating the essential role played by LPS as a determinant of virulence.

Ability to vary the expression of LPS may optimise the virulence potential of this organism by promoting the expression of the most appropriate LPS structures for a given environment, or stage in pathogenesis. This may be of particular significance in the development of invasive disease, which in the case of meningitis requires translocation of *H. influenzae* from the nasopharynx to the bloodstream and the central nervous system. This sequence involves the colonisation of several distinct environments and the interaction of *H. influenzae* with a variety of different host cells. Consistent with this idea, phase-variable LPS structures play a critical role in the translocation of *H. influenzae* from the nasopharynx to the bloodstream. Early studies on the role of LPS phase variation in disease pathogenesis focused on the phase-variable LPS structures expressed by *lic1* and *lic2*. Deletion mutations were introduced into both gene clusters and their effect on virulence assessed in the infant rat model (Weiser *et al.*, 1990b). Although this double mutant lacked expression of phase-variable oligosaccharide structures, it still generated levels of bacteraemia comparable with wild-type *H. influenzae* after intraperitoneal inoculation of infant rats. After

intranasal challenge, however, there was a significant reduction in the incidence of bacteraemia in rats infected with the mutant, suggesting that its ability to cause invasive disease was impaired. Since the mutant strain was able to colonise the nasopharynx of animals as efficiently as wild-type strains, it was concluded that phase-variable LPS structures play a critical role in the translocation of *H. influenzae* from the nasopharynx to the bloodstream.

Since these early experiments the role of *lic1A* and *lic2A* in invasive disease has been defined in more detail. It is now known that the product of *lic1A* has homology to eukaryotic choline kinases and plays a pivotal role in decorating *H. influenzae* LPS with phosphorylcholine. Incorporation of this structure into LPS influences colony morphology. Organisms that express choline in their LPS produce translucent colonies, while the colonies generated by choline-negative phase variants are opaque. Choline is a target for C-reactive protein and translucent organisms that express this LPS phenotype are rendered more susceptible to killing by innate immunity through the classical complement pathway. Expression of this structure is therefore detrimental to the intravascular survival of *H. influenzae* and such phase variants are rapidly killed. As a consequence *lic1A* is predominantly switched off in organisms isolated from the bloodstream of infected animals (Weiser *et al.*, 1998; Hosking *et al.*, 1999).

Although phosphorylcholine is deleterious during systemic infection, it appears to play a critical role in mediating long-term colonisation of the nasopharynx by *H. influenzae*. After intranasal inoculation, the number of 5'-CAAT-3' repeats in *lic1A* gradually changed. After 10 days a phenotypic switch had occurred and the majority of the organisms isolated from the nasopharynx of infected animals expressed choline-containing LPS structures (Weiser *et al.*, 1998). How the expression of choline enhances nasopharyngeal colonisation by *H. influenzae* is however unknown. In pneumococci, the presence of choline in the cell wall promotes adhesion to platelet-activating factor receptors and causes bacterial internalisation by nasopharyngeal epithelial cells (Cundell *et al.*, 1994). Whether choline triggers internalisation of *H. influenzae* has yet to be determined.

In contrast to Lic1A, expression of Lic2A plays a critical role in protecting *H. influenzae* from the bactericidal activity of serum (Weiser and Pan, 1998) and this is thought to be attributable to molecular mimicry. The expression of the αGal(1–4)βGal LPS structure, for the synthesis of which Lic2A is required, partially shields *H. influenzae* from antigen-specific host immune defences.

Other Phase-variable Virulence Determinants

Tetrameric Repeats Present in Virulence Genes

Analysis of the genome sequence of *H. influenzae* strain Rd revealed an additional 12 loci that contained multiple tandem repeats in potential ORFs (Hood *et al.*, 1996). Nine of these were previously uncharacterised virulence genes, which indicate an intimate association between phase variation and phenotypic traits implicated in pathogenesis. Of the tandem tetranucleotide repeats identified in this study four had novel sequences distinct from the previously described 5′-CAAT-3′ and 5′-GCAA-3′ motifs. Structural analysis of the amino acid sequence of these newly identified repeats indicated that, like 5′-CAAT-3′ and 5′-GCAA-3′, they encoded polypeptides predicted to form highly flexible random coiled structures. This observation supported the earlier hypothesis that repetitive intra-genic nucleotide sequences could only be retained in ORFs because the peptides that they encode did not impede the formation of tertiary structures necessary from catalytic activity.

Of the novel tetranucleotide repeats sequences identified in the Rd genome sequence, 5′-CAAC-3′ was found in association with five distinct loci, four of which were involved in the sequestration of iron. Iron uptake is an essential requirement for bacterial survival within the host and is thus an important determinant of virulence. Moreover, *H. influenzae* has an absolute requirement for iron and even during *in vitro* growth this component must be supplied in the form of haemin. It is likely that the presence of four distinct iron-uptake systems reflects the fundamental requirement for iron characteristic of this organism. The three other tetranucleotide repeats identified include 5′-AGTC-3′, which was present in a methyltransferase gene, 5′-TTTA-3′ located in an unidentified ORF with homology to hypothetical *Bacillus subtilis* proteins, and 5′-GACA-3′. The latter was found in *lgtC*, a gene that encodes a glycosyltransferase involved, in conjunction with *lic2A*, in the phase-variable expression of the αGal(1−4)βGal LPS structure. The ability of *H. influenzae* to vary the expression of so many virulence genes defines it as an organism with outstanding potential for undergoing adaptive mutagenesis.

Whilst the significance of the phase-variable expression of many of these novel genes in the pathogenesis of invasive disease is unknown, it is clear that this ability enables *H. influenzae* to express the most appropriate virulence determinants for a given stage in pathogenesis. This may be of particular significance in the development of invasive disease which involves translocation of *H. influenzae* from the nasopharynx to the bloodstream. Survival in these two distinct anatomical sites demands very different phenotypic attributes. The potential to vary the repertoire of phenotypic traits expressed is, therefore, likely to be essential for the development of invasive disease by *H. influenzae*. For example, whilst *lic1A* appears important for persistence within the nasopharynx, its expression in bloodstream is disadvantageous. Similar observations have recently been made for the product of *lic3A*, which appears to be a prerequisite for survival in the nasopharynx, but is counter-productive during systemic infection (Hosking *et al.*, 1999).

Are Tetrameric Repeats Important for Catalytic Function?

The amino acid sequences of Lic2A and Lic2B, encoded by the *lic2* gene cluster, are homologous to the galactosyltransferases LgtE and LgtB of *Neisseria* spp. and LpsA, an LPS biosynthesis gene from *Pasteurella haemolytica*. This homology suggests that they perform a related function in LPS biosynthesis. The major difference between this family of peptides is that only Lic2A contains the repetitive tetrapeptide motif SINQ. The absence of this amino acid sequence in all the other proteins provided the first indication that this domain might not be a prerequisite for the catalytic activity of Lic2A. Indeed, if this region was deleted the homology between Lic2A and the other members of this family of enzymes would be continuous throughout the length of the amino acid sequence.

To test this hypothesis the 5′-CAAT-3′ region was deleted from *lic2A* by fusing the regions 5′ and 3′ to the repeats (High *et al.*, 1996). Introduction of this constitutively expressed *lic2A* variant into the genome of *H. influenzae* had no observed effect on LPS biosynthesis. The overall molecular weight of the LPS molecule appeared unchanged and reactivity with mAb 4C4 was retained. The rate of phase variation of the αGal(1−4)βGal structure recognised by this monoclonal antibody was, however, markedly reduced as compared with wild-type *H. influenzae*, which was consistent with the constitutive expression of *lic2A*. Since deletion of the 5′-CAAT-3′ repeat region had no effect on the catalytic activity of the Lic2A protein, it was predicted that considerable variation in the length of the SINQ peptide could be tolerated in the protein, without loss of function.

In support of this hypothesis, the number of 5′-CAAT-3′ repeats in a series of clinical isolates, comprising strains from each capsular serotype and non-typable *H. influenzae*, varied considerably in both *lic1A* and *lic2A*. In the case of *lic2A* the number of 5'-CAAT-3' repeats ranged from 7 to 29 copies. The degree of variability observed in *lic1A* was much greater, with the number of repeats varying between 5 and 70 copies, and the latter represented over 20% of the coding sequence. The wide range of repeat units tolerated in *lic1A* and *lic2A* provided further circumstantial evidence that the repetitive SINQ tetrapeptide is not important for the catalytic activity of Lic proteins. It has been suggested that, because of its flexible nature, this peptide does not impede the interaction between the N- and C- terminal domains of the protein, which lie at either side, allowing a biologically active tertiary structure to be formed. The need for a structure with this characteristic predicts that only a limited number of combinations of nucleotides could be used as repeat units.

Conclusion

The pathogenesis of disease caused by non-typable *H. influenzae* and *H. influenzae* type b involves multiple steps. Although similar mechanisms are used to promote colonisation of the nasopharynx, both organisms express unique virulence determinants that dictate the spectrum of disease they are able to cause. A major difference is the expression of a PRP capsule by type b strains. This alone is sufficient to facilitate intravascular survival and the onset of invasive disease. Although it is a major determinant of virulence, the capsule has played a useful role as the basis of the highly successful *H. influenzae* vaccine. In contrast, no such highly conserved virulence determinant is common to non-typable strains of *H. influenzae*, which complicates the search for a candidate vaccine for this group of organisms. As a consequence, while disease due to type b stains is now rare, non-typable strains continue to be a significant cause of morbidity and often mortality in children.

References

Adhikari P, Kirby SD, Nowalk AJ, Veraldi KL, Schryvers AB, Mietzner TA (1995) Biochemical characterisation of a *Haemophilus influenzae* periplasmic iron transport operon. *J. Biol. Chem.* 270: 25142–25149.

Anderson P, Johnston RB, Smith DH (1972) Human serum activities against *Haemophilus influenzae* type b. *J. Clin. Invest.* 51: 39–45.

Barbour ML, Mayon-White RT, Coles C, Crook DW, Moxon ER (1995) The impact of conjugate vaccine on the carriage of *Haemophilus influenzae* type b. *J. Infect. Dis.* 171: 93–98.

Barenkamp SJ, Bodor FF (1990) Development of serum bactericidal activity following nontypable *Haemophilus influenzae* otitis media. *Pediatr. Infect. Dis. J.* 9: 333–339.

Barenkamp SJ, Leininger E (1992) Cloning, expression and DNA sequence analysis of genes encoding nontypable *Haemophilus influenzae* high-molecular-weight-surface exposed proteins related to filamentous hemagglutinin of *Bordetella pertussis. Infect. Immun.* 60: 1302–1313.

Barenkamp SJ, St. Geme JW III (1996) Identification of a second family of high-molecular-weight adhesion proteins expressed by nontypable *Haemophilus influenzae. Mol. Microbiol.* 18: 378–380.

Barenkamp SJ, Shurin PA, Marchant CD et al. (1984) Do children with recurrent otitis media become infected with a new organism or reaquire the original strain? *J. Pediatr.* 105: 533–537.

Beutler B, Krochin N, Milsark IW, Luedke C, Cerami A (1986) Control of cachetin (tumour necrosis factor) synthesis: mechanisms of endotoxin resistance. *Science* 232: 977–980.

Bock K, Breimer ME, Brignole A et al. (1985) The specificity of binding of a strain of uropathogenic *Escherichia coli* to Galα(1–4)βGal containing glycosphingolipids. *J. Biol. Chem.* 260: 8545–8551.

Booy R, Kroll JS (1997) Is *Haemophilus influenzae* finished? *J. Antimicrob. Chem.* 40: 149–153.

Boulnois GJ, Roberts IS, Hodge R, Hardy K, Jann K, Timmis KN (1987). Analysis of the K1 capsule biosynthesis genes of *Escherichia coli*: definition of three functional regions for capsule production. *Mol. Gen. Genet.* 208: 242–246.

Brandtzaeg P (1992) Humoral immune response patterns of human mucosae: inductin and relation to bacterial respiratory infections. *J. Infect. Dis.* 165: S167–S176.

Brinton CC Jr, Carter MJ, Derber DB et al. (1989) Design and development of pilus vaccines for *Haemophilus influenzae* diseases. *Pediatr. Infect. Dis.* 8: S54–S61.

Brophy LN (1991) Capsulation gene loss and 'rescue' mutations during the Cap+ to Cap-transition in *Haemophilus influenzae* type b. *J. Gen. Microbiol.* 137: 2571–2576.

Brown KB, Anderson SM, Young NS (1993) Erythrocyte P antigen: cellular receptor for B19 parvovirus. *Science* 262: 114–117.

Campagnari AA, Spinola SM, Lesse AJ, Abu Kwaik Y, Mandrell RE, Apicella MA (1990) Lipooligosaccharide epitopes shared among Gram-negative non-enteric mucosal pathogens. *Microb. Pathog.* 8: 353–362.

Camussi G, Albano E, Tetta C, Bussolino F (1991) The molecular action of tumour necrosis factor-α. *Eur. J. Biochem.* 202: 3–14.

Connor EM, Loeb MR (1983) A hemadsorption method for detecting colonies of *Haemophilus influenzae* type b expressing fimbriae. *J. Infect. Dis.* 148: 855–860.

Cope LD, Yogev R, Muller Ebhard U, Hansen EJ (1995) A gene cluster involved in the utilisation of both free heme and heme-hemopexin by *Haemophilus influenzae* type b. *J. Bacteriol.* 177: 2644–2653.

Corn P (1993) Genes involved in *Haemophilus influenzae* type b capsule expression are frequently amplified. *J. Infect. Dis.* 167: 356–364.

Crisel RM, Baker RS, Dorman DE (1975) Capsular polymer of *Haemophilus influenzae* type bI. Structural characterisation of the capsular polymer of the strain Eagan. *J. Biol. Chem.* 250: 4926–4930.

Cundell D, Gerard NP, Gerard C, Idanpaan-Heikkila I, Tuomanen EI (1995) *Streptococcus pneumoniae* anchor to activated human cells by the receptor for platelet-activating factor. *Nature* 377: 435–438.

Cundell D, Tuomanen E (1995) *Streptococcus pneumoniae* anchor to activated human cells by the receptor for platelet-activating factor. *Nature* 377: 435–438.

Dawid S, Barenkamp SJ, St Geme JW III (1999) Variation in expression of the *Haemophilus influenzae* HMW adhesins: a prokaryotic system reminiscent of eukaryotes. *Proc. Natl. Acad. Sci. USA* 96: 1077–1082.

Di Fabio JL, Michon F, Brisson JR, Jennings HJ (1990) Structures of the L1 and L6 core oligosaccharide epitopes of *Neisseria meningitidis. Can. J. Chem.* 68: 1029–1034.

Ernst JD, Hartiala KT, Goldstein IM, Sande MA (1984) Complement (C5)-derived chemotactic activity accounts for accumulation of polymorphonuclear leucocytes in cerebrospinal fluid of rabbits with pneumococcal meningitis. *Infect. Immun.* 46: 81–86.

Erwin AL, Brewah YA, Couchenour DA *et al.* (2000) Role of lipopolysaccharide phase variation in susceptibility of *Haemophilus influenzae* to bactericidal immunoglobulin M antibodies in rabbit sera. *Infect. Immun.* 68: 2804–2807.

Farley MM, Stephens DS, Mulks MH *et al.* (1986) Pathogenesis of IgA1 protease producing and non-producing *Haemophilus influenzae* in human nasopharyngeal organ cultures. *J. Infect. Dis.* 154: 752–759.

Flescher AR, Insel RA (1978) Characterisation of the lipopolysaccharide *of Haemophilus influenzae. J. Infect. Dis.* 138: 719–730.

Forsgren JA, Samuelson A, Ahlin A, Jonasson J, Rynnel-Gagoo B, Lindberg AA (1994) *Haemophilus influenzae* resides and multiplies intracellularly in human adenoid tissue as demonstrated by *in situ* hybridisation and bacterial viability assay. *Infect. Immun.* 60: 673–679.

Frangione B, Wolfenstein-Todel C (1972) Partial duplication in the 'hinge' region of Iga1 myeloma proteins. *Proc. Natl Acad. Sci. USA* 69: 3673–3676.

Gilligan PH (1991) Microbiology of airway disease in patients with cystic fibrosis. *Clin. Microbiol. Rev.* 4: 35–51.

Groenveld K, van Alphen L, Eijk PP, Jansen HM, Zanen HC (1988) Changes in outer membrane proteins of nontypable *Haemophilus influenzae* in patients with chronic obstructive pulmonary disease *J. Infect. Dis.* 158: 360–365.

Guerina NG, Langermann S, Clegg HW, Kessler TW, Goldmann DA, Gilsdorf JR (1982) Adherence of piliated *Haemophilus influenzae* type b to human oropharyngeal cells. *J. Infect. Dis.* 146: 564.

Guerina NG, Langermann S, Clegg HW, Schoolnik GK, Kessler TW, Goldman DA (1985) Purification and characterisation of *Haemophilus influenzae* pili and their structural and serological relatedness to *Escherichia coli* P-fimbriae. *J. Exp. Med.* 161: 145–159.

Hanson MS, Pelzel SE, Latimer JL, Mueller-Eberhard U, Hansen EJ (1992) Identification of a genetic locus of Haemophilus inlfuenzae type b necessary for the binding and utilisation of heme bound to human hemopexin. *Proc. Natl Acad. Sci. USA* 89: 979–987.

Hendrixson DR, St. Geme JW III (1998) The *Haemophilus influenzae* Hap serine protease promotes adherence and aggregation potentiated by soluble host protein. *Mol. Cell.* 2: 841–850.

Hers JFP, Mulder J (1953) The mucosal epithelium in muco-purulent bronchitis caused by *Haemophilus influenzae. J. Pathol. Bacteriol.* 62: 103–108.

High NJ, Deadman ME, Moxon ER (1993) The role of a repetitive DNA motif (5′-CAAT-3′) in the variable expression of the *Haemophilus influenzae* lipopolysaccharide epitope αGal$(1–4)\beta$Gal. *Mol. Microbiol.* 9: 1275–1282.

High NJ, Deadman MB, Moxon ER (1995) The role of a repetitive DNA motif (5′-CAAT-3′) in the variable expression of the *Haemophilus influenzae* lipopolysaccharide epitope αGal$(1-4)\beta$Gal. *Mol. Microbiol.* 9: 1275–1282.

High NJ, Jennings MP, Moxon ER (1996) Tandem repeats of the tetramer 5′-CAAT-3′ present in lic2A are required for phase variation but not lipopolysaccharide biosynthesis in *Haemophilus influenzae. Mol. Microbiol.* 20: 165–174.

Hoiseth SK, Moxon ER, Silver RP (1986) Genes involved in *Haemophilus influenzae* type b capsulae expression are part of an 18-kilobase tandem duplication. *Proc. Natl Acad. Sci. USA* 83: 1106–1110.

Holland J, Towner KJ, Williams P (1992) Tn916 insertion mutagenesis in *Escherichia coli* and *Haemophilus influenzae* following conjugative transfer. *J. Gen. Microbiol.* 138: 509–515.

Hood DW, Deadman ME, Jennings MP *et al.* (1996) DNA repeats identify novel virulence genes in *Haemophilus influenzae. Proc. Natl Acad. Sci. USA* 93: 11121–11125.

Hood DW, Makepeace K, Deadman ME *et al.* (1999) Sialic acid in the lipopolysaccharide of *Haemophilus influenzae*: strain distribution influence on serum resistance and structural characterisation. *Mol. Microbiol.* 33: 679–692.

Hosking SL, Craig JE, High NJ (1999) Phase variation of *lic1A, lic2A* and *lic3A* in colonisation of the nasopharynx, bloodstream and cerebrospinal fluid by *Haemophilus influenzae* type b. *Microbiology* 145: 3005–3011.

Jarosik GP, Hansen EJ (1994) Identifiation of a new locus involved in the expression of *Haemophilus influenzae* lipooligosaccharide. *Infect. Immun.* 62: 4861–4867.

Jarosik GP, Maciver I, Hansen EJ (1995) Utilisation of transferrin-bound iron by *Haemophilus influenzae* requires an intact tonB gene. *Infect. Immun.* 63: 710–713.

Johnson AP, Inzana TJ (1986) Loss of ciliary activity in organ cultures of rat trachea treated with lipo-oligosaccharide isolated from *Haemphilus influenzae*. *J. Med. Microbiol.* 22: 265–268.

Kilbourn JP, Haas H, Morris JF, Samson S (1983) *Haemophilus influenzae* biotypes and chronic bronchitis. *Am. Rev. Respir. Dis.* 128: 1093–1094.

Kilian M, Mestecky J, Kulhavy R, Tomana M, Butler WT (1980) Iga1 proteases from *Haemophilus influenzae*, *Streptococcus pneumoniae*, *Neisseria meningitidis* and *Streptococcus sanguis*: comparitive immunological studies. *J. Immunol.* 124: 2596–2600.

Kilian M, Reinholdt J, Lomholt H, Poulsen K, Frandsen EVG (1996) Biological significance of IgA proteases in bacterial colonisation and pathogenesis: critical evaluation of experimental evidence. *APMIS* 104: 321–338.

Kimura A, Hansen EJ (1987) Antigenic and phenotypic variants of *Haemophilus influenzae* lipopolysaccharide and their relationship to virulence. *Infect. Immun.* 51: 60–79.

Kroll JS, Moxon ER (1988) Capsulation and gene copy number at the cap locus of *Haemophilus influenzae* type b. *J. Bacteriol.* 170: 859–864.

Kroll JS, Hopkins I, Moxon ER (1988) Capsule loss in *Haemophilus influenzae* type b occurs by recombination-mediated disruption of a gene essential for polysaccharide export. *Cell* 53: 347–356.

Kroll JS, Zamze S, Loynds B, Moxon ER (1989) Common organization of chromosomal loci for production of different capsular polysaccharides in *Haemophilus influenzae*. *J. Bacteriol.* 171: 3343–3347.

Kroll JS, Loynds B, Brophy LN, Moxon ER (1990) The *bex* locus in encapsulated *Haemophilus influenzae*: a chromosomal region involved in capsule polysaccharide export. *Mol. Microbiol.* 4: 1853–1862.

Kroll JS, Loynds BM, Brophy LN, Moxon ER (1991) The *bex* locus in encapsulated *Haemophilus influenzae*: a chromosomal region involved in capsule polysaccharide export. *Mol. Microbiol.* 4: 1853–1862.

Kuehn MJ, Heuser J, Normark S, Hultgren SJ (1992) P pili in *E. coli* are composite fibres with distinct adhesive fibrillar tips. *Nature* 356: 252–255.

Lindberg AA, Brown E, Stromberg N, Westling-Ryd M, Schultz JE, Karlsson KA (1987) Identification of the carbohydrate receptor for Shiga toxin produced by *Shigella dysenteriae* type 1. *J. Biol. Chem.* 262: 1779–1785.

Lomholt HL van Alphen L, Kilian M (1993) Antigenic variation of immunoglobulin A1 proteases among sequential isolates of *Haemophilus influenzae* from healthy children and pateints with chronic obstructive pulmonary disease. *Infect. Immun.* 61: 4575–4581.

Maskell DJ, Szabo MJ, Deadman ME, Moxon ER (1992a) The *gal* locus from *Haemophilus influenzae*: cloning, sequencing and the use of *gal* mutants to study lipopolysaccharide. *Mol. Microbiol.* 20: 3051–3063.

Maskell DJ, Szabo MJ, Butler PD, Williams AE, Moxon ER (1992b) Molecular analysis of a complex locus from *Haemophilus influenzae* involved in phase variable lipopolysaccharide biosynthesis. *Mol. Microbiol.* 5: 1013–1022.

Masoud H, Moxon ER, Martin A, Krajcarski D, Richards JC (1997) Structure of variable and conserved oligosaccharide epitopes expressed by *Haemophilus influenzae* strain Eagan. *Biochemistry* 36: 2091–2103.

Moxon ER, Kroll JS (1988) Type b capsular poysaccharide as a virulence factor of *Haemophilus influenzae*. *Vaccine* 6: 113–115.

Moxon ER, Kroll JS (1990) The role of bacterial polysaccharide capsules as virulence factors. *Curr. Top. Microbiol. Immunol.* 150: 65–85.

Moxon ER, Murphy PA (1978) *Haemophilus influenzae* bacteraemia and meningitis resulting from survival of a single organism. *Proc. Natl Acad. Sci. USA* 75: 1534–1536.

Moxon ER, Vaughn KA (1981) The type b capsular polysaccharide as a virulence determinant of *Haemophilus influenzae*: studies using clinical isolates and laboratory transformants. *J. Infect. Dis.* 143: 517.

Moxon ER, Smith AL, Averill DR, Smith DH (1974) *Haemophilus influenzae* meningitis in infant rats after intranasal inoculation. *J. Infect. Dis.* 129: 154–162.

Mulks MH, Kornfeld SJ, Frangione B, Plaut AG (1982) Relationship between the specificity of IgA proteases and serotypes in *Haemophilus influenzae*. *J. Infect. Dis.* 146: 266–274.

Musser JM, Barenkamp SJ, Granoff DM, Selander RK (1986) Genetic relationships of serologically nontypable and serotype b strains of *Haemophilus influenzae*. *Infect. Immun.* 52: 183–191.

Musser JM, Kroll JS, Moxon ER, Selander RK (1988) Clonal population structure of encapsulated *Haemophilus influenzae*. *Infect. Immun.* 56: 1835–1845.

Otto BR, Verweij-van Vught AM, MacLaren DM (1992) Transferrins and heme-compounds as iron sources for pathogenic bacteria. *Crit. Rev. Microbiol.* 18: 217–233.

Patrick D, Betts J, Frey FA, Prameya R, Dorovini-zis K, Finlay BB (1992) *Haemophilus influenzae* lipopolysaccharide disrupts confluent monolayers of bovine brain endothelial cells via a serum dependent cytotoxic pathway. *J. Infect. Dis.* 165: 865–872.

Pfeiffer R (1892) Vorflaufige Mittheilungen uber die Erreger der Influenzae. *Dtsch Med. Wochenschr.* 18: 284.

Phillips NJ, Apicella MA, Griffis JM, Gibson BW (1992) Structural characterisation of the cell surface oligosaccharides from a nontypable strain of *Haemophilus influenzae*. *Biochemistry* 31: 4515–4526.

Phillips NJ, Apicella MA, Griffis JM, Gibson BW (1993) Structural studies of the oligosaccharides from *Haemophilus influenzae* type b strain A2. *Biochemistry* 32: 2003–2012.

Pichichero ME, Anderson P, Loeb M, Smith DH (1982) Do pili play a role in pathogenicity of *Haemophilus influenzae* type b? *Lancet* ii: 960–962.

Pietola H, Kahty H, Sivonen A *et al.* (1977) Haemophilus influenzae type b capsular polysaccharide vaccine in children: a double blind field study of 100 000 vaccines 3 months to 5 years of age in Finland. *Pediatrics* 60: 730.

Pittman M (1931) Variation and specificity in the bacterial species *Haemophilus influenzae*. *J. Exp Med.* 53: 471–492.

Plautt AG (1978) Microbial IgA proteases. *N. Engl. J. Med.* 298: 1459–1463.

Plautt AG, Wistar R Jr, Capra JD (1974) Differential susceptibility of human IgA immunoglobulins to streptococcal IgA protease. *J. Clin. Invest.* 54: 1295–1300.

Pohlner J, Halter R, Beyreuther K, Meyer TF (1987) Gene structure and extracellular secretion of *Neisseria gonnorrhoeae* IgA protease. *Nature* 325: 458–462.

Porra O, Cougant DA, Lagergard T, Svanborg-Eden C (1986) Application of multilocus enzyme gel electrophoresis to *Haemophilus influenzae*. *Infect. Immun.* 53: 71–78.

Poulsen K, Brandt J, Hjorth JP, Thorgersen HC, Kilian M (1989) Cloning and sequencing of the immunoglobulin A1 protease gene (*iga*) of *Haemophilus influenzae* serotype b. *Infect. Immun.* 57: 3097–3105.

Qiu J, Hendrixson DR, Baker EN, Murphy TF, St. Geme III JW, Plaut AG (1998) Human milk lactoferrin inactivates two putative colonisation factors expressed by *Haemophilus influenzae*. *Proc. Natl Acad. Sci. USA* 95: 12641–12646.

Read RC, Wilson R, Rutman A *et al.* (1991) Interaction of nontypeable *Haemophilus influenzae* with human respiratory mucosa *in vitro*. *J. Infect. Dis.* 163: 549–558.

Risberg A, Masoud H, Martin A, Richards JC, Moxon ER, Schweda EK (1999) Structural analysis of the lipopolysaccharide oligosaccharide epitopes expressed by a capsule deficient strain of *Haemophilus influenzae* Rd. *Eur. J. Biochem.* 261: 171–180.

Robbins JB, Scneerson R, Anderson P, Smith DH (1996) Prevention of systemic infections, especially meningitis, caused by *Haemophilus influenzae* type b-impact on public health, implications for other polysaccharide-based vaccines. *J. Am. Med. Assoc.* 50: 533–550.

Roberts M, Jacob RF, Haas JE, Smith AL (1984) Adherence of *Haemophilus influenzae* to monkey respiraory tissue in organ culture. *J. Gen. Microbiol.* 130: 1437–1447.

Rodriguez LP, Schneerson R, Robbins JB (1971) Immunity to *Haemophilus influenzae* type bI. The isolation and some physiochemical serologic and biologic properties of the capsular polysaccharides of *Haemophilus influenzae* type b. *J. Immunol.* 107: 1071.

Samuelson A, Freijd A, Jonasson J, Lindberg AA (1995) Turnover of nonencapsulated *Haemophilus influenzae* in the nasopharynges of otitis prone children: a longitudinal study. *J. Clin. Microbiol.* 33: 2027–2031.

Sanders JD, Cope LD, Hansen EJ (1994) Identification of a locus involved in the utilisation of iron by *Haemophilus influenzae*. *Infect. Immun.* 62: 4515–4525.

Schneerson R, Rodrigues LP, Parke JC Jr, Robbins JB (1971) Immunity to *Haemophilus influenzae* II: specificity and some biological characterisitics of 'natural' infection-acquired and immunisation-induced antibodies to the capsular polysaccharide of *Haemophilus influenzae* type b. *J. Immunol.* 107: 1081–1089.

Selander RK, Caugant DA, Ochmna H, Musser JM, Gilmour MN, Whittam TS (1986) Methods of multilocus enzyme electrophoresis for bacterial population genetics and systematics. *Appl. Environ. Microbiol.* 51: 873–884.

Smith AL, Smith DH, Averill DR, Moxon ER (1973) Production *of Haemophilus influenzae* type b meningitis in infant rats by intraperitoneal inoculation. *Infect. Immun.* 8: 278–290.

Smith DH, Peter G, Ingram DL, Anderson P (1973) Responses of children immunized with the capsular polysaccharide of *Haemophilus influenzae* type b. *Pediatrics* 52: 637–641.

Smith W, Andrewes CH, Laidlow PP (1933) A virus from influenza patients. *Lancet* ii: 66–69.

St. Geme JW III, Falkow S (1990) *Haemophilus influenzae* adheres to and enters cultured epithelial cells. *Infect. Immun.* 58: 403–4044.

St. Geme JW III, Falkow S, Barenkamp SJ (1993) High-molecular-weight proteins of nontypeable *Haemophilus influenzae* mediate attachment to human epithelial cells. *Proc. Natl Acad. Sci. USA* 90: 2875–2879.

St. Geme JW III, de la Morena ML, Falkow S (1994a) A *Haemophilus influenzae* IgA protease-like protein promotes intimate interaction with human epithelial cells. *Mol. Microbiol.* 14: 217–233.

St. Geme JW III, Takal A, Esko E, Falkow S (1994b) Evidence for capsule gene sequences among pharyngeal isolates of nontypeable *Haemophilus influenzae*. *J. Infect. Dis.* 169: 337–342.

St. Geme JW III, Cutter D, Barenkamp SJ (1996a) Characterisation of the genetic locus encoding *Haemophilus influenzae* type b surface fibrils. *J. Bacteriol.* 178: 6281–6287.

St. Geme JW III, Pinker JS, Krasan GP *et al.* (1996b) *Haemophilus influenzae* pili are composite structures assembled by the Hif B chaperone. *Proc. Natl Acad. Sci. USA* 93: 11913–11918.

St. Geme JW III, Kumar VV, Cutter D, Barenkamp SJ (1998) Prevalence and distribution of *hmw* and *hia* genes and the HMW and Hia proteins among genetically diverse strains of nontypeable *Haemophilus influenzae*. *Infect. Immun.* 66: 364–368.

Stool SE, Field MJ (1989) The impact of otitis media. *Pediatr. Infect. Dis.* 8: S11–S14.

Sutton A, Scheerson R, Kendall-Morris S, Robbins JB (1982) Differential complement resistance mediates virulence of *Haemophilus influenzae* type b. *Infect. Immun.* 35: 95–104.

Syrogiannopoulos GA, Hansen EJ, Erwin AL *et al.* (1988) *Haemophilus influenzae* type b lipooligosaccharide induces meningeal inflammation. *J. Infect. Dis.* 157: 237–244.

Teele DW, Klein JO, Rosner B (1989) Epidemiology of otitis media during the first seven years of life in children in greater Boston: a prospective cohort study. *J. Infect. Dis.* 160: 83–94.

Teele DW, Klein JO, Chase C, Menyuk P, Rosner B (1990) Otitis media in infancy and intellectual ability, school achievement, speech and language at age 7 years. *J. Infect. Dis.* 162: 685–694.

Tunkel AR, Scheld WM (1993) Pathogenesis and pathophysiology of bacterial meningitis. *Clin. Microb. Rev.* 6: 118–136.

Tuomanen E (1995) Mediators of inflammation and the treatment of bacterial meningitis. *Curr. Opin. Infect. Dis.* 8: 218–223.

Turk DC (1982) Clinical importance of *Haemophilus influenzae*. In: Sell SH, Wright PW (eds) *Biology of Haemophilus influenzae. Epidemiology, Immunology and prevention of disease*. New York: Elsevier North Holland, pp. 3–9.

Turk DC (1984) The pathogenicity of *Haemophilus influenzae*. *J. Med. Microbiol.* 18: 1–16.

Tunkel AR, Scheld WM (1995) Acute bacterial meningitis. *Lancet* 346: 1675–1680.

van Ham SM, van Alphen L, Mooi FR, van Putten JP (1993) Phase variation of *Haemophilus influenzae* fimbriae: transcriptional control of two divergent genes through a variable combined promoter region. *Cell* 73: 1187–1196.

van Ham SM, van Alphen L, Mooi FR, van Putten JP (1994) The fimbrial gene cluster of *Haemophilus influenzae* type b. *Mol. Microbiol.* 13: 673–684.

van Schilfgaarde ML, van Alphen L, Eijk P, Everts V, Dankert J (1995) Para cytosis of *Haemophilus influenzae* through cell layers of NCI-H292 lung epithelial cells. *Infect. Immun.* 63: 4729–4737.

Virji M, Weiser JN, Lindberg AA, Moxon ER (1989) Antigenic similarities in lipopolysaccharides of *Haemophilus* and *Neisseria* and expression of a digalactoside structure also present on human cells. *Microb. Pathog.* 9: 441–450.

Virji M, Kayhty H, Ferguson DJP, Alexandrescu C, Moxon ER (1992) Interactions with *Haemophilus influenzae* with human endothelial cells *in vitro*. *J. Infect. Dis.* 165 (Suppl. 1): S115–S116.

Wallace RJ Jr, Baker CJ, Quinones FJ, Hollis DG, Weaver RJ, Wiss K (1983) Nontypeable *Haemophilus influenzae* (biotype 4) as a neonatal, maternal and genital pathogen. *Rev. Infect. Dis.* 5: 123–136.

Weiser JN, Pan N (1998) Adaptation of *Haemophilus influenzae* to acquired and innate humoral immunity based on phase variation of lipopolysaccharide. *Mol. Microbiol.* 30: 767–775.

Weiser JN, Love JM, Moxon ER (1989a) The molecular mechanism of phase variation of *Haemophilus influenzae* lipopolysaccharide. *Cell* 59: 657–665.

Weiser JN, Lindberg AA, Manning J, Hansen EJ, Moxon ER (1989b) Identification of a chromosomal locus for expression of lipopolysaccharide epitopes in *Haemophilus influenzae*. *Infect. Immun.* 57: 3945–3052.

Weiser JN, Maskell DJ, Butler PD, Lindberg AA, Moxon ER (1990a) Characterisation of repetitive sequences controlling phase variation of *Haemophilus influenzae* lipopolysaccharide. *J. Bacteriol.* 172: 3304–3309.

Weiser JN, Williams A, Moxon ER (1990b) Phase-variable lipopolysaccharide structures enhance the invasive capacity of *Haemophilus influenzae*. *Infect. Immun.* 58: 3455–3457.

Weiser JN, Scchepetov M, Chong ST (1997) Decoration of lipopolysaccharide with phosphorylcholine: a phase-variable characteristic of *Haemophilus influenzae*. *Infect. Immun.* 65: 943–950.

Weiser JN, Pan N, McGowan KL, Musher D, Martin A, Richards J (1998) Phosphorylcholine on the lipopolysaccharide of *Haemophilus influenzae* contributes to persistence in the respiratory tract and sensitivity to serum killing mediated by C-reactive protein. *J. Exp. Med.* 187: 631–640.

Wenger JD, Hightower AW, Facklam RR, Gaventa S, Broome CV (1990) Bacterial Meningitis Study Group. Bacterial meningitis in the United States, 1986: report of a multistate surveillance study. *J. Infect. Dis.* 162: 1316–1323.

Wispelway B, Lesse AJ, Hansen EJ, Scheld WM (1988) *Haemophilus influenzae* lipooligosaccharide-induced blood-brain barrier permeability during experimental meningitis. *J. Clin. Invest.* 82: 1339–1346.

Zamze SE, Moxon ER (1987) Composition of the lipopolysaccharide from different capsular serotypes of *Haemophilus influenzae*. *J. Infect. Dis.* 138: 719–730.

Zwahlen A, Nydegger UE, Vaudaux P, Lambert P-H, Waldvogel FA (1982). Complement-mediated opsonic activity in normal and infected human cerebrospinal fluid: early response during bacterial meningitis. *J. Infect. Dis.* 145: 635–646.

Zwahlen A, Rubin LG, Connelly CJ, Inzana TJ, Moxon ER (1985) Alteration of the cell wall of *Haemophilus influenzae* type b by transformation with cloned DNA: association with attenuated virulence. *J. Infect. Dis.* 152: 485–492.

Zwahlen A, Kroll JS, Rubin LG, Moxon ER (1989) The molecular basis of pathogenicity in *Haemophilus influenzae*: comparitive virulence of genetically related capsular transformants and correlation with changes at the capsulation locus *cap*. *Microb. Pathog.* 7: 225–235.

PART
17
ANIMAL AND ECTOPARASITIC SOURCE INFECTIONS

94

Brucella

David O'Callaghan[1] and Alastair MacMillan[2]

[1]INSERM U431, Faculté de Médecine, Avenue Kennedy, 30900 Nîmes, France
[2]FAO/WHO Collaborating Centre for Brucella *Reference and Research, Veterinary Laboratories Agency, Weybridge, Surrey, UK*

The genus *Brucella* is composed of Gram-negative bacteria that produce characteristic intracellular infections of animals and humans resulting in the disease brucellosis. In humans the disease is also known as Malta fever or undulant fever. Brucellosis is an important zoonosis and the disease in humans is very severe and is contracted by direct contact with infected animals or after the ingestion of certain food products derived from infected animals. A wide range of wild and domestic animal species may be infected, but the disease is usually of public-health importance only when it occurs in domestic animal species such as cattle, sheep, goats, pigs, buffalo and camels. With rare exceptions, all human cases are derived from animals, and great efforts are made to control (and eventually eradicate) the disease in domestic animals to safeguard public health. The disease in animals is of worldwide distribution, and occurs everywhere except in northern Europe and Australasia where it has been successfully eradicated. In most animals the disease is mild, the main symptom being abortion, which usually occurs in mid to late pregnancy. Extremely large numbers of organisms are excreted in the products of abortion and, therefore, most danger is posed to humans who come into contact with animals at that time. Therefore, those most at risk from this route of exposure are farmers and veterinarians. The more general population, including those in urban areas, is at risk if dairy products continue to be prepared from unpasteurised milk.

The early history epidemiological investigation of the disease is fascinating and is reviewed by Hall (1989) and Williams (1989). The British physician Sir David Bruce (1887) isolated an organism from the spleen of a deceased patient while investigating an outbreak of an often fatal disease, known then as Mediterranean or Malta fever, affecting British soldiers stationed on the island of Malta. He named the organism *Micrococcus melitensis* because of the coccoid morphology of the organism. This was subsequently isolated from the milk of local goats and the epidemiological link was established. Professor Bernhard Bang (1897), working in Denmark, recovered what was described as *Bacterium abortus* from the intrauterine membranes of aborting cattle. The relationship between *Bacterium abortus* and *Micrococcus melitensis* was first suggested by Alice Evans (1918) and confirmed by Karl Meyer (Meyer and Shaw, 1920), who suggested the name *Brucella* in honour of Sir David Bruce.

Classification and Taxonomy

Brucella is included in the α_2 subclass of the class Proteobacteria on the basis of its 16S rRNA sequence (Moreno *et al.*, 1990). This subclass includes a variety

 doi:10.1006/bkmm.2001.0094

Table 1 Characters used in the differentiation of *Brucella* species and biovars

Species	Biovar	CO₂ requirement	H₂S production	Growth with		Agglutination with antisera			Lysis with phage at RTD			
				Thionin[a]	Basic fuchsin[a]	A	M	R	Tb	Bk₂	Wb	Fi
B. abortus[b]	1	(+)	+	−	+	+	−	−	L	L	L	L
	2	(+)	+	−	−	+	−	−	L	L	L	L
	3[c]	(+)	+	+	+	+	−	−	L	L	L	L
	4	(+)	+	−	+[d]	−	+	−	L	L	L	L
	5	−	−	+	+	−	+	−	L	L	L	L
	6[c]	−	(−)	+	+	+	−	−	L	L	L	L
	9	(−)	+	+	+	−	+	−	L	L	L	L
B. suis	1	−	+	+	−[e]	+	−	−	NL	L	L	PL
	2	−	−	+	−	+	−	−	NL	L	L	PL
	3	−	−	+	+	+	−	−	NL	L	L	PL
	4	−	−	+	(+)	+	+	−	NL	L	L	PL
	5	−	−	+	−	−	+	−	NL	L	L	PL
B. melitensis	1	−	−	+	+	−	+	−	NL	NL	NL	NL
	2	−	−	+	+	+	−	−	NL	NL	NL	NL
	3	−	−	+	+	+	+	−	NL	NL	NL	NL
B. ovis		+	−	+	(−)	−	−	+	NL	NL	NL	NL
B. canis		−	−	+	(−)	−	−	+	NL	NL	NL	NL
B. neotomae		−	+	−[f]	−	+	−	−	PL	L	L	L

(+), most strains positive; (−), most strains negative; L, confluent lysis; PL = partial lysis; NL, no lysis.

[a] Concentration = 1/50 000 w/v.

[b] *B. abortus* biotypes 7 and 8 are no longer recognised.

[c] Thionin at 1/25 000 (w/v) is used in addition. 3, +; 6, −.

[d] Some strains of this biotype are inhibited by basic fuchsin.

[e] Some isolates may be resistant to basic fuchsin.

[f] Not inhibited by thionin at 1/150 000 w/v.

B. abortus biotypes 7 and 8 are no longer recognised.

After Corbel and Brinley-Morgan (1984).

of plant and animal pathogens that have a characteristic pericellular or intracellular association with plant cells (*Agrobacterium* and the Rhizobiaceae) and as intracellular pathogens of mammals (*Brucella, Bartonella, Ochrobactrum* and the Rickettsiae). The importance of this genetic relationship is becoming increasingly apparent as common mechanisms in the way these bacteria interact with their respective eukaryotic hosts are discovered (Ugalde, 1999; Boschiroli *et al.*, 2001).

The official classification of *Brucella* is based solely on phenotypic characterisation with a range of bacteriological and biochemical tests. This subject is dealt with in considerable detail in a number of publications, most notably *Bergey's Manual* (Corbel and Brinley-Morgan, 1984). Under the nomenclature currently in use, six species are differentiated on the basis of their preferred host, their susceptibility to lysis by a number of specific phages and the pattern of oxidation of a variety of carbohydrate and amino acid substrates (**Table 1**). They may be further divided into a number of biovars on the basis of differences in their ability to grow on media that contain certain aniline dyes, their reactions in agglutination tests with monospecific sera and the production of hydrogen sulphide (**Table 1**).

The use of DNA–DNA hybridisation with a large number of strains representative of all six species and their biovars suggested that the genus is mono-specific (Verger *et al.*, 1985). Verger proposed that *Brucella* should comprise a single species, namely *B. melitensis*, with the current species reassigned as its biovars. The *Brucella* Taxonomic Sub-Committee is reluctant to abandon the classical nomenclature, which is of clinical, epidemiological and legislative value. The term 'nomen-species' rather than species is now preferred, to reflect the lack of convincing evidence for different genetic species. As will be discussed below, however, the classical divisions are not artificial because they reflect real conserved differences in genome organisation.

The *Brucella* Genome

The *Brucella* Genome is Complex

The information available about the genomes of the members of the *Brucella* genus is mostly derived from physical mapping by pulsed-field gel electrophoresis (PFGE). The *Brucella* genome is composed of two circular replicons. First found for *B. melitensis* 16M, which has two circular chromosomes, one of 1.1 Mb and the other of 2.2 Mb giving a genome of approximately 3.2 Mb (Michaux *et al.*, 1993), this was rapidly extended to the type strains all of the major species. The situation is slightly different in the four biovars of

B. suis. B. suis biovar 1 has two circular chromosomes of about 2.1 and 1.15 Mb, similar to those of the other *Brucella* species, but biovar 3 has a single circular chromosome of 3.3 Mb (Jumas-Bilak *et al.*, 1998). The genomes of biovars 2 and 4 differ further from the other *Brucella* species in that they have two circular chromosomes of 1.85 and 1.35 Mb, respectively. Jumas-Bilak *et al.* (1998) suggested that the different *Brucella* strains evolved from an ancestor with a single circular chromosome, since the products of recombination events between the three *rrn* loci are in a manner similar to the chromosomal rearrangements seen in *Salmonella typhi* (Liu and Sanderson, 1996).

Stability of the *Brucella* Genome

The view that the species is monospecific was confirmed by PFGE. This showed that the genomes of *Brucella* species are very stable and that a given species, or even a biovar within a species, has a particular genomic organisation, as shown by RFLP after digestion with a rare cutting restriction enzymes such as *Xba*I or *Spe*I (Allardet-Servent *et al.*, 1988; Michaux-Charachon *et al.*, 1997). Other than the rearrangements seen in the *B. suis* biovars, the major difference is the presence of a 600-kb inversion in the small chromosomes of biovars 1–4 of *B. abortus* (Michaux-Charachon *et al.*, 1997), a large inversion in a strain from a marine mammal (Boschiroli *et al.*, Unpublished), and numerous small deletions and insertions. This suggests that the genus consists of clonal lineages, each adapted to a specific, but not exclusive, mammalian host. This may be a result of the fact that individual bacteria are genetically isolated, because in nature they multiply principally within their preferential host and there is no evidence of transfer of genetic material in nature and natural plasmids and temperate phage are not known to occur.

Genome Sequencing

The complete sequence of the *B. melitensis* 16M genome has been determined and is being annotated (January 2001). It will be available over the internet through Genbank (www.ncbi.nlm.nih.gov). The sequencing of the *B. suis* 1330 genome has been begun by The Institute For Genomic Research (TIGR: www.TIGR.org). Sequencing of the *B. abortus* 2308 genome is also in progress (Ugalde, 1999) and preliminary data confirm the close relationship with the α-proteobacteria (Sanchez *et al.*, 2001). Access to the complete genome will be a quantum leap in understanding of the genetics, physiology, virulence and evolution of the genus *Brucella*.

Identification

Classical Bacteriology

Identification of *Brucella* should be performed with great care. These bacteria are class III pathogens and are highly infectious; they are the cause of one of the most commonly reported laboratory infections (Fiori *et al.*, 2000). Aerosols are the most common cause of laboratory infections. Cultures should be manipulated in a microbiological safety cabinet and centrifugation should always be carried out with sealed canisters in aerosol-free rotors. No single routine test is available by which an organism can be identified as belonging to the genus *Brucella*. They are very small Gram-negative coccobacilli that stain weakly with safranin. Slide agglutination with commercial antisera is positive for all strains except *B. canis* and *B. ovis*. Although *Brucella* is a strict aerobe, some strains require carbon dioxide, especially on primary isolation. *Brucella* is non-motile and generally oxidase-positive and urease-positive. The more complex methods of typing, based on a combination of growth characteristics, serological and bacteriological methods, have been comprehensively described by Alton *et al.* (1988) and Corbel and Brinley-Morgan (1984). They should be left to a qualified reference laboratory.

Molecular Methods

The classical typing methods are difficult and time-consuming, and present dangers to the laboratory staff who handle large quantities of virulent bacteria. The instability of several of the phenotypic characteristics used may also cause problems. The identification of stable molecular markers is, therefore, considered to be a high priority for taxonomic, diagnostic and epidemiological purposes. Several methods, mainly PCR–RFLP and Southern blot analysis of various genes or loci, have been used to find DNA polymorphism and facilitate the molecular identification and typing of the *Brucella* species and their biovars (reviewed by Vizcaino *et al.*, 2000). Southern blotting and probing for the IS6501/IS711 insertion sequence is an effective though time-consuming method, capable of the differentiation of the species of *Brucella* (Bricker *et al.*, 1994). Detection of polymorphisms by PCR–RFLP is easier to perform and can more easily be applied to large numbers of samples. The outer membrane protein genes of *Brucella* have so far proved the best candidates for investigation, since they show sufficient polymorphism to allow reasonable differentiation between *Brucella* strains (Cloeckaert *et al.*, 1996). By targeting the genes *omp25*, *omp2a* and *omp2b* it is possible to differentiate the six *Brucella* species and some of their biovars (Ficht *et al.*, 1990; Cloeckaert *et al.*, 1995; de Wergifosse *et al.*, 1995). The addition of primers for other genes such as *omp31* and *dnaK* improves the discrimination still further (Cloeckaert *et al.*, 1996; Vizcaino *et al.*, 1996, 1997). The advent of these new methods has assisted the taxonomic designation of newly discovered strains, such as those recently identified in marine mammals (Clavareau *et al.*, 1998; Bricker *et al.*, 2000). Among these methods, the detection of polymorphism by PCR–RFLP is considered to have an advantage over Southern blotting since it is easier to perform and apply to large numbers of samples. REP and ERIC PCR fingerprinting can be used to differentiate strains to the biovar level (Mercier *et al.*, 1996). We routinely use PFGE to compare *Spe*I restriction profiles for the typing of clinical isolates. It is a very sensitive and reliable method that exploits the stability of the genome discussed above and can distinguish between strains at the biovar level.

Structure

Lipopolysaccharide

Lipopolysaccharide Structure

As with other Gram-negative bacteria, the LPS is a major component of the *Brucella* outer membrane. Extensive studies on the enterobacteria (see Chapter 6) have shown that the LPS molecule has three parts. First, lipid A, a hydrophobic lipid moiety that is anchored in the membrane and is the toxic part of the molecule. Second, the core, a non-repeated phosphorylated polysaccharide, which plays a role in the non-permeability of the outer membrane. The core has two distinct regions; the inner core, which is characterised by sugars such as L-glycero-D-mannoheptose and the essential eight-carbon sugar acid 3-deoxy-D-manno-octulosonic acid (KDO); and the outer core which is a branched pentasaccharide that contains mainly glucose, galactose and *N*-acetyl-D-glucosamine. Third, the O antigen, which consists of repeating oligosaccharide that are often highly variable, even within strains of the same species. The LPS of *Brucella* shows several differences from those of enterobacteria and these are described below.

Lipid A The structure of the *Brucella* lipid A shows several differences to those of the enterobacteria found commonly in the α-proteobacteria. The presence of diaminoglucose in addition to glucosamine, suggests that there are two populations of core lipid A molecules. The fatty acid chains contain long saturated molecules (C16:0 to C18:0) and a very long-chain molecule 27-hydroxy-octacosanoate (27-OH-C28:0) (Moreno *et al.*, 1979, 1987a, b, 1990). Another characteristic is the absence of ethanolamine, neutral sugars and phosphate.

Core Although the structure of the core region has been only partially elucidated, it is clear that this is also different from the enterobacterial core (Bowser *et al.*, 1974; Kreutzer *et al.*, 1979; Moreno *et al.*, 1979, 1981, 1984). The major components include glucose, mannose, quinovosamine (2-amino-2,6-dideoxy-D-glucose), small amounts of other sugars not yet identified and small quantities of non-substituted KDO. Another major peculiarity is the absence of the heptose region.

O antigen The LPS of all *Brucella* species, except *B. canis* and *B. ovis*, is smooth. The O-chain structure of all species of *Brucella* has been elucidated and shown to be a linear homopolymer of 4,6-dideoxy-4-formamido-α-D-mannose (Caroff *et al.*, 1984; Bundle *et al.*, 1987b). Individual units are joined by either 1,2 or 1,3 glycosidic linkages, and it is the relative proportion of these linkages in the O-polysaccharide that differs between *Brucella* species. The particular arrangement of these linkages gives rise to three main epitopes recognised by monoclonal antibodies; the A epitope, the M epitope and the common epitope, all of which occur, albeit in different proportions, in all *Brucella* (Bundle *et al.*, 1987a, b, c; Meikle *et al.*, 1989). The importance of these three epitopes in diagnosis is discussed elsewhere in this chapter.

Genetics of Lipopolysaccharide Biosynthesis

The genetics of LPS biosynthesis were first analysed in *Salmonella* (see Chapter 6). Historically, the mutations that affect LPS biosynthesis were identified in three loci named *rfa*, *rfb*, and *rfc*. The nomenclature for LPS biosynthesis genes has been harmonised and the three loci have been renamed *waa*, *wb**, and *wzy*

respectively (Reeves *et al.*, 1996). This nomenclature is used in this chapter.

Little is known about the genetics of lipid A or core biosynthesis in *Brucella*. Analysis of attenuated mutants has led to the characterisation of the *manB/pmm* locus that encodes phosphomanomutase (Allen *et al.*, 1998; Foulongne *et al.*, 2000). The rough phenotype of the attenuated vaccine strain RB51 (see below) is due to the transposition of an insertion sequence in the *wboA* gene (McQuiston *et al.*, 1999; Vemulapalli *et al.*, 1999, 2000). The rough phenotype of a mutant in the phosphoglucomutase gene also shows the role of this enzyme in LPS biosynthesis (Ugalde *et al.*, 2000). Recent work has led to the identification and characterisation of the *wbk* region of *B. melitensis* 16M which encodes genes involved in O antigen biosynthesis (**Fig. 1**). The 13 828-bp region contains seven open reading frames (ORFs) of which six have homology with genes involved in O antigen synthesis in other bacteria (Godfroid *et al.*, 1998, 2000; Cloeckaert *et al.*, 2000). The G + C content of the seven ORFs in the region vary between 44% and 49%, which is significantly lower than the 56–59% of the total genome and suggests that this region has been acquired from a different bacteria by horizontal transfer, as has been suggested for other bacteria (Reeves, 1993). This hypothesis is supported by the presence of five insertion sequences in the region, another typical signature of a pathogenicity island.

Consequences of Lipopolysaccharide Structure

The peculiarities of the chemical structure of *Brucella* LPS have been shown to be responsible for several of the unusual features of the envelope of this bacterium, such as its permeability to hydrophobic compounds and its resistance to destabilisation by EDTA and cationic peptides such as polymyxin (Martinez de Tejada *et al.*, 1995; Freer *et al.*, 1996, 1999; Velasco *et al.*, 2000). Another important consequence of these structural differences is that *Brucella* lipid A is far less toxic than that of the enterobacteria.

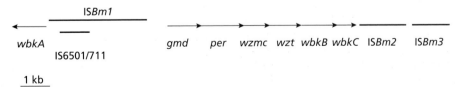

Fig. 1 O antigen biosynthesis in *B. melitensis*. The O antigen is a linear homopolymer of 1,2 and 1,3 linked 4,6-dideoxy-4-formamido-α-D-mannose. The genes for the biosynthesis of perosamine from mannose-6-phosphate are encoded by the *manB* and *manC* genes as well as genes from the *wbk* region. *wbkA* encodes a mannosyltransferase, *gmd*, GDP-mannose dehydratase, *per*, perosamine synthase, *wmz* and *wzt* are both ATPases, *wbkB* has no homologues and *wbkC* is similar to formyl transferases. There are also five insertion sequences in the region, a copy of the IS6501/711 insertion sequence and three new IS names ISBm1–4 (Adapted from Godfroid *et al.*, 2000).

Outer-Membrane Proteins

Analysis of *Brucella* outer-membrane proteins (OMPs) began in the 1980s with a view to developing vaccine candidates and diagnostic tools (see reviews by Cloeckaert *et al.*, 1996; Moriyon and Lopez-Goni, 1998; Vizcaino *et al.*, 2000). Seven surface-exposed OMPs have identified with monoclonal antibodies (Cloeckaert *et al.*, 1990) and the genes for these have been cloned and sequenced (Ficht *et al.*, 1989; de Wergifosse *et al.*, 1995; Lindler *et al.*, 1996; Vizcaino *et al.*, 1997; Tibor *et al.*, 1999). These include the genes for the Omp10, Omp16 and Omp19 lipoproteins, the Omp25 protein, the 36-kDa Omp2b protein and the Omp31 proteins. Passive immunisation with monoclonal antibodies to several of these proteins affords some protection in the mouse model, but attempts to vaccinate with purified protein have been unsuccessful (Bowden *et al.*, 1995, 2000).

Cyclic β(1–2) Glucan

Smooth and rough *Brucella* produce cyclic β(1–2) glucan, a low-molecular-weight polysaccharide also known as polysaccharide B. These cyclic β(1–2) glucans are also produced by *Agrobacterium* and *Rhizobium*, two other members of the α-proteobacteria, in which it plays a role in bacteria–plant interactions. The synthesis of cyclic β(1–2) glucan in *Agrobacterium*, *Rhizobium* and *Brucella* is a novel process (Inon de Iannino *et al.*, 1998, 2000). The synthesis depends on a 300-kDa inner-membrane synthetase that has the enzymatic activities required for initiation elongation and cyclisation and acts as a protein intermediate, with UDP-glucose as the sugar donor. Mutants in this gene have reduced virulence in mice. The cloned cyclic β(1–2) glucan synthetase from *Brucella* can functionally complement *Agrobacterium* and *Rhizobium* mutants.

Physiology

The classical work on the physiology and metabolism of *Brucella* has been referred to above. Here the more recent discoveries and their relevance to the virulence of *Brucella* are described.

Acid Resistance

Brucella encounter acid conditions in the stomach, duodenum, vagina and the phagosomes of infected macrophages of the host. Out of the host, fermenting milk and manure are also acid. Most *Brucella* strains are extremely acid-resistant and can survive exposure to

pH 4 for over 24 hours with little effect on viability. The bacteria are also far more resistant than the enterobacteria to extremely low pH (pH 3.2). Exposure to low pH completely changes protein synthesis in the bacteria. The expression of most proteins is repressed, but certain proteins, including a 24-kDa acid-shock protein, are strongly induced. As with the enterobacteria, exposure to mild acid (pH 5.8) induces an acid-tolerance response (ATR). This is far more evident in *B. canis* which, unlike *B. suis*, is extremely acid-sensitive.

Adaptation to the Intracellular Environment

Brucella must modulate its gene expression to adapt to the different environmental conditions encountered during the infectious process. Genes specifically induced in the host are often essential virulence factors (Handfield and Levesque, 1999). Biochemical analysis, by one- and two-dimensional electrophoresis, has been used to identify proteins induced in the macrophage and also under stress conditions *in vitro*, which mimic those thought to be encountered within the host (Lin and Ficht, 1995; Rafie-Kolpin *et al.*, 1996; Teixeira-Gomes *et al.*, 1997a, b, 2000). Western blotting with monoclonal antibodies and N-terminal sequencing have been used to identify the individual proteins. In bovine macrophages, 42 proteins are induced or up-regulated and over 100 are down-regulated (Rafie-Kolpin *et al.*, 1996). It is interesting that, although there is an overlap between the responses to stress and infection, the response to intracellular growth is more complex than simply a sum of the different stress responses. The hypothesis that *Brucella* encounters a harsh environment within the macrophage has led to the identification of several stress-related proteins including SOD, Kat, HtrA, RecA, DnaK, ClpA, ClpB and GroE. However, apart from DnaK, their role in virulence is only marginal (Liautard *et al.*, 1996; Sangari and Aguero, 1996; Ekaza *et al.*, 2000), which suggests that *Brucella* avoids situations in which it is exposed to harsh conditions.

Erythritol

Brucella is unusual in its ability to utilise a sugar alcohol, erythritol, which is used in preference to glucose and enhances the growth of some strains (Anderson and Smith, 1965; Meyer, 1967). The fact that ruminant placentas contain high concentrations of erythritol has been suggested as the basis for the predilection of *Brucella* for this organ. The biochemistry of the metabolic pathway used by *B. abortus* to degrade erythritol was investigated by Sperry and Robertson (1975a). Theoretically, the catabolism of erythritol produces 27 molecules of ATP, meaning that it is a more efficient carbon source than glucose for *Brucella*, which uses the inefficient pentose phosphate pathway rather than glycolysis to catabolise glucose. *B. abortus* S19, the live vaccine strain, is inhibited by erythritol, and Sperry and Robertson (1975b) found this to be due to the lack of D-erythrulose 1-phosphate dehydrogenase, so that a toxic intermediate, D-erythrulose 1-phosphate, accumulates and ATP is depleted. The genetics of erythritol catabolism have been analysed (Sangari *et al.*, 1994, 1996, 1998, 2000). The four genes of the *ery* locus (*eryA–D*) are arranged as an inducible operon (**Fig. 2a**). The first three genes encode enzymes involved in the transformation of erythritol into 3-keto-L-erythrose4-phosphate and the fourth a probable DNA-binding regulatory protein. These four genes are sufficient to allow *E. coli* to utilise erythritol, showing that *E. coli* possesses enzymes that can complete the final steps of erythritol catabolism. In *B. abortus* S19 there is a deletion of the 3′ end of the *eryC* gene and of the 5′ end of the *eryD* gene, which results in a hybrid gene that encodes a non-functional chimeric protein. The role of erythritol in virulence remains unclear, since an *ery*::Tn5 strain is virulent in mice, but information about its virulence in ruminants is not available.

Nickel Uptake

A region of the genome that encodes a periplasmic protein-dependent nickel uptake system (Nik) has been identified by the analysis of the genes induced in macrophages (see below). The five genes that encode proteins of the transport system (*nikA–E*) are divergently transcribed from the *nikR* (**Fig. 2b**) and the cloned genes complement *nik* mutants in *E. coli*. Analysis by fluorescence-activated cell sorting (FACS) was used to study the expression of green fluorescent protein gene, *gfp*, fusion under the control of the *nikA* promoter. This showed that, apart from its specific induction in macrophages, the promoter was activated by divalent metal ion deficiency and by low oxygen tension, which is interesting because *Brucella* is a strict aerobe; it was repressed by nickel chloride excess. Nickel is important for several metalloproteins, including urease. Reduced levels of urease were produced by a *nikA*::*kan* mutant when grown in nickel-depleted medium, and the *Brucella* and *E. coli* operons complemented this defect.

Pathogenesis

Pathology

Human brucellosis is responsible for a diverse range of clinical symptoms and signs, the most important of which are fever, chills, sweats, weakness, myalgia, arthralgia, depression and anorexia. The fever pattern is intermittent and a mild painless lymphadenopathy may be a subtle finding. A substantial proportion of patients present with splenomegaly and/or hepatomegaly. When the disease becomes chronic, a very wide range of pathological conditions may occur, including spondylitis (Gokhle *et al.*, 1999; Solera *et al.*, 1999), endocarditis (Cohen *et al.*, 1997; Uddin *et al.*, 1998) and meningo-encephalitis (Estevao *et al.*, 1995; Akdeniz *et al.*, 1998), related to infection of nearly all

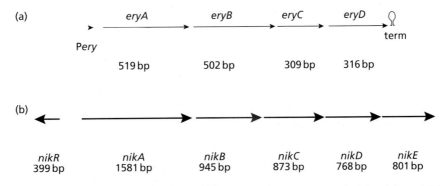

Fig. 2 (a) Erythritol utilisation operon in *B. abortus*. Closest homologues are: *eryA*, sugar (xylulose/glycerol) kinase, FGGY family; *eryE*, FAD-dependent glycerol-3-phosphate dehydrogenase; *eryC*, no homologues with extended similarity, limited homology to HupL hydrogenase; *eryD*, regulator, DNA-binding protein, helix–turn–helix motif. (Adapted from Sangari *et al.*, 2000), (b) Nickel uptake. Nickel uptake in *B. suis* is by a binding-protein dependent ABC transporter. The protein sequence shows high levels of similarity to the genes found in *E. coli*; *nikA* encodes a periplasmic binding-protein, *nicE* and *nicC* inner-membrane proteins, and *nikD* and *nikE* ATPases. The *nikR* gene, transcribed divergently from *nikA*, encodes a regulator (Adapted from Jubier-Maurin *et al.*, 2001).

the specific organ systems. Regardless of the infecting species, brucellosis has many common features, but the severity and the occurrence of complications vary to some extent with the organism responsible. In general, *B. melitensis* causes the most severe and acute symptoms. Biovars 1 and 3 of *B. suis* may produce severe infections but have a particular tendency to cause chronic suppurative lesions of the skeletal system. *B. abortus* and *B. canis* tend to produce milder disease with a higher proportion of subclinical infection (Corbel and MacMillan, 1998).

The pathogenesis of brucellosis in humans and animals is remarkably similar because of common events in the interaction of the bacteria with susceptible cells in their preferred host. In the days after the initial penetration an acute local inflammatory reaction occurs, but this is usually unsuccessful in preventing the further spread of bacteria to the regional lymph node and onwards by secondary haematogenous spread (Enright, 1990; Sutherland and Searson, 1990). *Brucella* may utilise both neutrophils and macrophages during the period of haematogenous spread, which probably occurs as early as 2 weeks after the initial challenge (Duffield *et al.*, 1984; Cheville *et al.*, 1992). As the result of the bacteraemia,

organisms spread throughout the body but they are most frequently isolated from lymphoid tissues and, in animals, from the mammary gland and reproductive tract.

Cell Biology

In animals *Brucella* infects the reproductive tract and extremely large numbers of organisms can be isolated from placental tissue. The bacteria are located in the rough endoplasmic reticulum of the chorionic trophoblasts, the epithelial cells of the foetal placenta (Anderson and Cheville, 1986; Detilleux *et al.*, 1988; Enright *et al.*, 1990; Samartino and Enright, 1993; Tobias *et al.*, 1993). The exact mechanisms that underlie the exquisite ability of *Brucella* to survive within macrophages (**Fig. 3**) and other cells are not yet fully understood, but this is an active area of research. It has long been thought that *Brucella* inhibit phagocyte functions by blocking phagosome-lysosome fusion in macrophages and by inhibiting degranulation in neutrophils (Oberti *et al.*, 1981; Canning *et al.*, 1986). Electron microscopic examination of the placentas from infected goats suggests that the bacteria multiply in the rough endoplasmic reticulum of the

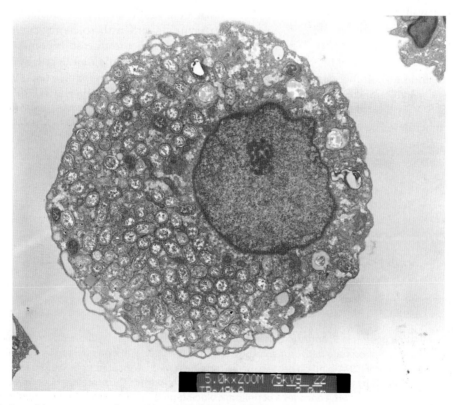

Fig. 3 *Brucella*-infected human macrophage. The THP-I macrophage-like cell line, 48 hours after infection with *B. suis* 1330 (Electron micrograph from Dr Chantal Cazevieille).

trophoblasts, and similar observations were made with Vero (monkey kidney) cells.

Recent developments in cellular microbiology have led to further investigation of these observations. *Brucella* have been shown to block and modify the process of phagosome maturation in a *Brucella*–HeLa cell infection model. Virulent strains of *Brucella* block phagosome–lysosome fusion and then pass through a novel intracellular compartment that has certain autophagosome-like characteristics. The bacteria are then targeted to, and replicate in, another novel compartment which has markers of the endoplasmic reticulum (Pizarro-Cerda *et al.*, 1998a, b, 2000), thus partly confirming the original observations. In macrophages, the situation is more complicated since a large proportion of bacteria that are taken up are killed (Arenas *et al.*, 2000). In the case of organisms that survive the initial bactericidal attack of the phagocyte, phagosome maturation is affected but, unlike the situation in Hela cells, only a small proportion of the bacteria co-localise with autophagosome or endoplasmic reticulum markers. By means of a model of phagosome–lysosome fusion reconstituted *in vitro* it was shown that inhibition of fusion is an active process and that phagocytosed *Brucella* exert a local effect on the phagosome in which they are contained rather than blocking all events in the cell (Naroeni *et al.*, 2001). Porte *et al.* (1999) showed that after uptake of *Brucella* the phagosome is rapidly acidified to a pH of 4. This acidification is an essential signal for the bacteria, since if acidification is blocked with the drugs, such as bafylomycin or monensin or if the pH is neutralised with ammonium chloride, intracellular multiplication of *B. suis* is inhibited. This suggests that acidification of the phagosome is a signal that induces the expression of virulence factors.

Virulence Factors

Type IV Secretion

A new class of secretion, named type IV secretion, in which bacteria transport macromolecular complexes across their envelope, has recently been described (Christie, 1997; Christie *et al.*, 2000; Lai and Kado, 2000). Members of this class of secretion systems transport macromolecular complexes and often transfer them directly into eukaryotic cells. The best studied is in the phytopathogen *Agrobacterium tumefaciens* which induces tumours in plants by transferring part of the 200-kbp Ti plasmid into the plant cell. The components of the transport machinery are encoded by the *virB* operon, which is also found on the Ti plasmid. Proteins of the *Agrobacterium* VirB system exhibit extensive sequence similarities with Tra proteins

encoded by, and involved in the transfer of, the broad-host-range plasmids of the IncP, N, and W incompatibility groups (Christie, 1997). Structure-function studies have shown that the VirB and Tra systems form a multi-component pore that spans both bacterial membranes and transports a single-stranded DNA protein complex across both bacterial membranes and into a recipient plant or bacterial cell. Components of the *Bordetella pertussis* Ptl transporter, which directs the secretion of the pertussis toxin to mammalian cells, have sequence similarities with the VirB and Tra systems (Covacci and Rappuoli, 1993; Weiss *et al.*, 1993). Furthermore, other homologues have been identified in the *cag* pathogenicity island of *Helicobacter pylori* (Censini *et al.*, 1996; Covacci *et al.*, 1997). Several *Legionella* Dot proteins are also homologues of VirB proteins and mediate the transfer of plasmid DNA from one bacterium to another (Segal *et al.*, 1998; Vogel *et al.*, 1998).

Brucella possesses a 12-kb region with 12 ORFs that encode a type IV secretion system homologous to the *Agrobacterium* VirB system (**Fig. 4**). Mutants in the genes in this region, called *virB*, have lost their ability to multiply in both macrophages and epithelial cells. Signature-tagged mutagenesis (see below) has also shown that this system is essential for virulence in mice. The 12 genes form an operon, the transcription of which is regulated by environmental signals and is specifically induced in the early stages after infection of macrophages and epithelial cells. The Dot/Icm type IV system of *Legionella pneumophila* is thought to export the macromolecules that affect the maturation of the phagosome, to allow the bacteria to develop intracellularly. In *H. pylori* the proteins encoded by genes of the *cag* pathogenicity island translocate the CagA protein (Odenbreit *et al.*, 2000; Stein *et al.*, 2000) which activates the NF-κB signalling pathway, to stimulate the production of interleukin 8 (IL-8) and a cascade of modifications to the host-cell cytoskeleton. The identification of the effector molecule exported by the *Brucella* type IV secretion system, and the mechanisms that regulate its expression, are now a priority.

Regulation of Virulence

Regulation of the expression of virulence factors in response to environmental signals is an essential aspect of bacterial virulence. Knowledge of the regulation of *Brucella* gene expression in general is still limited, but several two-component response regulators have been identified and characterised (Dorrell *et al.*, 1998, 1999, Sola-Landa *et al.*, 1998). A two-component response regulator, BvrRS, has been identified and is

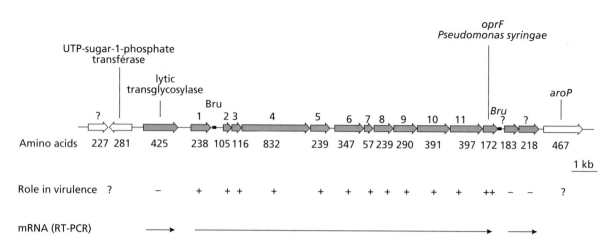

Fig. 4 The *virB* region of *B. suis* strain 1330. As shown by RT-PCR analysis the *B. suis virB* region encodes 12 transcriptionally-linked genes, *virB1–12*, which encode a type IV secretion system that is essential for virulence. The surrounding genes are not required for virulence.

essential for virulence in cells and mice (Sola-Landa *et al.*, 1998). These two proteins show homology with ChvIG in *A. tumefaciens* and ChvI-ExoS in *R. meliloti*, two-component systems essential for bacterium–plant interactions. Mutants in *bvrA* or *bvrS* are affected in their ability to enter HeLa cells and have lost the ability to modulate the endocytic pathway. After entry into cells they rapidly co-localise with the lysosomal marker cathepsin D. The environmental signals sensed by the system are still unknown, but the polymyxin-sensitive phenotype and analysis of proteins expressed by mutants suggest that it controls the expression of genes that encode components of the bacterial envelope.

A homologue of the *hfq* gene, which encodes integration host factor, an RNA-binding protein that participates in stationary phase stress resistance by enhancing the translation of RpoS, is essential for *Brucella* virulence (Robertson and Roop, 1999), but the factors that are controlled by this protein are unknown. Over-expression of the CcrM methyltransferase, a regulatory protein common in the α-proteobacteria, also attenuates the virulence of *Brucella*, possibly by perturbing the co-ordination of regulatory networks (Robertson *et al.*, 2000b). It has also been suggested that attenuation caused by a mutation in the gene that encodes the Lon protease is due to an accumulation of CcrM, which is normally degraded by Lon (Robertson *et al.*, 2000a).

Regulatory genes that control virulence have also been identified by the recently described signature-tagged transposon mutagenesis (STM). These include two two-component regulatory systems, a homologue of the *Ralstonia solanacearum vsrB* gene, which apparently controls LPS biosynthesis, and a homologue of NtrY, a regulator of nitrogen metabolism. A member of the LysR family has also been found to be essential for virulence in cell infection models.

Signature-tagged Transposon Mutagenesis

Knowledge of the genes involved in *Brucella* virulence has been vastly increased by STM. This permits the identification of attenuated mutants in pools of up to 96 mutants by negative selection and has been applied to a wide range of bacterial and fungal pathogens (Chiang and Meklanos, 1998). In the case of *Brucella*, the technique has been used in a mouse model (Hong *et al.*, 2000; Lestrate *et al.*, 2000) and it has also been adapted for use in cultured human macrophages (Foulongne *et al.*, 2000) (**Fig. 5**). Several mutants that affect the *virB* operon have been identified. Most of the genes are involved in basic metabolism or in metabolite transport and in shaping the bacterial envelope and are required for virulence in a cell culture model and in mice (reviewed by Boschiroli *et al.*, 2001).

Differential Fluorescent Induction

Differential fluorescent induction (DFI) is a technique in which green fluorescent protein (GFP) fusions are used to identify genes induced intracellularly (**Fig. 6**). The technique has been used to identify the *Brucella* genes that are expressed only in macrophages (Kohler *et al.*, 1999; Splitter, personal communication). It is interesting that, although there is little overlap, the genes identified fall into the same classes as those identified by STM (Boschiroli *et al.*, 2001). This shows how the metabolism of this bacterium changes to adapt to the intracellular environment. The role of these genes in virulence must now be assessed by

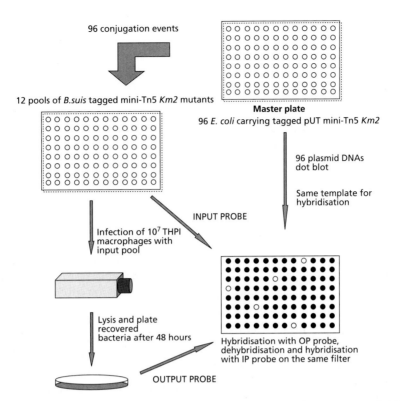

96 conjugation events

Master plate
96 *E. coli* carrying tagged pUT mini-Tn5 *Km2*

12 pools of *B.suis* tagged mini-Tn5 *Km2* mutants

96 plasmid DNAs
dot blot

Same template for
hybridisation

INPUT PROBE

Infection of 10⁷ THPI
macrophages with
input pool

Lysis and plate
recovered
bacteria after 48 hours

Hybridisation with OP probe,
dehybridisation and hybridisation
with IP probe on the same filter

OUTPUT PROBE

Fig. 5 Modification of signature-tagged mutagenesis for cell culture virulence assays (Foulongne *et al.*, 2000).

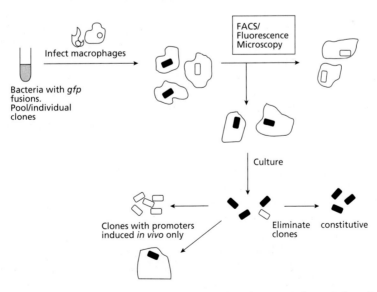

FACS/
Fluorescence
Microscopy

Infect macrophages

Bacteria with *gfp*
fusions.
Pool/individual
clones

Culture

Clones with promoters
induced *in vivo* only

Eliminate
clones

constitutive

Fig. 6 Differential fluorescence induction to identify intracellularly induced genes. Cells are infected with bacteria containing a plasmid bank of random transcriptional fusions to GFP. Cells may be infected with a pool of bacteria and containing fluorescent bacteria identified by FACS. An adaptation of the method, which avoids the aerosol infection risks of an FACS machine, is to infect cells with a single clone and identify fluorescent bacteria with a fluorescence microscope. When 96-well plates are used, large banks can be screened rapidly (Kohler *et al.*, 1999).

systematic mutagenesis. DFI has also been used to assess levels of gene expression within the cell and in the medium (Jubier-Maurin *et al.*, 2001; Boschiroli *et al.*, submitted).

Virulence Mechanisms Shared with the α-Proteobacteria

A remarkable feature of the genes that have been identified is that many encode proteins with homology or similarity to proteins involved in the interactions between α-proteobacterial plant pathogens or endosymbionts and their plant host. Clear examples are the VirB type IV secretion system, for which homologues are also found in other α proteobacterial pathogens of mammals, such as *Rickettsia* (Chapter 102) and *Bartonella* (Chapter 99), the BvrRS two-component system, CcrM, and the genes involved in the synthesis of cyclic $\beta(1-2)$glucan exopolysaccharide. A further example is the identification of a homologue of the *R. meliloti* BacA protein as an essential virulence factor (LeVier *et al.*, 2000).

Diagnosis

The problems associated with the diagnosis of human and animal brucellosis are different. In human disease, an individual patient presents with symptoms suggestive of the illness and diagnosis must be rapid to ensure timely and efficient treatment. In animal disease, diagnosis is generally a part of routine whole-herd sanitary monitoring by testing blood or milk samples. Such surveillance is universally governed by legislation. Infected animals are slaughtered, but the meat can be consumed. Although the general principles behind the tests used are similar, the two aspects are described separately.

Human Brucellosis

The first presumptive diagnosis is made by the Rose Bengal slide agglutination test, the results of which can then be confirmed by ELISA, complement fixation tests and indirect immunofluorescence for which commercial kits are available. The possibility of false-positive serological results because of cross-reaction with *Yersinia enterocolitica* O:9, *Vibrio cholerae*, especially in vaccinated individuals, and *Escherichia coli* O:157, should always be kept in mind. A definitive diagnosis is provided by a positive blood culture and the reader is referred to the excellent review by Yagupsky (1999). It should be stressed once again that all manipulation of *Brucella* should be performed with the greatest of care to avoid risk of laboratory-acquired infection.

Animal Brucellosis

Brucellosis in farm animals is a disease of international importance, and the tests used for its diagnosis and surveillance should be those recommended by the Office International des Epizooties (OIE) (MacMillan and Stack, 2001a, b). A definitive diagnosis can be made only by the recovery and identification of the organism. Blood testing is used, since testing is more usually carried out on a large scale rather than on an individual basis, as is the case in humans. Serological testing of serum or milk is widely used in the diagnosis of brucellosis and, in general, a strong and readily detectable humoral immune response can be detected, particularly against LPS epitopes. At present, tests with a sensitivity of 100% are not available, but the use of multiple tests increases the confidence in the diagnosis and sequential tests over time give a better insight than single tests.

The currently available and commonly used serological tests can be categorised as agglutination tests, buffered *Brucella* antigen tests, complement fixation tests, immunodiffusion tests and ELISAs. They have been reviewed by Alton *et al.* (1988) and MacMillan (1990).

Molecular Diagnostics

Molecular biology has yet to make its mark on the routine diagnosis of human brucellosis. PCR-based assays have been developed for human and animal brucellosis but are still not in routine use. The majority of the tests are based on the detection of species-specific insertion sites of a *Brucella*-specific insertion sequence (IS6501 or IS711) or the BCSP31 gene (reviewed by Ewalt and Bricker, 2000; Vizcaino *et al.*, 2000). A multiplex PCR assay for the detection of both *Brucella* and *M. bovis* in cattle has recently been validated in field trials in endemic areas. This assay has been proposed as a highly sensitive, cost-effective and economically viable alternative to serological testing (Sreevatsan *et al.*, 2000). The BCSP31 gene has also been used as a target for human diagnostic PCR assays and this has been successfully used with blood samples for the diagnosis of brucellosis (Morata *et al.*, 1999).

Antimicrobial Therapy

Before the antibiotic era, brucellosis was a serious chronic illness with frequent relapse and a mortality rate of about 2%. The intracellular nature of the infection means that, although the bacteria are sensitive *in vitro* to a wide range of antibiotics, many (ampicillin is a good example) are of no clinical value because they cannot reach the intracellular bacteria. Long courses of a combination of two antibiotics are required to reduce the risk of relapse. The WHO recommendation is a course of oral tetracycline (200 mg/day) in combination with rifampin (600–900 mg/day) for at least 6 weeks. Other successful combinations include oral tetracycline plus intramuscular streptomycin or tetracycline plus gentamicin. Antibiotic resistance is not a problem with *Brucella*. More detailed information about treatment can be found in the review by Young (1995).

Immunity

The mouse model of brucellosis was one of the first studied by Mackaness in his ground-breaking work that led to the discovery of cell-mediated immunity. The excellent review by Collins (1974) should be consulted for details of these classical studies. Although the relative merits of cell-mediated and humoral immunity in the control and protection against brucellosis has long been a subject for debate, it is clear that both arms of the immune response play a role.

The Humoral Response

The principal antigen recognised by the humoral response is the O antigen of smooth LPS. This response is the basis of the majority of the diagnostic tests for both human and animal brucellosis. In mice, *Brucella* LPS is a T-independent antigen to which the response is first principally IgM, which then switches to IgG3. *Brucella* LPS has unusual properties because it accumulates in lysosomes and associated with class II molecules in a haplotype-dependent manner and it inhibits class II but not class I presentation by macrophages (Forestier *et al.*, 1999, 2000). Protein antigens are also recognised by the humoral response, including many of the OMPs described above and also stress-response proteins such as GroEL, DnaK and HtrA (Spencer *et al.*, 1994; Zygmunt *et al.*, 1994; Lin *et al.*, 1996; Teixeira-Gomes *et al.*, 1997a). Antibodies against LPS give some protection against smooth strains in the mouse, whereas antibodies directed against certain proteins, including the OMPs, give some protection against rough strains. It is, however, generally accepted that the cellular response is also required for complete protection.

Cell-mediated Immunity

The macrophage is the central player in the immune response against *Brucella*. It is also the site of bacterial multiplication and not only produces cytokines that

direct the evolution of the response, but it is also influenced by the bacterium itself to modify the response. *Brucella* also inhibits the death of infected cells by apoptosis by up-regulating the expression of an anti-apoptotic factor of the *bcl-2* family (Gross *et al.*, 2000). Inhibition of apoptosis has also been seen in lymphocytes and monocytes from infected animals (Galdiero *et al.*, 2000). The role of these cell types in the control of brucellosis in mice has been clearly shown by depletion of these cell types *in vivo* anti-CD4 and anti-CD8 sera and by the use of knockout mice that do not express MHC class I (β2-m$^{-/-}$), Class II (A$\beta^{-/-}$) or any functional T or B cells (*Rag1*) (Oliveira and Splitter, 1995; Izadjoo *et al.*, 2000). These studies also showed that other cell types, such as NK cells, also play a part. These different cell types probably play roles as cytotoxic effectors that kill infected target cells, in the induction of antibody responses and in cytokine production.

Extensive attempts have been made to untangle the complicated network of cytokines that control brucellosis. Depletion with neutralising antibodies showed that interferon γ (IFNγ) is the key cytokine in the control of the infection by activating macrophages to kill the intracellular bacteria (Zhan and Cheers, 1993, 1995, 1998; Zhan *et al.*, 1996). Tumour necrosis factor α (TNFα) is one of the first cytokines produced by macrophages and acts on the macrophage itself to stimulate TNFα production. It also acts to stimulate IL-12 production, which then stimulates Th1 and NK cells to produce IFNγ. *Brucella* produces a soluble protease-sensitive factor that inhibits TNFα production by human, but not murine macrophages (Caron *et al.*, 1994, 1996). It is of interest that TNFα does not increase in cows after vaccination with live *Brucella* (Palmer *et al.*, 1998).

In patients with brucellosis there is an increase in the numbers of Vγ9Vδ2 T cells in the peripheral blood (Bertotto *et al.*, 1993). $\gamma\delta$ cells are non-MHC restricted T cells that are stimulated by non-peptide antigens. In humans, $\gamma\delta$ cells are now regarded as playing a key role in the early response to infection with intracellular pathogens. A non-peptide fraction from *Brucella* has been shown to activate human Vγ9Vδ2 cells and these inhibit *Brucella* multiplication in macrophages by the secretion of IFNγ and TNFα and also by a contact-dependent cytotoxicity (Ottones *et al.*, 2000a, b).

Epidemiology and Control

Brucellosis is a zoonosis that is almost always transmitted from animal reservoirs directly to humans by one of three routes: (1) direct contact of conjunctivae or abraded skin with excretions, secretions or tissues of infected animals or contaminated fomites; (2) inhalation of infectious aerosols followed by invasion through the mucosa of the respiratory tract; and (3) ingestion of tissues, foodstuffs or fluids containing organisms (Kaufmann *et al.*, 1980). Inter-human transmission of brucellosis is very rare. Examples of cases resulting from blood transfusion, the administration of bovine tissue for cosmetic purposes, shared needles and coitus are cited by Nicoletti (1989).

Public Health Significance

Infected farm animals, especially cows infected with *B. abortus*, effectively remain carriers for the rest of their lives, even though they may abort only once. During this time they excrete large numbers of organisms in their milk as well as in products of abortion and in those of subsequent, apparently normal, parturitions. Infection of humans can, therefore, occur by the ingestion of raw milk or milk products such as cream or cheese or by the handling of infected animals, especially around the time of parturition. Pasteurisation effectively protects the urban population in most regions, but the stockowners and their families often drink raw milk and are at risk from direct contact with infected animals.

By far the most important means of controlling the disease in humans is to reduce or eliminate the disease in farm animals. Prevention and control are best conducted on a regional or country basis, when they can be supported by an effective state veterinary infrastructure backed by legislation governing diagnosis, vaccination, surveillance, and the movement of animals and the compulsory slaughter of infected animals and their contacts. Effective education, information and advice to farmers and veterinarians is also an essential component. Such national campaigns must be compatible with international rules governing trade such as those defined by the OIE (MacMillan and Stack, 2001a, b).

Vaccination

No human vaccine for brucellosis is currently in use. The reasons are that current live vaccines used in animals remain virulent in humans, and no effective killed or subunit vaccine exists (reviewed by Nicoletti, 1989). Two strategies are in animals used for the control of brucellosis. In areas of low incidence a test and slaughter regime is used in which vaccination is not permitted, because it is impossible to distinguish between an infected and a vaccinated animal. Where the prevalence of brucellosis is high, test and slaughter

would not be feasible, since the costs of compensating farmers for their losses would be very high and insufficient brucellosis-free replacement stock would be available. In this situation, the widespread use of a licensed vaccine is a highly effective means of reducing the prevalence to a level at which test and slaughter would be feasible. Three live attenuated vaccines are currently used to vaccinate tens of millions of animals annually. For *B. melitensis* in goats and sheep Rev1 is used, which is fully virulent in humans and is a frequent cause of infection in veterinarians and farmers. For *B. abortus* in cattle, strain S19 has been used with great efficiency for well over half a century. Strain RB51 was developed and introduced into routine use in the USA in 1997. This strain is a rough mutant of *Brucella abortus* biovar 1 and as such does not give rise to antibodies detectable by conventional serological diagnostic tests (Jiminez de Bagues *et al.*, 1994).

Attempts are in progress to develop new, rationally attenuated, vaccine strains. Attenuation strategies include the creation of auxotrophic strains (Hoover *et al.*, 1999; Foulongne *et al.*, 2001), inactivation of genes encoding OMPs, inactivation of genes that encode proteins involved in the stress response and inactivation of virulence factor encoding genes. Other strategies include the development of an acellular vaccine with defined antigens and the use of naked DNA vaccination (Oliveira and Splitter, 1996; Kurar and Splitter, 1997; Cosivi and Corbel, 1998). All these projects are in early stages and the passage from the mouse model to the natural animal host and, perhaps, to humans will be the crucial step.

Acknowledgements

We would like to thank Steve Spencer for critical review of the manuscript, Vincent Foulongne for allowing us to use artwork from his PhD thesis and Amy Jennings for invaluable help with the proofreading.

References

Akdeniz H, Irmak H, Anlar O, Demiroz AP (1998) Central nervous system brucellosis: presentation, diagnosis and treatment. *J. Infect.* 36(3): 297–301.

Allardet-Servent A, Bourg G, Ramuz M *et al.* (1988) DNA polymorphism in strains of the genus *Brucella*. *J. Bacteriol.* 170(10): 4603–4607.

Allen CA, Adams LG, Ficht TA (1998) Transposon-derived *Brucella abortus* rough mutants are attenuated and exhibit reduced intracellular survival. *Infect. Immun.* 66(3): 1008–1016.

Alton GG, Jones LM, Angus RD, Verger JM (1988) *Techniques for the Brucellosis Laboratry.* Paris: Institut National de la Recherche Agronomique (INRA), p. 190.

Anderson JD, Smith H (1965) The metabolism of erythritol in *Brucella abortus. J. Gen. Microbiol.* 38: 109–124.

Anderson TD, Cheville NF (1986) Ultrastructural morphometric analysis of *Brucella abortus*-infected trophoblasts in experimental placentitis. Bacterial replication occurs in rough endoplasmic reticulum. *Am. J. Pathol.* 124(2): 226–237.

Arenas GN, Staskevich AS, Aballay A, Mayorga LS (2000) Intracellular trafficking of *Brucella abortus* in J774 macrophages. *Infect. Immun.* 68: 4255–4263.

Bang B (1897) The aetiology of epizootic abortion. *J. Comp. Pathol. Ther.* 10: 125.

Bertotto A, Gerli R, Spinozzi F *et al.* (1993) Lymphocytes bearing the gamma delta T cell receptor in acute *Brucella melitensis* infection. *Eur. J. Immunol.* 23(5): 1177–1180.

Boschiroli ML, Foulongne V, O'Callaghan D (2001) Brucellosis : a world wide zoonosis. *Curr. Opin. Microbiol.* 4(1): 58–64.

Bowden RA, Cloeckaert A, Zygmunt MS, Dubray G (1995) Outer-membrane protein- and rough lipopolysaccharide-specific monoclonal antibodies protect mice against *Brucella ovis. J. Med. Microbiol.* 43(5): 344–347.

Bowden RA, Estein SM, Zygmunt MS, Dubray G, Cloeckaert A (2000) Identification of protective outer membrane antigens of *Brucella ovis* by passive immunization of mice with monoclonal antibodies. *Microb. Infect.* 2(5): 481–488.

Bowser DV, Wheat RW, Foster JW, Leong D (1974) Occurrence of quinovosamine in lipopolysaccharides of *Brucella* species. *Infect. Immun.* 9(4): 772–774.

Bricker BJ, Halling SM (1994) Differentiation of *Brucella* abortus bv. 1, 2, and 4, *Brucella melitensis, Brucella ovis,* and *Brucella suis* bv. 1 by PCR. *J. Clin. Microbiol.* 32(11): 2660–2666.

Bricker BJ, Ewalt DR, MacMillan AP, Foster G, Brew S (2000) Molecular characterization of *Brucella* strains isolated from marine mammals. *J. Clin. Microbiol.* 38: 1258–1262.

Bruce D (1887) Note on the discovery of a microorganism in Malta fever. *Practitioner* 39: 161.

Bundle DR, Cherwonogrodzky JW, Caroff M, Perry MB (1987a) The lipopolysaccharides of *Brucella abortus* and *B. melitensis. Ann. Inst. Pasteur Microbiol.* 138(1): 92–98.

Bundle DR, Cherwonogrodzky JW, Perry MB (1987b) Structural elucidation of the *Brucella melitensis* M antigen by high-resolution NMR at 500 MHz. *Biochemistry* 26(26): 8717–8726.

Bundle DR, Cherwonogrodzky JW, Perry MB (1987c) The structure of the lipopolysaccharide O-chain (M antigen) and polysaccharide B produced by *Brucella melitensis* 16M. *FEBS Lett.* 216(2): 261–264.

Canning PC, Roth JA, Deyoe BL (1986) Release of 5'-guanosine monophosphate and adenine by *Brucella abortus* and their role in the intracellular survival of the bacteria. *J. Infect. Dis.* 154: 464–470.

Caroff M, Bundle DR, Perry MB, Cherwonogrodzky JW, Duncan JD (1984) Antigenic S-type lipopolysaccharide of *Brucella abortus* 1119–3. *Infect. Immun.* 46(2): 384–388.

Caron E, Peyrard T, Kohler S *et al.* (1994) Live *Brucella* spp. fail to induce tumour necrosis factor alpha excretion upon infection of U937-derived phagocytes. *Infect. Immun.* 62: 5267–5274.

Caron E, Gross A, Liautard JP, Dornand J (1996) *Brucella* species release a specific, protease-sensitive, inhibitor of TNF-alpha expression, active on human macrophage-like cells. *J. Immunol.* 156: 2885–2893.

Censini S, Lange C, Xiang Z *et al.* (1996) Cag, a pathogenicity island of *Helicobacter pylori*, encodes type I-specific and disease-associated virulence factors. *Proc. Natl Acad. Sci. USA* 93: 14648–14653.

Cheville NF, Jensen AE, Halling SM *et al.* (1992) Bacterial survival, lymph node changes, and immunologic responses of cattle vaccinated with standard and mutant strains of *Brucella abortus. Am. J. Vet. Res.* 53(10): 1881–1888.

Chiang SL, Mekalanos JJ (1998) Use of signature-tagged transposon mutagenesis to identify *Vibrio cholerae* genes critical for colonization. *Mol. Microbiol.* 27: 797–805.

Christie PJ (1997) *Agrobacterium tumefaciens* T-complex transport apparatus: a paradigm for a new family of multifunctional transporters in eubacteria. *J. Bacteriol.* 179: 3085–3094.

Christie PJ, Vogel JP (2000) Bacterial type IV secretion: conjugation systems adapted to deliver effector molecules to host cells. *Trends Microbiol.* 8(8): 354–360.

Clavareau C, Wellemans V, Walravens K *et al.* (1998) Phenotypic and molecular characterization of a *Brucella* strain isolated from a minke whale (*Balaenoptera acutorostrata*). *Microbiology* 144(12): 3267–3273.

Cloeckaert A, de Wergifosse P, Dubray G, Limet JN (1990) Identification of seven surface-exposed *Brucella* outer membrane proteins by use of monoclonal antibodies: immunogold labeling for electron microscopy and enzyme-linked immunosorbent assay. *Infect. Immun.* 58(12): 3980–3987.

Cloeckaert A, Verger JM, Grayon M, Grepinet O (1995) Restriction site polymorphism of the genes encoding the major 25 kDa and 36 kDa outer-membrane proteins of *Brucella. Microbiology* 141(9): 2111–2121.

Cloeckaert A, Verger JM, Grayon M, Vizcaino N (1996) Molecular and immunological characterization of the major outer membrane proteins of *Brucella. FEMS Microbiol. Lett.* 145(1): 1–8.

Cloeckaert A, Grayon M, Verger JM, Letesson JJ, Godfroid F (2000) Conservation of seven genes involved in the biosynthesis of the lipopolysaccharide O-side chain in *Brucella* spp. *Res. Microbiol.* 151(3): 209–216.

Cohen N, Golik A, Alon I *et al.* (1997) Conservative treatment for *Brucella* endocarditis. *Clin. Cardiol.* 20(3): 291–294.

Collins FM (1974) Vaccines and cell-mediated immunity. *Bacteriol. Rev.* 38(4): 371–402.

Corbel MJ, Brinley-Morgan WJ (1984) Genus *Brucella*. In: Krieg NR, Holt JG (eds) *Bergey's Manual of Systematic Bacteriology*, Baltimore, MD: Williams & Wilkins, Vol. 1, pp. 377–388.

Corbel MJ, MacMillan AP (1998) Brucellosis. In: Collier L, Balows A, Sussman M (eds) *Topley and Wilson's Microbiology and Microbial Infections*, 9th edn. London: Arnold, Vol. 3, pp. 819–847.

Cosivi O, Corbel MJ (1998) WHO consultation on the development of new/improved brucellosis vaccines. 17 December 1997, Geneva, Switzerland. *Biologicals* 26(4): 361–363.

Covacci A, Rappuoli R (1993) Pertussis toxin export requires accessory genes located downstream from the pertussis toxin operon. *Mol. Microbiol.* 8: 429–434.

Covacci A, Falkow S, Berg DE, Rappuoli R (1997) Did the inheritance of a pathogenicity island modify the virulence of *Helicobacter pylori? Trends Microbiol.* 5(5): 205–208.

de Wergifosse P, Lintermans P, Limet JN, Cloeckaert A (1995) Cloning and nucleotide sequence of the gene coding for the major 25-kilodalton outer membrane protein of *Brucella abortus. J. Bacteriol.* 177(7): 1911–1914.

Detilleux PG, Cheville NF, Deyoe BL (1988) Pathogenesis of *Brucella abortus* in chicken embryos. *Vet. Pathol.* 25(2): 138–146.

Dorrell N, Spencer S, Foulonge V *et al.* (1998) Identification, cloning and initial characterisation of FeuPQ in *Brucella* suis: a new sub-family of two-component regulatory systems. *FEMS Microbiol. Lett.* 162: 143–150.

Dorrell N, Guigue-Talet P, Spencer S *et al.* (1999) Investigation into the role of the response regulator NtrC in the metabolism and virulence of *Brucella suis. Microb. Pathog.* 27: 1–11.

Duffield BJ, Streeten TA, Spinks GA (1984) Isolation of *Brucella abortus* from supramammary lymph nodes of cattle from infected herds vaccinated with low dose strain 19. *Aust. Vet. J.* 61(12): 411–412.

Ekaza E, Guilloteau L, Teyssier J, Liautard JP, Kohler S (2000) Functional analysis of the ClpATPase ClpA of *Brucella suis*, and persistence of a knockout mutant in BALB/c mice. *Microbiology* 146: 1605–1616.

Enright FM, Araya LN, Elzer PH, Rowe GE, Winter AJ (1990) Comparative histopathology in BALB/c mice infected with virulent and attenuated strains of *Brucella abortus. Vet. Immunol. Immunopathol.* 26(2): 171–182.

Estevao MH, Barosa LM, Matos LM, Barroso AA, da Mota HC (1995) Neurobrucellosis in children. *Eur. J. Pediatr.* 154(2): 120–122.

Evans AC (1918) Further studies on *Bacterium abortus* and related bacteria. II. A comparison of *Bacterium abortus* with *Bacterium bronchi-septicus* and with the organism which causes Malta fever. *J. Infect. Dis.* 2: 580–593.

Ewalt DR, Bricker BJ (2000) Validation of the abbreviated *Brucella* AMOS PCR as a rapid screening method for differentiation of *Brucella abortus* field strain isolates and the vaccine strains, 19 and RB51. *J. Clin. Microbiol.* 38: 3085–3086.

Ficht TA, Bearden SW, Sowa BA, Adams LG (1989) DNA sequence and expression of the 36-kilodalton outer membrane protein gene of *Brucella abortus. Infect. Immun.* 57(11): 3281–3291.

Ficht TA, Bearden SW, Sowa BA, Marquis H (1990) Genetic variation at the omp2 porin locus of the Brucellae: species-specific markers. *Mol. Microbiol.* 4(7): 1135–1142.

Fiori PL, Mastrandrea S, Rappelli P, Cappuccinelli P (2000) *Brucella abortus* infection acquired in microbiology laboratories. *J. Clin. Microbiol.* 38(5): 2005–2006.

Forestier C, Moreno E, Meresse S *et al.* (1999) Interaction of *Brucella abortus* lipopolysaccharide with major histocompatibility complex class II molecules in B lymphocytes. *Infect. Immun.* 67(8): 4048–4054.

Forestier C, Deleuil F, Lapaque N, Moreno E, Gorvel JP (2000) *Brucella abortus* lipopolysaccharide in murine peritoneal macrophages acts as a down-regulator of T cell activation. *J. Immunol.* 165(9): 5202–5210.

Foulongne V, Bourg G, Cazevieille C, Michaux-Charachon S, O'Callaghan D (2000) Identification of *Brucella suis* genes affecting intracellular survival in an *in vitro* human macrophage infection model by signature-tagged transposon mutagenesis. *Infect. Immun.* 68: 1297–1303.

Foulongne V, Walravens K, Bourg G *et al.* (2001) Aromatic compound-dependent *Brucella suis* is attenuated in both cultured cells and mouse models. *Infect. Immun.* 69(1): 547–550.

Freer E, Moreno E, Moriyon I *et al.* (1996) *Brucella–Salmonella* lipopolysaccharide chimeras are less permeable to hydrophobic probes and more sensitive to cationic peptides and EDTA than are their native *Brucella* sp. counterparts. *J. Bacteriol.* 178(20): 5867–5876.

Freer E, Pizarro-Cerda J, Weintraub A *et al.* (1999) The outer membrane of *Brucella ovis* shows increased permeability to hydrophobic probes and is more susceptible to cationic peptides than are the outer membranes of mutant rough *Brucella abortus* strains. *Infect. Immun.* 67(11): 6181–6186.

Galdiero E, Romano Carratelli C, Vitiello M *et al.* (2000) HSP and apoptosis in leukocytes from infected or vaccinated animals by *Brucella abortus. New Microbiol.* 23(3): 271–279.

Godfroid F, Taminiau B, Danese I *et al.* (1998) Identification of the perosamine synthetase gene of *Brucella melitensis* 16M and involvement of lipopolysaccharide O side chain in *Brucella* survival in mice and in macrophages. *Infect. Immun.* 66(11): 5485–5493.

Godfroid F, Cloeckaert A, Taminiau B *et al.* (2000) Genetic organisation of the lipopolysaccharide O-antigen biosynthesis region of *Brucella melitensis* 16M. *Res. Microbiol.* 151(8): 655–668.

Gokhle YA, Bichile LS, Gogate A, Tillu AV (1999) Zamre. *Brucella* spondylitis: an important treatable cause of low backache. *J. Assoc. Physicians India* 47(4): 384–388.

Gross A, Terraza A, Ouahrani-Bettache S, Liautard JP, Dornand J (2000) *In vitro Brucella suis* infection prevents the programmed cell death of human monocytic cells. *Infect. Immun.* 68: 342–351.

Hall WH (1989) History of *Brucella* as a human pathogen. In: Young EJ, Corbel MJ (eds) *Brucellosis: Clinical and Laboratory Aspects.* Boca Raton, FL: CRC Press, pp. 53–72.

Handfield M, Levesque RC (1999) Strategies for isolation of *in vivo* expressed genes from bacteria. *FEMS Microbiol. Rev.* 23: 69–91.

Hong PC, Tsolis RM, Ficht TA (2000) Identification of genes required for chronic persistence of *Brucella abortus* in mice. *Infect. Immun.* 68: 4102–4107.

Hoover DL, Crawford RM, Van De Verg LL *et al.* (1999) Protection of mice against brucellosis by vaccination with *Brucella melitensis* WR201(16MDeltapurEK). *Infect. Immun.* 67(11): 5877–5884.

Inon de Iannino N, Briones G, Tolmasky M, Ugalde RA (1998) Molecular cloning and characterization of *cgs*, the *Brucella abortus* cyclic beta(1–2) glucan synthetase gene: genetic complementation of *Rhizobium meliloti ndvB* and *Agrobacterium tumefaciens chvB* mutants. *J. Bacteriol.* 180: 4392–4400.

Inon de Iannino N, Briones G, Iannino F, Ugalde RA (2000) Osmotic regulation of cyclic 1,2-beta-glucan synthesis. *Microbiology* 146: 1735–1742.

Izadjoo MJ, Polotsky Y, Mense MG *et al.* (2000) Impaired control of *Brucella melitensis* infection in Rag1-deficient mice. *Infect. Immun.* 68(9): 5314–5320.

Jimenez de Bagues MP, Elzer PH, Jones SM *et al.* (1994) Vaccination with *Brucella abortus* rough mutant RB51 protects BALB/c mice against virulent strains of *Brucella abortus*, *Brucella melitensis*, and *Brucella ovis*. *Infect. Immun.* 62(11): 4990–4996.

Jubier-Maurin V, Rodrigue A, Ouahrani-Bettache S *et al.* (2001) Identification of the *nik* Gene cluster of *Brucella suis*: regulation and contribution to urease activity. *J. Bacteriol.* 183(2): 426–434.

Jumas-Bilak E, Michaux-Charachon S, Bourg G, Ramuz M, Allardet-Servent A (1998) Unconventional genomic organization in the alpha subgroup of the Proteobacteria. *J. Bacteriol.* 180: 2749–2755.

Kaufmann AF, Fox MD, Boyce JM, Anderson DC (1980) Airborne spread of brucellosis. *Ann. NY. Acad. Sci.* 353: 105.

Kohler S, Ouahrani-Bettache S, Layssac M, Teyssier J, Liautard JP (1999) Constitutive and inducible expression of green fluorescent protein in *Brucella suis*. *Infect. Immun.* 67: 6695–6697.

Kreutzer DL, Buller CS, Robertson DC (1979) Chemical characterization and biological properties of lipopolysaccharides isolated from smooth and rough strains of *Brucella abortus*. *Infect. Immun.* 23(3):811–818.

Kurar E, Splitter GA (1997) Nucleic acid vaccination of *Brucella abortus* ribosomal L7/L12 gene elicits immune response. *Vaccine* 15(17/18): 1851–1857.

Lai EM, Kado CI (2000) The T-pilus of *Agrobacterium tumefaciens*. *Trends Microbiol.* 8(8): 361–369.

Lestrate P, Delrue RM, Danese I *et al.* (2000) Identification and characterization of *in vivo* attenuated mutants of *Brucella melitensis*. *Mol. Microbiol.* 38(3): 543–551.

LeVier K, Phillips RW, Grippe VK, Roop RM 2nd, Walker GC (2000) Similar requirements of a plant symbiont and a mammalian pathogen for prolonged intracellular survival. *Science* 287: 2492–2493.

Liautard JP, Gross A, Dornand J, Kohler S (1996) Interactions between professional phagocytes and *Brucella* spp. *Microbiologia* 12: 197–206.

Lin J, Ficht TA (1995) Protein synthesis in *Brucella abortus* induced during macrophage infection. *Infect. Immun.* 63: 1409–1414.

Lin J, Adams LG, Ficht TA (1996) Immunological response to the *Brucella abortus* GroEL homolog. *Infect. Immun.* 64(10): 4396–4400.

Lindler LE, Hadfield TL, Tall BD *et al.* (1996) Cloning of a *Brucella melitensis* group 3 antigen gene encoding Omp28, a protein recognized by the humoral immune response during human brucellosis. *Infect. Immun.* 64(7): 2490–2499.

Liu SL, Sanderson KE (1996) Highly plastic chromosomal organization in *Salmonella typhi*. *Proc. Natl Acad. Sci. USA* 93(19): 10303–10308.

MacMillan AP (1990) Conventional serological testing. In: **A,** Nielsen K, Duncan JR (eds). *Brucellosis*. Boca Raton, FL: CRC Press, pp. 153–197.

MacMillan AP, Stack J (2001a) Bovine brucellosis. In: *OIE Manual of Standards for Diagnostic Tests and Vaccines*. Paris: Office International des Epizooties, in press.

MacMillan AP, Stack J (2001b) Porcine brucellosis. In: *OIE Manual of Standards for Diagnostic Tests and Vaccines*. Paris: Office International des Epizooties, in press.

Martinez de Tejada G, Pizarro-Cerda J, Moreno E, Moriyon I (1995) The outer membranes of *Brucella* spp. are resistant to bactericidal cationic peptides. *Infect. Immun.* 63(8): 3054–3061.

McQuiston JR, Vemulapalli R, Inzana TJ *et al.* (1999) Genetic characterization of a Tn5-disrupted glycosyltransferase gene homolog in *Brucella abortus* and its effect on lipopolysaccharide composition and virulence. *Infect. Immun.* 67(8): 3830–3835.

Meikle PJ, Perry MB, Cherwonogrodzky JW, Bundle DR (1989) Fine structure of A and M antigens from *Brucella* biovars. *Infect. Immun.* 57(9): 2820–2828.

Mercier E, Jumas-Bilak E, Allardet-Servent A, O'Callaghan D, Ramuz M (1996) Polymorphism in *Brucella* strains detected by studying distribution of two short repetitive DNA elements. *J. Clin. Microbiol.* 34(5): 1299–1302.

Meyer KF, Shaw EB (1920) A comparison of the morphologic, cultural and biochemical characteristics of *B. abortus* and *B. melitensis*. Studies on the Genus *Brucella* nov. gen. I. *J. Infect. Dis.* 27: 173.

Meyer ME (1967) Metabolic characterization of the genus *Brucella*. VI. Growth stimulation by i-erythritol compared with strain virulence for guinea pigs. *J. Bacteriol.* 93(3): 996–1000.

Michaux S, Paillisson J, Carles-Nurit MJ *et al.* (1993) Presence of two independent chromosomes in the *Brucella melitensis* 16M genome. *J. Bacteriol.* 175: 701–705.

Michaux-Charachon S, Bourg G, Jumas-Bilak E *et al.* (1997) Genome structure and phylogeny in the genus *Brucella*. *J. Bacteriol.* 179: 3244–3249.

Morata P, Queipo-Ortuno MI, Reguera JM *et al.* (1999) Post-treatment follow-up of brucellosis by PCR assay. *J. Clin. Microbiol.* 37: 4163–4166.

Moreno E, Pitt MW, Jones LM, Schurig GG, Berman DT (1979) Purification and characterization of smooth and rough lipopolysaccharides from *Brucella abortus. J. Bacteriol.* 138(2): 361–369.

Moreno E, Speth SL, Jones LM, Berman DT (1981) Immunochemical characterization of Brucella lipopolysaccharides and polysaccharides. *Infect. Immun.* 31(1): 214–222.

Moreno E, Jones LM, Berman DT (1984) Immunochemical characterization of rough *Brucella* lipopolysaccharides. *Infect. Immun.* 43(3): 779–782.

Moreno E, Borowiak D, Mayer H (1987) *Brucella* lipopolysaccharides and polysaccharides. *Ann. Inst. Pasteur Microbiol.* 138(1): 102–105.

Moreno E, Mayer H, Moriyon I (1987) Characterization of a native polysaccharide hapten from *Brucella melitensis. Infect. Immun.* 55(11): 2850–2853.

Moreno E, Stackebrandt E, Dorsch M *et al.* (1990) *Brucella abortus* 16S rRNA and lipid A reveal a phylogenetic relationship with members of the alpha-2 subdivision of the class Proteobacteria. *J. Bacteriol.* 172(7): 3569–3576.

Moriyon I, Lopez-Goni I (1998) Structure and properties of the outer membranes of *Brucella abortus* and *Brucella melitensis. Int. Microbiol.* 1(1): 19–26.

Naroeni A, Jouy N, Ouahrani-Bettache S, Liautard JP, Porte F (2001) *Brucella suis*-impaired specific recognition of phagosomes by lysosomes due to phagosomal membrane modifications. *Infect. Immun.* 69(1): 486–493.

Nicoletti P (1989) Relationship between animal and human disease. In: Young EJ, Corbel MJ (eds) *Brucellosis: Clinical and Laboratory Aspects.* Boca Raton, FL: CRC Press, pp. 53–72.

Oberti J, Caravano R, Roux J (1981) Essai de détermination quantitative de la fusion phagolysosomiale au cours de l'infection de macrophages murins par *Brucella suis. Ann. Inst. Pasteur Immunol.* 132D: 201–206.

Odenbreit S, Puls J, Sedlmaier B *et al.* (2000) Translocation of *Helicobacter pylori* CagA into gastric epithelial cells by type IV secretion. *Science* 287(5457): 1497–1500.

Oliveira SC, Splitter GA (1995) CD8 + type 1 CD44hi CD45 RBlo T lymphocytes control intracellular *Brucella abortus* infection as demonstrated in major histocompatibility complex class I- and class II-deficient mice. *Eur. J. Immunol.* 25(9): 2551–2557.

Oliveira SC, Splitter GA (1996) Immunization of mice with recombinant L7/L12 ribosomal protein confers protection against *Brucella abortus* infection. *Vaccine* 14(10): 959–962.

Ottones F, Dornand J, Naroeni A, Liautard JP, Favero J (2000) Vgamma9Vdelta2 T cells impair intracellular multiplication of *Brucella suis* in autologous monocytes through soluble factor release and contact-dependent cytotoxic effect. *J. Immunol.* 165 (12): 7133–7139.

Ottones F, Liautard J, Gross A *et al.* (2000) Activation of human Vgamma9Vdelta2 T cells by a *Brucella suis* nonpeptidic fraction impairs bacterial intracellular multiplication in monocytic infected cells. *Immunology* 100(2): 252–258.

Palmer MV, Elsasser TH, Cheville NF (1998) Tumor necrosis factor-alpha in pregnant cattle after intravenous or subcutaneous vaccination with *Brucella abortus* strain RB51. *Am. J. Vet. Res.* 59(2): 153–156.

Pizarro-Cerda J, Meresse S, Parton RG *et al.* (1998a) *Brucella abortus* transits through the autophagic pathway and replicates in the endoplasmic reticulum of non-professional phagocytes. *Infect. Immun.* 66: 5711–5724.

Pizarro-Cerda J, Moreno E, Sanguedolce V, Mege JL, Gorvel JP (1998b) Virulent *Brucella abortus* prevents lysosome fusion and is distributed within autophagosome-like compartments. *Infect. Immun.* 66: 2387–2392.

Pizzaro-Cerda J, Moreno E, Gorvel JP (2000) Invasion and intracellular trafficking of *Brucella abortus* in nonphagocytic cells. *Microb. Infect.* 2(7): 829–835.

Porte F, Liautard JP, Kohler S (1999) Early acidification of phagosomes containing *Brucella suis* is essential for intracellular survival in murine macrophages. *Infect. Immun.* 67: 4041–4047.

Rafie-Kolpin M, Essenberg RC, Wyckoff JH 3rd (1996) Identification and comparison of macrophage-induced proteins and proteins induced under various stress conditions in *Brucella abortus. Infect. Immun.* 64: 5274–5283.

Reeves P (1993) Evolution of *Salmonella* O antigen variation by interspecific gene transfer on a large scale. *Trends Genet.* 9(1): 17–22.

Reeves PR, Hobbs M, Valvano MA *et al.* (1996) Bacterial polysaccharide synthesis and gene nomenclature. *Trends Microbiol.* 4(12): 495–503.

Robertson GT, Roop RM 2nd (1999) The *Brucella abortus* host factor I (HF-I) protein contributes to stress resistance during stationary phase and is a major determinant of virulence in mice. *Mol. Microbiol.* 34: 690–700.

Robertson GT, Kovach ME, Allen CA, Ficht TA, Roop RM (2000a) The *Brucella abortus* Lon functions as a generalized stress response protease and is required for wild-type virulence in BALB/c mice. *Mol. Microbiol.* 35: 577–588.

Robertson GT, Reisenauer A, Wright R *et al.* (2000b) The *Brucella abortus* CcrM DNA methyltransferase is essential for viability, and its overexpression attenuates intracellular replication in murine macrophages. *J. Bacteriol.* 182: 3482–3489.

Samartino LE, Enright FM (1993) Pathogenesis of abortion of bovine brucellosis. *Comp. Immunol. Microbiol. Infect. Dis.* 16: 95–101.

Sanchez DO, Zandomeni RO, Crevero S *et al.* (2001) Gene discovery through genomic sequencing of *Brucella abortus. Infect. Immun.* 69(2): 865–868.

Sangari FJ, Garcia-Lobo JM, Aguero J (1994) The *Brucella abortus* vaccine strain B19 carries a deletion in the erythritol catabolic genes. *FEMS Microbiol. Lett.* 121(3): 337–342.

Sangari FJ, Aguero J (1996) Molecular basis of *Brucella* pathogenicity: an update. *Microbiologia* 12: 207–218.

Sangari FJ, Aguero J, Garcia-Lobo JM (1996) Improvement of the *Brucella abortus* B19 vaccine by its preparation in a glycerol based medium. *Vaccine* 14(4): 274–276.

Sangari FJ, Grillo MJ, Jimenez De Bagues MP *et al.* (1998) The defect in the metabolism of erythritol of the *Brucella abortus* B19 vaccine strain is unrelated with its attenuated virulence in mice. *Vaccine* 16(17): 1640–1645.

Sangari FJ, Aguero J, Garcia-Lobo JM (2000) The genes for erythritol catabolism are organized as an inducible operon in *Brucella abortus*. *Microbiology* 146(2): 487–495.

Segal G, Purcell M, Shuman HA, Freeman PMC (1998) Host cell killing and bacterial conjugation require overlapping sets of genes within a 22-kb region of the *Legionella pneumophila* genome. *Proc. Natl Acad. Sci. USA* 95(4): 1669–1674.

Sola-Landa A, Pizarro-Cerda J, Grillo MJ *et al.* (1998) A two-component regulatory system playing a critical role in plant pathogens and endosymbionts is present in *Brucella abortus* and controls cell invasion and virulence. *Mol. Microbiol.* 29: 125–138.

Solera J, Lozano E, Martinez-Alfaro E *et al.* (1999) *Brucella* spondylitis: review of 35 cases and literature survey. *Clin. Infect. Dis.* 29: 1440–1449.

Spencer SA, Broughton ES, Hamid S, Young DB (1994) Immunoblot studies in the differential diagnosis of porcine brucellosis: an immunodominant 62 kDa protein is related to the mycobacterial 65 kDa heat shock protein (HSP-65). *Vet. Microbiol.* 39: 47–60.

Sperry JF, Robertson DC (1975a) Erythritol catabolism by *Brucella abortus*. *J. Bacteriol.* 121(2): 619–630.

Sperry JF, Robertson DC (1975b) Inhibition of growth by erythritol catabolism in *Brucella abortus*. *J. Bacteriol.* 124(1): 391–397.

Sreevatsan S, Bookout JB, Ringpis F *et al.* (2000) A multiplex approach to molecular detection of *Brucella abortus* and/or *Mycobacterium bovis* infection in cattle. *J. Clin. Microbiol.* 38: 2602–2610.

Stein M, Rappuoli R, Covacci A (2000) Tyrosine phosphorylation of the *Helicobacter pylori* CagA antigen after cag-driven host cell translocation. *Proc. Natl Acad. Sci. USA* 97(3): 1263–1268.

Sutherland SS, Searson J (1990) The immune response to *Brucella abortus*: the humoral response. In: Nielsen K, Duncan JR (eds) *Animal Brucellosis*. Boca Raton, FL: CRC Press, pp. 65–81.

Teixeira-Gomes AP, Cloeckaert A, Bezard G *et al.* (1997a) Identification and characterization of *Brucella ovis* immunogenic proteins using two-dimensional electrophoresis and immunoblotting. *Electrophoresis* 18: 1491–1497.

Teixeira-Gomes AP, Cloeckaert A, Bezard G, Dubray G, Zygmunt MS (1997b) Mapping and identification of *Brucella melitensis* proteins by two-dimensional electrophoresis and microsequencing. *Electrophoresis* 18: 156–162.

Teixeira-Gomes AP, Cloeckaert A, Zygmunt MS (2000) Characterization of heat, oxidative, and acid stress responses in *Brucella melitensis*. *Infect. Immun.* 68: 2954–2961.

Tibor A, Decelle B, Letesson JJ (1999) Outer membrane proteins Omp10, Omp16, and Omp19 of *Brucella* spp. are lipoproteins. *Infect. Immun.* 67(9): 4960–4962.

Tobias L, Cordes DO, Schurig GG (1993) Placental pathology of the pregnant mouse inoculated with *Brucella abortus* strain 2308. *Vet. Pathol.* 30(2): 119–129.

Uddin MJ, Sanyal SC, Mustafa AS *et al.* (1998) The role of aggressive medical therapy along with early surgical intervention in the cure of *Brucella* endocarditis. *Ann. Thorac. Cardiovasc. Surg.* 4(4): 209–213.

Ugalde JE, Czibener C, Feldman MF, Ugalde RA (2000) Identification and characterization of the *Brucella abortus* phosphoglucomutase gene: role of lipopolysaccharide in virulence and intracellular multiplication. *Infect. Immun.* 68(10): 5716–5723.

Ugalde RA (1999) Intracellular lifestyle of *Brucella* spp. Common genes with other animal pathogens, plant pathogens, and endosymbionts. *Microb. Infect.* 1: 1211–1219.

Velasco J, Bengoechea JA, Brandenburg K *et al.* (2000) *Brucella abortus* and its closest phylogenetic relative, *Ochrobactrum* spp., differ in outer membrane permeability and cationic peptide resistance. *Infect. Immun.* 68(6): 3210–3218.

Vemulapalli R, McQuiston JR, Schurig GG *et al.* (1999) Identification of an IS711 element interrupting the wboA gene of *Brucella abortus* vaccine strain RB51 and a PCR assay to distinguish strain RB51 from other *Brucella* species and strains. *Clin. Diagn. Lab. Immunol.* 6: 760–764.

Vemulapalli R, He Y, Buccolo LS *et al.* (2000) Complementation of *Brucella abortus* RB51 with a functional wboA gene results in O-antigen synthesis and enhanced vaccine efficacy but no change in rough phenotype and attenuation. *Infect. Immun.* 68(7): 3927–3932.

Verger JM, Grimont F, Grimont PAD, Grayon M (1985) *Brucella*, a monospecific genus as shown by deoxyribonucleic acid hybridization. *Int. J. Syst. Bacteriol.* 35: 292–295.

Vizcaino N, Cloeckaert A, Zygmunt MS, Dubray G (1996) Cloning, nucleotide sequence, and expression of the *Brucella melitensis* omp31 gene coding for an immunogenic major outer membrane protein. *Infect. Immun.* 64(9): 3744–3751.

Vizcaino N, Verger JM, Grayon M, Zygmunt MS, Cloeckaert A (1997) DNA polymorphism at the omp-31 locus of *Brucella* spp.: evidence for a large deletion in *Brucella abortus*, and other species-specific markers. *Microbiology* 143: 2913–2921.

Vizcaino N, Cloeckaert A, Verger J-M, Grayon M, Fernandez-Lago L (2000) DNA polymorphism in the genus *Brucella*. *Microb. Infect.* 2: 1089–1100.

Vogel JP, Andrews HL, Wong SK, Isberg RR (1998) Conjugative transfer by the virulence system of *Legionella pneumophila*. *Science* 279(5352): 873–876.

Weiss AA, Johnson FD, Burns DL (1993) Molecular characterization of an operon required for pertussis toxin secretion. *Proc. Natl Acad. Sci. USA* 90(7): 2970–2974.

Williams E (1989) The Mediterranean Fever Commission: Its origin and achievements In: Young EJ, Corbel MJ (eds) *Brucellosis: Clinical and Laboratory Aspects*. Boca Raton, FL: CRC Press, pp. 53–72.

Yagupsky P (1999) Detection of Brucellae in blood cultures. *J. Clin. Microbiol.* 37: 3437–3442.

Young EJ (1995) An overview of human brucellosis. *Clin. Infect. Dis.* 21: 283–289.

Zhan Y, Cheers C (1993) Endogenous gamma interferon mediates resistance to *Brucella abortus* infection. *Infect. Immun.* 61(11): 4899–4901.

Zhan Y, Cheers C (1995) Endogenous interleukin-12 is involved in resistance to *Brucella abortus* infection. *Infect. Immun.* 63(4): 1387–1390.

Zhan Y, Liu Z, Cheers C (1996) Tumor necrosis factor alpha and interleukin-12 contribute to resistance to the intracellular bacterium *Brucella abortus* by different mechanisms. *Infect. Immun.* 64(7): 2782–2786.

Zhan Y, Cheers C (1998) Control of IL-12 and IFN-gamma production in response to live or dead bacteria by TNF and other factors. *J. Immunol.* 161(3): 1447–1453.

Zygmunt MS, Cloeckaert A, Dubray G (1994) *Brucella melitensis* cell envelope protein and lipopolysaccharide epitopes involved in humoral immune responses of naturally and experimentally infected sheep. *J. Clin. Microbiol.* 32(10): 2514–2522.

95

Bacillus anthracis and Other *Bacillus* species

P.C.B. Turnbull[1], C.P. Quinn[2] and I. Henderson[3]

[1]*Arjemptur Technology Ltd, Porton Down Science Park, Salisbury, UK*
[2]*Centre for Applied Microbiology and Research, Porton Down, Salisbury, UK*
[3]*DynPort Vaccine Company LLC, Frederick, MD, USA*

The authors dedicate this chapter to the memory of their colleague Dr Brian McBride, who was to have joined them in co-authorship. Brian, who was an outstanding immunologist, died of cancer at the peak of his career and is sadly missed by his colleagues.

Taxonomy

Since the establishment of the science of bacteriology, aerobic, Gram-positive spore-forming rods have traditionally been placed in the genus *Bacillus*. Even before it became possible to examine relationships at the DNA level, however, it was clear that this group consists of a very diverse range of species. The first truly effective organisation of the genus was carried out by Gordon and colleagues in the 1960s and 1970s (Smith *et al.*, 1952; Gordon *et al.*, 1973). Their classification, based on traditional morphological, physiological and biochemical tests, is still used but identification is complicated by the need for special media and the problems of strain-to-strain variation (Logan and Turnbull, 1999).

The heterogeneity of the genus was confirmed in the late 1970s and early 1980s by determinations of DNA base composition and DNA–DNA re-association studies, which showed that the species of the genus vary between extremes of 31 and 69 mol% G+C (Priest, 1981). Furthermore, several examples were found in which the range of DNA base composition varied by more than 5% among isolates assigned to a particular species by phenotypic methods. At the beginning of the 1990s, with the introduction of 16S rRNA gene sequence analysis as an indicator of molecular evolution, the full extent of the variation became more apparent. The results of 16S rRNA sequence analysis of the type strains of 51 species led Ash *et al.* (1991) to conclude that the genus should be subdivided into at least five separate genera.

Apart from confirming the diversity of the species in the genus as a whole, 16S rRNA sequence analysis also served to highlight the very conserved nature of the so-called *Bacillus cereus* group – *B. cereus*, *B. thuringiensis*, *B. anthracis* and *B. mycoides*. Ash *et al.* (1991) and Ash and Collins (1992) showed that rRNA gene sequences are almost identical between members of these species. *B. anthracis* NCTC 8234 (the Sterne vaccine strain) and emetic *B. cereus* NCTC 11143, for example, differed from the *B. cereus* type strain (NCDO 1771T) by only a single base. Overall, sequence differences of 4–9 nucleotides were found between the strains of *B. mycoides*, *B. thuringiensis*, *B. anthracis* and *B. cereus*.

At least 11 phylogenetic lines of descent have been recognised (Priest, 1997). There is, however, currently no accepted level of rRNA divergence or set of criteria for the establishment of a genus and there is a danger of fragmentation of *Bacillus* into an unwieldy number of new genera (Priest, 1997). The genus currently comprises at least 50 validly published species (Logan and Turnbull, 1999), as compared with listings of 75, 95, 146, 33, 48, and 63 species in the *Bergey's Manuals* of 1923, 1934, 1939, 1957, 1974, and 1986 respectively (Priest, 1997). Some reorganisation has recently taken place, with the transfer of a number of species to 7 newly created genera. Of those that have been encountered in the clinical context, *B. alvei*, *B. macerans* and *B. polymyxa* have been included with about a dozen other former *Bacillus* species in the new genus *Paenibacillus*, and *B. brevis* and *B. laterosporus* are now placed within the genus *Brevibacillus* together with *Br. centrosporus* and *Br. agri*.

Medical and Public Health Importance

Not all *Bacillus* species are of medical or public health importance and of relevance in the context of this chapter. Although only a few species are associated to a greater or lesser extent with primary infections, a substantial number are, however, of clinical or health importance in a variety of other ways (Parry *et al.*, 1983; Zukowski, 1992) (Table 1).

The restriction endonucleases and DNA polymerases of several *Bacillus* species are of importance as research tools. Similarly, *Bacillus* species lend themselves well to host-vector systems for production of bio-engineered therapeutic products. A strain of *B. cereus* was the active ingredient of an antidiarrhoeal formulation prescribed in certain European countries (Parry *et al.*, 1983), but it was withdrawn in the early 1990s. Other *Bacillus* species are apparently the active ingredients of probiotics (Green *et al.*, 1999).

The resistance of *Bacillus* spores to heat, radiation and disinfectants makes them troublesome contaminants in operating theatres, surgical dressings and pharmaceutical products, and in foods and feeds. They constitute a challenge as normal components of dust, in the environment generally and in the routine diagnostic laboratory, where pathogens must be detected and identified in a background of normal environmental flora. Unsurprisingly, *Bacillus* species are common bacteriological contaminants of the injection paraphernalia of drug addicts (Tuazon, 1984).

Table 1 Some applications of *Bacillus* species

Species	Application
B. licheniformis	Bacitracin production
B. subtilis	Bacitracin production
	Antibiotic assay
	Biotin and riboflavin production
	Subtilisin-dependent assay for *Chlamydia*
	Phenylketonuria screening test
B. polymyxa	Polymyxin production
B. megaterium	Vitamins B_2 and B_{12} production
	Antibiotic assay
B. cereus	Probiotic for animals
	Antibiotic assay
	Disinfectant validation
B. circulans	Probiotic for animals
	Antibiotic assay
B. pumilus	Antibiotic assay
B. stearothermophilus	Antibiotic assay
	Heat sterilisation monitoring
B. globigii	Monitoring of fumigation
B. pumilus	Monitoring of radiation processes
B. fastidiosis	Uric acid assay
B. coagulans	Probiotic for animals

Bacillus species in Disease

The only obligate pathogen in the genus is *B. anthracis*, the agent of anthrax. This remains an important disease in several countries of Africa, Asia and central and southern Europe. In certain countries anthrax is a so-called re-emerging disease, because of changing lifestyles, breakdown in veterinary services, political upheavals or unrest. Other industrialised countries, including Britain and the USA, continue to record low numbers of cases in livestock in most years, with occasional human cases. Countries that are anthrax-free, or almost so, must maintain constant surveillance against resurgence of the disease associated with import of contaminated animal products from endemic countries.

Several *Bacillus* species have been associated with a range of infections, but in the majority of cases there was probably an underlying immunological, metabolic or other disorder, or drug abuse. Apart from *B. anthracis*, only *B. cereus* and its close relative, *B. thuringiensis*, have clearly identifiable virulence factors to account for pathogenicity. All *Bacillus* species produce various extracellular enzymes, however, and some of these presumably play a direct or indirect role in the syndromes with which the bacteria have been associated.

B. cereus, in particular, has long been associated with two distinct types of food poisoning and with other types of infection (see Chapter 57). In recent years, *B. subtilis* and *B. licheniformis* have also entered food-poisoning statistics too frequently to be dismissed as irrelevant. *B. subtilis*-associated food poisoning resembles the *B. cereus* emetic syndrome in its symptoms, but diarrhoea is relatively common and the association with rice dishes is less marked. *B. licheniformis* food poisoning, on the other hand, more resembles the diarrhoeal-type of *B. cereus* food poisoning. Two reports (Beattie and Williams, 1999; Salkinoja-Salonen *et al.*, 1999) suggests that *B. subtilis* and *B. licheniformis* can elaborate toxins similar or analogous to those of *B. cereus*, but this remains to be confirmed. Similarly, the pathogenicity mechanism and virulence factor(s) behind a well-recognised association of *B. licheniformis* with abortion in cattle and sheep and with severe vaginal infections in cows which bed down on maize silage (Logan, 1988; Blowey and Edmondson, 1995) have yet to be elucidated.

Identification and Diagnosis

Identification of clinical laboratory isolates of aerobic spore-forming bacteria is usually not undertaken, mainly because isolates of *Bacillus* encountered in the laboratory are generally dismissed as contaminants. As a result the 20% of species more commonly recognised are probably correctly identified, but some doubt arises with respect to the remaining 80%. Various attempts have been made in the past two decades to devise technology that would ensure much higher identification rates. The relatively elaborate approach of pyrolysis mass spectrometry was applied with some success to differentiation in the 1980s (Shute *et al.*, 1988), but initial hopes that this technique might supply the means for reliable differentiation between members of the *B. cereus* group, including emetic, diarrhoeal and environmental subgroups, did not materialise (Helyer *et al.*, 1997). The technique requires expensive equipment and problems of safety arise when it is used for the identification of *B. anthracis* (hazard level 3) in non-specialist centres.

Other approaches that have been applied with some degree of success to *Bacillus* differentiation are chemotaxonomic fingerprinting by fatty acid methyl ester (FAME) profiling, polyacrylamide gel electrophoresis (PAGE) analysis and Fourier-transform infra-red spectroscopy (Logan and Turnbull, 1999). Large databases of authentic strains are necessary for the success of these methods. A database for FAME analysis is commercially available (Microbial ID Inc., Newark, Delaware, USA).

The use of commercial kits based on miniaturised metabolic and enzymatic profiles, such as the API^T 20E and 50CHB combination systems (bioMérieux, Marcy l'Étoile, France) or $Biolog^T$ (Biolog Inc., Hayward, California, USA) is increasing, but their complete reliability depends on updated databases. It must also be stressed that their use should always be preceded by appropriate basic characterisation tests, principally spore and colony characteristics, motility, haemolysis and egg yolk reaction (Logan and Turnbull, 1999).

Molecular differentiation methods are beginning to appear, but developments in molecular speciation are biased towards species and strains of high commercial, military or emotive interest, such as *B. anthracis* and other species with which it might be confused, or *B. thuringiensis* or the other *Bacillus* insect pathogens used in insecticides or the *Bacillus* species that cause food spoilage and food poisoning. As in the case of metabolic and enzyme profiling methods, the value of these depends on large databases of authentic strains.

Among the potential molecular genetic techniques for speciation and sub-speciation, the most conventional is the detection of restriction fragment length polymorphism (RFLP) in the DNA of the organism under study. For a large number of bacterial species, the principal targets are the 16S ribosomal rRNA genes, which are evolutionarily representative because of their central role in gene expression. The application of this approach to speciate *Bacillus* has been considered above under taxonomy.

Differentiation of *Bacillus* species

The molecular differentiation of *B. anthracis* from *B. cereus* and *B. thuringiensis* has attracted most attention, probably because of defence-related interests. The identification of virulent *B. anthracis* in the conventional bacteriology laboratory is generally a simple procedure. The basic identification parameters are appropriate colonial morphology, lack of haemolysis and motility, sensitivity to diagnostic (γ) phage and penicillin, and production of the polypeptide capsule. For the most part, an isolate that meets these criteria *in vitro* is unarguably virulent *B. anthracis*. Virulence in test animals, with demonstration of the capsulated bacilli in the blood at death, has been a definitive diagnostic test since the beginning of the twentieth century (M'Fadyean, 1903). The use of animals is now rarely justifiable, but this is still the test depended on in some parts of the world. It may still be the approach of last resort when isolation *in vitro* is hampered in some way, such as when testing human or animal tissues collected after the start of antibiotic treatment and it is essential to establish anthrax as the cause of death (Turnbull *et al.*, 1998). The need for test animals has been further displaced by simple polymerase chain reaction (PCR) technology for demonstrating the presence of the genes that encode the toxin and capsule precursors (Turnbull *et al.*, 1998), but this technology is not yet available worldwide.

As a generalisation, there has always been a tendency in clinical laboratories where the occasional isolate with phenotypic characteristics of *B. anthracis* is encountered, but which fails to induce anthrax in mice or guinea-pigs or to produce a capsule in appropriate tests, to discard it as inconsequential, or recorded it as *B. cereus*. Names recognised in the 1960s, such as *B. anthracis similis*, *B. anthracoides*, *B. pseudoanthracis* and others (Turnbull, 1999), indicate that there were exceptions to this generalisation. It was not, however, until the 1990s that technology became available to show the natural existence of *B. anthracis* that had lost the smaller of the two plasmids, pXO2, or less frequently, both plasmids, pXO1 and pXO2. Such isolates were thought likely to be the progeny of virulent parent strains that had lost these plasmids as a result of environmental pressure (Turnbull *et al.*, 1992b).

The use of conventional metabolic identification schemes, or their miniaturised and expanded counterparts, such as the API or Biolog systems, does not increase the level of certainty in the identification of *B. anthracis*, but they are helpful to distinguish *B. anthracis*, *B. cereus* and *B. thuringiensis* from other morphological group 1 species (Gordon *et al.*, 1973) with which they can be confused, such as *B. megaterium*. *B. mycoides*, the fourth member of the '*B. cereus* group', is unlikely to be confused with anything else.

Largely as a result of close scrutiny by those concerned with the potential use of anthrax spores as agents of biological aggression after the Sverdlovsk incident in 1979 and the Gulf War in 1991, there has been a heightened interest in avirulent strains of *B. anthracis* and the increasing ease with which DNA can be moved from one member of the '*B. cereus* group' to another. In theory, this could reduce the reliability of the conventional phenotypic identification characters – haemolysis, motility, gamma phage and penicillin sensitivity and capsule production – when it comes to distinguishing *B. anthracis* from *B. cereus*. Similarly, the plasmid-encoded crystal toxins by which *B. thuringiensis* is conventionally identified may become less reliable for identification purposes.

Early DNA homology studies of the *B. cereus* group indicated that the level of similarity between individual *B. anthracis* isolates is greater than 90%, whereas homologies among strains of *B. cereus*, *B. thuringiensis* and *B. mycoides* were less than 60% (Böhm and Spath, 1990). This provided an early indication that schemes based on specific DNA fragments might be useful for species separation in the *B. cereus* group but would have limited value in *B. anthracis* strain differentiation. It became clear that the variation of overall *B. anthracis* chromosomal structure should be assessed. PCR with either the randomly amplified polymorphic DNA (RAPD) fingerprinting method of Welsh and McClelland (1990), or sequence-specific primers of strains of the *B. cereus* group, showed that none of the primers designed with random sequences detected differences between different isolates of *B. anthracis* (Henderson *et al.*, 1994). However, primers based on the sequences of the M13 fingerprinting probe and the 20-bp inverted repeat of IS231 variants highlighted not only the genetic coherency of *B. anthracis*, but also that these differences at the chromosomal level exist within the species (Henderson *et al.*, 1994, 1995). RAPD PCR is a rapid process but suffers from variability of oligonucleotide synthesis and equipment performance and, at least for the *B. cereus* group, it could only be seriously considered a means of species identification.

Pulsed-field electrophoresis (PFGE) is a powerful technique in molecular epidemiology, which allows construction of chromosomal maps and detection and identification of chromosomal rearrangements. Its application to *Bacillus* speciation has been confined to comparative maps based on *AscI*, *NotI* and *SfiI* restriction sites. These confirm the conserved nature of *B. cereus* and *B. thuringiensis* chromosomes (Carlson *et al.*, 1996) and the banding patterns in 16S–23S rRNA and *gyrB*–*gyrA* intergenic spacers sequence to differentiate between species and strains of *B. anthracis*, *B. cereus* and *B. mycoides* (Harrell *et al.*, 1995). The 16S–23S spacer PFGE banding patterns were identical in three *B. anthracis* strains, which differed by one base from *B. cereus*, and by nine from *B. mycoides*. Similarly, the gyrase spacers were identical in the *B. anthracis* strains, but *B. cereus* and *B. mycoides* differed from *B. anthracis* by one and two bases respectively, and from each other by just one nucleotide.

Multi-locus enzyme electrophoresis (MLEE) and multi-locus sequence typing (MLST) are automated nucleotide sequence determination methods, which make the typing of bacterial populations possible by determining the nucleotide sequence of the alleles present at multiple unlinked housekeeping loci and by assigning isolates to a sequence type on the basis of these data. The use of an automated nucleotide sequencing approach to speciate members of the genus *Bacillus* has been confined to demonstrating genetic diversity among a large group of isolates of *B. cereus* and *B. thuringiensis*, mostly from Norwegian soil (Hill *et al.*, 1999; Jackson *et al.*, 1999b).

Another approach to the detection of differences between organisms at the genetic level is to look for fragment-length polymorphisms. This is appropriate for a species as coherent as *B. anthracis* because it looks at sequence variation rather than chromosomal structure and, in the absence of other genetic markers on the chromosome, distinction of *B. cereus* group members depends on differences seen in RFLP (Henderson, 1996) and amplified fragment-length polymorphism (AFLP) (Keim *et al.*, 1997; Jackson *et al.*, 1999a) patterns. AFLP profiles allow unambiguous identification of isolates on the basis of a large number of independent genetic loci, but this method generates very large amounts of molecular information, which cannot be analysed rapidly by manual methods. The necessary electronic AFLP signature database together with the appropriate computational packages are being developed with a view to their practical use for the identification and phylogenetic characterisation of a broad range of medical, veterinary and environmental isolates (Jackson *et al.*, 1999a).

Very few chromosomal genes useful for differential or epidemiological purposes, that is those with evolutionary significance, have so far been sequenced. The *B. anthracis* virulence plasmids pXO1 and pXO2 have been sequenced (Okinaka *et al.*, 1999a, b), but the epidemiological value of these sequences is limited because of the potential loss of plasmids.

The complete *B. anthracis* genome sequence (Read and Peterson, 1999), which is to be made available through the TIGR microbial database site (www. tigr.org/tdb/mdb/mdb.html), is eagerly awaited. In addition to useful housekeeping genes, genetic markers of the known phenotypic distinguishing features will presumably be recognised. These will include markers such as metabolic and enzyme profiles, motility factors, phage receptors, β lactamase, phospholipase and enterotoxin activities, S-layer proteins and other specific vegetative cell and spore antigens, as well as virulence factors. These will allow a fuller understanding of the relationships between *Bacillus* species and their distinguishing features.

Strain and Subspecies Differentiation

In a molecular sense, *B. anthracis* has been described as one of the most monomorphic of known bacteria (Keim *et al.*, 1997). This may reflect the fact that, as an obligate pathogen, *B. anthracis* only multiplies in a host during an infection, but does so only rarely in comparison with most other pathogens. The extreme phenotypic coherency of *B. anthracis* strains, and their similarity to other closely related species, has long frustrated epidemiological and strategic tracking exercises. Conventional epidemiological methods, such as serotyping, biotyping and phage typing, have proved incapable of discriminating between individual strains of *B. anthracis*.

Approaches to classifying bacterial species have included the use of DNA fingerprinting techniques that rely on the detection of the RFLP revealed by DNA probes. M13 coliphage DNA can detect RFLPs in a variety of mammalian and bacterial species (Vassart *et al.*, 1987). This has been used to fingerprint *B. thuringiensis* strains (Miteva *et al.*, 1991) and Henderson *et al.* (1994) detected minor differences in a 400–500-bp band region of total DNA digests from 38 'strains' of *B. anthracis*, which constituted the first successful attempts to find molecular differences among *B. anthracis* strains (Keim *et al.*, 1999).

The selection of probes for the detection of RFLPs in *B. anthracis* is difficult, because few sequences of chromosomal origin are known, other than the rRNA genes and the *ermJ* gene that codes for erythromycin resistance (Harrell *et al.*, 1995). However, because of the homogeneity of the *rrn* genes in the *B. cereus* group, these readily lend themselves to assessing sequence variation in their flanking regions (Henderson, 1996). Probes directed at the 23S rRNA gene that detect RFLPs upstream to it were able to differentiate 50 *B. anthracis* isolates from a wide range of origins into 6 strain groups, whereas RFLPs downstream to the 16S gene reflected variations that result from laboratory manipulation (Henderson, 1996). A probe based on the 16S–23S intergenic region of *B. anthracis rrn* operons also sub-differentiated some of the *B. anthracis* isolates, but not isolates of other *B. cereus* group members.

The single molecular strain variation identified by Henderson *et al.* (1994) with the M13 phage primer was further characterised by Andersen *et al.* (1996), who showed that this variation was due to a 12-nucleotide variable number tandem repeat (VNTR) within a large open reading frame (ORF) designated as the *vrrA* locus (variable repeat region A). The molecular basis for the variation was due to differences in the number of repeats of 5'CAATATCAACAA3' encoding a putative glutamine-rich protein of 30 kDa (Andersen *et al.*, 1996; Jackson *et al.*, 1997). Five polymorphic types were ultimately identified in the genome of *B. anthracis*, with the VNTR varying from 2 to 6 repeats (Jackson *et al.*, 1997). Characterisation of VNTR in 198 *B. anthracis* isolates revealed some association between VNTR group and the geographical source of strain, but the majority (60%) fell into a single group (Jackson *et al.*, 1997). Only a few *B. cereus, B. thuringiensis* and *B. mycoides* strains have been characterised. The *uvrA* locus amplifies most frequently but the resulting alleles do not correspond to those of *B. anthracis* (Keim *et al.*, 2000). It is expected that this region will reveal many different types across these three species (Dr Paul Keim, personal communication).

Some 30 polymorphic loci have been identified by AFLP (Keim *et al.*, 1997). Of these marker loci, those with greater diversity, and hence high information content, represented restriction fragment-length mutations, rather than site changes attributable to the VNTRs. Seven chromosomal single-locus markers (*vvrB1, vvrB2, vvrC1, vvrC2*, etc.) and two plasmid markers with high information content in addition to *vrrA* have been developed from variable AFLP markers or from the DNA sequence (Keim *et al.*, 1999; Klevytska *et al.*, 1999; Price *et al.*, 1999; Zinser *et al.*, 1999). This combination of 10 highly informative single-locus markers and nearly 30 AFLP markers has provided great discriminatory power for the analysis of *B. anthracis* strains. These have been used to delineate 7 major geographical diversity groups from a study of several hundred isolates obtained from numerous countries (Keim *et al.*, 1997, 1999; Price *et al.*, 1999) and more local diversity groups in a single location (Smith *et al.*, 1999). The 7 geographical diversity groups may represent the only worldwide *B. anthracis* clones (Keim *et al.*, 1999).

These genetically variable regions have homologous variable loci in the genomes of *B. cereus* and *B. thuringiensis* (Keim *et al.*, 1999), and AFLP has been used to show the phylogenetic relationships between serovars of *B. thuringiensis* and between these and a range of *B. cereus* isolates (Jackson *et al.*, 1999a). Closer relationships were found between *B. thuringiensis* isolates in a given serovar than between those from different serovars, and between isolates of either species from one geographic location than from those from differing locations, but certain anomalies were highlighted. In particular, many Norwegian soil isolates, a *B. cereus* isolate from a periodontal infection and an environmental *Bacillus* spp. isolate from Iraq, all of which had phenotypic fits of *B. cereus/thuringiensis*, appeared more closely related to *B. anthracis* than to the *B. cereus* and *B. thuringiensis* control strains included in the study. A modified, more user-friendly, version of AFLP has been developed for species-level and strain-level differentiation of members of the *B. cereus* group (Velappan *et al.*, 1999).

Detection of *B. anthracis*

The focus of the molecular detection of *B. anthracis* has been on the environment. The needs of UN inspection teams for biological facilities in Iraq after the 1991 Gulf War also provided an impetus for the detection of *B. anthracis* by molecular means. The belated availability of human tissue specimens from casualties of the 1979 Sverdlovsk incident (Jackson *et al.*, 1998) has also seen the application of the most sophisticated molecular detection systems available to show that the victims had apparently been exposed to a mixture of at least four different *B. anthracis* strains.

The development of gene probes to detect and identify *B. anthracis* became possible after the publication of the DNA sequences of the virulence factor genes (Robertson *et al.*, 1988; Welkos *et al.*, 1988; Bragg and Robertson, 1989; Makino *et al.*, 1989). The development of probes specific for the toxin and capsule genes made it possible not only to distinguish virulent strains from non-virulent strains but also to differentiate *B. anthracis*, other than apparently rare pXO1$^-$/pXO2$^-$ isolates, from a heterologous background of closely related bacteria (Hutson *et al.*, 1993). Gene probes, however, require enrichment of target bacteria to generate sufficient copies of the target sequence and purification of the target DNA to ensure a signal. In the case of *B. anthracis*, this is frequently not achievable in environmental samples where *B. anthracis* is readily out-competed by other bacteria during enrichment. Furthermore, in the absence of chromosomal DNA sequences specific for *B. anthracis*, the use of probes is restricted to the plasmid-borne toxin and capsule gene sequences.

Attempts to develop gene probe protocols for the detection of *B. anthracis* were short-lived and rapidly overtaken by the more sensitive, rapid and convenient PCR-based methods. These too, however, depend on highly specific oligonucleotide primers to the genes encoding the four virulence factors. Claims have been made that single spores of *B. anthracis* can be detected (Carl *et al.*, 1992; Johns *et al.*, 1994; Reif *et al.*, 1994) but these were based on purified DNA and, for the detection of the organism in environmental samples, sensitivities greater than those of conventional methods (i.e. ~5 spores/g homogeneously dispersed in the soil) are hard to achieve. Carl *et al.* (1992) demonstrated that, in a nested PCR, the use of *O*-carbamyl-D-serine in the spore germination protocols radically improved detection limits by preventing the formation of L-alanine, an inhibitor of germination.

Two claims of detection sensitivities in soil at least equivalent to that of the best conventional methods have been published. Beyer *et al.* (1995) reported a nested PCR procedure with a sensitivity of less than 10 spores/100 g of soil, and Makino *et al.* (1993), who used a cap gene-based PCR method, claimed a detection sensitivity of one spore-forming unit in samples of contaminated soil (sample size not specified). Most reports have, however, recorded a limit of 1000 spores/g. These sensitivity limits are ascribed to potent PCR (Taq polymerase) inhibitors, sometimes identified as residing in the humic acids of the soil. This subject, including their own attempts to address it, has been reviewed by Sjöstedt *et al.* (1997). Makino *et al.* (1993) recorded a sensitivity of one 'spore-forming unit' in the spleens of infected mice (sample sizes not recorded). With a view to distinguishing virulent and avirulent strains of *B. anthracis* from *B. cereus*, Ramisse *et al.* (1996) published a multiplex PCR method to detect a chromosomal sequence they regarded as specific to *B. anthracis* in addition to the toxin and capsule genes. A rapid multiplex PCR system to screen for *B. anthracis* in the field is being designed (Robertson *et al.*, 1999). It is helpful to run multiplex PCR, which includes a chromosomal marker, when confirming the suspected absence of both pXO1 and pXO2 in cured strains or occasional natural isolates. A chromosomal sequence, *Ba813*, apparently specific for *B. anthracis*, was detected by Patra *et al.* (1966) in a genomic library but the sequence was subsequently also found in a small proportion of *B. cereus* and *B. thuringiensis* strains (Ramisse *et al.*, 1999). We have found primers to genes for S-layer proteins useful to supplement those to the virulence genes (Turnbull *et al.*, 1998) but since they do not have absolute specificity, their products must also be interpreted in the light of phenotypic information for the isolate being examined. A small acid-soluble protein (SASP-B) signature has, so far, shown absolute specificity in a TaqMan PCR assay (McKinney *et al.*, 1998). Impressive advances are being made towards rapid (<15 minutes), automated, multiplex PCR systems with fluorescent detection methods, as illustrated by the Idaho LightCycler (Lee *et al.*, 1999).

With the imminent completion of the *B. anthracis* genome sequence, it is to be hoped that universally accepted anthrax-specific chromosomal primers will be designed in the foreseeable future. Use of these is, however, mostly of academic importance for the diagnosis of infection, which is normally concerned with confirming the presence of the virulence genes. The need for chromosomal markers becomes of greater importance, however, in relation to taxonomic accuracy and possible concerns about cross-species engineering, such as that reported by Pomerantsev *et al.* (1997).

Immunological Diagnosis and Detection

The test devised by Ascoli (1911) to detect *B. anthracis* antigens in animal tissues and products must be one of the earliest immunological detection tests in the history of immunology. Its purpose was to supply rapid retrospective evidence of anthrax infection in animals. The test has played a major role in anthrax control and it remains in use, at least in eastern Europe. Since the thermostable antigens on which the test depends are common to other *Bacillus* species, the test depends on the assumption that only *B. anthracis* is likely to have proliferated sufficiently to deposit precipitating antigens throughout the tissues.

Serological techniques came into their own in the 1950s, as the basis of microbial species and strain differentiation. They resulted in the many systems that remain in place for the speciation and subspeciation of *Salmonella*, *Shigella*, *E. coli* and many other Gram-negative and some Gram-positive bacteria. In the case of *Bacillus*, serotyping systems based on flagellar antigens were effectively developed to subspeciate *B. thuringiensis* and for the strain typing of *B. cereus* isolates from food poisoning and other episodes. The number of serovars of *B. thuringiensis* based on this classification system currently stands at 82 (Lecadet *et al.*, 1999).

In the 1970s and early 1980s attempts were made to find species-specific spore antigens for detecting *B. anthracis* in the environment. Polyclonal antisera were absorbed for the differential detection of *B. anthracis* and *B. cereus* spores by immunofluorescence and immunoradiometry (Phillips and Martin, 1982a, b, c, 1983, 1984), lectin-based agglutination (Cole *et al.*, 1984) or enzyme-linked lectinosorbent assay (Graham *et al.*, 1984). Subsequently attempts were made to identify monoclonal antibodies specific for anthrax spore antigens (Phillips *et al.*, 1988). None of those systems was widely adopted and, after a period of abandonment, the subject has again come under examination (Kearney *et al.*, 1999; Longchamp and Leighton, 1999), with new monoclonal and recombinant antibody techniques and biotechnological protocols to examine spore coats and surfaces in isolation.

An attempt to develop a more sophisticated immunological detection system was made by Gatto-Menking *et al.* (1995) and Yu (1996) by capturing anthrax spores on antibody-conjugated micron-sized magnetic beads followed by the binding of ruthenium (II) trisbipyridal chelate (Ru(bpy)23+)-labelled reporter antibodies. The spores were magnetically captured and chemiluminescence brought about by applying an electrical potential. The detection limit on spore preparations was 10–100 spores, but it appears not to have been applied to environmental samples.

Enzyme immunoassay based on the toxin antigens, principally protective antigen (PA; see below), has proved useful to confirm the diagnosis in individual cases and for epidemiological investigations (Turnbull *et al.*, 1992a) but the test is confined to a few centres, because the purified antigens are not commercially available. Moreover, since treatment early in the course of the infection may prevent elaboration of sufficient antigen to induce a detectable antibody response, negative results must be interpreted with caution.

Anthraxin, a commercially produced heat-stable protein–polysaccharide–nucleic acid complex extracted chemically from an unencapsulated strain of *B. anthracis*, was licensed in 1962 in the former USSR. It has become widely used in the Russian sphere of influence as a skin test for the retrospective diagnosis of human and animal anthrax, and for vaccine evaluation (Shlyakhov, 1996; Shlyakhov and Rubinstein, 1996; Shlyakhov *et al.*, 1997). The test involves intradermal injection of 0.1 mL of Anthraxin and inspection of the site of injection after 24 hours for erythema and induration that lasts for 48 hours. This delayed-type hypersensitivity response reflects cell-mediated immunity and reportedly allows the retrospective diagnosis of anthrax in up to 72% of cases some 31 years after infection. It was successfully in a retrospective investigation of anthrax in a Swiss spinning mill, where synthetic fibres were combined with goat hair from Pakistan (Pfisterer, 1990). Anthraxin, like Ascoli antigen, depends on the nature of anthrax rather than on the specificity of the antigens involved.

Although a variety of chromosomally encoded anthrax-specific antigens, such as epitopes of the S-layer proteins EA1 or Sap, are believed to exist (Farchaus *et al.*, 1995; Mesnages *et al.*, 1997, 1999), specific immunological diagnostic and detection tests at present depend on the anthrax toxin antigens. These are the only truly specific anthrax antigens that can readily be isolated and purified. Burans *et al.* (1996) devised a highly sensitive, specific, rapid and simple-to-use immunochromatographic on-site diagnostic test based on antigen capture by monoclonal antibody to an epitope on PA. Complex biosensors based on similar antigen capture principles are under development (Long and O'Brien, 1999). Should they become commercially available, they will prove a great advantage, particularly where diagnostic laboratory resources and medical, veterinary or microbiological skills are limited.

Virulence Factors

Research on the pathogenesis of anthrax since World War II has elucidated the nature of two unique virulence factors of anthrax pathogenesis, the capsule and the tripartite protein toxin. Of these, the toxin has received by far the greatest attention.

Capsule

The role of the homopolymeric γ-linked poly-D-glutamic acid capsule in pathogenesis is not fully understood beyond its antiphagocytic function (Zwartou and Smith, 1956; Keppie *et al.*, 1963). Although it is undoubtedly an essential virulence factor, it is a poor immunogen and in the absence of toxin it appears to have no protective value for vaccine purposes (Ivins *et al.*, 1988). Its poor antigenicity may be a consequence of its polymeric structure although antisera against it have been raised. For example, Makino *et al.* (1988) used rabbit anticapsule antisera to study capsule gene expression. Ezzell and Abshire (1996) generated a monoclonal anticapsule antibody (FDF-1B9), which is the basis of a fluorescent antibody method for the routine diagnostic confirmation of capsule production.

Anthrax Toxin

The toxic effects of cell-free extracts from cutaneous anthrax lesions were first observed in 1904 after the use of oedema fluid as a vaccine (Watson *et al.*, 1947; Lincoln and Fish, 1970). Smith and Keppie (1954) showed conclusively that death in anthrax was due primarily by an intoxication resulting from a septicaemia. They noted that, if the terminal bacteraemia was aborted with streptomycin before it reached a critical level, the infection could be eliminated and the animals survived. If this critical level of micro-organisms was exceeded before antibiotic administration, however, the animals died even though the bacteraemia was cleared. Before this it was believed that death was due to capillary blockage, hypoxia and nutrient depletion by the exceedingly large number of bacilli.

Studies in primates in the 1960s indicated that the toxin depresses electrical activity in the cerebral cortex affecting the respiratory centre and thereby possibly contributing to anoxia, cardiac collapse, shock and sudden death (Walker *et al.*, 1967). Stanley and Smith (1961) demonstrated that anthrax toxin is a three-component toxin, but it was a further two decades before effective procedures became available to purify each of the components, protective antigen (PA), oedema factor (EF) and lethal factor (LF) (Leppla, 1982, 1984; Leppla *et al.*, 1985). Research continues into the many aspects of the pathology of anthrax that remain unresolved, such as the relationship between toxin production, the onset of bacteraemia and the pathological features visible in the lymph nodes and the spleen. The existence of such a relationship is clear from the protection afforded by PA in vaccinated animals.

Protective Antigen, Lethal Toxin and Oedema Toxin

PA had for some time been known to be the pivotal protein in the pathogenicity of anthrax toxin, and its name is derived from this fact (Watson *et al.*, 1947). Since purification procedures were established in the early 1980s, a considerable degree of understanding of the toxin and its mode of action has developed.

What was previously referred to as the three-component toxin is now regarded as two related toxins: the lethal toxin (LeTx), comprised of PA in combination with LF, and the oedema toxin, composed of PA in combination with EF. The three proteins are produced simultaneously during exponential growth of the organism. PA is an 83-kDa protein that binds to mammalian cell receptors and is cleaved to a 63-kDa fragment by furin, a host protease. This proteolytic cleavage leads to the formation of a secondary-binding site for which LF and EF compete with high-affinity (Leppla *et al.*, 1988). The possibility has been raised that PA and at least LF may interact before binding to the cell receptor (Brossier *et al.*, 2000). The protease activation site in PA83 is defined by the sequence --Arg-Lys-Lys-Arg- at residues 164–167 of the mature polypeptide (Singh *et al.*, 1989). This sequence is susceptible to a range of proteases, including trypsin and the furin-type enzymes (Gordon and Leppla, 1994; Gordon *et al.*, 1995). PA is also susceptible to a discrete cleavage by chymotrypsin at residues Phe-Phe-Asp315. This cleavage is considered to be cryptic in that the two molecules, PA37 and PA47, so created (Novak *et al.*, 1992) remain associated under non-denaturing conditions. The importance of this region to the translocation function of PA was identified by Singh *et al.* (1994). The residual 20-kDa fragment (PA20) has not, as yet, been assigned a biological or immunoprotective activity.

Activated PA63 forms heptamers at reduced pH, and each subunit of the heptamer can bind one LF molecule (Singh *et al.*, 1999). The water-soluble heptamer undergoes a substantial pH-induced conformational change that involves the creation of a 14-stranded β barrel. LF or EF binds to the cell-bound PA heptamer and, after endocytosis and acidification of the endosome, they translocate to the cytosol (Milne *et al.*, 1994; Liddington *et al.*, 1999).

Anthrax toxin is, therefore, regarded as fitting the binary or subunit toxin (A–B) model which describes the activity of several other bacterial protein toxins (Friedlander *et al.*, 1993). The crystal structure of PA has been elucidated and refined to 2.1Å by Petosa *et al.* (1997) and bears no resemblance to other bacterial toxins for which the three-dimensional structure is known (Liddington *et al.*, 1999). The PA83 monomer is organised into four domains that consist mainly of antiparallel β sheets. Domain 1 is an N-terminal, calcium-binding region within which the protease cleavage motif is situated. Domain 2 promotes heptamer formation and probably membrane insertion via a large flexible loop region. Domain 3 is small and of unknown function, and domain 4 comprises a C-terminal receptor-binding region (Little and Lowe, 1991; Singh *et al.*, 1991; Brossier *et al.*, 1999b; Varughese *et al.*, 1999). The formation of the heptamer after proteolytic cleavage is regarded as exposing a large hydrophobic surface for binding EF or LF. The efficiency and elegance of these molecules and their mode of action becomes increasingly evident as understanding of their mode of action increases.

Oedema Factor

EF is an 89-kDa protein which, in common with the other anthrax toxin components, does not contain the amino acid cysteine. When the underlying mechanisms of toxin activity at the cellular level were being elucidated, attention initially focused on the EF component as it became clear that the oedema-producing effect was due to raised levels of intracellular cyclic adenosine monophosphate (cAMP), in a manner analogous to the effects of cholera toxin (CT) and *Escherichia coli* heat-labile toxin (LT) (Leppla, 1982). Whereas CT and LT act indirectly by inducing irreversible conformational changes in membrane-bound guanidine diphosphate (GDP)-binding proteins and thereby stimulate constitutive adenylate cyclase activity (Gill and Coburn, 1987), EF is itself an adenylate cyclase. More remarkably, it proved to be dependent on calmodulin, the major intracellular calcium receptor in eukaryotic cells. Hence EF is functional only in eukaryotic cells (Leppla, 1982; Leppla *et al.*, 1985). By catalysing the abnormal production of cAMP, EF induces the altered water and ion movements that lead to the characteristic oedema of anthrax.

The role of oedema toxin in the anthrax disease process may be to prevent mobilisation and activation of polymorphonuclear leukocytes (PMNL) and thereby to prevent the phagocytosis of the bacteria (Leppla *et al.*, 1985; Leppla, 1991). O'Brien *et al.* (1985) demonstrated that combinations of PA and EF suppress PMNL activity and inhibit phagocytosis of *B. anthracis*. Unlike other cAMP-stimulating toxins, PA + EF (oedema toxin) enhances the migration of stimulated PMNL, and this is also the case for PA + LF (LeTx) (Wade *et al.*, 1985).

The only other bacterium known to produce a calmodulin-dependent adenylate cyclase is *Bordetella pertussis*, and antigenic cross-reaction has been demonstrated between the *B. anthracis* and the *Bo. pertussis* adenylate cyclases (Mock *et al.*, 1988). Three conserved regions of amino acid and nucleotide homology in the two molecules were subsequently reported (Escuyer *et al.*, 1988; Robertson, 1988). The overall sequences have little homology, however, and the enzymes also appear to have different mechanisms of cell internalisation. Cell entry by EF is blocked by a range of inhibitors of receptor-mediated endocytosis, whereas cell entry by pertussis adenylate cyclase is not (Gordon *et al.*, 1988). Escuyer *et al.* (1988) and Robertson (1988) identified putative calmodulin-binding regions on each molecule, and these exhibit a high degree of DNA homology.

Lethal Factor

On the basis of the predicted translation of its DNA sequence LF is the largest polypeptide (90 kDa) of the anthrax toxin complex, but on one-dimensional denaturing PAGE (SDS-PAGE) it migrates as a protein of approximately 87 kDa, between PA and EF (Leppla, 1988; Quinn *et al.*, 1988).

The binary combination of PA + LF (LeTx) is lethal to laboratory animals and causes lysis of certain eukaryotic cell lines, particularly those of monocyte/macrophage lineage (Friedlander, 1986). When LeTx is administered intravenously to laboratory animals, specifically Fischer 334 rats, death ensues within 60–90 minutes (Ezzell *et al.*, 1984; Singh *et al.*, 1989). The rat-lethality bioassay developed in the sixties has been superseded by an *in vitro* macrophage lysis test (Friedlander, 1986; Quinn *et al.*, 1991). The observations *in vitro* were directly linked with the lethality of anthrax in *in vivo* experiments which demonstrated that laboratory mice were rendered insensitive to anthrax lethal toxin when their native macrophage population was depleted by silica injection (Hanna *et al.*, 1993). Sensitivity to the toxin could be restored by co-injection of lethal toxin with the toxin-sensitive monocyte/macrophage cell line RAW 264, indicating

that macrophages mediate the action of lethal toxin *in vivo*. It was further revealing that levels of toxin sublytic for macrophages stimulated production of interleukin-1β (IL-1β) and tissue necrosis factor α (TNFα) and it was suggested that this may be responsible for the symptoms of secondary shock characteristic of fatal anthrax (Hanna *et al.*, 1993).

Studies on sensitive macrophages *in vitro* have demonstrated that lethal toxin causes membrane leakage before cell lysis and that small ions can pass through the cell membrane in an ATP-independent manner (Hanna *et al.*, 1992). The lethal effect on macrophages *in vitro* is also calcium-dependent and lysis is preceded by a large influx of extracellular calcium ions. Chelators of calcium ions, such as EGTA, or agents that block calcium channels, such as verapamil or nitrendipine, protect macrophages from the lytic effect of the toxin (Bhatnagar *et al.*, 1989). The lytic effect also requires concomitant protein synthesis, inhibition of which not only protects the cells *in vitro* but also abrogates the influx of calcium (Bhatnagar and Friedlander, 1994). Macrophages are also protected *in vitro* by osmotic stabilisation, indicating that lysis is probably the direct result of the unregulated cation pulse. Removal of the osmotic support allows the influx of calcium to proceed and the cells are lysed (Bhatnagar and Friedlander, 1994).

Protein sequence analysis indicates that LF contains one or more zinc-binding amino acid motifs. One of these is characteristic of the thermolysin family of zinc-dependent metalloproteases (Klimpel *et al.*, 1994, Koch *et al.*, 1994). Mutational analyses of the catalytic domain of LF have confirmed the relevance of these motifs in an *in vitro* macrophage lysis assay (Quinn *et al.*, 1991; Klimpel *et al.*, 1994).

The LF metalloprotease cleaves and inhibits several members of the mitogen-activated protein kinase-kinase family (MAPKKs or MEKs; Duesbery *et al.*, 1998). Tang and Leppla (1999) have shown that inhibitors of proteosome activity, such as acetyl-Leu-Leu-norleucinal, MG132 and lactacystin, inhibit lethal toxin activity without preventing LF cleavage of MEK-1, thus indicating that cell lysis events are proteosome-dependent and 'downstream' of MEK-1 hydrolysis. The correlation of these LeTx intoxication events with cell death is under investigation.

Genetics of Virulence

Both of the major virulence factors of *B. anthracis* are plasmid mediated. The genes for each of the toxin components and the transcriptional activator (*atxA*) are located on a single high-molecular-weight

(182 kb) plasmid designated pXO1 (Mikesell *et al.*, 1983; Okinaka *et al.*, 1999). The genes for capsule expression, assembly and degradation (*CapA, CapB, CapC* and *dep*) and their transcriptional activator, *acpA*, are located on the smaller (95 kb) pXO2 plasmid (Green *et al.*, 1985; Uchida *et al.*, 1985, 1993; Makino *et al.*, 1989; Vietri *et al.*, 1995). A wild-type strain may be differentially or fully cured of these plasmids by extended incubation at 43°C (pXO1 curing) or by culture in the presence of novobiocin (pXO2 curing) or both. Strains cured of one or both plasmids have reduced virulence. Toxigenic, non-capsulated strains form the basis of current live spore vaccines (see below). Strains selectively cured of pXO1 or pXO2 retain residual pathogenicity for certain strains of inbred mice (Welkos *et al.*, 1986, 1993). Unlike the phylogenetically related species *B. cereus* and *B. thuringiensis*, which have a very heterogeneous plasmid content, *B. anthracis* isolates of clinical and environmental origin have not yet been found with plasmids other than these two virulence plasmids.

The structural genes for each of the toxin components have been cloned (Vodkin and Leppla, 1983; Robertson and Leppla, 1986; Mock *et al.*, 1988; Tippets and Robertson, 1988) and sequenced (Escuyer *et al.*, 1988; Robertson, 1988; Robertson *et al.*, 1988; Welkos *et al.*, 1988; Bragg and Robertson, 1989). The 5′ regions of the LF and EF genes, *lef* and *cya*, share areas of high nucleotide homology (Escuyer *et al.*, 1988; Robertson, 1988; Bragg and Robertson, 1989). This is reflected in their deduced N-terminal amino acid sequences and is indicative of a common PA-binding and cell internalisation domain (Bragg and Robertson, 1989; Quinn *et al.*, 1991). The *lef* gene also contains a central region with perhaps five imperfect nucleotide sequence repeats, which are similarly reflected in the amino acid composition of the mature protein (Bragg and Robertson, 1989). The functions of these repeats have not been determined, but they may be involved in calcium-binding (Lowe *et al.*, 1990). Mutations in this region destabilise the LF protein, which probably indicates that it has an important structural role (Quinn *et al.*, 1991).

Analysis of the predicted LF protein sequence with the program 'Coils' (Pallen *et al.*, 1997) predicts that the N-terminal 200 amino acids and also the repeat region of LF are rich in coiled coils (C.P. Quinn, unpublished observations). This is indicative of regions that may be involved in promoting protein-protein interactions. The importance of the LF N-terminal domain (LF1-254) for the uptake of LF and fused polypeptides is well documented (Arora and Leppla, 1993, 1994). The repeat region is less well

characterised but, by extrapolation from the analyses of the N-terminal domain, it is tempting to speculate that this region of LF may have a key role in the interaction of the enzyme with its intracellular targets. Leppla (1999) suggested that this region has evolved to optimise the spatial separation of the PA-binding and catalytic domains.

Toxin and capsule expression *in vitro* are both modulated by ambient levels of carbon dioxide or hydrogen carbonate ion in the medium. Transcription of the PA structural gene (*pag*) is under the control of hydrogen carbonate ion in the medium and also by elements present on the pXO1 plasmid (Bartkus and Leppla, 1989), of which a *trans*-acting positive regulator *atxA* has been identified and cloned (Uchida *et al.*, 1993). This *trans*-acting gene product can up-regulate transcription from one of two promoters at the *pag* gene and, since all three genes of the anthrax toxin complex (*pag*, *lef* and *cya*) are co-ordinately regulated by both hydrogen carbonate ion and temperature, they may share a common activating element (Sirard *et al.*, 1994). A 300-bp sequence has been identified downstream from the *pag* structural gene (now designated *pagA*), which is co-transcribed with *pagA* and represses transcription of *pagA* and *atxA* (Hoffmaster and Koehler, 1999). This new regulatory element has been designated *pagR* and it has been proposed that it acts in concert with the positive transcriptional effect of *atxA*, as part of an autogenous *pagA* regulatory system. Uchida *et al.* (1997) demonstrated that the *atxA* gene product has a further role in the up-regulation of capsule expression and thereby identified the probable reason, at a genetic level, for the well established observation that, in order to produce wild-type levels of capsular material, pXO1$^-$/pXO2$^+$ strains of *B. anthracis* have an increased requirement for carbon dioxide.

Capsule production is only observed *in vitro* in atmospheres of 5–20% carbon dioxide or in serum or whole blood. The capsule-encoding region of pXO2 has been cloned, mapped and sequenced (Uchida *et al.*, 1987; Makino *et al.*, 1988, 1989). The genes essential for capsule formation encode three membrane-associated enzymes that mediate the polymerisation of D-glutamic acid via the cell membrane. They do not appear to have DNA sequence homology with the capsule-producing species *B. subtilis* (natto) and *B. megaterium*, but the capsular materials have some antigenic cross-reactivity (Makino *et al.*, 1989). Unlike the toxin genes, these elements are not functional when sub-cloned into *B. subtilis* (Makino *et al.*, 1989) but, in contrast, *B. anthracis* capsule material could apparently be expressed in *E. coli* (Uchida *et al.*, 1993). As with the regulation of expression of the toxin genes on pXO1, some aspect of regulation of capsule expression is modulated by a *trans*-acting element located on pXO2. Much remains to be understood about the regulation of capsule synthesis and the role this plays in the expression of virulence (Vietri *et al.*, 1995). It is becoming increasingly clear, however, that the two native *B. anthracis* plasmids have an integrated and interdependent role in the modulation of virulence. For example, the *trans*-acting regulator *atxA* located on pXO1 activates expression of *capB* on pXO2 (Guignot *et al.*, 1997; Uchida *et al.*, 1997) as well as the toxin genes on pXO1 (Uchida *et al.*, 1993; Koehler *et al.*, 1994; Dai *et al.*, 1995).

The sequence and organisation of pXO1 has been published (Okinaka *et al.*, 1999a, b). The toxin structural and regulatory genes reside on a distinct 44.8-kb region identified by inverted IS1627 elements, and only 61% of the plasmid potentially codes for protein products. Additionally, this IS1627-flanked 'pathogenicity island' includes genes identified as germination-response elements and a further 19 putative ORFs of unknown function. An intriguing observation was the lack of homology of pXO1 sequences to genes typically associated with the replication and stability of large plasmids in bacilli. It was proposed that pXO1 may have an, as-yet unidentified and perhaps unique, origin of replication.

Genetic elements on pXO1 and distinct from the toxin genes may therefore be of critical importance for the virulence of *B. anthracis*, possibly providing a selective advantage during the infection process. Evidence is available (Bowen and Quinn, 1999), that pXO1 can influence the segregational exclusion of certain non-native plasmid replicons. At significant metabolic cost, *B. anthracis* strives to retain pXO1 under negative selection pressure. This is, however, ameliorated by pXO2, which again indicates 'crosstalk' between the plasmids and that metabolic control elements governing the maintenance of the native plasmids are at least partly located on pXO2. It may be concluded, therefore, that collaborating regions of the chromosome and plasmids of *B. anthracis* constitute a 'virulence genome'.

Vaccines

Pasteur derived his original vaccine strains by subculturing *B. anthracis* at 42–43°C for 15–20 days; in doing so, he was inadvertently curing the bacilli of pX01. Strains that are cap$^+$/tox$^-$ strains are not protective, however, and the partial efficacy and residual virulence of Pasteur's vaccines are now explained in terms of mixed cultures with some uncured forms.

Welkos (1991) showed in inbred mice that cap$^+$/tox$^-$ strains possess a degree of virulence but, in the absence of toxin, their pathogenicity was unexplained.

Strains of cap$^-$/tox$^+$ *B. anthracis* are protective, and the best-known strain of this kind is strain 34F2 produced by Sterne (1937, 1939). This strain is the basis of the highly successful live-spore animal vaccine used in most countries outside the former USSR and China, where other cap$^-$/tox$^+$ strains are used. These also retain residual virulence. Certain livestock species, notably goats, appear to be particularly prone to vaccination casualties and a proportion of guinea-pigs injected with more than 10^6 spores can be expected to succumb. The relative susceptibilities of inbred mice to strain 34F2 has been studied in relation to the ability to synthesise the complement component C5 (Welkos *et al.*, 1986; Welkos and Friedlander, 1988). Live-spore vaccines are regarded as unsuitable for use in humans in most industrialised countries, because of their residual virulence and the many contra-indications to their use in certain individuals. Live-spore vaccines are, however, given to humans in China and countries of the former Soviet Union, allegedly with an excellent safety record (Shlyakhov and Rubinstein, 1994).

The two human anthrax vaccines used in developed countries are produced in the UK and the USA. The UK vaccine (licence Nos. 1511/0037 and 0058) is an alum-precipitated cell-free culture filtrate of statically incubated Sterne strain 34F2 cultures, and is still made essentially as first formulated in 1954 (Belton and Strange, 1954). It was introduced in 1965 for workers in at-risk occupations and licensed for human use in 1979 after biologicals first fell under the European Directive 75/319/EEC. The US human vaccine (product licence No. 99) is an alhydrogel-adsorbed cell-free culture filtrate of anaerobically fermenter-grown *B. anthracis* strain V770-NP1-R. It was licensed in 1972 for administration to those in at-risk occupations.

The UK and the US vaccines are both produced and administered under what is termed a 'licence of right', because they have been in existence for so long. Recently they have been the subject of criticism, for lacking the clinical, pharmacological, immunological and safety data that would now be required for the licensing of a new product. The efficacy of these vaccines has been the subject of debate for some years, because they are poorly defined, subject to batch variation and associated with unwanted reactions (Turnbull, 2000). Considerable reliance has long been placed on the results of protection tests in guinea-pigs (Ivins *et al.*, 1986; Little and Knudson, 1986; Turnbull *et al.*, 1986, 1988; Ivins and Welkos, 1988).

These led to the conclusion that the protective efficacies of these two vaccines are less than ideal (Turnbull, 1991). The thrust of research in the 1980s was, therefore, targeted at identifying ways of enhancing vaccine efficacy.

Passive protection tests indicate a role for anti-PA antibodies in protection (Little *et al.*, 1997). Direct evidence (Pezard *et al.*, 1995) and indirect evidence (Ivins and Welkos, 1988; Turnbull *et al.*, 1988) indicate that antibodies to other than PA play only a minor, possibly synergistic (Pezard *et al.*, 1995), role in protection. In this context it is worth noting that serum therapy was regarded as one of the only available treatments before the advent of antibiotics (Knudson, 1986) and until at least 1990 it was in use in China (Dong, 1990).

PA given with a suitable adjuvant elicits a high degree of protection (Ivins *et al.*, 1990). This has led to work aimed at designing a putative next-generation vaccine based on PA as the sole antigen either with a potent adjuvant (Ivins and Welkos, 1988; Turnbull *et al.*, 1990; Ivins *et al.*, 1992, 1995, 1998; Jones *et al.*, 1996; McBride *et al.*, 1998) or as expressed by a vector which itself served as the adjuvant (Ivins *et al.*, 1990, 1998; Iacono-Connors *et al.*, 1991; Coulson *et al.*, 1994; Barnard and Friedlander, 1999; Gupta *et al.*, 1999; Worsham and Sowers, 1999; Zegers *et al.*, 1999).

Guinea-pig protection studies with trypsin-generated fragments of PA, together with a selection of adjuvants, revealed that the protective epitopes reside in the PA63 species and that the purified PA20 species does not confer protection (Ivins and Welkos, 1988). In another modification of this approach (Singh *et al.*, 1998), a mutant version of PA unable to bind EF or LF was shown to be as effective as native PA, making it an attractive alternative from the standpoint of reduced potential toxicity. Studies on monoclonal antibodies directed against PA, which neutralise lethal toxin activity *in vivo* and *in vitro*, have identified at least three non-overlapping antigenic regions (Little *et al.*, 1996). Two of these regions are involved in the binding of LF to cell-bound PA, the first region mapping between amino acids Ile-581 and Asn-601 and the second within the PA17 fragment (amino acid residues Ser-168 and Phe-314) produced by limited chymotrypsin digestion of PA63. The third region is involved in the binding of PA to cells, and maps between amino acids Asp-671 and Ile-721 of the carboxy-terminus of PA.

The crystal structures of the monomeric protective antigen and the heptamer complex have been solved (Petosa *et al.*, 1997; Benson *et al.*, 1998). Progress is also being made towards an atomic resolution crystal structure for LF (Liddington *et al.*, 1999). These and other molecular structure-function studies on the anthrax lethal toxin (Varughese *et al.*, 1998; Wesche *et al.*, 1998; Beauregard *et al.*, 1999; Brossier *et al.*, 1999b; Collier, 1999; Duesbery and Van de Woude, 1999; Miller *et al.*, 1999; Singh *et al.*, 1999) are serving to design better prophylactic strategies.

It was noted in the 1980s and subsequently confirmed that, although antibodies to PA are essential for protection, the degree of protection was adjuvant-dependent and that there is no correlation between anti-PA antibody titre and degree of protection (Ivins *et al.*, 1986, 1990a, b, 1992, 1994, 1995, Ivins *et al.*, 1998; Little and Knudson, 1986; Turnbull *et al.*, 1986, 1988, 1990; Ivins and Welkos, 1988; McBride *et al.*, 1998). Some interest was even shown in the fact that enhanced efficacy can be obtained by supplementation of the licensed human vaccines with potent adjuvants, such as Freund's adjuvant, killed whole-cells of *Corynebacterium ovis* or *Bo. pertussis*, or complex cell-wall extracts, such as the Ribi adjuvants (Ribi ImmunoChem Research, Hamilton, Montana, USA), (Turnbull *et al.*, 1988, 1990; Jones *et al.*, 1996). These observations led to in-depth attempts to understand the nature of protective immunity and the roles in acquired protection of different antibody subclasses (McBride *et al.*, 1988; Baillie *et al.*, 1999; Williamson *et al.*, 1999; Gu *et al.*, 2000), delayed-type hypersensitivity and specific T-cell stimulation (Ivins *et al.*, 1998; McBride *et al.*, 1998; Beedham, 1999).

The current UK vaccine induces a mainly IgG1 response to PA in guinea-pigs and affords little protection from an aerosol challenge with the Ames strain of *B. anthracis*, whereas PA with a Ribi adjuvant induces a very high IgG2 response to PA and affords 100% protection from a similar challenge (McBride *et al.*, 1998). Passive transfer of an ammonium sulphate fraction of serum from such guinea-pigs into SCID-Beige mice affords better protection against intraperitoneal challenge with the Ames strain than a similar preparation from guinea-pigs vaccinated with the UK human vaccine (the late Dr B. McBride, unpublished observations). Interestingly, guinea-pig IgG2 is the only immunoglobulin subclass that induces K-cell-mediated lysis, suggesting that the Fc receptors on K-cells recognise only IgG2 (Ohlander *et al.*, 1978). Also, lymphokines increase expression of Fc receptors on guinea-pig macrophages for IgG2, but not for IgG1 (Limb *et al.*, 1988).

In spite of the observation of delayed-type hypersensitivity reactions after infection or vaccination (Pfisterer, 1990; Shlyakhov and Rubinstein, 1994; Shlyakhov, 1996; Shlyakhov *et al.*, 1997), no correlation between survival and lymphocyte proliferation responses could be demonstrated *in vitro* (Ivins *et al.*, 1998; McBride *et al.*, 1998; Beedham, 1999). Better tools in the form of cytokine assays may clarify the situation in the foreseeable future. In the mean time, the early reports by Henderson *et al.* (1956) and Darlow *et al.* (1956) that the vaccines, as made at that time, were effective in macaque monkeys have been revisited and found still to hold true (Friedlander *et al.*, 1993; Ivins *et al.*, 1996, 1998). This indication that the protective efficacy of the US vaccine in a non-human primate is far greater than that afforded to guinea-pigs has focused attention on species differences in immune response, which can result in misleading conclusions. The current view is that the differences in vaccine efficacy in guinea-pigs and primates is attributable to differences in Th1 and Th2 responses in these species to antigens delivered with aluminium-based adjuvants. A concerted effort is under way to understand the molecular basis of these responses, with a view to identifying markers for protection in humans in the absence of target populations in whom protective efficacy can be established by conventional clinical trials. This is an essential stepping-stone towards the development of a future generation of vaccines (McBride *et al.*, 1998; Baillie *et al.*, 1999).

The polypeptide capsule of *B. anthracis* is only weakly antigenic and little evidence to suggest it contributes to naturally acquired immunity or that it can produce artificial protection against anthrax. Other possible virulence-factor-based candidate supplementary vaccine antigens are being sought. Among these are the S-layer EA1 and Sap proteins (Mesnage *et al.*, 1998;, 1999; Fouet *et al.*, 1999). It can be expected that this search will be intensified when the *B. anthracis* genome sequence becomes available (Read and Peterson, 1999).

Acknowledgement

The authors thank Dr Paul Keim, Department of Biological Sciences, Northern Arizona University, Flagstaff, Arizona, USA, for kindly checking parts of the manuscript.

References

Andersen GL, Simchock JM, Wilson KH (1996) Identification of a region of genetic variability among *Bacillus anthracis* and related species. *J. Bacteriol.* 178: 377–384.

Arora N, Leppla SH (1993) Residues 1–254 of anthrax toxin lethal factor are sufficient to cause cellular uptake of fused polypeptides. *J. Biol. Chem.* 268: 3334–3341.

Arora N, Leppla SH (1994) Fusions of anthrax toxin lethal factor with shiga toxin and diphtheria toxin enzymatic domains are toxic to mammalian cells. *Infect. Immun.* 62: 4955–4961.

Ascoli A (1911) Die Präzipitindiagnose bei Milzbrand. *Zbl. Bact. Parasit. Infekt.* 58: 63.

Ash C, Farrow JAE, Dorsch M, Stackebrandt E, Collins MD (1991) Comparative analysis of *Bacillus anthracis*, *Bacillus cereus*, and related species on the basis of reverse transcriptase sequencing of 16S rRNA. *Int. J. Syst. Bacteriol.* 41: 343–346.

Ash C, Collins MD (1992) Comparative analysis of 23S ribosomal RNA gene sequences of *Bacillus anthracis* and emetic *Bacillus cereus* determined by PCR-direct sequencing. *FEMS Microbiol. Lett.* 94: 75–80.

Baillie LWJ, Fowler K, Turnbull PCB (1999) Human immune responses to the UK human anthrax vacccine. *J. Appl. Microbiol.* 87: 306–308.

Barnard JP, Friedlander AM (1999) Vaccination against anthrax with attenuated recombinant strains of *Bacillus anthracis* that produce protective antigen. *Infect. Immun.* 67: 562–567.

Bartkus JM, Leppla SH (1989) Transcriptional regulation of the protective antigen gene of *Bacillus anthracis*. *Infect. Immun.* 57: 2295–2300.

Beattie SH, Williams AG (1999) Detection of toxigenic strains of *Bacillus cereus* and other *Bacillus* spp. with an improved cytotoxicity assay. *Lett. Appl. Microbiol.* 28: 221–225.

Beauregard KE, Wimer-Mackin S, Collier RJ, Lencer WI (1999) Anthrax toxin entry into polarized epithelial cells. *Infect. Immun.* 67: 3026–3030.

Beedham RJ (1999) The role of cell-mediated immunity in protection against *Bacillus anthracis*. MPhil Thesis. Open University, UK.

Belton FC, Strange RE (1954) Studies on a protective antigen produced *in vitro* from *Bacillus anthracis*: medium and methods of production. *Br. J. Exp. Pathol.* 35: 144–152.

Benson EL, Huynh PD, Finkelstein A, Collier RJ (1998) Identification of residues lining the anthrax protective antigen channel. *Biochemistry* 37: 3941–3948.

Beyer W, Glockner P, Otto J, Böhm R (1995) A nested PCR method for the detection of *Bacillus anthracis* in environmental samples collected from former tannery sites. *Microbiol. Res.* 150: 179–186.

Bhatnagar R, Singh Y, Leppla SH, Friedlander AM (1989) Calcium is required for the expression of anthrax lethal toxin activity in the macrophage cell line J774A.1. *Infect. Immun.* 57: 2107–2114.

Bhatnagar R, Friedlander AM (1994) Protein synthesis is required for expression of anthrax lethal toxin cytotoxicity. *Infect. Immun.* 62: 2958–2962.

Blowey R, Edmondson P (1995) Mastitis Control in Dairy Herds. *An Illustrated and Practical Guide.* Ipswich: Farming Press Books, p. 41.

Böhm R, Spath G (1990) The taxonomy of *Bacillus anthracis* according to DNA–DNA hybridization and G + C content. Proceedings of the International Workshop on Anthrax, 11–13 April 1989, Winchester, UK. *Salisbury Med. Bull.* No. 68, Special suppl: 29–31.

Bowen JE, Quinn CP (1999) The native virulence plasmid combination affects the segregational stability of a theta-replicating shuttle vector in *Bacillus anthracis* var New Hampshire. *J. Appl. Microbiol.* 87: 270–278.

Bragg TS, Robertson DL (1989) Nucleotide sequence of the *Bacillus anthracis* lethal factor gene. *Gene* 81: 45–54.

Brossier F, Mock M, Sirard J-C (1999a) Antigen delivery by attenuated *Bacillus anthracis*: new prospects in veterinary vaccines. *J. Appl. Microbiol.* 87: 298–302.

Brossier F, Sirard J-C, Guidi-Rontani C, Duflot E, Mock M (1999b) Functional analysis of the carboxy-terminal domain of *Bacillus anthracis* protective antigen. *Infect. Immun.* 67: 964–967.

Brossier F, Weber-Levy M, Mock M, Sirard J-C (2000) Role of toxin functional domains in anthrax pathogenesis. *Infect. Immun.* 68: 1781–1786.

Burans J, Keleher A, O'Brien T *et al.* (1996) Rapid method for the diagnosis of *Bacillus anthracis* infection in clinical samples using a hand-held assay. Proceedings of the International Workshop on Anthrax, 19–21 September 1995, Winchester, UK. *Salisbury Med. Bull.* No. 87, Special suppl.: 36–37.

Carl M, Hawkins R, Coulson N *et al.* (1992) Detection of spores of *Bacillus anthracis* using the polymerase chain reaction. *J. Infect. Dis.* 165: 1145–1148.

Carlson CR, Johansen T, Koltsø A-B (1996) The chromosome map of *Bacillus thuringiensis* subsp. *canadensis* HD224 is highly similar to that of the *Bacillus cereus* type strain ATCC 14579. *FEMS Microbiol. Lett.* 141: 163–167.

Cole HB, Ezzell JW, Keller KF, Doyle RJ (1984) Differentiation of *Bacillus anthracis* and other *Bacillus* species by lectins. *J. Clin. Microbiol.* 19: 48–53.

Collier RJ (1999) Mechanism of membrane translocation by anthrax toxin: insertion and pore formation by protective antigen. *J. Appl. Microbiol.* 87: 283.

Coulson NM, Fulop M, Titball RW (1994) Effect of different plasmids on colonization of mouse tissues by the aromatic amino acid dependent *Salmonella typhimurium* SL 3261. *Microb. Pathog.* 16: 305–311.

Dai Z, Sirard J-C, Mock M, Koehler TM (1995) The *atxA* gene product activates transcription of the anthrax toxin genes and is essential for virulence. *Mol. Microbiol.* 16: 1171–1181.

Darlow HM, Belton FC, Henderson DW (1956) The use of anthrax antigen to immunise man and monkey. *Lancet* ii: 476–479.

Dong SL (1990) Progress in the control and research of anthrax in China. Proceedings of the International Workshop on Anthrax, 11–13 April 1989, Winchester, UK. *Salisbury Med. Bull*, No. 68, Special suppl: 104–105.

Duesbery NS, Woude GF Van de (1999) Anthrax lethal factor causes proteolytic inactivation of mitogen-activated protein kinase . *J. Appl. Microbiol.* 87: 289–293.

Escuyer V, Duflot E, Sezer O, Danchin A, Mock M (1988) Structural homology between virulence-associated bacterial adenylate cyclases. *Gene* 71: 293–298.

Ezzell JW, Ivins BE, Leppla SH (1984) Immunoelectrophoretic analysis, toxicity, and kinetics of *in vitro* production of the protective antigen and lethal factor components of *Bacillus anthracis* toxin. *Infect. Immun.* 45: 761–767.

Ezzell JW, Abshire TG (1996) Encapsulation of *Bacillus anthracis* spores and spore identification. Proceedings of the International Workshop on Anthrax, 9–21 September 1995, Winchester, UK. *Salisbury Med. Bull*, No. 87, Special suppl: 87.

Farchaus W, Ribot WJ, Downs MB, Ezzell JW (1995) Purification and characterization of the major surface array protein from the avirulent *Bacillus anthracis* strain Delta Sterne-1. *J. Bacteriol.* 177: 2481–2489.

Fouet A, Mesnage S, Tosi-Couture E, Gounon P, Mock M (1999) *Bacillus anthracis* surface: capsule and S-layer. *J. Appl. Microbiol.* 87: 251–255.

Friedlander AM (1986) Macrophages are sensitive to anthrax lethal toxin through an acid-dependent process. *J. Biol. Chem.* 261: 7123–7126.

Friedlander AM, Welkos SL, Pitt MLM *et al.* (1993) Postexposure prophylaxis against experimental inhalation anthrax. *J. Infect. Dis.* 167: 1239–1242.

Gatto-Menking DL, Yu H, Bruno JG *et al.* (1995) Sensitive detection of biotoxoids and bacterial spores using an immunomagnetic electrochemiluminescence sensor. *Biosensors Bioelectron.* 10: 501–507.

Gill DM, Coburn J (1987) ADP-ribosylation by cholera toxin: functional analysis of a cellular system that stimulates the enzymatic activity of cholera toxin fragment A1. *Biochemistry* 26: 6364–6371.

Gordon RE, Haynes WC, Pang CH-N (1973) The Genus *Bacillus. Agriculture Handbook No. 427.* Washington, DC: USDA, Agricultural Research Service, pp. 427.

Gordon VM, Leppla SH, Hewlett EL (1988) Inhibitors of receptor-mediated endocytosis block the entry of *Bacillus anthracis* adenylate cyclase toxin but not that of *Bordetella pertussis* adenylate cyclase toxin. *Infect. Immun.* 56: 1066–1069.

Gordon VM, Leppla SH (1994) Proteolytic activation of bacterial toxins: role of bacterial and host cell proteases. *Infect. Immun.* 62: 333–340.

Gordon VM, Klimpel KR, Arora N, Henderson MA, Leppla SH (1995) Proteolytic activation of bacterial toxins by eukaryotic cells is performed by furin and by additional cellular proteases. *Infect. Immun.* 63: 82–87.

Graham K, Keller K, Ezzell J, Doyle R (1984) Enzyme-linked lectinosorbent assay (ELLA) for detecting *Bacillus anthracis. Eur. J. Clin. Microbiol.* 3: 210–212.

Green BD, Battisti L, Koehler TM, Thorne CB, Ivins BE (1985) Demonstration of a capsule plasmid in *Bacillus anthracis. Infect. Immun.* 49: 291–297.

Green DH, Wakeley PR, Page A *et al.* (1999) Characterization of two *Bacillus* probiotics. *Appl. Environ. Microbiol.* 65: 4288–4291.

Gu ML, Leppla SH, Klinman DM (2000) Protection against anthrax toxin by vaccination with a DNA plasmid encoding anthrax protective antigen. *Vaccine* 17: 340–344.

Guignot J, Mock M, Fouet A (1997) *AtxA* activates the transcription of genes harbored by both *Bacillus anthracis* virulence plasmids. *FEMS Microbiol. Lett.* 147: 203–207.

Gupta P, Waheed SM, Bhatnagar R (1999) Expression and purification of the recombinant protective antigen of *Bacillus anthracis*. *Protein Expr. Purif.* 16: 369–376.

Hanna PC, Kochi S, Collier RJ (1992) Biochemical and physiological changes induced by anthrax lethal toxin in J774 macrophage-like cells. *Mol. Biol. Cell.* 3: 1269–1277.

Hanna PC, Acosta D, Collier RJ (1993) On the role of macrophages in anthrax. *Proc. Natl Acad. Sci. USA* 90: 10198–10201.

Harrell LJ, Andersen GL, Wilson KH (1995) Genetic variability of *Bacillus anthracis* and related species. *J. Clin. Microbiol.* 33: 1847–1850.

Helyer RJ, Kelley T, Berkeley RCW (1997) Pyrolysis mass spectrometry studies on *Bacillus anthracis, Bacillus cereus* and their close relatives. *Int. J. Med. Microbiol. Virol. Parasitol. Infect. Dis.* 285: 319–328.

Henderson DW, Peacock S, Belton FC (1956) Observations on the prophylaxis of experimental pulmonary anthrax in the monkey. *J. Hyg.* 54: 28–36.

Henderson I, Duggleby CJ, Turnbull PCB (1994) Differentiation of *Bacillus anthracis* from other *Bacillus cereus* group bacteria with the PCR. *Int. J. Syst. Bacteriol.* 44: 99–105.

Henderson I, Yu D, Turnbull PCB (1995) Differentiation of *Bacillus anthracis* and other *Bacillus cereus* group bacteria using IS231-derived sequences. *FEMS Microbiol. Lett.* 128: 113–118.

Henderson I (1996) Fingerprinting *Bacillus anthracis* strains. Proceedings of the International Workshop on Anthrax, 19–21 September 1995, Winchester, UK. *Salisbury Med. Bull.*, No. 87, Special suppl.: 55–58.

Hill KK, Kolstr A-B, Ticknor LO *et al.* (1999) AFLP analysis of 154 Norwegian soil isolates of *Bacillus cereus/B. thuringiensis*. Proceedings of the 2nd International Workshop on the Molecular Biology of *B. anthracis, B. cereus* and *B. thuringiensis*, 11–13 August 1999, Taos, New Mexico, p. 40.

Hoffmaster AR, Koehler TM (1999) Autogenous regulation of the *Bacillus anthracis pag* operon. *J. Bacteriol.* 181: 4485–4492.

Hutson RA, Duggleby CJ, Lowe JR, Manchee RJ, Turnbull PCB (1993) The development and assessment of DNA and oligonucleotide probes for the specific detection of *Bacillus anthracis*. *J. Appl. Bacteriol.* 75: 463–472.

Iacono-Connors LC, Welkos SL, Ivins BE, Dalrymple JM (1991) Protection against anthrax with recombinant virus-expressed protective antigen in experimental animals. *Infect. Immun.* 59: 1961–1965.

Ivins BE, Ezzell JW, Jemski J *et al.* (1986) Immunization studies with attenuated strains of *Bacillus anthracis*. *Infect. Immun.* 52: 454–458.

Ivins BE, Welkos SL (1988) Recent advances in the development of an improved human vaccine. *Eur. J. Epidemiol.* 4: 12–19.

Ivins BE, Welkos SL, Little SF, Knudson GB (1990a) Proceedings of the International Workshop on Anthrax. Winchester, 11–13 April 1989, UK. *Salisbury Med. Bull.* No. 68, Special Suppl.: 86–88.

Ivins BE, Welkos SL, Knudson GB, Little SF (1990b) Immunization against anthrax with aromatic compound-dependent (*Aro-*) mutants of *Bacillus anthracis* and with recombinant strains of *Bacillus subtilis* that produce anthrax protective antigen. *Infect. Immun.* 58: 303–308.

Ivins BE, Welkos SL, Little SF, Crumrine MH, Nelson GO (1992) Immunization against anthrax with *Bacillus anthracis* protective antigen combined with adjuvants. *Infect. Immun.* 60: 662–668.

Ivins BE, Fellows PF, Nelson GO (1994) Efficacy of a standard human anthrax vaccine against *Bacillus anthracis* spore challenge in guinea-pigs. *Vaccine* 12: 872–874.

Ivins B, Fellows P, Pitt L *et al.* (1995) Experimental anthrax vaccines: efficacy of adjuvants combines with protective antigen against an aerosol *Bacillus anthracis* spore challenge in guinea pigs. *Vaccine* 13: 1779–1784.

Ivins BE, Fellows PF, Pitt MLM *et al.* (1996) Efficacy of a standard human anthrax vaccine against *Bacillus anthracis* aerosol spore challenge in rhesus monkeys. Proceedings of the International Workshop on Anthrax, 9–21 September 1995, Winchester, UK. *Salisbury Med. Bull.*, No. 87, Special suppl.: 125–126.

Ivins BE, Pitt MLM, Fellows PF *et al.* (1998) Comparative efficacy of experimental anthrax vaccine candidates against inhalation anthrax in rhesus macaques. *Vaccine* 16: 1141–1148.

Jackson PJ, Walthers EA, Kalif AS *et al.* (1997) Characterization of the variable-number tandem repeats in *vrrA* from different *Bacillus anthracis* isolates. *Appl. Environ. Microbiol.* 63: 1400–1405.

Jackson PJ, Hugh-Jones ME, Adair DM *et al.* (1998) PCR analysis of tissue samples from the 1979 Sverdlovsk anthrax victims: The presence of multiple *Bacillus anthracis* strains in different victims. *Proc. Natl Acad. Sci. USA* 95: 1224–1229.

Jackson PJ, Hill KK, Laker MT, Ticknor LO, Keim P (1999a) Genetic comparison of *Bacillus anthracis* and its close relatives using amplified fragment length polymorphism and polymerase chain reaction analysis. *J. Appl. Bacteriol.* 87: 263–269.

Jackson PJ, Ticknor LO, Okinaka RT *et al.* (1999b) Phylogenetic relationships among different *B*. In: *B. anthracis, B. cereus and B. thuringiensis isolates based on AFLP and MLST analysis. Proceedings of the 2nd*

International Workshop on the Molecular Biology of B. anthracis, B. cereus *and* B. thuringiensis, 11-13 August 1999, Taos, New Mexico, p. 4.

Johns M, Harrington L, Titball RW, Leslie DL (1994) Improved methods for the detection of *Bacillus anthracis* spores by the polymerase chain reaction. *Lett. Appl. Microbiol.* 18: 236–238.

Jones MN, Beedham RJ, Turnbull PCB, Manchee RJ (1996) Efficacy of the UK human anthrax vaccine in guinea pigs against aerosolised spores of *Bacillus anthracis.* Proceedings of the International Workshop on Anthrax, 9–21 September 1995, Winchester, UK. *Salisbury Med Bull*, No. 87, Special suppl.: 123–124.

Kearney JF, Kushner N, Shu F (1999) Monoclonal antibodies to *Bacillus anthracis* spores. Proceedings of the 2nd International Workshop on the Molecular Biology of *B. anthracis, B. cereus* and *B. thuringiensis,* 11–13 August 1999, Taos, New Mexico, p. 41.

Keim P, Kalif A, Schupp J *et al.* (1997) Molecular evolution and diversity in *Bacillus anthracis* as detected by amplified fragment length polymorphism markers. *J. Bacteriol.* 179: 818–824.

Keim P, Klevytska AM, Price LB *et al.* (1999) Molecular diversity in *Bacillus anthracis. J. Appl. Microbiol.* 87: 215–217.

Keim P, Price LB, Klevytska AM *et al.* (2000) Multiple-locus variable-number tandem repeat analysis reveals genetic relationships within *Bacillus anthracis. J. Bacteriol.* 182: 2928–2936.

Keppie J, Harris-Smith W, Smith H (1963) The chemical basis of the virulence of *Bacillus anthracis.* IX. Its aggressins and their mode of action. *Br. J. Exp. Pathol.* 44: 446–453.

Klevytska AM, Schupp JM, Price LB, Adair DM, Keim P (1999) Multiple poymorphic regions within the *Bacillus anthracis vrrC* locus. Proceedings of the 2nd International Workshop on the Molecular Biology of *B. anthracis, B. cereus* and *B. thurringiensis,* 11–13 August 1999, Taos, New Mexico. p. 42.

Klimpel KR, Arora N, Leppla SH (1994) Anthrax lethal factor contains a zinc metalloprotease consensus sequence which is required for lethal toxin activity. *Mol. Microbiol.* 13: 1093–1100.

Knudson GB (1986) Treatment of anthrax in man: history and current concepts. *Mil. Med.* 151: 71–77.

Koch SK, Martin I, Schiavo G, Mock M, Cabiaux V (1994) The effects of pH on the interaction of anthrax toxin lethal and edema factors with phospholipid vesicles. *Biochemistry* 33: 2604–2609.

Koehler TM, Dai Z, Kaufman-Yarbray M (1994) Regulation of the *Bacillus anthracis* protective antigen gene: CO2 and a *trans*-acting element activate transcription from one of two promoters. *J. Bacteriol.* 176: 586–595.

Lecadet MM, Franchon E, Dumanoir VC *et al.* (1999) Updating the H-antigen classification of *Bacillus thuringiensis. J. Appl. Microbiol.* 86: 660–672.

Lee MA, Brightwell G, Leslie D, Bird H, Hamilton A (1999) Fluorescent detection techniques for real-time multiplex strand-specific detection of *Bacillus anthracis* using rapid PCR. *Appl. Microbiol.* 87: 218–223.

Leppla SH (1982) Anthrax toxin edema factor: a bacterial adenylate cyclase that increases cyclic AMP concentrations in eukaryotic cells. *Proc. Natl Acad. Sci. USA* 79: 3162–3166.

Leppla SH (1984) *Bacillus anthracis* calmodulin-dependent adenylate cyclase: chemical and enzymatic properties and interactions with eucaryotic cells. *Adv. Cyclic Nucleotide Res.* 17: 189–198.

Leppla SH, Ivins BE, Ezzell JW (1985) Anthrax toxin. In: Loretta L (ed.) *Microbiology – 1985.* Washington, DC: American Society for Microbiology, pp. 63–66.

Leppla SH (1988) Production and purification of anthrax toxin. In: Harshman S (ed.) *Methods in Enzymology,* Orlando, FL: Academic Press, Vol 165, pp. 103–116.

Leppla SH, Friedlander AM, Cora E (1988) Proteolytic activation of anthrax toxin bound to cellular receptors. In: Fehrenbach F, Alouf JE, Falmagne P *et al.* (ed.) *Bacterial Protein Toxins.* New York: Gustav Fischer Verlag, pp. 111–112.

Leppla SH (1991) The anthrax toxin complex. In: Alouf JE, Freer JH (eds) *Sourcebook of Bacterial Protein Toxins.* New York: Academic Press, pp. 277–302.

Leppla SH (1999) In: Alouf JE, Freer JH (ed.) *The Comprehensive Sourcebook of Bacterial Protein Toxins.* New York: Academic Press, pp. 243–263.

Liddington R, Pannifer A, Hanna P, Leppla S, Collier RJ (1999) Crystallographic studies of the anthrax lethal toxin. *J. Appl. Microbiol.* 87: 282.

Limb GA, Brown KA, Wolstencroft RA, Ellis BA, Dumonde DC (1988) Modulation of Fc and C3b receptor expression on guinea pig macrophages by lymphokines. *Clin. Exp. Immunol.* 74: 171–176.

Lincoln RE, Fish DC (1970) Anthrax toxin. In: Montie TC, Kadis S, Ajl SJ (ed.) *Microbial Toxins III.* New York: Academic Press, pp. 361–414.

Little SF, Knudson GB (1986) Comparative efficacy of *Bacillus anthracis* live spore vaccine and protective antigen vaccine against anthrax in the guinea pig. *Infect. Immun.* 52: 509–512.

Little SF, Lowe JR (1991) Location of receptor-binding region of protective antigen from *Bacillus anthracis. Biochem. Biophys. Res. Com.* 180: 531–537.

Little SF, Novak JM, Lowe JR *et al.* (1996) Characterization of lethal factor-binding and cell receptor-binding domains of protective antigen of *Bacillus anthracis* using monoclonal antibodies. *Microbiology* 142: 707–715.

Little SF, Ivins BE, Fellows PF, Friedlander AM (1997) Passive protection by polyclonal antibodies against *Bacillus anthracis* infection in guinea pigs. *Infect. Immun.* 65: 5171–5175.

Logan NA (1988) *Bacillus* species of medical and veterinary importance. *J. Med. Microbiol.* 25: 157–165.

Logan NA, Turnbull PCB (1999) *Bacillus* and recently derived genera. In: Murray RR, Baron EJ, Pfaller MA, Tenover FC, Yolken RH (ed.) *Manual of Clinical Microbiology, 6th edn*. Washington DC: ASM Press, pp. 357–369.

Long GW, O'Brien T (1999) Antibody systems for the detection of *Bacillus anthracis* in environmental samples. *J. Appl. Microbiol.* 87: 214.

Longchamp P, Leighton T (1999) Molecular recognition specificity of *Bacillus anthracis* spore antibodies. *J. Appl. Microbiol.* 87: 246–249.

Lowe JR, Salter O, Avina JA (1990) Evidence for a calcium-binding domain in the sequence of *Bacillus anthracis* lethal factor. Abstracts of the 90th Annual Meeting of the American Society for Microbiology, 13–17 May 1990, Annaheim, California. Washington, DC: ASM Press, B-320, p. 80.

M'Fadyean J (1903) A peculiar staining reaction of the blood of animals dead of anthrax. *J. Comp. Pathol.* 16: 35–41.

Makino S, Sasakawa C, Uchida I, Terakado N, Yoshikawa M (1988) Cloning and CO2-dependent expression of the genetic region for encapsulation from *Bacillus anthracis*. *Mol. Microbiol.* 2: 371–376.

Makino S, Uchida I, Terakado N, Sasakawa C, Yoshikawa M (1989) Molecular characterization and protein analysis of the *cap* region, which is essential for encapsulation in *Bacillus anthracis*. *J. Bacteriol.* 171: 722–730.

Makino S, Iinuma-Okado Y, Maruyama T *et al.* (1993) Direct detection of *Bacillus anthracis* in animals by polymerase chain reaction. *J. Clin. Microbiol.* 31: 547–551.

McBride BW, Mogg A, Telfer JL *et al.* (1998) Protective efficacy of a recombinant protective antigen against *Bacillus anthracis* challenge and assessment of immunological markers. *Vaccine* 16: 810–817.

McKinney N, Goldman S, Hunter-Cevera J, Leighton T, Long G, Nelson B (1998) Molecular phylogenetic analysis of the *B. anthracis* clade: spore structural proteins as evolutionary chronometers. Abstracts of the 3rd International Conference on Anthrax, 7–10 September 1998, p. 14. Bedford: Society for Applied Microbiology.

Mesnage S, Tosi-Couture E, Mock M, Gounon P, Fouet A (1997) Molecular characterisation of the *Bacillus anthracis* main S-layer component: evidence that it is the major cell-associated antigen. *Mol. Microbiol.* 23: 1147–1155.

Mesnage S, Tosi-Couture E, Gounon P, Mock M, Fouet A (1998) The capsule and S-layer: two independent and yet compatible macromolecular structures in *Bacillus anthracis*. *J. Bacteriol.* 180: 52–58.

Mesnage S, Tosi-Couture E, Mock M, Fouet A (1999) The S-layer homology domain as a means for anchoring heterologous proteins on the cell surface of *Bacillus anthracis*. *J. Appl. Microbiol.* 87: 256–260.

Mikesell P, Ivins BE, Ristroph JD, Dreier TD (1983) Evidence for plasmid-mediated toxin production in *Bacillus anthracis*. *Infect. Immun.* 39: 371–376.

Miller CJ, Elliott JL, Collier RJ (1999) Anthrax protective antigen: prepore-to-pore conversion. *Biochemistry* 38: 10432–10441.

Milne JC, Furlong D, Hanna PC, Wall JS, Collier RJ (1994) Anthrax protective antigen forms oligomers during intoxication of mammalian cells. *J. Biol. Chem.* 269: 20607–20612.

Miteva V, Abadjieva A, Grigorova R (1991) Differentiation among strains and serotypes of *Bacillus thuringiensis* by M13 DNA fingerprinting. *J. Gen. Microbiol.* 137: 593–600.

Mock M, Labruyere E, Glaser P, Danchin A, Ullman A (1988) Cloning and expression of the calmodulin-sensitive *Bacillus anthracis* adenylate cyclase in *Escherichia coli*. *Gene* 64: 277–284.

Novak JM, Stein M-P, Little SF, Leppla SH, Friedlander AM (1992) Functional characterization of protease-treated *Bacillus anthracis* protective antigen. *J. Biol. Chem.* 267: 17186–17193.

O'Brien J, Friedlander A, Dreier T, Ezzell J, Leppla S (1985) Effects of anthrax toxin components on human neutrophils. *Infect. Immun.* 47: 306–310.

Ohlander C, Larsson A, Perlmann P (1978) Specificity of Fc-receptors on lymphocytes and monocytes for guinea pig IgG1 and IgG2: phagocytosis of erythrocytes. *Scand. J. Immunol.* 7: 285–296.

Okinaka R, Cloud K, Hampton O *et al.* (1999) Sequence, assembly and analysis of pX01 and pX02. *J. Appl. Microbiol.* 87: 261–262.

Okinaka R, Cloud K, Hampton O *et al.* (1999) Sequence and organization of pXO1, the large *Bacillus anthracis* plasmid harboring the anthrax toxin genes. *J. Bacteriol.* 181: 6509–6515.

Pallen MJ, Dougan G, Frankel G (1997) Coiled-coil domains in proteins secreted by type III secretion systems. *Mol. Microbiol.* 25: 423–425.

Parry JM, Turnbull PCB, Gibson JR (1983) *A Colour Atlas of Bacillus Species*. Wolfe Medical Atlases No. 19 Edinburgh: Harcourt Health Services.

Patra G, Sylvestre P, Ramisse V, Therasse J, Guesdon J-L (1996) Isolation of a specific chromosomic DNA sequence of *Bacillus anthracis* and its possible use in diagnosis. *FEMS Microbiol. Lett.* 15: 223–231.

Petosa C, Collier RJ, Klimpel KR, Leppla SH, Liddington RC (1997) Crystal structure of the anthrax toxin protective antigen. *Nature* 385: 822–838.

Pezard C, Weber M, Sirard J-C, Berche P, Mock M (1995) Protective immunity by *Bacillus anthracis* toxin-deficient strains. *Infect. Immun.* 63: 1369–1372.

Pfisterer RM (1990) Retrospective verification of the diagnosis of anthrax by means of the intracutaneous skin test with the Russian allergen 'Anthraxin' in a recent epidemic in Switzerland. Proceedings of the International Workshop on Anthrax, 11–13 April 1989, Winchester, UK. *Salisbury Med. Bull.* No. 68, Special suppl.: 80.

Phillips AP, Martin KL (1982a) Assessment of immunofluorescence measurements of individual bacteria in direct and indirect assays for *Bacillus anthracis* and *Bacillus cereus* spores. *J. Appl. Bacteriol.* 53: 223–231.

Phillips AP, Martin KL (1982b) Evaluation of a microfluorometer in immunofluorescence assays of individual spores of *Bacillus anthracis* and *Bacillus cereus*. *J. Immunol. Meth.* 49: 271–282.

Phillips AP, Martin KL (1982c) Variations on the stainig method in quantitative indirect immunofluorescence assays for *Bacillus* spores, and the use of fluorescein-protein A. *J. Immunol. Meth.* 54: 361–369.

Phillips AP, Martin KL (1983) Comparison of direct and indirect immunoradiometric assays (IRMA) for *Bacillus anthracis* spores immobilised on multispot microscope slides. *J. Appl. Bacteriol.* 55: 315–324.

Phillips AP, Martin KL (1984) Radioactive labels for Protein A: evaluation in the indirect immunoradiometric assay (IRMA) for *Bacillus anthracis* spores. *J. Appl. Bacteriol.* 56: 449–456.

Phillips AP, Campbell AM, Quinn R (1988) Monoclonal antibodies against spore antigens of *Bacillus anthracis*. *FEMS Microbiol. Immunol.* 47: 169–178.

Pomerantsev AP, Staritsin NA, Mockov YV, Marinin LI (1997) Expression of cereolysine AB genes in *Bacillus anthracis* vaccine strain ensures protection against experimental hemolytic anthrax infection. *Vaccine* 15: 1846–1850.

Price LB, Klevytska AM, Smith KL *et al.* (1999) Molecular diversity in *B. anthracis*. Proceedings of the 2nd International Workshop on the Molecular Biology of *B. anthracis*, *B. cereus* and *B. thuringiensis*, 11–13 August 1999, Taos, New Mexico, p. 3.

Priest FG (1981) DNA homology in the genus *Bacillus*. In: Berkeley RCW, Goodfellow M (eds) *The Aerobic Endospore-forming Bacteria*. London: Academic Press, pp. 33–57.

Priest FG (1997) *Bacillus* – how many genera, how many species? Proceedings of the FEMS International Conference on Fundamental and Applied Aspects of Bacterial Spores, 8-11 July 1997, Girton College, Cambridge.

Quinn CP, Shone CC, Turnbull PC, Melling J (1988) Purification of anthrax-toxin components by high-performance anion-exchange, gel-filtration and hydrophobic-interaction chromatography. *Biochem. J.* 252: 753–758.

Quinn CP, Singh Y, Klimpel KR, Leppla SH (1991) Functinal mapping of anthrax toxin lethal factor by in-frame insertion mutagenesis. *J. Biol. Chem.* 266: 20124–20130.

Ramisse V, Patra G, Garrigue H, Guesdon J-L, Mock M (1996) Identification and characterisation of *Bacillus anthracis* by multiplex PCR analysis of sequences on plasmids pXO1 and pXO2 and chromosomal DNA. *FEMS Microbiol. Lett.* 145: 9–16.

Ramisse V, Patra G, Vaissaire J, Mock M (1999) The Ba813 chromosomal DNA sequence effectively traces the whole *Bacillus anthracis* community. *J. Appl. Bacteriol.* 87: 224–228.

Read T, Peterson S (1999) Whole genome sequencing of *Bacillus anthracis*. Abstracts of the 2nd International Workshop on the molecular biology of *Bacillus cereus*, *Bacillus anthracis* and *Bacillus thuringiensis*, 11–13 August 1999, Taos, New Mexico, p. 20.

Reif TC, Johns M, Pillai SD, Carl M (1994) Identification of capsule-forming *Bacillus anthracis* spores with the PCR and a novel dual-probe hybridization format. *Appl. Environ. Microbiol.* 60: 1622–1625.

Robertson DL (1988) Relationships between the calmodulin-dependent adenylate cyclases produced by *Bacillus anthracis* and *Bordetella pertussis*. *Biochem. Biophys. Res. Comm.* 157: 1027–1032.

Robertson DL, Leppla SH (1986) Molecular cloning and expression in *Escherichia coli* of the lethal factor gene of *Bacillus anthracis*. *Gene* 44: 71–78.

Robertson DL, Tippetts MT, Leppla SH (1988) Nucleotide sequence of the *Bacillus anthracis* edema factor gene (*cya*): a calmodulin-dependent adenylate cyclase. *Gene* 73: 363–371.

Robertson JM, Orandello KR, Mangold B, Long G (1999) A multiplex PCR system for *Bacillus anthracis*. Proceedings of the 2nd International Workshop on the Molecular Biology of *B. anthracis*, *B. cereus* and *B. thuringiensis*, 11–13 August 1999, Taos, New Mexico, p. 53.

Salkinoja-Salonen MS, Vuorio R, Andersson MA *et al.* (1999) Toxigenic strains of *Bacillus licheniformis* related to food poisoning. *Appl. Environ. Microbiol.* 65: 4637–4645.

Shlyakov E, Rubinstein E (1994) Human live anthrax vaccine in the former USSR. *Vaccine* 12: 727–730.

Shlyakov E (1996) Anthraxin – a skin test for early and retrospective diagnosis of anthrax and anthrax vaccination assessment. Proceedings of the International Workshop on Anthrax, 9–21 September 1995, Winchester, UK. *Salisbury Med. Bull.*, No. 87, Special suppl.: 109–110.

Shlyakov E, Rubinstein E (1996) Evaluation of the anthraxin skin test for diagnosis of acute and past human anthrax. *Eur. J. Clin. Microbiol. Infect. Dis.* 15: 242–245.

Shlyakov E, Rubinstein E, Novikov I (1997) Anthrax post-vaccinal cell-mediated immunity in humans: kinetics pattern. *Vaccine* 15: 631–636.

Shute LA, Gutteridge CS, Norris JR, Berkeley RCW (1988) Reproducibility of pyrolysis mass spectrometry: effect of growth medium and instrument stability on the differentiation of selected *Bacillus* species. *J. Appl. Bacteriol.* 64: 79–88.

Singh Y, Chaudhary VK, Leppla SH (1989) A deleted variant of *Bacillus anthracis* protective antigen is nontoxic and blocks anthrax toxin *in vivo*. *J. Biol. Chem.* 264: 19103–19107.

Singh Y, Klimpel KR, Quinn CP, Chaudhary VK, Leppla SH (1991) The carboxyl-terminal end of protective antigen is required for receptor-binding and anthrax toxin activity. *J. Biol. Chem.* 266: 15493–15497.

Singh Y, Klimpel KR, Arora N, Sharma M, Leppla SH (1994) The chymotrypsin-sensitive site, FFD315, in anthrax toxin protective antigen is required for translocation of lethal factor. *J. Biol. Chem.* 269: 29039–29046.

Singh Y, Ivins BE, Leppla SH (1998) Study of immunization against anthrax with the purified recombinant protective antigen of *Bacillus anthracis. Infect. Immun.* 66: 3447–3448.

Singh Y, Klimpel KR, Goel S, Swain PK, Leppla SH (1999) Oligomerization of anthrax toxin protective antigen and binding of lethal factor during endocytic uptake into mammalian cells. *Infect. Immun.* 67: 1853–1859.

Sirard J-C, Mock M, Fouet A (1994) The three *Bacillus anthracis* toxin genes are coordinately regulated by bicarbonate and temperature. *J. Bacteriol.* 176: 5188–5192.

Sjöstedt A, Eriksson U, Ramisse V, Garrigue H (1997) Detection of *Bacillus anthracis* spores in soil by PCR. *FEMS Microbiol. Ecol.* 23: 159–168.

Smith H, Keppie J (1954) Observations on experimental anthrax: demonstration of a specific lethal factor produced *in vivo* by *Bacillus anthracis. Nature* 173: 869–871.

Smith KL, Vos V de, Bryden HB *et al.* (1999) Meso-scale ecology of anthrax in southern Africa: a pilot study of diversity and clustering. *J. Appl. Microbiol.* 87: 204–207.

Smith NR, Gordon RE, Clark FE (1952) *Aerobic Spore-forming Bacteria. Agricultural Monograph No. 16.* Washington, DC: USDA.

Stanley JL, Smith H (1961) Purification of Factor I and recognition of a third factor of the anthrax toxin. *J. Gen. Microbiol.* 26: 49–66.

Sterne M (1937) The effects of different carbon dioxide concentrations on the growth of virulent anthrax strains. Pathogenicity and immunity tests on guinea-pigs and sheep with anthrax variants derived from virulent strains. *Onderstepoort J. Vet. Sci. Ind.* 9: 49–67.

Sterne M (1939) The use of anthrax vaccines prepared from avirulent (uncapsulated) varients of *Bacillus anthracis. Onderstepoort J. Vet. Sci. Ind.* 13: 307–312.

Tang G, Leppla SH (1999) Proteasome activity is required for anthrax lethal toxin to kill macrophages. *Infect. Immun.* 67: 3055–3060.

Tippetts MT, Robertson DL (1988) Molecular cloning and expression of the *Bacillus anthracis* edema factor toxin gene: a calmodulin-dependent adenylate cyclase. *J. Bacteriol.* 170: 2263–2266.

Tuazon CU, Murray HW, Levy C *et al.* (1974) Serious infections from *Bacillus* sp. *JAMA* 241: 1137–1140.

Turnbull PCB, Broster MG, Carman JA, Manchee RJ, Melling J (1986) Development of antibodies to protective antigen and lethal factor components of anthrax toxin in humans and guinea pigs and their relevance to protective immunity. *Infect. Immun.* 52: 356–363.

Turnbull PCB, Leppla SH, Broster MG, Quinn CP, Melling J (1988) Antibodies to anthrax toxin in humans and guinea pigs and their relevance to protective immunity. *Med. Microbiol. Immunol.* 177: 293–303.

Turnbull PCB, Doganay M, Lindeque PM, Aygen B, McLaughlin J (1992a) Serology and anthrax in humans, livestock and Etosha National Park wildlife. *Epidemiol. Infect.* 108: 299–313.

Turnbull PBC, Quinn CP, Hewson R, Stockbridge MC, Melling J (1990) Protection conferred by microbially-supplemented UK and purified PA vaccines. Proceedings of the International Workshop on Anthrax, 11–13 April 1989, Winchester, UK. *Salisbury Medical Bulletin*, No. 68, Special suppl: 89–91.

Turnbull PCB (1991) Anthrax vaccines: past, present and future. *Vaccine* 9: 533–539.

Turnbull PCB, Hutson RA, Ward MJ *et al.* (1992b) *Bacillus anthracis* but not always anthrax. *J. Appl. Bacteriol.* 72: 21–26.

Turnbull PCB, Böhm R, Cosivi O *et al.* (1998) Guidelines on surveillance and control of anthrax in humans and animals. World Health Organization/EMC/ZDI/98.6.

Turnbull PCB (1999) Definitive identification of *Bacillus anthracis* – a review. *J. Appl. Microbiol.* 87: 237–240.

Turnbull PCB (2000) Current status of immunization against anthrax: old vaccines may be here to stay for a while. *Curr. Opin. Infect. Dis.* 13: 113–120.

Uchida I, Sekizaki T, Hashimoto K, Terakado N (1985) Association of the encapsulation of *Bacillus anthracis* with a 60 megadalton plasmid. *J. Gen. Microbiol.* 131: 363–367.

Uchida I, Hashimoto K, Makino S *et al.* (1987) Restriction map of a capsule plasmid of *Bacillus anthracis. Plasmid* 18: 178–181.

Uchida I, Hornung JM, Thorne CB, Klimpel KR, Leppla SH (1993) Cloning and characterization of a gene whose product is a *trans*-activator of anthrax toxin synthesis. *J. Appl. Bacteriol.* 175: 5329–5339.

Uchida I, Makino S, Sekizaki T, Terakado N (1997) Cross-talk to the genes for *Bacillus anthracis* capsule synthesis by *atsA*, the gene encoding the *trans*-activator of anthrax toxin synthesis. *Mol. Microbiol.* 23: 1229–1240.

Varughese M, Chi A, Teixeira AV *et al.* (1998) Internalization of a *Bacillus anthracis* protective antigen-c-Myc fusion protein mediated by cell surface anti-c-Myc antibodies. *Mol. Med.* 4: 87–95.

Varughese M, Teixeira AV, Liu S, Leppla SH (1999) Identification of a receptor-binding region within domain 4 of the protective antigen component of anthrax toxin. *Infect. Immun.* 67: 1860–1865.

Vassart G, Georges M, Monsieur R, Brocas H *et al.* (1987) A sequence in M13 phage detects hypervariable mini-satellites in human and animal DNA. *Science* 235: 683–684.

Velappan N, Burde S, Grace K, Marrone BL (1999) *Bacillus* species and strain differentiation by a simple modified amplified fragment length polymorphism (AFLP) protocol. Proceedings of the 2nd International Workshop on the Molecular Biology of *B. anthracis, B. cereus* and *B. thuringiensis*, 11–13 August 1999, Taos, New Mexico, p. 59.

Vietri N, Marrero J, Hoover TA, Welkos SL (1995) Identification and characterization of a trans-activator involved in the regulation of encapsulation by *Bacillus anthracis*. *Gene* 152: 1–9.

Vodkin MH, Leppla SH (1983) Cloning of the protective antigen gene of *Bacillus anthracis*. *Cell* 34: 693–697.

Wade BH, Wright GG, Hewlett EL, Leppla SH, Mandell GL (1985) Anthrax toxin components stimulate chemotaxis of human polymorphonuclear neutrophils. *Proc. Soc. Exp. Biol. Med.* 179: 159–162.

Walker JS, Lincoln RE, Klein F (1967) Pathophysiology and biochemical changes in anthrax. *Fed. Proc.* 26: 1539–1544.

Watson DW, Cromartie WJ, Bloom WL, Kegeles G, Heckly RJ (1947) Studies on infection with *Bacillus anthracis*. III. Chemical and immunological properties of the protective antigen in crude extracts of skin lesions of *B. anthracis*. *J. Infect. Dis.* 80: 28–40.

Welkos SL, Keener TJ, Gibbs PH (1986) Differences in susceptibility of inbred mice to *Bacillus anthracis*. *Infect. Immun.* 51: 795–800.

Welkos SL, Friedlander AM (1988) Pathogenesis and genetic control of resistance to the Sterne strain of *Bacillus anthracis*. *Microb. Pathog.* 4: 53–69.

Welkos SL, Lowe JR, Eden-McCutchan F *et al.* (1988) Sequence and analysis of the DNA encoding protective antigen of *Bacillus anthracis*. *Gene* 69: 287–300.

Welkos SL (1991) Plasmid-associated virulence factors of non-toxigenic (pX01-) *Bacillus anthracis*. *Microb. Pathog.* 10: 183–198.

Welkos SL, Vietri NJ, Gibbs PH (1993) Non-toxigenic derivatives of the Ames strain of *Bacillus anthracis* are fully virulent for mice: role of plasmid pX02 and chromosome in strain-dependent virulence. *Microb. Pathog.* 14: 381–388.

Welsh J, McClelland M (1990) Fingerprinting genomes using PCR with arbitrary primers. *Nucleic Acids Res.* 18: 7213–7218.

Wesche J, Elliott JL, Falnes PØ, Olsnes S, Collier RJ (1998) Characterization of membrane translocation by anthrax protective antigen. *Biochemistry* 37: 15737–15746.

Williamson ED, Beedham RI, Bennett AM, Perkins SD, Miller J, Baillie LWJ (1999) Presentation of protective antigen to the mouse immune system: immune sequelae. *J. Appl. Microbiol.* 87: 315–317.

Worsham PL, Sowers MR (1999) Isolation of an asporogenic (spoOA) protective antigen-producing strain of *Bacillus anthracis*. *Can. J. Microbiol.* 45: 1–8.

Yu H (1996) Enhancing immunoelectrochemiluminescence (IECL) for sensitive bacterial detection. *J. Immunol. Meth.* 192: 63–71.

Zegers ND, Kluter E, Stap H van der *et al.* (1999) Expression of the protective antigen of *Bacillus anthracis* by *Lactobacillus casei*: towards the development of an oral vaccine against anthrax. *J. Appl. Microbiol.* 87: 309–314.

Zinser G, Schupp JM, Keim P (1999) *Identification of novel VNTR loci in B. anthracis*. Proceedings of the 2nd International Workshop on the Molecular Biology of *B. anthracis*, *B. cereus* and *B. thuringiensis*, 11–13 August 1999, Taos, New Mexico, p. 61.

Zukowski MM (1992) Production of commercially valuable products. In: Doi RH, McGloughlin M (eds) *Biology of Bacilli. Applications to Industry*. Oxford: Butterworth-Heinemann, pp. 311–337.

Zwartouw HT, Smith H (1956) Polyglutamic acid from *Bacillus anthracis* grown *in vivo*: structure and aggressin activity. *Biochem. J.* 63: 437–442.

96

Yersinia pestis

Robert R. Brubaker

Michigan State University, East Lansing, Michigan, USA

Yersinia pestis, the cause of bubonic plague, has three remarkable attributes. First, it causes the most severe of all human bacterial infections as judged by the occurrence of 20 million human deaths in mediaeval Europe during the 'Black Death' and perhaps 10 times that number since the evolution of humanity. Second, survival and multiplication of the plague bacillus within the host and vector are largely mediated by three plasmids and at least one chromosomal high-pathogenicity island. Removal of this 'imported' information from the genome converts the formidable plague bacillus to an utterly innocuous entity no longer capable of survival in its mammalian host or the flea. Third, the emergence of *Y. pestis* has occurred in recent geological times and its evolution may still be in progress. The purpose of this chapter is to integrate these three features and to define the molecular mechanisms that enable *Y. pestis* to promote uncompromisingly acute disease.

Two additional *Yersinia* species pathogenic to humans provide an excellent foil for identifying the unique determinants of acute disease utilised by plague bacilli. These closely related 'enteropathogenic' species are *Y. pseudotuberculosis*, which typically causes mesenteric lymphadenitis in humans (sometimes mistaken for appendicitis) and *Y. enterocolitica* that initiates outright enteric disease characterised by gastro-intestinal distress and diarrhoea (see Chapter 68). Considerable attention has been given to defining the mechanisms that enable the enteropathogenic yersiniae to colonise the gut after ingestion with contaminated food or water and thereby cause chronic infections. These processes involve the use of an adhesin (YadA) and host-cell invasins (Inv and Ail) not shared by *Y. pestis*. A comparison of infectivity by mouth between the enteropathogenic yersiniae and the plague bacillus is moot because the latter possesses a distinct plasmid-mediated mechanism that facilitates immediate penetration of any moist mucosal surface (**Table 1**).

Cells of *Y. pseudotuberculosis* are the more invasive of the two enteropathogenic species; all serovars are as lethal as *Y. pestis* in the mouse after intravenous administration but of reduced virulence by the subcutaneous or intradermal routes (**Table 2**). Most isolates of *Y. enterocolitica*, typified by the common European O:3 serovar, are not lethal in mice by any route of infection but, nevertheless, cause a chronic intestinal disease after oral administration. The North American serovar O:8 isolates of *Y. enterocolitica* are exceptions to this pattern in that they cause systemic complications in humans and acute infection in mice, especially by intravenous injection. The epidemiology of these uncommon isolates is mysterious; they tend to favour cold climates and may thus fill the same niche as is occupied by *Francisella tularensis* in Scandinavia. Unless stated otherwise, comparisons provided here between plague and experimental disease caused by *Y. enterocolitica* refer to the virulent O:8 serovar.

Table 1 Major phenotypic distinctions between typical wild-type strains of *Yersinia pestis*, *Yersinia pseudotuberculosis* and *Yersinia enterocolitica*

Characteristic	Location of structural gene(s)[a]	*Y. pestis*	*Y. pseudotuberculosis*	*Y. enterocolitica*
Pesticin (bacteriocin) plasminogen activator	pPCP	+	0	0
Fraction I (capsular antigen), murine toxin	pMT	+	0	0
YadA (fibrillar adhesin)	pYV	0	+	+
Absorption of exogenous pigments (haemin, Congo red)	Chromosome (within ΔPgm)	+[b]	0	0
Expression of flagella	Chromosome	0	+[b]	+[b]
Extensive LPS O-group structure	Chromosome	0	+[b]	+[b]
Invasin	Chromosome	0	+	+
Ail	Chromosome	0	+	+
Glucose 6-phosphate dehydrogenase	Chromosome	0	+	+
Aspartase	Chromosome	0	+	+
Biosynthesis of L-threonine (or glycine), L-isoleucine, L-valine, L-methionine,[c] and L-phenylalanine[c]	Chromosome	0	+	+
Urease[c]	Chromosome	0	+	+
Fermentation of rhamnose[c] and melibiose[c]	Chromosome	0	+	0
Assimilation of low levels of NH_3[c]	Chromosome	0	+	+

[a] Genetic element encoding structural gene(s) when present.
[b] Typically expressed at 26°C but not 37°C.
[c] Has yielded meiotrophic revertants.

Table 2 Influence of phenotypic traits on 50% lethal doses of *Yersinia pestis* and *Yersinia pseudotuberculosis* by intravenous, intraperitoneal and subcutaneous routes of injection in control mice and mice receiving sufficient injected iron to saturate serum transferrin

Phenotype	Injected Fe^{3+} (40 μg/mouse)[a]	Route of infection		
		Intravenous	Intraperitoneal	Subcutaneous
Y. pestis				
pCD⁺, pPCP⁺, Pgm⁺	0	8.1×10^0	9.9×10^0	6.1×10^0
	+	9.8×10^0	9.8×10^0	5.9×10^0
pCD⁻, pPCP⁺, Pgm⁺	0	$>5 \times 10^7$	$>5 \times 10^7$	$>5 \times 10^7$
	+	$>5 \times 10^7$	$>5 \times 10^7$	$>5 \times 10^7$
pCD⁺, pPCP⁻, Pgm⁺	0	7.1×10^1	3.8×10^5	$>5 \times 10^7$
	+	2.3×10^1	1.4×10^1	$>5 \times 10^7$
pCD⁺, pPCP⁺, Pgm⁻	0	1.5×10^1	$>5 \times 10^7$	$>5 \times 10^7$
	+	4.3×10^1	2.4×10^1	ND
Y. pseudotuberculosis				
pYV⁺	0	3.9×10^1	2.9×10^4	1.1×10^4
	+	3.1×10^1	2.0×10^2	1.4×10^2
pYV⁻	0	$>5 \times 10^7$	$>5 \times 10^7$	$>5 \times 10^7$
	+	$>5 \times 10^7$	$>5 \times 10^7$	$>5 \times 10^7$

[a] Iron was administered intraperitoneally as Fe^{2+}.
ND not determined.

During the course of its evolution *Y. pestis* has lost a number of structural and regulatory genes (**Table 1**) that are expressed by the enteropathogenic yersiniae and the more distantly related *Escherichia coli*. As noted by M. Schaechter (Neidhardt *et al.*, 1996), microbiologists work not only intentionally with the species of their choice but also inadvertently with *E. coli*, which always serves as a yardstick for comparison. This adage is especially pertinent to yersiniae and the reader is urged to consult Neidhardt *et al.* (1996) and Newman and Lin (1995) to evaluate the significance of these missing functions and contemplate the consequences of their absence.

Considerable progress has been made during the last two decades in defining the workings of the plague bacillus. D.J. Brenner (Communicable Disease Center, Atlanta), H.H. Mollaret (Institute Pasteur, Paris) and their colleagues have been instrumental in unravelling the taxonomy of the genus. G.R. Cornelis (Université Catholique de Louvain, Brussels), S.C. Straley (University of Kentucky, Lexington), and H. Wolf-Watz (University of Umeå, Sweden) and their colleagues have been paramount in defining the low-calcium response and attendant type III process of secretion. E. Carniel (Institute Pasteur, Paris), J. Heesemann (Ludwig Maximilian University, Munich), R.D. Perry (University of Kentucky, Lexington), and their collaborators largely defined mechanisms used by yersiniae to assimilate iron *in vivo*. Many others have made outstanding recent contributions to this field. This chapter relies heavily on the excellent comprehensive review of *Y. pestis* provided by Perry and Fetherston, 1997.

Epidemiology

Y. pestis is of nearly worldwide distribution and inhabitants of at least portions of all of the populated continents except Europe and Australia are at potential risk. The endurance of the bacteria in a given locale depends upon their maintenance in the resident rodent population (sylvatic plague). The organisms are believed to have emerged periodically from ancient reservoirs in eastern Russia and western China to occupy permanently territory in the Far East, Middle East and perhaps Africa, although, as noted below, the species may originally have evolved in Africa. Plague bacilli were transported from these foci by caravan and then ship to mediaeval Europe to initiate the Black Death. The last pandemic occurred at the end of the nineteenth century and resulted in serious outbreaks in India and the Far East, as well as permanently introducing the species into South America and western

North America. Europe is now free of plague because the population of domestic rats, which served as the primary host during the mediaeval epidemics, has been significantly reduced as opposed to other social rodents that exist independently of humans in remote endemic areas. As a consequence of its present widespread rural distribution, it is extremely unlikely that plague can be eradicated within the foreseeable future.

Theobald Smith noted that it is poor strategy for a parasite seriously to debilitate its host, since this act endangers the only environment capable of sustaining that parasite (Smith, 1934). This epidemiological tenet clearly applies to the enteropathogenic yersiniae, where prolonging the course of disease enhances the probability of transfer to new hosts by faecal contamination. However, *Y. pestis* is a rare exception to Smith's principle, in that the bacteria must kill the current host in order to assure departure of the flea vector required to infect new hosts. It is also significant in this context that infection by *Y. pestis* is also lethal to the flea vector. As a result of this novel mode of transmission, plague bacilli exist in a closed, fixed and protected life cycle, which eliminates any necessity to respond to changes in natural environments, a capability that undoubtedly favours survival of the enteropathogenic yersiniae. Instead, acquisition by *Y. pestis* of the wherewithal to colonise the flea resulted in its subjection to a different selective pressure, namely, the need to provoke a marked terminal septicaemia. As a result of further evolution in this direction, the typical rodent and primate hosts now succumb to an acute disease so severe that death typically occurs before the immune system can intervene.

Taxonomy

The yersiniae constitute genus XI of the family *Enterobacteriaceae*. It includes *Y. pestis* as the type species, enteropathogenic *Y. pseudotuberculosis* and *Y. enterocolitica*, a primary pathogen of fish (*Y. ruckeri*), and secondary invaders or saprophytes (*Y. aldovae*, *Y. bercovieri*, *Y. frederiksenii*, *Y. intermedia*, *Y. kristensenii*, *Y. mollaretii* and *Y. rohdei*). The three species pathogenic for humans share a $\sim 70 \, kb$ plasmid, which is termed pCD in *Y. pestis* and pYV in *Y. pseudotuberculosis* and *Y. enterocolitica*. The guanine plus cytosine content of 47–48.5 mol% of the DNA of *Y. enterocolitica* slightly exceeds that of *Y. pestis* and *Y. pseudotuberculosis* (46–47 mol%). The DNA of *Y. pestis* has 9%, 23% and 83% homology, under stringent conditions of reassociation, to the

DNA of *E. coli*, *Y. enterocolitica* and *Y. pseudotuberculosis*, respectively. There is no significant hybridisation with species classified outside the *Enterobacteriaceae* (Moore and Brubaker, 1975). These values accurately assess the relatedness of *Y. pestis* to the two enteropathogenic yersiniae and to *E. coli*. DNA hybridisation under less stringent conditions, as well as biochemical characterisation (Brenner *et al.*, 1976) provides the basis for defining the additional eight species of the genus. Analysis of the 16S rRNA gene of yersiniae shows that the genus is a coherent cluster, where *Y. pestis*, *Y. pseudotuberculosis*, and certain isolates of *Y. kristensenii*, are one subline and *Y. enterocolitica* formed another (Ibrahim *et al.*, 1993).

The Genome

The chromosome of a typical strain of *Y. pestis* (KIM10) is 4208.4 kb in size as judged by two-dimensional pulsed-field gel electrophoresis of DNA first digested with SpeI and then ApaI (Lucier and Brubaker, 1992). This is about 90% of that reported for *E. coli*. The total genome of this strain is thus 4389.5 kb or the sum of the chromosome plus three plasmids of 9.7 kb (pPCP1), 70.5 kb (pCD1) and 100.9 kb (pMT1) (Hu *et al.*, 1998). All three plasmids have been sequenced (Hu *et al.*, 1998), as have extensive regions of the chromosome (*Yersinia pestis* Sequencing Group, 1998; Buchrieser *et al.*, 1999; Garcia *et al.*, 1999). The nature of structural gene products associated with expression of virulence or survival in the vector is discussed more fully below.

The Chromosome

An interesting feature of the *Y. pestis* chromosome is the presence of active IS elements termed IS*100*, IS*200*-like and IS*285*. Of these, over 30 IS*100* inserts can exist in the genome, including at least one copy in each of the three plasmids typically carried by wild-type isolates (Portnoy and Falkow, 1981). In addition, this element flanks a 102-kb chromosomal pigmentation (Pgm) locus defined in **Fig. 1a** which, as a consequence of reciprocal recombination, undergoes precise deletion at a rate of 10^{-5} (Brubaker, 1970). These insertions may also account for at least some of the mutations that cause loss of important metabolic and structural functions in *Y. pestis* that are expressed in *Y. pseudotuberculosis* (**Table 1**) in which copies of IS elements are far less common (Perry and Fetherston, 1997).

Plasmids

An appreciation of pPCP (pesticin, coagulase, and plasminogen activator) is necessary to contrast the invasive nature of plague with the chronic disease caused by the enteropathogenic yersiniae that lack this element. A physical map of pPCP1 is provided in **Fig. 1b**. Mutants lacking this plasmid are conveniently isolated by repeated subculture at 5°C.

The discovery of pYV (yersiniae virulence) in *Y. enterocolitica* (Zink *et al.*, 1980) marked the beginning of the molecular era of yersiniae research. The versions carried by the enteropathogenic yersiniae are functionally interchangeable with, and largely homologous to, their analogue termed pCD (calcium dependency) from *Y. pestis* (**Fig. 2**). However, frameshift mutations in the latter prevent production of the fibrillar adhesin YadA (Rosqvist *et al.*, 1988) and lipoprotein YlpA (Hu *et al.*, 1998) encoded on pYV. Furthermore, the order of certain genes on pYV of *Y. enterocolitica* is distinct from that of pYV in *Y. pseudotuberculosis* (**Fig. 2**). This plasmid is central to the expression of both acute and chronic disease. Mutants cured of pCD arise spontaneously on Ca^{2+}-deficient media at a rate of 10^{-4} at 37°C (Perry and Fetherston, 1997).

pMT (murine exotoxin) encodes capsular or fraction 1 antigen and an exotoxin that is highly lethal for members of the rodent family *Muridae*. This plasmid is also unique to *Y. pestis* where these two functions facilitate acute disease and survival in the flea vector. pMT can also be eliminated by prolonged cultivation at 5°C, but cured mutants may still express fraction 1 and murine toxin from copies of structural genes residing on the chromosome (Perry and Fetherston, 1997). The remaining genes of pMT are largely plebeian and facilitate vegetative growth; they would normally be expected to reside on the chromosome (Hu *et al.*, 1998).

Genetic Systems

Most methods used for genetic analysis of *E. coli* are also suitable for *Y. pestis*. Examples are gene transfer mediated by certain conjugative plasmids or bacteriophage P1, Mu hP1, and λ which are unable to replicate in yersiniae. Standard procedures for insertion and fusion of transposons and bacteriophage have been instrumental in identifying genes encoding virulence effectors and regulators. The three plasmids can readily be isolated by standard techniques and transformed by electroporation (Perry and Fetherston, 1997).

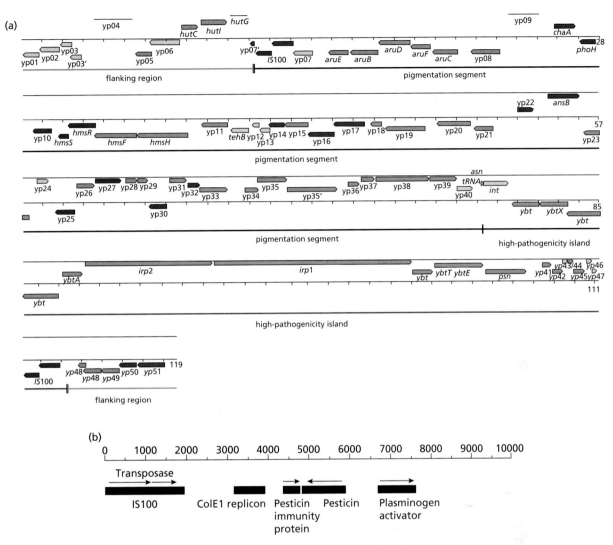

Fig. 1 (a) Map of the *pgm* locus and flanking regions showing the position and orientation of known genes and putative coding sequences: functional categories related to virulence (red), phage-related functions (light blue), transport proteins (dark blue), regulatory proteins (purple), nitrogen and carbon metabolism (brown), miscellaneous (yellow), insertion elements (black), and proteins of unknown function (green). The thick lines below the gene map indicate flanking regions (grey), pigmentation segment (black) and high-pathogenicity island (blue). Numbers at the right end of the map indicate the scale in kilobases. From Buchrieser *et al.* (1999). (b) Structural organisation of the 9610-bp plasmid pPCP1 derived from *Y. pestis* KIM10. The directions of transcription are indicated by the arrows. The single IS*100* element was used to define position 1 of this plasmid. The numbering above the line is the molecular size in base pairs. From Hu *et al.* (1999) (See also colour plate 23).

Ligand-fusion protein engineering has been especially useful in defining the relationship between immunity and repression of pro-inflammatory cytokines.

Physiology

Claims in the old literature that *Y. pestis* grows more rapidly at room temperature ($\sim 26°C$) than at host temperature (37°C) are incorrect and probably reflect ignorance of sluggish regulation, a temperature-dependent nutritional requirement for Ca^{2+} (see below), and the stimulatory effect of CO_2. If the latter is supplied at a concentration of 10%, the organisms form visible colonies in 24 hours, rivalling the enteropathogenic yersiniae. To assure steady-state kinetics in liquid cultures, two or three prior sub-cultures in the same medium are desirable before performing experiments. If these precautions are followed, *Y. pestis* and enteropathogenic yersiniae exhibit mid-exponential phase doubling times of about 70 minutes at 26°C and 37°C. Potassium

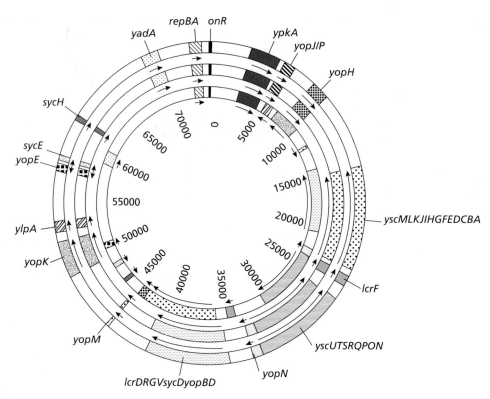

Fig. 2 Diagram comparing the organisation of selected genes and elements of pCD1 of *Y. pestis* KIM10 and pYV of *Y. pseudo-tuberculosis* (pIBI) and *Y. enterocolitica* (pYVe). The relative positions of selected loci with respect to the order of replication of pCD1 are shown: outer circle, pCD1; middle circle, pIBI; inner circle, pYVe. The genes and sequence features of pCD1 and the corresponding regions in pIBI and pYVe are depicted in the same colour to aid in their visualisation. Numbering inside the circles indicates the approximate sizes of the plasmids in nucleotides, measured from the start of their origins of replication. Arrows above each shaded segment representing a gene or gene group point to the direction of transcription. From Hu *et al.* (1999). (See also Colour Plate 24).

gluconate, which can be tolerated at concentrations of at least 0.04 M, serves as an excellent energy source by virtue of its ability to maintain neutral pH (Fowler and Brubaker, 1994; Perry and Fetherston, 1997).

Catabolism

Wild-type *Y. pestis* utilises mechanisms of glycolysis and terminal oxidation typical of *E. coli*, but lacks glucose 6-phosphate dehydrogenase and aspartase. The first is necessary for synthesis of pentose and reduction of $NADP^+$ via the hexose monophosphate pathway and the second is required for conversion of L-aspartate to fumarate, an intermediate of the tricarboxylic acid cycle. In addition, indirect evidence suggests that the tricarboxylic acid cycle may be blocked in *Y. pestis*, as judged by constitutive expression of the glyoxylate bypass (Hillier and Charnetzky, 1981). L-Serine undergoes unregulated deamination to pyruvate at a remarkable rate and L-glutamate is converted stoichiometrically to L-aspartate, which, because of the absence of aspartase, accumulates in the culture medium. It is

probably this lesion that accounts for the requirement for CO_2 to regenerate the oxaloacetate pool lost by transamination to L-aspartic acid (Dreyfus and Brubaker, 1978; Perry and Fetherston, 1997).

Anabolism

Y. pestis grows slowly at $\sim 5°C$ to $\sim 28°C$ on medium containing appropriate inorganic salts, including sulphite or thiosulphate as a source of sulphur, a fermentable carbohydrate, L-methionine, L-phenylalanine, and either glycine or L-threonine. Addition of L-isoleucine and L-valine stimulates growth but is generally not essential. The basis for an inability to assimilate low levels of NH_4^+ into organic linkage has not been defined. Nutritional requirements at 37°C are more complex and may depend on unresolved variables such as inorganic salt content, ionic strength and osmolarity. Variation in typically conserved processes of macromolecular synthesis is not known to occur in *Y. pestis* (Perry and Fetherston, 1997).

Global Regulation

The organisms are stringent as judged by the ability to express guanosine tetra- and pentaphosphate after starvation for L-phenylalanine. High-frequency RecA-dependent loss of the Pgm locus (Hare and McDonough, 1999), and ability of ultraviolet light or radiomimetic compounds to promote and induce pesticin, indicate the presence of typical DNA repair systems. Detection of a cAMP-binding protein sequence in regulation of iron assimilation by the Fur system has been reported (Staggs and Perry, 1992). The nutritional requirement for L-isoleucine, L-valine, glycine (or L-threonine) but not L-serine coupled with unrestrained deamination of the latter and inability to assimilate NH_4^+ would be expected if *Y. pestis* has a defective global leucine-responsive regulatory system. A chromosomal sequence corresponding to lux has been detected (*Yersinia pestis* Sequencing Group, 1998), suggesting that the organisms are capable of quorum sensing.

Structure

Cells of *Y. pestis* are small, stout but otherwise typical Gram-negative rods during growth in adequate culture media or *in vivo*. The genes of the enteropathogenic yersiniae that regulate or encode flagella and the lipopolysaccharide (LPS) O-group structure are generally expressed at 26°C but not at 37°C, and are either absent or entirely cryptic in *Y. pestis* (Federova and Devdariani, 1998). As a consequence, cells of the latter are always non-motile and morphologically rough at room temperature, although at host temperature the colonies may exhibit a smooth appearance because of the temperature-dependent production of fraction 1 antigen. Cell-free extracts of *Y. pestis* and *Y. pseudotuberculosis*, but not *Y. enterocolitica*, prepared after growth at 37°C have a marked red-brown colour because of the presence of KatY, a major protein with catalase-peroxidase activity (Garcia *et al.*, 1999).

Colonial morphology of *Y. pestis* is undistinguished after growth at 37°C, whereas incubation at 26°C permits absorption of certain exogenous small planar pigmented molecules including haemin and its analogue Congo red. This attribute, which confers the Pgm⁺ epithet (**Fig. 1a**), is encoded by a haemin storage locus (Hms) within the deletable sequence. It should be noted, however, that this locus also contains a distinct cluster of genes encoding high-affinity assimilation of iron, and its loss has pleiotropic consequences. It is now common practice to use the term Pgm⁻ for isolates that lack the entire 102-kb Pgm locus, and to designate strains lacking only the ability to absorb haemin or Congo red as Hms⁻ (Perry *et al.*, 1990) or Crb⁻ (Hare and McDonough, 1999).

Disease

Regardless of the host, plague is characterised by an initial phase of tissue invasion in which the bacteria gain access to the viscera from the dermis by way of the bloodstream after introduction by a flea bite. This may be traumatic as evidenced by extreme lymphadenopathy in humans, but proliferation in earnest does not commence until the organisms invade liver and spleen. Visceral invasion can be hastened by intravenous injection when the bacteria are immediately filtered from the vascular system by passage through the capillary beds of these organs. Similarly, the invasive process is bypassed on primary lung involvement, which leads to prompt death due to destruction of a critical organ. In the absence of pneumonic involvement, the terminal phase of the disease is always marked by pronounced septicaemia. This ominous symptom is not a result of multiplication within the vascular system. Instead, it reflects spillage of the bacteria into the circulation from massive necrotic lesions contained within the remnants of visceral organs.

Experimental Murine Plague

Yersiniae are recognised parasites of lymphoid tissue but in the acute disease the bacteria use lymphatic vessels primarily to reach internal organs that serve as major sites of multiplication. Thus, the virulence factors of *Y. pestis* function either to promote dissemination of the bacteria from peripheral sites of infection to the viscera or to facilitate bulk growth once these favoured niches are reached.

Invasion of Tissues

The remarkable infectivity of wild-type *Y. pestis* on peripheral administration is not shared by *Y. pseudotuberculosis* (Brubaker *et al.*, 1965) although, as already noted, the latter is of comparable virulence when injected intravenously (**Table 2**). The basis for this distinction between routes of administration is that, on intravenous injection, both organisms are immediately removed from the vascular system by the liver and spleen. Clearance in these organs is efficiently mediated by fixed macrophages, as well as physical filtration. Since *Y. pestis* and *Y. pseudotuberculosis* are facultative intracellular parasites (Perry and Fetherston, 1997), they promptly emerge within interstitial spaces to form localised foci of infection

(Une *et al.*, 1986; Straley and Cibull, 1989; Nakajima *et al.*, 1995). This sequence does not occur after intraperitoneal, subcutaneous or intradermal injection, where the enteropathogenic species can be immobilised and eliminated at the original site of infection. In contrast, peripherally injected cells of *Y. pestis* have a unique ability to invade adjacent tissues and eventually reach the viscera.

Inhibition of Inflammation

Mammalian hosts respond to bacterial invasion or the insult of injected foreign biological matter by the prompt up-regulation of major pro-inflammatory cytokines, especially interferon γ (IFNγ) and tumour necrosis factor α (TNFα). This generic inflammatory response serves as a non-specific mechanism of host defence designed to eliminate foreign matter by facilitating activation of professional phagocytes and formation of protective granulomas. In this regard pCD$^-$ yersiniae are the same as *E. coli* and other common bacteria, as judged by their ability promptly to induce IFNγ and TNFα on intravenous injection into mice (**Fig. 3a**). Intravenously injected pCD$^+$ cells, however, down-regulate host expression of IFNγ throughout the course of disease, but some TNFα is produced after mice become moribund (**Fig. 3b**). This astonishing ability of pCD to abrogate inflammation is reversible if the host is primed with trace levels of IFNγ plus TNFα before infection (**Fig. 3c**) or by passive immunisation with a polyclonal antiserum against pCD-encoded LcrV (**Fig. 3d**), a known protective antigen (see below). In both of cases, the animals survive the infection (Nakajima and Brubaker, 1993).

This shows that plague is not an uncompromising struggle between host and parasite, traditionally seen as analogous to trench warfare. Rather, it is a function of deception as shown by the ability of *Y. pestis* effectively to disguise ongoing tissue destruction and thereby maintaining the illusion that the host has no cause for alarm.

Favoured *in vivo* Niches

After invading the viscera indirectly via lymphatic vessels and solid tissue, or directly by intravenous injection, yersiniae assume residence in the spleen and liver, and begin to multiply (**Fig. 4**). Thereafter, the lungs are invaded and replication progresses until organ function is lost and the host becomes moribund. At this stage, there is a significant net accumulation of bacteria in the vascular system, which ensures that departing fleas are infected. Intravenously injected pCD$^-$ yersiniae are similarly filtered from the liver and spleen where they initiate growth. However, proliferation at these sites soon ceases, because the bacteria are eliminated without re-entering the vascular system (Une *et al.*, 1986; Nakajima *et al.*, 1995).

Examination of thin sections of liver and spleen reveals that pCD$^+$ isolate grow extracellularly within enclosed non-vascularised necrotic foci that progressively enlarge and then coalesce as the infection continue (**Fig. 5a**). At no time does the host mount a detectable inflammatory response against the bacteria

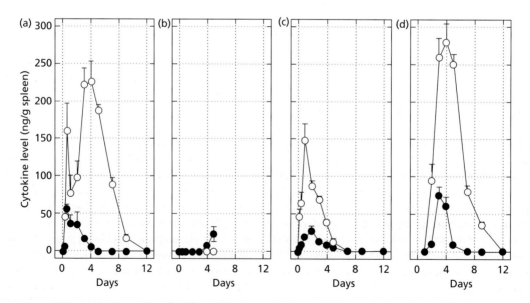

Fig. 3 Expression of the pro-inflammatory cytokines IFNγ (open circle) and TNFα (closed circle) in C57/BL/c mouse spleen after intravenous challenge with (a) 10^6 pCD$^-$ cells of *Y. pestis* KIM10; (b) 10^2 pCD$^+$ cells of *Y. pestis* KIM10; (c) 10^2 pCD$^+$ cells of *Y. pestis* KIM10 after priming with 20 μg of IFNγ and 20 ng of TNFα; and (d) 10^2 pCD$^+$ cells of *Y. pestis* KIM10 upon passive immunisation on post-infection day 1 with 100 μg of polyclonal rabbit anti-LcrV. Redrawn from Nakajima *et al.* (1993).

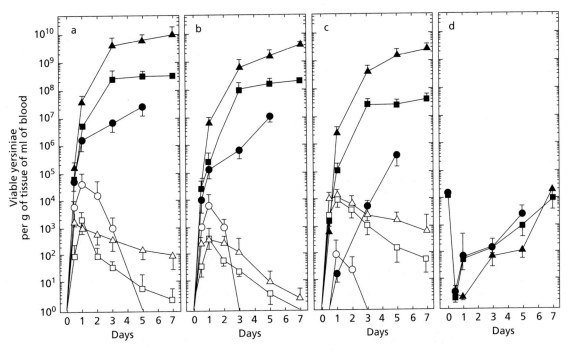

Fig. 4 Distribution of pCD⁺ (closed circles) and pCD⁻ (open circles) *Y. pestis* KIM10, pYV⁺ (closed squares) and pYV⁻ (open squares) *Y. pseudotuberculosis* PB1, or pCD⁺ (closed triangles) and pCD⁻ (open triangles) *Y. enterocolitica* WA in mouse (Swiss-Webster) spleen (a), liver (b), lung (c) and blood (d) following intravenous injection of ~ 10⁴ viable bacteria. All mice infected at this dose with *Y. pestis* expired on post-infection day 6 whereas those injected with *Y. pseudotuberculosis* or *Y. enterocolitica* typically survived for 7 and 11 days, respectively. Isolates of *Y. pestis* were *Pgm⁻* and thus unable to synthesise Ybt. From unpublished observations of T. Une and R.R. Brubaker.

contained within these lesions, which eventually haemorrhage, resulting in net accumulation within the vascular system. Intravenously injected pCD⁻ yersiniae, like the pCD⁺ parent, also assume residence in the visceral interstitium, but the resulting lesions are almost immediately surrounded by inflammatory cells that promptly facilitate the formation of protective granulomas (**Fig. 5b**) (Une *et al.*, 1986; Nakajima *et al.*, 1995).

Throughout the course of the disease, bulk vegetative growth occurs exclusively in spleen, liver, and then the lung. These organs become consumed and are almost stoichiometrically converted to bacterial mass. Accordingly, the ability of yersiniae to resist uptake by neutrophils and monocytes within the vascular system is of minimal importance as a mechanism of virulence, because significant septicaemia does not occur until the host has already become moribund. These observations do not, of course, preclude a role for professional phagocytes in delaying invasion of the viscera after peripheral administration or infection of the dermis by flea bite.

Mortality

The terminal population of enteropathogenic yersiniae in moribund mice is typically about 10 times that

observed for *Y. pestis* which, at comparable challenge doses, also causes an earlier death (**Fig. 4**). The possible role of activities unique to *Y. pestis* in assuring early death is discussed below.

Human Plague

It is sometimes suggested that *Y. pestis* has somehow become attenuated in modern times and is no longer the scourge accountable for the Black Death. This assumption is entirely fallacious, as observed by physicians confronted by recent epidemics, such as those that occurred during the Vietnam War. For example, Thomas Butler noted that, 'We were humbled by the swift progression of this disease that carried healthy individuals to death within three days' (Butler, 1983).

Other than pronounced fever and headache, the most typical symptom of untreated plague in humans is oedema and tenderness of the regional lymph nodes adjacent to the initial flea bite. These cervical, femoral/inguinal or axillary lesions are termed buboes; they delay but generally fail to prevent eventual invasion of the viscera, which is necessary for the subsequent development of the marked septicaemia characteristic of plague. Rodents typically expire before a comparable lymphadenopathy can develop, but the succession

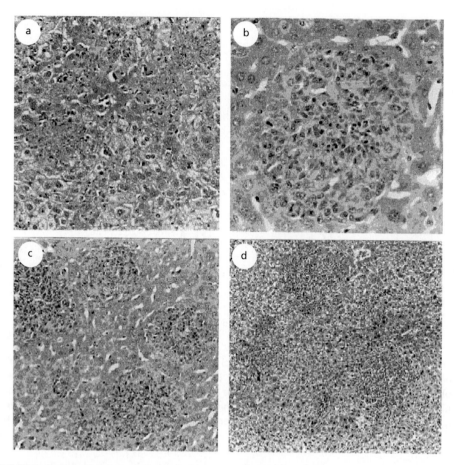

Fig. 5 Characteristic histopathological changes in mouse liver caused by *Y. pestis* KIM10 on post-infection day 3. Haematoxylin and eosin stain were used. (a) pCD$^+$ organisms in control mouse showing multiple necrotic focal lesions without inflammatory cell response (magnification ×140). (b) pCD$^-$ organisms in control mouse exhibiting granuloma formation (magnification ×280). (c) pCD$^+$ organisms in mouse actively immunised with PAV showing protective granulomatous lesions (magnification ×140). (d) pCD$^+$ organisms in mouse passively immunised with polyclonal rabbit anti-PAV showing pre-granulomatous lesions prompting infiltration of inflammatory (mononuclear) cells (magnification ×70). From Nakajima *et al.* (1995).

of events in both hosts is otherwise similar. Indeed, filtration of the invading bacteria by regional lymph nodes may sometimes entirely fail in humans, in which case visceral invasion and subsequent bacteraemia occurs promptly. This type of disease, termed septicaemic plague, is noted for a higher mortality rate (∼90%) than the bubonic form (∼50%).

An even more unfortunate and almost uniformly lethal variation of the human disease is pneumonic plague, which is characterised by fever, cough and chest pain. This syndrome is especially severe, and exposed individuals may become ill and die on the same day. The appearance of pneumonic plague during an epidemic is especially ominous, because this form of the disease can be transmitted directly by droplet without the flea vector. Even pneumonic plague, however, can be cured by antibiotics if appropriate prophylactic therapy is initiated 20 hours before the onset of symptoms (Butler, 1983).

Virulence Factors

As outlined above, plague bacilli first reach favoured visceral multiplication sites from the dermis and then undergo massive proliferation with eventual loss of organ function, accompanied by spillage into the vascular system. It is apparent that determinants of virulence facilitating the first invasive phase of the disease are not always required for the second vegetative phase and visa versa. It is also evident that the three pathogenic species of yersiniae are capable of causing acute disease after intravenous injection but only *Y. pestis* has the means efficiently to access internal organs from peripheral sites of infection. The journey from dermis to viscera is tumultuous as judged by the larger inocula of enteropathogenic yersiniae required to cause death by subcutaneous or intradermal administration (**Table 2**). Clearly, the plague bacillus easily defeats the non-specific host defence mechanisms that successfully waylay the enteropathogenic yersiniae *en route* or prevent their exit from the original site of injection. Indeed, the virulence factors that overcome these initial defensive measures largely account for the acute nature of bubonic plague.

Resistance to Non-specific Host Defence Mechanisms

Even bacterial pathogens of modest virulence are resistant to host extracellular antimicrobial activities such as lysozyme and β lysin, and yersiniae pathogenic to humans are no exception. Resistance to complement-mediated killing is constitutive in *Y. pestis* as opposed to *Y. pseudotuberculosis*, where it occurs in both pYV$^+$ and pYV$^-$ cells cultivated at 37°C but not 26°C. In contrast, only pYV$^+$ cells of *Y. enterocolitica* grown at 37°C are serum-resistant (Une and Brubaker, 1984a). The evidence for adhesin- and invasin-mediated resistance to complement in enteropathogenic yersiniae is compelling (Finlay and Falkow, 1997). These proteins are, however, not produced by *Y. pestis*, instead expression of rough LPS that lacks an O-group structure correlates with serum-resistance. This association has been confirmed by the inability of complement to lyse liposomes prepared with LPS from serum-resistant phenotypes of all three pathogenic species of *Yersinia* (Porat *et al.*, 1995).

Fraction 1 antigen of *Y. pestis* confers resistance to uptake by certain professional phagocytes, especially neutrophils and monocytes, but is evidently ineffective against fixed macrophages that line the capillary network of liver and spleen. This limited ability to provide resistance to phagocytosis *in vivo* probably explains why mutational loss of fraction 1 delays, but does not reduce, mortality in the mouse (Burrows, 1963). However, LcrV inhibits even neutrophil chemotaxis (Welkos *et al.*, 1998). Once phagocytosed, enteropathogenic yersiniae cured of pYV fare poorly, while those of *Y. pestis* lacking pCD survive at low multiplicities in macrophages just as well as the pCD$^+$ parent. At higher multiplicities, pCD/pYV-mediated cytotoxins assure prompt destruction of both professional and non-professional phagocytes (Perry and Fetherston, 1997).

Visceral Invasion

High-frequency deletion of the Pgm locus results in selective reduction of virulence by the intraperitoneal (Jackson and Burrows, 1956) but not intravenous route of injection (Une and Brubaker, 1984b). This lesion can be repaired by concomitant injection of sufficient Fe^{3+} to saturate serum transferrin,

suggesting that the locus encodes a mechanism required for iron assimilation *in vivo* (Jackson and Burrows, 1956). Similarly, loss of pPCP in Pgm$^+$ isolates results in reduced virulence by peripheral routes of injection (**Table 2**). This was interpreted as evidence that the plasminogen activator (Pla) encoded by pPCP was also necessary to provide rapid access from the dermis to liver and spleen (Brubaker *et al.*, 1965).

Plasminogen Activator

pPCP encodes the bacteriocin pesticin (Ben-Gurion and Hertman, 1958), its immunity protein (Pilsl *et al.*, 1996) and a particulate fibrinolysin (Madison, 1936) that activates plasminogen (Beesley *et al.*, 1967). The latter is translated as a transient 34.6-kDa protein that undergoes processing on insertion into the outer membrane to yield a 32.6-kDa product termed α Pla. This form is then converted, probably by shearing, in ~ 2 hours to a slightly smaller derivative (β Pla) (Mehigh *et al.*, 1989; Sodeinde and Goguen, 1988). Both forms bind tenaciously to the outer membrane of *Y. pestis*, whereas cells of *E. coli* transformed with pPCP produce only α Pla. Biologically active Pla can readily be extracted from pPCP$^+$ *E. coli* in soluble form with 1 M NaCl as a mixture of α Pla, β Pla, and smaller forms (Kutyrev *et al.*, 1999). The reason for this discrepancy is not fully understood but Pla possesses significant homology with OmpT of *E. coli* (Sodeinde and Goguen, 1989) and may, therefore, undergo faulty anchoring in this species. Tissue invasion is accomplished by Pla-dependent adherence to host basement membrane and extracellular matrix, where plasminogen activation facilitates bacterial metastasis (Lähteenmäki *et al.*, 1998).

Assimilation of Fe^{3+}

Of the many inorganic ions that are required for life, only iron is withheld during infection as a non-specific mechanism of host defence (see Chapter 32). This sequestration is associated with fever and is characterised by the removal and intracellular deposition of transferrin-bound iron, which reduces the extracellular level of Fe^{3+} to a concentration that is insufficient to permit bacterial growth (Weinberg and Weinberg, 1995). Jackson and Burrows (1956) correctly assumed that avirulence of Pgm$^-$ yersiniae administered by peripheral routes *in vivo* reflects a lesion in assimilation of iron, as judged by the ability of injected Fe^{+3} or haemin to enhance lethality in mice. Verification of this notion required considerable effort because Pgm$^-$ yersiniae grow well in marginally iron-deficient media. However, Pgm$^-$ isolates fail to sustain growth in an extremely iron-deficient medium after repeated passage at 37°C; curiously, multiplication at 26°C is not inhibited in this environment (Sikkema and Brubaker, 1987).

It is established that yersiniae pathogenic for humans can utilise at least three high-affinity mechanisms to scavenge Fe^{3+} in the febrile host. In *Y. pestis*, these processes consist of a siderophore-independent yersiniae ferric uptake (Yfu) system (Braun, 1997; R.D. Perry, personal communication), a siderophore-independent ABC transporter (Yfe) system (Bearden *et al.*, 1998), and a siderophore (yersiniabactin or Ybt)-dependent process. An additional ill-defined mechanism evidently functions in *Y. pestis* grown at 26°C but not 37°C (Sikkema and Brubaker, 1987; Lucier *et al.*, 1996). Finally, distinct genes encode products that enable *Y. pestis* to utilise haemin and haemoproteins directly as nutritional sources of iron (Hornung *et al.*, 1996). Loss of this system for utilising organic forms of iron does not reduce virulence, at least in the mouse (Thompson *et al.*, 1999).

It is the Ybt-dependent process of iron uptake discovered by Heesemann (1987) that is encoded within the Pgm locus (Guilvout *et al.*, 1993; Rakin *et al.*, 1994; Bearden *et al.*, 1997; Buchrieser *et al.*, 1998b). Members of this new class of siderophores generally have a lower affinity for Fe^{3+} (Ybt $K_d > 4 \times 10^{-36}$) than environmental siderophores (i.e. enterobactin $K_d > 10^{-52}$) (Gehring *et al.*, 1998). Indeed, Ybt is not detected by methods used to determine the stronger ligands (Perry and Brubaker, 1979). The discovery of two high-molecular-weight proteins, HMWP1 and HMWP2 (Carniel *et al.*, 1989; Guilvout *et al.*, 1993), provides the key for defining the synthesis of Ybt.

The 11 genes involved in the Fur-regulated (Staggs and Perry, 1992) ribosome-independent biosynthesis of Ybt constitute a ~23-kb high-pathogenicity island (Carniel *et al.*, 1996; Buchrieser *et al.*, 1998a, 1999; Hare and McDonough, 1999) located at the right-hand end of the Pgm locus (**Fig. 1a**). These genes comprise synthetic (*irp2–ybtE*), regulatory (*ybtA*), and transport (*ybtP–ybtS*) operons. *psn* encodes the unprocessed outer membrane receptor for Ybt, while *ybtP* and *ybtQ* encode the inner membrane permeases evidently needed for manipulation of the charged complex. YbtX may be involved in maintaining tertiary structure, as judged by its hydrophobicity and transmembrane domains. YbtS, YbtT and YbtE are concerned with salicylate, thioesterase, and salicyl-AMP metabolism. YbtA is an AraC-like regulator that promotes transcriptional activation of *psn*, *irp2* and *ybtP*. The role of YbtU is unknown, but this protein is necessary for biosynthesis (Gehring *et al.*, 1998). The construction of Ybt reflects modular passage of covalent intermediates from the N-terminus of the *irp2* product (HMWP2) to the C-terminus of the *irp1* product (HMWP1). *S*-Adenosylmethionine and malonyl-CoA, respectively, provide the methyl moieties and linker between the thiazoline and thiazolidine rings (Gehring *et al.*, 1998).

The phenotype of Pgm⁻ mutants lacking the entire 102-kb locus is mimicked, with respect to virulence, by isolates with point mutations in salient genes comprising the Ybt high-pathogenicity island (Bearden *et al.*, 1997). This confirms that the ability to express and utilise Ybt *per se* can account for the penchant of *Y. pestis* to invade the viscera from peripheral sites of infection. The reason why Ybt is dispensable following infection by intravenous injection is not fully resolved. One likely possibility is that host-cell cytoplasm, potentially rich in ferritin-bound and organic iron, is promptly accessed after intravenous injection. It is, therefore, significant that mutational loss of the siderophore-independent Yfe or ABC transporter system by Pgm⁻ yersiniae, which are unable to synthesize Ybt, results in complete loss of virulence ($LD_{50} > 10^7$ bacteria) by all routes of injection (Bearden and Perry, 1999).

The virulence in mice of mutants cured of pPCP, like those lacking the Pgm locus, is also restored on peripheral injection by the administration of sufficient iron to saturate serum transferrin (Brubaker *et al.*, 1965). This was unexpected, because loss of Pla is not known to influence the ability of yersiniae to assimilate Fe^{3+}. It is now recognised that injected, but not necessarily ingested, iron can seriously interfere with a number of non-specific mechanisms of host defence, and that this probably accounts for the ability of the cation to enhance the invasiveness of mutants lacking Pla. Indeed, injected iron restored virulence even to a mutant lacking otherwise essential pCD-encoded cytotoxic YopE (Mehigh *et al.*, 1989).

pPCP–Pgm Interaction

As noted above, pPCP encodes the bacteriocin pesticin and its immunity protein in addition to Pla. Pesticin is not directly involved in the expression of disease, but it provides important insights into the nature of iron transport and the role of the Pgm locus. Its mechanism of killing target bacteria is unusual in that it promotes the generation of enormous, osmotically stable spheroplasts by hydrolysing peptidoglycan (Hall and Brubaker, 1978). The conclusion that this cleavage reflects *N*-acetylglucosaminidase activity (Ferber and Brubaker, 1979) was fallacious; use of modern methods revealed that the bacteriocin is actually a muramidase (Vollmer *et al.*, 1997). Pesticin is active against *Y. pestis* that has lost pPCP, and thus its immunity protein, while retaining the Pgm locus. It can also kill serotype 1 strains of *Y. pseudotuberculosis*, serotype O:8 strains of *Y. enterocolitica*, and certain clinical isolates of *E. coli* (Brubaker and Surgalla, 1961; Perry and Fetherston, 1997) now known to synthesise Ybt. The common denominator conferring sensitivity to pesticin is, therefore, the ability to express Psn, the Ybt receptor (Buchrieser *et al.*, 1998a; Perry and Fetherston, 1997).

It is generally established that many *E. coli* bacteriocins (colicins) are unable to kill otherwise sensitive bacteria grown in excess iron. The reason for this inhibition is that Fe^{3+} represses Fur-mediated functions, including outer membrane siderophore receptors that are utilised by the colicins in question (Davis and Reeves, 1975). This phenomenon was first observed with pesticin (Brubaker and Surgalla, 1961) and, with the discovery of Ybt, was shown to reflect down-regulation of Psn by iron, which prevents absorption of the bacteriocin (Fetherston *et al.*, 1995). Selection for pesticin-resistance in the absence of iron would logically be expected to yield avirulent Psn^- mutants unable to assimilate Ybt. However, when this was tested, all of the recovered isolates had lost the entire Pgm locus within which *psn* resides (Brubaker, 1970).

While this provides a convenient method to establish the mutation rate to Pgm^-, the value was so high (10^{-5}) that survivors lacking only Psn could not be detected. This difficulty was resolved by co-selecting for spontaneous resistance to pesticin and retention of the Pgm locus-encoded Hms^+ trait defined above as the ability to absorb haemin at 26°C. One such isolate could still bind this pigment but was unable to grow at 37°C in iron-deficient medium (Sikkema and Brubaker, 1989). The DNA sequence of the Pgm locus of this mutant contained a point mutation in *psn* identified as a 5bp deletion (GACCT), which causes premature termination of translation (Lucier *et al.*, 1996). This sequence occurred immediately after another GACCT run, and its removal probably reflects a rare error in DNA replication.

This emphasises that spontaneous point mutations can occur in the Pgm locus, but that their presence is masked by its RecA-dependent high-frequency deletion (Hare and McDonough, 1999). This conclusion is, however, at variance with results obtained in another study where Hms^- mutants of laboratory strains that retained *irp2* in the high-pathogenicity island were readily isolated, although the reciprocal was not detected (Iteman *et al.*, 1993). Loss of the Hms^+ phenotype in these strains was attributed in part to smaller deletions encompassing all or portions of the Hms region or to point mutations therein (Buchrieser *et al.*, 1998a). Later work verified that

pesticin-resistant mutants of a $RecA^+$ Pgm^+ strain of *Y. pestis* cured of pPCP arise at very high frequency and lack the entire Pgm locus. Hms^- mutants, which usually retained at least portions of the high-pathogenicity island, arose in an isogenic $RecA^-$ isolate at a lower frequency, so that it would not be possible to detect in a $RecA^+$ background. The $RecA^-$ isolate also yields rare pesticin-resistant mutants that were typically Hms^+ (Hare and McDonough, 1999). One explanation to account for the difference between the KIM strain and laboratory stocks capable of more frequent mutation in the Hms locus (Buchrieser *et al.*, 1998a) is that the latter may have accumulated additional IS elements or homologous sequences during storage (Hare and McDonough, 1999). These findings illustrate that the Pgm locus, and surrounding DNA, possesses considerable plasticity. As will be considered below, portions of this chromosomal region may have been acquired recently during the evolution of *Y. pestis*.

Vegetative Growth

Studies of the pathology of murine and human plague outlined above provide three important insights. First, the course of disease is similar in both hosts, except that the mouse succumbs before comparable lymphadenopathy develops. Second, death in both cases occurs after the bacteria have undergone numerous doublings in visceral organs. Third, the ability of yersiniae to sustain this prolonged assault depends on carriage of pCD, as is shown by the observation that even massive doses of mutants cured of this plasmid, or lacking certain genes encoded on it, are rapidly cleared from liver and spleen (**Fig. 4**). This vegetative phase of the disease involves destruction of the tissues of these organs and subsequent metabolism of released nutrients. Cell disintegration is probably accomplished exclusively by a powerful pCD-encoded type III protein secretion system that translocates cytotoxic Yops from the cytoplasm of docked yersiniae into that of target cells. This process of 'injection' occurs without exposure to the extracellular environment and is tightly regulated by temperature, metal cations and ability successfully to form an intact structural unit required for translocation. Thus the type III secretion machine is the engine of virulence in all yersiniae pathogenic to mammals (Cornelis, 1998; Cornelis *et al.*, 1998).

PCD

This shared ~ 70 kb PCD plasmid encodes an 'anti-host genome' (Cornelis *et al.*, 1998) capable of mediating destruction of mammalian tissue while maintaining the deception that necrosis is under control and that the healing process may commence. *Y. pestis* that harbours pCD exhibits a 'low-calcium response' (LCR) during cultivation in typical culture media rendered modestly deficient in Ca^{2+} (~ 1.0 mM) at 37°C but not 26°C. The LCR phenotype is characterised by bacteriostasis, up-regulation of the type III secretion system, and release of virulence effectors (cytotoxic Yops and LcrV). Addition of sufficient Ca^{2+} to bring its concentration to that in mammalian plasma (2.5 mM), down-regulates these virulence effectors and their secretion system, while restoring vegetative growth. Equimolar Sr^{2+} or Zn^{2+} can replace Ca^{2+}, whereas the LCR is potentiated by addition of the concentration of Mg^{2+} present in mammalian cytoplasm (20 mM). *E. coli* transformed with pCD does not exhibit the complete LCR phenotype, indicating that as yet undefined chromosomal genes are necessary for its full expression (Perry and Fetherston, 1997).

Organisation and Regulation

Bacteriostasis at 37°C in Ca^{2+}-deficient medium and synthesis of virulence effectors are closely intertwined features of the LCR. Restriction of vegetative growth is associated with a reduction of adenylate energy charge that does not immediately influence viability, as is shown by initiation of cell division on return to permissive conditions, such as shift to 26°C or addition of Ca^{2+}. Ca^{2+} itself is not required for multiplication, because pCD^+ yersiniae exhibit almost full-scale growth at 37°C without the cation, provided that both Na^+ and L-glutamic acid are also eliminated (Fowler and Brubaker, 1994). The basis of this relationship is not clear, but it may be significant that Na^+ is a common anti-porter for L-glutamate which, as already noted, is stoichiometrically converted by *Y. pestis* to inert L-aspartate, causing loss of metabolic carbon (Dreyfus and Brubaker, 1978). The phenomenon of Ca^{2+}-dependence is, therefore, probably not directly associated with the expression of virulence *per se* but mutants exhibiting altered responses to the cation have provided powerful tools for probing the LCR.

pCD encodes about 50 structural genes that mediate the LCR and numerous transposases and non-coding elements (Hu *et al.*, 1998). Many of these genes were initially thought to promote regulatory (*lcr*), type III secretion (*ysc*), virulence effector (*yop*), or chaperone (*syc*) activities; additional categories exist. These epithets have often been retained in cases where the product was later found to fulfil some distinct or bifunctional role. Mutational loss of certain pCD-encoded genes causes a 'Ca^{2+}-independent'

phenotype similar to that of isolates cured of pCD, where growth at 37°C is constitutive and effectors of virulence are super-repressed. Examples of this Ca^{2+}-independent phenotype are isolates lacking *lcrV*, *lcrF* (*virF*), and core secretion genes such as *yscV* (*lcrD*). 'Ca^{2+}-blind' mutants are a second class of regulatory mutant phenotype in which bacteriostasis and production of virulence effectors are constitutive at 37°C in the presence or absence of Ca^{2+}. Some important mutants of this phenotype are those lacking *yopD*, *lcrG*, *lcrQ* or *lcrE* (*yopN*).

The transcription of genes that encode virulence effectors is activated by LcrF (Lambert de Rouvroit *et al.*, 1992) after its thermal induction on DNA melting within specific domains of pYV/pCD (Rohde *et al.*, 1999). The expression of these virulence factors at 37°C is down-regulated by Ca^{2+}, which indirectly prevents their secretion and thereby promotes a form of negative feedback inhibition. Removal of Ca^{2+} from the environment, especially in excess Mg^{2+}, or intimate contact with host cells in the presence of Ca^{2+} eliminates this block, whereupon LcrF-regulated determinants undergo prompt synthesis and secretion. Ca^{2+} may down-regulate their exit in the absence of host cell contact by stabilising an outer gate LcrE- and TyeA-mediated block of Ysc translocation channels. Elimination of this block by either withdrawal of Ca^{2+} or docking to host cells facilitates dissipation of negative control on secretion of LcrQ (a negative regulator) and YopD. The latter, in concert with YopB and LcrV, then forms a translocation pore modulated by YopK. It is through this pore, which is raised from the basic Ysc foundation, that pCD-encoded cytotoxic Yops undergo translocation to targets located in the host cell cytoplasm (Perry and Fetherston, 1997; Cornelis *et al.*, 1998; Fields and Straley, 1999). A diagram illustrating this model is provided in **Fig. 6** (Fields *et al.*, 1999).

lcrV is located in a pCD/pYV-encoded *lcrGVH–yopBD* operon and, as noted above, its non-polar loss results in acquisition of Ca^{2+}-independence, accompanied by down-regulation of LcrF-mediated functions. This effect does not reflect a direct role in regulating the LCR at the level of DNA, but rather results from feedback inhibition caused by the accumulation of free internal LcrG, a cytoplasmic inner gate protein (Nilles *et al.*, 1997). Removal of this impediment by titration with internal LcrV, clears the passage through which effector Yops enter the host cell (Nilles *et al.*, 1998). LcrV thus functions both as a structural component of the outer portion of the injection apparatus and as a cytoplasmic element required for retraction of LcrG from the inner gate of the secretion pore (**Fig. 6**). In addition, LcrV

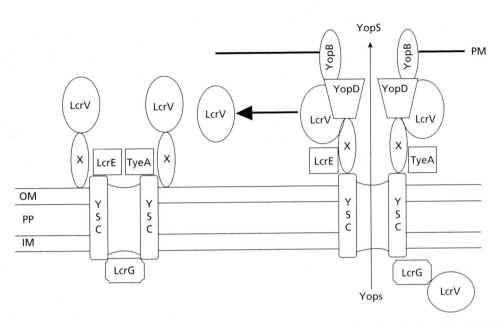

Fig. 6 Model for the role of surface-located pCD-encoded proteins in mediating translocation of effector Yops. The Ysc creates a secretion channel through the bacterial inner membrane (IM), across the periplasm (PP), and through the outer membrane (OM). These pores remain blocked in the absence of direct contact with the eukaryotic cell plasma membrane (PM) by LcrE, TyeA and LcrG. LcrV is associated with the extracellular surface of these closed pores directly or by association with another protein (X). Upon host cell contact, LcrE, TyeA and LcrG are displaced, and LcrV mediates formation of a delivery apparatus for transfer of Yops directly from the bacterium to the eukaryotic cell cytoplasm. From Fields *et al.* (1999).

is translocated to host cell cytoplasm by an entirely distinct process mediated by chromosomal genes (Fields and Straley, 1999). As a consequence of this process, as well as its dissipation from the injection apparatus, considerable extracellular LcrV accumulates *in vitro* (Lawton *et al.*, 1963) and *in vivo* (Smith *et al.*, 1960).

It is important to recognise that the literature abounds with reports demonstrating that the translocation of Yops by the enteropathogenic yersiniae is markedly enhanced if not dependent on expression of one or more of the surface ligands YadA, Inv or Ail. As already noted, none of these adhesins is expressed by *Y. pestis*, and thus comparison of Yop delivery between this species and the enteropathogenic yersiniae is tenuous. Further study may demonstrate that *Y. pestis* utilises Pla or some other adhesin to maintain analogous host cell attachment. For example, another prospect for this role is a putative adhesin encoded in the Pgm locus, but outside the high-pathogenicity island and the Hms locus (Buchrieser *et al.*, 1999).

Cytotoxic Yops are thus sequestered from blood, lymph and interstitial fluid during their translocation from docked yersiniae to host cell cytoplasm. This mechanism of transfer is efficient and insidious, in that it

assures partitioning from effectors of the immune system. Consequently, specific antibodies directed against these virulence effectors are unable to prevent cytotoxicity, even if they could be raised sufficiently early during the course of infection. This form of compartmentalisation provides a parameter of stealth not shared by pathogenic bacteria that release exotoxins directly into host extracellular fluids.

Yops

The term 'Yop' was initially coined to define major pYV-encoded outer membrane-associated proteins discovered in enteropathogenic yersiniae cultivated at 37°C in Ca^{2+}-deficient media (Portnoy *et al.*, 1981; Straley and Brubaker, 1981). Further study revealed, however, that Yops lack the typical leader sequences of generic outer membrane proteins and that their association with the bacterial surface reflects denaturation with fortuitous non-specific attachment. The term Yop is now reserved for the established major pCD/pYV-encoded proteins that are either virulence effectors or fill major roles in their translocation. Specific chaperones are required to mobilise Yops for type III secretion. However, convincing evidence has also accrued to indicate that at least YopE is recognised and then translocated on the basis of a unique non-coding mRNA configuration, rather than primary amino acid structure (Lee and Schneewind, 1999).

At least six Yops directly or indirectly promote symptoms of disease. These include YopE, which is essential for lethality in mice and YopT, which is not. Both of these proteins cause HeLa cell rounding associated with actin rearrangement. YopH is an essential virulence factor with protein-tyrosine phosphatase activity that prevents assembly of the macrophage focal adhesions required for phagocytosis. YopH also fosters visible host cell cytotoxicity. YopJ (YopP in *Y. enterocolitica*) is essential for full virulence of *Y. pseudotuberculosis* in mice where it binds directly to salient members of the mitogen-activated protein kinase superfamily (Orth *et al.*, 1999), thereby down-regulating TNFα with concomitant macrophage apoptosis (Monack *et al.*, 1997, 1998; Palmer *et al.*, 1998). However, at least in the murine system, YopJ of *Y. pestis* is dispensable for lethality (Straley and Bowmer, 1986). It is not clear whether this discrepancy reflects some difference in Yop translocation between the two species or whether plague bacilli cause death in mice before significant intervention by macrophages can occur. Further study will almost surely demonstrate that YopJ also fulfils a critical role during infection of the more resistant rodents indigenous to natural sylvatic reservoirs and humans. Mutants lacking YopM also retain significant virulence

in mice. The physiological role of this protein and the consequences of its loss in other hosts are unknown. YopO (YpkA in *Y. pseudotuberculosis*) is necessary for lethality in the murine model, and possesses serine/threonine kinase activity (Perry and Fetherston, 1997; Cornelis, 1998; Cornelis *et al.*, 1998).

Although YopB and YopD are established components of the type III secretion machine, they also have independent abilities to form pores in membranes. This effect is particularly striking in the case of YopB, which possesses haemolytic activity. The injection apparatus or needle shown in **Fig. 6** may exist in a dynamic state undergoing continuous reassembly (Fields *et al.*, 1999). Like LcrV, release of free YopB and YopD during this process may, therefore, contribute to the overall cytotoxic effect observed after receipt of the total package of delivered Yops.

LcrV

Since its discovery (Burrows, 1956), LcrV (V antigen) has remained in the forefront of yersiniae research. This 37.3-kDa pure protein, of 327 amino acid residues, has marked anti-inflammatory activity as judged by the ability of a homogeneous staphylococcal protein A–LcrV fusion protein (PAV) (Motin *et al.*, 1994) to mimic repression (**Fig. 7**) of the generic inflammatory response mediated by whole pCD⁺ yersiniae (**Fig. 3a**). PAV also prolongs the retention of *lcrV* yersiniae *in vivo* and enhances the virulence of

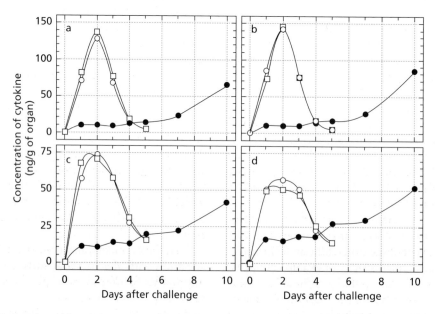

Fig. 7 Levels of IFNγ in spleen (a) and liver (b) or of TNFα in spleen (c) or liver (d) of Swiss-Webster mice following induction of cytokine synthesis on day 0 by intravenous injection of 10⁶ pCD⁻ cells of *Y. pestis* KIM. Mice received daily injections of 0.1 mL of 0.033 M potassium phosphate buffer, pH 7.0 (open squares), or 100 µg of either control truncated protein A (open circles) or PAV (closed circles) in 0.1 mL of the same buffer on post-infection days 0–5. Values represent averages determined for four mice. From Nakajima *et al.* (1995).

unrelated bacteria, especially *Listeria monocytogenes* (Nakajima *et al.*, 1995).

The retention of skin allografts by mice has been monitored in the presence and absence of injected PAV, in order to distinguish between pure anti-inflammatory activity resulting from down-regulation of pro-inflammatory cytokines and inhibition of specific immunity (Motin *et al.*, 1997). No significant difference was noted between treated and control mice, with regard to time of total allograft retention, but the period before onset of visible inflammation was doubled from 6 to 12 days ($p < 0.001$). This illustrates that LcrV is, indeed, anti-inflammatory but does not interfere with the onset of specific immunity *per se*, other than inhibiting cytokine-mediated processes common to both systems.

It is established that endotoxic shock is a consequence of tissue damage resulting from excessive stimulation by pro-inflammatory cytokines. An LPS-free polyhistidine-tagged LcrV fusion protein (Motin *et al.*, 1996) promotes an immediate resistance to LPS that is rapidly dissipated and then reappears after 48 hours (Nedialkov *et al.*, 1997). Down-regulation of IFNγ and TNFα occurs during both phases, as does up-regulation of anti-inflammatory cytokines, especially IL-10. The early phase cannot represent LPS tolerance, which requires 48 hours for full expression. Thus, immediate down-regulation of IFNγ and TNFα appears to reflect a novel LcrV-mediated ability to up-regulate anti-inflammatory cytokines that prevent inflammation.

These assumptions were verified to the extent that an analogous LcrV–polyhistidine fusion protein engineered from *Y. enterocolitica* also down-regulates LPS-induced TNFα in mouse spleen and liver (Schmidt *et al.*, 1999). Furthermore, this inhibition occurs at the level of mRNA synthesis and can be blocked by specific antibody. Antibodies directed against IL-10 do not, however, prevent repression of TNFα. Instead, inhibition by LcrV depends on the presence of both macrophages and activated T cells, which elaborate an as yet undefined soluble factor that accounts for down-regulation.

Further study will be necessary to determine whether this mechanism also accounts for the absolute repression of IFNγ, primarily synthesised by TH1 cells, during the generic inflammatory response (**Fig. 3a**). Absolute inhibition of this important cytokine is the hallmark of the anti-inflammatory umbrella maintained by pCD$^+$ yersiniae throughout the course of infection (**Fig. 3b**). Considered together, these results suggest that LcrV preserves this illusion of salubrity although other effectors of virulence, especially YopJ, may also contribute to down-regulation of TNFα.

pPCP–pCD Interaction

Curiously, Yops are absent from outer membrane preparations of pCD$^+$ *Y. pestis* grown at 37°C under conditions that favoured their expression in *Y. pseudotuberculosis* (20 mM Mg^{2+} and no added Ca^{2+}) (Straley and Brubaker, 1981). Further examination revealed that the Yops in question undergo normal synthesis but are promptly hydrolysed by pPCP-encoded plasminogen activator (Sample and Brubaker, 1987; Sodeinde *et al.*, 1988). LcrV was not vulnerable to hydrolysis by Pla. Net synthesis of salient Yops by *Y. pestis* obviously occurs *in vivo* because, as noted above, mutants lacking their structural genes are avirulent. A possible explanation for this difference is that Yops remain fully compartmentalised during translocation prompted by contact with the host cell surface, but they interact with Pla during the course of secretion induced by Ca^{2+}-deficiency. If correct, this notion also argues that Pla may fulfil a more active additional role during translocation of Yops, by serving as the *Y. pestis* adhesin that promotes docking to target cells.

In any event, Pla-mediated Yop degradation distinguishes *Y. pestis* from the enteropathogenic yersiniae, and may also facilitate expression of acute disease. In this regard, it should be recalled that the concentrations of Ca^{2+} (< 1 mM) and Mg^{2+} (20 mM) that induce secretion of Yops *in vitro* also exist in mammalian cytoplasm. Indeed, released host cell cytoplasm comprises the environment in the focal lesions of liver and spleen that supports bulk vegetative growth *in vivo*. Consequently, it would be expected that Yops are secreted and accumulate in lesions occupied by enteropathogenic yersiniae, but that they encounter post-translational hydrolysis in lesions generated by cells of *Y. pestis*. Since only translocated Yops facilitate disease, their extracellular existence would serve merely to signal the presence of foreign material, thereby compromising the anti-inflammatory umbrella. This difference between *Y. pestis* and the enteropathogenic yersiniae very probably accounts for the sharp necrotic foci that contain pCD$^+$ plague bacilli (**Fig. 5a**), in contrast to abscesses formed by pYV$^+$ *Y. pseudotuberculosis* that are surrounded by infiltrated neutrophils (Nakajima *et al.*, 1995).

Introduction of pPCP into pYV$^+$ *Y. pseudotuberculosis* results in comparable post-translational degradation of Yops, although the adhesin YadA is only marginally affected (Kutyrev *et al.*, 1999). These transformants did not show increased virulence by peripheral injection routes, as might be anticipated upon receipt of Pla, the major invasive activity of *Y. pestis*. Further study will be required to ascertain whether this reflects retention of other factors likely to impede dissemination in tissues but required for expression of chronic enteric disease (i.e. YadA, Inv and Ail).

Chromosomal Determinants

Certain chromosomally encoded determinants may also facilitate bulk multiplication in organs, although dual roles in accessing the viscera from peripheral sites of infection are also possible. Important examples are the temperature-dependent antigens 3, 4 and 5 of Crumpton and Davies (1956) now recognised as fraction 1, pH 6 antigen and KatY, respectively. As already noted, fraction 1 probably provides resistance to neutrophils and monocytes during the journey from the dermis to viscera. In contrast, pH 6 antigen serves as an outer membrane adhesin. Expression of this activity is favoured at the slightly reduced pH of host cell cytoplasm, suggesting a role in bacterial docking to the host cells that line the progressively expanding necrotic lesions of visceral organs. This function is not, however, critical in the mouse, as judged by only marginally decreased virulence of mutants lacking pH 6 antigen (Perry and Fetherston, 1997).

It is interesting that the specific activity of catalase-peroxidase in *Y. pestis* is among the highest reported in prokaryotes. This is due in part to production of a major protein, now termed KatY, that has marked homology with KatP, a virulence plasmid-encoded catalase-peroxidase of enterohaemorrhagic *E. coli* O157:H7 (Garcia *et al.*, 1999). The consequences on virulence of mutational loss of KatY have not been determined but, in view of its abundant temperature-conditional synthesis, it seems probable that the enzyme facilitates disease expression. A possible role for KatY is scavenging of hydrogen peroxide generated by professional phagocytes to defeat oxygen-dependent killing processes. If this mechanism can be verified, further work will be required to determine whether KatY accounts for initial survival in fixed liver and spleen macrophages or in free phagocytic cells encountered during the journey from dermis to viscera. It is of interest that KatY, like Yops, is at least partially hydrolysed by Pla, suggesting a transient association with the outer membrane (Garcia *et al.*, 1999).

As already noted, bulk cellular protein is not synthesised by pCD$^+$ yersiniae under conditions of strict bacteriostasis imposed by the LCR, although marked expression of Yops and LcrV occurs in this environment. Chromosomally encoded functions associated with virulence are also produced in this environment, including fraction 1, pH 6 antigen, KatY, Pla and murine toxin, as well as the ubiquitous chaperone GroEL (Mehigh and Brubaker, 1993). This suggests

that these activities are regulated by mechanisms not shared by proteins required for vegetative growth. It may, therefore, be significant in this context that the KatY promoter contains three homologues of the consensus recognition sequence of pCD-encoded transcriptional activator LcrF (Garcia *et al.*, 1999). In spite of the presence of these sequences, synthesis of KatY remains temperature-conditional in mutants that lack pCD (Crumpton and Davies, 1956), which suggests that a second LcrF-like transcriptional activator is encoded by the chromosome.

Immunity

Early work indicated that antisera raised against yersiniae now known to be pCD$^+$ provides better passive immunity to plague in mice than do anti-pCD$^-$ sera (Burrows and Bacon, 1958). This was attributed to LcrV, and antisera directed against crude preparations of this protein were later shown to be protective. Attempts to verify that LcrV, rather than contaminating Yops or other antigens, actually accounts for immunity were only partially successful because of the tendency of LcrV to undergo spontaneous hydrolysis during purification by conventional chromatographic methods (Brubaker *et al.*, 1987; Motin *et al.*, 1994). Nevertheless, rabbit polyclonal antisera against a highly purified mixture of degradation products give passive immunity in mice (Une and Brubaker, 1984b) and, as shown in **Fig. 1d**, enable infected mice to generate normal levels of IFNγ and TNFα (Nakajima and Brubaker, 1993).

Formal proof that anti-LcrV accounts for immunity, rather than antibodies directed against an undetectable antigen, was obtained by expressing PAV in *E. coli* and then purifying this ligand-fusion protein to homogeneity by affinity chromatography (Motin *et al.*, 1994). The passive immunity provided by rabbit anti-PAV could be removed by absorption with engineered truncated derivatives of LcrV, which showed that at least one essential protective epitope resides between amino acid residues 168 and 275. Curiously, anti-PAV or antibodies directed against other derivatives of LcrV from *Y. pestis* or *Y. pseudotuberculosis* fail to provide significant protection against *Y. enterocolitica*. This discrepancy reflects heterogeneity of LcrV in that homologous antiserum is required to confer immunity against virulent strains of this species (Roggenkamp *et al.*, 1997). As already noted, injected PAV blocks the generic inflammatory response in infected mice (**Fig. 3**), facilitates retention of *lcrV* mutants *in vivo*, enhances the virulence of unrelated pathogenic bacteria, and reduces the inflammation associated with allograft rejection (Nakajima *et al.*, 1995; Motin *et al.*, 1997). Of equal importance, mice either actively immunised with PAV (**Fig. 5c**) or passively immunised with anti-PAV (**Fig. 5d**) sequester intravenously injected pCD$^+$ yersiniae in granulomatous lesions in a manner entirely analogous to that observed for pCD$^-$ mutants (**Fig. 5b**). This shows that anti-LcrV protects against plague by abrogating the block that prevents expression of the generic inflammatory response. The ability of LcrV to serve as a protective antigen has been verified and extended by many workers (Perry and Fetherston, 1997).

These findings do not, of course, prove that LcrV exclusively maintains the anti-inflammatory umbrella characteristic of plague, or that its other roles in mediating expression and translocation of Yops are irrelevant to its ability to immunise against disease. However, anti-LcrV inhibits neutrophil chemotaxis (Welkos *et al.*, 1998). Furthermore, the finding that forms of anti-LcrV capable of providing passive immunity can re-establish the ability to produce pro-inflammatory cytokines *in vivo* (Nakajima and Brubaker, 1993) and in cultured cells (Schmidt *et al.*, 1999), strongly argues that it is this capability rather than inhibition of Yop translocation that accounts for immunity. The validity of this hypothesis is strengthened by the discovery that a protective antiserum against LcrV fails to prevent translocation of YopE into HeLa cell cytoplasm (Fields *et al.*, 1999).

Anti-Yops are either unable to immunise against plague or their effectiveness is modest in comparison to that of anti-LcrV (Andrews *et al.*, 1999). However, antibodies directed against other determinants of virulence can provide significant protection. For example, anti-fraction 1 can immunise effectively because of its ability to opsonise yersiniae en route to the viscera. Reliance upon this antiserum is venturesome, however, because in many hosts including humans, fraction 1 is not essential for expression of disease. It is, in fact, spontaneously lost at high frequency and thus, in the absence of selective pressure, fraction 1-deficient mutants may emerge as the predominant phenotype (Burrows, 1963). This probably accounts for their isolation in pure culture at human autopsy (Winter *et al.*, 1960). Solubilised Pla can also immunise against plague (V.V. Kutyrev and R.R. Brubaker, 1999, unpublished observations). It is not clear whether this form of protection also reflects opsonisation or whether, as in the case of LcrV, neutralisation of at least one of its many roles in facilitating disease. Further study will be required to evaluate the efficacy of antibodies directed against LPS from yersiniae.

The Flea Vector

Not only does the plague bacillus subvert all non-specific mechanisms of host defence, it has also acquired the ability to proliferate in the hostile environment of its vector. However, Pgm$^-$ mutants are unable to block the proventriculus of the flea (Kutyrev *et al.*, 1992), a lethal process that induces the repeated attempts at feeding, which assure transmission of the disease. Blockage is now known to require an intact Hms locus (**Fig. 1a**) in the Pgm locus (Hinnebusch *et al.*, 1996). As noted above, the Hms operon mediates the Hms$^+$ or Crb$^+$ phenotypes at 26°C but not at 37°C; it comprises *hmsHFRS* (Perry and Fetherston, 1997) but its expression is dependent on *hmsT* located outside the deletable region (Jones *et al.*, 1999). HmsH and HmsR are probable outer membrane proteins that facilitate haemin-binding by stacking and thereby accounting for the coloured appearance and brittle nature of colonies grown on solid medium containing haemin or Congo red. Similar binding undoubtedly promotes mechanical blockade of the flea gut, but the nuances of this reaction are not fully understood. Regulation of pigmentation may depend on the formation of a surface complex composed of multiple gene products. It is curious in this context, that the Hms$^+$/Crb$^+$ phenotypes are regulated by Fur in spite of the absence of salient-binding sites near *hmsHFRS*; a good potential-binding site was later mapped near *hmsT* (Jones *et al.*, 1999). Exhaustive attempts to define a role for the Hms$^+$/Crb$^+$ phenotypes in the expression of virulence in the mouse were unsuccessful, indicating that this function serves solely to induce transmission of flea-borne plague (Lillard *et al.*, 1999).

Work by Å. Forsberg (presented at the 1999 Chicago ASM Meetings) strongly implicated pMT-encoded murine toxin, a prototype phospholipase (Rudolph *et al.*, 1999), as an additional component required for colonisation of the flea. As opposed to blocking the proventriculus, murine toxin is evidently necessary for the survival of the bacteria in the gut. This was unexpected, in view of the ability of this toxin to assure a lethal outcome in infected mice and rats. As such, the phenomenon provides another illustration of dual functionality in the virulence determinants of *Y. pestis*.

Safety and Containment

Other than the obvious exception of identification in the field, wild-type *Y. pestis* should never be handled except within stringent conditions of enclosure (class 3 containment). In addition, workers lacking such facilities should always take precautions to assure that isolates received from others lack either pCD or the Pgm locus. Isolates cured of pCD have proved avirulent ($LD_{50} > 10^7$ bacteria) in all tested mammalian species, including mice and humans. Cells of this phenotype can thus be safely used in experiments concerned with functions not encoded by pCD. Pgm^-, pCD^+ mutants are also avirulent in humans, as shown by their long use as a live vaccine. As emphasised above, this phenotype is innocuous in the white mouse by peripheral routes of injection but remain fully virulent by intravenous injection. These responses enable investigators not equipped with class 3 containment facilities safely to undertake experiments concerned with pCD-mediated activities.

Identification

Demonstration of Gram-negative rods in buboes or from moribund rodents is sufficient to warrant a diagnosis of plague where the disease has already been recognised. Specific biochemical, cultural and immunological tests exist that permit identification of isolates in areas where plague is uncommon or unexpected. Once identified, the strain can be subjected to further analysis to ascertain its biotype, DNA macrorestriction pattern (Lucier and Brubaker, 1992), and ribotype (Guiyoule *et al.*, 1994).

Y. pestis can be identified with considerable certainty in laboratories lacking specific diagnostic reagents by observation of the Hms^+/Crb^+ phenotype at 26°C but not 37°C and ability to activate plasminogen and thus lyse fibrin plates. Presence of the phenotype in any serogroup I *Y. pseudotuberculosis*, serotype O:8 *Y. enterocolitica*, or the universal colicin indicator *E. coli* ϕ, permits identification on the basis of ability to express pesticin (Brubaker and Surgalla, 1961). Another simple test to distinguish *Y. pestis* from the enteropathogenic yersiniae in the field, is the ability of 0.1 mL of a filtered preparation of disrupted plague bacilli (~ 1 mg of protein) to kill mice within 24 hours of intraperitoneal injection. Similar preparations of *Y. pseudotuberculosis* or *Y. enterocolitica*, which lack the plague murine exotoxin, are not lethal at this concentration. If suitable antisera are available, more rapid identification by immunological methods is to be preferred. In this context, an immunofluorescence assay for the fraction 1 antigen of yersiniae in sputum or in fluid aspirated from buboes serves as a traditional tool for the diagnosis of plague.

Many laboratories are developing monoclonal antibodies capable of recognising *Y. pestis*-specific antigens, such as plasminogen activator, pesticin, murine exotoxin or fraction 1 for use in detecting the organisms *in vivo* or in natural environments. An alternative approach is use of sensitive methods based upon polymerase chain reactions to recognise hybridisation *in situ* between bacterial DNA and sequences comprising the structural genes of these specific antigens. Analogous procedures are use of branched DNA or TaqMan technology. It is reasonable to assume that these and similar methods will replace the traditional procedures currently in use for the identification of *Y. pestis* in the environment and the diagnosis of plague (Hinnebusch and Schwan, 1993; Norkina *et al.*, 1993; Tsukano *et al.*, 1996; Iqbal *et al.*, 1999, 2000).

As discussed below, the emergence of plague probably occurred very recently in evolutionary time. All isolates of *Y. pestis*, therefore, express lipopolysaccharide of identical or very similar composition. Consequently, strains cannot be distinguished by serotyping, and epidemiologists must rely on alternative methods. Biotyping utilises glycerol fermentation and nitrate reduction; isolates that do neither, ferment only glycerol or are positive for both are assigned, respectively, to varieties orientalis, medievalis or antiqua (Devignat, 1951). The validity of this scheme was verified by use of distinct determinants that map in the Pgm locus (Buchrieser *et al.*, 1999).

More detailed information regarding individual strains can be acquired by comparison of DNA macrorestriction patterns obtained by pulsed field gel electrophoresis. Although limited in scope, results from one study showed that, provided laboratory stocks are avoided, the chromosome of *Y. pestis* is remarkably homogeneous as would be expected for a recently evolved species (Lucier and Brubaker, 1992). Furthermore, the chromosome is stable if care was taken to avoid overgrowth of Pgm^- mutants which, as noted above, arise at high frequency by reciprocal recombination between IS*100* insertions. The close similarity observed between strains isolated from Java and western North America favour the hypothesis that plague was introduced into the United States at the turn of the present century from vessels arriving from Indochina. Similar data were obtained by ribotyping, where greatest variation was observed among strains isolated from Africa (Guiyoule *et al.*, 1994). These molecular techniques will eventually supersede biotyping and undoubtedly contribute to a better understanding of the evolution and current distribution of the plague bacillus.

Evolution

Y. pestis is a recently evolved species as judged by remarkably close similarities in ribotype (Guiyoule *et al.*, 1994), sequences of selected loci (Buchrieser *et al.*, 1999), and DNA macrorestriction patterns of typical strains (Lucier and Brubaker, 1992). The presence in *Y. pestis* of numerous pseudogenes that are functional in *Y. pseudotuberculosis*, indicates that the latter is the immediate progenitor. However, the precise sequence of events that led to the emergence of the plague bacillus is far from clear. pYV of *Y. enterocolitica* is organisationally distinct from that of *Y. pseudotuberculosis*, suggesting that it probably evolved before divergence into the line leading to *Y. pseudotuberculosis* and then *Y. pestis*. This relationship emphasises that lateral transfer of genes shared by pYV/pCD and the chromosomes of other pathogenic bacteria (Hsia *et al.*, 1997; Yahr *et al.*, 1997; Sawa *et al.*, 1999) occurred before the emergence of *Y. pseudotuberculosis*. The marked similarity between pCD of *Y. pestis* and pYV of *Y. pseudotuberculosis* (**Fig. 2**) further illustrates the close relationship between these two species, as does the existence of shared functions (i.e. KatY) not expressed by *Y. enterocolitica*.

The evolution of *Y. pestis* from *Y. pseudotuberculosis* required acquisition of new functions necessary for expression of acute disease in natural rodent hosts (i.e. Pla, murine toxin and fraction 1) and survival in the flea (i.e. the Hms$^+$ phenotype and murine toxin). Emergence of the plague bacillus was also dependent on mutational loss of determinants associated with chronic disease, including YadA, Inv and Ail. Acquisition of the high-pathogenicity island in the Pgm locus, which mediates Ybt biosynthesis, is probably unrelated to the evolution of *Y. pestis*, because this ability is also expressed by *Y. pseudotuberculosis* serotypes IA and IB, *Y. enterocolitica* serovar O:8, and certain pathogenic forms of *E. coli* (Heesemann, 1987; Guilvout *et al.*, 1993; Rakin *et al.*, 1994; Bearden *et al.*, 1997; Buchrieser *et al.*, 1998a). The high-pathogenicity island may have been acquired before the divergence of *Y. pestis*, or it may repeatedly have been transferred and lost. Evidence for the latter is that the high-pathogenicity island can insert into any of the three Asn tRNA loci of *Y. pseudotuberculosis* and is currently in a position not shared by *Y. pestis* (Buchrieser *et al.*, 1998a; Hare *et al.*, 1999). In view of this, it is possible that the ability to synthesise Ybt exists in nature as an infectious genetic unit capable of repeatedly invading new genomes from an unknown origin.

A functional Hms locus within the Pgm locus conferring the Hms$^+$ or Crb$^+$ phenotype was clearly necessary for emergence of *Y. pestis*. However, this trait is again widespread in nature as judged by the presence of cryptic *hms* genes or their homologues in other yersiniae or *E. coli* (Buchrieser *et al.*, 1998a, 1999; Jones *et al.*, 1999). Further study will be necessary to determine if these genes are no longer functional in these isolates or if their products mediate functions other than the binding of planar small molecules. The plague murine toxin, however, has not yet been detected elsewhere in nature. This activity was also implicated as an essential element for survival in the flea, thus its acquisition may have been sufficient to enable the organisms to abandon reliance upon transmission via the oral route. Murine toxin also assures the death of rats and mice, therefore its expression guarantees that resident fleas will depart in search of new hosts, thereby perpetuating the causative agent. In view of this relationship, it is significant that pMT encoding murine toxin is unique to *Y. pestis*. The other known virulence determinant encoded on this plasmid (fraction 1) is not essential in the murine system, although it is interesting that this capsular antigen is required for full lethality in certain non-muridae such as the guinea-pig (Burrows, 1963). Accordingly, the acquisition of pMT (or at least murine toxin and fraction 1) was evidently a necessary initial step in the emergence of plague.

Once the wherewithal to survive in the flea was realised, the organisms no longer needed to maintain the ability to compete for nutrients with saprophytes in natural environments, as must the enteropathogenic yersiniae. Consequently, mutations causing modest nutritional requirements or inability to ferment obscure carbohydrates could accumulate with impunity. Many of these mutations are typical of errors in DNA replication and often account for missing genes that remain functional in *Y. pseudotuberculosis* (**Table 1**). However, other mutational losses in *Y. pestis* reflect the insertion of IS sequences (Portnoy and Falkow, 1981). It is not always clear whether the observed attrition of chromosomally encoded functions in *Y. pestis* has occurred by accident or design. Clearly, mutational loss of adhesins expressed by enteropathogenic yersiniae (YadA, Inv and Ail) benefits expression of acute disease, because these activities serve no function after infection by flea bite, and may actually immobilise the organisms at peripheral sites of infection. The consequences of other losses are subtler but may, nevertheless, be detrimental to the host. For example, aspartase deficiency modifies the metabolism of L-glutamic acid by causing accumulation of extracellular L-aspartate (Dreyfus and Brubaker, 1978). As noted above, the effect of this metabolic loss, which may be rectified *in vivo* by regeneration of L-glutamate from L-aspartate, profoundly influences the LCR of *Y. pestis* by prompting an exacting *in vitro* nutritional requirement for Ca^{2+} in order to relieve toxicity caused by Na^+ (Fowler and Brubaker, 1994). In this context, it may be significant that L-glutamate metabolism is also critical in other intracellular parasites including *Brucella* spp. and *Rickettsia*, and that the latter also accumulate extracellular L-aspartate (Moulder, 1962).

Plague bacilli deposited in the dermis by flea bite would remain localised were it not for the invasive nature of Pla, which promotes dissemination through interstitial spaces (Lähteenmäki *et al.*, 1998). Since this pPCP-encoded activity is unnecessary for intestinal invasion, it seems reasonable to assume that its appearance in evolving *Y. pestis* probably awaited the acquisition of pMT required for survival in the flea. However, the established interactions between pPCP, the high-pathogenicity island and pCD, argue for a prior association between the products encoded by these elements. As described above, these interactions constitute the ability of pesticin to attach to the Ybt receptor Psn and the ability of Pla to eliminate secreted Yops that fail to undergo translocation into host cells. Pesticin itself probably has no role in causing disease, but rather assures that mutational loss of pPCP (encoding the pesticin immunity protein) is punished. However, Pla may have other functions, in addition to an established primary role as tissue invasin, such as serving as an adhesin and scavenger of undelivered Yops. Further experimental work and perhaps isolation of intermediates between *Y. pseudotuberculosis* and *Y. pestis* may be necessary to clarify the sequence of events that led to the emergence of plague.

Another feature of *Y. pestis* that complicates determination of its evolution and origin is the multiple copies of IS elements in its genome. These sequences effectively divide the chromosome into about 30 segments (Portnoy and Falkow, 1981) which, if of appropriate polarity, can undergo recombination resulting in large inversions and transpositions. These events could theoretically result in almost unlimited variation in the genome and possibly cause significant changes in virulence. This may very well account for known distinctions between plasmid size and number in certain isolates and explain the expression of murine toxin and fraction 1 in strains lacking detectable pMT. Marked differences in virulence have been noted among strains recovered from separate regions of the ancient plague reservoirs of central Asia. These distinctions are seldom observed in fresh isolates that represent the last pandemic, which are uniformly of high virulence. Perhaps study of these 'inter-epidemic' forms of *Y. pestis* resident in the ancient plague centres will shed more light on the evolution of acute disease.

Conclusion

Bubonic plague caused by *Yersinia pestis* ranks as the most severe bacterial disease known to humanity, as judged by historical records and the extent of pathology. In the mouse, this bleak process consists of an initial invasive stage characterised by colonisation of the host viscera from dermal sites of infection caused by flea bite. Thereafter the organisms undergo bulk vegetative growth in liver and spleen before spilling into the vascular system where they may again be ingested by fleas before the now imminent death of the host. A similar course occurs in humans except that lymphadenopathy is more severe, as judged by the formation of buboes in regional lymph nodes that drain the vicinity of the infected flea bite.

A potent outer membrane plasminogen activator capable of causing bacterial metastasis largely mediates the initial invasive phase of plague. In addition, the ability to express the siderophore yersiniabactin, encoded chromosomally within a high-pathogenicity island, is essential for *Y. pestis* to gain access to the viscera from the dermis. The second phase of vegetative growth involves destruction of organ tissue and its conversion to bacterial mass. The process is mediated by a type III protein secretion mechanism capable of delivering cytotoxins (Yops) directly from the cytoplasm of docked yersiniae to that of host cells. This mechanism provides an aspect of stealth by preventing effectors of the host immune system from recognising or neutralising these cytotoxins. In addition, the bacteria generate an anti-inflammatory umbrella by down-regulation of the major pro-inflammatory cytokines IFNγ and TNFα, which prevents infiltration of the inflammatory cells necessary for formation of protective granulomas. LcrV (V antigen) largely mediates this mechanism of deception. Reversal of IFNγ and TNFα repression by anti-LcrV provides immunity against plague.

Bubonic plague as opposed to the chronic enteropathogenic process caused by the closely related *Y. pseudotuberculosis* is mediated by two unique plasmids termed pPCP and pMT. The former encodes the plasminogen activator required for invasion of tissue and the latter encodes capsular fraction 1 antigen and murine toxin, which assure rapid death of certain rodent hosts. Murine toxin is also necessary for the survival of yersiniae in the flea vector. In contrast, the chromosomal Hms locus of *Y. pestis* assures that the bite of infected fleas initiates disease.

The vegetative phase of plague is largely mediated by a third plasmid that is shared with enteropathogenic *Y. pseudotuberculosis* and the more distantly related *Y. enterocolitica*. This common plasmid, termed pCD in *Y. pestis*, encodes the cytotoxic Yops and LcrV.

In addition to acquiring the unique plasminogen activator, murine toxin, fraction 1 antigen and the Hms locus, *Y. pestis* has lost genes that encode surface adhesins present in the enteropathogenic yersiniae (YadA, Inv and Ail). These activities are necessary for infection after oral administration but they are unnecessary even if not detrimental for expression of acute disease. In addition, the genome of *Y. pestis* contains numerous IS elements that facilitate reciprocal recombination resulting in high frequency deletion, plasmid integration and excision, and possible chromosomal transposition and inversion. The significance of these insertion sequences is discussed with respect to loss of metabolic function, alteration of virulence and evolution of acute disease. In short, the acute disease mounted by the plague bacillus is a campaign of molecular deception and stealth rather than a straightforward frontal assault on the host.

References

Andrews GP, Strachan ST, Benner GE *et al.* (1999) Protective efficacy of recombinant *Yersina* outer proteins against bubonic plague caused by encapsulated and nonencapsulated *Yersina pestis*. *Infect. Immun.* 67: 1533–1537.

Bearden SW, Fetherston JD, Perry RD (1997) Genetic organization of the yersiniabactin biosynthetic region and construction of avirulent mutants in *Yersinia pestis*. *Infect. Immun.* 65: 1659–1668.

Bearden SW, Staggs TM, Perry RD (1998) An ABC transporter system of *Yersinia pestis* allows utilization of chelated iron by *Escherichia coli* SAB11. *J. Bacteriol.* 180: 1135–1147.

Bearden SW, Perry RD (1999) The Yfe system of *Yersinia pestis* transports iron and manganese and is required for full virulence of plague. *Mol. Microbiol.* 32: 403–414.

Beesley ED, Brubaker RR, Janssen WA, Surgalla MJ (1967) Pesticins. III. Expression of coagulase and mechanism of fibrinolysis. *J. Bacteriol.* 94: 19–26.

Ben-Gurion R, Hertman I (1958) Bacteriocin-like material produced by *Pasteurella pestis*. *J. Gen. Microbiol.* 19: 289–297.

Braun V (1997) Avoidance of iron toxicity through regulation of bacterial iron transport. *Biol. Chem.* 378: 779–786.

Brenner DJ, Steingerwalt AG, Falcao DP, Weaver RE, Fanning GR (1976) Characterization of *Yersinia*

enterocolitica and *Yersinia pseudotuberculosis* by deoxyribonucleic acid hybridization and by biochemical reactions. *Int. J. Syst. Bacteriol.* 26: 180–194.

Brubaker RR, Surgalla MJ (1961) Pesticins I. Pesticin–bacterium interrelationships, and environmental factors influencing activity. *J. Bacteriol.* 82: 940–949.

Brubaker RR, Beesley ED, Surgalla MJ (1965) *Pasteurella pestis*: role of pesticin I and iron in experimental plague. *Science* 149: 422–424.

Brubaker RR (1970) Mutation rate to nonpigmentation in *Pasteurella pestis*. *J. Bacteriol.* 98: 1404–1406.

Brubaker RR, Sample AK, Yu D-Z, Zahorchak RJ, Hu PC, Fowler JM (1987) Proteolysis of V antigen from *Yersinia pestis*. *Microb. Pathog.* 2: 49–62.

Buchrieser C, Brosch R, Bach S, Guiyoule A, Carniel E (1998a) The high-pathogenicity island of *Yersinia pseudotuberculosis* can be inserted into any of the three chromosomal asn tRNA genes. *Mol. Microbiol.* 30: 965–978.

Buchrieser C, Prentice M, Carniel E (1998b) The 102-kilobase unstable region of *Yersinia pestis* comprises a high-pathogenicity island linked to a pigmentation segment which undergoes internal rearrangement. *J. Bacteriol.* 180: 2321–2329.

Buchrieser C, Rusniok C, Frangeul L *et al.* (1999) The 102-kilobase *pgm* locus of *Yersinia pestis*: sequence analysis and comparison of selected regions among different *Yersinia pestis* and *Yersinia pseudotuberculosis* strains. *Infect. Immun.* 67: 4851–4861.

Burrows TW (1956) An antigen determining virulence in *Pasteurella pestis*. *Nature* 177: 426–427.

Burrows TW, Bacon GA (1958) The effects of loss of different virulence determinants on the virulence and immunogenicity of strains of *Pasteurella pestis*. *Br. J. Exp. Pathol.* 39: 278–291.

Burrows TW (1963) Virulence of *Pasteurella pestis* and immunity to plague. *Ergebn. Mikrobiol.* 37: 59–113.

Butler T (1983) *Plague and other Yersinia Infections*. New York: Plenum Press, pp. 220.

Carniel E, Antoine JC, Guiyoule A, Guiso N, Mollaret HH (1989) Purification, location, and immunological characterization of the iron-regulated high-molecular weight-proteins of the highly pathogenic yersiniae. *Infect. Immun.* 57: 540–545.

Carniel E, Guilvout I, Prentice M (1996) Characterization of a large chromosomal 'high-pathogenicity island' in biotype 1B *Yersinia enterocolitica*. *J. Bacteriol.* 178: 6743–6751.

Cornelis GR (1998) The *Yersinia* deadly kiss. *J. Bacteriol.* 180: 5495–5504.

Cornelis GR, Boland A, Boyd AP *et al.* (1998) The virulence plasmid of *Yersinia*, an antihost genome. *Microbiol. Mol. Biol. Rev.* 62: 1315–1352.

Crumpton MY, Davies DAL (1956) An antigenic analysis of *Pasteurella pestis* by diffusion of antigens and antibodies in agar. *Proc. R. Soc. Lond. Ser. B* 145: 109–134.

Davis JK, Reeves P (1975) Genetics of resistance to colicins in *Escherichia coli* K-12: cross resistance among colicins of group B. *J. Bacteriol.* 123: 96–101.

Devignat R (1951) Varietes de l'espece *Pasteurella pestis*. Nouvelle hypothese. *Bull. WHO* 4: 247–263.

Dreyfus LA, Brubaker RR (1978) Consequences of aspartase deficiency in *Yersinia pestis*. *J. Bacteriol.* 136: 757–764.

Federova VA, Devdariani ZL (1998) Study of antigenic determinants of *Yersinia pestis* lipopolysaccharide using monoclonal antibodies. *Mol. Gen. Mikrobiol. Virusol.* 18: 22–26.

Ferber DM, Brubaker RR (1979) Mode of action of pesticin: *N*-acetylglucosaminidase activity. *J. Bacteriol.* 139: 495–501.

Fetherston JD, Lillard JW Jr, Perry RD (1995) Analysis of the pesticin receptor from *Yersinia pestis*: role in iron deficient growth and possible regulation by its siderophore. *J. Bacteriol.* 177: 1824–1833.

Fields KA, Straley SC (1999) LcrV of *Yersinia pestis* enters infected eukaryotic cells by a virulence plasmid-independent mechanism. *Infect. Immun.* 67: 4801–4813.

Fields KA, Nilles ML, Cowan C, Straley SC (1999) Virulence role of V antigen of *Yersinia pestis* at the bacterial surface. *Infect. Immun.* 67: 5395–5408.

Finlay RB, Falkow S (1997) Common themes in microbial pathogenicity revisited. *Microbiol. Mol. Biol. Rev.* 61: 136–169.

Fowler JM, Brubaker RR (1994) Physiological basis of the low calcium response in *Yersinia pestis*. *Infect. Immun.* 62: 5234–5241.

Garcia E, Nedialkov YA, Elliott J, Motin VL, Brubaker RR (1999) Molecular characterization of KatY (Antigen 5), a thermoregulated chromosomally encoded catalase-peroxidase of *Yersinia pestis*. *J. Bacteriol.* 181: 3114–3122.

Gehring AM, DeMoll E, Fetherston JD *et al.* (1998) Iron acquisition in plague: modular logic in enzymatic biogenesis of yersiniabactin by *Yersinia pestis*. *Chem. Biol.* 5: 573–586.

Guilvout I, Mercereau-Pujalon O, Bonnefey S, Pugsley AP, Carniel E (1993) High-molecular weight protein 2 of *Yersinia enterocolitica* is a homolog to AngR of *Vibrio anguillarum* and belongs to a family of proteins involved in nonribosomal peptide synthesis. *J. Bacteriol.* 175: 5488–5504.

Guiyoule A, Grimont F, Iteman I, Grimont PA, Lafevre M, Carniel E (1994) Plague pandemics investigated by ribotyping of *Yersinia pestis* strains. *J. Clin. Microbiol.* 32: 634–641.

Hall PJ, Brubaker RR (1978) Pesticin-dependent generation of osmotically stable spheroplast-like structures *J. Bacteriol.* 136: 786–789.

Hare JM, McDonough KA (1999) High-frequency RecA-dependent and -independent mechanisms of Congo red binding mutations in *Yersinia pestis*. *J. Bacteriol.* 181: 4896–4904.

Hare JM, Wagner AK, McDonough KA (1999) Independent acquisition and insertion into different chromosomal locations of the same pathogenicity island in *Yersinia pestis*

and *Yersinia pseudotuberculosis*. *Mol. Microbiol.* 31: 291–304.

Heesemann J (1987) Chromosomal-encoded siderophores are required for mouse virulence of enteropathogenic *Yersinia* species. *FEMS Microbiol. Lett.* 48: 229–233.

Hillier S, Charnetzky WL (1981) Glyoxalate bypass enzymes in *Yersinia* species and multiple forms of isocitrate lyase in *Yersinia pestis*. *J. Bacteriol.* 145: 452–458.

Hinnebusch J, Schwan TG (1993) New method for plague surveillance using polymerase chain reaction to detect *Yersinia pestis* in fleas. *J. Clin. Microbiol.* 31: 1511–1514.

Hinnebusch BJ, Perry RD, Schwan TG (1996) Role of the *Yersinia pestis* hemin storage (*hms*) locus in the transmission of plague by fleas. *Science* 273: 367–370.

Hornung JM, Jones HA, Perry RD (1996) The *hmu* locus of *Yersinia pestis* is essential for utilization of free haemin and haem-protein complexes as iron sources. *Mol. Microbiol.* 20: 725–739.

Hsia RC, Panneekoek Y, Ingerowski E, Bavoil PM (1997) Type III secretion genes identify a putitive virulence locus of *Chlamydia*. *Mol. Microbiol.* 25: 351–359.

Hu P, Elliott J, McCready P *et al.* (1998) Structural organization of virulence-associated plasmids of *Yersinia pestis*. *J. Bacteriol.* 180: 5192–5202.

Ibrahim A, Goebel BM, Liesack W, Griffiths M, Stackebrandt E (1993) The phylogeny of the genus *Yersinia* based on 16S rDNA sequences. *FEMS Microbiol. Lett.* 114: 173–177.

Iqbal SS, Chambers JP, Brubaker RR, Goode MT, Valdez JJ (1999) Detection of *Yersinia pestis* using branched DNA. *Mol. Cell. Probes* 13: 315–320.

Iqbal SS, Chambers JP, Goode MT, Valdes JJ, Brubaker RR (2000) Detection of *Yersinia pestis* by pesticin fluorogenic probe coupled PCR. *Mol. Cell. Probes* 14: 109–114.

Iteman I, Guiyole A, De Almeida AMP, Guilvot I, Barabton G, Carniel E (1993) Relationship between loss of pigmentation and deletion of the chromosomal iron-regulated *irp2* gene in *Yersinia pestis*: evidence for separate but related events. *Infect. Immun.* 61: 2717–2722.

Jackson S, Burrows TW (1956) The virulence enhancing effect of iron on non-pigmented mutants of virulent strains of *Pasteurella pestis*. *Br. J. Exp. Pathol.* 37: 577–583.

Jones HA, Lillard JW, Perry RD (1999) HmsT, a protein essential for expression of the haemin storage (Hms+) phenotype of *Yersinia pestis*. *Microbiology* 145: 2117–2128.

Kutyrev VV, Filippov AA, Oparina OS, Protsenko OA (1992) Analysis of *Yersinia pestis* chromosomal determinants Pgm+ and Pst^s associated with virulence. *Microb. Pathog.* 12: 177–186.

Kutyrev V, Mehigh RJ, Motin VL, Pokrovskaya MS, Smirnov GB, Brubaker RR (1999) Expression of the plague plasminogen activator in *Yersinia pseudotuberculosis* and *Escherichia coli*. *Infect. Immun.* 67: 1359–1367.

Lähteenmäki K, Virkola R, Sarén A, Emödy E, Korhonen TK (1998) Expression of the plasminogen activator Pla of *Yersinia pestis* enhances bacterial enhancement to the mammalian extracellular matrix. *Infect. Immun.* 66: 5755–5762.

Lambert de Rouvroit C, Sluiters C, Cornelis GR (1992) Role of the transcriptional activator, VirF, and temperature in the expression of the pYV plasmid genes of *Yersinia enterocolitica*. *Mol. Microbiol.* 6: 395–409.

Lawton WD, Erdman RL, Surgalla MJ (1963) Biosynthesis and purification of V and W antigen in *Pasteurella pestis*. *J. Immunol.* 91: 179–184.

Lee VT, Schneewind O (1999) Type III secretion machines and the pathogenesis of enteric infections caused by *Yersinia* and *Salmonella* spp. *Immunol. Rev.* 168: 241–255.

Lillard JW Jr, Bearden SW, Fetherston JD, Perry RD (1999) The haemin storage (Hms$^+$) phenotype of *Yersinia pestis* is not essential for the pathogenesis of bubonic plague in mammals. *Microbiology* 145: 197–209.

Lucier TS, Brubaker RR (1992) Determination of genome size, macrorestriction pattern polymorphism and nonpigmentation-specific deletion in *Yersinia pestis* by pulsed-field gel electrophoresis *J. Bacteriol.* 174: 2078–2086.

Lucier TS, Fetherston JD, Brubaker RR, Perry RD (1996) Iron uptake and iron-repressible polypeptides in *Yersinia pestis*. *Infect. Immun.* 64: 3023–3031.

Madison RR (1936) Fibrinolytic specificity of *Bacillus pestis*. *Proc. Soc. Exp. Biol. Med.* 34: 301–302.

Mehigh RJ, Sample AK, Brubaker RR (1989) Expression of the low-calcium response in *Yersinia pestis*. *Microb. Pathog.* 6: 203–217.

Mehigh RJ, Brubaker RR (1993) Major stable peptides of *Yersinia pestis* synthesized during the low-calcium response. *Infect. Immun.* 61: 13–22.

Monack DM, Mecsas J, Ghori N, Falkow S (1997) *Yersinia* signals macrophages to undergo apoptosis and YopJ is necessary for this cell death. *Proc. Natl Acad. Sci. USA* 94: 10385–10390.

Monack DM, Mecsas J, Bouley D, Falkow S (1998) *Yersinia*-induced apoptosis *in vivo* aids in the establishment of a systemic infection of mice. *J. Exp. Med.* 188: 2127–2137.

Moore RL, Brubaker RR (1975) Hybridization of deoxyribonucleotide sequences of *Yersinia enterocolitica* and other selected members of *Enterobacteriaceae*. *Int. J. Syst. Bacteriol.* 25: 336–339.

Motin VL, Nakajima R, Smirvov GB, Brubaker RR (1994) Passive immunity to yersiniae mediated by anti-recombinant V antigen and Protein A-V antigen fusion peptide. *Infect. Immun.* 62: 4192–4201.

Motin VM, Nedialkov YA, Brubaker RR (1996) V antigen-polyhistidine fusion peptide: binding to LcrH and active immunity against plague. *Infect. Immun.* 64: 4313–4318.

Motin VL, Kutas SM, Brubaker RR (1997) Suppression of mouse skin allograft rejection by Protein A-yersiniae V antigen fusion peptide. *Transplantation* 63: 1040–1042.

Moulder JW (1962) *The Biochemistry of Intracellular Parasitism.* Chicago: The University of Chicago Press.

Nakajima R, Brubaker RR (1993) Association between virulence of *Yersinia pestis* and suppression of gamma interferon and tumor necrosis factor alpha. *Infect. Immun.* 61: 23–31.

Nakajima R, Motin VL, Brubaker RR (1995) Suppression of cytokines in mice by Protein A-V antigen fusion peptide and restoration of synthesis by active immunization. *Infect. Immun.* 63: 3021–3029.

Nedialkov YA, Motin VL, Brubaker RR (1997) Resistance to lipopolysaccharide mediated by the *Yersinia pestis* V antigen-polyhistidine fusion peptide: amplification of interleukin-10. *Infect. Immun.* 65: 1196–1203.

Neidhardt FC, Curtiss RI, Ingraham JL *et al.* (1996) *Escherichia coli and Salmonella. Cellular and Molecular Biology.* Washington, DC: ASM Press.

Newman EB, Lin R (1995) Leucine-responsive regulatory protein: a global regulator of gene expression in *E. coli. Annu. Rev. Microbiol.* 49: 747–775.

Nilles ML, Williams AW, Skrzypek E, Straley SC (1997) *Yersinia pestis* LcrV forms a stable complex with LcrG and may have a secretion-related regulatory role in the low-Ca^{2+} response. *J. Bacteriol.* 179: 1307–1316.

Nilles ML, Fields KA, Straley SC (1998) The V antigen of *Yersinia pestis* regulates Yop vectorial targeting as well as Yop secretion through effects on YopB and LcrG. *J. Bacteriol.* 180: 3410–3420.

Norkina OV, Kulichenko AN, Gintsburg AL *et al.* (1993) Development of a diagnostic test for *Yersinia pestis* by the polymerase chain reaction. *J. Appl. Bacteriol.* 76: 240–245.

Orth K, Palmer LE, Bao ZQ *et al.* (1999) Inhibition of the mitogen-activated protein kinase superfamily by a *Yersinia* effector. *Science* 285: 1920–1923.

Palmer LE, Hobbie S, Galán JE, Bliska JB (1998) YopJ of *Yersinia pseudotuberculosis* is required for the inhibition of macrophage TNF-α production and downstream regulation of the MAP kinase p38 and JNK. *Mol. Microbiol.* 27: 953–965.

Perry RD, Brubaker RR (1979) Accumulation of iron by yersiniae. *Infect. Immun.* 137: 1290–1298.

Perry RD, Pendrak ML, Schuetze P (1990) Identification and cloning of a hemin storage locus involved in the pigmentation phenotype of *Yersinia pestis. J. Bacteriol.* 172: 5929–5937.

Perry RD, Fetherston JD (1997) *Yersinia pestis*-etiologic agent of plague. *Clin. Microbiol. Rev.* 10: 35–66.

Pilsl H, Killmann H, Hantke K, Braun V (1996) Periplasmic location of the pesticin immunity protein suggests inactivation of pesticin in the periplasm. *J. Bacteriol.* 178: 2431–2435.

Porat R, McCabe WR, Brubaker RR (1995) Lipopolysaccharide-associated resistance to killing of yersiniae by complement. *J. Endotox. Res.* 2: 91–97.

Portnoy DA, Falkow S (1981) Virulence-associated plasmids from *Yersinia enterocolitica* and *Yersinia pestis*. *J. Bacteriol.* 148: 877–883.

Portnoy DA, Moseley SL, Falkow S (1981) Characterization of plasmids and plasmid-associated determinants of *Yersinia enterocolitica* pathogenesis. *Infect. Immun.* 31: 775–782.

Rakin A, Saken E, Harmsen D, Heesemann J (1994) The pesticin receptor of *Yersinia enterocolitica*: a novel virulence factor with dual function. *Mol. Microbiol.* 13: 253–263.

Roggenkamp A, Geiger AM, Leitritz L, Kessler A, Heeseman J (1997) Passive immunity to infection with *Yersinia* spp. mediated by anti-recombinant V antigen is dependent on polymorphism of V antigen. *Infect. Immun.* 65: 446–451.

Rohde JR, Luan X-S, Rohde H, Fox JM, Minnich SA (1999) The Yersinia enterocolitica pYV virulence plasmid contains multiple intrinsic DNA bends which melt at 37°C. *J. Bacteriol.* 181: 4198–4204.

Rosqvist R, Skurnik M, Wolf-Watz H (1988) Increased virulence of *Yersinia pseudotuberculosis* by two independent mutations. *Nature* 334: 522–525.

Rudolph AE, Stuckey JA, Zhao Y *et al.* (1999) Expression, characterization, and mutagenesis of the *Yersinia pestis* murine toxin, a phospholipase D superfamily member. *J. Biol. Chem.* 274: 11824–11831.

Sample AK, Brubaker RR (1987) Post-translational regulation of Lcr plasmid-mediated peptides in pesticinogenic *Yersinia pestis*. *Microb. Pathog.* 3: 239–248.

Sawa T, Yahr TL, Ohara M *et al.* (1999) Active and passive immunization with the Pseudomonas V antigen protects against type III intoxication and lung injury. *Nature Med.* 5: 392–398.

Schmidt A, Rollinghoff M, Beuscher HU (1999) Suppression of TNF by V antigen of *Yersinia* spp. involves activated T cells. *Eur. J. Immunol.* 29: 1149–1157.

Sikkema DJ, Brubaker RR (1987) Resistance to pesticin, storage of iron and invasion of HeLa cells by yersiniae. *Infect. Immun.* 55: 572–578.

Sikkema DJ, Brubaker RR (1989) Outer-membrane peptides of *Yersinia pestis* mediating siderophore-independent assimilation of iron. *Biol. Metals* 2: 174–184.

Smith H, Keppie J, Cocking EC, Witt K (1960) The chemical basis of the virulence of *Pasteurella pestis* I. The isolation and the aggressive properties *of Past. pestis* and its products from infected guinea pigs. *Br. J. Exp. Pathol.* 41: 452–459.

Smith T (1934) *Parasitism and Disease.* Princeton, NJ: Princeton University Press.

Sodeinde OA, Goguen JD (1988) Genetic analysis of the 9.5-kilobase virulence plasmid of *Yersinia pestis*. *Infect. Immun.* 56: 2743–2748.

Sodeinde OA, Sample AK, Brubaker RR, Goguen JD (1988) Plasminogen activator/coagulase gene of *Yersinia pestis* is responsible for degradation of plasmid-encoded outer membrane proteins. *Infect. Immun.* 56: 2749–2752.

Sodeinde OA, Goguen JD (1989) Nucleotide sequence of the plasminogen activator gene of *Yersinia pestis*: relationship to *ompT* of *Escherichia coli* and gene E of *Salmonella typhimurium*. *Infect. Immun.* 57: 1517–1523.

Staggs TM, Perry RD (1992) Fur regulation in *Yersinia* species. *Mol. Microbiol.* 6: 2507–2516.

Straley SC, Brubaker RR (1981) Cytoplasmic and membrane proteins of yersiniae cultivated under conditions simulating mammalian intracellular environment. *Proc. Natl Acad. Sci. USA* 78: 1224–1228.

Straley SC, Bowmer WS (1986) Virulence genes regulated at the transcriptional level by Ca^{2+} in *Yersinia pestis* include structural genes for outer membrane proteins. *Infect. Immun.* 51: 445–454.

Straley SC, Cibull ML (1989) Differential clearance and host-pathogen interactions of YopE⁻ and YopK⁻ YopL⁻ *Yersinia pestis* in BALB/c mice. *Infect. Immun.* 57: 1200–1210.

Thompson JM, Jones HA, Perry RD (1999) Molecular characterization of the hemin uptake locus (*hmu*) from *Yersinia pestis* and analysis of *hmu* mutants for hemin and hemoprotein utilization. *Infect. Immun.* 67: 3879–3892.

Tsukano H, Itoh K, Suzuki S, Watanabe H (1996) Detection and identification of *Yersinia pestis* by polymerase chain reaction (PCR) using multiplex primers. *Microbiol. Immunol.* 40: 773–775.

Une T, Brubaker RR (1984a) *In vivo* comparison of avirulent Vwa⁻ and Pgm⁻ or Pstʳ phenotypes of yersiniae. *Infect. Immun.* 43: 895–900.

Une T, Brubaker RR (1984b) Roles of V antigen in promoting virulence and immunity in yersiniae. *J. Immunol.* 133: 2226–2230.

Une T, Nakajima R, Brubaker RR (1986) Roles of V antigen in promoting virulence in *Yersinia*. *Contrib. Microbiol. Immunol.* 9: 179–185.

Vollmer V, Pilsl H, Hantke K, Holtje J-V, Braun V (1997) Pesticin displays muramidase activity. *J. Bacteriol.* 179: 1580–1583.

Weinberg ED, Weinberg GA (1995) The role of iron in infection. *Curr. Opin. Infect. Dis.* 8: 164–169.

Welkos S, Friedlander A, McDowell D, Weeks J, Tobery S (1998) V antigen of *Yersinia pestis* inhibits neutrophil chemotaxis. *Microb. Pathol.* 24: 185–196.

Winter CC, Cherry WB, Moody MD (1960) An unusual strain of *Pasteurella pestis* isolated from a fatal case of human plague. *Bull. WHO* 23: 408–409.

Yahr TL, Mende-Mueller LM, Friese MB, Frank DW (1997) Identification of type III secreted products of the *Pseudomonas aeruginosa* exoenzyme S regulon. *J. Bacteriol.* 179: 7165–7168.

Yersinia pestis Sequencing Group (1998) *Y. pestis* genome database [Online.]. Sanger Centre Cambridge, England, http://www.sanger.ac.uk/Projects/Y_pestis/.

Zink DL, Feeley JC, Wells JG *et al.* (1980) Plasmid-mediated tissue invasiveness in *Yersinia enterocolitica*. *Nature* 283: 224–226.

97

Borrelia burgdorferi

Guiqing Wang, Gary P. Wormser and Ira Schwartz

New York Medical College, Valhalla, New York, USA

Borrelia burgdorferi is the cause of Lyme borreliosis (LB) or Lyme disease (Johnson, 1984). It possesses some genetic and phenotypic features that are unique among prokaryotes, such as a linear chromosome and multiple linear and circular plasmids in a single cell. *B. burgdorferi* is maintained in nature mainly by means of a rodent/bird–tick cycle and is occasionally transmitted to humans by the bite of an infected *Ixodes* tick. Clinical manifestations of Lyme borreliosis depend upon the stage of the infection and may affect dermatological, neurological, cardiac and musculoskeletal systems (Steere, 1989; Nadelman and Wormser, 1998). Several *B. burgdorferi* species associated with Lyme borreliosis have been described and are referred to collectively as '*B. burgdorferi* sensu lato' (Baranton *et al.*, 1992; Wang *et al.*, 1999a). In this chapter, general aspects regarding *B. burgdorferi* sensu lato, hereafter referred to *B. burgdorferi* unless otherwise noted, are presented, with a focus on its genome and molecular characteristics.

Lyme borreliosis was first described as a multisystem disease in the late 1970s when Steere *et al.* (1977a, b, 1978) investigated the clinical and epidemiological features of an arthritic illness with geographic clustering in the communities surrounding Old Lyme, Connecticut. In 1981, Burgdorfer and his colleagues isolated a spirochaete from an adult *Ixodes dammini* tick collected on Shelter Island, a known focus for Lyme borreliosis in New York State. The spirochaete reacted immunologically with the sera of patients with Lyme borreliosis, and was termed *I. dammini* spirochaete (Burgdorfer *et al.*, 1982). By 1983, the *I. dammini* spirochaete was cultured directly from patients with Lyme borreliosis and was conclusively demonstrated to be the aetiological agent of human disease (Benach *et al.*, 1983; Steere *et al.*, 1983). In 1984 this spirochaete was designated *Borrelia burgdorferi*, a new member of the genus *Borrelia* (Johnson, 1984).

In Europe, most of the clinical syndromes of Lyme borreliosis had been reported since 1883, mainly independently of each other, before the systematic description of Lyme borreliosis in the United States (Weber and Pfister, 1993). *B. burgdorferi* DNA has now been detected in museum specimens of *Ixodes* ticks and rodents collected on the European and North American continents more than 100 years ago (Marshall *et al.*, 1994; Matuschka *et al.*, 1996).

Classification

The spirochaetes represent a phylogenetically ancient and distinct group among the prokaryotes. They are currently categorised into a single order 'Spirochaetales' in the class 'Spirochaetes' of the phylum 'Spirochaetes phy. nov.'. In the second edition of *Bergey's Manual of Systematic Bacteriology*, the order 'Spirochaetales' is re-classified into three families: *Spirochaetaceae* (family I), *Serpulinaceae* (family II), and

Leptospiraceae (family III) (Garrity and Holt, 2000). The family *Spirochaetaceae* consists of nine genera: *Spirochaeta*, *Borrelia*, *Brevinema*, *Clevelandina*, *Cristispira*, *Diplocalyx*, *Hollandina*, *Pillotina* and *Treponema*. Family *Serpulinaceae* contains two genera *Brachyspira* and *Serpulina*; and family *Leptospiraceae* encompasses the genera *Leptonema* and *Leptospira*.

A total of 28 species that include the Lyme borreliosis borreliae and the relapsing fever borreliae have been officially recognised in the genus *Borrelia*. Based on DNA–DNA homology analysis and phylogeny inferred from 16S rRNA gene sequences, 10 *B. burgdorferi* sensu lato species have been described: *B. burgdorferi* sensu stricto, *Borrelia garinii*, *Borrelia afzelii*, *Borrelia japonica*, *Borrelia andersonii*, *Borrelia valaisiana*, *Borrelia lusitaniae*, *Borrelia tanukii*, *Borrelia turdi* and *Borrelia bissettii* (Wang *et al.*, 1999a). All of these species, except *B. andersonii* and *B. bissettii*, are also validated by the International Union of Microbiological Societies. The major features of these *B. burgdorferi* sensu lato species are summarised in **Table 1**. In addition, a number of *B. burgdorferi* isolates with genetic and phenotypic characteristics distinct from those of the described species have been reported (Casjens *et al.*, 1995; Postic *et al.*, 1998; Wang *et al.*, 1999b), but their taxonomic status remains to be determined.

Identification

Traditionally, the identification of spirochaetes has been based mainly on their morphological, physiological, biochemical and ecological characteristics. *B. burgdorferi* has the basic phenotypic features of spirochaetes and of the genus *Borrelia*. Therefore, it can be easily distinguished from other eubacteria (Barbour and Hayes, 1986). Typically, *Borrelia* cells are helical with dimensions of 0.2–0.5 μm by 10–30 μm and motile with three modes of movement. An outer membrane surrounds the protoplasmic cylinder complex. The flagella of the spirochaetes are entirely endocellular, and are located in the periplasmic space between the outer membrane and the protoplasmic cylinder.

Further, *B. burgdorferi* can be differentiated from other pathogenic treponemes and leptospires on the basis of morphological traits. These include the wavelength of the cell coils, the presence or absence of terminal hooks, the shape of the cell poles and the number of periplasmic flagella. For instance, the cells of *Borrelia* are generally larger but with fewer and looser coils than treponemes, whereas the cells of leptospires are thinner than *Borrelia* but tightly coiled with a hook at one or both ends. Other morphological details for these three pathogenic spirochaetes are

Table 1 Characteristics of *B. burgdorferi* sensu lato species associated with Lyme borreliosis[a]

Taxon	Vector	Animal host	Human disease	Distribution	References
B. burgdorferi sensu stricto[b]	*I. scapularis*, *I. pacificus* *I. ricinus*	Mammals, birds	EM, arthritis, carditis, neuroborreliosis	USA Europe	Baranton *et al.* (1992)
B. garinii	*I. ricinus*, *I. persulatus*	Birds, small mammals	EM, arthritis, neuroborreliosis	Europe, Asia	Baranton *et al.* (1992)
B. afzelii	*I. ricinus*, *I. persulatus*	Small mammals	EM, arthritis, neuroborreliosis, ACA	Europe, Asia	Canica *et al.* (1993) Baranton *et al.* (1992)
B. japonica	*I. ovatus*	Small mammals	No	Japan	Kawabata *et al.* (1993)
B. valaisiana	*I. ricinus* *I. nipponensis*, *I. columnae*	Birds	Unclear	Europe Japan, Korea	Wang *et al.* (1997)
B. lusitaniae	*I. ricinus*	Unknown	No	Central Europe	Le Fleche *et al.* (1997)
B. andersonii	*I. dentatus*	Rabbit	No	USA	Marconi *et al.* (1995)
B. bissettii[c]	*I. pacificus*, *I. neotomae*, *I. scapularis*	Rodents, birds	Unclear	USA	Postic *et al.* (1998)
B. takunii	*I. takunus*	Small mammals	No	Japan	Fukunaga *et al.* (1996a)
B. turdi	*I. turdus*	Small mammals	No	Japan	Fukunaga *et al.* (1996a)

[a] Modified from Wang *et al.* (1999a).

[b] The presence of *B. burgdorferi* sensu stricto in Asia is controversial (Wang *et al.*, 1999a).

[c] Nine isolates with genetic and phenotypic similarity to *B. bissettii* isolate 25 015 were cultured from patients with EM and lymphocytoma in Slovenia.

described in the *Bergey's Manual of Systematic Bacteriology* (Canale-Parola *et al.*, 1989) and the *Bergey's Manual of Determinative Bacteriology* (Holt, 1994).

It is almost impossible to discriminate *B. burgdorferi* from the relapsing fever borreliae, and to distinguish different species of *B. burgdorferi* sensu lato, by their morphology. Lyme borreliosis spirochaetes do not have a specific parasite–vector relationship involving a single tick species. Therefore, the identification of *B. burgdorferi* species largely depends on analyses of genetic characteristics. For this purpose, several genetic approaches, such as DNA–DNA homology analysis, DNA sequencing of 16S rRNA and other conserved genes, ribosomal DNA restriction fragment analysis, pulsed-field gel electrophoresis and randomly amplified polymorphic DNA analysis have been developed and used widely for the identification and classification of *B. burgdorferi* (Wang *et al.*, 1999a). As will be discussed below, isolates belonging to different *B. burgdorferi* sensu lato species can easily be distinguished from each other by most of the genotyping methods.

Molecular Identification of *B. burgdorferi* Species

DNA–DNA Homology Analysis

DNA homology analysis is currently used as a reference method for species delineation of *B. burgdorferi* sensu lato. As recommended by the Ad Hoc Committee on reconciliation of approaches to bacterial systematics, the phylogenetic definition of a species generally would include strains with approximately 70% or more DNA–DNA relatedness and with a ΔT_m of 5°C or less (Wayne *et al.*, 1987). Usually, *B. burgdorferi* isolates within a species have 70–100% DNA homology, while isolates from different species exhibit only 30–65% DNA homology. For example, the type strain of *B. afzelii*, VS461, has 48% DNA homology to the type strain of *B. burgdorferi* sensu stricto B31, 65% to *B. garinii* strain 20047, 54% to *B. japonica* HO14, 64% to *B. valaisiana* VS116 and 58% to *B. lusitaniae* PotiB2 (Baranton *et al.*, 1992; Kawabata *et al.*, 1993; Postic *et al.*, 1994).

Ribosomal DNA Restriction Fragment Analysis (Ribotyping)

Grouping of bacteria by ribotyping is based on the profiles obtained after restriction enzyme digestion of chromosomal DNA and hybridisation with a probe derived from a highly conserved ribosomal RNA. Almost all *B. burgdorferi* isolates studied had a 3.2-kb *Eco*RV, a 3.2-kb *Hpa*I or a 2.2-kb *Hind*III restriction

fragment (Baranton *et al.*, 1992; van Dam *et al.*, 1993; Postic *et al.*, 1996). These bands can be used as genetic markers for the identification of *B. burgdorferi*. Moreover, species-specific *Hin*dIII fragments have been reported for *B. burgdorferi* sensu stricto (1.45 kb), *B. garinii* (1.2 kb) and *B. afzelii* (4.0 and 1.45 kb) (van Dam *et al.*, 1993).

Pulsed-field Gel Electrophoresis and Plasmid Fingerprinting

Discrimination between *B. burgdorferi* species has been achieved by analyses of the large restriction-fragment length polymorphism (LRFP) and the presence of species-specific fragments as obtained by pulsed-field gel electrophoresis (PFGE) (Busch *et al.*, 1996a). Analysis of the plasmid profiles of *B. burgdorferi* isolates also provides a potential for strain and species identification (Xu and Johnson, 1995).

Randomly Amplified Polymorphic DNA Fingerprinting

Randomly amplified polymorphic DNA (RAPD) fingerprinting (Williams *et al.*, 1990), or arbitrarily primed PCR uses low-stringency PCR amplification with a single primer of arbitrary sequence to generate strain-specific arrays of anonymous DNA fragments. It is reported that RAPD is a powerful tool for *B. burgdorferi* strain and species delineation (Welsh *et al.*, 1992; Wang *et al.*, 1998). Different *B. burgdorferi* species and isolates within each species have been distinguished by this technique.

Ribosomal DNA PCR-RFLP

The rRNA gene cluster of *B. burgdorferi* contains a single copy of a 16S rRNA sequence (*rrs*) and tandemly repeated 23S (*rrlA* and *rrlB*) and 5S rRNA sequences (*rrfA* and *rrfB*) (**Fig. 1**) (Fukunaga *et al.*, 1992; Schwartz *et al.*, 1992a; Gazumyan *et al.*, 1994). Based on this unique rRNA gene structure, a PCR-based RFLP analysis has been developed and used widely for species identification of *B. burgdorferi* (Postic *et al.*, 1994). PCR amplification of the variable *rrfA*–*rrlB* inter-genic spacer usually yields a 225–266-bp amplicon among strains from different *B. burgdorferi* species. Digestion of this amplicon with *Mse*I results in different restriction fragments with species-differentiating characteristics (Wang *et al.*, 1999a).

The *rrs*–*rrlA* inter-genic spacer is about 3.0 kb in *B. burgdorferi* sensu stricto and 5 kb in *B. garinii* and *B. afzelii* (Schwartz *et al.*, 1992a; Ojaimi *et al.*, 1994). Amplification of a portion of this *rrs*–*rrlA* spacer, followed by digestion with *Hin*fI and *Mse*I, produces both species and subspecies-specific RFLP patterns (Liveris *et al.*, 1995). This method has been applied to the typing of *B. burgdorferi* directly in human tissue specimens and field-collected ticks, thus obviating the need for culture isolation (Liveris *et al.*, 1996, 1999).

DNA Sequence Analysis

DNA sequence analysis of some highly conserved gene loci, such as *rrs* (Fukunaga *et al.*, 1996a; Le Fleche

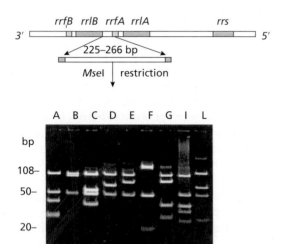

Fig. 1 Ribosomal RNA gene structure and PCR-based RFLP analysis of *B. burgdorferi*. The eight *Borrelia* species included are *B. burgdorferi* sensu stricto (lane A), *B. garinii* (lanes B and C), *B. afzelii* (lane D), *B. japonica* (lane E), *B. valaisiana* (lane F), *B. lusitaniae* (lane G), *B. bissettii* (lane I) and *B. andersonii* (lane L). Modified from Postic *et al.* (1996) with permission.

et al., 1997; Wang *et al.*, 1997) and *fla* (Fukunaga *et al.*, 1996b), is useful for species identification of *B. burgdorferi*. Different *B. burgdorferi* species can also be distinguished from each other by their sequences of *ospA* (Dykhuizen *et al.*, 1993; Will *et al.*, 1995), *hbb* (Valsangiacomo *et al.*, 1997), *p93* (Dykhuizen *et al.*, 1993) and *p39* (Roessler *et al.*, 1997). Further details of these genotyping methods for the differentiation of *B. burgdorferi* sensu lato have been reviewed elsewhere (Wang *et al.*, 1999a).

Structure

Morphology

B. burgdorferi is a Gram-negative helically shaped bacterium with periplasmic flagella. The cells, composed of 3–10 loose coils, are 10–30 μm in length and 0.2–0.5 μm in width (Barbour and Hayes, 1986). *B. burgdorferi* can be easily recognised by dark-field or phase-contrast microscopy; the cultured organisms are motile and swim in freshly prepared slides (**Fig. 2a**). The spirochaetes can also be observed by light microscopy after Gram, Giemsa or silver staining. Recognition of *B. burgdorferi* by microscopy in biological fluids and tissues is usually difficult and challenging, because of the small numbers of organisms, the apparent lack of colony formation in tissues, and the considerable variations in length and structural

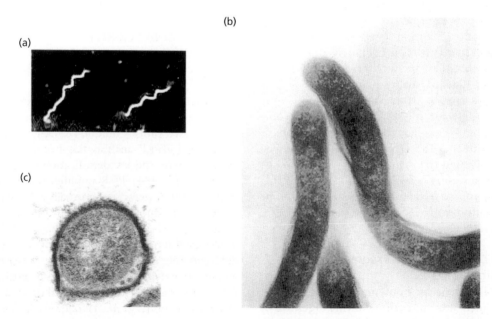

Fig. 2 Micrographs of *B. burgdorferi* by light microscopy (a) and high-voltage electron microscopy (b). The cell diameter of *B. burgdorferi* strain 297 shown in electron microscopy is 0.33 μm. The bundle of periplasmic flagella is clearly visible in the cell on the right (the EM micrograph was provided by K. Buttle, S.F. Goldstein and N.W. Charon). (c) Cross section of *B. burgdorferi* strain B31. The periplasmic flagella are visible in the lower right of the cell (micrograph provided by N.W. Charon, West Virginia University).

morphology of individual spirochaetes (Aberer and Duray, 1991). The end-shape profile of *B. burgdorferi* varies from blunt to pointed. Blebs, which may contain plasmid DNA, and a cyst form of *B. burgdorferi* have been observed (Hayes and Burgdorfer, 1993; Alban *et al.*, 2000). The variability and nature of these structures have not yet been well determined.

Ultrastructure

B. burgdorferi possesses a slime surface layer (S-layer), a trilaminar outer membrane surrounding the periplasmic space that usually contains 7–11 periplasmic flagella and an innermost compartment, the protoplasmic cylinder (Hayes and Burgdorfer, 1993). The S-layer is in immediate apposition to the outer leaflet of the outer membrane. It appears as a homogeneous electron-lucent 4- to 30-nm-thick layer after negative staining. The chemical components of the S-layer have not been determined, partly because it is easily lost in aqueous solutions.

The outer membrane of *B. burgdorferi* is similar to the trilaminar cytoplasmic membrane of eukaryotic cells. This highly flexible membrane possesses an electron-dense outer leaflet and a less electron-dense inner leaflet as determined by thin-section transmission electron microscopy. The membrane architecture and biochemical characteristics of *B. burgdorferi* have been investigated by analysis of the isolated outer membrane (Radolf *et al.*, 1995), by freeze-fracture and high-voltage electron microscopy (Radolf *et al.*, 1994; Goldstein *et al.*, 1996), and immunofluorescence microscopy (Cox *et al.*, 1996). It is proposed that two classes of integral membrane proteins, the surface-exposed lipoproteins and the non-lipidated transmembrane proteins, are embedded in the lipid bilayer of the outer membrane of *B. burgdorferi* (Radolf *et al.*, 1994). A number of abundant outer surface proteins (Osp), such as OspA, OspB and OspC, may be anchored in this outer membrane by the lipid moieties at their N-terminus. In contrast to the enteric Gram-negative bacteria that contain a high density of proteins with extensive β-pleated sheet structure (e.g. porins), the outer membrane of *B. burgdorferi* contains a greater abundance of membrane lipoproteins, relatively few transmembrane proteins and little or no lipopolysaccharide (LPS).

The periplasmic flagella or endoflagella of *B. burgdorferi* are entirely endocellular in the periplasmic space beneath the outer membrane (**Fig. 2**). Genetic evidence has shown that these flagella play a role in spirochaetal cell shape and motility and are important for infection of the mammalian host (Goldstein *et al.*,

1996). In *B. burgdorferi* 7–11 flagella are attached at each end of the periplasmic cylinder and extend toward the opposite end, usually overlapping in the centre of the cell. The flagella of *B. burgdorferi* resemble one another and have the same architecture as flagella of other eubacteria, consisting of a basal disk, neck, hook and filament portion. At the point of insertion, the basal disks at the apical end of the flagella become an integral part of the periplasmic cylinder. Structural analysis with high-voltage electron microscopy reveals that the flagella of *B. burgdorferi* are wrapped around the periplasmic cylinder in a right-handed sense, but these flagella form a left-handed helix in space and have a wavelength (2.83 μm) identical to the cell body (Goldstein *et al.*, 1996).

The cell wall of *B. burgdorferi* consists of a peptidoglycan layer that shows as an electron-dense layer just external to the cytoplasmic membrane of the protoplasmic cylinder by scanning electron microscopy. The peptidoglycan layer of *B. burgdorferi* is known to contain ornithine as the diamino acid. Few data are currently available about the biochemical, immunological and structural features of the peptidoglycan of *B. burgdorferi* and other *Borrelia* species. The innermost compartment of *B. burgdorferi* cells is a protoplasmic cylinder surrounded by a trilaminar cytoplasmic membrane. The protoplasmic cylinder of the spirochaetes contains ground substance, nucleoplasm, ribosomes, mesosomal components and occasional crystalline elements.

Physiology

Growth Conditions *in vitro*

B. burgdorferi sensu lato can be grown routinely in liquid culture under micro-aerophilic conditions in Barbour–Stoenner–Kelly II (BSK II) medium (Barbour, 1984), or variations of it (Preac-Mursic *et al.*, 1986; Pollack *et al.*, 1993). The medium contains *N*-acetylglucosamine, yeast extract, amino acids, vitamins, nucleotides, serum and selective components such as kanamycin and 5-fluorouracil. The optimum growth temperature for *B. burgdorferi* is 30–34°C. The bacterium grows slowly, dividing only every 8–12 hours during the exponential growth phase *in vitro* (Barbour, 1984). Usually, culture-adapted isolates can reach a cell density of 10^6–10^{-8} per mL after cultivation *in vitro* for 5–7 days. The lack of biosynthetic pathways may explain why growth of *B. burgdorferi in vitro* requires serum-supplemented mammalian tissue-culture medium (Fraser *et al.*, 1997).

B. burgdorferi can also be grown on the surface of semi-solid BSK II medium in 1.5% agarose. This makes it possible to obtain individual, pure clones for precise studies of spirochaete genetics and molecular characteristics. Two distinct types of colonies with different diameters are observed after incubation of *B. burgdorferi* on BSK II plates under micro-aerophilic conditions for 3–6 weeks. Colony morphology and plating efficiency vary between strains (Kurtti *et al.*, 1987).

Growth Conditions and Dynamics *in vivo*

Dynamics of Growth and Dissemination in Ticks

Growth of *B. burgdorferi* in ticks has been assessed systematically by feeding nymphal *I. scapularis* on infected mice (Piesman *et al.*, 1990; de Silva and Fikrig, 1995). In one of these studies, *B. burgdorferi* was detected only in the midgut of unfed nymphs and each nymph had a mean of 496 spirochaetes. On attachment of nymphs to the host, the bacteria grew and reached a mean of 7848 spirochaetes per tick 15 hours after attachment. The bacteria appeared to be restricted to the gut during the first 36 hours of rapid growth, but disseminated to the salivary glands in the majority of nymphs after 48 hours. Thus, a critical event that allows the spirochaetes to disseminate and infect the salivary glands takes place 36–48 hours after attachment. A maximum number of 166 575 spirochaetes per nymph was noted 72 hours after attachment. Soon after completion of feeding and detachment from the host, the mean number of spirochaetes decreased to 95 410 per nymph and the spirochaetes appeared to be cleared from organs other than the midgut (de Silva and Fikrig, 1995). Thus, dissemination of spirochaetes within the vector appears to be a transient phenomenon.

Dissemination and Localisation in Animal Models

The laboratory mouse is a useful animal model to study the biological parameters of *B. burgdorferi in vivo* (Moody and Barthold, 1998). By intra-dermal inoculation of 10^4 spirochaetes into C3H/HeN mice, Barthold *et al.* (1991) found that the spirochaetes could be cultured from the blood and spleen of most mice by day 3 and from ear tissue by day 10. The number of spirochaetes in mouse tissues peaked on day 15. Distribution of spirochaetes in the infected mice was multi-focal, with a predilection for collagenous connective tissue of joints, heart, arteries, nerves, muscle and skin. Predominant locations for the spirochaete in heart tissues were observed (Pachner *et al.*, 1995).

Physiology and Biochemical Characteristics

B. burgdorferi lacks genes that encode enzymes required for the synthesis of most amino acids, fatty acids, enzymes co-factors and nucleotides. It also shows limited metabolic capacity. Growth of *B. burgdorferi* depends largely on the availability of nutrients provided in the culture medium or from the host during tick feeding. Analysis of the genome and reconstruction of metabolic pathways suggests that *B. burgdorferi* uses glucose as a primary carbon and energy source, but other carbohydrates such as glycerol, glucosamine, fructose and maltose may be used in glycolysis. *B. burgdorferi* does not contain genes for tricarboxylic acid cycle enzymes or for components of the electron transport system. Thus, it is assumed that lactic acid is the main end product of glycolysis, which is consistent with the micro-aerophilic nature of this spirochaete. A total of 52 open reading frames (ORFs) encode transport and binding proteins which would contribute to 16 distinct membrane transport system for amino acids, carbohydrates, anions and cations (Fraser *et al.*, 1997). The lipid composition of the outer membrane and whole cell are very similar, suggesting that bulk transfer of lipid occurs between the cytoplasmic and outer membranes. Since genes for the respiratory electron transport chain were not identified, ATP production must be accomplished by substrate-level phosphorylation.

The isolated outer membrane of *B. burgdorferi* had a specific gravity of $1.12–1.19 \, \text{g/cm}^2$, depending on the purification procedure used (Bledsoe *et al.*, 1994; Radolf *et al.*, 1995), and comprises approximately 16.5% of the whole spirochaete by dry weight. Chemical analysis of the outer envelope of *B. burgdorferi* reveals a composition of 45.9% protein, 50.8% lipid and 3.3% carbohydrate (Coleman *et al.*, 1986). The purified periplasmic membrane has a lower specific gravity ($1.12 \, \text{g/cm}^2$), but a higher percentage of proteins (56%) by dry weight of the membrane. Since *B. burgdorferi* lacks the ability to elongate long-chain fatty acids, the fatty acid composition of the cells reflects that present in the growth medium. The presence of LPS in the *B. burgdorferi* cell wall is controversial (Beck *et al.*, 1985; Takayama *et al.*, 1987).

Genome

B. burgdorferi is the first spirochaete whose complete genome was determined (Fraser *et al.*, 1997). The genome size of the type strain *B. burgdorferi* sensu stricto B31 is 1 521 419 bp. This consists of a linear chromosome of 910 725 bp, with a $G + C$ content of 28.6%, and 21 plasmids (9 circular and 12 linear) with

Table 2 Genome features of *B. burgdorferi* B31[a]

Replicon	Geometry	Size (bp)	G + C (%)	Coding (%)	Gene[b]
Chromosome	Linear	910 725	28.6	93	853
Plasmid[c]		610 694	27.6	81	837
cp9	Circular	9 386	23.9	75	11
cp26	Circular	26 498	26.5	88	29
cp32-1	Circular	30 750	29.4	92	42
cp32-3	Circular	30 223	28.9	92	44
cp32-4	Circular	30 299	29.3	92	43
cp32-6	Circular	29 838	29.3	92	41
cp32-7	Circular	30 800	29.1	93	42
cp32-8	Circular	30 885	29.1	92	43
cp32-9	Circular	30 651	29.3	92	42
lp5	Linear	5 228	23.8	73	7
lp17	Linear	16 928	23.1	64	28
lp21	Linear	18 901	20.7	32	12
lp 25	Linear	24 177	23.4	66	38
lp28-1	Linear	28 250	32.3	79	49
lp28-2	Linear	29 766	31.6	92	34
lp28-3	Linear	28 601	25.0	66	40
lp28-4	Linear	27 323	24.5	62	31
lp36	Linear	36 849	26.9	76	50
lp38	Linear	38 829	26.1	67	37
lp54	Linear	53 541	28.2	82	76
lp56	Linear	52 971	27.3	87	78
Total		1 521 419	28.2	88	1690

[a] Adapted from Fraser *et al.* (1997), Casjens *et al.* (2000).

[b] Number includes 168 pseudogenes (1 chromosome and 167 plasmid-derived); of the 837 plasmid-encoded genes, 535 of these are larger than 300 bp.

[c] About 2000 bp of undetermined telomere sequences from linear plasmids included.

a combined size of 610 694 bp (Fraser *et al.*, 1997; Casjens *et al.*, 2000). The genome features are summarised in **Table 2**.

Genome analysis has revealed that *B. burgdorferi* possesses certain genetic structures that are unique among prokaryotes. These include (1) a linear chromosome and multiple linear and circular plasmids in a single bacterium, (2) a unique organisation of the rRNA gene cluster, consisting of a single 16S rRNA and tandemly repeated 23S and 5S rRNA genes, (3) a frequency of lipoprotein-encoding genes significantly higher than that of any other bacterial genome sequenced to date, that is 4.9% of the chromosomal genes and 14.5% of the plasmid genes as against only 1.3% in *Helicobacter pylori*, and 2.1% in *T. pallidum*, (4) a substantial fraction of plasmid DNA that appears to be in a state of evolutionary decay, and (5) evidence for numerous, and potentially recent, DNA re-arrangements among the plasmid genes.

Chromosome

The genomes of pathogenic micro-organisms and parasites that have been sequenced vary in size from approximately 580 kb in *Mycoplasma genitalium* to 35 000 kb in *Trypanosoma brucei*. The chromosomes of various *B. burgdorferi* isolates have been found to be approximately 935–955 kb and are among the smallest of the bacterial genomes sequenced to date (Casjens and Huang, 1993; Casjens *et al.*, 1995). The chromosome of *B. burgdorferi* is a linear molecule as identified by pulsed-field electrophoresis and DNA sequence analysis (Ferdows and Barbour, 1989; Casjens *et al.*, 1995; Fraser *et al.*, 1997). The physical and genetic maps of the linear chromosomes of several species, including the pathogenic species *B. burgdorferi* sensu stricto, *B. garinii* and *B. afzelii*, have been determined. Comparison of their physical maps indicates that the gene order on the chromosome in different *B. burgdorferi* sensu lato species is highly conserved (Casjens and Huang, 1993; Ojaimi *et al.*, 1994; Casjens *et al.*, 1995).

In *B. burgdorferi* B31 the chromosome contains 853 predicted ORFs with an average size of 992 bp. Of these, 59% have been assigned biological functions, and 12% matched hypothetical coding sequences of functions unknown from other organisms. A large number of lipoprotein genes and genes unique to

Borrelia spp. were recognised; 29% of the ORFs were unique to *B. burgdorferi*.

The chromosome of *B. burgdorferi* contains a basic set of genes that encode proteins for DNA replication, transcription, translation, solute transport and energy metabolism, but not for most known cellular biosynthetic reactions. The location of the origin of replication for the linear chromosome is not yet known, but it is most likely to be near the centre, as there is a striking break in the GC skew (G–C/G+C) curve near the middle of the chromosome (Fraser *et al.*, 1997). All 61 triplet codons, and 31 tRNAs with specificity for all 20 amino acids, have been identified in the ORFs of *B. burgdorferi*. The most frequently used codons are AAA (Lys), AAU (Asn), AUU (Ile), UUU (Phe), GAA (Glu), GAU (Asp) and UUA (Leu), which account for approximately 40% of triplet codons used (Fraser *et al.*, 1997).

It is estimated that a single copy of the chromosome is present in each cell during cultivation *in vitro*, since the yield of cultured *B. burgdorferi* DNA per bacterium (0.002 pg) is comparable to the calculated DNA content based on its genome size (1.5×10^6 bp, equal to approximately 0.0015 pg). A recent study has shown, however, that *B. burgdorferi* B31 cultured *in vitro* contains 1–17 copies of the chromosome-encoded *gyrB* gene per organism (Shang *et al.*, 2000). This is similar to the observation in *B. hermsii* that it contains 13–18 copies of the chromosome per cell after isolation from infected mice (Kitten and Barbour, 1992). Further studies are required to clarify how many copies of the chromosomes are present per *B. burgdorferi* cell grown *in vivo* and *in vitro*.

Ribosomal RNA Gene Organisation

The rRNA genes are highly conserved among eubacteria. Most are arranged into single operons in the order of 16S (*rrs*)–23S (*rrl*)–5S (*rrf*) rRNA. The number and location of the rRNA operons varies from species to species with, for example, seven copies present in *Escherichia coli* but only one copy in *M. genitalium*. In *B. burgdorferi* B31, the rRNA gene cluster is located between 434 and 447 kb on the chromosome; the gene order is *rrs*–tRNA^{Ala}–tRNA^{Ile}–*rrlA*–*rrfA*–*rrlB*–*rrfB* (Fraser *et al.*, 1997). The sizes of the 16S, 23S and 5S rRNA genes are 1527, 2926 and 112 nucleotides, respectively (Gazumyan *et al.*, 1994). Compared with the rRNA operons of most eubacteria, the rRNA gene cluster in *B. burgdorferi* has a number of unusual features. These include (1) a single copy of 16S rRNA and tandemly repeated 23S and 5S rRNA genes, (2) a large 16S–23S rRNA inter-genic spacer which is 3052 bp in *B. burgdorferi* B31 and approximately 5.0 kb in *B. garinii* and *B. afzelii* (Gazumyan

et al., 1994; Ojaimi *et al.*, 1994), which are significantly larger than those typically found in other eubacteria (400–500 bp), (3) a spacer tRNA^{Ile} gene in the opposite orientation from the other genes in this cluster. Based on the above, it is speculated that the rRNA genes in *B. burgdorferi* are not expressed as a whole operon but as at least three separate transcripts, *rrs*–tRNA^{Ala}, tRNA^{Ile} and *rrlA*–*rrfA*–*rrlB*–*rrfB*.

The 16S rRNAs of *B. burgdorferi* sensu lato have 70–75% sequence similarity to those of *E. coli* and *Leptospira interrogans*, and 90.0–95.6% similarity to the 16S rRNAs of relapsing fever *Borreliae* (Gazumyan *et al.*, 1994). The DNA sequence similarity among different species of *B. burgdorferi* sensu lato ranges from 95.3% to 99.6% (Le Fleche *et al.*, 1997). Analysis of the 16S rRNA gene sequences has been used to compare the evolutionary history of *B. burgdorferi* with other eubacteria and to discriminate different species and subspecies of *B. burgdorferi* sensu lato (Paster *et al.*, 1991; Wang *et al.*, 1999a).

Motility and Chemotaxis Operons

In *B. burgdorferi* B31 a total of 54 genes, or 6.3% of the chromosomal genes, encode proteins involved in motility and chemotaxis. Most of these genes are arranged in eight operons containing from 2 to 25 genes (Fraser *et al.*, 1997). At least four motility and chemotaxis operons (*flaA/che*, *flaB*, *flgB* and *flgK*) have also been described in *B. burgdorferi* isolate 212 (Ge *et al.*, 1997a, b). With few exceptions, the location of the motility operons and the gene order in each operon among *B. burgdorferi* sensu lato species are highly conserved and similar to those of other spirochaetes (Ojaimi *et al.*, 1994; Casjens *et al.*, 1995).

Transcription of the motility and chemotaxis operons in *B. burgdorferi* is initiated by σ^{70}-like promoters (Ge *et al.*, 1997b). This is different from the situation in enteric bacteria and *Bacillus subtilis*, in which the well-conserved σ^{28}-like promoter elements are involved. It is also unlike that in other spirochaetes, such as *Spirochaeta aurantia*, *T. pallidum* and *Serpulina hyodysenteriae* in which the outer layer of flagellar polypeptides are transcribed from σ^{70}-like promoters, whereas the core polypeptide genes are transcribed from σ^{28}-like promoters.

B. burgdorferi maintains duplicate copies of several chemotaxis genes (*cheR*, *cheW*, *cheA*, *cheY* and *cheB*) on a small chromosome, but the reason is not clear. It is possible that multiple *che* genes provide redundancy so that the spirochaetes can migrate to distant sites in the tick and mammalian host by differential gene expression under varied physiological conditions. Alternatively, the flagellar motors at the two ends of the *B. burgdorferi* cell may be different and require distinct *che* systems for more efficient invasion and dissemination in mammalian hosts (Fraser *et al.*, 1997).

Plasmids

Plasmids occur in all studied *B. burgdorferi* isolates (Xu and Johnson, 1995). In *B. burgdorferi* B31, 21 plasmids (12 linear and 9 circular) have been identified, and in size account for approximately one-third of the total genome. The plasmids of *B. burgdorferi* B31 contain 535 predicted ORFs that are larger than 300 bp. Only 8% of these ORFs were assigned biological roles (Fraser *et al.*, 1997; Casjens *et al.*, 2000). Genes encoding the abundant major outer membrane proteins, such as OspA and OspC, as well as genes encoding proteins with important biosynthetic functions, are located on plasmids. The copy number of plasmids relative to the chromosome has been determined in *B. burgdorferi* isolates B31 and Sh-2-82 by comparison of the relative hybridisation of each plasmid to replicon-specific DNA probes (Hinnebusch and Barbour, 1992; Casjens and Huang, 1993; Casjens *et al.*, 1995). All plasmids tested were present in low copy number of about 1–3 per chromosome equivalent, and remained the same even after approximately 7000 generations in continuous culture *in vitro* (Hinnebusch and Barbour, 1992). The number and size of plasmids varies between *B. burgdorferi* isolates, ranging from 5 to 21 and from 9 to 57.7 kb, respectively (Xu and Johnson, 1995; Casjens *et al.*, 2000). Plasmids with different sizes may contain related sequences, while plasmids with the same size may contain different sequences.

One of the unusual features of *B. burgdorferi* plasmids is the co-existence of multiple linear and circular DNA molecules in a single organism. Linear plasmids are linear duplex DNA molecules with covalently closed ends (Barbour and Garon, 1987). This form of DNA occurs only in some animal viruses and was first described in prokaryotic organisms in the relapsing fever spirochaete *B. hermsii* (Plasterk *et al.*, 1985). In *B. burgdorferi* B31, 12 linear plasmids with sizes ranging from 5 to 56 kb were identified (Casjens *et al.*, 2000). A study of the distribution of these linear plasmids among 15 *B. burgdorferi* sensu stricto isolates indicated that 5 of the 12 linear plasmids found in isolate B31 were present in all of the isolates tested; four additional linear plasmids were present in 10–14 of these isolates; only three plasmids appeared to have cognates in fewer than 25% of the isolates examined (Palmer *et al.*, 2000). This suggests that the B31 linear plasmid sequences are usually present in other *B. burgdorferi* isolates.

The mechanisms of replication and segregation for borrelial plasmids are undefined. Telomeric structures at each end of the linear plasmids have been identified that contain a series of phased, short direct repeats and a palindrome adjacent to a highly AT-rich sequence. These are similar to those of a eukaryotic virus (Hinnebusch *et al.*, 1990; Hinnebusch and Barbour, 1991). This finding implies that the novel linear plasmids of *B. burgdorferi* may have originated through a horizontal genetic transfer across kingdoms. In addition, linear plasmid dimers in different species of *B. burgdorferi* (Marconi *et al.*, 1996) and conversion from a linear to a circular plasmid in *B. hermsii* have been reported (Ferdows *et al.*, 1996), suggesting that these alternate plasmid forms may be utilised for linear DNA replication.

Another striking feature of *B. burgdorferi* plasmids is the presence of a large number of paralogous gene families and pseudogenes. A total of 161 paralogous gene families containing 2–41 members have been identified in *B. burgdorferi* B31; 107 of these contain at least one plasmid-borne member (Fraser *et al.*, 1997; Casjens *et al.*, 2000). The significance of the large number of paralogous plasmid-encoded genes is not understood. They may be expressed differentially in diverse environments or undergo homologous recombination to generate antigenic variation. Relatively few pseudogenes have been found in other bacteria. However, 167 pseudogenes have been recognised on the 21 plasmids of *B. burgdorferi* B31. These pseudogenes are defined as 'any region of DNA similar in sequence to a paralogous *Borrelia* predicted gene or to a gene from other organism, but which is obviously truncated and/or does not have full open reading frames relative to its homologues' (Casjens *et al.*, 2000). It is interesting that 87% of these pseudogenes are located on 10 linear plasmids of isolate B31. Most pseudogenes appear to be damaged by deletions, insertions, inversions and frame-shift mutations (Casjens *et al.*, 2000).

Nine circular plasmids in the 29–32 kb range (cp32-1 through cp32-9), which have a high degree of similarity with one another, are carried by *B. burgdorferi* isolate B31 (Stevenson *et al.*, 1996; Casjens *et al.*, 1997, 2000). Truncated forms of these supercoiled plasmids (designated cp18) have been described in strains N40 and 297 (Stevenson *et al.*, 1997; Akins *et al.*, 1999). Multiple plasmids of about 32 kb with sequences related to those of B31 are also present in all other *B. burgdorferi* isolates (Casjens *et al.*, 2000). The ubiquity of these plasmids suggests that they may be important in the natural life cycle of these organisms. Recent evidence indicates that cp32 plasmids may be conjugative plasmids or prophages, since the integration of a cp32-like circular DNA into a linear progenitor to form the linear plasmid lp56 might have occurred (Casjens *et al.*, 2000). Furthermore, cp32 plasmid-related DNA has been identified within the capsids of bacteriophage-like particles released from *B. burgdorferi* isolate CA-11.2A (Eggers and Samuels, 1999).

In *B. burgdorferi*, some plasmids appear to be relatively unstable and can be lost during cultivation *in vitro* (Schwan *et al.*, 1988; Xu *et al.*, 1996). Plasmid loss does not appear to affect the growth of spirochaetes in culture, indicating that they are not essential for cellular housekeeping functions (Sadziene *et al.*, 1995; Tilly *et al.*, 1997). Nevertheless, several studies have correlated the plasmid profile of *B. burgdorferi* with infectivity of the studied isolate. The loss of one or more plasmids in *B. burgdorferi*, such as lp27.5 and lp28-1, may result in significant decreases in pathogenicity in laboratory mice (Xu *et al.*, 1996; Zhang *et al.*, 1997a).

The *ospAB* Operon

The *ospA* and *ospB* genes, located on a 49- to 57-kb linear plasmid (Samuels *et al.*, 1993), encode the major outer membrane proteins OspA and OspB. These were the first plasmid-encoded genes to be characterised in *B. burgdorferi* (Bergstrom *et al.*, 1989). DNA sequence and RT-PCR analysis indicate that these two genes are transcribed from a common promoter and constitute an operon. The deduced translation products from *ospA* and *ospB* in *B. burgdorferi* B31 are 273 amino acids with a molecular weight of 29 334, and 296 amino acids with a molecular weight of 31 739, respectively (Bergstrom *et al.*, 1989). OspA and OspB are very similar in sequence (53%), indicating a recent evolutionary event in which a duplication of the *osp* genes may have occurred. Sequence analysis of *ospA* genes from a large collection of *B. burgdorferi* sensu lato showed homogeneity in *B. burgdorferi* sensu stricto and *B. afzelii* but revealed major subgroups within the *B. garinii* species (Wilske *et al.*, 1993; Will *et al.*, 1995).

The Vmp-like Sequence (*vls*) Locus

In *B. hermsii*, the switch between expression of two variable major protein (Vmp) types (7 and 21) is associated with a DNA re-arrangement between linear plasmids, which results in notable antigenic variation (Plasterk *et al.*, 1985). Recently, a Vmp-like sequence (*vls*) locus has been described, which consists of one expressed *vlsE* gene and 15 silent *vls* cassettes on a 28-kb linear plasmid (lp28-1) in *B. burgdorferi* B31 (**Fig. 3**) (Zhang *et al.*, 1997a). The Vls proteins of *B. burgdorferi* are variable and expressed in experimentally infected C3H/HeJ mice (Zhang *et al.*, 1997a) and in patients with Lyme borreliosis (Kawabata *et al.*, 1998; Lawrenz *et al.*, 1999). A cassette-specific, segmental gene conversion mechanism may be involved in yielding the extensive antigenic variation of these VlsE proteins (Zhang and Norris, 1998). More recently, Vmp-like sequences were also reported to be present in more than 20 *B. burgdorferi* sensu stricto isolates cultured from patients with Lyme borreliosis in North America (Iyer *et al.*, 2000). Comparison of the *vls* sequences from low-passage clinical isolates indicates that both highly conserved and heterogeneous subgroups exist among these *B. burgdorferi* sensu stricto isolates.

The *Borrelia* Direct Repeat (Bdr)

Multiple circular and linear plasmids of *B. burgdorferi* sensu lato and relapsing fever borreliae carry genes for members of the Bdr protein family (Zuckert *et al.*, 1999; Roberts *et al.*, 2000; Zuckert and Barbour, 2000). These plasmid-carried *bdr* genes encode polymorphic, acidic proteins with predicted sizes ranging from 10.7 to 30.6 kDa and with putative phosphorylation sites and transmembrane domains. Several Bdr paralogues with sizes from 19.5 to 30.5 kDa are expressed by cultured strain B31 in a temperature-independent manner. Distinct patterns of cross-reacting proteins of the Bdr family were also detected in other *B. burgdorferi* sensu stricto, *B. garinii* and *B. afzelii* strains as well as in the relapsing fever

(a)

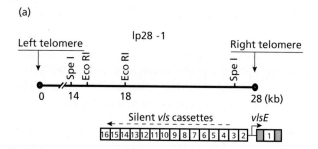

(b)

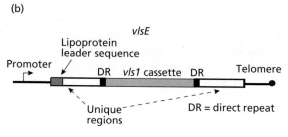

Fig. 3 The Vmp-like sequence (*vls*) locus of *B. burgdorferi* B31. (a) Diagram of the overall arrangement of the *vls* locus in plasmid lp28-1. (b) Structure of *vlsE*. Adapted from Zhang *et al.* (1997a) with permission.

spirochaetes *B. hermsii* and *B. turicatae*. Approximately 50% of sera from patients with Lyme borreliosis had detectable antibodies to one or more of the Bdr proteins (Zuckert *et al.*, 1999). Proteolysis *in situ*, immunofluorescence and growth inhibition assays indicate that Bdr proteins are not surface exposed, but may be located in the periplasmic space. It is speculated that Bdr sequences may play regulatory roles in regulation of gene expression in *Borrelia*.

Genetic Manipulation of the Genome

The genome sequence of *B. burgdorferi* has provided a wealth of data about the genetic composition of this bacterium. Relatively little is known, however, about the function of most of the deduced proteins encoded by the genome. Moreover, molecular studies of the biology of *B. burgdorferi* and the pathogenesis of Lyme borreliosis are hindered by the lack of suitable genetic tools that have been used in other well-characterised bacteria. A selectable marker, *gyrB*(r), a mutated form of the chromosomal *gyrB* gene, which confers resistance to the antibiotic coumermycin A(1), has been developed (Rosa *et al.*, 1996) and used successfully to disrupt the *ospC*, *guaB* and *oppA*IV genes (Tilly *et al.*, 1997; Stevenson *et al.*, 1998). The utility of this coumermycin-resistant *gyrB*(r) gene for targeted gene disruption is limited by a high frequency of recombination with the endogenous *gyrB* gene.

A kanamycin-resistance gene (*kan*) has been introduced into *B. burgdorferi*, which allowed for more efficient direct selection of mutants and hence significantly improved the ability to construct isogenic mutant strains (Bono *et al.*, 2000). In addition, a plasmid that is able to propagate in both Gram-positive and Gram-negative bacteria and that confers erythromycin resistance has been propagated extra-chromosomally in *B. burgdorferi* B31 after electroporation and has been used to express enhanced green fluorescent protein (*egfp*) in strain B31 under the control of the *flaB* promoter (Sartakova *et al.*, 2000). These advances should provide a basis for further development of more efficient genetic systems for *B. burgdorferi*.

Bacteriophage are widely used as tools for genetic manipulation in bacteria. Bacteriophage have been detected in *B. burgdorferi* isolates cultured under normal conditions (Hayes *et al.*, 1983) or treated with the DNA gyrase inhibitor ciprofloxacin (Neubert *et al.*, 1993). Eggers and Samuels (1999) reported the recovery of a new bacteriophage with a polyhedral head and a diameter of 55 nm that appears to have a simple 100-nm-long tail. Molecular analysis of this bacteriophage reveals that its DNA is a 32-kb double-stranded linear molecule, possibly derived from the 32-kb circular plasmids. Since they are neither routinely observed nor reproducibly induced in most borrelial cultures, phage particles of *B. burgdorferi* have not been well characterised.

Genome Comparison between *B. burgdorferi* and *T. pallidum*

The complete genome of another pathogenic spirochaete, *Treponema pallidum* has been determined (Fraser *et al.*, 1998). The genome of *T. pallidum* subsp. *pallidum* (Nichols) is a circular chromosome of 1 138 006 bp with a G + C content of 52.8%. There are a total of 1041 predicted ORFs, with an average size of 1023 bp, representing 92.9% of total genomic DNA. The number of genes that have been assigned biological roles in *T. pallidum* (55% of the 1041 ORFs) is comparable with that in *B. burgdorferi* (59% of the 853 chromosomal genes) (Fraser *et al.*, 1997, 1998).

Comparison of the *T. pallidum* genome sequence with that of *B. burgdorferi* suggests that less than one-half of the genes show readily detectable orthologous relationships and this substantiates the considerable diversity among pathogenic spirochaetes. Among the 476 (46% of total) ORFs in *T. pallidum* that have orthologues in *B. burgdorferi*, 76% have a predicted biological function. More than 40% of these genes are also highly conserved in other bacteria (Fraser *et al.*, 1998). In general, the genes encode proteins for the flagellar apparatus and those involved in core biological function, such as genome replication and expression and have a high orthology coefficient value (OC > 0.8) between *B. burgdorferi* and *T. pallidum*. In contrast, marked variability was found among proteins involved in specific processes, such as nutrient transport, metabolism, gene-specific transcription regulation, signal transduction and host response (Subramanian *et al.*, 2000). Some of the genes conserved in these two spirochaetes are not recognised in other bacterial genomes. They are likely to represent 'spirochaete-specific' genes that contribute to the unusual structural properties of these bacteria. Of the 54% of *T. pallidum* ORFs that are not shared with *B. burgdorferi*, more than 80% are of unknown biological function.

Population Genetics

Clonality of *B. burgdorferi* Population

Population genetic analysis can provide valuable insights for epidemiological tracking of the spread and stability of microbe genotypes over space and time, to define and delimit currently described taxa and to search for new genetic subdivisions within species and to evaluate the impact of microbial genetic diversity on their biological properties (Tibayrenc, 1995). Analysis by multi-locus enzyme electrophoresis has revealed a clonal population structure of *B. burgdorferi* based upon the linkage disequilibrium of allele distributions (Boerlin *et al.*, 1992), indicating that in the absence of recombination all genes in strains belonging to the same species share a common evolutionary history. This conclusion has also been drawn from RAPD fingerprinting and comparison of the sequences of some highly conserved chromosomal genes such as *fla* and *p93* (Welsh *et al.*, 1992; Dykhuizen *et al.*, 1993). In addition, clonality is supported by the conserved chromosomal gene order in different *B. burgdorferi* species (Casjens *et al.*, 1995; Ojaimi *et al.*, 1994).

It is not clear whether such populations are strictly clonal, since localised horizontal gene transfer of the outer membrane protein-encoding genes has been reported in *B. burgdorferi* sensu lato. For example, a close examination of the highly variable *ospC* gene indicates that lateral gene transfer may have occurred quite frequently, not only between members of the same species, but also between strains from different species (Livey *et al.*, 1995; Ras *et al.*, 1997). As a result of such gene transfer and subsequent re-combination, the linkage disequilibrium among different species may be disturbed. This would lead to a different topology of phylogenetic relationship based on *ospC* sequence analysis as compared with that based on *rrs* gene analysis. Nevertheless, *ospC* genes from strains of the same species appear to be more closely related to each other than to *ospC* genes from different species. Furthermore, species-specific motifs recognised in the conserved N-terminal of the OspC proteins indicate that *ospC* is still clonally inherited (Livey *et al.*, 1995; Fukunaga *et al.*, 1996c).

It is assumed that speciation among *B. burgdorferi* sensu lato is possibly a recent phenomenon, since the *rrs* gene sequences are highly conserved among different *B. burgdorferi* species. Given the greater diversity of *B. burgdorferi* in Eurasia than in North America, it is likely that an ancestor of this complex originated in Eurasia. This hypothesis may, however, be challenged by the increasing awareness of the genetic diversity of the *B. burgdorferi* population in North America (Mathiesen *et al.*, 1997; Ras *et al.*, 1997; Postic *et al.*, 1998).

Genetic Diversity of *B. burgdorferi* Population

B. burgdorferi was originally thought to be a single species that causes human Lyme borreliosis. It was recognised to be divergent when differential reactivity against two monoclonal antibodies (H6831 and H5TS) was observed among North American strains and when the antigenic properties of North American and European isolates were compared (Barbour *et al.*, 1985). Subsequently, the phenotypic and genetic properties of a large number of *B. burgdorferi* isolates from various geographic and biological sources have been investigated with different molecular approaches (reviewed by Wang *et al.*, 1999a). Thereafter considerable genetic and phenotypic diversity among *B. burgdorferi* was documented (Baranton *et al.*, 1992; van Dam *et al.*, 1993; Wilske *et al.*, 1993; Casjens *et al.*, 1995; Picken *et al.*, 1995; Mathiesen *et al.*, 1997; Postic *et al.*, 1998; Saint Girons *et al.*, 1998; Wang *et al.*, 1998), and this led to the delineation of 10 different species in the *B. burgdorferi* sensu lato complex (Wang *et al.*, 1999a). The observed genetic diversity of *ospC* in a single population circulating on Shelter Island, New York may reflect the high genetic heterogeneity of the *B. burgdorferi* population in nature (Wang *et al.*, 1999d).

B. burgdorferi is maintained in nature mainly by a tick–rodent cycle. A variety of observational and experimental data demonstrate substantial genotypic or genetic polymorphism in cultured isolates or PCR-amplified *B. burgdorferi*-specific DNAs from natural sources or experimentally infected animals (Picken *et al.*, 1995; Ryan *et al.*, 1998; Hofmeister *et al.*, 1999). The mechanisms for generation of this genetic diversity remain unknown. Selection of distinct genotypes by reservoir hosts and ticks, or adaptive evolution may contribute to the genetic diversity of *B. burgdorferi* populations (Theisen *et al.*, 1995; Kurtenbach *et al.*, 1998). Lateral gene transfer and recombination between *Borrelia* isolates belonging to the same species and between different species in hosts and vectors with mixed infections may also play a role in the maintenance of the genetic heterogeneity of *B. burgdorferi* in nature. As reported recently, the genetic diversity of *B. burgdorferi* among clinical isolates, as measured directly in patient tissues, may be much higher than was assumed based on analysis of cultured isolates (Liveris *et al.*, 1999). Further studies are necessary to compare the genetic heterogeneity and conservation of *B. burgdorferi* populations by the direct typing of samples collected from reservoirs, hosts, ticks and patients in well-defined endemic regions.

DNA Exchange between and within Species

Comparative analysis of the spirochaete genomes has provided evidence of gene exchange with other bacteria, archaea and eukaryotic hosts. These events are assumed to have occurred at different points in evolution and at different taxon levels of the spirochaetes (Subramanian *et al.*, 2000). Notably, DNA re-arrangement, presumably by lateral gene transfer and recombination, has been reported among *B. burgdorferi* sensu stricto, *B. garinii*, *B. afzelii* and *B. valaisiana* and between isolates within each species (Rosa *et al.*, 1992; Marconi *et al.*, 1994; Livey *et al.*, 1995; Ras *et al.*, 1997; Zhang *et al.*, 1997a; Wang *et al.*, 1999c; Casjens *et al.*, 2000; Sung *et al.*, 2000). In *B. burgdorferi* B31, extensive genetic similarities have been observed among various linear and circular plasmids, suggesting possibly inter-plasmid recombination in a single organism. Additional intra-plasmid recombinations, such as the *vls* locus, and possible DNA exchange between the chromosome and plasmids have also been noted (Casjens *et al.*, 2000).

Proteins

Protein profile analysis shows that whole-cell lysates of *B. burgdorferi* comprise more than 50 visible protein bands in silver-stained polyacrylamide gels (**Fig. 4**). In

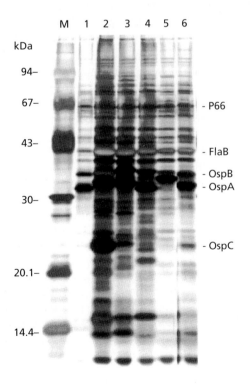

Fig. 4 Silver-stained SDS-PAGE gel of whole cell lysates of five *B. burgdorferi* sensu lato species: *B. burgdorferi* sensu stricto strain B31 (lane 1), *B. garinii* 20047 (lane 2), *B. afzelii* VS461 (lane 3), *B. japonica* HO14 (lane 4) and *B. valaisiana* VS116 and M53 (lanes 5 and 6).

the past two decades considerable effort has been devoted to identifying and characterising these proteins with focus on (1) molecular determinants unique to *B. burgdorferi*, (2) *B. burgdorferi*-specific antigens that are abundant and immunogenic with potential use in improving the specificity and sensitivity of serologic tests, (3) target antigens that elicit a protective immune response and which could be used as candidates for development of an effective vaccine, and (4) proteins that may be involved in the interaction of the spirochaetes with their arthropod and vertebrate hosts through differential gene expression.

Protein expression in *B. burgdorferi* is differentially regulated by the surrounding environment. For example, OspA is expressed in the tick, but is down-regulated in the mammalian host. Conversely, OspC synthesis is up-regulated on tick feeding. A number of proteins are not expressed during culture *in vitro*, but are synthesised *in vivo*. These include EppA, OspE/F paralogues (Erps), Bbk32, Bbk50 and DbpA (de Silva and Fikrig, 1997; Sung *et al.*, 1998; Anguita *et al.*, 2000). The functions of most of these conditionally expressed proteins are not known. Genes selectively synthesised in the vector may play a role in nutrient uptake and transmission from the ticks to vertebrate hosts, whereas genes expressed only *in vivo* may be needed by the spirochaetes to sense and adapt to diverse host environments, evade the host immune response and disseminate from the site of deposition to distant tissues (de Silva and Fikrig, 1997).

Global regulatory mechanisms for gene expression in *B. burgdorferi* are not well understood. Transcription of *B. burgdorferi* genes appears to be initiated at σ^{70}-like promoters (Ge *et al.*, 1997b). In *B. burgdorferi* B31, the *ospAB* promoter is comprised of a core region containing typical '-35' and '-10' elements and a unique T-rich region (Sohaskey *et al.*, 1999), while the transcription of *ospC* may be under the control of multiple promoters that are highly similar in nucleotide sequence (Marconi *et al.*, 1993). The expression of OspA and OspB proteins may be regulated at both the transcriptional and translational level (Jonsson and Bergstrom, 1995). Furthermore, the inverse relationship between the transcription of *ospC* and the *ospAB* operon may indicate co-regulation of these separately encoded operons. It has been observed that several factors, such as culture temperature (Schwan *et al.*, 1995), pH (Carroll *et al.*, 1999), serum concentration in the medium (Alban *et al.*, 2000), cell density (Indest *et al.*, 1997), expression of a DNA-binding protein (Indest and Philipp, 2000) and other as yet unidentified environmental signals may affect *B. burgdorferi* gene expression (de Silva and Fikrig, 1997; Yang *et al.*, 1999). Molecular approaches to identification of environmentally regulated genes are currently not available for *B. burgdorferi*.

Major Outer Surface Proteins

Several surface-exposed membrane proteins designated outer surface protein OspA to OspF, have been identified and characterised. Many of these Osp proteins are expressed during human infection and some are abundant and strong immunogens. They have, therefore, been used as targets for serologic tests and vaccine development. All Osp proteins share some common structural and biochemical features (Bergstrom *et al.*, 1989; Fuchs *et al.*, 1992; Norris *et al.*, 1992; Lam *et al.*, 1994). Analysis of the leader sequences reveals a hydrophobic domain and a consensus cleavage sequence recognised by signal peptidase I or II at its N-terminus, and [³H]palmitate labelling confirms that they are lipoproteins. Proteinase sensitivity and immunofluorescence-labelling analysis demonstrate that these proteins are outer surface-exposed. Finally, genes that encode the Osp proteins map on linear or circular plasmids rather than on the chromosome.

OspA and OspB

OspA and OspB were the first outer surface proteins to be described and are the most intensively studied Osp lipoproteins (Bergstrom *et al.*, 1989). The molecular mass of OspA varies from 31 to 33.5 kDa between isolates belonging to different *B. burgdorferi* sensu lato species (**Fig. 4**). OspA is expressed in the tick vector and its expression is down-regulated on mammalian infection. The fraction of spirochaetes that express OspA in the midgut of an infected nymphal tick may fall from 100% to 30% within 24 hours of feeding. OspA appears to be involved in several aspects of the pathogenesis and immune protection of Lyme borreliosis. For example, it serves as a plasminogen receptor that may facilitate the dissemination of *B. burgdorferi* in infected hosts (Coleman *et al.*, 1997). OspA is capable of inducing nuclear translocation of nuclear factor κB (NF-κB) in human endothelial cells (Wooten *et al.*, 1996) and of stimulating production of IL-1 and IL-6 by human macrophages (reviewed in Szczepanski and Benach, 1991), indicating that it may play a role in the pathogenesis of Lyme borreliosis. Because OspA is abundant in spirochaetal cell lysates and is highly immunogenic, it has been used widely as target antigen for vaccine development.

OspC

OspC is a lipoprotein of 20–25 kDa that is heterogeneous in size and expression among different *B. burgdorferi* isolates. It is an early immunodominant protein that elicits a strong antibody response in the majority of infected patients. Thus, OspC has been used as an important diagnostic antigen. OspC protein is also regarded as a possible vaccine candidate, because active immunisation with recombinant OspC protects mice against subsequent challenge with the homologous strain and inhibits migration of *B. burgdorferi* from the tick midgut to the salivary glands (Gilmore and Piesman, 2000). Expression of OspC is up-regulated on initiation of tick feeding and may be involved in transmission of spirochaetes from tick to mammal (Masuzawa *et al.*, 1994; Gilmore and Piesman, 2000; Schwan and Piesman, 2000).

The gene that encodes OspC is located on a 26 to 28-kb circular plasmid (cp26 in strain B31) (Tilly *et al.*, 1997). Sequences of *ospC* from a large number of *B. burgdorferi* isolates representing different species have been determined (Jauris-Heipke *et al.*, 1995; Livey *et al.*, 1995; Theisen *et al.*, 1995; Ras *et al.*, 1997; Wang *et al.*, 1999c). There is significant sequence heterogeneity among *B. burgdorferi* isolates and species at the antigenic and genetic levels. The amino acid sequence identity of the OspC proteins among different *B. burgdorferi* sensu lato species ranges from 63% to 82% (Jauris-Heipke *et al.*, 1995). Homology between *B. burgdorferi* OspC and the Vmp proteins of *B. hermsii* has been reported (Margolis *et al.*, 1994), but because of its genetic stability in OspC-immunised mice, antigenic variation in OspC among *B. burgdorferi* isolates and species during chronic infection is not likely to be an important mechanism for immune evasion (Hodzic *et al.*, 2000).

OspD

OspD is a 28-kDa surface-exposed protein described as a low-passage and virulence-associated protein (Norris *et al.*, 1992). Its gene (*ospD*) is located on a 38-kb linear plasmid (lp38) in *B. burgdorferi* B31. The plasmid that carries *ospD* varies from 36 to 40 kb in size among different *B. burgdorferi* isolates and is not universally present in all *B. burgdorferi* sensu lato isolates (e.g. only 8 of 33 *B. burgdorferi* sensu stricto, 6 of 12 *B. afzelii* and 18 of 20 *B. garinii* examined had *ospD* gene) (Marconi *et al.*, 1994). Thus, OspD protein appears not to be an essential determinant of *B. burgdorferi* infectivity.

OspEF-related Proteins

OspE and OspF are outer surface lipoproteins first characterised from *B. burgdorferi* isolate N40 (Lam *et al.*, 1994). They contain 171 and 230 amino acids with a calculated molecular mass of 19 and 26 kDa, respectively. The *ospE* and *ospF* genes are arranged in tandem as one transcriptional unit and are located on a 45 kb plasmid in *B. burgdorferi* N40. Molecular and evolutionary analyses have shown that a paralogous family of proteins in *B. burgdorferi* is related to OspE and OspF and family members have been designated OspEF-related proteins (Erps) (Marconi *et al.*, 1997). Several studies suggest that Erp proteins are selectively expressed during infection *in vivo* and may play a role in immune evasion (Akins *et al.*, 1995; Sung *et al.*, 2000).

Flagellin

Flagellin is an important structural and functional protein. The flagellar filaments of *B. burgdorferi* are composed of two proteins, FlaB and FlaA, which are assumed to constitute the core filament and the outer sheath of the periplasmic flagella, respectively. FlaB is a 41-kDa protein encoded by *flaB* that is located in a motility operon on the chromosome of *B. burgdorferi* (Gassmann *et al.*, 1991; Ge *et al.*, 1997b). This protein is abundant, highly immunogenic and promotes one of the earliest antibody responses during human infection. Comparative analysis of the amino acid sequences reveals a high degree of sequence conservation with the flagellins from phylogenetically related and unrelated bacteria, e.g. the flagellin sequences of *B. burgdorferi* and *B. hermsii* have a 95% amino acid similarity (Gassmann *et al.*, 1991; Fukunaga *et al.*, 1996b). A *Borrelia* genus-specific monoclonal antibody (H9724) developed to an epitope of the FlaB has been used widely to distinguish *Borrelia* species from other bacteria. FlaA has only recently been found to be associated with the flagellar filaments of *B. burgdorferi* (Ge and Charon, 1997). The putative FlaA has a molecular mass of 37–38 kDa. It is antigenic and expressed *in vivo* at a lower level. Thus, FlaA is not an immunodominant protein. The ability of *B. burgdorferi* to move through extremely viscous fluids that approach the viscosity of skin may be critical for the dissemination and persistence of the spirochaete during infection.

Porin Proteins Oms28 and Oms66 (P66)

Approximately 5- to 10-fold less of outer membrane spanning (Oms) proteins is observed in the outer membrane of *B. burgdorferi* as compared with enteric Gram-negative bacteria. Electrophysiological analysis indicates that the outer membrane of *B. burgdorferi* contains two distinct porin activities with single-channel conductance of 0.6 and 1.2 nanosiemens (nS) (Skare *et al.*, 1995). Two Oms proteins with porin activity, Oms28 and Oms66, have been identified in

B. burgdorferi B31 (Skare *et al.*, 1996, 1997). These two, as well as other unidentified, porin proteins would presumably allow for the passive diffusion of low-molecular-weight compounds across the outer membrane of *B. burgdorferi*.

Oms28, encoded by a 771-bp ORF on a 28-kb linear plasmid (lp28-1) in *B. burgdorferi* B31, has a 0.6-nS porin activity. The deduced amino acid sequence of Oms28 predicts a 257-amino-acid precursor protein with a calculated molecular mass of 28 kDa. A putative 24-amino-acid leader peptidase I signal sequence could be recognised at its N-terminus (Skare *et al.*, 1996). Oms28 is possibly a *B. burgdorferi* sensu stricto-specific porin protein, since it was observed only in *B. burgdorferi* sensu stricto isolates, but not in other pathogenic species such as *B. garinii*, *B. hermsii* and *T. pallidum*.

The 1.2-nS porin activity resides in Oms66 in *B. burgdorferi* B31. The purified Oms66 is non-ion selective, but voltage-dependent (Skare *et al.*, 1997). Genetic and immunological evidence indicates that Oms66 is identical to the previously described P66 protein and is chromosomally encoded (Bunikis *et al.*, 1995). The deduced amino acid sequences of Oms66/ P66 are conserved, showing 92–94% identity among different *B. burgdorferi* species. Native Oms66 can elicit potent bactericidal activity and significant protective immunity against host-adapted organisms (Exner *et al.*, 2000). The immune response to this surface-exposed protein during natural infection is directed against a predicted immunoreactive surface domain (Bunikis *et al.*, 1996). In addition, recombinant Oms66/P66 binds specifically to β_3-chain integrins and inhibits attachment of intact *B. burgdorferi* to the same integrins, which makes it an attractive candidate bacterial ligand for integrins $\alpha_{IIb}\beta_3$ and $\alpha_v\beta_3$ (Coburn *et al.*, 1999).

Decorin-binding Proteins DbpA and DbpB

Decorin is a collagen-associated extracellular matrix proteoglycan found in the skin and many other tissues of humans. *B. burgdorferi* expresses two decorin-binding proteins on its surface, designated DbpA and DbpB (Guo *et al.*, 1998; Hagman *et al.*, 1998). The genes that encode DbpA and DbpB are organised as an operon on a 54-kb linear plasmid (lp54) in *B. burgdorferi* B31. Molecular analysis reveals that the DbpA proteins are heterogeneous among *B. burgdorferi* isolates between and within species (58.3–100% similarity among 29 isolates studied). In contrast, the predicted DbpB sequences are highly conserved (96.3–100% similarity among 15 isolates analysed) (Roberts *et al.*, 1998).

Biochemical and molecular analysis suggests that DbpA and DbpB are surface-exposed lipoproteins and act as adhesins of the microbial surface component-recognising adhesive matrix molecule family, which mediate *B. burgdorferi* attachment to the extracellular matrix of the host (Guo *et al.*, 1998). As demonstrated by site-directed mutation analysis, three lysine residues (Lys82, Lys163 and Lys170) of DbpA are thought to be crucial for its binding to decorin (Brown *et al.*, 1999a). Both DbpA and DbpB are immunogenic and expressed *in vivo*. Early studies showed that active and passive immunisation of mice with DbpA or DbpA antiserum protects mice against needle-inoculated challenge of *B. burgdorferi* isolates cultured *in vitro*. A recent study, however, found that DbpA may not be protective when immunised mice are challenged by tick infestation, because of the lack of DbpA expression by *B. burgdorferi* in ticks (Hagman *et al.*, 2000).

Basic Membrane Protein BmpA

BmpA (P39) is a 39-kDa conserved, species-specific protein that is highly immunogenic in mice and humans infected with *B. burgdorferi* (Simpson *et al.*, 1990; Brunet *et al.*, 1995). Mice exposed to *B. burgdorferi*-infected ticks (*I. scapularis*) usually produce an antibody response to P39, but not to OspA, within one week post-infection. This indicates that P39 is an effective immunogen in natural infections. IgG antibody against P39 is detectable in 15–64% of Lyme borreliosis patients, depending on the clinical stage at which the sample is taken (Aguero-Rosenfeld *et al.*, 1996; Roessler *et al.*, 1997). P39 has 27% amino acid identity to the TmpC protein of *T. pallidum* (Fraser *et al.*, 1998). It may be a major periplasmic membrane protein, but the location of P39 on the spirochaetal cell and its biological roles are still unknown.

P39 is encoded by *bmpA* of the *bmp* gene cluster which includes four chromosomal genes in tandem in the order *bmpD–bmpC–bmpA–bmpB* (Aron *et al.*, 1996). DNA sequence homology between *bmpA* and other *bmp* genes ranges from 56% to 64%. Molecular analysis indicates that BmpA of *B. garinii* is relatively heterogeneous, with an amino acid sequence identity ranging from 91% to 97%, whereas the BmpAs of *B. afzelii* and *B. burgdorferi* sensu stricto appear to be highly conserved (>98.5% intra-species identity). Inter-species identity of the BmpA sequences ranges from 86% to 92% (Roessler *et al.*, 1997).

Other Proteins

The molecular and immunological properties of several other proteins identified in whole-cell lysates of *B. burgdorferi* have been characterised. These include an immunogenic protein with a molecular weight of 83–100 kDa (p83/p100) (Dykhuizen *et al.*, 1993; Rossler *et al.*, 1995), a 60-kDa heat-shock protein (HSP60) (Scopio *et al.*, 1994), a virulent strain-associated repetitive protein A (VraA) (Skare *et al.*, 1999), a 37-kDa arthritis-related protein (Arp) (Feng *et al.*, 2000) and several multi-copy lipoproteins (Mlp) (Yang *et al.*, 1999; Porcella *et al.*, 2000).

The properties of the abundant or potentially pathogenic proteins of *B. burgdorferi* which have been characterised are presented in **Table 3**.

Pathogenesis

A variety of factors contribute to the pathogenesis of Lyme borreliosis, including the virulence of the *B. burgdorferi* strain, the host–pathogen interaction, inflammatory immune and possibly auto-reactive responses and specific host-associated factors (Coleman and Belach, 1993; Nordstrand *et al.*, 2000). Although the precise means of the interaction between the organism, the immune response elicited by the organism and the host are not well understood, the presence of spirochaetes, or spirochaetal components, in targeted tissues is thought to be crucial in the pathogenesis of Lyme borreliosis.

Attachment, Direct Cell Damage and Dissemination

Strong evidence suggests that *B. burgdorferi* is present at the site of inflammation in Lyme borreliosis patients and infected animals. The spirochaetes residing at these sites may invade, damage and/or kill the host cells (Aberer *et al.*, 1997; Dorward *et al.*, 1997) or induce host cells to release cytokines that may amplify inflammation in the affected tissues.

Attachment and adherence to host tissues is presumably the first step in the establishment of infection by microbes. DbpA and DbpB, two bacterial adhesins that bind to a collagen-associated extracellular matrix proteoglycan of the skin and other tissues, have been identified in *B. burgdorferi* (Guo *et al.*, 1998). Another candidate adhesin is a 26-kDa glycosaminoglycan (GAG)-binding protein (Bgp), which can potentially promote the colonisation of spirochaetes in diverse tissues through binding to various GAG molecules, such as heparin, heparan sulphate and dermatan sulphate (Parveen and Leong, 2000). In addition, *B. burgdorferi* cells can bind to human platelets by the platelet-specific integrin $\alpha_{\mathrm{IIb}}\beta_3$ and to endothelial cells by the integrins $\alpha_v\beta_3$ and $\alpha_5\beta_1$ (Coburn *et al.*,

Table 3 Some characterised *B. burgdorferi* proteins

Protein	Molecular mass (kDa)	Gene location	Expression *in vivo*	Immuno-protective	Description	Reference
Chromosome						
FlaA	38–39	BB668	+	+/−	Outer sheath of flagella	Ge and Charon (1997)
FlaB	41	BB147	+	−	Core filament of flagella Cell shape and motility	Ge *et al.* (1997b)
BmpA (P39)	39	BB383	+	−	*Borrelia* membrane protein	Aron *et al.* (1996)
Oms66 (P66)	66	BB603	+	+	Outer membrane spanning protein Porin, integrin ligand	Skare *et al.* (1997); Parveen and Leong (2000)
Bgp	26	BB588	+	?	*Borrelia* GAG-binding protein Adhesin/ haemagglutinin	Parveen and Leong (2000)
HSP60	60		+	−	Heat-shock protein, chaperone	Scopio *et al.* (1994)
P83/100	83–100	BB744	+	?	Periplasmic membrane protein	Rossler *et al.* (1995)
Plasmid						
OspA	31–33.5	lp54	+/−	+	Outer surface protein, plasminogen receptor, cytokine inducer	Bergstrom *et al.*, (1989)
OspB	33–34	lp54	−	+ or −	Outer surface protein	Bergstrom *et al.*, (1989)
OspC	20–25	cp26	+	+	Outer surface protein	Fuchs *et al.* (1992)
OspD	28	lp38	?	−	Outer surface protein	Norris *et al.* (1992)
OspE	19	lp45	+	+/−	Outer surface protein	Lam *et al.* (1994)
OspF	26	lp45	+	+/−	Outer surface protein	Lam *et al.* (1994)
ErpT	30	lp28-1	+	−	OspE-related protein	Fikrig *et al.* (1999)
DbpA	18–20	lp54	+	+	Decorin-binding protein Adhesin	Guo *et al.* (1998)
DbpB	19–20	lp54	+	−	Adhesin	Guo *et al.* (1998)
BBK32	47	lp36	+	+	Fibronectin-binding protein Adhesin (fibronectin receptor)	Probert and Johnson (1998)
BlyA	7.4	cp30	+	?	Haemolysin	Guina and Oliver (1997)
Oms28	28	lp28-3	+	?	Porin	Skare *et al.* (1996)
Mlp	17–24	cp32/18	+	?	Multi-copy lipoprotein	Yang *et al.* (1999)
VlsE	35	lp28-1	+	?	Vmp-like sequence protein Antigenic variation	Zhang *et al.* (1997a)

1994, 1998). Although toxins have not been identified in *B. burgdorferi*, direct toxic and cytopathic effects on various host cells as a result of spirochaetal attachment, as well as spirochaete-induced inflammatory reactions, have been observed (Garcia-Monco *et al.*, 1991; Habicht, 1992; Aberer *et al.*, 1997; Dorward *et al.*, 1997).

Relatively few spirochaetes are present at the lesion sites, as shown by clinical and histological observations of patients and experimentally infected mice. This indicates that the Lyme borreliosis spirochaetes have a mechanism for causing or amplifying the inflammatory process. It has been reported that *B. burgdorferi* can cause monocytes to aggregate and release pro-inflammatory cytokines, including IL-1β, IL-6 and TNFα. *B. burgdorferi* is also a potential inducer of a variety of chemokines, including IL-8, which may, by the attraction and activation of leucocytes, contribute to inflammation and tissue damage observed in Lyme borreliosis.

B. burgdorferi may invade the bloodstream from the site of deposition and spread to several tissues, resulting in disseminated infection. Invasive infection and haematogenous dissemination in Lyme borreliosis has been associated with specific genotypes of the infecting *B. burgdorferi* (Wormser *et al.*, 1999a; Seinost *et al.*, 1999). Strain variations in recognition and attachment to different types of glycosaminoglycan and proteoglycans, or interaction with host cells, are possibly one of the determinants for distinct tissue tropisms of *B. burgdorferi* (Leong *et al.*, 1998; Parveen *et al.*, 1999; Grab *et al.*, 1999). Since it does not appear to produce endogenous enzymes for digesting extracellular matrix components, *B. burgdorferi* may be capable of utilising host factors to facilitate its dissemination and establishment of infection in various distant organs. The plasminogen-activation system (PAS) plays a role in the spread of spirochaetes *in vivo*. *B. burgdorferi* has a plasminogen receptor, possibly OspA, and can assemble host plasminogen onto its surface (Hu *et al.*, 1995; Coleman *et al.*, 1997). The activation of *B. burgdorferi*-bound plasmin(ogen), a broad-spectrum serine protease, would enhance the ability of the spirochaetes to penetrate endothelial cell layers and to degrade components of the extracellular matrix. In fact, plasminogen is required for efficient dissemination of *B. burgdorferi* in ticks and for high-grade spirochaetaemia in mice (Coleman *et al.*, 1997).

Immunopathogenesis

The host immune response and autoimmunity may play an important role in the pathogenesis of late Lyme borreliosis, such as Lyme arthritis and neuroborreliosis (Hu and Klempner, 1997; Sigal, 1997a). It has been suggested that both B-cell- and T-cell-mediated humoral and cellular immune responses are involved in the pathology of Lyme borreliosis. In patients resistant to antibiotic treatment, Lyme arthritis may be associated with immune reactivity to OspA and the major histocompatibility complex (MHC) class II alleles (Chen *et al.*, 1999). The identification of an immunodominant epitope of OspA, which is homologous to the human leucocyte function-associated antigen-1 (hLFA-1), highlights the potential of autoimmunity in the development of chronic Lyme arthritis (Gross *et al.*, 1998a; Hemmer *et al.*, 1999). Molecular mimicry between other *B. burgdorferi*

antigens and host components has also been reported and may be involved in the immunopathogenesis of late Lyme borreliosis (Aberer *et al.*, 1989; Sigal and Williams, 1997).

Additional evidence indicates that T-cell-mediated cellular immune responses modulate the susceptibility and severity of Lyme arthritis, carditis and neuro-borreliosis in human and animal models (Habicht *et al.*, 1991; Keane-Myers and Nickell, 1995; Busch *et al.*, 1996b). The course of Lyme borreliosis in immunocompetent and B-cell-deficient mice has been compared (McKisic *et al.*, 2000). It was found that immunocompetent mice can resolve both carditis and arthritis, whereas B-cell-deficient mice developed myocarditis and severe destructive arthritis by 8 weeks of inoculation of *B. burgdorferi* N40. Cell transfer experiments with infected B6-*Rag1* knockout mice, which are devoid of T and B cells, demonstrated that the transfer of both naïve T cells and B cells induced resolution of carditis and arthritis. Infected mice reconstituted with T cells, however, developed myocarditis and severe destructive arthritis, and CD4 + T cells were the primary mediator of the observed immune-mediated pathology. This demonstrates directly the deleterious effect of T cells in the pathogenesis of Lyme borreliosis on an animal system (McKisic *et al.*, 2000).

Mechanisms of Persistence

B. burgdorferi may persist for months to years in various tissues of Lyme borreliosis patients and infected mice (Steere, 1989; Barthold *et al.*, 1993). Several studies suggest that multiple strategies may be utilised by the spirochaetes to evade host immunity. Among these are differential gene expression (de Silva and Fikrig, 1997; Anguita *et al.*, 2000), limited exposure of the major surface lipoproteins (Cox *et al.*, 1996), modulation of surface molecule structure (Bunikis and Barbour, 1999), immune mimicry to host tissue components (Sigal, 1997b) and antigenic variation (Zhang *et al.*, 1997a). Experimentally, *B. burgdorferi* is able to invade fibroblast cells and is protected from killing by exposure to an antibiotic (Georgilis *et al.*, 1992). Since spirochaetes are usually localised in the intracellular matrix of the infected tissues, it is unlikely that *B. burgdorferi* escapes immune recognition as a result of intracellular localisation. It remains to be determined whether *B. burgdorferi* can 'hide out' in some immunologically privileged sites in the host.

As mentioned above, the development of chronic Lyme arthritis may be associated with specific host MHC class II alleles (Kalish *et al.*, 1993; Gross *et al.*, 1998a). In Europe, patients infected with *B. afzelii* may have only a mild skin lesion (erythema migrans, EM) or develop the severe chronic skin disorder, acrodermatitis chronica atrophicans (ACA). *B. afzelii* isolates cultured from patients with EM and ACA are genetically indistinguishable by means of various molecular methods (Busch *et al.*, 1996c; Wang *et al.*, 1998). These findings indicate that genetic susceptibility and other undefined host factors are critical, in some circumstances, for the development of persistent infection by *B. burgdorferi*.

Immunity

Innate Defences

Innate or non-specific defences play an important role in the early stage of *B. burgdorferi* infection. This is evident from the presence of macrophages at the sites of local infection. When spirochaetes invade the bloodstream and begin the process of dissemination to other sites, they are exposed to the complement system, and come into contact with the circulating polymorphonuclear leucocytes and monocytes. *B. garinii* (OspA serotypes 5 and 6) and *B. valaisiana* are sensitive to complement-mediated killing, *B. burgdorferi* sensu stricto displays intermediate sensitivity, whereas *B. afzelii* is complement-resistant (van Dam *et al.*, 1997).

Specific Humoral and T-cell Immunity

Humoral immunity appears to be the most effective defence against *B. burgdorferi* infection. The development of *B. burgdorferi*-specific antibodies follows the classic pattern of IgM preceding IgG and IgA and is characterised by the initial recognition of a limited number of antigens followed by a marked expansion in the number of antigens recognised later in the disease. The peak of IgM usually occurs 3–6 weeks after infection. A study of the evolution of the serologic response to *B. burgdorferi* in 46 antibiotic-treated patients with culture-proven EM revealed that 91% of patients develop detectable IgM antibodies 8–14 days after the first sample (baseline) was collected. Peak IgM antibody levels were seen at this time in patients with localised or disseminated disease. The most frequent IgM at baseline and peak were antibodies against the 41-kDa flagellin (p41), the 24-kDa OspC, and a 37-kDa protein. IgM to antigens of 39, 58, 60, 66 or 93 kDa were most often detected in sera obtained within 1 month. Approximately 89% of the culture-confirmed patients developed IgG antibodies as determined at a follow-up examination. The most frequently observed IgG antibodies were those against p41 (in 100% of patient sera), p18, p21, p39, p45, p60 and p24 (57–79% of patient sera) (Aguero-Rosenfeld *et al.*, 1996). High levels of IgG and occasionally of IgM may persist for several years after adequate treatment and resolution of symptoms.

Protective borreliacidal antibody and growth-inhibiting antibody have been detected and characterised from sera of Lyme borreliosis patients or individuals vaccinated with an OspA vaccine (Callister *et al.*, 1993; Padilla *et al.*, 1996; Luke *et al.*, 2000). Passive immunisation with serum from patients or infected animals protects mice against challenge by *B. burgdorferi* and may even resolve an established infection (Fikrig *et al.*, 1990; Zhong *et al.*, 1999; Feng *et al.*, 2000). Antibody-mediated killing may or may not be mediated by complement (Kochi *et al.*, 1991; Aydintug *et al.*, 1994; van Dam *et al.*, 1997).

B. burgdorferi-specific cellular immune responses, primarily T-lymphocyte-mediated, arise 1–3 weeks after infection and may persist during the entire course of disease. As described earlier, activated T cells are capable of releasing pro-inflammatory and anti-inflammatory cytokines, or assist B lymphocytes in developing *Borrelia*-specific antibodies. The T-cell-mediated immune response may have dual roles in immune protection and in pathogenesis of Lyme borreliosis. Both TH1-type (e.g. IL-1β, IL-2, IL-6, TNFα, IL-11, IL-12, IFNγ) and TH2-type (e.g. IL-4 and IL-10) cytokines are induced in T lymphocytes, monocytes, macrophages or mast cells by *B. burgdorferi* lipoproteins *in vitro* and are detected in the synovial fluid, cerebrospinal fluid (CSF) and cardiac lesions of patients and infected mice (Habicht, 1992; Zeidner *et al.*, 1997; Pachner *et al.*, 1997; Gross *et al.*, 1998b; Kelleher *et al.*, 1998; Giambartolomei *et al.*, 1998; Talkington and Nickell, 1999; Anguita *et al.*, 1999). The activation and induction of cytokines in monocytic cells by *Borrelia* surface-exposed lipoproteins is mediated by a CD14-dependent pathway distinct from that used by LPS (Sellati *et al.*, 1998; Giambartolomei *et al.*, 1999). In mice, arthritis susceptibility and severity were correlated with the presence of CD8 + T cells and/or TH1 cells, whereas resistance was associated with a TH2 response (Keane-Myers and Nickell, 1995; Anguita *et al.*, 1998).

Diagnosis

Clinical Diagnosis and Case Definition

Infection with *B. burgdorferi* may result in multisystem involvement, including the dermatological, neurological, cardiac and musculoskeletal systems. EM is the clinical hallmark of early infection. It is recognised in up to 90% of the patients with objective clinical evidence of *B. burgdorferi* infection, that is those who fulfil the CDC surveillance criteria for Lyme borreliosis (Steere, 1989; Nadelman and Wormser, 1998). The rash typically begins at the site of the tick bite as a red macule or papule, sometimes with central clearing. At this stage, patients may be either asymptomatic or present with flu-like symptoms, such as headache, myalgia, arthralgias and fever.

Dissemination of *B. burgdorferi* sensu lato to the nervous system, joints, heart or other skin areas, and occasionally to other organs may give rise to a wide spectrum of manifestations in the involved systems. Usually, patients at this stage experience one or more of the following syndromes: multiple EM lesions, Lyme carditis, arthritis, and early neuroborreliosis including meningo-radiculoneuritis (Bannwarth's syndrome), meningitis, meningo-encephalitis, and/or facial palsy (van der Linde, 1991; Strle *et al.*, 1996; Oschmann *et al.*, 1998).

Late Lyme borreliosis may develop among some untreated or antibiotic treatment-resistant patients months to years after the onset of the disease. The major manifestations of late Lyme borreliosis include Lyme arthritis, chronic neuroborreliosis and ACA. More detailed description of the clinical manifestations of Lyme borreliosis can be found in some excellent published reviews (Steere, 1989; Stanek *et al.*, 1996; Nadelman and Wormser, 1998).

The case definition of Lyme borreliosis either developed by the CDC (1997) for epidemiological purposes (**Table 4**) or proposed by the European Union Concerted Action on Risk Assessment in Lyme borreliosis (Stanek *et al.*, 1996) is clinically useful. The diagnosis of Lyme borreliosis can easily be made in an endemic area when the characteristic skin lesions, i.e. EM, borrelial lymphocytoma at typical sites (nipple or ear lobe) or ACA are present. A history of a recent tick bite may be supportive, but only 30–54% of the culture-confirmed EM patients could recall a tick bite (Strle *et al.*, 1999). In patients with EM, laboratory testing is neither necessary nor recommended. Diagnosis of Lyme borreliosis on the basis of the presence of manifestations other than EM is usually less indicative. Thus laboratory testing is necessary for the diagnosis of Lyme borreliosis in these patients and elevated specific antibody to *B. burgdorferi* titres can usually be demonstrated.

Cultivation of *B. burgdorferi*

Cultivation of *B. burgdorferi* from clinical specimens remains the 'gold standard' for the diagnosis of Lyme borreliosis. The recovery rate varies depending on a number of factors, including the stage of disease (early or late infection), when the sample is taken, the use of antibiotics (before or after antimicrobial therapy), the type of specimen used (whole blood, serum, plasma, skin, CSF, synovial fluid or synovial tissue), the site of the skin biopsy (centre or peri-lesional skin of EM), and the cultural technique applied. *B. burgdorferi* can be cultured from 4–25% of blood samples from patients with early Lyme borreliosis (Nadelman *et al.*, 1990; Goodman *et al.*, 1995; Wormser *et al.*, 1998), 11–20% of CSF specimens from patients with neuro-borreliosis (Karlsson *et al.*, 1990; Oschmann *et al.*, 1998), and 57–86% of skin biopsies from patients with EM (Strle *et al.*, 1996, 1999; Picken *et al.*, 1997). The yield of blood culture may be improved up to 50% by using a large volume of plasma rather than serum samples from patients with early Lyme borreliosis (Wormser *et al.*, 2000a).

Detection of *B. burgdorferi* by PCR

Detection of spirochaetes in tissue specimens by histochemical staining is not reliable, because spirochaetes cannot be distinguished easily from elastic tissue or procollagen fibres. Cultivation *in vitro* demonstrates directly the presence of *B. burgdorferi* in clinical samples, but it is a difficult, time-consuming procedure. Except for skin biopsies, the yield of culture from clinical specimens is usually low. *B. burgdorferi* has been rarely cultured from the synovial fluid and myocardium of patients. Therefore, polymerase chain reaction (PCR) detection of *B. burgdorferi* DNA represents an alternative and practical approach for laboratory diagnosis of Lyme borreliosis (Schmidt, 1997). PCR has been used to detect the DNA of *B. burgdorferi* sensu lato in various clinical specimens, including skin biopsy (Schwartz *et al.*, 1992b; Picken *et al.*, 1997; Liveris *et al.*, 1999), blood (Goodman *et al.*, 1995; Picken *et al.*, 1997), urine (Goodman *et al.*, 1991; Priem *et al.*, 1997), CSF (Jaulhac *et al.*, 1991; Priem *et al.*, 1997), and synovial fluid (Nocton *et al.*, 1994; Priem *et al.*, 1997). In a large retrospective analysis, *B. burgdorferi* DNA was detected in 75 of 88 (85%) synovial fluid specimens of patients with arthritis. It is reported that the diagnostic sensitivity of a nested PCR with primer sets targeting the plasmid-located *ospA* gene and a chromosomal *p66* gene is 91% with paired synovial fluid–urine samples from patients with Lyme arthritis ($n = 35$) and 87% with paired CSF–urine specimens from neuroborreliosis patients ($n = 22$) (Priem *et al.*, 1997). Since PCR inhibitors are present in host blood, skin and possibly other tissues, appropriate

Table 4 Lyme disease surveillance case definition (modified by CDC in 1996)

Clinical case definition	Erythema migrans, or at least one late manifestation, as defined below, and laboratory confirmation of infection
General definitions	
A. Erythema migrans (EM)	A skin lesion that typically begins as a red macule or papule and expands over a period of days to weeks to form a large round lesion, often with partial central clearing. A solitary lesion must reach at least 5 cm in size. Secondary lesions may also occur. Annular erythematous lesions occurring within several hours of a tick bite represent hypersensitivity reactions and do not qualify as EM. The expanding EM lesion may be accompanied by other acute symptoms, particularly fatigue, fever, headache, mild stiff neck, arthralgia or myalgia, which are typically intermittent. Diagnosis of EM must be made by a physician. Laboratory confirmation is recommended for persons with no known exposure.
B. Late manifestations	Late manifestations include any of the following when an alternate explanation is not found: *Musculoskeletal system* Recurrent, brief attacks (weeks or months) of objective joint swelling in one or a few joints, sometimes followed by chronic arthritis in one or a few joints. Manifestations not considered as criteria for diagnosis include chronic progressive arthritis not preceded by brief attacks and chronic symmetrical polyarthritis. Additionally, arthralgia, myalgia, or fibromyalgia syndromes alone are not criteria for musculoskeletal involvement. *Nervous system* Any of the following, alone or in combination: Lymphocytic meningitis; cranial neuritis, particularly facial palsy (may be bilateral); radiculoneuropathy; or, rarely, encephalomyelitis. Encephalomyelitis must be confirmed by showing antibody production against *B. burgdorferi* in the cerebrospinal fluid (CSF), demonstrated by a higher titre of antibody in CSF than in serum. Headache, fatigue, paraesthesia or mild stiff neck alone are not criteria for neurological involvement. *Cardiovascular system* Acute onset, high-grade (II° or III°) atrioventricular conduction defects that resolve in days to weeks and are sometimes associated with myocarditis. Palpitations, bradycardia, bundle branch block, or myocarditis alone are not criteria for cardiovascular involvement.
C. Exposure	Exposure is defined as having been in wooded, brushy or grassy areas (potential tick habitats) in a county in which Lyme disease is endemic no more than 30 days before onset of EM. A history of tick bite is not required.

Table 4 *Continued*

D. Endemic to county	A county in which Lyme disease is endemic is one in which at least two definite cases have been previously acquired or in which a known tick vector has been shown to be infected with *B. burgdorferi*.
E. Laboratory confirmation	Isolates the spirochaete from tissue or body fluid, detects diagnostic levels of IgM or IgG antibodies to the spirochaete in serum or CSF, or detects a significant change in antibody levels in paired acute- and convalescent-phase serum samples. States may determine the criteria for laboratory confirmation and diagnostic levels of antibody. Syphilis and other known causes of biologic false-positive serologic test results should be excluded when laboratory confirmation has been based on serologic testing alone.

preparation of template DNA is important for successful amplification of *B. burgdorferi* DNA (Cogswell *et al.*, 1996; Schwartz *et al.*, 1997a). In general, the sensitivity of PCR and culture for detection of *B. burgdorferi* in patient and infected animal specimens is comparable (Pachner *et al.*, 1993; Lebech *et al.*, 1995; Picken *et al.*, 1997).

Serodiagnosis

The most common methods for detection of antibodies to *B. burgdorferi* are enzyme-linked immunosorbent technology (ELISA or EIA) or indirect immunofluorescence assay (IFA) (Tugwell *et al.*, 1997; Brown *et al.*, 1999b). The antigen for these assays is typically whole-cell lysate, which is a complex mixture containing several immunogenic proteins, lipoproteins and carbohydrates from a cultured strain. The specificity of ELISA for detection of *B. burgdorferi* antibodies is 72–94% (Tugwell *et al.*, 1997). Current serological tests may be limited by the delay in development of an antibody response, cross-reactivities with other organisms, high background sero-prevalence in asymptomatic persons residing in endemic areas, difficulty in distinguishing past from present infections and lack of standardisation (Norman *et al.*, 1996; Bakken *et al.*, 1997; Brown *et al.*, 1999b; Wormser *et al.*, 1999b). In addition, antigenic heterogeneity of *B. burgdorferi* sensu lato may influence the sensitivity and specificity of serological tests for Lyme borreliosis, especially in Europe where three pathogenic species of *B. burgdorferi* sensu lato and at least eight OspA serotypes are well-documented (Wilske *et al.*, 1993; Hauser *et al.*, 1998). In an effort to improve the performance of these serological tests, a two-step approach was proposed by the Centers for Disease Control and Prevention (CDC, 1995) and was recommended by the American College of Physicians

(1997). In this so-called two-step testing approach, Western blotting of sera with positive and indeterminate results by ELISA or IFA is required in order to increase specificity. The protocol, as well as its simplified approach (Trevejo *et al.*, 1999), improve the specificity of serodiagnosis for Lyme borreliosis (Craven *et al.*, 1996; Johnson *et al.*, 1996; Ledue *et al.*, 1996). A limitation of the two-step test is that the first- and second-step assays are not independent (Wormser *et al.*, 2000b). Interpretive criteria have been suggested for standardised Western blotting for the three species of *B. burgdorferi* sensu lato that cause human disease in Europe (Hauser *et al.*, 1997, 1999). The significance of other serological tests, such as detection of indirect haemagglutination antibody (IHA) (Pavia *et al.*, 2000) and specific immune complexes (Schutzer *et al.*, 1999) for the serodiagnosis of Lyme borreliosis requires confirmation.

Serological tests for *B. burgdorferi* antibodies should be used only to support a clinical diagnosis of Lyme borreliosis, and not as the primary basis for making a diagnosis or for treatment decisions. The positive predictive value (the ability to diagnose disease) of a test depends on both the sensitivity and specificity of the assay, and the prevalence of the disease in the population to which the test is being applied.

Antimicrobial Therapy

The susceptibility of *B. burgdorferi* to several classes of antibiotics *in vitro* has been tested by determining the minimal inhibitory concentrations (MICs) and minimal borreliacidal concentrations (MBCs) (Baradaran-Dilmaghani and Stanek, 1996; Preac-Mursic *et al.*, 1996; Hunfeld *et al.*, 2000). The tested *B. burgdorferi* isolates are susceptible *in vitro* to penicillins (e.g.

penicillin G, amoxycillin and piperacillin), second- and third-generation cephalosporins (e.g. ceftriaxone and cefotaxime), tetracyclines (e.g. tetracycline, doxycycline and minocycline), macrolides (e.g. azithromycin, erythromycin, roxithromycin), streptogramins (e.g. quinupristin-dalfopristin) and glycopeptides (e.g. vancomycin), but they are relatively resistant to aminoglycosides, trimethoprim, rifampicin and quinolones. The MIC_{90} of the most active antibiotics, such as ceftriaxone, azithromycin and roxithromycin, is typically between 0.0015 and 0.15 µg/mL (Hunfeld *et al.*, 2000).

The treatment of patients with EM with tetracycline, doxycycline, amoxicillin or cefuroxime axetil is highly effective. Patients with advanced cardiac or neurological manifestations can be treated with ceftriaxone or cefotaxime (Nadelman and Wormser, 1995; Wormser, 1997; Dotevall and Hagberg, 1999; Wormser *et al.*, 2000c). Individualised duration of antimicrobial therapy is recommended, and varies from 2 to 3 weeks for localised and early disseminated infection and 2–4 weeks for late symptoms. Treatment of patients with early manifestations of Lyme borreliosis can prevent late sequelae (Seltzer *et al.*, 2000). Antimicrobial treatment failures of patients with EM, ACA and Lyme arthritis have, however, been noted. In patients with Lyme arthritis, such failures may be associated with immune responses to specific epitopes of OspA and the presence of the MHC II allele, DRB1-0401, in patients (Gross *et al.*, 1998a; Chen *et al.*, 1999). The current recommendations for the antimicrobial treatment of Lyme borreliosis are summarised in guidelines from the Infectious Diseases Society of America (Wormser *et al.*, 2000c). A cost-effectiveness analysis indicated that oral therapy was as effective as intravenous therapy and was less costly for patients with either early Lyme borreliosis or uncomplicated Lyme arthritis (Eckman *et al.*, 1997). Prophylactic treatment with antibiotics after tick bite is not recommended (Wormser *et al.*, 2000c).

Epidemiology and Control

Lyme borreliosis has been documented in patients residing in North America, Europe and Asia. In North America and Europe, it is the most common vector-borne disease. Between 1992 and 1998, a total of 88 967 cases of Lyme disease were reported to the CDC by 49 states and the District of Columbia. Of these cases, 92% were reported from eight north-eastern and mid-Atlantic states and two north-central states (Orloski *et al.*, 2000). In 1998, a total of 16 461 cases of Lyme borreliosis were reported by 45 states,

resulting in a national incidence of 6.2 per 100 000 population, and a 69% increase of cases compared with those reported in 1992. Endemic areas of Lyme borreliosis in Europe include the British Isles, Scandinavia, Western Europe and states of the former USSR, from the Baltic States east through Russia to the Pacific Coast. It is estimated that annually more than 50 000 Lyme borreliosis cases occur in the European countries (O'Connell *et al.*, 1998). Lyme borreliosis is also well documented in Far Eastern Russia (Sato *et al.*, 1996), Japan (Nakao and Miyamoto, 1995) and China (Ai *et al.*, 1990; Zhang *et al.*, 1997b). Although there have been reports of Lyme borreliosis-like cases in Australia, Africa and South America, *B. burgdorferi* has not been isolated from patients, tick vectors or vertebrate hosts from those areas.

The frequency of Lyme borreliosis in males and females is similar (Steere *et al.*, 1986; Berglund *et al.*, 1995), and seasonal variations and a bimodal age distribution have been noted. In the United States, children aged 5–9 years and adults aged 45–54 years have the highest mean annual incidence (Orloski *et al.*, 2000). Outdoor occupations and activities appear to influence the sero-prevalence and incidence.

B. burgdorferi is transmitted among reservoirs and hosts by ticks of the family *Ixodidae*, mainly ticks in the *Ixodes ricinus* complex. *I. scapularis* and *I. pacificus* in the United States, and *I. ricinus* and *I. persulcatus* in Eurasia are the principal vectors of the spirochaetes (Burgdorfer *et al.*, 1991). The prevalence rate of *B. burgdorferi* in *I. scapularis* collected from endemic areas of the United States varies from 12% to almost 100% (Anderson, 1989), and in *I. ricinus* in Europe from 0 to 58% (Hubalek and Halouzka, 1998). In a summarised survey across the European continent, the overall mean proportion of unfed *I. ricinus* larvae ($n = 5699$) infected with *B. burgdorferi* was 1.9% (range 0–11%), of nymphs ($n = 48 804$) was 10.8% (range 2–43%) and of adults ticks ($n = 41 666$) was 17.4% (range 3–58%) (Hubalek and Halouzka, 1998). Although the presence of *B. burgdorferi* in other tick species and insects has been reported, their role as vectors for Lyme borreliosis spirochaetes remains unknown (Burgdorfer *et al.*, 1991). Co-infection with *B. burgdorferi* and other tick-borne pathogens such as *Babesia* and *Ehrlichia* in both *I. scapularis* and *I. ricinus* ticks has been reported (Cinco *et al.*, 1997; Schwartz *et al.*, 1997b).

All *Ixodes* species transmitting *B. burgdorferi* sensu lato are three-host ticks, i.e. each individual tick feeds on three different host animals during its life. The white-footed mouse (*Peromyscus leucopus*) in the United States, and the wood mouse (*Apodemus sylvaticus*) and bank vole (*Clethrionomys glareolus*) in

Europe are among the most important reservoirs for *B. burgdorferi* (Anderson, 1989; Gern *et al.*, 1998). Migrating birds may be reservoirs of certain *B. burgdorferi* species and therefore may play a role in the geographic dissemination of *B. burgdorferi* (Olsen *et al.*, 1993; Nicholls and Callister, 1996). Usually, larval ticks acquire *B. burgdorferi* by feeding on infected hosts. The spirochaetes survive through the moults and remain present in all subsequent stages of the vectors (Burgdorfer *et al.*, 1991). Trans-ovarial transmission of *B. burgdorferi* sensu lato has been reported in various tick species, but its occurrence is thought to be very low.

Strategies for prevention and control of Lyme borreliosis include the reduction of tick host populations, ecological and chemical control of tick vectors, and personal protection of individuals who have a high risk for Lyme borreliosis. Practical and environmental approaches for reservoir and tick control have not, however, been well developed. Moreover, neither the reservoir hosts nor the tick vectors are readily accessible to measures that could reduce the risk of infection for the human population. Therefore, personal protection is currently the most practical option in high-risk areas. Personal protection includes the wearing of light-coloured clothes to make crawling ticks visible, tucking trousers into socks to prevent ticks from gaining access to exposed skin, using tick or insect repellents and vaccination.

Several whole-cell and subunit vaccines have been demonstrated to protect against infection of *B. burgdorferi* in laboratory animals (reviewed in Wormser, 1995; Thanassi and Schoen, 2000). The efficacy and safety of recombinant OspA vaccines in humans has been evaluated by two large clinical trials (Sigal *et al.*, 1998; Steere *et al.*, 1998). Each involved over 10 000 volunteers and showed that OspA vaccines are safe and effective for prevention of Lyme borreliosis. One of these is approved by the FDA for use in the United States in persons aged 15–70 years (CDC, 1999). OspA immunity appears to be effective in the tick vector by blocking the transmission of spirochaetes from the tick gut to the host (de Silva *et al.*, 1996). In Europe, use of polyvalent vaccine may be required because of a genetically heterogeneous population of *B. burgdorferi* sensu lato.

Acknowledgements

The authors thank Dr Nyles W. Charon for providing the electron micrographs of *B. burgdorferi* and Dr Steven J. Norris for providing the adapted diagram of the *vls* structure. Research in the authors' laboratories was supported by grants for the National Institute of Health (AR41511 and AI45601).

References

Aberer E, Duray PH (1991) Morphology of *Borrelia burgdorferi*: structural patterns of cultured borreliae in relation to staining methods. *J. Clin. Microbiol.* 29: 764–772.

Aberer E, Brunner C, Suchanek G *et al.* (1989) Molecular mimicry and Lyme borreliosis: a shared antigenic determinant between *Borrelia burgdorferi* and human tissue. *Ann. Neurol.* 26: 732–737.

Aberer E, Koszik F, Silberer M (1997) Why is chronic Lyme borreliosis chronic? *Clin. Infect. Dis.* 25: S64–S70.

American College of Physicians (1997) Guidelines for laboratory evaluation in the diagnosis of Lyme disease. *Ann. Intern. Med.* 127: 1106–1108.

Aguero-Rosenfeld ME, Nowakowski J, Bittker S *et al.* (1996) Evolution of the serologic response to *Borrelia burgdorferi* in treated patients with culture-confirmed erythema migrans. *J. Clin. Microbiol.* 34: 1–9.

Ai CX, Hu RJ, Hyland KE *et al.* (1990) Epidemiological and aetiological evidence for transmission of Lyme disease by adult *Ixodes persulcatus* in an endemic area in China. *Int. J. Epidemiol.* 19: 1061–1065.

Akins DR, Porcella SF, Popova TG *et al.* (1995) Evidence for *in vivo* but not *in vitro* expression of a *Borrelia burgdorferi* outer surface protein F (OspF) homologue. *Mol. Microbiol.* 18: 507–520.

Akins DR, Caimano MJ, Yang X *et al.* (1999) Molecular and evolutionary analysis of *Borrelia burgdorferi* 297 circular plasmid-encoded lipoproteins with OspE- and OspF-like leader peptides. *Infect. Immun.* 67: 1526–1532.

Alban PS, Johnson PW, Nelson DR (2000) Serum-starvation-induced changes in protein synthesis and morphology of *Borrelia burgdorferi*. *Microbiology* 146: 119–127.

Anderson JF (1989) Epizootiology of *Borrelia* in *Ixodes* tick vectors and reservoir hosts. *Rev. Infect. Dis.* 11 (Suppl. 6): S1451–S1459.

Anguita J, Rincon M, Samanta S *et al.* (1998) *Borrelia burgdorferi*-infected, interleukin-6-deficient mice have decreased Th2 responses and increased Lyme arthritis. *J. Infect. Dis.* 178: 1512–1515.

Anguita J, Barthold SW, Samanta S *et al.* (1999) Selective anti-inflammatory action of interleukin-11 in murine Lyme disease: arthritis decreases while carditis persists. *J. Infect. Dis.* 179: 734–737.

Anguita J, Samanta S, Revilla B *et al.* (2000) *Borrelia burgdorferi* gene expression *in vivo* and spirochete pathogenicity. *Infect. Immun.* 68: 1222–1230.

Aron L, Toth C, Godfrey HP *et al.* (1996) Identification and mapping of a chromosomal gene cluster of *Borrelia burgdorferi* containing genes expressed *in vivo*. *FEMS Microbiol. Lett.* 145: 309–314.

Aydintug MK, Gu Y, Philipp MT (1994) *Borrelia burgdorferi* antigens that are targeted by antibody-dependent, complement-mediated killing in the rhesus monkey. *Infect. Immun.* 62: 4929–4937.

Bakken LL, Callister SM, Wand PJ *et al.* (1997) Inter-laboratory comparison of test results for detection of Lyme disease by 516 participants in the Wisconsin State Laboratory of Hygiene/College of American Pathologists Proficiency Testing Program. *J. Clin. Microbiol.* 35: 537–543.

Baradaran-Dilmaghani R, Stanek G (1996) *In vitro* susceptibility of thirty Borrelia strains from various sources against eight antimicrobial chemotherapeutics. *Infection* 24: 60–63.

Baranton G, Postic D, Saint Girons I *et al.* (1992) Delineation of *Borrelia burgdorferi* sensu stricto, *Borrelia garinii* sp. nov, and group VS461 associated with Lyme borreliosis. *Int. J. Syst. Bacteriol.* 42: 378–383.

Barbour AG (1984) Isolation and cultivation of Lyme disease spirochetes. *Yale J. Biol. Med.* 57: 521–525.

Barbour AG, Hayes SF (1986) Biology of *Borrelia* species. *Microbiol. Rev.* 50: 381–400.

Barbour AG, Garon CF (1987) Linear plasmids of the bacterium *Borrelia burgdorferi* have covalently closed ends. *Science* 237: 409–411.

Barbour AG, Heiland RA, Howe TR (1985) Heterogeneity of major proteins in Lyme disease borreliae: a molecular analysis of North American and European isolates. *J. Infect. Dis.* 152: 478–484.

Barthold SW, Persing DH, Armstrong AL *et al.* (1991) Kinetics of *Borrelia burgdorferi* dissemination and evolution of disease after intradermal inoculation of mice. *Am. J. Pathol.* 139: 263–273.

Barthold SW, de Souzas MS, Janotka JL *et al.* (1993) Chronic Lyme borreliosis in the laboratory mouse. *Am. J. Pathol.* 143: 959–971.

Beck G, Habicht GS, Benach JL *et al.* (1985) Chemical and biologic characterization of a lipopolysaccharide extracted from the Lyme disease spirochete (*Borrelia burgdorferi*). *J. Infect. Dis.* 152: 108–117.

Benach JL, Bosler EM, Hanrahan JP *et al.* (1983) Spirochetes isolated from the blood of two patients with Lyme disease. *N. Engl. J. Med.* 308: 740–742.

Berglund J, Eitrem R, Ornstein K *et al.* (1995) An epidemiologic study of Lyme disease in southern Sweden. *N. Engl. J. Med.* 333: 1319–1327.

Bergstrom S, Bundoc VG, Barbour AG (1989) Molecular analysis of linear plasmid-encoded major surface proteins, OspA and OspB, of the Lyme disease spirochaete *Borrelia burgdorferi*. *Mol. Microbiol.* 3: 479–486.

Bledsoe HA, Carroll JA, Whelchel TR *et al.* (1994) Isolation and partial characterization of *Borrelia burgdorferi* inner and outer membranes by using isopycnic centrifugation. *J. Bacteriol.* 176: 7447–7455.

Boerlin P, Peter O, Bretz A-G *et al.* (1992) Population genetic analysis of *Borrelia burgdorferi* isolates by multilocus enzyme electrophoresis. *Infect. Immun.* 60: 1677–1683.

Bono JL, Elias AF, Kupko JJ III *et al.* (2000) Efficient targeted mutagenesis in *Borrelia burgdorferi*. *J. Bacteriol.* 182: 2445–2452.

Brown EL, Guo BP, O'Neal P (1999a) Adherence of *Borrelia burgdorferi*. Identification of critical lysine residues in DbpA required for decorin binding. *J. Biol. Chem.* 274: 26272–26278.

Brown SL, Hansen SL, Langone JJ (1999b) Role of serology in the diagnosis of Lyme disease. *JAMA* 282: 62–66.

Brunet LR, Sellitto C, Spielman A *et al.* (1995) Antibody response of the mouse reservoir of *Borrelia burgdorferi* in nature. *Infect. Immun.* 63: 3030–3036.

Bunikis J, Barbour AG (1999) Access of antibody or trypsin to an intergral outer membrane protein (P66) of *Borrelia burgdorferi* is hindered by Osp lipoprotein. *Infect. Immun.* 67: 2874–2883.

Bunikis J, Noppa L, Bergstrom S (1995) Molecular analysis of a 66-kDa protein associated with the outer membrane of Lyme disease *Borrelia*. *FEMS Microbiol. Lett.* 131: 139–145.

Bunikis J, Noppa L, Ostberg Y *et al.* (1996) Surface exposure and species specificity of an immunoreactive domain of a 66-kilodalton outer membrane protein (P66) of the *Borrelia* spp. that cause Lyme disease. *Infect. Immun.* 64: 5111–5116.

Burgdorfer W, Barbour AG, Hayes SF *et al.* (1982) Lyme disease – a tick-borne spirochetosis? *Science* 216: 1317–1319.

Burgdorfer W, Anderson JF, Gern L *et al.* (1991) Relationship of *Borrelia burgdorferi* to its arthropod vectors. *Scand. J. Infect. Dis. Suppl.* 77: 35–40.

Busch U, Hizo-Teufel C, Boehmer R *et al.* (1996a) Three species of *Borrelia burgdorferi* sensu lato (*B. burgdorferi* sensu stricto, *B afzelii*, and *B. garinii*) identified from cerebrospinal fluid isolates by pulsed-field gel electrophoresis and PCR. *J. Clin. Microbiol.* 34: 1072–1078.

Busch DH, Jassoy C, Brinckmann U *et al.* (1996b) Detection of *Borrelia burgdorferi*-specific CD8 + cytotoxic T cells in patients with Lyme arthritis. *J. Immunol.* 157: 3534–3541.

Busch U, Hizo-Teufel C, Bohmer R *et al.* (1996c) *Borrelia burgdorferi* sensu lato strains isolated from cutaneous Lyme borreliosis biopsies differentiated by pulsed-field gel electrophoresis. *Scand. J. Infect. Dis.* 28: 583–589.

Callister SM, Schell RF, Case KL *et al.* (1993) Characterization of the borreliacidal antibody response to *Borrelia burgdorferi* in humans: a serodiagnostic test. *J. Infect. Dis.* 167: 158–164.

Canale-Parola E, Johnson RC, Faine S (1989) Spirochetes. In: Staley JT (ed.) *Bergey's Manual of Systematic Bacteriology*. Baltimore: Williams and Wilkins, pp. 38–70.

Canica MM, Nato F, du Merle L *et al.* (1993) Monoclonal antibodies for identification of *Borrelia afzelii* sp. nov. associated with late cutaneous manifestations of Lyme borreliosis. *Scand. J. Infect. Dis.* 25: 441–448.

Carroll JA, Garon CF, Schwan TG (1999) Effects of environmental pH on membrane proteins in *Borrelia burgdorferi*. *Infect. Immun.* 67: 3181–3187.

Casjens S, Huang WM (1993) Linear chromosomal physical and genetic map of *Borrelia burgdorferi*, the Lyme disease agent. *Mol. Microbiol.* 8: 967–980.

Casjens S, Delange M, Ley HLI II *et al.* (1995) Linear chromosomes of Lyme disease agent spirochetes: genetic diversity and conservation of gene order. *J. Bacteriol.* 177: 2769–2780.

Casjens S, van Vugt R, Tilly K *et al.* (1997) Homology throughout the multiple 32-kilobase circular plasmids present in Lyme disease spirochetes. *J. Bacteriol.* 179: 217–227.

Casjens S, Palmer N, van Vugt R *et al.* (2000) A bacterial genome in flux: the twelve linear and nine circular extrachromosomal DNAs in an infectious isolate of the lyme disease spirochete *Borrelia burgdorferi. Mol. Microbiol.* 35: 490–516.

CDC (Centers for Disease Control and Prevention) (1995) Recommendations for test performance and interpretation from the Second National Conference on Serologic Diagnosis of Lyme Disease. *MMWR* 44: 590–591.

CDC (1997) Case definitions for infectious conditions under public health surveillance: Lyme disease. *MMWR* 46: 20–21.

CDC (1999) Recommendations for the use of Lyme disease vaccine. *MMWR* 48: 1–5.

Chen J, Field JA, Glickstein L *et al.* (1999) Association of antibiotic treatment-resistant Lyme arthritis with T cell responses to dominant epitopes of outer surface protein A of *Borrelia burgdorferi. Arthritis Rheum.* 42: 1813–1822.

Cinco M, Padovan D, Murgia R *et al.* (1997) Coexistence of *Ehrlichia phagocytophila* and *Borrelia burgdorferi* sensu lato in *Ixodes ricinus* ticks from Italy as determined by 16S rRNA gene sequencing. *J. Clin. Microbiol.* 35: 3365–3366.

Coburn J, Barthold SW, Leong JM (1994) Diverse Lyme disease spirochetes bind integrin $\alpha_{IIb}\beta_3$ on human platelets. *Infect. Immun.* 62: 5559–5567.

Coburn J, Magoun L, Bodary SC *et al.* (1998) Integrins $\alpha_v\beta_3$ and $\alpha_5\beta_1$ mediate attachment of Lyme disease spirochetes to human cells. *Infect. Immun.* 66: 1946–1952.

Coburn J, Chege W, Magoun L *et al.* (1999) Characterization of a candidate *Borrelia burgdorferi* beta3-chain integrin ligand identified using a phage display library. *Mol. Microbiol.* 34: 926–940.

Cogswell FB, Bantar CE, Hughes TG *et al.* (1996) Host DNA can interfere with detection of *Borrelia burgdorferi* in skin biopsy specimens by PCR. *J. Clin. Microbiol.* 34: 980–982.

Coleman JL, Benach JL (1993) Pathogenesis of Lyme disease. In: Coyle PK (ed.) *Lyme Disease.* St Louis: Mosby Year Book, pp. 179–183.

Coleman JL, Benach JL, Beck G *et al.* (1986) Isolation of the outer envelope from *Borrelia burgdorferi. Zbl. Bakteriol. Mikrobiol. Hyg. A* 263: 123–126.

Coleman JL, Gebbia JA, Piesman J *et al.* (1997) Plasminogen is required for efficient dissemination of *B. burgdorferi* in

ticks and for enhancement of spirochetemia in mice. *Cell* 89: 1111–1119.

Cox DL, Akins DR, Bourell KW *et al.* (1996) Limited surface exposure of *Borrelia burgdorferi* outer surface lipoproteins. *Proc. Natl Acad. Sci. USA* 93: 7973–7978.

Craven RB, Quan TJ, Bailey RE *et al.* (1996) Improved serodiagnostic testing for Lyme disease: results of a multicenter serologic evaluation. *Emerg. Infect. Dis.* 2: 136–140.

de Silva AM, Fikrig E (1995) Growth and migration of *Borrelia burgdorferi* in *Ixodes* ticks during blood feeding. *Am. J. Trop. Med. Hyg.* 53: 397–404.

de Silva AM, Fikrig E (1997) Arthropod- and host-specific gene expression by *Borrelia burgdorferi. J. Clin. Invest.* 99: 377–379.

de Silva AM, Telford SR III, Brunet LR *et al.* (1996) *Borrelia burgdorferi* OspA is an arthropod-specific transmission-blocking Lyme disease vaccine. *J. Exp. Med.* 183: 271–275.

Dorward DW, Fischer ER, Brooks DM (1997) Invasion and cytopathic killing of human lymphocytes by spirochetes causing Lyme disease. *Clin. Infect. Dis.* 25: S2–8.

Dotevall L, Hagberg L (1999) Successful oral doxycycline treatment of Lyme disease-associated facial palsy and meningitis. *Clin. Infect. Dis.* 28: 569–574.

Dykhuizen DE, Polin DS, Dunn JJ *et al.* (1993) *Borrelia burgdorferi* is clonal: implications for taxonomy and vaccine development. *Proc. Natl Acad. Sci. USA* 90: 10163–10167.

Eckman MH, Steere AC, Kalish RA *et al.* (1997) Cost effectiveness of oral as compared with intravenous antibiotic therapy for patients with early Lyme disease or Lyme arthritis. *N. Engl. J. Med.* 337: 357–363.

Eggers CH, Samuels DS (1999) Molecular evidence for a new bacteriophage of *Borrelia burgdorferi. J. Bacteriol.* 181: 7308–7313.

Exner MM, Wu X, Blanco DR *et al.* (2000) Protection elicited by native outer membrane protein oms66 (p66) against host-adapted *Borrelia burgdorferi*: conformational nature of bactericidal epitopes. *Infect. Immun.* 68: 2647–2654.

Feng S, Hodzic E, Barthold SW (2000) Lyme arthritis resolution with antiserum to a 37-kilodalton *Borrelia burgdorferi* protein. *Infect. Immun.* 68: 4169–4173.

Ferdows MS, Barbour AG (1989) Megabase-sized linear DNA in the bacterium *Borrelia burgdorferi*, the Lyme disease agent. *Proc. Natl Acad. Sci. USA* 86: 5969–5973.

Ferdows MS, Serwer P, Griess GA *et al.* (1996) Conversion of a linear to a circular plasmid in the relapsing fever agent *Borrelia hermsii. J. Bacteriol.* 178: 793–800.

Fikrig E, Barthold SW, Kantor FS *et al.* (1990) Protection of mice against the Lyme disease agent by immunizing with recombinant OspA. *Science* 250: 553–556.

Fikrig E, Chen M, Barthold SW *et al.* (1999) *Borrelia burgdorferi* erpT expression in the arthropod vector and murine host. *Mol. Microbiol.* 31: 281–290.

Fraser CM, Casjens S, Huang WM *et al.* (1997) Genomic sequence of a Lyme disease spirochaete, *Borrelia burgdorferi. Nature* 390: 580–586.

Fraser CM, Norris SJ, Weinstock GM *et al.* (1998) Complete genome sequence of *Treponema pallidum*, the syphilis spirochete. *Science* 281: 375–388.

Fuchs R, Jauris S, Lottspeich F *et al.* (1992) Molecular analysis and expression of a *Borrelia burgdorferi* gene encoding a 22 kDa protein (pC) in *Escherichia coli*. *Mol. Microbiol.* 6: 503–509.

Fukunaga M, Yanagihara Y, Sohnaka M (1992) The 23S/5S ribosomal RNA genes (*rrl/rrf*) are separate from the 16S ribosomal RNA gene (*rrs*) in *Borrelia burgdorferi*, the aetiological agent of Lyme disease. *J. Gen. Microbiol.* 138: 871–877.

Fukunaga M, Hamase A, Okada K *et al.* (1996a) *Borrelia tanukii* sp. nov. and *Borrelia turdae* sp. nov. found from ixodid ticks in Japan: rapid species identification by 16S rRNA gene-targeted PCR analysis. *Microbiol. Immunol.* 40: 877–881.

Fukunaga M, Okada K, Nakao M *et al.* (1996b) Phylogenetic analysis of *Borrelia* species based on flagellin gene sequences and its application for molecular typing of Lyme disease borreliae. *Int. J. Syst. Bacteriol.* 46: 898–905.

Fukunaga M, Hamase A, Okada K *et al.* (1996c) Characterization of spirochetes isolated from ticks (*Ixodes tanuki*, *Ixodes turdus*, and *Ixodes columnae*) and comparison of the sequences with those of *Borrelia burgdorferi* sensu lato strains. *Appl. Environ. Microbiol.* 62: 2338–2344.

Garcia-Monco JC, Fernandez Villar B, Szczepanski A *et al.* (1991) Cytotoxicity of *Borrelia burgdorferi* for cultured rat glial cells. *J. Infect. Dis.* 163: 1362–1366.

Garrity GM, Holt JG (2000) An overview of the road map to the manual Bergey's Manual of Systematic Bacteriology. In: *Bergey's Manual of Systematic Bacteriology*, 2nd edn. New York: Springer-Verlag, pp. 1–19.

Gassmann GS, Jacobs E, Deutzmann R *et al.* (1991) Analysis of the *Borrelia burgdorferi* GeHo *fla* gene and antigenic characterization of its gene product. *J. Bacteriol.* 173: 1452–1459.

Gazumyan A, Schwartz JJ, Liveris D *et al.* (1994) Sequence analysis of the ribosomal RNA operon of the Lyme disease spirochete, *Borrelia burgdorferi*. *Gene* 146: 57–65.

Ge Y, Charon NW (1997) An unexpected *flaA* homolog is present and expressed in *Borrelia burgdorferi*. *J. Bacteriol.* 179: 552–556.

Ge Y, Old IG, Saint Girons I *et al.* (1997a) The *flgK* motility operon of *Borrelia burgdorferi* is initiated by a sigma 70-like promoter. *Microbiology* 143: 1681–1690.

Ge Y, Old IG, Saint Girons I *et al.* (1997b) Molecular characterization of a large *Borrelia burgdorferi* motility operon which is initiated by a consensus sigma70 promoter. *J. Bacteriol.* 179: 2289–2299.

Georgilis K, Peacocke M, Klempner MS (1992) Fibroblasts protect the Lyme disease spirochete, *Borrelia burgdorferi*, from ceftriaxone *in vitro*. *J. Infect. Dis.* 166: 440–444.

Gern L, Estrada-Pena A, Frandsen F *et al.* (1998) European reservoir hosts of *Borrelia burgdorferi* sensu lato. *Zbl. Bakteriol.* 287: 196–204.

Giambartolomei GH, Dennis VA, Philipp MT (1998) *Borrelia burgdorferi* stimulates the production of interleukin-10 in peripheral blood mononuclear cells from uninfected humans and rhesus monkeys. *Infect. Immun.* 66: 2691–2697.

Giambartolomei GH, Dennis VA, Lasater BL *et al.* (1999) Induction of pro- and anti-inflammatory cytokines by *Borrelia burgdorferi* lipoproteins in monocytes is mediated by CD14. *Infect. Immun.* 67: 140–147.

Gilmore RD Jr, Piesman J (2000) Inhibition of *Borrelia burgdorferi* migration from the midgut to the salivary glands following feeding by ticks on OspC-immunized mice. *Infect. Immun.* 68: 411–414.

Goldstein SF, Buttle KF, Charon NW (1996) Structural analysis of the Leptospiraceae and *Borrelia burgdorferi* by high-voltage electron microscopy. *J. Bacteriol.* 178: 6539–6545.

Goodman JL, Jurkovich P, Kramber JM *et al.* (1991) Molecular detection of persistent *Borrelia burgdorferi* in the urine of patients with active Lyme disease. *Infect. Immun.* 59: 269–278.

Goodman JL, Bradley JF, Ross AE *et al.* (1995) Bloodstream invasion in early Lyme disease: results from a prospective, controlled, blinded study using the polymerase chain reaction. *Am. J. Med.* 99: 6–12.

Grab DJ, Lanners H, Martin LN *et al.* (1999) Interaction of *Borrelia burgdorferi* with peripheral blood fibrocytes, antigen-presenting cells with the potential for connective tissue targeting. *Mol. Med.* 5: 46–54.

Gross DM, Forsthuber T, Tary-Lehmann M *et al.* (1998a) Identification of LFA-1 as a candidate autoantigen in treatment-resistant Lyme arthritis. *Science* 281: 703–706.

Gross DM, Steere AC, Huber BT (1998b) T helper 1 response is dominant and localized to the synovial fluid in patients with lyme arthritis. *J. Immunol.* 160: 1022–1028.

Guina T, Oliver DB (1997) Cloning and analysis of a *Borrelia burgdorferi* membrane-interactive protein exhibiting haemolytic activity. *Mol. Microbiol.* 24: 1201–1213.

Guo BP, Brown EL, Dorward DW *et al.* (1998) Decorin-binding adhesins from *Borrelia burgdorferi*. *Mol. Microbiol.* 30: 711–723.

Habicht GS (1992) Cytokines in *Borrelia burgdorferi* infection. In: Schutzer SE (ed.) *Lyme Disease: Molecular and Immunologic Approaches*. Cold Spring Harbor: Cold Spring Harbor Laboratory Press, pp. 149–168.

Habicht GS, Katona LI, Benach JL (1991) Cytokines and the pathogenesis of neuroborreliosis: *Borrelia burgdorferi* induces glioma cells to secrete interleukin-6. *J. Infect. Dis.* 164: 568–574.

Hagman KE, Lahdenne P, Popova TG *et al.* (1998) Decorin-binding protein of *Borrelia burgdorferi* is encoded within a two-gene operon and is protective in the murine model of Lyme borreliosis. *Infect. Immun.* 66: 2674–2683.

Hagman KE, Yang X, Wikel SK *et al.* (2000) Decorin-binding protein A (DbpA) of *Borrelia burgdorferi* is not protective when immunized mice are challenged via tick infestation and correlates with the lack of DbpA expression by *B. burgdorferi* in ticks. *Infect. Immun.* 68: 4759–4764.

Hauser U, Lehnert G, Lobentanzer R *et al.* (1997) Interpretation criteria for standardized Western blots for three European species of *Borrelia burgdorferi* sensu lato. *J. Clin. Microbiol.* 35: 1433–1444.

Hauser U, Krahl H, Peters H *et al.* (1998) Impact of strain heterogeneity on Lyme disease serology in Europe: comparison of enzyme-linked immunosorbent assays using different species of *Borrelia burgdorferi* sensu lato. *J. Clin. Microbiol.* 36: 427–436.

Hauser U, Lehnert G, Wilske B (1999) Validity of interpretation criteria for standardized Western blots (immunoblots) for serodiagnosis of Lyme borreliosis based on sera collected throughout Europe. *J. Clin. Microbiol.* 37: 2241–2247.

Hayes SF, Burgdorfer W (1993) Ultrastructure of *Borrelia burgdorferi*. In: Weber K, Burgdorfer W (eds) *Aspects of Lyme Borreliosis*. Berlin: Springer-Verlag, pp. 29–43.

Hayes SF, Burgdorfer W, Barbour AG (1983) Bacteriophage in the *Ixodes dammini* spirochete, etiological agent of Lyme disease. *J. Bacteriol.* 154: 1436–1439.

Hemmer B, Gran B, Zhao Y *et al.* (1999) Identification of candidate T-cell epitopes and molecular mimics in chronic Lyme disease. *Nature Med.* 5: 1375–1382.

Hinnebusch J, Barbour AG (1991) Linear plasmids of *Borrelia burgdorferi* have a telomeric structure and sequence similar to those of a eukaryotic virus. *J. Bacteriol.* 173: 7233–7239.

Hinnebusch J, Barbour AG (1992) Linear- and circular-plasmid copy numbers in *Borrelia burgdorferi*. *J. Bacteriol.* 174: 5251–5257.

Hinnebusch J, Bergstrom S, Barbour AG (1990) Cloning and sequence analysis of linear plasmid telomeres of the bacterium *Borrelia burgdorferi*. *Mol. Microbiol.* 4: 811–820.

Hodzic E, Feng S, Barthold SW (2000) Stability of *Borrelia burgdorferi* outer surface protein C under immune selection pressure. *J. Infect. Dis.* 181: 750–753.

Hofmeister EK, Glass GE, Childs JE *et al.* (1999) Population dynamics of a naturally occurring heterogeneous mixture of *Borrelia burgdorferi* clones. *Infect. Immun.* 67: 5709–5716.

Holt JG (1994) The spirochetes. In: Holt JG, Krieg NR, Sneath PHA, Staley JT, Williams ST (eds). *Bergey's Manual of Determinative Bacteriology*. Baltimore: Williams and Wilkins, pp. 27–36.

Hu LT, Klempner MS (1997) Host–pathogen interactions in the immunopathogenesis of Lyme disease. *J. Clin. Immunol.* 17: 354–365.

Hu LT, Perides G, Noring R *et al.* (1995) Binding of human plasminogen to *Borrelia burgdorferi*. *Infect. Immun.* 63: 3491–3496.

Hubalek Z, Halouzka J (1998) Prevalence rates of *Borrelia burgdorferi* sensu lato in host-seeking *Ixodes ricinus* ticks in Europe. *Parasitol. Res.* 84: 167–172.

Hunfeld K, Kraiczy P, Wichelhaus TA *et al.* (2000) Colorimetric *in vitro* susceptibility testing of penicillins, cephalosporins, macrolides, streptogramins, tetracyclines, and aminoglycosides against *Borrelia burgdorferi* isolates. *Int. J. Antimicrob. Agents* 15: 11–17.

Indest KJ, Philipp MT (2000) DNA-binding proteins possibly involved in regulation of the post-logarithmic-phase expression of lipoprotein P35 in *Borrelia burgdorferi*. *J. Bacteriol.* 182: 522–525.

Indest KJ, Ramamoorthy R, Sole M *et al.* (1997) Cell-density-dependent expression of *Borrelia burgdorferi* lipoproteins *in vitro*. *Infect. Immun.* 65: 1165–1171.

Iyer R, Hardham JM, Wormser GP *et al.* (2000) Conservation and heterogeneity of *vlsE* among human and tick isolates of *Borrelia burgdorferi*. *Infect. Immun.* 68: 1714–1718.

Jaulhac B, Nicolini P, Piemont Y *et al.* (1991) Detection of *Borrelia burgdorferi* in cerebrospinal fluid of patients with Lyme borreliosis. *N. Engl. J. Med.* 324: 1440.

Jauris-Heipke S, Liegl G, Preac-Mursic V *et al.* (1995) Molecular analysis of genes encoding outer surface protein C (OspC) of *Borrelia burgdorferi* sensu lato: Relationship to *ospA* genotype and evidence of lateral gene exchange of *ospC*. *J. Clin. Microbiol.* 33: 1860–1866.

Johnson BJ, Robbins KE, Bailey RE *et al.* (1996) Serodiagnosis of Lyme disease: accuracy of a two-step approach using a flagella-based ELISA and immunoblotting. *J. Infect. Dis.* 174: 346–353.

Johnson RC (1984) *Borrelia burgdorferi* sp. nov.: etiological agent of Lyme disease. *Int. J. Syst. Bacteriol.* 34: 496–497.

Jonsson M, Bergstrom S (1995) Transcriptional and translational regulation of the expression of the major outer surface proteins in Lyme disease *Borrelia* strains. *Microbiology* 141: 1321–1329.

Kalish RA, Leong JM, Steere AC (1993) Association of treatment-resistant chronic Lyme arthritis with HLA-DR4 and antibody reactivity to OspA and OspB of *Borrelia burgdorferi*. *Infect. Immun.* 61: 2774–2779.

Karlsson M, Hovind-Hougen K, Svenungsson B *et al.* (1990) Cultivation and characterization of spirochetes from cerebrospinal fluid of patients with Lyme borreliosis. *J. Clin. Microbiol.* 28: 473–479.

Kawabata H, Masuzawa T, Yanagihara Y (1993) Genomic analysis of *Borrelia japonica* sp. nov. isolated from *Ixodes ovatus* in Japan. *Microbiol. Immunol.* 37: 843–848.

Kawabata H, Myouga F, Inagaki Y *et al.* (1998) Genetic and immunological analyses of Vls (VMP-like sequences) of *Borrelia burgdorferi*. *Microb. Pathog.* 24: 155–166.

Keane-Myers A, Nickell SP (1995) T cell subset-dependent modulation of immunity to *Borrelia burgdorferi* in mice. *J. Immunol.* 154: 1770–1776.

Kelleher DM, Telford SR, Criscione L *et al.* (1998) Cytokines in murine lyme carditis: Th1 cytokine expression follows expression of proinflammatory cytokines in a susceptible mouse strain. *J. Infect. Dis.* 177: 242–246.

Kitten T, Barbour AG (1992) The relapsing fever agent *Borrelia hermsii* has multiple copies of its chromosome and linear plasmids. *Genetics* 132: 311–324.

Kochi SK, Johnson RC, Dalmasso AP (1991) Complement-mediated killing of the Lyme disease spirochete *Borrelia burgdorferi*. Role of antibody in formation of an effective membrane attack complex. *J. Immunol.* 146: 3964–3970.

Kurtenbach K, Peacey M, Rijpkema SG *et al.* (1998) Differential transmission of the genospecies of *Borrelia burgdorferi* sensu lato by game birds and small rodents in England. *Appl. Environ. Microbiol.* 64: 1169–1174.

Kurtti TJ, Munderloh UG, Johnson RC *et al.* (1987) Colony formation and morphology in *Borrelia burgdorferi*. *J. Clin. Microbiol.* 25: 2054–2058.

Lam TT, Nguyen TP, Montgomery RR *et al.* (1994) Outer surface proteins E and F of *Borrelia burgdorferi*, the agent of Lyme disease. *Infect. Immun.* 62: 290–298.

Lawrenz MB, Hardham JM, Owens RT *et al.* (1999) Human antibody responses to VlsE antigenic variation protein of *Borrelia burgdorferi*. *J. Clin. Microbiol.* 37: 3997–4004.

Le Fleche A, Postic D, Girardet K *et al.* (1997) Characterization of *Borrelia lusitaniae* sp. nov. by 16S ribosomal DNA sequence analysis. *Int. J. Syst. Bacteriol.* 47: 921–925.

Lebech AM, Clemmensen O, Hansen K (1995) Comparison of *in vitro* culture, immunohistochemical staining, and PCR for detection of *Borrelia burgdorferi* in tissue from experimentally infected animals. *J. Clin. Microbiol.* 33: 2328–2333.

Ledue TB, Collins MF, Craig WY (1996) New laboratory guidelines for serologic diagnosis of Lyme disease: evaluation of the two-test protocol. *J. Clin. Microbiol.* 34: 2343–2350.

Leong JM, Wang H, Magoun L *et al.* (1998) Different classes of proteoglycans contribute to the attachment of *Borrelia burgdorferi* to cultured endothelial and brain cells. *Infect. Immun.* 66: 994–999.

Liveris D, Gazumyan A, Schwartz I (1995) Molecular typing of *Borrelia burgdorferi* sensu lato by PCR-restriction fragment length polymorphism analysis. *J. Clin. Microbiol.* 33: 589–595.

Liveris D, Wormser GP, Nowakowski J *et al.* (1996) Molecular typing of *Borrelia burgdorferi* from Lyme disease patients by PCR-restriction fragment length polymorphism analysis. *J. Clin. Microbiol.* 34: 1306–1309.

Liveris D, Varde S, Iyer R *et al.* (1999) Genetic diversity of *Borrelia burgdorferi* in Lyme disease patients as determined by culture versus direct PCR with clinical specimens. *J. Clin. Microbiol.* 37: 565–569.

Livey I, Gibbs CP, Schuster R *et al.* (1995) Evidence for lateral transfer and recombination in OspC variation in Lyme disease *Borrelia*. *Mol. Microbiol.* 18: 257–269.

Luke CJ, Marshall MA, Zahradnik JM *et al.* (2000) Growth-inhibiting antibody responses of humans vaccinated with recombinant outer surface protein A or infected with *Borrelia burgdorferi* or both. *J. Infect. Dis.* 181: 1062–1068.

Marconi RT, Samuels DS, Garon CF (1993) Transcriptional analyses and mapping of the *ospC* gene in Lyme disease spirochetes. *J. Bacteriol.* 175: 926–932.

Marconi RT, Samuels DS, Landry RK *et al.* (1994) Analysis of the distribution and molecular heterogeneity of the *ospD* gene among the Lyme disease spirochetes: evidence for lateral gene exchange. *J. Bacteriol.* 176: 4572–4582.

Marconi RT, Liveris D, Schwartz I (1995) Identification of novel insertion elements, restriction fragment length polymorphism patterns, and discontinuous 23S rRNA in Lyme disease spirochetes: phylogenetic analyses of rRNA genes and their intergenic spacers in *Borrelia japonica* sp. nov. and genomic group 21038 (*Borrelia andersonii* sp. nov.) isolates. *J. Clin. Microbiol.* 33: 2427–2434.

Marconi RT, Casjens S, Munderloh UG *et al.* (1996) Analysis of linear plasmid dimers in *Borrelia burgdorferi* sensu lato isolates: implications concerning the potential mechanism of linear plasmid replication. *J. Bacteriol.* 178: 3357–3361.

Marconi RT, Sung SY, Hughes CA *et al.* (1997) Molecular and evolutionary analyses of a variable series of genes in *Borrelia burgdorferi* that are related to *ospE* and *ospF*, constitute a gene family, and share a common upstream homology box. *J. Bacteriol.* 178: 5615–5626.

Margolis N, Hogan D, Cieplak W Jr *et al.* (1994) Homology between *Borrelia burgdorferi* OspC and members of the family of *Borrelia hermsii* variable major proteins. *Gene* 143: 105–110.

Marshall WF.3, Telford SR.3, Rys PN *et al.* (1994) Detection of *Borrelia burgdorferi* DNA in museum specimens of *Peromyscus leucopus*. *J. Infect. Dis.* 170: 1027–1032.

Masuzawa T, Kurita T, Kawabata H *et al.* (1994) Relationship between infectivity and OspC expression in Lyme disease *Borrelia*. *FEMS Microbiol. Lett.* 123: 319–324.

Mathiesen DA, Oliver JH Jr, Kolbert CP *et al.* (1997) Genetic heterogeneity of *Borrelia burgdorferi* in the United States. *J. Infect. Dis.* 175: 98–107.

Matuschka FR, Ohlenbusch A, Eiffert H *et al.* (1996) Characteristics of Lyme disease spirochetes in archived European ticks. *J. Infect. Dis.* 174: 424–426.

McKisic MD, Redmond WL, Barthold SW (2000) Cutting edge: T cell-mediated pathology in murine Lyme borreliosis. *J. Immunol.* 164: 6096–6099.

Moody KD, Barthold SW (1998) Lyme borreliosis in laboratory mice. *Lab. Anim. Sci.* 48: 168–171.

Nadelman RB, Wormser GP (1995) Erythema migrans and early Lyme disease. *Am. J. Med.* 98: 15S–23S.

Nadelman RB, Wormser GP (1998) Lyme borreliosis. *Lancet* 352: 557–565.

Nadelman RB, Pavia CS, Magnarelli LA *et al.* (1990) Isolation of *Borrelia burgdorferi* from the blood of seven patients with Lyme disease. *Am. J. Med.* 88: 21–26.

Nakao M, Miyamoto K (1995) Mixed infection of different *Borrelia* species among *Apodemus speciosus* mice in Hokkaido, Japan. *J. Clin. Microbiol.* 33: 490–492.

Neubert U, Schaller M, Januschke E *et al.* (1993) Bacteriophages induced by ciprofloxacin in a *Borrelia burgdorferi* skin isolate. *Zbl. Bakteriol.* 279: 307–315.

Nicholls TH, Callister SM (1996) Lyme disease spirochetes in ticks collected from birds in midwestern United States. *J. Med. Entomol.* 33: 379–384.

Nocton JJ, Dressler F, Rutledge BJ *et al.* (1994) Detection of *Borrelia burgdorferi* DNA by polymerase chain reaction in synovial fluid from patients with Lyme arthritis. *N. Engl. J. Med.* 330: 229–234.

Nordstrand A, Barbour AG, Bergstrom S (2000) *Borrelia* pathogenesis research in the post-genomic and post-vaccine era. *Curr. Opin. Microbiol.* 3: 86–92.

Norman GL, Antig JM, Bigaignon G *et al.* (1996) Sero-diagnosis of Lyme borreliosis by *Borrelia burgdorferi* sensu stricto, *B. garinii*, and *B. afzelii* western blots (immunoblots). *J. Clin. Microbiol.* 34: 1732–1738.

Norris SJ, Carter CJ, Howell JK *et al.* (1992) Low-passage-associated proteins of *Borrelia burgdorferi* B31: characterization and molecular cloning of OspD, a surface-exposed, plasmid-encoded lipoprotein. *Infect. Immun.* 60: 4662–4672.

O'Connell S, Granstrom M, Gray JS *et al.* (1998) Epidemiology of European Lyme borreliosis. *Zbl. Bakteriol.* 287: 229–240.

Ojaimi C, Davidson BE, Saint Girons I *et al.* (1994) Conservation of gene arrangement and an unusual organization of rRNA genes in the linear chromosomes of the Lyme disease spirochaetes *Borrelia burgdorferi*, *B. garinii* and *B. afzelii*. *Microbiology* 140: 2931–2940.

Olsen B, Jaenson TG, Noppa L *et al.* (1993) A Lyme borreliosis cycle in seabirds and *Ixodes uriae* ticks. *Nature* 362: 340–342.

Orloski KA, Hayes EB, Campbell GL *et al.* (2000) Surveillance for Lyme disease – United States, 1992–1998. *MMWR* 49: 1–11.

Oschmann P, Dorndorf W, Hornig C *et al.* (1998) Stages and syndromes of neuroborreliosis. *J. Neurol.* 245: 262–272.

Pachner AR, Ricalton N, Delaney E (1993) Comparison of polymerase chain reaction with culture and serology for diagnosis of murine experimental Lyme borreliosis. *J. Clin. Microbiol.* 31: 208–214.

Pachner AR, Basta J, Delaney E *et al.* (1995) Localization of *Borrelia burgdorferi* in murine Lyme borreliosis by electron microscopy. *Am. J. Trop. Med. Hyg.* 52: 128–133.

Pachner AR, Amemiya K, Delaney E *et al.* (1997) Interleukin-6 is expressed at high levels in the CNS in Lyme neuroborreliosis. *Neurology* 49: 147–152.

Padilla ML, Callister SM, Schell RF *et al.* (1996) Characterization of the protective borreliacidal antibody response in humans and hamsters after vaccination with a *Borrelia burgdorferi* outer surface protein A vaccine. *J. Infect. Dis.* 174: 739–746.

Palmer N, Fraser C, Casjens S (2000) Distribution of twelve linear extrachromosomal DNAs in natural isolates of Lyme disease spirochetes. *J. Bacteriol.* 182: 2476–2480.

Parveen N, Leong JM (2000) Identification of a candidate glycosaminoglycan-binding adhesin of the Lyme disease spirochete *Borrelia burgdorferi*. *Mol. Microbiol.* 35: 1220–1234.

Parveen N, Robbins D, Leong JM (1999) Strain variation in glycosaminoglycan recognition influences cell-type- specific-binding by Lyme disease spirochetes. *Infect. Immun.* 67: 1743–1749.

Paster BJ, Dewhirst FE, Weisburg WG *et al.* (1991) Phylogenetic analysis of the spirochetes. *J. Bacteriol.* 173: 6101–6109.

Pavia CS, Wormser GP, Bittker S *et al.* (2000) An indirect hemagglutination antibody test to detect antibodies to *Borrelia burgdorferi* in patients with Lyme disease. *J. Microbiol. Methods* 40: 163–173.

Picken RN, Cheng Y, Han D *et al.* (1995) Genotypic and phenotypic characterization of *Borrelia burgdorferi* isolated from ticks and small animals in Illinois. *J. Clin. Microbiol.* 33: 2304–2315.

Picken MM, Picken RN, Han D *et al.* (1997) A two year prospective study to compare culture and polymerase chain reaction amplification for the detection and diagnosis of Lyme borreliosis. *Mol. Pathol.* 50: 186–193.

Piesman J, Oliver JR, Sinsky RJ (1990) Growth kinetics of the Lyme disease spirochete (*Borrelia burgdorferi*) in vector ticks (*Ixodes dammini*). *Am. J. Trop. Med. Hyg.* 42: 352–357.

Plasterk RH, Simon MI, Barbour AG (1985) Transposition of structural genes to an expression sequence on a linear plasmid causes antigenic variation in the bacterium *Borrelia hermsii*. *Nature* 318: 257–263.

Pollack RJ, Telford SR III, Spielman A (1993) Standardization of medium for culturing Lyme disease spirochetes. *J. Clin. Microbiol.* 31: 1251–1255.

Porcella SF, Fitzpatrick CA, Bono JL (2000) Expression and immunological analysis of the plasmid-borne *mlp* genes of *Borrelia burgdorferi* strain B31. *Infect. Immun.* 68: 4992–5001.

Postic D, Assous MV, Grimont PAD *et al.* (1994) Diversity of *Borrelia burgdorferi* sensu lato evidenced by restriction fragment length polymorphism of *rrf* (5S)-*rrl* (23S) intergenic spacer amplicons. *Int. J. Syst. Bacteriol.* 44: 743–752.

Postic D, Assous M, Belfaiza J *et al.* (1996) Genetic diversity of *Borrelia* of Lyme borreliosis. *Wien Klin. Wochenschr.* 108: 748–751.

Postic D, Ras NM, Lane RS *et al.* (1998) Expanded diversity among Californian *Borrelia* isolates and description of *Borrelia bissettii* sp. nov (Formerly *Borrelia* group DN127). *J. Clin. Microbiol.* 36: 3497–3504.

Preac-Mursic V, Wilske B, Schierz G (1986) European *Borrelia burgdorferi* isolated from humans and ticks culture conditions and antibiotic susceptibility. *Zbl. Bakteriol. Mikrobiol. Hyg. A* 263: 112–118.

Preac-Mursic V, Marget W, Busch U *et al.* (1996) Kill kinetics of *Borrelia burgdorferi* and bacterial findings in relation to the treatment of Lyme borreliosis. *Infection* 24: 9–16.

Priem S, Rittig MG, Kamradt T *et al.* (1997) An optimized PCR leads to rapid and highly sensitive detection of *Borrelia burgdorferi* in patients with Lyme borreliosis. *J. Clin. Microbiol.* 35: 685–690.

Probert WS, Johnson BJ (1998) Identification of a 47 kDa fibronectin-binding protein expressed by *Borrelia burgdorferi* isolate B31. *Mol. Microbiol.* 30: 1003–1015.

Radolf JD, Bourell KW, Akins DR *et al.* (1994) Analysis of *Borrelia burgdorferi* membrane architecture by freeze-fracture electron microscopy. *J. Bacteriol.* 176: 21–31.

Radolf JD, Goldberg MS, Bourell K *et al.* (1995) Characterization of outer membranes isolated from *Borrelia burgdorferi*, the Lyme disease spirochete. *Infect. Immun.* 63: 2154–2163.

Ras NM, Postic D, Foretz M *et al.* (1997) *Borrelia burgdorferi* sensu stricto, a bacterial species 'made in the USA'? *Int. J. Syst. Bacteriol.* 47: 1112–1117.

Roberts DM, Carlyon JA, Theisen M *et al.* (2000) The *bdr* gene families of the Lyme disease and relapsing fever spirochetes: potential influence on biology, pathogenesis, and evolution. *Emerg. Infect. Dis.* 6: 110–122.

Roberts WC, Mullikin BA, Lathigra R *et al.* (1998) Molecular analysis of sequence heterogeneity among genes encoding decorin binding proteins A and B of *Borrelia burgdorferi* sensu lato. *Infect. Immun.* 66: 5275–5285.

Roessler D, Hauser U, Wilske B (1997) Heterogeneity of BmpA (P39) among European isolates of *Borrelia burgdorferi* sensu lato and influence of interspecies variability on serodiagnosis. *J. Clin. Microbiol.* 35: 2752–2758.

Rosa PA, Schwan T, Hogan D (1992) Recombination between genes encoding major outer surface proteins A and B of *Borrelia burgdorferi*. *Mol. Microbiol.* 6: 3031–3040.

Rosa P, Samuels DS, Hogan D *et al.* (1996) Directed insertion of a selectable marker into a circular plasmid of *Borrelia burgdorferi*. *J. Bacteriol.* 178: 5946–5953.

Rossler D, Eiffert H, Jauris-Heipke S *et al.* (1995) Molecular and immunological characterization of the p83/100 protein of various *Borrelia burgdorferi* sensu lato strains. *Med. Microbiol. Immunol. (Berl.)* 184: 23–32.

Ryan JR, Levine JF, Apperson CS *et al.* (1998) An experimental chain of infection reveals that distinct *Borrelia burgdorferi* populations are selected in arthropod and mammalian hosts. *Mol. Microbiol.* 30: 365–379.

Sadziene A, Thomas DD, Barbour AG (1995) *Borrelia burgdorferi* mutant lacking Osp: biological and immunological characterization. *Infect. Immun.* 63: 1573–1580.

Saint Girons I, Gern L, Gray JS *et al.* (1998) Identification of *Borrelia burgdorferi* sensu lato Species in Europe. *Zbl. Bakteriol.* 287: 190–195.

Samuels DS, Marconi RT, Garon CF (1993) Variation in the size of the ospA-containing linear plasmid, but not the linear chromosome, among the three *Borrelia* species associated with Lyme disease. *J. Gen. Microbiol.* 139: 2445–2449.

Sartakova M, Dobrikova E, Cabello FC (2000) Development of an extrachromosomal cloning vector system for use in *Borrelia burgdorferi*. *Proc. Natl Acad. Sci. USA* 97: 4850–4855.

Sato Y, Miyamoto K, Iwaki A *et al.* (1996) Prevalence of Lyme disease spirochetes in *Ixodes persulcatus* and wild rodents in far eastern Russia. *Appl. Environ. Microbiol.* 62: 3887–3889.

Schmidt BL (1997) PCR in laboratory diagnosis of human *Borrelia burgdorferi* infections. *Clin. Microbiol. Rev.* 10: 185–201.

Schutzer SE, Coyle PK, Reid P *et al.* (1999) *Borrelia burgdorferi*-specific immune complexes in acute Lyme disease. *JAMA* 282: 1942–1946.

Schwan TG, Piesman J (2000) Temporal changes in outer surface proteins A and C of the Lyme disease-associated spirochete, *Borrelia burgdorferi*, during the chain of infection in ticks and mice. *J. Clin. Microbiol.* 38: 382–388.

Schwan TG, Burgdorfer W, Garon CF (1988) Changes in infectivity and plasmid profile of the Lyme disease spirochete, *Borrelia burgdorferi*, as a result of *in vitro* cultivation. *Infect. Immun.* 56: 1831–1836.

Schwan TG, Piesman J, Golde WT *et al.* (1995) Induction of an outer surface protein on *Borrelia burgdorferi* during tick feeding. *Proc. Natl Acad. Sci. USA* 92: 2909–2913.

Schwartz JJ, Gazumyan A, Schwartz I (1992a) rRNA Gene organization in the Lyme disease spirochete, *Borrelia burgdorferi*. *J. Bacteriol.* 174: 3757–3765.

Schwartz I, Wormser GP, Schwartz JJ *et al.* (1992b) Diagnosis of early Lyme disease by polymerase chain reaction amplification and culture of skin biopsies from erythema migrans lesions. *J. Clin. Microbiol.* 30: 3082–3088.

Schwartz I, Varde S, Nadelman RB *et al.* (1997a) Inhibition of efficient polymerase chain reaction amplification of *Borrelia burgdorferi* DNA in blood-fed ticks. *Am. J. Trop. Med. Hyg.* 56: 339–342.

Schwartz I, Fish D, Daniels TJ (1997b) Prevalence of the rickettsial agent of human granulocytic ehrlichiosis in ticks from a hyperendemic focus of Lyme disease. *N. Engl. J. Med.* 337: 49–50.

Scopio A, Johnson P, Laquerre A *et al.* (1994) Subcellular localization and chaperone activities of *Borrelia burgdorferi* Hsp60 and Hsp70. *J. Bacteriol.* 176: 6449–6456.

Seinost G, Dykhuizen DE, Dattwyler RJ *et al.* (1999) Four clones of *Borrelia burgdorferi* sensu stricto cause invasive infection in humans. *Infect. Immun.* 67: 3518–3524.

Sellati TJ, Bouis DA, Kitchens RL *et al.* (1998) *Treponema pallidum* and *Borrelia burgdorferi* lipoproteins and synthetic lipopeptides activate monocytic cells via a CD14-dependent pathway distinct from that used by lipopolysaccharide. *J. Immunol.* 160: 5455–5464.

Seltzer EG, Gerber MA, Cartter ML *et al.* (2000) Long-term outcomes of persons with Lyme disease. *JAMA* 283: 609–616.

Shang ES, Champion CI, Wu XY *et al.* (2000) Comparison of protection in rabbits against host-adapted and cultivated *Borrelia burgdorferi* following infection-derived immunity or immunization with outer membrane vesicles or outer surface protein A. *Infect. Immun.* 68: 4189–4199.

Sigal LH (1997a) Lyme disease: a review of aspects of its immunology and immunopathogenesis. *Annu. Rev. Immunol.* 15: 63–92.

Sigal LH (1997b) Immunologic mechanisms in Lyme neuroborreliosis: the potential role of autoimmunity and molecular mimicry. *Semin. Neurol.* 17: 63–68.

Sigal LH, Williams S (1997) A monoclonal antibody to *Borrelia burgdorferi* flagellin modifies neuroblastoma cell neuritogenesis *in vitro*: a possible role for autoimmunity in the neuropathy of Lyme disease. *Infect. Immun.* 65: 1722–1728.

Sigal LH, Zahradnik JM, Lavin P *et al.* (1998) A vaccine consisting of recombinant *Borrelia burgdorferi* outer-surface protein A to prevent Lyme disease. *N. Engl. J. Med.* 339: 216–222.

Simpson WJ, Schrumpf ME, Schwan TG (1990) Reactivity of human Lyme borreliosis sera with a 39-kilodalton antigen specific to *Borrelia burgdorferi*. *J. Clin. Microbiol.* 28: 1329–1337.

Skare JT, Shang ES, Foley DM *et al.* (1995) Virulent strain associated outer membrane proteins of *Borrelia burgdorferi*. *J. Clin. Invest.* 96: 2380–2392.

Skare JT, Champion CI, Mirzabekov TA *et al.* (1996) Porin activity of the native and recombinant outer membrane protein Oms28 of *Borrelia burgdorferi*. *J. Bacteriol.* 178: 4909–4918.

Skare JT, Mirzabekov TA, Shang ES *et al.* (1997) The Oms66 (p66) protein is a *Borrelia burgdorferi* porin. *Infect. Immun.* 65: 3654–3661.

Skare JT, Foley DM, Hernandez SR *et al.* (1999) Cloning and molecular characterization of plasmid-encoded antigens of *Borrelia burgdorferi*. *Infect. Immun.* 67: 4407–4417.

Sohaskey CD, Zuckert WR, Barbour AG (1999) The extended promoters for two outer membrane lipoprotein genes of *Borrelia* spp. uniquely include a T-rich region. *Mol. Microbiol.* 33: 41–51.

Stanek G, O'Connell S, Cimmino M *et al.* (1996) European union concerted action on risk assessment in Lyme borreliosis: clinical case definitions for Lyme borreliosis. *Wien Klin. Wochenschr.* 108: 741–747.

Steere AC (1989) Lyme disease. *N. Engl. J. Med.* 263: 201–205.

Steere AC, Malawista SE, Hardin JA *et al.* (1977a) Erythema chronicum migrans and Lyme arthritis: the enlarging clinical spectrum. *Ann. Intern. Med.* 86: 685–698.

Steere AC, Malawista SE, Snydman DR *et al.* (1977b) Lyme arthritis: an epidemic of oligoarticular arthritis in children and adults in three Connecticut communities. *Arthritis Rheum.* 20: 7–17.

Steere AC, Broderick TF, Malawista SE (1978) Erythema chronicum migrans and Lyme arthritis: epidemiologic evidence for a tick vector. *Am. J. Epidemiol.* 108: 312–321.

Steere AC, Grodzicki RL, Kornblatt AN *et al.* (1983) The spirochetal etiology of Lyme disease. *N. Engl. J. Med.* 308: 733–740.

Steere AC, Taylor E, Wilson ML *et al.* (1986) Longitudinal assessment of the clinical and epidemiological features of Lyme disease in a defined population. *J. Infect. Dis.* 154: 295–300.

Steere AC, Sikand VK, Meurice F *et al.* (1998) Vaccination against Lyme disease with recombinant *Borrelia burgdorferi* outer-surface lipoprotein A with adjuvant. *N. Engl. J. Med.* 339: 209–215.

Stevenson B, Tilly K, Rosa PA (1996) A family of genes located on four separate 32-kilobase circular plasmids in *Borrelia burgdorferi* B31. *J. Bacteriol.* 178: 3508–3516.

Stevenson B, Casjens S, van Vugt R *et al.* (1997) Characterization of cp18, a naturally truncated member of the cp32 family of *Borrelia burgdorferi* plasmids. *J. Bacteriol.* 179: 4285–4291.

Stevenson B, Bono JL, Elias A *et al.* (1998) Transformation of the Lyme disease spirochete *Borrelia burgdorferi* with heterologous DNA. *J. Bacteriol.* 180: 4850–4855.

Strle F, Nelson JA, Ruzic-Sabljic E *et al.* (1996) European Lyme borreliosis: 231 culture-confirmed cases involving patients with erythema migrans. *Clin. Infect. Dis.* 23: 61–65.

Strle F, Nadelman RB, Cimperman J *et al.* (1999) Comparison of culture-confirmed erythema migrans caused by *Borrelia burgdorferi* sensu stricto in New York State and by *Borrelia afzelii* in Slovenia. *Ann. Intern. Med.* 130: 32–36.

Subramanian G, Koonin EV, Aravind L (2000) Comparative genome analysis of the pathogenic spirochetes *Borrelia burgdorferi* and *Treponema pallidum*. *Infect. Immun.* 68: 1633–1648.

Sung SY, Lavoie CP, Carlyon JA *et al.* (1998) Genetic divergence and evolutionary instability in ospE-related members of the upstream homology box gene family in *Borrelia burgdorferi* sensu lato complex isolates. *Infect. Immun.* 66: 4656–4668.

Sung SY, McDowell JV, Carlyon JA *et al.* (2000) Mutation and recombination in the upstream homology box-flanked *ospE*-related genes of the Lyme disease spirochetes result in the development of new antigenic variants during infection. *Infect. Immun.* 68: 1319–1327.

Szczepanski A, Benach JL (1991) Lyme borreliosis: host responses to *Borrelia burgdorferi*. *Microbiol. Rev.* 55: 21–34.

Takayama K, Rothenberg RJ, Barbour AG (1987) Absence of lipopolysaccharide in the Lyme disease spirochete, *Borrelia burgdorferi*. *Infect. Immun.* 55: 2311–2313.

Talkington J, Nickell SP (1999) *Borrelia burgdorferi* spirochetes induce mast cell activation and cytokine release. *Infect. Immun.* 67: 1107–1115.

Thanassi WT, Schoen RT (2000) The Lyme disease vaccine: conception, development, and implementation. *Ann. Intern. Med.* 132: 661–668.

Theisen M, Borre M, Mathiesen MJ *et al.* (1995) Evolution of the *Borrelia burgdorferi* outer surface protein OspC. *J. Bacteriol.* 177: 3036–3044.

Tibayrenc M (1995) Population genetics of parasitic protozoa and other microorganisms. *Adv. Parasitol.* 36: 47–115.

Tilly K, Casjens S, Stevenson B *et al.* (1997) The *Borrelia burgdorferi* circular plasmid cp26: conservation of plasmid structure and targeted inactivation of the *ospC* gene. *Mol. Microbiol.* 25: 361–373.

Trevejo RT, Krause PJ, Sikand VK *et al.* (1999) Evaluation of two-test serodiagnostic method for early Lyme disease in clinical practice. *J. Infect. Dis.* 179: 931–938.

Tugwell P, Dennis DT, Weinstein A *et al.* (1997) Laboratory evaluation in the diagnosis of Lyme disease. *Ann. Intern. Med.* 127: 1109–1123.

Valsangiacomo C, Balmelli T, Piffaretti JC (1997) A phylogenetic analysis of *Borrelia burgdorferi* sensu lato based on sequence information from the *hbb* gene, coding for a histone-like protein. *Int. J. Syst. Bacteriol.* 47: 1–10.

van Dam AP, Kuiper H, Vos K *et al.* (1993) Different genospecies of *Borrelia burgdorferi* are associated with distinct clinical manifestations of Lyme borreliosis. *Clin. Infect. Dis.* 17: 708–717.

van Dam AP, Oei A, Jaspars R *et al.* (1997) Complement-mediated serum sensitivity among spirochetes that cause Lyme disease. *Infect. Immun.* 65: 1228–1236.

van der Linde MR (1991) Lyme carditis: clinical characteristics of 105 cases. *Scand. J. Infect. Dis. Suppl.* 77: 81–84.

Wang G, van Dam AP, Le Fleche A *et al.* (1997) Genetic and phenotypic analysis of *Borrelia valaisiana* sp. nov. (*Borrelia* genomic groups VS116 and M19). *Int. J. Syst. Bacteriol.* 47: 926–932.

Wang G, van Dam AP, Spanjaard L *et al.* (1998) Molecular typing of *Borrelia burgdorferi* sensu lato by randomly amplified polymorphic DNA fingerprinting analysis. *J. Clin. Microbiol.* 36: 768–776.

Wang G, van Dam AP, Schwartz I *et al.* (1999a) Molecular typing of *Borrelia burgdorferi* sensu lato: taxonomic, epidemiological, and clinical implications. *Clin. Microbiol. Rev.* 12: 633–653.

Wang G, van Dam AP, Dankert J (1999b) Phenotypic and genetic characterization of a novel *Borrelia burgdorferi* sensu lato isolate from a patient with Lyme borreliosis. *J. Clin. Microbiol.* 37: 3025–3028.

Wang G, van Dam AP, Dankert J (1999c) Evidence for frequent *ospC* gene transfer between *Borrelia valaisiana* sp. nov. and other Lyme disease spirochetes. *FEMS Microbiol. Lett.* 177: 289–296.

Wang IN, Dykhuizen DE, Qiu W *et al.* (1999d) Genetic diversity of *ospC* in a local population of *Borrelia burgdorferi* sensu stricto. *Genetics* 151: 15–30.

Wayne LG, Brenner DJ, Colwell RR *et al.* (1987) Report of the Ad Hoc Committee on reconciliation of approaches to bacterial systematics. *Int. J. Syst. Bacteriol.* 37: 463–464.

Weber K, Pfister HW (1993) History of Lyme borreliosis in Europe. In: Weber K, Burgdorfer W (eds) *Aspects of Lyme Borreliosis.* Berlin: Springer-Verlag, pp. 1–20.

Welsh J, Pretzman C, Postic D *et al.* (1992) Genomic fingerprinting by arbitrarily primed polymerase chain reaction resolves *Borrelia burgdorferi* into three distinct phyletic groups. *Int. J. Syst. Bacteriol.* 42: 370–377.

Will G, Jauris-Heipke S, Schwab E *et al.* (1995) Sequence analysis of *ospA* genes shows homogeneity within *Borrelia burgdorferi* sensu stricto and *Borrelia afzelii* strains but reveals major subgroups within the *Borrelia garinii* species. *Med. Microbiol. Immunol. (Berl.)* 184: 73–80.

Williams JG, Kubelik AR, Livak KJ *et al.* (1990) DNA polymorphisms amplified by arbitrary primers are useful as genetic markers. *Nucleic Acids Res.* 18: 6531–6535.

Wilske B, Preac-Mursic V, Göbel UB *et al.* (1993) An OspA serotyping system for *Borrelia burgdorferi* based on reactivity with monoclonal antibodies and OspA sequence analysis. *J. Clin. Microbiol.* 31: 340–350.

Wooten RM, Modur VR, McIntyre TM *et al.* (1996) *Borrelia burgdorferi* outer membrane protein A induces nuclear translocation of nuclear factor-kappa B and inflammatory activation in human endothelial cells. *J. Immunol.* 157: 4584–4590.

Wormser GP (1995) Prospects for a vaccine to prevent Lyme disease in humans. *Clin. Infect. Dis.* 21: 1267–1274.

Wormser GP (1997) Treatment and prevention of Lyme disease, with emphasis on antimicrobial therapy for neuroborreliosis and vaccination. *Semin. Neurol.* 17: 45–52.

Wormser GP, Nowakowski J, Nadelman RB *et al.* (1998) Improving the yield of blood cultures for patients with early Lyme disease. *J. Clin. Microbiol.* 36: 296–298.

Wormser GP, Liveris D, Nowakowski J *et al.* (1999a) Association of specific subtypes of *Borrelia burgdorferi* with hematogenous dissemination in early Lyme disease. *J. Infect. Dis.* 180: 720–725.

Wormser GP, Aguero-Rosenfeld ME, Nadelman RB. (1999b) Lyme disease serology: problems and opportunities. *JAMA* 282: 79–80.

Wormser GP, Bittker S, Cooper D *et al.* (2000a) Comparison of the yields of blood cultures using serum or plasma from patients with early lyme disease. *J. Clin. Microbiol.* 38: 1648–1650.

Wormser GP, Carbonaro C, Miller S *et al.* (2000b) A limitation of 2-stage serological testing for Lyme disease: enzyme immunoassay and immunoblot assay are not independent tests. *Clin. Infect. Dis.* 30: 545–548.

Wormser GP, Nadelman RB, Dattwyler RJ *et al.* (2000c) Practice guidelines for the treatment of Lyme disease. *Clin. Infect. Dis.* 31: S1–S14.

Xu Y, Johnson RC (1995) Analysis and comparison of plasmid profiles of *Borrelia burgdorferi* sensu lato strains. *J. Clin. Microbiol.* 33: 2679–2685.

Xu Y, Kodner C, Coleman L *et al.* (1996) Correlation of plasmids with infectivity of *Borrelia burgdorferi* sensu stricto type strain B31. *Infect. Immun.* 64: 3870–3876.

Yang X, Popova TG, Hagman KE *et al.* (1999) Identification, characterization, and expression of three new members of the *Borrelia burgdorferi mlp* (2.9) lipoprotein gene family. *Infect. Immun.* 67: 6008–6018.

Zeidner N, Mbow ML, Dolan M *et al.* (1997) Effects of *Ixodes scapularis* and *Borrelia burgdorferi* on modulation of the host immune response: induction of a Th2 cytokine response in Lyme disease-susceptible (C3H/HeJ) mice but not in disease-resistant (BALB/c) mice. *Infect. Immun.* 65: 3100–3106.

Zhang JR, Norris SJ (1998) Genetic variation of the *Borrelia burgdorferi* gene *vlsE* involves cassette-specific, segmental gene conversion. *Infect. Immun.* 66: 3698–3704.

Zhang JR, Hardham JM, Barbour AG *et al.* (1997a) Antigenic variation in Lyme disease borreliae by promiscuous recombination of VMP-like sequence cassettes. *Cell* 89: 275–285.

Zhang ZF, Wan KL, Zhang JS (1997b) Studies on epidemiology and etiology of Lyme disease in China. *Chin. J. Epidemiol.* 18: 8–11.

Zhong W, Gern L, Stehle T *et al.* (1999) Resolution of experimental and tick-borne *Borrelia burgdorferi* infection in mice by passive, but not active immunization using recombinant OspC. *Eur. J. Immunol.* 29: 946–957.

Zuckert WR, Barbour AG (2000) Stability of *Borrelia burgdorferi bdr* loci *in vitro* and *in vivo*. *Infect. Immun.* 68: 1727–1730.

Zuckert WR, Meyer J, Barbour AG (1999) Comparative analysis and immunological characterization of the *Borrelia* Bdr protein family. *Infect. Immun.* 67: 3257–3266.

98

Relapsing Fever *Borrelia*

Sally J. Cutler

Imperial College School of Medicine, London, UK

Relapsing fever is the term used to describe recurrent fevers with similar clinical presentation caused by different members of the genus *Borrelia* that are transmitted by different arthropod vectors. During the first half of the twentieth century, relapsing fever was a disease of great worldwide importance with major epidemics which affected 50 million of the population with a 10–40% mortality. During the 1930s, approximately one-third of the population in Africa were wiped out by a major epidemic described as relapsing fever.

Since 1967 the epidemic form of louse-borne relapsing fever has been largely confined to areas of extreme poverty in East Africa, with most cases arising in Ethiopia, and the Peruvian Andes. A recent outbreak in Sudan is estimated to have affected 20 000 individuals of the Dinka tribe during 1998–1999, with a mortality of 10–14%. Despite the reappearance elsewhere in the world of other louse-borne diseases, such as epidemic typhus in Bosnia and trench fever in the vagrant populations, louse-borne relapsing fever has not re-emerged. Furthermore, molecular analysis of lice collected from around the world including France, Peru, Russia and the African countries of Burundi, Congo and Zimbabwe, has failed to produce evidence of louse-borne relapsing fever (Roux and Raoult, 1999). Tick-borne relapsing fevers tend to be more sporadic, but still cause major health problems

in Africa where, in areas such as Central Tanzania, it is one of the major causes of child mortality (Talbert *et al.*, 1998). Although present in some European countries and America, tick-borne relapsing fever tends to be more of a rarity. It is usually associated with camping out in rural locations in close proximity to animal reservoirs of the spirochaete and their associated *Ornithodoros* (*Ornithodorus*) tick vectors.

Discovery of the Disease

Although descriptions of the clinical disease have been recorded since the days of Hippocrates, the term 'relapsing fever' was first used by David Craigie to describe an outbreak of the disease in Edinburgh in 1843. A spirochaete cause of relapsing fever was first demonstrated by Otto Obermeier during the 1867–1868 outbreak in Berlin. He recorded spirochaetes in the blood of patients with clinical relapsing fever, but his inability to reproduce the disease in animal models, and indeed in himself, delayed the publication of these findings until 1873 (Obermeier, 1873). In 1879, Moczutkowsky confirmed transmission of the disease by inoculating blood from cases of relapsing fever into healthy individuals. Later, Ross and Milne (1904) discovered the causative agent in the African variety of relapsing (tick) fever. The same finding was also made independently by Dutton and Todd (1905) who demonstrated relapsing fever in monkeys that had been transmitted by infected *Ornithodorus moubata* ticks. The role of the human body louse in the transmission of relapsing fever was reported by MacKie (1907).

Classification and General Characteristics

Members of the genus *Borrelia* belong to the order *Spirochaetales*. The type species is *Borrelia anserina*,

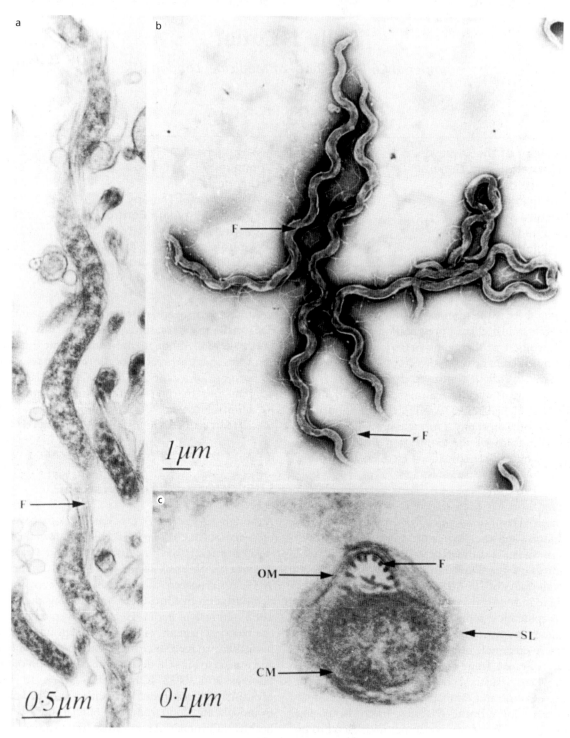

Fig. 1 (a)–(c)

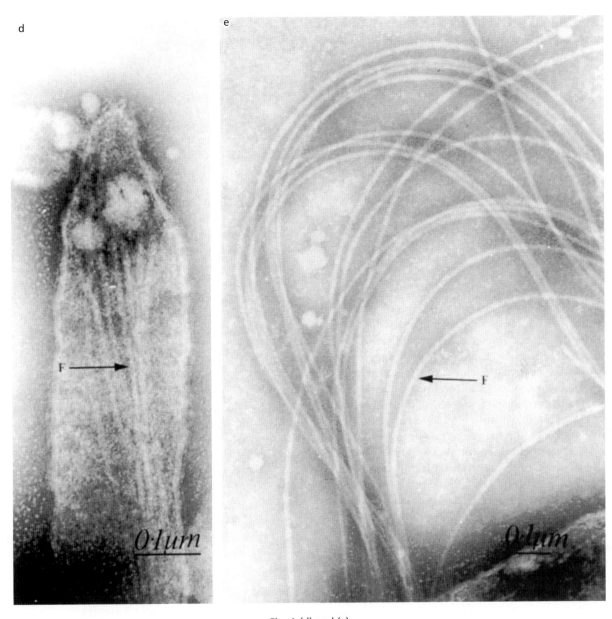

Fig. 1 (d) and (e)

Fig. 1 Electron micrographs of *Borrelia recurrentis* strain A1. (a) The spirochaete morphology of cells; (b) negatively stained spirochaete cells showing the tapering ends (phosphotungstic acid); (c) transverse section of cell showing the periplasmic flagella; (d) negatively stained preparation showing sub-terminally associated flagella (ammonium molybdate); (e) negatively stained preparation showing flagella associated with the central portion of a cell (ammonium molybdate). F, flagella; SL, surface layer; OM, outer membrane; CM, cytoplasmic membrane. With permission from Cutler *et al.* (1997).

the cause of avian borreliosis. *Borrelia* are one of the few groups where traditional taxonomic studies based on morphological criteria have been fully supported by phylogenetic analysis based on *rrs* gene sequence data.

The borreliae are helical in shape (**Fig. 1a, b**), and are highly motile with an outer membrane surrounding unsheathed flagella in the periplasmic space outside the protoplasmic cylinder (**Fig. 1c**). The protoplasmic contents of the cell are surrounded by a cytoplasmic membrane and cell wall. Cultivation *in vitro* requires the medium to be supplemented with long-chain fatty acids together with glucose which is fermented to produce lactic acid.

A G + C content ranging from 27 to 32 mol% has been reported for members of the genus *Borrelia*

(Hyde and Johnson, 1984). The most recently determined are for *B. recurrentis* (A1) and *B. duttonii* (strain Ly) with G + C contents of 28.4 mol% (Cutler *et al.*, 1997) and 27.6 mol% (Cutler *et al.*, 1999) respectively, both in the reported range for the genus.

All members of the genus are arthropod-borne; most are transmitted by a variety of ticks, and one by the human body louse, *Pediculus humanis* var. *corporis*. The *Borrelia* species are associated with diseases of both veterinary and clinical importance, including avian borreliosis, epizootic bovine abortion, Lyme borreliosis, tick and louse-borne relapsing fever. The members of the genus associated with relapsing fevers form the basis of this chapter.

Biodiversity

Differences have been noted between relapsing fever-associated strains, including numbers of clinical relapses, disease severity, mammalian hosts and arthropod vector preferences. Traditionally these spirochaetes have been classified according to their arthropod vector, pathogenicity for laboratory animals and on their geographical location. The species associated with relapsing fever are listed in **Table 1**.

Only recently, with the development of molecular taxonomic methods has the taxonomic organisation of *Borrelia* been addressed. Studies based on the DNA base content and the relationship between three American strains by DNA hybridisation techniques suggest that these three 'species' should be classified as one (Hyde and Johnson, 1984). Others have suggested that *B. parkeri* and *B. turicatae* should constitute a single species. This is supported by the observation that a 790-bp *Hin*dIII fragment of *B. hermsii* DNA hybridises with DNA from *B. parkeri* (Schwan *et al.*, 1989).

Phylogenetic Identification of Relapsing Fever *Borrelia* by Comparison of *rrs* Genes

In a phylogenetic study of 20 relapsing fever borreliae based on the *rrs* 16S RNA gene, similarity values of between 97.7 and 99.9% were demonstrated (Ras *et al.*, 1996). Cluster analysis divided the relapsing fever strains into three distinct groups. The first group includes *B. crocidurae*, *B. duttonii*, *B. recurrentis* and *B. hispanica*; *B. persica* appeared on the periphery of this cluster. The second group included *B. coriaceae*, *B. lonestari* and *B. miyamotoi*, while the final cluster is composed of *B. hermsii*, *B. parkeri* and *B. turicatae* (see **Fig. 2**) (Ras *et al.*, 1996). Comparison of the DNA sequence of *B. recurrentis* with a strain of African tick-borne relapsing fever, *B. duttonii*,

has revealed that the strains are closely related, but differ in four base pairs in their *rrs* genes (Ras *et al.*, 1996).

Phylogeny by Flagellin Sequencing

A 984 amino acid portion of the flagellin of *B. recurrentis* has been compared with similar sections of the flagellin from other relapsing fever spirochaetes and subjected to phylogenetic analysis. The results of this are displayed in **Fig. 3**. Phylogenetic trees based on the flagellin gene sequence gave cluster patterns similar to those with the *rrs* gene, confirming the relationship between *B. recurrentis*, *B. duttonii* and *B. crocidurae*, which are more distantly related to relapsing fever strains of Spanish origin. American relapsing fever strains fall into a separate cluster together with *B. miyamotoi* and *B. lonestari*, while the Lyme disease-associated borreliae cluster separately. In the flagellin gene only a single base difference was found between *B. recurrentis* and *B. duttonii*. The divergence seen in *rrs* genes and the flagellin gene sequences of *B. recurrentis* and *B. duttonii* suggest that there is a selective pressure to maintain the flagellin gene without mutation. Whether the similarity between *B. recurrentis* and *B. duttonii* reflects adaptation of the strain to a louse vector, or results from geographical or circumstantial differences remains to be resolved. Some evidence suggests that *B. duttonii* may be transmitted to a louse vector, but this remains to be substantiated (see Spirochaete–Vector Interactions, p. 2108). What is certain is that the sporadic tick-borne relapsing fever is not as severe and does not seem to have the epidemic potential of the louse-borne disease.

Alternative Phylogenetic Targets

Another potential phylogenetic target has recently been described. A surface-exposed outer membrane protein known as P66 is present in all *Borrelia* species examined so far. Analysis of sequence data reveals two conserved hydrophobic regions flanking a highly variable surface exposed loop (Bunikis *et al.*, 1998). Though the function of this protein is unknown, it provides an additional phylogenetic tool for comparative analysis of these spirochaetes.

Possible Association of Relapsing Fever Species with *Borrelia miyamotoi*

The isolation of *B. miyamotoi* from ticks in Japan (Fukunaga *et al.*, 1995) may have uncovered an

Table 1 Traditional classification of relapsing fever *Borrelia* using vectors and geographical distribution

Species	Arthropod vector	Distribution
B. recurrentis	*Pediculus humanus*	Ethiopia/Sudan (now)
B. duttonii	*Ornithodoros moubata*	Central/East Africa
B. hispanica	*O. erraticus*	Spain, Portugal, Greece, Cyprus
Borrelia spp. (unnamed)		Spain
Crocidurae group	*O. sonrai*	Africa, Middle East, Iran
B. crocidurae		
B. merionesi		
B. microti		
B. dipodilli		
B. persica and related	*O. tholozani*	S. Russia, Iran, Middle East, Cyprus
B. uzbekhistanica		
B. sogdiana		
Borrelia spp.	*O. tholozani crossi*	
B. babylonensis	*O. asperus*	
B. turkmenica	*O. cholodkovskyi*	
B. latyshewii	*O. tartakowskyi*	Central Asia
B. caucasica	*O. verrucosus*	Armenia, Caucasus, Georgia
B. hermsii	*O. hermsi*	Western USA, British Columbia
B. parkeri	*O. parkeri*	Western USA
B. turicatae	*O. turicata*	Texas, Kansas, Mexico
B. venezuelensis	*O. venezuelensis*	Panama, Colombia, Venezuela, Ecuador, Paraguay
B. mazzottii	*O. talaje*	Central/South America
B. dugesii	*O. dugesi*	Central/South America

Goubau (1984); Anda *et al.* (1996).

evolutionary link between the relapsing fever and Lyme-associated *Borrelia*. Phylogenetic analysis of the *rrs* gene and flagellin of *B. miyamotoi* has shown that this spirochaete, which has not been associated with clinical disease, falls within the relapsing fever-associated strains rather than those causing Lyme borreliosis. This is surprising because this spirochaete is carried by *Ixodes persulcatus* ticks, which belong to the *Ixodes ricinus* complex. They are hard-bodied ticks only distantly related to the *Ornithodoros* ticks that transmit relapsing fever. It is possible that this species may represent an important link between the

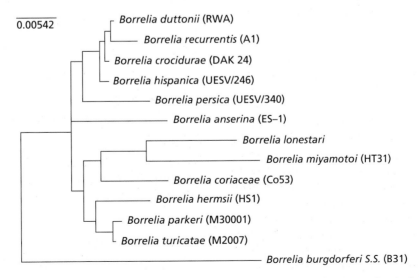

Fig. 2 Phylogenetic tree based on a comparison of the 16S rRNA sequences of relapsing fever *Borrelia*. The branching pattern was generated by the neighbour-joining method. With permission from Ras *et al.* (1996).

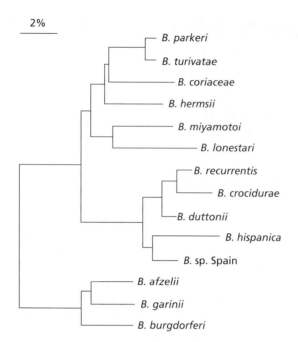

2%

- B. parkeri
- B. turivatae
- B. coriaceae
- B. hermsii
- B. miyamotoi
- B. lonestari
- B. recurrentis
- B. crocidurae
- B. duttonii
- B. hispanica
- B. sp. Spain
- B. afzelii
- B. garinii
- B. burgdorferi

Fig. 3 Phylogenetic tree of *Borrelia* constructed by the neighbour-joining method of flagellin sequences. With permission from Cutler *et al.* (1997).

evolution of *Borrelia* and their adaptation to arthropod vectors.

Identification

Classical

Diagnosis of relapsing fever has traditionally been based on the demonstration of spirochaetes in blood smears, usually by Wright's or Giemsa stains. Serological identification is inappropriate, because of the cross-reactivity and antigenic variation of surface proteins in these species. Attempts to improve identification have been made by the use of 'species-specific' monoclonal antibodies against *B. hermsii* (Schwan *et al.*, 1992), but the identity of non-reactive strains would remain unresolved.

Speciation of strains has largely depended on knowledge of the geographically prevalent species and identification of the arthropod vector. Indeed, many species of relapsing fever *Borrelia* have been named after the arthropod vector responsible for their transmission. The susceptibility of small mammals to infection by the spirochaete was at one time used in conjunction with arthropod identification to speciate isolates. Similarly, serological techniques have been used to identify spirochaetes and provide serotype information.

These methods all have their constraints. Geographical information is often absent, especially for the more

sporadic tick-borne relapsing fevers. Similarly, identification of the vector is often not possible, because ticks usually feed for only a short period during the night. Serological identification is plagued by poor specificity resulting from the similarity among *Borrelia* species and the ability of strains to undergo antigenic variation of their major outer membrane proteins. These methods also require cultivation or animal passage to provide sufficient spirochaetes for identification.

Molecular Methods

Difficulties with antibody-specificity and cultivation of relapsing fever spirochaetes to obtain sufficient material for diagnosis and strain differentiation can be largely overcome with molecular techniques. PCR-based strategies are ideally suited for this purpose. Relapsing fever-specific primer sets for the amplification of the *rrs* gene have been described (Ras *et al.*, 1996). Speciation of relapsing fever-associated strains can be achieved by restriction enzyme digestion of the *rrs* gene product with *Mse*I, *Bfa*I and *Hph*I, but restriction sites have not been identified that can distinguish between *B. duttonii*, *B. crocidurae* or *B. hispanica* (Ras *et al.*, 1996).

Structure

The cell structure of relapsing fever spirochaetes is typical of the genus *Borrelia* (**Fig. 1a**). The cells have tapering ends (**Fig. 1b**), typical of those described for the genus, but slightly less pointed than those described for *B. burgdorferi* (Hovind Hougen, 1984; Hulínská *et al.*, 1989). The length is variable, but usually 10–25 μm (Hovind Hougen, 1974), while the width of cells is 0.2–0.5 μm. An average wavelength of 1.8 μm with an amplitude of 0.8 μm has been described for *B. recurrentis* (Cutler *et al.*, 1997). Flagella reside in the periplasmic space between the outer membrane and protoplasmic cylinder (**Fig. 1c**). They are sub-terminal and attached to opposite ends of the cell (**Fig. 1d**), overlapping in the central portion of the spirochaete, and they mediate both motility and shape of cells. The cell wall peptidoglycan contains both muramic acid and ornithine. Cholesterol and lecithin have been demonstrated in *B. turicatae*, while muramic acid has been found in *B. duttonii*. The outer membrane contains heterogeneous lipoproteins described in more detail below.

Outer Membrane

Constituents of the outer membrane have been studied in *B. hermsii* (Shang *et al.*, 1998). Outer membrane vesicles were found to contain approximately

60 proteins, including a porin-like protein with activities ranging from 0.2 to 7.2 nS, glycerophosphodiester phosphodiesterase and variable membrane proteins (VMP).

Flagella

The flagella are typically unsheathed, which distinguishes them from those of treponemes, with attachment points at either end of the cell (**Fig. 1c**). Numbers of 15–30 flagella have been reported for the tick-borne varieties. Early reports of the numbers of flagella in louse-borne relapsing fever spirochaetes were in keeping with their tick-borne relatives. This has, however, been questioned in *B. recurrentis* strain A1, which has 8–10 periplasmic flagella with a diameter of 0.27–0.34 µm (**Fig. 1c–e**). Similar observations were made for *B. duttonii* (Cutler *et al.*, 1999). The *B. recurrentis* strains previously studied had 15–30 unsheathed flagella, and had been multiply passaged, mouse-adapted and the original country of origin was not stated. Higher numbers of flagella (15–26), have been observed in a newly described species of relapsing fever *Borrelia* isolated in Spain (Anda *et al.*, 1996).

The flagellin protein has a molecular weight of 39 kDa in isolates of *B. hermsii*, *B. turicatae* and *B. parkeri*. In contrast, a band located at 41 kDa was detected by immunoblotting with the genus-specific flagellin monoclonal antibody H9724 (Barbour *et al.*, 1986) in isolates of *B. recurrentis* (Cutler *et al.*, 1997) and *B. duttonii* (Cutler *et al.*, 1999). A 41 kDa flagellin is also found in Lyme-associated borreliae.

Other Membrane-associated Proteins

A plasmid-encoded family of proteins known as the Bdr protein family have been analysed in *B. turicatae*. Though their function has not yet been elucidated, a possible regulatory function has been predicted. Comparable proteins have also been reported in the Lyme borreliosis-associated borreliae (Carlyon *et al.*, 2000).

Lipids

Palmitate is the principal long-chain fatty acid in spirochaetes. It forms the anchor by which membrane lipoproteins are attached. Radio-labelling with palmitate has been used successfully to locate predominant membrane-associated lipoproteins (Vidal *et al.*, 1998). The lipoproteins present in *B. hermsii* and *B. turicatae* are mostly in the form of α helix with a conserved secondary structure. The proposed pathway for lipid metabolism is considered in the section on Physiology (p. 2103).

Variable Membrane Proteins

The majority of membrane proteins in other spirochaetes, namely *Treponema pallidum* and *Borrelia burgdorferi* are covalently modified with lipids. These lipoproteins induce both specific cellular and humoral

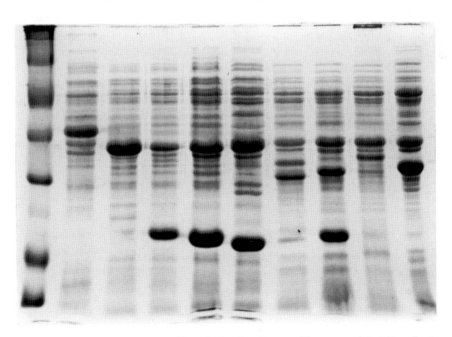

Fig. 4 Coomassie blue-stained SDS-PAGE protein profiles of relapsing fever and Lyme-associated *Borrelia*. Lane 1, *B. recurrentis* strain A1; lanes 2–4, *B. hermsii* strains HS1, Man1 and Con1; lane 5, *B. turicatae*; lane 6, *B. burgdorferi* sensu lato; lane 7, *B. garinii*; lane 8, *B. afzelii*; lane 9, *B. valaisiana*. With permission from Cutler *et al.* (1994).

immune responses and have been established as principal cytokine inducers in syphilis and Lyme borreliosis (Radolf *et al.*, 1995). Activation of macrophages depends on protein acylation and this activity believed to reside in the N-termini of native molecules (Radolf *et al.*, 1995).

Detergent-soluble variable major lipoproteins (VMPs) have also been detected in relapsing fever spirochaetes. These antigenic proteins reside in the outer membrane and can be labelled with palmitate

(Shang *et al.*, 1998; Vidal *et al.*, 1998; Scragg *et al.*, 2000). They vary in molecular weight, depending on the species and serotype of the spirochaete. Two groups containing at least 26 different Vmp serotypes have been described for *B. hermsii*: the large Vmps of 34–39 kDa, also known as variable large proteins (VLPs), and the small Vmps containing Vmps of about 20 kDa, also known as variable small proteins (VSPs) (Barbour, 1990; Hinnebusch *et al.*, 1998; **Fig. 4**). Varying degrees of homology have been reported for

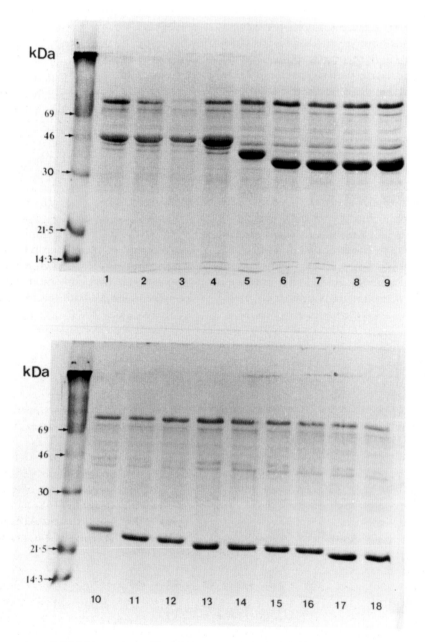

Fig. 5 Coomassie blue-stained SDS-PAGE protein profiles of *B. recurrentis* strains A1–A18. Lanes 1–4, protein group 1 strains; lane 5, protein group 1 strains; lanes 6–9, protein group 3; lane 10, protein group 4a; lanes 11 and 12, protein group 4b; lanes 13–16, group 4c; lanes 17 and 18, protein group 4d. With permission from Cutler *et al.* (1997).

these small Vmps, ranging from 80% down to 40% (Barstad *et al.*, 1985). They have an isoelectric point of 4.4–5 (Barbour, 1990).

DNA sequencing to compare 23 Vmp genes from *B. hermsii* disclosed five groups, in which members of the Vlp group divided into four subgroups, while Vsps formed a single group (Hinnebusch *et al.*, 1998). Lipoproteins of around 40 kDa have also been described in *B. turicatae* and *B. parkeri* (**Fig. 4**), while for *B. recurrentis* and *B. duttonii*, both high-molecular-weight (35.5–45 kDa) and low-molecular-weight (21.5–24 kDa) proteins have been described (**Fig. 5**; Cutler *et al.*, 1997, 1999). Protein sequence analysis has disclosed 72% similarity between one of these major proteins of *B. recurrentis* and the Vmp12 of *B. hermsii* (Vidal *et al.*, 1998), which places this protein in the Vlp beta family proposed by Hinnebusch *et al.* (1998).

These proteins of *B. recurrentis* have been shown to be potent stimulators of TNF, and anti-TNF therapy has indeed been shown to benefit patients with louse-borne relapsing fever (Fekade *et al.*, 1996). It is probable that these lipoproteins have an essential role in the pathogenesis of relapsing fever and this will be discussed further below, under Pathogenesis.

Analysis by quadrupole orthogonal time-of-flight mass spectrometry has shown a triacyl lipid moiety located at the N-terminal cysteine of the native polypeptide in *B. recurrentis*. These studies revealed that *Borrelia* lipoproteins are structural homologues of *Escherichia coli* lipoproteins. Rather than being a single moiety, three different attachments of 789, 815 and 817 daltons have been found in strain A1 of *B. recurrentis*, each attached to identical polypeptides. These are consistent with N-terminal glyceryl cysteine with the following fatty acid substitutions: (1) C16:0, C16:0, C16:0; (2) C16:0, C16:0, C18:1; (3) C16:0, C16:0, C18:0 (Scragg *et al.*, 2000).

Genome Organisation

Analysis of the genomic organisation among the relapsing fever *Borrelia* has revealed that these spirochaetes have segmented genomes (Hayes *et al.*, 1988). This was later supported by the work of Margolis *et al.* (1994b) who located the *guaA* and *guaB* genes required for purine biosynthesis in *B. hermsii* to a linear plasmid. It is possible that the *Borrelia* have adopted this strategy to enable rapid switches between arthropod and mammalian environments. Genomic organisation has been examined in the relapsing fever *Borrelia* by pulsed-field methods (**Fig. 6**). Profiles

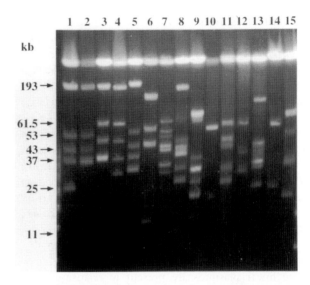

Fig. 6 Pulsed-field gel electrophoresis patterns to compare each plasmid type of *B. recurrentis* with tick-borne and Lyme *Borrelia*. Lane 1, *B. recurrentis* type 1; lane 2, *B. recurrentis* type 2; lane 3, *B. recurrentis* type 3; lane 4, *B. recurrentis* type 4; lane 5, *B. recurrentis* type 5; lane 6, *B. hermsii*; lane 7, *B. parkeri*; lane 8, *B. turicatae*; lane 9, *B. miyamotoi*; lane 10, *B. burgdorferi* sensu stricto; lane 11, *B. garinii*; lane 12, *B. afzelii*; lane 13, *B. japonica*; lane 14, *B. valaisiana*; lane 15, *B. lusitaniae*. With permission from Cutler *et al.* (1997).

appear to be heterogeneous even among individual species, as seen with the profiles for *B. recurrentis* isolates in **Fig. 6** and also with *B. duttonii* (Cutler *et al.*, 1999) and *B. coriaceus* (Hendson and Lane, 2000), but these studies have been restricted by the numbers of available isolates.

Arrangement and Transcription Regulation of Genes

Linear Chromosomes

Relapsing fever *Borrelia*, in common with other members of the genus, are polyploid, with linear chromosomes that migrate into the gel with a size of approximately one megabase (Baril *et al.*, 1989; Bergström *et al.*, 1991; Kitten and Barbour, 1992; Ferdows *et al.*, 1996; Cutler *et al.*, 1997). The telomeres have covalently closed hairpins (Hinnebusch and Barbour, 1991). It is thought that these linear DNA molecules may form circular intermediates during replication (Ferdows *et al.*, 1996). Physical mapping of genes onto the chromosome for relapsing fever *Borrelia* has shown similarity with other members of the genus, but the restriction sites are different between species (Takahashi and Fukunaga, 1996). Relapsing fever strains studied to date fail to show the tandem repeat of *rrl/rrf* genes seen in the Lyme-associated *Borrelia* (Schwartz *et al.*, 1992).

Mini-chromosomes

Several relapsing fever strains, including *B. turicatae*, *B. parkeri*, *B. duttonii* and *B. recurrentis* possess a large linear plasmid just under 200 kb in size (**Fig. 6**), and is considered by some to be a mini-chromosome (Barbour, 1994). Isolates of *B. recurrentis* carry a large plasmid of approximately 192 kb (**Fig. 6**, range 183–194 kb). High-molecular-weight mini-chromosomes of 180 and 170 kb, respectively, have previously been described in *B. hermsii* and *B. turicatae* (Ferdows *et al.*, 1996). A variant containing a circular form of this plasmid showed no adverse effects with regard to its protein profile or generation time (Ferdows *et al.*, 1996).

Linear Plasmids

Large linear plasmids are present in relapsing fever *Borrelia*. Like the chromosome, they have covalently closed hairpins at their telomeres. The copy number of these plasmids approximates to the chromosome copy number, suggesting that they are linked (Hinnebusch and Tilly, 1993). The plasmid patterns observed for *B. recurrentis* are distinct from those of tick-borne relapsing fever and Lyme *Borrelia* (**Fig. 6**). The largest group of relapsing fever isolates so far studied are 18 isolates of *B. recurrentis* (Cutler *et al.*, 1997). These fall into one of six plasmid patterns (**Fig. 6, Table 2**). Plasmid bands of between 25 and 194 kb resolved as sharp bands during pulsed-field electrophoresis and thus probably represent linear DNA (Casjens *et al.*, 1995; Ferdows *et al.*, 1996). A plasmid of 11 kb was present in all strains, but whether this is linear or circular was not determined. This band was not always detectable with ethidium bromide-stained gels, but it could be seen after hybridisation.

Multiple Copies of Chromosomes and Linear Plasmids

Comparison of *B. hermsii* grown in mice with those cultured *in vitro* demonstrated multiple chromosome copies and linear plasmids. The copy number of expression plasmids was 14 (range 12–17), while that of silent plasmids was 8 (range 7–9). In mouse-derived spirochaetes, the chromosome copy number was 16 (range 13–18). Those cells grown in BSK II possessed only a quarter to one-half of this number of plasmids or chromosomes (Kitten and Barbour, 1992).

Linear plasmids and chromosomes have both been detected in relapsing fever borreliae. Replicons with similar structures are more frequently associated with eukaryotic cells than with prokaryotes. The molecular characterisation of these structures may provide clues to the evolutionary relationship between prokaryotes and eukaryotes and may additionally provide evidence for genetic exchange between these kingdoms (Hinnebusch and Tilly, 1993). Similarity has been noted between the AT-rich sequences at the ends of the specific linear plasmids of *B. burgdorferi* and tick-borne strains of relapsing fever, and African swine fever (iridovirus), another tick-borne disease (Hinnebusch and Barbour, 1991; Bergström *et al.*, 1992), suggesting that their linear replicons may have a common origin and, again, may have been subject to trans-kingdom exchange. This is more likely because the tick is a common vector of both iridovirus and relapsing fever. Since *B. recurrentis* is louse-borne, the possible presence of these common sequences would have taxonomic and evolutionary implications that may particularly illuminate the relationship to tick-borne disease.

Genome Sequencing

With the genome sequencing of *Treponema pallidum* and *B. burgdorferi* now complete (Fraser *et al.*, 1997, 1998), the case strengthens for the sequencing of relapsing fever spirochaete genomes. Both *B. hermsii* and *B. recurrentis* would be likely candidates, with the possibility of identification of genes that determine vector competence and differences in clinical features of these relapsing fever borreliae. Availability of genomic sequences would allow the application of microarray technology to examine patterns of gene expression under different environmental stimuli. Knowledge of the genomic sequence would pave the

Table 2 Molecular sizes of plasmids for *B. recurrentis*. With permission from Cutler *et al.* (1997)

Plasmid group	Plasmid size in kilobases							
Type 1	193		53		43	37		25	11
Type 2	192		53		43	37			11
Type 3	187	61.5		46		37			11
Type 4	188	61.5		46		37	29		11
Type 5	200		53		43	37	31		11
Type 6	192		53		43	37	31	25	11

way to assigning probable gene functions, which in turn may provide insights into the host–parasite and vector–parasite relationships of these spirochaetes.

Physiology

The physiology of relapsing fever *Borrelia* is largely unstudied, because attempts over the years to cultivate them have been plagued with problems. Cultivation of certain *Borrelia* strains in Kelly's medium A and B (Kelly, 1976) became possible, and was further improved by the introduction of BSK II medium (Barbour, 1984), but the physiological requirements of these strains have yet to be fully determined. Similarly, the requirements of strains *in vivo* has yet to be established. Certain strains, in particular *B. recurrentis*, are not easily adapted to growth in any mammal other than the human host (Kelly, 1976). Only monkeys have been used successfully to propagate *B. recurrentis* and to produce a relapsing clinical picture similar to that seen in humans (Judge *et al.*, 1974a, b, c).

Estimations of growth rates have been made for a few strains. A mean generation time *in vitro* of 7.1 hours (± 0.2 hours) for *B. hermsii* (Ferdows *et al.*, 1996), while generation times of 8–9 hours *in vitro* have been reported for *B. recurrentis* (Cutler *et al.*, 1994). This contrasts to the approximately 3 hours (2.8–4.2 hours) reported for *B. hermsii* in mice (Stoenner *et al.*, 1982), but others have reported 6 hours for *B. hermsii* in rats or mice (Barbour, 1990).

Glucose is metabolised by the glycolytic pathway, and multiplication becomes restricted as lactic acid accumulates. Lipids are essential for the growth of *Borrelia*. The spirochaetes obtain essential fatty acids from the albumin component of medium or through phospholipase B-mediated hydrolysis of lysolecithin

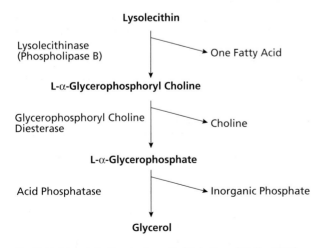

Fig. 7 Lipid metabolism in *Borrelia* (Pickett and Kelly, 1974).

present in the rabbit serum also included as a medium component. The pathway proposed for lipid catabolism in *B. hermsii*, *B. parkeri* and *B. turicatae* is illustrated in **Fig. 7** (Pickett and Kelly, 1974).

Pathogenesis

During major epidemics a range of clinical severity is observed, with mortality rates varying from 70% down to 10% in untreated patients with louse-borne relapsing fever.

Spirochaetes tend to be confined to the lumen of blood vessels, but tangled masses may also reside in the characteristic splenic miliary abscesses and infarcts, or in the central nervous system associated with haemorrhages. Perivascular, histiocytic, interstitial myocarditis may be seen in many cases of the louse-borne disease, resulting in conduction defects, dysrhythmia or sudden heart failure. Splenic rupture, cerebral haemorrhage and liver failure are other causes of death. The liver may also show hepatitis, haemorrhages and necrosis. Meningitis may be a feature of the disease with petechial haemorrhages affecting the surfaces of viscera.

After antimicrobial treatment of relapsing fever, pathophysiological changes may be seen which are typical of an 'endotoxin reaction', but classical endotoxin has not been detected in these spirochaetes. They do however possess major membrane lipoproteins which are potent inducers of TNFα, but also stimulate IL-1β, IL-6 and IL-12 (Vidal *et al.*, 1998). These lipoproteins at concentrations of 10–20 μM produce a TNFα stimulation equivalent to 30–40 ng/mL of lipopolysaccharide (40 ng/mL LPS = 10 nM) (Radolf *et al.*, 1995). A different mechanism of macrophage activation has been postulated on the basis that spirochaete lipoproteins are able to activate LPS-resistant macrophages (Radolf *et al.*, 1995). A TNFα-inducing toxin from *B. recurrentis* strain A1 is approximately half as potent as LPS on a molar basis (Vidal *et al.*, 1998). Comparison of TNFα-inducing activity has been shown to vary at least 50-fold, depending on which size of *B. recurrentis* Vmp is used (Vidal *et al.*, 1998). The ability of Vmp to induce TNFα to varying degrees may have exciting implications for the virulence of strains and pathogenesis of the disease.

A central role of the Vmp lipoprotein in disease presentation has been documented for *B. turicatae* infection in mice. Those that express Vmp type A produce neurological sequelae, while type B results in larger numbers of blood-borne spirochaetes (Cadavid *et al.*, 1997).

Certain species, namely *B. crocidurae* and *B. duttonii* are able to envelope themselves in host red blood cells in a phenomenon known as rosetting. This may conceal the spirochaetes from the host immune response, and thus prolong survival (Burman *et al.*, 1998).

Benzylpenicillin attaches to the penicillin-binding proteins of the spirochaete, resulting in large surface blebs. Phagocytosis of damaged cells by neutrophils may be enhanced by complement, but is not essential for their elimination. Contact of the spirochaete with mononuclear leukocytes induces production of pyrogen and thromboplastin, which may have a role in the fever and disseminated intravascular coagulation that typifies relapsing fever.

Clearance of spirochaetes from the host appears to be B cell-dependent, and irradiated or cyclophosphamide-treated animal models show a delay in the clearance of the spirochaetes. T cell-deficient animals showed no difference from controls in their ability to clear spirochaetes from their blood (Barbour, 1990).

Clinically, the disease starts suddenly with the onset of rigors and fever. Other early signs are headache, dizziness, generalised aches and pains, anorexia, nausea, vomiting and diarrhoea. Later, abdominal pain, cough and epistaxis may be seen. Jaundice or petechial rashes, particularly affecting the trunk, may also be present. Signs of meningism occur in about 40% of cases. In louse-borne relapsing fever the first attack of fever resolves by crisis 4 to 10 days after initial onset, while the duration of the initial fever for tick-borne relapsing fever is only about 3 days. Then there is a period of afebrile remission of 5–9 days followed by up to 4 relapses of fever in louse-borne disease or 13 relapses in tick-borne disease. The relapsing nature of the disease is due to antigenic variation in the outer membrane proteins of the spirochaete. The presumed mechanisms involved are detailed below.

Molecular Mechanisms of Antigenic Variation

Variable Membrane Proteins

The sero-specificity of *Borrelia* in each clinical relapse is determined by their outer membrane proteins. The serotypes differ in the molecular weights of their major outer membrane protein, their peptide maps and reactivity with monoclonal antibodies. In *B. hermsii* 24 different serotypes were identified with sero-specific FITC-labelled antibodies (Stoenner *et al.*, 1982), and later two further serotypes were described (Meier *et al.*, 1985). It was noted that sero-conversion is not random but appears in a programmed manner (Stoenner *et al.*, 1982). Conversions were also shown

to occur in the absence of antibody in fortified Kelly's medium at a rate of 10^{-4} to 10^{-3} cells per generation and were frequently noted to contain mixtures of serotypes (Stoenner *et al.*, 1982). More recently, spontaneous antigenic variation has been observed with a cloned isolate of *B. turicatae* originally of serotype B. After 50 passages *in vitro* a further three Vsp serotypes were observed (A, B, E and F) (Ras *et al.*, 2000).

The genetic mechanism of this antigenic variation has been studied primarily in *B. hermsii*. It is suggested that there is a 'mini-chromosome' shuttle mechanism similar to that found in sleeping sickness trypanosomes (Meier *et al.*, 1985; Burman *et al.*, 1990). In *B. hermsii* the variable membrane lipoprotein (Vmp) is the antigen responsible (Restrepo *et al.*, 1994). The variant genes are carried on linear plasmids, mostly as promoterless silent copies, which are expressed after transposition into a single telomeric expression site on another linear plasmid. This event leads to duplication of the silent copy and loss of the original active gene. Recombination is mediated through similar telomeric sequences with an approximate 300 bp upstream recombination site and a downstream recombination site about 1000 bp from the end of the gene (Barbour, 1990). Studies of *B. hermsii* have suggested three possible mechanisms of Vmp variation: (1) a complete silent *vmp* gene may replace a complete expressed *vmp* using intermolecular recombination; (2) a silent complete *vmp* gene located on another plasmid may partially replace expressed *vmp* (the donor *vmp* sequence may be an incomplete pseudo-gene), and (3) the donor gene may be on the same replicon as the active *vmp* and the serotype switch is achieved by intramolecular recombination in the expressed plasmid (Restrepo *et al.*, 1994).

The mechanism of antigenic variation in *B. turicatae* differs somewhat from that described for *B. hermsii*. The switching mechanism involves intraplasmid recombination between those linear plasmids that carry silent and expressed genes, with the substitution of the active gene by a replacement gene at a sub-telomeric expression site. This antigenic variation results in alteration of the linear plasmid profile in *B. turicatae* (Ras *et al.*, 2000) as has been previously postulated for *B. recurrentis* (Cutler *et al.*, 1997).

Areas of homology exist at the telomeres of the different *vmp* genes of *B. hermsii* and, indeed, they are homologous with other relapsing fever species, including *B. turicatae* and *B. crocidurae* (A.G. Barbour, 1997, personal communication) and with *B. recurrentis* (Cutler, unpublished observations). Similarity has also been demonstrated between *B. hermsii* Vsp sequences and those of other borreliae, namely the OspC protein of *B. burgdorferi* sensu lato (Marconi *et al.*, 1993;

Margolis *et al.*, 1994a) and a *vmp*-like gene in *B. myamotoi* (Hamase *et al.*, 1996). Primers devised against these conserved telomeric sequences could be used to identify and sequence the *vmp* genes of cultivable strains. Primers could then be used to examine *vmp* genes present in blood samples from animal models, patients with relapsing fever and their lice or tick vectors. Identification of sequences located downstream of promoters and northern blotting and reverse transcription PCR should establish which *vmp*s are expressed, and this would establish whether certain 'culture-adapted' variants are selected by *in vitro* methods. A culture/tick-associated Vsp33, formerly serotype C (Carter *et al.*, 1994), has been described for *B. hermsii* (Barbour, 1990). This type of study would establish the range of Vmps expressed in primary disease. Relapse variants would require the use of animal models, as well as those expressed in the louse or tick vector. Indeed, a serotype switch to a possible tick-associated phenotype, Vsp33, is observed when *O. hermsi* ticks are allowed to feed on mice infected with *B. hermsii* Vmp types 7 or 8. A reverse switch is observed when the spirochaete are returned to a mammalian host (Schwan and Hinnebusch, 1998).

The switch in Vmps from those associated with mammalian hosts and that seen in culture or arthropod-adapted variants (see Physiology, p. 2103), *B. hermsii vsp33*, appears to be the result of a major genetic reorganisation. This particular serotype shows only low-frequency *vmp* switches. In addition, the expression promoter for *vsp33* is different from that used for other *vmp* genes so far characterised. A surprising finding is that this Vsp shows greater homology with Vmps of similar molecular weight and OspC of *B. burgdorferi* than it does with other 35- to 39-kDa Vmps derived from the same isolate (Carter *et al.*, 1994). Thus, the *B. hermsii vsp33* 'C' serotype, was derived from serotype 7 after passage *in vitro*. It has a molecular weight of 20 kDa and lacks antigenic relatedness to types 7 or 21, which have molecular weights of 39 and 38 kDa, respectively (Meier *et al.*, 1985). Further investigations suggested that *vsp33* is a tick-associated variant, similar to the OspA/B expressed proteins in *B. burgdorferi* during its carriage in a tick vector. On transfer into a mammalian host, *B. hermsii* cells switch their *vmp* expression to disease-associated serotypes (Stoenner *et al.*, 1982; Schwan and Hinnebusch, 1998). A homologue of this protein has been described in multiply passaged *B. duttonii* (Takahashi *et al.*, 2000).

Analysis of a collection of strains from patients with louse-borne relapsing fever also showed that isolates had either high-molecular-weight major outer membrane proteins of 35.5–45 kDa or low-molecular-weight proteins of 21.5–24 kDa. A significant difference was found between strains with low-molecular-weight proteins and certain plasmid profiles (Cutler *et al.*, 1997). The expression site of one group of *B. recurrentis* strains has been mapped to a plasmid of 24 kb, with a silent copy of the gene residing on a plasmid of 53 kb (S.J. Cutler *et al.*, unpublished observations). *B. recurrentis* strains of other groups that do not express a 45-kDa Vlp, all lacked the 25-kb plasmid (see **Table 2** and **Fig. 6**). Though both silent and expressed copies of the gene have identical mature protein-coding regions, they possess different 5′ regions, both containing a potential promoter and lipoprotein leader sequence. Antigenic variation may result from a recombination event at the junction between the lipoprotein leader sequence and mature protein-coding region. It is possible that a repressor mechanism may suppress the expression of this silent *vlp* (V.I. Vidal *et al.*, personal communication).

Antigenic variation has also been identified in *B. turicatae*, in which two different serotypes are associated with very different clinical outcomes. In a mouse model, infection with serotype *vmpB* results in a severe arthritis, while serotype *vmpA* results in more extensive central nervous system involvement. The number of borreliae present in joints or blood of mice infected with serotype B was far greater than that seen with serotype A, suggesting a possible relationship between *vmp* serotype and disease severity (Pennington *et al.*, 1997, 1999). Antigenic variation in *B. turicatae*, unlike *B. hermsii*, involves recombination at a non-telomeric site. Expressed *vspA* and *vspB* are preceded, respectively, by an identical promoter sequence on a plasmid of 43 or 53 kb. Archived copies of *vspA* are located on both 43- and 51-kb linear plasmids, while silent *vspB* is located on a plasmid of 37 kb (Pennington *et al.*, 1999). Antigenic variation is associated with large genomic rearrangements, including extensive duplications and deletions. It has been postulated that during a switch from serotype A to B the entire right end of the expression plasmid is replaced (Pennington *et al.*, 1999). If a similar non-telomeric site is used by *B. recurrentis*, this may account for the differences observed in pulsed-field gel plasmid profiles seen among the 18 clinical isolates (**Table 2**).

It was believed that only antigenic drift occurs in the Lyme borreliosis-associated *Borrelia*, but recently promiscuous recombination has been identified at the *vlsE* site on a virulence-associated plasmid of 28 kb. This plasmid codes for outer membrane lipoprotein VlsE. Variation in this lipoprotein has been associated with decreased reactivity to antiserum directed against the parental Vls1 cassette region, and thus may contribute to evasion of the host immune response, which allows persistence of the spirochaete. The predicted amino acid sequence of VlsE shares a high degree of similarity to Vmps of *B. hermsii*, in particular, Vmp17 (27.2% identity and 56.8% similarity; Zhang *et al.*, 1997). These similarities suggest that both the *vls* and *vmp* systems may have evolved from a common borrelial ancestral gene.

Several important differences can be seen between these two means of antigenic variation. *B. hermsii* possesses multiple *vmp*-containing linear plasmids, while only one is present in the Lyme-associated borreliae. Second, the silent *vmp* genes in *B. hermsii* are separated by intragenic non-coding regions and may be arranged in either orientation, while the silent *vls* cassettes are organised in a 'head-to-tail' single open reading frame. The silent *vmp* genes in *B. hermsii* lack promoters, but are otherwise complete with their own ribosome-binding sites, but those in the *vls* cassettes represent only the central third of the expression site. Another important difference is that in the antigenic variation of *B. hermsii* the spirochaetes express a predominant single *vmp* allele, while with the *vlsE* allelic variation numerous variants may be found in a single sample during infection.

Immunity

It is thought that in areas where *Ornithodoros* ticks are prevalent a certain degree of immunity against relapsing fever is acquired during childhood (Felsenfeld, 1965). This is reinforced by the local custom of planting ticks 'in specially prepared holes of newly built dwellings to assure a good tick infestation and lasting immunity to relapsing fever as a result of frequent exposure' (Burgdorfer, 1976).

Antibodies destroy the population of spirochaetes for which they are specific, but they do not influence the rate of antigenic variation. Persistence of spirochaetes may occur during afebrile periods, with the spirochaete residing in immunologically protected sites such as the brain and eye, or persisting in sites including the spleen, liver, kidney or as a low level spirochaetaemia. Cerebrospinal fluid taken between febrile episodes is infectious for laboratory animals (Felsenfeld, 1965). Furthermore, *B. hispanica* can be recovered from brain material collected from infected guinea-pigs 2.5 months to 3 years later (cited in Felsenfeld, 1965). Variability has been noted between the survival of various strains in different animal models. For example, mouse brains have been reported to be poor for the survival of relapsing fever

Borrelia (Felsenfeld, 1965). These sites may provide a source for future relapses of the disease or for subclinical spirochaetaemia which would extend the reservoir of infection.

Diagnosis

Relapsing fever has traditionally been diagnosed by the demonstration of spirochaetes in the blood of patients during febrile episodes. Detection is either by combining Giemsa or Wright's staining with examination by light microscopy, or by use of dark-ground or phase-contrast microscopy of wet preparations. Alternatively, intraperitoneal injection of patient's blood into mice of rats usually produces large numbers of spirochaetes in the animals' blood within 3–5 days for tick-borne borreliae. A variety of relapsing fever spirochaetes has now been successfully cultivated in Barbour-Stoenner-Kelly II (BSK II) medium (Barbour, 1984) including *B. hermsii*, *B. parkeri*, *B. turicatae*, *B. duttonii* and *B. recurrentis*. Freshly prepared medium with a neutral pH, a micro-aerophilic environment and a temperature of 30–37°C gives best results with growth of most strains reaching 10^7 after about 3 days. It is of paramount importance before use to evaluate the medium for its ability to support the growth of these fastidious organisms, especially with regard to the bovine serum albumin fraction V.

Borreliae are resistant to rifampicin, phosphomycin, sulphonamides, trimethoprim and 5-fluorouracil (Barbour and Hayes, 1986; Morshed *et al.*, 1993). These antimicrobial agents have been used in attempts to isolate relapsing fever spirochaetes from contaminated material. The combination of phosphomycin 400 μg/mL, 5-fluorouracil 100 μg/mL, trimethoprim 10 μg/mL and sulphamethoxazole 50 μg/mL has been proposed to give broad antimicrobial coverage, yet permit the selective isolation of *Borrelia* (Morshed *et al.*, 1993). The sensitivity of culture is not yet sufficient for diagnosis. In addition, the poor success to date with isolating relapsing fever strains suggests that selective pressures are forced on the organism, and that these may be further increased by the inclusion of selective antimicrobial agents.

Animal inoculation provides an alternative means for the isolation and presumptive identification of relapsing fever *Borrelia*, but with the range of alternative methods now available this method is infrequently used. Certain strains, have still not been cultivated *in vitro*, and for these animals passage, usually in mice, is the only means of propagating the organism.

Serological methods are not generally used. The currently available serological approaches, ELISA and immunoblotting, fail to discriminate between different *Borrelia* species, resulting in probable misidentification of patients as cases of Lyme borreliosis. Differential titres against selected species of *Borrelia* have been used in an attempt to distinguish between different borrelial infections (Rawlings, 1995), but are probably unreliable. It would be difficult to match the species of infecting *Borrelia* with the serotype to which the patient would have antibodies. Much heterogeneity has been reported among Lyme borreliosis patients with respect to their serological reactivity to their own isolates and other strains of Lyme *Borrelia* (Berger *et al.*, 1988).

Progress has been made in the field of serodiagnosis by the identification of an immunoreactive 39-kDa protein homologue of GlpQ, a glycerophosphodiester phosphodiesterase isolated from *B. hermsii* (Schwan *et al.*, 1996). This glycoprotein has been cloned and reactivity demonstrated with over 90% of convalescent phase sera from patients with relapsing fever from an area endemic for *B. hermsii*. In addition, reactivity has been demonstrated with *B. parkeri*, *B. turicatae*, *B. crocidurae*, *B. coriaceae*, *B. anserina*, *B. duttonii* and *B. recurrentis*. Sera collected from patients with Lyme borreliosis or syphilis did not react with GlpQ, so that this antigen could be used in a serodiagnostic test to distinguish between Lyme and other borrelial infections.

Molecular approaches for the diagnosis of relapsing fever will overcome several of the problems outlined above. PCR-based strategies with *Borrelia*-specific primers followed by hybridisation with speciating probes should provide a useful approach. Hybridisation probes able to differentiate between *B. burgdorferi* and *B. hermsii* have been described (Schwan *et al.*, 1989; Picken, 1992). The major problems are the identification of past infection and the collection of specimens during afebrile periods.

Antimicrobial Therapy

Though tick-borne relapsing fever is usually milder than the louse-borne disease, it is often more difficult to treat because spirochaetes tend to persist in tissues such as the central nervous system and eye. Treatment is usually with tetracyclines or penicillin. Numerous studies have compared the efficacy of both regimes: tetracyclines produce a more rapid clearance of spirochaetes from the blood, but with the concomitant increased risk of Jarisch–Herxheimer reactions, whereas with penicillin the disappearance of *Borrelia* is slower. Usual therapeutic regimes are intramuscular penicillin or oral tetracycline or doxycycline for 5–10 days for tick-borne relapsing fever, while a single dose will cure the louse-borne disease. Oral erythromycin can be used as an alternative treatment for pregnant women or children. Chloramphenicol has also been used for 10 days to treat tick-borne relapsing fever or as a single dose for the treatment of louse-borne relapsing fever.

In order to prevent the Jarisch–Herxheimer reaction, some prefer to combine penicillin and tetracycline, with penicillin given first, followed on the next day by tetracycline. Others have used single low doses of penicillin or tetracycline, but the low dose of penicillin is associated with unacceptable levels of relapse (Seboxa and Rahlenbeck, 1995).

Studies *in vitro* with an isolate of *B. recurrentis* have confirmed its exquisite susceptibility to penicillin (MIC 0.2 µg/mL), tetracycline (MIC and MBC 0.06 µg/mL) and erythromycin (MIC 0.04 µg/mL; MBC 0.02 µg/mL) (Cutler *et al.*, 1994), which supports the clinical observation that louse-borne relapsing fever can successfully be cured with single-dose treatment. Slightly higher inhibitory values were found for tick-borne relapsing fever strains (three strains of *B. hermsii* and one of *B. turicatae*), with MICs of 1–4 µg/mL and MBCs of 2–4 µg/mL for tetracycline.

Jarisch–Herxheimer Reaction

The Jarisch–Herxheimer reaction, frequently seen as a complication of the therapy of relapsing fever, may be a consequence of the release of abundant spirochaete lipoproteins and other constituents from dead or dying organisms. Patients characteristically show rigors, fever and a rise in blood pressure, followed by a sudden fall in blood pressure to near shock levels. Typically, the Jarisch–Herxheimer reaction follows 1–2 hours after treatment with penicillin or tetracycline in approximately half of those treated. Jarisch–Herxheimer reactions in louse-borne relapsing fever patients have been associated with a fatality rate of approximately 5% (Fekade *et al.*, 1996). Immediately before onset of the Jarisch–Herxheimer reaction an increase in TNFα, IL-6 and IL-8 is detectable. A significant reduction in the severity and incidence of Jarisch–Herxheimer reactions in louse-borne relapsing fever patients can be achieved by the infusion of polyclonal ovine anti-TNFα Fab 30 minutes before antibiotic treatment (Fekade *et al.*, 1996).

The lipid moiety of Vmp of these spirochaetes has been directly linked with TNFα induction. By analogy with other spirochaetes, it is believed that the TNFα-inducing component is presented through the toll-like receptor 2 and signalling is facilitated by CD14 (Scragg *et al.*, 2000). The central role of lipoproteins in the pathogenicity of relapsing fever spirochaetes and its host cell interactions deserve to be pursued.

Epidemiology and Control

With the exceptions of *B. recurrentis* and *B. duttonii*, the relapsing fever borreliae are maintained by a cycle through rodent reservoirs and the tick vectors which naturally feed on them. *B. recurrentis* and *B. duttonii*, however, appear respectively to be maintained solely by humans and either the human body louse or *O. moubata* ticks.

Since tick-borne relapsing fever is sporadic in its occurrence, very little is known about the incidence of this disease on a worldwide basis. The disease is more frequently encountered in Jordan, Rwanda, Tanzania and Iran. A prevalence of 1% has been reported in children in western Senegal, while at one health centre in Rwanda 1650 proven cases were treated in a year. In southern Zaire 4.3–7.4% of new out-patients seen each day have relapsing fever (Dupont *et al.*, 1997). In some areas of central Tanzania infection with *B. duttonii* is the sixth most common cause of admission and the seventh most common cause of death among children (Talbert *et al.*, 1998). Certain traditions, such as the introduction of *O. moubata* ticks into specially prepared burrows in new dwellings to guarantee a good and long-lasting immunity to infection with *B. duttonii* (Burgdorfer, 1976) may have a role in the incidence of this disease. More typically the disease is associated with customs of living in close proximity with livestock and their parasites, either from necessity as in Africa where more than 85% of traditional style dwellings are infested with *O. moubata* ticks, or by choice in an attempt to 'get back to nature' in the National Parks of America. During 1973, 62 cases of tick-borne relapsing fever occurred among people staying in log cabins in the Grand Canyon National Park (Boyer *et al.*, 1977). Later, in 1995 48% of 23 family members developed tick-borne relapsing fever after a log cabin holiday in Colorado (Trevejo *et al.*, 1998). A total of 280 cases of tick-borne relapsing fever have been reported in the USA in the past 25 years. Increased occurrence of *B. turicatae* infection has been reported in cavers in Texas, possibly as a result of bats serving as reservoirs of the disease (Rawlings, 1995).

Tick-borne relapsing fever in endemic areas is probably under-recognised. If the clinical presentation is not recognised and blood not drawn at the appropriate febrile period, the diagnosis may be overlooked. In addition, exposure to *Ornithodoros* ticks is often not reported, probably as a result of their preferred short nocturnal feeding habits. Relapsing fever appears to be serologically indistinguishable from Lyme borreliosis with currently available tests (Cutler and Wright, 1994); although this will result in misdiagnosis, nevertheless correct therapeutic management of patients will be used.

Spirochaete–Vector Interactions

Lice

The human body louse has long been established as the vector of louse-borne relapsing fever (MacKie, 1907). Spirochaetes have been detected in the haemolymph of infected lice, but not in their salivary glands or ovaries. Transmission requires the body louse to be crushed into abraded skin, usually by scratching. The custom of crushing lice between the host's teeth has not been proven as a route of infection. The louse life span is only 10–61 days. During this time they prefer to be maintained at body temperature, but they can survive for up to 6 days away from the human host (S.J. Cutler, personal observation). During a blood meal a louse will only take approximately 1 mg of blood (cited in Felsenfeld, 1965). Observations on lice have shown that only 12–17% of lice feeding on a patient with relapsing fever acquire the ability to transmit the disease (cited in Felsenfeld, 1965).

B. recurrentis is believed to be maintained solely by a louse–human cycle, but the above facts have led researchers to question the extent of the *B. recurrentis* reservoir. The specificity of this cycle may reside with the host specificity of the human body louse. The body louse probably evolved from the human head louse. The role of this species, and indeed the more distantly related pubic louse, has yet to be evaluated with sensitive currently available technology. Early reports suggest that neither *Pediculus humanis* var. *capitis* nor *Phthirus pubis* serve as vectors for *B. recurrentis* (cited in Felsenfeld, 1965).

The time between seasonal epidemics is longer than the life span of lice, which suggests that the reservoir of louse-borne relapsing fever infection remains in humans. Although louse-borne relapsing fever was widespread during the last century, it now persists in a few isolated foci, particularly the Ethiopian highlands, but the reason for this remains an enigma.

When collecting samples from Ethiopia, the author observed that the close contacts of clinical cases were also heavily infested with lice, but without showing overt signs of disease. The asymptomatic louse-carrying patient may represent either immunity to the disease, or adaptation of the spirochaete to persistence. Confirmation of borrelial DNA in lice from these individuals or detection of the spirochaete in asymptomatic carriers will support this supposition. The question then arises whether it is the louse or human that constitutes the reservoir of infection. Furthermore, the infected human body louse (*Pediculus humanis* var. *corporis*) loses its capacity to transmit the disease after a single generation of approximately 3 weeks (Salih *et al.*, 1977). It is believed that transovarial transmission does not occur. Since the louse is the only vector of infection, and has an ecological niche confined to humans, it appears logical to assume that humans, or possibly primates, may represent the only reservoir for *B. recurrentis*, the causative organism of louse-borne relapsing fever.

This problem may be resolved by considering the possibility of a secondary reservoir of *B. recurrentis* residing in East African ticks. Nicolle and Anderson in 1926 commented that 'the epidemiology of relapsing fever would ultimately be clarified by the discovery of some reservoir – most probably a tick – in which the louse-borne virus is able to persist for long periods independently of the human host' (cited in Felsenfeld, 1965). For many years a variety of tick vectors have been assessed for their ability to transmit the disease. The potential role of *Ornithodoros* ticks in the transmission of *B. recurrentis* has been addressed, usually by allowing the ticks to feed on a patient during a febrile episode, and assessing their ability to transmit the disease to susceptible hosts either by allowing them to bite, injection of coxal fluid or inoculation of total tick contents (Brumpt, 1908, Nicolle *et al.*, 1913, Ingram, 1924 all cited in Felsenfeld, 1965), however, without success. With the advent of suitably sensitive molecular tools, such as PCR, it is hoped that the true extent of the reservoir for louse-borne relapsing fever can at last be identified.

Studies of the phylogeny of relapsing fever *Borrelia* have highlighted the similarity in DNA sequences between *B. recurrentis* and *B. duttonii* (Ras *et al.*, 1996; Cutler *et al.*, 1997). Estimates from looking at the evolutionary clock show that the 'change' from the tick-borne spirochaete *B. duttonii*, to the louse-borne *B. recurrentis* may have occurred approximately one million years ago. One can speculate that at that time humans were first infected with *B. recurrentis*. East Africa contained the ancestor of *Homo sapiens* and again at that time, humans may have begun to wear beads and, perhaps, rudimentary clothing, which would encourage the persistence of body lice.

B. recurrentis, or perhaps a precursor, may still persist in *O. moubata* ticks. Use of molecular probes to investigate whether ticks contain *B. recurrentis* DNA and to evaluate the efficiency of *O. moubata* ticks to serve as vectors will resolve this problem. The finding of *B. recurrentis* in a tick vector would then explain the persistence of this disease during periods of low clinical incidence and why East Africa is the last remaining reservoir of the disease.

Ticks

Tick-borne relapsing fevers are either transmitted by infected saliva or wound contamination with coxal fluid. Regurgitation of gut material while feeding may also have a role, since *O. moubata* is reported to regurgitate up to 1.3% of its midgut contents during a blood meal. In the tick spirochaetes penetrate the gut wall and disseminate to the ovaries, central ganglion, salivary glands and occasionally the coxal glands. During this dissemination, the spirochaetes may undergo antigenic variation with expression of Vmp types not encountered in the mammalian host (Schwan and Hinnebusch, 1998). The preferred route of transmission varies depending on the particular tick and spirochaete involved. Adult ticks feed for 20–30 minutes approximately every 6 weeks, whereas nymphal stages feed every 2 weeks (Felsenfeld, 1965). This rapid feeding reduces the likelihood that the tick moves from its feeding site to an alternative site where hosts may not be so plentiful. Marked ticks of *O. turicata* in gopher tortoise burrows fail to show any movement away when observed for periods of more than a year (Korch, 1994). In addition, *O. moubata* ticks can survive for over 20 months without a blood meal (S.J. Cutler, personal observation).

Transovarial transmission has been demonstrated to varying degrees among the tick-borne relapsing fever species with a range from 0.29% for *O. hermsi* to 100% for *O. turicata* (Felsenfeld, 1965). It is unlikely that this is sufficient to sustain borrelial infection of ticks beyond a generation or two.

Taxonomic studies of *B. duttonii* and *B. recurrentis* have highlighted the similarity between these species. It is possible that *B. recurrentis* is a louse-adapted variant of *B. duttonii*, and there is some evidence that *B. duttonii* survives in the body louse (Baltazard *et al.*, 1947 cited in Felsenfeld, 1965), but this remains to be substantiated. Further support for this comes from the ability of the monkey louse, *P. longiceps*, to transmit *B. duttonii* (Heisch, 1950 cited in Felsenfeld, 1965). Whether this cycle represents an alternative reservoir for *B. duttonii* has yet to be addressed.

Global Prospects for Control

Louse-borne Relapsing Fever

If louse-borne relapsing fever is a disease of humans with only the human louse as a vector, and there is no evidence of transovarial transmission, the prospects for control and eventual eradication are excellent. Two approaches could be used. First, the elimination of body louse carriage, which could be tackled either by improvements in living conditions, especially the provision of adequate facilities for the washing of clothing, or by application of insecticides. Alternatively, agents which would prevent the breeding of lice or a vaccine targeting any specific vector-associated serotypes could be used. The latter approach would be adequate if all cases were treated together with close family contacts, to cover the possibility of the asymptomatic carriage of spirochaetes.

Past experience has shown that treatment of the clinical disease with antibiotics alone fails to control outbreaks, whereas widespread use of topical insecticides controls epidemics but does not prevent seasonal recurrences. Typically, outbreaks of louse-borne relapsing fever have been controlled by dusting every member of the community with insecticides as quickly as possible. This usually used to be with 10% DDT, but this has largely been replaced by 1% malathion or 0.5% permethrin. The dusting must spread insecticide evenly over the inner surface of garments next to the skin, with special attention to seams and folds. Mass delousing of large groups of people is possible with motor-driven air compressors with up to 10 duster heads (Busvine, 1980).

If less host-specific ticks are also identified as carrying *B. duttonii*, it is probable that alternative reservoirs may exist. Eradication of the disease would then become more complicated to achieve. Political difficulties, namely the war in Eritrea, have prevented further assessment of these strategies.

Tick-borne Relapsing Fever

B. duttonii disease is believed to be a restricted to humans and transmitted by *O. moubata*. Survival of these spirochaetes in ticks is possible over prolonged periods of time. In addition, *O. moubata* may feed from a variety of alternative hosts, which may be potential reservoirs of infection. Application of sensitive molecular approaches to determine the extent of transovarial transmission in these ticks would also need to be determined before a mathematical model can be constructed from which the effects of various disease elimination strategies could be determined.

Control and/or elimination of other tick-borne relapsing fever *Borrelia* remains more problematic.

A diverse range of *Ornithodoros* ticks carry borreliae and each of these tick species has its own ecological niche and preferred host-range. This makes the implementation of any form of control strategy difficult. Limited tick control may be achieved by spraying buildings with 2% benzene hexachloride or 0.5% malathion in combination with reducing the numbers of rodent reservoirs. More recently, use of a synthetic pyrethroid (lambda-cyhalothrin) in a durable formulation has kept dwellings tick-free for up to 2 years and resulted in significant reductions in clinical cases of tick-borne relapsing fever (Talbert *et al.*, 1998). Alternatively, biting by these nocturnal feeders may be prevented by use of bed nets.

Vaccines

The role of vaccines for future control of relapsing fever is uncertain. What is clear is that the inherent variability in Vmps between strains and species would make use of these in a polyvalent vaccine unlikely. A conserved immunogenic component in relapsing fever spirochaetes has yet to be identified and evaluated for its potential to elicit protective immunity.

References

Anda P, Sánchez-Yebra W, del Mar Vitutia M *et al.* (1996) A new *Borrelia* species isolated from patients with relapsing fever in Spain. *Lancet* 348: 162–165.

Baltazard M, Bahmanyar M, Mofidi C (1947) Fiévres récurrentes transmises a la fois par ornithodores et par poux. *Ann. Inst. Pasteur* 73: 1066–1071.

Barbour AG (1984) Isolation and cultivation of Lyme disease spirochetes. *Yale J. Biol. Med.* 57: 521–523.

Barbour AG (1990) Antigenic variation of a relapsing fever *Borrelia* species. *Annu. Rev. Microbiol.* 44: 155–171.

Barbour AG, Hayes SF (1986) Biology of Borrelia species. *Microbiol. Rev.* 50: 381–400.

Barbour AG, Hayes SF, Heiland RA, Schrumpf ME, Tessier SL (1986) A *Borrelia*-specific monoclonal antibody binds to a flagellar epitope. *Infect. Immun.* 52: 549–554.

Baril C, Richaud C, Baranton G, Saint-Girons IS (1989) Linear chromosome of *Borrelia burgdorferi*. *Res. Microbiol.* 140: 507–516.

Barstad PA, Coligan JE, Raum MG, Barbour AG (1985) Variable major proteins of *Borrelia hermsii*. Epitope mapping and partial sequence analysis of CNBr peptides. *J. Exp. Med.* 161: 1302–1314.

Berger BW, MacDonald AB, Benach JL (1988) Use of an autologous antigen in the serologic testing of patients with erythema migrans of Lyme disease. *J. Am. Acad. Dermatol.* 18: 1243–1246.

Bergström S, Barbour AG, Garon CF, Hindersson P, Saint-Girons I, Schwan TG (1991) Genetics of *Borrelia burgdorferi*. *Scand. J. Infect. Dis.* Suppl. 77: 102–107.

Bergström S, Garon CF, Barbour AG, MacDougall J (1992) Extrachromosomal elements of spirochetes. *Res. Microbiol.* 143: 623–628.

Boyer KM, Munford RS, Maupin GO *et al.* (1977) Tick-borne relapsing fever: an interstate outbreak originating at Grand Canyon National Park. *Am. J. Epidemiol.* 105: 469–479

Burgdorfer W (1976) The diagnosis of the relapsing fevers. In: Johnson RC (ed.) *The Biology of Parasitic Spirochetes.* New York: Academic Press, pp. 225–234.

Burman N, Bergstrom S, Restrepo BI, Barbour AG (1990) The variable antigens Vmp7 and Vmp21 of the relapsing fever bacterium *Borrelia hermsii* are structurally analogous to the VSG proteins of the African trypanosome. *Mol. Microbiol.* 4: 1715–1726.

Burman N, Shamaei-Tousi A, Bergström S (1998) The spirochete *Borrelia crocidurae* causes erythrocyte rosetting during relapsing fever. *Infect. Immun.* 66: 815–819.

Busvine JR (1980) *Insects and Hygiene. The Biology and Control of Insect Pests of Medicat and Domestic Importance* 3rd edn. London: Chapman and Hall, pp. 256–272.

Bunikis J, Luke CJ, Bunikiene E, Bergström S, Barbour AG (1998) A surface-exposed region of a novel outer membrane protein (P66) of *Borrelia* spp. is variable in size and sequence. *J. Bacteriol.* 180: 1618–1623.

Cadavid D, Pennington PM, Kerentseva TA, Bergström S, Barbour AG (1997) Immunologic and genetic analyses of VmpA of a neurotropic strain of *Borrelia turicatae. Infect. Immun.* 65: 3352–3360.

Carlyon JA, Roberts DM, Theisen M, Sadler C, Marconi RT (2000) Molecular and immunological analyses of the *Borrelia turicatae* Bdr protein family. *Infect. Immun.* 68: 23269–23273.

Carter CJ, Bergström S, Norris SJ, Barbour AG (1994) A family of surface-exposed proteins of 20 kilodaltons in the genus *Borrelia. Infect. Immun.* 62: 2792–2799.

Casjens S, Delange M, Ley III HL, Rosa P, Huang WM (1995) Linear chromosomes of Lyme disease agent spirochetes: genetic diversity and conservation of gene order. *J. Bacteriol.* 177: 2769–2780.

Cutler SJ, Wright DJM (1994) Predictive value of serology in diagnosing Lyme borreliosis. *J. Clin. Pathol.* 47: 344–349.

Cutler SJ, Fekade D, Hussein K *et al.* (1994) Successful *in vitro* cultivation of *Borrelia recurrentis* [letter]. *Lancet* 343: 242.

Cutler SJ, Moss J, Fukunaga M, Wright DJM, Fekade D, Warrell D (1997) *Borrelia recurentis* characterization and comparison with relapsing-fever, Lyme-associated, and other *Borrelia* spp. *Int. J. Syst. Bacteriol.* 47: 958–968.

Cutler SJ, Akintunde COK, Moss J *et al.* (1999) Successful *in vitro* cultivation of *Borrelia duttonii* and its comparison with *Borrelia recurrentis. Int. J. Syst. Bacteriol.* 49: 1793–1799.

Dupont HT, La Scola B, Williams R, Raoult D (1977) A focus of tick-borne relapsing fever in southern Zaire. *Clin. Infect. Dis.* 25: 139–144.

Dutton JE, Todd JL (1905) The nature of tick fever in the eastern part of the Congo Free State. *BMJ* ii: 1259–1260.

Fekade D, Knox K, Hussein K *et al.* (1996) Prevention of Jarisch-Herxheimer reactions by treatment with anti-bodies against tumor necrosis factor alpha. *N. Engl. J. Med.* 335: 311–315.

Felsenfeld O (1965) Borreliae, human relapsing fever, and parasite-vector-host relationships. *Bacteriol. Rev.* 29: 46–74.

Ferdows MS, Serwer P, Griess GA, Norris SJ, Barbour AG (1996) Conversion of a linear to a circular plasmid in the relapsing fever agent *Borrelia hermsii. J. Bacteriol.* 178: 793–800.

Fraser C, Casjens S, Huang WM *et al.* (1997) Genomic sequence of a Lyme disease spirochaete, *Borrelia burgdorferi. Nature* 390: 580–586.

Fraser CM, Norris SJ, Weinstock GM *et al.* (1998) Complete genome sequence of *Treponema pallidum*, the syphilis spirochete. *Science* 281: 375–388.

Fukunaga M, Takahashi Y, Tsuruta Y *et al.* (1995) Genetic and phenotypic analysis of *Borrelia miyamotoi* sp. nov, isolated from the Ixodid tick *Ixodes persulcatus*, the vector for Lyme disease in Japan. *Int. J. Syst. Bacteriol.* 45: 804–810.

Goubau PF (1994) Relapsing fevers. A review. *Ann. Soc. Belge Méd. Trop.* 64: 335–364.

Hamase A, Takahashi Y, Nohgi K, Fukunaga M (1996) Homolog of variable major protein genes between *Borrelia hermsii* and *Borrelia miyamotoi*. *FEMS Microbiol. Lett.* 140: 131–137.

Hayes LJ, Wright DJ, Archard LC (1988) Segmented arrangement of *Borrelia duttonii* DNA and location of variant surface antigen genes. *J. Gen. Microbiol.* 134: 1785–1793.

Hendson M, Lane RS (2000) Genetic characteristics of *Borrelia coriaceae* isolates from the soft tick *Ornithodoros coriaceus* (Acari: Argasidae). *J. Clin. Microbiol.* 38: 2678–2682.

Hinnebusch J, Barbour AG (1991) Linear plasmids of *Borrelia burgdorferi* have a telomeric structure and sequence similar to those of a eukaryotic virus. *J. Bacteriol.* 173: 7233–7239.

Hinnebusch J, Tilly K (1993) Linear plasmids and chromosomes in bacteria. *Mol. Microbiol.* 10: 917–922.

Hinnebusch BJ, Barbour AG, Restrepo BI, Schwan TG (1998) Population structure of the relapsing fever spirochete *Borrelia hermsii* as indicated by polymorphism of two multigene families that encode immunogenic outer surface lipoproteins. *Infect. Immun.* 66: 432–440.

Hovind Hougen K (1974) Electron microscopy of *Borrelia merionesi* and *Borrelia recurrentis. Acta Pathol. Microbiol. Scand. Sect. B* 82: 799–809.

Hovind Hougen K (1984) Ultrastructure of spirochetes isolated from *Ixodes ricinus* and *Ixodes dammini. Yale J. Biol. Med.* 57: 543–548.

Hulínská D, Jirous J, Valesová M, Herzogová J (1989) Ultrastructure of *Borrelia burgdorferi* in tissues of patients with Lyme disease. *J. Basic Microbiol.* 29: 73–83.

Hyde JF, Johnson RC (1984) Genetic relationship of Lyme disease spirochetes to *Borrelia*, *Treponema* and *Leptospira* spp. *J. Clin. Microbiol.* 20: 151–154.

Judge DM, La-Croix JT, Perine PL (1974a) Experimental louse-borne relapsing fever in the grivet monkey, *Cercopithecus aethiops*. I. Clinical course. *Am. J. Trop. Med. Hyg.* 23: 957–961.

Judge DM, La-Croix JT, Perine PL (1974b) Experimental louse-borne relapsing fever in the grivet monkey, *Cercopithecus aethiops*. II. Pathology. *Am. J. Trop. Med. Hyg.* 23: 962–968.

Judge DM, La-Croix JT, Perine PL (1974c) Experimental louse-borne relapsing fever in the grivet monkey, *Cercopithecus aethiops*. III. Crisis following therapy. *Am. J. Trop. Med. Hyg.* 23: 969–973.

Kelly RT (1976) Cultivation and physiology of relapsing fever borreliae. In: Johnson RC (ed.) *The Biology of Parasitic Spirochetes*. New York: Academic Press, pp. 87–94.

Kitten T, Barbour AG (1992) The relapsing fever agent *Borrelia hermsii* has multiple copies of its chromosome and linear plasmids. *Genetics* 132: 311–324.

Korch GW (1994) Geographic dissemination of tick-borne zoonoses. In: Sonenshine DE, Mather TN *Ecological Dynamics of Tick-borne Zoonoses*. New York: Oxford University Press, pp. 139–197.

MacKie FP (1907) The part played by *Pediculus corporis* in the transmission of relapsing fever. *BMJ* 2: 1706–1709.

Marconi RT, Samuels DS, Schwan TG, Garon CF (1993) Identification of a protein in several *Borrelia* species which is related to OspC of the Lyme disease spirochetes. *J. Clin Microbiol.* 31: 2577–2583.

Margolis N, Hogan D, Cieplak W, Schwan TG, Rosa PA (1994a) Homology between *Borrelia burgdorferi* OspC and members of the family of *Borrelia hermsii* variable major proteins. *Gene* 143: 105–110.

Margolis N, Hogan D, Tilly K, Rosa PA (1994b) Plasmid location of *Borrelia* purine biosynthesis gene homologs. *J. Bacteriol.* 176: 6427–6432.

Meier JT, Simon MI, Barbour AG (1985) Antigenic variation is associated with DNA rearrangements in a relapsing fever *Borrelia*. *Cell* 41: 403–409.

Moczutkowsky OO (1879) Materialien zur Pathologie und Therapie des Rückfalltyphus. *Dtsch Arch. klin. Med.* 24: 80–97.

Morshed MG, Konishi H, Nishimura T, Nakazawa T (1993) Evaluation of agents for use in medium for selective isolation of Lyme disease and relapsing fever *Borrelia* species. *Eur J. Clin. Microbiol. Infect. Dis.* 12: 512–518.

Obermeier OHF (1873) Vorkommen feinster, eine Eigenbewequng zeigender Fäden im Blute von Recurrenskranken. *Zbl. Med. Wissenschaft.* 11: 145–147.

Pennington PM, Alfred CD, West CS, Alvarez R, Barbour AG (1997) Arthritis severity and spirochete burden are determined by serotype in the *Borrelia turicatae*-mouse model of Lyme disease. *Infect. Immun.* 65: 285–292.

Pennington PM, Cadavid D, Bunikis J, Norris SJ, Barbour AG (1999) Extensive interplasmidic duplications change the virulence phenotype of the relapsing fever agent *Borrelia turicatae*. *Mol. Microbiol.* 34: 1120–1132.

Picken RN (1992) Polymerase chain reaction primers and probes derived from flagellin gene sequences for specific detection of the agents of Lyme disease and North American relapsing fever. *J. Clin. Microbiol.* 30: 99–114.

Pickett J, Kelly R (1974) Lipid catabolism of relapsing fever borreliae. *Infect. Immun.* 9: 279–285.

Radolf JD, Arndt LL, Akins DR *et al.* (1995) *Treponema pallidum* and *Borrelia burgdorferi* lipoproteins and synthetic lipopeptides activate monocytes/macrophages. *J. Immunol.* 154: 2866–2877.

Ras N, Lascola B, Postic D *et al.* (1996) Phylogenesis of relapsing fever *Borrelia*. *J. Int. Syst. Bacteriol.* 46: 859–865.

Ras NM, Postic D, Ave P *et al.* (2000) Antigenic variation of *Borrelia turicatae* vsp surface lipoproteins occurs *in vitro* and generates novel serotypes. *Res. Microbiol.* 151: 5–12.

Rawlings JA (1995) An overview of tick-borne relapsing fever with emphasis on outbreaks in Texas. *Texas Med.* 91: 56–59.

Restrepo BI, Carter CJ, Barbour AG (1994) Activation of a vmp pseudo gene in *Borrelia hermsii*: an alternative mechanism of antigenic variation during relapsing fever. *Mol. Microbiol.* 13: 287–299.

Ross PH, Milne AD (1904) 'Tick fever'. *BMJ* ii: 1453–1454.

Roux V, Raoult D (1999) Body lice as tools for diagnosis and surveillance of reemerging diseases. *J. Clin. Microbiol.* 37: 596–599.

Salih SY, Mustafa D, Abdel Wahab SM, Ahmed MAM, Omer A (1977) Louse-borne relapsing fever: a clinical and laboratory study of 363 cases in the Sudan. *Trans. R. Soc. Trop. Med. Hyg.* 71: 43–51.

Schwan TG, Hinnebusch BJ (1998) Bloodstream- versus tick-associated variants of a relapsing fever bacterium. *Science* 280: 1938–1940.

Schwan TG, Simpson WJ, Schrumpf ME, Karstens RH (1989) Identification of *Borrelia burgdorferi* and *B. hermsii* using DNA hybridization probes. *J. Clin. Microbiol.* 27: 1734–1738.

Schwan TG, Gage KL, Karstens RH, Schrumpf ME, Hayes SF, Barbour AG (1992) Identification of the tick-borne relapsing fever spirochete *Borrelia hermsii* by using a species-specific monoclonal antibody. *J. Clin. Microbiol.* 30: 790–795.

Schwan TG, Schrumpf ME, Hinnebusch BJ, Anderson Jr DE, Konkel ME (1996) GlpQ: an antigen for serological discrimination between relapsing fever and Lyme borreliosis. *J. Clin. Microbiol.* 34: 2483–2492.

Schwartz JJ, Gazumyan A, Schwartz I (1992) rRNA gene organization in the Lyme disease spirochete, *Borrelia burgdorferi*. *J. Bacteriol.* 174: 3757–3765.

Scragg IG, Kwiatkowski D, Vidal V *et al.* (2000) Structural characterization of the inflammatory moiety of a variable major lipoprotein of *Borrelia recurrentis*. *J. Biol. Chem.* 275: 937–941.

Seboxa T, Rahlenbeck SI (1995) Treatment of louse-borne relapsing fever with low dose penicllin or tetracycline: a clinical trial. *Scand. J. Infect. Dis.* 27: 29–31.

Shang ES, Skare JT, Exner MM *et al.* (1998) Isolation and characterization of the outer membrane of *Borrelia hermsii*. *Infect. Immun.* 66: 1082–1091.

Stoenner HG, Dodd T, Larsen C (1982) Antigenic variation of *Borrelia hermsii*. *J. Exp. Med.* 156: 1297–1311.

Takahashi Y, Cutler SJ, Fukunaga M (2000) Size conversion of a linear plasmid in the relapsing fever agent *Borrelia duttoni*. *Microbiol. Immunol.* 44: 1071–1074.

Takahashi Y, Fukunaga M (1996) Physical mapping of the *Borrelia miyamotoi* HT31 chromosome in comparison with that of *Borrelia turicatae*, an etiological agent of tick-borne relapsing fever. *Clin. Diagn. Lab. Immunol.* 3: 533–540.

Talbert A, Nyange A, Molteni F (1998) Spraying tick-infested houses with lambda-cyhalothrin reduces the incidence of tick-borne relapsing fever in children under five years old. *Trans. R. Soc. Trop. Med. Hyg.* 92: 251–253.

Trevejo RT, Schriefer ME, Gage KL *et al.* (1998) An interstate outbreak of tick-borne relapsing fever among vacationers at a Rocky Mountain cabin. *Am. J. Trop. Med. Hyg.* 58: 743–747.

Vidal VI, Scragg IG, Cutler SJ *et al.* (1998) Variable major lipoprotein is a principal TNF-inducing factory of louse-borne relapsing fever. *Nature Med.* 4: 1416–1420.

Zhang JR, Hardham JM, Barbour AG, Norris SJ (1997) Antigenic variation in Lyme disease borreliae by promiscuous recombination of vmp-like sequence cassettes. *Cell* 89: 275–285.

99

Bartonella

Michael F. Minnick

The University of Montana, Missoula, Montana, USA

The *Bartonella* group has recently undergone a taxonomic renaissance. For nearly 85 years, *Bartonella bacilliformis* was the only representative of an enigmatic bacterial genus first described by A.L. Barton (Barton, 1909). However, the bacteria that once constituted the genera *Rochalimaea* and *Grahamella* were re-classified as bartonellae (Brenner *et al.*, 1993; Birtles *et al.*, 1995), thus increasing the number of potential species to 14. In addition, the perennial belief that the genus was a member of the order *Rickettsiales* was discarded, and a closer relationship to the family *Rhizobiaceae* was recognised (Birtles *et al.*, 1991; Brenner *et al.*, 1991, 1993; O'Connor *et al.*, 1991; Relman *et al.*, 1992). Five *Bartonella* species are aetiological agents of emerging infectious disease in humans (Table 1). The potential impact of this group on human health has fuelled an interest in the basic biology of these bacteria and how genetic adaptation contributes towards their emergence and pathogenicity.

Bartonella spp. share many general characteristics. They are small (approximately $0.3\,\mu m \times 1\,\mu m$), Gram-negative, pleiomorphic coccobacilli with a chromosomal guanine + cytosine (G + C) content of approximately 40 mol%. The bacteria are facultative intracellular pathogens that employ haemotrophy (infection of erythrocytes) as a parasitic strategy. All members of the genus are notoriously fastidious and grow slowly *in vitro*. Vector-mediated transmission is another common theme in the genus. Bartonellae are typically transmitted between mammalian hosts by arthropods, and each species is transmitted by a particular insect vector (Table 1). The reservoirs of bartonellae include mammals and arthropods (Tables 1 and 2).

Table 1 *Bartonella* species that are pathogenic for humans

Species	Common manifestation(s)	Vector(s)	Reservoir(s)
B. bacilliformis	Oroya fever, verruga peruana	Sandflies	Humans
B. clarridgeiae	Cat-scratch disease	Cats	Cats
B. elizabethae	Endocarditis	Unknown	Unknown
B. henselae	Cat-scratch disease, endocarditis, bacillary angiomatosis, bacillary peliosis, bacteraemia syndrome	Cats, fleas	Cats
B. quintana	Trench fever, endocarditis, bacillary angiomatosis, bacteraemia syndrome	Body louse	Body louse

Table 2 *Bartonella* species that have not been demonstrated to be pathogenic for humans

Species	Host(s)	Reference(s)
B. alsatica	Rabbits	Heller *et al.* (1999)
B. doshiae	Rodents	Birtles *et al.* (1995)
B. grahamii	Rodents, insectivores	Birtles *et al.* (1995)
B. koehlerae	Domestic cat	Droz *et al.* (1999)
B. peromysci	Rodents	Birtles *et al.* (1995)
B. talpae	Moles	Birtles *et al.* (1995)
B. taylorii	Rodents	Birtles *et al.* (1995)
B. tribocorum	Rats	Heller *et al.* (1998)
B. vinsonii	Domestic dog, rodents	Weiss and Dasch (1982), Kordick *et al.* (1996)

Bartonellosis

Bartonellosis, originally a term referring only to Oroya fever, now includes a variety of human diseases (**Table 1**). Although a wide array of manifestations can occur during infection, the bartonelloses present with bacteraemia, a low-grade fever and malaise, vascular lesions (haemangiomas, papules or peliosis) and lymphadenopathy. Immunodeficient patients are particularly at risk of contracting an opportunistic bartonellosis, but immunocompetent individuals can be infected.

Oroya Fever

Oroya fever, also referred to as Carrion's disease or verruga peruana, is endemic to South America (Alexander, 1995). It is particularly prominent in Peru, Colombia and Ecuador (Kreier and Ristic, 1981), and has afflicted travellers to this region (Matteelli *et al.*, 1994). The course of disease is very unusual and has two sequentially disparate stages. The primary (haematic) phase is characterised by an acute bacteraemia that occurs within 4 weeks of a bite by a contaminated phlebotamine sandfly. From the inoculation site, *B. bacilliformis* efficiently colonises the entire circulatory system. Nearly every erythrocyte becomes parasitised and approximately 80% are lysed (Hurtado *et al.*, 1938), presumably by culling in the spleen (Reynafarje and Ramos, 1961). Without antibiotic therapy the case fatality rate can reach 40–88% (Weinman, 1965; Gray *et al.*, 1990). The acute-phase can be accompanied by anorexia, headache and, in fatal cases, coma (Roberts, 1995). Secondary infectious diseases such as salmonellosis, shigellosis or recurrence of toxoplasmosis or

tuberculosis, are not uncommon during the haematic phase and can complicate the prognosis (Urteaga and Payne, 1955; Cuadra, 1956; Gray *et al.*, 1990). Patients present with symptoms of the secondary (tissue) phase, approximately 4 weeks after resolution of the primary phase. Tissue involvement results from bacterial invasion of the capillary endothelium, and generates bacteria-filled vacuoles, termed rocha lima inclusions, and localised cellular proliferation leading to the formation of nodule or papule lesions, termed verruga peruana (Arias-Stella *et al.*, 1986). Verruga eruptions are usually cutaneous but may involve mucous membranes and viscera (Ricketts, 1949). Cutaneous lesions are found on the skin of the head and extremities and can persist for several weeks to months.

While the secondary phase is rarely fatal, verrugas can bleed and cause scarring (Weinman, 1965). Anaemia is no longer present during the tissue stage (Ricketts, 1949), but viable bacteria can be isolated from the blood, marrow and haemangioma tissue of patients with verruga peruana.

Trench Fever

Trench fever, or shinbone fever, is a louse-borne disease caused by *B. quintana* and is distributed throughout the world. The illness was first reported during World War I and was responsible for several epidemics among troops (McNee and Renshaw, 1916). The morbidity of trench fever was so severe that only influenza caused a greater loss of man-hours during World War I (Strong, 1918). After a brief period of quiescence, trench fever re-appeared during World War II (Kostrzewski, 1950), and has occurred sporadically ever since. Currently, infection with *B. quintana*, the so-called 'urban trench fever', is re-emerging in homeless inner-city populations of the United States (Jackson and Spach, 1996).

Trench fever occurs approximately 9 days after inoculation by the bite of an infected louse (Byam, 1919). Although symptoms of trench fever vary significantly between patients, the disease usually presents with mild to moderately severe fever, chills, malaise, myalgia and bone pain that is especially prominent in the tibia (hence shinbone fever) (Varela *et al.*, 1969). Occasionally, patients develop splenomegaly and a maculopapular rash resembling the rose spots of typhoid fever (Strong, 1918). The illness usually lasts about a week, but febrile episodes may be recurrent and the bacteraemia may be protracted.

In addition to trench fever, *B. quintana* has been implicated as an emerging aetiological agent of bacillary angiomatosis and endocarditis (see below).

Cat-scratch Disease

The aetiological agent of cat-scratch disease (CSD) was a mystery until 1988, when the putative CSD bacillus was finally isolated from the lymph nodes of a patient (English *et al.*, 1988) and later named *Afipia felis* (Brenner *et al.*, 1991). However, confusion quickly developed because other CSD patients were not sero-positive for *A. felis*, and novel isolates of the bacterium from other cases were not obtained (Regnery and Tappero, 1995). Later, the preponderance of evidence pointed to *B. henselae* as the causative agent of CSD (Perkins *et al.*, 1992; Welch *et al.*, 1992; Regnery *et al.*, 1992a; Anderson *et al.*, 1993; Dolan *et al.*, 1993; Zangwill *et al.*, 1993; Bergmans *et al.*, 1995). More recently, *B. clarridgeiae* was also found to cause CSD (Kordick *et al.*, 1997). Both *B. henselae* and *B. clarridgeiae* are physically transmitted to humans by scratches or bites of cats (Margileth, 1993; Kordick *et al.*, 1997) and they may be transmitted by cat fleas (Zangwill *et al.*, 1993).

CSD typically manifests itself as granulomatous skin lesions (papules or pustules) that develop within one week after injury inflicted by an infected cat. Skin lesions contain areas of necrosis bordered by histiocytes, lymphocytes and giant cells (Johnson and Helwig, 1969). Lymphadenitis is characteristic, with proximal draining lymphadenopathy 2–3 weeks after infection. Like the skin papules, lymph node lesions are granulomatous microabcesses with lymphocyte infiltration and giant cells, together with follicular hyperplasia (Carithers, 1985). In addition to skin and lymph node involvement, patients with CSD may present with mild fever, malaise and gastro-intestinal distress. Complications involving the central nervous system, bone, lung, liver, spleen and eyes have also been reported (Carithers, 1985; Milam *et al.*, 1990; Golden, 1993; Doyle *et al.*, 1994; McCrary, 1994; Caniza *et al.*, 1995). CSD typically resolves spontaneously within 8–12 weeks. *B. henselae* has also been implicated as an emerging infectious agent of bacillary angiomatosis and endocarditis (see below).

Bacillary Angiomatosis

Bacillary angiomatosis (BA) results from infection with *B. henselae* or *B. quintana* (Relman *et al.*, 1990, 1992; Koehler *et al.*, 1992; Welch *et al.*, 1992), and was first observed in AIDS patients in the 1980s (Stoler *et al.*, 1983; Cockerell *et al.*, 1987; Berger *et al.*, 1989). Although it usually affects immunodeficient individuals (Stoler *et al.*, 1983; Koehler and Tappero, 1993), it has been reported in immunocompetent patients (Cockerell *et al.*, 1990; Lucey *et al.*, 1992; Tappero *et al.*, 1993a). The course of the disease can be subacute and insidious in immunodeficient patients, whereas it is sudden in immunocompetent individuals (Schwartzman, 1992).

As with CSD, risk factors for BA probably include exposure to infected cats or cat fleas (Koehler *et al.*, 1994). It is characterised by pseudo-neoplastic cutaneous or subcutaneous vascular lesions, and a lack of granulomatous tissue in BA lesions distinguishes them from CSD. The papule or nodule-like lesions of BA contain extensive vascular channels bordered by cuboidal, protuberant endothelium and a multi-cellular inflammatory infiltrate with polymorphonuclear leucocytes which display leucocytoclastic characteristics (LeBoit *et al.*, 1989; Cockerell *et al.*, 1990). In addition, when stained by Warthin-Starry silver stain, the lesions usually contain aggregates of bartonellae (Angritt *et al.*, 1988; LeBoit *et al.*, 1988). Cutaneous lesions of BA are similar to verruga peruana and superficially resemble vascular neoplasms such as Kaposi's sarcoma (LeBoit *et al.*, 1988; Webster *et al.*, 1992) or pyogenic granulomas (Koehler and Tappero, 1993). The lesions can last for several months (Koehler *et al.*, 1992). Subcutaneous or visceral BA can involve a host of organ systems including the brain, bone, lymph nodes and eyes (Kemper *et al.*, 1990; Koehler *et al.*, 1992; Spach *et al.*, 1992; Koehler and Tappero, 1993; Tappero *et al.*, 1993a, b; Golnik *et al.*, 1994; Slater *et al.*, 1994; Waldvogel *et al.*, 1994).

Bacillary Peliosis

B. henselae is also the agent of bacillary peliosis (BP) (Marullo *et al.*, 1992; Slater *et al.*, 1992; Welch *et al.*, 1992; Tappero *et al.*, 1993b). BP lesions are characterised by cystic blood-filled cavities (Perkocha *et al.*, 1990; Garcia-Tsao *et al.*, 1992). Peliosis may be accompanied by gastro-intestinal distress, fever and chills, and the syndrome may occur alone or in combination with bacillary angiomatosis or bacteraemia syndrome. BP can involve single or multiple organs such as the liver, spleen and lymph nodes and, when stained by Warthin-Starry silver stain, the lesions contain bartonellae (Perkocha *et al.*, 1990). BP in the liver can produce hepatomegaly, and patients typically show elevated serum levels of liver enzymes (e.g. γ-glutamyltransferase and alkaline phosphatase) (Perkocha *et al.*, 1990). Death from peliosis-induced liver failure has also been reported (Perkocha *et al.*, 1990).

Endocarditis

Although many micro-organisms can cause endocarditis, evidence suggests that blood culture-negative cases often involve *Bartonella* spp. (Raoult *et al.*,

1996). Endocarditis has been linked to infection with *B. henselae* (Hadfield *et al.*, 1993; Drancourt *et al.*, 1996a; Raoult *et al.*, 1996), *B. quintana* (Spach *et al.*, 1993; Drancourt *et al.*, 1995; Jalava *et al.*, 1995; Raoult *et al.*, 1996) and *B. elizabethae* (Daly *et al.*, 1993). Endocarditis caused by bartonellae have been reported in immunocompetent and immunosuppressed individuals.

Bacteraemia Syndrome

Cases of relapsing fever with persistent bacteraemia have been reported in patients infected with *B. henselae* (Slater *et al.*, 1990; Welch *et al.*, 1992, 1993; Lucey *et al.*, 1992; Regnery *et al.*, 1992a) and *B. quintana* (Koehler *et al.*, 1992; Maurin *et al.*, 1994).

Classification

The family *Bartonellaceae* is no longer a member of the order *Rickettsiales,* a group that originally included the *Anaplasmataceae, Rickettsiaceae* and *Bartonellaceae.* The earlier taxonomy was undoubtedly based on rickettsia-like characteristics such as the small size of *B. bacilliformis* ('virus-like particles'), its transmission to humans by an arthropod vector (sandflies), and its ability to live within host cells (Moulder, 1974).

Although molecular means of distinguishing bartonellae from rickettsiae are a fairly recent development, classification based upon superficial characteristics was tenuous at best. For instance, *Bartonella, Grahamella* and *Rochalimaea* were the only members of the *Rickettsiales* that could be cultured *in vitro*, while true rickettsiae are strictly obligate intracellular parasites of eukaryotic cells. In addition, *B. bacilliformis* was the only flagellated member in the order (Weinman, 1974).

Only recently have DNA-relatedness data, mainly based on 16S rRNA sequences, been employed to elucidate the correct phylogeny of *Bartonella* and *Rochalimaea*. DNA studies clearly showed that *Rochalimaea* and *Bartonella* are more closely related to each other than to any rickettsiae. In addition, the evidence indicated that *Bartonella* and *Rochalimaea* are most closely related to members of the α2 subgroup of the *Proteobacteria* class, especially the *Rhizobiaceae* (Birtles *et al.*, 1991; Brenner *et al.*, 1991; O'Connor *et al.*, 1991; Relman *et al.*, 1992). On the basis of these findings, *Rochalimaea* species were subsequently re-classified as bartonellae and the family *Bartonellaceae*, containing the genera *Grahamella* and *Bartonella*, was removed from the *Rickettsiales* (Brenner *et al.*, 1993). A subsequent study, based upon DNA relatedness data and phenotypic characteristics, re-classified five *Grahamella* species as

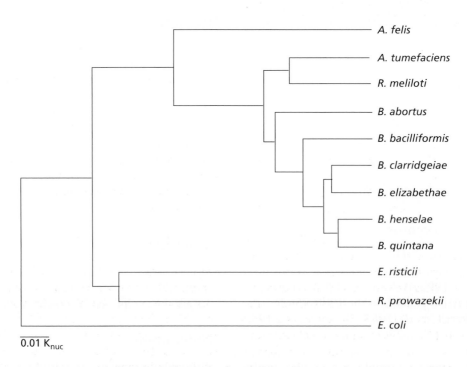

Fig. 1 Phylogenetic tree showing relationships between the five pathogenic *Bartonella* species and various α-*Proteobacteria*. *E. coli* (γ-*Proteobacteria*) is included for comparison. The tree was constructed using 16S rDNA and rRNA sequences from GenBank and Clustal software (PCGene 6.0, Intelligenetics Corp.). K_{nuc}, evolutionary distance.

bartonellae and eliminated the genus (Birtles *et al.*, 1995).

At present, the genus *Bartonella* contains the 14 species listed in **Tables 1** and **2**. The phylogenetic relationship of *Bartonella* to other α-*Proteobacteria* is shown in **Fig. 1**. The close relatives of *Bartonella* include *Agrobacterium*, *Rhizobium* and *Brucella* spp., the latter being the most closely related (Brenner *et al.*, 1993). At first glance this family portrait seems diverse, but it does show natural history resemblance. Specifically, all four genera have evolved towards a parasitic or mutualistic association with eukaryotic cells. Bartonellae and brucellae are both intracellular parasites of mammalian cells, while *Agrobacterium* and *Rhizobium* species can parasitise or symbiotically associate with plant cells, respectively. None of these bacteria are obligate parasites, and all can be cultivated *in vitro*.

Interestingly, human diseases caused by *Brucella* and *Agrobacterium* share superficial similarities with bartonelloses. Like bartonellae, opportunistic infections with *Agrobacterium radiobacter* occur mainly in immunocompromised patients and can cause bacteraemia and endocarditis (Plotkin, 1980; Edmond *et al.*, 1993; Freney *et al.*, 1985; Southern, 1996). In addition, some manifestations of brucellosis resemble the symptoms associated with infection by *B. henselae* or *B. quintana*. For instance, brucellosis is a febrile illness characterised by granulomas in the lymph nodes, liver, spleen and bone, and is accompanied by lymphadenopathy, bacteraemia and occasionally endocarditis (Wilfert, 1992).

Identification

Growth Requirements

When bartonellosis is suspected clinically, attempts should be made to isolate the bacterium by plating blood samples prepared by the lysis-centrifugation method (Slater *et al.*, 1990) (Wampole Laboratories, Cranbury, NJ, USA). Alternatively, homogenised tissue can serve as the inoculum for blood plates or endothelial cell monolayers (Koehler *et al.*, 1992). Since all *Bartonella* spp. require a 5% blood- or haemin-supplemented growth medium, blood plates with a complex base medium (trypticase soy, heart infusion or Columbia agar) are used. For optimal growth, the medium pH should be 7.0–7.5. Excluding *B. bacilliformis*, the yield of bartonellae is increased by the lysis of erythrocytes by freezing or osmotic shock, before plating on blood or heated-blood agar. Bartonellae also require high humidity and, with the exception of *B. bacilliformis*, prefer an atmosphere of 5% CO_2.

The slow growth rate of these bacteria can complicate identification and requires that the isolate be cultured for several days (*B. bacilliformis* and *B. elizabethae*) or up to 4 weeks (*B. henselae* and *B. quintana*) to visualise colonies. The combination of a slow growth rate and their fastidious nature can make isolation and axenic culture of bartonellae a challenging undertaking. Indeed, successful isolation of *B. henselae* from endocarditis has only recently been achieved (Drancourt *et al.*, 1996a). After isolation and pure culture, identification of the suspected *Bartonella* can be accomplished by the techniques to be described below.

Colonial Morphology

Bartonella colonies are typically small (1–3 mm diameter), round and lenticular, and range from translucent to opaque (white or cream) in colour. With repeated passage, colonies may display a dry-to-mucoid phase variation. Colonies obtained from low-passage isolates typically adhere to, and pit, the medium, but this phenotype disappears with repeated passage. Colonies of *B. bacilliformis* freely interconvert between a small, translucent round colony (T1) to a larger colony with an irregular edge (T2) (Walker and Winkler, 1981). Colonies from primary isolates can take up to 4 weeks to become visible, but growth is much more rapid (2–5 days) in subsequent passages.

Cell Morphology

Bartonella are Gram-negative, non-acid fast, pleiomorphic rods. They stain poorly with safranin, but stain well with Giemsa or Gimenez stains. Cells are typically coccobacilli or slightly curved rods with a polar enlargement(s). Cells can also be coccoid, beaded, filamentous or in chains. Annular (ring forms) and auto-aggregates can also occur. Cell size is uniformly less than 3 μm in the greatest dimension, with most cells measuring 0.5 μm × 1.0 μm. Electron micrographs of the four pathogenic bartonellae are shown in **Fig. 2**.

Biochemical Tests

Since bartonellae are non-fermentative aerobes with unremarkable physiology (see below), biochemical tests are not usually conclusive for species identification. Exceptions are given in **Table 3** to facilitate the presumptive identification of *Bartonella* isolates. In addition to these characteristics, RapID ANA panels (Innovative Diagnostic Systems, Inc., Norcross, GA, USA) (Daly *et al.*, 1993; Clarridge *et al.*, 1995), DNA hybridisation (Regnery *et al.*, 1991; Welch *et al.*,

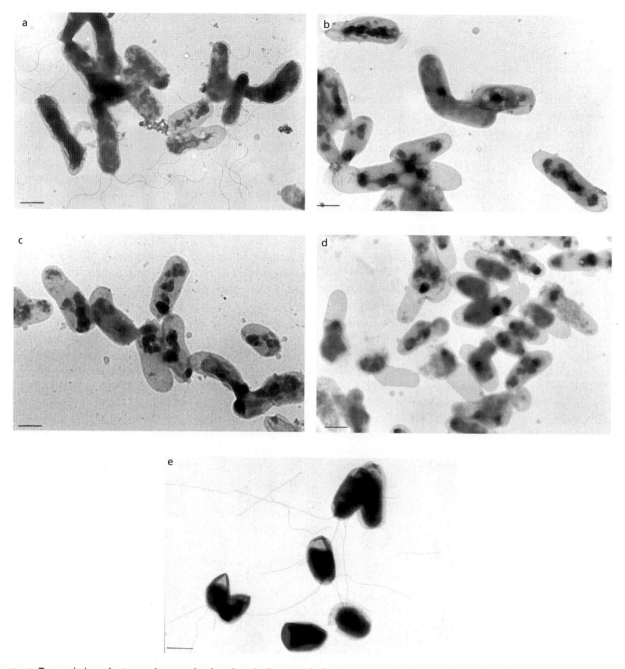

Fig. 2 Transmission electron micrographs showing similar morphologies among the *Bartonella* species that are pathogenic for humans. Bacteria were grown by standard methods and stained with uranyl acetate. (a) *B. bacilliformis*; (b) *B. elizabethae*; (c) *B. henselae*; (d) *B. quintana* and (e) *B. clarridgeiae*. Bar marker 0.5 μm.

1992) and pre-formed peptidases (Welch *et al.*, 1992) have also been used with some success.

Polymerase Chain Reaction

The polymerase chain reaction (PCR) is a sensitive and specific tool for the identification of 'non-culturable' bartonellae in human samples, or to confirm presumptive identification of isolates. PCR has frequently been used to amplify all or part of the 16S rDNA gene from suspected bartonellae. The 16S rDNA PCR product is subsequently sequenced, and the data compared with known 16S rDNA sequences in GenBank to confirm the presumptive identification

Table 3 Differential characteristics of pathogenic *Bartonella* species using standard tests

Species	Flagella	Temperature (°C)[a]	Hemolysis[b]	Phe[c]	Trypsin[c]	Oxidase
B. bacilliformis	+	28	+	−	−	−
B. clarridgeiae	+	37	−	+	+	−
B. elizabethae	−	35	+	+	+	−
B. henselae	−	37	−	+	+	−
B. quintana	−	37	−	+	−	V[d]

[a] Optimal temperature for growth.
[b] Incomplete β-hemolysis on blood plates after 4 days' growth.
[c] β-Naphthylamide-conjugated substrate for aminopeptidase activity.
[d] Variable positive (may be weak or absent).

(Relman *et al.*, 1990; Koehler *et al.*, 1992; Hadfield *et al.*, 1993). To streamline identification, PCR strategies have also been devised from *Bartonella* genus-specific or species-specific targets. These include the 16S rDNA gene (Dauga *et al.*, 1996), the citrate synthase gene (*gltA*) (Regnery *et al.*, 1991; Birtles and Raoult, 1996), the *ftsZ* gene (Kelly *et al.*, 1998), repetitive extragenic palindromic (REP) or enterobacterial repetitive intergenic consensus (ERIC) DNA sequences (Clarridge *et al.*, 1995; Rodriguez-Barradas *et al.*, 1995b) and the 16S–23S rDNA intergenic spacer (Minnick and Barbian, 1997).

Restriction-fragment Length Polymorphism

Restriction-fragment length polymorphism (RFLP) has been used to identify bartonellae by employing rare-cutting restriction endonucleases together with genomic DNA (Slater *et al.*, 1990; Maurin *et al.*, 1994; Roux and Raoult, 1995), with PCR fragments of *gltA* (Norman *et al.*, 1995), or PCR products containing the 16S-23S rDNA *ITS* (Matar *et al.*, 1993; Roux and Raoult, 1995; Bergmans *et al.*, 1996).

Cellular Fatty Acid Analysis

Cellular fatty acid (CFA) analysis has been used to identify bartonellae to the genus level. The unusual fatty acid composition of bartonellae when compared with other bacteria makes this technique particularly useful. The predominant fatty acids of all bartonellae include *cis*-11-octadecanoate $(C_{18:1\omega7c})$ (about 54%) and hexadecanoate $(C_{16:0})$ (about 20%). *B. elizabethae*, *B. henselae*, *B. clarridgeiae* and *B. quintana* contain considerable amounts of octadecanoate $(C_{18:0})$ (about 23%), whereas *B. bacilliformis* contains very little (about 2%). *B. bacilliformis* also contains an unusually high amount (about 20%) of *cis*-11-hexadecanoate $(C_{16:1\omega7c})$ in comparison with most *Bartonella* species

(<1%) (Westfall *et al.*, 1984; Slater *et al.*, 1990; Daly *et al.*, 1993; Clarridge *et al.*, 1995; Kordick *et al.*, 1995).

Structure

Cell

Bartonella cells walls are similar to those of other Gram-negative bacteria (Kreier *et al.*, 1991). The cellular fatty acids are unusual in composition, with over 50% being *cis*-11-octadecanoate $(C_{18:1\omega7c})$ (see above). Motility has been observed in two pathogenic species (**Table 3**), and is conferred by unipolar lophotrichous flagella (*B. bacilliformis* and *B. clarridgeiae*) (Scherer *et al.*, 1993; Clarridge *et al.*, 1995) or type IV pili (*B. henselae*) (Batterman *et al.*, 1995). Capsules or spore-like structures are not present in any species. Spheroplasts of *Bartonella* can be made with lysozyme, and outer membrane fractions can be prepared from these by cell lysis and collection by sucrose step-gradient centrifugation. The outer membrane of *B. bacilliformis* contains 14 outer membrane proteins (OMPs) ranging in molecular mass from 11.2 to 75.3 kDa when analysed by SDS-PAGE (Minnick, 1994). Further analysis of outer membrane fractions by two-dimensional SDS-PAGE suggests that the actual number of OMPs is closer to 50 (M.F. Minnick, unpublished). Nine surface proteins have been identified in *B. henselae* (Burgess and Anderson, 1998). The flagellin subunit (42 kDa) and a phage coat protein (31.5 kDa) are the major OMPs of *B. bacilliformis*. The outer membrane of *B. bacilliformis* contains a uniform lipopolysaccharide (LPS) molecule of approximately 5 kDa on SDS-PAGE (Knobloch *et al.*, 1988; Minnick, 1994). The absence of multiple LPS bands suggests that only subtle differences exist in the O side-chains of *B. bacilliformis*. Early work on purified LPS from *B. quintana* showed that it contained 2-keto-3-deoxyoctonate and heptose, and that it is lethal for the chick embryo, fixes complement and is positive in *Limulus* amoebocyte lysate tests (Hollingdale *et al.*, 1980).

Genome

The G + C content for *Bartonella* spp. is 40 ± 1 mol%, with values ranging from 38.5 mol% for *B. quintana* (Tyeryar *et al.*, 1973) to 41.1 mol% for *B. vinsonii* (Daly *et al.*, 1993). In addition to a highly conserved G + C content, DNA hybridisation indicates a high degree of sequence homology (32–67% relatedness at 55°C) between the genomes of *Bartonella* species (Welch *et al.*, 1992; Daly *et al.*, 1993). The genome sizes of *B. bacilliformis* and *B. quintana* are approximately 1.6 and 1.5 million base pairs, respectively (Myers *et al.*, 1979; Krueger *et al.*, 1995). Thus, the typical *Bartonella* genome is less than half the size of that of *E. coli*. An extrachromosomal, linear DNA band of 14 kbp is frequently observed in genomic DNA preparations from *B. henselae* and *B. bacilliformis*. It is heterogeneous in nature and derived from a putative defective bacteriophage (Anderson *et al.*, 1994). Plasmids have not been described from any bartonellae.

Gene Structure

Genes in bartonellae are biased for A or T in the second and third positions of their codons, because of their modest G + C content. For example, codon usage analysis of the 2275-bp *gyrB* gene (gyrase B subunit) of *B. bacilliformis* (Battisti *et al.*, 1998) and the 1512-bp *htrA* gene (heat-shock antigen) of *B. henselae* (Anderson *et al.*, 1994), shows an A/T of 60% in the second position and 77% in the third position of their respective codons. A clear bias is not, however, observed in the first position of the codons. Although promoters have not so far been fully characterised from a *Bartonella* species, all genes characterised are preceded by putative -35 and -10 hexamers with homology to the *E. coli* consensus promoter (McLure, 1985). Similarly, these genes are immediately preceded by a ribosomal-binding site similar to that of *E. coli* (Gold *et al.*, 1981). Indeed, the recent complementation of *E. coli* mutants with a cloned *ppA* gene (inorganic pyrophosphatase) and *gyrB* (gyrase B) from *B. bacilliformis* suggests that there is conservation of promoter sequence and function between the two bacteria (Mitchell and Minnick, 1997b; Battisti *et al.*, 1998). The majority of the characterised *Bartonella* ORFs end with a TAA stop codon, and putative transcriptional terminators have been observed downstream of the *ialA*, *ialB*, *ctpA* and *ppA* genes of *B. bacilliformis* (Mitchell and Minnick, 1995, 1997a, b).

Physiology

Haemotrophy is the most striking physiological aspect of *Bartonella*. Parasitism of erythrocytes is very unusual for bacteria, but is seen in the species from two other genera, *Anaplasma* and *Haemobartonella* (Kreier and Ristic, 1981). Haemotrophy undoubtedly fulfils the growth requirement for blood or haemin by all bartonellae. Haem uptake is used by several pathogenic bacteria to acquire iron and porphyrin (Reidl and Mekalanos, 1996). Work with *B. quintana* shows that a high concentration of haem (20–40 µg/mL), but not protoporphyrin, is essential for growth. In addition, serum is not a required growth factor (Myers *et al.*, 1969). The molecular basis for haemin uptake by bartonellae remains to be determined.

Apart from haemotrophy, the physiology of bartonellae is not particularly exciting. One potential problem with standard biochemical tests is that they do not include haemin for *Bartonella* growth, and therefore test results must be judged cautiously. If haemin is added to physiological test media, test results for acid production from carbohydrates (lactose, maltose, saccharose), hippurate hydrolysis, pyrazinamidase and Voges-Proskauer can be used to differentiate *B. henselae* from *B. quintana* (Drancourt and Raoult, 1993). With standard tests, however, bartonellae are strictly aerobic and do not utilise carbohydrates (no acid or gas) by preformed or de-novo enzymes.

The oxidase test is usually negative for bartonellae, but variable and weak positive reactions have been reported for *B. quintana* and *B. vinsonii* (Daly *et al.*, 1993; Kordick *et al.*, 1996). Tests for catalase, indole production, nitrate reduction and urease activity are all negative (Birtles *et al.*, 1995; Clarridge *et al.*, 1995). In addition, tests for hippurate hydrolysis, alkaline phosphatase, tetrathionate reductase, pyrazinamidase, tributyrin, *o*-nitrophenyl-β-D-galactoside, esculin hydrolysis and arginine dihydrolase are all negative (Birtles *et al.*, 1995). Voges-Proskauer tests are negative for all species pathogenic for humans, but positive for bartonellae previously designated as *Grahamella* species (**Table 2**; Birtles *et al.*, 1995). Aminopeptidase hydrolysis of a variety of amino acids and peptides has been observed in the pathogenic *Bartonella* species, and for phenylalanine or trypsin, and may be useful for differentiation of some species (**Table 3**) (Birtles *et al.*, 1995; Clarridge *et al.*, 1995; Kordick *et al.*, 1995). Utilisation of succinate has been demonstrated for *B. quintana* (Weiss and Moulder, 1984).

Bartonellae are slow-growing bacteria with generation times of approximately 6 hours (5.8–6.7 hours) under optimal conditions *in vitro* for *B. quintana*

(Weiss and Dasch, 1982) and 6–8 hours for *B. bacilliformis* (Benson *et al.*, 1986). Because of long generation times, maximum colony size is reached only after several days' growth on plates. They can also be cultivated in embryonated chicken eggs and tissue culture.

Products

Only three extracellular products from bartonellae have been described, including deformation factor (deformin), a defective bacteriophage and haemolysin. Very few enzymes have been characterised and exotoxins have not been described for any *Bartonella* species. A putative ABC transporter of *B. bacilliformis* has been cloned and sequenced (*txpA*; GenBank accession no. U68242); it shares homology with the ABC glucan exporter of *Agrobacterium tumefaciens*. The TxpA protein may be involved in export from, or import into, bartonellae (for a review of ABC transporters see Fath and Kolter, 1993).

Enzymes

Five putative *Bartonella* enzymes have been deduced from nucleotide sequence. These include several citrate synthase species (GltA) (Birtles and Raoult, 1996), FtsZ (Padmalayam *et al.*, 1997), GroEL (Haake *et al.*, 1997), alanyl tRNA synthetase (AlaS), leucyl tRNA synthetase (LeuS) and orotidine monophosphate decarboxylase (PyrF). Only four *Bartonella* enzymes have been characterised at the nucleotide sequence and protein level. An autolytic C-terminal protease (CtpA) (Mitchell and Minnick, 1997a), the invasion-associated locus A protein (IalA) – a nudix hydrolase (Cartwright *et al.*, 1999), the gyrase B subunit (GyrB) (Battisti *et al.*, 1998), and inorganic pyrophosphatase (PPase) (Mitchell and Minnick, 1997b). In keeping with PPases from other prokaryotes, the *B. bacilliformis* enzyme displays maximal activity at pH 8.0 and is high thermostable in the presence of Mg^{2+} (highest activity at 55°C).

Deformation Factor (Deformin)

B. bacilliformis produces an extracellular protein called deformin that can independently generate indentations and trenches in erythrocyte membranes. The pits and trenches produced with purified deformin are morphologically similar to those observed in infected cells (Benson *et al.*, 1986). The protein is actively secreted during bacterial growth and is a 130-kDa homodimer in its native state. Deformin is sensitive to heat (70–80°C) and proteases. Its activity is enhanced by pre-treatment of the erythrocytes with trypsin or neuraminidase, and abrogated if erythrocytes are pre-treated with phospholipase D. Deformin-induced invaginations are reversible by vanadate, DLPC, or if intracellular Ca^{2+} levels are increased with ionophores (Mernaugh and Ihler, 1992; Xu *et al.*, 1995). Whether other bartonellae produce deformin, or if deformin is active against other cell types, is not known. The gene for deformin has not been characterised.

Defective Bacteriophage

Uniform bacteriophage-like particles containing icosahedral heads (~ 40 nm diameter) and a filamentous sheath-like tail structure (16 nm in length) were first observed in *B. bacilliformis* (Umemori *et al.*, 1992). The authors suggested that the bacteriophage was temperate and that the *Bartonella* strains (KC583 and KC584) were lysogenic. In addition, observed decreases in *B. bacilliformis* yields with increased passage were thought to be due to bacteriophage infection. Subsequent work showed that phage-like particles can also be obtained from *B. henselae*, but not *B. elizabethae* or *B. quintana*. Phage from *B. henselae* and *B. bacilliformis* consist of a heterogeneous, linear DNA of 14 kbp and at least three proteins (Anderson *et al.*, 1994). Two of the phage proteins, Pap31 and PapA, have been characterised and are 31 and 36 kDa, respectively (Anderson *et al.*, 1997; Bowers *et al.*, 1998).

Haemolysins

Haemolysins are produced by *B. bacilliformis* and *B. elizabethae*. Until recently, all literature on *B. bacilliformis* claimed that the bacterium was non-haemolytic. However, if *B. bacilliformis* cultures are plated on thin blood agar plates and cultured for at least 4 days, incomplete β-haemolysis can be observed. The haemolysin is extracellular and can pass through 0.2 μm filters into the underlying medium to produce a zone of haemolysis which is limited to the outline of the colony (Minnick, 1997). Likewise, an incomplete β haemolysin with delayed appearance (4 days' growth) has been reported for *B. elizabethae* (Daly *et al.*, 1993). The molecular nature of these haemolysins is not known.

Angiogenic Factor

Pathogenic bartonellae produce a protein, angiogenic factor, which stimulates angiogenesis, and probably facilitates the generation of vascular lesions during bartonellosis. Fractions from *B. bacilliformis* containing this protein induce vascularisation *in vivo* and can induce human umbilical vein endothelial cells (HUVECs) to proliferate *in vitro* (Garcia *et al.*, 1990). Live *B. bacilliformis* (Garcia *et al.*, 1992), *B. henselae* or *B. quintana* (Conley *et al.*, 1994) stimulate endothelial cell proliferation when co-cultured with HUVECs. The latter also migrate towards bartonellae in co-cultures (Conley *et al.*, 1994). Angiogenic protein is actively mitogenic for endothelial cells, but fibroblasts, smooth muscle and mesenchyme cells are unaffected (Garcia *et al.*, 1990; Conley *et al.*, 1994). The molecular nature of the protein and its mechanism of action are unknown.

Pathogenesis

Bartonellae are adapted to infect a variety of invertebrate and vertebrate hosts and several cell types, including erythrocytes, epithelial cells and endothelial cells. Thus, these pathogens provide a splendid opportunity to investigate host-parasite interactions at the cellular and molecular level. Research on molecular pathogenesis has been done mainly with *B. bacilliformis* as a model system for the genus.

Host Colonisation

Bartonellae are inoculated directly into the bloodstream by the bite or scratch of a contaminated arthropod or cat. Subsequent colonisation of the host lymphatics and circulatory system is probably enhanced by blood flow and, if possible, bacterial motility. *B. bacilliformis* and *B. clarridgeiae* (previously 94-F40), a novel *Bartonella* species isolated from a veterinarian who had been bitten by a cat (Kordick *et al.*, 1997), are highly motile by peritrichous flagella. Twitching motility has been observed in low-passage isolates of *B. henselae* and is conferred by type IV bundle-forming pili (Batterman *et al.*, 1995). Putative-type IV pili have also been observed in *B. bacilliformis*, but probably play a subordinate role to flagella in motility (Minnick *et al.*, 1996). *B. bacilliformis* flagella have been biochemically characterised and consist of multiple 42-kDa flagellin subunits that are highly resistant to protease treatment (Scherer *et al.*, 1993). The N-terminus of flagellin (Scherer *et al.*, 1993), and the nucleotide sequence of the *fla* gene (GenBank no. L20677), have been characterised. The *fla* locus was recently used for the first time to demonstrate site-directed mutagenesis of a *Bartonella* spp. (Battisti and Minnick, 1999).

Adherence

B. bacilliformis and *B. henselae* show a correlation between colony morphology and adherence to host cells in assays *in vitro*. The adherence of *B. bacilliformis* obtained from colony type T2 (see above) was nearly twice that of bacteria derived from colony type T1 (Walker and Winkler, 1981). Similarly, high-passage phase variants of *B. henselae* from mucoid colonies showed a decrease in adherence to Hep-2 cells when compared with low-passage bacteria obtained from dry and embedded colonies (Batterman *et al.*, 1995). Reduction in *B. henselae* adherence is believed to be a result of repeated-passage phase variation, which causes the loss of expressed type IV bundle-forming pili (BFP) on the surface of the pathogen (Batterman *et al.*, 1995). The recent discovery of a putative type IV BFP on low-passage *B. bacilliformis* suggests that other bartonellae also use these appendages (Minnick *et al.*, 1996).

B. bacilliformis can associate with HUVECs and epithelial cells (Hep-2) with equal binding efficiency, and the majority of adherence occurs within the first 60 minutes of a 3-hour incubation period (Hill *et al.*, 1992). This is in contrast to erythrocyte adherence, where maximal complexing occurs at approximately 6 hours after incubation (Benson *et al.*, 1986). Disparate-binding kinetics suggests that unique receptor–ligand interactions occur between bartonellae and each host cell type. Adhesion to erythrocytes can be inhibited by reagents that inactivate proton-motive force (*N*-ethylmaleimide) or respiration (KCN), but it is unaffected by pre-treating the pathogen with glycolysis inhibitors (NaF) or proton-motive force. These observations suggest that adhesion is energy-dependent (Walker and Winkler, 1981). It is believed, however, that the red blood cell is passive during the process and cannot contribute to energy-dependent adhesion. The identity of the host cell receptor is not known. However, pre-treating the erythrocyte with pronase or subtilisin enhances adhesion, whereas α or β glucosidase treatment decreases adhesion. Presumably, protease treatment of red cells exposes a glycolipid receptor that can subsequently be destroyed by glucosidase. It is also known that human erythrocytes are preferentially bound by *Bartonella* as compared with red blood cells from rabbits or sheep, suggesting that human red cells possess a more appropriate receptor or greater receptor density/accessibility than other erythrocyte types (Walker and Winkler, 1981). Given the ability of *B. henselae* to infect cats and humans, it would be interesting to compare its binding efficiency with cat and human erythrocytes. More recently, it has been shown that

B. bacilliformis and *B. henselae* recognise five or six proteins, respectively, from human erythrocyte membranes (Iwaki-Egawa and Ihler, 1997).

Flagella may also serve as adhesins. A polar tuft of fibrous projections on *B. bacilliformis* was observed to make contact with the erythrocyte membrane during adhesion (Walker and Winkler, 1981). The fibrous tuft closely resembles the peritrichous flagella of the bacterium. It is also known that non-motile bacteria bind poorly to erythrocytes, suggesting that an adhesin (flagella?) is missing in non-motile bacteria, or that non-motile mutants have fewer bacteria–erythrocyte collisions (Benson *et al.*, 1986). If *B. bacilliformis* is treated with rabbit antiflagellin antiserum, bacterial association with red cells is significantly reduced as compared with pre-immune rabbit serum controls (Scherer *et al.*, 1993). This suggests that flagella may possess adhesive qualities and/or they increase the number of bacteria–host cell collisions.

The endothelial cell receptor for bartonellae has not been characterised. The observation that binding of *B. bacilliformis* to epithelial cells and endothelial cells displays a similar degree of efficiency suggests that the apparent predilection of the pathogen for endothelial cells may actually be due to tissue site (e.g. circulatory system), rather than receptor-mediated, constraints (McGinnis-Hill *et al.*, 1992). In addition, binding data also suggest that both cell types contain a suitable, if not the same, receptor(s). The role of surface-exposed *Bartonella* polypeptides in adhesion is largely unknown, but recent work with *B. henselae* has identified five biotinylated proteins (28–58 kDa) capable of binding to intact HUVECs. Of these, a 43-kDa polypeptide was identified as the major adhesin of the pathogen. It is significant that the 43-kDa protein was also recognised by reciprocal probing with biotinylated HUVEC surface proteins (Burgess and Anderson, 1998). The exact nature of the 43-kDa adhesin is currently under investigation.

Invasion

Invasion of erythrocytes has been documented for *B. bacilliformis* (Benson *et al.*, 1986), *B. henselae* (Kordick and Breitschwerdt, 1995; Mehock *et al.*, 1998) and bartonellae previously classified as *Grahamella* species (Table 2) (Birtles *et al.*, 1995). Host cell invasion by *B. quintana* and *B. elizabethae* has not been demonstrated, although both require blood or haemin for growth. In fact, *B. quintana* is believed to associate epicellularly with erythrocytes (Merrell *et al.*, 1978). In addition to red blood cells, invasion of other host cell types (epithelial and endothelial cells) has been demonstrated for *B. henselae* (Batterman *et al.*,

1995; Dehio *et al.*, 1997), *B. bacilliformis* (Garcia *et al.*, 1992; McGinnis-Hill *et al.*, 1992) and *B. quintana* (Brouqui and Raoult, 1996).

Since a variety of cells can serve as hosts for bartonellae, molecular mechanisms for entry probably vary depending upon the type of cell being invaded. Virulence studies with *B. bacilliformis*, together with cultured epithelial or endothelial monolayers, demonstrate that host cells can be induced by *Bartonella* to re-configure the cytoskeleton, thereby enhancing bacterial uptake. Internalisation is significantly reduced ($\sim 30\%$ of controls) when actin filament formation is inhibited with cytochalasin D, or when bacteria are pre-treated with anti-*Bartonella* antiserum (McGinnis-Hill *et al.*, 1992). These inhibition studies suggest that the bacterium is not passive during internalisation, and that the process involves a surface-borne molecule(s) accessible to antibody. Erythrocyte invasion by bartonellae is very different, because red cells are necessarily passive (non-endocytotic) and cannot contribute to bacterial uptake. *B. henselae* enters endothelial cells by a novel structure termed an invasome (Dehio *et al.*, 1997). In models of infection *in vitro*, *B. henselae* cells are contacted and moved rearward to form an aggregate on the leading lamella of the endothelial cell. The 'clumps' of bacteria are subsequently engulfed by membrane protrusions rich in cortical F-actin, ICAM-1 and phosphotyrosine. By chemical inhibition studies, it has been shown that invasome activity is actin-dependent and microtubule-independent. Although most clinical strains of *B. henselae* were internalised by the formation of the invasome, a natural mutant of *B. henselae* was internalised by an alternative process and was ultimately located in a perinuclear phagosome. The authors suggested that invasome-mediated internalisation may somehow interfere with the perinuclear phagosome formation. It is also possible that the mutant lacked the appropriate surface ligand to trigger formation of the structure.

B. quintana invades epithelial cells in classical *in vitro* experiments with Hep-2 cells (Vinson and Fuller, 1961). More recently it was shown that the pathogen is also invasive for human endothelial cells *in vitro* and, more importantly, in cardiac valve tissue from endocarditis patients (Brouqui and Raoult, 1996). Studies *in vitro* with *B. quintana* and human endothelial cells show that bacteria can become internalised within just 1 minute of co-incubation. Concurrently, host cells show ruffling and the bacterial cell wall becomes modified to produce surface appendages (20–40 nm wide by up to 500 nm length). The appendages are apparently lost after adherence or internalisation of the pathogen, and are similar to those observed in *Salmonella typhimurium*. The authors suggest that the appendage is mediating or directing host cell adherence and endocytosis. After host cell uptake, *B. quintana* multiplies in a vacuole, culminating in the formation of morulae resembling those seen during infection by ehrlichiae or chlamydiae. Older *B. quintana* and endothelial cell co-cultures showed that morulae contain bacteria and vesicle-like blebs presumably derived from the bacterial membrane. However, membrane blebs were not observed in cardiac tissue samples. It is interesting to note that *B. bacilliformis* (M.F. Minnick, unpublished data), *B. henselae* and *B. quintana* (Brouqui and Raoult, 1996) all produce blebs during growth *in vitro*, but their potential role in pathogenesis remains a mystery.

Three virulence determinants are implicated in *Bartonella* invasiveness, including deformin, flagella and proteins encoded by the invasion-associated locus of *B. bacilliformis*. Deformin-induced invaginations in the plasma membrane of erythrocytes (see above) probably produce entry portals for red cell colonisation. However, even with deformin, bacteria cannot enter the red cell cytosol unless they are motile (Mernaugh and Ihler, 1992). Perhaps *B. bacilliformis* swims into the trenches produced by deformin. If *B. bacilliformis* cells are treated with monospecific antibodies against the flagellin subunit, invasiveness of human erythrocytes is almost abrogated in virulence assays *in vitro* (Scherer *et al.*, 1993).

B. bacilliformis possesses an invasion-associated locus that contains two genes, *ialAB* (Mitchell and Minnick, 1995). The locus is approximately 1500 bp and contains two ORFs (*ialA* and *ialB*) that confer an invasive phenotype on minimally invasive strains of *E. coli* (strains HB101 and DH5α) when combined with human red cells *in vitro*. Both genes are required to produce the invasiveness phenotype. The *ialA* gene codes for a (di)nucleoside polyphosphate hydrolase that may function to reduce stress-induced

dinucleotide 'alarmones' during host cell invasion and therefore enhance pathogen survival (Cartwright *et al.*, 1999). The *ialB* gene codes for a protein with similar molecular mass and approximately 60% amino acid sequence similarity to the adhesion and invasion locus (Ail) protein of *Yersinia enterocolitica* (Miller *et al.*, 1990) and the resistance to complement killing (Rck) protein of *Salmonella typhimurium* (Heffernan *et al.*, 1994) (**Fig. 3**). Both Ail and Rck are implicated in host cell attachment, invasion and serum resistance. Whether the *Bartonella* invasion-associated locus mediates or facilitates the invasion of other cell types is unknown.

Characterisation of genes that flank the *ialAB* locus suggests that *ialA* and *ialB* are components of a larger pathogenicity gene cluster. A gene encoding a C-terminal protease, *ctpA*, lies upstream of *ialA* and *ialB* (**Fig. 4**) (Mitchell and Minnick, 1997a). The encoded protease may be involved in *B. bacilliformis* virulence, especially because the *prc* C-terminal protease of *S. typhimurium* is involved in intracellular survival. The *prc* protease is thought to degrade abnormally folded stress-response proteins generated in the intracellular environment of the macrophage, thereby enhancing survival of *S. typhimurium* (Baumler *et al.*, 1994). While no evidence was found for CtpA involvement in invasion of human erythrocytes (Mitchell and Minnick, 1997a), the possible role that CtpA plays in *B. bacilliformis* intracellular survival is unknown.

Immediately upstream of the *ctpA* gene is a 1200 bp ORF termed *filA* (**Fig. 3**) (M.F. Minnick and S.J. Mitchell, unpublished data). The characteristics of the predicted FilA polypeptide include: (1) a typical secretory signal sequence that follows the $(-3, -1)$ rule; (2) a 60% α-helical secondary structure; (3) a C-terminal hydrophobic domain that could anchor the protein in a membrane; (4) leucine-rich composition

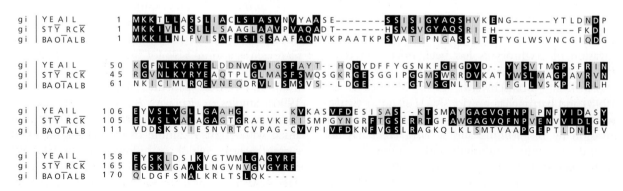

Fig. 3 Multiple sequence alignment of the IalB protein of *B. bacilliformis* (BAOIALB) with the Rck protein of *Salmonella typhimurium* (STY_RCK) and the Ail protein of *Yersinia enterocolitica* (YE_AIL) (GenBank accession nos. M29945, M76130 and L25276, respectively). Identical amino acids are indicated in black, conserved amino acids in grey and introduced gaps by dashed lines.

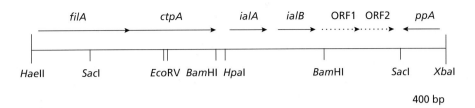

Fig. 4 Linkage and partial restriction map of the invasion-associated locus of *B. bacilliformis* plus flanking sequences characterised to date. Relative positions of the genes are indicated by arrows. Gene abbreviations: *filA*, filament A gene; *ctpA*, C-terminal processing protease; *ialA* and *B*, invasion-associated locus genes A and B, respectively; ORFs 1 and 2, open reading frames 1 and 2, respectively; and *ppa*, inorganic pyrophosphatase. (GenBank accession nos. U73652, L37094, L25276 and L46591, respectively).

(12%) with numerous leucine repeats, a characteristic found in proteins that engage in protein–protein interactions (Kobe and Deisenhofer, 1995), and (5) amino acid similarity to a variety of filamentous proteins, including smooth muscle myosin and the M1 protein of *Streptococcus pyogenes*, a virulence determinant involved in adhesion and invasion (Fischetti, 1989; Kehoe, 1994). The similarity of FilA to M1 protein, the potential surface location and filamentous nature, plus the possibility that FilA provides for protein–protein interaction, suggests that the protein is a virulence determinant. The 3′ end of the virulence gene cluster is presumably demarcated by an inorganic pyrophosphatase (PPase) gene (*ppa*) located 1022 bp downstream of the locus (Mitchell and Minnick, 1997b). PPase is not involved in virulence, since the enzyme plays an essential metabolic housekeeping role in many organisms.

DNA hybridisation data suggest that *B. quintana* and *B. henselae* possess homologues of *ialA*, *ialB* and *ctpA*, while *B. vinsonii*, an agent that does not infect humans, contains *ialA* and *ctpA* but may not contain an *ialB* homologue. *B. elizabethae* hybridises poorly to any of the probes for the three genes (Mitchell and Minnick, 1997a). The homologue of the *ialAB* locus depicted in **Fig. 4** has been cloned from *B. henselae* and was found to confer an increase in invasive phenotype on *E. coli* to approximately 100-fold over controls, and it possesses identical gene linkage with approximately 70–85% sequence identity to the *B. bacilliformis* locus (Murakawa *et al.*, 1997). Likewise, invasiveness by *B. tribocorum* for rodent erythrocytes required the activity of IalB (Gille *et al.*, 1999). Collectively, these data suggest that the invasion locus is conserved in most pathogenic *Bartonella* species.

Immunity

The general course of bartonellosis begins with transient or persistent bacteraemia that may involve erythrocyte parasitism, depending upon bacterial species.

The pathogen subsequently infects the vascular bed in a variety of tissues, where vascular lesions are produced. Tissue involvement can be chronic, and presumably the pathogen is shed from the affected tissues into the bloodstream. Lymphadenopathy is frequently present in bartonellosis. The immune status of the host is a risk factor for infection. Typically, immunocompromised patients are at higher risk of opportunistic infection by bartonellosis than healthy individuals. As a rule of thumb, trench fever, CSD, endocarditis and Oroya fever take place regardless of the immune status of the patient, while bacillary angiomatosis (BA), bacillary peliosis and bactraemia syndrome occur mainly in immunocompromised individuals. The symptoms of bartonellosis are generally more pronounced in immunodeficient patients. In BA, many patients display a marked $CD4^+$ lymphocyte leukocytopenia, with counts of less than 100 (LeBoit *et al.*, 1989; Koehler and Tappero, 1993). The human immune response against *Bartonella* is poorly characterised, and for some species (*B. elizabethae* and *B. clarridgeiae*) has never been studied. Since bartonellae are found in a variety of body fluids and cells, involvement of both humoral and cellular arms of the immune system needs to be invoked for resolution of infection.

B. bacilliformis

Life-long humoral immunity supposedly results from an active infection with *B. bacilliformis* (Ricketts, 1949; Weinman, 1965). Population seropositivity (mainly IgM isotype) in endemic areas of Peru can reach 60%, and many of these individuals are healthy (Knobloch *et al.*, 1985). However, in spite of circulating antibody, many individuals who are asymptomatic or post-eruptive for verruga peruana are often blood culture-positive for *B. bacilliformis* (Howe, 1943), and carrier rates can reach 15% in some areas of Peru (Herrer, 1953). Immunosuppression is characteristic, and probably results from the acute anaemia and leukocytosis. Although a variety of proteins from *B. bacilliformis* are immunogenic (Knobloch, 1988),

2128 ANIMAL AND ECTOPARASITIC SOURCE INFECTIONS

the most prominent antigen include a GroEL homo-logue (BB65) (Knobloch and Schreiber, 1990), flagellin protein (42 kDa), a phage coat protein (31.5 kDa) and an uncharacterised protein of 45 kDa (Minnick, 1994).

B. quintana

Although during *B. quintana* infections symptoms are highly variable, patients generally produce low-titre, complement-fixing antibody during the early phase of convalescence (Varela *et al.*, 1969). In spite of the antibodies, however, patients frequently display per-sistent and protracted bacteraemia (Varela *et al.*, 1969; Koehler *et al.*, 1992; Maurin *et al.*, 1994). In addition, *B. quintana* is markedly resistant to complement kill-ing by the classical or alternative pathways (Myers and Wisseman, 1978). Infection is frequently accompanied by leukocytosis (Vinson *et al.*, 1969), possibly enhancing bacteraemia.

B. henselae

Patients infected with *B. henselae* are generally sero-positive by indirect fluorescence antibody tests (Regnery *et al.*, 1992a). Antibodies apparently opso-nise the pathogen and enhance production of oxygen radicals after phagocytosis, but do not activate com-plement by the classical pathway, which proceeds mainly by the alternative pathway (Rodriguez-Barradas *et al.*, 1995a). Persistent bacteraemia with *B. henselae* is common in humans and cats (Koehler *et al.*, 1994), and respective antibody titres do not necessarily correlate with protection. An immunodom-inant protein of approximately 83-kDa is recognised by immune serum from cats and humans (K. Karem, personal communication). Challenge by the live agent, but not killed bacteria, generates protective immunity in cats (K. Karem, personal communication).

Diagnosis

Since bartonellae are difficult to culture from clinical isolates, indirect identification of the pathogen is usually based on the individual patient history and symptoms. If culture of the agent is possible, identifi-cation of the bacterium can be made using the tech-niques described above.

Oroya Fever

Patients give a history of travel or residence in South America and of having been bitten by a sandfly.

Diagnostic symptoms of the primary stage include a dramatically low erythrocyte count ($\sim 500\,000\,\mathrm{mm}^{-3}$) (Hurtado *et al.*, 1938; Reynafarje and Ramos, 1961; Kreier and Ristic, 1981) and numerous infected eryth-rocytes when blood smears are stained with eosin/thiazine stain (Knobloch *et al.*, 1985). The secondary (tissue) phase presents with cutaneous angiomatous lesions (verruga peruana) that contain bacteria when sectioned and stained with Warthin-Starry silver stain.

Trench Fever

Patients will probably give a history of infestation with body louse (*Pediculus humanus*), and chronic alcohol abuse, homelessness and non-Caucasian descent are risk factors (Jackson and Spach, 1996). Diagnostic characteristics include pain in the bones (especially the tibia), fever and persistent bacteraemia for several weeks to months after the resolution of symptoms (Vinson *et al.*, 1969). Seroconversion to *B. quintana* can be tested (Regnery *et al.*, 1992a).

Cat-scratch Disease

Patients are generally immunocompetent children or young adults. There is a history of cat bite or scratch resulting in a papule or pustule-like lesion. Diagnostic characteristics include a positive CSD-skin test, an unexplainable lymphadenopathy and characteristic lesion histopathology. Diagnosis depends on tests for seroconversion (Regnery *et al.*, 1992a; Zangwill *et al.*, 1993; Patnaik *et al.*, 1992) and a more sensi-tive ELISA system (Barka *et al.*, 1993). A titre of ≥ 64 by immunofluorescence assay is considered positive for *B. henselae* by the Centers for Disease Control, USA (Regnery *et al.*, 1992b).

Bacillary Angiomatosis

Patients are usually immunocompromised. Cutaneous and/or subcutaneous non-granulomatous vascular lesions affect a variety of organs. Bartonellae can be observed in sections of affected tissue with Warthin-Starry silver stain. Diagnostic reagents include an anti-*B. henselae* IgG testing kit (Specialty Laboratories, Santa Monica, CA, USA), tests for seroconversion (Regnery *et al.*, 1992a; Zangwill *et al.*, 1993; Patnaik *et al.*, 1992) and an ELISA system (Barka *et al.*, 1993).

Bacillary Peliosis

Patients are usually immunocompromised. Diagnosis based on symptoms is difficult and may require biopsy. Cystic blood-filled liver lesions that are similar to those of BA, together with bartonellae in sections of affected tissue, are diagnostic. Hepatomegaly and elevated serum levels of liver enzymes are characteristic. Bacillary peliosis should be suspected in patients with BA or bactraemia syndrome. Detection reagents used for BA diagnosis can also be used for bacillary peliosis.

Endocarditis

Bartonellae should be suspected in blood culture-negative cases of endocarditis. Patients usually have predisposing heart conditions or are at high risk for contracting infection with *B. henselae* or *B. quintana* (see below). Vegetative lesions are common in cases of bartonella endocarditis (Daly *et al.*, 1993; Raoult *et al.*, 1996). The difficulty of culture and serological cross-reactivity between chlamydiae, bartonellae and *Coxiella burnetii* can complicate diagnosis (Drancourt *et al.*, 1995; Knobloch *et al.*, 1988; La-Scola and Raoult, 1996). Seropositivity for bartonellae can be tested with the detection reagents described for diagnosis of BA. Molecular techniques involving PCR should also be used (see above).

Bacteraemia Syndrome

Patients are usually immunocompromised and persistent bacteraemia and recurring fevers together with chronic fatigue and malaise are diagnostic. Diagnosis can be made by blood cultures with subsequent identification protocols (see above) or by testing for seropositivity to *B. henselae* or *B. quintana* as described above for BA.

Antimicrobial Therapy

A variety of antibiotics have been successfully used to treat bartonelloses. Antimicrobial therapy varies depending upon *Bartonella* species, and in some cases it is syndrome-dependent. CSD and trench fever are generally self-limiting illnesses that do not require therapy unless the infection is systemic. In contrast, BA can be life threatening and requires prompt antibiotic therapy. The immune state of the patient is also an important consideration for the type and duration of antimicrobial therapy for bartonellosis. For example, BA therapy for healthy patients is approximately 3 weeks, whereas immunocompromised patients may require several weeks to months, or possibly a lifetime, of therapy to resolve the infection (Adal *et al.*, 1994). Relapse of bartonellosis is common even with antimicrobial therapy. Bartonellae are typically sensitive to most antibiotics *in vitro* (Maurin and Raoult, 1996), but are frequently resistant to them during therapy. This may reflect the inability of the chemotherapeutic agent to gain access to the intracellular pathogen. Aminoglycosides are particularly bactericidal for bartonellae *in vivo* (Musso *et al.*, 1995), and are recommended for therapy. Although optimal antibiotic regimens are still unclear, a list of antimicrobials that have successfully been used in treating various manifestations of bartonellosis are given in **Table 4**.

Table 4 Antimicrobial agents used to treat bartonelloses

Syndrome	Antimicrobial agent(s)	Reference(s)
Adenomegaly	Gentamicin	Drancourt *et al.* (1996b)
Bacillary angiomatosis	Erythromycin or doxycycline. Also chloramphenicol, sulphatrimethoprim, aminoglycosides, azithromycin, ciprofloxacin	Bartlett (1996)
	Clarithromycin	Milde *et al.* (1995)
Bacteraemia syndrome	Erythromycin or doxycycline	Adal *et al.* (1994)
Bacillary peliosis	Erythromycin (alternatives as in BA)	Perkocha *et al.* (1990); Bartlett (1996)
Cat-scratch disease [a]	Ciprofloxacin, sulphatrimethoprim or gentamicin	Bartlett (1996)
	Erythromycin or doxycycline	Smith (1997)
Endocarditis	Nafcillin + gentamicin	Daly *et al.* (1993)
Oroya fever	β-Lactams, tetracycline + streptomycin, aminoglycosides, chloramphenicol if co-infected by enterics	Weinman (1965)
Pulmonary nodules	Doxycycline	Caniza *et al.* (1995)
Retinal vasculitis and vitreitis	Ciprofloxacin	Soheilian *et al.* (1996)
Trench fever [a]	Tetracycline or chloramphenicol	Moulder (1974)

[a] Usually self-limiting (antibiotic therapy not required unless systemic).

Natural antibiotic resistance has not been reported for bartonellae, but coumermycin A-resistant mutants have been generated and the *gyrB* gene (for gyrase B) lesions that confer coumerin-resistance have been characterised (Battisti *et al.*, 1998).

Epidemiology and Control

Bartonellosis has been reported in rodents, insectivores, dogs, cats and humans (**Tables 1** and **2**). The pathogen is usually transmitted to mammals by arthropods, and once infected the host may serve as a reservoir if the infection is chronic. Transmission may also be enhanced by persistent infection of the vector. For example, *Bartonella* (previously *Grahamella*) species that are endemic in rodent populations (**Table 2**) also infect the flea, thereby rendering the insect a reservoir (Kreier and Ristic, 1981). In nature bartonelloses are usually sporadic and epidemic, suggesting that infection of the vector or reservoir is cyclical, rather than continuous.

B. bacilliformis

The only known risk factor for *B. bacilliformis* infection is exposure to bites from phlebotamine sandflies (*Lutzomyia verrucarum*) in South America. The bacterium was once thought to be endemic to the high-altitude regions of the Andes, because its vector was restricted to that habitat. However, a recent review reports that related sandflies may serve as vectors for the agent and that numerous cases of Oroya fever have occurred in areas of much lower altitude (Alexander, 1995). Because of this, *B. bacilliformis* may be an emerging infectious agent in South America. The high carrier rate and seropositivity of individuals in endemic areas strongly suggests that these individuals serve as a reservoir for the pathogen. In addition, the sandfly vector may be infected by the bacterium (Hertig, 1942; Kreier and Ristic, 1981). The best control measure is insecticide application to eradicate sandflies. The use of antibiotics has produced a dramatic decline in the number of deaths caused by Oroya fever. No vaccine is currently available.

B. henselae

The evidence suggests that exposure to cats is the most significant risk factor for contracting bartonellosis due to *B. henselae* (Zangwill *et al.*, 1993; Koehler *et al.*, 1994). A very high incidence (89%) of prolonged bacteraemia with this pathogen has been observed in cats belonging to CSD patients (Kordick *et al.*, 1995).

In addition, a high incidence of this agent (9–41%) has been seen in control cats of all ages in the United States and Japan (Koehler *et al.*, 1994; Kordick *et al.*, 1995; Maryuyama *et al.*, 1996). Transmission routes include cat scratches and bites, and possibly by infected fleas (Zangwill *et al.*, 1993; Koehler *et al.*, 1994). Epidemiological evidence suggests that fleas might also serve as a vector (Lucey *et al.*, 1992; Welch *et al.*, 1992). *B. henselae* is currently the only *Bartonella* spp. known to be transmitted to humans by non-arthropod means. Control of the disease includes reduced exposure to cats during immunodepressed states, antibiotic therapy of infected cats and humans, when available vaccination of the cat reservoir, and flea control. A vaccine is not yet available, but is being developed for use in cats.

B. quintana

Trench fever is transmitted to humans by exposure to infected human body lice (*Pediculus humanus*), and the insects can transmit the pathogen for long periods of time (Vinson and Fuller, 1961). Infected lice perennially shed *B. quintana* in their faeces, and humans are subsequently exposed through breaches in the skin, such as bites or scratches, and contact with the vector is increased by conditions of overcrowding and poor hygiene. However, lice have not been clearly implicated in the re-emergence of *B. quintana* in inner cities. Overt risk factors for 'urban trench fever' include homelessness, non-Caucasian descent and alcoholism (Spach *et al.*, 1995; Jackson and Spach, 1996). Although the arthropod vector of urban trench fever is still unclear, it is probably a blood-sucking insect such as a louse, mite or flea. Spread of trench fever has historically been controlled by de-lousing and hygienic measures. Vaccines are not currently available.

B. elizabethae

The epidemiology of *B. elizabethae* is a mystery. The endocarditis patient from whom the species was isolated had no known exposure to small mammals, nor did he have pre-disposing factors, such as cardiac valvular abnormalities, immunosuppression, HIV infection, etc., which could have contributed to bartonellosis (Daly *et al.*, 1993).

B. clarridgeiae

This organism is transmitted to humans by cat bite (Kordick *et al.*, 1997) and it is likely that the pathogen can be transmitted to humans by other forms of cat trauma and ectoparasites such as fleas. The co-incidental infection of house cats with *B. henselae* and *B. clarridgeiae* suggests that co-transmission of these pathogens to humans may also occur (Gurfield *et al.*, 1997).

References

Adal KA, Cockerell CJ, Petri WA (1994) Cat scratch disease, bacillary angiomatosis and other infections due to *Rochalimaea*. *N. Engl. J. Med*. 330: 1509–1515.

Alexander B (1995) A review of bartonellosis in Ecuador and Colombia. *Am. J. Trop. Med. Hyg*. 52: 354–359.

Anderson B, Kelley C, Threlkel R, Edwards K (1993) Detection of *Rochalimaea henselae* in cat-scratch disease skin test antigens. *J. Infect. Dis*. 168: 1034–1036.

Anderson B, Goldsmith C, Johnson A, Padmalayam I, Baumstark B (1994) Bacteriophage-like particle of *Rochalimaea henselae*. *Mol. Microbiol*. 13: 67–73.

Anderson B, Scotchlas D, Jones D, Johnson A, Tzianabos T, Baumstark B (1997) Analysis of 36-kilodalton protein (PapA) associated with the bacteriophage particle of *Bartonella henselae*. *DNA Cell Biol*. 16: 1223–1229.

Angritt P, Tuur SM, Macher AM *et al*. (1988) Epithelioid angiomatosis in HIV infection: neoplasm or cat-scratch disease? *Lancet* 1: 996.

Arias-Stella J, Lieberman PH, Erlandson RA, Arias-Stella J Jr (1986) Histology, immunohistochemistry, and ultrastructure of the verruga in Carrion's disease. *Am. J. Surg. Pathol*. 10: 595–610.

Barka NE, Hadfield T, Patnaik M, Schwartzman WA, Peter JB (1993) EIA for detection of *Rochalimaea henselae*-reactive IgG, IgM and IgA antibodies in patients with suspected cat-scratch disease. *J. Infect. Dis*. 167: 1503–1504.

Bartlett JG (1996) *Pocket Book of Infectious Disease Therapy*. Baltimore: Williams and Wilkins, p. 22.

Barton A (1909) Descripcion de elementos endoglobulares en los enfermos de Fiebre de Verruga. *Cron. Med. Lima* 26: 7.

Batterman HJ, Peek JA, Loutit JS, Falkow S, Tompkins LS (1995) *Bartonella henselae* and *Bartonella quintana* adherence to and entry into cultured human epithelial cells. *Infect. Immun*. 63: 4553–4556.

Battisti JM, Smitherman LS, Samuels DS, Minnick MF (1998) Mutations in *Bartonella bacilliformis gyrB* confer resistance to coumermycin A1. *Antimicrob. Agents Chemother*. 42: 2906–2913.

Battisti JM, Minnick MF (1999) Development of a system for genetic manipulation of *Bartonella bacilliformis*. *Appl. Environ. Microbiol*. 65: 3441–3448.

Baumler AJ, Kusters JG, Stojiljkovic I, Heffron F (1994) *Salmonella typhimurium* loci involved in survival within macrophages. *Infect. Immun*. 62: 1623–1630.

Benson LA, Kar S, McLaughlin G, Ihler GM (1986) Entry of *Bartonella bacilliformis* into erythrocytes. *Infect. Immun*. 54: 347–353.

Berger TG, Tappero JW, Kaymen A, LeBoit PE (1989) Bacillary (epithelioid) angiomatosis and concurrent Kaposi's sarcoma in acquired immune deficiency syndrome. *Arch. Dermatol*. 125: 1543–1547.

Bergmans AM, Groothedde JW, Schellekens JF, van Embden JD, Ossewaarde JM, Schouls LM (1995) Etiology of cat scratch disease: comparison of polymerase chain reaction detection of *Bartonella* (formerly *Rochalimaea*) and *Afipia felis* DNA with serology and skin tests. *J. Infect. Dis*. 171: 916–923.

Bergmans AM, Schellekens JF, VanEmbden JD, Schouls LM (1996) Predominance of two *Bartonella henselae* variants among cat-scratch disease patients in the Netherlands. *J. Clin. Microbiol*. 34: 254–260.

Birtles RJ, Harrison TG, Fry NK, Saunders NA, Taylor AG (1991) Taxonomic considerations of *Bartonella bacilliformis* based on phylogenetic and phenotypic characteristics. *FEMS Microbiol. Lett*. 67: 187–191.

Birtles RJ, Harrison TG, Saunders NA, Molyneux DH (1995) Proposals to unify the genera *Grahamella* and *Bartonella*, with descriptions of *Bartonella talpae* comb. nov., *Bartonella peromysci* comb. nov., and three new species, *Bartonella grahamii* sp. nov., *Bartonella taylorii* sp. nov., and *Bartonella doshiae* sp. nov. *Int. J. Syst. Bacteriol*. 45: 1–8.

Birtles RJ, Raoult D (1996) Comparison of partial citrate synthase gene (*gltA*) sequences for phylogenetic analysis of *Bartonella* species. *Int. J. Syst. Bacteriol*. 46: 891–897.

Bowers TJ, Sweger D, Jue D, Anderson B (1998) Isolation, sequencing and expression of the gene encoding a major protein from the bacteriophage associated with *Bartonella henselae*. *Gene* 206: 49–52.

Brenner DJ, O'Connor SP, Hollis DG, Weaver RE, Steigerwalt AG (1991) Molecular characterization and proposal of a neotype strain for *Bartonella bacilliformis*. *J. Clin Microbiol*. 29: 1299–1302.

Brenner DJ, O'Connor SP, Winkler HH, Steigerwalt AG (1993) Proposals to unify the genera *Bartonella* and *Rochalimaea*, with descriptions of *Bartonella quintana* comb. nov., *Bartonella vinsonii* comb. nov., *Bartonella henselae* comb. nov., and *Bartonella elizabethae* comb. nov., and to remove the family *Bartonellaceae* from the order *Rickettsiales*. *Int. J. Syst. Bacteriol*. 43: 777–786.

Brouqui P, Raoult D (1996) *Bartonella quintana* invades and multiplies within endothelial cells *in vitro* and *in vivo* and forms intracellular blebs. *Res. Microbiol*. 147: 719–731.

Burgess AW, Anderson BE (1998) Outer membrane proteins of *Bartonella henselae* and their interaction with human endothelial cells. *Microb. Pathog*. 25: 157–164.

Byam W (1919) Trench fever. In: Loyd LL (ed.) *Lice and their Menace to Man*. Oxford: Oxford University Press, pp. 120–130.

Caniza MA, Granger DL, Wilson KH, Washington MK, Kordick DL, Frush DP, Blitchington RB (1995) *Bartonella* (*Rochalimaea*) *henselae*: etiology of pulmonary nodules in a patient with depressed cell-mediated immunity. *Clin. Infect. Dis*. 20: 1505–1511.

Carithers HA (1985) Cat-scratch disease: an overview based on a study of 1,200 patients. *Am. J. Dis. Child.* 139: 1124–1133.

Cartwright JL, Britton P, Minnick MF, McLennan AG (1999) The *ialA* invasion gene of *Bartonella bacilliformis* encodes a (di)nucleoside polyphosphate hydrolase of the MutT motif family and has homologs in other invasive bacteria. *Biochem. Biophys. Res. Commun.* 256: 474–479.

Clarridge JE, Raich TJ, Pirwani D *et al.* (1995) Strategy to detect and identify *Bartonella* species in routine clinical laboratory yields *Bartonella henselae* from human immunodeficiency virus-positive patient and unique *Bartonella* strain from his cat. *J. Clin. Microbiol.* 33: 2107–2113.

Cockerell CJ, Whitlow MA, Webster GF, Friedman-Kien AE (1987) Epithelioid angiomatosis: a distinct vascular disorder in patients with the acquired immunodeficiency syndrome or AIDS-related complex. *Lancet* 2: 654–656.

Cockerell CJ, Bergstresser PR, Myrie-Williams C, Tierno PM (1990) Bacillary epithelioid angiomatosis occurring in an immunocompetent individual. *Arch. Dermatol.* 126: 787–790.

Conley T, Slater L, Hamilton K (1994) *Rochalimaea* species stimulate human endothelial cell proliferation and migration *in vitro*. *J. Lab. Clin. Med.* 124: 521–528.

Cuadra MS (1956) Salmonellosis complication in human bartonellosis. *Texas Rep. Biol. Med.* 14: 97–113.

Daly JS, Worthington MG, Brenner DJ *et al.* (1993) *Rochalimaea elizabethae* sp. nov. isolated from a patient with endocarditis. *J. Clin. Microbiol.* 31: 872–881.

Dauga C, Miras I, Grimont PA (1996) Identification of *Bartonella henselae* and *B. quintana* 16S rDNA sequences by branch-, genus- and species-specific amplification. *J. Med. Microbiol.* 45: 192–199.

Dehio C, Meyer M, Berger J, Schwarz H, Lanz C (1997) Interaction of *Bartonella henselae* with endothelial cells results in bacterial aggregation on the cell surface and the subsequent engulfment and internalisation of the bacterial aggregate by a unique structure, the invasome. *J. Cell Sci.* 110: 2141–2154.

Dolan MJ, Wong MT, Regnery RL *et al.* (1993) Syndrome of *Rochalimaea henselae* adenitis suggesting cat scratch disease. *Ann. Intern. Med.* 118: 331–336.

Doyle D, Eppes SC, Klein JD (1994) Atypical cat-scratch disease: diagnosis by a serologic test for *Rochalimaea* species. *South. Med. J.* 87: 485–487.

Drancourt M, Raoult D (1993) Proposed tests for the routine identification of *Rochalimaea* species. *Eur. J. Clin. Microbiol. Infect. Dis.* 12: 710–713.

Drancourt M, Mainardi JL, Brouqui P *et al.* (1995) *Bartonella (Rochalimaea) quintana* endocarditis in three homeless men. *N. Engl. J. Med.* 332: 419–423.

Drancourt M, Birtles R, Chaumentin G, Vandenesch F, Etienne J, Raoult D (1996a) A new serotype of *Bartonella henselae* in endocarditis and cat-scratch disease. *Lancet* 347: 441–443.

Drancourt M, Moal V, Brunet P, Dussol B, Berland Y, Raoult D (1996b) *Bartonella (Rochalimaea) quintana* infection in a seronegative hemodialyzed patient. *J. Clin. Microbiol.* 34: 1158–1160.

Droz S, Chi B, Horn E, Steigerwalt AG, Whitney AM, Brenner DJ (1999) *Bartonella koehlerae* sp. nov., isolated from cats. *J. Clin. Microbiol.* 37: 1117–1122.

Edmond MB, Riddler SA, Baxter CM, Wicklund BM, Pasculle AW (1993) *Agrobacterium radiobacter*: a recently recognised opportunistic pathogen. *Clin. Infect. Dis.* 16: 388–391.

English CK, Wear DJ, Margileth AM, Lissner CR, Walsh GP (1988) Cat-scratch disease. Isolation and culture of the bacterial agent. *JAMA* 259: 1347–1352.

Fath MJ, Kolter R (1993) ABC transporters: bacterial exporters. *Microbiol. Rev.* 57: 995–1017.

Fischetti VA (1989) Streptococcal M protein: molecular design and biological behavior. *Clin. Microbiol. Rev.* 2: 285–314.

Freney J, Gruer LD, Bornstein N *et al.* (1985) Septicemia caused by *Agrobacterium* sp. *J. Clin. Microbiol.* 22: 683–685.

Garcia FU, Wojta J, Broadley KN, Davidson JM, Hoover RL (1990) *Bartonella bacilliformis* stimulates endothelial cells *in vitro* and is angiogenic *in vivo*. *Am. J. Pathol.* 136: 1125–1135.

Garcia FU, Wojta J, Hoover RL (1992) Interactions between live *Bartonella bacilliformis* and endothelial cells. *J. Infect. Dis.* 165: 1138–1141.

Garcia-Tsao G, Panzini L, Yoselevitz M, West AB (1992) Bacillary peliosis hepatis as a cause of acute anemia in a patient with the acquired immune deficiency syndrome. *Gastroenterology* 102: 1065–1070.

Gille C, Lanz C, Dehio C (1999) Site-specific mutagenesis of the *ialB* locus of *Bartonella tribocorum* by single-crossover gene disruption. In: *Abstracts of the First International Conference on Bartonella as Emerging Pathogens*, Max Planck Institute, Tübingen, Germany, p. 56, poster 28.

Gold L, Pribnow D, Schneider T, Shinedling S, Singer BS, Stormo G (1981) Translation initiation in prokaryotes. *Annu. Rev. Microbiol.* 35: 365–403.

Golden SE (1993) Hepatosplenic cat-scratch disease associated with elevated anti-*Rochalimea* antibody titers. *Pediatr. Infect. Dis. J.* 12: 868–871.

Golnik KC, Marotto ME, Fanous MM *et al.* (1994) Opthalmic manifestations of *Rochalimaea* species. *Am. J. Opthalmol.* 118: 145–151.

Gray GC, Johnson AA, Thornton SA *et al.* (1990) An epidemic of Oroya fever in the Peruvian Andes. *Am. J. Trop. Med. Hyg.* 42: 215–221.

Gurfield AN, Boulouis HJ, Chomel BB *et al.* (1997) Coinfection with *Bartonella clarridgeiae* and *Bartonella henselae* and with different *Bartonella henselae* strains in domestic cats. *J. Clin. Microbiol.* 35: 2120–2123.

Haake DA, Summers TA, McCoy AM, Schwartzman W (1997) Heat shock response and *groEL* sequence of *Bartonella henselae* and *Bartonella quintana*. *Microbiology* 143: 2807–2815.

Hadfield TL, Warren R, Kass M, Brun E, Levy L (1993) Endocarditis caused by *Rochalimaea henselae*. *Hum. Pathol.* 24: 1140–1141.

Heffernan EJ, Wu L, Louie J, Okamoto S, Fierer J, Guiney DG (1994) Specificity of the complement resistance and cell association phenotypes encoded by the outer membrane protein genes *rck* from *Salmonella typhimurium* and *ail* from *Yersinia enterocolitica*. *Infect. Immun.* 62: 5183–5186.

Heller R, Riegel P, Hansmann Y *et al.* (1998) *Bartonella tribocorum* sp. nov., a new *Bartonella* species isolated from the blood of wild rats. *Int. J. Syst. Bacteriol.* 48: 1333–1339.

Heller R, Kubina M, Mariet P *et al.* (1999) *Bartonella alsatica* sp. nov., a new *Bartonella* species isolated from the blood of wild rabbits. *Int. J. Syst. Bacteriol.* 49: 283–288.

Herrer A (1953) Carrion's disease. II. Presence of *Bartonella bacilliformis* in the peripheral blood of patients with the benign tumor form. *Am. J. Trop. Med.* 2: 645–649.

Hertig M (1942) Phlebotomus and Carrion's disease. *Am. J. Trop. Med.* 22: 1–80.

Hill E, Raji A, Valenzuela MS, Garcia F, Hoover R (1992) Adhesion to and invasion of cultured human cells by *Bartonella bacilliformis*. *Infect. Immun.* 60: 4051–4058.

Hollingdale MR, Vinson JW, Herrmann JE (1980) Immunochemical and biological properties of the outer membrane-associated lipopolysaccharide and protein of *Rochalimaea quintana*. *J. Infect. Dis.* 141: 672–679.

Howe C (1943) Carrion's disease. Immunologic studies. *Arch. Intern. Med.* 72: 147–167.

Hurtado A, Musso JP, Merino C (1938) La anemia en la enfermedad de Carrion (verruga peruana). *Ann. Fac. Med. Lima* 28: 154–168.

Iwaki-Egawa S, Ihler GM (1997) Comparison of the abilities of proteins from *Bartonella bacilliformis* and *Bartonella henselae* to deform red cell membranes and to bind to red cell ghost proteins. *FEMS Microbiol. Lett.* 157: 207–217.

Jackson LA, Spach DH (1996) Emergence of *Bartonella quintana* infection among homeless persons. *Emerg. Infect. Dis.* 2: 141–144.

Jalava J, Kotilainen P, Nikkari S *et al.* (1995) Use of the polymerase chain reaction and DNA sequencing for detection of *Bartonella quintana* in the aortic valve of a patient with culture-negative infective endocarditis. *Clin. Infect. Dis.* 21: 891–896.

Johnson WT, Helwig EB (1969) Cat-scratch disease. Histopathologic changes in the skin. *Arch. Dermatol.* 100: 148–154.

Kehoe MA (1994) Cell wall-associated proteins in Gram-positive bacteria. *New Compr. Biochem.* 27: 217–261.

Kelly TM, Padmalayam I, Baumstark BR (1998) Use of the cell division protein FtsZ as a means of differentiating among *Bartonella* species. *Clin. Diagn. Lab. Immunol.* 5: 766–772.

Kemper CA, Lombard CM, Deresinski SC, Tompkins LS (1990) Visceral bacillary epithelioid angiomatosis: possible manifestations of disseminated cat scratch disease in the immunocompromised host: a report of two cases. *Am. J. Med.* 89: 216–222.

Knobloch J, Solano L, Alvarez O, Delgado E (1985) Antibodies to *Bartonella bacilliformis* as determined by fluorescence antibody test, indirect haemagglutination and ELISA. *Trop. Med. Parasitol.* 36: 183–185.

Knobloch J (1988) Analysis and preparation of *Bartonella bacilliformis* antigens. *Am. J. Trop. Med. Hyg.* 39: 173–178.

Knobloch J, Bialek R, Muller G, Asmus P (1988) Common surface epitope of *Bartonella bacilliformis* and *Chlamydia psittaci. Am. J. Trop. Med. Hyg.* 39: 427–433.

Knobloch J, Schreiber M (1990) BB65, a major immunoreactive protein of *Bartonella bacilliformis. Am. J. Trop. Med. Hyg.* 43: 373–379.

Kobe B, Deisenhofer J (1995) Proteins with leucine-rich repeats. *Curr. Opin. Struct. Biol.* 5: 409–416.

Koehler JE, Quinn FD, Berger TG, LeBoit PE, Tappero JW (1992) Isolation of *Rochalimaea* species from cutaneous and osseous lesions of bacillary angiomatosis. *N. Engl. J. Med.* 327: 1625–1631.

Koehler JE, Tappero JW (1993) Bacillary angiomatosis and bacillary peliosis in patients infected with human immunodeficiency virus. *Clin. Infect. Dis.* 17: 612–624.

Koehler JE, Glaser CA, Tappero JW (1994) *Rochalimaea henselae* infection. A new zoonosis with the domestic cat as reservoir. *JAMA* 271: 531–535.

Kordick DL, Breitschwerdt EB (1995) Intraerythrocytic presence of *Bartonella henselae. J. Clin. Microbiol.* 33: 1655–1656.

Kordick DL, Wilson KH, Sexton DJ, Hadfield TL, Berkhoff HA, Breitschwerdt EB (1995) Prolonged *Bartonella* bacteremia in cats associated with cat-scratch disease patients. *J. Clin. Microbiol.* 33: 3245–3251.

Kordick DL, Swaminathan B, Greene CE *et al.* (1996) *Bartonella vinsonii* subsp. *berkhoffii* subsp.nov., isolated from dogs; *Bartonella vinsonii* subsp. *vinsonii;* and emended description of *Bartonella vinsonii. Int. J. Syst. Bacteriol.* 46: 704–709.

Kordick DL, Hilyard EJ, Hadfield TL *et al.* (1997) *Bartonella clarridgeiae*, a newly recognised zoonotic pathogen causing inoculation papules, fever, and lymphadenopathy (cat scratch disease). *J. Clin. Microbiol.* 35: 1813–1818.

Kostrzewski J (1950) The epidemiology of trench fever. *Med. Dosw. Mikrobiol.* 11: 233–263.

Kreier JP, Ristic M (1981) The biology of hemotrophic bacteria. *Annu. Rev. Microbiol.* 35: 325–338.

Kreier JP, Gother R, Ihler GM, Krampitz HE, Mernaugh G, Palmer GH (1991) The hemotrophic bacteria: the families *Bartonellaceae* and *Anaplasmataceae*. In: Balows A, Truper HG, Dworkin M, Harder W, Schleifer K-H (ed.) *The Prokaryotes*, New York: Springer-Verlag, 2nd edn. pp. 3994–4022.

Krueger CM, Marks KL, Ihler GM (1995) Physical map of the *Bartonella bacilliformis* genome. *J. Bacteriol.* 177: 7271–7274.

La-Scola B, Raoult D (1996) Serological cross-reactions between *Bartonella quintana, Bartonella henselae,* and *Coxiella burnetii. J. Clin. Microbiol.* 34: 2270–2274.

LeBoit PE, Berger TG, Egbert BM, Beckstead JH, Yen TS, Stoler MH (1988) Epithelioid haemangioma-like vascular proliferation in AIDS: manifestation of cat-scratch disease bacillus infection? *Lancet* 1: 960–963.

LeBoit PE, Berger TG, Egbert BM, Beckstead JH, Yen TS, Stoler MH (1989) Bacillary angiomatosis. The histopathology and differential diagnosis of a pseudoneoplastic infection in patients with human immunodeficiency virus disease. *Am. J. Surg. Pathol.* 13: 909–920.

Lucey D, Dolan MJ, Moss CW *et al.* (1992) Relapsing illness due to *Rochalimaea henselae* in immunocompetent hosts: implication for therapy and new epidemiological associations. *Clin. Infect. Dis.* 14: 683–688.

Margileth AM (1993) Cat scratch disease. *Adv. Pediatr. Infect. Dis.* 8: 1–21.

Marullo S, Jaccard A, Roulot D, Mainquene C, Clauvel JP (1992) Identification of the *Rochalimaea henselae* 16S rRNA sequence in the liver of a French patient with bacillary peliosis hepatis. *J. Infect. Dis.* 166: 1462.

Maryuyama S, Nogami S, Inoue I, Namba S, Asanome K, Katsube Y (1996) Isolation of *Bartonella henselae* from domestic cats in Japan. *J. Vet. Med. Sci.* 58: 81–83.

Matar GM, Swaminathan B, Hunter SB, Slater LN, Welch DF (1993) Polymerase chain reaction-based restriction fragment length polymorphism analysis of a fragment of the ribosomal operon from *Rochalimaea* species for subtyping. *J. Clin. Microbiol.* 31: 1730–1734.

Matteelli A, Castelli F, Spinetti A, Bonetti F, Graifenberghi S, Carosi G (1994) Short report: verruga peruana in an Italian traveler from Peru. *Am. J. Trop Med. Hyg.* 50: 143–144.

Maurin M, Roux V, Stein A, Ferrier F, Viraben R, Raoult D (1994) Isolation and characterization by immunofluorescence, sodium dodecyl sulfate-polyacrylamide gel electrophoresis, western blot, restriction fragment length polymorphism-PCR, 16S rRNA gene sequencing, and pulsed-field gel electrophoresis of *Rochalimaea quintana* from a patient with bacillary angiomatosis. *J. Clin. Microbiol.* 32: 1166–1171.

Maurin M, Raoult D (1996) *Bartonella (Rochalimaea) quintana* infections. *Clin. Microbiol. Rev.* 9: 273–292.

McClure WR (1985) Mechanism and control of transcription initiation in prokaryotes. *Annu. Rev. Biochem.* 54: 171–204.

McCrary B, Cockerham W, Pierce P (1994) Neuroretinitis in cat-scratch disease associated with the macular star. *Pediatr. Infect. Dis. J.* 13: 838–839.

McNee JW, Renshaw A (1916) 'Trench fever': a relapsing fever occurring with the British forces in France. *BMJ* 1: 225–234.

Mehock JR, Greene CE, Gherardini FC, Hahn TW, Krause DC (1998) *Bartonella henselae* invasion of feline erythrocytes *in vitro. Infect. Immun.* 66: 3462–3466.

Mernaugh G, Ihler GM (1992) Deformation factor: an extracellular protein synthesised by *Bartonella bacilliformis* that deforms erythrocyte membranes. *Infect. Immun.* 60: 937–943.

Merrell BR, Weiss E, Dasch GA (1978) Morphological and cell association characteristics of *Rochalimaea quintana*: comparison of the vole and Fuller strains. *J. Bacteriol.* 135: 633–640.

Milam MW, Balerdi MJ, Toney JF, Foulis PR, Milam CP, Behnke RH (1990) Epithelioid angiomatosis secondary to disseminated cat scratch disease involving the bone marrow and skin in a patient with acquired immune deficiency syndrome: a case report. *Am. J. Med.* 88: 180–183.

Milde P, Brunner M, Borchard F *et al.* (1995) Cutaneous bacillary angiomatosis in a patient with chronic lymphocytic leukemia. *Arch. Dermatol.* 131: 933–936.

Miller VL, Bliska JB, Falkow S (1990) Nucleotide sequence of the *Yersinia enterocolitica ail* gene and characterization of the Ail protein product. *J. Bacteriol.* 172: 1062–1069.

Minnick MF (1994) Identification of outer membrane proteins of *Bartonella bacilliformis. Infect. Immun.* 62: 2644–2648.

Minnick MF, Barbian KD (1997) Identification of *Bartonella* using PCR: genus- and species-specific primer sets. *J. Microbiol. Methods* 31: 51–57.

Minnick MF, Mitchell SJ, McAllister SJ (1996) Cell entry and the pathogenesis of *Bartonella* infections. *Trends Microbiol.* 4: 343–347.

Minnick MF (1997) Virulence determinants of *Bartonella bacilliformis*. In: Anderson B, Bendinelli M, Friedman H (eds) *Rickettsial Infection and Immunity*. New York: Plenum Press, pp. 197–211.

Mitchell SJ, Minnick MF (1995) Characterization of a two-gene locus from *Bartonella bacilliformis* associated with the ability to invade human erythrocytes. *Infect. Immun.* 63: 1552–1562.

Mitchell SJ, Minnick MF (1997a) A carboxy-terminal processing protease gene is located immediately upstream of the invasion-associated locus from *Bartonella bacilliformis. Microbiology* 143: 1221–1233.

Mitchell SJ, Minnick MF (1997b) Cloning, functional expression, and complementation analysis of an inorganic pyrophosphatase from *Bartonella bacilliformis. Can. J. Microbiol.* 43: 734–743.

Moulder JW (1974) Order I. Rickettsiales Giewszczkiewicz (1939), 25. In: Buchanan RE, Gibbons NE (eds) *Bergey's Manual of Determinative Bacteriology*, 8th edn. Baltimore: Williams & Wilkins.

Murakawa GJ (1997) Pathogenesis of *Bartonella henselae* in cutaneous and systemic disease. *J. Am. Acad. Dermatol.* 37: 775–776.

Musso D, Drancourt M, Raoult D (1995) Lack of bactericidal effect of antibiotics except aminoglycosides on *Bartonella (Rochalimaea) henselae. J. Antimicrob. Chemother.* 36: 101–108.

Myers WF, Cutler LD, Wisseman CL (1969) Role of erythrocytes and serum in the nutrition of *Rickettsia quintana. J. Bacteriol.* 97: 663–666.

Myers WF, Wisseman CL (1978) Effect of specific antibody and complement on the survival of *Rochalimaea quintana* in vitro. *Infect. Immun.* 22: 288–289.

Myers WF, Wisseman CL, Fiset P, Oaks EV, Smith JF (1979) Taxonomic relationship of vole agent to *Rochalimaea quintana. Infect. Immun.* 26: 976–983.

Norman AF, Regnery R, Jameson P, Greene C, Krause DC (1995) Differentiation of *Bartonella*-like isolates at the species level by PCR-restriction fragment length polymorphism in the citrate synthase gene. *J. Clin. Microbiol.* 33: 1797–1803.

O'Connor SP, Dorsch M, Steigerwalt AG, Brenner DJ, Stackebrandt E (1991) 16S rRNA sequences of *Bartonella bacilliformis* and cat scratch disease bacillus reveal phylogenetic relationships with the alpha-2 subgroup of the class *Proteobacteria. J. Clin. Microbiol.* 29: 2144–2150.

Padmalayam I, Anderson B, Kron M, Kelly T, Baumstark B (1997) The 75-kilodalton antigen of *Bartonella bacilliformis* is a structural homolog of the cell division protein FtsZ. *J. Bacteriol.* 179: 4545–4552.

Patnaik M, Schwartzman WA, Barka NE, Peter JB (1992) Possible role of *Rochalimaea henselae* in pathogenesis of AIDS encephalopathy. *Lancet* 340: 971.

Perkins BA, Swaminathan B, Jackson LA, Brenner DJ, Wenger JD, Regnery RL, Wear DJ (1992) Case 22-1992-pathogenesis of cat scratch disease [letter]. *N. Engl. J. Med.* 327: 1599–1600.

Perkocha LA, Geaghan SM, Yen TS *et al.* (1990) Clinical and pathological features of bacillary peliosis hepatis in association with human immunodeficiency virus infection. *N. Engl. J. Med.* 323: 1581–1586.

Plotkin GR (1980) *Agrobacterium radiobacter* prosthetic valve endocarditis. *Ann. Intern. Med.* 93: 839–840.

Raoult D, Fournier PE, Drancourt M *et al.* (1996) Diagnosis of 22 new cases of *Bartonella* endocarditis. *Ann. Intern. Med.* 125: 646–652.

Reidl J, Mekalanos JJ (1996) Lipoprotein e(P4) is essential for hemin uptake by *Haemophilus influenzae. J. Exp. Med.* 183: 621–629.

Ricketts WE (1949) Clinical manifestations of Carrion's disease. *Arch. Intern. Med.* 84: 751–781.

Regnery R, Tappero J (1995) Unraveling mysteries associated with cat-scratch disease, bacillary angiomatosis, and related syndromes. *Emerg. Infect. Dis.* 1: 16–21.

Regnery RL, Spruill CL, Plikaytis BD (1991) Genotypic identification of rickettsiae and estimation of intraspecies sequence divergence for portions of two rickettsial genes. *J. Bacteriol.* 173: 1576–1589.

Regnery RL, Anderson BE, Clarridge JE, Rodriguez-Barradas MC, Jones DC, Carr JH (1992a) Characterization of a novel *Rochalimaea* species, R. *henselae* sp. nov., isolated from blood of a febrile, human immunodeficiency virus-positive patient. *J. Clin. Microbiol.* 30:265–274.

Regnery RL, Olson JG, Perkins BA, Bibb W (1992b) Serological response to '*Rochalimaea henselae*' antigen in suspected cat-scratch disease. *Lancet* 339: 1443–1445.

Relman DA, Loutit JS, Schmidt TM, Falkow S, Tompkins LS (1990) The agent of bacillary angiomatosis: an approach to the identification of uncultured pathogens. *N. Engl. J. Med.* 323: 1573–1580.

Relman DA, Lepp PW, Sadler KN, Schmidt TM (1992) Phylogenetic relationships among the agent of bacillary angiomatosis, *Bartonella bacilliformis*, and other alpha proteobacteria. *Mol. Microbiol.* 6: 1801–1807.

Reynafarje C, Ramos J (1961) The hemolytic anemia of human bartonellosis. *Blood* 17: 562–578.

Roberts NJ Jr (1995) Bartonella bacilliformis (bartonellosis). In: Mandell GL, Bennett JE, Dolin R (eds) *Principles and Practice of Infectious Diseases*, New York: Livingstone Press, 4th edn. pp. 2209–2210.

Rodriguez-Barradas MC, Bandres JC, Hamill RJ *et al.* (1995a) *In vitro* evaluation of the role of humoral immunity against *Bartonella henselae. Infect. Immun.* 63: 2367–2370.

Rodriguez-Barradas MC, Hamill RJ, Houston ED *et al.* (1995b) Genomic fingerprinting of *Bartonella* species using repetitive-element PCR for distinguishing species and isolates. *J. Clin. Microbiol.* 33: 1089–1093.

Roux V, Raoult D (1995) Inter- and intraspecies identification of *Bartonella (Rochalimaea)* species. *J. Clin. Microbiol.* 33: 1573–1579.

Scherer DC, DeBuron-Connors I, Minnick MF (1993) Characterization of *Bartonella bacilliformis* flagella and effect of antiflagellin antibodies on invasion of human erythrocytes. *Infect. Immun.* 61: 4962–4971.

Schwartzman WA (1992) Infections due to *Rochalimaea*: The expanding clinical spectrum. *Clin. Infect. Dis.* 15: 893–900.

Slater LN, Welch DF, Hensel D, Coody DW (1990) A newly recognised fastidious gram-negative pathogen as a cause of fever and bacteremia. *N. Engl. J. Med.* 323: 1587–1593.

Slater LN, Welch DF, Min KW (1992) *Rochalimaea henselae* causes bacillary angiomatosis and peliosis hepatis. *Arch. Intern. Med.* 152: 602–606.

Slater LN, Pitha JV, Herrera L, Hughson MD, Min KW, Reed JA (1994) *Rochalimaea henselae* infection in acquired immunodeficiency syndrome causing inflammatory disease without angiomatosis or peliosis. Demonstration by immunochemistry and corroboration by DNA amplification. *Arch. Pathol. Lab. Med.* 118: 33–38.

Smith DL (1997) Cat-scratch disease and related clinical syndromes. *Am. Fam. Physician* 55: 1783–1794.

Soheilian M, Markomichelakis N, Foster CS (1996) Intermediate uveitis and retinal vasculitis as manifestations of cat scratch disease. *Am. J. Ophthalmol.* 122: 582–584.

Southern PM (1996) Bacteremia due to *Agrobacterium tumefaciens (radiobacter)*. Report of infection in a pregnant woman and her stillborn fetus. *Diagn. Microbiol. Infect. Dis.* 24: 43–45.

Spach DH, Panther LA, Thorning DR, Dunn JE, Plorde JJ, Miller RA (1992) Intracerebral bacillary angiomatosis in a patient infected with human immunodeficiency virus. *Ann. Intern. Med.* 116: 740–742.

Spach DH, Callis KP, Paauw DS *et al.* (1993) Endocarditis caused by *Rochalimaea quintana* in a patient infected with human immunodeficiency virus. *J. Clin. Microbiol.* 31: 692–694.

Spach DH, Kanter AS, Dougherty MJ *et al.* (1995) *Bartonella* (*Rochalimaea*) *quintana* bacteremia in inner-city patients with chronic alcoholism. *N. Engl. J. Med.* 332: 424–428.

Stoler MH, Bonfiglio TA, Steigbigel RT, Pereira M (1983) An atypical subcutaneous infection associated with acquired immune deficiency syndrome. *Am. J. Clin. Pathol.* 80: 714–718.

Strong RP (ed.) (1918) *Trench Fever. Report of Commission: Medical Research Committee, American Red Cross*, Oxford: Oxford University Press, pp. 40–60.

Tappero JW, Koehler JE, Berger TG *et al.* (1993a) Bacillary angiomatosis and bacillary splenitis in immunocompetent adults. *Ann. Intern. Med.* 118: 363–365.

Tappero JW, Mohle-Boetani J, Koehler JE *et al.* (1993b) The epidemiology of bacillary angiomatosis and bacillary peliosis. *JAMA* 269: 770–775.

Tyeryar FJ, Weiss E, Millar DB, Bozeman FM, Ormsbee RA (1973) DNA base composition of rickettsiae. *Science* 180: 415–417.

Umemori E, Sasaki Y, Amano K-I, Amano Y (1992) A phage in *Bartonella bacilliformis*. *Microbiol. Immunol.* 36: 731–736.

Urteaga BO, Payne EH (1955) Treatment of the acute febrile phase of Carrion's disease with chloramphenicol. *Am. J. Trop. Med.* 4: 507–511.

Varela G, Vinson JW, Molina-Pasquel C (1969) Trench fever II. Propagation of *Rickettsia quintana* on cell-free medium from the blood of two patients. *Am. J. Trop. Med. Hyg.* 18: 708–712.

Vinson JW, Fuller HS (1961) Studies on trench fever I. Propagation of rickettsia-like micro-organisms from a patient's blood. *Pathol. Microbiol.* 24: 152–166.

Vinson JW, Varela G, Molina-Pasquel C (1969) Trench fever. III. Induction of clinical disease in volunteers inoculated with *Rickettsia quintana* propagated on blood agar. *Am. J. Trop. Med. Hyg.* 18: 713–722.

Waldvogel K, Regnery RL, Anderson BE, Caduff R, Caduff J, Nadal D (1994) Disseminated cat-scratch disease: detection of *Rochalimaea henselae* in affected tissue. *Eur. J. Pediatr.* 153: 23–27.

Walker TS, Winkler HH (1981) *Bartonella bacilliformis*: colonial types and eyrthrocyte adherence. *Infect. Immun.* 31: 480–486.

Webster GF, Cockerell CJ, Friedman-Kien AE (1992) The clinical spectrum of bacillary angiomatosis. *Br. J. Dermatol.* 126: 535–541.

Weinman D (1965) The *bartonella* group. In: Dubos RJ, Hirsch JG (eds) *Bacterial and mycotic infections of man*. Philadelphia: Lippincott, pp. 775–785.

Weinman D (1974) Family II. *Bartonellaceae* Giewszczkiewicz (1939), 25. In: Buchanan RE, Gibbons NE (eds) *Bergey's Manual of Determinative Bacteriology*, 8th edn. Baltimore: Williams & Wilkins.

Weiss E, Dasch GA (1982) Differential characteristics of strains of *Rochalimaea*: *Rochalimaea vinsonii* sp. nov., the Canadian vole agent. *Int. J. Syst. Bacteriol.* 32: 305–314.

Weiss E, Moulder JW (1984) Order I. *Rickettsiales*. In: Krieg NR (ed.) *Bergey's Manual of Systematic Bacteriology*. Baltimore: Williams & Wilkins, pp. 687–719.

Welch DF, Pickett DA, Slater LN, Steigerwalt AG, Brenner DJ (1992) *Rochalimaea henselae* sp. nov., a cause of septicemia, bacillary angiomatosis, and parenchymal bacillary peliosis. *J. Clin. Microbiol.* 30: 275–280.

Welch DF, Hensel DM, Pickett DA, San-Joaquin VH, Robinson A, Slater LN (1993) Bacteremia due to *Rochalimaea henselae* in a child: practical identification of isolates in the clinical laboratory. *J. Clin. Microbiol.* 31: 2381–2386.

Westfall HN, Edman DC, Weiss E (1984) Analysis of fatty acids of the genus *Rochalimaea* by electron capture gas chromatography: detection of nonanoic acid. *J. Clin Microbiol.* 19: 305–310.

Wilfert CM (1992) *Brucella*. In: Joklik WK, Willett HP, Amos DB, Wilfert CM (eds) *Zinsser Microbiology*. Norwalk, CT: Appleton and Lange, pp. 609–614.

Xu Y-H, Lu Z-Y, Ihler GM (1995) Purification of deformin, an extracellular protein synthesized by *Bartonella bacilliformis* which causes deformation of erythrocyte membranes. *Biochim. Biophys. Acta* 1234: 173–183.

Zangwill KM, Hamilton DH, Perkins BA *et al.* (1993) Cat scratch disease in Connecticut: epidemiology, risk factors, and evaluation of a new diagnostic test. *N. Engl. J. Med.* 329: 8–13.

100

Leptospira

Martin J. Woodward

Veterinary Laboratories Agency, Addlestone, Surrey, UK

It is a common misconception that *Leptospira* infections of humans equate only with Weil's disease. Weil first gave a precise clinical description in 1886 of a severe form of leptospirosis in humans, probably caused by *Leptospira interrogans* serovar *icterohaemorrhagiae*. However, the symptoms of leptospiral infection vary considerably in severity depending on which serovar causes infection. Importantly, in the early literature at the turn of the twentieth century, a number of descriptions were published of the clinical manifestation of leptospiral infection worldwide. Leptospiral infections in humans have been described variously as rice-harvest jaundice, 7-day or autumn fever, cane cutters' disease, mud fever and swineherd's disease. Similar clinical descriptions were also noted for soldiers affected during trench warfare and by coal miners. As these examples suggest, similar clinical syndromes have been associated with particular occupations and environments, the common link being warm water and association with various animals. In developed countries today, this disease is associated with outdoor sporting activities such as swimming, canoeing and camping in which individuals are exposed to leptospires in natural water sources. Commonly, the cause of infection is the excretion by infected rats of large numbers of viable organisms in their urine into watercourses.

Leptospirosis is an acute, generalised infectious disease characterised by headache and fever, myalgia and chills, frequently with nausea but not always associated with vomiting, and conjunctivitis. The muscular pain in the abdomen, the spinal region and notably in the calf muscles is frequently severe and patients are usually prostrate. The onset is usually rapid and very frequently diagnosed as an influenza-like infection on initial presentation, particularly in non-endemic areas. This early, bacteraemic phase is never fatal and lasts some 3–7 days with concomitant reduction of fever. In some cases the disease does not develop further after this phase. More often than not, however, one of two secondary phases follows depending on the host and the serovar responsible. One phase is a severe and potentially life-threatening stage lasting up to a further 30 days in which fever is renewed, often with greater intensity and accompanied with jaundice, haemorrhage, renal failure and myocarditis. This is the classical picture first described by Weil, and is more correctly defined as icteric leptospirosis. Mortality rates approaching 10% are common. The other phase is a less severe secondary phase, often associated with the recurrence of fever. This is less intense, of shorter duration and sometimes absent altogether, but associated with a rash, uveitis and meningitis. Delirium is noted in some patients with anicteric leptospirosis.

This variable clinical outcome of leptospiral infection of humans is readily confused in the early presentation as a generalised influenza-like infection. Whilst Weil's 1886 description was the first accurate account of one form of leptospirosis, it was over a quarter of a century before the causative agent was identified and even then there was considerable confusion in its definition. Hecker and Otto in 1910 were the first to take blood from infected patients and inoculate surrogate animal models to test for transmission of disease. Of the many animal species tested, however, symptoms were produced only in the ape and these were not truly typical of Weil's disease. Surprisingly, these researchers failed to undertake appropriate microbiological analyses and it was not until 1916 that a microbiological basis for Weil's disease was demonstrated. Inada and co-workers in 1916 transmitted infection to the guinea-pig and demonstrated the amplification of spirochaetes (spirochaetes; *spiro* = coiled, *chaete* = hair) in the blood. Inada coined the term 'Spirochaeta icterohaemorrhagiae' to describe these organisms. At about the same time, similar confirmatory experiments were performed in Europe.

In World War I, blood taken from affected soldiers from the trenches who presented with clinical symptoms typical of Weil's disease, was inoculated into guinea-pigs. Spirochaetes were observed in Geimsa-stained guinea-pig blood smears. Significantly, Inada and co-workers were the first to culture these organisms *in vitro* in a complex medium containing 10% rabbit serum, and they then described their biological properties in detail. Perhaps the most compelling study of Inada and co-workers to confirm the role of these spirochaetes in pathogenesis was the demonstration of passive immunity by administration of specific immune serum.

Before the definitive work of Inada and others, Stimson in 1907 was the first to observe coiled hair-like organisms in tissue sections, notably large clumps of bacterial spiral-shaped cells in the lumen of renal tubules of the kidney, from a patient dying of yellow fever. Subsequent morphological analysis of Stimson's material by Noguchi in the 1920s led to the detailed morphological description of these spirochaetes but he perpetuated the notion that these spirochaetes were associated with yellow fever, and as a consequence, much of this work was overlooked. For a more detailed discussion of the early history of leptospirosis the reader is referred to Faine (1994) and Faine *et al.* (1999).

Nutrition and Cultivation of *Leptospira*

The first successful culture of a number of leptospires was in the second decade of the twentieth century by Noguchi, Inada, Ito and others, who made use of pig kidney in ascitic fluid sealed under a layer of paraffin oil. Later, simpler media essentially comprised 10% rabbit or guinea-pig serum with various additions. However,

Ellinghausen and McCullogh (1965) and Johnson and Gary (1962) demonstrated by fractionation of existing media that long-chain fatty acids, notably oleic acid, were the primary nutritional source. Thus it was that EMJH (Ellinghausen McCullogh Johnson Harris) medium was developed and it is still used extensively, usually in the form of a semi-solid medium gelled with about 0.2% agar. Leptospires prefer subsurface growth.

Various forms of oleic acid (OA)-based media are described in laboratory manuals and these are used for primary isolation of pathogenic leptospires from clinical and natural sources. Leptospires may be selected by the addition to OA medium of 5-fluoro-uracil in combination with various antibiotics such as nalidixic acid, vancomycin and polymixin, bacitracin or neomycin to which they are innately insensitive. The antibiotics are added primarily to prevent rapid over-growth by contaminants. However, transfer to fresh non-selective medium after 48 hours is usually essential for successful culture. Pathogenic leptospires are sensitive to streptomycin, erythromycin, tetracycline, second- and third-generation β lactams but they are occasionally resistant to penicillins, and most are resistant to chloramphenicol.

Pathogenic leptospires have an optimum growth temperature *in vitro* in the range 28–30°C and, unlike saprophytic leptospires, will not grow at 13°C. Considering the pathogenicity of these organisms in animal hosts this is surprising indeed. It would be interesting to pursue studies into stress responses for there may be an association between the ability to cause to infection at a non-optimal growth temperature with upregulation of virulence determinants under temperature stress.

All leptospires are aerobic chemo-organotrophic, but oxygen may be a growth-limiting factor and CO_2 is essential. The optimum pH for growth is in the range of 7.2–7.6 and mean doubling times of 14–18 hours are common, although a considerable lag phase is often noted on transfer from clinical samples to laboratory media. Cultures maintained for laboratory diagnosis in such tests as the microscopic agglutination test (MAT) are laboratory-adapted and grow well in a matter of 1–2 weeks to a density appropriate for these tests. It is possible that cultivation of fresh clinical isolates is hampered by the slow adaptation to laboratory media, so that it is particularly dependent on the provision of essential fatty acids (see below).

The carbon and energy sources are long-chain fatty acids, and sugars are not fermented. Energy is derived by β oxidation of fatty acids (Henneberry and Cox, 1970) with oxygen as the terminal electron acceptor.

Although long-chain fatty acids may be shortened and modified for incorporation, leptospires cannot synthesise long-chain fatty acids from pyruvate or acetate. Thus, their chemical composition reflects the medium in which they grow and, indeed, the fatty acids are themselves toxic and should be provided in the presence of a detoxifier or slow-release mechanism such as bovine serum albumen. A characteristic component of leptospires is *cis*-11-hexadecanoic acid, which is very rarely found in other bacteria. However, detailed chemical analysis (Vinh *et al.*, 1986, 1989) showed serovar variation in lipid composition of the lipopolysaccharide component of the bacterial cells. Some differences were striking, such as the presence of 3-hydroxy-dodecanoic acid in *L. interrogans* serovar *copenhagi* and its absence from *L. interrogans* serovar *hardjo-prajitno* and *L. borgspetersenii* serovar *hardjo-bovis*. The converse was true for tridecanoic acid and there were many other variations both in terms of type and concentration of lipid. Significantly, *L. interrogans* serovar *hardjo-prajitno* and *L. borgs-petersenii* serovar *hardjo-bovis* share similar lipid composition and are regarded serologically as very closely related, despite belonging to quite distinct genomospecies. This complex of problems relating to classification of these bacteria will be more fully discussed below.

Not all leptospires produce lipases while many pathogenic leptospires produce phospholipases and sphingomyelinases (Real *et al.*, 1989; Segers *et al.*, 1990). Whether these enzymes are true virulence determinants is open to debate. As yet, little is known of the metabolism of lipids by leptospires other than that long-chain fatty acids are an essential requirement and shorter chain fatty acids may be utilised. There is no evidence for the incorporation of acetate and pyruvate into long fatty acid biosynthesis.

Ammonia is an essential nutrient for leptospires and several pathogens produce urease, which may be suppressed by the presence of ammonia, whereas non-pathogenic leptospires do not. Nitrogen may be obtained from amino acids, notably asparagine, by auto-deamination caused by the presence of asparaginases in sera, whilst many amino acids are readily incorporated into protein as shown by pulse chase experiments. Work on nitrogen metabolism has shown unusual purine and pyrimidine metabolism by leptospires (Johnson and Rogers, 1964a, 1967; Johnson and Harris, 1968). Indeed, sensitivity and resistance to the presence of 8-azaguanine in a growth medium differentiates between pathogenic and saprophytic leptospires, respectively (Johnson and Rogers, 1964b). Thus, it can be seen that leptospires are fastidious and difficult to cultivate.

Cellular Morphology and Chemical Composition

Leptospires are Gram-negative and the typical cellular morphology is that of a slender helical coil of the order of 250–500 nm in diameter and 10–20 μm in length. Laboratory-adapted strains may, however, be significantly longer (**Fig. 1**). The bacteria are highly motile, propelled by a periplasmic flagella enclosed entirely between the inner and outer membranes.

Chemotaxis towards haemoglobin has been noted, but this and the biomolecular mechanics of locomotion in this bacterium has been little studied. The major subunit of the flagellin is immunogenic and the gene encoding this protein has been cloned (see below). The standard method for observation is by dark-field microscopy of wet mounts. The helical bacteria constantly spin along their long axis and the tips are often 'hooked'. It is amusing to recall that the initial naming of the organism as *Spirochaeta interrogans* arose because of its hook-like question mark shape, and the confusion over its association with yellow fever. So-called straight variants occur, and possibly arise because of atypical attachment of the flagella to the inner membrane at the very tip of the bacterium. Straight variants are considerably less motile and form denser colonies on solid media.

Although laboratory cultivation gives rise to avirulent straight forms, the correlation between virulence and helical or straight morphology is not absolute. Indeed, freshly isolated pathogenic leptospires tend to be shorter and more highly coiled. The typical helical structure is susceptible to alterations caused by ageing,

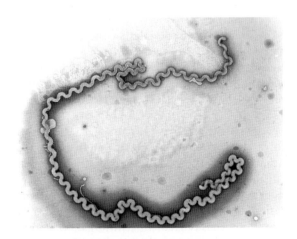

Fig. 1 *Leptospira* serovar Saxkoebing 766V grown in EMJH semi-solid medium (magnification × 6600). Electron micrograph kindly provided by Bill Coolley, Charlie Dalley and Juan Gallego-Beltran.

growth conditions and exogenous chemical agents. Within tissues or phagocytes the bacteria adopt a spherical (1.5–2 μm) granular form.

The outer membrane contains various antigenic components, including protein, lipids and lipopolysaccharide (LPS), the latter being highly immunogenic. The LPS is typical of Gram-negative bacteria but, because of its relative lack of toxicity compared with other Gram-negative LPS pyrogens, it was until recently considered atypical even without biochemical confirmation. The sugars identified in LPS include arabinose, fucose, galactosamine, glucose, glucosamine, glucose-6-phosphate, N-acetylglucosamine, mannose, mannosamine, rhamnose, ribose and xylose in various proportions according to serovar. Interestingly, 4-o-methyl mannose is characteristic for leptospires (Yanagihara et al., 1983). The detailed chemistry of lipopolysaccharides will not be given here because they are considered in Chapter 6. Of significance is the fact that leptospires do not ferment carbohydrates but possess many of the enzymes associated with the tricarboxylic acid cycle, the pentose-phosphate shunt and other sugar-specific hydrolases, which suggests that these enzymes enable synthesis and bioconversions of sugars as part of the metabolic capability of these organisms.

The outer membrane is easily removed by plasmolysis, gentle disruption and detergents and such preparations have been used as vaccine candidates (Auran et al., 1972). The structure of the outer membrane is little understood but various outer membrane proteins (OMPs) and heat-shock proteins (HSPs) have been identified (see below) and the overall charge is negative, which may contribute to cellular adhesion (Tsuchimoto et al., 1984). A number of common OMPs are conserved among Leptospira (Alves et al., 1999).

The contents of the periplasmic space are unknown other than the presence of flagella. The inner membrane or true cell wall contains peptidoglycan and microfibrils comprising glycoprotein that span the coils of the helical cellular structure (Yanagihara et al., 1975).

Each bacterium has two axial flagella, one at each end of the bacterium, arising through the peptidoglycan layer from a complex hook structure similar to that described for other Gram-negative bacteria. Flagella may be extracted more or less intact by physical disruption of protoplasts or sphaeroplasts followed by density gradient centrifugation, and each flagellum is found to be about half the length of the bacterium. The flagellum is immunogenic, and sera from naturally infected animals react in western blot preparations with a heat-stable protein doublet of about 34 kDa present in whole cell and flagella preparations (Kelson et al., 1988). Each flagellum appears to comprise an inner core composed of one protein of about 34 kDa and an outer sheath composed of another protein of about 36 kDa (Trueba et al., 1992). The flaB gene from L. borgspetersenii serovar hardjo was cloned and expressed in E. coli to yield a 32-kDa FlaB protein (Mitchison et al., 1991) and the flaB gene was found to be highly conversed within the genus Leptospira (Woodward and Redstone, 1994; Lin et al., 1997).

Identification of immunoreactive proteins (B-cell subset only) has been studied by western blots of whole-cell preparations of leptospires probed with immune or convalescent sera. A number of proteins in the size range 34–95 kDa have been observed, of which the flagella proteins were prominent, and, not unsurprisingly, temperature shift demonstrated many heat-shock proteins, including GroEL and DnaK homologues (Stamm et al., 1991). Lipids represent about 15–25% of the dry weight of leptospires, of which about 50% are phosphatidylethanolamine and others are complex in structure. The lipopolysaccharide content is not as well-characterised as, for example, that of salmonellas and E. coli, and although it is regarded as relatively similar in composition, it is unusual in being relatively non-toxic when compared with other Gram-negative LPS.

Classification

Serology and Classification

According to Solomon Faine (Faine, 1994):

> Following the development of genetic taxonomy, nomenclature is confused. The serologically based classification of Kmety and Dikken (1988) still recognises only the species interrogans and biflexa and serovars are sub-specific designations within the species.
>
> Many isolates are the only ones of their serovar. Partly because the difficulties of serotyping tend to inhibit attempts at full identification, partly because the strains may be unique, there are no others serologically identical.
>
> Conventional wisdom treats all leptospires as if they were the same, except for serology and pathogenicity. Increasingly it becomes apparent that there are several species, if not genera, within which members share distinct metabolic as well as genetic characteristics.

This author does not intend to embark on this debate. Suffice it to say, commonly employed serological methods have served the medical and veterinary communities well and meaningfully for some considerable time. However, the lack of readily identifiable phenotypic traits has resulted in the almost exclusive use of serology for identification and classification with its inherent problems. The advent of contemporary genetic methods has given new insights into this old problem and, as in Faine's comments above, there are moves to agree the creation of several genera, or genomospecies, to be recognised within the leptospires.

Serology

The clinical manifestations of leptospiral infections vary, whilst when isolates are compared, the physiological and biochemical characteristics *in vitro* are similar. The lack of sufficient phenotypic differences to support a realistic taxonomic scheme led inevitably to the construction of a differentiation scheme based primarily on serological features. Thus, isolates were tested for agglutination with patient convalescent serum, or sera from animals known to have been infected with characterised isolates, or immune serum from rabbits dosed with characterised isolates. Refinements were made by testing for agglutination after absorption of sera with characterised isolates. There are internationally accepted protocols for the derivation of sera and for undertaking cross-agglutination studies that are not relevant here. Suffice it to say, two species were accepted, namely *L. interrogans*, which includes the pathogenic types, and *L. biflexa*, which includes the saprophytes. Within *L. interrogans* there are 24 defined serogroups that comprise clusters of individual serovars, of which there are over 200, which share antigenic cross-reactivity (Kmety and Dikken, 1988). Another dozen or so less well-defined serovars have yet to be placed into serogroups. It would, however, be unwise to consider such a monothetic scheme as having any taxonomic relevance. Nomenclature follows the convention of the assumed serologically designated genus, and species names are followed by the specific serovar name and often the epithet of the type or individual isolate. Thus *Leptospira interrogans* serovar *icterohaemorrhagiae* strain Verdun is the form commonly used. Recent suggestions favour a simplified form of nomenclature such as *Leptospira* Icterohaemorrhagiae (verdun). It should be pointed out the family name, *Leptospiraceae*, comprises *Leptospira* (*L. interrogans* (all pathogenic), *L. biflexa* (all non-pathogenic)), *Leptonema* with its single species *Leptonema illini*, and *Turneria parva* (formerly *Leptospira parva*) (all non-pathogenic) (Brenner *et al.*, 1999).

Genetic Classification

Genome Structure

With the advent over the past three decades of sophisticated molecular analytical methods, the genome of representative leptospires have been subjected to a number of whole-genome and targeted gene studies. Paster *et al.* (1991) stated that the spirochaetes are an ancient branch of the eubacteria with *Leptospira* representing the deepest division. Leptospires are characterised by a low guanine and cytosine content (G + C ratio) of 35–41%. Significant observations have been made regarding the genome structure largely by pulsed-field gel electrophoresis (PGFE) and gene probing (Saint-Girons *et al.*, 1992; Zuerner *et al.*, 1991, 1993). Two replicons of about 4400 kb and 350 kb in *L. interrogans* serovars *icterohaemorrhagiae* and *pomona* were identified. The smaller was considered to be a large plasmid, but the presence of an essential household gene encoding β-semialdehyde dehydrogenase has led to the recognition that this replicon is a true chromosome.

Simple whole-genome analytical systems have been investigated in order to aid differentiation of leptospires. Various forms of whole-genome restriction enzyme analysis (REA) and PFGE are valuable tools in the description and confirmation of the identity of isolates. Ellis *et al.* (1988) were the first to demonstrate by REA approaches that isolates defined serologically as serovar *hardjo* were readily differentiated into two quite discrete groups, *hardjo-bovis* and *hardjo-prajitno*. Later cross-hybridisation studies confirmed these findings (see below). Interestingly, Ralph *et al.* (1993) showed that the restriction endonuclease *Rsa*I did not cut the genomic DNA from a number of *Leptospira* serovars, and it was suggested that the DNA may be modified by methylation at the recognition sequence, 5′-GTAC-3′. Other analytical procedures have been applied such as arbitrarily primed PCR in order to differentiate serovars.

Cross-hybridisation

The genetic relatedness amongst leptospires has been determined at 55°C and 70°C by DNA–DNA hybridisation of total genomic DNA from a total of 47 serovars, representative of both pathogenic and saprophytic serovars (Yasuda *et al.*, 1987). The hybridisation data indicated relatedness between serovars that in many cases showed little if any correlation with traditional serological classification. Seven new genomospecies were proposed over and above *icterohaemorrhagiae*, and they were designated *L. noguchii*, *L weilii*, *L. santorosa*, *L. borgspetersenii*, *L. meyeri*, *L. wolbachii* and *L. inadai*. Similar studies were undertaken by Ramadass *et al.* (1992) with 66 sero-vars. Broadly the same conclusions as those of Yasuda *et al.* (1987) were reached, and a further genomo-species, *L. kirschneri*, was proposed. Considerable debate continues regarding the classical serological typing of leptospires and the development of a true phylogenetic taxonomy based on DNA similarities. The genomospecies are now accepted but medical and veterinary clinicians continue to use the serologically based nomenclature system. This may well reflect the ability of laboratories to undertake serological rather than genetic analysis of isolates.

Recently, Brenner *et al.* (1999) extended the cross-hybridisation studies to include 303 individual isolates of pathogenic and non-pathogenic lepto-spires and concluded that five new genomospecies were discernible, over and above those defined by Yasuda *et al.* (1987) and Ramadass *et al.* (1992). These are as yet unnamed and listed as genomo-species 1–5; *L. alexanderi* is the proposed name for genomospecies 2.

Ribosomal RNA, Ribosomal Proteins and their Genes

The rRNA scaffold of ribosomes is regarded as highly homologous throughout the eubacteria, and the genes that encode these molecules have been studied intensively for evolutionary relationships between bacteria. Indeed, universal primers based on the 16S rRNA gene are used routinely for the specific detection and differentiation of bacteria by direct sequence PCR approaches. Pathogenic leptospires are unique in that there is a single copy of the 5S rRNA gene and two highly homologous copies of each of the 23S and 16S rRNA genes (Takahashi *et al.*, 1998). These genes are dispersed rather than co-located on the genome. Evidence has been presented that at least in one pathogenic serovar, *L. interrogans* serovar *balcanica*, the 23S rRNA molecule is uniquely pro-cessed into 17S and 14S moieties (Hsu *et al.*, 1990). Ribotyping patterns have been used to differentiate between serovars (Perolat *et al.*, 1993).

The genetic organisation of the genes encoding ribosomal proteins have been studied by Zuerner *et al.* (2000). Commencing with *secY*, a pre-protein translocase gene usually co-located in the *spc* ribo-somal cluster, a chromosomal walking exercise identified a large contiguous region that encodes 32 open reading frames. All shared significant homology with *Borrelia burgdorferi* and *Treponema pallidum* orthologues and have been annotated as the S10-*spc*-α ribosomal protein region. Comparisons between this region in the three genera indicated a significant evolutionary change between *Leptospira*

on the one hand and *Borrelia* and *Treponema* on the other.

Insertion Elements

Repetitive DNA elements have been identified in a number of leptospires. Two quite distinct elements, IS1500 and IS1533, have been characterised to sequence level by a number of authors. Using these elements separately as gene probes in Southern hybrid-isations of whole *Leptospira* genomes, evidence has been gained for discrete hybridisation patterns for these elements in terms of genomic location and number. In very broad terms, the distribution of these elements between serovars follows the probable evolutionary history of leptospires. Woodward and Sullivan (1991) demonstrated the presence of about 40 copies of IS1533 in a cattle isolate originally determined as *L. borgspetersenii* serovar *hardjo-bovis* by serology, and showed considerable degeneracy in the DNA sequence of many of the copies of these elements (**Fig. 2**).

It was speculated that the IS elements may have been acquired as a single burst of multiplication and integration in a primordial *Leptospira* early in the evolution of the genus. Southern hybridisation has shown that the genomic location of the IS1533 ele-ments in *L. borgspetersenii* serovar *hardjo-bovis* and a number of other leptospires are highly consistent between isolates even from widely separated geo-graphical locations. This suggests that the IS elements may no longer be transposable and do not contribute to genomic rearrangements, but this concept is deba-ted (see below). It is possible that specific serovars and IS elements have co-evolved and that this has been beneficial in terms of natural selection and possibly even host adaptation. Similar studies by Zuerner and Bolin (1997) made use of IS1500 to differentiate between closely related isolates and to define new genetic groups. They also proposed the use of IS typing for epidemiological studies.

PCR and ligase-mediated PCR making use of IS element sequences and adjacent chromosomal geno-mic sequences have also been used to develop rapid diagnostic and differentiation tools (Woodward and Redstone, 1993; Zuerner and Bolin, 1997). Whilst the sequences of some IS1533 elements were degen-erate, analysis of IS1500 suggested that this element is related to IS5 and may have active transposase activity. The possibility has been suggested that IS elements may be hot spots for recombination which may induce chromosomal plasticity (see below).

Gene Analysis

A number of contemporary approaches have been taken to clone, sequence and analyse specific genes of

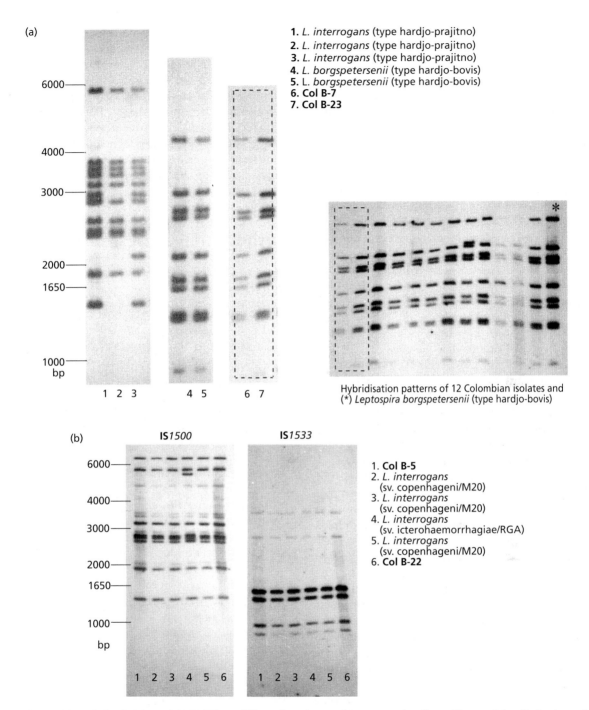

(a)

1. *L. interrogans* (type hardjo-prajitno)
2. *L. interrogans* (type hardjo-prajitno)
3. *L. interrogans* (type hardjo-prajitno)
4. *L. borgspetersenii* (type hardjo-bovis)
5. *L. borgspetersenii* (type hardjo-bovis)
6. **Col B-7**
7. **Col B-23**

Hybridisation patterns of 12 Colombian isolates and (*) *Leptospira borgspetersenii* (type hardjo-bovis)

(b) IS*1500* IS*1533*

1. **Col B-5**
2. *L. interrogans* (sv. copenhageni/M20)
3. *L. interrogans* (sv. copenhageni/M20)
4. *L. interrogans* (sv. icterohaemorrhagiae/RGA)
5. *L. interrogans* (sv. copenhageni/M20)
6. **Col B-22**

Fig. 2 (a) Southern hybridisation with IS*1500* to differentiate *Leptospira* serovars *hardjo-prajitno* and *hardjo-bovis*, and to facilitate classification of clinical isolates from Columbian cattle. (b) Southern hybrisation patterns with IS*1500* and IS*1533* to facilitate classification of clinical isolates from Columbian cattle. Data were kindly provided by Juan Gallego-Beltran.

leptospires. These include immunoscreening of gene expression libraries and complementation of selected autotrophic markers by screening libraries in *E. coli* autotrophs. Many genes have been described, and one of the first genes to be analysed was that encoding the endoflagellin major subunit. Antigen analysis by western blot demonstrated similarities between leptospires (Doherty *et al.*, 1989; Brown *et al.*, 1991) and one protein of approximately 32 kDa was identified as the highly conserved endoflagellin major subunit (Kelson *et al.*, 1988). The endoflagellin was unique in being within the outer sheath and therefore of intrinsic scientific interest to compare with the flagella structures of other well-characterised bacterial species. It was highly immunogenic and a recombinant form was considered of potential use for diagnosis and possibly vaccine development.

Motility *per se* is considered a major virulence determinant in many bacterial pathogens, whilst immunogenic variation within the flagellin has been reported as playing a role in immune evasion. Thus, Mitchison *et al.* (1991) cloned and sequenced the *flaB* gene that encodes the endoflagellin from *L. borgspetersenii* serovar *hardjo-bovis* strain L171. Woodward and Redstone (1994) showed by analysis of the gene from 23 serovars representing 10 serogroups that the gene is highly conserved at the sequence level. In addition, restriction-fragment length polymorphism (RFLP) of *flaB* PCR products was useful in serovar discrimination, whilst the intensity of cross-hybridisation of *flaB* PCR products was in line with the differentiation of leptospires into genomospecies. Similar confirmatory observations were made by Lin *et al.* (1997). Similar findings were shown in a study of another putative virulence determinant, namely the gene encoding the sphingomyelinase, *sphA* (Segers *et al.*, 1990).

Numerous other genes (e.g. *asd*, *aroA*, *dapD*, *groEL*, *leuA*, *proA*, *rpsL*, *rpoC*, *trpE*, *trpG*, etc.) have been cloned and sequenced, but a concerted effort to raise the profile of leptospires is awaited in the near future, so that a complete genome sequence is established of at least one serovar. Boursaux-Eude *et al.* (1998) undertook a preliminary whole-genome study of *L. interrogans* strain RZ11 and *L. interrogans* serovar *icterohaemorrhagiae* Verdun, which are considered to be closely related. Total genomic DNA was digested to completion and resolved by pulsed-field gel electrophoresis and hybridised with probes prepared from characterised genes and also from products generated by random hexanucleotide priming. The latter products were sequenced and compared with the genetic sequence databases to assign putative designations. The larger of the two chromosomes was mapped in both strains and gene locations assigned. Two general conclusions arose from this study. First, genes common to a single metabolic pathway or a particular function are frequently dispersed about the genome. Secondly, the number and location of IS*1500* and IS*1533* elements showed a correlation between clusters of genes that were differentially located on the two genomes. These results point to plasticity of the genome, with IS elements potentially acting as sites for chromosomal rearrangements.

Genetic Approaches to Serovar Differentiation

The elegant work of Ellis and others used whole-genome restriction enzyme analysis (REA) to differentiate between and within serovars (Robinson *et al.*, 1982; Ellis *et al.*, 1989, 1991; Zuerner *et al.*, 1993). Interestingly, direct PCR sequencing of conserved and variable genes and gene regions is a valid and widely accepted tool to tackle both detection and differentiation, but it has not been used widely for *Leptospira*. Other approaches to rapid serovar differentiation have included the development and use serovar-specific gene probes (Van Eys *et al.*, 1988) and Southern hybridisation with IS elements (Woodward and Sullivan, 1991; Paccarini *et al.*, 1992; Zuerner and Bolin, 1997).

Pathogenesis

Source, Routes and the Course of Infection

Animal and Environmental Factors

Man is an incidental or accidental host for many different serovars of *Leptospira* that are primarily host-adapted animal pathogens. The source of leptospire infection is ultimately from animals in which these pathogens are host-adapted and replicate. For all animal species, the organisms localise to the renal tubules and are therefore readily excreted in large numbers in the urine. With the exception of venereal transmission, which is considered an important route for *L. interrogans* serovar *bratislava* in the pig, transmission between animals to humans of all leptospires is most probably by urine. Although leptospires are fastidious, they may survive well in warm water and neutral soils for many days. Numerous studies have identified leptospires in rice (paddy) field water and other surface waters where buffalo are work animals and in which rodents abound. Specifically, for example, Smith and Self (1955) showed that *L. australis* survives in paddy fields for up to 42 days. Many local names are used in rice-growing communities in Asia for leptospirosis acquired by field workers, including rice-harvest jaundice, autumn fever, seven-day fever and so on, and some fields are described as haunted. Other occupations at risk are sewerage workers and miners, who are often in contact with contaminated water sources, whilst veterinarians and animal handlers are at high risk through direct contact with animals. Human-to-human transmission is very infrequent, but routes of infection may include breast-feeding, urine splash and possibly sexual transmission.

Pathogenesis – An Overview

Much of the early work on pathogenesis was carried out by Faine and others from the late 1950s onwards

Table 1 The stages of the pathogenesis of leptospirosis

Entry	Mechanisms unknown	Skin permeability Skin abrasion Oral Conjunctival route	Chemotactic responses to haemoglobin identified but role of motility unknown
Dissemination	Blood circulation	Free-living state in blood with rapid multiplication of pathogenic bacteria Macrophage engulfment	Relative serum resistance *in vivo* not demonstrated *in vitro* Intracellular survival and possible induction of host cell apoptosis
Cell targets and adhesion	Immune evasion	Kidney Heart Meninges Ocular aqueous humour	Tropism for specific tissues well characterised Active cellular adhesion mechanism possibly aided by opsonisation Cellular invasion noted with probable role for motility Up-regulation of LPS and LipL41 and down-regulation of LipL36 during cellular association
Immune responses	Humoral	Rapid IgM responses Variable IgG responses (often absent) Role for cell-mediated immune responses unclear	LPS is the primary immunogen Other immunogens include endoflagellin, the major OMP and other minor proteins
Sequelae	Renal failure	Haemorrhage of renal tubules	Toxin (lipases and haemolysins identified)

with small animal surrogate models, but still relatively little work has been done on the detailed analysis of the pathogenicity mechanisms of this fascinating genus. Pathogenesis may be considered as a series of discrete stages (**Table 1**). Leptospires enter the host and circulate to all tissues via the bloodstream and the lymphatics. Non-pathogenic leptospires are very sensitive to rapid clearance by phagocytosis, whereas pathogenic serovars survive and multiply rapidly. The role of complement and other factors such as innate immunoglobulins is unclear *in vivo*, but leptospires are sensitive to complement-dependent lysis *in vitro*. By definition, these pathogens must possess mechanisms for survival in serum and within phagocytes. Direct animal inoculation has been used to separate leptospires from other agents on the basis of their innate survival and rapid growth. Recent data suggest that leptospires not only survive in macrophages but also mediate macrophage apoptosis. In surrogate animal models, pathogenic leptospires generally do not elicit an inflammatory response at the inoculation site, but a small abscess may be observed at the site if very high doses are administered. Such reactions may be due to leucocyte infiltration or tissue damage; there is no evidence for macrophage taxis toward leptospires.

There is, however, evidence of kidney damage in models such as the hamster in which low doses (10^2 cfu) of specific serovars cause rapid death due largely to renal failure, after intra-peritoneal inoculation. Lesions are also noted in the liver, but before renal damage. Initially the leptospires may be localised in liver, spleen and lung tissues, before surviving organisms colonise and grow rapidly at preferred tissue sites, of which the best described are the renal tubules (Marshall, 1976). Other sites are the heart, meninges and ocular aqueous humour, where the immune response is less efficient. The bacteria form very large multi-cellular intimately attached clumps in the renal tubules and, with time, the organisms penetrate to be excreted in high numbers in the urine. In humans, an inflammatory response generates interstitial nephritis in which lymphocytes, monocytes and plasma cells infiltrate, leading to acute and chronic kidney damage with loss of renal function (Arean et al., 1964). Although there is clear evidence that ocular disorders occur, recovery from short-term visual disturbance associated with the acute disease is generally good (Martins et al., 1998).

Entry in the Human Host

The exact mode of entry remains unclear, but it is thought that leptospires gain access through the skin barrier either through cuts and abrasions or through thoroughly wetted skin. The latter concept is compatible with many occupational forms of the disease, but intra-nasal and intra-ocular routes of infection by spray, droplets and aerosols may also occur. Whether ingestion of contaminated water contributes significantly is also unclear, but a number of epidemiological surveys have suggested that ingestion of contaminated water is the most likely cause of infection in certain circumstances. The implication is that entry may be by both passive and active mechanisms, the later requiring adhesion to and subsequent penetration of target cells. A number of studies have shown that leptospires may enter phagocytes and neutrophils, and adhere and invade a variety of epithelial cells including kidney-derived Vero cells. Whether entry requires an active, as yet undefined, mechanism is unclear and there is no firm evidence that motility aids entry or mediates tissue penetration. Formalin-killed cells are less invasive, but it is open to question whether this constitutes evidence that motility plays a role. Motility contributes to invasiveness in many pathogens, and it is not unreasonable to assume that motility and flagella are specific virulence determinants of leptospires.

Clearance

In the longer term, clearance is achieved by the development of an immune response in which opsonisation of the organisms leads to engulfment by phagocytes and neutrophils. Immunity seems only to be humoral and responses against a number of surface antigens have been described, notably including LPS and a number of specific proteins such as flagellin and OmpL1 (see below). However, a number of lines of evidence point to LPS being the only antigen of significance in inducing immunity, since immunity is limited to homologous serovars.

There is much debate about the use of inappropriate vaccines prepared against heterologous serovars to those endemic in areas where disease is being treated. In serological terms LPS alone contributes to the classical serotyping scheme, and LPS as well as LPS-derived oligosaccharides can induce immunity, whilst monoclonal antibodies against LPS may mediate passive immunity. An oligosaccharide containing rhamnose, ribose, glucose and glucosamine derived by endoglycosidase H digestion and column chromatography from *L. interrogans* serovar *pomona* has been purified (Midwinter *et al.*, 1990). This oligosaccharide, conjugated with diphtheria toxoid, was inoculated into BALB/c mice to raise immune antisera. These agglutinated not only to the specific antigen but also to heterologous serovars such as *hardjo-bovis* and *copenhagi*, but the agglutination titres were lower. Opsonised bacteria of serovars *pomona* and *hardjo-bovis*

were readily phagocytosed by mouse peritoneal macrophages.

Interestingly, Vinh *et al.* (1994) showed that phosphorylated mannose was part of an opsonising epitope of *L. interrogans* serovar *hardjo-prajitno*. The role of other cellular components in immunity remains unclear.

Molecular Biology of Putative Virulence Determinants

Lipopolysaccharide Biosynthesis

Many lines of evidence point to a key role for LPS in the pathogenicity of leptospires, but only recently has LPS attracted detailed molecular-based research. Genes that encode the machinery for LPS biosynthesis are conserved across Gram-negative bacteria. Based on this information, Mitchison *et al.* (1997) prepared a gene probe, *rfbA* and *rfbB*, from *Salmonella enterica* serotype Typhimurium that encodes the dTDO-D-glucose-1-phosphate thimydylyltransferase and the glucose-4,6-dehydratase. The probe was used to hybridise genomic DNA from *L. interrogans* serovar *copenhagi* at low stringency. A genomic region of some 12 kb was identified that shared homology with the probe. Sequence analysis identified six open reading frames (ORF) of which four, designated *rfbC*, *rfbD*, *rfbB* and *rfbA* and transcribed in this order, shared a high degree of amino acid homology with equivalent orthologues of *S. enterica*, *E. coli* and *Sh. flexneri*. Furthermore, these genes were shown by RT-PCR to be expressed in *L. interrogans* serovar *copenhagi* and they complemented a *rfbB* mutation in *Sh. flexneri*, thereby demonstrating functionality.

As discussed earlier, *L. interrogans* serovar *hardjo-prajitno* and *L. borgspetersenii* serovar *hardjo-bovis* were once considered to belong to a single serovar, *L. interrogans* serovar *hardjo*, because of identical serological cross-reactions based on the LPS structure. Recent work by Surujballi *et al.* (1999) suggests, however, that differences in their LPS structures may be detected by monoclonal antibodies and suggests that these may be used in differentiation by blocking ELISA. Genetic approaches proved these serovars to belong to two quite discrete species (see above) and interest in the organisation of their respective LPS biosynthetic machinery was viewed as important in gaining an understanding of the evolution of this major virulence determinant.

Kalambaheti *et al.* (1999) used the rhamnose biosynthesis genes of *L. interrogans* serovar *copenhageni* as a gene probe to clone and sequence the *rfb* cluster of *L. borgspetersenii* serovar *hardjo-bovis*. Thirty-one ORFs, all transcribed in the same direction, were identified as a contiguous series of genes located within a 36.7-kb region. Interestingly the entire region was flanked by IS*1533* elements, while in one intergenic region adjacent to *orfT* there were a number of IS elements described as an IS mosaic. Using degenerate oligonucleotide primers based on this sequence, the homologous region from *L. interrogans* serovar *hardjo-prajitno* was amplified successfully and the sequences of the contiguous 38-kb region was determined (Kalambaheti *et al.*, 1999). Thirty-two ORFs were identified, of which all except one shared significant homology (about 85%) at the peptide level. Not unexpectedly, inter-genic regions showed sequence diversity and the IS element mosaic was absent from *L. interrogans* serovar *hardjo-prajitno*. Comparisons were also made between these sequences and extended sequence data of the *rfb* region of *L. interrogans* serovar *copenhageni*. The two *L. interrogans* serovars shared 98.6% homology for 17 ORFs, between 49% and 82% homology for a further six ORFs, while no homologues were identified for nine ORFs in *L. interrogans* serovar *copenhageni*. Comparisons with database entries have made it possible to identify putative designations of gene function for about 27 of the ORFs and provide clues regarding the hypothetical assembly and structure of LPS, but functional analysis is awaited. In addition, these data give interesting insights into the possible evolutionary pathways for LPS biosynthesis in these organisms, but further analysis of other serovars is required.

Expression of Outer Membrane Proteins during Infection

Whilst LPS is still considered the major antigen for agglutination and opsonisation, the presence of a number of outer membrane proteins prompted their analysis both in terms of immune response and possible functionality in pathogenesis. It should be pointed out that the outer membrane of leptospires contains relatively few proteins as compared with other Gram-negative bacteria. However, Brown *et al.* (1991) showed that nine serovars representative of five serogroups possessed very similar outer membrane protein profiles as determined by SDS-PAGE. In similar studies, about six common proteins were identified, of which a flagella doublet at 34–35.5 kDa was prominent (Nicholson and Prescott, 1993). These data may indicate conservation across the genus of proteins and of their respective functions.

Doherty *et al.* (1989) screened an *L. interrogans* serovar *pomona* strain L10 gene library expressed in *E. coli* with polyclonal rabbit antisera and identified two leptospiral proteins designated p12 and p20. Subsequent analysis showed these proteins to be present in the outer membrane of all 23 pathogenic serovars tested, but in only two of six non-pathogenic serovars. Neither protein had agglutinating or opsonising abilities. Zuerner *et al.* (1991) extended the work on this serovar with a very elegant technique to identify surface-exposed immunogenic antigens. Bacteria grown in a medium supplemented with [^{35}S]methionine were adsorbed with specific rabbit antiserum and the outer membranes were then solubilised. The antibody-bound proteins were precipitated with staphylococcal protein A and six prominent proteins of 22, 26, 31, 36, 42 and 63 kDa were identified. The 31-kDa protein was heat-labile and highly immunogenic. Radioisotope-labelled proteins were also secreted into the medium and it was postulated that one of the secreted proteins was a haemolysin, because there was a correlation between lysis of sheep red blood cells and secretion of this protein.

A 31-kDa outer membrane protein of *L. alstoni* serovar *grippotyphosa* has been sequenced by N-terminal amino acid sequencing, and a degenerate oligonucleotide was designed to probe genomic DNA in order to clone the encoding gene. The deduced amino acid sequence indicated that the mature product (31 113 kDa) contains 10 transmembrane loops. OmpL1 was shown by immunelectron microscopy to be surface located in all 10 pathogenic leptospires tested and it was absent from four non-pathogenic serovars, but agglutination and opsonisation studies for this protein were not reported. However, Shang *et al.* (1995) showed by chemical cross-linkage analysis that OmpL1 formed multimers and, when reconstituted into planar lipid bilayers, generated a channel conductance similar to that of the native leptospiral membrane. OpmL1 is therefore probably a porin.

Shang *et al.* (1996) and Haake *et al.* (1998) used similar cloning approaches to clone and sequence other leptospire outer membrane protein. Significantly, two proteins were demonstrated by pulse-chase experiments to be post-translationally modified to incorporate lipid moieties, and these were designated LipL41 and LipL36. LipL41 is surface exposed and remains constant in amount throughout the life cycle, whereas LipL36, which is abundant in early exponential phase, decreases in concentration with the age of the batch culture. These lipoproteins were present only on pathogenic leptospires. LipL36 was originally cloned from *L. kirschneri* strain RM52 and this serovar was used to inoculate hamsters – a very sensitive animal model – and sera from survivors were tested by ELISA for responses to LipL36. Interestingly, the data indicated that LipL36 may be down-regulated during infection.

Secreted Antigens

Haemolysin activity of leptospires has been well known for many years. Early detailed descriptions were made by Alexander *et al.* (1956). The ability to cause haemolysis has been used to characterise and differentiate isolates (Hathaway and Marshall, 1980) and loss of haemolysin activity is equated with attenuation (Volina *et al.*, 1986). The first to suggest that haemolysin and sphingomyelinase activity were synonymous and that there were likely to be a variety of different types of haemolysin within the genus was Real *et al.* (1989). The gene was cloned in the mid-1980s.

An open reading frame that encoded a 63-kDa protein, which shares homology with orthologous products from *Bacillus cereus* and *Staphylococcus aureus*, was identified by detailed sequence analysis of the haemolysin (sphingomyelinase) gene of *L. borgspetersenii* serovar *hardjo-bovis* strain sponselee (Segers *et al.*, 1990). Secretion was associated with significant post-translational modification, and the secreted haemolysins from both leptospires and recombinant *E. coli* are in the order of 37–41 kDa. Hybridisation and cloning studies by Gaastra *et al.* (1994) showed that seven discrete sphingomyelinases encoded by the *sphA–sphG* genes may be identified within the genus, and some serovars possessed more than one type. Interestingly, the distribution of these genes amongst leptospires is in line with current thinking on genetic classification. Detailed analysis of other secreted lipases and other products is awaited.

Expression of Leptospiral Antigens *In Vivo*

LPS and certain outer membrane proteins are expressed *in vivo* as determined by their response to humoral antibodies, and there are clues that certain antigens, notably LipL36, were down-regulated in the hamster model. Since leptospires adhere to the renal tubules as a site of preferred colonisation, histological and immunocytochemical studies of infected kidney have been undertaken by various researchers (Sterling and Thierman, 1981; Sitprija, 1984; Alves *et al.*, 1987, 1992; Scanziani *et al.*, 1991; Pereira *et al.*, 1998; Barnett *et al.*, 1999). The elegant work of Barnett *et al.* (1999) showed conclusively by immunocytochemical methods that in the hamster model inoculated with *L. kirschneri* strain RM52, LPS, OmpL1 and LipL41 but not LipL36 are elaborated by bacteria that colonise the renal tubules. Ten days after inoculation the kidney structure was entire but the glomeruli were contracted, and though the glomerular spaces were enlarged, an inflammatory infiltrate was not observed. At 28 days, however, an inflammatory infiltrate was prominent, and all the glomeruli were contracted with a protein exudate present in the glomerular spaces.

Leptospires were readily identified in the tubules and in the luminal spaces, as well as in some phagocytes. It is significant that LPS and OmpL1 were also identified in phagocytes.

Adherence to Host Cells and Invasion

Leptospires colonise favoured sites, such as the renal tubules and the aqueous humour of the eye via blood and the reticuloendothelial system. Cinco *et al.* (1980) demonstrated that leptospires invade, survive and escape from within macrophages, an indication that leptospires probably possess specific uptake and survival mechanisms. It is well established that leptospires adhere to mouse fibroblast L929 cells (Vinh *et al.*, 1984), renal epithelial cells (Ballard *et al.*, 1986) and MDCK cells (Thomas and Higbie, 1990). In addition, Ito and Yanagawa (1987) also showed specific adherence to the extracellular matrix of L929 cells, and certain opsonising antibodies enhanced adherence to these cells (Vinh *et al.*, 1984). Merien *et al.* (1997) used double fluorescence labelling to investigate the interactions between *L. interrogans* serovar *icterohaemorrhagiae* strain Verdun and both Vero cells and monocyte–macrophage-like J774.1 cells. Virulent live bacteria adhered to both cell types and almost 100% of bacteria were rapidly internalised, and significant numbers were adherent and internalised within 20 minutes. Laboratory-passaged live avirulent organisms adhered to a similar extent to both cell types, but they were not internalised by Vero cells and poorly internalised by J774.1 cells (about 12% by 2 hours post-infection). Formalin-killed bacteria adhered less efficiently to both cell types, and uptake by J774.1 cells was only 5%. This confirms that virulent leptospires actively invade permissive host cells, but the precise mechanisms of adherence and cell–cell signalling remain to be elucidated.

Uptake is inhibited by monodansylcadaverine but not by cytochalasin D, which indicates that uptake is by endocytosis in coated pits and does not depend on the host cell microfilament system. Significantly, in time course DNA fragmentation studies of J774.1 cells that contain internalised leptospires, apoptosis is induced by the bacteria. In tests *in vivo*, guinea-pig hepatocytes infected with *L. interrogans* serovar *icterohaemorrhagiae* show signs of leptospira-induced apoptosis 48 hours after inoculation (Merien *et al.*, 1998). These elegant studies provide evidence that leptospires have an intracellular stage in their life cycle, and it is possible that this is associated with persistent infection, resistance to antimicrobial treatments and immune evasion. Furthermore, induced macrophage apoptosis may modulate the host immune response, and act early in infection as a mechanism for release from non-specific phagocytosis to facilitate escape to targeted colonisation sites.

Peptidoglycans from leptospires activate complement and stimulate leucocyte phagocytosis and lymphocyte mitogenesis, activities in common with the LPS of Gram-negative bacteria (Cinco *et al.*, 1993). Subsequently it was shown that viable *Leptospira* organisms and leptospire peptidoglycan induce a pro-adhesive effect in human umbilical vein endothelial cells (HUVECs) for neutrophils (Dobrina *et al.*, 1995). Antibodies to neutrophil CD11/CD18 moieties and inhibition of bacterial protein and RNA synthesis inhibited host cell adhesiveness. *Leptospira*-derived LPS stimulates platelet aggregation and adherence of polymorphonuclear neutrophils (PMNs) to HUVECs, both of which depend on platelet-activating factor (PAF) (Isogai *et al.*, 1997).

Motility

All leptospires are motile and the structure of the flagellar apparatus is assumed to be common to all, but this remains unconfirmed. In ordinary liquid media at room temperature the bacteria remain rigid, spin on their longitudinal axes and move at about 10 μm/s in straight lines or wide arcs. They may pause, shudder and rapidly change direction. Towards the end of the exponential phase the bacteria become less motile. Motility is sensitive to reagents that disrupt the outer membrane. Increased viscosity of the medium results in bacterial cells with an altered shape and, indeed, it is postulated that the bacteria are attracted to higher viscosity. In addition, chemotaxis towards haemoglobin has been demonstrated for *L. interrogans* serovar *copenhageni* (Yuri *et al.*, 1993). This may be advantageous for survival in the environment or invasion of host tissue. As described above, formalin-killed cells were less invasive for tissue culture cells, but whether this reflects chemical denaturation of specific ligand–receptors or inactivation of flagella is unknown. The gene, *flaB*, that encodes the major flagellar subunit protein has been cloned and shown to be highly conserved in the genus *Leptospira* (see above).

Immunity and Vaccination

Protective Responses are Humoral

The study of leptospirosis has traditionally been undertaken in susceptible animal hosts (Alston and Broom, 1958) such as the hamster and guinea-pig, and also in the gerbil in which the histopathological responses are similar to those in acute human infections. These models are, however, not as well characterised as the many in-bred genetic mouse lines that have been used extensively to study other bacterial infections. Adler and Faine (1977) used a number of different *Leptospira* serovars including *pomona, copenhageni, icterohaemorrhagiae, hardjo* and *ballum*, which cause acute leptospirosis in guinea-pigs and hamsters, to infect mice. Six-week-old BALB/c mice given up to 10^9 cfu showed no signs of infection and cleared the bacteria within 3 days, but if treated with cyclophosphamide, the mice died by day 5 with increasing bacterial blood counts. The results with nude athymic mice given serovar *pomona* were similar, but they survived if actively immunised or if antileptospiral antibodies were given passively.

The evidence is compelling that in the mouse model protection against lethal dosing with leptospires is due to humoral antibodies. Indeed, there are many examples of passive protection by inoculation with IgG and monoclonal antibodies against LPS and LPS-derived oligosaccharides. Masuzawa *et al.* (1996) has shown that anti-idiotypic antibodies specific for a monoclonal antibody against the main protective surface antigen of *L. interrogans* serovar *lai* are protective in hamster challenge experiments.

The question remains, to what extent, if at all, cell-mediated responses play a part in immunity to *Leptospira* infection. Pereira *et al.* (1998) showed that C3H/HeJ mice, which are hyporesponsive to LPS, are highly susceptible to *L. interrogans* serovar *icterohaemorrhagiae* infection if dosed no later than 21 days of age. Typical kidney damage with haemorrhage was observed, and if a dose of 10^7 cfu was given the lethality was 100%. An interesting observation was made when older mice, with and without CD4 +- and CD8 +-specific depleting antibodies, were dosed, however. Although no significant difference in survival was observed, the extent of vascular damage, particularly in the lung, was significantly increased in CD4 +- and CD8 +-depleted mice.

The Major Immunogens and Killed-cell Vaccines (Bacterins)

As we have seen above, LPS and a number of outer membrane proteins and lipoproteins are major immunogens of leptospires (Masuzawa *et al.*, 1990). Whole-cell preparations killed, for example by heat or formalin, and various extracts, such as EDTA, chloroform/methanol and surface extracts, have been used as bacterins to understand the immune responses and to develop appropriate vaccines. Many of these

Table 2 *Leptospira* serovars[a]

Serovar	Genomospecies	Animal host	Distribution	Disease in host	Disease in humans
Pathogens					
bratislava	interrogans	Pigs (horses)	World-wide	Abortion, stillbirth and infertility	Anicteric disease
canicola	interrogans	Dogs	World-wide	Acute and chronic nephritis	Acute icteric disease
grippotyphosa	kirschneri	Rodents	World-wide	Chronic	Anicteric > icteric disease
hardjo-bovis	borgspetersenii	Cattle	World-wide	Milk drop syndrome and abortion	Anicteric disease
hardjo-prajitno	interrogans	Cattle	World-wide	Milk drop syndrome and abortion	Anicteric disease
icterohaemorrhagiae	interrogans	Rodents (pets; dogs)	World-wide		Acute > icteric disease
pomona	interrogans	Pigs (cattle and sheep)	North America	Acute leptospirosis in piglets, abortion and acute nephritis in adults	Anicteric > icteric disease
tarrasovi	borgspetersenii	Pigs	Europe and Australasia	Abortion	Anicteric disease
lyme	inadai	Mongoose	World-wide		Low pathogenicity
rio	santorosai	Cattle and rodents	South America		Low pathogenicity
panama	noguchii	Rodents (opossum)	Americas		
celledoni	weilii	Rodents	Australasia		
sofia	meyeri	Rodents (cattle)	Europe and Indonesia		
manhao	alexanderi	?	China		
hualin	Species 4	?	China		
Non-pathogens					
patoc	biflexa				
gent	wolbachii				
parva	parva				
pingchan	Species 1	Frog			
holland	Species 3	Water			
saopaulo	Species 5	Water			
illini	Leptonema illini				

[a]A partial list of pathogenic and widespread *Leptospira* serovars. A single serovar of the newly constructed genomospecies of pathogenic and non-pathogenic serovars is listed. A detailed description of the current state of the *Leptospira* phylogeny is given by Brenner *et al.* (1999).

preparations are well-established and commercially available for vaccination of domestic and farmed animal species. The method of preparation significantly alters the potential immunogenicity of a bacterin. Boiled *L. interrogans* serovar *pomona* is not immunogenic when given intra-dermally to humans, rabbits or hamsters, whereas formalin-killed bacterins are immunogenic (Adler and Faine, 1980).

Leptospirosis is regarded largely as a chronic infection of farmed animal species, companion animals and wild animal reservoirs, notably rodents. There is a close relationship between infecting serovar and host (**Table 2**), and humans are an accidental or incidental host, susceptible to many if not all pathogenic serovars. Vaccine development has, therefore, been targeted at protecting farmed and companion animal species with specific regard to the endemic and/or prevalent serovars in the particular host.

Vaccines for Use in Humans

Protection of populations at risk has been a goal for many years, and several human vaccine trials have been undertaken. Vaccines against leptospira have been established in Japan, Russia and China. In a trial supported by the World Health Organization, for example, Torten *et al.* (1973) developed a killed vaccine with *L. interrogans* serovar *grippotyphosa* and *L. interrogans* serovar *szwajizak* for use with 700 irrigation workers. Previous studies had suggested that bacteria grown in a chemically defined protein-free growth medium were highly immunogenic and might be well-tolerated. The vaccine was innocuous and seroconversion was noted in 56% of volunteers and in 100% after boosting. Chapman *et al.* (1990) monitored the response of human volunteers to a bivalent vaccine prepared from killed *L. interrogans* serovar *hardjo* and *L. interrogans* serovar *pomona*. The response was rapid, and specific IgM to both serovars was produced by 6 days after immunisation. Specific IgG production followed, but it was variable and less pronounced against serovar *pomona*. As expected, western blot analysis showed that LPS was recognised as an antigen, but another was flagellin.

Human antiserum protected hamsters against acute infection after challenge with *L. interrogans* serovar *pomona*. According to Masuzawa *et al.* (1991), a complex glycolipid extract of a number of serovars was not only efficacious but had very low endotoxicity, and these would be suitable candidates for further development for use in humans.

A field study with a 'Zhejiang type-D' vaccine in China gave promising results in that the number of cases of leptospirosis during an epidemic season was none in 3260 vaccinated individuals and nine in 3970 unvaccinated controls. Similar studies with other vaccine types have also been reported from Russia, but in certain endemic areas passive immunisation was regarded as the only effective treatment. Trials of a trivalent vaccine (serovars *icterohaemorrhagiae*, *canicola* and *pomona*) are in progress in Cuba. It is interesting that Perolat *et al.* (1992) described severe leptospirosis in a breeding colony of squirrel monkeys and demonstrated the efficacy of treatment with homologous killed vaccine and antibiotic treatment. They suggested that squirrel monkeys might be a useful model for the development of human vaccines.

A number of general conclusions can be drawn from the many studies thus far carried out. First, success with killed vaccines is not consistent and depends on the pre-exposure state of the population. Second, homologous vaccine is essential. The choice of vaccine candidates is important, in the light of the observation by Skilbeck *et al.* (1992), that isolates of *L. interrogans* serovar *pomona* that cause disease in the field show significant genotypic differences from reference strains used to make a vaccine. Third,

immunological responses are generally rapid and detectable as IgM, followed by low levels of IgG that are not persistent or even undetectable. There are, however, examples where as few as 10% of a vaccinated population showed any IgM and none showed IgG (Stoinanova *et al.*, 1982). No easy method is available to monitor either for successful vaccination or for the likely success of killed vaccines in individuals.

Vaccines for Use in Animals

The literature on the development and use of vaccines for animals is extensive. In principle, bacterins have been prepared with endemic and prevalent serovars relevant to the carrier animal host. Vaccines for cattle, pigs, sheep and dogs are commercially available and widely used throughout the world, and a few examples will be discussed to highlight certain problems. Infection with *L. interrogans* serovar *bratislava* is common in pigs worldwide, and *bratislava* bacterin administered intramuscularly protects from renal infection after challenge at 6 weeks but not at 26 weeks (Ellis *et al.*, 1989). A herd suspected of endemic infection with serovar *bratislava* was vaccinated with a hexavalent vaccine that included the homologous serovar and showed reproductive improvement with 96% of sows farrowing. In a similar herd given a pentavalent vaccine lacking the homologous serovar, only 75% of sows farrowed (Frantz *et al.*, 1989). The immune competence of pigs may be related to T-helper cell profiles in particular breeds (Nguyen *et al.*, 1998). The serovars associated with milk drop syndrome and abortion in cattle and prevalent worldwide are *L. interrogans* serovar *hardjo-prajitno* and *L. borgspetersenii* serovar *hardjo-bovis*. Monovalent and multi-valent whole-cell bacterin vaccines and outer membrane subfractions derived from these and other common serovars have been developed and used commercially.

IgM is often readily assayed after vaccination but IgG levels are generally negligible. Evidence for prevention of infection or improvement of reproductive productivity has been mixed (Tripathy *et al.*, 1976; Bey and Johnson, 1986; Bolin *et al.*, 1989a, b; Goddard *et al.*, 1991; Dhaliwal *et al.*, 1994; Samina *et al.*, 1997), but there is some indication that heterologous protection may be afforded (Hancock *et al.*, 1984). The presence of maternal antibody at 4 weeks post-partum did not compromise calf vaccination and was highly protective against homologous challenge (Palit *et al.*, 1991). A closed-herd policy and annual vaccination for 5 years completely eliminated *Leptospira* infection from a previously infected herd (Little *et al.*, 1992). These data suggest that vaccination, as part of a total control policy, may be effective in eliminating *Leptospira* infection from cattle and may thereby reduce the risk of infection to farm workers.

Since cattle and sheep often co-graze and cattle-adapted serovars are maintained in sheep, cattle vaccines are often used in sheep (Gerritsen *et al.*, 1994). Bacterins of serovars *canicola* and *icterohaemorrhagiae* and other serovars have been prepared for use as dog vaccines. Gitton *et al.* (1994) showed that, as is the case in other animals, dogs respond by producing humoral antibodies to the key major antigens of LPS and a number of proteins, notably the flagellin.

Novel Approaches to the Development of Leptospiral Vaccines

Vaccine development has been largely empirical and has followed traditional routes for preparation and delivery. Protective responses are generated in all host species tested to a limited number of surface antigens, and attempts to define these are now being made. The development of live vaccines by attenuation by laboratory culture or by genetic manipulation is under consideration. Toward this goal, Baril *et al.* (1992) cloned the *dapD*, *aroD* and *asd* genes of *L. interrogans* serovar *icterohaemorrhagiae*. In other bacterial species deletions in these genes lead to attenuation without compromising immunogenicity, and vaccines with these deletions are strongly protective. However, the lack of genetic systems for the manipulation of leptospires hinders this approach, although purified recombinant proteins are protective. Thus, vaccination with recombinant OmpL1 and LipL41 presented together in the context of the host *E. coli* cloning vector membrane, give significant levels of protection against acute leptospirosis in a golden Syrian hamster model, but either protein alone is ineffective (Haake *et al.*, 1999). Clearly, immunogen presentation is important.

Recombinant FlaB flagellin of *L. interrogans* serovar *pomona* has been expressed in large quantities in a host *E. coli* cloning vector and shown to react in western blots with convalescent serum (Lin *et al.*, 1999). Its use in vaccination, diagnosis and functional studies is awaited.

Diagnosis

It should be borne in mind that humans are an accidental or incidental host for leptospires, which are adapted to certain specific hosts in which infection is usually chronic and may even be inapparent, whereas infections of humans may be acute. Nevertheless, many of the diagnostic approaches are common to humans and animals, and they will be discussed together. A septicaemic phase is common to icteric and anicteric human leptospirosis. It is usually of short duration and notably presents with fever and a variety of other potentially clinically relevant diagnostic symptoms (Lecour *et al.*, 1989). In geographic areas where leptospirosis is endemic, seasonality and occupation coupled with relatively non-specific signs of fever, headache, light sensitivity, myalgia with notable abdominal and calf muscular pain and nausea possibly with vomiting are a clear pointer to leptospiral infection and should presumptively be treated as such. In non-endemic areas, however, such symptoms may be associated with any number of infections and confirmation of the diagnosis usually requires laboratory investigations.

Direct Detection of *Leptospira*

Cultural Methods

In humans, leptospires invade and circulate in the blood, aggregating and multiplying at a number of key sites, most notably in the renal tubules of the kidney. If antibiotics have not been given organisms may persist in the circulation for about 10 days and may be isolated by direct culture from small blood samples (0.1–0.2 mL) inoculated into 10-mL volumes of EMJH medium. If penicillin has been administered, penicillinase may be added to the medium. Cerebrospinal fluid (CSF) may be positive for organisms for about 5 days from mid-way through the septicaemic phase. The kidney may be infected during the second phase, but the numbers of organisms excreted and their viability vary considerably between samples and serovars. Culture is rarely successful except in very experienced hands and alternative diagnostic strategies must be adopted. Furthermore, slow growth rates preclude culture as a first-line diagnostic tool.

Direct Visualisation

Leptospires may be observed by direct microscopy of blood, CSF and urine, but their presence depends on the stage of infection. Blood is commonly used and may be stained, either non-specifically with stains such as Giemsa or with various silver stains, or preferably with polyvalent antisera conjugated with an appropriate reporter. Of many such reagents a fluorochrome, such as fluorescein isothiocyanate, is widely used (fluorescent antibody test, FAT) in many contexts, notably for the detailed examination of tissues such as kidney and aborted animal fetuses. Generally, and because leptospires are not regarded primarily as intracellular pathogens, tissue samples are rarely examined routinely by FAT. For example, however, in the context of an animal herd with a suspected chronic infection that manifests itself as an abortion outbreak, FAT examination of aborted material would be required (Ellis *et al.*, 1982; Ellis, 1986).

DNA-based Methods and the Polymerase Chain Reaction

The polymerase chain reaction (PCR) is an exquisitely sensitive method for the detection of pathogens in the clinical setting, and gene probes have been used to detect leptospires in a number of clinical samples (Terpstra *et al.*, 1986; LeFebvre, 1987; Millar *et al.*, 1987). Many PCR tests have been developed and are based on conserved genes, such as *flaB* of the major flagellin subunit of the endoflagellin, *ompL1* of the outer membrane protein, regions of the 16S rRNA genes and certain repetitive genetic elements (Van Eys *et al.*, 1989; Woodward *et al.*, 1991). There are, however, many reports that PCR methods are subject to inhibition by components in clinical specimens.

Immunomagnetic separation combined with PCR has been considered as a means to capture organisms and to remove inhibitory substances. Similarly, many commercially available methods for the extraction and clean-up of DNA from clinical specimens are suitable for PCR use, but PCR has yet to be widely adopted. DNA gene probes and oligonucleotide probes have been developed for the confirmation of other test results, in particular to probe smears, impressions and thin sections.

Animal Inoculation

Although, in the early history of leptospirosis study guinea-pig inoculation was used to demonstrate the bacterial aetiology of the disease, this approach to diagnosis is inappropriate because of the many other available strategies and animal inoculation will not here be considered. Such an approach is, however, of value for primary isolation where the identity of the organism must be verified, as in the case of atypical symptoms or failure of culture by other means.

Radioimmunoassay

Assay for the presence of specific antigen has been considered an appropriate approach to the detection of leptospires, and radioimmunoassays for urine samples have been developed, but these have not used extensively except in a limited number of specialist laboratories.

Indirect Detection of *Leptospira*

Biochemical Changes as Sequelae of Infection

Kidney dysfunction is common in the second phase of icteric leptospirosis and biochemical analyses of urine and blood are undertaken. The urine protein concentration is increased, and haemoglobin and all types of blood cells also are frequently found. The blood urea, uric acid, phosphate and creatinine levels are increased. In severe cases there is hepatocellular damage with raised serum levels of certain indicator enzymes, notably glutamine oxaloacetate transaminase and glutamate pyruvate transaminase. The bilirubin is increased and, free haemoglobin is present in the blood, in association with haemorrhage. Jaundice is sometimes encountered in mild forms of the disease, but is common in severe cases.

Microscopic Agglutination Tests (MAT)

Leptospires are immunogenic and an immune response is present early in infection. IgM is usually produced during the first or second week of infection and the titre will gradually rise, later to be replaced by IgG. In serological tests at least, IgM seems less specific than IgG. Titres in humans and animals vary considerably over time and usually fall over months and even years. The titres of late acute phase sera are often variable and reflect the infecting serovar and the immunocompetence of the host. In humans infected with *L. icterohaemorrhagiae* titres may reach 20 000, while in *L. borgspetersenii* serovar *hardjo-bovis* infections may be only about 100. As infection continues, the likelihood of the specific direct detection of the causative organism progressively diminishes as it is cleared by the host, and assay of the immune response becomes the favoured approach to diagnosis.

There are some 212 recognised serovars of pathogenic leptospires that fortunately share common epitopes and are therefore cross-reactive in related serogroups. Diagnostic laboratories maintain a library of live antigens representative of serovars common to the endemic status. Reference laboratories may maintain representatives of all serogroups and these may be used with dilutions of test sera, singly or pooled in simple agglutination tests. Live antigen is generally regarded as generating greater sensitivity and specificity in the MAT, but killed antigen is safer to handle in the routine diagnostic laboratory. Quality control of this diagnostic is a contentious subject, since many laboratories fail to validate their antigens! Another problem is definition of a cut-off point, which will vary from host to host, the serovar involved and whether the infection is or is not endemic in the particular geographical location. Provision of validated negative and positive control serum is also necessary to support these tests.

The MAT retains a favoured status as a front-line diagnostic tool, largely for veterinary reasons. The test is well understood and, if reference strains are available, it is suitably sensitive and relatively easy assay to perform. Skilled workers can cope with a high throughput and accurately interpret the results. Indeed, international trade in live domesticated farm animals and international movement of companion animals is governed by test requirements laid down by individual importing countries. MAT is the gold standard for all leptospiral assays.

Enzyme-linked Immunosorbent and Related Assays

The major B-cell immunogens of leptospires have been identified by whole-cell fractionation and western blot with convalescent sera.

As we have seen above, LPS and various protein antigens including OmpL1 and FlaB, the major subunit protein of the endoflagella, are highly immunogenic and, therefore, are potential diagnostic antigens. Various antigenic fractions and various ELISA/EIA formats have been developed for use in diagnostic and screening programmes. For example in bovine infection, ELISAs have been used routinely with serum and genital discharges (Trueba et al., 1990; Dhaliwal et al., 1996a, b). Attempts to generate specific tests with purified antigen fractions have largely failed because of cross-reactivity. This has in part been overcome by the use of indirect blocking ELISA formats with monoclonal serovar-specific antibodies. The author has shown that MAT and two commercially available immunodiagnostic tests are comparable and share similar sensitivities and specificities for the detection of leptospiral infections in cattle (Woodward et al., 1997). While at first sight this may be regarded as good news for the replacement of the labour-intensive MAT with potentially automated ELISA or related tests, a significant problem remains for the diagnosis of chronic carrier animals, where the titres may be negligible.

Novel serological methods have been developed in recent years for detecting infection in humans. These include, for example, an IgM-based 'immuno-dipstick' (Gussenhoven et al., 1997), a proteinase-K-resistant antigen dot ELISA (Ribiero et al., 1995), an erythrocyte activation ELISA (Shamardin et al., 1995), and a one-point microcapsule agglutination test (MCAT) (Armitsu et al., 1994). ELISAs for veterinary diagnosis and surveillance have also been developed. An ELISA for the diagnosis of L. hardjobovis infection in cattle has been compared with MAT (Bercovich et al., 1990). There was a 90% correlation between the two tests with positive and negative field sera from 704 adult cattle from 90 separate farms. The ELISA was specific, and reacted with only three of 227 seropositive sera from heterologous Leptospira

infections. Comparisons of complement fixation, MAT and ELISA with a limited selection of test sera by Staak et al. (1990) gave less favourable results. There was a close correlation between CFT and MAT, but the correlation between MAT and ELISA was only 74%.

Other ELISA formats have also been developed and evaluated. For example, Palit et al. (1991) used MAT and ELISA to monitor the fate of maternal antibody and the responses of calves to vaccination. To monitor the immune responses in vaccinated bovines, Goddard et al. (1991) compared the detection of circulating IgM and IgG by ELISA with MAT. Sting and Dura (1994) used 2% sodium taurocholate at 50°C to extract antigens from eight Leptospira species for use in ELISA. These extracts performed well in comparative studies with the MAT and the immunofluorescence test on sera from cattle, pigs, horses and dogs. An ELISA system for mass screening of cattle against 14 infectious agents including Leptospira was used by Behymer et al. (1991).

References

Adler B, Faine S (1977) Host immunological mechanisms in the resistance of mice to leptospiral infections. *Infect. Immun.* 17: 67–72.

Adler B, Faine S (1980) Immunogenicity of boiled compared with formalised leptospiral vaccines in rabbits, hamsters and humans. *J. Hyg. (Lond.)* 84: 1–10.

Alexander AD, Smith OH, Hiatt CW, Gleiser CA (1956) Presence of haemolysin in cultures of pathogenic leptospires. *Proc. Soc. Exp. Biol. Med.* 91: 205–211.

Alston JM, Broom JC (1958) *Leptospirosis in Man and Animals.* Edinburgh: E and S Livingstone.

Alves VAF, Vianna MR, Yasuda PH, De Brito T (1987) Detection of leptospiral antigen in the human liver and kidney using an immunoperoxidase staining procedure. *J. Pathol.* 151: 125–131.

Alves VAF, Gayotto LCC, De Brito T et al. (1992) Leptospiral antigens in the liver of experimentally infected guinea pig and their relation to morphogenesis of liver damage. *Exp. Toxicol. Pathol.* 44: 425–434.

Alves VA, LeFevbre RB, Probert W (1999) Identification of outer envelope proteins conserved among *Leptospira* serovars. *Rev. Med. Vet.* 150: 877–884.

Arean VM, Sasarin G, Green JH (1964) The pathogenesis of leptospirosis: toxin production by *Leptospira icterohaemorrhagiae. Am. J. Vet. Res.* 25: 836–843.

Armitsu Y, Kmety E, Ananyina Y et al. (1994) Evaluation of the one-point microcapsule agglutination test for diagnosis of leptospirosis. *Bull. WHO* 72: 395–399.

Auran NE, Johnson RC, Ritzi DM (1972) Isolation of the outer sheath of Leptospira and its immunogenic properties in hamsters. *Infect. Immun.* 5: 968–975.

Ballard SA, Williamson M, Adler B, Vinh T, Faine S (1986) Interactions of virulent and avirulent leptospires with primary cultures of renal epithelial cells. *J. Med. Microbiol.* 21: 59–67.

Baril C, Richaud C, Fournie E, Baranton G, Saint-Girons I (1992) Cloning of *dapD*, *aroD* and *asd* genes of *Leptospira interrogans* serovar *icterohaemorrhagiae*, and nucleotide sequence of the *asd* gene. *J. Gen. Microbiol.* 138: 47–53.

Barnett JK, Barnett D, Bolin CA *et al.* (1999) Expression and distribution of leptospiral outer membrane components during renal infection in hamsters. *Infect. Immun.* 67: 853–861.

Behymer DE, Riemann HP, Utterback W, D-Elmi C, Franti CE (1991) Mass screening of cattle sera against 14 infectious disease agents, using an ELISA system for monitoring health in livestock. *Am. J. Vet. Res.* 52: 1699–1705.

Bercovich Z, Taaijke R, Bokhout BA (1990) Evaluation of an ELISA for the diagnosis of experimentally induced and naturally occurring *Leptospira hardjo* infections in cattle. *Vet. Microbiol.* 21: 255–262.

Bey RF, Johnson RC (1986) Current status of leptospiral vaccines. *Progr. Vet. Microbiol. Immunol.* 2: 175–197.

Bolin CA, Thiermann AB, Handsaker AL, Foley JW (1989a) Effect of vaccination with a pentavalent leptospiral vaccine on *Leptospira interrogans* serovar *hardjo* type *hardjo-bovis* infection in pregnant cattle. *Am. J. Vet. Res.* 50: 161–165.

Bolin CA, Zuerner RL, Trueba G (1989b) Effect of vaccination with a pentavalent leptospiral vaccine on *Leptospira interrogans* serovar *hardjo* type *hardjo-bovis* infection of cattle. *Am. J. Vet. Res.* 50: 2004–2008.

Boursaux-Eude C, Saint-Girons I, Zuerner R (1998) *Leptospira* genomics. *Electrophoresis* 19: 589–592.

Brenner DJ, Kaufman AF, Sulzer KR, Steigerwalt AG, Rogers FC, Weyant RS (1999) Further determination of DNA relatedness between serogroups and serovars in the family *Leptospiraceae* with a proposal for *Leptospira alexanderi* sp. nov, four new *Leptospira* genomospecies. *Int. J. Syst. Bacteriol.* 49: 839–858.

Brown JA, LeFebvre RB, Pan MJ (1991) Protein and antigen profiles of prevalent serovars of *Leptospira interrogans*. *Infect. Immun.* 59: 1772–1777.

Chapman AJ, Faine S, Adler B (1990) Antigens recognised by the human immune response to vaccination with a bivalent hardjo/pomona leptospiral vaccine. *FEMS Microbiol. Immunol. Lett.* 2: 111–118.

Cinco M, Banfi E, Soranzo MR (1980) Studies on the interaction between macrophages and leptospires. *J. Gen. Microbiol.* 124: 409–413.

Cinco M, Perticari S, Presani G, Dobrina A, Liut F (1993) Biological properties of a peptidoglycan extracted from *Leptospira interrogans*: in vitro studies. *J. Gen. Microbiol.* 139: 2959–2964.

Dhaliwal GS, Murray RD, Downham DY, Dobson H (1994) The effect of vaccination against *Leptospira interrogans* serovar *hardjo* infection at the time of service on pregnancy rates in dairy cows. *Vet. Res.* 25: 271–274.

Dhaliwal GS, Murray RD, Dobson H, Montgomery J, Ellis WA, Baker J (1996a) Presence of antigen and antibodies in serum and genital discharges of heifers after experimental intrauterine inoculation with *Leptospira interrogans* serovar hardjo. *Res. Vet. Sci.* 60: 157–162.

Dhaliwal GS, Murray RD, Dobson H, Montgomery J, Ellis WA (1996b) Presence of antigen and antibodies in serum and genital discharges of cows from dairy herds naturally infected with *Leptospira interrogans* serovar hardjo. *Res. Vet. Sci.* 60: 163–167.

Dobrina A, Nardon E, Vecile E, Cinco M, Patrarca P (1995) *Leptospira icterohaemorrhagiae* and leptospire peptidoglycans induce endothelial cell adhesiveness for polymorphonuclear leukocytes. *Infect. Immun.* 63: 2995–2999.

Doherty JP, Adler B, Rood JI, Billington SJ, Faine S (1989) Expression of two conserved leptospiral antigens in *Escherichia coli*. *J. Med. Microbiol.* 28: 143–149.

Ellinghausen HC, McCullogh WG (1965) Nutrition of '*Leptospira pomona*' and growth of 13 other serotypes: fractionation of oleic albumin complex and a medium of bovine albumin and polysorbate 80. *Am. J. Vet. Res.* 26: 45–51.

Ellis WA (1986) The diagnosis of leptospirosis in farm animals. In: Ellis WA, Little TWA (eds) *The Present State of Leptospirosis Diagnosis and Control*. Dordrecht: Martinus Nijhoff, pp. 13–31.

Ellis WA, O'Brien JJ, Neill SD (1982) Bovine leptospirosis: microbiological and serological findings in aborted foetuses. *Vet. Rec.* 110: 147.

Ellis WA, Thiermann AB, Montgomery JM, Handsaker A, Winter PJ, Marshall BB (1988) Restriction endonuclease analysis of *Leptospira interrogans* serovar hardjo. *Res. Vet. Sci.* 44: 375–379.

Ellis WA, Montgomery JM, McParland PJ (1989) An experimental study with a *Leptospira interrogans* serovar *bratislava* vaccine. *Vet. Rec.* 125: 319–321.

Ellis WA, Montgomery JM, Thiermann AB (1991) Restriction endonuclease analysis as a taxonomic tool in the study of pig isolates belonging to the Australis serogroup of *Leptospira interrogans*. *J. Clin. Microbiol.* 29: 957–961.

Faine S (1994) *Leptospira and Leptospirosis*. Ann Arbor: CRC Press.

Faine S, Adler B, Bolin CA, Perolat P (1999) *Leptospira and Leptospirosis*, 2nd edn. Melbourne, Australia: MediSci.

Frantz JC, Hanson LE, Brown AL (1989) Effect of vaccination with a bacterin containing *Leptospira interrogans* serovar *bratislava* on the breeding performance of swine herds. *Am. J. Vet. Res.* 50: 1044–1047.

Gaastra W, van der Zeijst BAM, Segers RPAM (1994) Sphingomyelinase genes of Leptospira. In: Freer JH (ed.) *Bacterial Protein Toxins, Zbl. Bakteriol.* suppl. 24: 88–89.

Gerritsen MJ, Koopmans MJ, Peterse D, Olyhoek T (1994) Sheep as a maintenance host for *Leptospira interrogans* serovar *hardjo* subtype hardjo-bovis. *Am. J. Vet. Res.* 55: 1232–1237.

Gitton X, Daubie MB, Andre F, Ganiere JP, Andre-Fountaine G (1994) Recognition of *Leptospira interrogans* antigens by vaccinated or infected dogs. *Vet. Microbiol.* 41: 87–97.

Goddard RD, Luff PR, Thornton DH (1991) The serological response of calves to Leptospira interrogans serovar hardjo vaccines and infection measured by the microscopic agglutination test and anti-IgM and anti-IgG enzyme-linked immunosorbent assay. *Vet. Microbiol.* 26: 191–201.

Gussenhoven GC, Van Der Hoorn MA, Goris MG *et al.* (1997) LEPTO dipstick, a dipstick assay for detection of Leptospira-specific immunoglobulin M antibodies in man. *J. Clin. Microbiol.* 35: 92–97.

Haake DA, Martinich C, Summers TA *et al.* (1998) Characterisation of leptospiral outer membrane lipoprotein LipL36: downregulation associated with late-log-phase growth and mammalian infection. *Infect. Immun.* 66: 1579–1587.

Haake DA, Mazel MK, McCoy AM *et al.* (1999) Leptospiral outer membrane proteins OmpL1 and LipL41 exhibit synergistic immunoprotection. *Infect. Immun.* 67: 6572–6582.

Hancock GA, Wilks CR, Kotiw M, Allen JD (1984) The long term efficacy of a hardjo-pomona vaccine preventing leptospiruria in cattle exposed to natural challenge with *Leptospira interrogans* serovar *hardjo*. *Austr. Vet. J.* 61: 54–56.

Hathaway SC, Marshall RB (1980) Haemolysis as a means of distinguishing between *Leptospira interrogans* serovars *balcanica and hardjo*. *J. Med. Microbiol.* 13: 477–481.

Henneberry RC, Cox CD (1970) β-oxidation of fatty acids by *Leptospira*. *J. Can. Microbiol.* 16: 41–45.

Hsu D, Pan MJ, Zee YC, LeFebvre RB (1990) Unique ribosome structure of *Leptospira interrogans* is composed of four rRNA components. *J. Bacteriol.* 172: 3478–3480.

Isogai E, Hirose K, Kimura K *et al.* (1997) Role of platelet-activating-factor (PAF) on cellular responses after stimulation with leptospire lipopolysaccharide. *Microbiol. Immunol.* 41: 271–275.

Ito T, Yanagawa R (1987) Leptospiral attachment to extracellular matrix of mouse fibroblast (L929) cells. *Vet. Microbiol.* 15: 89–96.

Johnson RC, Gary ND (1962) Nutrition of 'Leptospira pomona'. I. Studies on a chemically defined substitute for the rabbit serum ultrafiltrate. *J. Bacteriol.* 85: 976–982.

Johnson RC, Harris VG (1968) Purine analogue sensitivity and lipase activity of leptospires. *Appl. Microbiol.* 16: 1584–1590.

Johnson RC, Rogers P (1964a) 5-fluorouracil as a selective agent for growth of Leptospirae. *J. Bacteriol.* 88: 422–426.

Johnson RC, Rogers P (1964b) Differentiation of pathogenic and saprophytic leptospires by 8-azaguanidine. *J. Bacteriol.* 88: 1618–1623.

Johnson RC, Rogers P (1967) Utilisation of amino acids and purine and pyrimidine bases. *Arch. Biochem. Biophys.* 107: 459–470.

Kalambaheti T, Bulach DM, Rajakumar K, Adler B (1999) Genetic organisation of the lipopolysaccharide O-antigen biosynthetic locus of *Leptospira borgspetersenii* subtype hardjo-bovis. *Microb. Pathog.* 27: 105–117.

Kelson JS, Adler B, Chapman AJ, Faine S (1988) Identification of flagellar antigens by gel electrophoresis and immunoblotting. *J. Med. Microbiol.* 26: 47–53.

Kmety E, Dikken H (1988) *Revised List of Leptospira Serovars*. International Committee on Systematic Bacteriology of IUMS. Groningen, The Netherlands: University Press.

Lecour H, Miranda M, Magro C, Rocha A, Goncalves V (1989) Human Leptospirosis – a review of 50 cases. *Infection* 17: 8–12

LeFebvre RB (1987) DNA probe for the detection of *Leptospira interrogans* serovar *hardjo* type hardjo-bovis. *J. Clin. Microbiol.* 25: 2236–2238.

Lin M, Surujballi O, Nadin-Davis S, Randall G (1997) Identification of a 35-kilodalton serovar-cross-reactive flagellar protein, FlaB from Leptospira interrogans by N-terminal sequencing, gene cloning, and sequence analysis. *Infect. Immun.* 65: 4355–4359.

Lin M, Bughio N, Surujballi O (1999) Expression in *Escherichia coli* of *flaB*, the gene encoding for a periplasmic flagellin of *Leptospira interrogans* serovar *pomona*. *J. Med. Microbiol.* 48: 977–982.

Little TWA, Hathaway SC, Broughton ES, Seawright D (1992) Control of *Leptospira hardjo* infection in beef cattle by whole-herd vaccination. *Vet. Rec.* 131: 90–92.

Marshall RB (1976) Ultrastructural changes in renal tubules of sheep following experimental infection with *Leptospira interrogans* serotype *pomona*. *J. Med. Microbiol.* 7: 505–508.

Martins MG, Matos KT, da Silva MV, Abreu MT (1998) Ocular manifestations in the acute phase on leptospirosis. *Ocular Immunol. Inflamm.* 6: 75–79.

Masuzawa T, Nakamura R, Shimizu T, Yanagihara Y (1990) Biological activities and endotoxic activities of protective antigens (PAgs) of *Leptospira interrogans*. *Zbl. Bakteriol.* 274: 109–117.

Masuzawa T, Suzuki R, Yanagihara Y (1991) Comparison of protective effects with tetra-valent glycolipid antigens and whole cell-inactivated vaccine in experimental infection of Leptospira. *Microbiol. Immunol.* 35: 199–208.

Masuzawa T, Suzuki R, Yanagihara Y (1996) Protective activity of rabbit polyclonal anti-idiotype antibody against *Leptospira interrogans* infection in hamsters. *Biol. Pharm. Bull.* 19: 613–615.

Merien F, Baranton G, Perolat P (1997) Invasion of Vero cells and induction of apoptosis in macrophages by pathogenic *Leptospira interrogans* are correlated with virulence. *Infect. Immun.* 65: 729–738.

Merien F, Trucculo J, Rougier Y, Baranton G, Perolat P (1998) *In vivo* of apoptosis of hepatocytes in guinea pigs infected with *Leptospira interrogans* serovar *icterohaemorrhagiae*. *FEMS Microbiol. Immunol. Lett.* 169: 95–102.

Midwinter A, Adler B, Faine S (1990) Vaccination of mice with lipopolysaccharide (LPS) and LPS-derived immuno-conjugates from *Leptospira interrogans*. *J. Med. Microbiol.* 33: 199–204.

Millar BD, Chappel RJ, Adler B (1987) Detection of leptospires in biological fluids using DNA hybridisation. *Vet. Microbiol.* 15: 71–78.

Mitchison M, Rood JI, Faine S, Adler B (1991) Molecular analysis of a *Leptospira borgspetersenii* gene encoding an endoflagellar subunit protein. *J. Gen. Microbiol.* 137: 1529–1536.

Mitchison M, Bulach DM, Vinh T, Rajakumar K, Faine S, Adler B (1997) Identification and characterisation of the dTDP-Rhamnose biosynthesis and transfer genes of lipopolysaccharide-related *rfb* locus in *Leptospira interrogans* serovar *copenhagi*. *J. Bacteriol.* 179: 1262–1267.

Nicholson VM, Prescott JF (1993) Outer membrane proteins of three pathogenic *Leptospira* species. *Vet. Microbiol.* 36: 123–138.

Nguyen VP, Wong CW, Hinch GN, Singh D, Colditz IG (1998) Variation in the immune status of two Australian pig breeds. *Austr. Vet. J.* 76: 613–617.

Paccarini ML, Savio ML, Tagliabue S, Rossi C (1992) Repetitive sequences cloned from *Leptospira interrogans* serovar hardjo genotype hardjo-prajitno and their application to serovar identification. *J. Clin. Microbiol.* 30: 1243–1249.

Palit A, Middleton H, Sheers J, Basilone C (1991) The influence of maternal antibody and age of calves on effective vaccination against *Leptospira interrogans* hardjo. *Austr. Vet. J.* 68: 299–303.

Paster BJ, Dewhirst FE, Weisberg WG *et al.* (1991) Phylogenetic analysis of the spirochaetes. *J. Bacteriol.* 173: 6101–6109.

Pereira MM, Andrade J, Marchevsky RS, Ribiero dos Santos R (1998) Morphological characterisation of lung and kidney lesions in C3H/HeJ mice infected with *Leptospira interrogans*: defect of CD4 and CD8 T-cells are prognosticators of the disease progression. *Exp. Toxicol. Pathol.* 50: 191–198.

Perolat P, Poingt JP, Vie JC, Jouaneau C, Baranton G, Gysin J (1992) Occurrence of severe leptospirosis in a breeding colony of squirrel monkeys. *Am. J. Trop. Med. Hyg.* 46: 538–545.

Perolat P, Merien F, Ellis WA, Barranton G (1993) Characterisation of *Leptospira* isolates from serovar hardjo using ribotyping, arbitrarily primed PCR and mapped restriction site polymorphisms. *Res. Microbiol.* 144: 5–15.

Ralph D, Que Q, van Etten JL, McCelland M (1993) *Leptospira* genomes are modified at 5′ GTAC. *J. Bacteriol.* 175: 3913–3915.

Ramadass P, Jarvis BDW, Corner RJ, Penny D, Marshall RB (1992) Genetic characterisation of pathogenic Leptospira species by DNA hybridisation. *Int. J. Syst. Bacteriol.* 42: 215–219.

Real G del, Segers RPAM, van der Zeijst BAM, Gaastra W (1989) Cloning of a haemolysin gene from *Leptospira interrogans* serovar hardjo. *Infect. Immun.* 57: 2588–2590.

Ribiero MA, Souza CC, Almeida SH (1995) Dot-ELISA for human leptospirosis employing immunodominant antigen. *J. Trop. Med. Hyg.* 98: 452–456.

Robinson AJ, Ramadas P, Lee A, Marshall RB (1982) Differentiation of subtypes within *Leptospira interrogans* serovars *hardjo*, *balcanica*, and *tarrasovi* by bacterial restriction-endonuclease DNA analysis (BRENDA). *J. Med. Microbiol.* 15: 331–338.

Saint-Girons I, Norris SJ, Gobel U, Meyer J, Walker EM, Zuerner R (1992) Genome structure of spirochaetes. *Res. Microbiol.* 143: 615–621.

Samina I, Brenner J, Moalem U, Berenstein M, Cohen A, Peleg BA (1997) Enhanced antibody response in cattle against *Leptospira hardjo* by intradermal vaccination. *Vaccine* 15: 1434–1436.

Scanziani E, Luinin M, Fabbi M, Pizzocara P (1991) Comparison between specific immunoperoxidase staining and bacteriological culture in the diagnosis of renal leptospirosis in pigs. *Res. Vet. Sci.* 50: 229–232.

Segers RPAM, van der Drift A, de Nijs A, Corcione P, van der Zeijst BAM, Gaastra W (1990) Molecular analysis of sphingomyelinase C gene from *Leptospira interrogans* serovar hardjo. *Infect. Immun.* 58: 2177–2185.

Shamardin VA, Ruud NV, Il'iasov BK (1995) The effectiveness of immunological methods in the diagnosis of leptospirosis. *Z. Mikrobiol. Epidemiol. Immunobiol.* 2: 84–86.

Shang ES, Exner MM, Summers TA *et al.* (1995) The rare outer membrane protein, OmpL1, of pathogenic Leptospira species is a heat modifiable porin. *Infect. Immun.* 63: 3174–3181.

Shang ES, Summers TA, Haake DA (1996) Molecular cloning and sequence analysis of the gene encoding LipL41: a surface exposed lipoprotein of pathogenic *Leptospira* species. *Infect. Immun.* 64: 2322–2330.

Sitprija V (1984) Renal involvement in leptospirosis. In: Robinson RR (ed.) *Nephrology*. New York: Springer-Verlag, pp. 1041–1052.

Skilbeck NW, Lyon M, Stallman N (1992) Genetic diversity among Australian and New Zealand isolates of *Leptospira interrogans* serovar *pomona*. *Austr. Vet. J.* 69: 29–30.

Smith DJW, Self HRM (1955) Observations on the survival of *Leptospira australis* in soil and water. *J. Hyg.* 53: 436–444.

Staak C, Mekapratreep M, Kampe U, Schonberg A (1990) Serological reactions of leptospirosis-positive (MAR and CFT) bovine sera in ELISA. *Zbl. Veterinarmed.* 37: 581–589.

Stamm LV, Gherardini FC, Parrish EA, Moomaw CR (1991) Heat shock response of spirochaetes. *Infect. Immun.* 8: 1572–1575.

Sterling CR, Thierman AB (1981) Urban rats as chronic carriers of leptospirosis: an ultrastructural investigation. *Vet. Pathol.* 18: 628–637.

Sting R, Dura U (1994) Isolation of serovar-specific leptospiral antigens for use in an enzyme-linked immunosorbent assay (ELISA) compared with the microscopic agglutination test and immunofluorescence. *Zbl. Veterinarmed.* 41: 166–175.

Stoinanova NA, Popova EM, Sergio LM, Potachev AF (1982) Nature of the antibodies in persons inoculated with a leptospirosis vaccine. *Z. Mikrobiol. Epidemiol. Immunol.* 4: 93–95.

Surujballi O, Howlett C, Henning D (1999) Production and characterisation of monoclonal antibodies specific for *Leptospira borgspetersenii* serovar *hardjo* type *hardjo-bovis* and *Leptospira interrogans* serovar *hardjo* type *hardjo-prajitno*. *Can. J. Vet. Res.* 63: 62–68.

Takahashi Y, Akase K, Hirano H, Fukunaga M (1998) Physical and genetic maps of the *Leptospira interrogans* serovar *icterohaemorrhagiae* strain ictero no.1 chromosome and sequencing of a 19-kb region of the genome containing the 5S rRNA gene. *Gene* 215: 37–45.

Terpstra WJ, Schoone GJ, Ter-schegget J (1986) Detection of leptospiral DNA by nucleic acid hybridisation with 32P and biotin labeled probes. *J. Med. Microbiol.* 22: 23–28.

Thomas DD, Higbie LM (1990) *In vitro* association of leptospires with host cells. *Infect. Immun.* 58:581–585.

Torten M, Shenberg E, Gerichter CB, Neuman P, Klingberg MA (1973) A new leptospiral vaccine for use in man. II. Clinical and serologic evaluation of a field trial with volunteers. *J. Infect. Dis.* 128: 647–651.

Tripathy DN, Hanson LE, Mansfield ME (1976) Evaluation of the immune response of cattle to leptospiral bacterins. *Am. J. Vet. Med.* 37: 51–55.

Trueba GA, Bolin CA, Thoen CO (1990) Evaluation of an enzyme immunoassay for diagnosis of the bovine leptospirosis caused by *Leptospira interrogans* serovar *hardjo* type *hardjo-bovis*. *J. Vet. Diagn. Invest.* 2: 323–329.

Trueba G, Bolin C, Zuerner R (1992) Characterisation of the periplasmic flagellum proteins of *Leptospira interrogans*. *J. Bacteriol.* 174: 4761–4768.

Tsuchimoto M, Niikura M, Ono E, Kida H, Yanagawa R (1984) Leptospiral attachment to cultured cells. *Zbl. Bakteriol. Microbiol. Hyg. Reihe* A258: 268–274.

Van Eys GJJM, Zaal J, Schoone GJ, Terpstra WJ (1988) DNA hybridisation with *hardjo-bovis* specific recombinant probes as a method for type discrimination of *L. interrogans* serovar *hardjo*. *J. Gen. Microbiol.* 134: 567–574.

Van Eys GJJM, Gravenkamp G, Gerritsen MJ et al. (1989) Detection of leptospires in urine by polymerase chain reaction. *J. Clin. Microbiol.* 27: 2258–2262.

Vinh T, Faine S, Adler B (1984) Adhesion of leptospires to mouse fibroblasts (L929) and its enhancement by specific antibody. *J. Med. Microbiol.* 18: 73–85.

Vinh T, Adler B, Faine S (1986) Ultrastructure and chemical composition of lipopolysaccharide extracted from *Leptospira interrogans* serovar *copenhagi*. *J. Gen. Microbiol.* 132: 103–109.

Vinh T, Shi MH, Adler B, Faine S (1989) Glycoprotein cytotoxin from *Leptospira interrogans* serovar *copenhagi*. *J. Gen. Microbiol.* 132: 111–123.

Vinh T, Faine S, Handley CJ, Adler B (1994) Immunochemical studies of epitopes of opsonic epitopes of

the lipopolysaccharide of *Leptospira interrogans* serovar *hardjo*. *FEMS Immunol. Med. Microbiol.* 8: 99–107.

Volina EG, Levina LF, Soboleva GL (1986) Phospholipase activity and virulence of pathogenic *Leptospiraceae*. *J. Hyg. Epidemiol. Microbiol. Immunol.* 30: 163–169.

Woodward MJ, Redstone JS (1993) Differentiation of *Leptospira* serovars by the polymerase chain reaction and restriction fragment length polymorphism. *Vet. Rec.* 132: 325–326.

Woodward MJ, Redstone JS (1994) Deoxynucleotide sequence conservation of the endoflagellin subunit protein gene, *flaB*, within the genus *Leptospira*. *Vet. Microbiol.* 40: 239–251.

Woodward MJ, Sullivan GJ (1991) Nucleotide sequence of a repetitive element isolated from *Leptospira interrogans* serovar hardjo type hardjo-bovis. *J. Gen. Microbiol.* 137: 1101–1109.

Woodward MJ, Sullivan GJ, Palmer NMA, Woolley JC, Redstone JS (1991) Development of a PCR test specific for *L. hardjo-bovis*. *Vet. Rec.* 128: 281–283.

Woodward MJ, Swallow C, Kitching A, Dalley C, Sayers AR (1997) *Leptospira hardjo* serodiagnosis: a comparison of MAT, ELISA and Immunocomb. *Vet. Rec.* 141: 603–604.

Yanagihara Y, Hattori Y, Mifuchi I (1975) Purification of microfibres of *Leptospira* and their chemical and immunological properties. In: *Proceedings of the National Symposium on Leptospirosis, Leptospira and other Spirochaeta, Bucharest, Romania*, pp. 347–353.

Yanagihara Y, Kamisango K, Takeda K, Mifuchi I, Azuma I (1983) Identification of 4-*o*-methylmannose in cell wall polysaccharide of *Leptospira*. *Microbiol. Immunol.* 27: 711–715.

Yasuda PH, Steigerwalt AG, Snizer KR, Kaufmann AF, Rogers F, Brenner DJ (1987) Deoxyribonucleic acid relatedness between serogroups and serovars in the family *Leptospiraceae* with proposals for seven new *Leptospira* species. *Int. J. Syst. Bacteriol.* 37: 407–415.

Yuri K, Takamoto Y, Okada M, Hiramune T, Kikuchi N, Yanagawa R (1993) Chemotaxis of leptospires to haemoglobin in relationship to virulence. *Infect. Immun.* 61: 2270–2272.

Zuerner RL, Bolin CA (1997) Differentiation of *Leptospira interrogans* isolates by IS1500 hybridisation and PCR assays. *J. Clin. Microbiol.* 35: 2612–2617.

Zuerner RL, Knudtson W, Bolin CA (1991) Characterisation of outer membrane and secreted proteins of *Leptospira interrogans* serovar *pomona*. *Microb. Pathog.* 10: 311–322.

Zuerner RL, Ellis WA, Bolin CA, Montgomery JM (1993) Restriction fragment length polymorphisms distinguish *Leptospira borgspetersenii* serovar *hardjo* type *hardjo-bovis* from different geographical locations. *J. Clin. Microbiol.* 31: 578–583.

Zuerner RL, Hartskeerl RA, van de Kemp H, Bal AE (2000) Characterization of the Leptospira interrogans S10-spc-α operon. *FEMS Microbiol. Lett.* 182: 303–308.

101

Francisella

Karen L. Elkins[1] and Francis E. Nano[2]

[1]Center for Biologics Evaluation and Research, FDA, Rockville, MD, USA
[2]University of Victoria, Victoria, British Columbia, Canada

Francisella tularensis is the best-known member of the genus *Francisella*, and is the causative agent of tularaemia. *F. tularensis* is responsible for a relatively small number of cases of disease in North America, Europe and northern Asia, but remains a notable public health concern in Russia. It is a highly contagious bacterium that can infect most mammals by a variety of routes.

The name of the genus properly honours Edward Francis, who contributed classic studies on this pathogen (Francis, 1925, 1928). *F. tularensis* was originally isolated by McCoy and Chapin (1912) after they observed a plague-like illness among ground squirrels in Tulare County, California. It soon became clear that the agent responsible for this disease in rodents, first called *Bacterium tularense*, was also responsible for human illnesses variously called lemming fever, deer-fly fever, rabbit fever, tick fever and Ohara's disease (Francis, 1925, 1928; Ohara, 1954). As several of these early names imply, *F. tularensis* can be carried by both arthropod vectors, such as ticks and deer flies, and by over 150 species of vertebrates. The animal groups most often observed to carry *F. tularensis* are lagomorphs and rodents. The pathogen is transmitted to humans by insect bites or by contact with animals who are themselves infected by insects. Thus, the populations most often afflicted with tularaemia include hunters, hikers, and others who are exposed to wild animals and their arthropod parasites.

Although this chapter will focus primarily on *F. tularensis*, two other species of *Francisella* are recognised and deserve mention. *F. novicida* ('new killer'), originally isolated from water collected near dead muskrats (Larson *et al.*, 1955), has been recognised as a rare human pathogen; in two cases, *F. novicida* caused a disease similar to tularaemia (Hollis *et al.*, 1989). The other species, *F. philomiragia*, was originally isolated from muskrats and water during an attempt to re-isolate *F. novicida* (Jensen *et al.*, 1969). This bacterium, named because of frequent mirages seen in the area, is associated with an acute febrile disease in near-drowning victims and in patients with chronic granulomatous disease (Wenger *et al.*, 1989).

F. tularensis is a facultative intracellular bacterium, and wild-type strains and the attenuated variant denoted LVS (live vaccine strain) have been extensively studied as model representatives of this important class of pathogens (Anthony and Kongshavn, 1987; Fortier *et al.*, 1991; Tärnvik *et al.*, 1992; Yee *et al.*, 1996). Knowledge about the molecular biology of *Francisella* and its virulence mechanisms are, however, at an early stage. Furthermore, much of the data concerning its pathogenesis and immunology has been gathered over a period of decades, and there has been little consistency in the use of strains, cell cultures and animal models.

Classification and Identification

Since it was originally designated as *Bacterium tularense*, the aetiological agent of tularaemia has been re-classified several times. It was placed for a time in

Table 1 Characteristics of *Francisella* species

Species/biotype	Distinguishing characteristics	Geographic distribution	Usual vectors of human disease	Disease
F. tularensis type A (biotype tularensis or nearctica)	Has citrulline ureidase, ferments glycerol and maltose	North America (90%)	Rabbits, ticks	Severe acute febrile disease with non-specific symptoms
F. tularensis type B, including LVS (biotype palaearctica or holarctica)	Ferments maltose only	North America (10%), Scandinavia, lower Europe, Russia, Japan	Water-related animals and water	Mild to moderate acute febrile disease
F. novicida	Faster growth, less dependent on cysteine; ferments glycerol and sucrose; no capsule	United States	Ground water	Mild acute febrile illness
F. philomiragia	Distinguished from *F. tularensis* by DNA relatedness; oxidase positive	United States, Switzerland	Muskrats and water	Acute febrile illness in near-drowning victims and CGD patients

the genus *Pasteurella* (*P. tularensis*), and then provisionally also in the genus *Brucella*. In 1947 the separate genus *Francisella* was proposed, but this did not come into common usage until the 1960s. Similarly, *F. novicida* was originally placed in the genus *Pasteurella*, but it was subsequently recognised as a member of the genus *Francisella*. *F. tularensis* and *F. novicida* have many properties that indicate a close relationship, including a similar and unusual fatty acid composition, inter-species genetic transformation, DNA relatedness of 75–98% (Hollis *et al.*, 1989), and near identity of 16S ribosomal DNA sequences (Forsman *et al.*, 1994). Phylogenetic analysis of the 16S RNA sequences places the genus *Francisella* in the γ-subclass of the *Proteobacteria*. Although *F. novicida* is somewhat less fastidious than *F. tularensis* in its cysteine requirement for growth, and somewhat less virulent for mammals, the proposal to consider *F. novicida* as a biotype, that is *F. tularensis* biotype *novicida*, rather than a separate species (Hollis *et al.*, 1989) is well justified.

It has long been recognised that in nature there are two biotypes of *F. tularensis* associated with different virulence. *F. tularensis* type A, also known as biotype *tularensis* or *nearctica*, is highly virulent for both animals and humans, and is found only in North America. This biotype produces the disease syndrome most commonly associated with the name 'tularaemia', and exposure is often linked to contact with ticks and rabbits. In contrast, *F. tularensis* type B, also known as biotype *palaearctica* or *holarctica*, is found in North America,

Europe and Asia, and is usually less virulent. The disease produced by the type B biotype is milder and often not recognised as classical tularaemia; exposure is typically by contact with rodents or with contaminated water. The two *F. tularensis* biotypes can usually be distinguished from each other biochemically in that biotype B is generally unable to produce acid from glycerol and lacks a citrulline ureidase system. Biochemical characteristics do not, however, always lead to unambiguous typing, and classification by genetic relationships is currently being explored (Eigelsbach and McGann, 1984; Sandström *et al.*, 1992a; Forsman *et al.*, 1994) (**Table 1**).

Identification of *F. tularensis* can be aided by some of its unusual morphological and growth characteristics. The organisms are non-motile, encapsulated and generally coccobacilli, but they are pleiomorphic in late exponential phase cultures. The bacteria are strictly aerobic and grow optimally at 37°C. With some notable exceptions (Bernard *et al.*, 1994), growth is cysteine-dependent, and the doubling time is about 2 hours in enriched artificial media. Most importantly for identification, *Francisella* are quite small, and typically less than 0.4 μm wide. On polychrome staining they are weakly Gram-negative bipolar rods. *Francisella* have a characteristic lipid composition that can be useful in identification. All species analysed so far have long chain C_{18}–C_{26} unsaturated and saturated fatty acids, as well as two long-chain hydroxy fatty acids that may be part of the bacterial lipid A (Jantzen *et al.*, 1979).

Although *Francisella* can be identified in the clinical laboratory by standard methods, its isolation and identification in clinical settings is often not successful. *Francisella* cannot be cultured on routine laboratory media, and the relatively slow growth rate dictates that clinical isolates must be cultured in the presence of carefully chosen antibiotics to prevent overgrowth by normal flora or typical contaminants. Alternatively, in reference laboratories, biopsy material is often inoculated into mice in an attempt to recover the pathogen from the liver of the moribund animal. Microscopic identification is difficult because of the small size and weak staining characteristics of *F. tularensis*, but identification by immunofluorescence is possible if the necessary reagents are available. Given the difficulties in culturing *F. tularensis* and its highly infectious nature, physicians often choose to make a diagnosis of tularaemia on the basis of the case history and serology (see below).

Surface, Secreted, and Virulence-associated Components

The most thoroughly studied structural features of *Francisella* are the capsule and lipopolysaccharide (LPS). The classical papers of Eigelsbach *et al.* (1951) and Eigelsbach and Downs (1961) described colony opacity variants of the highly virulent Schu S4 strain and of the LVS strain that were almost certainly LPS or capsule production variants (**Fig. 1**). Since the opacity

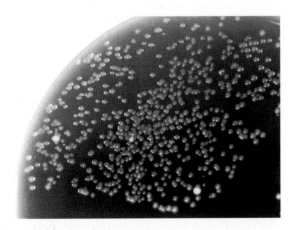

Fig. 1 Reversion of opacity variants of *F. tularensis* LVS to wild type. Small, translucent colonies are opacity variants of LVS that possess less of the *F. tularensis* O antigen and express *F. novicida*-type O antigen. Larger, opaque colonies are phenotypically identical to the parent strain, LVS, expressing exclusively the *F. tularensis* O antigen. The revertants were amplified by passage of the opacity variant through rat macrophages *in vitro*.

variants are avirulent and non-immunogenic, monitoring their presence became important in production of LVS for human vaccination. Surprisingly, the inclusion of significant numbers of more than a few per cent of the opacity variants in a vaccine lot renders the preparation non-immunogenic.

The molecular basis for the opacity variation became apparent when it was discovered that *F. tularensis* undergoes a phase-variation in its LPS O antigen, a structure that is probably shared with the capsule (Cowley *et al.*, 1996). Western immunoblot analysis of the LPS from *F. tularensis* colony opacity variants showed that a shift from the wild type to the opacity variant was due to a change in LPS composition. The O antigen of wild-type *F. tularensis* was greatly diminished in the opacity variant, and another O-antigen form was newly expressed. This second O-antigen form is the one found in *F. novicida*, which apparently does not undergo phase-variation or possess a capsule. In *F. tularensis* a reverse phase-variation also occurs.

An unusual feature of the O antigen phase-variation is that it is accompanied by a change in the properties of the lipid A portion of the LPS. Unlike the lipid A of most Gram-negative bacteria, the one found in *F. tularensis* LPS has apparently evolved to be poorly reactive with inflammatory cells (Sandström *et al.*, 1992b; Cowley *et al.*, 1996). Surprisingly, the stimulatory effect of the lipid A associated with the *F. novicida* LPS O antigen on rat macrophages is much greater than that of the wild-type *F. tularensis* LPS, and results in the stimulation of nitric oxide by macrophages. Nitric oxide production in turn limits the growth of *F. tularensis* in rat macrophages. Hence, phase-variation modulates the ability of *F. tularensis* to grow in rat macrophages, a phenomenon that has not been observed in any other pathogen.

An apparent two-cistron operon, *mglAB*, is required for virulence and intra-macrophage growth (Baron and Nano, 1998). Changing Thr54 to lysine in *mglA*, or insertion of mutations in either *mglA* or *mglB*, results in the creation of a mutant strain unable to grow in macrophages and unable to cause disease in mice. The deduced amino acid sequence of MglA and MglB shows significant identity with SspA/SspB, proteins thought to be global regulators that interact with RNA polymerase during stationary phase. Inactivation of *mglAB* results in the loss of expression of numerous proteins, including a secreted 70-kDa protein.

An analysis of the *F. tularensis* proteins that are preferentially expressed during growth in macrophages has been performed by two-dimensional gel separation of labelled proteins (Golovliov *et al.*, 1997).

Surprisingly, only a few proteins are significantly induced by intracellular growth. The most prominently induced protein of 23 kDa was analysed further by cloning and sequencing the encoding gene, but the deduced amino acid sequence shows no significant similarity to other proteins, and protein motifs were not apparent.

Only a few proteins in the outer membrane of *Francisella* have been studied. One of these, FopA, migrates at an apparent molecular mass of 34 kDa when solubilised below 80°C, but as a doublet at 41 and 43 kDa when outer membrane extracts were solubilised above 95°C (Nano, 1988). DNA sequence analysis of the encoding gene revealed a deduced protein of 313 amino acids of 32 808 kDa after cleavage of the putative signal sequence (Leslie *et al.*, 1993). The deduced amino acid sequence suggests that FopA is composed largely of β sheets, typical of outer membrane proteins, and comparison of the amino acid sequence of FopA shows a weak similarity to outer membrane proteins of other Gram-negative bacteria. A typical Gram-negative lipoprotein, TUL4, has also been described for *F. tularensis* (Sjöstedt *et al.*, 1989, 1990a). The DNA sequence of the encoding gene predicts a protein of 15 772 kDa with consensus sequences for a lipoprotein signal peptidase and acylation. Studies of TUL4 in *F. tularensis* and of the recombinant form of the protein in *E. coli* demonstrated the incorporation of radiolabel when cells are supplied with [^{3}H]palmitate. Globomycin, a specific inhibitor of the lipoprotein signal peptidase, also prevents maturation of the protein (Sjöstedt *et al.*, 1991). The TUL4 proteins do not show strong similarity to other lipoproteins.

A number of strains of *Francisella* possess an acid phosphatase that inhibits the respiratory burst of neutrophils (Reilly *et al.*, 1996). The acid phosphatase, AcpA, is easily released from *Francisella* cells. AcpA has broad specificity, and responds to phosphatase inhibitors in a fashion similar to that of other burst-inhibiting acid phosphatases. The deduced amino acid sequence of AcpA is similar to that of some bacterial phopholipases but unlike that of acid phosphatases. AcpA has phospholipase activity of similar specific activity to other bacterial phospholipases, but this activity was at least 1000-fold less than the phosphomonoesterase-specific activity.

An operon that encodes an ABC transporter has been identified in *F. novicida* (Mdluli *et al.*, 1994). The encoded proteins show high homology with *msbA/orfE* of *E. coli* (Karow and Georgopoulos, 1993), proteins thought to be involved in secretion of LPS. Homologues of the *E. coli* heat-shock proteins GroEL, GroES and DnaK are induced in *F. tularensis*

in response to elevated heat or hydrogen peroxide exposure (Ericsson *et al.*, 1994b), and the locus that encodes the DnaK proteins has been cloned and sequenced (Zuber *et al.*, 1995). Finally, a siderophore-like compound has been identified that appears to be necessary to initiate growth of low-density cultures of *F. tularensis* (Halmann *et al.*, 1967).

Genetics

The genetic analysis of *Francisella* is still in its early stages, but it is clear that the basis for rapid progress is in place. DNA has been introduced into *F. tularensis* and integrated by electroporation into the chromosome (Anthony *et al.*, 1991b). A 'cryotransformation' technique has been used to introduce autonomous replicating plasmids into *F. tularensis* (Mokrievich *et al.*, 1994). DNA can be introduced into *F. novicida* after treating cells with a calcium-containing buffer (Tyeryar and Lawton, 1970). Transformation of *F. novicida* with recombinant clones that carry transposon-interrupted *F. novicida* DNA, leads to the integration of the DNA, with frequent gene replacement events. As expected, gene duplication events also occur. Exploitation of the high transformation frequency allows the generation of insertional mutants by virtue of gene replacement (Berg *et al.*, 1992), and the generation of complement mutations *in cis* by virtue of gene duplications (Mdluli *et al.*, 1994; Baron and Nano, 1998). Autonomous replicating plasmids should allow *in trans* complementation of mutations, but this has not yet been demonstrated. A plasmid from a clinical isolate of *F. novicida* has been used to generate two cloning vectors that can replicate in *F. tularensis* (Pavlov *et al.*, 1996). This plasmid hybridises to several restriction fragments of the *Francisella* chromosome, suggesting frequent integration of the plasmid or the presence of repeated elements (S. Myltseva and F.E. Nano, unpublished data). Broad host-range plasmids of incompatibility group W have also been shown to replicate in *F. novicida* (Anthony *et al.*, 1991b).

A mini-transposon that encodes erythromycin-resistance, TnMax2 (Hass *et al.*, 1993), is useful to create random transposon mutant banks of *F. novicida* (S. Cowley *et al.*, unpublished data). Modifications of TnMax2 have led to the construction of a mini-cassette and a mini-transposon that carry a promoterless chloramphenicol acetyltransferase (*cat*) gene and an erythromycin-selectable marker (Baron and Nano, 1999). These two elements can be used to make transcriptional fusions in *F. novicida* that produce CAT under control of an *F. novicida* promoter.

Pathogenesis

The only verified virulence factor of *Francisella* so far identified is the bacterial capsule (Hood, 1977). He developed methods to remove the capsule from the virulent *F. tularensis* Schu S4 strain, and found that the resulting bacteria are viable but have a 100-fold reduction in LD_{50} for guinea-pigs, but not mice. Initial presence of a capsule and its removal after treatment were demonstrated by electron microscopy. Although the conclusion that the capsule is necessary for virulence is reasonable, it was not possible to demonstrate that other bacterial factors were not affected. However, the capsule itself was neither toxic nor immunogenic, which indicates that its role in virulence is related to protection of the intact bacteria against host defence. Similarly, loss of capsule by *F. tularensis* LVS, by manipulation of culture conditions, results in reduced virulence for mice (Cherwonogrodzky *et al.*, 1994). More directly, a capsule-deficient mutant of *F. tularensis* LVS designated LVSR, derived by chemical mutagenesis, was found to be much less virulent in out-bred mice than the parent strain (Sandström *et al.*, 1988). This was probably because of its increased sensitivity to killing by serum, by activation of the classical complement pathway. Interestingly and paradoxically, LVSR is also less susceptible to intracellular killing by human polymorphonuclear leucocytes (PMNs), which suggests that recognition of the capsule by the host is also responsible for activation of oxygen-dependent killing by PMNs. Other research (Cowley *et al.*, 1996) demonstrated that LVSR also lacks much of its normal amount of LPS O antigen, which may be the basis of serum-sensitivity. *F. novicida*, which apparently does not have a capsule, is nonetheless resistant to the bactericidal action of serum.

Secreted toxins or other protein virulence factors have not so far been identified in *F. tularensis*, and although all strains have LPS, for all practical purposes this lacks traditional endotoxic activity. Recent reports indicate that purified LVS LPS is not endotoxic in D-galactosamine-sensitised mice, and fails to activate *Limulus* amoebocyte lysate (Sandström *et al.*, 1992b). Furthermore, LVS LPS fails to stimulate human monocytes or peripheral blood lymphocytes to proliferate, produce TNFα, or produce IL-1 (Sandström *et al.*, 1992b). Similarly, mouse peritoneal exudate macrophages treated with LVS LPS do not produce TNFα or nitric oxide, and a mouse pre-B cell line does not respond to LVS LPS with increases in surface immunoglobulin expression. LVS LPS also fails to act as an antagonist to block stimulation of TNFα or nitric

oxide production by other LPS (Ancuta *et al.*, 1996). Since *Francisella* is a facultative intracellular bacterium, its ability to cause disease is instead probably related to its ability to infect macrophages and macrophage-like cells in the reticuloendothelial system. Thus, pathogenicity is best considered in relation to the interaction of the bacterium with host cells.

As with most bacterial infections, PMNs are recruited to the site of *Francisella* infection within hours of entry. Attenuated strains such as LVS are, however, much more susceptible to killing by PMNs than are wild-type strains (Löfgren *et al.*, 1983). In mice infected intravenously with *F. tularensis* LVS, bacteria initially infect Kupffer cells in the liver, but can also infect adjacent hepatocytes. It appears that PMNs recognise and kill infected hepatocytes, thus releasing bacteria for conventional phagocytosis and direct killing by PMNs (Conlan and North, 1992). If PMNs are not available, unrestricted bacterial growth continues in hepatocytes, which lack intracellular killing mechanisms and rapid death of the animal results. These observations, although made with an attenuated strain of *F. tularensis* in an animal model, emphasise that pathogenesis of disease caused by *Francisella* may more reflect interaction between host and bacteria, than production of particular bacterial virulence factors. Specific mechanisms of susceptibility or resistance to PMN-mediated killing have not, however, been identified in *Francisella*.

Francisella cells that escape early killing by PMNs infect and grow readily in resident mammalian macrophages. *F. tularensis* LVS and *F. novicida* both multiply in macrophages isolated from mice and guinea-pigs; unlike LVS, however, *F. novicida* does not grow in rat macrophages (Anthony *et al.*, 1991a), since LPS stimulation of rat macrophages induces production of nitric oxide, which in turn ablates intracellular growth of the bacterium (see above). Both LVS and *F. novicida* occupy a membrane-bound compartment that appears to be a phagosome that is not fused with lysosomes. Growth of LVS in mouse macrophages requires an acidified compartment that permits availability of iron (Fortier *et al.*, 1995).

The killing of *F. tularensis* LVS by activated macrophages has been well characterised. Macrophages deliberately activated by interferon γ (IFNγ) readily control the intracellular growth of *F. tularensis* LVS by both nitric oxide-dependent and nitric oxide-independent means (Anthony *et al.*, 1992; Fortier *et al.*, 1992; Polsinelli *et al.*, 1994), but it is not completely clear whether the activity is actually bactericidal or simply bacteriostatic. Strategies that limit availability of IFNγ or nitric oxide production in animals clearly increase the virulence of LVS, while those which activate macrophages increase host resistance (Leiby *et al.*, 1992; Green *et al.*, 1993). Thus, the state of host phagocyte activation has a clear effect on the severity of disease produced by *Francisella*.

Disease progression and pathology have been described in monkey models of aerosol infection with both type A (White *et al.*, 1964) and type B (Schricker *et al.*, 1972) *F. tularensis*. Studies of this kind have established that, while monkeys are somewhat more susceptible than humans for infection, virulence and pathogenesis parallel that in humans. Infected monkeys have an acute febrile illness with pneumonia, since bacteria were introduced by aerosol, in which bacteria multiply readily on the lungs, disseminate to regional lymph nodes, and then spread to spleen and liver; animals that survive clear bacteria within 2 months. The cause of death in fatal cases has not, however, been specifically determined, but is generally assumed to be due to an overwhelming bacterial burden and liver failure.

Immunity

Specific T-cell-mediated proliferative and cytokine responses, and activation of macrophages for intracellular killing, are probably of most importance in immunity to *Francisella*. It is now generally accepted that the specific antibody response has little or no, or a very minor, role in protection against virulent infection. Resolution of natural infection probably leads to long-lived immunity to subsequent infection, and documented cases of re-infection are rare (Burke, 1977). The immunology of *Francisella* infection has been reviewed in detail elsewhere (Tärnvik, 1989), and the focus here will be on more recent developments.

Over the last decade, studies of several intracellular pathogens such as *Listeria monocytogenes*, *Salmonella typhimurium* and *Leishmania* have indicated that mammals have an initial early, days to weeks, lymphocyte-independent phase and a later, weeks to months, lymphocyte-dependent phase of resistance to primary infection. *Francisella* appears to be no exception. Thus, both T-cell-deficient, athymic *nu/nu* mice (Elkins *et al.*, 1993) and total lymphocyte-deficient *scid* mice (Elkins *et al.*, 1996) are able to control and survive primary intra-dermal infection with LVS for 3–4 weeks before succumbing to overwhelming infection. The activity of PMNs is necessary for initial survival of LVS infection in mice (see above; Sjöstedt *et al.*, 1994; Elkins *et al.*, 1996). Similar to other intracellular infections, this initial survival also clearly depends on the production of the so-called TH1 cytokines IFNγ and tumour necrosis factor α (TNFα)

(Anthony *et al.*, 1989; Leiby *et al.*, 1992; Elkins *et al.*, 1993;, 1996). TNFα, IFNγ, IL-10 and IL-12 messenger RNA and the corresponding cytokine proteins, but not IL-2, IL-3 or IL-4, are detectable in livers from BALB/cJ mice within 24–48 hours of subcutaneous infection with LVS (Golovliov *et al.*, 1995), and also after lethal infection with virulent tularaemia (Golovliov *et al.*, 1996). This suggests that these cytokines may be necessary, but are not sufficient, for survival of *Francisella* infection. Macrophages and natural killer cells, and not lymphocytes, probably produce these cytokines, but direct demonstration of the cellular sources awaits further study. Furthermore, the stimulus for activation of macrophages and natural killer cells does not proceed through specific receptor-mediated recognition of *Francisella*-related molecules, but presumably requires recognition of determinants common to pathogens. Cell wall polysaccharides, capsular polysaccharides and bacterial LPS have been proposed as ligands for common 'pattern recognition' receptors on cells, such as macrophages and natural killer cells of the innate immune system (Janeway, 1992). Elucidation of such putative *Francisella* structures awaits further study.

Long-term resolution of primary *Francisella* infection clearly requires the activity of conventional T lymphocytes, since neither *nu/nu* mice (which have B cells) nor α/β T-cell knockout mice survive more than a month after LVS infection (Elkins *et al.*, 1993; Yee *et al.*, 1996). On the other hand, B-cell knockout mice appear only slightly impaired in their ability to survive and clear a primary intradermal infection with LVS (Elkins *et al.*, 1999a). Either CD4+ T cells or CD8+ T cells are apparently sufficient to effect survival, since knockout mice deficient in only one T-cell subset are readily able to clear primary intradermal LVS infection (Yee *et al.*, 1996; see also Conlan *et al.*, 1994). In total T-cell-deficient mice, bacterial numbers in organs quickly reach a plateau level several days after infection. This steady-state level is maintained over several weeks in apparently asymptomatic mice, until bacterial numbers begin a steep rise shortly before death of the animal (Elkins *et al.*, 1993).

The specific effector function contributed by T lymphocytes, over and above the cytokine production provided by cells of the innate immune system, remains to be revealed. It is unlikely that T helper cell function for specific antibody production is critical, since CD4− knockout mice that fail to make specific IgG anti-LVS antibodies easily resolve primary, and secondary, infection (Yee *et al.*, 1996). In other systems, T cells function as cytotoxic cells against bacteria-infected macrophages (Kaufmann *et al.*, 1988).

Since macrophages activated by IFNγ appear to control, but not necessarily completely clear, bacterial numbers (Fortier *et al.*, 1992), the concept of final elimination of bacteria by T-cell-mediated killing is attractive. Indeed, a recent study described CD4+ cytotoxic T cells, re-stimulated from peripheral blood lymphocytes of humans vaccinated with LVS, that were specific for *F. tularensis*-pulsed monocytes (Surcel *et al.*, 1991b).

The contribution of $\gamma/\delta+$ T cells to primary infection may be minimal, since knockout mice deficient for the δ chain of the T-cell receptor, and thus lacking $\gamma/\delta+$ T cells, appeared comparable to normal mice in their ability to survive and clear primary infection (Yee *et al.*, 1996). Some deficiency in their ability to handle secondary infection has, however, been described (see below). This is in contrast to the currently popular hypothesis that $\gamma/\delta+$ T cells recognise common pathogenic antigens such as heat-shock proteins, and in this way are important as very early first-line defenders against infection (Kaufmann, 1990). On the other hand, dramatic polyclonal expansion of Vγ9/Vδ2+ T cells in the peripheral blood cells of a tularaemia patient within 10 days of apparent infection has been described. This was interpreted as suggesting that *Francisella* may possess a superantigen responsible for stimulation of a particular $\gamma/\delta+$ T-cell receptor without regard to antigen specificity (Sumida *et al.*, 1992). Indeed, other tularaemia patients, but not LVS vaccinees, exhibit a dramatic increase in Vγ9/Vδ2+ T-cell levels, and such cells respond to non-peptide phosphorylated antigen stimulation (Poquet *et al.*, 1998). This is consistent with other attempts, with a comprehensive set of *Francisella* proteins, to stimulate human peripheral blood lymphocytes (PBLs) in which such a molecule was not detected (Surcel, 1990).

Humans infected with wild-type *Francisella* exhibit a T-cell-dependent delayed-type hypersensitivity response within about a week of infection (Buchanan *et al.*, 1971), a property that has been exploited to develop a diagnostic skin test. Early responses of cells of the innate human immune system, and corresponding cytokine production, to primary infection with either wild-type *Francisella* or vaccination with LVS have not been comprehensively studied. Primary T-cell responses in human volunteers vaccinated with LVS have been described. By days 10–14 after vaccination, a proliferative response *in vitro* by PBLs to killed antigen could be detected, and by day 14 IL-2, IFNγ and TNFα production, but not IL-4 production, was observed (Karttunen *et al.*, 1991). Similarly, proliferative responses to killed antigen by PBLs obtained from tularaemia patients were demonstrated

2 weeks after appearance of symptoms. Production of TNFα was detected within one week, and IL-2 and IFNγ, but not IL-4, were found from 2 weeks onwards (Surcel *et al.*, 1991a). The T-cell or monocyte contribution to either proliferation or cytokine production was not, however, specifically determined.

Effector mechanisms of the innate immune system may be necessary prerequisites to the development of specific secondary immunity. Thus, macrophage activation, natural killer cell function and cytokine production may be required not only to control initial infection but also to promote development of specific, lymphocyte-mediated immunity. In other models of intracellular infection such as with *Leishmania*, IL-12 has been proposed as a cytokine responsible for bridging between innate and specific immunity (Scott, 1993). Data on the role of IL-12 in survival of primary *Francisella* infection and secondary protection is, however, limited and conflicting. IL-12-depleted wild-type mice and IL-12 knockout mice show a chronic infection but do not die (Elkins *et al.*, 2001). Another potential bridging mechanism between innate and specific immunity in *Francisella* infection has been proposed (Culkin *et al.*, 1997). Normal mice given a sublethal infection with LVS are, very surprisingly, strongly resistant to a second lethal LVS infection that is introduced only 2–3 days later. This resistance mechanism appears to depend primarily on lymphocytes and especially B cells. It is present in athymic *nu/nu* mice, T-cell-depleted mice and all T-cell knockout mice, but it is absent altogether in *scid* mice, and it is largely diminished in B-cell knockout mice (Elkins *et al.*, 1992, 1993; Conlan *et al.*, 1994; Culkin *et al.*, 1997). It is not, however, related to B-cell-mediated production of specific antibodies, and thus is more likely to derive from another B-cell function, such as cytokine production. This mechanism is non-specific and also operates in *Listeria monocytogenes* infection (Elkins *et al.*, 1998). Stimulation of protection may, at least in part, be due to recognition of bacterial DNA that contains unmethylated CpG motifs (Elkins *et al.*, 1999b).

The secondary immune response to *F. tularensis* has been studied for many years, in animal models and in humans who were either vaccinated with LVS or suffering from natural infection. The animals used as models include the mouse, rat, guinea-pig, rabbit and monkey. By far the most information is available from studies with mice, which have been considered reasonable models, since the histopathology of *Francisella* infection is similar in mice and humans (Anthony and Kongshavn, 1987; Tärnvik, 1989; Fortier *et al.*, 1991; Tarnvik *et al.*, 1992). Indeed,

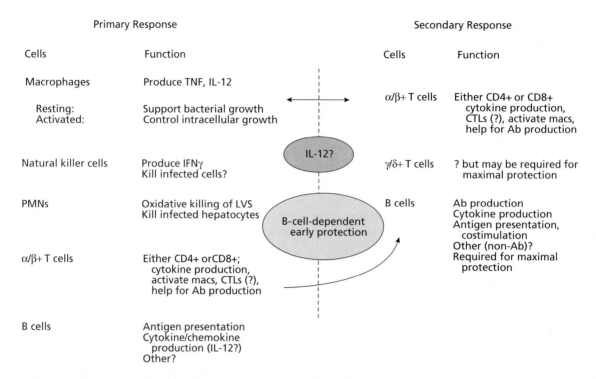

Fig. 2 Schematic illustration of the relationship between components of the immune response to *F. tularensis*. Elements of the early primary antibacterial response are illustrated on the left, elements of the later secondary response on the right, and possible bridging components (necessary for intermediate control of infection as well as optimal development of a secondary response) in the centre. All components of the primary response participate again in secondary responses, but (except for T cells) have no memory amplification. Primary and secondary T-cell functions are probably identical qualitatively, but differ quantitatively in that specific anti-*Francisella* T cells have been clonally expanded and captured as memory cells.

results from human studies generally correlate well with information derived from studies in mice, and they are consistent with concepts of secondary immunity to intracellular pathogens developed from studies of other pathogens.

In general, resolution of natural infection leads to strong protective immunity against secondary challenge. It has, however, proved very difficult to stimulate strong protection by immunisation with killed bacteria, both in *Francisella* infection (Foshay, 1950; Claflin and Larson, 1972; Hambleton *et al.*, 1974) and in other intracellular infections. Since introduction of killed bacteria results in a specific antibody response that is very similar to the antibody response after live infection, it has been inferred that antibody is not sufficient for successful protection. Further, in experimental situations where little or no specific antibody is present, excellent protection can be demonstrated (see below). Rather, specific T-cell-mediated immunity is probably central in protection, and stimulation of a vigorous T-cell-mediated response is the usual goal of vaccination strategies. It should be remembered, however, that in any second exposure to the bacteria a concomitant innate immune

response is elicited, just as in primary infection, and these effector mechanisms may be quite important in both initiating and augmenting specific T-cell-mediated responses (**Fig. 2**). For example PMNs contribute significantly to survival of second LVS infections in mice (Sjöstedt *et al.*, 1994).

A variety of studies have attempted to characterise the relative contribution of T-cell subsets by transfer of immune cells to naïve animals, but these systems have been inconsistent, technically difficult, and often directed to examine effects on bacterial burdens in organs rather than survival (see Allen, 1962; Thorpe and Marcus, 1965; Eigelsbach *et al.*, 1975). The most direct information in animal models comes from use of either T-cell-depleted or genetically created T-cell knockout mice. Thus, while mice that have resolved initial LVS infection and are then depleted of total $\alpha/\beta+$ T cells have some ability to survive a second lethal challenge, probably because of the early B-cell-mediated mechanism discussed above (Conlan *et al.*, 1994; Yee *et al.*, 1996), such mice never completely resolve infection. Further, mice that lack mature B cells and antibodies (B-cell knockout mice) resolve sublethal primary LVS infection, but they are 100-fold less well-protected against a second lethal challenge than their normal counterparts (Elkins *et al.*, 1999a). This defect in optimal specific protective immunity is readily reconstituted by transfer of primed, and to a lesser degree unprimed, B cells, but not by transfer of specific antibodies. Thus, there appears to be a significant role for B cells in secondary immunity to *Francisella* by a function other than antibody production, such as cytokine production, antigen presentation and/or regulation of appropriate cell trafficking. Mice either depleted of CD4+ or CD8+ T cells individually, and CD4 knockout or β_2 microglobulin/CD8 knockout mice, readily survive and clear a maximal second lethal challenge (Conlan *et al.*, 1994; Yee *et al.*, 1996). Therefore, either CD4+ or CD8+ T cells are sufficient in mice to provide maximal protection, at least when B cells and other cells of the innate immune system are present.

The role of $\gamma/\delta+$ T cells in secondary immunity deserves further study. *Francisella* immune TcRδ knockout mice have difficulty in surviving maximal lethal challenge doses, and make a reduced anti-LVS antibody response (Yee *et al.*, 1996). The relationship between these two related observations remains to be further explored, as do the exact effector mechanisms utilised by all classes of T cells in providing protection against a second lethal challenge.

The nature of the LVS antigens recognised by mouse T cells has not been completely studied. At least one lipoprotein, designated TUL4, affects both proliferation and cytokine secretion in T cells from immune mice (Sjöstedt *et al.*, 1992b). This protein does not appear to be immunodominant, but it causes proliferation by T cells from vaccinated individuals (Sjöstedt *et al.*, 1989).

Similarly, the human T-cell response to *Francisella* appears to be quite long-lived and heterogeneous. Peripheral blood lymphocytes or purified T cells, in the presence of monocytes, from LVS-vaccinated individuals proliferate *in vitro* when incubated with heat-killed bacteria (Tärnvik and Löfgren, 1975; Tärnvik and Holm, 1978). This proliferative response is primarily due to recognition of a wide variety of proteins, at least some of which are outer membrane proteins (Sandström *et al.*, 1987; Surcel, 1990; Sjöstedt *et al.*, 1992a). Subjects who had recovered from natural tularaemia infection between 2 weeks and 25 years earlier, also had PBLs that proliferate in response to heat-killed bacteria (Koskela and Herva, 1980; Ericsson *et al.*, 1994a), and also a variety of bacterial proteins (Sjöstedt *et al.*, 1990a). T-cell responses to at least four of these proteins are observed in cells from naturally infected subjects and LVS-vaccinated subjects, which suggests that antigenic proteins important

in human responses to wild-type infection are also expressed by LVS. Furthermore, an obvious immunodominant antigen has not been found. Although both *Francisella*-specific proliferative and TH1 cytokine production responses by the CD45RO+ (memory)/CD4+ T-cell subset can readily be demonstrated, responses from the CD8+ subset can also be detected (Sjöstedt *et al.*, 1992a). On antigen stimulation *in vitro*, T cells from vaccinated or naturally infected individuals produce IL-2 and IFNγ, and exhibited up-regulated IL-2 receptor expression (Karttunen *et al.*, 1987). T-cell clones established from PBLs of naturally infected subjects and LVS-vaccinated individuals have been reported, and are class II-restricted, CD4+, produce IL-2 and IFNγ, and may provide help for antibody production *in vitro* (Surcel *et al.*, 1989; Sjöstedt *et al.*, 1990b, 1992a; Surcel, 1990). T cells from LVS-immune mice also proliferate in response to heat-killed bacteria. As may be expected, the proliferative response is macrophage-dependent, and the interaction between macrophages and T cells is H-2, and largely class II-restricted (Anthony and Kongshavn, 1988). Although such observations suggest that CD4+ T cells may be especially important in immunity to *Francisella*, it must be remembered that mice totally lacking in CD4+ T cells readily resolve primary and secondary infection.

Since antibody responses are useful for diagnosis (see below), the antibody response to *Francisella* of animals and humans has been studied in some detail. In mice, IgM anti-LVS antibody responses are detectable 5 days after infection, reach a peak by 2 weeks after infection, and then decline slowly but persist for more than 4 months. IgG anti-LVS antibody responses are detectable 10 days after infection, reach a peak by about one month, and are maintained at a plateau level for at least 4 months (Rhinehart-Jones *et al.*, 1994). The overwhelming majority of the specific IgG response is of the IgG2a subclass, as might be expected for an infection that results in production of high levels of IFNγ. Limited protection can be transferred to naïve mice with anti-LVS IgG antibodies, but the strength of this protection is quite small compared with that afforded by immunisation with live bacteria, and protection cannot be transferred with IgM anti-LVS antibodies (Rhinehart-Jones *et al.*, 1994). Anti-LVS antibodies affected the relative distribution of bacteria in organs but not overall mortality (Anthony and Kongshavn, 1987). In contrast, CD4− knockout mice infected with LVS produce only small amounts of IgM anti-LVS antibodies, but are able to survive a maximal second lethal challenge (Yee *et al.*, 1996).

The older literature demonstrated that serum from animals infected with wild-type *F. tularensis* is unable to transfer protection against wild-type challenge, and at best slightly extends the mean time to death (Allen, 1962; Thorpe and Marcus, 1965). In addition, anti-*Francisella* antibodies appear to have little functional activity, in that neither bactericidal action nor opsonisation for uptake and subsequent killing by macrophages or PMNs can readily be demonstrated *in vitro* (Löfgren *et al.*, 1983; Sandström *et al.*, 1988; Rhinehart-Jones *et al.*, 1994). Taken together, these results indicate that anti-LVS antibodies are neither necessary nor sufficient for optimal secondary immunity.

In humans naturally infected with *F. tularensis*, specific IgM, IgG and IgA serum antibodies appear almost simultaneously about 6–10 days after the onset of symptoms, or about 2 weeks after infection. They all peak about 1–2 months after infection, and can still be detected up to 11 years later (Koskela and Salminen, 1985). However, most subjects who recovered from natural infection did not have significant amounts of specific anti-*Francisella* antibodies in their serum 25 years later (Ericsson *et al.*, 1994a). Similarly, specific IgM, IgA and IgG antibodies can be detected in the serum of humans vaccinated with LVS, about 2 weeks after vaccination and they persist for at least 1.5 years (Koskela and Herva, 1982). No correlation is apparent between levels of serum antibodies and protection against subsequent infection. For example, two vaccinated laboratory workers known to have high titres of agglutinating antibodies at the time of infection, nonetheless suffered severe tularaemia after a laboratory exposure (Overholt *et al.*, 1961).

Presentation and Diagnosis

Disease caused by *F. tularensis* usually presents in humans as a febrile illness with an acute onset, but can be accompanied by a constellation of non-specific symptoms such as malaise, fatigue, headache, muscle pain, lymphadenopathy and gastro-intestinal disturbances (Tärnvik, 1989). The specific form of the disease acquired usually depends on the route of infection. The majority of cases in the United States are ulcero-glandular, in which infection begins with a skin scratch or insect bite that leads to ulceration at the site of entry, followed by systemic symptoms. Oculo-glandular disease is similar, except that the infection begins at the conjunctiva and may include a granulomatous conjunctivitis. The site of infection in the typhoidal form of tularaemia is uncertain, but may result from ingestion of undercooked infected meat, and gastro-intestinal symptoms may be the first manifestation. Although pneumonia may result

after dissemination of bacteria in any form of the disease, primary pneumonic tularaemia follows aerosol exposure.

A simpler classification system that describes tularaemia as ulcero-glandular or typhoidal has been proposed (Evans *et al.*, 1985). In cases described in this manner, it is evident that typhoidal tularaemia has a much poorer prognosis, with a mortality rate of 6–50%, than ulcero-glandular tularaemia, which has a mortality rate of 1.5–11%. Usually the disease must be differentiated from other generalised bacterial diseases by clinical history, in which contact with potential sources, such as ticks, rodents or rabbits, is specifically questioned.

Although originally isolated from animals or water and not patients, both *F. novicida* and *F. philomiragia* have been reported to be associated with human diseases. Two isolates of *F. novicida* were obtained from men suffering from lymphadenopathy and fever (Hollis *et al.*, 1989). Fourteen isolates of *F. philomiragia* have also been characterised from patients with pneumonia and fever. All but two of these patients were either near-drowning victims or had an underlying immune dysfunction, usually chronic granulomatous disease (Wenger *et al.*, 1989), and none had documented contact with the vector animals usually associated with *Francisella*.

Diagnosis of tularaemia is most often made by serology, in which end-point titres of specific agglutinating antibodies are determined. Serum agglutinating antibodies are, however, not usually detectable until at least the second week of disease, and diagnosis by conventional serology may not be appropriate when recent infection is suspected. A titre of 1 : 160 is usually considered to indicate previous infection, but since the antibody response to *Francisella* infection is long-lived, a single positive titre does not discriminate between recent and past infection. If pre-infection or acute and convalescent serum samples are available, a 4-fold rise in titre is generally accepted as evidence of recent infection. *F. tularensis* is rarely isolated from blood, except in the latest stages of disease, presumably because the bulk of bacterial replication takes place in tissue macrophages; culture from sputum, pharyngeal secretions or ulcers is more likely to be successful. A number of rapid diagnostic tests for tularaemia, including ELISA (Viljanen *et al.*, 1983; Bevanger *et al.*, 1988), skin tests (Buchanan *et al.*, 1971) and PCR have been described, but none of them is yet available for clinical use. Although little genetic information is so far available for *F. novicida*, and sequence data have not been published for *F. philomiragia*, PCR primers have been designed based on a *F. tularensis* common lipoprotein (Long *et al.*, 1993; Junhui *et al.*, 1996) and an outer membrane protein (Fulop *et al.*, 1996). At present these appear to be specific for *Francisella*, and do not give positive results with clinical isolates of other related bacteria such as *Yersinia*. It remains to be determined, however, whether these assays will be generally applicable to the detection of all *Francisella* isolates.

Antimicrobial Therapy

All strains of *Francisella* appear to be susceptible to aminoglycoside antibiotics, and streptomycin has long been the drug of choice when tularaemia is suspected or diagnosed (Enderlin *et al.*, 1994). The use of chloramphenicol and tetracyclines has met with variable success, in spite of apparent susceptibility to these classes of antibiotics *in vitro*, and relapses are common. Gentamicin treatment is also reasonably effective for adults and children (Cross *et al.*, 1995). Quinolones, such as ciprofloxacin and norfloxacin, have occasionally been used with some success. However, in spite of the observation that minimal inhibitory concentrations (MICs) of third-generation cephalosporins determined *in vitro* are quite low, there is evidence that ceftriaxone is ineffective in treatment (Cross and Jacobs, 1993). Penicillin and related drugs have no effect on *Francisella*, and most type B strains are resistant to erythromycin.

Epidemiology and Vaccination

Understanding the distribution of *Francisella* in nature, and its ecology, is key to appreciating its biology and its potential as a human clinical problem. Virtually all available epidemiological information concerns *F. tularensis*. *F. novicida* has only been isolated in the United States, once from water in Utah (Larson *et al.*, 1955) and twice from patients with a tularaemia-like illness (Hollis *et al.*, 1989), and 13 of the 14 *F. philomiragia* isolates reported to date were from the US, the other being from Switzerland (Wenger *et al.*, 1989).

Since its original description in the US, *F. tularensis* has been isolated regularly in Canada, several states of the former Soviet Union, Sweden, Finland, Norway, Italy, Austria, Czechoslovakia, Hungary and Japan. Virtually all isolates have been found between the latitudes 30° and 71° north. In the US, over 22 000 cases of tularaemia were diagnosed between 1927 and 1948, with a case fatality rate of about 8% (Sanford, 1983). In the last two decades, however, only 100–300 cases are reported per year and, since the introduction of antibiotic treatment, the mortality rate has dropped dramatically. The reasons for the decline in the overall number of cases in the US during the 1950s are not clear, but they probably include a reduction in rural housing patterns and the amount of rabbit hunting.

Tularaemia was recognised in the Soviet Union in the late 1920s and it became a significant public health problem during the war years. Approximately 100 000 cases occurred between 1940–1945, as sanitation and population movements changed and the disease became endemic in voles and field mice. In the United Kingdom, tularaemia has been reported only in isolated cases as an imported travel-related infection (Blomley and Pearson, 1972; Wood *et al.*, 1976), which suggests that *F. tularensis* is not enzootic in animals of the British Isles.

F. tularensis has been isolated from over 100 types of wildlife, and the most important animal reservoirs include rabbits, hares, mice, rats, muskrats, beavers and voles. Tularaemia has also been diagnosed in, and acquired from, domestic cats (Capellan and Fong, 1993). Further, *F. tularensis* has been isolated from over 50 arthropod species, including a wide variety of ticks that live in association with wildlife and domestic animals. The organism causes disease in virtually all its animal hosts, but incubation times and severity of symptoms vary widely. Not surprisingly, infected wild animals that are too ill to avoid human contact or capture become vectors for the transmission to humans. Person-to-person transmission, even by aerosol in cases of primary pneumonic tularaemia, is very rare.

There are two seasonal peaks of *Francisella* infections. Disease recognised in the spring or early summer is often associated with tick or mosquito bites, when human outdoor activity and insect exposure is at its highest. Bursts of disease in the winter are most often associated with rabbit hunting.

A particularly well-documented epidemic of tularaemia serves to illustrate the epidemiology of the disease (Brachman, 1969; Young *et al.*, 1969). Tularaemia has been a reportable disease in the United States for several decades, but it never occurred in Vermont until the spring of 1968. Vermont then identified a cluster of tularaemia cases in residents who had influenza-like symptoms with hand sores and lymphadenopathy; all had handled muskrats trapped from three streams in Addison County. Once the outbreak was recognised, however, thorough epidemiological investigation revealed a somewhat unexpected clinical picture. Thirty-nine symptomatic cases were identified, but the distribution between clinically mild, moderate and severe disease was approximately even, respectively, with 11, 13 and 14 cases. This is in contrast to the typical concept of tularaemia as a severe systemic illness in all persons infected with the same virulent strain, which was assumed in this localised outbreak, and suggested marked differences in host susceptibility to disease. In addition, serological screening of residents of the area who had known contact with muskrats resulted in detection of eight asymptomatic cases. *F. tularensis* was subsequently cultured from about 5% of captured muskrats in the area, and from mud and water in the affected stream, but not from rabbits or from the ectoparasites combed from the muskrats. Subsequent studies of isolates from dead muskrats and mud suggested that these were type B bacteria (Hornick and Eigelsbach, 1969). The means by which *F. tularensis*, which was clearly not endemic in the area, was transported to Vermont could not be determined. Thus, this investigation demonstrates the importance of identified contact with host animals, as well as the broad range of clinical outcomes possible following infection.

Well-investigated outbreaks of pneumonic tularaemia in Martha's Vineyard (Teutsch *et al.*, 1979) and in Sweden (Dahlstrand *et al.*, 1971) also emphasise that tularaemia can be readily transmitted by aerosols. In Martha's Vineyard a cluster of cases occurred in holiday-makers who were all exposed to the same indoor fire, while in Sweden a large outbreak occurred among farmers who inhaled dust from infected hay.

The extent of disease, and its effect on military operations, spurred interest in development of a tularaemia vaccine in the former Soviet Union in the 1940s; in the west, efforts at vaccination began with development of killed vaccines, but it soon became clear that killed vaccine or antigen preparations were relatively ineffective in the prevention of disease (Foshay, 1950). Scientists at the Gamaleia Institute in the former Soviet Union developed attenuated strains from *F. tularensis* type B clinical isolates (Tigertt, 1962), mass vaccination campaigns were undertaken in the late 1940s, with several million people being vaccinated between 1946 and 1955 (Sandström, 1994). By 1960, after several decades of mass vaccinations with various attenuated strains, only a few hundred cases of tularaemia occurred annually. More recent data are not readily available.

In 1956, an ampoule of mixed tularaemia vaccine from the Gamaleia Institute was transferred to the United States. This culture was expanded and serially passaged through mice five times with an accompanying increase in virulence, and finally isolated from the blood of moribund animals. The final isolate was designated LVS, for live vaccine strain (Eigelsbach and Downs, 1961). LVS is thus significantly more virulent for mice than strains used in Soviet mass vaccination programmes. The genetic basis of attenuation of LVS remains uncharacterised, but LVS is encapsulated, and shares a number of similarities with wild-type *F. tularensis* in overall carbohydrate, lipid and outer membrane protein composition.

The success of LVS as a vaccine in humans is uncertain. In Europe and North America, a number of laboratory workers and other 'at risk' populations have been vaccinated. Several somewhat limited studies have been offered in support of the efficacy of LVS as a vaccine against either type A or type B tularaemia. Remarkably, aerosol challenge studies in human volunteers were performed in the late 1950s, in which volunteers were given LVS by scarification; they were subsequently challenged by inhalation with the virulent Schu S4 strain, a clinical type A isolate (McCrumb, 1961). Results in this limited population suggested that some protection against development of symptoms was afforded, at least at lower challenge doses. Similar observations in a monkey model have also been published (Eigelsbach *et al.*, 1962). An analysis of laboratory-acquired cases at one research facility, before and after LVS was routinely used for vaccination, was interpreted as indicating that LVS prevented typhoidal tularaemia, including pneumonic tularaemia, fairly well and at least ameliorated symptoms of ulcero-glandular tularaemia (Burke, 1977).

References

Allen WP (1962) Immunity against tularemia: passive protection of mice by transfer of immune tissues. *J. Exp. Med.* 115: 411–420.

Ancuta P, Pedron T, Girard R, Sandström G, Chaby R (1996) Inability of the *Francisella tularensis* lipopolysaccharide to mimic or antagonize the induction of cell activation by endotoxins. *Infect. Immun.* 64: 2041–2046.

Anthony LSD, Kongshavn PAL (1987) Experimental murine tularemia caused by *Francisella tularensis*, live vaccine strain: a model of acquired cellular resistance. *Microb. Pathog.* 2: 3–14.

Anthony LSD, Kongshavn PAL (1988) H-2 restriction in acquired cell-mediated immunity to infection with *Francisella tularensis* LVS. *Infect. Immun.* 56: 452–456.

Anthony LSD, Ghadirian E, Nestel FP, Kongshavn PAL (1989) The requirement for gamma interferon in resistance of mice to experimental tularemia. *Microb. Pathog.* 7: 421–428.

Anthony LSD, Burke RD, Nano FE (1991a) Growth of *Francisella* spp. in rodent macrophages. *Infect. Immun.* 59: 3291–3296.

Anthony LSD, Gu M, Cowley SC, Leung WWS, Nano FE (1991b) Transformation and allelic replacement in *Francisella* spp. *J. Gen. Microbiol.* 137: 2697–2703.

Anthony LSD, Morrissey PJ, Nano FE (1992) Growth inhibition of *Francisella tularensis* live vaccine strain by IFN-gamma-activated macrophages is mediated by reactive nitrogen intermediates derived from L-arginine metabolism *J. Immunol.* 148: 1829–1834.

Baron GS, Nano FE (1998) *MglA* and *MglB* are required for the intramacrophage growth of *Francisella novicida. Mol. Microbiol.* 29: 247–259.

Baron GS, Nano FE (1999) An erythromycin resistance cassette and mini-transposon for constructing transcriptional fusions to *cat. Gene* 229: 59–65.

Berg JM, Mdluli KE, Nano FE (1992) Molecular cloning of the *recA* gene and construction of a *recA* strain of *Francisella novicida. Infect. Immun.* 60: 690–693.

Bernard K, Tessier S, Winstanley J, Chang D, Borczyk A (1994) Early recognition of atypical *Francisella tularensis* strains lacking a cysteine requirement. *J. Clin. Microbiol.* 32: 551–553.

Bevanger L, Maeland JA, Naess AI (1988) Agglutinins and antibodies to *Francisella tularensis* outer membrane antigens in the early diagnosis of disease during an outbreak of tularemia. *J. Clin. Microbiol.* 26: 433–437.

Blomley DJ, Pearson AD (1972) Case of tularaemia in England. *BMJ* 4: 235.

Brachman PS (1969) *Francisella tularensis* in New England. *N. Engl. J. Med.* 280: 1296–1297.

Buchanan TM, Brooks GF, Brachman PS (1971) The tularemia skin test. 325 skin tests in 210 persons: serologic correlation and review of the literature. *Ann. Intern. Med.* 74: 336–343.

Burke DS (1977) Immunization against tularemia: analysis of the effectiveness of live *Francisella tularensis* vaccine in prevention of laboratory-acquired tularemia. *J. Infect. Dis.* 135: 55–60.

Capellan J, Fong IW (1993) Tularemia from a cat bite: case report and review of feline-associated tularemia. *Clin. Infect. Dis.* 16: 472–475.

Cherwonogrodzky JW, Knodel MH, Spence MR (1994) Increased encapsulation and virulence of *Francisella tularensis* live vaccine strain (LVS) (LVS) by subculturing on synthetic medium. *Vaccine* 12: 773–775.

Claflin JL, Larson CL (1972) Infection-immunity in tularemia: specificity of cellular immunity. *Infect. Immun.* 5: 311–318.

Conlan JW, North RJ (1992) Early pathogenesis of infection in the liver with the facultative intracellular bacteria *Listeria monocytogenes, Francisella tularensis*, and *Salmonella typhimurium* involves lysis of infected hepatocytes by leukocytes. *Infect. Immun.* 60: 5164–5171.

Conlan JW, Sjöstedt A, North RJ (1994) CD4+ and CD8+ T-cell-dependent and -independent host defense mechanisms can operate to control and resolve primary and secondary *Francisella tularensis* LVS infection in mice. *Infect. Immun.* 62: 5603–5607.

Cowley S, Myltseva S, Nano FE (1996) Phase variation in *Francisella tularensis* affecting intracellular growth, lipopolysaccharide antigenicity, and nitric oxide production. *Mol. Microbiol.* 20: 867–874.

Cross JT, Jacobs RF (1993) Tularemia: treatment failures with outpatient use of ceftriaxone. *Clin. Infect. Dis.* 17: 976–980.

Cross JT Jr, Schutze GE, Jacobs RF (1995) Treatment of tularemia with gentamicin in pediatric patients. *Pediatr. Infect. Dis. J.* 14: 151–152.

Culkin SJ, Rhinehart-Jones T, Elkins KL (1997) A novel role for B cells in early protective immunity to an intracellular pathogen, *Francisella tularensis* strain LVS. *J. Immunol.* 158: 3277–3284.

Dahlstrand S, Ringertz O, Zetterberg B (1971) Airborne tularemia in Sweden. *Scand. J. Infect. Dis.* 3: 7–16.

Eigelsbach HT, Downs CM (1961) Prophylactic effectiveness of live and killed tularemia vaccines. *J. Immunol.* 87: 415–425.

Eigelsbach HT, McGann VG (1984) Genus *Francisella* Dorofe'ev (1947), 176[AL], pp. 394–399. In: Kreig NR (ed.) *Bergey's Manual of Systematic Bacteriology.* Baltimore: Williams and Wilkins.

Eigelsbach HT, Braun W, Herring RD (1951) Studies on the variation of *Bacterium tularense. J. Bacteriol.* 61: 557–569.

Eigelsbach HT, Hunter DH, Janssen WA, Dangerfield HG, Rabinowitz SG (1975) Murine model for study of cell-mediated immunity: protection against death from fully virulent *Francisella tularensis* infection. *Infect. Immun.* 12: 999–1005.

Eigelsbach HT, Tulis JJ, McGavran MH, White JD (1962) Live tularemia vaccine. I Host–parasite relationship in monkeys vaccinated intracutaneously or aerogenically. *J. Bacteriol.* 84: 1020–1027.

Elkins KL, Cooper AC, Kieffer TL, Colombini S (2001) *In vivo* clearance of an intracellular bacterium, *F. tularensis* LVS, is dependent of the p4.0 subunit of IL 12. Submitted for publication.

Elkins KL, Leiby DA, Winegar RK, Nacy CA, Fortier AH (1992) Rapid generation of specific protective immunity to *Francisella tularensis. Infect. Immun.* 60: 4571–4577.

Elkins KL, Rhinehart-Jones T, Nacy CA, Winegar RK, Fortier AH (1993) T-cell-independent resistance to infection and generation of immunity to *Francisella tularensis. Infect. Immun.* 61: 823–829.

Elkins KL, Rhinehart-Jones TR, Culkin SJ, Yee D, Winegar RK (1996) Minimal requirements for murine resistance to infection with *Francisella tularensis* LVS. *Infect. Immun.* 64: 3288–3293.

Elkins KL, MacIntyre AT, Rhinehart-Jones TR (1998) Nonspecific early protective immunity in *Francisella* and *Listeria* infections can be dependent on lymphocytes. *Infect. Immun.* 66: 3467–3469.

Elkins KL, Bosio CM, Rhinehart-Jones TR (1999a) Importance of B cells, but not specific antibodies, in primary and secondary protective immunity to the model intracellular bacterium, *Francisella tularensis* LVS. *Infect. Immun.* 67: 6002–6007.

Elkins KL, Rhinehart-Jones TR, Stibitz C, Conover JS, Klinman DM (1999b) Bacterial DNA containing CpG motifs stimulates lymphocyte-dependent protection of mice against lethal infection with intracellular bacteria. *J. Immunol.* 162: 2291–2298.

Enderlin G, Morales L, Jacobs RF, Cross JT (1994) Streptomycin and alternative agents for the treatment of tularemia: review of the literature. *Clin. Infect. Dis.* 19: 42–47.

Ericsson M, Sandström G, Sjöstedt A, Tärnvik A (1994a) Persistence of cell-mediated immunity and decline of humoral immunity to the intracellular bacterium *Francisella tularensis* 25 years after natural infection. *J. Infect. Dis.* 170: 110–114.

Ericsson M, Tärnvik A, Kuoppa K, Sandström G, Sjöstedt A (1994b) Increased synthesis of DnaK, GroEL, and GroES homologs by *Francisella tularensis* LVS in response to heat and hydrogen peroxide. *Infect. Immun.* 62: 178–183.

Evans ME, Gregory DW, Schaffner W, McGee ZA (1985) Tularemia: a 30-year experience with 88 cases. *Medicine* 64: 251–269.

Forsman M, Sandström G, Sjöstedt A (1994) Analysis of 16S ribosomal DNA sequences of *Francisella* strains and utilization for determination of the phylogeny of the genus and for identification of strains by PCR. *Int. J. Syst. Bacteriol.* 44: 38–46.

Fortier AH, Slayter MV, Ziemba R, Meltzer MS, Nacy CA (1991) Live vaccine strain of *Francisella tularensis*: infection and immunity in mice. *Infect. Immun.* 59: 2922–2928.

Fortier AH, Polsinelli T, Green SJ, Nacy CA (1992) Activation of macrophages for destruction of *Francisella tularensis*: identification of cytokines, effector cells, and effector molecules. *Infect. Immun.* 60: 817–825.

Fortier AH, Leiby DA, Narayan RB *et al.* (1995) Growth of *Francisella tularensis* LVS in macrophages: the acidic intracellular compartment provides essential iron required for growth. *Infect. Immun.* 63: 1478–1483.

Foshay L (1950) Tularaemia. *Annu. Rev. Microbiol.* 4: 313–330.

Francis E (1925) Tularemia. *JAMA* 84: 1243–1250.

Francis E (1928) A summary of the present knowledge of tularemia. *Medicine* 7: 411–426.

Fulop M, Leslie D, Titball R (1996) A rapid, sensitive method for the detection of *Francisella tularensis* in clinical samples using the polymerase chain reaction. *Am. J. Trop. Med. Hyg.* 54: 364–366.

Golovliov I, Sandström G, Ericsson M, Sjöstedt A, Tärnvik A (1995) Cytokine expression in the liver during the early phase of murine tularemia. *Infect. Immun.* 63: 534–538.

Golovliov I, Kuoppa K, Sjöstedt A, Tärnvik A, Sandström G (1996) Cytokine expression in the liver of mice infected with a highly virulent strain of *Francisella tularensis*. *FEMS Immunol. Med. Microbiol.* 13: 239–244.

Golovliov I, Ericsson M, Sandström G, Tärnvik A, Sjöstedt A (1997) Identification of proteins of *Francisella tularensis* induced during growth in macrophages and cloning of the gene encoding a prominently induced 23-kilodalton protein. *Infect. Immun.* 65: 2183–2189.

Green SJ, Nacy CA, Schreiber RD *et al.* (1993) Neutralization of gamma interferon and tumor necrosis factor alpha blocks *in vivo* synthesis of nitrogen oxides from L-arginine and protection against *Francisella tularensis* infection in *Mycobacterium bovis* BCG-treated mice. *Infect. Immun.* 61: 689–698.

Hass R, Kahrs AF, Facius D, Allmeier H, Schmitt R, Meyer TF (1993) TnMax – a versatile min-transposon for the analysis of cloned genes and shuttle mutagenesis. *Gene* 130: 23–31.

Halmann M, Benedict M, Mager J (1967) An endogenously produced substance essential for growth initiation of *Pasteurella tularensis*. *J. Gen. Microbiol.* 49: 461–468.

Hambleton P, Evans CGT, Hood AM, Strange RE (1974) Vaccine potencies of the live vaccine strain of *Francisella tularensis* and isolated bacterial components. *Br. J. Exp. Pathol.* 55: 363–373.

Hollis DG, Weaver RE, Steigerwalt AG, Wenger JD, Moss CW, Brenner DJ (1989) *Francisella philomiragia* comb. nov. (formerly *Yersinia philomiragia* and *Francisella tularensis* biogroup novicida (formerly *Francisella novicida*) associated with human disease. *J. Clin. Microbiol.* 27: 1601–1608.

Hood AM (1977) Virulence factors of *Francisella tularensis*. *J. Hyg.* 79: 47–65.

Hornick RB, Eigelsbach HT (1969) Tularemia epidemic – Vermont, 1968. *N. Engl. J. Med.* 281: 1310.

Janeway CA Jr (1992) The immune system evolved to discriminate infectious nonself from noninfectious self. *Immunol. Today* 13: 11–16.

Jantzen E, Berdal BP, Omland T (1979) Cellular fatty acid composition of *Francisella tularensis*. *J. Clin. Microbiol.* 10: 928–930.

Jensen WI, Owen DR, Jellison WJ (1969) *Yersinia philomirogia* sp. n., a new member of the *Pasteurella* group of bacteria, naturally pathogenic for the muskrat (*Ondatra zibethica*). *J. Bacteriol.* 100: 1237–1241.

Junhui Z, Jianchun L, Songle Z, Meiling C, Fengxiang C, Hong C (1996) Detection of *Francisella tularensis* by the polymerase chain reaction. *J. Med. Microbiol.* 45: 477–482.

Karow M, Georgopoulos C (1993) The essential *Escherichia coli msbA* gene, a multicopy suppressor of null mutations in the *htrB* gene, is related to the universally conserved family of ATP-dependent translocators. *Mol. Microbiol.* 7: 69–79.

Karttunen R, Andersson G, Hans PT *et al.* (1987) Interleukin 2 and gamma interferon production, interleukin 2 receptor expression, and DNA synthesis induced by tularemia antigen *in vitro* after natural infection or vaccination. *J. Clin. Microbiol.* 25: 1074–1078.

Karttunen R, Surcel HM, Andersson G, Ekre HPT, Herva E (1991) *Francisella tularensis*-induced *in vitro* gamma interferon, tumor necrosis factor alpha, and interleukin 2 responses appear within 2 weeks of tularemia vaccination in human beings. *J. Clin. Microbiol.* 29: 753–756.

Kaufmann SHE (1990) Heat shock proteins and the immune response. *Immunol. Today* 11: 129–136.

Kaufmann SHE, Rodewald HR, Hug E, DeLibero G (1988) Cloned *Listeria monocytogenes* specific non-MHC-restricted Lyt-2$^+$ T cells with cytolytic and protective activity. *J. Immunol.* 140: 3173–3179.

Koskela P, Herva E (1980) Cell-mediated immunity against *Francisella tularensis* after natural infection. *Scand. J. Infect. Dis.* 12: 281–287.

Koskela P, Herva E (1982) Cell-mediated and humoral immunity induced by a live *Francisella tularensis* vaccine. *Infect. Immun.* 36: 983–989.

Koskela P, Salminen A (1985) Humoral immunity against *Francisella tularensis* after natural infection. *J. Clin. Microbiol.* 22: 973–979.

Larson CL, Wicht W, Jellison WL (1955) An organism resembling *P. tularensis* from water. *Public Health Rep.* 70: 253–258.

Leiby DA, Fortier AH, Crawford RM, Schreiber RD, Nacy CA (1992) *In vivo* modulation of the murine immune response to *Francisella tularensis* LVS by administration of anticytokine antibodies. *Infect. Immun.* 60: 84–89.

Leslie DL, Cox J, Lee M, Titball RW (1993) Analysis of a cloned *Francisella tularensis* outer membrane protein gene and expression in attenuated *Salmonella typhimurium*. *FEMS Microbiol. Lett.* 111: 331–336.

Löfgren S, Tärnvik A, Bloom GD, Sjöberg W (1983) Phagocytosis and killing of *Francisella tularensis* by human polymorphonuclear leukocytes. *Infect. Immun.* 39: 715–720.

Long GW, Oprandy JJ, Narayanan RB, Fortier AH, Porter KR, Nacy CA (1993) Detection of *Francisella tularensis* in blood by polymerase chain reaction. *J. Clin. Microbiol.* 31: 152–154.

McCoy GW, Chapin CW (1912) Further observations on a plague like disease of rodents with a preliminary note on the causative agent *bacterium tularense*. *J. Infect. Dis.* 10: 61–72.

McCrumb FR (1961) Aerosol infection of man with *Pasteurella tularensis*. *Bacteriol. Rev.* 25: 262–267.

Mdluli KE, Anthony LSD, Baron GS, McDonald MK, Myltseva SV, Nano FE (1994) Serum-sensitive mutation of *Francisella novicida*: association with an ABC transporter gene. *Microbiology* 140: 3309–3318.

Mokrievich AN, Fursov VV, Pavlov VM (1994) Plasmid cryotransformation. *Problemy Osobo Opasnyh Infecktsy Saratov* 4: 186–190.

Nano FE (1988) Identification of a heat-modifiable protein of *Francisella tularensis* and molecular cloning of the encoding gene. *Microb. Pathog.* 5: 109–119.

Ohara S (1954) Studies on Yato-Byo (Ohara's disease, tularemia in Japan), report I. *Jpn J. Exp. Med.* 24: 69–85.

Overholt EL, Tigertt WD, Kadull PJ *et al.* (1961) An analysis of forty-two cases of laboratory-acquired tularemia: treatment with broad-spectrum antibiotics. *Am. J. Med. Sci.* 30: 785–791.

Pavlov VM, Mokrievich AN, Volkovoy K (1996) Cryptic plasmid pFNL10 from *Francisella novicida*-like

F6168: the base of plasmid vectors for *Francisella tularensis*. *FEMS Immunol. Med. Microbiol.* 13: 253–256.

Polsinelli T, Meltzer MS, Fortier AH (1994) Nitric oxide-independent killing of *Francisella tularensis* by IFNγ-stimulated murine alveolar macrophages. *J. Immunol.* 153: 1238–1245.

Poquet Y, Kroca M, Halary F *et al.* (1998) Expansion of Vgamma9Vdelta2 T cells is triggered by *Francisella tularensis* derived-phosphoantigens in tularemia but not after tularemia vaccination. *Infect. Immun.* 66: 2107–2114.

Reilly TJ, Baron GS, Nano FE, Kuhlenschmidt MS (1996) Characterization and sequencing of a respiratory burst-inhibiting acid phosphatase from *Francisella tularensis*. *J. Biol. Chem.* 271: 10973–10983.

Rhinehart-Jones TR, Fortier AH, Elkins KL (1994) Transfer of immunity against lethal murine *Francisella* infection by specific antibody depends on host gamma interferon and T cells. *Infect. Immun.* 62: 3129–3137.

Sandström G (1994) The tularemia vaccine. *J. Chem. Tech. Biotechnol.* 59: 315–320.

Sandström G, Tärnvik A, Wolf-Watz H (1987) Immuno-specific T-lymphocyte stimulation by membrane proteins from *Francisella tularensis*. *J. Clin. Microbiol.* 25: 641–644.

Sandström G, Löfgren S, Tärnvik A (1988) A capsule-deficient mutant of *Francisella tularensis* LVS exhibits enhanced sensitivity to killing by serum but diminished sensitivity to killing by polymorphonuclear leukocytes. *Infect. Immun.* 56: 1194–1202.

Sandström G, Sjöstedt A, Forsman M, Pavlovic NV, Mishankin BN (1992a) Characterization and classification of strains of *Francisella tularensis* isolated in the central asian focus of the Soviet Union and in Japan. *J. Clin. Microbiol.* 30: 172–175.

Sandström G, Sjöstedt A, Johansson T, Kuoppa K, Williams JC (1992b) Immunogenicity and toxicity of lipopolysaccharide from *Francisella tularensis* LVS. *FEMS Microbiol. Immunol.* 105: 201–210.

Sanford JP (1983) Tularemia. *JAMA* 250: 3225–3226.

Schricker RL, Eigelsbach HT, Mitten JQ, Hall WC (1972) Pathogenesis of tularemia in monkeys aerogenically exposed to *Francisella tularensis* 425. *Infect. Immun.* 5: 734–744.

Scott P (1993) IL-12: initiation cytokine for cell-mediated immunity. *Science* 260: 496–497.

Sjöstedt A, Sandström G, Tärnvik A, Jaurin B (1989) Molecular cloning and expression of a T-cell stimulating membrane protein of *Francisella tularensis*. *Microb. Pathog.* 6: 403–414.

Sjöstedt A, Sandström G, Tärnvik A, Jaurin B (1990a) Nucleotide sequence and T cell epitopes of a membrane protein of *Francisella tularensis*. *J. Immunol.* 145: 311–317.

Sjöstedt A, Sandström G, Tärnvik A (1990b) Several membrane polypeptides of the live vaccine strain *Francisella tularensis* LVS stimulate T cells from naturally infected individuals. *J. Clin. Microbiol.* 28: 43–48.

Sjöstedt A, Tärnvik A, Sandström G (1991) The T-cell-stimulating 17-kilodalton protein of *Francisella tularensis* LVS is a lipoprotein. *Infect. Immun.* 59: 3163–3168.

Sjöstedt A, Eriksson M, Sandström G, Tärnvik A (1992a) Various membrane proteins of *Francisella tularensis* induce interferon-γ production in both CD4[+] and CD8[+] T cells of primed humans. *Immunology* 76: 584–592.

Sjöstedt A, Sandström G, Tärnvik A (1992b) Humoral and cell-mediated immunity in mice to a 17-kilodalton lipoprotein of *Francisella tularensis* expressed by *Salmonella typhimurium*. *Infect. Immun.* 60: 2855–2862.

Sjöstedt A, Conlan JW, North RJ (1994) Neutrophils are critical for host defense against primary infection with the facultative intracellular bacterium *Francisella tularensis* in mice and participate in defense against reinfection. *Infect. Immun.* 62: 2779–2783.

Sumida T, Maeda T, Takahashi H *et al.* (1992) Predominant expansion of Vγ9/Vδ2 T cells in a tularemia patient. *Infect. Immun.* 60: 2554–2558.

Surcel H-M (1990) Diversity of *Francisella tularensis* antigens recognized by human T lymphocytes. *Infect. Immun.* 58: 2664–2668.

Surcel H-M, Ilonen J, Poikonen K, Herva E (1989) *Francisella tularensis*-specific T-cell clones are human leukocyte antigen class II restricted, secrete interleukin-2 and gamma interferon, and induce immunoglobulin production. *Infect. Immun.* 57: 2906–2908.

Surcel H-M, Syrjälä H, Karttunen R, Tapaninaho S, Herva E (1991a) Development of *Francisella tularensis* antigen responses measured as T-lymphocyte proliferation and cytokine production (tumor necrosis factor alpha, gamma interferon, and interleukin-2 and -4) during human tularemia. *Infect. Immun.* 59: 1948–1953.

Surcel HM, Tapaninaho S, Herva E (1991b) Cytotoxic CD4[+] T cells specific for *Francisella tularensis*. *Clin. Exp. Immunol.* 83: 112–115.

Tärnvik A (1989) Nature of protective immunity to *Francisella tularensis*. *Rev. Infect. Dis.* 11: 440–451.

Tärnvik A, Holm SE (1978) Stimulation of subpopulations of human lymphocytes by a vaccine strain of *Francisella tularensis*. *Infect. Immun.* 20: 698–704.

Tärnvik A, Löfgren S (1975) Stimulation of human lymphocytes by a vaccine strain of *Francisella tularensis*. *Infect. Immun.* 12: 951–957.

Tärnvik A, Eriksson M, Sandström G, Sjöstedt A (1992) *Francisella tularensis* – a model for studies of the immune response to intracellular bacteria in man. *Immunology* 76: 349–354.

Teutsch SM, Martone WJ, Brink EW *et al.* (1979) Pneumonic tularemia on Martha's Vineyard. *N. Engl. J. Med.* 301: 826–828.

Thorpe BD, Marcus S (1965) Phagocytosis and intracellular fate of *Pasteurella tularensis*. III. *In vivo* studies with passively transferred cells and sera. *J. Immunol.* 94: 578–585.

Tigertt WD (1962) Soviet viable *Pasteurella tularensis* vaccines. A review of selected articles. *Bacteriol. Rev.* 26: 354–373.

Tyeryar FJ, Lawton WD (1970) Factors affecting transformation of *Pasteurella novicida*. *J. Bacteriol.* 104: 1312–1317.

Viljanen MK, Nurmi T, Salminen A (1983) Enzyme-linked immunosorbent assay (ELISA) with bacterial sonicate antigen for IgM, IgA, and IgG antibodies to *Francisella tularensis*: comparison with bacterial agglutination test and ELISA with lipopolysaccharide antigen. *J. Infect. Dis.* 148: 715–720.

Wenger JD, Hollis DG, Weaver RE *et al.* (1989) Infection caused by *Francisella philomiragia* (formerly *Yersinia philomiragia*) A newly recognized human pathogen. *Ann. Intern. Med.* 110: 888–892.

White JD, Rooney JR, Prickett PA, Derrenbacher EB, Beard CW, Griffith WR (1964) Pathogenesis of experimental respiratory tularemia in monkeys. *J. Infect. Dis.* 114: 277–283.

Wood JB, Valteris K, Hardy RH (1976) Imported tularaemia. *BMJ* 76: 811–812.

Yee D, Rhinehart-Jones TR, Elkins KL (1996) Loss of either CD4[+] or CD8[+] T cells does not affect the magnitude of protective immunity to an intracellular pathogen, *Francisella tularensis* strain LVS. *J. Immunol.* 157: 5042–5048.

Young LS, Bicknell DS, Archer BG *et al.* (1969) Tularemia epidemic: Vermont, 1968. *N. Engl. J. Med.* 280: 1253–1260.

Zuber M, Hoover TA, Dertzbaugh MT, Court DL (1995) Analysis of the DnaK molecular chaperone system of *Francisella tularensis*. *Gene* 164: 149–152.

Further Reading

Baron GS, Nano FE (1998) *MglA* and *MglB* are required for the intramacrophage growth of *Francisella novicida*. *Mol. Microbiol.* 29: 247–259.

Bosio CM, Elkins KL (2001) Susceptibility to secondary *Francisella tularensis* LVS infection in B cell deficient mice is associated with neutrophilia, not defects in specific T cell mediated immunity. *Infect. Immun.* 69: 194–302.

Cowley S, Myltseva S, Nano FE (1996) Phase variation in *Francisella tularensis* affecting intracellular growth, lipopolysaccharide antigenicity, and nitric oxide production. *Mol. Micro.* 20: 867–874.

Enderlin G, Morales L, Jacobs RF, Cross JT (1994) Streptomycin and alternative agents for the treatment of tularemia: review of the literature. *Clin. Inf. Dis.* 19: 42–47.

Evans ME, Gregory DW, Schaffner W, McGee ZA (1985) Tularemia: a 30-year experience with 88 cases. *Medicine* 64: 251–269.

Hornick RB, Eigelsbach HT (1969) Tularemia epidemic – Vermont, 1968. *N. Engl J. Med.* 281: 1310.

Poquet V, Kroca M, Halary F *et al.* (1998) Expansion of Vgamma9 Vdelta2 T cells is triggered by *Francisella tularensis*-derived phosphoantigens in tularemia but not after tularemia vaccination. *Infect. Immun.* 66: 2107–2114.

Sanford JP (1983) Tularemia. *JAMA* 250: 3225–3226.

Tärnvik A, Eriksson M, Sandström G, Sjöstedt A (1992) *Francisella tularensis* – a model for studies of the immune response to intracellular bacteria in man. *Immunol.* 76: 349–354.

Tärnvik A (1989) Nature of protective immunity to *Francisella tularensis*. *Rev. Intect. Dis.* 11: 440–451.

102

Rickettsia and *Orientia*

Marina E. Eremeeva[1] and Gregory A. Dasch[2]

[1]University of Maryland at Baltimore, Baltimore, Maryland, USA
[2]Naval Medical Research Center, Bethesda, Maryland, USA

The term 'rickettsia' has for many years been loosely applied to a very wide range of Gram-negative bacteria simply because of their obligate association with arthropods or other hosts, their size and their intracellular habitat. These micro-organisms are, however, now known to be a highly polyphyletic group. Molecular approaches to the phylogeny of rickettsiae have been used to demonstrate that most species in the genera *Rickettsia*, *Orientia*, *Ehrlichia*, *Anaplasma*, *Wolbachia*, *Cowdria* and *Neorickettsia* have similar evolutionary origins and that these genera all belong to the α-subdivision of Proteobacteria. However, *Coxiella burnetii* and *Rickettsiella grylli*, which are closely related to *Legionella*, and *Francisella (Wolbachia) persica* belong to the γ-subdivision of Proteobacteria. Furthermore, rickettsia-like species formerly classified in the genera *Rochalimaea* and *Grahamella* were renamed and combined with *Bartonella* species based on their genomic characteristics (see Chapter 99). Bartonellas are not only cultivable on axenic media but they belong in a different lineage (α-2) of the α-Proteobacteria from *Rickettsia*, *Orientia* and the ehrlichial genera. The other *Bartonella*-like genera, *Haemobartonella* and *Eperythrozoon*, are mycoplasmas (Neimark and Kocan, 1997; Rikihisa *et al.*, 1997). The ehrlichiae are now included in the family *Anaplasmataceae* (Dumler *et al.*, 2001), so the present chapter is restricted to *Rickettsia* and *Orientia*.

The name 'Rickettsia' was first used in the specific sense by the Brazilian Henrique da Rocha-Lima, who identified the agent of epidemic typhus in infected body lice (reviewed by K. Weiss, 1988 and E. Weiss, 1988). The name honours Dr Howard Taylor`Ricketts, who discovered the aetiological agent of Rocky Mountain spotted fever (RMSF) in 1909 and established the role of ticks in its transmission (reviewed by Harden, 1990). Ricketts first recognised that the microbial agents that cause Rocky Mountain spotted fever and epidemic typhus are similar, but distinct, micro-organisms. In 1909 Charles Nicolle established that epidemic typhus is transmitted by the human body louse. He received the Nobel Prize in 1928 for his contributions. The specific epithet of the aetiological agent of epidemic typhus, *Rickettsia prowazekii*, honours Stanislav von Prowazek, who greatly contributed to our understanding of this disease with his studies of the epidemics in Serbia in 1913 and Hamburg in 1914.

Orientia (formerly *Rickettsia*) *tsutsugamushi* was originally implicated by Hayashi as the cause of tsutsugamushi fever in Japan in 1920, but he mistakenly described the agent as a protozoan, *Theileria tsutsugamushi* (Kawamura *et al.*, 1995). The agent was correctly characterised as a bacterium in 1930 and given the name *Rickettsia orientalis* by Nagayo and co-workers. Later, the name *Rickettsia tsutsugamushi* was commonly accepted in the literature until 1995, when it was placed in the new genus *Orientia* based on numerous phenotypic and genetic differences which distinguish it from the species of *Rickettsia* (Tamura *et al.*, 1995). Because *Rickettsia* and *Orientia* are more closely related to each other than to any other currently known bacteria, for present purposes they will be discussed in this chapter together. This is also justified by historical aspects of the discovery of the species in these genera, their similar association with arthropod vectors, common aspects of the clinical symptoms and pathogenesis of the diseases they cause, and their similar interactions with eukaryotic host cells and metabolic characteristics. For the purposes of this chapter, the term 'rickettsiae' will be used to refer to both *Rickettsia* and *Orientia*.

Classification

Micro-organisms in the genera *Rickettsia* and *Orientia* are obligately intracellular, Gram-negative bacteria. Based on analysis of their 16S rRNA and groEL (60 kDa heat-shock chaperonin protein) genes, *Rickettsia* and *Orientia* represent very closely related evolutionary lineages of micro-organisms in the α-subdivision of *Proteobacteria* (Stover *et al.*, 1990;

Viale and Arakaki, 1994; Ohashi *et al.*, 1995). According to current concepts *Rickettsia* and *Orientia* are the sole genera belonging to the family *Rickettsiaceae*, order Rickettsiales (Dumler *et al.*, 2001).

Rickettsia

The genus *Rickettsia* is classically separated into the typhus group (TG) and the spotted fever group (SFG) based on the presence of distinct group lipopolysaccharide (LPS) antigens. The typhus group includes two human pathogens, *Rickettsia prowazekii* and *R. typhi*. *R. canada* has been isolated from ticks only but is traditionally considered a member of TG since it contains LPS which is cross-reactive with LPS from these species and shares some of their biological traits. Molecular data, however, indicate that it is a deeply diverging separate lineage within the genus *Rickettsia* (Roux and Raoult, 1995, 1999b; Stothard and Fuerst, 1995; Roux *et al.*, 1997). The SFG contains more than 30 diverse genotypes which cause at least 13 different diseases or pathological syndromes in humans (**Fig. 1, Table 1**). Improvements in isolation techniques for rickettsiae and widespread application of molecular tools for the identification of these fastidious micro-organisms have resulted in recognition of some new

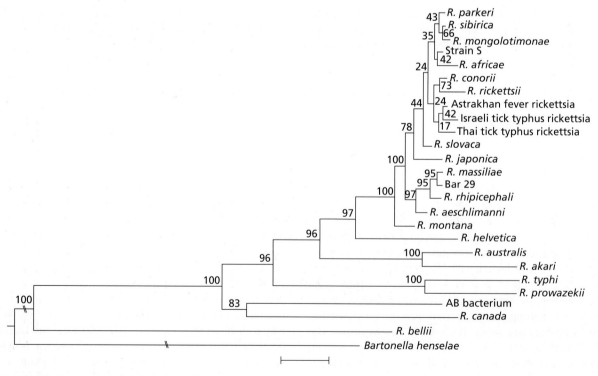

Fig. 1 Phylogenetic analysis of genotypes in the genus *Rickettsia* derived from *gltA* DNA sequences. The evolutionary distance values were determined by the method of Kimura, and the tree constructed by the neighbour-joining method. The scale bar represents 1% difference in nucleotide sequences. The numbers at the nodes are the proportions of 100 bootstrap re-samplings that support the topology shown. Reproduced from Roux *et al.* (1997) with permission of the General Society for Microbiology.

Table 1 Diseases and habitats of pathogenic *Rickettsia* and *Orientia*

Species	Disease or pathogenic effect	Principal arthropod vector	Principal vertebrate reservoirs[a,b]	Geographic distribution
Typhus group				
R. prowazekii	Epidemic typhus	Human body louse	**Human**	Worldwide
	Brill-Zinsser (recrudescent typhus)	None	**Human**	Worldwide
	Sylvatic typhus	Squirrel flea, louse	**Flying squirrels**	Eastern USA
R. typhi	Endemic (murine) typhus	Rat flea, louse	**Rats, other rodents**	Worldwide
R. canada	Unknown	*Haemaphysalis* ticks	Rabbits, hare, birds	Ontario, California
Spotted fever group				
R. akari	Rickettsialpox	Mouse mite	**Rodents**	Worldwide
R. felis	Murine typhus-like	Cat flea	**Opossum**, rats	Texas, California, Yucatan
R. rickettsii	Rocky Mountain SF	*Dermacentor* ticks, *Amblyomma cajennense* tick	**Rodents, lagomorphs, canines**, birds	North America, South America
R. amblyommii	Suspected mild SFG rickettsiosis	*Amblyomma americanum* tick	Rodents, birds, ruminants	Southern USA
R. conorii	Boutonneuse fever, Mediterranean SF	*Rhipicephalus* and *Haemaphysalis* ticks	**Rodents, dogs**, lagomorphs	Africa, Southern Europe to India
R. slovaca	Suspected in meningitis Tick-borne lymphadenopathy	*Dermacentor* ticks	**Rodents, lagomorphs**, ruminants	Europe, Asia
R. mongolotimonae	Suspected mild SF	?	?	Europe, Asia
R. sibirica	North Asian tick typhus	*Dermacentor*, *Haemaphysalis* ticks	**Rodents**, canines, ruminants	Europe, Asia
R. africae	African tick bite fever	*Amblyomma* ticks	**Ruminants**, rodents	Sub-Saharan Africa
R. australis	Queensland tick typhus	*Ixodes* ticks	Rodents, marsupials	Australia
R. honei	Flinders Island SF	*Aponomma* tick?	Rodents, canines	Australia, Thailand
R. japonica	Oriental SF	*Haemaphysalis*, *Dermacentor* ticks	Rodents	Japan
R. sharonii	Israeli tick typhus	*Rhipicephalus* ticks	Dogs, rodents, hedgehog	Israel
R. helvetica	Chronic perimyocarditis	*Ixodes ricinus* tick	Rodents, canines, ruminants	Europe
Rickettsia sp.	Astrakhan SF	*Rhipicephalus pumilio* tick	Canines, hedgehog	Europe
Other species				
Rickettsia sp.	Papaya bunchy top disease	Leafhopper *Empoasca papayae* Oman	None	Puerto Rico, Costa Rico
R. sp. AB bacterium	Sex ratio distortion in insect host	Ladybird beetle *Adalia bipunctata*	None	Europe, Asia
O. tsutsugamushi	Scrub typhus, tsutsugamushi fever	Trombiculid mites	**Rodents, marsupials**	Southern Asia, Australia

a Primary vectors are indicated. Many vectors, but not all, serve as the primary rickettsial reservoir.

b Vertebrate hosts that have been shown to be rickettsial reservoirs are in **bold print**. The other vertebrates listed serve as hosts for the vectors and frequently have antibodies to rickettsiae, but rickettsiae have not yet been isolated from their tissues or blood. SF, spotted fever.

species and other as yet unnamed genotypes, particularly within the SFG. However, firm criteria for the naming of new species and the propriety of the nomenclature existing within the genus *Rickettsia* have been a subject of considerable debate. The difficulty is that while different rickettsial genotypes can be readily distinguished, the differences detected between rickettsial species are smaller than those often used to distinguish other bacterial species. In part this is counterbalanced by the relatively small amount of genetic variation found to date within most named rickettsial species (Fuerst *et al.*, 1990; Fuerst and Poetter, 1991).

A polyphasic approach to the taxonomy of the genus *Rickettsia* has been suggested (Roux and Raoult, 1999b, 2000). It proposes to integrate phenotypic, genotypic and phylogenetic criteria in order to establish an improved classification of *Rickettsia*, comparable with that used for other bacterial genera. However, the obligate intracellular growth of *Rickettsia* makes classic phenotypic characteristics difficult to employ for most species and impossible for others as continuous culture outside the arthropod host has not been established for them. Consequently, comparative analysis of gene sequences is now the major criterion for the creation of new species of *Rickettsia*. This molecular systematics approach separates *R. prowazekii* and *R. typhi* from *R. canada* as two different lineages. The *R. canada* cluster also contains *R. bellii* and several other insect-associated bacteria (AB bacterium, pea aphid rickettsia and PTB bacterium), whose cultural requirements are not established and for which only minimal phenotypic information is available. With the polyphasic approach the SFG rickettsiae can be divided into four lineages: the *R. rickettsii*, *R. massiliae*, *R. helvetica* and *R. akari* groups.

Orientia

Orientia (formerly *Rickettsia*) *tsutsugamushi* is the aetiological agent of scrub typhus, also known as tsutsugamushi disease. It is a single species that contains many antigenic and genotypic variants. Immunological techniques such as complement fixation and fluorescent antibody tests with guinea-pig or rabbit antisera, and biological tests, including *in vivo* and *in vitro* neutralisation assays and cross-vaccination protection tests were originally used to define antigenic prototypes. The most extensively used set of prototype isolates includes the Karp (New Guinea), Gilliam (Assam), and Kato (Japan) strains and five isolates from Thailand (TA763, TA716, TA678, TA686 and TH1817). Cross-adsorbed rabbit sera against these eight prototypes were later used to determine the antigen properties of numerous isolates of *Orientia* obtained throughout the endemic region for scrub typhus by an indirect fluorescent antibody test (Shirai *et al.*, 1982). Recent studies of *Orientia* isolates with both monoclonal antibodies and genetic typing have demonstrated both that other divergent antigenic types exist (notably, Kuruoki, Kawasaki, Shimokoshi in Japan and Yonchon in Korea) and that isolates related antigenically to the classic prototypes by antibody typing need not be identical and can be further divided into genetic subtypes (Enatsu *et al.*, 1999; Tamura, 1999). The isolate variability of *Orientia* is not restricted to the major variable surface protein (50–62 kDa), but also occurs in the 22 kDa, 47 kDa and 110 kDa antigens and GroESL operon (Oaks *et al.*, 1989; Kelly *et al.*, 1994; Dasch *et al.*, 1996). The extent to which individual genotypes of *Orientia* may be associated with specific trombiculid mite vectors, different disease epidemiologies and atypical clinical presentations of scrub typhus, including refractory responses to standard antibiotic therapies, is presently unknown.

Identification

Staining Procedures

Direct identification of rickettsiae is most often accomplished with specific differential staining procedures that depend on their cell wall structure. *Rickettsia* is stained most satisfactorily with Gimenez stain (Gimenez, 1964) while Giemsa staining is preferable for *Orientia*. The cell walls of typhus and SFG rickettsiae resemble those of other Gram-negative bacteria, except for the presence of an S-layer. However, species of *Rickettsia* are not stained well by the classical Gram stain. The method of Gimenez is based on the selective retention of carbol fuchsin by rickettsiae, which are seen as bright red slender coccobacillary forms against a pale greenish-blue background produced by the counter-stain, typically malachite green. Both the cell wall composition and protein antigens of *O. tsutsugamushi* are very different from those of typhus and SFG rickettsiae. *O. tsutsugamushi* stains a dark-purple colour with Giemsa stain or Dif-Quick blood stain.

Alternatively, *Orientia* and *Rickettsia* can be visualised by ultraviolet (fluorescent) microscopy following acridine orange staining. The dye binds preferentially to RNA and permits differentiation of physiologically active and dead micro-organisms (Lauer *et al.*, 1981). Viable micro-organisms fluoresce a bright orange to red colour, while dead microbes appear green like the stained protein in the cytoplasm of their host cells.

Growth and Plaquing Characteristics

Rickettsia and *Orientia* are obligate intracellular bacteria that are unable to grow axenically. Both *Rickettsia* and *Orientia* can be cultivated in the yolk sacs of embryonated chicken eggs (Dasch and Weiss, 1991). The optimal growth temperature for SFG rickettsiae is 32–34°C. The embryo is killed early in the course of infection but the SFG rickettsiae continue to multiply for the next 2–3 days. In contrast, TG rickettsiae and *Orientia* grow better at 35°C. The peak titres of TG rickettsiae and *Orientia* are achieved just before embryo death, after which their viability rapidly declines. Yields of TG rickettsiae from yolk sacs are excellent, while tissue culture is more suitable for the abundant growth of SFG rickettsiae and *Orientia*. *Rickettsia* and *Orientia* can be grown in primary chicken fibroblasts or human umbilical vein endothelial cells (HUVECs) or in a variety of established permanent cell lines including Vero green monkey kidney cells, L929 mouse fibroblasts, RAW264.6 mouse macrophage-like cells, and HMEC-1, ESV40 and EA.hy 926 human endothelial cells. Additional growth properties of *Rickettsia* and *Orientia* are discussed later in this chapter.

Different rickettsiae exert markedly different effects on eukaryotic cells and these can be quantified by comparison of their plaque-forming activity (Wike *et al.*, 1972a, b). This technique has been suggested as a distinguishing method for pathogenic and non-pathogenic rickettsiae (Hackstadt, 1996). Plaque-forming activity is also an important taxonomic characteristic (Weiss and Moulder, 1984). Among SFG, *R. rickettsii* forms clearly distinguishable plaques of 2–3 mm in diameter at 4–7 days after inoculation, although some variability in size and plaque morphology exists depending on the cell used, diluent, incubation temperature, and the density of host cells (Wike *et al.*, 1972a; Walker and Cain, 1980; Walker *et al.*, 1982). The plaque-forming ability of SFG rickettsiae also varies in different cell lines (Johnson and Pedersen, 1978) and some species typically require a few days more than *R. rickettsii* to produce visible plaques. *R. prowazekii* forms plaques within 10–13 days and of significantly smaller size (0.5–1.5 mm in diameter) and less defined morphology than those of SFG. Interestingly, diverse strains from the three typhus group rickettsiae, *R. prowazekii*, *R. typhi* and *R. canada*, exhibit very similar plaque-forming properties in primary chicken embryo cells, even though their growth characteristics in cell culture differ

(Woodman *et al.*, 1977). Supplementation of the agarose overlay with 2 µg/mL of emetine and 40 µg/mL NaF, or with dextran sulphate results in formation of more distinct plaques and increases the plaquing efficiency of typhus rickettsiae (Policastro *et al.*, 1996). *O. tsutsugamushi* may take 12–20 days to produce extremely small lytic plaques of 0.4–1.2 mm in diameter in different cell types (Hanson, 1987a). Treatment of host cells with 400 ng/mL daunomycin significantly increases the efficiency of plaque formation by *O. tsutsugamushi* (Hanson, 1987b). The molecular or genetic basis for these phenotypic differences among *Rickettsia* and *Orientia* is not yet known.

Immunological Identification

Non-specific and Group Assays

Rickettsial LPS is the group-specific antigen which differentiates typhus and SFG rickettsiae. The LPS of both rickettsial groups share epitopes with the LPS of *Proteus* OX19 (TG to a greater extent than SFG) and *Proteus* OX2 (SFG to a greater extent than TG) and with LPS of *Legionella bozemanii* and *L. micdadei*. The cross-reactive epitopes are detected with human IgM that is present early in the course of most typhus and spotted fever group rickettsial infections (Amano *et al.*, 1993a, 1998; Sompolinsky *et al.*, 1986; Raoult and Dasch, 1995). This LPS IgM cross-reactivity may lead to misdiagnosis while human anti-rickettsial IgG antibodies are more specific for rickettsial LPS. An ELISA for antibodies to SFG LPS has been described (Jones *et al.*, 1993). Both latex agglutination and passive agglutination assays as well as complement fixation tests with soluble antigen fractions containing LPS are useful for detecting group-specific antibodies against both TG and SFG rickettsiae (Shirai *et al.*, 1975; Hechemy *et al.*, 1981, 1986; La Scola and Raoult, 1997). Scrub typhus infection elicits antibodies against *Proteus* OXK LPS as detected by Weil-Felix reaction and Western blotting (Amano *et al.*, 1993b); however, the *O. tsutsugamushi* antigen responsible for eliciting these antibodies has not been identified. Heat-shock proteins of *Rickettsia* and *Orientia* contain both genus and widely conserved epitopes (Eremeeva *et al.*, 1998).

Specific Assays

The species-specific antigens used to serotype TG and SFG rickettsiae are the rOmpA and rOmpB proteins, also known as SPAs (Dasch, 1981; Dasch *et al.*, 1981; Anacker *et al.*, 1986; Raoult and Dasch, 1989a; Ching *et al.*, 1990). Specific monoclonal and polyclonal antibodies may be used to identify rickettsial cells by

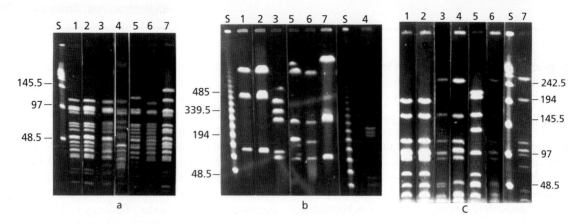

Fig. 2 Identification of genotypes of *Rickettsia* in the spotted fever group by restriction enzyme digestion of chromosomal DNA and separation of fragments by pulsed-field gel electrophoresis. (a) *Sma*I fragments separated for 24 hours at 190 V and a ramped pulse time of 3–10 s. (b) *Bss*HII fragments separated for 48 hours at 150 V and 5–120 s ramped pulse time. (c) *Eag*I fragments separated for 35 hours at 180 V and 5–20 s ramped pulse time. Lane 1. Strain Astrakhan; Lane 2. Tick isolate A-167 from Astrakhan; Lane 3. Israeli tick typhus; Lane 4. *R. conorii* Moroccan; Lane 5. *R. africae*; Lane 6. *R. sibirica* K-1; Lane 7. *R. slovaca* 13-B; Lane S. Bacteriophage lambda DNA PFG ladder. Values are in base pairs. Reproduced from Eremeeva *et al.* (1994d) with permission of Allen Press.

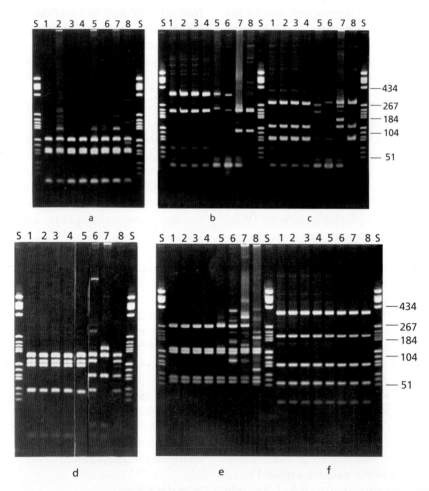

Fig. 3 Differentiation of genotypes of *Rickettsia* in the spotted fever group by PCR/RFLP analysis of amplicons from multiple genes. (a) PCR with primer pairs RpCS.877p and RpCS.1258n from *gltA* digested with *Alu*I. (b,c) PCR with primer pairs Rr190.70p and Rr190.602n from *romp*A digested with *Rsa*I and *Pst*I, respectively. (d,e,f) PCR with primer pairs BG-1 and BG-2, BG-3 and BG-4, BG-5 and BG-6, from *romp*B, respectively, each digested with *Rsa*I. Lane 1. Strain Astrakhan; Lane 2. Tick isolate A-108 from Astrakhan; Lane 3. Tick isolate A-167 from Astrakhan; Lane 4. Israeli tick typhus; Lane 5. *R. conorii* Moroccan; Lane 6. *R. africae*; Lane 7. *R. sibirica* K-1; Lane 8. *R. slovaca* 13-B; Lane S. DNA size markers with values indicated on the right in base pairs. Reproduced from Eremeeva *et al.*, 1994d with permission of Allen Press.

direct micro-immunofluorescence (MIF) and immunoperoxidase techniques (Pickens *et al.*, 1965; Philip *et al.*, 1978; Dumler *et al.*, 1990; Paddock *et al.*, 1999). Western blotting of antigens separated by SDS-PAGE can distinguish closely related isolates of rickettsiae, particularly when specific polyclonal mouse typing sera or monoclonal antibodies are available, but this procedure requires laborious cultivation and purification of the unknown agent and its comparison with standard isolates. Specific monoclonal antibodies are available for many typhus, SFG and *Orientia* isolates and they have been used in blocking assays, capture assays, MIF and Western blotting procedures to identify rickettsiae (Dobson *et al.*, 1989; Raoult and Dasch, 1989a, b; Uchiyama *et al.*, 1990; Tange *et al.*, 1991; Moree and Hanson, 1992; Park *et al.*, 1993; Radulovic *et al.*, 1993, 1994). Monoclonal antibodies useful for identification are mostly directed against the major serotyping proteins rOmpA and rOmpB and LPS of *Rickettsia* (Xu and Raoult, 1997, 1998), and 47, 56 and 110 kDa protein antigens of *Orientia*.

Restriction-fragment Length Polymorphism Analysis of PCR Amplicons and Total Chromosomal DNA

Restriction-fragment length polymorphism (RFLP) analysis of chromosomal DNA with rare cutting restriction endonucleases followed by pulsed-field gel electrophoresis has been used to identify new genotypes, which then can be cultivated for further characterisation (**Fig. 2**) (Roux and Raoult, 1993; Eremeeva *et al.*, 1993a, b, 1995; Babalis *et al.*, 1994).

16S rDNA sequences have also been employed for agent identification, but they have relatively low variability so they are not as useful as other genes for distinguishing among closely related genotypes (Roux and Raoult, 1995; Stothard and Fuerst, 1995). The simple method of PCR-RFLP analysis of the rOmpA, rOmpB, 120-kDa protein, 17-kDa protein and citrate synthase (*gltA*) genes and/or sequencing of these PCR amplicons has, however, become the primary means for rapid characterisation of new isolates of

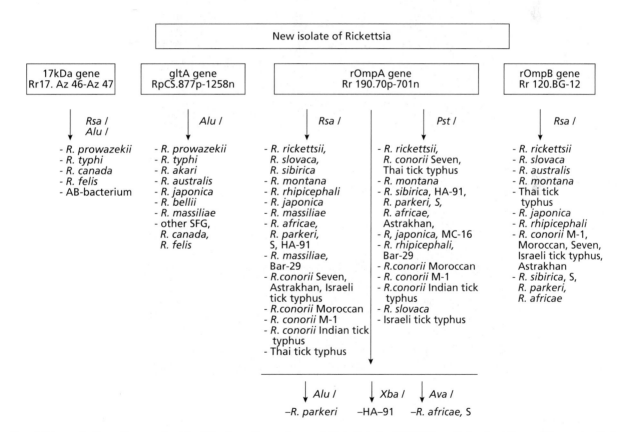

Fig. 4 Schematic flow diagram for identification of genotypes of *Rickettsia* by PCR/RFLP analysis of portions of the genes that encode 17-kDa protein, citrate synthetase, rOmpA and rOmpB. Restriction enzymes used to obtain unique identifying fragmentation patterns and genotypes that can be identified by the combination of fragmentation patterns obtained to that point in the flow diagram are shown. Abbreviations of genotypes: Astrakhan, Astrakhan fever isolate; HA-91, Chinese isolate; S, Armenian tick isolate; MC-16, Moroccan tick isolate MC-16; Bar29, Barcelona tick isolate Bar29. Redrawn from Regnery *et al.*, 1991; Eremeeva *et al.*, 1994d; Roux *et al.*, 1996.

typhus and SFG rickettsiae (**Fig. 3**) (Regnery *et al.*, 1991; Azad *et al.*, 1992; Eremeeva *et al.*, 1993a, 1994b, c, 1996; Walker *et al.*, 1995; Roux *et al.*, 1996, 1997; Dasch and Jackson, 1998; Fournier *et al.*, 1998; Roux and Raoult, 2000). A combination of analysis of *gltA* and r*OmpA*, r*OmpB* and 17-kDa protein gene amplicons by PCR/RFLP can be used to identify the different species and genotypes of typhus and spotted fever rickettsiae (**Fig. 4**). Scrub typhus isolates are best typed by PCR-RFLP or sequencing methods employing the conserved groESL operon and the more variable 22- and 56-kDa antigen genes (Kelly *et al.*, 1994; Dasch *et al.*, 1996; Seong *et al.*, 1997; Enatsu *et al.*, 1999; Tamura *et al.*, 1999).

PCR/RFLP methods are ideal for rickettsiae which are refractory to cultivation in cell culture and are thus available only in primary clinical or arthropod samples (Webb *et al.*, 1990; Beati *et al.*, 1992; Balayeva *et al.*, 1995; Higgins and Azad, 1995). PCR assays have proven to be quite sensitive for clinical use, requiring a minimum of about 10 rickettsiae in an assay sample for a positive result. Field samples, crushed arthropod vectors, whole blood, buffy coats and biopsy samples can be used as sources of DNA for PCR. To increase the reliability and specificity of PCR amplification with low numbers of template copies, specific nested PCR procedures have been employed for both *Rickettsia* and *Orientia* (Anderson *et al.*, 1993; Furuya *et al.*, 1993; Kawamori *et al.*, 1993; Kelly *et al.*, 1994; Tay *et al.*, 1996; Casleton *et al.*, 1998).

Structure

Morphology: Cell Wall Structure and Surface Proteins

Members of the genus *Rickettsia* are small Gram-negative bacterial rods about 0.3–0.5 µm in diameter and 0.8–2.0 µm in length. When cell division is impaired by adverse growth conditions, they may be considerably longer. Rickettsiae do not have morphologically identifiable flagella or pili.

In most respects the cell walls of typhus and SFG rickettsiae are typical of other Gram-negative bacteria: they contain peptidoglycan, LPS, a cytoplasmic membrane with numerous proteins and an outer membrane with a much simpler protein composition and some proteins which are present in high concentrations (Winkler, 1990; Pang and Winkler, 1994). The rickettsiae also contain a microcapsular protein layer, often called a crystalline surface protein layer or S-layer (Dasch, 1981; Ching *et al.*, 1990). In different rickettsiae, it is composed of either one or two

multimeric proteins, rOmpB and rOmpA, also known as the species-specific surface protein antigens or SPAs, which are arrayed in a regular periodic layer that is detectable by electron microscopy with appropriate fixation and staining methods (Ching *et al.*, 1990; Uchiyama *et al.*, 1994b).

O. tsutsugamushi differs from *Rickettsia* spp. in its morphology in that the external layer of its outer membrane is much thicker than the inner leaflet, while the opposite is true of other rickettsial species (Silverman and Wisseman, 1978). *Orientia* also has a diameter about twice as large (1.2–2.5 µm), because it is more coccobacillary than species of *Rickettsia* (Tamura *et al.*, 1990). The cell wall structure and antigens of *O. tsutsugamushi* are very different from those of typhus and SFG rickettsiae. The microcapsule, peptidoglycan and LPS found in *Rickettsia* have not been detected in *Orientia* (Hanson, 1985; Amano *et al.*, 1987). *O. tsutsugamushi* lacks such typical constituents of peptidoglycan and LPS as muramic acid, hydroxy fatty acids and 2-keto-3-deoxyoctonic acid (KDO) (Amano *et al.*, 1987). It is particularly fragile to osmotic shock, is sensitive to digestion with phospholipase A_2, and its growth is more resistant to penicillin than that of other rickettsiae, which may reflect the lack of a peptidoglycan sacculus (reviewed in Kawamura *et al.*, 1995). In contrast, *R. prowazekii* can be induced to form unstable spheroplasts by penicillin (Wisseman *et al.*, 1974). The electron-lucent halo zones ('slime layer') that occur in *Rickettsia* are not observed around growing *O. tsutsugamushi* cells (Silverman *et al.*, 1978; Tamura *et al.*, 1990).

Genomic Sequence of *R. prowazekii*

The complete genomic sequence of the circular chromosome of the *R. prowazekii* attenuated vaccine strain Madrid E has been determined (Andersson *et al.*, 1998). It has 1 111 523 bp with an average $G + C$ content of 29.1 mol%. The genome contains 834 putative complete open reading frames (ORFs) with an average length of 1005 bp. Protein-coding genes represent 75.4% of the genome, and 0.6% of the genome encodes stable RNA. Thirty-three genes encode the 32 different isoacceptor-tRNA species. Each species of rRNA gene, 16S, 23S and 5S, is present as a single copy in the genome and none is adjacent to a tRNA gene (Andersson *et al.*, 1995). The rRNA operons of *R. prowazekii* and the other typhus and SFG rickettsiae are unusual since the 16S rRNA gene and 23S-5S rRNA genes are not contiguous, so this dislocation probably occurred before the divergence of these two groups of rickettsiae (Andersson *et al.*, 1995, 1999).

Functional analysis of the genome of *R. prowazekii* can now be based primarily on deduced sequence homologies to characterised genes and proteins in other bacteria. It confirms earlier observations on the biology and physiology of rickettsiae that were derived from classical biochemical tests. Genome analysis suggests that, as compared with free-living proteo-bacteria, only a very small proportion of rickettsial genes are involved in the biosynthesis of amino acids and nucleosides and their regulation. On the other hand, some putative *R. prowazekii* genes resemble functional homologues found in mitochondria. ATP production in *R. prowazekii* also resembles that of mitochondria; the identified rickettsial genes encode a complete tricarboxylic acid cycle, respiratory-chain complexes, ATP-synthase gene complexes, and the ATP/ADP translocases, while the genes that encode proteins which mediate anaerobic glycolysis are absent (Andersson *et al.*, 1998). Altogether, these results suggest that the functional roles of the genes absent from the rickettsial genome were probably replaced by homologous functions encoded by the host cell genome during the evolutionary adaptation of rickettsiae to a strict intracellular lifestyle. As a reflection of this fact, *R. prowazekii* has only a small number of regulatory genes, particularly two-component signal transduction systems. About 25% of the genome of *R. prowazekii* is represented by non-coding sequences, which are thought to be ORFs that were degraded because of a high spontaneous rate of deletion mutations in the rickettsial genome (Andersson and Andersson, 1997, 1999). Conversely, essential roles may be inferred for retained genes of *R. prowazekii* such as those that encode a complete *sec*-dependent secretory system and the bacterial chaperones, *dnaK*, *dnaJ*, *hslU*, *hslV*, *groEL*, *groES*, *htpG* and *htrA*.

Gene Structure, Codon Usage and Origin of Replication

The genome of *R. prowazekii* is extremely A + T-rich (Andersson and Sharp, 1996). Consequently, rickettsial genes are characterised by a relatively low G + C content of their coding sequences. The 27 genes that have been analysed have a G + C content of 43%, 34%, and 20% for the first, second and third positions of codons, with a strong bias toward U- and A-ending codons, so that UAA is the most common termination codon (Andersson *et al.*, 1998). The codon usage pattern varies randomly among the genes, while the

frequency of codon usage (U versus A or C versus G) also differs in different codon families. As a consequence of the AT-biased codon usage, *R. prowazekii* proteins have a high frequency of AT-coded amino acids (30%) and a low frequency of GC-coded amino acids (16.8%). Tyr, Phe, Asn, Lys and Ile are found the most often, while Pro, Gly and Ala occur less frequently in *R. prowazekii* proteins. The inter-genic spacer regions of *R. prowazekii* have an even lower G + C content of 21.4%. There is no translational selection because of biased codon usage in the *R. prowazekii* genome, since there is very little difference in codon usage patterns between genes, such as the SPA gene, 17-kDa lipoprotein gene and ribosomal genes *rpsG*, *rpsJ* and *rpsL*, which are expressed at different rates.

The *dnaA* gene found near the putative origin of replication is located at approximately 750 kb in the *R. prowazekii* genome (Andersson *et al.*, 1998). The genes that flank *dnaA* differ from the conserved gene arrangement found in *E. coli* and *B. subtilis* (*rnpA*-*rpmH*-*dnaA*-*dnaN*-*recF*-*gyrB*). In particular, while the genes *rnpA* and *rpmH* of *R. prowazekii* are located in the vicinity of *dnaA*, their orientation is reversed as compared with the consensus motif, and *dnaN*, *recF* and *gyrB* are found elsewhere on the chromosome. The functional activity of the putative *dnaA* gene and the exact site of the origin of replication of *R. prowazekii* remain to be demonstrated.

The Genomes of *Rickettsia* and *Orientia*

DNA sequence analysis of the typhus and SFG rickettsiae beside *R. prowazekii* is still fragmentary and for many species has been done only for a few genes including the 16S and 23S rRNAs, 16S–23S intergenic spacer region, *gltA*, *rompA*, *rompB*, the 17-kDa genus-specific protein antigen, the 120-kDa cytoplasmic protein antigen and the *rpoB* subunit of rifampin resistance genes (Anderson and Tzianabos, 1989; Roux and Raoult, 1995; Stothard and Fuerst, 1995; Roux *et al.*, 1997; Fournier *et al.*, 1998; Andersson *et al.*, 1999; Drancourt and Raoult, 1999; Uchiyama, 1999a, b). The genes found in different rickettsial isolates have very little sequence divergence, typically within the range of 0.5%–2%. Phylogenetic analysis of these gene sequences confirms that all *Rickettsia* belong to a closely related cluster of microorganisms and suggests that their metabolic properties may be very similar as well.

Before DNA sequences were obtained so easily, the genomic information about *Rickettsia* that was garnered reflected the limitations of the extant methodology and its applicability to obligate intracellular bacteria. As determined by pulsed-field gel electrophoresis the genome sizes of *R. typhi* and *R. prowazekii* are about 1.1 Mbp, while those of 14 species of SFG rickettsiae are about 1.2–1.3 Mbp (Eremeeva *et al.*, 1993b; Roux and Raoult, 1993). *R. massiliae* and *R. helvetica* genomes are larger (1.3–1.4 Mbp) than those of the other SFG species, while the genome size of the more ancestral species, *R. bellii*, is about 1.66 Mbp (Roux and Raoult, 1993). The G + C content of typhus rickettsiae and *R. bellii* DNA is 29–30 mol% while the composition of DNA from six species of SFG is 31–33 mol% (Schramek, 1972; Tyeryar *et al.*, 1980; Philip *et al.*, 1983).

Homologies established by DNA–DNA hybridisation have been determined for typhus group rickettsiae and a few isolates of SFG. Homology between *R. prowazekii* and *R. typhi* is 70–72%, while their respective homologies to *R. canada* and *R. rickettsii* are 44–52% and 36–53% (Myers and Wisseman, 1980, 1981). Within the SFG group, *R. rickettsii* shares 91–94% DNA homology with *R. conorii*, 70–74% with *R. sibirica* and *R. montana*, 53% with *R. australis* and 46% with *R. akari*. Homologies with other species have not been determined. Electroporation and transformation of *R. prowazekii* and *R. typhi* have been achieved, and will for the first time permit genetic manipulation of *Rickettsia* (Rachek *et al.*, 1998, 2000; Troyer *et al.*, 1999).

O. tsutsugamushi contains a chromosome of 2.4–2.7 Mbp (Tamura *et al.*, 1995). Whether *Orientia* has a more extensive metabolic repertoire than *Rickettsia* or its genome contains significantly more non-coding or paralogous sequences remains unknown. The G + C contents determined for six prototype isolates varied between 28 and 30.5 mol% (Kumura *et al.*, 1991).

Plasmids have not been described in *Rickettsia* or *Orientia*.

Physiology

Rickettsia and *Orientia* are obligate intracellular bacteria that do not grow in axenic medium. The spectrum of their known biochemical and physiological capacities suggests that adaptation has occurred to the physiology of the eukaryotic host cell and to survival in its cytoplasmic and nuclear compartments. However, the precise basis of rickettsial dependence on the host cell for replication remains undefined.

Metabolic Characteristics

Rickettsiae possess a wide range of metabolic activities, but classical biochemical tests are not used for their routine identification since the growth of rickettsiae is strictly eukaryotic host cell-dependent. The biochemical properties of *R. prowazekii* and *R. typhi* have been characterised most completely because they can be purified in much greater quantities than SFG rickettsiae. In addition, the availability of the complete genome sequence of *R. prowazekii* has permitted predictions about the presence of many metabolic pathways based on the homology of putative ORFs to proteins characterised in other bacteria (Andersson *et al.*, 1998). The recent completion of the *R. conorii* genome sequence should shed light on the differences in the biochemical properties of the typhus and spotted fever groups of rickettsiae. Although this comparison will probably confirm their great similarity, there are significant limitations in their applicability to *Orientia* because of its much larger genome size. Typhus rickettsiae are non-glycolytic and utilise citric acid cycle intermediates, notably derived from pyruvate, glutamine and glutamate, and aspartic acid and asparagine, to drive oxidative phosphorylation and ATP synthesis (Austin and Winkler, 1988; Winkler, 1990). They also actively transport activated compounds such as ATP, ADP and NAD, sugar compounds such as UDP-glucose, amino acids and mono-, di- and triphosphate nucleotides but not purine and pyrimidine bases or nucleosides (Winkler *et al.*, 1999). AMP is neither synthesised nor degraded by rickettsiae, as is probably true for the other nucleotide monophosphates. Purified rickettsiae can readily synthesise RNA and proteins from defined nucleotide and amino acid precursors, but only small amounts of DNA (Winkler, 1987). Peak metabolic activities of rickettsiae occur at pH 7.3–7.6, optimally at high concentrations of K^+ and phosphate, and physiological levels of Mg^{2+}. *Orientia* has metabolic properties similar to those of *Rickettsia* but it is more osmotically fragile and less sensitive to potassium ion levels. Its sugar and nucleotide transport properties have not been reported.

Bioenergetics

Rickettsiae utilise the tricarboxylic acid cycle to generate their own metabolic energy by coupling the phosphorylation of ADP to ATP primarily with the oxidation of glutamate and glutamine (Bovarnick, 1956; Williams and Weiss, 1978). The *R. prowazekii* genome contains many genes with homology to characterised enzymes of the electron transport system, particularly the NADPH dehydrogenase complex, the respiratory chain, including cytochrome b and c_1, the cytochrome oxidase complex, and the ubiquinone system (Andersson *et al.*, 1998). Furthermore, *R. prowazekii* possesses its own ATP synthase. The presence of these proteins in other rickettsiae has not been demonstrated to this extent, but there is direct biochemical and indirect physiological evidence for similar functional activities in other species (Austin and Winkler, 1988).

In addition to oxidative phosphorylation of ADP to ATP, rickettsiae can obtain high-energy phosphate directly from the host cell cytoplasm by exchange of rickettsial ADP for host cell ATP. Rickettsiae have an unusual transport system for ATP and ADP, the ATP/ADP translocase (Tlc), which has also been described in *Chlamydia* (see Chapter 87) and plant plastids, but is not known in free-living bacteria (Winkler, 1976; Plano and Winkler, 1991; Alexeyev and Winkler, 1999a, b). A functionally similar transport system exists in mitochondria, but it exchanges internally produced ATP for cytosolic ADP. The rickettsial ATP/ADP exchange transporter system has no sequence similarity with the mitochondrial ATP/ADP system. Furthermore, the rickettsial ATP/ADP translocase is insensitive to carboxyatractyloside and bongkrekic acid, which specifically inhibit the mitochondrial enzyme. The 56-kDa Tlc of *R. prowazekii* has 12 transmembrane domains that traverse the cytoplasmic membrane. Based on membrane topology studies of rickettsial Tlc expressed in *E. coli*, its N- and C-termini probably both lie in the cytoplasm. The *R. prowazekii* genome contains five related genes that encode for putative ATP/ADP translocases, which may reflect their biological importance (Andersson *et al.*, 1998). All genes of this family are probably actively transcribed and translated, based on detection of mRNA transcribed from four of these homologues. Duplication and divergence of the ATP/ADP translocase genes occurred before the divergence of typhus and SFG rickettsiae.

Products

Purified and Recombinant Proteins

Very little work has been done on proteins purified directly from rickettsiae because of the difficulties inherent in obtaining sufficient mass from any species except the TG. More recently a number of proteins have been expressed and characterised as recombinant proteins in *E. coli*.

A major characteristic feature of all typhus and spotted fever group rickettsiae, except *R. bellii*, is the presence of an outermost crystalline surface protein layer composed of either one or two multimeric proteins, rOmpA and rOmpB. These proteins are also known as the species-specific surface protein antigens (SPA). *R. prowazekii* and *R. typhi* contain only one type of SPA, rOmpB, while *R. canada* and the majority of SFG rickettsiae express two different SPAs, rOmpA and rOmpB. The rOmpBs of *R. prowazekii* and *R. typhi* contain two and one cysteine residues, respectively, that are oxidised and form disulphide-linked multimers (Dasch, 1985). rOmpB is processed from a 168-kDa precursor by cleavage of a conserved carboxypeptide domain resulting in a mature N-terminal-encoded 120-kDa surface layer protein and a 32-kDa β protein derived from the C-terminus, which contains a membrane-anchoring domain (Gilmore *et al.*, 1991; Hackstadt *et al.*, 1992). The 32-kDa protein sequence is highly conserved among rOmpB genes from different species. A homologous sequence is present in some rOmpA protein genes which have been sequenced completely (Fournier *et al.*, 1998), but it is not known whether rOmpA genes also undergo carboxy cleavage. One peculiarity of rOmpA is the presence of tandemly repeated 72 and 75 amino acid sequences which vary in number, sequence and order among different SFG isolates (Anderson *et al.*, 1990; Gilmore and Hackstadt, 1991; Gilmore, 1993; Croquet-Valdes *et al.*, 1994). rOmpA and rOmpB are very rich in β-sheet structure and are heat-labile antigens. rOmpA and rOmpB contain strain, species, group-specific and group-cross-reactive epitopes which can be demonstrated with monoclonal antibodies and hyperimmune animal and patient sera (Anacker *et al.*, 1987a, b; Raoult and Dasch, 1989a, b; Ching *et al.*, 1990; Uchiyama *et al.*, 1990, 1994a, 1995; Vishwanath, 1991). Some monoclonal antibodies as well as human immune sera also recognise domains that contain methylated lysines on the rOmpB of typhus group rickettsiae (Ching *et al.*, 1993, 1999a). The rOmp proteins are the dominant protein antigens which elicit anti-rickettsial responses in humans and experimental animals (Eremeeva *et al.*, 1994a, b). They have been used as protective sub-unit vaccines for typhus and SFG rickettsiae (Dasch *et al.*, 1984, 1999; Sumner *et al.*, 1995). A recombinant antigen containing the repeat region of rOmpA provided protection in guinea-pigs against *R. rickettsii* (McDonald *et al.*, 1987, 1988) and a *R. conorii* recombinant provided both homologous and heterologous protection against *R. rickettsii* (Vishwanath *et al.*, 1990). The rOmpA proteins exhibit differences in the number and the arrangement of these repeats as well as minor differences in sequence (Gilmore, 1993) but the degree to which this may affect cross-protective immunity is not known.

A major outer membrane protein of 29.5 kDa from *R. prowazekii* has significant protein sequence homology with peptidyl-prolyl *cis/trans* isomerase and related proteins of the parvulin family (Emelyanov *et al.*, 1996). Cloned genes of 29.5-kDa protein from the virulent strains Breinl and EVir and avirulent strain E were found to be identical by RFLP analysis, and in the electrophoretic mobility of corresponding recombinant proteins encoded by these genes. Proteins of the parvulin family have a wide variety of biological functions, including protein maturation, trafficking and arrangement in the outer membrane, adhesion to the host cell, and binding of cyclosporins. The biological function of the rickettsial parvulin is unknown.

The genus-specific 17-kDa outer membrane protein is another rickettsial antigen that undergoes post-translational modification by acquiring a lipid moiety (Anderson *et al.*, 1988). *Rickettsia* also possesses several widely conserved heat-shock proteins, among which DnaK and GroEL share epitopes found widely in other bacteria, while DnaJ, GrpE and GroES possess more genus-specific epitopes (Eremeeva *et al.*, 1998).

A rickettsial intracytoplasmic protein of 120 kDa (PS120) is conserved among SFG rickettsiae (Schuenke and Walker, 1994; Uchiyama *et al.*, 1994b; Uchiyama, 1997, 1999a). This protein is heat-stable and comigrates in SDS-polyacrylamide gels with rOmpB when the organisms are solubilised without heating before electrophoresis (Uchiyama *et al.*, 1994b). Heat denaturation before electrophoresis permits separation of rOmpB and PS120 at 135 and 120 kDa, respectively. rOmpB and PS120 can also be distinguished by peptide mapping and Western immunoblotting (Uchiyama *et al.*, 1994a). PS120 genes from *R. conorii* and *R. japonica* have 97% homology but share only 68% homology to PS120 from *R. prowazekii* (Schuenke and Walker, 1994; Uchiyama, 1999a).

The major variable immunogenic surface protein, designated 56 kDa antigen or vOmp, that differs between serotypes of *Orientia* varies considerably in size from 52 to 62 kDa and contains serotype-specific and conserved epitopes (Kawamura *et al.*, 1995). Other antigenic proteins in the range of 110 to 22 kDa are also located on the cell surface. The 22- and 110-kDa *Orientia*-specific proteins are also type-specific antigens while the 47-, 60- and 70-kDa protein antigens are more conserved and are homologous to the HtrA, GroEL and DnaK heat-shock proteins, respectively. However, the DnaJ and GroES proteins of *Orientia* do not exhibit cross-reactivity with antisera against the corresponding homologues from *E. coli*. The 56-kDa proteins elicit very early IgM and IgG serologic responses but convalescent sera also contain antibodies against the 72, 60, 47 and 22–35 kDa proteins of *O. tsutsugamushi*. Scrub typhus infection also elicits antibodies against *Proteus* OXK LPS, as detected by

the Weil–Felix reaction (Amano *et al.*, 1993b). While the rickettsial antigen responsible for this reaction has not yet been purified, it appears to be a heat-stable carbohydrate (Lim *et al.*, 2000).

Lipopolysaccharide

Lipopolysaccharides (LPS) are thermostable, group-specific antigens of the genus *Rickettsia* which are absent in *Orientia* (Amano *et al.*, 1987). Some details of the chemical structure have been established for LPS purified from *R. prowazekii* and *R. typhi*, and, among SFG group organisms, for LPS from *R. japonica* and Thai tick typhus rickettsia TT118 (Amano *et al.*, 1993a, 1998). TG LPS migrates in a ladder-like pattern on SDS-PAGE. Major components include glucose, 3-deoxy-D-manno-octulosonic acid, glucosamine, quinovosamine, phosphate and fatty acids, β-hydroxylmyristic acid and heneicosanoic acid (Amano *et al.*, 1998). The O-side-chain polysaccharides contain glucose, glucosamine, quinovosamine and phosphorylated hexosamine. The first three sugars are also found in the O-polysaccharide chains of the LPS from *Proteus vulgaris* OX19, so they probably represent the cross-reacting antigens recognised in the Weil–Felix test (see below under Diagnosis). LPS from *R. japonica* and TT118 contains KDO, glucosamine, quinovosamine, ribose, phosphate and palmitic acid (Amano *et al.*, 1993a). SFG LPS does not have the β-hydroxy fatty acids found in TG LPS. Sera from patients infected with Japanese spotted fever reacted with SFG LPS, with the polysaccharide moiety of *Proteus* OX2 LPS, and with core saccharide or lipid A moiety of OX19 LPS. Neither TG LPS nor SFG LPS contain heptose, which is a component of enterobacterial LPS. Rickettsial LPS is also relatively resistant to acid hydrolysis, which suggests that the linkage between the core polysaccharide and lipid A moieties of rickettsial LPS differs from that of enterobacterial LPSs. The relationship between the chemical moieties of rickettsial LPS and its biological activities are not known.

Pathogenesis

The pathogenic effects of rickettsiae have been attributed to their successful adherence, penetration and intracellular multiplication in the cytoplasm of eukaryotic cells. Each of these processes has been examined in great detail using electron, light and fluorescence microscopy, and by biochemical analyses, but much is still not understood (reviewed in Eremeeva *et al.*,

2000). The following steps of rickettsial entry can be differentiated by electron microscopy: adhesion of the rickettsia to the host cell membrane, uptake by engulfment of the rickettsia by host cell membrane extensions or by invagination of the host cell membrane, inclusion of the rickettsia in a phagocytic vacuole, lysis of the vacuole and escape of the rickettsia into the cytoplasm and intracytosolic growth.

Invasion of Host Cells

Adhesion

Attachment of rickettsiae to the cell surface of eukaryotic cells is the first event during rickettsial invasion. Rickettsial adhesion is a very rapid process immediately followed by uptake of the bound microorganisms (Walker and Winkler, 1978). Adhesion of viable *R. prowazekii* to either fibroblasts or HUVECs is not influenced by pre-treatment of cells with *N*-ethylmaleimide (NEM), NaF, cytochalasin B or *p*-chloromercuriphenylsulphonic acid, but it is greatly dependent on temperature, time of incubation and the number of rickettsiae per cell (Walker, 1984; Walker and Winkler, 1978). Treatment of rickettsiae with the calcium ionophore A23187 stimulated their binding to HUVECs (Walker, 1984), indicating the importance of calcium levels for invasion by *R. prowazekii*. In contrast, inactivation of rickettsiae with NEM, formalin or UV light significantly reduced adhesion. Adhesion of *R. prowazekii* to the host cell is irreversible, and once it has occurred, it is no longer sensitive to the inhibitory effect of NEM (Winkler and Miller, 1984; Winkler and Turco, 1988). Since similar observations were made in experiments with *R. prowazekii* and *R. conorii* infection of fibroblasts and endothelial cells (Winkler and Miller, 1984; Winkler and Turco, 1988; Li and Walker, 1992), unique cell-specific surface receptors are probably not involved in adhesion of rickettsiae to different types of eukaryotic cells. Active roles for both rickettsiae and host cells are important in adhesion, since binding does not occur following heat inactivation of the rickettsiae, or treatment of either host cells or rickettsiae with 1% paraformaldehyde or with 0.25% trypsin for 15 minutes (Li and Walker, 1992). Since TG rickettsiae express haemolytic activity *in vitro* but do not invade erythrocytes (Winkler, 1985), this interaction has been used as a model system for studying adhesion (Winkler, 1977; Winkler and Miller, 1980).

Entry of Rickettsiae into the Host Cell

Viable rickettsiae that bind to the external cell surface appear to perturb the host cell plasma membrane in a way that results in their internalisation at the site of adhesion (Walker and Winkler, 1978). The site of *R. rickettsii* adhesion and entry exhibits remarkable changes in its electron density (Eremeeva *et al.*, 1999). The uptake of *R. prowazekii* by fibroblasts and HUVECs is sensitive to treatment with cytochalasins, NEM, NaF, requires metabolically active rickettsiae, and is probably calcium-dependent (Walker and Winkler, 1978; Walker, 1984). Accordingly, this process has been called induced phagocytosis (Walker and Winkler, 1978) in contrast with two other possible models for rickettsial entry: direct penetration of a passive host cell by active rickettsiae or phagocytosis of passive rickettsiae by an active host cell (Winkler and Turco, 1988). The current hypothesis, developed primarily for *R. prowazekii*, suggests that induced phagocytosis is coupled with rickettsial release from the phagosome by means of a rickettsial phospholipase (Winkler and Turco, 1988; Winkler and Daugherty, 1989). The general applicability of this concept has been tested and confirmed only for *R. rickettsii* invasion of HUVECs (Eremeeva *et al.*, 1999), *R. conorii* in Vero cells (Teysseire *et al.*, 1995), *R. japonica* in L929 mouse fibroblasts (Iwamasa *et al.*, 1992) and *O. tsutsugamushi* in mouse fibroblasts, HUVECs and phagocytic cells (Urakami *et al.*, 1982, 1984; Ng *et al.*, 1985; Tamura, 1988).

Both internalisation and escape of *R. rickettsii* from the phagocytic vacuole in HUVECs are very rapid processes, since 10 to 30 minutes after internalisation was begun, many rickettsiae could be observed in the cytoplasm. During *R. conorii* infection of Vero cells between 40% and 90% of the bacteria were internalised by 3 and 12 minutes or later, respectively. By 12 minutes, about 45% of the intracellular rickettsiae were already seen free in the cytoplasm without being surrounded with phagosomal membrane (Teysseire *et al.*, 1995). Entry of *R. rickettsii* into EA.hy 926 endothelial cells is associated with an immediate significant increase in the polymeric F-actin and decrease in the monomeric G-actin content of infected cells, as compared with uninfected control cells (Eremeeva *et al.*, 1999). Unipolar nucleation of actin tails and replication occur immediately after release of *R. rickettsii* from the phagocytic vacuole (Heinzen *et al.*, 1993; Eremeeva *et al.*, 1999).

Rickettsial Components Involved in Invasion

Surface Proteins   Our understanding of the rickettsial structures that bind to the host cell and mediate rickettsial invasion is very limited. Since adhesion of *R. conorii* to cells is sensitive to heat and protease treatment, surface-exposed protein(s) may play an important role in this interaction (Li and Walker,

1992, 1998). It is possible that different SFG species recognise the same host cell receptor, since when they are mixed in equal numbers, *R. rickettsii* competitively inhibits the attachment of *R. conorii* to L929 cells by 51%. The species-specific *rOmpA* and *rOmpB* proteins have been suggested as the rickettsial ligands (adhesins) because they are surface-exposed and thermo-modifiable. Recombinant *rOmpA* and *rOmpB* expressed in *E. coli* also inhibited the attachment of *R. japonica* to Vero cells (Uchiyama, 1999b). Surface array proteins have weak sequence homology with flagella and some adhesins and have been proposed to have similar structural and functional properties (Kuen and Lubitz, 1996). If this hypothesis is correct, it may be expected that *R. prowazekii* and *R. typhi* will not display the same type of adhesive characteristics as SFG rickettsiae. These typhus rickettsiae only have rOmpB proteins that form intermolecular disulphide bonds and which exhibit high resistance to trypsin treatment, while SFG rickettsiae possess non-disulphide-linked rOmpA and rOmpB, which appear to be more susceptible to proteases. Differences in the interaction of TG and SFG rickettsiae with erythrocytes are consistent with these molecular differences in surface composition.

Phospholipase A$_2$ Since release of free fatty acids and lysophosphatides occurs during both the haemolysis of erythrocytes and the entry of L929 cells by *R. prowazekii*, a role for rickettsia-associated PLA$_2$ in invasion is also very likely (Winkler and Miller, 1980, 1982; Winkler, 1985). Phospholipase activity also occurs during the infection of different types of eukaryotic cells by the non-haemolytic rickettsia, *R. rickettsii* (Walker *et al.*, 1983; Silverman *et al.*, 1992). Proteins of 15, 16 and 60 kDa from *R. rickettsii* can be detected by Western blotting using anti-eukaryotic PLA$_2$ and anti-phospholipase C antisera, respectively (Manor *et al.*, 1994). Structures on the *R. rickettsii* cells that react with these antibodies have also been localised by electron microscopy with immuno-gold labelling (Manor *et al.*, 1994).

Attributes of the rickettsial PLA$_2$ and the haemolytic activity of viable *R. prowazekii in vitro* have been compared (Ojcius *et al.*, 1995). As expected for an extracellular phospholipase, the rickettsial PLA$_2$ activity was primarily active on negatively charged phospholipids. PLA$_2$ was most active at pH 6–7 and its activity declines at alkaline and acidic pH, whether or not calcium is present. *Rickettsia*-associated PLA$_2$ activity is sensitive to reduction with dithiothreitol and is magnesium-dependent. The haemolytic activity of viable *R. prowazekii* is affected similarly by pH in the absence of calcium. However, in contrast to PLA$_2$

activity, haemolytic activity is enhanced significantly upon calcium supplementation at either acidic or basic pH. Similarly, *R. typhi* expresses haemolytic activity at both neutral and acidic pH, but its activity is somewhat lower than that observed in *R. prowazekii* preparations. These findings indicate that rickettsiae may produce more than one membranolytic factor. Although a gene(s) for rickettsial PLA$_2$ has not yet been identified, recent genome sequencing of *R. prowazekii* has detected two genes, *tlyA* and *tlyC*, that may be haemolysins (Andersson *et al.*, 1998). The TlyC homologue from *R. typhi* is very similar to TlyC of *R. prowazekii*, since they have 97% homology (Radulovic *et al.*, 1999). Following transformation with *tlyC* cloned from *R. typhi*, a non-haemolytic strain of *Proteus mirabilis* exhibited a haemolytic phenotype (Radulovic *et al.*, 1999), but the precise function of TlyC in host cell invasion by rickettsiae remains unknown.

Intracellular Growth of *Rickettsia* and *Orientia*

Growth Patterns and Release from Cells

Endothelial cells of small and middle-sized vessels are the primary sites for replication of *Rickettsia* in vertebrate hosts. *R. rickettsii* can also invade underlying smooth muscle cells. In arthropods, rickettsiae colonise salivary gland cells, midgut cells and intestinal and gonadal cells. After penetration and lysis of the phagocytic vacuole membrane, rickettsiae start their intracellular growth cycle, which is unique for different species and sometimes even for the strain of rickettsia (Wisseman, 1986).

Differences in the growth patterns of some typhus group and SFG are well described (Silverman *et al.*, 1980; Wisseman, 1986). After infection, *R. prowazekii* starts to multiply in the cytoplasm of its host cells and undergoes clearly distinguished and measurable lag, exponential and stationary growth phases with a generation time of 9 hours at 34°C. The infected cells become overfilled with rickettsiae, sometimes reaching more than 800 micro-organisms per cell. These cells finally lyse 2–4 days after infection and release rickettsiae into the intracellular medium. From there they can initiate a new cycle of infection in nearby viable cells. Rickettsial phospholipase A may also be active during growth and release of the rickettsiae from cells as well as the initial infection (Winkler and Daugherty, 1989).

This type of infection was observed for both virulent and non-virulent strains of *R. prowazekii* in non-phagocytic cells (Wisseman, 1986). In contrast, macrophages and macrophage-like cells support the growth of virulent strains, but inhibit the intracellular growth of the attenuated Madrid E strain (Gambrill and Wisseman, 1973a; Gudima, 1979; Turco and Winkler, 1982; Winkler and Daugherty, 1983; Penkina *et al.*, 1995). The nature of this phenomenon is not completely understood. Both attenuated E and virulent Breinl strains exhibit serine and glycine auxotrophy in Vero cells. The growth of the E strain in Vero cells is more severely inhibited than that of Breinl strain unless these amino acids are added to the medium, even though their growth is similar in irradiated L929 cells (Austin *et al.*, 1987). The non-virulent Madrid E strain also has an alteration in its rOmpB lysine methylation patterns as compared with virulent strains (Rodionov *et al.*, 1991; Ching *et al.*, 1993), which may influence its sensitivity to cytokines (Turco and Winkler, 1994a). The underlying metabolic basis for these differences in growth properties may be due to a defective *metK* gene in the E strain (Andersson and Andersson, 1997).

Different isolates of *R. typhi* exhibit very similar growth characteristics, which differ from those of *R. prowazekii* Madrid E and Breinl strains (Wisseman, 1986). Growth of *R. typhi* is also confined to the cytoplasm, but shortly after invasion, the rickettsiae begin to escape from the host cell and to infect other cells in the culture, causing a rapid increase in the number of infected cells. *R. canada* grows in the cytoplasm, but it can also invade and replicate in the nucleus like SFG rickettsiae (Burgdorfer and Brinton, 1970).

The growth of *R. rickettsii* has been studied in detail as a paradigm for other SFG rickettsiae, although different isolates from this group display a variety of growth characteristics. *R. rickettsii* infection in chicken embryo and L929 cell cultures starts exponentially without a measurable lag phase (Wisseman *et al.*, 1976). However, very soon after infection, *R. rickettsii* begins to escape from the host cell into the medium in large numbers, without concomitant pathological changes in host cell structure (Wisseman *et al.*, 1976; Silverman, 1984; Silverman and Bond, 1984). In some cells, SFG rickettsiae invade the nucleus where they multiply to form compact masses (Wisseman *et al.*, 1976). Later stages of infection are characterised by typical changes in cellular structure, notably the dilatation of the rough endoplasmic reticulum and outer nuclear membrane and accumulation of electron-dense material within the cisternae of intracellular membranes (Walker and Cain, 1980; Silverman, 1984).

Ultrastructural changes similar to those observed in *R. rickettsii*-infected cells were also found in L929 cells infected with *R. japonica* Katayama strain (Iwamasa *et al.*, 1992) and *R. sibirica* 246 strain (Popov *et al.*,

1986), indicating that common cytopathic mechanisms are likely to occur during SFG rickettsial growth in cell cultures. The appearance and accumulation of pathological changes in the cultures infected with *R. japonica* or *R. sibirica* are merely delayed compared with those occurring in *R. rickettsii*-infected cultures, with the greatest effects observed at 5–6 days after infection. *R. conorii* infection also produces fewer necrotic effects in Vero cells when compared with *R. rickettsii*. These cytopathic effects are not due to a secreted rickettsial toxin, since it fails to pass a membrane barrier (Walker *et al.*, 1984). Less virulent isolates of *R. rickettsii* cause mild and delayed cytopathic effects resembling those caused by other SFG agents (M.E. Eremeeva, unpublished observations), but the molecular basis for the differences among isolates are not understood.

O. tsutsugamushi grows to high density in the cytoplasm of its infected cells and may also occasionally invade the nucleus (Urakami *et al.*, 1982). To effect its release from infected cells, *Orientia* evaginates from the cell plasma membrane in a process that resembles the budding process of enveloped viruses (Ewing *et al.*, 1978). The budding rickettsiae adhere tenaciously to host cell components and accumulate to high density at the surface of infected cells. Like the other rickettsiae, *O. tsutsugamushi* grows slowly in vertebrate cells at 37°C with a doubling time between 9 and 18 hours (Hanson, 1987b).

Interaction with the Actin Cytoskeleton

The cell cytoskeleton has an active role in the course of rickettsial infection, since uptake from and release of rickettsiae into the extracellular space is inhibited by cytochalasins B and D (Winkler and Miller, 1982; Walker, 1984; Heinzen *et al.*, 1993). Microfilament-associated intracellular movement is another aspect of the rickettsia–cell interaction (Todd *et al.*, 1983; Teysseire *et al.*, 1992; Heinzen *et al.*, 1993 (**Fig.5**). Intracellular movement was first reported for *R. rickettsii* in cultured rat fibroblasts (Schaechter *et al.*, 1957), and later it was documented for *R. conorii* (Kokorin, 1968) and *O. tsutsugamushi* (Ewing *et al.*, 1978).

Filamentous actin tails of up to 70 μm length were detected at one pole of intracellular *R. rickettsii* and *R. conorii* in Vero cells and HUVECs, respectively, following staining with fluorescent-labelled phallotoxins (Teysseire *et al.*, 1992; Heinzen *et al.*, 1993). Nucleation of actin monomers occurs very rapidly after rickettsiae escape from the phagocytic vacuole into the cytoplasm. Micro-organisms coated with F-actin are seen by 15 minutes after entry, and formation of fully formed actin tails can be detected 15–30 minutes after

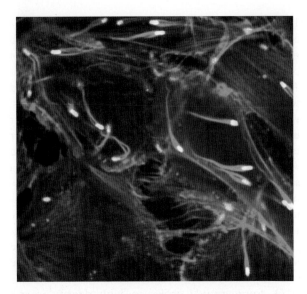

Fig. 5 Confocal fluorescent microscope image of *R. rickettsii* and associated actin tails in Vero cells. Rickettsiae (strain HIp) were stained at 60 hours post-infection of Vero cells with monoclonal antibody mAb13-2 against rOmpA followed by FITC-labelled anti-mouse immunoglobulin. The actin tails are stained with rhodamine phalloidin. The rickettsiae are approximately 1 μm in length. Courtesy of Robert A. Heinzen.

infection in both Vero cells and HUVECs (Heinzen *et al.*, 1993; Eremeeva *et al.*, 1999). Formation of actin tails is arrested following chloramphenicol treatment of infected cells, suggesting that de-novo rickettsial protein synthesis is required.

Rickettsia is one of three different genera of bacteria known to have actin-based motility, but the mechanisms of actin polymerisation it uses are poorly understood when compared with those employed by *Shigella flexneri* and *Listeria monocytogenes* (see Chapters 61 and 69). Recent studies demonstrated that actin-based movement of both *R. rickettsii* and *R. conorii* differs from that of *Shigella* and *Listeria* (Gouin *et al.*, 1999; Heinzen *et al.*, 1999). Intracellular *Rickettsia* move 2.5–3 times slower than *Listeria* and *Shigella*. The *Rickettsia*-induced actin comet tails are organised in a different structure, which is more stable since its half-life is three times longer. Rickettsiae display longer tails in which non-cross-linked actin filaments form on both side of the cell, while *Listeria* and *Shigella* actin tails form branching networks. *Rickettsia* also interacts with a different set of mammalian cytoskeleton proteins (Gouin *et al.*, 1999). VASP (vasodilator-stimulated phosphoprotein) and α-actinin, particularly, are co-localised with actin in the actin tails of *R. conorii*, but neither the Arp2/3 nor the neural Wiskott–Aldrich syndrome protein (N-WASP) complex, which is detected in *Shigella* actin tails, are detected in

rickettsial actin tails. Both *R. rickettsii* and *Listeria monocytogenes* actin tails have been shown to contain the cytoskeletal proteins VASP, profilin, vinculin and filamin, but rickettsial tails lack ezrin, paxillin and tropomyosin, proteins associated with actin tails of *Listeria* (Van Kirk *et al.*, 2000). These findings suggest that *Rickettsia* may employ a unique mechanism of actin polymerisation.

The importance of actin polymerisation in rickettsial virulence is uncertain, since formation of actin tails occurs in both virulent and avirulent strains of *R. rickettsii* and several other isolates of SFG rickettsiae of varying pathogenicity, including *R. montana*, *R. australis* and *R. parkeri* (Heinzen *et al.*, 1993). Moreover, F-actin tails have not been detected in association with pathogenic *R. prowazekii*, or with *R. canada*, whose pathogenicity is uncertain (Heinzen *et al.*, 1993). Only a few *R. typhi* cells display actin tails which are generally much shorter than the long tails produced by SFG rickettsiae (Teysseire *et al.*, 1992; Heinzen *et al.*, 1993).

R. rickettsii infection of EA.hy 926 cells results in dose-dependent changes in the intracellular actin pools of infected cells by 48 hours, including accumulation of intracellular polymeric F-actin and progressive decreases in the monomeric G-actin fractions. This process correlates with a very fast decline in the number of viable cells in infected monolayers. These changes can be prevented by treatment of infected cells with 200 μM of α-lipoic acid (Eremeeva *et al.*, 1999). Intracellular polymeric F-actin pools increase between 24 and 48 hours after inoculation with *R. rickettsii*, and then levels decline dramatically almost to zero by 96 hours after infection. *R. sibirica*, which is less cytotoxic than *R. rickettsii*, does not affect intracellular actin pools for up to 72 hours after infection. At 96 hours and 144 hours after infection, F-actin levels in *R. sibirica*-infected cells were elevated 2-fold and 3-fold, respectively, as compared with uninfected cells. Indeed, 54% of cells infected with *R. sibirica* are still viable at day 7 of experiment. The mRNA pool for actin decreases in human endothelial cells after *ex vivo* infection of umbilical cords with *R. rickettsii* for 4 hours (Courtney *et al.*, 1996), while *in vitro* experiments indicate accumulation of actin mRNA by 24 hours after inoculation.

Rickettsia-induced Changes in Infected Cells

Vascular injury is the major initial lesion observed in rickettsial diseases. The injury results from penetration and replication of the rickettsiae in endothelial cells. The damage inflicted on infected endothelial cells and resulting dissemination of rickettsiae into the surrounding and underlying tissues can be detected in histological studies of severe cases. This damage results in enhanced permeability of tissue fluids into the infection foci, infiltration by monocytes and polymorphonuclear neutrophils and inflammatory reactions, and the formation of micro-thrombi. Since multiple events are involved in this damage *in vivo*, *in vitro* models have been used to understand individual aspects of this complex process. The changes observed in rickettsia-infected endothelium include up-regulation of cellular adhesion receptor expression (E- and P-selectin, ICAM, VCAM), increased production of pro-inflammatory cytokines (IL-1α, IL-6, IL-8 and TNFα) and prostaglandins, and activation of pro-coagulant (tissue factor, thrombomodulin and von Willebrand factor) and fibrinolytic systems (**Table 2**). Some of these perturbances are transient in character, while other changes accrue during the course of rickettsial infection. These changes may be due to increased transcriptional activation of the corresponding operons, and increased expression and/or secretion of protein factor(s). The relevance of the responses to rickettsial infection *in vitro* is supported by detection of similar pathological changes in patients suffering from RMSF, Mediterranean spotted fever and epidemic typhus (Walker *et al.*, 1990; Elghetany and Walker, 1999; Vitale *et al.*, 1999).

Transcriptional Activation and Apoptosis in Rickettsial Infection

R. rickettsii infection of endothelial cells activates the transcriptional factor NF-κb (Sporn *et al.*, 1997), which regulates tissue factor expression. The kinetics of activation of NF-κB is biphasic with the highest levels of activation at 3–7 hours and again later at 24 hours after infection. Two activated NF-κB species, the p50 homodimer and the p50–p65 heterodimer, are present after *R. rickettsii* infection. *R. rickettsii* infection also results in an increase in the steady level of IκBα mRNA, which encodes the inhibitory subunit of NF-κB. The mechanism of *R. rickettsii*-induced NF-κB activation is clearly independent of protein kinase C (Sahni *et al.*, 1999), and probably involves a direct interaction between intra-cytoplasmic rickettsiae and NF-κB. Induction of a subset of chemokine genes by still unidentified heat-stable molecules of *O. tsutsugamushi* results in the nuclear translocation and transcriptional activation of NF-κB in murine macrophages (Cho *et al.*, 2000).

The early steps of *R. rickettsii*-induced NF-κB activation, particularly in the intact cellular system, are likely to involve a proteasome-mediated mechanism, since it can be inhibited by treatment with the specific proteasome inhibitors TPCK and MG 132 (Clifton

et al., 1998; Shi *et al.*, 1998). Surprisingly, such treatment results in the death of infected cells, which undergo apoptotic changes accompanied by evident nuclear DNA fragmentation (Clifton *et al.*, 1998. Apoptotic changes after *R. rickettsii* infection also occur in human embryonic fibroblasts transfected with a suppressor mutant inhibitory subunit IκBα, but non-transfected cells are not killed. Accordingly, *R. rickettsii* appears to inhibit host cell apoptosis by a mechanism which depends on NF-κB activation and thus modulates the host cell response to its own advantage, allowing the targeted cell to continue as the site of infection (Clifton *et al.*, 1998). It is unknown whether inhibition of apoptosis is utilised by all species of the genus *Rickettsia*, or how inhibition of apoptosis is effected and the infected cell begins the sequence of events which lead eventually to its necrosis. Interaction of *O. tsutsugamushi* with their host cells contrasts significantly with the behaviour of *Rickettsia* species. *O. tsutsugamushi* induces apoptosis in lymphocytes in spleen and regional lymph nodes of infected mice (Kasuya *et al.*, 1996), and cultured J774 macrophage-like cells and the endothelial cell line ECV340 (Kee *et al.*, 1999). Heavily infected ECV340 cells showed early disruption of focal adhesions and decreased polymerisation of actin stress fibres. These events may contribute to *Orientia*-induced apoptosis.

Oxidant-associated Injury of *Rickettsia*-infected Cells

Toxic reactive oxygen species (ROS) are typically associated with killing of bacteria by phagocytic cells. However, accumulation of ROS also appears to be induced during rickettsial infection in non-phagocytic cells, particularly vascular endothelial cells. Hydrogen peroxide may be responsible for killing *R. conorii* in cytokine-treated human microvascular endothelial cells (HMEC-1) since rickettsial growth increases significantly in response to catalase treatment (Walker *et al.*, 1996). On the other hand, ROS, particularly superoxide radicals and peroxides, appear to cause the cytopathic effect elicited by *R. rickettsii* in cultured HUVECs (Hong *et al.*, 1998; Silverman, 1997). Four to five hours after exposure of HUVECs to *R. rickettsii*, elevated levels of peroxide were detected in the cell culture medium and infected monolayer (Hong *et al.*, 1998). Accumulation of superoxide is first detected within 1 hour in endothelial cells, when attachment and internalisation of rickettsiae are taking place. At later stages of infection, intensive intracellular multiplication of rickettsiae correlates with a dramatic accumulation of intracellular peroxides in the monolayer, and to irreversible necrotic changes in the

Table 2 Non-immune eukaryotic cell responses induced *in vitro* by *Rickettsia* and *Orientia*

Character of changes	Micro-organism	Cell type tested	Changes in mRNA level	Changes in protein expression		Clinical observation	References
				Technique	Maximum expression		
Expression of cellular receptors							
1. E-selectin	*R. rickettsii*	HUVEC	ND	Flow cytometry Neutrophil adherence inhibition assay	6–8 h		Sporn *et al.*, 1993
	R. conorii	HUVEC	ND	Flow cytometry Monocyte adherence assay	4–8 h		Dignat-George *et al.*, 1997
2. Intercellular adhesion molecule (ICAM)-1[a]	*R. conorii*	HUVEC	ND	Flow cytometry Monocyte adherence assay	4–24 h	RMSF[a]	Dignat-George *et al.*, 1997
3. Vascular adhesion molecule [VCAM]-1	*R. conorii*	HUVEC	ND	Flow cytometry Monocyte adherence assay	0–8 h		Dignat-George *et al.*, 1997
Synthesis of pro-inflammatory cytokines							
1. IL-1α	*R. rickettsii*	HUVEC	4 h	ELISA	18 h		Sporn and Marder, 1996
	R. conorii	HUVEC	ND	ELISA	24 h	MSF	Kaplanski *et al.*, 1995
		HEp-2	ND	ELISA	48–96 h		Manor, 1992
		BGM	ND	ELISA	48–96 h		Manor, 1992
2. IL-6	*R. conorii*	HUVEC	ND	ELISA	24 h		Kaplanski *et al.*, 1995
3. IL-8	*R. conorii*	HUVEC	ND	ELISA	6–48 h		Kaplanski *et al.*, 1995
4. TNFα	*R. conorii*	Human monocyte-derived macrophages	ND	ELISA	24 h		Manor and Sarov, 1990
		Murine peritoneal macrophages, P388D1 cell line, HEp-2	ND	MTT bioassay	24–72 h (linear)		Jerrells and Geng, 1994
		BGM	ND	ELISA	48 h		Manor, 1992
			ND	ELISA	96 h		Manor, 1992
	R. sharonii	Human monocyte-derived macrophages	ND	ELISA	24 h		Manor and Sarov, 1990

Changes in protocoagulant system							
1. Tissue factor	R. conorii	HUVEC	Clotting assay Flow cytometry	ND	4–24 h	MSF[b]	Teysseire et al., 1992a
	R. rickettsii	HUVEC	Clotting assay ELISA	3–4 h	4–24 h (transient)		Sporn et al., 1994; Shi et al., 1998
2. Thrombomodulin[b]	R. conorii	HUVEC	Flow cytometry, Protein C activation	ND	4–24 h (steady declining level)		Teysseire et al., 1992
3. von Willebrand factor[b]	R. conorii	HUVEC	ELISA, MIF	ND	8–12 h	MSF[b]	Teysseire et al., 1992
	R. rickettsii	HUVEC	MIF, protein electrophoresis and immunoprecipitation	ND	24 h		Sporn et al., 1991
4. Platelet-activating factor	R. prowazekii, E	HUVEC		ND	24–72 h		Walker and Mellot, 1993
Changes in fibrinolytic system							
1. Prostaglandin I$_2$ (PGI$_2$)[c]	R. prowazekii, E	HUVEC	ND	ND		RMSF[c]	Walker et al., 1990
2. Prostaglandin E$_2$ (PGE$_2$)	R. prowazekii, E	HUVEC	ND	ND			Walker et al., 1990
	R. conorii and R. sharonii	Monocyte-derived macrophages	ND	ND			Manor and Sarov, 1990
3. Leukotriene B$_4$	R. prowazekii	Mouse polymorpho-nuclear leucocytes Peritoneal and alveolar macrophages	ND	ND			Walker et al., 1991 Walker and Hoover, 1991

[a] Changes in the circulatory concentration of ICAM-1 in the blood of a patient with a lethal case of RMSF has been reported (Sessler et al., 1995).

[b] Patients with MSF exhibited elevated levels of plasma thrombomodulin and vWF, which both progressively decreased during the treatment of the disease (George et al., 1993).

[c] Changes in PGI$_2$ plasma levels could be detected during acute RMSF and convalescent period (Walker and Triplett, 1991).

structure of infected cells. The origin and timing of the appearance and accumulation of necrotic changes, and the basis of the failure of the cellular antioxidant system to neutralise ROS are not fully understood. Whether this mechanism of cytotoxicity is typical for other species of *Rickettsia*, and whether different species have similar abilities to induce and resist the conditions associated with oxidative injury are unknown.

After treatment of *R. rickettsii*-infected human endothelial cells with 100–500 µM of the antioxidant α-lipoic acid, a dose-dependent reduction of intracellular peroxide levels and increase of cellular antioxidant pools (reduced glutathione and glutathione peroxidase) occurs (Eremeeva and Silverman, 1998a, b). α-Lipoic acid treatment of the endothelial cells results in an increase in their viability at 72 and 96 hours after infection with *R. rickettsii* as compared with untreated infected cells, and prevents the occurrence of morphological changes in cells infected for 72 hours. Since α-lipoic acid does not directly alter the functional activity of rickettsiae, its protective effect probably involves complex changes in the metabolism of the infected host cell. Further studies are required to identify these mechanisms and to evaluate the potential of α-lipoic acid as an adjunct treatment for RMSF in humans.

Immunity

The intracellular lifestyle of *Rickettsia* and *Orientia* presents some unusual challenges to the immune system. These organisms inhabit an environment sequestered from some compartments of the immune system. The majority of natural and experimental rickettsial infections are, however, resolved by specific anti-rickettsial immune responses, including specific T lymphocyte-mediated responses, effected by T-helper and T-cytotoxic lymphocytes, cytokine-mediated activation of macrophages and lymphokine-activated killer cells (Carl *et al.*, 1988), and synthesis of specific anti-rickettsial antibodies with a variety of functions (reviewed in Jerrells, 1988, 1997; Turco and Winkler, 1997).

T-cell-mediated Immune Responses to *Rickettsia* and *Orientia*

Recovery from infection of humans and experimentally infected animals with rickettsiae in the genera *Rickettsia* and *Orientia* is associated with the development of a delayed-type hypersensitivity (Wisseman, 1967; Murphy *et al.*, 1980; Jerrells and Osterman, 1982; Crist *et al.*, 1984) mediated by T lymphocytes. These

observations were obtained from the results of two major types of studies: passive transfer of immune lymphocytes or cell populations enriched for T cells and lymphocyte proliferation assays *in vitro*.

The importance of antigen-responsive T cells in rickettsial immunity has also been established *in vitro* and *in vivo*. In particular, CD4 and CD8 subpopulations of T cells develop in mice in response to infection with *O. tsutsugamushi* Karp strain, and their presence is required for resistance of immune mice to a secondary challenge with this rickettsia (Feng *et al.*, 1997; Jerrells, 1997). Furthermore, class I MHC-restricted CD8 lymphocytes can also be detected and their specific responsiveness to rickettsial antigens determined.

Although understanding of this process is still very limited, it is probably not species-specific but rather represents a conserved eukaryotic response to *Rickettsia* and especially to SFG isolates. This conclusion is based on the observation that serologically distant *R. rhipicephali* has the full ability to cross-stimulate immune T cells, resulting in establishment of protective immunity in guinea-pigs to challenge with the virulent *R. rickettsii* strain (Gage and Jerrells, 1992).

Whether there is any group specificity in the way cytotoxic lymphocytes recognise SFG or TG rickettsiae is not known. As shown for *R. typhi*, however, they possess a unique mechanism of presenting their antigens to the host immunocompetent cells (Rollwagen *et al.*, 1985). *R. typhi* antigens increasingly appear on the surface of infected L929 fibroblasts 1–3 days after infection without the concomitant presence of adherent micro-organisms on the surface of the infected cells. B cells transfected with Epstein–Barr virus are also able to present TG rickettsial antigens to cytotoxic T cells (Carl *et al.*, 1987). The mouse endothelial cell line SVEC-10 infected with *R. conorii* strain Malish for 48 hours can also activate CD8 T lymphocytes, but the expression of rickettsial antigen on the infected cell surface was not measured (Diaz *et al.*, 1999). The mechanism of such antigen presentation and its contribution to host survival is not understood.

Antibody Responses to *Rickettsia* and *Orientia*

In infected humans and animals rickettsiae elicit vigorous antibody responses, which result in production of IgM and IgG, but not IgA, which is directed mostly against the large crystalline surface protein antigens and LPS (Anacker *et al.*, 1985, 1986, 1987a, b; Ching *et al.*, 1990; Eremeeva *et al.*, 1994a, b). Antibodies against *Orientia* are largely directed to the 56, GroEL (60 kDa), 47, 70 and 110-kDa antigens, but antibodies to other surface antigens present in lower quantities can also be detected (Ohashi *et al.*, 1988; Oaks *et al.*, 1989). A direct role for antibody in the clearance of rickettsiae from infected cells is dubious since immunoglobulins cannot enter infected cells. This conclusion is supported by data showing that the development of *O. tsutsugamushi*, *R. akari* and *R. conorii* infection in athymic mice occurs in spite of production of significant levels of antibodies (Murata and Kawamura, 1977; Kenyon and Pedersen, 1980; Kokorin *et al.*, 1982). Moreover, passive transfer of immune serum does not protect against infection whether it is given prior to or after the inoculation with viable rickettsiae (Murphy *et al.*, 1979, 1980; Crist *et al.*, 1984). On the other hand, pre-treatment of rickettsiae with immune sera or monoclonal antibodies prevents attachment and entry of rickettsiae into the host cell and therefore reduces the chance for the development of infection (Turco and Winkler, 1982; Hanson, 1983; Lange and Walker, 1984; Anacker *et al.*, 1985, 1987b). Similarly, rickettsiae pre-treated with convalescent human serum are efficiently phagocytosed and killed by professional macrophages and macrophage-like cells (Gambrill and Wisseman, 1973b; Beaman and Wisseman, 1976).

Role of Activated Macrophages and NO

Resting macrophages and macrophage-like cells are destroyed by rickettsial infection. In contrast, lymphokine-activated macrophages are important effector cells in anti-rickettsial immunity (Winkler and Turco, 1994). Their activation by IFNγ results in the inhibition of the growth and/or killing of *Rickettsia* and *Orientia*. Infection with SFG rickettsiae and *Orientia* also up-regulates the production of tumour necrosis factor (TNFα) by human and mouse macrophages, which in its turn inhibits rickettsial replication acting alone or in concert with IFNγ (Manor and Sarov, 1990; Geng and Jerrells, 1994; Jerrells and Geng, 1994).

Human THP-1 monocytic macrophages, activated with the chemokine RANTES alone or with a variety of cytokines, including TNFα, IFNγ, or IL-1β, significantly inhibit growth of *R. akari* (Walker *et al.*, 1999a). The effector mechanism is probably mediated by a peroxide-dependent system, since treatment with catalase increases the rickettsial growth over that found in infected untreated cells. The depletion of the intracellular tryptophan pools by the cytokine-induced expression of indoleamine 2,3-dioxygenase (IDO), the tryptophan-degradation enzyme, may also be involved in systemic anti-rickettsial responses of activated macrophages. Nitric oxide-dependent killing is another potential anti-rickettsial mechanism found in human macrophages, since the nitric oxide synthase

inhibitor, N^G-monomethyl-L-arginine (N^GMMLA) can reverse the RANTES-induced anti-rickettsial activity of THP-1 cells.

Nitric oxide production is also an effector mechanism(s) of anti-rickettsial activity that occurs in mouse L929 fibroblasts (Feng and Walker, 1993; Turco and Winkler, 1993), mouse macrophage-like RAW264.7 cells (Turco and Winkler, 1994b) and mouse endothelial cells (Walker *et al.*, 1997b). In these cellular systems NO synthesis occurs as a result of treatment of cultured cells with IFNγ alone, or in combination with TNFα, prior to infection with *R. prowazekii* or *R. conorii*. Accumulation of NO in these cells is probably a result of the activation of inducible nitric oxide synthase (*i*-NOS), since *i*-NOS mRNA was detected as early as 4 hours after cytokine stimulation, it increased at 8 hours and it slowly decreased by 72 hours. The stimulatory effect of cytokines can be mimicked by addition of sodium nitroprusside, a source of NO. In contrast, treatment with aminoguanidine and N^GMMLA inhibits NO production and relieves its suppression of the rickettsial infection. Extracellularly released NO inhibits the ability of *R. prowazekii* to invade mouse cytokine-activated L929 fibroblasts and RAW264.7 cells by killing them directly (Turco *et al.*, 1998).

NO-mediated killing has been implicated in clearance of *R. conorii* infection in immunocompetent C3H/HeN mice, since depletion of IFNγ and TNFα results in animal mortality due to an overwhelming rickettsial infection (Feng *et al.*, 1994). Consequently, one may hypothesise that IFNγ secreted by T lymphocytes and natural killer cells and TNFα secreted by macrophages act in a synergistic fashion to stimulate synthesis of NO in rickettsia-infected cells, and this in turn kills the intracellular rickettsiae.

Effect of Lymphokines on Rickettsial Infection in Non-immune Cells

The details of lymphokine effects on rickettsial infection are known most completely for the interferons and tumour necrosis factor. These observations have been summarised in several reviews (Winkler and Turco, 1988; Turco and Winkler, 1993, 1997). Cytokines can modify the interactions between rickettsiae and their host cells in several ways: inhibition of rickettsial growth; direct cell killing of rickettsiae; cytotoxic effects on infected host cells; and the inhibition of initial rickettsial infection (Turco and Winkler, 1997). All these effects are found in model systems involving TG, SFG rickettsiae or *Orientia* (**Table 3**). As a rule, however, infection with each examined species and even strains of *Rickettsia*

Table 3 Role of cytokines in infection with *Rickettsia* and *Orientia*

Response and inducing organism	Cytokine					
	IFN α/β		IFN γ		TNFα	
	Cell type in vitro	Organism in vivo	Cell type in vitro	Organism in vivo	Cell type in vitro	Organism in vivo
1. Induction of cytokine secretion						
R. prowazekii	PBL[a]	Mice			RAW264.7 MDM	
R. typhi	Primary chick embryonic cells					
R. conorii		Mice		Mice	MDM MPM, immune mouse splenocytes HEp-2 P388D1	Immune mouse, human
Israeli tick typhus rickettsia					MDM HEp-2	
R. australis						Mouse
O. tsutsugamushi			Monkey PBM	Monkey	Immune mouse splenocytes	Immune mouse
2. Inhibition of growth of rickettsiae						
R. prowazekii	L929[b], MRC-5 human fibroblasts		All tested human and mouse cell lines		L929	
R. akari	L292		?			
R. rickettsii	Mouse fibroblasts from BALB/3T3 clone A31 and C3H/10T1/2 clone 8		Mouse fibroblasts from BALB/3T3 clone A31 and C3H/10T1/2 clone 8			

				Mouse
R. conorii	?	L929, J774A.1, mouse macrophages	L929, HEp-2	
O. tsutsugamushi	Mouse fibroblasts from BALB/3T3 clone A31 and C3H/10T1/2 clone 8	Mouse macrophages from C57BL/6 and BALB/3T3 clone A31	Mouse macrophages from C3H/10T1/2 clone 8	

3. Cytotoxicity for the host cell

R. prowazekii	L929	L929, RAW264.7, human fibroblasts	L929
R. conorii		HEp-2	
O. tsutsugamushi	?	BALB/3T3 clone A31	

---

[a] Referred to a secondary response of monocytes obtained from animals and humans recovered from rickettsial infection.
[b] Only avirulent strain Madrid E.
PBL, peripheral blood leucocytes; MDM, monocyte-derived macrophages; PBM, peripheral blood monocytes.

and *Orientia* causes unique patterns of responses to lymphokines, therefore the observations obtained with a particular rickettsia should not be regarded as a general rule.

Rickettsia and *Orientia* infections induce production of both IFNα/β and IFNγ in many cell lines, and induced interferon activity correlates with an anti-rickettsial effect. Several inhibitory mechanisms are involved to produce IFN-induced anti-rickettsial effects, including those also identified in macrophages: up-regulation of indoleamine 2,3-dioxygenase and depletion of tryptophan pools in infected cells, and activation of NO-synthase-dependent and -independent mechanisms. Virulent strains of *R. prowazekii*, including strain Breinl, differ from a virulent Madrid E in their susceptibility to IFN-mediated inhibitory effects (Turco and Winkler, 1991, 1994a). In particular, virulent strains are resistant to IFNγ and IFNα/β, and to TNFα, and this characteristic correlates well with their ability to grow in macrophage-like cells. Among SFG rickettsiae, effects of lymphokines have been examined for infection with *R. conorii* and *R. akari*, which apparently do not respond to the same anti-rickettsial mechanism effective in killing *R. prowazekii* (Feng and Walker, 1993; Walker *et al.*, 1997b).

Combined cytotoxic effects of IFNγ treatment and infection with *Rickettsia* or *Orientia* vary significantly according to the type of host cell line. The exact mechanism of cytotoxic effect has not been determined, but the proper entry and the initial metabolism of rickettsiae are mostly required for its induction.

The anti-rickettsial effects of TNFα may be mediated by nitric oxide synthase, since its inhibition restores the growth of *R. conorii* Malish strain (Feng and Walker, 1993) and *R. prowazekii* strain E in L929 cells (Turco and Winkler, 1993). The inhibitory effect of TNFα on rickettsial growth is enhanced by pre-treatment with IFNγ. A cytotoxic, NO-synthase-independent effect of TNFα on *R. prowazekii* infection has been described in L929 cells only, while a similar effect could not be detected in RAW264.7 cells (Turco and Winkler, 1993, 1994b). Host cell-directed cytotoxicity mediated by TNFα treatment was not observed with different types of cells infected with *R. conorii* (Manor and Sarov, 1990; Feng and Walker, 1993).

Diagnosis

During the first few days of illness an accurate clinical diagnosis of rickettsial disease is very difficult and is strongly dependent on epidemiological clues (Dumler and Walker, 1994). A rash, the hallmark of rickettsial infection, typically appears only after several days of illness, a history of arthropod exposure is not always elicited, and a primary eschar or 'tache noire' is not always noticed or present. Physicians must, therefore, often evaluate rather non-specific clinical symptoms, including onset of malaise, chills, myalgia, fever and severe headache. Definitive diagnosis requires differentiation of the rickettsioses from various infectious diseases, including typhoid fever, dengue fever, malaria and meningitis. More than one rickettsial infection may also occur in the same geographic areas. The patient history, such as louse infestation, recreational activity, contacts with animals and their ectoparasites and exposure to natural or disturbed habitats, is important for diagnosis. Three major approaches are currently used for rapid laboratory diagnosis of rickettsioses: detection of micro-organisms in blood or biopsy samples by PCR-based techniques, detection of specific antibodies with a wide variety and complexity of serological tests, and direct immunohistochemical detection of organisms in biopsies. These approaches provide reliable alternatives to isolation of micro-organisms from patients, which requires specialised handling, is laborious, and is generally too slow to be clinically useful. PCR detection of rickettsial DNA in arthropods, including lice, collected from patients or in suspected foci, has been suggested as an additional alternative tool to help in the diagnosis of rickettsioses (Roux and Raoult, 1999a).

PCR

PCR is a rapid, sensitive and specific tool for detecting bacterial DNAs in clinical samples. Several authors have reported that PCR can detect rickettsial infection during the phase of acute rickettsiaemia, which occurs before specific antibody can be detected (Furuya *et al.*, 1991, 1993; Tay *et al.*, 1996). Nested, hemi-nested or re-amplified PCR tests make it possible to detect *Orientia* (Tay *et al.*, 1996; Horinouchi *et al.*, 1997; Pai *et al.*, 1997) and *Rickettsia* (Sexton *et al.*, 1994; Breitschwerdt *et al.*, 1999; Nilsson *et al.*, 1999) in clinical specimens more reliably than primary PCR but at the expense of additional time and chance of false-positives. PCR has been used to follow the efficacy of antibiotic treatment in eradicating *Orientia* and *Rickettsia* from the blood of patients (Murai *et al.*, 1995; Breitschwerdt *et al.*, 1999).

Serology

Many techniques are available for the serological diagnosis of rickettsial diseases, including the non-specific Weil–Felix test (WF). Specific tests with rickettsial antigens include micro-agglutination (MA), complement fixation (CF), indirect haemagglutination (HA), latex agglutination (LA), enzyme-linked immunosorbent assay (ELISA), micro-immunofluorescence assay (MIF), immunoperoxidase assay (IPA) and immunoblot assays (**Table 4**). The variable advantages, drawbacks and sensitivities of these assays have been recently reviewed (La Scola and Raoult, 1997). The difficulty of serological diagnosis in rickettsial diseases is the high cross-reactivity of human sera between different species of rickettsiae. Differentiation between TG, SFG and *Orientia* is usually easy, since differences in titre are typically large enough to distinguish between these types of infection. Serological differentiation between infections caused by the members of the same group, either TG or SFG, is more difficult, since the differences may often be within only one serum dilution. This analysis is particularly difficult in areas where more than one pathogenic rickettsia is known. In such situations, definitive serological diagnosis can be effected by cross-adsorbing the sera with different antigens and then re-assaying these sera against antigens from potential aetiologic agents (La Scola *et al.*, 2000). The recent development of rapid dot immunoassays as well as capillary flow chromatographic assays holds great promise for simplifying the rapid detection of anti-rickettsial antibodies in laboratories with limited apparatus (Weddle *et al.*, 1995; Ching *et al.*, 1999b). The availability of purified recombinant rickettsial antigens has also permitted the development of more standardised commercial tests (Land *et al.*, 2000).

Immunohistochemistry

Immunohistochemical (IHC) detection of rickettsiae in patient tissue provides a powerful alternative tool for diagnosis of rickettsioses (Walker and Cain, 1978; Dumler *et al.*, 1990; Walker *et al.*, 1997a). It is usually performed on a skin biopsy, and may combine routine haematoxylin and eosin staining to detect pathological changes and to visualise rickettsiae with specific antibodies and alkaline phosphatase-labelled conjugates. The immunohistochemical technique is particularly valuable in the early stages of disease when serological tests are not very reliable, and samples stained using conventional histochemical methods are often difficult to interpret. The IHC staining of skin biopsy specimens is the method of choice to investigate the pathophysiology of rickettsioses, to evaluate the anatomic distribution of rickettsiae in specific organs, and for retrospective diagnosis of fatal cases from autopsy specimens (Paddock *et al.*, 1999).

Susceptibility to Antimicrobial Agents

Physical Stability and Chemical Inactivation

Rickettsiae have stringent requirements for maintaining their physiology outside their host cells (Winkler, 1990). High ionic strength buffers or buffers supplemented with sucrose to provide osmolarity are essential for long-term maintenance of isolated microorganisms in a cell-free medium. Even in media designed to mimic the internal host cell environment, metabolic activity or infectivity cannot be preserved longer than a few days. However, the stability of *Rickettsia* in dried arthropod faeces or in properly lyophilised cultures is well known (Dasch and Weiss, 1991). Chemical disinfectants and heating above 60°C are highly effective for inactivation of rickettsiae.

O. tsutsugamushi is more sensitive to chemical disinfectants and temperature changes and more labile than typhus group rickettsiae under physiological conditions, while *R. rickettsii* and *R. conorii* can continue to divide after the death of the host cell.

Susceptibility to Antibiotics

The susceptibility of rickettsiae to antibiotics and the suitability of these for treatment have been evaluated in experimental models of infection, which include cell culture, embryonated chicken eggs, and animals (reviewed by Maurin and Raoult, 1999). When efficacy has been demonstrated in these models with a new drug, clinical trials are necessary further to evaluate its value for prophylaxis and treatment.

The susceptibility of rickettsiae to antibiotics *in vitro* is commonly evaluated with various modifications of the plaque assay (Wisseman *et al.*, 1974; Maurin and Raoult, 1999). The plaque assay is suitable for determination of the minimum inhibitory concentration (MIC), and the rate of killing of rickettsiae by a single antibiotic concentration and the reduction in number of plaques is measured with respect to the time of exposure of infected cells to the antibiotic. Alternatively, a colorimetric assay in 96-well plates is used to test the antibiotic susceptibility of rickettsiae. In this assay, cells infected with rickettsiae for 4 days are stained with neutral red or methylthiazol tetrazolium dye, and reductions in host cell cytopathology due to rickettsiae are evaluated spectrophotometrically by measuring the optical density of extracts from treated and untreated infected cells. Finally, a variation of the shell vial procedure has been used for antibiotic susceptibility tests. Treated and untreated infected cultures are examined by direct

Table 4 Serologic diagnosis of rickettsial infections

Technique	Diagnosed disease	Type of detected antibodies	Minimum positive titre	Time after onset of disease to antibody detection	Sensitivity	Specificity	Correlation with other assays	Recommended applications
Weil-Felix test (WF)	TG and SFG rickettsioses,[a] scrub typhus	Agglutinins (IgM)	40–320 depending on investigator	1–2 weeks	Poor	Non-specific	MIF	Not recommended
Microagglutination test (MA)	All rickettsioses	Agglutinins	≥8	1–2 weeks	Low	High	CF, HA, MIF	Not recommended
Complement fixation test (CF)	All rickettsioses	IgM, IgG	8–16	3–4 weeks	High with convalescent sera	Species-specific for Rickettsia, strain-specific for Orientia	MIF (for convalescent sera)	Detection of antibodies in late convalescent sera
Indirect haemagglutination (HA)	All rickettsioses	IgM > IgG	40	1–2 weeks	High with acute sera	High group specificity	MIF (for acute sera)	Diagnosis of acute infection
Latex agglutination test (LA)	RMSF, MSF, epidemic typhus, endemic typhus	IgM > IgG	64	1–2 weeks	High	High group specificity	MIF	Diagnosis of acute infection
Enzyme-linked immunosorbent assay (ELISA)	RMSF, epidemic typhus, endemic typhus, scrub typhus	IgM, IgG	OD 0.25 > control	1 week	Very high	Species-specific	MIF, IPA	Diagnosis, sero-epidemiological study
Micro-immuno-fluorescence assay (MIF)	All rickettsioses	IgM, IgG	16–64 depending on investigator	1–2 weeks	High	Species-specific[b]	WF, MA, CF, LA, ELISA, IPA, WB	Diagnosis, sero-epidemiological study
Immunoperoxidase assay (IPA)	Scrub typhus MSF epidemic typhus	IgM, IgG	20	1 week	High	Species-specific	MIF	Diagnostic field study

Immunoblotting based tests

Western blotting (WB)	All rickettsioses	IgM, IgG, IgA	1–2 weeks	High	Species-specific	MIF	Diagnosis of acute and convalescent diseases, differential diagnosis, seroepidemiological study
Line blot assay	All rickettsioses	IgM, IgG		High	Species-specific	WB, MIF	Diagnosis of acute and convalescent diseases, differential diagnosis, seroepidemiological study
Dot-blot assay	Scrub typhus, endemic typhus, MSF, RMSF	IgM, IgG	2 weeks	Moderate	Group-specific	MIF, ELISA	Diagnostic in field conditions, serosurvey

[a] Except rickettsialpox.
[b] Detection of IgM is complicated by cross-reactivity of rheumatoid factor, serum absorption is recommended before IgM detection.

staining of rickettsiae or indirect detection of rickettsiae with fluorescent antibodies (Ives *et al.*, 1997). To measure the effect of antibiotics in embryonated eggs, rickettsiae are inoculated into the yolk sac with and without antibiotics and the difference in embryo mean survival time is determined. Subculture of rickettsiae from yolk sacs of surviving embryos permits evaluation of the rickettsiacidal or rickettsiastatic activity of antibiotics. Reduction of mortality and the duration of fever in susceptible animals after intraperitoneal inoculation with viable rickettsiae have been used as measures of antibiotic effectiveness. Although the animal model may be considered a more physiological test of the anti-rickettsial activity of the drug, since it acts in concert with specific and non-specific defence mechanisms, the reliability of animal models for predicting human responses is uncertain for many drugs because their pharmacokinetics and the physiology of animals and humans may differ.

Tetracyclines, chloramphenicol, rifampicin and fluoroquinolones can all be used to treat infections with *Rickettsia*, but adverse effects may limit their administration. Treatment of pregnant women and young children is the most problematic, since the tetracyclines are toxic, and the use of chloramphenicol exposes these patients to the risk of aplasia.

The susceptibility of the rickettsiae to some antibiotics is summarised in **Table 5** (Maurin and Raoult, 1999; Rolain *et al.*, 1999, 2000). *R. prowazekii* and *R. typhi* are susceptible to all macrolides tested to date. *R. canada* and SFG rickettsiae are susceptible to josamycin, but exhibit more resistance to other macrolides. Several tick-restricted SFG rickettsiae are more resistant to rifampin than are typhus and other SFG rickettsiae. Although the nature of this resistance is not completely understood, a single point mutation resulting in a phenylalanine-to-leucine change was found at position 973 of the *R. conorii rpoB* sequence

Table 5 Susceptibility of *Rickettsia* and *Orientia* to antibiotics *in vitro*

Antibiotic	Minimal inhibitory concentration (μg /mL)[a]		
	R. prowazekii, R. typhi[b]	SFG rickettsiae, *R. canada*[b]	*Orientia*
Doxycycline	0.06–0.125	0.06–0.25	0.0625
Tetracycline	0.1	0.25	0.15–0.31, 1
Minocycline			
Thiamphenicol	1–2	0.5–4	
Chloramphenicol	1	0.25–0.5	1.25–2.5
Rifampin	0.06–0.25	0.03–1	0.31
Erythromycin	0.06–1	1–16	2.5–10
Clarithromycin	0.125–1	0.5–4	
Josamycin	0.5–1	0.5–2	
Roxithromycin	1	8–16	
Pristinamycin	2–4	1–8	
Spiramycin		16–32	
Azithromycin	0.25	8–16[c]	0.0078
Ciprofloxacin	0.5–1	0.25–1	
Ofloxacin	1	0.5–2	
Pefloxacin	1	0.5–2	
Sparfloxacin		0.125–0.25	
PII		2.0	
PI		256	
PD 127,391		0.125–0.25	
PD 131,628		0.25–0.5	
Co-trimoxazole	> 8/2	> 8/2	
Gentamicin	16	4–16	
Amoxicillin	128	128–256	
Ampicillin	500		R
Penicillin G	> 20,100		R

[a] Unless otherwise specified, the MIC of each antibiotic is the lowest antibiotic concentration, resulting in complete inhibition of bacterial growth or plaque formation. R, resistant to the antibiotic.

[b] MIC determined by both plaque and dye uptake assays.

[c] MIC for *R. akari* is 0.25 μg/mL.

(Drancourt and Raoult, 1999). This mutation has been detected in all tested rifampin-resistant species and is absent in the rifampin-susceptible species. Some groups of antibiotics, including β lactams, aminoglycosides and co-trimoxazole are not effective at all for treatment of rickettsiae.

The susceptibility of many genotypes of *O. tsutsugamushi* to antibiotics has not been examined. Consistent with its apparent lack of peptidoglycan layer, *Orientia* is insensitive to penicillin (Amano *et al.*, 1987). Chloramphenicol and tetracyclines are active when assayed in eggs and mice and effective in the treatment of scrub typhus in humans. Rifampicin, ciprofloxacin, doxycycline and erythromycin are all potentially useful drugs according to tests in cell cultures or in mice *in vitro*. Different susceptibilities to antibiotics of distinct isolates of *O. tsutsugamushi* have been documented. Possible antibiotic resistance has been described in clinical cases of scrub typhus that occur in some endemic areas of Northern Thailand (Watt *et al.*, 1996). Numerous genetic types of *Orientia* exist in this region as well as a previously undescribed infected chigger species, *L. chiangraiensis*.

Epidemiology and Control

Rickettsiae include a wide range of agents with varying pathogenicity for humans (**Table 1**). Historical and contemporary aspects of the epidemiology and ecology of *Rickettsia* and *Orientia* have been reviewed extensively (Rehacek and Tarasevich, 1988; Dasch and Weiss, 1991, 1998; Kawamura *et al.*, 1995; Azad *et al.*, 1997; Raoult and Roux, 1997; Azad and Beard, 1998).

R. prowazekii is the aetiological agent of epidemic typhus, which is transmitted by the human body louse. In the history of humankind epidemic typhus typically accompanies wars and catastrophes. Typhus epidemics are often associated with cold seasons of the year or cold climate and poor hygienic conditions suitable for body louse infestation. Active foci of epidemic typhus are known in the Andes regions of South America and in Burundi and Ethiopia (Raoult *et al.*, 1997; WHO Working Group on Rickettsial Diseases, 1982; Walker *et al.*, 1999b). The louse becomes infected by ingesting the blood of diseased humans. Infection is typically acquired by scarification of skin bites with infected faeces. Recovery from epidemic typhus is thought to result in non-sterile immunity, permitting persistence of *R. prowazekii* in a human reservoir for decades between epidemics (Zinsser and Castaneda, 1933; Murray *et al.*, 1950). Occasionally, these individuals suffer from a relapsed or recrudescent form of typhus called Brill–Zinsser disease, the symptoms of which are similar, but usually milder, than those of primary epidemic typhus.

Sylvatic cycles of *R. prowazekii* have also been described with the American flying squirrel, *Glaucomys volans* (Bozeman *et al.*, 1975). Several human cases of flying squirrel-associated infections have been found that are typically associated with invasion of houses by squirrels during the cold seasons. Epidemic typhus rickettsiae are believed to be transmitted by the specific louse of the squirrel between animals and to humans by the fleas, which are not host-specific. Epidemic typhus is most easily controlled by elimination of body lice with improved hygienic conditions and use of insecticides. Vaccines can be used to protect people in infected foci and health workers who are exposed to the typhus agent in their working environment. Epidemic typhus vaccines include the Cox vaccine from World War II, the attenuated Madrid E strain vaccine, and modern recombinant and chemical vaccines (Zdrodovskyi and Golinevitch, 1972; Dasch *et al.*, 1984; Weiss, 1992). Some of these vaccines may provide only partial protective effect, but they clearly lower the risk of rickettsial infection and reduce the severity of acute disease (Ignatovich *et al.*, 1999). In epidemics, doxycycline can be used both therapeutically and prophylactically.

R. typhi causes endemic typhus, also called murine typhus. It is often a milder disease in humans than the infection caused by *R. prowazekii* (Woodward, 1988). It has a worldwide distribution that corresponds to the distribution of its principal reservoirs, the rats *Rattus norvegicus* and *R. rattus*, although other vertebrate hosts may be involved (Adams *et al.*, 1970; Azad, 1988; Azad *et al.*, 1997). *R. typhi* is transmitted by rat lice among rodent populations, but the oriental rat flea, *Xenopsylla cheopis*, is the chief transmitter of the disease to humans. *R. typhi* grows in the midgut epithelial cells of the flea and is excreted in its faeces. The life span of the flea is not affected by the rickettsiae, which are maintained by trans-ovarial transmission in infected populations of fleas (Farhang-Azad *et al.*, 1985). Humans usually become infected by inoculation of infected flea tissues and faeces into the scarified flea bite site, and in rare cases, by inhalation of aerosolised faeces. Preventive measures include rat control and application of insecticides. Historically, vaccines against *R. typhi* were developed contemporarily with epidemic typhus vaccines, and have also been used as a bivalent vaccine (Dasch *et al.*, 1984). *R. canada*, serologically and biologically another typhus-like rickettsia, has only been isolated from *Haemaphysalis* ticks from Richmond, Canada and from Mendocino County, California. Its role in human disease has not been proven.

R. felis (ELB agent) causes cat flea rickettsiosis (Sorvillo *et al.*, 1993). It was originally characterised as a typhus group-like rickettsia based on its serological characteristics and flea association, but phylogenetic analysis has placed it in a distinct clade among the SFG rickettsiae. *R. felis* has been difficult to isolate and cultivate. It has been maintained in a laboratory population of cat fleas, *Ctenocephalides felis*, where it is inherited trans-ovarially. *R. felis* has been identified in the tissues of opossums and their fleas, and from several patients originally thought to have murine typhus. *R. felis* is known from California and Texas in the United States and has been described in Yucatan, Mexico (Zavala-Velazquez *et al.*, 2000).

The majority of SFG rickettsiae are typified by their association with specific ticks, whose ecology limits the geographic distribution of each genotype of rickettsiae. Ticks acquire rickettsiae by feeding on infected vertebrates and remain infected throughout the rest of their lives, transmitting the bacteria trans-stadially and trans-ovarially to their offspring. Rocky Mountain spotted fever, caused by *R. rickettsii*, is found throughout the western hemisphere (Walker, 1989). It is transmitted by the wood tick, *Dermacentor andersonii*, in the western USA and the dog tick, *D. variabilis*, in the eastern USA. *Rhipicephalus sanguineus* and *Amblyomma cajennense* are thought to be vectors of *R. rickettsii* in Mexico and South America, respectively. Several other SFG rickettsiae have been identified in North America, including *R. bellii*, *R. parkeri*, *R. montana*, *R. amblyommii*, *R. peacockii* (the former East Side agent), and several other as yet unnamed distinct serotypes. The area of distribution for some of these SFG rickettsiae overlaps with that of *R. rickettsii* and they may be transmitted by the same tick vectors. Whether they interfere with maintenance of *R. rickettsii* in ticks and whether they are pathogenic for humans is unknown. However, these SFG agents may cause seroconversion without apparent disease in some endemic areas.

R. conorii is responsible for boutonneuse fever, which is also known as Mediterranean spotted fever or Marseilles fever in Mediterranean countries. It also occurs from Eurasia to India and throughout Africa. Its geographical range correlates with the distribution of its principal vector, the brown dog tick, *Rhipicephalus sanguineus*. *R. sibirica* causes North Asian tick typhus which is endemic in western and eastern Siberia of the former Soviet Union and northern China. *R. sibirica* has been detected in more than 20 different species of ixodid ticks. Individual isolates have been

found far from the limits of known foci of disease, so the distribution of disease caused by *R. sibirica* may actually exist from Siberia and Pakistan through central Europe to Portugal. *R. slovaca* is primarily transmitted by *Dermacentor marginatus* in European countries, and it has been associated with sporadic cases of meningoencephalitis and tick-borne lymphoadenopathy (Mittermayer *et al.*, 1980; Lakos and Raoult, 1999). *R. helvetica* is found in association with *Ixodes ricinus* ticks, which is widely distributed through all Europe; it has been implicated in lethal cases of myocarditis (Nilsson *et al.*, 1999). *R. africae*, the agent of *Amblyomma*-transmitted tick typhus in sub-Saharan Africa and another spotted fever agent causing the disease known as Israeli tick typhus, is also found in the region endemic for *R. conorii*. Another endemic focus of spotted fever group rickettsiosis, Astrakhan spotted fever, was recently discovered in the northern part of the Caspian Sea region of Russia (Eremeeva *et al.*, 1994c). *R. australis*, *R. honei* and *R. japonica* are the agents of human diseases in Australia, Flinders Island and Japan, respectively. Other tick-borne spotted fever agents are also known from Europe and Asia, but their pathogenicity for humans has not been established (**Table 1**).

Like *R. felis*, *R. akari* differs markedly from the other tick-transmitted SFG rickettsiae in that its main vector is a mite, *Liponysoides sanguineus*, which parasitises domestic mice, *Mus musculus*. Mites are the main reservoir of *R. akari*, where it is maintained trans-ovarially and trans-stadially. Typically, mites circulate in rodent populations, but when their host animal's temperature is raised by disease, the mites may leave them, infest available humans, and transmit *R. akari* to them. The distribution of *R. akari* is probably worldwide, since outbreaks of rickettsialpox have occurred in the USA, Europe and Asia.

Two plant-associated arthropods have been found to contain rickettsiae closely related to *R. bellii* and one of these causes disease in plants (**Table 1**). Another recently discovered unusual *Rickettsia*, AB-bacterium, is found in some populations of the ladybird beetle, *Adalia bipunctata*. AB-agent is responsible for sex ratio distortion in its insect hosts and has been detected in beetles from England, European regions of Russia, and Japan. Although the AB-bacterium is also not yet cultivable, it is more closely related to typhus and SFG rickettsiae than the more prevalent arthropod-associated *Wolbachia pipientis* and its relatives.

The natural foci of *O. tsutsugamushi* typically consist of transitional forms of vegetation that are associated with changing ecological conditions. It is transmitted by the bite of trombiculid mites, particularly those of the genus *Leptotrombidium*. As with many other rickettsiae, *O. tsutsugamushi* is transmitted vertically in its invertebrate hosts; it also causes sex ratio distortion to a female bias in some species of trombiculid mites. The larval stage or chigger feeds on a wide range of vertebrate hosts, including humans, and the bite is responsible for the transmission of tsutsugamushi disease to them. Scrub typhus occurs throughout the Orient from Afghanistan and Tadzhikistan to Korea and Japan, from the Maritime provinces of Russia to northern Australia and to the western islands of the Pacific Ocean.

Acknowledgements

This work was supported by the Naval Medical Research and Development Command, Research Task 61102A.001.01.BJX.1293. The opinions and statements contained herein are the private ones of the authors and are not to be construed as official or reflecting the views of the Naval Department or the Naval Service at large.

References

Adams WH, Emmons RW, Brooks JE (1970) The changing ecology of murine (endemic) typhus in Southern California. *Am. J. Trop. Med. Hyg.* 19: 311–318.

Alexeyev MF, Winkler HH (1999a) Membrane topology of the *Rickettsia prowazekii* ATP/ADP translocase revealed by novel dual pho-lac reporters. *J. Mol. Biol.* 285: 1503–1513.

Alexeyev MF, Winkler HH (1999b) Gene synthesis, bacterial expression and purification of the *Rickettsia prowazekii* ATP/ADP translocase. *Biochim. Biophys. Acta* 1419: 299–306.

Amano K, Tamura A, Ohashi N *et al.* (1987) Deficiency of peptidoglycan and lipopolysaccharide components in *Rickettsia tsutsugamushi*. *Infect. Immun.* 55: 2290–2292.

Amano KI, Fujita M, Suto T (1993a) Chemical properties of lipopolysaccharides from spotted fever group rickettsiae and their common antigenicity with lipopolysaccharides from *Proteus* species. *Infect. Immun.* 61: 4350–4355.

Amano KI, Suzuki N, Fujita M, Nakamura Y, Suto T (1993b) Serological reactivity of sera from scrub typhus patients against Weil-Felix test antigens. *Microbiol. Immunol.* 37: 927–933.

Amano KI, Williams JC, Dasch GA (1998) Structural properties of lipopolysaccharides from *Rickettsia typhi* and *Rickettsia prowazekii* and their chemical similarity to the lipopolysaccharide from *Proteus vulgaris* OX19 used in the Weil-Felix test. *Infect. Immun.* 66: 923–926.

Anacker RL, List RH, Mann RE, Hayes SF, Thomas LA (1985) Characterization of monoclonal antibodies protecting mice against *Rickettsia rickettsii*. *J. Infect. Dis.* 151: 1052–1060.

Anacker RL, List RH, Mann RE, Wiedbrauk DL (1986) Antigenic heterogeneity in high- and low-virulence strains of *Rickettsia rickettsii* revealed by monoclonal antibodies. *Infect. Immun.* 51: 653–660.

Anacker RL, Mann RE, Gonzales C (1987a) Reactivity of monoclonal antibodies to *Rickettsia rickettsii* with spotted fever and typhus group rickettsiae. *J. Clin. Microbiol.* 25: 167–171.

Anacker RL, McDonald GA, List RH, Mann RE (1987b) Neutralizing activity of monoclonal antibodies to heat-sensitive and heat-resistant epitopes of *Rickettsia rickettsii* surface proteins. *Infect. Immun.* 55: 825–827.

Anderson B (1993) PCR detection of *Rickettsia rickettsii* and *Ehrlichia chaffeensis*. In: Persing DH, Smith TF, Tenover FC, White TJ (eds) *Diagnostic Molecular Microbiology, Principles and Applications*. Washington, DC: American Society for Microbiology, pp. 197–202.

Anderson BE, Tzianabos T (1989) Comparative sequence analysis of a genus-common rickettsial antigen gene. *J. Bacteriol.* 171: 5199–5201.

Anderson BE, Baumstark BR, Bellini WJ (1988) Expression of the gene encoding the 17-kilodalton antigen from *Rickettsia rickettsii*: transcription and posttranslational modification. *J. Bacteriol.* 170: 4493–4500.

Anderson BE, McDonald GA, Jones DC, Regnery RL (1990) A protective protein antigen of *Rickettsia rickettsii* has tandemly repeated, near-identical sequences. *Infect. Immun.* 58: 2760–2769.

Andersson JO, Andersson SGE (1997) Genomic rearrangements during evolution of the obligate intracellular parasite *Rickettsia prowazekii* as inferred from an analysis of 52015 bp nucleotide sequence. *Microbiology* 143: 2783–2795.

Andersson JO, Andersson SG (1999) Genome degradation is ongoing process in *Rickettsia*. *Mol. Biol. Evol.* 16: 1178–1191.

Andersson SG, Sharp PM (1996) Codon-usage and base composition in *Rickettsia prowazekii*. *J. Mol. Evol.* 42: 525–536.

Andersson SGE, Zomorodipour A, Winkler HH, Kurland CG (1995) Unusual organization of the rRNA genes in *Rickettsia prowazekii*. *J. Bacteriol.* 177: 4171–4175.

Andersson SGE, Zomorodipour A, Andersson JO *et al.* (1998) The genome sequence of *Rickettsia prowazekii* and the origin of mitochondria. *Nature* 396: 133–140.

Andersson SGE, Stothard DR, Fuerst P, Kurland CG (1999) Molecular phylogeny and rearrangement of rRNA genes in *Rickettsia* species. *Mol. Biol. Evol.* 16: 987–995.

Austin FE, Winkler HH (1988) Relationship of rickettsial physiology and composition to the rickettsia-host cell interaction. In: Walker DH (ed.) *Biology of Rickettsial Diseases*, Vol. 2. Boca Raton, FL: CRC Press, pp. 29–50.

Austin FE, Turco J, Winkler HH (1987) *Rickettsia prowazekii* requires host cell serine and glycine for growth. *Infect. Immun.* 55: 240–244.

Azad AF (1988) Relationship of vector biology and epidemiology of louse- and flea-borne rickettsioses. In: Walker DH (ed.) *Biology of Rickettsial Diseases*, Vol. 2. Boca Raton, FL: CRC Press, pp. 51–61.

Azad AF, Beard CB (1998) Rickettsial pathogens and their arthropod vectors. *Emerg. Infect. Dis.* 4: 179–186.

Azad AF, Sacci JB, Nelson WM Jr *et al.* (1992) Genetic characterization and transovarial transmission of a typhus-like rickettsia found in cat fleas. *Proc. Natl Acad. Sci. USA* 89: 43–46.

Azad AF, Radulovic S, Higgins JA, Noden BH, Troyer JM (1997) Flea-borne rickettsiosis: ecologic considerations. *Emerg. Infect. Dis.* 3: 319–327.

Babalis T, Tselentis Y, Roux V, Psaroulaki A, Raoult D (1994) Isolation and identification of a rickettsial strain related to *Rickettsia massiliae* in Greek ticks. *Am. J. Trop. Med. Hyg.* 50: 365–372.

Balayeva NM, Eremeeva ME, Tissot-Dupont H, Zakharov IA, Raoult D (1995) Genotype characterization of the bacterium expressing the male-killing trait in the ladybird beetle *Adalia bipunctata* with specific rickettsial molecular tools. *Appl. Environ. Microb.* 61: 1431–1437.

Beaman L, Wisseman CL Jr (1976) Mechanisms of immunity in typhus infections. VI. Differential opsonizing and neutralizing action of human typhus *Rickettsia*-specific antibodies in cultures of human macrophages. *Infect. Immun.* 14: 1071–1076.

Beati L, Finidori JP, Gilot B, Raoult D (1992) Comparison of serologic typing, sodium dodecyl sulfate-polyacrylamide gel electrophoresis protein analysis, and genetic restriction fragment length polymorphism analysis for identification of rickettsiae: characterization of two new rickettsial strains. *J. Clin. Microbiol.* 30: 1922–1930.

Bovarnick MR (1956) Phosphorylation accompanying the oxidation of glutamate by the Madrid E strain of typhus rickettsiae. *J. Biol. Chem.* 220: 353–361.

Bozeman FM, Masiello SA, Williams MS, Elisberg BL (1975) Epidemic typhus rickettsiae isolated from flying squirrels. *Nature* 255: 545–547.

Breitschwerdt EB, Papich MG, Hegarty BC *et al.* (1999) Efficacy of doxycycline, azithromycin, or trovafloxacin for treatment of experimental Rocky Mountain spotted fever in dogs. *Antimicrob. Agents Chemother.* 43: 813–821.

Burgdorfer W, Brinton LP (1970) Intranuclear growth of *Rickettsia canada*, a member of the typhus group. *Infect. Immun.* 2: 112–114.

Carl M, Robbins FM, Hartzman RJ, Dasch GA (1987) Lysis of cells infected with typhus group rickettsiae by a human cytotoxic T cell clone. *J. Immunol.* 139: 4203–4207.

Carl M, Ching WM, Dasch GA (1988) Recognition of typhus group rickettsia-infected targets by human lymphokine-activated killer cells. *Infect. Immun.* 56: 2526–2529.

Casleton BG, Salata K, Dasch GA, Strickman D, Kelly DJ (1998) Recovery and viability of *Orientia tsutsugamushi* from packed red cells and the danger of acquiring scrub typhus from blood transfusion. *Transfusion* 38: 680–689.

Ching WM, Dasch GA, Carl M, Dobson ME (1990) Structural analyses of the 120-kDa serotype protein antigens of typhus group rickettsiae. Comparison with other S-layer proteins. *Ann. N.Y. Acad. Sci.* 590: 334–351.

Ching WM, Wang H, Davis J, Dasch GA (1993) Amino acid analysis and multiple methylation of lysine residues in the surface protein antigen of *Rickettsia prowazekii*. In: Angelotti RH (ed.) *Techniques in Protein Chemistry*, Vol. IV, San Diego: Academic Press, pp. 307–314.

Ching WM, Wang H, Dasch GA (1999a) Identification and characterization of linear epitopes recognized by mouse monoclonal antibodies on the surface protein antigen of *Rickettsia prowazekii*. In: Raoult D, Brouqui P (eds) *Rickettsiae and Rickettsial Diseases at the Turn of the Third Millennium*. Paris: Elsevier, pp. 16–22.

Ching WM, Zhang Z, Dasch GA, Rowland D, Devine PL (1999b) Early diagnosis of scrub typhus by a rapid flow assay using recombinant 56 kDa antigen. Abstract 23. In: *American Society for Rickettsiology. 15th Sesquiannual Meeting. Program and Abstracts*, South Seas Plantation, Captiva Island, Florida.

Cho NH, Seong SY, Huh MS *et al.* (2000) Expression of chemokine genes in murine macrophages infected with *Orientia tsutsugamushi*. *Infect. Immun.* 68: 594–602.

Clifton DR, Gross RA, Sahni SK (1998) NF-κB-dependent inhibition of apoptosis is essential for host cell survival during *Rickettsia rickettsii* infection. *Proc. Natl. Acad. Sci. USA* 95: 4646–4651.

Courtney MA, Haidaris PJ, Marder VJ, Sporn LA (1996) Tissue factor mRNA expression in the endothelium of an intact umbilical vein. *Blood* 87: 174–179.

Crist AE Jr, Wisseman CL Jr, Murphy JR (1984) Characteristics of lymphoid cells that adoptively transfer immunity to *Rickettsia mooseri* infection in mice. *Infect. Immun.* 44: 55–60.

Croquet-Valdes PA, Weiss K, Walker DH (1994). Sequence analysis of the 190-kDa antigen-encoding gene of *Rickettsia conorii* (Malish strain). *Gene* 140: 115–119.

Dasch GA (1981) Isolation of species-specific protein antigens of *Rickettsia typhi* and *Rickettsia prowazekii* for immunodiagnosis and immunoprophylaxis. *J. Clin. Microbiol.* 14: 333–341.

Dasch GA (1985) Distinctive properties of components of the cell envelope of typhus group rickettsiae. In: Kazar J (ed.) *Rickettsiae and Rickettsial Diseases. Proceedings of the Third International Symposium*. Bratislava: Publishing House of the Slovak Academy of Sciences, pp. 54–61.

Dasch GA, Jackson LM (1998) Genetic analysis of isolates of the spotted fever group of rickettsiae belonging to the *R. conorii* complex. *Ann. N.Y. Acad. Sci.* 849: 11–20.

Dasch GA, Weiss E (1991) The Genera *Rickettsia, Rochalimaea, Ehrlichia, Cowdria,* and *Neorickettsia*. In: Balows A, Truper HG, Dworkin M, Harder W, Schleifer

KH (eds) *The Prokaryotes. A Handbook on the Biology of Bacteria: Ecophysiology, Isolation, Identification, Applications,* Vol. III, 2nd edn. New York: Springer-Verlag, pp. 2408–2470.

Dasch GA, Weiss E (1998) The Rickettsiae. In: Collier L, Balows A, Sussman M (eds) *Topley & Wilson's Microbiology and Microbial Infections,* Vol. 2, 9th edn. London: Arnold; New York: Oxford University Press, pp. 853–876.

Dasch GA, Samms JR, Williams JC (1981) Partial purification and characterization of the major species-specific protein antigens of *Rickettsia typhi* and *Rickettsia prowazekii* identified by rocket immunoelectrophoresis. *Infect. Immun.* 31: 276–288.

Dasch GA, Burans JP, Dobson ME, Rollwagen FM, Misiti J (1984) Approaches to subunit vaccines against the typhus rickettsiae, *Rickettsia typhi* and *Rickettsia prowazekii.* In: Leive L, Schlessinger D (eds) *Microbiology 1984.* Washington, DC: American Society for Microbiology, pp. 251–256.

Dasch GA, Strickman D, Watt G, Eamsila C (1996) Measuring genetic variability in *Orientia tsutsugamushi* by PCR/RFLP analysis: a new approach to questions about its epidemiology, evolution, and ecology. In: Kazar J, Toman R (eds) *Rickettsiae and Rickettsial Diseases. Proceedings of the Vth International Symposium.* Bratislava: Veda, pp. 79–84

Dasch GA, Bourgeois AL, Rollwagen FM (1999) The surface protein antigen of *Rickettsia typhi: in vitro* and *in vivo* immunogenicity and protective efficacy in mice. In: Raoult D, Brouqui P (eds) *Rickettsiae and Rickettsial Diseases at the Turn of the Third Millennium.* Paris: Elsevier, pp. 116–122.

Diaz M, Feng HM, Walker DH (1999) *Rickettsia conorii* antigen presentation to CD8 T-lymphocytes by a murine endothelial cell line. In: Raoult D, Brouqui P (eds) *Rickettsiae and Rickettsial Diseases at the Turn of the Third Millennium.* Paris: Elsevier, pp. 123–127.

Dignat-George F, Teysseire N, Mutin M *et al.* (1997) *Rickettsia conorii* infection enhances vascular cell adhesion molecule-1- and intercellular adhesion molecule-1-dependent mononuclear cell adherence to endothelial cells. *J. Infect. Dis.* 175: 1142–1152.

Dobson ME, Azad AF, Dasch GA, Webb L, Olson JG (1989) Detection of murine typhus infected fleas with an enzyme-linked immunosorbent assay. *Am. J. Trop. Med. Hyg.* 40: 521–528.

Drancourt M, Raoult D (1999) Characterization of mutations in the *rpo*B gene in naturally rifampin-resistant rickettsia species. *Antimicrob. Agents Chemother.* 43: 2400–2403.

Dumler JS, Walker DH (1994) Diagnostic tests for Rocky Mountain spotted fever and other rickettsial diseases. *Dermatol. Clin.* 12: 25–36.

Dumler JS, Gage WR, Pettis GL, Azad AF, Kuhadja FP (1990) Rapid immunoperoxidase demonstration of *Rickettsia rickettsii* in fixed cutaneous specimens from patients with Rocky Mountain spotted fever. *Am. J. Clin. Pathol.* 93: 410–414.

Dumler JS, Barbet AF, Bekker GPJ *et al.* (2001) Reorganization of genera in the families *Rickettsiaceae* and *Anaplasmataceae* in the order *Rickettsiales;* unification of some species of *Ehrlichia* and *Anaplasma, Cowdria* with *Ehrlichia,* and *Ehrlichia* with *Neorickettsia;* descriptions of five new species combinations; and designation of *Ehrlichia equi* and 'HE' agent as subjective synonyms of *Ehrlichia phagocytophila. Int. J. Syst. Evol. Microbiol.* In press.

Elghetany MT, Walker DH (1999) Hemostatic changes in Rocky Mountain spotted fever and Mediterranean spotted fever. *Am. J. Clin. Pathol.* 112: 159–168.

Emelyanov VV, Demyanova NG, Kalmyrzayev BB, Krasnova MA (1996) Probable nature of *Rickettsia prowazekii* virulence. *Mol. Genet. Mikrobiol. Virusol.* 2: 39–40.

Enatsu T, Urakami H, Tamura A (1999) Phylogenetic analysis of *Orientia tsutsugamushi* strains based on the sequence homologies of 56-kDa type-specific antigen genes. *FEMS Microbiol. Lett.* 180: 163–169.

Eremeeva ME, Silverman DJ (1998a) Effects of the antioxidant α-lipoic acid on human umbilical vein endothelial cells infected with *Rickettsia rickettsii. Infect. Immun.* 66: 2290–2299.

Eremeeva ME, Silverman DJ (1998b) *Rickettsia rickettsii* infection of the EA.hy 926 endothelial cell line: morphological response to infection and evidence for oxidative injury. *Microbiology* 144: 2037–2048.

Eremeeva ME, Balayeva NM, Ignatovich VF, Raoult D (1993a) Proteinic and genomic identification of spotted fever group rickettsiae isolated in the former USSR. *J. Clin. Microbiol.* 31: 2625–2633.

Eremeeva ME, Roux V, Raoult D (1993b) Determination of genome size and restriction pattern polymorphism of *Rickettsia prowazekii* and *Rickettsia typhi* by pulsed field gel electrophoresis. *FEMS Microbiol. Lett.* 112: 105–112.

Eremeeva ME, Balayeva NM, Raoult D (1994a) Serological response of patients suffering from primary and recrudescent typhus: comparison of complement fixation reaction, Weil-Felix test, microimmunofluorescence, and immunoblotting. *Clin. Diagn. Immunol.* 1: 318–324.

Eremeeva ME, Balaeva NM, Genig VA *et al.* (1994b) The protective activity of individual proteins and immunogenic fractions isolated from *Rickettsia prowazekii. Zh. Mikrobiol. Epidemiol. Immunobiol.* 1: 95–100.

Eremeeva ME, Beati L, Makarova LA *et al.* (1994c) Astrakhan fever rickettsiae: antigenic and genotypic analysis of isolates obtained from human and *Rhipicephalus pumilio* ticks. *Am. J. Trop. Med. Hyg.* 51: 697–706.

Eremeeva M, Yu XJ, Raoult D (1994d) Differentiation among spotted fever group rickettsiae species by analysis of restriction fragment length polymorphism of PCR-amplified DNA. *J. Clin. Microbiol.* 32: 803–810.

Eremeeva M, Balayeva N, Ignatovich V, Raoult D (1995) Genomic study of *Rickettsia akari* by pulsed-field gel electrophoresis. *J. Clin. Microbiol.* 33: 3022–3024.

Eremeeva M, Ignatovich V, Dasch G, Raoult D, Balayeva N (1996) Genetic, biological and serological differentiation of *Rickettsia prowazekii* and *Rickettsia typhi*. In: Kazar J, Toman R (eds) *Rickettsiae and Rickettsial Diseases. Proceedings of the Vth International Symposium.* Bratislava: Veda, pp. 43–50.

Eremeeva ME, Ching WM, Wu Y, Silverman DJ, Dasch GA (1998) Western blotting analysis of heat shock proteins of *Rickettsiales* and other eubacteria. *FEMS Microbiol. Lett.* 167: 229–237.

Eremeeva ME, Santucci LA, Popov VL, Walker DH, Silverman DJ (1999) *Rickettsia rickettsii* infection of human endothelial cells: oxidative injury and reorganization of the cytoskeleton. In: Raoult D, Brouqui P (eds) *Rickettsiae and Rickettsial Diseases at the Turn of the Third Millennium*. Paris: Elsevier, pp. 128–144.

Eremeeva ME, Dasch GA, Silverman DJ (2000) Interaction of rickettsiae with eukaryotic cells: adhesion, entry, intracellular growth, and host cell responses. In: Oelschlaeger TA, Hacker J (eds) *Subcellular Biochemistry Bacterial Invasion into Eukaryotic Cells.* Vol. 33: New York: Kluwer Academic/Plenum Press, pp. 479–516.

Ewing EP Jr, Takeuchi A, Shirai A, Osterman JV (1978) Experimental infection of mouse peritoneal mesothelium with scrub typhus rickettsiae: an ultrastructural study. *Infect. Immun.* 19: 1068–1075.

Farhang-Azad A, Traub R, Baqar S (1985) Transovarial transmission of murine typhus rickettsiae in *Xenopsylla cheopis* fleas. *Science* 227: 543–545.

Feng HM, Walker DH (1993) Interferon-gamma and tumor necrosis factor-alpha exert their antirickettsial effect via induction of synthesis of nitric oxide. *Am. J. Pathol.* 143: 1016–1023.

Feng HM, Popov VL, Walker DH (1994) Depletion of gamma interferon and tumor necrosis factor alpha in mice with *Rickettsia conorii*-infected endothelium: impairment of rickettsicidal nitric oxide production resulting in fatal overwhelming rickettsial disease. *Infect. Immun.* 62: 1952–1960.

Feng H-M, Popov VL, Yuoh G, Walker DH (1997) Role of T lymphocyte subsets in immunity to spotted fever group rickettsiae. *J. Immunol.* 158: 5314–5320.

Fournier PE, Roux V, Raoult D (1998) Phylogenetic analysis of the spotted fever group rickettsiae by study of the outer surface protein rOmpA. *Int. J. Syst. Bacteriol.* 48: 839–849.

Fuerst PA, Poetter KF (1991) DNA sequence differentiation in North American spotted fever group species of *Rickettsia*. In: Kazar J, Raoult D (eds) *Rickettsiae and Rickettsial Diseases. Proceedings of the IVth International Symposium.* Bratislava: Publishing House of the Slovak Academy of Sciences, pp. 162–169.

Fuerst PA, Poetter KP, Pretzman C, Perlman PS (1990) Molecular genetics of populations of intracellular bacteria: the spotted fever group rickettsiae. *Ann. N.Y. Acad. Sci.* 590: 430–438.

Furuya Y, Yoshida Y, Katayama T *et al.* (1991) Specific amplification of *Rickettsia tsutsugamushi* DNA from clinical specimens by polymerase chain reaction. *J. Clin. Microbiol.* 29: 2628–2630.

Furuya Y, Yoshida Y, Katayama T, Yamamoto S, Kawamura A Jr (1993) Serotype-specific amplification of *Rickettsia tsutsugamushi* DNA by nested polymerase chain reaction. *J. Clin. Microbiol.* 31: 1637–1640.

Gage KL, Jerrells TR (1992) Demonstration and partial characterization of antigens of *Rickettsia rhipicephali* that induce cross-reactive cellular and humoral immune responses to *Rickettsia rickettsii*. *Infect. Immun.* 60: 5099–5106.

Gambrill MR, Wisseman CL Jr (1973a) Mechanisms of immunity in typhus infections. II. Multiplication of typhus rickettsiae in human macrophage cell cultures in the nonimmune system: influence of virulence of rickettsial strains and of chloramphenicol. *Infect. Immun.* 8: 519–527.

Gambrill MR, Wisseman CL Jr (1973b) Mechanisms of immunity in typhus infections. III. Influence of human immune serum and complement on the fate of *Rickettsia mooseri* within human macrophages. *Infect. Immun.* 8: 631–640.

Geng P, Jerrells TR (1994) The role of tumor necrosis factor in host defense against scrub typhus rickettsiae. I. Inhibition of growth of *Rickettsia tsutsugamushi*, Karp strain, in cultured murine embryonic cells and macrophages by recombinant tumor necrosis factor-alpha. *Microbiol. Immunol.* 38: 703–711.

George F, Brouqi P, Boffa MC *et al.* (1993) Demonstration of *Rickettsia conorii* induced endothelial injury *in vivo* by measuring circulating endothelial cells, thrombomodulin and von Willebrand factor in patients with Mediterranean spotted fever. *Blood* 82: 2109–2116.

Gilmore RD Jr (1993) Comparison of the rompA gene repeat regions of rickettsiae reveals species-specific arrangements of individual repeating units. *Gene* 125: 97–102.

Gilmore RD Jr, Hackstadt T (1991) DNA polymorphism in the conserved 190 kDa antigen gene repeat region among spotted fever group rickettsiae. *Biochim. Biophys. Acta* 1097: 77–80.

Gilmore RD Jr, Cieplak W Jr, Policastro PF, Hackstadt T (1991) The 120 kilodalton outer membrane protein (rOmpB) of *Rickettsia rickettsii* is encoded by an unusually long open reading frame: evidence for protein processing from a large precursor. *Mol. Microbiol.* 5: 2361–2370.

Gimenez DF (1964) Staining rickettsiae in yolk-sac cultures. *Stain Technol.* 39: 135–140.

Gouin E, Gantelet H, Egile C (1999) A comparative study of the actin-based motilities of the pathogenic bacteria *Listeria monocytogenes*, *Shigella flexneri* and *Rickettsia conorii*. *J. Cell Sci.* 112: 1697–1708.

Gudima OS (1979) Quantitative study on the reproduction of virulent and vaccine *Rickettsia prowazeki* strains in cells of different origin. *Acta Virol.* 23: 421–427.

Hackstadt T (1996) The biology of rickettsiae. *Infect. Agents Dis.* 5: 127–143.

Hackstadt T, Messer R, Cieplak W, Peacock MG (1992) Evidence for proteolytic cleavage of the 120-kilodalton outer membrane protein of rickettsiae: identification of an avirulent mutant deficient in processing. *Infect. Immun.* 60: 159–165.

Hanson BA (1983) Effect of immune serum on infectivity of *Rickettsia tsutsugamushi*. *Infect. Immun.* 42: 341–349.

Hanson BA (1985) Identification and partial characterization of *Rickettsia tsutsugamushi* major protein immunogens. *Infect. Immun.* 50: 603–609.

Hanson BA (1987a) Improved plaque assay for *Rickettsia tsutsugamushi*. *Am. J. Trop. Med. Hyg.* 36: 631–638.

Hanson BA (1987b) Factors influencing *Rickettsia tsutsugamushi* infection of cultured cells. *Am. J. Trop. Med. Hyg.* 36: 621–630.

Harden VA (1990) *Rocky Mountain Spotted Fever: History of a Twentieth-century Disease.* Baltimore: Johns Hopkins University Press.

Hechemy KE, Osterman JV, Eisemann CS, Elliott LB, Sasowski SJ (1981) Detection of typhus antibodies by latex agglutination. *J. Clin. Microbiol.* 13: 214–216.

Hechemy KE, Raoult D, Eisemann C, Han Y, Fox JA (1986) Detection of antibodies to *Rickettsia conorii* with a latex agglutination test in patients with Mediterranean spotted fever. *J. Infect. Dis.* 153: 132–135.

Heinzen RA, Hayes SF, Peacock MG, Hackstadt T (1993) Directional actin polymerization associated with spotted fever group rickettsia infection of Vero cells. *Infect. Immun.* 61: 1926–1935.

Heinzen RA, Grieshaber SS, Van Kirk LS, Devin CJ (1999) Dynamics of actin-based movement by *Rickettsia rickettsii* in Vero cells. *Infect. Immun.* 67: 4201–4207.

Higgins JA, Azad AF (1995) Use of polymerase chain reaction to detect bacteria in arthropods: a review. *J. Med. Entomol.* 32: 213–222.

Hong JE, Santucci LA, Tian X, Silverman DJ (1998) Superoxide dismutase-dependent, catalase-sensitive peroxides in human endothelial cells infected by *Rickettsia rickettsii*. *Infect. Immun.* 66: 1293–1298.

Horinouchi H, Murai K, Okayama A *et al.* (1997) Prevalence of genotypes of *Orientia tsutsugamushi* in patients with scrub typhus in Miyazaki Prefecture. *Microbiol. Immunol.* 41: 503–507.

Ignatovich VF, Kekcheeva NG, Genig NG *et al.* (1999) Necessity of revising the tactics of the prophylactic immunization of employees working with *Rickettsia prowazekii*. *Zh. Mikrobiol. Epidemiol. Immunobiol.* 2: 52–56.

Ives TJ, Manzewitsch P, Regnery RL, Butts JD, Kebede M (1997) *In vitro* susceptibilities of *Bartonella henselae*, *B. quintana*, *B. elizabethae*, *Rickettsia rickettsii*, *R. conorii*, *R. akari*, and *R. prowazekii* to macrolide antibiotics as determined by immunofluorescent-antibody analysis of infected Vero cell monolayers. *Antimicrob. Agents Chemother.* 41: 578–582.

Iwamasa K, Okada T, Tange Y, Kobayashi Y (1992) Ultrastructural study of the response of cells infected *in vitro* with causative agent of spotted fever group rickettsiosis in Japan. *APMIS* 100: 535–542.

Jerrells TR (1988) Mechanisms of immunity to *Rickettsia* species and *Coxiella burnetii*. In: Walker DH (ed.) *Biology of Rickettsial Diseases*, Vol. II. Boca Raton, FL: CRC Press, pp. 79–100.

Jerrells TR (1997) Immunity to rickettsiae (redux). In: Anderson B, Friedman H, Bendinelli M (eds) *Rickettsial Infection and Immunity*. New York: Plenum Press, pp. 15–28.

Jerrells TR, Osterman JV (1982) Host defenses in experimental scrub typhus: delayed-type hypersensitivity responses of inbred mice. *Infect. Immun.* 35: 117–123.

Jerrells TR, Geng P (1994) The role of tumor necrosis factor in host defense against scrub typhus rickettsiae. II. Differential induction of tumor necrosis factor-alpha production by *Rickettsia tsutsugamushi* and *Rickettsia conorii*. *Microbiol. Immunol.* 38: 713–719.

Johnson JW, Pedersen CE Jr (1978) Plaque formation by strains of spotted fever rickettsiae in monolayer cultures of various cell types. *J. Clin. Microbiol.* 7: 389–391.

Jones D, Anderson B, Olson J, Greene C (1993) Enzyme-linked immunosorbent assay for detection of human immunoglobulin G to lipopolysaccharide of spotted fever group rickettsiae. *J. Clin. Microbiol.* 31: 138–141.

Kaplanski G, Teysseire N, Farnarier C *et al.* (1995) IL-6 and IL-8 production from cultured human endothelial cells stimulated by infection with *Rickettsia conorii* via a cell-associated IL-1α-dependent pathway. *J. Clin. Invest.* 96: 2839–2844.

Kasuya S, Nagano I, Ikeda T, Goto C, Shimokawa K, Takahashi Y (1996) Apoptosis of lymphocytes in mice induced by infection with *Rickettsia tsutsugamushi*. *Infect. Immun.* 64: 3937–3941.

Kawamori F, Akiyama M, Sugieda M (1993) Two-step polymerase chain reaction for diagnosis of scrub typhus and identification of antigenic variants of *Rickettsia tsutsugamushi*. *J. Vet. Med. Sci.* 55: 749–755.

Kawamura A Jr, Tanaka H, Tamura A (eds) (1995) *Tsutsugamushi Disease – An Overview*. Tokyo: University of Tokyo Press.

Kee SH, Cho KA, Kim MK, Lim BU, Chang WH, Kang JS (1999). Disassembly of focal adhesions during apoptosis of endothelial cells ECV304 infected with *Orientia tsutsugamushi*. *Microb. Pathogen.* 27: 265–271.

Kelly DJ, Dasch GA, Chye TC, Ho TM (1994) Detection and characterization of *Rickettsia tsutsugamushi* (*Rickettsiales: Rickettsiaceae*) in infected *Leptotrombidium* (*Leptotrombidium*) *fletcheri* chiggers (Acari: Trombiculidae) with the polymerase chain reaction. *J. Med. Entomol.* 31: 691–699.

Kenyon RH, Pedersen CE Jr (1980) Immune responses to *Rickettsia akari* infection in congenitally athymic nude mice. *Infect. Immun.* 28: 310–313.

Kokorin IN (1968) Biological peculiarities of the development of rickettsiae. *Acta Virol.* 12: 31–35.

Kokorin IN, Kabanova EA, Shirokova EM (1982) Role of T lymphocytes in *Rickettsia conorii* infection. *Acta Virol.* (Praha) 26: 91–97.

Kuen B, Lubitz W (1996) Analysis of S-layer proteins and genes. In: Sleytr UB, Messner P, Pum D, Sara M (eds) *Crystalline Bacterial Cell Surface Proteins*. Austin, Texas: RG. Landes, pp. 77–102.

Kumura K, Minamishima Y, Yamamoto S, Ohashi N, Tamura A (1991) DNA base composition of *Rickettsia tsutsugamushi* determined by reversed-phase high-performance liquid chromatography. *Int. J. Syst. Bacteriol.* 41: 247–248.

Lakos A, Raoult D (1999) Tick-borne lymphadenopathy (TIBOLA) a *Rickettsia slovaca* infection? In: Raoult D, Brouqui P (eds) *Rickettsiae and Rickettsial Diseases at the Turn of the Third Millennium*. Paris: Elsevier, pp. 258–261.

Land MV, Ching WM, Dasch GA *et al.* (2000) Evaluation of a commercially available recombinant-protein enzyme-linked immunosorbent assay for detection of antibodies produced in scrub typhus rickettsial infections. *J. Clin. Microbiol.* 38: 2701–2705.

Lange JV, Walker DH (1984) Production and characterization of monoclonal antibodies to *Rickettsia rickettsii*. *Infect. Immun.* 46: 289–294.

La Scola B, Raoult D (1997) Laboratory diagnosis of rickettsiosis: current approaches to diagnosis of old and new rickettsial diseases. *J. Clin. Microbiol.* 35: 2715–2727.

La Scola B, Rydkina L, Ndihokubwayo JB, Vene S, Raoult D (2000) Serological differentiation of murine typhus and epidemic typhus using cross-adsorption and Western blotting. *Clin. Diagn. Lab. Immunol.* 7: 612–616.

Lauer BA, Reller LB, Mirrett S (1981). Comparison of acridine orange and Gram stains for detection of microorganisms in cerebrospinal fluid and other clinical specimens. *J. Clin. Microbiol.* 14: 205–210.

Li H, Walker DH (1992) Characterization of rickettsial attachment to host cells by flow cytometry. *Infect. Immun.* 60: 2030–2035.

Li H, Walker DH (1998) rOmpA is a critical protein for the adhesion of *Rickettsia rickettsii* to host cells. *Microb. Pathog.* 24: 289–298.

Lim BU, Kim MK, Kang JS (2000) Heat stable component of *Orientia tsutsugamushi* stimulates the production of tumor necrosis factor alpha by murine macrophages. Abstract 96. In *American Society for Rickettsiology. 15th Sesquiannual Meeting. Program and Abstracts*. South Seas Plantation, Captiva Island, Florida.

Manor E (1992) The effect of monocyte-derived macrophages on the growth of *Rickettsia conorii* in permissive cells. *Acta Virol.* (Praha) 36: 13–18.

Manor E, Sarov I (1990) Inhibition of *Rickettsia conorii* growth by recombinant tumor necrosis factor alpha: enhancement of inhibition by gamma interferon. *Infect. Immun.* 58: 1886–1890.

Manor E, Carbonetti NH, Silverman DJ (1994) *Rickettsia rickettsii* has proteins with cross-reacting epitopes to eukaryotic phospholipase A2 and phospholipase C. *Microb. Pathog.* 17: 99–109.

Maurin M, Raoult D (1999) Antimicrobial therapy of rickettsial diseases. In: Raoult D, Brouqui P (eds) *Rickettsiae and Rickettsial Diseases at the Turn of the Third Millennium*. Paris: Elsevier, pp. 330–342.

McDonald GA, Anacker RL, Garjian K (1987) Cloned gene of *Rickettsia rickettsii* surface antigen: candidate vaccine for Rocky Mountain spotted fever. *Science* 235: 83–85.

McDonald GA, Anacker RL, Mann RE, Milch LJ (1988) Protection of guinea pigs from experimental Rocky Mountain spotted fever with a cloned antigen of *Rickettsia rickettsii*. *J. Infect. Dis.* 158: 228–231.

Mittermayer T, Brezina R, Urvolgyi J (1980) First report of an infection with *Rickettsia slovaca*. *Folia Parasitol.* 27: 373–376.

Moree MF, Hanson B (1992) Growth characteristics and proteins of plaque-purified strains of *Rickettsia tsutsugamushi*. *Infect. Immun.* 60: 3405–3415.

Murai K, Okayama A, Horinouchi H, Oshikawa T, Tachibana N, Tsubouchi H (1995) Eradication of *Rickettsia tsutsugamushi* from patient's blood by chemotherapy, as assessed by the polymerase chain reaction. *Am. J. Trop. Med. Hyg.* 52: 325–327.

Murata M, Kawamura A Jr (1977) Restoration of the infectivity of *Rickettsia tsutsugamushi* to susceptible animals by passage in athymic nude mice. *Jpn. J. Exp. Med.* 47: 385–391.

Murphy JR, Wisseman CL Jr, Fiset P (1979) Mechanisms of immunity in typhus infection: adoptive transfer of immunity to *Rickettsia mooseri*. *Infect. Immun.* 24: 387–393.

Murphy JR, Wisseman CL Jr, Fiset P (1980) Mechanisms of immunity in typhus infection: analysis of immunity to *Rickettsia mooseri* infection of guinea pigs. *Infect. Immun.* 27: 730–738.

Murray ES, Baehr G, Schwartzman G *et al.* (1950) Brill's disease. I. Clinical and laboratory diagnosis. *JAMA* 142: 1059–1066.

Myers WF, Wisseman CL Jr (1980) Genetic relatedness among the typhus group rickettsiae. *Int. J. Syst. Bacteriol.* 30: 143–150.

Myers WF, Wisseman CL Jr (1981) The taxonomic relationship of *Rickettsia canada* to the typhus and spotted fever groups of the genus *Rickettsia*. In: Burgdorfer W, Anacker RL (eds) *Rickettsiae and Rickettsial Diseases*. New York: Academic Press, pp. 313–325.

Neimark WL, Kocan KM (1997) The cell wall-less rickettsia *Eperythrozoon wenyonii* is a *Mycoplasma*. *FEMS Microbiol. Lett.* 156: 287–291.

Ng FKP, Oaks SC Jr, Lee M, Groves MG, Lewis GE Jr (1985) A scanning and transmission electron microscopic examination of *Rickettsia tsutsugamushi*-infected human endothelial, MRC-5, and L-929 cells. *Jpn. J. Med. Sci. Biol.* 38: 125–139.

Nilsson K, Lindquist O, Pahlson C (1999) Association of *Rickettsia helvetica* with chronic perimyocarditis in sudden cardiac death. *Lancet* 354: 1169–1173.

Oaks EV, Rice RM, Kelly DJ, Stover CK (1989) Antigenic and genetic relatedness of eight *Rickettsia tsutsugamushi* antigens. *Infect. Immun.* 57: 3116–3122.

Ohashi N, Tamura A, Suto T (1988) Immunoblotting analysis of antirickettsial antibodies produced in patients of tsutsugamushi disease. *Microbiol. Immunol.* 32: 1085–1092.

Ohashi N, Fukuhara M, Shimada M, Tamura A (1995) Phylogenetic position of *Rickettsia tsutsugamushi* and the relationship among its antigenic variants by analyses of 16S rRNA gene sequences. *FEMS Microbiol. Lett.* 125: 299–304.

Ojcius DM, Thibon M, Mounier C, Dautry-Varsat A (1995) pH and calcium dependence of hemolysis due to *Rickettsia prowazekii*: comparison with phospholipase activity. *Infect. Immun.* 63: 3069–3072.

Paddock CC, Greer PW, Ferebee TL (1999) Hidden mortality attributable to Rocky Mountain spotted fever: immunohistochemical detection of fatal, serologically unconfirmed disease. *J. Infect. Dis.* 179: 1469–1476.

Pai H, Sohn S, Seong Y, Kee S, Chang WH, Choe KW (1997) Central nervous system involvement in patients with scrub typhus. *Clin. Infect. Dis.* 24: 436–440.

Pang H, Winkler HH (1994) Analysis of the peptidoglycan of *Rickettsia prowazekii*. *J. Bacteriol.* 176: 923–926.

Park CS, Kim IC, Lee JB *et al.* (1993) Analysis of antigenic characteristics of *Rickettsia tsutsugamushi* Boryong strain and antigenic heterogeneity of *Rickettsia tsutsugamushi* using monoclonal antibodies. *J. Korean Med. Sci.* 8: 319–324.

Penkina GA, Ignatovich VF, Balaeva NM (1995) Interaction of *Rickettsia prowazekii* strains of different virulence with white rat macrophages. *Acta Virol.* 39: 205–209.

Philip RN, Casper EA, Burgdorfer W (1978) Serologic typing of rickettsiae of the spotted fever group by micro-immunofluorescence. *J. Immunol.* 121: 1961–1968.

Philip RN, Casper EA, Anacker RL *et al.* (1983) *Rickettsia bellii* sp. nov.: a tick-borne rickettsia, widely distributed in the United States, that is distinct from the spotted fever and typhus biogroups. *Int. J. Syst. Bacteriol.* 33: 94–106.

Pickens EG, Bell EJ, Lackman DB, Burgdorfer W (1965) Use of mouse serum in identification and serologic classification of *Rickettsia akari* and *Rickettsia australis*. *J. Immunol.* 94: 883–889.

Plano GV, Winkler HH (1991) Identification and initial topological analysis of the *Rickettsia prowazekii* ATP/ADP translocase. *J. Bacteriol.* 173: 3389–3396.

Policastro PF, Peacock MG, Hackstadt T (1996) Improved plaque assays for *Rickettsia prowazekii* in Vero76 cells. *J. Clin. Microbiol.* 34: 1944–1948.

Popov VL, Dyuisalieva RG, Smirnova NS, Tarasevich IV, Rybkina NN (1986) Ultrastructure of *Rickettsia sibirica* during interaction with the host cell. *Acta Virol.* 30: 494–498.

Rachek LO, Tucker AM, Winkler HH, Wood DO (1998) Transformation of *Rickettsia prowazekii* to rifampin resistance. *J. Bacteriol.* 180: 2118–2124.

Rachek LO, Hines A, Tucker AM, Winkler HH, Wood DO (2000) Transformation of *Rickettsia prowazekii* to erythromycin resistance encoded by the *Escherichia coli ereB* gene. *J. Bacteriol.* 182: 3289–3291.

Radulovic S, Speed R, Feng HM, Taylor C, Walker DH (1993) EIA with species-specific monoclonal antibodies: a novel seroepidemiologic tool for determination of the etiologic agent of spotted fever rickettsiosis. *J. Infect. Dis.* 168: 1292–1295.

Radulovic S, Feng HM, Crocquet-Valdes P *et al.* (1994) Antigen-capture enzyme immunoassay: a comparison with other methods for the detection of spotted fever group rickettsiae in ticks. *Am. J. Trop. Med. Hyg.* 50: 359–364.

Radulovic S, Troyer JM, Beier MS, Lau AOT, Azad AF (1999) Identification and molecular analysis of the gene encoding *Rickettsia typhi* hemolysin. *Infect. Immun.* 67: 6104–6108.

Raoult D, Dasch GA (1989a) Line blot and Western blot immunoassays for diagnosis of Mediterranean spotted fever. *J. Clin. Microbiol.* 27: 2073–2079.

Raoult D, Dasch GA (1989b) The line blot: an immunoassay for monoclonal and other antibodies; its application to the serotyping of gram-negative bacteria. *J. Immunol. Meth.* 125: 57–65.

Raoult D, Dasch GA (1995) Immunoblot cross-reactions among *Rickettsia*, *Proteus* spp. and *Legionella* spp. in patients with Mediterranean spotted fever. *FEMS Immun. Med. Microbiol.* 11: 13–18.

Raoult D, Roux V (1997) Rickettsioses as paradigms of new or emerging infectious diseases. *Clin. Microbiol. Rev.* 10: 694–719.

Raoult D, Roux V, Ndihokubwayo JB (1997) Jail fever (epidemic typhus) outbreak in Burundi. *Emerg. Infect. Dis.* 3: 357–359.

Regnery RL, Spruill CL, Plikaytis BD (1991) Genotypic identification of rickettsiae and estimation of intraspecies sequence divergence for portions of two rickettsial genes. *J. Bacteriol.* 173: 1576–1589.

Rehacek J, Tarasevich IV (1988) *Acari-borne Rickettsiae and Rickettsioses in Eurasia*. Bratislava: Veda Publishing House of the Slovak Academy of Sciences. 343 pp.

Rikihisa Y, Kawahara M, Wen B *et al.* (1997) Western immunoblot analysis of *Haemobartonella muris* and comparison of 16S rRNA gene sequences of *H. muris*, *H. felis*, and *Eperythrozoon suis*. *J. Clin. Microbiol.* 35: 823–829.

Rodionov AV, Eremeeva ME, Balayeva NM (1991) Isolation and partial characterization of the M(r) 100 kD protein from *Rickettsia prowazekii* strains of different virulence. *Acta Virol.* 35: 557–565.

Rolain JM, Maurin M, Vestris G, Raoult D (1999) *In vitro* susceptibilities of 27 rickettsiae to 13 antimicrobials. *Antimicrobial. Agents Chemother.* 36: 1342–1344.

Rolain JM, Maurin M, Bryskier A, Raoult D (2000) *In vitro* activities of telithromycin (HMR 3647) against *Rickettsia rickettsii*, *Rickettsia conorii*, *Rickettsia africae*, *Rickettsia typhi*, *Rickettsia prowazekii*, *Coxiella burnetii*, *Bartonella henselae*, *Bartonella quintana*, *Bartonella bacilliformis*, and *Ehrlichia chaffeensis*. *Antimicrobial. Agents Chemother.* 44: 1391–1393.

Rollwagen FM, Bakun AJ, Dorsey CH, Dasch GA (1985) Mechanisms of immunity to infection with typhus rickettsiae: infected fibroblasts bear rickettsial antigens on their surfaces. *Infect. Immun.* 50: 911–916.

Roux V, Raoult D (1993) Genotypic identification and phylogenetic analysis of the spotted fever group rickettsiae by pulsed-field gel electrophoresis. *J. Bacteriol.* 175: 4895–4904.

Roux V, Raoult D (1995) Phylogenetic analysis of the genus *Rickettsia* by 16S rDNA sequencing. *Res. Microbiol.* 146: 385–396.

Roux V, Raoult D (1999a) Body lice as tools for diagnosis and surveillance of reemerging diseases. *J. Clin. Microbiol.* 37: 596–599.

Roux V, Raoult D (1999b) Phylogenetic analysis and taxonomic relationships among the genus *Rickettsia*. In: Raoult D, Brouqui P (eds) *Rickettsiae and Rickettsial Diseases at the Turn of the Third Millennium*. Paris: Elsevier, pp. 52–66.

Roux V, Raoult D (2000) Phylogenetic analysis of members of the genus *Rickettsia* using the gene encoding the outer-membrane protein rOmpB (*ompB*). *Int. J. Syst. Evol. Microbiol.* 50: 1449–1455.

Roux V, Fournier PE, Raoult D (1996) Differentiation of spotted fever group rickettsiae by sequencing and analysis of restriction fragment length polymorphism of PCR-amplified DNA of the gene encoding the protein rOmpA. *J. Clin. Microbiol.* 34: 2058–2065.

Roux V, Rydkina E, Eremeeva M, Raoult D (1997) Citrate synthase gene comparison, a new tool for phylogenetic analysis, and its application for rickettsiae. *Int. J. Syst. Bacteriol.* 47: 252–261.

Sahni SK, Turpin LC, Brown TL, Sporn LA (1999) Involvement of protein kinase C in *Rickettsia rickettsii*-induced transcriptional activation of the host endothelial cells. *Infect. Immun.* 67: 6418–6423.

Schaechter M, Bozeman FM, Smadel JE (1957) Study on the growth of rickettsiae. II. Morphologic observations of living rickettsiae in tissue culture cells. *Virology* 3: 160–172.

Schramek S (1972) Deoxyribonucleic acid base composition of members of the typhus group of rickettsiae. *Acta Virol.* 16: 447.

Schuenke K, Walker DH (1994) Cloning, sequencing, and expression of the gene coding for an antigenic 120-kilodalton protein of *Rickettsia conorii*. *Infect. Immun.* 62: 904–909.

Seong SY, Park SG, Huh S *et al.* (1997) T-track PCR fingerprinting for the rapid detection of genetic polymorphism. *FEMS Microbiol. Lett.* 152: 37–44.

Sessler CN, Schwartz M, Windsor AC, Fowler AA, 3rd (1995) Increased serum cytokines and intercellular adhesion molecule-I in fulminant Rocky Mountain spotted fever. *Crit. Care Med.* 23: 973–976.

Sexton DJ, Kanj SS, Wilson K *et al.* (1994) The use of a polymerase chain reaction as a diagnostic test for Rocky Mountain spotted fever. *Am. J. Trop. Med. Hyg.* 50: 59–63.

Shi RJ, Simpson-Haidaris PJ, Lerner NB, Marder VJ, Silverman DK, Sporn LA (1998) Transcriptional regulation of endothelial cell tissue factor expression during *Rickettsia rickettsii* infection: involvement of the transcription factor NF-κB. *Infect. Immun.* 66: 1070–1075.

Shirai A, Dietel JW, Osterman JV (1975) Indirect hemagglutination test for human antibody to typhus and spotted fever group rickettsiae. *J. Clin. Microbiol.* 2: 430–437.

Shirai A, Coolbaugh JC, Gan E, Chan TC, Huxsoll DL, Groves MG (1982) Serologic analysis of scrub typhus isolates from the Pescadores and Philippine Islands. *Jpn. J. Med. Sci. Biol.* 35: 255–259.

Silverman DJ (1984) *Rickettsia rickettsii*-induced cellular injury of human vascular endothelium *in vitro*. *Infect. Immun.* 44: 545–553.

Silverman DJ (1997) Oxidative cell injury and spotted fever group rickettsiae. In: Anderson BE, Friedman H, Bendinelli M (eds) *Rickettsial Infection and Immunity*. New York: Plenum, pp. 79–98.

Silverman DJ, Bond SB (1984) Infection of human vascular endothelial cells by *Rickettsia rickettsii*. *J. Infect. Dis.* 149: 201–206.

Silverman DJ, Wisseman CL Jr (1978) Comparative ultrastructural study on the cell envelopes of *Rickettsia prowazekii*, *Rickettsia rickettsii*, and *Rickettsia tsutsugamushi*. *Infect. Immun.* 21: 1020–1023.

Silverman DJ, Wisseman CL Jr, Waddell AD, Jones M (1978) External layers of *Rickettsia prowazekii* and *Rickettsia rickettsii*. Occurrence of a slime layer. *Infect. Immun.* 22: 233–246.

Silverman DJ, Wisseman CL Jr, Waddell A (1980) *In vitro* studies of rickettsia–host cell interactions: ultrastructural study of *Rickettsia prowazekii*-infected chicken embryo fibroblasts. *Infect. Immun.* 29: 778–790.

Silverman DJ, Santucci LA, Meyers N, Sekeyova Z (1992) Penetration of host cells by *Rickettsia rickettsii* appears to be mediated by a phospholipase of rickettsial origin. *Infect. Immun.* 60: 2733–2740.

Sompolinsky D, Boldur I, Goldwasser RA *et al.* (1986) Serological cross-reactions between *Rickettsia typhi*, *Proteus vulgaris* OX19 and *Legionella bozemanii* in a series of febrile patients. *Isr. J. Med. Sci.* 22: 745–752.

Sorvillo FJ, Gondo B, Emmons R (1993) A suburban focus of endemic typhus in Los Angeles County: association with seropositive domestic cats and opossums. *Am. J. Trop. Med. Hyg.* 48: 269–273.

Sporn LA, Marder VJ (1996) Interleukin-1α a production during *Rickettsia rickettsii* infection of cultured endothelial cells: potential role in autocrine cell stimulation. *Infect. Immun.* 64: 1609–1613.

Sporn LA, Shi RJ, Lawrence SO *et al.* (1991) *Rickettsia rickettsii* infection of cultured endothelial cells induces release of large von Willebrand factor multimers from Wiebel-Palade bodies. *Blood* 78: 2595–2602.

Sporn LA, Lawrence SO, Silverman DJ, Marder VJ (1993) E-selectin-dependent neutrophil adhesion to *Rickettsia rickettsii*-infected endothelial cells. *Blood* 81: 2406–2412.

Sporn LA, Haidaris PJ, Shi RJ *et al.* (1994) *Rickettsia rickettsii* infection of cultured human endothelial cells induces tissue factor expression. *Blood* 83: 1527–1534.

Sporn LA, Sahni SK, Lerner NB *et al.* (1997) *Rickettsia rickettsii* infection of cultured human endothelial cells induces NF-kappaB activation. *Infect. Immun.* 65: 2786–2791.

Stothard DR, Fuerst PA (1995) Evolutionary analysis of the spotted fever and typhus groups of *Rickettsia* using 16S rRNA gene sequences. *Syst. Appl. Microbiol.* 18: 52–61.

Stover CK, Marana DP, Dasch GA, Oaks EV (1990) Molecular cloning and sequence analysis of the Sta58 major antigen gene of *Rickettsia tsutsugamushi*: sequence homology and antigenic comparison to the 60-kilodalton family of stress proteins. *Infect. Immun.* 58: 1360–1368.

Sumner JW, Sims KG, Jones DC, Anderson BE (1995) Protection of guinea pigs from experimental Rocky Mountain spotted fever by immunization with baculovirus-expressed *Rickettsia rickettsii* rOmpA protein. *Vaccine* 13: 29–35.

Tamura A (1988) Invasion and intracellular growth of *Rickettsia tsutsugamushi. Microbiol. Sci.* 5: 228–232.

Tamura AA (1999) Genetic diversity of *Orientia tsutsugamushi.* In: Raoult D, Brouqui P (eds) *Rickettsiae and Rickettsial Diseases at the Turn of the Third Millennium.* Paris: Elsevier, pp. 67–73.

Tamura A, Urakami H, Ohashi N (1990) A comparative view of *Rickettsia tsutsugamushi* and the other groups of rickettsiae. *Eur. J. Epidemiol.* 7: 259–269.

Tamura A, Ohashi N, Urakami H, Miyamura S (1995) Classification of *Rickettsia tsutsugamushi* in a new genus, *Orientia* gen. nov., as *Orientia tsutsugamushi* comb. nov. *Int. J. Syst. Bacteriol.* 45: 589–591.

Tange Y, Kanemitsu N, Kobayashi Y (1991) Analysis of immunological characteristics of newly isolated strains of *Rickettsia tsutsugamushi* using monoclonal antibodies. *Am. J. Trop. Med. Hyg.* 44: 371–381.

Tay ST, Nazma S, Rohani MY (1996) Diagnosis of scrub typhus in Malaysian aborigines using nested polymerase chain reaction. *Southeast Asian J. Trop. Med. Public Health* 27: 580–583.

Teysseire N, Amoux D, George F *et al.* (1992a) von Willebrand factor release and thrombomodulin and tissue factor expression in *Rickettsia conorii*-infected endothelial cells. *Infect. Immun.* 60: 4388–4393.

Teysseire N, Chiche-Portiche C, Raoult D (1992b) Intracellular movements of *Rickettsia conorii* and *R. typhi* based on actin polymerization. *Res. Microbiol.* 143: 821–829.

Teysseire N, Boudier JA, Raoult D (1995) *Rickettsia conorii* entry into Vero cells. *Infect. Immun.* 63: 366–374.

Todd WJ, Burgdorfer W, Wray GP (1983) Detection of fibrils associated with *Rickettsia rickettsii. Infect. Immun.* 41: 1252–1260.

Troyer JM, Radulovic S, Azad AF (1999) Green fluorescent protein as a marker in *Rickettsia typhi* transformation. *Infect. Immun.* 67: 3308–3311.

Turco J, Winkler HH (1982) Differentiation between virulent and avirulent strains of *Rickettsia prowazekii* by macrophage-like cell lines. *Infect. Immun.* 35: 783–791.

Turco J, Winkler HH (1991) Comparison of properties of virulent, avirulent, and interferon-resistant *Rickettsia prowazekii* strains. *Infect. Immun.* 59: 1647–1655.

Turco J, Winkler HH (1993) Role of the nitric oxide synthase pathway in inhibition of growth of interferon-sensitive and interferon-resistant *Rickettsia prowazekii* strains in L929 cells treated with tumor necrosis factor alpha and gamma interferon. *Infect. Immun.* 61: 4317–4325.

Turco J, Winkler HH (1994a) Cytokine sensitivity and methylation of lysine in *Rickettsia prowazekii* EVir and interferon-resistant *R. prowazekii* strains. *Infect. Immun.* 62: 3172–3177.

Turco J, Winkler HH (1994b) Relationship of tumor necrosis factor alpha, the nitric oxide synthase pathway, and lipopolysaccharide to the killing of gamma interferon-treated macrophagelike RAW264.7 cells by *Rickettsia prowazekii. Infect. Immun.* 62: 2568–2574.

Turco J, Winkler HH (1997) Cytokines influencing infections by *Rickettsia* species. In: Anderson BE, Friedman H, Bendinelli M (eds) *Rickettsial Infection and Immunity.* New York: Plenum, pp. 29–52.

Turco J, Liu H, Gottlieb SF, Winkler HH (1998) Nitric oxide-mediated inhibition of the ability of *Rickettsia prowazekii* to infect mouse fibroblasts and mouse macrophagelike cells. *Infect. Immun.* 66: 558–566.

Tyeryar FJ Jr, Weiss E, Millar DB, Bozeman FM, Ormsbee RA (1980) DNA base composition of rickettsiae. *Science* 180: 415–417.

Uchiyama T (1997) Intracytoplasmic localization of antigenic heat stable 120- to 130-kilodalton proteins (PS120) common to spotted fever group rickettsiae demonstrated by immunoelectron microscopy. *Microbiol. Immunol.* 41: 815–818.

Uchiyama T (1999a) Sequence analysis of the gene encoding a spotted fever group-specific intracytoplasmic protein PS120 of *Rickettsia japonica. Microbiol. Immunol.* 43: 983–987.

Uchiyama T (1999b) Role of major surface antigens of *Rickettsia japonica* in the attachment to host cells. In: Raoult D, Brouqui P (eds) *Rickettsiae and Rickettsial Diseases at the Turn of the Third Millennium.* Paris: Elsevier, pp. 182–188.

Uchiyama T, Uchida T, Walker DH (1990) Species-specific monoclonal antibodies to *Rickettsia japonica*, a newly identified spotted fever group rickettsia. *J. Clin. Microbiol.* 28: 1177–1180.

Uchiyama T, Uchida T, Walker DH (1994a) Analysis of major surface polypeptides of *Rickettsia japonica. Microbiol. Immunol.* 38: 575–579.

Uchiyama T, Uchida T, Wen JW, Walker DH (1994b) Demonstration of heat-labile and heat-stable epitopes of *Rickettsia japonica* on ultrathin sections. *Lab. Invest.* 71: 432–437.

Uchiyama T, Zhao L, Yan Y, Uchida T (1995) Cross-reactivity of *Rickettsia japonica* and *Rickettsia typhi* demonstrated by immunofluorescence and Western immunoblotting. *Microbiol. Immunol.* 39: 951–957.

Urakami H, Tsuruhara T, Tamura A (1982) Intranuclear *Rickettsia tsutsugamushi* in cultured mouse fibroblasts (L cells). *Microbiol. Immunol.* 26: 445–447.

Urakami H, Tsuruhara T, Tamura A (1984) Electron microscopic studies on intracellular multiplication of *Rickettsia tsutsugamushi* in L cells. *Microbiol. Immunol.* 28: 1191–1201.

Van Kirk LS, Hayes SF, Heinzen RA (2000) Ultrastructure of *Rickettsia rickettsii* actin tails and localization of cytoskeletal proteins. *Infect Immun.* 68: 4706–4713.

Viale AM, Arakaki AK (1994) The chaperone connection to the origins of the eukaryotic organelles. *FEBS Lett.* 341: 146–151.

Vishwanath S (1991) Antigenic relationships among the rickettsiae of the spotted fever and typhus groups. *FEMS Microbiol. Lett.* 65: 341–344.

Vishwanath S, McDonald GA, Watkins NG (1990) A recombinant *Rickettsia conorii* vaccine protects guinea pigs from experimental boutonneuse fever and Rocky Mountain spotted fever. *Infect. Immun.* 58: 646–653.

Vitale G, Mansueto S, Gambino G (1999) Differential up-regulation of circulating soluble selectins and endothelial adhesion molecules in Sicilian patients with Boutonneuse fever. *Clin. Exp. Immunol.* 117: 304–308.

Walker DH (1989) Rocky Mountain spotted fever: a disease in need of microbiological concern. *Clin. Microbiol. Rev.* 2: 227–240.

Walker DH, Cain BG (1978) A method for specific diagnosis of Rocky Mountain spotted fever on fixed, paraffin-embedded tissue by immunofluorescence. *J. Infect. Dis.* 137: 206–209.

Walker DH, Cain BG (1980) The rickettsial plaque. Evidence for direct cytopathic effect of *Rickettsia rickettsii*. *Lab. Invest.* 43: 388–396.

Walker DH, Firth WT, Edgell CJS (1982) Human endothelial cell culture plaques induced by *Rickettsia rickettsii*. *Infect. Immun.* 37: 301–306.

Walker DH, Firth WT, Ballard JG, Hegarty BC (1983) Role of phospholipase-associated penetration mechanism in cell injury by *Rickettsia rickettsii*. *Infect. Immun.* 40: 840–842.

Walker DH, Firth WT, Hegarty BC (1984) Injury restricted to cells infected with spotted fever group rickettsiae in parabiotic chambers. *Acta Trop. Basel* 41: 307–312.

Walker DH, Feng HM, Saada JI *et al.* (1995) Comparative antigenic analysis of spotted fever group rickettsiae from Israel and other closely related organisms. *Am. J. Trop. Med. Hyg.* 52: 569–576.

Walker DH, Popov VL, Welsh CJR, Feng HM (1996) Mechanisms of rickettsial killing within cytokine-stimulated endothelial cells. In: Kazar J, Toman R (eds) *Rickettsiae and Rickettsial Diseases. Proceedings of the Vth International Symposium.* Bratislava: Veda, pp. 51–56.

Walker DH, Feng HM, Ladner S (1997a) Immunohistochemical diagnosis of typhus rickettsioses using an anti-lipopolysaccharide monoclonal antibody. *Mod. Pathol.* 10: 1038–1042.

Walker DH, Popov VL, Crocquet-Valdes PA, Welsh CJR, Feng HM (1997b) Cytokine-induced, nitric oxide-dependent, intracellular antirickettsial activity of mouse endothelial cells. *Lab. Invest.* 76: 129–138.

Walker DH, Crocquet-Valdes PA, Feng HM (1999a) Intracellular anti-rickettsial mechanisms of chemokine- and cytokine-activated human macrophages and hepatocytes. In: Raoult D, Brouqui P (eds) *Rickettsiae and Rickettsial Diseases at the Turn of the Third Millennium.* Paris: Elsevier, pp. 189–194.

Walker DH, Zavala-Velazquez JE, Ramirez G, Olano JP (1999b) Emerging infectious diseases in the Americas. In: Raoult D, Brouqui P (eds) *Rickettsiae and Rickettsial Diseases at the Turn of the Third Millennium.* Paris: Elsevier, pp. 274–278.

Walker TS (1984) Rickettsial interactions with human endothelial cells *in vitro*: adherence and entry. *Infect. Immun.* 44: 205–210.

Walker TS, Winkler HH (1978) Penetration of cultured mouse fibroblasts (L cells) by *Rickettsia prowazeki*. *Infect. Immun.* 22: 200–208.

Walker TS, Hoover CS (1991) Rickettsial effects on leukotriene and prostaglandin secretion by mouse polymorphonuclear leukocytes. *Infect. Immun.* 59: 351–356.

Walker TS, Tripplet DA (1991) Serologic characterization of Rocky Mountain spotted fever. Appearance of antibodies reactive with endothelial cells and phospholipids, and factors that alter protein C activation and prostacyclin secretion. *Am. J. Clin. Pathol.* 95: 725–732.

Walker TS, Mellott GE (1993) Rickettsial stimulation of endothelial platelet-activating factor synthesis. *Infect. Immun.* 61: 2024–2029.

Walker TS, Brown JS, Hoover CS, Morgan DA (1990) Endothelial prostaglandin secretion: effects of typhus rickettsiae. *J. Infect. Dis.* 162: 1136–1144.

Walker TS, Dersch MW, White WE (1991) Effects of typhus rickettsiae on peritoneal and alveolar macrophages: rickettsiae stimulate leukotriene and prostaglandin secretion. *J. Infect. Dis.* 163: 568–573.

Watt G, Chouriyagune C, Ruangweerayud R *et al.* (1996) Scrub typhus infections poorly responsive to antibiotics in Northern Thailand. *Lancet* 348: 86–89.

Webb L, Carl M, Malloy DC, Dasch GA, Azad AF (1990) Detection of murine typhus infection in fleas by using the polymerase chain reaction. *J. Clin. Microbiol.* 28: 530–534.

Weddle JR, Chan TC, Thompson K (1995) Effectiveness of a dot-blot immunoassay of anti-*Rickettsia tsutsugamushi* antibodies for serologic analysis of scrub typhus. *Am. J. Trop. Med. Hyg.* 53: 43–46.

Weiss E (1988) History of Rickettsiology. In: Walker DH (ed.) *Biology of Rickettsial Diseases*, Vol. I. Boca Raton, FL: CRC Press, pp. 15–32.

Weiss E (1992) Rickettsias. In: Lederberg J (ed.) *Encyclopedia of Microbiology*, Vol. 3. San Diego, CA: Academic Press, pp. 585–610.

Weiss E, Moulder JW (1984) The rickettsias and chlamydias. In: Krieg NR, Holt JG (eds) *Bergey's Manual of Systematic Bacteriology*, Vol. 1. Baltimore: Williams and Wilkins, pp. 687–739.

Weiss K (1988) The role of rickettsioses in history. In: Walker DH (ed.) *Biology of Rickettsial Diseases*, Vol. I. Boca Raton, FL: CRC Press, pp. 1–14.

WHO Working Group on Rickettsial Diseases (1982) Rickettsioses: a continuing disease problem. *Bull. WHO* 60: 157–164.

Wike DA, Ormsbee RA, Tallent G, Peacock MG (1972a) Effects of various suspending media on plaque formation by rickettsiae in tissue culture. *Infect. Immun.* 6: 550–556.

Wike DA, Tallent G, Peacock MG, Ormsbee RA (1972b) Studies of the rickettsial plaque assay technique. *Infect. Immun.* 5: 715–722.

Williams JC, Weiss E (1978) Energy metabolism of *Rickettsia typhi*: pools of adenine nucleotides and energy charge in the presence and absence of glutamate. *J. Bacteriol.* 134: 884–892.

Winkler HH (1976) Rickettsial permeability: an ADP-ATP transport system. *J. Biol. Chem.* 251: 389–396.

Winkler HH (1977) Rickettsial hemolysis: adsorption, desorption, readsorption, and hemagglutination. *Infect. Immun.* 17: 607–612.

Winkler HH (1985) Rickettsial phospholipase A activity. In: Kazar J (ed.) *Rickettsiae and Rickettsial Diseases.* Bratislava: Publishing House of the Slovak Academy of Sciences, pp. 185–194.

Winkler HH (1987) Protein and RNA synthesis by isolated *Rickettsia prowazekii*. *Infect. Immun.* 55: 2032–2036.

Winkler HH (1990) Rickettsia species (as organisms). *Annu. Rev. Microbiol.* 44: 131–153.

Winkler HH, Daugherty RM (1983) Cytoplasmic distinction of avirulent and virulent *Rickettsia prowazekii*: fusion of infected fibroblasts with macrophage-like cells. *Infect. Immun.* 40: 1245–1247.

Winkler HH, Daugherty RM (1989) Phospholipase A activity associated with the growth of *Rickettsia prowazekii* in L929 cells. *Infect. Immun.* 57: 36–40.

Winkler HH, Miller ET (1980) Phospholipase A activity in the hemolysis of sheep and human erythrocytes by *Rickettsia prowazekii*. *Infect. Immun.* 29: 316–321.

Winkler HH, Miller ET (1982) Phospholipase A and the interaction of *Rickettsia prowazekii* and mouse fibroblasts (L-929 cells). *Infect. Immun.* 38: 109–113.

Winkler HH, Miller ET (1984) Activated complex of L-cells and *Rickettsia prowazekii* with *N*-ethylmaleimide-insensitive phospholipase A. *Infect. Immun.* 45: 577–581.

Winkler HH, Turco J (1988) *Rickettsia prowazekii* and the host cell: entry, growth and control of the parasite. *Curr. Top. Microbiol. Immunol.* 138: 81–107.

Winkler HH, Turco J (1994) Rickettsiae and macrophages. *Immunol. Series* 60: 401–414.

Winkler HH, Daugherty R, Hu F (1999) *Rickettsia prowazekii* transports UMP and GMP, but not CMP, as building blocks for RNA synthesis. *J. Bacteriol.* 181: 3238–3241.

Wisseman CL Jr (1967) The present and future of immunization against the typhus fevers. In: *International Conference on Vaccines against Viral and Rickettsial Diseases of Man,* Scientific Publication No. 147. Washington, DC: Pan American Health Organization, pp. 523–527.

Wisseman CL Jr (1986) Selected observations on rickettsiae and their host cells. *Acta Virol.* 30: 81–95.

Wisseman CL Jr, Waddell AD, Walsh WT (1974) *In vitro* studies of the action of antibiotics on *Rickettsia prowazekii* by two basic methods of cell culture. *J. Infect. Dis.* 130: 564–574.

Wisseman CL Jr, Edlinger EA, Waddell AD, Jones MR (1976) Infection cycle of *Rickettsia rickettsii* in chicken embryo and L-929 cells in culture. *Infect. Immun.* 14: 1052–1064.

Woodman DR, Weiss E, Dasch GA, Bozeman FM (1977) Biological properties of *Rickettsia prowazekii* strains isolated from flying squirrels. *Infect. Immun.* 16: 853–860.

Woodward TE (1988) Murine typhus fever: its clinical and biological similarity to epidemic typhus. In: Walker DH (ed.) *Biology of Rickettsial Diseases*, Vol. 2. Boca Raton, FL: CRC Press, pp. 79–92.

Xu W, Raoult D (1997) Distribution of immunogenic epitopes on the two major immunodominant proteins (rOmpA and rOmpB) of *Rickettsia conorii* among the other rickettsiae of the spotted fever group. *Clin. Diagn. Lab. Immunol.* 4: 753–763.

Xu W, Raoult D (1998) Taxonomic relationships among spotted fever group rickettsiae as revealed by antigenic analysis with monoclonal antibodies. *J. Clin. Microbiol.* 36: 887–896.

Zavala-Velazquez JE, Ruiz-Sosa JA, Sanchez-Elias RA, Becerra-Carmona J, Walker DH (2000) *Rickettsia felis* rickettsiosis in Yucatan. *Lancet* 356: 1079–1080.

Zdrodovskyi PF, Golinevitch EM (1972) *Rickettsiae and Rickettsial Diseases.* Moscow: Meditzina.

Zinsser H, Castaneda MR (1933) On the isolation from a case of Brill's disease of a typhus strain resembling the European type. *N. Engl. J. Med.* 209: 815–819.

103
Identification of Uncultured Pathogens

James Flexman

Royal Perth Hospital, Perth, Western Australia

Of the micro-organisms found in terrestrial and aquatic environments, more than 99% cannot be cultivated in the laboratory by conventional techniques (Relman, 1999). The significance of this is illustrated by the large percentage of uncultivated bacteria from a selected range of bacterial divisions (**Fig. 1**). Indeed, 13 bacterial divisions have been noted exclusively to contain uncultivated organisms (Hugenholtz *et al.*, 1998). It follows that a significant number of organisms that cause human disease may not have been cultivated. The failure to grow micro-organisms on artificial media or in tissue culture can be due to a variety of reasons, including unidentified growth factors, incorrect incubation temperatures, inadequate growth times and dependence on other organisms or the host to support their growth. Cultivatable organisms may also fail to grow because of exposure to antibiotics or poor specimen handling in the laboratory. It is, therefore, not surprising that a culture-confirmed diagnosis is sometimes not made in the diagnostic laboratory. Also, the slow growth of human pathogens, including mycobacteria and fungi, may result in a delayed diagnosis.

In many cases, patients need to be treated with prolonged antibiotic therapy to treat these organisms and a specific diagnosis, such as of *Mycobacterium tuberculosis*, can result in rationalisation of therapy and prompt attention to infection control issues. Therefore, molecular methods have the potential to

Table 1 Some human diseases identified or confirmed by molecular techniques

Disease	Organism identified
Cat-scratch disease	*Bartonella hensellae*
Bacillary angiomatosis	*B. hensellae, B. quintana*
Non-A non-B hepatitis	Hepatitis C
Whipple's disease	*Tropheryma whippelii*
Hantavirus pulmonary syndrome	Sin Nombre virus
Kaposi's sarcoma	Kaposi's sarcoma-associated virus

provide a rapid diagnosis for known pathogens and to identify new microbes associated with human disease. Some examples of the latter are shown in **Table 1**, and the focus of this chapter is on the utilisation of molecular methods for detecting previously unrecognised pathogens.

Genotypic Detection of Unrecognised Pathogens

Several proven methods are available to investigate the possible causative agent of a disease considered to have an infective aetiology. These include broad range primers (e.g. to a gene that encodes the small subunit

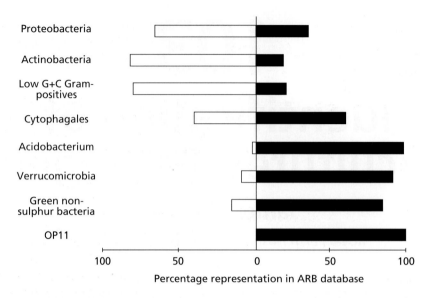

Fig. 1 Percentage of cultivated and uncultivated bacteria from several bacterial divisions present in the ARB sequence database. From Hugenholtz *et al.* (1998) wih permission.

ribosomal RNA), representational difference analysis and expression library screening.

Broad-range polymerase chain reaction (PCR) primers can be selected to amplify genetic loci conserved among a broad group of organisms (**Fig. 2**). Primers that bind to all bacterial small subunit rDNAs and amplify-specific sequences are commonly used for this purpose. By means of this approach, previously unrecognised bacteria can be identified by comparing the sequence data from the amplified product with an extensive database. The disadvantage of this method is that contamination can occur with DNA from bacteria not associated with the underlying disease.

When multiple organisms are present in a test sample, direct sequencing of the amplification product is not possible because of the mixed amplification products. This is less of a problem when material from a usually sterile site (e.g. blood) is used as the source of tissue to be tested, since in this situation a single causative agent is most common. However, there remains a risk of contamination during handling. If there are multiple sequences from different organisms in the specimen, these must be distinguished by other methods, including cloning, single-stranded conformational polymorphism analysis or group-specific probe hybridisation. Thus, a combination of techniques, such as broad-range PCR and cloning can be used to study bacterial diversity in a particular environment.

By means of this approach, Kroes *et al.* (1999) examined the bacterial diversity in the human

subgingival crevice. They compared 264 small subunit rDNA sequences from 21 clone libraries created with products amplified directly from subgingival plaque, with sequences obtained from bacteria cultivated from the same specimen. The results showed that direct amplification of 16S rDNA yielded a more diverse bacterial flora than cultivation.

In addition to studying biodiversity in a bacterial environment, the technology can also be used to study the evolutionary relationships between organisms. For example, *Cyclospora* are intestinal pathogens of humans, but their likely reservoirs and host range are not well understood. Lopez *et al.* (1999) analysed the 18S ribosomal DNA from human and baboon *Cyclospora* species. The baboon *Cyclospora* were defined as a clade within the diverse group of *Eimera* species. Phylogenetic analysis indicated that the baboon-associated *Cyclospora* are distinct from the human *Cyclospora*, but are the closest known relatives of human *Cyclospora*. The authors concluded that the findings raised important questions about the evolutionary relationships of the eimeriids and could lead to improved PCR-based diagnostic methods.

Representational difference analysis (RDA) is another method used to identify unrecognised pathogens. RDA involves subtractive hybridisation to identify genomic fragments present in infected, but not in uninfected control tissue. The use of a control uninfected specimen is designed to identify sequences due to contaminants or endogenous flora common to the control and infected tissue, thus allowing unique genomic fragments to be identified. The

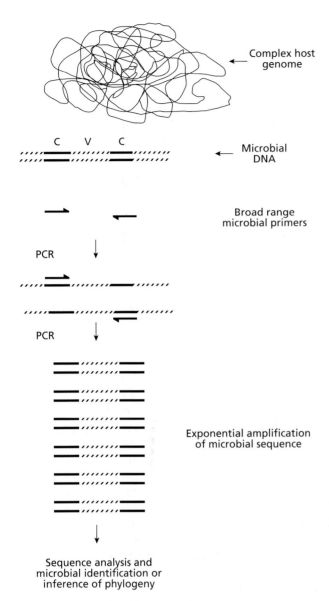

Fig. 2 Diagram showing a broad-range consensus PCR approach to microbial detection and identification. C, conserved; V, variable regions of microbial DNA. Primers anneal to conserved regions of microbial DNA and the polymerase generates new DNA strands, including the variable region that may contain a phylogenetically informative sequence. From Fredricks and Relman (1999) with permission.

unique genomic fragments are, however, enriched in a non-selective manner and not all of the fragments identified by this method will be from the putative pathogen.

Expression library screening involves the use of host antisera to detect antigens in infected tissue. Therefore, this technique identifies only genomic fragments that encode antigenic products. Furthermore, the method can be hindered by the presence of auto-antibodies.

Problems of Molecular Approaches to Identification of New Pathogens

The success of molecular methods used to identify a new pathogen depend on the tissue sample that contains the infective agent. Failure to detect the organism may be due to a sampling error. Targeting specific areas of tissue, such as granulomas, by laser-capture micro-dissection can help to deal with this problem. The tissue may also have been sampled after the agent has been eliminated by the host immune response. In cat-scratch disease the lymph node often contains granulomas and micro-abscesses but organisms are not seen on microscopy, including silver staining, and culture negative for *Bartonella henselae*. The lymph node often contains DNA from *B. henselae* and Anderson *et al.* (1994) found it in 21 of 25 specimens. *B. henselae* DNA was, however, absent in 16% of the lymph nodes. Therefore, to attempt to identify unrecognised pathogens with limited tissue samples may result in failure to identify the aetiological agent. Similarly, a range of control tissues may be necessary to reduce the likelihood of identifying organisms that are not involved in the underlying pathological process.

Another potential problem arises if the pathology is related to the quantity of the infecting agent and not simply on its presence in the infected tissue. The probability of Cytomegalovirus (CMV) causing disease in immunocompromised patients is related to the viral load (Cope *et al.*, 1997). Consequently the more sensitive the PCR test the less specific is the positive result for the disease. In this case both control and pathological tissue can be positive for CMV, but only tissue with a high virus content may show histopathology consistent with disease. This concept can also be applied to other agents where the level of the organism influences pathogenicity, such as *Pneumocystis carinii*.

Interestingly, molecular methods may also identify new agents not associated with a known disease. Examples of this are hepatitis G virus (Linnen *et al.*, 1996) and TT virus (Nishizawa *et al.*, 1997). These agents were identified with molecular methods, including RDA, to investigate possible new viruses that cause hepatitis. However, despite periods of prolonged viraemia subsequent investigations of cases and controls found no evidence that either of these viruses causes chronic hepatitis (Alter *et al.*, 1997; Cossart, 1998). Indeed, this has led to the idea that 'is it time to consider the concept of a commensal virus' (Griffiths, 1999). Therefore, although new pathogens may be sought in patients with a specific disease, unrecognised agents may be identified that are commensals or non-pathogens.

Another example was the discovery of *Afipia felis* as a possible causative agent of cat-scratch disease (English *et al.*, 1988). Subsequent investigations, including the use of serological markers (Regnery *et al.*, 1992; Flexman *et al.*, 1997), and further molecular studies (Anderson *et al.*, 1994), confirmed that *B. hensellae* is the agent responsible for the majority of cases of cat-scratch disease. A dual role has, however, been suggested with *Afipia* responsible for some cases of cat-scratch disease (Alkan *et al.*, 1995). Thus, sensitive molecular methods may identify new agents that seem to be convincing as a possible cause of a disease, but which do not stand up to further scrutiny when other investigators attempt to reproduce the results. Thus, the weakness of the 'case report' can be seen in the context of the search for new pathogens.

Even in a series of cases and controls the possibility has to be considered that immunosuppressive drugs enhance viruses to levels of detectability in the cases but not the controls (Griffiths, 1999). The susceptibility of the host must also be considered when a possible new pathogen is investigated. Because of a number of factors, including innate resistance, previous immunity and the current level of immune competence, active infection does not result in disease in all of the infected individuals. The virulence of the infecting organism introduces a further complexity into the identification of a new pathogen. Disease defined by histopathological changes in tissues may only be evident with virulent organisms. A further consideration is that a dual role or polymicrobial infection is required for the development of disease, such as the involvement of anaerobes in the aetiology of some abscesses or hepatitis D virus requiring the presence of hepatitis B to replicate.

Establishing Disease Causation

Robert Koch introduced his ideas of how a causal relationship between a micro-organism and a disease may be proved. Koch's postulates created a scientific standard for establishing a microbe as the causative agent of a particular disease. However, the link between an organism and disease is not always linear and unidirectional (Relman, 1999). This has resulted in a reconsideration of Koch's postulates, with a new set of guidelines for establishing disease causation with sequenced-based technology (Fredericks and Relman, 1997).

Many of the problems of trying to establish that an infectious agent is the cause of a particular disease have been considered above. If the organism is highly pathogenic, easily cultivated, causes disease regardless of inoculum size and is pathogenic in all hosts, the link is easy to establish. In essence, Koch's third postulate are that the organism is isolated in pure culture, inoculated into an animal model and the disease studied. However, if the organism can also be a commensal, an opportunistic pathogen or a pathogen depending on the environment or host, the link is more difficult to establish. Furthermore, the identification of new pathogens that resist attempts at cultivation present a new challenge when it comes to establishing causation.

There are, however, new ways in which the link between an organism and disease can be studied with molecular methods. For example PCR may be more useful than histology to monitor the response of *Tropheryma whippelii* to antibiotic therapy, since histological resolution of intestinal lesions may take months to years. In contrast, PCR-based evidence of infection with *T. whippelii* correlates with disease resolution and relapse after treatment (Ramzan *et al.*, 1997).

In conclusion, organisms that are uncultured present problems in attempts to establish their role as causes of a particular disease, but some organisms that initially resisted cultivation, such as *B. hensellae* (Koehler *et al.*, 1992; Dolan *et al.*, 1993; Flexman *et al.*, 1995) and *T. whippelii* (Fredericks and Relman, 1997; Relman, 1997), have subsequently been cultured. Since many associations between organisms and disease are non-causal, a combination of approaches is necessary to establish a new agent as the cause of a particular disease. Well-controlled epidemiology, clinical and histological studies, comparisons of molecular with other methods and repeated observations by independent researchers are required.

References

Alkan S, Morgan MB, Sandin RL *et al.* (1995) Dual role for *Afipia felis* and *Rochalimae henselae* in cat scratch disease. *Lancet* 345: 385.

Alter HJ, Nakatsuji Y, Melpolder J *et al.* (1997) The incidence of transfusion-associated hepatitis G virus infection and its relation to liver disease. *N. Engl. J. Med.* 336: 747–754.

Anderson B, Sims K, Regnery RL *et al.* (1994) Detection of *Rochalimaea* DNA in specimens from cat scratch disease patients by PCR. *J. Clin. Microbiol.* 2: 942–948.

Cope AV, Sabin C, Burroughs A *et al.* (1997) Interrelationships among quantity of human cytomegalovirus

(HCMV) DNA in blood, donor-recipient serostatus, and administration of methylprednisolone as risk factors for HCMV disease following liver transplantation. *J. Infect Dis.* 176: 1484–1490.

Cossart Y (1998) TTV a common virus, but pathogenic? *Lancet* 352: 164.

Dolan MJ, Wong MT, Regnery RL *et al.* (1993) Syndrome of *Rochalimaea henselae* adenitis suggesting cat scratch disease. *Ann. Intern. Med.* 118: 331–336.

English LK, Wear DJ, Margileth AM *et al.* (1988) Cat scratch disease. Isolation and culture of the bacterial agents. *JAMA* 259: 1347–1352.

Flexman JP, Lavis NJ, Kay ID *et al.* (1995) *Bartonella henselae* is a causative agent of cat scratch disease in Australia. *J. Infect.* 31: 241–245.

Flexman JP, Chen S, Dickeson DJ *et al.* (1997) Detection of antibodies to *Bartonella henselae* in clinically diagnosed cat scratch disease. *Med. J. Aust.* 166: 532–535.

Fredericks DN, Relman DA (1997) Cultivation of Whipple bacillus: the irony and the ecstasy. *Lancet* 350(9087): 1262–1263.

Fredericks DN, Relman DA (1999) Application of polymerase chain reaction to the diagnosis of infectious diseases. *Clin. Infect. Dis.* 29: 475–488.

Griffiths P (1999) Time to consider the concept of a commensal virus. *Rev. Med. Virol.* 9: 73–74.

Hugenholtz P, Goebel BM, Pace NR (1998) Impact of culture-independent studies on the emerging phylogenetic view of bacterial diversity. *J. Bacteriol.* 180: 4765–4774.

Koehler JE, Quinn FD, Berger TG *et al.* (1992) *Rochalimaea* species from cutaneous and osseous lesions of bacillary angiomatosis. *N. Engl. J. Med.* 327: 1625–1631.

Kroes I, Lepp PW, Relman D (1999) Bacterial diversity within the human subgingival space. *Proc. Natl Acad. Sci. USA* 96(25): 14 547–14 552.

Linnen J, Wages JJ, Zhang-Keck ZY *et al.* (1996) Molecular cloning and disease association of hepatitis G virus: a transfusion-transmissible agent. *Science* 271: 505–508.

Lopez FA, Mangliemot J, Schmidt TM *et al.* (1999) Molecular characterisation of *Cyclospora*-like organisms from baboons. *J. Infect. Dis.* 179(3): 670–676.

Nishizawa T, Okamoto H, Konishi K *et al.* (1997) A novel DNA virus (TTV) associated with elevated transaminase levels in post-transfusion hepatitis of unknown etiology. *Biochem. Biophys. Res. Commun.* 241: 92–97.

Ramzan NN, Loftus E, Burgart LJ *et al.* (1997) Diagnosis and monitoring of Whipple disease by polymerase chain reation. *Ann. Intern. Med.* 126: 520–527.

Regnery RL, Olsen JG, Perkins BA *et al.* (1992) Serologic response to *Rochalimaea hensellae* antigen in suspected cat scratch disease. *Lancet* 339: 1443–1445.

Relman DA (1997) The Whipple bacillus lives (*ex vivo*). *J. Infect. Dis.* 176(3): 752–754.

Relman D (1999) The search for unrecognised pathogens. *Science* 284: 1308–1310.

Index

protective antigen (PA) 2018–19
  antibodies 2023
  binding 2021
  structure 2019
protein
  biosynthesis inhibition 606–8
  contact-mediated transport 389
  secretion pathways 388–90
protein A, staphylococcal 732, *733*, 842, 849, 862
  anti-phagocytic 865
  atopy 1024
protein C, thrombin formation inhibition 702
protein C/thrombomodulin 701
protein electrophoresis 539–41
  typing 540
protein F1, *Streptococcus pyogenes* 660–1
protein I/IIf 1001
protein III, *Neisseria gonorrhoeae* 1789
protein kinase A (PKA) 766–7
  activation 1197–8
protein kinase C (PKC) 745, 1103–5
  activation 1105, 1177, 1475
  mitogen-activated pathway 1177
protein pump 20
protein synthesis 321, 323–38
  antimicrobial agent targets 586
  control 334–7
  error-proneness with streptomycin 606
  inhibitors 586, 587
  transcription 323, *324*, 325–8
    initiation 334–6
  transcriptional control 334–7
  translation 323, *324*, 328, *329*, 330–4
protein toxins 566, 740–5
  adenylate cyclase deregulation 742
  ADP-ribosylation 741
  cholesterol-binding 745
  covalent modification 741
  cytoskeleton disruption 743, *744*
  glycosylation 741, 743
  intracellular targets 740–1
  membrane-damaging 743–4
    enzymes 744–5
  pore-forming 745–6
    toxin 747
  protein synthesis inhibition 742
  proteolytic 743
    cleavage 741
  ribonucleases 743
  RTX 745–6
  superantigens 747–8
  translocation 740–1
protein tyrosine kinases 1094, 1104–5
protein tyrosine phosphorylation 645, 1417
protein–protein interactions 630
proteins
  antigens 809
  group B streptococci 904–6
  post-translational modification 813
  synthesis 602
    regulation 321
  tandem repeat-containing 904–5
    group B streptococci 910
proteomes 528
proteomics 2–3, 344
Proteus
  ascending urinary tract infection 1510
  capsular polysaccharides 67
  L-forms 591
  meningitis 1954
  motility 155
  stone formation 1511
Proteus mirabilis
  adherence 1542
  adhesins 1542
  fimbriae 1542
  flagella 1543

Proteus mirabilis (*continued*)
  flagellin 1543
  haemolysin 1543
  host response evasion 1543–4
  IgA protease 1525
  motility 1543
  P-like pilus (PMP) 1542
  stone formation 1524, 1543
  swarmers 1543
  urease 1543
  urinary tract infection *1516*, 1517, 1541–4
Proteus vulgaris, biofilms 141
prothrombin activation 731
proto-porphyrin 1974
proton electrochemical gradient 244
proton electrochemical potential difference 245
proton motive force (PMF) 155, 157, 233–4
  DNA transport 380
  transmembrane 244, 249–50
  transport-coupled membrane energisation 250–1
proton motive loops 245
proton pump 234, 245
  inhibitors in *Helicobacter pylori* 1333
proton-translocating ATPase 234
proton translocation, transmembrane 245–7
protonophores 244
protons, vectorial translocation 245
protoplast transformation systems 522
protoplasts 141
protoporphyrin 1929
protozoal infection, cell-mediated immunodeficiency 835
Prt proteases 40
prtV gene 1202
PrtV protein 1202
PsaA permease 1637, 1644
pseudo-appendicitis syndrome, *Yersinia enterocolitica* 1405
pseudo-membranous colitis 1153
Pseudomonas
  genus 961
  meningitis 1954
  O antigens 120
  skin infections 1024, 1029–30
Pseudomonas aeruginosa
  adherence 1553
  adhesion blocking by sugars 640
  algB gene 62
  alginate capsule 48, 731
    gene cluster 55
    polymerisation 59
  alginate regulation system 62–3
  algR gene 62
  antimicrobial resistance traits *571–2*
  arginine deiminase (ADI) pathway 238
  colistin treatment 597
  cystic fibrosis 1551, 1559
  defensin expression 1553
  elastase 675, 750
  exotoxin A 742
  exotoxin S 749
  extracellular polysaccharide 48
  fibronectin receptors 1553
  genetic potential for biosynthesis of individualities *571–2*
  hydroxyphenazine 1552
  integrin receptors 1553
  iron acquisition 967
  lectins 631
  microscopy *30*
  motility 155
  neutropenia 790, 835
  non-clonal populations 471
  phenylalanine formation 278
  porins 35

Pseudomonas aeruginosa (*continued*)
  pyocyanin 1552
  RND family 578
  secreted proteins 1545
  skin infections 1029
  slime production 730–1
  small-colony variants (SCV) 732
  toxic injection 749
  type 4 pili 187, 189
  tyrosine formation 278
  urinary tract infection *1516*, 1517
  virulence factors 742
    expression regulation 433
Pseudomonas cepacia 1029
Pseudomonas syringae, hypersensitive response 186
psiB genes 368
psittacosis 1851
psoriasis 1025
  guttate 1025
  staphylococcal enterotoxins 1098, 1099
PspA protein 1625–6, 1637, 1644
ptl gene 1584, 1586
Ptl proteins 1585–6, 1999
pTLC 1193
ptx gene 1584
puberty, acne 1045–6
public health, lipopolysaccharide genetic polymorphism 131–2
PulD protein 1241
pullulanase 1241, 1243
pulmonary defences, secretory 1641–3
pulmonary disease, chronic 1691
pulmonary nocardiosis 944–5
pulmonary vasculature endothelium 912
pulse-field gel electrophoresis (PFGE) 474–5, 536, 542
  applications 543
purA gene mutant 529
purine
  base interconversion 289–91, *292*
  biosynthesis 282–3, *285*
    regulation 286–91
  biosynthetic pathway 281, *284*
  operator sites for transcription 288
  pathway genes 286–7
  salvage pathway 289–91
puromycin, translation inhibitors 337
PurP protein 286–7
PurR gene 287–8
putrescine 267, 1258
pXO2 2014
pyelonephritis 177, 180, 1516, 1524
  acute 1509–10
  chronic 1510
  MR/P fimbriae 1542
  O-antigens 1523
  P fimbriae 1525
  Proteus mirabilis 1542
  S fimbriae 1525
pyelonephritis-associated pili (Pap) 434
pyocyanin 1552
pyoderma 1026
pyosalpinx 1788
pyrazinamide
  resistance 1742, 1775
  tuberculosis 1561, 1741
    susceptibility testing 1742
pyrethroids, synthetic 2110
pyridoxalphosphate-requiring enzymes 576
pyridoxaminephosphate-requiring enzymes 576
pyrimethamine 292
  dihydrofolate reductase inhibition 604
pyrimidine
  base interconversion 291, *292*
  biosynthesis 283–6, *288*
    regulation 286–91
  ring 288, *290*